Mutation Breeding for Sustainable Food Production and Climate Resilience

Suprasanna Penna • S. Mohan Jain
Editors

Mutation Breeding for Sustainable Food Production and Climate Resilience

Springer

Editors
Suprasanna Penna
Nuclear Agriculture and Biotechnology Division
Bhabha Atomic Research Centre
Mumbai, India

Amity Centre for Nuclear Biotechnology & Amity Institute of Biotechnology
Amity University
Mumbai, Maharashtra, India

S. Mohan Jain
Department Agricultural Sciences
University of Helsinki
Helsinki, Finland

ISBN 978-981-16-9719-7 ISBN 978-981-16-9720-3 (eBook)
https://doi.org/10.1007/978-981-16-9720-3

This Springer imprint is published by the registered company Springer Nature Singapore Pte Ltd.
The registered company address is: 152 Beach Road, #21-01/04 Gateway East, Singapore 189721, Singapore

Foreword

Agricultural growth is important to ensure food and nutritional security for the growing population and also is a vital to boost economic prosperity. Globally, scientists are challenged to achieve higher food production per unit of both land and water from shrinking farm size, and biodiversity erosion facing climate change. Currently, climate resilience forewarns significant threats to food security, ecosystems, economic stability, and water resources worldwide. To overcome such challenges, sustained agricultural research is needed through the judicious blend of both conventional and innovative methodologies for sustainable food production, nutritional food security, and socio-economic aspects. Crop improvement is a continuous process for the evolution of promising and improved varieties with specifically targeted traits by exploiting a wide range of spontaneous/natural and induced greater genetic variability. Plant breeders are continuously striving to sustain crop production and look for innovative technologies including mutagenesis and molecular tools such as genome editing and transgenic plants.

Among many methodologies that have been and are being employed, mutation breeding has made significant contributions and is poised to create useful genetic variations to be employed in breeding strategies in different crop plants. Although first natural mutant was reported about 2300 years ago in China, the use of irradiations to induce mutations was pioneered in the early twentieth century by Prof. Lewis John Stadler (1928), after the discovery of X-rays by W.K. Rontgen. In 1964, the Joint FAO/IAEA Division of Nuclear Techniques in Food and Agriculture was established in Vienna, and thereafter mutation breeding received greater attention as a tool by plant breeders around the globe. I had an opportunity to visit FAO/IAEA division when Dr. Mohan Jain was working there. Application of radiation (X-rays, Gamma rays, ion beams, fast neutrons) and use of chemical mutagens (EMS) have played an important role in inducing novel genetic variability, improving existing popular varieties, and developing promising varieties in different crop plants. The application of mutation breeding has since shown great potential to evolve climate-resilient crop varieties to mitigate the effects of climate change. Various agricultural institutes have been engaged in induced mutagenesis research in different crop plants including vegetatively propagated crops, ornamental, medicinal, and aromatic plants.

It is noteworthy to mention that mutation breeding has significantly contributed to developing more than 3365 improved, mutant crop varieties globally. These varieties have been cultivated extensively by millions of farmers and immensely benefitted them by enhancing their farm income and the nation's food security. New technologies have also been developed in the last decade to create and identify useful genetic variation. These include in vitro mutagenesis, heavy ion beam and space mutation breeding, mutagenomics, and TILLING (Targeting Induced Local Lesions in Genomes).

In view of all the new developments and an important role of mutation breeding in agriculture, horticulture, floriculture, etc., it is very pertinent to bring out a book on this topic. This book presents the techniques of creating mutants for utilization towards developing improved crop varieties. It also emphasizes on the basic mutation methodologies and success stories in different countries in food crops, horticultural plants, and plantation crops. In this context, it is timely that the editors of this illustrious volume, Drs. P. Suprasanna and S. Mohan Jain have made dedicated efforts for the compilation of this book that covers the latest information on basic and applied mutation breeding in various crops. Their efforts are highly commendable to kindle thoughts among researchers and students to encourage them to exploit mutation technologies and molecular tools in crop improvement. I compliment all the scientists for their valuable contributions in this book, which will serve as an excellent source of reference material to researchers, commercial seed companies, teachers, young breeders in new genomic era to undertake induced mutagenesis in respective crops in their future programs aimed at crop improvement.

International Centre for Genetic Engineering and Biotechnology
New Delhi, India

Sudhir K. Sopory

Preface

Food security continues to be on the center stage of agriculture development. Several factors including climate change, growing population, and dwindling natural resources continue to mount pressure on finding sustainable solutions. In this backdrop, it is imperative to enhance and sustain food production in the present and future. Genetic variation is the mainstay for plant breeding and methods to enhance genetic variability are essentially needed for developing new improved crop varieties. Breeding of crop plants has been practiced over several decades and considerable progress has been made to evolve high yielding varieties with enhanced quality and stress tolerance. The success in the development of high-yielding crops using different breeding methods like selection, hybridization, and associated breeding strategies led to the Green Revolution. However, since induction of new and novel genetic variability has to be exploited for improving a wide range of self and cross-pollinated plants including vegetatively propagated plant species, there has been a continued implementation of alternative breeding strategies including induction of mutations.

Since the initial experiments with ionizing radiations in 1920s, several studies have been conducted to isolate useful mutations for plant traits culminating into improved mutant varieties for cultivation. More than 3365 mutant varieties are officially registered in a database curated by the FAO/IAEA. The success can be very well appreciated in the economic impact that these varieties have created globally in different countries. There has been tremendous progress in fine tuning the crucial processes of mutation induction, mutant selection, and mutant characterization. Innovative breeding strategies are in place to tackle the issues surrounding food and nutritional security. Novel genomics tools are opening up new avenues for the induction and characterization of genetic variation.

This book has covered different facets of mutation breeding in crop plants and has chapters right from relevance of mutation breeding, historical account, mutagens, mutant selection, molecular tools, and targeted mutagenesis techniques. The book also presents successful examples of "application of induced mutations" in different crop plants, horticultural crops, ornamentals, commercial crops including tuber crops. It is also an attempt to collate seminal contributions of pioneering researchers of international repute who are engaged in the trade of art of mutation breeding. Each chapter has an introduction with the background of the respective mutation breeding area and gives in-depth account of the topic.

The book will be an excellent reference source for plant biologists, breeders, biotechnologists, and geneticists engaged in mutation breeding programs as well as for seed companies and policy makers. We profusely thank and appreciate the contributions from all authors, for their cooperation and patience during the publication process. Their contributions have been a big source of success and quality of this book. We are happy that this collaborative work has chapters from 74 scientists and 11 countries. Big thanks to Springer and their team for continuous support and for the opportunity to bring out this book.

Mumbai, India — Suprasanna Penna
Helsinki, Finland — S. Mohan Jain

Contents

1 **Mutation Breeding to Promote Sustainable Agriculture and Food Security in the Era of Climate Change** 1
Suprasanna Penna, Siamak Shirani Bidabadi, and Shri Mohan Jain

2 **History of Plant Mutation Breeding and Global Impact of Mutant Varieties** 25
M. C. Kharkwal

3 **Physical and Chemicals Mutagenesis in Plant Breeding** 57
S. Bado, B. P. Forster, and F. Maghuly

4 **Mutagenesis and Selection: Reflections on the In Vivo and In Vitro Approaches for Mutant Development** 99
Suprasanna Penna and Suresh G. Bhagwat

5 **Haploid Mutagenesis: An Old Concept and New Achievements** . . . 129
Monika Gajecka and Iwona Szarejko

6 **Strategies for Screening Induced Mutants for Stress Tolerance** 151
Fatma Sarsu, Suprasanna Penna, and Ganesh C. Nikalje

7 **Induced Mutagenesis for Developing Climate Resilience in Plants** 177
Mithlesh Kumar, Kirti Rani, and Suprasanna Penna

8 **Molecular Markers for Mutant Characterization** 205
R. S. Bhat, M. P. Brijesh Patil, I. S. Tilak, and K. Shirasawa

9 **Application of TILLING as a Reverse Genetics Tool to Discover Mutation in Plants Genomes for Crop Improvement** 233
M. I. Siddique, A. Younis, M. A. Gururani, and J. Venkatesh

10 **Plant Mutagenesis Tools for Precision Breeding: Conventional CRISPR/Cas9 Tools and Beyond** 269
Rahul Mahadev Shelake, Dibyajyoti Pramanik, Rajesh Ramdas Waghunde, and Jae-Yean Kim

11 Crop Improvement Through Induced Genetic Diversity and Mutation Breeding: Challenges and Opportunities 293
Sobhana Sivasankar

12 Induced Mutations for Development of New Cultivars and Molecular Analysis of Genes in Japan 301
Hitoshi Nakagawa

13 Role of Mutation Breeding in Crop Improvement with Special Reference to Indian Subcontinent 355
M. C. Kharkwal

14 Success of Mutation Breeding of Sorghum to Support Food Security in Indonesia 429
Soeranto Human, Wijaya M. Indriatama, and Sihono

15 Potential of Mutation Breeding in Genetic Improvement of Pulse Crops 445
Souframanien Jegadeesan and Dhanasekar Punniyamoorthy

16 Advances in Mutation Breeding of Groundnut (*Arachis hypogaea* L.) 487
Anand M. Badigannavar and Suvendu Mondal

17 Mutation Breeding for Sustainable Food Production in Latin America and the Caribbean Under Climate Change 521
María Caridad González Cepero, Luz Gómez Pando, Alexander De Andrade, Juan Felix Arguello Delgado, Héctor Nakayama Nakashima, Alejandra Landau, Alberto Prina, Mayra Rodríguez García, Aura Suchini Farfan, and Eulogio de la Cruz Torres

18 Mutation Breeding Studies in the Indian Non-basmati Aromatic Rice: Success and Outlook 539
N. B. Gaikwad and V. A. Bapat

19 Induced Mutagenesis in Chrysanthemum 565
S. K. Datta

20 Mutation Breeding Research in Sweet Pepper 599
A. Popatanasov, O. Timina, and N. Tomlekova

21 Induced Mutation Technology for Sugarcane Improvement: Status and Prospects 645
Suprasanna Penna, S. J. Mirajkar, M. V. Purankar, A. A. Nikam, S. G. Dalvi, E. R. Vaidya, and R. M. Devarumath

22 Induced Mutations for Developing New Ornamental Varieties 669
M. Tütüncü, K. Y. Kantoğlu, B. Kunter, and Y. Y. Mendi

23 Improvement of Fruit Crops Through Radiation-Induced Mutations Facing Climate Change 693
S. S. Maan, V. Sharma, and J. S. Brar

24 Induced Mutations for Genetic Improvement of Banana 719
H. Tak, S. Bhakta, S. Negi, and T. R. Ganapathi

25 Mutation Breeding in Date Palm (*Phoenix dactylifera* L.) 735
Djamila Yatta El Djouzi, Hafida Khelafi, Shri Mohan Jain, and Jameel M. Al-Khayri

26 Mutation Breeding in Tropical Root and Tuber Crops 779
A. V. V. Koundinya, Arindam Das, and Vivek Hegde

Editors and Contributors

About the Editors

Suprasanna Penna Ph.D. (Genetics), is formerly Head of Nuclear Agriculture and Biotechnology Division, Bhabha Atomic Research Centre, Mumbai, and Professor at Homi Bhabha National Institute, Mumbai. Currently he is Professor and the Director, Amity Institute of Biotechnology, Amity University of Maharashtra, Bhatan, Mumbai, India. He has made significant contributions to plant science covering plant biotechnology, mutation breeding, stress biology, genomics, and novel biopolymers. His work on technology development for radiation depolymerized oligo-chitosan has been successfully used in enhancing plant productivity in crop plants and vegetables. He has also led research on molecular understanding of plant abiotic stress tolerance in crop plants including salinity-adaptive mechanisms in halophytes. He has served as an expert with International Atomic Energy Agency (IAEA), Vienna (Austria). Dr Penna is on the editorial board of national and international journals and published over 350 publications and has edited books on salinity tolerance, plant nutrition and climate change, plant genetic diversity, and plant-metal interactions. He is the guest editor with various publishers.

Shri Mohan Jain is a Consultant and Plant Biotechnologist, Department of Agricultural Sciences, University of Helsinki, Helsinki, Finland. He received his M.Phil. in 1973 and Ph.D. in 1978 from Jawaharlal Nehru University, New Delhi, India. He was a postdoctoral fellow in Israel, USA, and visiting scientist/Professor in Japan, Malaysia, Germany, and Italy. He was a Technical Officer, Plant Breeding and Genetics, International Atomic Energy Agency (IAEA), Vienna, Austria, 1999–2005. He is a member of International Association of Plant Tissue Culture and Biotechnology and an editorial board member of *Euphytica*, *In Vitro*, *Propagation of Ornamental Plants*, *Emirates Journal of Food and Agriculture*, and a series on Forest Biotechnology. His has more than 160 publications in peer-reviewed journals, book chapters, and conference proceedings, and edited 55 books, is an invited speaker, and acted as a chairperson in several international conferences worldwide. He was awarded Nobel Peace Prize, in commemoration of the awarding to IAEA of the Nobel Peace Prize for 2005; He was also former consultant to IAEA, the European Union, the Government of Grenada and the Egyptian Government.

Currently his research interests are somatic embryogenesis, organogenesis, haploidy, somatic cell hybridization, somaclonal variation and mutagenesis mainly in medicinal plants, date palm, and banana genetic improvement, genetic diversity, erosion, conservation, and utilization in the context of climate change and food and nutritional security.

Contributors

Jameel M. Al-Khayri Department of Agricultural Biotechnology, College of Agriculture and Food Sciences, King Faisal University, Al-Ahsa, Saudi Arabia

Anand M. Badigannavar Nuclear Agriculture and Biotechnology Division, Bhabha Atomic Research Centre, Mumbai, India

S. Bado Department of Plant Science, PHYTOPRISE GmbH, Oberwart, Austria
Département de Biochimie-Microbiologie, Unité de Formation et de Recherche en Sciences de la Vie et de la Terre, Université Ouaga I Professeur Joseph KI-ZERBO, Ouagadougou, Burkina Faso

V. A. Bapat Department of Biotechnology, Shivaji University, Kolhapur, India

S. Bhagwat Nuclear Agriculture and Biotechnology Division, Bhabha Atomic Research Centre, Mumbai, India

S. Bhakta Plant Cell Culture Technology Section, Nuclear Agriculture and Biotechnology Division, Bhabha Atomic Research Center, Trombay, Mumbai, India

R. S. Bhat Department of Biotechnology, University of Agricultural Sciences, Dharwad, India

J. S. Brar Department of Fruit Science, College of Horticulture and Forestry, Punjab Agricultural University, Ludhiana, Punjab, India

M. P. Brijesh Patil Department of Biotechnology, University of Agricultural Sciences, Dharwad, India

Alexander de Andrade Company of Agricultural Investigations and Rural Extension (Epagri), Experimental Station of Itajai, Itajai, Brazil

María Caridad González Cepero National Institute of Agricultural Science (INCA), Mayabeque, Cuba

Eulogio de la Cruz Torres National Institute of Nuclear Investigations (ININ), Mexico City, Mexico

S. G. Dalvi Plant Tissue Culture Section, Vasantdada Sugar Institute, Pune, India

Arindam Das Department of Vegetable Science, Uttara Banga Krishi Viswavidyalaya, Cooch Behar, West Bengal, India

S. K. Datta CSIR-National Botanical Research Institute, Lucknow, India
Bose Institute, Kolkata, India

Juan Felix Arguello Delgado National University of Costa Rica, Heredia, Costa Rica

R. D. Devarumath Molecular Biology and Genetic Engineering Division, Vasantdada Sugar Institute, Pune, India

Dhanasekar Punniyamoorthy Nuclear Agriculture and Biotechnology Division, Bhabha Atomic Research Centre, Trombay, Mumbai, India

Djamila Yatta El Djouzi I.N.R.A. Biotechnology and Plant Breeding Division CRP Mehdi Boualem, Alger, Algeria

Aura Suchini Farfan Science and Agricultural Technology Institute (ICTA), Guatemala, Mexico

B. P. Forster Biohybrids Ltd, Reading, UK

N. B. Gaikwad Department of Botany, Shivaji University, Kolhapur, India

Monika Gajecka Institute of Biology, Biotechnology and Environmental Protection, Faculty of Natural Sciences, University of Silesia in Katowice, Katowice, Poland

T. R. Ganapathi Plant Cell Culture Technology Section, Nuclear Agriculture & Biotechnology Division, Bhabha Atomic Research Center, Mumbai, India
Homi Bhabha National Institute, Mumbai, India

Mayra Rodríguez García Genetic and Engineering and Biotechnology Center (CIGB), Havana, Cuba

M. A. Gururani Department of Biology, College of Science, UAE University, Al Ain, United Arab Emirates

Vivek Hegde Division of Crop Improvement, ICAR-Central Tuber Crops Research Institute, Thiruvananthapuram, Kerala, India

Soeranto Human Research Center for Isotopes and Radiation Application, Indonesian Research and Innovation Agency (BRIN), Jakarta, Indonesia

Wijaya Murti Indriatama Research Center for Isotopes and Radiation Application, Research Organization of Nuclear Technology (BATAN), Indonesian Research and Innovation Agency (BRIN), Jakarta, Indonesia

S. Mohan Jain Department of Agricultural Sciences, PL-27, University of Helsinki, Helsinki, Finland

K. Y. Kantoğlu Turkish Energy Nuclear and Mineral Research Agency, Nuclear Energy Research Institute, Ankara, Turkey

M. C. Kharkwal Division of Genetics, ICAR-Indian Agricultural Research Institute, New Delhi, India

Hafida Khelafi I.N.R.A. Biotechnology and Plant Breeding Division CRP Mehdi Boualem, Alger, Algeria

J. -Y. Kim Division of Applied Life Science (BK21 Four Program), Plant Molecular Biology and Biotechnology Research Center, Gyeongsang National University, Jinju, South Korea

A. V. V. Koundinya Division of Crop Improvement, ICAR-Central Tuber Crops Research Institute, Thiruvananthapuram, Kerala, India

Mithlesh Kumar AICRN on Potential crops, ARS Mandor Jodhpur Agriculture University, Jodhpur, Rajasthan, India

B. Kunter Turkish Energy Nuclear and Mineral Research Agency, Nuclear Energy Research Institute, Ankara, Turkey

Alejandra Landau Genetic Institute "Edwal A. Favret", Buenos Aires, Argentina

S. S. Maan Department of Fruit Science, College of Horticulture and Forestry, Punjab Agricultural University, Ludhiana, Punjab, India

F. Maghuly Plant Functional Genomics, Institute of Molecular Biotechnology, Department of Biotechnology, BOKU-VIBT, University of Natural Resources and Life Sciences, Vienna, Austria

Y. Y. Mendi Horticulture Department, Agriculture Faculty, Çukurova University, Adana, Turkey

S. J. Mirajkar School of Nanoscience and Biotechnology, Shivaji University, Kolhapur, India

Suvendu Mondal Nuclear Agriculture and Biotechnology Division, Bhabha Atomic Research Centre, Mumbai, India

Hitoshi Nakagawa Hamamatsu Photonics K.K., Central Research Laboratory, Hamamatsu, Shizuoka, Japan
Institute of Radiation Breeding, National Institute of Agrobiological Sciences, Tsukuba, Ibaraki, Japan

Héctor Nakayama Nakashima Multidisciplinary Center of Technological Investigations, Asuncion National University, San Lorenzo, Paraguay

S. Negi Department of Biotechnology, University of Mumbai, Mumbai, India

Ganesh C. Nikalje Seva Sadan's R. K. Talreja College of Arts, Science and Commerce, Mumbai, India

A. A. Nikam Plant Tissue Culture Section, Vasantdada Sugar Institute, Pune, India

Luz Gómez Pando Agrarian National University, Lima, Perú

A. Popatanasov Institute of Neurobiology, Bulgarian Academy of Sciences, Sofia, Bulgaria

D. Pramanik Division of Applied Life Science (BK21 Four Program), Plant Molecular Biology and Biotechnology Research Center, Gyeongsang National University, Jinju, Korea

Alberto Prina Genetic Institute "Edwal A. Favret", Buenos Aires, Argentina

M. V. Purankar Molecular Biology and Genetic Engineering Division, Vasantdada Sugar Institute, Pune, India

Kirti Rani Division of Crop Improvement, ICAR-Directorate of Groundnut Research, Junagadh, Gujarat, India

Fatma Sarsu Joint FAO/IAEA Centre of Nuclear Techniques in Food and Agriculture, International Atomic Energy Agency Vienna, Vienna, Austria

V. Sharma Faculty of Applied Sciences and Biotechnology, Shoolini University, Bhajol, Solan (HP), Himachal Pradesh, India

R. M. Shelake Division of Applied Life Science (BK21 Four Program), Plant Molecular Biology and Biotechnology Research Center, Gyeongsang National University, Jinju, Korea

S. Shirani Bidabadi Department of Horticulture, College of Agriculture, Isfahan University of Technology, Isfahan, Iran

K. Shirasawa Department of Frontier Research and Development, Kazusa DNA Research Institute, Chiba, Japan

M. I. Siddique Vegetable Research Division, National Institute of Horticulture and Herbal Science, Rural Development Administration, Jeonju, Republic of Korea

Sihono Research Center for Isotopes and Radiation Application, Indonesian Research and Innovation Agency (BRIN), Jakarta, Indonesia

S. Singh Department of Fruit Science, College of Horticulture and Forestry, Punjab Agricultural University, Ludhiana, Punjab, India

Sobhana Sivasankar Plant Breeding and Genetics Section, Joint FAO/IAEA Centre of Nuclear Techniques in Food and Agriculture, Vienna International Centre, Vienna, Austria

Souframanien Jegadeesan Nuclear Agriculture and Biotechnology Division, Bhabha Atomic Research Centre, Mumbai, India

Suprasanna Penna Nuclear Agriculture and Biotechnology Division, Bhabha Atomic Research Centre, Mumbai, India
Amity Centre for Nuclear Biotechnology & Amity Institute of Biotechnology, Amity University, Mumbai, Maharashtra, India

Iwona Szarejko Institute of Biology, Biotechnology and Environmental Protection, Faculty of Natural Sciences, University of Silesia in Katowice, Katowice, Poland

H. Tak Plant Cell Culture Technology Section, Nuclear Agriculture & Biotechnology Division, Bhabha Atomic Research Center, Mumbai, India

I. S. Tilak Department of Biotechnology, University of Agricultural Sciences, Dharwad, India

O. Timina Faculty of Natural Sciences and Geography, Transdnistrian State University, Tiraspol, Transdnistria, Moldova

N. Tomlekova Agricultural Academy, Maritsa Vegetable Crops Research Institute, Plovdiv, Bulgaria

M. Tütüncü Horticulture Department, Agriculture Faculty, Ondokuz Mayıs University, Samsun, Turkey

E. R. Vaidya Pulses Research Unit, Dr. Panjabrao Deshmukh Krishi Vidyapeeth, Akola, India

J. Venkatesh Department of Plant Science and Plant Genomics and Breeding Institute, Seoul National University, Seoul, Korea

R. R. Waghunde Department of Plant Pathology, College of Agriculture, Navsari Agricultural University, Bharuch, India

A. Younis Institute of Horticultural Sciences, University of Agriculture, Faisalabad, Pakistan

Mutation Breeding to Promote Sustainable Agriculture and Food Security in the Era of Climate Change

1

Suprasanna Penna, Siamak Shirani Bidabadi, and Shri Mohan Jain

Abstract

Globally, sustainable agriculture and food production systems face unprecedented challenges from population expanse, declining land, and water resources besides climate change. Climate change-driven factors like variation in temperature and precipitation patterns, the spread of drought and salinity on farmlands, increasing soil and water pollution, global warming due to rising carbon dioxide concentrations and extreme temperatures have shown tremendous impact on plant growth, development, and productivity of crop plants. Utmost attention is needed to develop and adopt effective, novel plant breeding technologies including mutation breeding, molecular marker-assisted selection, and genetically modified and gene-edited crops. Induced mutations have played an influential role in meeting challenges of food security through the development and adoption of new mutant varieties by the farming community in several countries. As on date, 3402 mutant varieties have been released worldwide as registered on the IAEA mutant variety database. A wide range of genetic variability for attributes related to plant type, yield, and stress tolerance has been induced for use in plant breeding and crop improvement programs. The mutant resources are now continually developed for creating new mutant germplasm for use in cro improvement

S. Penna (✉)
Nuclear Agriculture and Biotechnology Division, Bhabha Atomic Research Centre, Mumbai, India

Amity Centre for Nuclear Biotechnology & Amity Institute of Biotechnology, Amity University, Mumbai, Maharashtra, India

S. Shirani Bidabadi
Department of Horticulture, College of Agriculture, Isfahan University of Technology, Isfahan, Iran

S. M. Jain
Department of Agricultural Sciences, University of Helsinki, Helsinki, Finland
e-mail: mohan.jain@helsinki.fi

S. Penna, S. M. Jain (eds.), *Mutation Breeding for Sustainable Food Production and Climate Resilience*, https://doi.org/10.1007/978-981-16-9720-3_1

and, in functional genomics studies. In the past few years, advanced mutagenic tools of gene discovery and genome editing have emerged as the novel plant breeding technologies. The introduction of precise genetic changes in the target genes using genome editing has been fairly successful in the improvement of crop plants. Continued research and prioritization will foster further development of efficient mutagenesis, screening protocols and high-throughput phenotyping methods for use in crop improvement programs aimed at sustainable agriculture and climate resilience.

Keywords

Food security · Climate resilience · Mutation breeding · Induced mutations · Plant stress

1.1 Introduction

Since prehistoric time, more than 10,000 years ago, plant cultivation and agriculture have been associated with human evolution. Food security has become a priority as the human population continued to grow with the land and water becoming inadequate. Global population is anticipated to increase from the present 7.6 billion to 9.7 billion by 2050 demanding massive increase in the yield of the major food crops (FAO 2017). An estimate from FAO (2020) states that 8.9% of the world population is facing hunger, three billion people are facing under nutrition problems and two billion people are facing food insecurity. Thus, both global food security and nutritional demands have to be met with innovative agricultural practices. Plant breeding practices have significantly contributed for enhancing plant growth and productivity of several major crops (Cai et al. 2017); however, it has also become a challenge to meet the global food demands and crop yields without increasing the cultivatable area. Additionally, climate change has posed a serious challenge to food security including food availability, food accessibility, food utilization, and food system stability (Fig. 1.1).

In this scenario, supply chains are disrupted, market prices increase, assets and livelihood opportunities are lost, purchasing power falls, human health is endangered, and affected people are unable to cope. Therefore, finding a way to reduce the risks of food security caused by climate change is a prime concern (Lobell et al. 2011). The lack of knowledge about the climate impact presents a formidable challenge that needs to be overcome by designing, developing, and adopting appropriate technologies to prevent the adverse impact of climate change on agriculture (Campbell et al. 2016). In a recent study, Fedick and Santiago (2022) argued that during the drought periods of Mayan period (ca. 750–900/1000 CE) although drought situation impacted the obtainability of plant food, plant diversity could have prevented famine. This emphasizes the essentiality of "plant genetic diversity" to cope with climate change-driven stress factors like drought. Therefore, the limitations on climate forecasting as well as the reduction of harmful factors related to climate change such as drought, salinity, and heat stress are among the important

Fig. 1.1 The fundamental tenets of food security

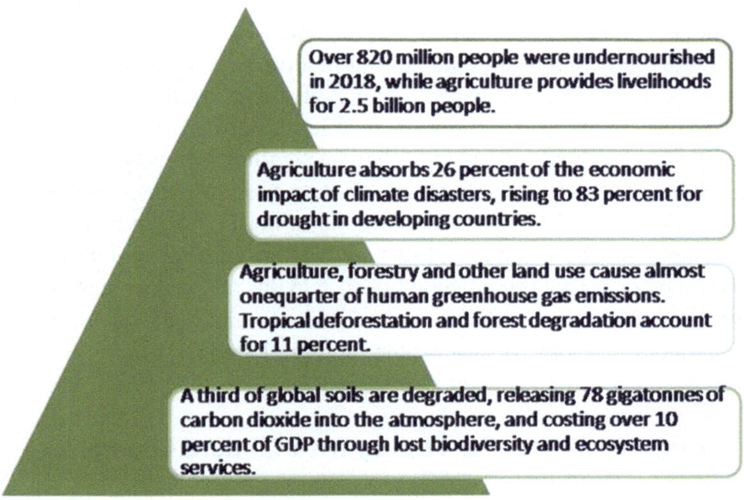

Fig. 1.2 Some of the climate change-driven challenges for agriculture and food systems (FAO 2020)

challenges that must be addressed in the future (Vermeulen et al. 2013; Raza et al. 2019). One of the most important challenges is to identify and assess "options and systems" that will ensure the production of food using sustainable approaches in a cost-effective, environment friendly, and feasible routes. Figure 1.2 has enlisted some of the climate change-driven challenges for agriculture and food systems (FAO 2020).

1.2 Climate Change and Food Security

The major factors affecting food production are occurrence of severe salinity, drought and temperature stresses and environmental pollutants. Furthermore, issues related to crop, livestock and fisheries production, as well as global trade, population, and human behavior, all of which are economic and socioeconomic aspects of agricultural systems, are affected by climate change, which ultimately affects food security (Tirado et al. 2010; Oppenheimer et al. 2014; Campbell et al. 2016). Climate change-associated factors such as high levels of CO_2, methane, and N_2O, extreme temperature, varying precipitation patterns, and erratic weather occurrences can have a greater impact on the incidence of biotic and abiotic stresses which include pests, pathogens, weeds, water paucity, etc. (Myers et al. 2014). Climate change enhances the prevalence of pests and diseases which risk agricultural crop performance (Bebber et al. 2013) and likely to reduce food safety due to the higher rate of microbial growth as temperatures rise, particularly regarding fresh fruit and vegetable crops (Hammond et al. 2015).

The genetic resources, climatic factors, diseases, pollination, soil formation as well as water and nutrient cycling that are directly or indirectly related to the ecosystem, are very closely linked to food security (Cabell and Oelofse 2012). Oppenheimer et al. (2014) reported that an increase in temperature and environmental stresses caused a significant reduction in biodiversity and shifts in relationships within communities that threatened productivity and resilience of current food systems (Khoury et al. 2014). Thornton and Herrero (2015) suggested that interaction of plants with climate change will have a significant impact on managing the risk on sustainable food production and nutrition security which increase the efficiency of inputs and reduced food losses in agroforestry. The interactive effects of climate change on food security and relevance of sustainable agricultural practices and mutation breeding are presented in Fig. 1.3. In this chapter, we have provided an overview of food security, climate change, and relevance of mutation breeding to improve agricultural crops for sustainable food production.

1.2.1 Traits/Responses for Coping with Climate Change

Climate change is a very complex process and consists of several factors, including changes in the concentration of greenhouse gases in the atmosphere, increasing temperatures, changes in precipitation patterns and increasing the frequency of severe weather events (Gray and Brady 2016). Plant responses to climate change can be studied from a developmental perspective. For example, the initiation of lateral roots in response to water scarcity can be considered as a developmental response (Babe et al. 2012; Gray and Brady 2016). Carbon dioxide (CO_2) is one of the gases in the atmosphere that is increasing and affects plant development and morphology both directly by affecting the process of photosynthesis and indirectly through the production of greenhouse gases that cause global warming induced changes in plants (Ainsworth and Long 2005; Meehl et al. 2007). One of the

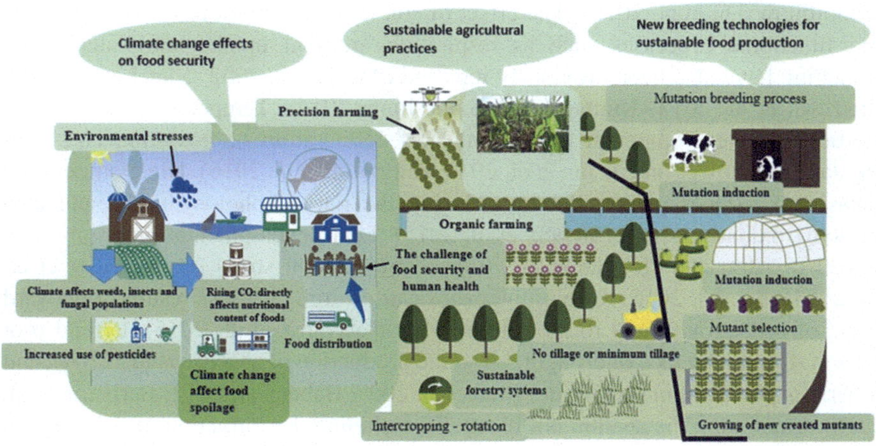

Fig. 1.3 Consequences of climate change on food security and using mutation breeding technique along with sustainable agricultural practices to help improve food security for humans

important developmental responses of plants to the enhanced level of CO_2 in the environment is increasing plant fresh and dry weight (Hatfield et al. 2011; Madhu and Hatfield 2013; Bishop et al. 2014; Gray et al. 2016). Kühn et al. (2021) analyzed several global studies and found eight plant functional traits of plant performance which can be considered significant to thrive under conditions of climate change. These include, specific leaf area, plant height, water-use efficiency, resprouting ability, growth rate, clonality/bud banks/below-ground storage, wood density, and rooting depth (Kühn et al. 2021). Such information on plant traits and associated attributes will help in predictive pre-breeding decision(s) to identify and select varieties/mutants that exhibit resilience or susceptibility to climatic change-associated impact (Garnier et al. 2016).

Plants show different physiological, biochemical, and molecular responses to increasing CO_2 level in the atmosphere, depending on whether they are plant species with C3, C4, and crassulacean acid metabolic (CAM) pathways. A noticeable enhancement in photosynthesis can be used as an important strategic adaptation to alleviate the global rise in atmospheric CO_2 (Reddy et al. 2010). However, the mechanisms related to the developmental processes in plants that interact with CO_2 are still unclear and will need to be addressed in the future. In addition to causing physiological and developmental changes in plants, temperature also affects their yield (Hatfield et al. 2011). Enhanced temperature could change plant physiology via modification of photosynthetic carbon assimilation in which the enzyme responsible for CO_2 fixation from the atmosphere into organic compounds (Rubisco) in the plants (Crafts-Brandner and Salvucci 2000; Sage et al. 2008; Raza et al. 2019). Developmental processes in leaf such as leaf initiation ratio, leaf expansion as well as duration of expansion are also influenced by temperature (Granier et al. 2002; Du et al. 2018). Leaf initiation requires auxin-induced expression of a KNOX gene (KNOTTED-LIKE HOMEOBOX transcription factor) in the incipient leaf primordium (Du et al. 2018). The effect of soil temperature on root developmental and functional processes such as respiration, nutrient uptake, and lateral root branching have also been previously reported (Awal et al. 2003; Nagel et al. 2009; Hanzawa et al. 2013; Kumagai and Sameshima 2014). Temperature also affects reproductive developmental process such as acceleration of flowering and even regulation of flower time (Zinn et al. 2010; Burghardt et al. 2016). High temperature stress also reduces plant yield by reducing the rate of photosynthesis and also inducing oxidative stress (Siebers et al. 2015). Leaf developmental programs such as cell division and elongation are affected by drought stress (Baerenfaller et al. 2012; Clauw et al. 2015). One of the major causes drought stress results downward angle of root growth and role of auxin (Rellán-Álvarez et al. 2015). Lateral root growth stops in the face of drought due to the production of abscisic acid (ABA), and suberin is produced in root tissues (Seo et al. 2009; Babe et al. 2012; Barberon et al. 2016). Changes in the anatomy of root cells, including the formation of aerenchyma as well as reduced levels of cortical cells, is associated with an improved plant performance when exposed to drought leading to maintain optimal root growth in the absence of water (Lynch 2015). Reproductive development in the plants is also the stage that is most sensitive to drought stress (Saini and Westgate 1999; Ma et al. 2014). Su et al. (2013) asserted that reproductive development responses of plants to drought include early detention of floral development. They reported that water stress during reproductive development leads to abnormal anther phenotypes. Tolerance to some abiotic stresses such as drought and heat during reproductive development is very critical for high yield potential and better yield stability of wheat and soybean under climate change (Stratonovitch and Semenov 2015; Ma et al. 2017; Tricker et al. 2018; Senapati et al. 2019; Du et al. 2020). This necessitates that further research is needed in the focus on inducing desirable variability for novel traits and trait associations for plant responses to climate change-driven stresses.

1.3 Induced Genetic Variation

Plant genetic diversity, both natural and induced, has contributed to improvement of crop plants. Natural plant populations have rich genetic diversity, but it is opined that the process of domestication has rather reduced genetic diversity relative to the wild ancestor(s) across the genome, and severely reduced genetic diversity for genes targeted by domestication (Flint-Garcia 2013). Moreover, it is possible that domestication has opted for favorable haplotypes for selected genes leading to low genetic diversity (Smýkal et al. 2018). In such a scenario, these populations can be analyzed to study and/or induce genetic diversity to sustain the impeding climate change and environmental stresses. Studies have shown that microevolutionary factors like mutation, genetic drift, and gene flow can influence existing genetic variation in plant populations over generations, and that this could configure genetic responses to climate change (Fu et al. 2019). The new breeding approaches in the past several decades have contributed to the induction of novel genetic variability, albeit in a random mode, through tools of in vitro-induced variation and mutagenesis

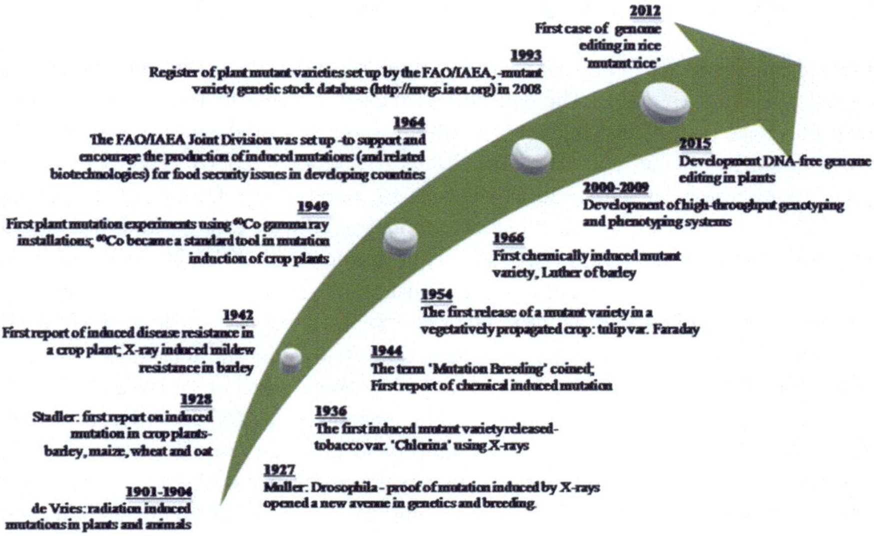

Fig. 1.4 Development of induced mutagenesis technology-chronological view. (Modified after Suprasanna et al. 2015)

(Songstad et al. 2017). It has been viewed that the generation of climate-smart varieties will require novel plant breeding technologies including induced mutations; biotechnological interventions and plant genetic engineering (Ahloowalia et al. 2004; Shu 2009; Suprasanna et al. 2021), and such genetically enhanced germplasm should contribute to climate-smart agriculture.

The use of mutation techniques by plant breeders has a long history, dating back to 1920, which peaked from 1950 to 1960 (Oladosu et al. 2016). Several experimental advances have emerged in the area of induced mutagenesis (Fig. 1.4). During this period, a significant number of mutants have been obtained, and the chronological evolution of the advances is shown in Fig. 1.4. New tools mutation induction such as, ion beam mutagenesis has been successfully employed for the development of 70 new crop varieties in Japan and China (Abe 2021). Mutation breeding also had a great impact on the economies of countries to maintain genetic diversity by providing new alleles to the genome and preventing gene erosion caused by the replacement of local varieties by modern varieties, environmental degradation, urbanization, and land clearing through deforestation and brush fires (Oerke 2006; Kozjak and Meglic 2012; Ahuja and Jain 2015). Some notable examples include, highly fertile aromatic rice mutants, lettuce mutants having low polyphenol oxidase, peanut mutants lacking major allergens, good shelf-life mandarin oranges, cherry mutants without cold vernalization, high oil productivity mutants of microalgae, mosses with high bioremediation ability for heavy metals.

1.3.1 Mutation Breeding for Sustainable Food Production

Induced mutations are being efficiently used to create genetic variability to improve various traits in crop plants. The data collected clearly shows that induced mutations have played a very significant role in solving food and food security problems in many countries (Kharkwal and Shu 2009; FAO/IAEA 2020). The purpose of inducing mutations in agricultural crops is to increase the frequency of genetic variability in order to more easily select plant species with desired traits for use in plant breeding programs (Jain 2010, 2012; Suprasanna et al. 2012, 2014; Suprasanna and Jain 2021). Mutation breeding offers considerable scope for sustainable agricultural development and food security (Jain and Suprasanna 2011; Mir et al. 2020). It can help to further increase yields, while addressing the shortcomings of green revolution technologies in landscape, water resources and environment towards agrochemicals, agricultural practices, and management (Aguilar-Rivera et al. 2019). Mutation breeding has also the ability to enhance crop diversity, provide better resistance to pests and diseases, increase nutrient use efficiency, make crops more resilient to climate shocks, and improve nutritional quality (Jain and Maluszynski 2004; Jain 2010; Suprasanna et al. 2017; Suprasanna and Jain 2017). Therefore, the problems of yield gap, low vigor, narrow gene pool, complex genomes, reduced fertility, and the lengthy breeding/selection cycle could very well be addressed using induced mutations (Sharma and Sharma 2004; Suprasanna et al. 2012).

According to FAO report (FAO 2019), more than 800 million people worldwide are in dire need of food. Considering that fertile land and water are becoming scarce, fundamental strategic changes, and novel breeding technologies will have to be implemented to address problems facing agriculture and food production systems (Jain 2010; Cai et al. 2017; Meemken and Qaim 2018; FAO 2019; Qaim 2020). Over the past century, plant breeding has played a major role in enhancing food production (Evenson and Gollin 2003). However, climate change is having negative impact on sustainable food production as evident in Asian and African countries mostly dependent on agriculture and livestock (Jain 2010; Wheeler and von Braun 2013). Qaim (2020) suggested that, sustainable agriculture and food security cannot be realized without new breeding technologies (NBTs) as these could contribute to higher crop yields, lower use of chemical fertilizers and pesticides, better crop resilience to climate stress, reduced postharvest losses, and more nutritious foods (Bailey-Serres et al. 2019; Eshed and Lippman 2019; Zaidi et al. 2019). The NBTs offer considerable scope for transforming crop improvement with more precision and resolution and, developing highly productive and improved crop varieties for achieving food security under climate resilience (Suprasanna et al. 2021).

Mutagenesis introduces new, novel variation into crop plants for use in plant breeding (Holme et al. 2019), and the mutant germplasm developed in the past few decades is still part of the gene pool used in modern plant breeding (Bradshaw 2017). Irradiation and treatment with chemical mutagens are the two major methods used to induce mutations in plants (Mba et al. 2011). X-rays and gamma radiation cause a mixture of bigger chromosome deletions and point mutations, i.e., single base

Table 1.1 Some of the major plant varieties developed by the use of induced mutations (FAO/IAEA 2020)

Variety name	Latin name	Common name	Country	Registration
IIyou 623 and IIyou 673	*Oryza sativa* L.	Rice	China	2007 and 2008, respectively
ABHIMANYU	*Bougainvillea* sp.	Paper flower	India	2010
AL-BEELY	*Musa* sp.	Banana	Sudan	2007
ALDAMLA	*Prunus avium* L.	Cherry	Turkey	2014
Albisoara	*Glycine max* L.	Soybean	Republic of Moldova	2010
Ana Delia	*Hibiscus* sp.	Hibiscus	Cuba	2013
ARTIpurple and ARTIqueen	*Dendranthema grandiflorum* (Ramat.) Kitamura	Chrysanthemum	Republic of Korea	2011
Beinong 103	*Glycine max* L.	Soybean	China	2009
Binadhan-19	*Oryza sativa* L.	Rice	Bangladesh	2017
Binachinabadam-7	*Arachis hypogaea* L.	Groundnut	Bangladesh	2014
Binamasur-11	*Lens culinaris* Medik.	Lentil	Bangladesh	2017
Binagom-1	*Triticum aestivum* L.	Wheat	Bangladesh	2016
Binasarisha-9	*Brassica napus* L.	Rapeseed	Bangladesh	2013
Binamoog-9	*Vigna radiata* (L.) Wil.	Mung bean	Bangladesh	2017
Binapiaz-1 and 2	*Allium cepa* L.	Onion	Bangladesh	2018
Binatomato-13	*Lycopersicon esculentum* M.	Tomato	Bangladesh	2018
Binatil-4	*Sesamum indicum* L.	Sesame	Bangladesh	2016
Binasola-10	*Cicer arietinum* L.	Chickpea	Bangladesh	2016
BURAK	*Prunus avium* L.	Sweet cherry	Turkey	2014

substitutions or deletions, whereas the most commonly used chemical mutagens (e.g., NaN3, EMS, MNU) almost exclusively cause single base substitutions (transitive, i.e., from G/C to A/T; see Table 1.1). Selection of stable mutants with good agronomic traits in a short period of time is highly beneficial for developing new improved plant varieties. However, one of the benefits of mutation breeding is that many mutants for different traits can be identified and can be screened under conditions of climate change (Jain 2010). Jain et al. (2010) considered nuclear technology is more cost-effective than any new technique. In vitro mutagenesis enhances genetic variability which is highly beneficial in plant breeding (Suprasanna et al. 2012; Sarsu et al. 2018). The rate of spontaneous mutations is very low and can't be used by breeders for crop improvement. Hence it is necessary to enhance mutation rate with physical and chemical mutagen treatment for generating genetic diversity to utilize in plant breeding programs (Jain 2010, 2012). The advantage of mutation breeding depends on efficient selection of premier variants in the second or third or further generation (Priyadarshan 2019). Mutated plants can be caused by errors that occur internally in cellular function or as a result of either external factors

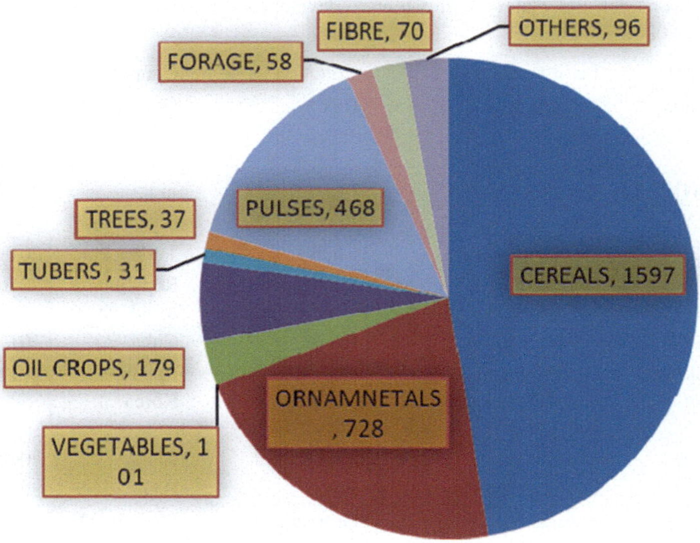

Fig. 1.5 Distribution of officially released mutant crop varieties (IAEA mutant database 2021)

Fig. 1.6 Officially released mutant varieties—major cereals (IAEA mutant database 2021)

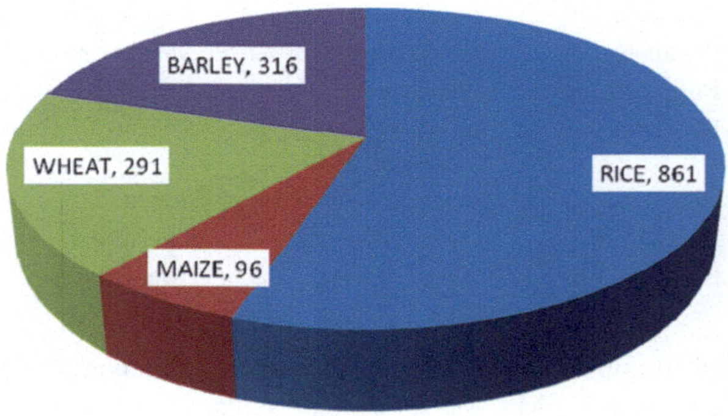

such as ultraviolet radiation or chemical compounds. As a result of these mutations, very important traits include increased crop yield, ease of harvesting, and so on can be selected. Figures 1.5 and 1.6 shows the success of mutation breeding in different agricultural crops and in major cereals (IAEA mutant database 2021).

Using mutations, traits for yield, quality, taste and disease, and pest resistance have been improved in several horticultural crops such as legumes, peppermint, peanut, citrus, banana, and cassava. Mutation-breeding is generally suitable in species with a narrow genetic base or those that are recalcitrant to traditional breeding methods (Jain and Suprasanna 2011; Suprasanna et al. 2014; Krasileva et al. 2017). Breeding of many agricultural crops by mutation induction has been widely used successfully in most species (Suprasanna et al. 2015). According to the database of Food and Agriculture Organization of the United Nations (FAO)/IAEA, many plant species of agricultural crops and ornamentals (more than 3365 mutant varieties) such as wheat, rice, grapefruit, rapeseed, sunflower, cotton, and banana have been included in breeding programs through induction of mutations (Jain 2005; Jain 2010; Suprasanna et al. 2015; Suprasanna and Jain 2021). These mutant varieties have also been categorized based on the basis of their development viz. (1) direct use of mutant lines, (2) indirect use of mutant lines, (3) use of a mutant gene allele (trait), and (4) use of wild species' genes translocated into plant genome via irradiation-enabled translocations (Oladosu et al. 2016).

New mutant germplasm generation must be continued in crop improvement programs in order to better cope with environmental stress conditions in the era of climate change (Suprasanna et al. 2014). Among all agricultural crops, the highest number of mutants is produced in rice, followed by barley, wheat, maize, durum, wheat, oat, millet, sorghum, and rye. By induced mutations, useful mutants have been produced in *Solanaceae* family plants with traits including high yields, increased vitamin C, and also higher antioxidant capacity as well as increased resistance to stressful conditions (Ozturk et al. 2012; Tomlekova et al. 2009; FAO/IAEA 2020). In vegetable crops, new mutants of pepper, tomato, potato, cucumber, eggplant, lettuce, onion, radish, and watermelon enhanced total production (Maluszynski et al. 2009; Anonymous 2013). Successful breeding programs by inducing mutations in ornamental plants also include breeding traits such as increased tolerance against abiotic stresses (Datta 2014). Among the tropical ornamentals, mutation breeding has been successfully attempted mainly in orchids, red ginger, torch ginger, and Anthurium (Sheela and Sheena 2014).

Induced mutations have contributed to global food security by increasing global food production, yield potential, and stability. A large number of mutant varieties in rice, banana, wheat, and cotton have been released in developing countries (Table 1.1). These varieties have been cultivated over hundreds of millions of ha of agricultural land, while the impact on national economies of these countries is measured based on billions of dollars (Jain 2005, 2010; Shu 2009; Hertel et al. 2010; Rosegrant 2011; Mba 2013; Suprasanna and Jain 2017; Priyadarshan 2019; FAO/IAEA 2020; Mir et al. 2020; Sarsu et al. 2021). Successful impact of mutation breeding includes the case of rice in China, Thailand, Vietnam, and the USA; barley in European countries and Peru; durum wheat in Bulgaria and Italy; wheat in China; soybean in China and Vietnam; and some other food legumes in India and Pakistan. Accordingly, mutant crop varieties continue to contribute significantly to addressing the food and nutritional securities of many countries in different continents across the globe (Fig. 1.7) especially in view of the potentials for harnessing novel traits in enhancing the adaptabilities of crops to climate change and variations (Mba 2013). China covers more than 30 million hectares of mutant cultivars and earns US$ 4.9 billion to boost the country's economy. In India and Bangladesh, 80% of the area under rice cultivation is being covered by the mutant varieties. In Indonesia, an approximate amount of US$ 2 billion has been received from a single top rice variety. In Peru, mutant varieties of barley have enabled farmers to earn up to seven million anodes, providing them with food and livelihoods through food security and economic benefits. In Vietnam, mutant varieties of rice and soybean helping poor farmers to improve their livelihood and a top rive mutant cultivar

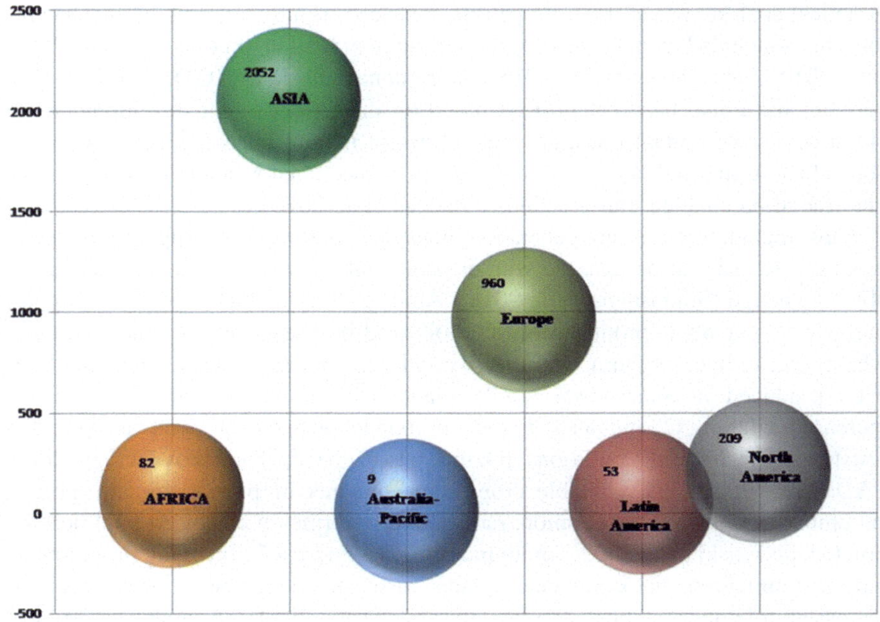

Fig. 1.7 Development of mutant varieties in different continents

earning US$ 3.3 billion with an increase of US$ 537.6 million over old varieties. While soybean mutant varieties bring about US$ 3 billion with 3.5 million farmers get a 30% increase in the economy. In Pakistan, 43 mutant varieties showed an economic impact with earnings of US dollars 6 billion during 2018 (Mir et al. 2020). In India, mutation breeding of groundnut has contributed through some notable high yielding cultivars like TG 39, TPG 41, TLG 45, TG 47, TAG 24, TG 37A, and TG 51 which have gained popularity and cultivation in several groundnut growing regions of the country.

Several successful examples of the effect of mutated genes on commercial crops have been reported (Table 1.1) which have significantly increased the nutritional value of those crops as well as enhanced biotic and abiotic stress tolerance. Increasing the quality of protein in corn, low phytic acid (LPA) barley, rice, wheat, and soybean production and oilseeds with optimized fatty acid compositions are examples of successful applications of mutations in plant breeding programs (Shu 2009; Mba 2013). According to the FAO/IAEA database, 1825 mutants have been successfully created to improve agronomic and botanical traits; of which, 577 mutants (18%) are created for increase in yield, 321 mutants (10%) for better quality and nutritional content, 200 mutants (6%) for biotic, and 125 mutants (4%) for abiotic stress tolerance (Suprasanna et al. 2015) (Fig. 1.8).

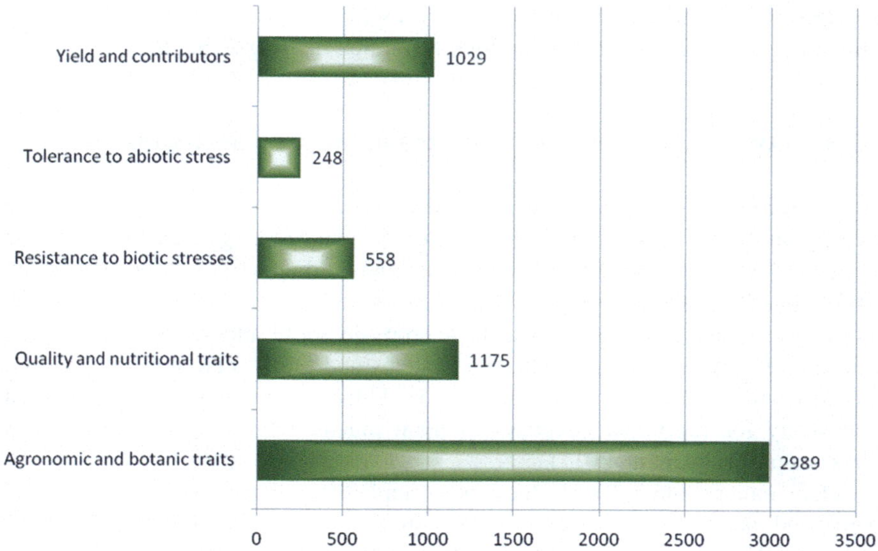

Fig. 1.8 Mutant varieties for major agronomic traits

1.3.2 Plant Mutant Resources

Plant mutant repositories serve as platforms for basic and applied research in crop improvement (Suprasanna and Jain 2017). Plant mutant libraries which are generally developed through physical or chemical mutagenesis are now available and are being maintained as mutant genetic resource hub for multitude of roles in functional genomics, gene mapping and understanding the nature of mutation, the physiological and metabolic alterations (Lundqvist 2009). Mutant libraries possessing high mutation density are desirable for studies on functional genomics and mutant gene discovery (Tsuda et al. 2015). For example, TOMATOMA (http://tomatoma.nbrp.jp/) is a tomato mutant database offers phenotypic data of ethylmethane sulfonate (EMS) γ-ray irradiation-derived tomato mutant lines (Shikata et al. 2015). EMS mutant libraries were developed in tobacco for a wide range of phenotypes and for the production of secondary metabolites. Jiao et al. (2016) established an EMS mutagenized large-scale genomics resource for functional genomics studies for drought tolerance in sorghum and observed that induced genetic variation was distinct from natural variation. Using EMS mutagenesis, a mutant resource was established in var. Nagina 22 for application in rice improvement (Amitha Mithra et al. 2016). In wheat, Guo et al. (2017) developed an EMS based-mutant resource with high-frequency phenotypic and genotypic variation for plant traits. Recently, 1504 FN-derived rice mutant lines were whole-genome sequenced and this resulted in the identification of 91,513 mutations affecting 32,307 genes (Li et al. 2017). In cotton, Lian et al. (2020) developed EMS mutant library for analyzing different mutant phenotypes for screening traits related to leaf, flower, fruit, fiber, and plant architecture. The mutant library was very useful to unravel EMS-induced site mutation using high-throughput and high-resolution melting technology.

1.3.3 Current Progress on Advances in Induced Mutagenesis

Crop breeding programs are more focused on achieving sustainability, nutritional security and local environmental fitness to help reduce energy demands for crop production while still providing adequate amounts of high-quality food (Fess et al. 2011; Mba 2013; Cowling et al. 2018; Roberts and Mattoo 2019). However, the currently available cultivars of most staple crops do not fit into this expected highly efficient yet low-input crop production systems and thus a new portfolio of crop varieties will need to be bred (Mba 2013). Induced mutations have been used extensively for functional genomics in plant mutation breeding of model crops (Jankowicz-Cieslak et al. 2017).

"Plant mutagenomics" tools have been applied to characterize the mutational events and mutant traits (Suprasanna and Jain 2017). Several high-throughput tools such as cDNA-amplified fragment length polymorphism, single strand conformational polymorphism, serial analysis of gene expression, microarray, differential display, TILLING, high-resolution melt analysis and next-generation sequencing have been used in the detection and molecular detailing of induced mutations in different crop plants (Gady et al. 2011; Li et al. 2017). Such mutagenomics investigations are useful in the analysis of gene functions using induced mutations emphasizing the utility of mutations induced by physical and chemical mutagens (Nawaz and Shu 2014). The techniques of forward and reverse genetics are now used in such analyses (Fig. 1.9).

Molecular techniques such as targeting induced local lesions in genomes (TILLING) are being used to detect specific sequential genomic changes in mutant plants (Mir et al. 2020). TILLING is used for the production of stable genetic stocks of mutant plant populations such as *Arabidopsis*, barley, soybean, tomato, and wheat. These strategies can be used for many years and screened repeatedly for different traits (Jankowicz-Cieslak et al. 2017; Suprasanna and Jain 2017; Taheri et al. 2017). In this regard, TILLING as a reverse genetics strategy can be beneficial to any species, regardless of genome size and ploidy level, to identify single nucleotide mutations in a specific region of a gene of interest with a powerful detection method resulted from chemical-induced mutagenesis, it could be a very promising technique (Taheri et al. 2017). However, due to the laborious and time-consuming process of TILLING and also slow development of suitable TILLING targets, another faster, simpler, and less expensive technique with non-enzymatic screening system known as high-resolution melting (HRM), has come up to assist TILLING in mutation detection (Taheri et al. 2017). Efficient procedures are required for the strict integration of induced mutations in breeding and functional genomics studies (Jankowicz-Cieslak et al. 2017). Szurman-Zubrzycka et al. (2018) developed a large TILLING population for barley referred to as HorTILLUS (Hordeum—TILLING—University of Silesia), following mutagenic treatment

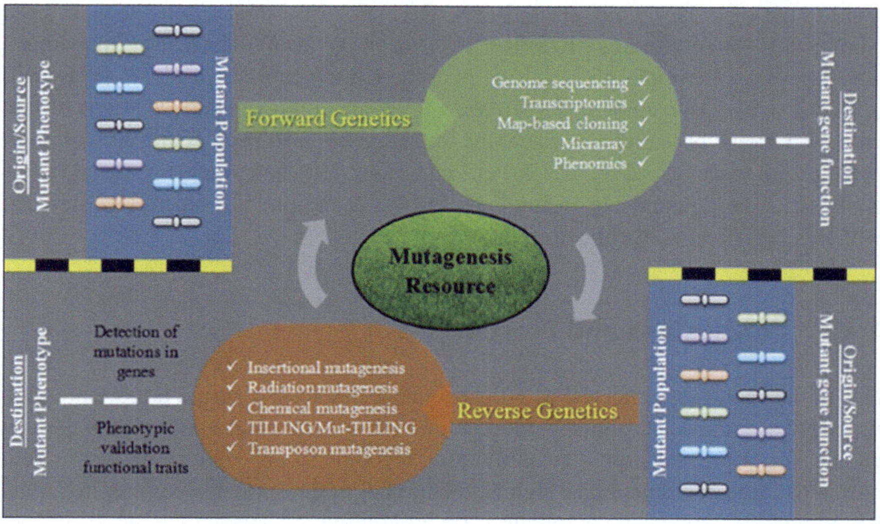

Fig. 1.9 A roadmap for mutagenomics studies in plants (Suprasanna and Jain 2017)

with sodium azide and *N*-methyl-*N*-nitrosourea. Authors suggested that this resource will be useful for studies on functional genomics and selection of barley mutants. T-DNA insertional mutagenesis projects are also very beneficial to support gene discovery projects. The mutagenesis-based T-DNA insertion may also reveal gene function via a gene knock-out and a gene knock-up (over-expression and miss-expression) or through expression patterns revealed by modified insertion elements (Pan et al. 2003). Discovery of mutant genes by using molecular and genomic characterization of induced mutations has successfully contributed to understanding the molecular nature of induced mutations and mutant traits. Further, genomic prediction for agronomic traits can be useful in the speedy selection of superior genotypes and step-up breeding gains for climate resilience in plants (Shirani Bidabadi et al. 2012a, b, 2021). In this regard, the advent of advanced genomics approaches and plant genome sequencing could benefit mutation discovery and molecular regulation of plant traits.

Biotechnological and genomics advancements have made the way for targeted modification of plant genomes and thus the new era of "genome editing" began (Zhang et al. 2018; Sharma and Vakhlu 2021) by using several tools such as ZFNs (zinc-finger nucleases), TALENs (transcription activator-like effector nucleases), and the CRISPR/Cas (clustered regularly interspaced short palindromic repeats) system. These molecular editors employ sequence-specific nucleases (SSNs) to recognize specific DNA sequence and create double-stranded breaks (DSBs). The DSBs are then repaired by the endogenous repairing machinery through addition or deletion of nucleotides resulting in mutants. Since the first report of genome editing of bacterial blight susceptibility gene OsSWEET14 in rice (Li et al. 2018), several plant species have been successfully modified to create "genome-edited plants" for a wide range of traits (Wada et al. 2020). Recently, new innovations such as base editing, DNA-free editing, CRISPR/Cpf1 system, and multiplex editing have been developed to extend the potentials of genome editing tool-kit in plants (Pramanik et al. 2021; El-Mounadi et al. 2020). Most traits may involve changes in one or few base alterations, base editing of adenine, cytosine has also been achieved with high editing efficiency (Mao et al. 2019). Therefore, the new tools of mutagenesis continue to be a promising technology for the development of novel germplasm with desired, precise modification in future.

1.4 Conclusions and Prospects

Due to the challenges imposed by climate change and food security, utmost attention is needed to establish sustainable food production systems and suitable technological choices for improving crop plants. Mutation breeding has become an important field in sustainable agriculture and a significant number of new mutants with desired traits have been developed and the mutant varieties continue to have a tremendous impact on improving crop productivity. A large majority of mutant varieties in cereals, pulses, oil, root and tuber crops, and ornamentals have been released across the globe contributing to significant economic impact. It could be predicted that induced mutagenesis will remain a useful tool for isolating the desired mutants and developing resistance to biotic and abiotic stresses in various crops as it is a rapid and relatively low-cost approach to create new alleles and phenotypes. Plant functional traits such as specific leaf area, plant height, water-use efficiency, resprouting ability, growth rate, clonality/bud banks/below-ground storage, wood density, and rooting depth are considered important for climate resilience and hence, these can be useful in pre-breeding decision(s) to identify and select varieties/mutants. In addition, mutation breeding is complemented by new advanced technologies to enable the detection of mutated alleles to create successful mutants, which in turn will elucidate gene function and improvement. The novel techniques in breeding new cultivars which are referred to as new breeding technologies (NBTs) such as genome editing and could allow for the introduction of targeted, specific mutations into the genomes for higher crop productivity and climate resilience. Plant mutation breeding technology will undoubtedly play a greater role in plant improvement for diverse functional traits for food security and sustainable agriculture in the coming years.

References

Abe T (2021) Ion beam mutagenesis for crop improvement. Plant Breed Genet Newsl 47:17–19

Aguilar-Rivera N, Michel-Cuello C, Cárdenas-González JF (2019) Green revolution and sustainable development. In: Leal Filho W (ed) Encyclopedia of sustainability in higher education. Springer, Cham. https://doi.org/10.1007/978-3-030-11352-0_87

Ahloowalia BS, Maluszynski M, Nichterlein K (2004) Global impact of mutation-derived varieties. Euphytica 135:187–204

Ahuja MR, Jain SM (2015) Genetic diversity and erosion in plants, vol 438. Springer International Publishing, Switzerland. https://doi.org/10.1007/978-3-319-25954-3

Ainsworth EA, Long SP (2005) Tansley review: what have we learned from 15 Years of Free-Air CO_2 Enrichment (FACE)? A meta-analytic review of the responses of photosynthesis, canopy properties and plant production to rising CO_2. New Phytol 165:351–371

Amitha Mithra SV, Kar MK, Mohapatra T, Robin S, Sarla N, Seshashayee M, Singh K, Singh AK, Singh NK, Sharma RP (2016) DBT propelled national effort in creating mutant resource for functional genomics in rice. Curr Sci 110:543–548

Anonymous (2013) FAO/IAEA mutant variety database. http://www-mvd.iaea.org/

Awal MA, Ikeda T, Itoh R (2003) The effect of soil temperature on source-sink economy in peanut (*Arachis hypogaea*). Environ Exp Bot 50:41–50

Babe A, Lavigne T, Severin JP, Nagel K, Walter A, Chaumont F, Batoko H, Beeckman T, Draye X (2012) Repression of early lateral root initiation events by transient water deficit in barley and maize. Philos Trans R Soc Lond B Biol Sci 367:1534–1541. https://doi.org/10.1098/rstb.2011.0240

Baerenfaller K, Massonnet C, Walsh S, Baginsky S, Bühlmann P, Hennig L, Hirsch-Hoffmann M, Howell KA, Kahlau S, Radziejwoski A, Russenberger D, Rutishauser D, Small I, Stekhoven D, Sulpice R, Svozil J, Wuyts N, Stitt M, Hilson P, Granier C, Gruissem W (2012) Systems-based analysis of Arabidopsis leaf growth reveals adaptation to water deficit. Mol Syst Biol 8:606. https://doi.org/10.1038/msb.2012.39

Bailey-Serres J, Parker JE, Ainsworth EA, Oldroyd GED, Schroeder JI (2019) Genetic strategies for improving crop yields. Nature 575:109–188

Barberon M, Vermeer JEM, DeBellis D, Wang P, Naseer S, Andersen TG, Humbel BM, Nawrath C, Takano J, Salt DE, Geldner N, Barberon M, Engelbertus J (2016) Adaptation of root function by nutrient-induced plasticity of endodermal differentiation article adaptation of root function by nutrient-induced plasticity of endodermal differentiation. Cell 164:447–459. https://doi.org/10.1016/j.cell.2015.12.021

Bebber DP, Ramotowski MAT, Gurr SJ (2013) Crop pests and pathogens move polewards in a warming world. Nat Clim Chang 3:985–988

Bishop KA, Betzelberger AM, Long SP, Ainsworth EA (2014) Is there potential to adapt soybean (*Glycine max* Merr.) to future [CO_2]? An analysis of the yield response of 18 genotypes in free-air CO_2 enrichment. Plant Cell Environ 38:1765–1774

Bradshaw JE (2017) Plant breeding: past, present and future. Euphytica 213:60. https://doi.org/10.1007/s10681-016-1815-y

Burghardt LT, Runcie DE, Wilczek AM, Cooper MD, Roe JL, Welch SM, Schmitt J (2016) Fluctuating, warm temperatures decrease the effect of a key floral repressor on flowering time in *Arabidopsis thaliana*. New Phytol 210:564–576. https://doi.org/10.1111/nph.13799

Cabell JF, Oelofse M (2012) An indicator framework for assessing agroecosystem resilience. Ecol Soc 17(1):18. https://doi.org/10.5751/ES-04666-170118

Cai Y, Golub AA, Hertel TW (2017) Agricultural research spending must increase in light of future uncertainties. Food Policy 70:71–83

Campbell BM, Vermeulen SJ, Aggarwal PK, Corner-Dolloff C, Girvetz E, Loboguerrero AM, Ramirez-Villegas J, Rosenstock T, Sebastian L, Thornton PK, Wollenberg E (2016) Reducing risks to food security from climate change. Glob Food Sec 11:34–43

Clauw P, Coppens F, DeBeuf K, Dhondt S, VanDaele T, Maleux K, Storme V, Clement L, Gonzalez N, Inzé D (2015) Leaf responses to mild drought stress in natural variants of *Arabidopsis*. Plant Physiol 167:800–816. https://doi.org/10.1104/pp.114.254284

Cowling WA, Li L, Siddique KHM, Banks RG, Kinghorn BP (2018) Modeling crop breeding for global food security during climate change. Food Energy Secur 8(2):e00157. https://doi.org/10.1002/fes3.157

Crafts-Brandner SJ, Salvucci ME (2000) Rubisco activase constrains the photosynthetic potential of leaves at high temperature and CO_2. Proc Natl Acad Sci U S A 97:13430–13435

Datta S (2014) Induced mutagenesis: basic knowledge for technological success. In: Tomlekova NB, Kozgar MI, Wani MR (eds) Mutagenesis: exploring genetic diversity of crops. Wageningen Academic, Netherlands, pp 97–139

Du F, Guan C, Jiao Y (2018) Molecular mechanisms of leaf morphogenesis. Mol Plant 11:1117–1134

Du Y, Zhao Q, Chen L, Yao X, Xie F (2020) Effect of drought stress at reproductive stages on growth and nitrogen metabolism in soybean. Agronomy 10:302. https://doi.org/10.3390/agronomy10020302

El-Mounadi K, Morales-Floriano ML, Garcia-Ruiz H (2020) Principles, applications, and biosafety of plant genome editing using CRISPR-Cas9. Front Plant Sci 11:56. https://doi.org/10.3389/fpls.2020.00056

Eshed Y, Lippman ZB (2019) Revolutions in agriculture chart a course for targeted breeding of old and new crops. Science 366:eaax0025. https://doi.org/10.1126/science.aax0025

Evenson RE, Gollin D (2003) Assessing the impact of the green revolution, 1960 to 2000. Science 300:758–762. https://doi.org/10.1126/science.1078710

FAO (2017) The future of food and agriculture—trends and challenges. Rome, Italy, FAO

FAO (2019) The State of Food and Agriculture 2019. Moving forward on food loss and waste reduction, Rome. Licence: CC BY-NC-SA 3.0 IGO

FAO-IAEA (2020). https://mvd.iaea.org/. Accessed 26 Aug 2020

Fedick SL, Santiago LS (2022) Large variation in availability of Maya food plant sources during ancient droughts. Proc Natl Acad Sci U S A 119(1):e2115657118. https://doi.org/10.1073/pnas.2115657118

Fess TL, Kotcon JB, Benedito VA (2011) Crop breeding for low input agriculture: a sustainable response to feed a growing world population. Sustainability 3:1742–1772. https://doi.org/10.3390/su3101742

Flint-Garcia SA (2013) Genetics and consequences of crop domestication. J Agric Food Chem 61: 8267–8276

Fu Y-B, Peterson GW, Horbach C, Konkin DJ, Beiles A, Nevo E (2019) Elevated mutation and selection in wild emmer wheat in response to 28 years of global warming. Proc Natl Acad Sci U S A 116(40):20002–20008

Gady ALF, Vriezen WH, Van de Wal MHBJ, Huang P, Bovy AG, Visser RGF, Bachem CWB (2011) Induced point mutations in the phytoene synthase 1 gene cause differences in carotenoid content during tomato fruit ripening. Mol Breed 29(3):801–812

Garnier E, Navas L, Grigulis K (2016) Plant functional diversity. Organism traits, community structure, and ecosystem properties. Oxford University Press, Oxford, UK

Granier C, Massonnet C, Turc O, Muller B, Chenu K, Tardieu F (2002) Individual leaf development in *Arabidopsis thaliana*: as table thermal-time-based programme. Ann Bot 89:595–604

Gray SB, Brady SM (2016) Plant developmental responses to climate change. Dev Biol 419:64–77

Gray SB, Siebers M, Locke AM, Rosenthal DR, Strellner RS, Paul RE, Klein SP, Ruiz-Vera UM, McGrath J, Dermody O, Ainsworth EA, Bernacchi CJ, Long SP, Ort DR, Leakey ADB (2016) Intensifying drought eliminates the expected benefits of elevated [CO_2] for soybean. Nat Plants. https://doi.org/10.1038/NPLANTS.2016.132

Guo H, Yan Z, Li X, Xie Y, Xiong H, Liu Y, Zhao L, Gu J, Zhao S, Liu L (2017) Development of a high-efficient mutation resource with phenotypic variation in hexaploid winter wheat and identification of novel alleles in the TaAGP.L-B1 gene. Front Plant Sci 8:1404

Hammond ST, Brown JH, Burger JR, Flanagan TP, Fristoe TS, Mercado-Silva N, Nekola JC, Okie JG (2015) Food spoilage, storage and transport: implications for a sustainable future. Bioscience 65(8):758–768

Hanzawa T, Shibasaki K, Numata T, Kawamura Y, Gaude T, Rahman A (2013) Cellular auxin homeostasis under high temperature is regulated through a SORTING NEXIN1–dependent endosomal trafficking pathway. Plant Cell 25:3424–3433. https://doi.org/10.1105/tpc.113.115881

Hatfield JL, Boote KJ, Kimball BA, Ziska LH, Izaurralde RC, Ort D, Thomson AM, Wolfe D (2011) Climate impacts on agriculture: implications for crop production. Agron J 103:351–370. https://doi.org/10.2134/agronj2010.0303

Hertel TW, Burke MB, Lobell DB (2010) The poverty implications of climate-induced crop yield changes by 2030. Glob Environ Change 20:577–585

Holme IB, Gregersen PL, Brinch-Pedersen H (2019) Induced genetic variation in crop plants by random or targeted mutagenesis: convergence and differences. Front Plant Sci 10:1468. https://doi.org/10.3389/fpls.2019.01468

IAEA mutant database (2021). https://mvd.iaea.org/

Jain SM (2005) Major mutation-assisted plant breeding programs supported by FAO/IAEA. Plant Cell Tissue Organ Cult 82:113–123

Jain SM (2010) Mutagenesis in crop improvement under the climate change. Rom Biotechnol Lett 15(2):88–106

Jain SM (2012) *In vitro* mutagenesis for improving date palm (*Phoenix dactylifera* L.). Emir J Food Agric 24(5):400–407

Jain SM, Maluszynski M (2004) Induced mutations and biotechnology on improving crops. In: Mujib A, Cho M, Predieri S, Banerjee S (eds) *In vitro* applications in crop improvement: recent progress. IBH-Oxford, India, pp 169–202

Jain SM, Suprasanna P (2011) Induced mutations for enhancing nutrition and food production. Gene Conserv 40:201–215

Jain SM, Ochatt SJ, Kulkarni VM, Predieri S (2010) In vitro culture for mutant development. In: Prakash J (ed) Proc. IVth IS on acclim. and establt. of micropropagated plants, Acta Hort, vol 865. ISHS, pp 59–67

Jankowicz-Cieslak J, Mba C, Till BJ (2017) Mutagenesis for crop breeding and functional genomics. In: Jankowicz-Cieslak J, Tai T, Kumlehn J, Till B (eds) Biotechnologies for plant mutation breeding. Springer, Cham, pp 3–18. https://doi.org/10.1007/978-3-319-45021-6_1

Jiao Y, Burke J, Chopra R, Burow G, Chen J, Wang B et al (2016) A sorghum mutant resource as an efficient platform for gene discovery in grasses. Plant Cell 28:1551–1562

Kharkwal MC, Shu QY (2009) The role of induced mutations in world food security. In: Shu QY (ed) Induced plant mutations in the genomics era. Food and Agriculture Organization of the United Nations, Rome, pp 33–38

Khoury CK, Bjorkman AD, Dempewolf H, Ramirez-Villegas J, Guarino L, Jarvis A, Rieseberg LH, Struik PC (2014) Increasing homogeneity in global food supplies and the implications for food security. Proc Natl Acad Sci U S A 111(11):4001–4006. https://doi.org/10.1073/pnas.1313490111

Kozjak P, Meglic V (2012) Mutagenesis in plant breeding for disease and pest resistance. In: Mishra R (ed) Mutagenesis. IntechOpen, Rijeka, Croatia. https://doi.org/10.5772/50332

Krasileva KV, Vasquez-Gross HA, Howell T, Bailey P, Paraiso F, Clissold L et al (2017) Uncovering hidden variation in polyploid wheat. Proc Natl Acad Sci U S A 114:E913–E921. https://doi.org/10.1073/pnas.1619268114

Kühn N, Carolina TI, Carretero J, Vandvik V, Enquist B, Willis K (2021) Globally important plant functional traits for coping with climate change. Front Biogeogr 13:e53774. https://doi.org/10.21425/F5FBG53774

Kumagai E, Sameshima R (2014) Genotypic differences in soybean yield responses to increasing temperature in a cool climate are related to maturity group. Agric For Meteorol 198:265–272

Li C, Zong Y, Wang Y, Jin S, Zhang D, Song Q, Zhang R, Cao C (2018) Expanded base editing in rice and wheat using a Cas9-adenosine deaminase fusion. Genome Biol 19:59. https://doi.org/10.1186/s13059-018-1443-z

Li G, Jain R, Chern M, Pham NT, Martin JA, Wei T, Schackwitz WS, Lipzen AM, Duong PQ, Jones KS, Jiang L, Ruan D, Bauer D, Peng Y, Barry KW, Schmutz J, Ronald PC (2017) The sequences of 1504 mutants in the model rice variety kitaake facilitate rapid functional genomic studies. Plant Cell 29(6):1218–1231

Lian X, Liu Y, Guo H, Fan Y, Wu J, Guo H, Jiao C, Tang Z, Zhang L, Fan Y, Gou Z, Zhang C, Li T, Zeng F (2020) Ethyl methanesulfonate mutant library construction in Gossypium hirsutum L. for allotetraploid functional genomics and germplasm innovation. Plant J 103:858–868

Lobell DB, Schlenker W, Costa-Roberts J (2011) Climate trends and global crop production since 1980. Science 333(6042):616–620. https://doi.org/10.1126/science.1204531

Lundqvist U (2009) Eighty years of Scandinavian barley mutation research and breeding. In: Shu QY (ed) Induced mutations in the genomic era. Food and Agriculture Organization of the United Nations, Rome, pp 39–43

Lynch JP (2015) Root phenes that reduce the metabolic costs of soil exploration: opportunities for 21st century agriculture. Plant Cell Environ 38:1775–1784. https://doi.org/10.1111/pce.12451

Ma X, Sukiran NL, Ma H, Su Z (2014) Moderate drought causes dramatic floral transcriptomic reprogramming to ensure successful reproductive development in Arabidopsis. BMC Plant Biol 14:164

Ma J, Li R, Wang H, Li D, Wang X, Zhang Y, Zhen W, Duan H, Yan G, Li Y (2017) Transcriptomics analyses reveal wheat responses to drought stress during reproductive stages under field conditions. Front Plant Sci 8:592. https://doi.org/10.3389/fpls.2017.00592

Madhu M, Hatfield JL (2013) Dynamics of plant root growth under increased atmospheric carbondioxide. Agron J 105:657–669. https://doi.org/10.2134/agronj2013.0018

Maluszynski M, Szarejko I, Bhatia CR, Nichterlein K, Lagoda PJL (2009) Methodologies for generating variability part 4: mutation techniques. In: Ceccarelli S, Guimaraes EP, Weltzien E (eds) Plant breeding and farmer participation. FAO, Rome, Italy, pp 159–194

Mao Y, Botella JR, Liu Y, Zhu JK (2019) Gene editing in plants: progress and challenges. Natl Sci Rev 6:421–437

Mba C (2013) Induced mutations unleash the potentials of plant genetic resources for food and agriculture. Agronomy 3:200–231. https://doi.org/10.3390/agronomy3010200

Mba C, Afza R, Shu QY (2011) Plant mutation breeding and biotechnology. In: Shu QY, Forster BP, Nakagawa H (eds) Mutagenic radiations: X-rays, ionizing particles and ultraviolet. CABI, Wallingford, pp 83–90

Meehl GA, Stocker TF, Collins WD, Friedlingstein P, Gaye AT, Gregory JM, Kitoh A, Knutti R, Murphy JM, Noda A, Raper SCB, Watterson IG, Weaver AJ, Zhao ZC (2007) Global climate projections. In: Solomon S, Qin D, Manning M, Chen Z, Marquis M, Averyt KB, Tignor M, Miller HL (eds) Climate change 2007: the physical science basis. Contribution of Working Group I to the fourth assessment report of the Intergovernmental Panel on Climate Change. Cambridge University Press, Cambridge, UK, pp 747–845

Meemken EM, Qaim M (2018) Organic agriculture, food security, and the environment. Ann Rev Resour Econ 10(1):39–63

Mir AS, Maria M, Muhammad S, Ali SM (2020) Potential of mutation breeding to sustain food security. IntechOpen, Rijeka, Croatia. https://doi.org/10.5772/intechopen.94087

Myers SS, Zanobetti A, Kloog I, Huybers P, Leakey ADB, Bloom AJ, Carlisle E, Dietterich LH, Fitzgerald G, Hasegawa T, Holbrook NM, Nelson RL, Ottman MJ, Raboy V, Sakai H, Sartor KA, Schwartz J, Seneweera S, Tausz M, Usui Y (2014) Increasing CO_2 threatens human nutrition. Nature 510:139–142

Nagel KA, Kastenholz B, Jahnke S, Van Dusschoten D, Aach T, Mühlich M, Truhn D, Scharr H, Terjung S, Walter A, Schurr U (2009) Temperature responses of roots: impact on growth, root system architecture and implications for phenotyping. Funct Plant Biol 36:947–959. https://doi.org/10.1071/FP09184

Nawaz Z, Shu Q (2014) Molecular nature of chemically and physically induced mutants in plants: a review. Plant Genet Resour 12:S74–S78

Oerke EC (2006) Crop losses to pests. J Agric Sci 144:31–43

Oladosu Y, Rafii MY, Abdullah N, Hussin G, Ramli A, Rahim HA, Miah G, Usman M (2016) Principle and application of plant mutagenesis in crop improvement: a review. Biotechnol Biotechnol Equip 30:1–16

Oppenheimer M, Campos M, Warren R, Birkmann J, Luber G, O'Neill B, Takahashi K (2014) Emergent risks and key vulnerabilities. In: Field CB, Barros VR, Dokken DJ, Mach KJ, Mastrandrea MD, Bilir TE, Chatterjee M, Ebi KL, Estrada YO, Genova RC, Girma B, Kissel ES, Levy AN, MacCracken S, Mastrandrea PR, White LL (eds) Climate change 2014: impacts, adaptation, and vulnerability. Part A: Global and sectoral aspects. Contribution of Working Group II to the fifth assessment report of the Intergovernmental Panel on Climate Change. Cambridge University Press, Cambridge, UK, pp 1039–1099

Ozturk S, Kantoglu O, Ozmen D, Ic E, Kantoglu KY (2012) Phytonutrient analysis of gamma rays mutated tomato. In: International conference nuclear science and its application abstracts book, September 25–28, Samarkand, Uzbekistan, pp 398–399

Pan X, Liu H, Clarke J, Jones J, Bevan M, Stein L (2003) ATIDB: Arabidopsis thaliana insertion database. Nucleic Acids Res 31:1245–1251

Pramanik D, Shelake RM, Kim MJ, Kim JY (2021) CRISPR-mediated engineering across the central dogma in plant biology for basic research and crop improvement. Mol Plant 14(1): 127–150

Priyadarshan PM (2019) Plant breeding: classical to modern. Springer Nature Singapore Pte Ltd, p 570. https://doi.org/10.1007/978-981-13-7095-3

Qaim M (2020) Role of new plant breeding technologies for food security and sustainable agricultural development. Appl Econ Perspect Policy 42:129–150. https://doi.org/10.1002/aepp.13044

Raza A, Razzaq A, Mehmood SS, Zou X, Zhang X, Lv Y, Xu J (2019) Impact of climate change on crops adaptation and strategies to tackle its outcome: a review. Plants 8:34. https://doi.org/10.3390/plants8020034

Reddy AR, Rasineni GK, Raghavendra AS (2010) The impact of global elevated CO_2 concentration on photosynthesis and plant productivity. Curr Sci 99(1):46–57

Rellán-Álvarez R, Lobet G, Lindner H, Pradier PL, Sebastian J, Yee MC, Geng Y, Trontin C, Larue T, Schrager-Lavelle A, Haney CH, Nieu R, Maloof J, Vogel JP, Dinneny JR (2015) GLO-Roots: an imaging platform enabling multidimensional characterization of soil-grown root systems. eLife 4:1–26. https://doi.org/10.7554/eLife.07597

Roberts DP, Mattoo AK (2019) Sustainable crop production systems and human nutrition. Front Sustain Food Syst 3:72. https://doi.org/10.3389/fsufs.2019.00072

Rosegrant MW (2011) Impacts of climate change on food security and livelihoods. In: Solh M, Saxena MC (eds) Food security and climate change in dry areas, Proceedings of an international conference, Amman, Jordan, 1–4 February 2010. International Center for Agricultural Research in the Dry Areas, Aleppo, Syria, pp 24–26

Sage RF, Way DA, Kubien DS (2008) Rubisco, Rubisco activase, and global climate change. J Exp Bot 59:1581–1595. https://doi.org/10.1093/jxb/ern053

Saini HS, Westgate ME (1999) Reproductive development in grain crops during drought. Adv Agron 68:59–96. https://doi.org/10.1016/S0065-2113(08)60843-3

Sarsu F, Spencer-Lopes MM, Sangwan R, Suprasanna P (2018) Specific techniques for increasing efficiency of mutation breeding a) *in vitro* methods in plant mutation breeding In: Spencer-Lopes MM, Forster BP, Jankuloski L (eds) Manual on mutation breeding, 3rd edn. FAO/IAEA, Rome, Italy, pp 205–226

Sarsu F, Kofi IB, Jankuloski L (2021) Contribution of induced mutation in crops to global food security. Aplicaciones Nucleares 12(22):2–11. https://doi.org/10.18272/aci.v12i3.2031

Senapati N, Stratonovitch P, Paul MJ, Semenov MA (2019) Drought tolerance during reproductive development is important for increasing wheat yield potential under climate change in Europe. J Exp Bot 70(9):2549–2560

Seo PJ, Xiang F, Qiao M, Park JY, Lee YN, Kim SG, Lee YH, Park WJ, Park CM (2009) The MYB96 transcription factor mediates abscisic acid signalling during drought stress response in *Arabidopsis*. Plant Physiol 151:275–289. https://doi.org/10.1104/pp.109.144220

Sharma B, Sharma SK (2004) Induced mutations and selection techniques for quantitative traits. In: Jain HK, Kharkwal MC (eds) Plant breeding. Springer, Dordrecht, pp 647–656. https://doi.org/10.1007/978-94-007-1040-5_27

Sharma S, Vakhlu J (2021) Evolution and biology of CRISPR system: a new era tool for genome editing in plants. Bot Rev 87:496–517

Sheela VL, Sheena A (2014) Novel trends and achievements in breeding of tropical ornamental crops especially orchids and anthuriums: the mutation breeding approach. In: Tomlekova NB, Kozgar MI, Wani MR (eds) Mutagenesis: exploring genetic diversity of crops. Wageningen Academic, Netherlands, pp 141–157

Shikata M, Hoshikawa K, Ariizumi T, Fukuda N, Yamazaki Y, Ezura H (2015) TOMATOMA update: phenotypic and metabolite information in the micro-tom mutant resource. Plant Cell Physiol 57(1):e11

Shirani Bidabadi S, Sariah M, Zakaria W, Subramaniam S, Maziah M (2012a) Induced mutations for enhancing variability of banana (Musa spp.) shoot tip cultures using ethyl methanesulphonate (EMS). Aust J Crop Sci 6(3):391–401

Shirani Bidabadi S, Sariah M, Zakaria W, Subramaniam S, Maziah M (2012b) *In vitro* selection and characterization of water stress tolerant lines among ethyl methanesulphonate (EMS) induced variants of banana (Musa spp., with AAA genome). Aust J Crop Sci 6(3):567–575

Shirani Bidabadi S, Sharifi P, Jain SM (2021, 2021) Plant breeding integrated with genomic-enabled prediction. OBM Genet 5(3). https://doi.org/10.21926/obm.genet.2103137

Shu QY (2009) A summary of the international symposium on induced mutations in plants. In: Shu QY (ed), Induced plant mutations in the genomics era. Food and Agriculture Organization of the United Nations, Rome, 15–18

Siebers MH, Yendrek CR, Drag D, Locke AM, Rios Acosta L, Leakey ADB, Ainsworth E, Bernacchi CJ, Ort DR (2015) Heat waves imposed during early pod development in soybean (*Glycine max*) cause significant yield loss despite a rapid recovery from oxidative stress. Glob Chang Biol 21(8):3114–3125. https://doi.org/10.1111/gcb.12935

Smýkal P, Nelson MN, Berger JD, Von Wettberg EJB (2018) The impact of genetic changes during crop domestication. Agronomy 8(7):119. https://doi.org/10.3390/agronomy8070119

Songstad DD, Petolino JF, Voytas DF, Reichert NA (2017) Genome editing of plants. Crit Rev Plant Sci. https://doi.org/10.1080/07352689.2017.1281663

Stratonovitch P, Semenov MA (2015) Heat tolerance around flowering in wheat identified as a key trait for increased yield potential in Europe under climate change. J Exp Bot 66:3599–3609

Su Z, Ma X, Guo H, Sukiran NL, Guo B, Assmann SM, Ma H (2013) Flower development under drought stress: morphological and transcriptomic analyses reveal acute responses and long-term acclimation in *Arabidopsis*. Plant Cell 25:3785–3807. https://doi.org/10.1105/tpc.113.115428

Suprasanna P, Jain SM (2017) Mutant resources and mutagenomics in crop plants. Emirates J Food Agric 29(9):651–657. https://doi.org/10.9755/ejfa.2017.v29.i9.86

Suprasanna P, Jain SM (2021) Biotechnology and induced mutations in ornamental plant improvement. Acta Hortic. https://doi.org/10.17660/ActaHortic.2022.1334.1

Suprasanna P, Jain SM, Ochatt SJ, Kulkarni VM, Predieri S (2012) Applications of *in vitro* techniques in mutation breeding of vegetatively propagated crops. In: Shu QY, Forster BP, Nakagawa H (eds) Plant mutation breeding and biotechnology. CABI Publishing, Wallingford, pp 371–385

Suprasanna P, Mirajkar SJ, Patade VY, Jain SM (2014) Induced mutagenesis for improving plant abiotic stress tolerance. In: Tomlekova NB, Kozgar MI, Wani MR (eds) Mutagenesis: exploring novel genes and pathways. Academic, Wageningen, pp 347–376

Suprasanna P, Mirajkar SJ, Bhagwat SG (2015) Induced mutations and crop improvement. In: Bahadur B, Rajam VM, Sahijram L, Krishnamurthy KV (Eds), Plant biology and biotechnology. Vol. I. Plant diversity, organization, function and improvement Springer, India, pp. 593–617

Suprasanna P, Ganapathi TR, Ghag SB, Jain SM (2017) Genetic modifications of horticultural plants by induced mutations and transgenic approach. Acta Hortic 1187:219–232. https://doi.org/10.17660/ActaHortic.2017.1187.22

Suprasanna P, Saddhe A, Ghuge SP, Ingle K (2021) New and novel genetic tools for improving crops. CAB Rev 16:028. https://doi.org/10.1079/PAVSNNR202116028

Szurman-Zubrzycka ME, Zbieszczyk J, Marzec M, Jelonek J, Chmielewska B, Kurowska MM, Krok M, Daszkowska-Golec A, Guzy-Wrobelska J, Gruszka D, Gajecka M, Gajewska P, Stolarek M, Tylec P, Sega P, Lip S, Kudełko M, Lorek M, Gorniak-Walas M, Malolepszy A, Podsiadlo N, Szyrajew KP, Keisa A, Mbambo Z, Todorowska E, Gaj M, Nita Z, Orlowska-Job W, Maluszynski M, Szarejko I (2018) HorTILLUS—a rich and renewable source of induced mutations for forward/reverse genetics and pre-breeding programs in barley (*Hordeum vulgare* L.). Front Plant Sci 9(216). https://doi.org/10.3389/fpls.2018.00216

Taheri S, Lee Abdullah T, Jain SM, Sahebi M, Azizi P (2017) TILLING, high-resolution melting (HRM), and next-generation sequencing (NGS) techniques in plant mutation breeding. Mol Breed 37:40. https://doi.org/10.1007/s11032-017-0643-7

Thornton PK, Herrero M (2015) Adapting to climate change in the mixed crop and livestock farming systems in sub-Saharan Africa. Nat Clim Chang 5:830–836

Tirado MC, Clarke R, Jaykus LA, McQuatters-Gollop A, Frank JM (2010) Climate change and food safety: a review. Food Res Int 43:1745–1765

Tomlekova NB, Timina OO, Timin OY (2009) Achievements and perspectives of sweet pepper breeding towards high beta-carotene. In: Krasteva L, Panayotov N (eds) IV Balkan symposium on vegetables and potatoes, Acta Horticulturae, vol 830, pp 205–212

Tricker PJ, ElHabti A, Schmidt J, Fleury D (2018) The physiological and genetic basis of combined drought and heat tolerance in wheat. J Exp Bot 69:3195–3210

Tsuda M, Kaga A, Anai T, Shimizu T, Sayama T, Takagi K, Machita K, Watanabe S, Nishimura M, Yamada N, Mori S, Sasaki H, Kanamori H, Katayose Y, Ishimoto M (2015) Construction of a high-density mutant library in soybean and development of a mutant retrieval method using amplicon sequencing. MC Genomics 16:1014

Vermeulen SJ, Challinor AJ, Thornton PK, Campbell BM, Eriyagama N, Vervoort JM, Kinyangi J, Jarvis A, Läderach P, Ramirez-Villegas J, Nicklin KJ, Hawkins E, Smith DR (2013) Addressing uncertainty in adaptation planning for agriculture. Proc Natl Acad Sci U S A 110(21): 8357–8362. https://doi.org/10.1073/pnas.1219441110

Wada N, Ueta R, Osakabe Y, Osakabe K (2020) Precision genome editing in plants: state-of-the-art in CRISPR/Cas9-based genome engineering. BMC Plant Biol 20:234. https://doi.org/10.1186/s12870-020-02385-5

Wheeler T, von Braun J (2013) Climate change impacts on global food security. Science 341(6145): 508–513

Zaidi SS, Vanderschuren H, Qaim M, Mahfouz MM, Kohli A, Mansoor S, Tester M (2019) New plant breeding technologies for food security. Science 363(6434):1390–1391

Zhang Y, Massel K, Godwin ID (2018) Applications and potential of genome editing in crop improvement. Genome Biol 19:210. https://doi.org/10.1186/s13059-018-1586-y

Zinn KE, Tunc-Ozdemir M, Harper JF (2010) Temperature stress and plant sexual reproduction: uncovering the weakest links. J Exp Bot 61:1959–1968. https://doi.org/10.1093/jxb/erq053

History of Plant Mutation Breeding and Global Impact of Mutant Varieties

2

M. C. Kharkwal

Abstract

Hugo de Vries coined the term 'Mutation' in his classical hypothesis known as 'Mutationtheorie'. Immediately after the discovery of mutagenic effects of X-rays on Drosophila by Muller and barley, maize by Stadler, about nine decades ago extensive experiments on induced mutations were initiated. In the three decades that followed the pioneering work of Muller and Stadler to understand the nature of mutations, induction techniques and their role in evolution, genetics and plant breeding, a great deal of work on basic aspects of induced mutation technique in understanding the mechanism of gene mutations, mode of action of physical and chemical mutagens was done world over. Several countries took up the task of crop improvement through the use of mutation technique in their classical breeding programmes as well as through molecular approaches. During 1950–1960, several countries took up the task of crop improvement through mutation breeding approaches, particularly after the establishment of the International Atomic Energy Agency (IAEA) which started coordinated programmes on the use of mutation breeding technique in a large number of crops in several countries of the world. Over 3500 mutant varieties belonging to >240 plant species including cereals, pulses, oilseeds, vegetables, fruits, fibres and ornamentals that have been developed and released by 2022 are evidence of the successful use of mutation technique in plant breeding. A wide range of characters including yield, flowering and maturity duration, plant architecture, quality and tolerance to biotic and abiotic stresses have been improved in the mutant varieties developed so far. As per the IAEA records, majority of these mutant varieties were developed and released as direct mutants, the rest were released through cross-breeding with mutants. Most of the mutant varieties have

M. C. Kharkwal (✉)
Division of Genetics, ICAR-Indian Agricultural Research Institute, New Delhi, India

S. Penna, S. M. Jain (eds.), *Mutation Breeding for Sustainable Food Production and Climate Resilience*, https://doi.org/10.1007/978-981-16-9720-3_2

been developed using physical mutagens, with gamma rays alone accounting for the development of majority of the mutant varieties. Since induced mutagenesis is gaining importance in plant molecular biology as a tool to identify and isolate genes and to study their structure and function, interest in mutation techniques and mutation breeding has increased recently in several area of biological research. These studies have an enormous potential for future crop improvement programmes. To redesign our crops by placing important traits on genetic maps and to equip them with the genes and attributes that could meet the huge food production challenges, there is an urgent need to use a combination of molecular and induced mutation techniques. Large-scale use of mutation breeding methods have made a significant contribution to the national economies of several countries.

Keywords

Mutation · Mutagen · Mutation breeding · Mutant variety · Gamma rays · X-rays · Crop improvement

2.1 Introduction

A heritable change in a genetic characteristic of an organism caused by mutation is a natural process that creates new variants (alleles) of genes. Mutation is the ultimate source of all genetic variations existing in any organism, including plants. The variation so created by mutation provides the raw material for natural selection and is a driving force in evolution. This variation created by mutation is further amplified by recombination of alleles on homologous chromosomes and their independent assortment. No mutations-no evolution is an established fact, as mutations create new genes and, in this way, provides the raw material for natural selection to act on. Natural selection operates to bring about evolution of new races and species through the variability created by spontaneous natural mutations and amplified by subsequent recombination of genes during sexual reproduction. Mutations arising spontaneously are random events in terms of the time of their occurrence and the gene in which they occur. In this way, mutant forms showing both large and small effects on the phenotype arise for all kinds of traits. Many of the lethal mutations may make the organism less adapted to its environment. Others may confer no immediate advantage but may help to generate a wide range of useful recombinant genotypes through the subsequent process of independent segregation and crossing over of genes. Besides natural mutations that occur spontaneously due to various kinds of radiations and cosmic rays received from the sun and also emitted by several radioactive elements on the earth, mutation can also be artificially induced by a number of physical agents like gamma rays and X-rays and several types of chemical agents known as chemical mutagens.

Historically, it was in 300 BC in China that the first story of mutants in crop plants was described in the book *Lula*. However, the first natural mutant plant in cereals was found about 2317 years ago in China (Van Harten 1998). The earliest recorded history of a practical example of dominant germinal mutation in domestic animals and its use in cross-breeding is traced back to 1791, when an English farmer Seth Write noted an unusually short-legged male sheep in his herd at his farm by the Charles River in Dover, Massachusetts in the USA. This mutant sheep was further used to develop a short-legged breed of sheep named 'Ancon' (Van Harten 1998). Charles Darwin (1859) in his historical book *Origin of Species* postulated that variations are created spontaneously in nature, and these naturally created variations are responsible for the development and origin of different species from a single common ancestor. Darwin reported that only those genotypes that create variations and adapt themselves to the environmental condition prevailing will alone survive, and he called this phenomenon as *survival of the fittest by natural selection*. He further reported that these variations are heritable from one generation to another. He called these factors that cause variation as *sports* or *bud variation* produced *spontaneously* by unknown causes that play a major role in speciation. Darwin did recognize that mutant forms arose suddenly and spontaneously, but he was doubtful about their role in evolution. Although Darwin originally used the word *transmutation*, he however, narrowly missed the discovery of the concept of *mutation* in his theories as he largely rejected *discontinuous variations* as significant factors in evolution.

The fact that natural selection was a necessary condition for evolution was recognized by Bateson (1894), but he also did not believe that it was a sufficient explanation, primarily because so little was known about the facts of heritable variation. Bateson published his monumental 598-page treatise *Materials for the Study of Variation, treated with special regard to Discontinuity in the Origin of Species* in 1894. Bateson's *Materials* treatise bears a curious resemblance to Darwin's Origin of Species. Bateson tended to assume that there was a one-to-one relationship between unitary genetic factors and their associated heritable traits. In 1905, Bateson coined the word 'genetics' to describe and establish a new scientific academic discipline.

The genetic behaviour of variegation in the flowers of *Antirrhinum*, the snapdragon was studied by Hugo de Vries (1889). Here the flowers are white or yellow, with red stripes, the variegation being recessive to the self red condition. On a variegated plant, the size of the red areas is variable, and sometimes includes an entire branch. He showed that the flowers on those red branches behave like the red flowers on a wholly red F_1, that is, on self-pollination they give 3 red:1 variegated. The results were not given a Mendelian interpretation because, at the time the experiments were done, de Vries did not know Mendelism. He did, however, suggest that some kind of segregation was occurring, so that self colour was somehow split off from the variegated element and that no true whites were produced.

2.2 Discovery of Mutations

While experimenting for rediscovery of Mendel's laws of inheritance, Hugo de Vries in the late nineteenth century found variation in evening primrose (*Oenothera lamarckiana*) that doesn't follow mendelian pattern of inheritance (3:1), but it was heritable. Mutation as a mechanism of creating variability was first identified by Hugo de Vries in 1901, and he considered them as heritable changes by mechanisms distinct from recombination and segregation. Credited with the discovery of mutations, de Vries (1901, 1903) described them as suddenly arising changes in the organisms, which were inherited and produced relatively large effects on the phenotype. He coined the term 'mutation' and presented an integrated concept concerning the occurrence of sudden, shock-like changes (leaps) of existing traits, which led to the origin of new species and variation. He also clearly included in this term mutations with 'small effects'. In his experiments with evening primrose, de Vries observed many aberrant types, which he called 'mutants'. The notion of mutation which was used by de Vries to indicate sudden genetic changes as a major cause of evolution quickly became established after the publication of his great 'Mutationstheorie' (1901–1903) in German. de Vries's work was soon translated in English and appeared as 'Species and Varieties, Their Origin by Mutation'. A great service was rendered by de Vries in 1900 in the Netherlands, along with Carl Correns (1900) in Germany and Erich Tshermak (1900) in Austria, as they simultaneously and independently re-discovered Mendel's long forgotten work. de Vries also had suggested that the new types of radiations like X-rays and gamma rays, discovered earlier by scientists like Conrad von Roentgen (1895), Bequerel Pierre and Marie Curie (1898) might be applied to induce mutations artificially. Because of the 'Mutation Theory of Evolution' based on the concept of mutations as the source of genetic variation and his early ideas about their potential use and value for plant breeding, de Vries's work around the turn of the century may be marked as the starting point of the history of mutation techniques in the discipline of plant breeding. The role of mutations in breeding was, however, first realized by H. J. Muller (1927) and Stadler (1928) through their X-rays experiments with fruit fly (*Drosophila melanogaster*) and maize and barley, respectively, for producing genetic changes.

In the early nineteenth century, there was little to be said about the cause of mutations except that they were rare, sudden, and discrete events that cause 'genes' as Johannsen (1909) called them to pass from one stable state to another. In his experiments with common bean, Johannsen (1913) described not only spontaneous drastic mutations, but also small mutations affecting the seed index (seed width versus seed length proportion). This is a character which closely falls into the class of continuous variation, and hence Johannsen may be regarded as the first who really proved the existence of spontaneous small mutations. Baur (1924) also emphasized repeatedly the importance of small mutations, which he called 'Kleinmutationen', majority of which exert only slight effects. The difference in Darwinian evolution in plant and animal breeding are not so different as the natural selection is substituted by human selection in the hybrid progeny in the course of breeding.

2.2.1 Muller's Discovery of Induction of Mutations on Drosophila

The early work of Thomas Hunt Morgan (1910, 1912) and C. B. Bridges (1916) on mutations on Drosophila, furnished several instances of the occurrence of new dominant genes, and many new sex-linked recessives, in pedigree material. These examples confirmed the conclusion that mutation occurs in a single gene in a single cell, and that it can occur at any stage of development. However, as these results were merely qualitative, since the frequencies were too low for a quantitative study and were also strongly influenced by the personal equation of the observer, an objective index was needed so that one would recognize a class of mutations that was frequent enough to give significant numerical values. Both of these requirements were met in the elegant mutation techniques devised by Muller for the study of newly arisen sex-linked lethals in Drosophila.

Muller's mutation technique depended on the study of the sex ratios from individual females that were heterozygous for sex-linked 'marker' genes. This technique made possible an objective and unambiguous determination of the frequency of the sex-linked lethal. Muller published a brief report in Science (Muller 1927) on the use of mutation technique to establish priority for his discovery of X-ray-induced mutagenesis on fruit fly (*Drosophila melanogaster*) before leaving for Berlin in Germany to attend the fifth International Congress of Genetics (ICG) to present the very first paper on X-ray-induced mutagenesis technique. It appeared with the sensational title 'Artificial transmutation of the gene'. This was an important discovery with far reaching implications both for genetics and for crop improvement. The first paper on the use of mutation technique and the discovery of induced mutagenesis presented at the International Congress of Genetics, Berlin was Muller's first comprehensive report on his X-ray work on fruit fly (*Drosophila melanogaster*). His pioneering work, which opened a new era in genetics and plant breeding, was greatly appreciated. Muller presented the elegant mutation techniques that he had developed for scoring mutations in Drosophila and in determining their rates. He reported that mutation rates of sex-linked recessive lethals could be greatly increased in Drosophila following treatment of sperms in male flies with high doses of X-rays. It was his technique, which showed that genes generally show a mutation rate of 10^{-5} to 10^{-6} per locus per generation. This means that one out of a 100,000 copies of the wild type gene can mutate in the course of one generation. He also introduced the concept of generation time while considering mutation rates in different organisms. Muller's discovery of induction of mutation and rates of mutations, a breakthrough in the history of genetics, has contributed enormously to our understanding of genes and their evolution.

2.2.1.1 *ClB*: An Elegant Technique

Muller (1930) then improved the mutation technique by using the 'ClB' chromosome that he found in his experiments. This is an X chromosome that carries a crossover reducer, a lethal, and the dominant mutant gene 'Bar'. As is well known now, this chromosome made it possible to detect new sex-linked lethals without anaesthetizing the flies or make counts—merely by rapid examination of individual culture bottles. Through this technique one could test many more chromosomes than possible ever before and thus could obtain adequate data on the lethal frequencies. Crosses were made in the earlier mutation experiments between flies containing a different collection of genes in their X-chromosomes. In most cases, only one parent was subjected to the X-ray treatment. The female offspring from these crosses would contain any new mutant gene that might have arisen in only one of their X-chromosomes, and hence they would manifest no abnormality if, as usual, the mutant gene were recessive.

A major difference between the control and treated series was soon observed when the offspring ('F_2') from the above cultures were examined. The controls showed a very low frequency of lethal mutations (1 in 947 fertile cultures), like that usually found, whereas the treated series showed a surprisingly high frequency of lethals (88 in 758 cultures), and all but three of these lethals were confined to the chromosomes derived from the treated progenitor. Similar results were found with regard to visible mutations and semi-lethals in these cultures though of course they were not as numerous as the lethals. In later generations ('F_3' and 'F_4'), only a small number of new mutations were found, not significantly more in the treated than in the control series. When Muller applied his 'ClB' technique to the study of gamete treated with X-rays, it was apparent that there was a marked increase in the frequency of newly arising lethals. This is the clear example of the use of mutation technique for artificial induction of mutations.

2.2.2 Stadler's Discovery of Induction of Mutations in Plant Systems

The discovery of mutagenic action of X-rays demonstrated in maize, barley and wheat by Lewis John Stadler in 1928 and 1930 started the use of mutation technique for generating novel genetic variability through radiations in plants also. Stadler had begun his studies with barley at about the same time that Muller's work with X-rays on Drosophila began, but since he was using an annual plant, his results were not available until after Muller's papers were published. His first paper did not furnish independent confirmation of Muller's result, on a different material studied with a different mutation technique. Although Stadler also published data on induced mutations in barley, he described them with much pessimism, stating that they would be of no use for plant breeding (Stadler 1928). By 1930, Stadler had obtained X-ray mutations by hundreds. Stadler reported the production of solitary mutations and an increase in lethality by X-rays. His experimental designs were elegant and critical, and the dimensions most impressive. Many geneticists believe the induction of mutation in plants as a breakthrough in the history of genetics and plant breeding.

2.3 Early Experiments with Induced Mutations

The exhaustive results of the mutation technique used by Muller and of Stadler at last opened the way to an extensive experimental study on induced mutations that is still being actively practiced. A great deal of work on basic aspects of induced mutation technique in understanding the mechanism of gene mutations, mode of action of physical and chemical mutagens was done world over to understand the nature of mutations, induction techniques and their role in evolution, genetics and plant breeding.

Hanson and Heys (1929) studied the induction of lethals in Drosophila by using radium mutation technique. They interposed lead shields of different thickness and recorded the ionization in each treatment. The curves for ionization and for mutation rate were superposable, leading to the conclusion that ionization is responsible for the mutations and that the relation is a simple, direct one—a 'one hit' phenomenon. Oliver (1930) confirmed this conclusion by showing that varying the dose by varying the duration of exposure to a constant X-ray source also gave a linear curve relating dosage to mutation rate. Muller showed that in the same year that the amount of background radiation is far less than would be required to produce the normal 'spontaneous' frequency of mutations. It came to be very generally accepted that the total ionization is all that need be considered in connection with radiation-induced mutations, at least within a strain.

Altenberg (1934) discovered the mutagenic effect of ultraviolet light (UV) through irradiation of the polar cap cells of fruit fly eggs. Stadler (1941) introduced a new mutagenic agent, ultraviolet light, in plants into his laboratory. The mutagenic potential of these rays has since been confirmed in many organisms in which germ tissue could be easily exposed to the low-penetrating UV. While the gamma radiation has become a very popular mutagen since the 1950s that has been used extensively to induce mutations, various forms of neutron had also been studied (Muller 1954) for their use in mutagenesis in the 1960–1970s though their application in induced mutagenesis has been limited.

2.3.1 Classic Examples of Early Application of Induced Mutations

The two major discoveries of the induction of mutations made by Muller (1927) in *Drosophila* and Stadler (1928) in maize and barley plants demonstrated that with the help of these physical mutagenic agents, it was possible to obtain mutation rates that were much higher than spontaneous rates. These discoveries lead to extensive works on induced mutations and showed the practical potential of radiation as a plant breeding tool and resulted almost immediately in the practical recovery of some economically useful mutants in wheat (Delaunay 1931; Sapehin 1930, 1936). Stubbe (1934) described 'small mutations' (now known as 'point mutations' or 'micromutations') for the first time in higher plants such as *Antirrhinum.* Knapp (1950) presented clear suggestions about how to utilize the micromutations in barley.

Tollenaar (1934) was the first worker to isolate a light green 'chlorina' mutant in tobacco which was released for commercial cultivation (Kharkwal 2012). The lecture delivered by Nilsson-Ehle on polymorphic factors in barley in Halle in the year 1939 opened a new era in the use of induced mutation in Germany. Timofeeff-Ressovsky (1941) reviewed his extensive work on X-ray mutagenesis techniques in Drosophila and his pioneering interpretation of mutagenicity data in terms of the classical target theory (Timofeeff-Ressovsky and Zimmer 1947; Zimmer 1961). Later, Freisleben and Lein (1942) reported the induction of mildew resistance in barley by X-irradiation. Mutation is not a directed process as is evident from the DNA structure which also suggested that mutations would occur in a random manner because basically they are unpredictable mistakes during the process of replication. Luria and Delbruck's mutation technique in 1943 provided convincing evidence in support of random, non-adaptive nature of mutations in an experiment. According to their mutation technique, they counted the number of mutant individuals showing drug resistance in different cultures of wild type bacteria.

2.3.2 Gustafsson Builds Up the Momentum for Mutation Breeding

Swedish research on induced mutations in barley started in 1928 on a small scale at Svalöf, Sweden, initiated by the eminent Swedish geneticists Herman Nilsson-Ehle and Åke Gustafsson. The first treatments with irradiation commenced using the Svalöf cultivar 'Gull', which was the most common barley cultivar grown in Sweden at that time (Lundqvist 2009, 2021). A major stimulus for much of later work in further practical refinement of mutation technique and the usefulness of mutation breeding in crop improvement was demonstrated first in Sweden by the classic paper of Gustafsson (1947) which demonstrated practical plant breeding of agricultural plants by means of X-rays and ultraviolet rays induced mutations. Swedish plant breeders found many distinct categories of chlorophyll mutations like *albina*, *viridis* and *xantha* in barley by the use of X-rays. These chlorophyll mutations were always the first indication of treatment success (Gustafsson 1938, 1940), and their abundance served as the standard method for measuring the induced mutagenic effects. In the mid-1930s, the first viable mutations appeared, and it was possible to distinguish two subgroups: 'Morphological' and 'Physiological' mutations. They discovered some morphological mutants characterized by dense heads, late maturity and very stiff taller straw and resembled the *erectum* barleys, in comparison with the normal spikes in most of the barley cultivars. These mutants named *Erectoides* were found to yield higher and produce more straw than the maternal variety. A very large number of mutations of barley were shown to respond in a variety of ways in different genotypes (Lundqvist 2014). A variety of barley called 'Pallas' developed from stiff strawed and early mutants of the variety 'Bonus' was released for commercial cultivation in Sweden (Gustafsson 1963). These mutants represented the first actual accomplishments of the production of superior varieties by the use of radiation. Similar useful induced mutants reported included stem rust resistance caused by *Puccinia graminis* in wheat (MacKey 1954) and in oats (Frey 1955) and dwarf mutants in rice (Beachell 1957). The development of dwarf wheat and rice varieties that led to the green revolution are classic examples of mutation breeding achieved through successful exploitation of the natural mutant genes *Rht-1* and *Rht-2—Norin* in case of wheat and *sd1—dee-gee-woo-gen* in rice, which effect a large constellation of characters responsible for their superior agronomic responses and are called Green Revolution Genes.

2.4 Techniques for Detection and Analysis of Induced Quantitative Variation

Gaul (1961, 1965) classified mutations into 'macro' and 'micromutations' and used a systematic approach for the selection of micromutations in barley. Swaminathan (1963, 1965) in India broadly grouped induced mutations into four major types according to the degree of change in the phenotype and method of detection: (1) macro mutations, (2) mutations affecting single characters, (3) systematic mutations, and (4) micromutations. While estimating the induced variability due to micromutations, an increase in variance irrespective of character, symmetrical or skewed, has been the general observation while the mean mostly remained unchanged and sometimes even decreased. The hypothesis forwarded by Brock (1965) to explain such behaviour of induced mutations in quantitatively inherited traits is that random mutations are expected to increase the variance and shift the mean away from the direction of previous selection history. However, Gaul and Aastveit (1966) concluded that the mutations for quantitative characters are not related to the genotype, and with random mutation a change in the mean value of almost any quantitative character is to be expected. As a consequence of the nature of phenotypic manifestation of quantitative characters, the only method available to detect the induction of new variation for quantitative traits following mutagenic treatment is the estimation of mean and variance by statistical methods. The mean values of quantitative traits in populations derived from irradiated gametes (pollen) or embryos (seeds) are in most instances lower in the treated than in untreated populations. In a very extensive study, Scossiroli et al. (1966) reported a similar effect in the same population for a large number of characters. The effect of radiation on the means has been interpreted to be due to detrimental mutations which can be removed through selection in subsequent generations.

2.5 Discovery of Chemical Mutagenesis

After establishing X-rays mutagenesis in both animals and plants, numerous new mutagens, both physical like gamma rays, alpha and beta particles, neutrons, protons, ultraviolet (UV) radiation and chemicals like ethyl methane sulphonate (EMS), ethylineimine (EI), *N*-nitroso-*N*-methyl urea (NMU), sodium azide (SA), etc. were found to be effective in generating genetic variability. The technique of induction of mutations by chemical agents was attempted by many people over a long period, but until 1941 there were no clear and convincing positive results until it became established that certain chemicals could produce mutations. Auerbach (1941) was the first to report that mustard gas had a mutagenic effect on *Drosophila*, which was similar to that of X-rays on plants. Auerbach and Robson (1946) later obtained clear evidence that mustard gas is mutagenic.

The chemical mutagens were found to be highly effective in inducing true gene mutations and this specificity of action could be analysed in terms of the mechanism of their reaction with different DNA bases. The question, whether chemical agents do indeed produce mutations with the same frequency as the physical mutagens, was settled after the first paper published by Auerbach and Robson (1946). They used the standard mutation technique devised by Muller to score recessive and visible gene mutations in Drosophila following exposures of flies to a predetermined dose of the gas. Their most important observation was that mustard gas is highly mutagenic and capable of producing lethal and visible mutations at rates, which were comparable with the effect of X-rays. Besides gene mutations, chromosomal aberrations in the form of deletions, inversions and translocations were produced. In these diverse ways, mustard gas was found to be highly mutagenic. Auerbach (1949) presented her discovery of the mutagenicity of the 'radiomimetic' alkylating agent mustard gas at the ICG held in Stockholm in 1948. Oehlker (1943), and Gustafsson and Mackey (1948) also proved that mustard gas was mutagenic in barley as well.

Rapoport (1946, 1948) in Russia also discovered and demonstrated mutagenic effects of mustard gas and several other chemicals such as formaldehyde, diethylsulphate (dES), diazomethane, and other compounds and established that alkylating agents are the most important group of chemical mutagens.

2.5.1 Mechanism of Gene Mutation

The future direction of work on chemical mutagens was determined mainly by rapid advances in the understanding of gene structure and function following the demonstration of DNA as the genetic material. Watson and Crick (1953) while proposing their model for the structure of DNA pointed out that both replication and mutation of genes can be understood in terms of the new structure. Based on chemical studies nitrous acid was expected to convert cytocine to uracil and adenine to hypoxanthine, two of the known analogues of the nucleic acid bases. The modified bases mispaired during replication of the treated DNA leading to mutation through transitions. Freese (1959) reported extensive studies on reverse mutations resulting in wild type phenotype in rII mutants by two analogues—5-bromouracil (BU) and 2-amino-purine (AP), found to be highly mutagenic in viruses of the T4 series through their action in inducing A:T to G:C transitions.

The simplest kinds of base pair changes are due to transitions and transversions, but they can result in a phenotypically visible mutation. There are no restrictions on the different kinds of sequence changes in the DNA of a gene following different types of misprints during replication. Addition or deletion of a nucleotide base pair is another common error, when one of the bases manages to pair with two bases or fails to pair at all. These kinds of sequence changes resulting in an alteration in the reading frame of the gene's DNA are known as frameshift mutations. Some of the mutations occur from rearrangement of bases in the DNA. A small or large sequence of bases may be inverted as a result of chromosome breakage, and reunion of the broken ends may involve different DNA molecules in a reciprocal rearrangement or in loss of a fragment. Duplication of a DNA sequence is another common mechanism for change in the structure of a gene leading to gene mutation.

2.6 Application of Mutation Technique in Crop Improvement

A great deal of work was done world over on basic aspects of induced mutation technique in understanding the mechanism of gene mutations, mode of action of physical and chemical mutagens to understand the nature of mutations, induction techniques and their role in evolution, genetics and plant breeding (Kharkwal 2012). By now extensive use of mutation breeding has already established itself as a powerful plant breeding tool in its own right and made a significant contribution to food and fibre production and to farmer's incomes and national economies worldwide. Several countries took up the task of crop improvement through the use of induced mutagenesis technique and mutation breeding approaches during 1950–1970 and reported spectacular accomplishments of mutation breeding technique in evolving several superior crop varieties. Coordinated programmes were initiated by IAEA in a large number of crops in several countries of the world on the use of mutation breeding technique.

Significant contributions in understanding basic and applied nature of mutation phenomenon and techniques on cereal and legume crops were successfully made in several countries like Sweden (Gustafsson 1947, 1967 on barley), Germany (Gaul 1961, 1965; Gaul et al. 1969 on barley; Gottashalk and Wolff 1983 on peas), the USA (Konzak 1954 and Frey 1955 on oats, and by Gregory 1955; Gregory et al. 1960 on peanuts, Konzak et al. 1984), and India (Swaminathan et al. 1968 on wheat; Patil et al. 1995 on groundnut, and Kharkwal et al. 1988, 2005; Kharkwal and Shu 2009, Kharkwal 1998b, c on chickpea, lentil, pea and cowpea). Gregory et al. (1960, 1965) produced the first groundnut mutant variety NC 4x, about 5% better in total production in the USA by exposing about 200 pounds (90 kg) of peanuts to high doses of X-rays and planted one million M_2 plants in 64 acres in the university's crop development fields. Mutant var. NC 4x possesses between 400% and 500% fewer cracked pods than NC 2 and was released in January 1959 in the USA.

2.6.1 Ultra Modern Techniques of Mutation Breeding in Crop Improvement

The discovery of structure of DNA by Watson and Crick (1953) had brought a fundamental change in our understanding of gene mutations—transitions, transversions, frameshift mutations and misrepair mutagenesis, particularly caused by chemical mutagens. Recent advances in molecular biology have led to the establishment of several ultra modern techniques, which have helped researchers to redefine the scope, nature and applications of mutagenesis techniques. Induced mutagenesis technique is gaining importance as a tool in the identification of plant genes using molecular approaches. The main important established techniques are: (a) in vitro mutagenesis, (b) double haploid (DH) techniques (c) transposon-mediated mutagenesis, (d) site-directed mutagenesis (SDM), (e) random mutagenesis, (f) insertional mutagenesis, (g) systematic mutational analysis, and (h) targeting induced local lesions IN genomes (TILLING) (Colbert et al. 2001). With the recent advances in genomics, it has been documented that the use of high-throughput platforms, such as TILLING (Targeting Induced Local Lesions in Genomes) and EMAIL (Endonucleolytic Mutation Analysis by Internal Labelling) (Cross et al. 2008) in the rapid evaluation of mutant stocks for specific genomic sequence alterations can be very much helpful in studying the genetic variability at molecular level. With the advancement of molecular techniques, the mutation breeding comes into the era that is called the 'Molecular Mutation Breeding' in which molecular or genomic information and tools are used in the development of breeding strategies, screening, selection and verification of induced mutants, and in the utilization of mutated genes in the breeding process (Shu 2009). The latest rapid, high-throughput method in this field 'Mutagenomics'—the discovery of genes using the combination of mutants and genome sequencing (Hodgens et al. 2020)—is becoming increasingly useful in screening for point mutations. Some of the recently introduced ultra modern innovative techniques presently used for induction and study of mutation breeding parameters are briefly listed and introduced below.

2.6.1.1 High Hydrostatic Pressure (HHP)

Defined as an extreme thermo-physical factor that affects the multiple cellular processes like synthesis of DNA, RNA, proteins, cell survival (Ishii et al. 2004) and high hydrostatic pressure (HHP) has been very effective in inducing mutagenesis in microorganisms. The usage of HHP, a cost-effective technique in mutation breeding for creation of new mutant varieties has started since last few years only and one of the examples is the creation of mutant varieties of rice (Zhang et al. 2013).

2.6.1.2 Ion Beam Technology (IBT)

Ion Beam Technology was found to show high relative biological effectiveness (RBE) as compared to low linear energy transfer (LET) radiations, such as gamma rays, X-rays and electrons (Feng et al. 2009). Mutation breeding with heavy ion beams is unique technology wherein heavy ion beams are generated by accelerating atomic ions using a particle accelerator. Ion beams have been used to develop approximately 70 new crop varieties in Japan in the last two decades, and beneficial mutants have been identified in a variety of species (Abe et al. 2007, 2012, 2015, 2021).

2.6.1.3 Space Breeding Technology (SBT)

Space breeding technology, in which the growth cycle of the seeds could be shortened (Li 2013), has now become a proven way and can be applicable in modern mutation breeding strategies. The two parameters in combination i.e., presence of cosmic rays and microgravity, affect the genetic diversity of the crops in space and thus they are the main causes of the changes in breeding new crop varieties (Mei et al. 1998; Gu and Shen 1989). China is the leader in the field of space breeding, having more than 66 new varieties developed in the space breeding programme (Liu et al. 2005, 2021; Liu 2021).

2.6.1.4 Targeting Induced Local Lesions IN Genomes (TILLING)

TILLING is a method that allows directed identification of the mutations in a very specific gene. It is one of the most high-throughput, non-transgenic reverse-genetic approaches, which combines mutagenesis with a sensitive DNA screening technique and enables the recovery of individuals carrying allelic variants of candidate genes (Sato et al. 2006). It is an efficient early-screening tool for specific point mutations in genes of interest from a small population and enables geneticists to analyse gene function and associate genotype with phenotype. TILLING combines traditional mutagenesis followed by high-throughput mutation discovery, which can improve the efficiency of using induced mutations and nucleotide polymorphism discovery methods for a reverse genetic strategy that is high in throughput, low in cost and applicable to most organisms, and in less than a decade will help to develop crops with improved traits (Colbert et al. 2001; Till et al. 2009).

2.6.1.5 Endonucleolytic Mutation Analysis by Internal Labelling (EMAIL)

EMAIL has been developed by Cross et al. (2008) for detecting rare mutations in specific genes in pooled samples using capillary electrophoresis. EMAIL, is an alternate approach to mismatch detection, in which amplicon labelling is achieved by incorporating fluorescently labelled deoxynucleotides. This technique which is highly improved over TILLING, offers the plant breeder a new tool for efficient screening of induced point mutation at an early stage for variants in genes of specific interest before taking plants to field trials, offers increased sensitivity in gene-specific mutant detection in pooled samples, enabling enlarged pool sizes and improving throughput and efficiency (Lee et al. 2009).

2.7 Role of Mutation Breeding in Crop Improvement

The standard technique of creating variability by means of altering genes through induction of mutations by physical or chemical mutagens and using the same effectively through elaborate methods of selection techniques in various generations for improvement of a particular crop species for desired objectives is called mutation breeding and is frequently practiced by plant breeders all over the world for crop improvement. Results of mutation breeding were more often useful in self-pollinating plant species. Success has also been tremendous in ornamental plants and in vegetatively propagated crops, which usually are heterozygous. Today, mutation breeding for crop improvement is not based only upon classical physical mutagens like X- or gamma rays or classical chemical mutagens such as EMS or NMU, but also upon variation that occurs during in vitro culture (Suprasanna and Nakagawa 2012) and has been termed 'somaclonal variation'. Use of haploids derived from anther culture has also found its best application in the 'doubled-haploids-technique', which leads faster to homozygosity for more effective selection (Shu et al. 2012).

The impact of induced mutation on crop improvement is reflected in the 3365 mutant varieties officially registered by FAO/IAEA (http://mvd.iaea.org) carrying novel-induced variation. Moreover, about three-quarters of these are direct mutant varieties derived from treatment with gamma rays, thus highlighting the importance of peaceful usage of radiations that belong to the group of physical mutagens. All this translates into a tremendous economic impact on world agriculture, poverty alleviation, food security and food production that is currently valued in billions of dollars and millions of cultivated hectares (Kharkwal and Shu 2009). A detailed description of the economic impact of a large number of prominent mutant varieties released in several countries of the world is given in Chap. 13 contributed by this author.

2.8 Development of Crop Varieties Through Mutation Breeding: Global Scenario

Ever since the discoveries of induction of mutations made by Muller and Stadler, a large number of breeders in several countries followed mutation breeding, induced and generated significant amount of genetic variability through various mutagens. Because of the popularity, easy and safe access, majority of the genetic variability has been induced through radiations, particularly gamma rays that have contributed significantly to modern plant breeding. Among the mutant varieties released, the majority are food crops and ornamentals. A detailed database on the mutation-derived varieties developed and released in major crops all over the world is being maintained by FAO/IAEA (http://mvd.iaea.org). The number of mutant varieties officially released at the beginning of the year 2022 is >3365 belonging to >240 plant species. However, as the Mutant Variety Database of IAEA is perhaps not being updated very regularly, as is evident from the number of mutant varieties released in China (817), India (341) and Pakistan (56) reported in the MVD, whereas China has already reported >1000 mutant varieties (Liu 2021), India has already reported 542 (Table 2.1) and Pakistan has released 79 mutant varieties (Table 2.1—T. M. Shah personal communication), the number of mutant varieties released in the world from >240 plant species is expected to be >3500. Majority of the mutant varieties have been released during the last three decades. The cumulative number of officially released mutant varieties in six continents of the world indicates that Asia tops the regional list with 2052 closely followed by Europe (960) and North America (209). With more than 100 mutant varieties each, China, India, Japan, Russian

Table 2.1 Number of mutant cultivars released in top 20 countries and top 20 crops of the world

No.	Country	No. of mutants	No.	Crop	No. of mutants
1	China	817	1	Rice	853
2	India	542[a]	2	Barley	311
3	Japan	479	3	Chrysanthemum	266
4	Russian Federation	216	4	Wheat	265
5	Netherlands	176	5	Soybean	181
6	Germany	171	6	Maize	96
7	USA	139	7	Groundnut	79
8	Pakistan	79[b]	8	Rose	67
9	Bulgaria	76	9	Common bean	57
10	Bangladesh	76	10	Cotton	48
11	Vietnam	58	11	Mung bean	39
12	Canada	40	12	Dahlia	36
13	South Korea	40	13	Durum wheat	31
14	France	39	14	Pea	34
15	Italy	35	15	Sesame	30
16	United Kingdom	34	16	Chickpea	27
17	Poland	31	17	Tomato	25
18	Sweden	26	18	Rapeseed	24
19	Guyana	26	19	Oat	23
20	Thailand	24	20	Faba bean	20

http://mvd.iaea.org (accessed in March, 2022)
[a] Detailed lists in Chap. 12 by this author
[b] Personal communication from T. M. Shah, Director, NIAB, Faisalabad, Pakistan

Federation, the Netherlands, Germany and the USA are the leading countries engaged in the development and release of mutant varieties. A perusal of the data on specific crops and number of mutant varieties released in the world (Table 2.1) indicates that the top 20 ranks are occupied by some of the most important food (cereals, pulses, oilseeds), fibre and ornamental plant species in the world agriculture and economics.

During the last five decades, several countries took up extensive crop improvement programmes through the use of induced mutagenesis and mutation breeding and made spectacular accomplishments in evolving several superior mutant varieties in large number of important agricultural crop species including cereals, pulses, oilseeds, vegetables, fruits, fibres and ornamentals. A wide range of characters including yield, maturity, quality and tolerance to biotic and abiotic stresses have been improved in the mutant varieties developed so far. Although an exact estimate of the total area covered by commercially released mutant cultivars in a large number of countries is not readily available, but they are being cultivated in millions of hectares and have made a very significant contribution worth billions of dollars in global agriculture addressing food and nutritional security problems in many countries of the world (Kharkwal and Shu 2009; Kharkwal 2017).

Most of the released induced mutant varieties belong to seed-propagated crop plants species and nearly 25% are of the ornamentals. Of the total mutant varieties, majority mutant varieties were developed 'directly' after mutagenic treatment and selection in the subsequent generations. The remaining new mutant varieties were developed 'indirectly' through cross-breeding of mutants or already released mutant varieties as sources of desired characters in cross-breeding programmes. Among the various mutagenic agents used for developing varieties, a great majority were obtained with the use of physical mutagens like radiations, particularly gamma rays as the mutagen.

Mutation breeding in rice has been very successful worldwide, probably due to its diploid nature and self-fertilizing character. Out of 1616 mutant varieties of cereals, rice alone accounts for 853 varieties. The first widely known rice mutant cultivar has been cv. Reimei, a short straw (semi-dwarf) mutant from cv. Fujiminori, released in Japan in 1966. Several important cultivars were developed thereafter through cross-breeding with this cultivar (Van Harten 1998; Nakagawa 2021). The mutant cultivar carried an allele that was allelic to the well-known *sd1* [semi-dwarf allele in cultivar *Dee-geo-wu-gen* (DGWG), a spontaneous dwarf mutant discovered by Chinese scientists]. This mutant DGWG possessing dwarfness, stiff straw, fertilizer responsiveness, non-lodging and day length insensitivity as its most important traits, was the forerunner of green revolution. It was used in crossing programmes for developing important cultivars like Taichung Native-1 and IR 8. In barley, 311 mutant varieties superior to best control lines for various plant characters have been isolated and released by various countries, including Sweden (Gustafsson 1963). Some of the important mutant varieties of barley are 'Pallas' (Sweden), 'Balder J' (Finland), 'Diamant' (Czech Republic), 'Trumpf' (Germany), 'Goldspear' (UK), 'Pennrad' and 'Luther' (USA) and 'Betina' (France) (Van Harten 1998).

2.8.1 Mutants in Recombination Breeding

Although majority of mutant varieties out of the total of 3365 reported were released as direct mutants and one third through crosses with various mutants, the trend of late is changing towards more use of induced mutants in recombination breeding. Mutants of no value often give promising recombinants when inter-crossed. Nichterlein (1999) enlisted a large number of new and better varieties of common bean (*Phaseolus vulgaris* L.) that were developed during 1960–1988 by using X-ray-induced bush type mutants in the pedigree. Mutation breeding programme for durum wheat in Italy involving extensive selection and hybridization work led to 11 registered varieties, five of which resulted from direct mutant selection and six from the cross-breeding procedures (Scarascia-Mugnozza et al. 1991). Mung bean variety NIAB Mung 98 was developed through hybridization between an induced mutant and an exotic accession (Siddique et al. 1999). Most of the mutant varieties of pulses, oilseeds, cereals and fibre crops released by Bhabha Atomic Research Centre, Mumbai in India are developed primarily by using mutants in cross-breeding.

2.9 Major Plant Traits Improved by Induced Mutations

2.9.1 Yield and Yield Components Improvement

Stable and high yield potential over a range of environmental conditions is probably the most important objective of most plant breeding programmes. One of the most important characters for judging agronomic value of mutants is their yield potential. Therefore, improvement in grain/seed yield has always been the main objective in almost all the crops. The improvement in grain yield through induced mutations has been brought about by alterations in yield contributing traits (Ahloowalia et al. 2004). Isolation of micromutations for higher yield coupled with some other desirable attributes like disease and pest resistance has been reported in chickpea (Kharkwal 2001, 2003; Kharkwal et al. 2001, 2005; Kharkwal and Shu 2009). Examples of important mutant cultivars with improved yield traits are listed in Table 2.2.

2.9.2 Tolerance/Resistance to Abiotic and Biotic Stresses

2.9.2.1 Tolerance to Abiotic Stresses

Abiotic stresses encompass several unfavourable environmental conditions such as drought, soil salinity, extreme pH, flooding and temperature. Induction of mutation for abiotic stresses has been successfully attempted in several crops and mutants for aluminium tolerance in banana (Matsumoto and Yamaguchi 1991), chlorate tolerance in barley and field pea (Kleinhofs et al. 1978), and salt stress in rice (Abe et al. 2007; Do et al. 2009), heat tolerance in rice (Poli 2013) and cold (harsh weather) tolerance in Amaranth (Gómez-Pando et al. 2009) and several other abiotic stress conditions have been obtained (Kharkwal and Shu 2009). Examples of important mutant cultivars with improved abiotic stress tolerance traits are listed in Table 2.3.

2.9.2.2 Tolerance/Resistance to Biotic Stress

Some of the crop varieties improved for yield or yield components through induced mutations have also shown improved tolerance to biotic and abiotic stresses. High yielding chickpea mutant varieties, Pusa 408, Pusa 413, Pusa 417 and Pusa 547 with resistance to Ascochyta blight, Fusarium wilt and other diseases and pests have been released for commercial cultivation in India (Kharkwal et al. 1988, 2005; Kharkwal and Shu 2009; Kharkwal 2017); similarly CM-72, CM-88, NIFA-95 and CM 1918 were released in Pakistan (Haq et al. 1988; Haq 2009). A number of mutants resistant to specific diseases or insect pests have also been isolated and released in several other crops. Examples of important mutant cultivars with improved biotic stress tolerance/resistance are listed in Table 2.4.

Table 2.2 Examples of mutants with improved yield traits

Crop	Latin name	Improved trait	Mutagen	Cultivar	Country	Reference
Rice	*Oryza sativa*	High yielding	Gamma rays	Zhefu—8	China	MBNL[a] Nos. 25 and 26, 1985
Bread wheat	*Triticum aestivum*	High yielding		Jauhar—78	Pakistan	Ahloowalia et al. (2004)
Barley	*Hordeum vulgare*	Semi-dwarf, malting quality	Gamma rays	Golden Promise	United Kingdom	Sigurbjornsson and Micke (1974)
Barley	*Hordeum vulgare*	Semi-dwarf	X-rays	Diamant	Czech Republic	Ahloowalia et al. (2004)
Groundnut	*Arachis hypogaea*	High yielding	Gamma ray	TAG-24	India	Ahloowalia et al. (2004)
Chickpea	*Cicer arietinum*	High yielding	Gamma rays	Pusa-547	India	Kharkwal et al. (2005)
Black gram	*Vigna mungo*	High yielding	Gamma ray	TAU-1	India	Ahloowalia et al. (2004)
Cotton	*Gossypium* sp.	High yielding	Gamma ray	NIAB-78	Pakistan	Ahloowalia et al. (2004)
Banana	*Musa* sp.	High yielding	Gamma ray	Al Beely	Sudan	PBGNL[b] Nos. 16 and 17, 2006

[a] *MBNL* Mutation Breeding Newsletter
[b] *PBGNL* Plant Breeding and Genetics Newsletter

Table 2.3 Examples of mutants with improved abiotic stress tolerance traits

Crop	Latin name	Improved trait	Mutagen	Cultivar	Country	Reference
Rice	*Oryza sativa*	Salt tolerance	Gamma rays	NIAB-IRRI-9	Pakistan	MBNL[a] No. 45, 2001
Rice	*Oryza sativa*	Salt tolerance	Gamma rays	VND95-20	Viet Nam	Do et al. (2009)
Rice	*Oryza sativa*	Tolerance to cold	Gamma rays	Kashmir Basmati	Pakistan	Ahloowalia et al. (2004)
Rice	*Oryza sativa*	Tolerance to heat	Gamma rays	Nagina-22	India	Poli (2013)
Rice	*Oryza sativa*	Salt tolerance	Ion beam irradiation		Japan	Abe et al. (2007)
Bread wheat	*Triticum aestivum*	Drought tolerance	Gamma rays	Njoro BW1	Kenya	IAEA Bulletin, 50-1
Barley	*Hordeum vulgare*	High altitude, harsh weather	Gamma rays	UNA La Molina	Peru	MBNL No. 43, 1997 Gómez-Pando et al. (2009)
Maize	*Zea mays*	Drought tolerance	Gamma rays	Kneja 698W	Bulgaria	PMR[b], 2012
Amaranth	*Amaranthus caudatus* L.	High altitude, harsh weather	Gamma rays	Centenario	Peru	Gómez-Pando et al. (2009)
Soybean	*Glycine max*	Tolerance to cold, drought and water logging	Gamma rays	Heinong-26	China	Khan and Tyagi (2013)

[a] *MBNL* Mutation Breeding Newsletter
[b] *PMR* Plant Mutation Report

Table 2.4 Examples of mutants with improved biotic stress tolerance/resistance traits

Crop	Latin name	Improved trait	Mutagen	Cultivar	Country	Reference
Rice	*Oryza sativa*	Resistance to blast and virus diseases	Gamma rays	Camago-8	Costa Rica	MBNL[a] Nos. 43, 1997
Bread wheat	*Triticum aestivum*	Resistance to black stem rust (Ug99)	Gamma rays	EldoNgano-1	Kenya	PBGNL Nos. 32 and 33, 2014
Barley	*Hordeum vulgare*	Powdery mildew	X-rays	Comtesse	Germany	MBNL No. 33, 1989 and No. 36, 1990
Barley	*Hordeum vulgare*	Mildew resistance	EMS	Betina	France	Sigurbjornsson and Micke (1974)
Chickpea	*Cicer arietinum*	Blight and wilt resistance	Gamma rays	Pusa-408, Pusa-413, Pusa-417	India	Kharkwal et al. (1988)
Chickpea	*Cicer arietinum*	Blight resistance	Gamma ray	Hassan-2K	Pakistan	Hassan et al. (2001)
Lentil	Lens	Blight resistance	Gamma ray	NIAB-Masoor	Pakistan	Sadiq et al. (2008)
Japanese pear	*Pyrus pyrifolia*	Black spot resistance	Gamma ray	Gold Nijisseiki	Japan	Saito (2016)
Peppermint	*Mentha piperita*	Wilt disease resistance	Neutron irradiation	Murray Mitcham	USA	Todd et al. (1977)

[a] *MBNL* Mutation Breeding Newsletter

2.9.3 Grain Quality and Nutrition

In recent years, there has been greater emphasis on the improvement of protein content and specific amino acids seed quality of cereals, pulses and oil content and fatty acid composition in oilseeds through mutation breeding. The alteration of seed colour and seed size for better acceptability has been achieved through induced mutations in several crops. The change of wheat seed coat colour from red to amber by gamma radiation resulting in the development of 'Sharbati Sonora' (Swaminathan et al. 1968) is a classical example. An increase in seed size and protein content associated with high yield in several chickpea mutant cultivars has been reported by Kharkwal (1998a). Chickpea variety Pusa 547 with large seed size, attractive colour and thin testa has been developed through radiation induced mutation (Kharkwal et al. 2005). Similarly, a high yielding early maturing chickpea mutant with high protein content, named as Hyprosola has been released in Bangladesh (Sheikh et al. 1982). Mutants with altered fatty acid composition have been isolated in soybean, rapeseed, sunflower, linseed and minor oil crops (Dribnenki et al. 1996). Examples of important mutant cultivars with improved quality traits are listed in Table 2.5.

2.9.4 Mutation Breeding for Improvement of Agronomic Traits

2.9.4.1 Plant Type, Growth Habit and Architecture

Dwarf and semi-dwarf mutants with reduced plant height having positive effect on yield via improved fertilizer response, increased tillering and lodging resistance was selected in cereals. In grain legumes, several examples of improved plant architecture, including dwarf or bushy mutants with increased lodging resistance, improved agronomic traits like plant architecture with erect growth habit and higher harvest index have been isolated (Kharkwal 1996, 1999, 2000) and released (Kharkwal et al. 2004, 2005; Kharkwal and Shu 2009). Semi-dwarfness and earliness are the characters most frequently described in released rice, wheat and barley mutant cultivars.

2.9.4.2 Flowering and Ripening Time

Changes in maturity period, leading to earliness, have been brought about by induced mutations in several crops. Rice variety IIT 60, an EMS-induced mutant of IR-8 matures 1 month earlier than IR-8 with same yield potential (Kharkwal 1996) and the improved rice variety Yuanfengzao, released in China, matures 45 days earlier than the original variety IR-8 (Wang 1991). In castor, the mutant variety 'Aruna' developed in India through neutron irradiation matures in 120 days compared to 270 days of the parent cultivar, HC-6 (Ankineedu et al. 1968). Several mutants showing extra-early maturity have been isolated and used in cross-breeding in various crops. Examples of important mutant cultivars with improved agronomic traits are listed in Table 2.6.

Table 2.5 Examples of mutants with improved quality traits

Crop	Latin name	Improved trait	Mutagen	Cultivar	Country	Reference
Rice	*Oryza sativa*	Grain quality	Gamma rays	VND95-20	Viet Nam	Do et al. (2009)
Rice	*Oryza sativa*	Grain quality	Gamma rays	ShweWartun	Myanmar	MBNL Nos. 11 and 12, 1978, Ahloowalia et al. (2004)
Rice	*Oryza sativa*	Glutinous endosperm	Gamma rays	RD-6	Thailand	Ahloowalia et al. (2004)
Bread wheat	*Triticum aestivum*	Grain colour	Gamma rays	Sharbati sonora	India	Swaminathan et al. (1968)
Bread wheat	*Triticum aestivum*	Grain colour	Gamma rays	Jauhar-78	Pakistan	MBNL No. 2, 1973
Sorghum	*Sorghum* sp.	Grain colour	Gamma rays	Djeman	Mali	MBNL No. 44, 1999
Chickpea	*Cicer arietinum*	Grain colour and cooking quality	Gamma rays	Pusa-547	India	Kharkwal et al. (2005)
Sunflower	*Helianthus annuus*	High oleic acid	Gamma rays	NuSun	USA	Ahloowalia et al. (2004)
Cassava	*Manihot esculentus*	Cooking quality	Gamma rays	Tekbankye	Ghana	MBNL No. 44, 1999
Grapefruit	*Citrus paradise*	Red fruit flesh and juice	Retrotransposon	Rio Star	USA	MBNL No. 37, 1991

Table 2.6 Examples of mutants with improved agronomic traits

Crop	Latin name	Improved trait	Mutagen	Cultivar	Country	Reference
Rice	*Oryza sativa*	Semi-dwarf	Gamma rays	Reimei	Japan	Das et al. (2017)
Rice	*Oryza sativa*	Short stature and semi-dwarf	Gamma rays	Calrose 76	USA	Rutger et al. (1977), Letsari et al. (2016)
Rice	*Oryza sativa*	Short stature, early maturity	Gamma rays	TNDB 100	Viet Nam	MBNL No. 45, 2001
Basmati Rice	*Oryza sativa*	Short stature and non-lodging	Gamma rays	CRM 2007-1	India	PMR No. 1 & 2, 2006
Bread wheat	*Triticum aestivum*	Short, resistant to lodging	X-rays	Creso	Italy	MBNL No. 6, 1973
Bread wheat	*Triticum aestivum*	Resistant to lodging	Gamma rays	Gergana	Bulgaria	MBNL No. 37, 1991
Barley	*Hordeum vulgare*	Semi-dwarf	X-rays	Diamante	Czech Republic	Ahloowalia et al. (2004)
Barley	*Hordeum vulgare*	Early flowering	Gamma rays	Mari	Sweden	Sigurbjörnsson (1975)
Castor	*Ricinus communis*	Early maturity	Gamma rays	Aruna	India	Ankineedu et al. (1968)
Rye	*Secale cereale*	Short lie cycle	Gamma rays	Soron	Peru	Gómez-Pando et al. (2009)
Banana	*Musa acuminata*	Short stature	Gamma rays	Novaria	Malaysia	MBNL No. 44, 1999; Mak et al. (1996)
Rice	*Oryza sativa*	Herbicide resistance	Gamma ray	Rice	USA	Maluszynski et al. (1995)
Rice	*Oryza sativa*	Herbicide resistance	Gamma ray	IRAT 239	Guyana	MBNL No. 33, 1981
Corn	*Zea mays*	Herbicide resistance	Gamma ray	Corn	USA	Mabbett (1992)

2.10 Social and Economic Impact of Mutation Breeding Technique in Crop Improvement

As the knowledge accumulated, it became clear that mutation technique was not a magic wand that would create anything at any time. Therefore, following mutagenic treatments, a mixed bag of induced variants is found and many of these may not be of any value. Because of these reasons, it will be unrealistic to expect miracles out of programmes using induced mutagenesis technique. Nevertheless, having already developed and released more than 3000 mutant varieties in a wide variety of crop species (Maluszynski et al. 2000), the accomplishments of induced mutagenesis technique in evolving superior crop varieties and its role in basic studies confer on it an honourable niche in the ever-growing crop improvement programmes and also contributing significantly in providing food security in several countries (Ahloowalia et al. 2004; Kharkwal and Shu 2009; Kharkwal 2017).

The following selected examples illustrate the role that mutation breeding has played and continues to play in plant breeding programmes for crop improvement. The rapeseed cultivar 'Regina II' was developed by mutation in Sweden and was released in Canada in 1953. 'Redwood 65' flax was derived and registered from a mutation programme at the University of Saskatchewan (Larter et al. 1965) and is present in the pedigrees of many western Canadian flax cultivars. Elsewhere in the world, almost 70% of the durum wheat in Italy was mutant varieties and there are 200 rice cultivars derived from mutagenesis programmes. The economic impact of mutant varieties of rice released with improved characters such as semi-dwarfness, earliness, improved grain yield, and disease tolerance and improved grain quality have been reviewed by Rutger (1992) and Maluszynski et al. (1995). Two improved rice mutant varieties, 'TNDB-100' and 'THDB', with earliness and improved grain yield were released within only 6 years after mutagenic treatment in Mekong Delta of Vietnam. These varieties grown in millions of hectares have maintained tolerance to acid sulphate soil or soil salinity (Do et al. 2009; Le et al. 2021). The economic impact of short straw mutants in barley that are in the pedigrees of many cultivars grown today derive from mutation programmes and have been of immense economic value in several countries. Michelite, an X-ray-induced white bean mutant with altered plant type, is in the pedigrees of most of the white beans grown in North America (Ahloowalia et al. 2004; Kharkwal and Shu 2009; Kharkwal 2017).

Mutation-breeding programmes in China for crop improvement was a profitable approach as the estimated output of cereals, fibre and oilseeds had substantially increased and contributed to national economy of China (Wang 1991). High yielding Chinese mutant varieties of wheat (Luyuan-502), rice (Ilyou-D069 and Zhefu-802), soybean (Hefeng-25), groundnut (Luhua-11) and other crops like cotton, maize and forage legumes are being cultivated in millions of hectares and their breeders have won National Invention Awards (Liu 2021).

Commercial success and economic impact of mutant varieties of grapefruit in Texas and pear in Japan has been highlighted by Nichterlein et al. (2000). Grapefruit mutant varieties 'Star Rubi' and 'Rio Red' are grown in almost 75% of the total grapefruit area in Texas, USA (Sauls 1999). The mutant variety 'Gold Nijiseeiki' of Japanese pear is more resistant to black spot disease caused by *Alternaria alternata* than its parent. The additional annual income by growing this variety is estimated to be about US$ 50 million (Amano 1997; Nakagawa 2021). In peppermint, the mutant variety, 'Todd's mitcham' forms bulk of the world's production of mint oil. Several high yielding and disease-resistant mutant varieties of chickpea, mung bean, wheat, rice and cotton released for commercial cultivation in Pakistan and grown in millions of hectares have made significant impact on the total production and productivity of these crops (Ahloowalia et al. 2004; Haq 2009; Kharkwal and Shu 2009; Kharkwal 2017).

2.11 Conclusion

Spontaneous mutations were the only source of genetic diversity available for the domestication and breeding of plants until the dawn of the twentieth century. It was nine decades ago that Stadler first reported induction of mutation in plants like barley and maize by X-rays to generate novel genetic diversity for crop improvement. The successful application of radiation for the induction of plant mutations prompted a flurry of research and development in its use to generate novel genetic diversity for crop improvement in several countries. This also prompted and lead to discovery of several new physical as well as chemical mutagenic agents. Since then, induced mutations have been used to generate genetic variation in many food, feed, ornamental and cash crops, leading to the release of large numbers of mutant varieties for cultivation across the globe. Global food and nutrition security in the coming years will also depend heavily on the availability and cultivation of improved crop varieties that can perform well under the pressures of increasing frequencies and intensities of drought, flooding and coastal salinity, warming temperatures and the transboundary spread of intensifying plant pests and diseases triggered by warming temperatures. Recent years are witness to the increasing application of newer mutagen sources and techniques such as High Hydrostatic Pressure, Ion Beams, Electron beams, Targeting Induced Local Lesions in Genomes, Cosmic rays for space mutagenesis and Endonucleolytic Mutation Analysis by Internal Labelling for the improvement of plant species, and of functional genomics technologies that establish gene-to-phenotype relationships for the discovery of molecular variants underlying mutations to be used in marker-assisted breeding and gene editing. The target of mutation breeding, therefore, will now get transferred from yield traits to quality traits and stress tolerance for adapting to climate change. The glorious history and outstanding achievements of mutation breeding during the past nine decades indicates that it is going to play a major role in shaping the future research on improvement of crop plants in twenty-first century and beyond.

References

Abe T et al (2007) Plant breeding using the ion beam irradiation in RIKEN. In: Cyclotrons and their applications, pp 222–224

Abe T et al (2012) Ion beam mutagenesis. In: Shu QY et al (eds) Plant mutation breeding and biotechnology. CABI, Wallingford, UK, pp 99–106

Abe T et al (2015) Ion beam breeding and gene discovery for function analyses using mutants. Nucl Phys News 25:30–34

Abe T et al (2021) Ion beam mutagenesis – an innovative and effective method for plant breeding and gene discovery. In: Sivasankar S et al (eds) Mutation breeding, genetic diversity and crop adaptation to climate change. CABI, Wallingford, UK, pp 411–424

Ahloowalia B et al (2004) Global impact of mutation derived varieties. Euphytica 135(2):187–204
Altenberg E (1934) The artificial production of mutations by ultraviolet light. Am Naturalist 68: 491–507
Amano E (1997) Mutation breeding in Japan and contribution to the region. In: Strategy paper on application of mutation techniques for crop improvement in East Asia and Pacific Region. IAEA, Vienna
Ankineedu G, Sharma KD, Kulkarni LG (1968) Effect of fast neutrons and gamma-rays on castor. Indian J Genet 28:31–39
Auerbach C (1941) The effect of sex on the spontaneous mutations rate in *Drosophila melanogaster*. J Genet 11:255–265
Auerbach C (1949) Chemical induction of mutations. In: Bonnier G, Larsson R (eds) Proc. Eighth Intl. Cong. Genetics. Berlingska Boktryckeriet, Lund, Sweden, pp 128–147
Auerbach C, Robson JM (1946) Chemical production of mutations. Nature 157:302
Bateson W (1894) Materials for the study of variation, treated with especial regard to discontinuity in the origin of species. Macmillan, London
Baur E (1924) Untersuchungen uber das Wesen, die Entstehung und die Vererbung von Rassenunterschieden bei *Anthirrinum majus*. Bibl Genet 4:1–170
Beachell HM (1957) The use of X-ray and thermal neutrons in producing mutations in rice. Int Rice Comm Newsl 6(1):18–22
Bridges CB (1916) Nondisjunction as a proof of chromosome theory of heredity. Genetics 1:107
Brock RD (1965) Induced mutations affecting quantitative characters. In: The use of induced mutations in plant breeding (Rep. FAO/IAEA Tech. Meeting Rome, 1964). Pergamon Press, Oxford, pp 443–450
Colbert T et al (2001) High-throughput screening for induced point mutations. Plant Physiol 126(2): 480–484
Correns C (1900) G. Mendel's Regal uber das Verhalten der Nach-kommenschaft der Rassenbasterde. Ber Deutsch Bot Ges 18:158–168
Cross MJ, Waters DL, Lee LS et al (2008) Endonucleolytic mutation analysis by internal labeling (EMAIL). Electrophoresis 29(6):1291–1301
Darwin C (1859) On the origin of species by means of natural selection, or the preservation of favoured races in the struggle for life. John Murrey, London
Das G et al (2017) Corrigendum: Insight into MAS: a molecular tool for development of stress resistant and quality of rice through gene stacking. Front Plant Sci 8:1321
De Vries H (1889) Intracellulare pangenesis. Gustav Fischer Jena. (English translation, 1910. The Open Court, Chicago)
De Vries H (1901) Die Mutationstheorie. I. Veit & Co., Leipzig, Germany. (English translation, 1910. The Open Court, Chicago)
De Vries H (1903) Die Mutationstheorie. II. Veit & Co., Leipzig, Germany. (English translation, 1910. The Open Court, Chicago)
Delaunay LN (1931) Resultate eines dreijahrigen Rontgen Versuch mit Weitzen. Der Zuchter 3: 129–137
Do K et al (2009) Socio-economic impacts of mutant rice varieties in Southern Vietnam. Induced plant mutations in the genomics era. FAO, Rome, pp 65–68
Dribnenki JCP et al (1996) LinolaTM 989 low linolenic flax. Can J Plant Sci 76:329–331
FAO/IAEA Mutant Variety Database (2022). http://mvd.iaea.org. Accessed Mar 2022
Feng HY et al (2009) Mutagenic mechanisms of ion implantation in plants. In: Shu QY (ed) Induced plant mutations in the genomics era. Food and Agriculture Organization of the United Nations, Rome, pp 220–222
Freese E (1959) On the molecular explanation of spontaneous and induced mutations. Brookhaven Symp Biol 12:63–73
Freisleben R, Lein A (1942) Uber die Auffindung einer mehltauresistenten mutante nach Roentgenbestrahlung anfaelligen reinen Linnie von Sommergerste. Naturwissenschafen 30:608
Frey KJ (1955) Agronomic mutations in oats induced by X Ray treatment. Agron J 47:207–210

Gaul H (1961) Use of induced mutations in seed-propagated species. In: Mutation and plant breeding. National Academy of Science, National Research Council, USA, pp 206–252

Gaul H (1965) The concept of macro- and micro-mutations and results on induced micro-mutations in barley. In: The use of induced mutations in plant breeding (Rep. FAO/IAEA Tech. Meeting Rome, 1964). Pergamon Press, Oxford, pp 408–426

Gaul H, Aastveit K (1966) Induced variability of culm length in different genotypes of hexaploid wheat, following X-irradiation and EMS-Treatment. Savremena Poljopr 14:253–276

Gaul H et al (1969) Micromutations influencing yield in barley-studies over nine-generations. In: Induced mutations in plants. Proc. Symp. Pullman. IAEA, Vienna, pp 375–398

Gómez-Pando L, Eguiluz A et al (2009) Barley (*Hordeun vulgare*) and kiwicha (*Amaranthus caudatus*) improvement by mutation induction in Peru. In: Induced plant mutations in the genomics era. Food and Agriculture Organization of the United Nations, Rome, pp 371–374

Gottashalk W, Wolff G (1983) Induced mutations in plant breeding, Monograph on theoretical and applied genetics. Springer, Berlin, pp 323–327

Gregory WC (1955) X-ray breeding of peanuts (*Arachis hypogaea*). Agron J 47:396–399

Gregory WC (1965) Mutation frequency, magnitude of change and the probability of improvement in adaptation. In: The use of induced mutations in plant breeding. Pergamon Press, Oxford, pp 429–441

Gregory W et al (1960) The peanut NC 4x, a milestone in crop breeding. Crops Soils 12(8):12–13

Gu and Shen (1989) Effects of space flight on the growth and some cytological characteristics of wheat seedlings. Acta Photophysiol Sin 15(4):403–407

Gustafsson A (1938) Studies on the genetic basis of chlorophyll formation and the mechanism of induced mutating. Hereditas 24:33–93

Gustafsson A (1940) The mutation system of the chlorophyll apparatus. Acta Universitatis Lundensis 36:1–40

Gustafsson A (1947) Mutations in agricultural plants. Hereditas 33:1–100

Gustafsson A (1963) Productive mutations induced in barley by ionizing radiations and chemical mutagens. Hereditas 50:211–263

Gustafsson A, Mackey J (1948) The genetic effects of mustard gas substances and neutrons. Hereditas 34:371–386

Gustafsson A et al (1967) Yield reactions and rates of origin of Chromosome mutations in barley. Hereditas 56:200–206

Hanson FR, Heys F (1929) An analysis of the effects of the different types of radium in *Drosophila*. Am Nat 63:201–213

Haq MA (2009) Development of mutant varieties of crop plants at NIAB and the impact on the agricultural production in Pakistan. In: Shu QY (ed) Induced mutations in the genomic era. Food and Agriculture Organization of the United Nations, Rome, Italy, pp 61–64

Haq MA et al (1988) Improvement of chickpea through induced mutations. In: Proc. FAO/IAEA workshop on improvement of grain legume production using induced mutations, 1–5 July, 1986, Pullman, Washington (USA). IAEA, Vienna, pp 75–88

Hassan S et al (2001) Gamma ray induced high yielding Kabuli type chickpea mutant variety "Hassan-2K". Pak J Bot 33:703–707

Hodgens C et al (2020) Mutagenomics: a rapid high-throughput method to identify causative mutations from a genetic screen. Plant Physiol 184(4):1658–1673

Ishii A, Sato T, Wachi M et al (2004) Effects of high hydrostatic pressure on bacterial cytoskeleton FtsZ polymers *in vivo* and *in vitro*. Microbiology 150:1965–1972

Johannsen W (1909) Elemente der Exakten Erhlichkeitenlehre. Gustav Fischer, Jena, p 516

Johannsen W (1913) Elements der exakten Erblichkeitslehre. Gustav Fischer, Jena, p 723

Khan MH, Tyagi SD (2013) A review on induced mutagenesis in soybean. J Cereals Oilseeds 4(2): 19–25

Kharkwal MC (1996) Accomplishments of mutation breeding in crop improvement in India. In: Sachdev MS et al (eds) Isotopes & radiations in agriculture and environment research. Indian Society for Nuclear Techniques in Agriculture and Biology, New Delhi, pp 196–218

Kharkwal MC (1998a) Induced mutations for improvement of protein in chickpea (*Cicer arietinum* L.). Indian J Genet 58(1):61–68

Kharkwal MC (1998b) Induced mutations in chickpea (*Cicer arietinum* L.). I. Comparative mutagenic effectiveness and efficiency of physical and chemical mutagens. Indian J Genet Plant Breed 58(2):159–167

Kharkwal MC (1998c) Induced mutations in chickpea (*Cicer arietinum* L.). II. Frequency and spectrum of chlorophyll mutations. Indian J Genet Plant Breed 58(4):465–474

Kharkwal MC (1999) Induced mutations in chickpea (*Cicer arietinum* L.). III. Frequency and spectrum of viable mutations. Indian J Genet Plant Breed 59(4):451–464

Kharkwal MC (2000) Induced mutations in chickpea (*Cicer arietinum* L.) IV. Types of macromutations induced. Indian J Genet 60:305–320

Kharkwal MC (2001) Induced mutations in chickpea (*Cicer arietinum* L.). V. Evaluation of micromutations. Indian J Genet Plant Breed 61(2):115–124

Kharkwal MC (2003) Induced mutations in chickpea (*Cicer arietinum* L.). VI. Significance of induced altered correlations. Indian J Genet Plant Breed 63(3):219–224

Kharkwal M (2012) A brief history of plant mutagenesis. Plant mutation breeding and biotechnology. CABI, Wallingford, pp 21–30

Kharkwal MC (2017) Mutation breeding for crop improvement. Geography You 17(102):26–32

Kharkwal MC, Shu QY (2009) Role of induced mutations in world food security. In: Induced plant mutations in the genomics era. FAO, Rome, pp 33–38

Kharkwal MC et al (1988) Induced mutations for improvement of chickpea, lentil, pea and cowpea. In: Improvement of grain legume production using induced mutations, Proc. FAO/IAEA Workshop, 1–5 July 1986, Pullman, Washington, USA. IAEA, Vienna, pp 89–109

Kharkwal MC et al (2001) Seventy five years of research on induced mutations with special reference to crop improvement in India. In: Ramamurty N et al (eds) Proceedings of NAARRI international conference on applications of radioisotopes and radiation technology in the 21st century, Mumbai, pp 230–235

Kharkwal MC et al (2004) Mutation breeding in crop improvement. In: Jain HK, Kharkwal MC (eds) Plant breeding – Mendelian to molecular approaches. Narosa Publishing House, New Delhi, India, pp 601–645

Kharkwal MC et al (2005) Pusa-547, a high yielding chickpea (*Cicer arietinum* L.) mutant variety for late sown conditions of North Western Plains Zone of India. Indian J Genet 65:229–230

Kleinhofs A et al (1978) Induction and selection of specific gene mutations in *Hordeum* and *Pisum.* Mutat Res 51:29–35

Knapp E (1950) Grundfragen der experimentallen mutations anslosung und bedautung fur die praktische pflanzenzuchtung. Vortrag auf der Pflazen Zuchtertagung Einbeck:1–20

Konzak CF (1954) Stem rust resistance in oats induced by nuclear radiation. Agron J 46:538–540

Konzak CF et al (1984) Induced mutations in seed propagated crops. In: Janick J (ed) Plant breeding reviews, vol 2. AVI Publishing Company, Westport, CT, pp 13–72

Larter EN et al (1965) Redwood 65, an improved flax variety. Can J Plant Sci 45(5):515–516. https://doi.org/10.4141/cjps65-100

Le TD et al (2021) Soybean breeding through induced mutation in Vietnam. In: Sivasankar S et al (eds) Mutation breeding, genetic diversity and crop adaptation to climate change. CABI, Wallingford, UK, pp 40–46

Lee LS et al (2009) EMAIL – a highly sensitive tool for specific mutation detection in plant improvement programmes. In: Shu QY (ed) Induced plant mutations in the genomics era. Food and Agriculture Organization of the United Nations, Rome, Italy, pp 243–244

Letsari P. et al (2016) Genetic Diversity of Japonica Rice (Oryza sativa L.) based on Markers corresponding to Starch Synthesizing Genes. Makara Journal of Science: 20:2, 49-54. DOI: 10.7454mss.v20i2.5947

Li S (2013) Space breeding seeds to bring benefits to TCM. http://news.xinhuanet.com/english/china/2013-07/04/c_132512624.htm. Accessed 5 July 2013

Liu L (2021) New breakthroughs in plant mutation breeding prominent among top ten scientific research advances in China in 2019. FAO/IAEA Plant Breed Genet Newsl 46:20–22
Liu F, Cao M et al (2005) Screening a RAPD marker related to the maize male sterility gene obtained by space flight. J Sichuan Agric Univ 23(1):19–23
Liu L et al (2021) New mutation techniques for crop improvement in China. In: Sivasankar S et al (eds) Mutation breeding, Genetic Diversity and Crop Adaptation to Climate Change. CABI, Wallingford, UK, pp 47–52
Lundqvist U (2009) Eighty years of Scandinavian barley mutation genetics and breeding. In: Shu QY (ed) Induced mutations in the genomic era. Food and Agriculture Organization of the United Nations, Rome, Italy, pp 39–43
Lundqvist U (2014) Scandinavian mutation research in barley – a historical review. Hereditas 151(6):123–131
Lundqvist U (2021) Scandinavian mutation research during the past 90 years – a historical review. In: Sivasankar S et al (eds) Mutation breeding, genetic diversity and crop adaptation to climate change. CABI, Wallingford, UK, pp 10–23
Luria SE, Delbruck M (1943) Mutation of bacteria from virus sensitivity to virus resistance. Genetics 28:491
Mabbett T (1992) Herbicide tolerant crops – ICI seeds leads the way. Int Pest Control 34(2):49–50
MacKey J (1954) Neutron and X-ray experiments in wheat and a revision of the speltoid problem. Hereditas 40:65–180
Mak C et al (1996) Novaria – a new banana mutant induced by gamma irradiation. InfoMusa 5(1): 35–36
Maluszynski M et al (1995) Mutation techniques in plant breeding. In: Proc. IAEA/FAO Symp., on induced mutations and molecular techniques for crop improvement, Vienna, June 19–23, 1995, pp 489–504
Maluszynski M et al (2000) Officially released mutant varieties – the FAO/IAEA database. Mutat Breed Rev 12:1–84
Matsumoto K, Yamaguchi H (1991) Induction and selection of aluminium tolerance in banana. In: Proc FAO/IAEA Symp on plant mutation breeding for crop improvement, Vienna, 1990, vol 2. IAEA, Vienna, pp 249–256
Mei M, Qin Y, Sun Y (1998) Morphological and molecular changes of maize plants after seeds been flown on recoverable satellite. Adv Space Res 22:1691–1697
Morgan TH (1910) Sex limited inheritance in *Drosophila.* Science 32:120–122
Morgan TH (1912) Further experiments with mutations in eye-colour of *Drosophila*: the loss of the orange factor. J Acad Nat Sci Phila 5:321–346
Muller HJ (1927) Artificial transmutation of gene. Science 66:84–87
Muller HJ (1930) Types of visible variations induced by X-rays in *Drosophila.* J Genet 22:299–333
Muller HJ (1954) The nature of genetic effects produced by radiation. In: Holleander A (ed) Radiation biology, vol 1. McGraw Hill, New York, pp 351–473
Nakagawa H (2021) History of mutation breeding and molecular research using induced mutations in Japan. In: Sivasankar S et al (eds) Mutation breeding, genetic diversity and crop adaptation to climate change. CABI, Wallingford, UK, pp 24–39
Nichterlein K (1999) The role of induced mutations in the improvement of common beans (*Phaseolus vulgaris* L.). Mutat Breed Newsl 44:6–9
Nichterlein K et al (2000) Achievements and trends of using induced mutations in crop improvement. In: Proc., DAE-BRNS Symp., Dec. 6–8, 2000, Mumbai, India, pp 27–35
Oehlker F (1943) Chromosome mutation in meiosis by chemicals. In: Auerbach C (ed) Mutation research – problems, results and prospects. Chapman and Hall, UK
Oliver CP (1930) The effect of varying the duration of X-ray treatment upon the frequency of mutation. Science 71:44
Patil SH et al (1995) Semi-dwarf early maturing and high yielding new groundnut variety, TAG-24. J Oilseed Res 12:254–257

Poli Y (2013) Characterization of a Nagina-22 rice mutant for heat tolerance and mapping of yield traits. Rice 6(1):36
Rapoport IA (1946) Carbonyl compounds and the chemical mechanism of mutation. C R Doklady Acad Sci USSR 54:65
Rapoport IA (1948) Alkylation of gene molecule. C R Doklady Acad Sci USSR 59:1183–1186
Roentgen W (1895) Uber eine neue Art von Strahlen. Vorlaufige Mitteilung. In: Aus denSitzungsberichten der Würzburger Physik-Medic. Gesellschaft Wurzburg, pp 137–147
Rutger JN (1992) Impact of mutation breeding in rice. A review. Mutation breeding review, vol 8. IAEA, Vienna
Rutger JN, Peterson ML, Hu C (1977) Registration of Calrose 76 Rice1 (Reg. No. 45). Crop Sci 17(6):978–978
Sadiq MS et al (2008) A high yielding and disease resistant mutant of lentil developed through seed irradiation of an exotic germplasm. Can J Pure Appl Sci 2:411
Saito T (2016) Advances in Japanese pear breeding in Japan. Breed Sci 66(1):46–59
Sapehin AA (1930) Rontgen-mutationen beim Weizen (*Triticum vulgare*). Der Zuchter 2:257–259
Sapehin AA (1936) X-ray mutants in soft wheat. Bull Appl Bot Genet Plant Breed Ser II 9:3–37
Sato Y, Shirasawa K, Takahashi Y et al (2006) Mutant selection from progeny of gamma-ray irradiated rice by DNA Heteroduplex Cleavage using Brassica petiole extract. Breed Sci 56: 179–183
Sauls JW (1999) Texas citrus - root stock and scion varieties. http://aggie-horticulture.tamu.edu/citrus/12304.htm. pp 1–7
Scarascia-Mugnozza GT, D'Mato F et al (1991) Mutation breeding programme for durum Wheat (*Triticum turgidum* ssp. durum Desf.) improvement in Italy. In: Proc. FAO/IAEA Symp. on plant mutation breeding for crop improvement, June 18–22, 1991
Scossiroli RE et al (1966) Studies on the induction of new genetic variability for quantitative traits by seed irradiation and its use for wheat improvement. In: Mutations in plant breeding (Proc. Panel Vienna, 1966). IAEA, Vienna, pp 197–229
Sheikh MAQ et al (1982) A high-yielding and high-protein mutant of chickpea (*Cicer arietinum* L.) derived through mutation breeding. Environ Exp Bot 22:483–389
Shu QY (2009) Turning plant mutation breeding into a new era: molecular mutation breeding. In: Shu QY (ed) Induced plant mutations in the genomics era. Food and Agriculture Organization of the United Nations, Rome
Shu QY, Forster BP, Nakagawa H (2012) Plant mutation breeding and biotechnology. CABI FAO, Oxfordshire, UK, pp 1–608
Siddique SM et al (1999) Development of mungbean variety 'NIAB Mung-98' involving induced mutants through conventional breeding. Mutat Breed Newsl 44:11–13
Sigurbjörnsson B (1975) Methods of mutation induction, including efficiency, and utilization of induced genetic variability. Barley Genet III:84–95
Sigurbjornsson B, Micke A (1974) Philosophy and accomplishments of mutation breeding. In: Polyploidy & induced mutations in plant breeding proceedings
Stadler LJ (1928) Genetic effect of X-rays in maize. Proc Natl Acad Sci U S A 14:69–75
Stadler LJ (1930) Some genetic effects of X-rays in plants. J Hered 21:3–19
Stadler LJ (1941) Genetic studies with ultraviolet radiation. In: Punnet RC (ed) Proc. Seventh Intl. Congress of Genetics. Cambridge University Press, Cambridge, pp 269–276
Stubbe H (1934) Einige Kleinmutationen von *Anthirrinum majus' L.* Zuechter 6:299–303
Suprasanna P, Nakagawa H (2012) Mutation breeding of vegetatively propagated crops. Plant mutation breeding and biotechnology. Food and Agriculture Organization of the United Nations, Rome, pp 347–358
Swaminathan MS (1963) Evaluation of the use of induced micro- and macro-mutations in the breeding of polyploid crop plants. In: Symp application of nuclear energy in agriculture, Rome, 1961, pp 241–277

Swaminathan MS (1965) A comparison of mutation induction in diploids and polyploids. In: The use of induced mutations in plant breeding (Rep. FAO/IAEA Tech. Meeting, Rome, 1964). Pergamon Press, Oxford, pp 619–641

Swaminathan MS et al (1968) Mutations in plant breeding II. IAEA, Vienna, p 233

Till BJ, Afza R, Bado S et al (2009) Global TILLING Projects. In: Shu QY (ed) Induced plant mutations in the genomics era. Food and Agriculture Organization of the United Nations, Rome, pp 237–239

Timofeeff-Ressovsky NW (1941) Mechanismus der punktmutation. In: Punnet RC (ed) Proc. Seventh Intl. Cong. Cambridge University Press, Cambridge, pp 281–294

Timofeeff-Ressovsky NW, Zimmer KG (1947) Das Treffer prinzip in der Biologie. S. Hirzel, Leipzig, Germany

Todd W, Green R, Horner C (1977) Registration of Murray Mitcham Peppermint1 (Reg. No. 2). Crop Sci 17(1):188–188

Tollenaar D (1934) Untersuchungen uber Mutation bei Tabak: I. Entstechungsweise und Wesen Kntslich erzeugter Gene-Mutanten. Genetica 16:111–152

Van Harten AM (1998) Mutation breeding: theory and practical applications. Cambridge University Press, Cambridge

Von Tshermak E (1900) Uber Kunstliche Kreuzung bei *Pisum sativum*. Ber der Bot Ges 18:232–239

Wang LQ (1991) Induced mutation for crop improvement in China. In: Proc. IAEA Symp. on plant mutation breeding for crop improvement, June 18–22, 1990. IAEA, Vienna, pp 9–32

Watson JD, Crick FHC (1953) A structure for deoxyribose nucleic acid. Nature 171:964

Zhang W, Liu X, Zheng F et al (2013) Induction of rice mutations by high hydrostatic pressure. Plant Physiol Biochem 70:182–187

Zimmer KG (1961) Studies on quantitative radiation biology. Oliver and Boyd, London

Physical and Chemicals Mutagenesis in Plant Breeding

3

S. Bado, B. P. Forster, and F. Maghuly

Abstract

World agriculture faces the challenge of food security because of the rapid growth of human population, reduced arable land area, plant productivity and climate change. Since the pioneering work on mutation of Thomas Hunt Morgan in 1910 (Am Nat 44:449–496) and the first release of an improved cultivar in tobacco in the 1930s, plant mutation breeding has become an effective breeding method. It has produced direct mutant cultivars and provided materials for further breeding. It has contributed over 3330 cultivars in more than 220 plant species. Major advantages of plant mutation breeding are: (1) mutation induction in elite materials thus little or no additional breeding is necessary, (2) fastest way of developing new improved lines, (3) applicability to all plant species and (4) generation novel traits. Today, plant mutation breeding is a much-needed weapon to combat new challenges in agriculture such as direct and indirect effects of climate change. This chapter outlines and compares methods used for physical and chemical mutagenesis for crop plant improvement. The benefits of mutation induction needed to be assessed and placed in context with respect to other options (available resources, facilities, costs, pragmatism, etc.). In this chapter,

S. Bado (✉)
Department of Plant Science, PHYTOPRISE GmbH, Oberwart, Austria

Département de Biochimie-Microbiologie, Unité de Formation et de Recherche en Sciences de la Vie et de la Terre, Université Ouaga I Professeur Joseph KI-ZERBO, Ouagadougou, Burkina Faso
e-mail: souleymane.bado@phytoniq.com

B. P. Forster
Biohybrids Ltd, Reading, UK

F. Maghuly
Plant Functional Genomics, Institute of Molecular Biotechnology, Department of Biotechnology, BOKU-VIBT, University of Natural Resources and Life Sciences, Vienna, Austria

S. Penna, S. M. Jain (eds.), *Mutation Breeding for Sustainable Food Production and Climate Resilience*, https://doi.org/10.1007/978-981-16-9720-3_3

the future of mutation breeding in the light of new and exciting advances in plant sciences and technologies are discussed.

Keywords

Mutagens · Physical and chemical · Mutation induction · Mutation detection · Mutagenesis · Plant mutation breeding · Crop improvement

3.1 Introduction

A mutation is a heritable change in the genetic material of an organism, which is independent of genetic segregation or genetic recombination. Mutation may be natural or induced. Deliberate mutation induction and selection has had a massive impact on agriculture in providing improved cultivars. Plant mutation breeding follows three strategies: induced mutagenesis, mutation detection and selection. Mutations can be classified into three main categories based on where they occur in: (1) genome, (2) chromosome and (3) gene mutation. Other types of mutations also occur such as non-nuclear mutations and gene copy number mutations, there are also mutation mimics, (reviewed by Lundqvist et al. 2012). Mutation is a natural process and has been a primary driver for speciation and evolution (Darwin 1868). Natural mutations have had a significant impact on the domestication and cultivation of wild species by man's selection of favoured traits in plants and animals, and thereby the evolution of mankind (Bronowski 1973). Mutation has been responsible for the abundant diversity of life on Earth.

Present agriculture is challenged by a massive world population, which is estimated to peak at nine billion people by 2050, and declining areas for farming. This bleak situation is further exacerbated by new environmental challenges: pests, diseases, drought, salinity, water-logging and abnormal temperatures (Rosenzweig et al. 2001; Roy et al. 2011; Islam et al. 2012). Thus, major crop plants have suffered a decrease in production and productivity with yield reduction of over 50% resulting in enormous economic losses worldwide and food shortages (Oerke et al. 1999; Rai et al. 2011).

Classical plant breeding, the improvement of crop plants through hybridization, transgressive segregation and heterosis to produce better genotypes has been reviewed recently by Makay et al. (2020). However, conventional approaches are slow and laborious, for annual crops it can take over 10 years to produce a new cultivar, for perennial crops this is much longer (Maluszynski et al. 1995). In addition, all plant breeding programmes are limited by the genetic variation available. Traditionally, this has led to germplasm collections which may be classed into elite, breeding, landrace and wild species groups. The further a breeder has to go to access a trait the longer and more complicated the breeding. Therefore, high genetic variability within the elite and contemporary breeding materials is preferred, and thus mutation induction in these groups is attractive.

Plant mutation breeding is a form of accelerated plant breeding (Forster et al. 2015). In addition to the benefits listed above, plant mutation breeding is considered a safe and accepted breeding method, which has been practiced since 1927 (Gager and Blakeslee 1927; Muller 1927). It has led to new traits that have provided yield and quality improvements and has revolutionized farming. Examples include: (1) semi-dwarf small grain cereals crops allowed mechanical harvesting (combine harvesters) and provided a yield advantage (Mlcochova et al. 2004; Rutger 2009), (2) rust-resistant wheat (Bado et al. 2015), (3) salt tolerance rice (Do 2009; Vinh et al. 2009), (4) high altitude farming of barley in Peru (Gómez-Pando et al. 2009), (5) seedless citrus fruit (Spiegel-Roy 1990; Vardi et al. 2008; Sutarto et al. 2009), (6) self-compatibility in apple, cherry and almond (Matthews and Lapins 1967; Monastra et al. 1998; Spiegel-Roy 1990; Tehrani and Brown 1992), (7) resistance to wilt disease and high yielding peppermint to sustain chewing gum production (Sigurbjörnsson and Micke 1974), (8) red grapefruit (FAO/IAEA 1991), (9) 742 improved cultivars in vegetatively propagated crops (ornamentals, food and fruit trees, Bado et al. 2017), (10) non-plant crops—high yielding and good quality fungi (Sathesh-Prabu and Lee 2011) and high yielding extracellular polysaccharides from the algae (Liu et al. 2015).

Currently, the stringent regulations imposed on genetically modified crops (that use recombinant DNA technology) do not apply to cultivars produced by mutation breeding. Mutation induction is particularly relevant to crops that do not have a sexual reproduction system, i.e. cannot be improved by normal breeding. Induced mutation has been an effective approach to broaden the genetic diversity to improve vegetatively propagated crops (VPCs) and to develop successful cultivars in food crops (banana, potato, sweet potato, cassava and wasabi), fruit trees (cherry, citrus, apple, almond, peach, apricot), ornamentals (chrysanthemum, rose, dahlia and many other flowers) and other crops (Bermuda grass, field mint, peppermint and tea) (van Harten and Broertjes 1989; Jankowicz-Cieslak et al. 2012; Bado et al. 2017; Chepkoech et al. 2019). An advantage of mutation breeding is that the starting (target) genotype is usually a top-performing or popular line or cultivar; any improvement is readily accepted, and the improved line may be developed into a new cultivar directly.

This chapter provides an overview of the current knowledge of physical and chemical mutagens and induced mutants, emphasizing their effectiveness in improved traits for agriculture, food production and meeting market demands. The principles of mutation breeding and genetics are not new though new methods in mutation induction and detection have been developed. There are several enabling biotechnologies, especially tissue culture that broaden the scope and make mutation breeding more effective and targeted. These are described and discussed with respect to need, available resources and the emergence of new and exciting plant biotechnologies.

3.2 History of Physical and Chemicals Mutagens in Plant Breeding

Plant mutation breeding has a long history in technology development and optimization, and the development and release of mutant cultivars (Table 3.1). Currently, over 3330 cultivars have been released worldwide with over 88% being derived from physical and chemical mutagens (https://mvd.iaea.org/).

3.2.1 Physical and Chemical Mutagens

Since discovering the mutagenic effects of X-rays (Muller 1927; Stadler 1928a, b), a large number of physical and chemical mutagens have been assessed in plant breeding. From the mutant cultivars (officially) released worldwide and registered in the IAEA database (>3330, https://mvd.iaea.org), the vast majority involved physical mutagenesis (>77%) followed by chemical mutagens (>11%) and their combinations (>1%) (Fig. 3.1). Other (>9%) mutant cultivars have been released through further breeding using these induced mutants (9.85%).

3.2.1.1 Physical Mutagens

Physical mutagens have had the greatest impact in plant breeding (Table 3.2). They can be classified into two groups: (1) Ionizing radiations and (2) Non-ionizing radiation. The ionizing radiations are composed of particulate or densely ionizing radiation (alpha ray, beta ray, heavy-ion beam and fast, slow and thermal neutrons), and electromagnetic or sparsely ionizing radiation (X-rays, gamma rays, cosmic rays, electron beams). This classification is mainly based on radiation capacity in producing ions and their ability (enough energy) to affect the structure of atoms of impacted materials, such as plant propagules (Mba et al. 2012).

With respect to mutation breeding, ionizing radiation causes biological injuries through two main interactions (physical and chemical) with DNA. Physical action is the result of the energy deposited which cause direct DNA damage. Chemical reactions are mainly due to highly reactive free radicals that are generated (OH^- and H^+) which cause indirect DNA damage (Rao et al. 1965; Reisz et al. 2014). Light of various wavelengths can also cause photochemical damage, typically purine or pyrimidine dimers, that result in point mutations in the DNA sequence (Kurowska et al. 2012).

In the presence of oxygen, radiation-induced free radicals form peroxy radicals that are highly damaging (Sparrow and Singleton 1953; Sparrow 1961; Esnault et al. 2010; Lagoda 2012). Thus, plant materials with high water content and/or some biochemical composition (e.g. oil) are more sensitivity to radiation. This may result in small to large mutations, from point mutations to deletion, single and double-strand breaks, and chromosomal rearrangement.

Physical mutagens have been applied to a wide range of plant materials from whole plants (in gamma fields) to single cells (in gamma cells). Soft materials such as in vivo (excluding seed), in vitro materials (e.g. embryogenic callus and embryos)

Table 3.1 Historical milestones in physical and chemical mutagenesis in plant breeding

Date	Milestone		References
	Physical mutagenesis	Chemical mutagenesis	
1838–1839	Description of the cell theory and suggestion of totipotency of cells		Scheiden (1838)
1868	Introduction of the 'histogen theory' to explain shoot apex behaviour in plants		Darwin (1868), van Harten (1998)
1895–1900	The discovery of various kinds of radiation		Forster and Shu (2012)
1901–1905	Suggestions and promotion of radiation in plant and animal mutation breeding		de Vries (1901, 1903, 1905)
1907	Proposition of the word 'chimera'		Cramer (1907)
1927	Attempts of induced mutations in seed propagated crops using gamma and X-rays		Gager and Blakeslee (1927), Muller (1927), von Sengbush (1927)
1928	Successful mutation induction and discovery of mutagenic effects of X-rays in barley and maize		Stadler (1928a, b)
1930s	Continued of deliberated mutation in plants The first official mutation breeding programmes: in Sweden, Germany, United States		Stadler (1929), Goodspeed (1929), Kharkwal (2012)
1931	Deliberate mutation induction on potato		Asseyeva (1931)
1932		Attempts to induce mutations using chemicals	Sakharov (1932), Klein (1932)
1934	Release of physically (X-rays) improved cultivar tabak 'Chlorina' cv. (Indonesia)		Rana (1965)
1941–1943		Description of the ability of chemicals to induce mutations, i.e. mustard gas	Auerbach and Robson (1942), Oehlkers (1943)
1942	Report of induced disease resistance in a crop plant; X-ray-induced powdery mildew resistance in barley		Freislebe and Lein (1942)
1944		First report on the deliberate use of chemical mutagens	Auerbach and Robson (1944)
1944–1946		Continuation of the demonstration of the mutagenic effects of chemicals	Auerbach and Robson (1946), Oehlkers (1946), Rapoport (1946)

(continued)

Table 3.1 (continued)

Date	Milestone		References
	Physical mutagenesis	Chemical mutagenesis	
1949	Start of the gamma irradiation in plant mutation breeding Released of mutant in vegetatively propagated crop by X-rays: Tulip cv. 'Faraday' in Netherlands		Sparrow and Singleton (1953), https://mvd.iaea.org/
1954	Officially improved food crop plant by X-rays, pea mutant 'Stral-art' cv.		https://mvd.iaea.org/
1961	Officially released introgressed mutant cultivar Antirrhinum cv. 'Juliva'		https://mvd.iaea.org/
1963	Release of mutant from gamma ray: mutant variety 'Mori-hou-fu 3A' cv. of apple		https://mvd.iaea.org/
1964	Release of mutant cultivar in sweet cherry cv. 'Compact Lambert' in Canada		Sigurbjörnsson and Micke (1974)
1964	Set up of the FAO/IAEA Joint Division with a mandate to support the production of induced mutations for food security issues in developing countries		Forster and Shu (2012)
1966	UN Geneva Conference on 'Peaceful Application of Atomic Energy'		Donini and Sonnino (1998)
1966		Official release of the chemically (diethyl sulphate, dES) improved cultivar of barley, cv. Luther in the United States	Nilan and Muir (1967)
1968	Release of the mutant cultivar (X-ray) potato cv. 'Mariline 2' in Belgium		van Harten (1989)
1969–1970	The first FAO/IAEA international training course on crop mutation breeding Pullman Symposium on plant mutation breeding • Publication of first classified list of mutant cultivars • Publication of the first Manual on Mutation Breeding		IAEA (1970), Kharkwal (2012)
1972	Start of 'Mutation Breeding Newsletter' published by FAO/IAEA		Forster and Shu (2012)

(continued)

Table 3.1 (continued)

Date	Milestone		References
	Physical mutagenesis	Chemical mutagenesis	
1975	Initiation of coordination research programme organized by FAO and IAEA on 'Improvement of vegetatively propagated plants through induced mutations'		FAO/IAEA (1975)
1977	Publication of the second edition of Manual on Mutation Breeding		IAEA (1977)
1978	Start of space mutation breeding in China		Xianfang et al. (2004)
1980–1981	• Introduction of tissue culture (biotechnology) for in vitro mutation induction • First mutation breeding report on VPCs • FAO/IAEA first symposium on 'Use of induced mutations as a tool in plant research' in Vienna, Austria		Broertjes (1982), Kharkwal (2012)
1989–1990	Use of RIKEN RI-Beam Factory (ion beam) for radiation biology research and plant breeding		Abe et al. (2012)
1993	FAO/IAEA Mutant Variety Database (https://mvd.iaea.org/)		Forster and Shu (2012)
1998	Release of the rice mutant cv. 'Hangyu' 1 from cosmic irradiation		https://mvd.iaea.org/
2000–2005	TILLING (Targeting Induced Local Lesion in Genome) to induce and study mutation-phenotype association		McCallum et al. (2000), Colbert et al. (2001)
2002	Release of ion beam commercial mutant cv. 'Temari Bright Pink'		Abe et al. (2012)
2005–2010	Biotechnologies for targeted mutation discovery and wide spectrum establishment: • Direct genomic selection (DGS) method • Exome capture sequencing for mutation screening and functional genomic analysis • High-throughput sequencing and mutation discovery methods based on massive parallel sequencing • Using next-generation sequencing (NGS) for rapid detection of rare mutation in targeted gene		Basiardes et al. (2005), Okou et al. (2007), Mamanova et al. (2010), Gilchrist and Haughn (2010), Tsai et al. (2011)
2008	International Symposium on Induced Mutation in Plants in Vienna, Austria. Induced Plant Mutation in the Genomics Era.		Shu (2009)
2018	• IAEA/FAO International Symposium on Plant Mutation Breeding and Biotechnology • Publication of the third edition of the Manual on Mutation Breeding • New breeding techniques (e.g. genome editing) are not exempted from the current EU GMO legislation • Physical or chemical treatments are explicitly exempted from the EU GMO legislation		FAO/IAEA (2018), Court-of-Justice-of-the-European-Union (2018), Holme et al. (2019)
2020	Emmanuelle Charpentier and Jennifer A. Doudna won the 2020 Nobel Prize in Chemistry for the development of a method for genome editing (CRISPR/Cas)		Wu et al. (2020)

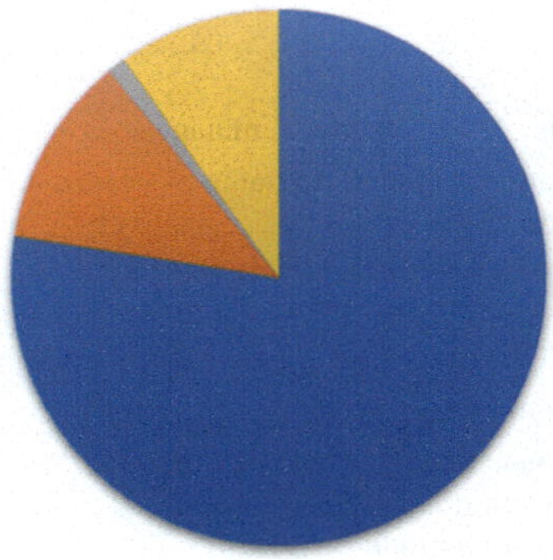

Fig. 3.1 Distribution of improved mutant crop cultivars developed by using different mutagenic agents. (Data from IAEA mutants' database: https://mvd.iaea.org/, accessed on 13 July 2020)

require relatively lower dosage of physical mutagens compared to seeds because of their high water content. Other variables include: storage time, applied mutagen dose, activity level of radioactive source and temperature influence physical mutagenesis (Mba et al. 2010). The material treated and the treatment itself are critical factors determining success.

The efficiency and effectiveness of physical mutagens have been described and are well known (Snyder et al. 1961; Lagoda 2012). A key feature is linear energy transfer (LET) of various mutagens, such as fast neutrons, ion beams, X-rays, gamma rays and UV light (in order of high to low LET) (Magori et al. 2010; Tanaka et al. 2010). The effects from the same dosage of different types of mutagens can be evaluated as relative biological effects (RBEs), which is a more pragmatic indicator for plant breeding (e.g. post-treatment survival and growth).

Gamma rays have been the most widely used physical mutagens in plant mutagenesis (Beyaz and Yildiz 2017; Li et al. 2019). Gamma rays are emitted from radioactive isotopes, mostly commonly Cobalt, ^{60}Co (with a half-life of 5.3 years) and Caesium, ^{137}Cs (with a half-life of 30 years). Because of their shorter wavelength (3×10^{-14} m), and uniform penetrating power, they are capable of reaching and changing DNA in many biological samples. Gamma rays gained popularity (after 1970s) during the promotion of peaceful uses of radioactive elements (following US President Dwight D. Eisenhower 'Atoms for Peace' speech on December 8, 1953), with an emphasis on world hunger concerns. Recently, gamma sources have become a security risk, and strict international regulations are imposed on the shipment and installation of new gamma sources and the refurbishment of old gamma irradiators. However, interest in induced mutation has not waned and other methods have been developed.

There has been a resurgence in the use of X-rays. X-ray irradiation does not involve radioactive isotope sources. X-rays represent the second most commonly used physical mutagen applied in plant mutation breeding, with over 21% of

Table 3.2 Comparison of various physical mutagens with respect to their respective advantages and disadvantages in plant breeding and genetics

Mutagen	Advantages	Disadvantages	Comments
Gamma rays	• Low LET radiation • Optimal procedures published and available • Regional, national and international services available • Highly reproducibility • Deep penetration in multi-cellular systems • Choice of high or low DNA damage • Cause single nucleotide substitution, inversion and deletion	• Requires a radioactive source • Many restrictions for new source and refurbished sources • Problem of high dose rate of new (highly reactive) sources • Requires specialized physical structure/laboratories and technicians	Most successful, has resulted in 66.91% of officially released mutant cultivars
X-rays	• Non-radioactive • Low LET radiation • Easy accessibility • Penetrates tissues from a few millimetres to a few centimetres	• Requires optimal settings for uniform irradiation • Optimized procedures are generally not available	Second most popular physical mutagen with 21.89% of officially released mutant cultivars
Neutron rays (fast, slow, thermal)	• High LET radiation • Penetrates tissues from a few millimetres to a few centimetres • Saturate genome for genetic studies • Cause point mutation, A/T to G/C transition, insertion, inversion, translocations tandem duplication and deletion • Gene knock-outs	• Requires nuclear reactor or accelerator • Relative high cost • Can create large deletions (>1 Mb) • Can delete multiple genes at a time • Difficulty of absorbed dosage estimation due to surface contamination	Third most popular physical mutagen with over 5.23% of officially released mutant cultivars
UV light	• Very low LET radiation • Available • Activation of transposable elements (indirect gene mutation) • Effective in pollen and fungal mycelium • Relative low cost	• Low penetration • Limited to single cell layer or sensitive material	Very limited effectiveness and use in plant mutation breeding
Ion beam	• High LET radiation • Large biological effects • High mutation rates • High survival rates of M_1/M_1V_1 • Wide and new	• Deposit high energy/damage • Not tested on a large number of species • Laborious sample preparation	• High installation cost of the accelerator • Highly effectiveness in ornamentals • Good for genetics studies

(continued)

Table 3.2 (continued)

Mutagen	Advantages	Disadvantages	Comments
	phenotypic variation • Targeted trait specificity	• Low accessibility of ion beam irradiators/facilities • Anatomy of seed and meristematic tissue can be limiting	
Cosmic rays	• High-energy ion radiation • High LET radiation • Tested on various plant species	• Requires space flight • Massive investment costs and intensive technological support • Very limited accessibility • Mechanism of space-induced mutation not fully understood • Limited to plant materials in dormant state (not actively growing)	• Access to space programme • Possibility to induce novel variation • Possibility to obtain rare mutants • Effective ground simulation of space environment factors may increase this approach

released mutant cultivars worldwide (https://mvd.iaea.org). It should be noted that X-rays were the first physical mutagen proven to induce mutation in crop plants in the early plant mutation era with mutational effects on barley described in 1928 (Stadler 1928a).

Today, ion beam is emerging as an effective physical mutagen (after its biological effects were discovered in the mid-1980s, Abe et al. 2012). Ion beam treatments have resulted in the release of several improved cultivars in rice, wheat, potato, tomato, soybean, tree crops (ginkgo, Chinese chestnut, sea buckthorn), and many ornamental plants (e.g. chrysanthemum, carnation, dahlia, lily, petunia, rose and cyclamen) (Tanaka 2009; Abe et al. 2012; Feng and Yu 2012; Yamaguchi 2018).

Ultrasound, UV light, accelerated ageing and combined ^{32}P and gamma rays were used for plant mutagenesis with limited effectiveness recorded (http://mvd.iaea.org).

3.2.1.2 Chemical Mutagens

Several early researchers attempted to induce mutations in animals and plants by chemical mutagens (Morgan 1911; Sakharov 1932; Klein 1932). Frequently used chemical mutagens in plants include, alkylating agents, e.g. mustard gas, ethyl methane sulphonate (EMS), ethyleneimine (EI), *n*-ethyl-nitrosourea (NEU), *n*-methyl-nitrosourea (NMU), 1,4-bis-diazoacetyl butane (DAB), hydroxylamine (HA), diethyl sulphate (DES), dimethyl sulphate (DMS), methyl methane sulphonate (MMS), maleic hydrazide (MH), hydrogen fluoride (HF) and sodium azide (SA) (Shkvarnikov and Morgun 1975; Leitão 2012). These cause point mutations, and their effects are often silent or missense mutations (50%), with only about 5% nonsense mutations (Viana et al. 2019). Alkylating agents have one, two, or more reactive groups known as mono, di, or polyfunctional that react with DNA (Davies 1966). When the chemical agents interact with genetic materials, they generate alkylated O^6guanine, N^3adenine and N^3cytosine. During the repair of the alkylated base, O^6guanine tends to convert G:C to A:T base-pairs, N^3adenine gives rise to A:T to T:A transversions and N^3cytosine often results in C:G to T:A transitions and C:G to G:C transversions (Minocha and Arnason 1962; Leitão 2012). These single nucleotide changes generate truncations and missense mutations and can broaden allelic variation. Nonsense mutations may cause premature stop codons and splice site deletions (Tadele et al. 2009). The chemical mutagens are specific in their action, whereas radiation activity is random (Kempner 2011).

Ethyl methane sulphonate (EMS) is one of the most popular alkylating agents in plant mutation breeding and genetics. EMS produced over 25% of the chemically improved mutant cultivars due to its effectiveness and ease of handling, especially its detoxification by hydrolysis for disposal (Pathirana 2011). Other effective chemical mutagens include: *n*-ethyl-nitrosourea (NEU), nitroso methyl urea (NMU), colchicine, ethyleneimine (EI) and dimethyl sulphate (DMS) with over 14%, 10%, 9%, 7% and 4% of total released mutant cultivars, respectively. Over 70% of mutant cultivars produced through chemical mutagenesis are from these six chemical agents (http://mvd.iaea.org). Other non-alkylating mutagens include (1) nitrous acid and nitric oxide, (2) base analogues and related compounds, (3) some antibiotics, e.g. streptozotocin, mitomycin C or azaserine and (4) intercalating agents and topoisomerase. Their mutagenic effects and their applications in plant mutation induction have been reviewed by Leitão (2012).

From the non-alkylating mutagens, sodium azide (SA) (a common bactericide, pesticide and industrial nitrogen gas generator) is one of the most commonly used mutagens in plant breeding (Kleinhofs et al. 1974; Gruszka et al. 2012). Common chemical mutagens with their mode of action are listed in Table 3.3. However, it is essential to note that most of those chemical agents are hazardous to health and are carcinogens. An exhaustive list of the chemical mutagens used in mutation breeding with their effectiveness can be found in IAEA mutant database (http://mvd.iaea.org).

The following parameters need to be considered to establish optimal procedures for the chemical mutagenesis:

- Dose (concentration × time)
- pH
- Physical and chemical properties of the agents
- Interaction with the culture media
- Post-treatment condition

Table 3.3 Classes of chemicals with some examples of chemical mutagens and their mode of action

Chemical class	Chemical mutagens	Mode of action
Base analogues and related compounds	5-Bromouracil, 5-bromo-deoxyuridine, 2-aminopurine, maleic hydrazide, 8-ethoxy caffeine	Deletion, addition, frameshift, Chromosome breakage transitions (purine to purine or pyrimidine to pyrimidine) and tautomerization
Acridines	Acridine orange, acriflavine, ethyleneimine, proflavine	Transitions AT to GC, GC to AT
Mustards	Sulphur mustard, nitrogen mustard	Alkylate various sites in DNA
Nitrosamines	Dimethyl nitrosamine, nitrosoguanidine, nitroso methyl urea	Alkylate various sites in DNA
Alkylating agents	1-Methyl-1-nitrosourea, 1-ethyl-1-nitrosourea, methyl methane sulphonate, ethyl methane sulphonate, dimethyl sulphate, diethyl sulphate; 1-methyl-2-nitro-1-nitrosoguanidine, 1-ethyl-2-nitro-1-nitrosoguanidine, *N,N*-dimethyl nitrous amide, *N,N*-diethyl nitrous amide and others	Alkylate various sites in DNA
Epoxides	Ethylene oxide, diepoxybutane	Alkylate various sites in DNA
Azide	Sodium azide	Generates azidoalanine causing G/C to A/T transitions. Gene mutation
Miscellaneous	Nitrous acid, maleic hydrazide, hydrazine, hydroxylamine, antibiotics, free radicals	Transitions AT to GC, GC to AT, Chromosomal aberrations
Intercalating agents and topoisomerase	Ethidium bromide and proflavine daunorubicin, 4′,6-diamidino-2-phenylindole (DAPI)	Block transcription and DNA replication
Colchicine	–	Chromosome doubling affects spindle microtubule formation and promotes chromosome doubling

Some other influencing factors need to be considered for effective treatments:

- Pre-soaking
- Treatment duration
- The temperature during treatment
- Hydrogen ion concentration
- Dry back
- Post washing
- Storage of treated materials

Table 3.4 Advantages and disadvantages of physical and chemical mutagens

Mutagens	Advantages	Disadvantages
Physical	• Tested, proven, robust, ubiquitous application, non-hazardous, environmentally friendly, transferrable, cost-effective, non-regulated, produce rare mutant events and low mutational load (with specific doses) • Reproducible • High and uniform penetration in multi-cellular plant materials (except UV)	• Mutagenesis is random, thus selection methods for desired mutations are required • Mutation spectra are not well known, except for fast neutrons • High degree of sterility in treated plants • Dose rate needs to be determined for each genotype • Chromosomal and gene mutations can occur simultaneously
Chemical	• Tested, proven, robust • Point mutation predominant • Relatively less chromosomal damage • High mutation rates and densities are useful for gene-phenotype association studies • Different mutation spectra • Identification of multiple mutant alleles	• Mutagenesis is random and widespread • Chemicals are very hazardous, non-environmentally friendly and regulated • Penetration difficulties in multi-cellular plant parts • Difficulties in reproducibility • Sensitivity of some target tissues and organs • Difficult to assess effective dose and reproducibility • The dose rate needs to be determined for each genotype • Stability of the mutagen

3.2.1.3 Advantages and Disadvantages of Physical and Chemical Mutagens

The choice of physical or chemical mutagen depends on the objective and all have advantages and disadvantages (Table 3.4). Ionizing radiations have benefits that they can be used for every plant material, and protocols have been developed that provide a degree of accuracy and reproducibility. Irradiation services are available at local, regional, national and international levels, with many supported by the International Atomic Energy Agency or national atomic energy institutes.

In addition to the random action and the ability to produce small mutagenic effects in an otherwise unaltered genetic background, there are some limitations and challenges. Limiting factors for gamma ray emitters (typically ^{60}Co and ^{137}Cs) is that new and refurbished sources are highly radioactive, so shields need to be installed to allow low dose treatments recommended for plant materials with a high-water content. More importantly, there are now restrictive international regulations on the transport of radioactive isotopes, and this complicates or prevents the setting up of new gamma sources or the refurbishment of old ones.

A concern for fast neutron treatments is they cause a low level of temporary radioactivity in the irradiated materials that may pose a safety problem in post-irradiation handling and absorbed dose assessment. X-rays being non-radioactive are more user-friendly and accessible as they are used widely in other uses worldwide (e.g. blood irradiator, sterilization and radiotherapy in hospitals). However, there is a problem with uniformity using large samples. Thus, X-ray irradiators need to be equipped with sample rotating devices to provide even irradiation exposure.

Ion beam mutagenesis suffers from laborious sample preparation, limiting treated sample size, e.g. a plastic box with 5 × 7.5 × 1.25 cm, or plastic Petri dishes of 3, 6 and 9 cm in automatic ion beam irradiation system (Abe et al. 2012) and results in severe damage to treated material DNA. Despite the effectiveness of ion beams in generating beneficial and a new spectrum of mutations (e.g. Khazaei et al. 2018), it has a major disadvantage in that there are a limited facilities compared to gamma and X-rays.

Chemical mutagens are known to be effective in gene mutation and permit the identification of new alleles, which is of high interest to plant geneticists. Their major drawback for plant breeding is they produce a heavy mutational load. The chemicals used are extremely hazardous and require suitable infrastructure/laboratories for safety use. Furthermore, these chemicals have an impact on the survival of the treated plant materials and sensitive materials such as in vitro cells (e.g. microspores and protoplasts), tissues consisting of single cell layers (somatic and gametic cells) and immature embryos.

3.3 Determination of Optimum Dosage for Mutation Induction

Cells of higher plants respond to physical and chemical mutagens with various biological changes influenced by the developmental stage, environment and biological activity of the plant materials to be targeted. These factors significantly affect the efficiency and effectiveness of the mutation treatment and are defined as:

Efficiency—the ratio of desirable mutagenic changes to undesired effects
Effectiveness—the number of mutations produced per unit of dose

Damaging effects are proportional to the dose (physical) or concentration (chemical) received. However, mutagens can directly or indirectly damage cellular components and can cause perturbations in metabolic and physiological functions as well as genetic changes. Therefore, before developing mutagenic populations or mutation studies, it is recommended to conduct radio-sensitivity (physical) or toxicity (chemical) tests to determine suitable treatments. The optimum irradiation dose or chemical treatment should balance the required mutation frequency and plant injury. The ultimate goal is to induce enough interesting new variations in a viable population and in a population size that is manageable with breeding or genetic resources. In general, the lower the density of mutations the more mutant lines that need to be screened. To determine optimal treatments, it is advisable to evaluate the effects of a range of treatments. The type of plant propagule to be treated has a strong effect; some examples follow. In ionizing irradiation for seeds higher doses are used (e.g. 50 to >600 Gy for dry small grain cereal seeds). In contrast, low dose ranges are recommended (10–100 Gy, or less) for tissues with high-water content, e.g. stems, buds, embryos, anthers, microspores and floral tissues. For fast neutrons, typical doses are between 5 and 50 Gy for sunflower seed (Gvozdenic et al. 2009). When using chemical mutagens, such as EMS, for seed mutagenesis, the concentration ranges from 0.05 to 0.1 M (0.25% to >2%) for a duration of 0.5–24 h with a temperature of 25–35 °C whereas for in vitro plant materials a concentration up to 1% is normally applied (Kodym and Afza 2003; Mba et al. 2010). In sodium azide (SA) toxicity, a range of 1–50 mM for duration 2–24 h, e.g.: 1–4 mM at pH 3 for 3–6 h was effective for dry barley seed, but for in vitro plant materials the concentration ranged from 0.1 to 1.0 mM for 10 min for leaf callus of sugarcane and 6 h for barley anthers, and 0.1–10 mM for 1 h for immature maize embryos (Gruszka et al. 2012). Maghuly et al. (2016) reported that an EMS range of 0.4–1.6% for 1.5 h was optimal for jatropha in vivo cuttings.

The acidity of SA and the alkalinity of EMS solutions give rise to different post-treatment response in germination and seedling development. SA treatments also affect cellular metabolism and mitosis (Gruszka et al. 2012). In general, vegetative materials with active growth (high mitotic rate) and reproductive cells (meiotic cells) are more susceptible to mutagen treatment than dormant or quiescent materials such as seed (Lagoda 2012).

Plant materials for testing should be: clean, genetically and morphologically uniform, disease and pest free, homogenous with high viability/vigour and health. Radio-sensitivity or toxicity tests (as for mutation treatments) are influenced by biological activity (e.g. mitotic rate), genotype, and by environmental factors such as oxygen, water content, post-irradiation storage, and temperature (Mba et al. 2010). The suitable dose and choice of mutagen also depends on the aim, chiefly whether this is for breeding or genetic studies. The effects of mutation treatments are also subject to the variability in plant genera, between and within species, and the genotype.

Response to treatment is assessed in the first (treated) generation, i.e. M_1 of seed propagated plants and M_1V_1 of vegetatively propagated plants. Typically, measurements are taken on biomass (germination and growth rate, multiplication rate, seedling height, hypocotyl, and root length, reviewed Bado et al. 2015). One measure of sensitivity is the lethal dose (LD) for 50% of treated materials (LD_{50}) or 50% growth reduction (GR_{50} or RD_{50}) though alternative LD_{30} or GR_{30} may also be useful. For seed, early stages in germination and seedling growth are often measured, such as germination/growth rate, seedling height (for monocots, Figs. 3.2 and 3.3a), or hypocotyl length (for dicots, Figs. 3.2 and 3.3b). This allows treatment effects in the M_1 samples to be distinguished from delayed germination and seedling biomass (Kodym et al. 2012). These data are then expressed as a percentage over untreated samples (control) to estimate the sensitivity parameters LD_{50} or GR_{50}, or their alternatives at 30%, for mutation induction. The plotted data is typically a curve (Fig. 3.2) with stimulated growth at low doses, however for practical purposes a straight line is often fitted (Mba et al. 2010) or else a probit analysis which is a function of the inverse cumulative distribution function or quantile function associated with the standard normal distribution (Surakshitha et al. 2017). Examples

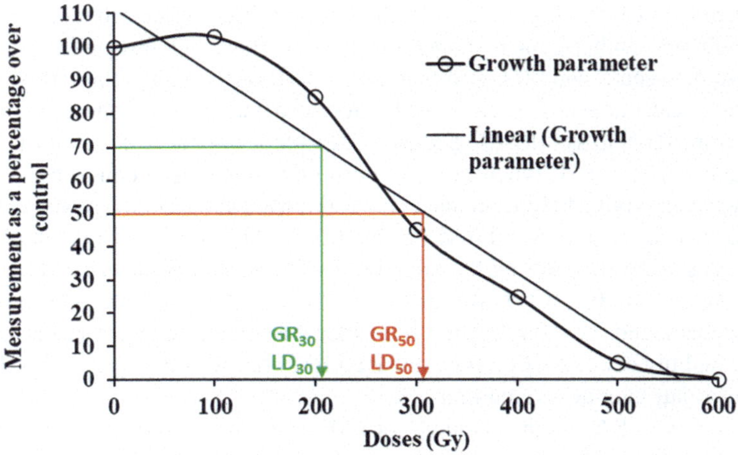

Fig. 3.2 Theoretical curve showing effects of irradiation on plant growth parameters (e.g. reduction of germination rate, seedling height, hypocotyl length, fresh weigh), note low doses can stimulate plant growth whereas optimal dosage for mutation induction causes reduced growth. Example of the GR_{30} or LD_{30} (green) and GR_{50} or LD_{50} (red)

of radio-sensitivity tests are given in Fig. 3.3, showing a theorical radiosensitivity test of monocot versus dicot seed to physical or chemical mutagens.

3.4 Mutagenesis

To improve crop plants, breeders and researchers have until recently relied on three main options: crossing, mutation and transformation. The choice is mainly guided by the following considerations: (1) is the target trait under genetic control, (2) can desired alleles be manipulated easily by crossing, (3) does linkage drag results in unwanted traits and (4) does crossing result in the break-up of refined linkage groups required for successful cultivars. If the choice is plant mutation breeding, then issues focus on the ability to produce a desired mutation with little or no effects on the background genome of the elite line selected. Other considerations include: space to grow on mutant generations (this is particularly problematic for tree crops that require large areas), time, resources and facilities for screening large mutant populations (either phenotypically or genotypically). The target genotype is normally a top performing cultivar or elite breeding line (that is deficient for a particular trait). Target genotypes are often popular with respect to market demands and adapted to local environments or favoured non-field growing conditions (e.g. hydroponics and indoor farming).

In principle, any live plant material can be used for physical and chemical mutagenesis; however, the plant material must have the ability to grow. Thus, most materials include a meristem (apical and side meristems or embryo), or tissue

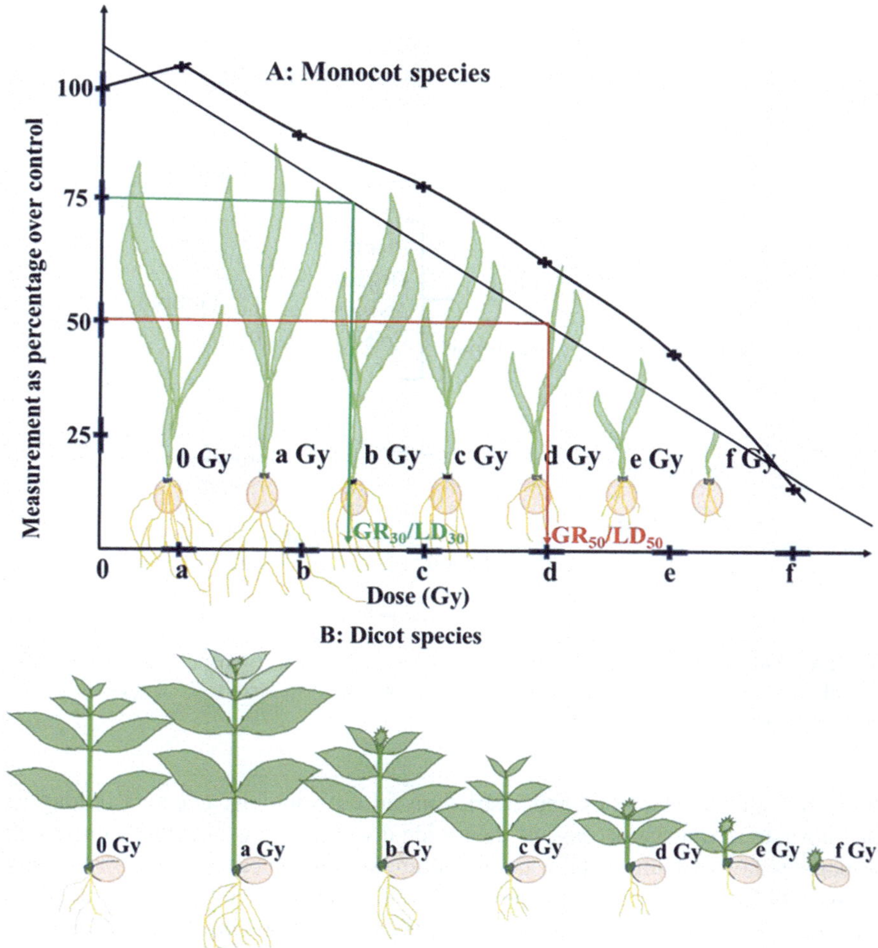

Fig. 3.3 Radio-sensitivity or toxicity test of a seed plant crops: (**a**) Monocot species with measurements of seedlings, (e.g. rice, wheat, maize, sorghum) and in vitro (e.g. banana, ginger and cocoyam); and (**b**) dicot species assessed by hypocotyl or epicotyl length, inter node (e.g. peanut, Bambara groundnut, pea, sunflower) and in vitro (e.g. potato, sweet potato, yam and cassava). Example of a, b, c, d, e and f for monocot species (0, 100, 200, 200, 300, 400, 500 and 600 Gy) and dicot species (0, 50, 100, 150, 200, 300, 400 Gy) and in vitro materials (0, 5, 10, 15, 20, 40 and 60 Gy)

culture methods in which cells, tissues or organs can be cultured. Totipotency of plant cells can be exploited to regenerate plants after mutagenesis via a range of techniques (Jankowicz-Cieslak et al. 2012; Chen et al. 2013; Bado et al. 2016; López et al. 2017; Suprasanna et al. 2012). A summary of live plant materials in plant mutation breeding is given in Fig. 3.4.

The major decisions for adopting tissue culture-based approaches for a mutation breeding are given in a flow diagram (Fig. 3.5). These include access to local

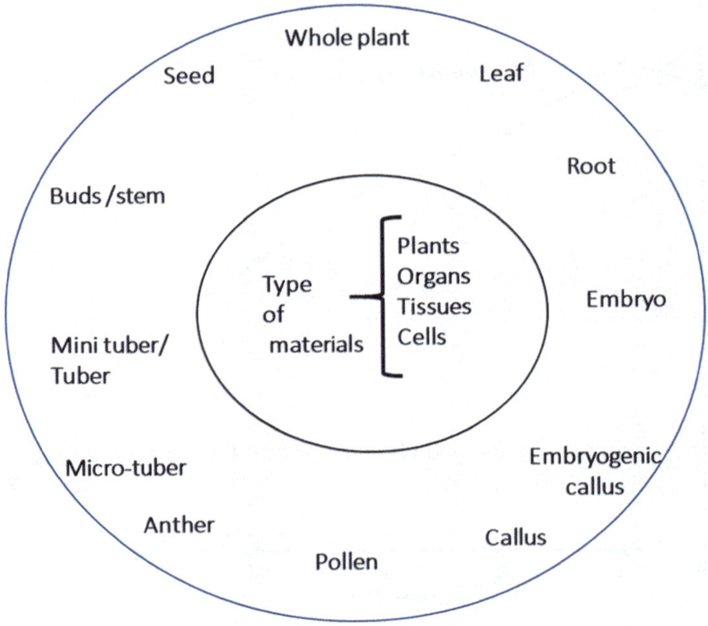

Fig. 3.4 Plant materials used in mutation breeding: single cells (somatic and gametic cells, protoplast, leaf disc, etc.) and multi-cellular systems (whole plant, seed, bud, stem cutting, graft, tuber, mini- and micro-tubers, etc.)

mutation induction facilities and restrictions in out-sourcing, chemical mutagens may have an advantage because of their wide diversity and availability. Single-cell propagules favour multi-cellular tissue culture systems because in vitro screening can be done easily and on large numbers of callus, embryogenic callus and plantlets. Gametic cells, being haploid, have an advantage in that on chromosome doubling, e.g. in doubled haploid production mutation events are made homozygous instantly and thus the mutant phenotype can be seen relatively quickly (in other materials mutations occur in one allele and since most induced mutations are recessive these materials need to be selfed to produced individuals that carry homozygous mutant genes which show the changed phenotype). However, mutation breeding using single-cell propagules is often laborious with low regeneration into plantlets and requires skills, resources, and facilities; often the target genotype is one which is responsive to tissue culture but not one that is in demand by breeders and end-users. This explains the limited adoption and success of single-cell systems in plant mutation breeding, which have generated less than 9% of improved cultivars in vegetatively propagated crops (Bado et al. 2017). However, tissue culture offers great potential especially for vegetatively propagated plants.

In developing a mutation breeding programme, one must be aware of the likelihood of success and the numbers of individuals and progenies that need to be screened. The most common mutant traits induced are late flowering, pale green

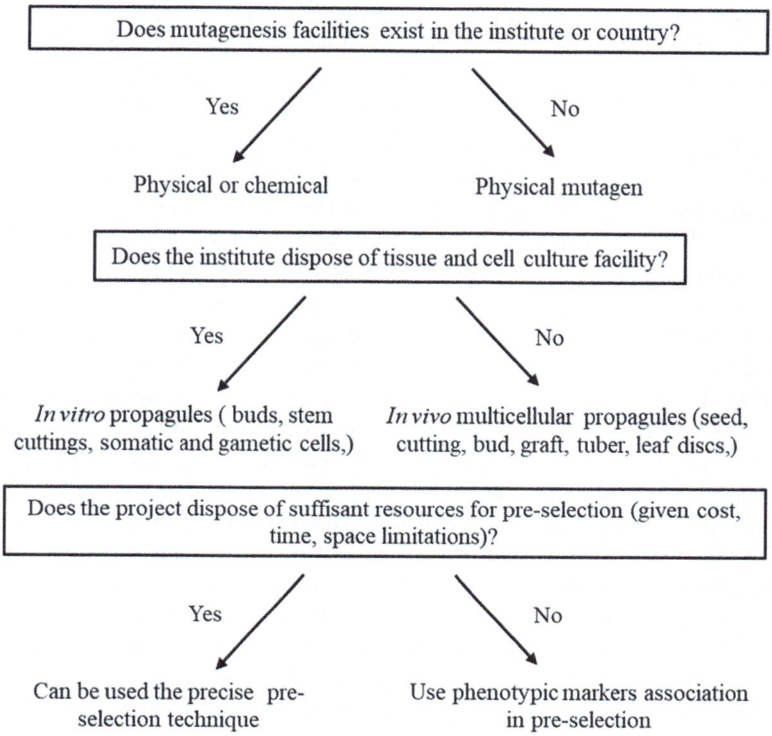

Fig. 3.5 Decision diagram for the development of mutation breeding and mutation genetics programmes

leaf colour, short stature and sterility, collectively these can make up over 50% of all mutants generated (Forster et al. 2012). To induce lodging-resistant mutants in oat, barley, bread and durum wheat with physical and chemical mutagens about 1200 seeds are required, whereas in the production of pest and disease resistance about 10,000–20,000 seeds are needed for mutation treatment (Donini and Sonnino 1998). Since induced mutation is a random process, large mutant populations need to be developed for screening. The success of mutation breeding depends on multiple factors, either external (i.e. resources, space, time, and facilities) and internal (i.e. choice of plant materials and mutagens, cost and space in laboratories, nurseries and field). The mode of mutagen application involves internal factors (breeder's choice) and external factors (availability of the mutation induction facility).

3.4.1 Mutagenesis of Seed

Seeds are the most commonly used targets for mutation treatment. This is because they are a common material used by breeders, they are in a dormant/sessile state and can be easily transported, treated, returned, stored and grown on. As stated above a radiosensitivity or toxicity test is recommended as a first step (Sect. 3.3). The seed for treatment should be healthy and uniform (both genetically and physiologically). For physical mutagenesis water content must be controlled, this is normally done in a desiccator. General pre-treatment procedures (parchment removal, storage desiccation, pre-soaking, etc.) can be found in Mba et al. (2010 and references therein).

After treatment, the parental M_0 seeds is referred to as M_1 seeds, i.e. the first mutant generation. In addition to carrying mutations, these seeds are also physiologically weak due to treatment and require to be grown in optimal conditions. Physiological imbalance can promote out-crossing and isolation (bagging) of floral parts may be required to prevent this (Ukai and Nakagawa 2012). In addition to physiological aberrations, M_1 plants are often chimeric and since most mutations are recessive and in a heterozygous state, thus M_1 plants cannot be used for phenotypic screening. It has become a dogma in plant mutation breeding that the earliest selection for mutant traits cannot be carried out until the M_2 at the earliest (Fig. 3.6, Black arrow scheme). However, with the advent of DNA analysis, mutations in specific genes can now be screened for and individual M_1 plants carry mutations in target genes can be selected and grown on. This is a major change in philosophy, as selection can be done a generation earlier, thus (1) eliminating the vast number of M_1 plants of no interest and (2) saving time (to produce subsequent generations), costs and subsequent laboratory, nursery and field space (Fig. 3.6, red arrow scheme). These factors make mutation breeding in perennial crops, e.g. trees more attractive, as an example mutation breeding in oil palm is now a reality (Nur et al. 2018). A limitation in genotypic selection is that DNA needs to be extracted and analysed, but there are an increasing number of service providers and such analyses can be out-sourced.

In mutagenesis of haploids (e.g. gametic cells), any interesting and exciting observations in M_1 materials can be fixed at an early stage through doubled haploid technologies (Fig. 3.6, green scheme, Zhao et al. 1994). M_1 plants (either the whole population or selected individual mutant genotypes) are grown up till maturity. Depending on the breeding system of the species (in-breeder or out-breeder), the M_1 plants may be selfed or crossed to produce the M_2. Although genotypic selection can be practiced in the M_1, it is the M_2 generation that provides the earliest display of the mutant phenotype, which is the ultimate target for breeders and geneticists. Space is therefore required to grown up the M_2 plants, and since many traits of interest are related to yield (or impact on yield indirectly) this requires growing to maturity and harvest. M_2 individuals may be grown on to produce M_3 families (and above) containing specific mutant traits (a common practice in gene-phenotype studies, e.g. Targeting Induced Local Lesions in Genomes (TILLING) or used directly in plant breeding).

3.4.2 Mutagenesis of Vegetative Propagules

The number of mutant cultivars in vegetatively propagated crops lags way behind seed propagated crops by 793 (23.80%) to 2539 (76.20%). However, the picture is

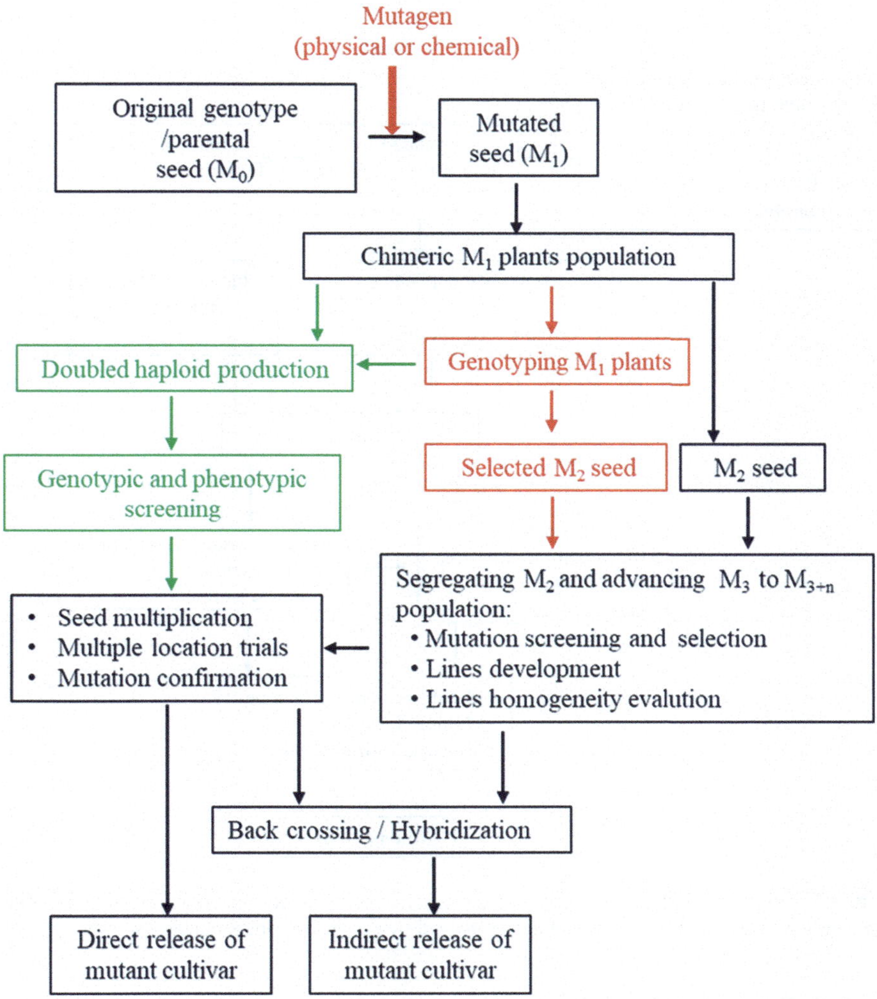

Fig. 3.6 Scheme of mutation breeding using seeds as plant propagules for mutation induction, showing: the commonly adopted strategy (previous dogma in black arrow), accelerated scheme mode through doubled haploidy (green arrow) and M_1 genotypic screening and selection

changing as advancements in cell/tissue culture techniques enable new approaches. Vegetative propagules, for mutation induction can be classified into two categories: (1) single-cell propagules (somatic and gametic calli, embryogenic calli, cell suspensions, microspores and protoplasts) and (2) multi-cellular propagules (stem cuttings, buds, mini-tubers, micro-tubers, tubers, whole plants, scions, grafts, bulbs, etc.) (Fig. 3.7). Vegetatively propagated plants usually have no or a very limited breeding system. Therefore, a major objective in producing mutations (the vast majority of which are recessive) is to knock-out the dominant allele at heterozygous loci of interest, thus producing a double recessive locus. This can be done with

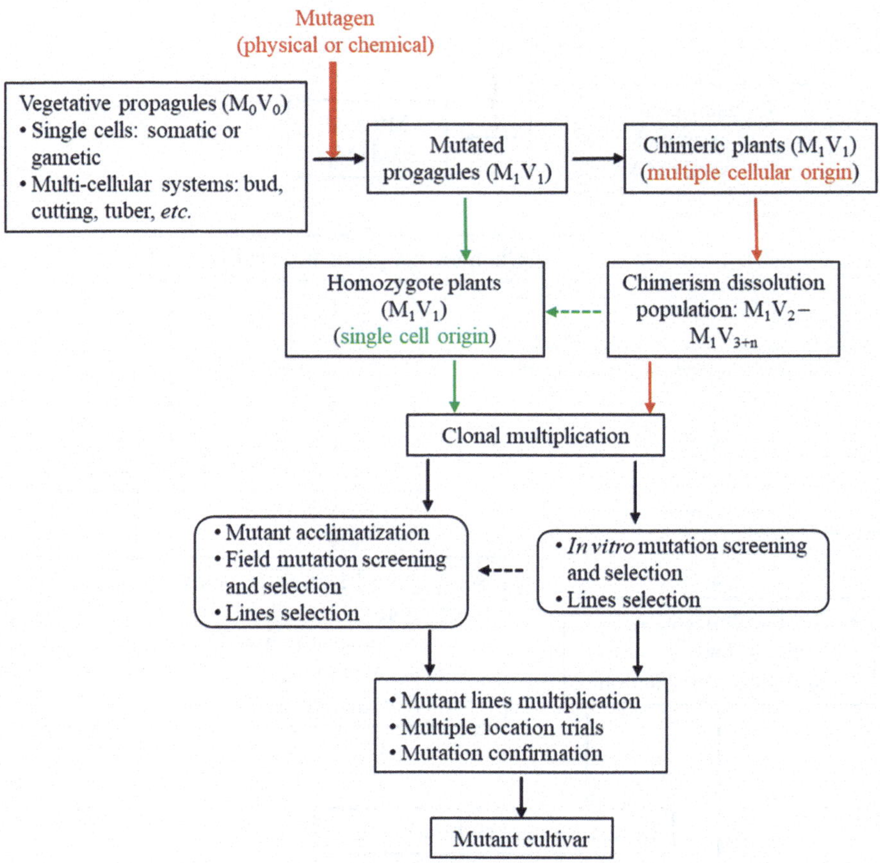

Fig. 3.7 Scheme of mutation breeding using vegetative materials for mutation induction showing the commonly adopted strategies and an accelerated mode through double haploidy

gamma and X-ray irradiation, and chemical mutagenic treatments using LD_{50} or LD_{30} treatments. Alternatively, high dose treatments and ion beam irradiation can be used to delete parts of the genome. As with seed treatments, mutagenesis of multi-cellular explants produces chimeric plants (M_1V_1). Outer layer cells/tissues may constitute a barrier to the target meristematic cells in both physical and chemical mutagenesis. A comparison of recommended/commonly used treatments is given by Bado et al. (2017) for several vegetatively propagated crop species.

The starting materials for treatment are termed M_0V_0, which become M_1V_1 after mutagenesis (Fig. 3.7). Treated multi-cellular explants (M_1V_1) normally have to undergo multiple sub-cultures to produce homohistont explants (solid mutants in which chimeras have been dissolved) which are uniform for the mutation and phenotype. At least three sub-cultures or propagation cycles are recommended in treated buds or stem explants to stabilize the mutated sectors before mutant selection (Broertjes and van Harten 1988; Roux et al. 2003; Sutarto et al. 2009; Jankowicz-

Cieslak et al. 2012). The more heterozygous are genotypes the more phenotypic mutations will be (Spiegel-Roy 1990). An alternative method for mutation event isolation is the detachment of the variegated sectors of mutated explants (possible in the M_1V_1 generation, but more normally practiced in higher generations, e.g. M_1V_{3+n}) to develop homohistonts through regeneration or somatic embryogenesis (Fig. 3.7, dash arrow), this has been a successful technique mutation breeding in ornamental plants (Dwivedi et al. 2000; Datta et al. 2005; Datta 2020).

Single cells have the advantage of generating solid mutant plants directly. The regenerated plants (M_1V_1) may be screened early and desired mutants selected (Suprasanna et al. 2012). The advanced and selected lines can then be micropropagated to produce enough plants for acclimatization and multi-locational field trialling. Stabilized mutants with suitable yield may then be released as direct mutant cultivars.

Other propagules of vegetative propagated plants can include seed (in facultative species) as well as somatic or gametic cell cultures of seed propagated plants (Saleem et al. 2005; Serrat et al. 2014). For facultative species cross breeding is possible to introgress, the mutation into other cultivars.

3.5 Combined Mutagenic Treatments

Combining two or more mutagenic treatments date back to the 1950s (Oster 1958; Moutschen 1960; Konzak et al. 1961; Merz et al. 1961; Moutschen and Moutschen-Dahme 1961; Soriano 1963; Favret 1963; Valeva 1964; Wallace 1965; Aastveit 1968; Doll and Sandfaer 1969). Two improved ornamental cultivars of streptocarpus, 'Cobalt Nymph' and 'Purple Nymph' were produced in The Netherlands in 1969 by combining physical (X-ray) and chemical (colchicine) treatments (http://mvd.iaea.org). Combined mutagenic treatments have not been as attractive in plant mutation breeding as single-acting mutagens. To date, only 37 improved cultivars have been documented out of a total of over 3330 (>1%) registered mutant cultivars (http://mvd.iaea.org, Fig. 3.2). Reports of combined mutagenic treatments in plant mutation breeding have declined with only three improved cultivars released after 2000. However, there are many recent reports on the effects of combined treatments by researchers and breeders (Mehandjiev et al. 2001; Bhat et al. 2007; Goyal and Khan 2010; Khan and Tyagi 2010; Makeen and Babu 2010; Srinivas and Veerabadhiran 2010; Channaoui et al. 2019). The highest number of combined mutagens resulting in a successful mutant cultivar involved five chemical mutagens (NEU, NMU, DES, DMS, and EI) resulting in the high yielding cultivar 'Prikarpatskie 4' of field bean in Russian Federation in 1986 (http://mvd.iaea.org). Konzak et al. (1975) were among the first to report a synergistic effect of SA on mutation yields following MNH treatment. Synergistic increases in the frequency of chimeras, chlorophyll-deficient mutations, and morphological mutations were observed in both SA post-irradiation treatments and pre-EMS treatments with gamma irradiation of barley seed by Cheng and Gao (1988). Wani (2009) also recorded synergism by combining gamma rays and EMS treatments in chickpea.

The efficiency and effectiveness of the combined mutagens, as with single treatments is influenced by the applied doses/treatments (Olejniczak and Patyna 1981; Srivastav and Raina 1981; Chauhan and Patra 1993; Olejniczak 1994), the mode of application (pre, post, or simultaneous), and the plant materials (seed and vegetative propagules). For instance, in seed mutagenesis with ionizing radiation, water content is an influencing factor affecting efficiency and effectiveness. Jafri et al. (2013) recorded a better response inducing chromosomal aberrations with combined gamma ray and EMS treatments of coriander. The combined mutagens (fast neutrons + NEU) for mutation induction in pea were the most efficient and produced the highest number and the widest of spectrum mutations. Other examples include barley (Aastveit 1968; Mohan 1972), black gram (Lal et al. 2009) and rapeseed (Channaoui et al. 2019).

However, non-synergistic effects have been observed in several other studies of seed treatments (e.g. Olejniczak 1994; Girija and Dhanavel 2009). In vegetative propagules, combined mutagenic treatments were applied to detached leaves of *Streptocarpus* sp. and rooted cuttings of *Bougainvillea* sp. As previously mentioned, the combination of X-rays and colchicine produced two mutant cultivars 'Cobalt Nymph' and 'Purple Nymph' (http://mvd.iaea.org). In bougainvillea, the combined treatment of gamma rays and colchicine resulted in two mutant cultivars, 'Silver Top' and Lady Hudson, released in 1978 and 1979, respectively (http://mvd.iaea.org). With these four exceptions there are no other published reports (to our knowledge) of this kind for other vegetatively propagated plant or vegetative propagules.

3.6 Mutations Screening

Mutagenic populations for desired traits and alleles can be selected for by classic phenotypic and/or more modern genotypic screening. The previous dogma was that mutation selection could only be performed at the M_2. Screening in the M_1 has been considered heresy. This is because M_1 plants contain chimeras, physiological disorders and usually do not express the mutant phenotype (as the vast majority, >90%, are genetically recessive). However, with the meteoric rise in DNA analytical methods and an increasing focus on DNA sequencing in plant breeding and genetics, high-throughput and affordable genotypic screening methods are now available (e.g. High Resolution DNA Melt and Diversity Arrays Technologies) that can be applied in mutation detection. Genotypic screening of mutants can be applied to M_1 generations as it is independent of plant health, physiology, phenotype and environmental effects. This presents a massive step forward for plant mutation breeding as selection of (usually rare) plants carrying mutations of interest can be selected earlier, and the vast majority of plants of no interest rejected, thus saving not only time, but future investment in development. This opportunity can be applied to all plant species, but is of particular relevance to perennial crops, e.g. trees that have long generation times, and which require a lot of time and space in nurseries and in the field. A limitation is that the target gene for mutation is known and sequence data are available, but the genomes of most major crop species and important genes are being revealed and other crops will follow suit. It is true that some selected M_1 plants will carry chimeras, but these can be dissolved in producing the next generation (by sexual or vegetative propagation), the most important issue for the breeder is to capture and exploit as early as possible the mutant alleles of interest.

Mutants derived from gametic cells and somatic cells can also be screened at the first generation, and these do not produce chimeras. Gametic cells and doubled haploidy have been of interest as mutation produced in haploid gametic cells are fixed immediately as homozygotes in the doubled haploid and thus the mutant phenotype can be observed (Fig. 3.6, green scheme, Castillo et al. 2001). In addition, cell culture allows the production of large population sizes in a short time for mass in vitro mutation induction and screening. Early examples of doubled haploid mutation screening in vitro include salt tolerance in wheat (Zhao et al. 1994, 1995). However, it is essential to note that many agronomically useful traits (e.g. resistance to biotic and abiotic stresses, and notably yield) are not expressed in single cells or callus or young plants (Maluszynski et al. 1995; van Harten 1998).

Regarding phenotypic screening of traits, some may require simple procedures such as visual selection, others need more elaborated investigations (see Sects. 3.4.1 and 3.4.2). It is important to point out that although mutation detection can now be performed in the M_1 via genotypic selection, the ultimate test is the observation of the desired phenotype and no detrimental effects. Thus, plant performance of M_2 individuals and M_3 families remain key assessment stages in plant mutation breeding.

Phenotypic traits may be qualitative or quantitative traits (Donini and Sonnino 1998; Makay et al. 2020). For qualitative (single gene) traits (e.g. morphological and physiological traits, response to daylength, vernalisation), the mutation can be detected on a single plant based on visual inspection or combined with appropriate (e.g. physiological) screening procedures. These screens are often destructive and can be carried out only on generations above the M_2 for seed crops or M_1V_3 for vegetatively propagated crops. For phenotypic quantitative trait screening (e.g. yield, quality, biochemical content, mineral content, seed or fruit size, weight, and density), mature plants are generally required though biochemical, biometric, physical, or mechanical methods may be applied as indicators. Here the traditional practice is to screen a mutant family usually the M_3 of seed propagated plants and the M_1V_3 of vegetatively propagated plants, but if the target gene is known the M_1 plants may be screened directly (see above).

Most mutant cultivars have been developed from forward-genetic (phenotypic screening) (Pathirana 2011). However, DNA technologies now allow the evaluation of mutant individuals and populations genotypically. One of the first examples has been the identification of a novel allele ($waxy_{E1100}$) in the *granule-bound starch synthase I* gene (waxy) from seed EMS mutagenesis. The novel allele was introgressed into elite breeding lines lacking granule-bound starch synthase I protein activity to produce genotypes with high-amylopectin starch (Muth et al. 2008).

Mutation screening based on heritable phenotypes remains useful and also essential for performance testing of potential cultivars, and this does not require genetic information. It is a useful and traditional method (Oladosu et al. 2016) and this has been the traditional selection method of breeders for decades. However, the specific mutation(s) mediating the phenotype may remain undiscovered. Therefore, screening a mutagenized population by phenotype should be followed by characterization of the molecular event underlying the modified character if possible (Sikora et al. 2011, see below for methods).

Several methods have been developed to identify a mutation in a gene of interest, but all have their limitations. Techniques rely on mobility change during gel electrophoresis of DNA, such as single-strand conformation polymorphism (SSCP) and denaturing gradient gel electrophoresis (DGGE), these provide a rapid screening method but they do not show the location nor the type of variation in the DNA fragment (Gilchrist et al. 2006). Methods that are based on polymerase chain reaction (PCR) require the knowledge of the target DNA sequence and only work for small fragments of DNA. Array hybridization techniques are efficient but require a large amount of genetic material and have low sensitivity (discovering approximately 50% of the single nucleotide polymorphism (SNPs)). Conventional sequencing (Sanger) is the most accurate approach for candidate genes from multiple genotypes; however, efforts and cost for detecting multiple loci in large numbers of individuals are relatively high (Gilchrist et al. 2006).

A number of rapid and sensitive reverse genetic screening methods, like denaturing high-pressure liquid chromatography (dHPLC), a post PCR method such as high-resolution melting (HRM), and single strand-specific endonucleases like CEL I, have been used for mutation detection (Shu et al. 2012a, b). However, changes in DNA base modification after the initial screening for the HRM and CEL I methods are required and normally confirmed by re-sequencing (Szurman-Zubrzycka et al. 2017).

TILLING (Targeting Induced Local Lesions in Genomes) as a reverse genetic technique holds much potential for examining natural and induced variation and bridging the phenotype-genotype gap (Gilchrist and Haughn 2005; Till et al. 2003; Comai et al. 2004). Many of the species for which complete genome sequence data are available, including the tree crops such as papaya, cocoa, peach, apple, avocado and oil palm *are* amenable to classical mutagenesis and genetic analysis. For these organisms, identifying natural genetic variants can provide much information about gene function and can also be useful for association mapping and linkage disequilibrium analysis. TILLING techniques are able to detect small deletions, insertions, and microsatellite (SSR) polymorphisms in addition to single-bp changes in the DNA sequence. New methods were introduced by Albert et al. (2007), who enriched genomic loci for NGS using Roche NimbleGen oligonucleotide arrays, and Hodges et al. (2007), who applied the arrays to capture the full human exome. Since then, methods requiring less hands-on work and smaller amounts of input DNA have been developed (Sulonen et al. 2011). A solution-based oligonucleotide hybridization and capture method based on Agilent's biotinylated RNA baits was described by Gnirke et al. (2009) and developed further (Hoischen et al. 2010; Harbour et al. 2010). These methods were used in soybean (Haun et al. 2011; Bolon et al. 2011) to find the genomic variation among individuals of reference cultivar Williams 82 and of a Fast Neutron mutant population, respectively. To characterize large genomes like wheat

(Winfield et al. 2012) and pine (Neves et al. 2013; Henry et al. 2014; Hussain et al. 2018) exome capture sequencing (ECS) helped to identify a large number of novel SNPs. In general, most of the functional variation resides in the coding regions of the genome. Thus, it is logical, to begin with, exons for the discovery of the causative genetic variants of important traits (Cosart et al. 2011; Bamshad et al. 2011).

Significant efforts towards large-scale characterization, screening, identification, and discovery of SNPs-mutants and short InDels were made with success in crop plants using TILLING combined with Next Generation Sequencing (NGS) (Gupta et al. 2017; Tsai et al. 2011). Fortunately, recent advances in genomic technology and computational methods make it possible to identify the genetic basis of variation using genome-wide approaches more cost-effectively (Hillier et al. 2008; Ng et al. 2009), such as WG (whole-genome) sequencing (Hillier et al. 2008), RNA-seq (Hansey et al. 2012), methylated DNA (Brunner et al. 2009), genotyping-by-sequencing (GBS) (Huang et al. 2009), and restriction-site associated DNA (RAD) sequencing (seq) (Baird et al. 2008). Compared to WG sequencing, the approaches mentioned above are more cost-efficient to provide genetic information through: (a) minimizing representation of uninformative genomic regions and (b) enabling efficient sample pooling strategies (Beissinger et al. 2013; Schlötterer et al. 2014). While RAD-seq involves sequencing DNA fragments with high coverage, GBS's focus is to interrogate DNA sequences with low target coverage and minimize reads in repetitive sequences (Beissinger et al. 2013). Both RAD-seq and GBS identify genetic markers that are usually not functionally relevant (Maghuly et al. 2018). Since RNA-seq involves the sequencing of mRNA, it targets only expressed transcripts. ECS focuses on sequencing and analyses of the genomic regions most likely to be functionally relevant (Schlötterer et al. 2014). However, identifying specific genome modification demands further reference genome, which is commonly not given (Grohmann et al. 2019).

In recent years, due to rapid advancement in NGS and bioinformatic tool kits, several novel mutation mapping strategies such as a forward genetic systems for pinpointing causal mutations have been developed including, mutational mapping (MutMap), MutMap-Gap, and mutant chromosome sequencing (MutChromSeq) (Periyannan 2018; Kumawat et al. 2019). They provide enormous information about the function of most of the unknown genes. While the MutMap method is suitable for identifying a causative mutation in the plants with small genomes with available high-quality reference genome sequences, the MutMap-gap method is helpful in the absence of a reference genome sequence, where gaps or rearrangement at the targeted sequence will be filled by de novo assembly. In plants with complex genomes or high ploidy levels, DArTseq (Diversity Arrays Technology) and MutChromSeq provide better alternatives to detect mutations (Baloch et al. 2017; Periyannan 2018). These techniques reduce genome complexity and facilitate rapid gene isolation by a combination of chromosome purification with mutagenesis. Although MutChromSeq is a powerful method for large complex genomes, it requires preliminary knowledge of targeted genes' map position on the chromosome and a reference genome.

All the methods mentioned above rely on short-reads, which are cost-effective, provide accurate data, and are supported by various bioinformatic tools; however, this platform is not suitable for long insertion sequence and structural variation detection. In this case, long reads of third- and fourth-generation sequencing can offer more accurate genome assemblies, detect structural variations, and preserve base modification by eliminating amplification bias in the case of native DNA and RNA. But the error rate of long-read sequencing platforms is high. Therefore, a combination of short- and long-read sequencing can overcome the weakness of both techniques (Grohmann et al. 2019; Zhao et al. 2019; Amarasinghe et al. 2020).

As stated above, it is recommended that mutants detected via phenotypic screening should be investigated to discover their underlying genetic modification. This is not simply for academic understanding but is of practical significance. Genetic markers can be developed from the identified mutational event, and these can be used in selection in subsequent cycles of plant breeding (marker-assisted selection). Although field trialling remains an essential stage in the development of new cultivars, the number of lines that are advanced to this stage can be focused on those carrying elite (proven) genetic backgrounds (defensive breeding) and which carry new mutant alleles (targeted breeding) to meet new market demands, and for this there is an increasing emphasis on breeding by genotype.

3.7 Impacts of Plant Mutation Breeding in Crop Plant Improvement

A major objective of mutation induction has been to improve well-adapted plant cultivars/elite lines with a deficiency in one or two traits, either directly or through the production of pre-breeding materials. Mutation has resulted in new and now established ideotypes, e.g. semi-dwarfs for mechanical harvesting, and such mutations are now present in many crops. The original mutant cultivar is termed a foundation cultivar (Mlcochova et al. 2004; Rutger 2009). Significant mutant traits include yield, plant morphotype (growth habit, semi-dwarf), flowering time, abiotic and biotic resistance, physiological traits, nutritional value, or processing and sexual reproduction (autogamy or allogamy) mutants (Konzak et al. 1984; Bado et al. 2015; Oladosu et al. 2016).

Food security has received particular attention from the joint FAO/IAEA division of the United Nations which has provided services, training and capacity building in plant mutation breeding and genetics for member states. There have been successes in a range of crops: cereals (>47% of induced mutant cultivars), ornamentals and flowers (>21%), legumes and pulses (>13%), oil crops (>3%) and others including medicinal plants, fodder crops, (>13%). Achievements can also be classed by mutagen treatments. Single physical treatments account for 77.51% of mutant cultivars, single chemical treatments for 11.53%, combined mutagens (1.11%) and somaclonal variations (spontaneous mutation, about 1%) (Fig. 3.8).

Mutation induction by physical and chemical mutagens has resulted in the widening of genetic variability in over 220 plant species (Bado et al. 2015,

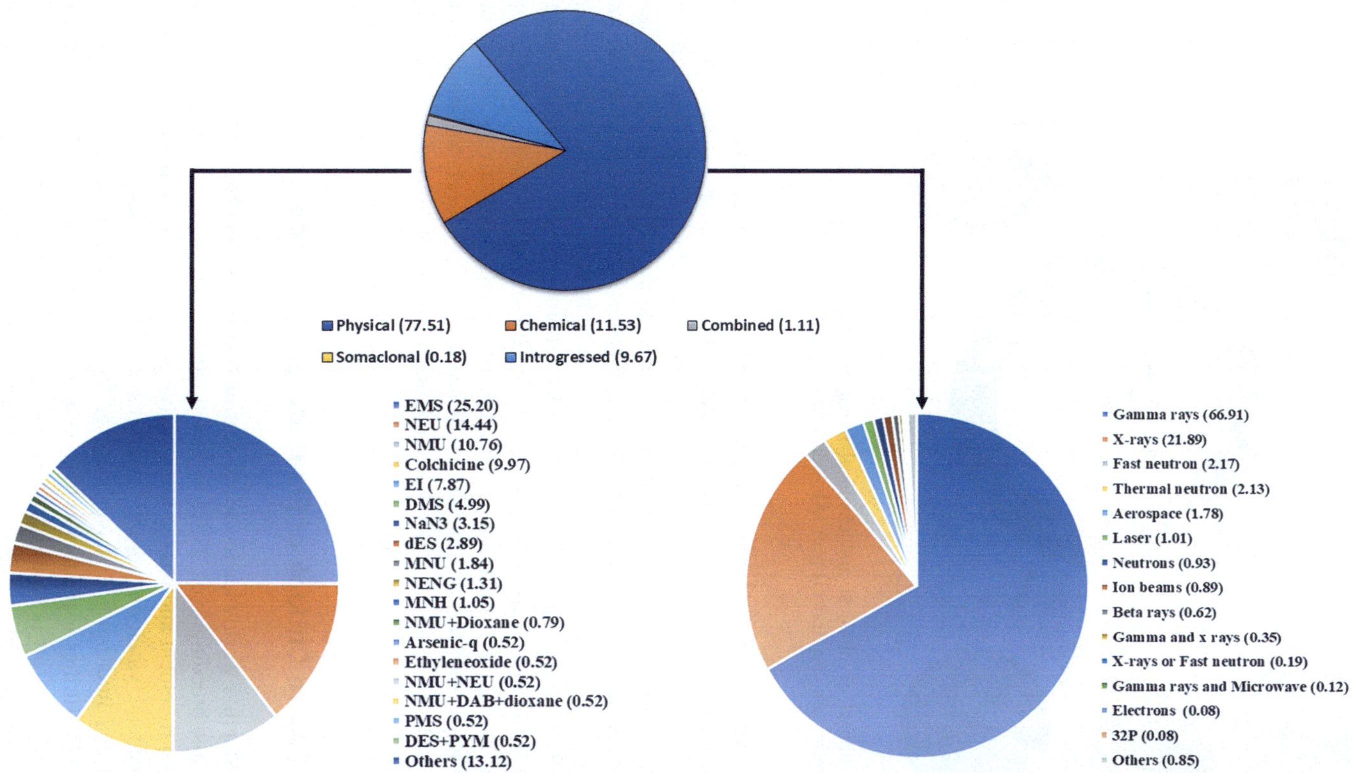

Fig. 3.8 Distribution of mutant cultivars as a percentage of the total number if mutant cultivars (From IAEA data base, accessed July 2020)

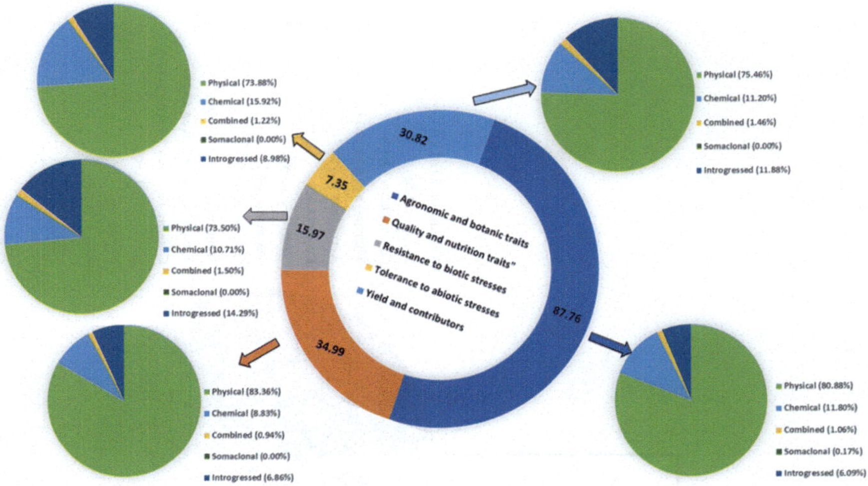

Fig. 3.9 Impact of physical and chemical mutagens (single and combination treatments) in inducing important crop traits (From IAEA data base, accessed July 2020)

Fig. 3.9). Agronomic and botanic traits modified include yield, quality, morphotype, life cycle, seed shape and size, seedling vigour, high biomass, flower size, shape and colour, fruit morphology and anatomy, shattering, early flowering and maturity, late maturity, winter hardiness, resistance/tolerance to abiotic stressors, wide adaptability, improved taste, elimination of toxins, etc. Physical mutagens have been the most effect in producing new traits, followed by chemical mutagens, combined treatments and finally somaclonal variation (Fig. 3.9). Improved mutant cultivars have a presence worldwide with over 60%, 28%, 7%, 2% and about 1% in Asia, Europe, America, Africa and Australia and Pacific continents, respectively. These new cultivars have contributed to wealth at all levels, from producer, to farmer, to the end user and nationally (see reviews from Kume et al. 2002; Ahloowalia et al. 2004; Bado et al. 2015, 2017; Oladosu et al. 2016).

3.8 Future Outlook of Plant Mutation Breeding

Plant production worldwide has led to narrow bottleneck in genetic variation among cultivars and breeding lines. This low genetic diversity represents a severe limitation to crop improvement and has led to extensive gene capture expeditions to centres of natural genetic variation among landraces and wild species. Various consortia (national and international) have been set up in germplasm collection of various crops, but the further a breeder has to go genetically to find a trait/allele of interest the more breeding effort is required. Moreover, countries housing wild germplasm are becoming increasingly protective about their natural resources and strict treaties have been signed in this respect (Moore et al. 2007).

Mutation induction is an attractive alternative to broadening genetic variability in elite lines as it permits the improvement of plant genotypes quickly (Roychowdhury and Tah 2013). This chapter as focused on the induction of mutations for plant breeding using physical and chemical mutagens. These approaches are relatively fast, cheap and simple and have served plant breeding well and will continue to do so for some considerable time. New methods in mutation detection, via genotyping, are set to have an impact not only on more rapid and specific mutation detection, but also application to wider groups of plants, notably perennial crops which have lagged behind annual crops especially in genotypic mutation selection. There are also exciting developments in mutation induction via gene editing, e.g. Clustered Regularly Interspaced Short Palindromic Repeats (CRISPR/Cas) (Deltcheva et al. 2011; Jinek et al. 2012) for which the inventors Emmanuelle Charpentier and Jennifer A. Doudna won the 2020 Nobel Prize for Chemistry (Wu et al. 2020). Advantages of plant mutation breeding include: (1) use of any type of plant material (Fig. 3.5) (2) can induce novel alleles in a gene and does not require knowledge of the gene controlling the trait, which is prerequisite when considering other methods; and (3) plant mutation breeding is an accelerated form of breeding (e.g. especially relevant to perennial crops). However, typical of innovation, new methods are normally initially expensive and require specialized facilities.

3.9 Conclusion

Traditional plant mutation breeding has had significant impacts worldwide. The effectiveness of physical and chemical mutagenesis has increased through enabling technologies, particularly tissue culture for vegetatively propagated plants and mutation detection and selection via high-throughput phenotyping and genotyping. Thus, more plant species can benefit from these approaches. Considerations for plant mutation breeding are:

1. Mutation induction: Greater use of locally accessible mutagen facilities. The resurgence of X-ray irradiation for mutation induction as an alternative to gamma irradiation. Less toxic and more environmentally friendly chemical mutagens will improve the diversity in the choice of mutagens by breeders and researchers.
2. Plant propagule: The increasing availability of improved methods in cell, tissue and organ culture of crop species which are not genotype dependent.
3. Selection: High-throughput phenotyping and genotyping and the development of marker-assisted selection. In vitro selection, especially for vegetatively propagated crops.

Mutation breeding is and remains a relevant and practical breeding option. Today, a major challenge for breeders is to respond rapidly to effects of climate change, which is bringing new diseases and pests, and changing environmental conditions. In general, induced mutation and its exploitation in plant breeding aims to achieve resilient crops with improved quantitative and qualitative traits within a short time. Due to its relative simplicity, flexibility, practicability, low cost, long safety record and universal acceptance, chemical and physical mutagens continue to provide solutions for plant breeding objectives.

Although physical and chemical mutagenesis have been used in many crop species, they still face challenges, preliminary in terms of (1) precision and efficiency, (2) high-throughput of whole-genome functional screening of modified genes or DNA sequences, (3) high-throughput phenotypic screening of a large population (accessible facilities), and (4) effective tissue culture systems for perennial crop species. Various new approaches are still needed to overcome the challenges of today and tomorrow to enhance mutation induction and mutation detection in generating desired effects. Physical and chemical mutagenesis remains an effective low-cost weapon in the armoury of plant breeders, and methods continue to improve.

Dedication The authors would like to dedicate this chapter to Udda Lundqvist who died recently. Udda Lundqvist was a pioneer in plant mutation genetics, especially in barley, which became a model crop for mutation studies. She inspired many and will be greatly missed.

References

Aastveit K (1968) Effects of combinations of mutagens on mutations frequency in barley. In: Mutations in plant breeding II. IAEA, Vienna, pp 5–14

Abe A, Kosugi S et al (2012) Genome sequencing reveals agronomically important loci in rice using MutMap. Nat Biotechnol 30:174–178

Ahloowalia BS, Maluszynski M, Nichterlein K (2004) Global impact of mutation-derived varieties. Euphytica 135:187–204

Albert TJ, Molla MN, Muzny DM et al (2007) Direct selection of human genomic loci by microarray hybridization. Nat Methods 4:903–905. https://doi.org/10.1038/nmeth1111

Amarasinghe SL, Su S, Dong X et al (2020) Opportunities and challenges in long-read sequencing data analysis. Genome Biol 21:30. https://doi.org/10.1186/s13059-020-1935-5

Asseyeva T (1931) Bud mutations in the potato. Bull Appl Bot (Leningrad) 27:135–217

Auerbach C, Robson JM (1942) Experiments on the action of mustard gas in Drosophila, production of sterility and mutation. Report to the Ministry of Supply 3979

Auerbach C, Robson JM (1944) Production of mutations by allyl isothiocyanate. Nature 154:81

Auerbach C, Robson J (1946) Chemical production of mutations. Nature 157:302. https://doi.org/10.1038/157302a0

Bado S, Forster BP, Nielen S, Ghanim A, Lagoda PJL, Till BJ, Laimer M (2015) Plant mutation breeding: current progress and future assessment. Plant Breed Rev 39:23–88

Bado S, Rafiri MA, El-Achouri K, Sapey E, Nielen S, Ghanim AMA, Forster BP, Laimer M (2016) In vitro methods for mutation induction in potato (Solanum tuberosum L.). Afr J Biotechnol 15(39):2132–2145

Bado S, Yamba NGG, Sesay JV, Laimer M, Forster BP (2017) Plant mutation breeding for the improvement of vegetatively propagated crops: successes and challenges. CAB Rev 12(028). https://doi.org/10.1079/PAVSNNR201712028

Baird NA, Etter PD, Atwood TS, Currey MC et al (2008) Rapid SNP discovery and genetic mapping using sequenced RAD markers. PLoS One 3:e3376

Baloch FS, Alsaleh A, Shahid MQ et al (2017) A whole genome DArTseq and SNP analysis for genetic diversity assessment in durum wheat from central fertile crescent. PLoS One 12(1): e0167821. https://doi.org/10.1371/journal.pone.0167821

Bamshad MJ, Ng SB, Bigham AW et al (2011) Exome sequencing as a tool for Mendelian disease gene discovery. Nat Rev Genet 12:745–755

Basiardes S, Veile R, Helms C, Mardis ER, Bowcock AM, Lovett M (2005) Direct genomic selection. Nat Methods 1(2):63–69. https://doi.org/10.1038/nmeth0105-63

Beissinger TM, Hirsch CN, Sekhon RS et al (2013) Marker density and read depth for genotyping populations using genotyping-by-sequencing. Genetics 193:1073–1081

Beyaz R, Yildiz M (2017) The use of gamma irradiation in plant mutation breeding. In: Plant engineering. Intech, Rijeka, Croatia, pp 33–46. https://doi.org/10.5772/intechopen.69974

Bhat TA, Khan AH, Parveen S (2007) Spectrum and frequency of chlorophyll mutations induced by MMS, gamma rays and their combination in two varieties of *Vicia faba* L. Asian J Plant Sci 6(3):558–561

Bolon YT, Haun WJ, Xu WW et al (2011) Phenotypic and genomic analyses of a fast neutron mutant population resource in soybean. Plant Physiol 156:240–253

Broertjes C (1982) Significance of *in vitro* adventitious bud techniques for mutation breeding of vegetatively propagated crops. In: Induced mutations in vegetatively propagated plants, vol II. IAEA, Vienna, pp 1–9

Broertjes C, van Harten AM (1988) Applied mutation breeding for vegetatively propagated crops. Developments in crop science No. 12. Elsevier, Amsterdam

Bronowski J (1973) The ascent of man. British Broadcasting Corp, London

Brunner AL, Johnson DS, Kim SW et al (2009) Distinct DNA methylation patterns characterize differentiated human embryonic stem cells and developing human fetal liver. Genome Res 19: 1044–1056

Castillo AM, Cistué L, Vallés JM, Romagosa I, Molina-Cano JL (2001) Efficient production of androgenic doubled-haploid mutants in barley by the application of sodium azide to anther and microspore cultures. Plant Cell Rep 20:105–111

Channaoui S, Labhilili M, Mouhib M, Mazouz H, El Fechtali M, Nabloussi A (2019) Development and evaluation of diverse promising rapeseed (*Brassica napus* L.) mutants using physical and chemical mutagens. OCL 26:35. https://doi.org/10.1051/ocl/2019031

Chauhan SP, Patra NK (1993) Mutagenic effects of combined and single doses of gamma rays and EMS in *Opium Poppy*. Plant Breed 110(4):342–345. https://doi.org/10.1111/j.1439-0523.1993.tb00600.x

Chen YL, Liang HL, Ma XL, Lou SL, Xie YY, Liu ZL, Chen LT, Liu YG (2013) An efficient rice mutagenesis system based on suspension-cultured cells. J Integr Plant Biol 55:122–130

Cheng X, Gao M (1988) Biological and genetic effects of combined treatments of sodium azide, gamma rays and EMS in barley. Environ Exp Bot 28(4):281–288

Chepkoech E, Kinyua MG, Kiplagat O, Ochuodho J, Bado S, Kimno S, Chelulei M (2019) Assessment of the ploidy level diversity by chloroplast counts in stomatal guard cells of potato mutants. Asian J Res Crop Sci 4(3):1–7

Colbert T, Till BJ, Tompa R, Reynolds S, Tteaine MN et al (2001) High-throughput screening for induced point mutations. Plant Physiol 126:480–484

Comai L, Young K, Till BJ, Reynolds SH, Greene EA, Codomo CA, Enns LC, Johnson JE, Burtner C, Odden AR, Henikoff S (2004) Efficient discovery of DNA polymorphisms in natural populations by Eco-tilling. Plant J 37:778–786

Cosart T, Beja-Pereira A, Chen S et al (2011) Exome-wide DNA capture and next generation sequencing in domestic and wild species. BMC Genomics 12:347

Court-of-Justice-of-the-European-Union (2018) Organisms obtained by mutagenesis are GMOs and are, in principle, subject to the obligations laid down by the GMO Directive. Press Release

Cramer PJS (1907) Kristische Übersicht der bekannten Fälle von Knospenvariation. Natuurkundige Verhandelingen der Hollandsche Maatschappij van Wetenschappen, Haarlem 3:6

Darwin C (1868) The variation of animals and plants under domestication. 10th imp. of the 2nd ed, 1 Vol I (1921). Murray, London

Datta SK (2020) Induced mutations: technological advancement for development of new ornamental varieties. Nucleus. https://doi.org/10.1007/s13237-020-00310-7

Datta S, Misra P, Mandal A (2005) *In vitro* mutagenesis – a quick method for establishment of solid mutant in chrysanthemum. Curr Sci 88(1):155–158. http://www.jstor.org/stable/24110108

Davies DR (1966) The comparative effects of a mono- and a bi-functional alkylating agent on recombination in *Chlamydomonas reinhardi*. Z Vererbungsl 98:61–70

de Vries H (1901) Die Mutationstheorie I. Veit & Co. Germany, Leipzig. (English transl., 1910. The Open Court, Chicago)

de Vries H (1903) Die Mutationstheorie II. Veit & Co. Germany, Leipzig. (English transl., 1910. The Open Court, Chicago)

de Vries H (1905) Species and varieties: their origin by mutation. The Open Court Publ. Co, Chicago, IL

Deltcheva E, Chylinski K, Sharma CM, Gonzales K, Chao Y, Pirzada ZA, Eckert MR, Vogel J, Charpentier E (2011) CRISPR RNA maturation by trans-encoded small RNA and host factor RNase III. Nature 471(7340):602–607. https://doi.org/10.1038/nature09886

Do KT (2009) Socio-economic impacts of mutant rice varieties in southern Vietnam. In: Shu QY (ed) Induced plant mutations in the genomics era. FAO, Rome, pp 65–68

Doll H, Sandfaer J (1969) Mutagenic effect of gamma rays, diethyl sulfate, ethyl methane sulfonate and various combinations of gamma rays and the chemicals. In: Induced mutations in plants. IAEA, Vienna, pp 195–206

Donini P, Sonnino A (1998) Induced mutation in plant breeding: current status and future outlook. In: Jain SM, Brar DS, Ahloowalia BS (eds) Somaclonal variation and induced mutations in crop improvement. Kluwer Academic, Dordrecht, pp 255–291

Dwivedi AK, Banerji BK, Chakraborty D, Mandal AKA, Datta SK (2000) Gamma ray induced new flower colour chimera and its management through tissue culture. Indian J Agric Sci 70(12): 853–855

Esnault MA, Legue F, Chenal C (2010) Ionizing radiation: advances in plant response. Environ Exp Bot 68:231–237

FAO/IAEA (1975) Improvement of vegetatively propagated plants through induced mutations. In: Proceedings of a research co-ordination meeting held at Tokai, 30 September–4 October 1974 organized by the Joint FAO/IAEA Division of Atomic Energy in Food and Agriculture, vol 173. IAEA, Vienna

FAO/IAEA (1991) Mutation Breed Newslett 37. Intl Atomic Energy Agency, Vienna

FAO/IAEA (2018) In: Spencer-Lopes MM, Forster BP, Jankuloski L (eds) Manual on mutation breeding, 3rd edn. Food and Agriculture Organization of the United Nations, Rome, Italy

Favret EA (1963) Genetic effects of single and combined treatment of ionizing radiations and ethyl methane sulfonate on barley seeds. In: Barley Genetics, 1'01. I, Proc. 1st Int Barley Genet. Symp., Wageningen, pp 26–3r

Feng HY, Yu ZL (2012) Ion beam implantation mutagenesis. In: Shu QY, Forster BP, Nakagawa H (eds) Plant mutation breeding and biotechnology. CABI, FAO, Oxfordshire, UK, Rome, pp 107–121

Forster BP, Shu QY (2012) Plant mutagenesis in crop improvement: basic terms and applications. In: Shu QY, Forster BP, Nakagawa H (eds) Plant mutation breeding and biotechnology. CABI, FAO, Oxfordshire, UK, Rome, pp 9–20

Forster BP, Franckowiak JD, Lundqvist U, Thomas WTB, Leader D, Shaw P, Lyon J, Waugh R (2012) Mutant phenotyping and pre-breeding in barley. In: Shu QY, Forster BF, Nakagawa H (eds) Plant mutation breeding and biotechnology. CABI, FAO, Oxfordshire, UK, Rome, pp 327–346

Forster BP, Till BJ, Ghanim AMA, Huynh HOA, Burstmayr H, Caligari PDS (2015) Accelerated plant breeding. CAB International, Oxfordshire, UK. (Online ISSN 1749-8848)

Freislebe R, Lein A (1942) Uber die Auffindung einer mehltauresistenten Mutante nach Röntgenbestrahlung einer anfalligen Linie von Sommergerste. Naturwissenschaften 30:608

Gager CS, Blakeslee AF (1927) Chromosome and gene mutations in Datura following exposure to radium rays. Proc Natl Acad Sci U S A 13:75–79

Gilchrist EJ, Haughn GW (2005) TILLING without a plough: a new method with applications for reverse genetics. Curr Opin Plant Biol 8:211–215

Gilchrist E, Haughn G (2010) Reverse genetics techniques: engineering loss and gain of gene function in plants. Brief Funct Genomics 9:103–110

Gilchrist EJ, Haughn GW, Ying CC, Otto SP, Zhuang J et al (2006) Use of eco-TILLING as an efficient SNP discovery tool to survey genetic variation in wild populations of *Populus trichocarpa*. Mol Ecol 15:1367–1378

Girija M, Dhanavel D (2009) Mutagenic effectiveness and efficiency of gamma rays, ethyl methane sulphonate and their combined treatment in cowpea (*Vigna unguiculata* L. Walp). Glob J Mol Sci 4:68–75

Gnirke A, Melnikov A, Maguire J, Rogov P et al (2009) Solution hybrid selection with ultra-long oligonucleotides for massively parallel targeted sequencing. Nat Biotechnol 27:182–189

Gómez-Pando L, Eguiluz A, Jiménez J, Falconí J, Heros E (2009) Barley (*Hordeun vulgare*) and kiwicha (*Amaranthus caudatus*) improvement by mutation induction in Peru. In: Shu QY (ed) Induced plant mutations in the genomics era. Food and Agriculture Organization of the United Nations, Rome, pp 330–332

Goodspeed TH (1929) The effects of X-rays and radium on species of the genus *Nicotiana*. J Hered 20:243–259

Goyal S, Khan S (2010) Differential response of single and combined treatment in moist seeds of urdbean. Indian J Bot Res 6:183–188

Grohmann L, Keilwagen J, Duensing N, Dagand E, Hartung F, Wilhelm R, Bendiek J, Sprink T (2019) Detection and identification of genome editing in plants: challenges and opportunities. Front Plant Sci 10:236. https://doi.org/10.3389/fpls.2019.00236

Gruszka D, Szarejiko I, Maluszynski M (2012) Sodium azide as mutagen. In: Shu QY, Forster BF, Nakagawa H (eds) Plant mutation breeding and biotechnology. CABI, FAO, Oxfordshire, UK, Rome, pp 159–165

Gupta P, Reddaiah B, Salava H et al (2017) Next-generation sequencing (NGS)-based identification of induced mutations in a doubly mutagenized tomato (*Solanum lycopersicum*) population. Plant J 92(3):495–508. https://doi.org/10.1111/tpj.13654

Gvozdenic S, Bado S, Afza R, Jocic S, Mba C (2009) Intervarietal differences in response of sunflower (*Helianthus annuus* L.) to different mutagenic treatments. In: Shu QY (ed) Induced plant mutations in the genomics era. FAO, Rome, pp 358–360

Hansey CN, Vaillancourt B, Sekhon RS et al (2012) Maize (*Zea mays* L.) genome diversity as revealed by RNA-sequencing. PLoS One 7(3):e33071

Harbour JW, Onken MD, Roberson ED, Duan S et al (2010) Frequent mutation of BAP1 in metastasizing uveal melanomas. Science 330:1410–1413

Haun WJ, Hyten DL, Xu WW et al (2011) The composition and origins of genomic variation among individuals of the soybean reference cultivar Williams 82. Plant Physiol 155:645–655

Henry IM, Nagalakshmi U, Lieberman MC, Ngo KJ, Krasileva KV et al (2014) Efficient genome-wide detection and cataloging of EMS induced mutations using exome capture and next-generation sequencing. Plant Cell 26:1382–1397

Hillier LW, Marth GT, Quinlan AR et al (2008) Whole-genome sequencing and variant discovery in *C. elegans*. Nat Methods 5:183–188

Hodges E, Xuan Z, Balija V, Kramer M et al (2007) Genome-wide in situ exon capture for selective resequencing. Nat Genet 39:1522–1527

Hoischen A, van Bon BW, Gilissen C, Arts P et al (2010) De novo mutations of SETBP1 cause Schinzel-Giedion syndrome. Nat Genet 42:483–485

Holme IB, Gregersen PL, Brinch-Pedersen H (2019) Induced genetic variation in crop plants by random or targeted mutagenesis: convergence and differences. Front Plant Sci 14:1468. https://doi.org/10.3389/fpls.2019.01468

Huang X, Feng Q, Qian Q et al (2009) High-throughput genotyping by whole-genome resequencing. Genome Res 19:1068–1076

Hussain M, Iqbal MA, Till BJ, Rahman M-u (2018) Identification of induced mutations in hexaploid wheat genome using exome capture assay. PLoS One 13(8):e0201918. https://doi.org/10.1371/journal.pone.0201918

IAEA (1970) Manual on mutation breeding, Tech. Rep. Ser, vol 119, 1st edn. IAEA, Vienna

IAEA (1977) Manual on mutation breeding, Tech. Rep. Ser, vol 119, 2nd edn. IAEA, Vienna

Islam MS, Rahman MA, Sultana N, Nath B, Paul A (2012) Using geospatial techniques to assess the salinity impact on agricultural land use: a study on Shyamnagar Upazila, Satkhira. J Agric Environ Int Dev 106(2):157–169

Jafri IF, Khan A, Gulfishan M (2013) Genomic damage induced by individual and combination treatment of gamma rays and ethyl methane sulphonate in Coriandrum sativum L var. Larishma. Int J Bot Res 3(2):79–86

Jankowicz-Cieslak J, Huynh OA, Brozynska M, Nakitandwe J, Till BJ (2012) Induction, rapid fixation and retention of mutations in vegetatively propagated banana. Plant Biotechnol J 10: 1056–1066

Jinek M, Chylinski K, Fonfara I, Hauer M, Doudna JA, Charpentier E (2012) A programmable dual-RNA–guided DNA endonuclease in adaptive bacterial immunity. Science 337(6096): 816–821. https://doi.org/10.1126/science.1225829

Kempner ES (2011) Direct effects of ionizing radiation on macromolecules. J Polym Sci B Polym Phys 49(12):827–831. https://doi.org/10.1002/polb.22250

Khan MH, Tyagi SD (2010) Studies on effectiveness and efficiency of gamma rays, EMS and their combination in soybean (*Glycine max* (L.) Merrill). J Plant Breed Crop Sci 2(3):55–58

Kharkwal MC (2012) A brief history of plant mutagenesis. In: Shu QY, Forster BP, Nakagawa H (eds) Plant mutation breeding and biotechnology. CABI, Wallingford, pp 21–30

Khazaei H, Mäkelä PSA, Stoddard FL (2018) Ion beam mutagenesis in rye (Secale cereale L.) and faba bean (*Vicia faba* L.). Agric Food Sci 27:146–151

Klein G (1932) Handbuch der Pflanzenanalyse, Part 2, Chapter 26. Springer, Vienna

Kleinhofs A, Sander C, Nilan RA, Konzak CF (1974) Azide mutagenicity - mechanism and nature of mutants produced. In: Proc. Polyploidy and Induced Mutations in Plant Breeding, Oct. 1972. IAEA, Vienna, pp 159–199

Kodym A, Afza R (2003) Physical and chemical mutagenesis. Methods Mol Biol 236:189–204. https://doi.org/10.1385/1-59259-413-1:189

Kodym A, Afza R, Forster BP, Ukai Y, Nakagawa H (2012) Methodology for physical and chemical mutagenic treatments. In: Shu QY, Forster BF, Nakagawa H (eds) Plant mutation breeding and biotechnology. CABI, FAO, Oxfordshire, UK, Rome, pp 169–180

Konzak CF, Nilas RA, Harle JR, Heiner RE (1961) Control of factors affecting the response of plants to mutagens. Brookhaven Symp Biol 14:128–157

Konzak CF, Niknejad M, Wickham I, Donaldson E (1975) Mutagenic interaction of sodium azide on mutations induced in barley seeds treated with diethyl sulfate or M-methyln′-N′-Nitrosourea. Mutat Res 3:55–62

Konzak CF, Kleinhofs A, Ullrich SE (1984) Induced mutations in seed-propagated crops. Plant Breed Rev 2:13–72

Kumawat S, Rana N, Bansal R, Vishwakarma G et al (2019) Expanding avenue of fast neutron mediated mutagenesis for crop improvement. Plants 8:164

Kume T, Amano E, Nakanishi TM, Chino M (2002) Economic scale of utilization of radiation (II): agriculture comparison between Japan and the USA. J Nucl Sci Technol 39:1106–1113

Kurowska M, Labocha-Pawłowska A, Gnizda D, Maluszynski M, Szarejko I (2012) Molecular analysis of point mutations in a barley genome exposed to MNU and gamma rays. Mutat Res Fundam Mol Mech Mutagen 738–739:52–70

Lagoda PJL (2012) Effects of radiation on living cells and plants. In: Shu QY, Forster BF, Nakagawa H (eds) Plant mutation breeding and biotechnology. CABI, FAO, Oxfordshire, UK, Rome, pp 123–134

Lal GM, Toms B, Smith S (2009) Induced chlorophyll mutations in black gram. Asian J Agric Sci 1:1–3

Leitão JM (2012) Chemical mutagenesis. In: Shu QY, Forster BF, Nakagawa H (eds) Plant mutation breeding and biotechnology. CABI, FAO, Oxfordshire, UK, Rome, pp 135–158

Li F, Shimizu A, Nishio T, Tsutsumi N, Kato H (2019) Comparison and characterization of mutations induced by gamma-ray and carbon-ion irradiation in rice (*Oryza sativa* L.) using whole-genome resequencing. G3 (Bethesda) 9(11):3743–3751. https://doi.org/10.1534/g3.119.400555

Liu B, Sun Z, Ma X, Yang B, Jiang Y, Wei D, Chen F (2015) Mutation breeding of extracellular polysaccharide-producing microalga *Crypthecodinium cohnii* by a novel mutagenesis with atmospheric and room temperature plasma. Int J Mol Sci 16:8201–8212. https://doi.org/10.3390/ijms16048201

López J, Rayas A, Santos A, Medero V, Beovides Y, Basail M (2017) Mutation induction using gamma irradiation and embryogenic cell suspensions in plantain (Musa spp.). In: Jankowicz-Cieslak J, Tai T, Kumlehn J, Till B (eds) Biotechnologies for plant mutation breeding. Springer, Cham. https://doi.org/10.1007/978-3-319-45021-6_4

Lundqvist U, Franckowiak JD, Forster BP (2012) Mutation categories. In: Shu QY, Forster BF, Nakagawa H (eds) Plant mutation breeding and biotechnology. CABI, FAO, Oxfordshire, UK, Rome, pp 47–55

Maghuly F, Bado S, Jankowicz-Cieslak J, Laimer M (2016) Protocols for induced mutagenesis in *Jatropha curcas*. In: Jankowicz-Cieslak J et al (eds) Biotechnologies for plant mutation breeding. Springer International Publishing AG, Switzerland, pp 21–38. https://doi.org/10.1007/978-3-319-45021-6_2

Maghuly F, Pabinger S, Krainer J, Laimer M (2018) The pattern and distribution of induced mutations in *J. curcas* using reduced representation sequencing. Front Plant Sci 9:524. https://doi.org/10.3389/fpls.2018.00524

Magori S, Tanaka A, Kawaguchi M (2010) Physically induced mutation: ion beam mutagenesis. In: Meksem K, Kahl G (eds) The handbook of plant mutation screening. Wiley-VCH, Weinheim, Germany, pp 1–16

Makay IJ, Cockram J, Howell P, Powell W (2020) Understanding the classics: the unifying concepts of transgressive segregation, inbreeding depression and heterosis and their central relevance for crop breeding. Plant Biotechnol J 19:26–34. https://doi.org/10.1111/pbi.13481

Makeen K, Babu GS (2010) Mutagenic effectiveness and efficiency of gamma rays, sodium azide and their synergistic effects in urd bean (*Vigna mungo* L.). World J Agric Sci 6(2):234–237

Maluszynski M, Ahloowalia BS, Sigurbjornsson B (1995) Application of *in-vivo* and *in-vitro* mutation techniques for crop improvement. Euphytica 85:303–315

Mamanova L, Coffey AJ, Scott CE, Kozarewa I, Turner EH, Kumar A, Howard E, Shendure J, Turner DJ (2010) Target-enrichment strategies for next-generation sequencing. Nat Methods 7:111–118. https://doi.org/10.1038/nmeth.1419

Matthews P, Lapins K (1967) Self-fertile sweet cherries. Fruit Var Hort Dig 21:36–37

Mba C, Afza R, Bado S, Jain SM (2010) Induced mutagenesis in plants using physical and chemical agents. In: Davey MR, Anthony P (eds) Plant cell culture: essential methods. Wiley, Chichester, UK, pp 111–130

Mba C, Afza R, Shu QY (2012) Mutagenic radiations: X-rays, ionizing particles and ultraviolet. In: Shu QY, Forster BF, Nakagawa H (eds) Plant mutation breeding and biotechnology. CABI, FAO, Oxfordshire, UK, Rome, pp 83–106

McCallum CM, Comai L, Greene EA, Henikoff S (2000) Targeting induced local lesions in genomes (TILLING) for plant functional genomics. Plant Physiol 123:439–442

Mehandjiev A, Kosturkova G, Mihov M (2001) Enrichment of *Pisum sativum* gene resources through combined use of physical and chemical mutagens. Israel J Plant Sci 49(4):280–284

Merz T, Swanson CP, Cohn NS (1961) Interaction of chromatid breaks produced by X-rays and radiomimetic compounds. Science I33:703–705
Mlcochova L, Chlourpek O, Uptmoor R, Ordon F, Friedt W (2004) Molecular analysis of the barley cv. 'Valticky and its X-ray-derived semidwarf-mutant 'Diamant'. Plant Breed 123:421–427
Minocha JL, Arnason TJ (1962) Mutagenic effectiveness of ethyl methane sulphonate and methyl methane sulphonate in barley. Nature 196:499
Mohan RPK (1972) Biological effects of combination treatments with ionizing radiations and diethyl sulfate (dES) in barley. Mutat Res 16:322–327
Monastra F, Strada GD, Fideghell C, Quarta R (1998) Supernova: Une nouvelle variété d'amandier obtenue par mutagenèse. 7e Colloque GREMPA. Rapport EUR 11557:3–7
Moore G, Engels J, Fowler C (2007) The international treaty on plant genetic resources for food and agriculture: access, benefit sharing and conservation. Acta Horticulturae 760:27–32. https://doi.org/10.17660/ActaHortic.2007.760.1
Morgan TH (1910) Chromosomes and heredity. Am Nat 44:449–496. http://www.jstor.org/stable/pdf/2455783.pdf
Morgan TH (1911) An attempt to analyze the constitution of the chromosomes on the basis of sex-limited inheritance in Drosophila. J Exp Zool 11:365–413
Moutschen J (1960) Action combinée du Mylcran et des rayons X et son importance au point de vue genetique. Hereditas 46:471–472
Moutschen J, Moutschen-Dahme M (1961) Effets de traitements combinés du Myleran vec d'autres poisons radiomimetiques et avec rayons X. In: Effects of ionizing radiations on seeds. IAEA, Vienna, pp 333–344
Muller HJ (1927) Artificial transmutation of gene. Science 66:84–87
Muth J, Hartje S, Twyman RM, Hofferbert H-R, Tacke E, Prüfer D (2008) Precision breeding for novel starch variants in potato. Plant Biotechnol J 6:576–584. https://doi.org/10.1111/j.1467-7652.2008.00340.x
Neves LG, Davis JM, Barbazuk WB (2013) Whole-exome targeted sequencing of the uncharacterized pine genome. Plant J 75:146–156
Ng SB, Turner EH, Robertson PD et al (2009) Targeted capture and massively parallel sequencing of 12 human exomes. Nature 461:272–276
Nilan RA, Muir CE (1967) Registration of Luther barley. Crop Sci 7:278
Nur F, Forster BP, Osei SA, Amiteye A, Ciomas J et al (2018) Mutation breeding in oil palm: a manual. In: Forster BP, Caligari PDS (eds) Techniques in plantation science. CAB International, Wallingford UK
Oehlkers F (1943) Die Auslösung Von Chromosomenmutationen in der Meiosis durch Einwirkung von Chemikalien. Z Indukt Abstammungs Vererbungs 81:313–341
Oehlkers F (1946) Weitere Versuche zur Mutationsauslösung durch Chemikalien. Biol Zentralbl 65:176–186
Oerke ECC, Weber A, Dehne WH, Schonbeck F (1999) Crop production and crop protection. Elsevier, Amsterdam
Okou DT, Steinberg KM, Middle C, Cutler DJ, Albert TJ, Zwick ME (2007) Microarray-based genomic selection for high-throughput resequencing. Nat Methods 4:907–909. https://doi.org/10.1038/nmeth1109
Oladosu Y, Rafii MY, Abdullah N, Hussin G, Ramli A, Rahim HA, Miah G, Usman M (2016) Principle and application of plant mutagenesis in crop improvement: a review. Biotechnol Biotechnol Equip 30(1):1–16. https://doi.org/10.1080/13102818.2015.1087333
Olejniczak J (1994) Genetic variability induced through mutations in maize. In: Bajaj YPS (ed) Maize, Biotechnology in agriculture and forestry, vol 25. Springer, Berlin, pp 355–365. https://doi.org/10.1007/978-3-642-57968-4_24
Olejniczak J, Patyna H (1981) Mutagenic effect of N-methyl-N-nitrosourea (MNUA) and sodium azide (SA) in maize line S-75 (*Zea mays* L.). Genet Polon 22(3):289–294
Oster II (1958) Interactions between ionizing radiations and chemical mutagens. Z Verebungslehre 89:1–6

Pathirana R (2011) Plant mutation breeding in agriculture. CAB Rev Perspect Agric Vet Sci Nutr Nat Resour 6:1–20

Periyannan S (2018) Sustaining global agriculture through rapid detection and deployment of genetic resistance to deadly crop diseases. New Phytol 219:45–51. https://doi.org/10.1111/nph.14928

Rai MK, Kalia RK, Singh R, Gangola MP, Dhawan AK (2011) Developing stress tolerant plants through in vitro selection-an overview of the recent progress. Environ Exp Bot 71:89–98

Rana RS (1965) Radiation-induced variation in ray-floret characteristics of annual chrysanthemum. Euphytica 14:296–300. https://doi.org/10.1007/BF00149515

Rao MPK, Konzak CF, Nilan RA, Dhesi SS (1965) The influence of hydrogen ion concentration on radiation induced damage in barley. Radiat Bot 5:455–463

Rapoport IA (1946) Carbonyl compounds and the chemical mechanism of mutations. Dok Acad Nauk SSSR Ser Biol 54:65–67

Reisz JA, Bansal N, Qian J, Zhao W, Furdui CM (2014) Effects of ionizing radiation on biological molecules—mechanisms of damage and emerging methods of detection. Antioxid Redox Signal 21(2):260–292. https://doi.org/10.1089/ars.2013.5489

Rosenzweig C, Iglesias A, Yang XB, Epstein PR, Chivian E (2001) Climate change and extreme weather events. Glob Change Hum Health 2(2):90–104

Roux N, Toloza A, Radecki Z, Zapata-Arias FJ, Dolezel J (2003) Rapid detection of aneuploidy in Musa using flow cytometry. Plant Cell 21:483–490

Roy SJ, Tucker EJ, Tester M (2011) Genetic analysis of abiotic stress tolerance in crops. Curr Opin Plant Biol 14(3):232–239

Roychowdhury R, Tah J (2013) Mutagenesis a potential approach for crop improvement. In: Hakeem KR, Ahmad P, Ozturk M (eds) Crop improvement: new approaches and modern techniques. Springer, New York, pp 149–187

Rutger JN (2009) The induced sd1 mutant and other useful mutant genes in modern rice varieties. In: Shu QY (ed) Induced plant mutations in the genomics era. FAO, Rome, pp 44–47

Sakharov VV (1932) Iodine as a chemical factor, affecting mutational process in Drosophila melanogaster. Biol Zh 1(3–4):1–8

Saleem MY, Mukhtar Z, Cheema AA, Atta BM (2005) Induced mutation and in vitro techniques as a method to induce salt tolerance in Basmati rice (*Oryza sativa* L.). Int J Environ Sci Technol 2(2):141–145. https://doi.org/10.1007/BF03325868

Sathesh-Prabu C, Lee Y-K (2011) Mutation breeding of mushroom by radiation. J Radiat Ind 5(4): 285–295

Scheiden MJ (1838) Beiträge zur phytogenesis. In: Müller J (ed) Archiv für Anatomie Pysiologgie Wissenschaftliche Medicin. Viet, Berlin, pp 137–176

Schlötterer C, Tobler R, Kofler R et al (2014) Sequencing pools of individuals - mining genome-wide polymorphism data without big funding. Nat Rev Genet 15:749–763

Serrat X, Esteban R, Guibourt N, Moysset L, Nogués S, Lalanne E (2014) EMS mutagenesis in mature seed-derived rice calli as a new method for rapidly obtaining TILLING mutant populations. Plant Methods 10(1):5. https://doi.org/10.1186/1746-4811-10-5

Shkvarnikov PK, Morgun VV (1975) Mutations in maize, induced by chemical mutagens. Proc Indian Nat Sci Acad 41B(3):177–187

Shu QY (2009) Induced plant mutation in the genomics era. FAO & IAEA, Rome, Italy. 458p

Shu QY, Shirasawa K, Hoffmann M, Hurlebaus J, Nishio T (2012a) Molecular techniques and methods for mutation detection and screening in plants. In: Shu QY, Forster BP, Nakagawa H (eds) Plant mutation breeding and biotechnology. CABI, FAO, Oxfordshire, UK, Rome, pp 241–268

Shu QY, Forster BP, Nakagawa H (2012b) Principles and applications of plant mutation breeding. In: Shu QY, Forster BP, Nakagawa H (eds) Plant mutation breeding and biotechnology. CABI, FAO, Oxfordshire, UK, Rome, pp 301–325

Sigurbjörnsson B, Micke A (1974) Philosophy and accomplishments of mutation breeding. In: Polyploidy and induced mutations in plant breeding. IAEA, Vienna, pp 303–343

Sikora P, Chawade A, Larsson M, Olsson J, Olsson O (2011) Mutagenesis as a tool in plant genetics, functional genomics, and breeding. Int J Plant Genomics 2011:314829., 13 pages. https://doi.org/10.1155/2011/314829

Snyder EB, Grigsy HC, Hidalgo JU (1961) X-radiation of southern pine seed at various moisture contents. Silvae Genetica Band 10 Heft 5

Soriano JD (1963) Some effects of combining X-rays and ethyl methane sulfonate on barley seeds (Abstr.). Genetics 48:91I

Sparrow AH (1961) Types of ionizing radiation and their cytogenetic effects. Symposium on mutation and plant breeding by National Academy of Sciences Council Publication 891, pp 55–129

Sparrow AH, Singleton WR (1953) The use of radio cobalt as a source of gamma-rays and some effects of chronic irradiation on growing plants. Am Nat 87:29–48

Spiegel-Roy P (1990) Economic and agriculture impact of mutation breeding in fruit trees. Mutat Breed Rev Vienna 5:215–235

Srinivas RT, Veerabadhiran P (2010) Efficiency and effectiveness of physical and chemical mutagens and their combination in inducing chlorophyll mutations in M2 generation of Lablab [*Lablab purpureus* (l.) Sweet var. Typicus.]. Electron J Plant Breed 813(1):752–757

Srivastav PK, Raina SN (1981) Cytogenetics of Tephrosia II. Mutagenic effect of single, pre- and post-irradiation treatments with EMS and MMS in *Tephrosia purpurea*. Cytologia 46:709–721

Stadler LJ (1928a) Genetic effects of x-rays in maize. Proc Natl Acad Sci U S A 14:69–75

Stadler LJ (1928b) Mutations in barley induced by X-rays and radium. Science 68:186–187

Stadler LJ (1929) Chromosome number and the mutation rate in Auena and Triticum. Proc Natl Acad Sci Washington 15:876–4381

Sulonen AM, Ellonen P, Almusa H et al (2011) Comparison of solution-based exome capture methods for next generation sequencing. Genome Biol 12:R94. https://doi.org/10.1186/gb-2011-12-9-r94

Suprasanna P, Jain SM, Ochatt SJ, Kulkarni VM, Predieri S (2012) Applications of *in vitro* techniques in mutation breeding of vegetatively propagated crops. In: Shu QY, Forster BP, Nakagawa H (eds) Plant mutation breeding and biotechnology. CABI, FAO, Oxfordshire, UK, Rome, pp 371–385

Surakshitha NC, Soorianathasundaram K, Ganesan NM (2017) Determination of mutagenic sensitivity of hardwood cuttings of grapes 'Red Globe' and 'Muscat' (*Vitis vinifera* L.) to gamma rays. Sci Hortic 226:152–156. https://doi.org/10.1016/j.scienta.2017.08.040

Sutarto I, Agisimanto D, Supriyanto A (2009) Development of promising seedless citrus mutants through gamma irradiation. In: Shu QY (ed) Induced plant mutations in the Genomics Era. FAO, Rome, Italy, pp 306–308

Szurman-Zubrzycka M, Chmielewska B, Gajewska P, Szarejko I (2017) Mutation detection by analysis of DNA heteroduplexes in TILLING populations of diploid species. In: Jankowicz-Cieslak J, Tai T, Kumlehn J, Till B (eds) Biotechnologies for plant mutation breeding. Springer, Cham. https://doi.org/10.1007/978-3-319-45021-6_18

Tadele Z, Mba C, Till BJ (2009) TILLING for mutations in model plants and crops. In: Jain SM, Brar DS (eds) Molecular techniques in crop improvement, 2nd edn. Springer, Netherlands, pp 307–332

Tanaka A (2009) Establishment of ion beam technology for breeding. In: Shu QY (ed) Induced plant mutation in the genomics era. Food and Agriculture Organizations of the United Nations, Rome, pp 216–219

Tanaka A, Shikazono N, Hase Y (2010) Studies on biological effects of ion beams on lethality, molecular nature of mutation, mutation rate, and spectrum of mutation phenotype for mutation breeding in higher plants. J Radiat Res 51:223–233

Tehrani G, Brown SK (1992) Pollen-incompatibility and self-fertility in sweet cherry. Plant Breed Rev 9:367–388

Till BJ, Reynolds SH, Greene EA, Christine AC, Enns LC, Johnson JE, Burtner C, Odden AR, Young K, Taylor NE, Henikoff JG, Comai L, Henikoff S (2003) Large-scale discovery of induced point mutations with high-throughput TILLING. Genome Res 13:524–530

Tsai H, Howell T, Nitcher R et al (2011) Discovery of rare mutations in populations: TILLING by sequencing. Plant Physiol 156(3):1257–1268. https://doi.org/10.1104/pp.110.169748

Ukai Y, Nakagawa H (2012) Strategies and approaches in mutant population development for mutant selection in seed propagated crops. In: Shu QY, Forster BF, Nakagawa H (eds) Plant mutation breeding and biotechnology. CABI, FAO, Oxfordshire, UK, Rome, pp 209–221

Valeva SA (1964) Cytogenetic analysis of the combined action of chemicals and irradiation on barley seeds. Radiobiologiya 4:451–456

van Harten AM (1998) Mutation breeding: theory and practical applications. Cambridge University Press, Cambridge

van Harten AM, Broertjes C (1989) Induced mutations in vegetatively propagated crops. Plant Breed Rev 6:55–91

Vardi A, Levin I, Carmi N (2008) Induction of seedlessness in citrus: from classical techniques to emerging biotechnological approaches. J Am Soc Hortic Sci 133:117–126

Viana VE, Pegoraro C, Busanello C, de Oliveira AC (2019) Mutagenesis in rice: the basis for breeding a new super plant. Front Plant Sci 10:1326. https://doi.org/10.3389/fpls.2019.01326

Vinh MQ, Thinh DK, Bang DT, At DH, Ham LH (2009) Current status and research directions of induced mutation application to seed crops improvement in Vietnam. In: Shu QY (ed) Induced plant mutations in the genomics era. FAO, Rome, pp 341–345

von Sengbush R (1927) Süsslupinen und öllupinen. Lanswirtschaftliches Jahrbuch 91:723–880

Wallace AT (1965) Increasing the effectiveness of ionizing radiations in induced mutations at the vital locus controlling resistance to fungus *Helminthosporium victoriae* in oats. Radiation Botany (Suppl) 5:237–250

Wani AA (2009) Mutagenic effectiveness and efficiency of gamma rays, ethyl methane sulphonate and their combination treatments in chickpea (*Cicer arietinum* L.). Asian J Plant Sci 8:318–321

Winfield MO, Wilkinson PA, Allen AM et al (2012) Targeted re-sequencing of the allohexaploid wheat exome. Plant Biol J 10:733–742

Wu KJ, Zimmer C, Peltier E (2020) Nobel Prize in chemistry awarded to 2 scientists for work on genome editing. The New York Times 7. https://www.nytimes.com/2020/10/07/science/nobel-prize-chemistry-crispr.html

Xianfang W, Long Z, Weixu D, Chunhua L (2004) Study of space mutation breeding in China. Acta Agric Nucl Sin 18(4):241–246

Yamaguchi H (2018) Mutation breeding of ornamental plants using ion beams. Breed Sci 68(1): 71–78. https://doi.org/10.1270/jsbbs.17086

Zhao R, Gao S, Qiao Y, Zhu H, Bi Y (1994) A new approach of screening salt-tolerance variants by anther culture to cultivate salt-tolerance wheat variety. Acta Agric Boreali Sin 9:34–38

Zhao R, Gao S, Qiao Y, Zhu H, Bi Y (1995) Studies on the application of anther culture in salt-tolerance breeding in wheat (*Triticum aestivum* L.). Acta Agron Sin 21:230–234

Zhao L, Zhang H, Kohnen MV, Prasad KVSK, Gu L, Reddy ASN (2019) Analysis of transcriptome and epitranscriptome in plants using PacBio Iso-Seq and nanopore-based direct RNA sequencing. Front Genet 10:253. https://doi.org/10.3389/fgene.2019.00253

Mutagenesis and Selection: Reflections on the In Vivo and In Vitro Approaches for Mutant Development

4

Suprasanna Penna and Suresh G. Bhagwat

Abstract

Plant mutation breeding has become a major tool for the improvement of crop plants. Mutation breeding of crop plants is achieved by using physical and/or chemical mutagens in plants propagated via vegetative and sexual modes of propagation. Optimized protocols for the isolation of mutants, their screening and handling of mutant population are now fairly established in several crop plants. Plant improvement has become feasible for desirable traits like plant type, flowering, higher yield, disease, pest resistance, abiotic stress tolerance, nutritional quality including traits of industrial importance. Plant type, propagation method, ploidy and genetics of traits are some of the critical factors to be considered before undertaking mutagenesis as the sole method to isolate desired mutants. Despite the feasibility of undertaking mutation induction by using a wide variety of mutagens, isolation of desired mutant using irradiation techniques, in a given time will depend on the radiation facilities and treatment regime, besides the plant species. Methods of in vitro mutagenesis and in vitro selection have added another dimension for the development of mutants, especially in the case of vegetatively propagated crops using in vitro shoot cultures, embryogenic callus, cell suspensions and protoplast cultures. In vitro *cultures* offer potential benefits to select for biotic and abiotic tolerance besides nutritional quality. Success is often achieved based on optimization of appropriate in vitro culture system(s), high frequency plant regeneration, post-mutagenesis handling

S. Penna (✉)
Nuclear Agriculture and Biotechnology Division, Bhabha Atomic Research Centre, Mumbai, India

Amity Centre for Nuclear Biotechnology & Amity Institute of Biotechnology, Amity University, Mumbai, Maharashtra, India

S. G. Bhagwat
Nuclear Agriculture and Biotechnology Division, Bhabha Atomic Research Centre, Mumbai, India

S. Penna, S. M. Jain (eds.), *Mutation Breeding for Sustainable Food Production and Climate Resilience*, https://doi.org/10.1007/978-981-16-9720-3_4

of mutant population and evaluation of mutants. Extensive research input is required to create and characterize mutation-induced genetic diversity followed by large-scale evaluation of the induced mutants for various agronomic, yield and nutritional quality characters for use in crop improvement.

Keywords

Mutants · In vitro selection · Biotic stress tolerance · Abiotic stress tolerance

4.1 Introduction

Food security and sustainable agriculture are emerging as a priority research theme in the changing scenario of climate change and population growth (Calicioglu et al. 2019). This necessitates that the agricultural productivity should also rise to meet the demand of plant growth and sustenance under the climate change associated biotic and abiotic stresses. There is a constant competition between human beings and plant pathogens for the production of crops (Harlan 1992; Breseghello and Coelho 2013; Fróna et al. 2019). Earlier, crop varieties were derived from germplasm with resistance to diseases which were acquired during the course of evolution. Intensive breeding for yield and quality has diminished the natural resistance due to elimination of resistance trait or intensive agriculture has provided opportunity for the pathogen to develop new types which breaks down the resistance. Consequently, improvement for resistance to disease has become very significant objective in crop breeding programmes. To breed a new resistant variety, a source of resistance is needed. Such sources may or may not be available in germplasm or they may not be suitable for use due to wild traits in the donor or linkages with undesirable traits. Hence, there is a need to explore new and novel plant breeding methods for enhancing genetic variability for different agronomic traits for use in crop improvement (Holme et al. 2019; Qaim 2020). Mutations provide good scope for generating variation for a wide range of desirable traits including biotic and abiotic resistance (Datta 2020).

Mutation is an event which occurs in nature, however, rarely. Mutation can be made to occur at higher frequency by treatment with certain agents called mutagens. With the availability of mutagens, mutation frequency can be increased to enable induced mutations to be applied in crop improvement. Crop improvement using induced mutations is referred to as mutation breeding. Freisleben and Lein (1944) coined the term for the mutant lines which are artificially induced through physical or chemical mutagens and are used in crop improvement programmes. Mutation breeding has become an important part of the genetic improvement of crops, and it has been an important component for assuring global food security and nutrition (Ahloowalia et al. 2004). Of all the mutagens that have been employed for mutation induction, physical agents and, especially gamma rays are most popular due to ease and their ability to penetrate deep into biological tissue (Mba et al. 2010). Gamma rays obtained from radioactive isotope of Cobalt ^{60}Co are widely used. The isotope has a half-life of 5.3 years and emits radiations of energies 1.33 and 1.17 MeV. The material does not have added or induced radioactivity and can be handled after treatment. Where neutrons are used as a mutagen, the sample post-irradiation may have induced radioactivity and hence special precautions are required.

4.2 Choice of the Starting Material

Most plant materials that are used in plant propagation can be used as starting material for mutation breeding experiments. The dose to be used depends upon the type of material used. Along with induction of mutations, the radiation treatment also causes radiation injury or radiation damage which is undesirable. Radiation sensitivity is seen more in the metabolically active plant or those with more water content. Also, the cuttings of stems or rhizomes or bulbs that are larger in size and large number of propagules to be irradiated become a limiting factor. Seeds on the other hand are compact, have lower water content and can be stored before or after irradiation for some time. Transportation is much easier when seeds are used. Due to these reasons, seeds are commonly used wherever possible. However, tissue culture-derived explants, stem cuttings, bulbs, rhizomes and tubers can be used as the target material in mutation breeding experiments.

Despite the feasibility of mutation induction by using a wide variety of mutagens in plant species of interest, isolation of desired mutant in a given time will depend on the radiation facilities and treatment regime, besides the plant species (Pathirana 2011; Bado et al. 2015). Plant species that have short generation time are self-pollinating or that grow in different seasons are more suitable owing to the possibility to grow them in big numbers in a small field plots and generation time is shorter. Thus, mutants can be identified and confirmed in shorter time periods. Cross-pollinated plants, large size trees or plants with long generation time are not suitable; however, they can be used for mutation breeding. Similarly, plants with higher levels of ploidy may continue to segregate before the mutant phenotype is identified and stabilized.

4.3 Doses to Be Used

Radiation dose should be sufficiently high to increase the probability of inducing a mutation; however, very high doses can result in tissue injury and lethality. In general, lethal dose 50 (LD_{50}) has been used in mutagenesis experiments, and it refers to the dose which results in 50% lethality or 50% survival of the tissue. This dose may vary with the plant species, target material and physiological/genetic nature of the tissue and developmental stage (Kodym et al. 2012). In mutagenesis experiments, it is often a practice to work with plant population that are generated through LD_{50} dose treated, through treatment with a dose lesser than LD_{50} and that generated through treatment with higher than LD_{50} dose. Mutation is a random event and the isolation of desirable mutation is often dependent on survival of irradiated material, and hence it is advised to use three treatment doses to enhance chance of recovering a desirable mutant. In some cases where previous published literature is available, one can obtain information on LD_{50} to be adopted as a guideline. It is suggested however to establish a baseline data by treatment of tissue/seeds with a range of low to high doses and record the plant survival. Another criterion that is used to determine the dose to be used in mutagenesis experiments is growth reduction in treated plants, termed GR50 which means mutagen dose required to reduce the growth parameter such as seedling height by 50% of the control. Both parameters LD_{50} and GR50 are formulated based on the premise that there will be low genetic impact of the mutagen at lower doses rarely resulting in a phenotypic change (Álvarez-Holguín et al. 2019). Several reports suggest that GR50 offers a good scope of producing effective mutations and mutagenic events (Wanga et al. 2020). Mutation induction is a chance event, and hence it is desirable to work with large plant population in the M_1 and M_2 generations. If the objective is to isolate a mutant controlled by a single gene in a non-polyploid plant species, mutant isolation becomes easier than working with polygenic traits and polyploid species which require large experimental population. It is also recommended to consider higher initial material (plant/seed/propagules) for irradiation. To facilitate the isolation of several mutants with desirable traits, initial material for irradiation should be adequately large and further populations will have to be high enough to be evaluated.

4.4 Objective for Mutation Breeding

It is essential to have a specific aim for the mutation breeding experiment, for example, resistance to a disease, plant height, yield or early maturity which can help to narrow down the plants with the specific mutant trait and to carry forward the population in subsequent generations. An ideal situation is where a high yielding and otherwise acceptable variety has one defect such as susceptibility to a disease. Other common traits to be modified are change in plant height, days to flowering, non-shattering pods. A single starting parent variety with a single defect to be eliminated makes an excellent case for mutation breeding. Simultaneous improvement in more than one trait is possible if sufficiently large populations are handled in the experiment. Also, starting with more than one variety as a parent for mutagenesis is possible where there is no constraint on the land for raising populations and trained manpower for screening the segregating populations. Purity of the parental seeds is extremely important, if the starting material is contaminated, the contaminant will be wrongly identified as mutation confusing the matters. In crops which show seed dormancy, use of plot in which the same crop has been grown in the previous season should be avoided. Consideration must be given to the possibility that the earlier seeds may germinate and contaminate the population. The M_1 generation population should be grown in a constraint-free environment to ensure maximum survival and good seed filling.

4.5 Generations

The first generation raised by planting irradiated seeds in field is referred to as M_1. In the population belonging to the M_1 generation, plants often show variations in growth and lethality due to several physiological disturbances. The mutational events are generally recessive in nature and hence remain unidentified in the first generation. However, the M_1 population is very important and forms the basis and source to raise the M_2 generation and, for this reason maximum survival of M_1 plants is very important. The plants in population M_1 are given identification numbers and harvested individually. At least 20 seeds from each M_1 plant harvest are sown as a row to rise the M_2 plant generation (plant to row method). It is a good practice to store the remaining seeds carefully. These can be used to recover more mutants as and when required. To ensure complete expression of traits in the plant population in M_2, the plants are often space planted. The M_2 generation contains mutants of different kinds which will require examining morphological traits (seedling colour, leaf shape and size, stem height, branching habit, flowering time, fruit shape, size or colour, etc.), for identification of mutants for physiological or biochemical traits other screening methods are required. It may take examination of large number of plants to identify a mutant. The produce from an M_2 generation plant is sown to obtain the M_3 generation population. The mutation expressed in M_2 generation is confirmed in M_3 generation. Produce from M_3 generation plants is used to raise the M_4 generation. Although the mutant is identified for a trait of interest, it may carry other unidentified mutations and a plant to row method is followed till all the segregation is complete and a uniform population is obtained.

4.6 Screening Methodology

To conduct a mutation breeding programme successfully, most important requirement is an appropriate and efficient screening method (Brown 2013). The work requires looking at a large number of individual plants over a few generations and hence a rapid and economical method is required. Phenotypic traits that can be assessed visually, for example, change in plant height, date of flowering or change in pigmentation, are relatively easy to handle. Such traits are assessed by visual inspection while traits which require actual counting, measurement or biochemical analysis are more difficult to be selected easily. In each generation, for a proper comparison the original parent should be available in sufficient numbers, this is particularly important in case of quantitative traits which are affected by the environmental conditions. To facilitate taking data and valid comparison, the parent variety rows may be planted after certain number of rows of the M_2 population. In a programme for selection of a disease-resistant mutant, it is necessary that the population in M_2 generation is grown at a hot spot where the disease is expected to be present and at sufficient intensity. Rows of susceptible parent variety not only allow proper comparison but also help in spreading of the disease in the entire M_2 population. If the disease intensity is not sufficient and disease is not spreading on its own, spray of pathogen inoculum may be carried out. For assessing the status of a plant for a disease, rating scales are available (Kranz 1988). The disease symptoms may be observed and recorded as per the rating scale for the particular disease. All plants which show the absence of disease or resistant reaction should be tagged. If the mutant phenotype involves change only at a single locus, it may segregate in a Mendelian monogenic ratio (3:1) and out of the 20 plants in the row one or more mutant plants may be observed. The M_1 plant corresponding to the row in the M_2 population is marked, the remaining seeds of the same M_1 plant may be sown in later season to recover more mutants. All plants with resistant reaction are putative mutants. The plants which are marked as resistant reaction are allowed to self-pollinate. All the seeds from resistant plants are collected individually and given an identification number. In the subsequent season, some of the seeds from each resistant plant are sown in a plant to row fashion. A putative mutant may confirm the presence of resistance mutation by showing an entire row which is true breeding for resistance, while another mutant may produce a row which exhibits the presence of both resistant and susceptible phenotype showing segregation for resistance. Often the mutation events are complex and segregation may not exhibit mutant phenotype in the M_2 generation. It is possible that in the subsequent generations such as M_3 or M_4 the resistant phenotype may be observed.

The newly identified mutant with resistance is grown for few generations at a location where disease occurs and single plant selection is practised in every generation till a completely uniform looking (for resistance and other visual parameters) plant population is obtained. The resistant mutant may or may not have the ability to out yield the parent variety in absence of the disease; however, it must show significant yield advantage in the presence of the disease for it to qualify for further advancement. When the mutant does not meet the criteria of agronomic performance, it is used for crossing with the original parent variety or as a source of resistance in a cross-breeding programme. Once stabilized, the mutant breeds true over generations. The mutant will be effective in keeping the disease at bay till a time when the pathogen mutates to develop a virulent type. The resistant variety may lose its advantage in commercial cultivation because of the accumulation of undesirable attributes, etc. In general, the cultivated varieties deteriorate with time or become irrelevant because of changes in the cultivation technology or consumer preference. Development of new crop varieties is thus a continuous process. Mutation breeding is one of the very effective options under certain circumstances.

4.7 In Vitro Mutagenesis and Selection

Genetic variability is the foundation for crop improvement which is generally practised through selection, recombination, mutation and hybridization. Induced mutations have played a central place in the improvement and productivity of crop plants. However, there are certain limitations such as less-effective screening techniques, mode of propagation, large mutated population and time for mutant development. In this context, it has become a necessity to develop effective and alternate methods. In vitro plant cell and tissue culture techniques offer immense scope for plant propagation, germplasm maintenance and genetic manipulation (Jain 2000; Loyola-Vargas and Ochoa-Alejo 2018). The combination of mutagenesis and in vitro culture (referred as, in vitro mutagenesis) has been successfully applied for induction and the selection of desirable mutations. There are several advantages such as (a) propagation and or multiplication of plant material prior to mutagenic treatment to generate a large plant population, (b) separation of chimera and increased recovery of mutations through decreased somatic competition through altered culture conditions and (c) selection of mutants for desirable traits. Thus, the technology has significant potential for speeding up the production of mutants since one can achieve high propagation and a high number of generations per unit time and space under in vitro environment (Suprasanna et al. 2012a, b). A general scheme of mutagenesis using in vivo/in vitro propagules for mutant development and subsequent stages of handling mutagenized plant population and development of a variety is illustrated in Fig. 4.1.

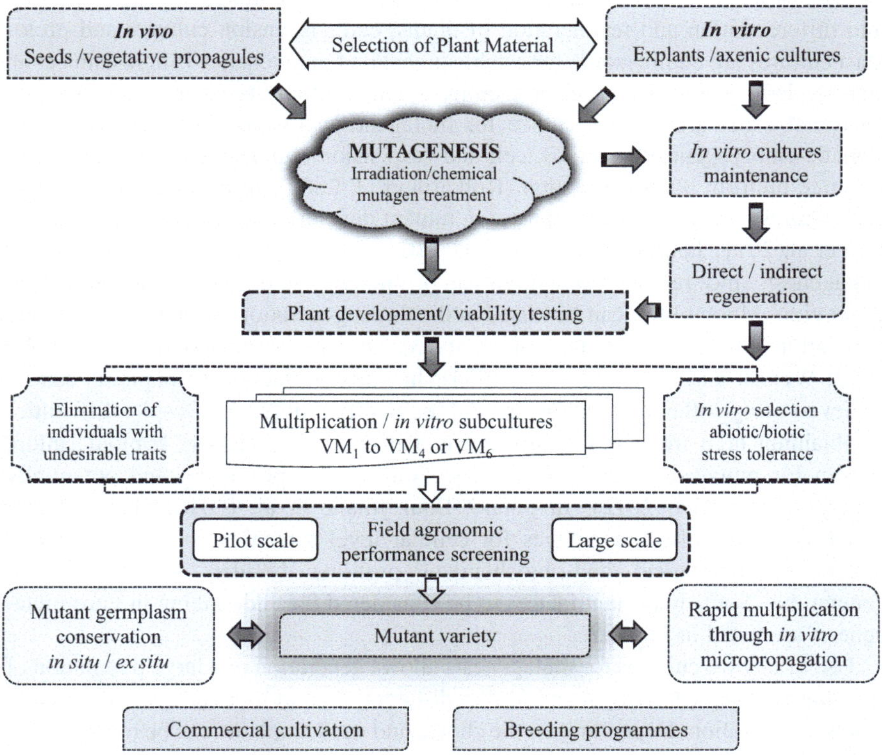

Fig. 4.1 Sequence of events in the development of a mutant variety. (Source: Suprasanna et al. 2015)

4.8 In Vitro Cultures

Plant cells owing to their totipotency can be cultured to generate whole plants. Somaclonal variation is the genetic variability that is expressed in cell and tissue cultures, and this variability can be enhanced through application of in vitro mutagenesis (Novak et al. 1989; Negrutiu 1990). Application of in vitro cultures in mutation breeding has potential benefits such as raising new culture systems for mutagenesis from actively growing in vitro cultures, recovery of mutants and their large-scale multiplication in a controlled environment. In vitro mutagenesis technology has advantages especially in case of vegetative propagated crops (VPC) and has been successfully adopted in several VPCs (Novak 1991; Maluszynski et al. 1995; Suprasanna et al. 2012a).

Mutagenesis can be performed with different culture systems like tissue, organ, callus, cell suspension or protoplasts depending upon the plant system and availability of the efficient regeneration system. The culture systems have been very well established in some model plants, but in other crop plants, few of them have been used for propagation. While shoot tip cultures and callus cultures are easily induced into differentiation and regeneration of plants, cell suspension cultures and protoplast cultures are cumbersome for establishment and maintenance (Suprasanna et al. 2012a). However, cell suspension cultures and protoplasts (being analogous to microbial cells) are a good choice for mutagenesis. In this regard, induction of somatic embryogenesis in callus, cell and suspension cultures can offer benefits to generate uniform plant population (Kulkarni et al. 2007). In banana, embryogenic cell cultures can be developed for use in mutant development (Kulkarni et al. 2007; Jain et al. 2011; Jankowicz-Cieslak et al. 2021). Haploids have a definite gain for mutagenesis since recessive mutations can be induced and detected in the immediate generation. Mutant lines can be raised in the direct generation from any segregating population, thus reducing the time required for varietal development (Szarejko 2012; Wenzel and Foroughi-Wehr 1994). In vitro production of haploids can be achieved using anther and microspore culture in different plants. Despite difficulties in obtaining high frequency androgenesis and plant regeneration, haploid culture system for mutagenesis and mutant selection is quite promising and potentially rewarding (Szarejko 2012). In potato, Souleymane et al. (2016) observed that in vitro cultures offer advantages for cellular level screening for different stress factors, morphological and biochemical profiling besides advancement of generations. Some important factors to be considered for undertaking in vitro mutagenesis are listed in Fig. 4.2.

Use of in vitro cultures for mutagenesis allows generation of a large population of cells for mutagenic treatment and further differentiation of mutated cells/tissues into plants regeneration. Depending on the choice and need, explants can be mutagenized prior to in vitro establishment under culture conditions or, the in vitro cultures (multiple shoot cultures/callus/cell suspensions/protoplasts) can be treated with mutagen. Cell suspension cultures involving rapidly dividing suspension of single cells and cell aggregates are an ideal choice for mutation induction followed by selection. Novak (1991) has described optimum mutagenic treatment conditions for

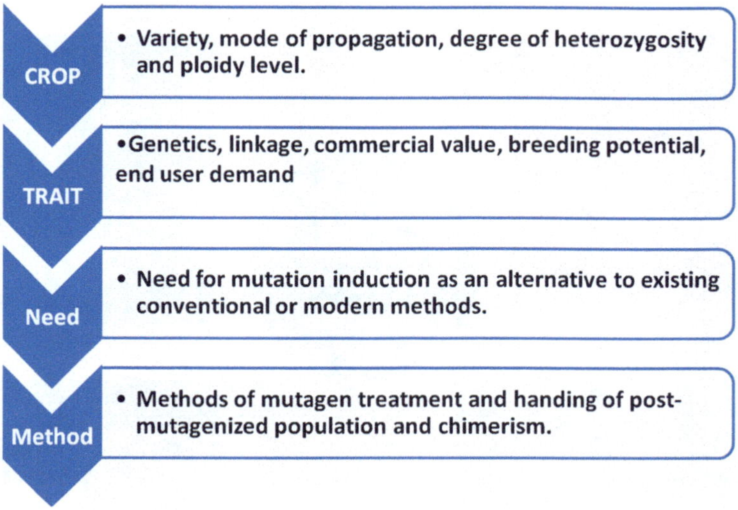

Fig. 4.2 Important considerations for in vitro mutagenesis

several plant species and for different in vitro culture systems. A successful in vitro mutagenesis strategy can depend on factors like reproducible plant regeneration, mutagenic treatment and selection of mutant lines for desired traits (Jain 2007; Suprasanna et al. 2012a). Somatic embryogenesis can be very useful in this context, because of the single cell origin of somatic embryos and any formation of chimera can be circumvented (Jain et al. 2010). For example, in banana, embryogenic suspensions containing single-cell originated embryos can be a good system for generating non-chimeric progeny and early separation of chimeric sectors (Roux et al. 2001). Banana cell suspensions have now become the widely used system for mutation induction (Lopez et al. 2017). In vitro mutation breeding has important components such as mutagenesis, propagation of selected mutants, selection of mutant lines and chimera dissociation (Fig. 4.3). All the four are very important considering the scope and success of mutation induction for a desirable trait. Several successful examples are available in vegetatively propagated plant species such as banana, sugarcane, potato, ornamental plants, which illustrate that such strategies can be helpful in the isolation and development of useful mutants (Roux et al. 2009; Ganapathi et al. 2016; Suprasanna and Negi 2020; Zia et al. 2018; Ibrahim et al. 2018).

4.9 Mutagens, Dose Optimization and Other Considerations

Among the mutagens, physical mutagens (gamma rays, X rays, and UV light) and particle radiations are preferred as there is least residual presence of mutagen and less contact of mutagen, whereas in case of using chemical mutagens, safety measures are essential for handling and disposal of the spent mutagen after

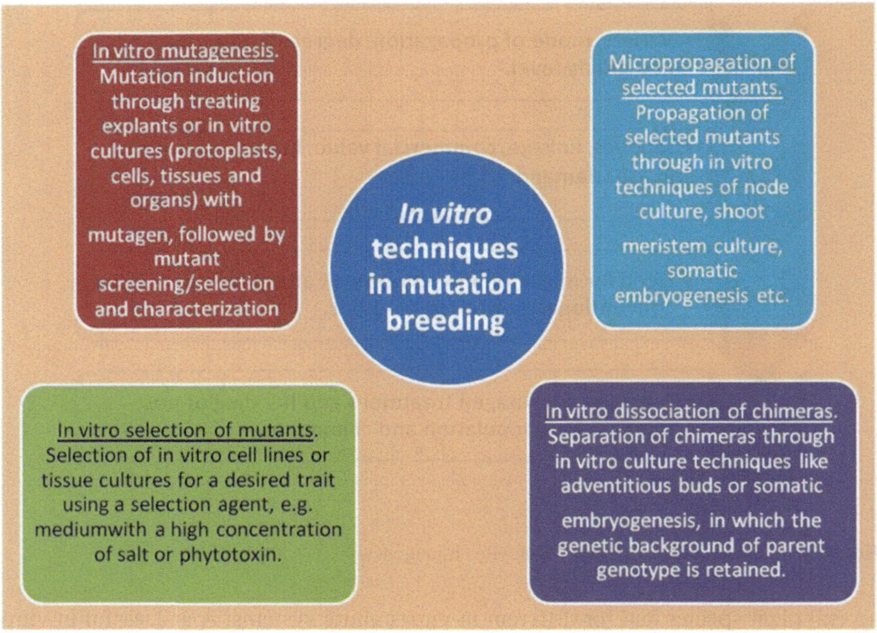

Fig. 4.3 Different components of in vitro mutation breeding

treatment. Different radiation sources are now available for mutation induction (Çelik and Atak 2017). Ion beams are high LET (Linear Energy Transfer) radiations compared with low LET radiations, such as gamma- and X-rays. Plant material can be treated with mutagens in several modes of treatment such as acute, chronic, recurrent and combined (Datta 2020; Hase et al. 2020). Ion beam-based mutation breeding has been successfully accomplished in different plant species, especially ornamentals and several mutants have been generated in Japan and other countries (Yamaguchi 2018). Chemical mutagens generally induce point mutations whereas physical mutagens induce chromosomal aberrations or rearrangements. Among the chemical mutagens, EMS has been used for in vitro cultures in concentration range of 0.2–1% V/V. In most of the annual and biennial ornamental plants, EMS has been widely used for mutant development (Melsen et al. 2021). Other chemical mutagens like MNNG, sodium azide, diethyl sulphate, ethylenimine, ethyl nitroso urethane, ethyl nitroso urea and methyl nitroso urea are also used in other plants (Leitão 2012; Kashtwari et al. 2018).

Sensitivity of a plant tissue to mutation induction often depends upon mutagen type, applied dose and dose rate, besides the plant-based factors such as ploidy, developmental stage and physiological condition. In case of radiations, this is the radiosensitivity to indicate the magnitude of identifiable effects on the irradiated plant part(s). Percentage of surviving cells, cell viability, cell division, growth inhibition, can be used as the indices for checking radiosensitivity. Lethal dose 50 (LD_{50}) is the dose at which 50% of the irradiated/mutagenized material

Table 4.1 Radiosensitivity of some important crops for in vitro mutagenesis

Plant	Explant in culture	Physical mutagen dose
Avocado	Zygotic embryos	27–28 Gy
Banana	Shoot tips	Carbon ion beam 0.5–128 Gy
Banana	Shoot tips	Gamma rays 60 Gy
Banana	In vitro plantlets	Ion beams, 0.5–128 Gy
Banana sp.	Embryogenic cell suspension	Gamma rays, 10–40 Gy
Banana var. Lakatan Latundan	Shoot tips	Gamma rays, 25–40 Gy
Begonia	In vitro cultured leaflets	Gamma rays, 100 Gy
Carnation	Leaf segments	Ion beams 15 Gy
Cassava	Somatic embryos	Gamma rays, 50 Gy
Cassava	In vitro stem cuttings with two nodes	Gamma rays, 25–30 Gy
Chrysanthemum	Rooted cuttings	Gamma rays, 25 Gy
Dendrobium orchids	Protocorm-like bodies	Gamma rays, 35 Gy
Eustoma grandiflorum	Leaf segments	Ion beams 3–10 Gy
Gerbera	In vitro shoots	Ion beams, 5–10 Gy
Orchids	Protocorm-like bodies	Ion beams, 50 Gy
Pear	In vitro shoots	Gamma rays, 3.5 Gy
Peppermint	Stolons and rhizomes	Gamma rays, 30–40 Gy
Perennial gentian	Segment nodes	Ion beams 2–4 Gy
Pineapple var. Queen	Crowns	Gamma rays, 0–45 Gy
Potato	Callus cultures	Gamma rays, 30–50 Gy
Potato	Microtubers	Gamma rays, 10–30 Gy
Potato	Microtuber sprouting	Gamma rays 20–54 Gy
Sugarcane	Buds/callus cultures	Gamma rays, 20–25 Gy
Sweet potato	Embryogenic suspensions	Gamma rays, 80 Gy
Sweet potato	In vitro plantlets	Ion beams 20 Gy
Tangerine	In vitro plantlets	4–24 krad
Yam	Stem cutting	Gamma rays, 20–50 Gy
Yam	Microtuber	Gamma rays, 40 Gy

(vegetative propagules, cells, tissues, pollen or other in vitro cultures) don't survive. Depending on the optimized dose, number of explants/vegetative propagules required for mutagenesis can be calculated as suggested by Predieri and Di Virgilio (2007). Radiosensitivity of some important crop plants is presented in Table 4.1. It is advisable to plan for mutagenic treatments with higher number of initial starting material (cuttings, leaves, tubers, shoot apices, buds, calli, shoots, etc.) for irradiation to obtain sufficient number of plants for screening and selection.

In vitro mutagenesis protocols should be optimized with high regeneration/proliferation rate, number of subcultures post-irradiation, rooting response of the shoots/other propagules and survival of plants. The number of subcultures after irradiation can range from 3 to 6 depending on the plant species and incidence of

Table 4.2 Some practical in vivo techniques used for chimera dissociation (Suprasanna and Nakagawa 2012)

Method	Description	Plant species
Adventitious buds	Regeneration of plants through adventitious buds from leaf cuttings	Chrysanthemum
Induced axillary branching	Induced axillary shoot production from a chimera cultivar by cytokinin application	*Cordyline terminalis* L.
Leaf cuttings	Leaf cuttings regenerate shoots and the plants are mostly non-chimeral (possibly arising from single cells or cells descended from a single apical layer)	*Peperomia*
Adventitious roots from stem cuttings	Generation of shoots from root cuttings (non-chimeral, originating mostly from L3)	Potato, Pelargonium
Disbudding of tubers	Regeneration of shoots from disbudded tubers (stems)	Potato
In vitro organogenesis	Regeneration through organogenesis from sectored chimeras	Begonia and several dicots and monocots
Shoot proliferation through micropropagation	Shoot cultures of chimera plants are proliferated on high cytokinin medium	Banana
Shoot-tip culture technique	Shoot tips (consisting of a meristematic dome with 2–5 leaf primordial) are cultured on cytokinin medium	Banana
Multi-apexing culture technique	A combination of a high proliferation rate and production of adventitious buds is used. With increased number of apical initials, and/or number of cell divisions, the chance are good for accelerating loss of mutations and the apical meristem becomes non-chimeral	Banana
Corm slice culture technique	Shoot-tip explants isolated from in vitro plantlets with a pseudostem are isolated and necrotic tissues were removed before making 3–5 transverse sections below the meristem tip are made. Slices of the corm tissue 0.5–1 mm thick are then placed on the semi-solid multiplication medium	Banana

chimera formation. Thus, rapid proliferation in vitro through subculture passages facilitate the separation of mutated from non-mutated sectors (Ahloowalia 1998). In vitro techniques including adventitious buds, induced axillary branching, leaf cuttings, in vitro organogenesis, shoot proliferation through micropropagation, fragmentation of shoot apices, shoot-tip culture technique, multi-apexing culture technique, corm slice culture technique offer as a good system for mutagenesis in case of vegetatively propagated plants for the production of non-chimera mutants (Table 4.2). When shoot apices are used, multiplication and proliferation of multiple shoots is induced after mutagenic treatment to support further growth and chimera

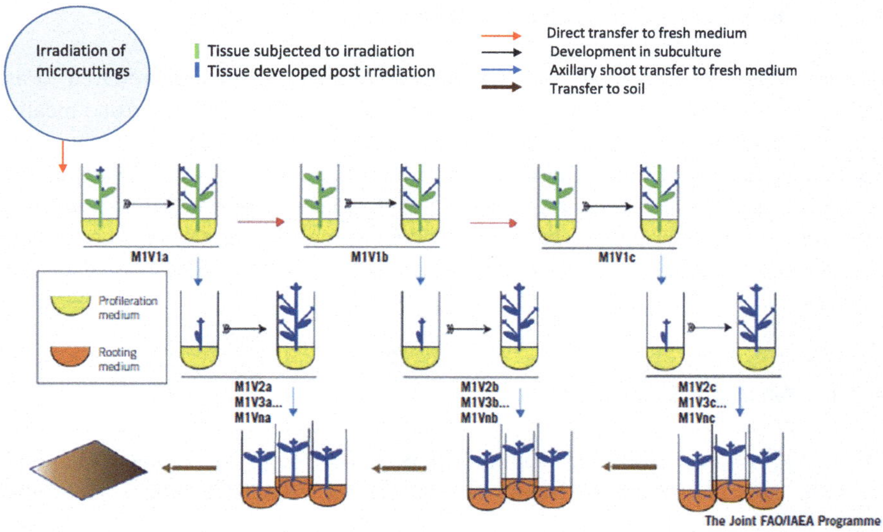

Fig. 4.4 Post-irradiation culture management (Suprasanna et al. 2012a, b)

separation (Kulkarni et al. 2007). Using micro-cuttings, Predieri and Di Virgilio (2007) proposed that post-irradiation management in vitro should necessarily include induction of meristems proliferation and shoot development (Fig. 4.4). Irradiated/in vitro/shoots are allowed to pass through in vitro multiplication passages (M_1V_1, M_1V_2, M_1V_3, ... M_1V_n) so that at every passage, each axillary shoot from irradiated meristems (blue) is excised and cultured onto fresh medium. Originally irradiated shoots (green) after the removal of the first axillary shoots (a) are submitted to two more cultures (b, c) to have more axillary shoots to develop. The post-irradiation culture multiplication scheme can be followed three to four subcultures since mutations occurring in the first vegetative generation after mutagenesis (VM_1) may remain undetected and hence subsequent VM_4 or VM_6 generations will have to be maintained in vitro (van Harten 1998).

During mutagenesis, there is a competition between the mutagenized and non-mutagenized cells in a plant tissue, and this, referred to as intrasomatic selection (Kaplan 1948) or diplontic selection (Gaul 1964), can be a challenge to isolate mutations in vegetatively propagated plant species. To circumvent this, chronic irradiation, neutron irradiation, chemical mutagenesis and mutagenesis of isolated single cells are practised to reduce intrasomatic competition (Nayar 1969). There could also be radiation-induced lethality and higher tissue damage which often results in less competitiveness among the cells and possibility of low regeneration rate among the higher dose mutagenized cells. This is often observed during in vitro mutagenesis experiments (Van Harten 1998; Patade and Suprasanna 2009). Partial desiccation treatments or media manipulations can be employed to invigorate regeneration response in gamma-irradiated in vitro cultures.

4.10 Somaclonal Mutant Varieties

In vitro culture is a potential tool to create novel genetic variation, referred to as 'somaclonal variation', and using the methods of in vitro selection agronomically useful somaclones can be isolated for specific, desirable traits such as yield, improved plant type, flower colour, disease resistance. The somaclonal mutants for distinct morphological traits represent a novel genetic resource for use in crop improvement programmes. These tissue culture-induced mutations have been demonstrated in many plant species and are catalogued in the IAEA mutant database (Table 4.3).

4.11 Mutant Selection

Plant cell, tissue and organ culture offers good scope for mutagenesis and in vitro mutant selection (Novak 1991; Rai et al. 2011; Suprasanna and Rao 1997; Suprasanna et al. 2012a, 2015). There has been considerable success for the application of in vitro mutagenesis and selection of useful agronomic traits like disease resistance, drought, cold and salt tolerance and nutritional quality for their utility in breeding and plant improvement (Suprasanna et al. 2012b). Selection scheme is based on the premise that the desirable trait(s) are selectable at the cellular level and expressed in the regenerated plants. Some of the agronomic characters can be selected in plant cell cultures (Fig. 4.5). Selection of mutants can be done by the addition of a selection agent into culture medium such that non-mutant cells without tolerance or resistance do not survive and only tolerant or resistant cells grow and proliferate. Generally, a dose-response curve for growth inhibition or survival can be established prior to selection studies. Duncan and Widholm (1990) have outlined detailed methodologies for the isolation of mutant cells in vitro. Protocols have been developed for selecting amino acid analogue-resistant cell lines (Miao et al. 1983), herbicide tolerance (Chaleff and Parsons 1978), lysine + threonine resistance (Hibberd and Green 1982), salt tolerance (Nabors 1990) and disease resistance (Gengenbach et al. 1977; Rines and Luke 1985). Generally, phenotypic assessment is done for useful mutants among the mutagenized population, and this constitutes as an essential step for successful mutation breeding. Since phenotyping selections can be often time-consuming and labour-intensive, high-throughput phenotyping (HTP) systems are now being used (Chang et al. 2020).

4.11.1 Selection for Biotic Stress Tolerance

Crop plants are threatened by several diseases and pests and often farmers incur heavy yield losses result in substantial economic losses (Li et al. 2021). In vitro cellular level selection is now used for the selection of disease-resistant plants and the selection criterion includes pathogen itself, pathogen metabolites, toxins or culture filtrates (El Hadrami et al. 2005). Table 4.3 contains some reports on the

Table 4.3 Somaclonal mutant varieties developed in important plants

Mutant variety	Country of origin	Common and scientific name	Improved features
Shanyou 371	China	*Oryza sativa* L.	The mutant variety Shanyou 371 was officially approved in 1998. It was developed by irradiation with gamma rays (10 Gy). Main improved attributes of mutant variety are high grain yield and improved plant structure. The first hybrid rice developed by using somaclonal variation technique in the world.
Hatsu-akane	Japan	*Oryza sativa* L.	The mutant variety Hatsu-akane was officially approved in 1990. Somaclonal mutant by protoplast culture. Main improved attribute of mutant variety is short stem.
Hatsu-yume	Japan	*Oryza sativa* L.	The mutant variety Hatsu-yume was officially approved in 1990. Somaclonal mutant through protoplast culture. Main improved attribute of mutant variety is shorter stem.
Sumi-takara	Japan	*Oryza sativa* L.	The mutant variety Sumi-takara was officially approved in 1991. Somaclonal mutation through anther culture. The real mutant characteristic is hairless on leaves.
Yume-kaori	Japan	*Oryza sativa* L.	The mutant variety Yume-kaori was officially approved in 1993. It is a somaclonal mutant developed through protoplast culture. Main improved attributes of mutant variety are short stem and lower 1000-kernel weight.
Hareyaka	Japan	*Oryza sativa* L.	The mutant variety Hareyaka was officially approved in 1995. Somaclonal mutant through protoplast culture. Main improved attribute of mutant variety is high tolerance to lodging.
Yume-gokochi	Japan	*Oryza sativa* L.	The mutant variety Yume-gokochi was officially approved in 1995. Somaclonal mutant from protoplast culture. Main improved attribute of mutant variety is reduced amylose content.
Ohita 3 Gou	Japan	*Oryza sativa* L.	The mutant variety Ohita 3 Gou was officially approved in 1997. Somaclonal mutant through protoplast culture. Main improved attribute of mutant variety is 10% shorter stem, but slightly smaller grain.
Yume-ekubo	Japan	*Oryza sativa* L.	The mutant variety Yume-ekubo was officially approved in 2000. Somaclonal mutant from protoplast culture. Main improved attributes of mutant variety are short stem and late maturity.

(continued)

Table 4.3 (continued)

Mutant variety	Country of origin	Common and scientific name	Improved features
Iwata 15 Gou	Japan	*Oryza sativa* L.	The mutant variety Iwata 15 Gou was officially approved in 2001. Somaclonal mutant through seed callus. Main improved attribute of mutant variety is low amylose content.
White Baron	Japan	*Solanum tuberosum* L.	The mutant variety White Baron was officially approved in 1997. Somaclonal mutant from protoplast culture. Main improved attribute of mutant variety is tuber tolerant to browning.
Jagakids Purple	Japan	*Solanum tuberosum* L.	The mutant variety Jagakids Purple was officially approved in 1994. Somaclonal mutant from protoplast culture. Main improved attribute of mutant variety is suitable for brewing.
Chikugo-midori	Japan	*Juncus effusus* L.	The mutant variety Chikugo-midori was officially approved in 2001. Somaclonal mutant from meristem culture. Main improved attributes of mutant variety are more long stems, higher weight of total long stems.
Mrs. Elegant	Japan	*Dianthus caryophyllus* L.	The mutant variety Mrs. Elegant was officially approved in 2001. Somaclonal mutation by tissue culture. The real mutant characteristic is small flowered; smaller number of flowers per flower stalk.
Boh-red	Japan	*Dianthus caryophyllus* L.	The mutant variety Boh-red was officially approved in 2002. Somaclonal mutation through tissue culture. The real mutant characteristic is vivid red flower colour.
Kirikami Red	Japan	*Dianthus caryophyllus* L.	The mutant variety Kirikami Red (Mrs. Red) was officially approved in 2002. Somaclonal mutation through tissue culture. The real mutant characteristic is deep orange red surface of flower.
Yua-red	Japan	*Dianthus caryophyllus* L.	The mutant variety Yua-red was officially approved in 2005. Somaclonal mutation through meristem culture. The real mutant characteristics are vivid red flower colour, suspended flower and large flower.
Pearl Prism	Japan	*Chrysanthemum* sp.	The mutant variety Sumi-takara was officially approved in 1991. Somaclonal mutation through anther culture. The real mutant characteristics are erect flower petal and more transparent petal.

(continued)

Table 4.3 (continued)

Mutant variety	Country of origin	Common and scientific name	Improved features
Royal Wedding	Japan	*Chrysanthemum* sp.	The mutant variety Royal Wedding was officially approved in 1998. Somaclonal mutation through tissue culture. The real mutant characteristic is deep purple pink flower colour.
Etenraku	Japan	*Chrysanthemum* sp.	The mutant variety Etenraku was officially approved in 2001. Somaclonal mutation through tissue culture. The real mutant characteristic is shorter stem.
My Comfort	Japan	*Agrostis* sp. (*Creeping bent grass*)	The mutant variety My Comfort was officially approved in 2001. Somaclonal mutant from tissue culture. Main improved attribute of mutant variety is taller plants.
Yume-ippai	Japan	*Oryza sativa* L.	The mutant variety Yume-ippai was officially approved in 2003. It was developed by hybridization with two mutant varieties Yume-gokochi and Yume-kaori obtained by somaclonal mutation of in vitro culture. Main improved attribute of mutant variety is short culm.
Yume-no-hana	Japan	*Oryza sativa* L.	The mutant variety Yume-no-hana was officially approved in 2004. It was developed by hybridization with two mutant varieties Yume-gokochi and Yume-kaori obtained by somaclonal mutation of in vitro culture. Main improved attributes of mutant variety are low amylose content and good taste.
Miyagi VWD 1 Gou	Japan	*Cryptotaenia japonica* (*Japanese parsley*)	The mutant variety Miyagi VWD 1 Gou was officially approved in 2002. Somaclonal mutant from protoplast culture. Main improved attribute of mutant variety is resistance to leaf diseases.
Okayama Nohshi B1 Gou	Japan	*Petasites* sp. (*Butterbur*)	The mutant variety Okayama Nohshi B1 Gou was officially approved in 1996. Somaclonal mutant through meristem callus. Main improved attribute of mutant variety is less anthocyanin in leaf stem.
Shin-Nyohou	Japan	*Fragaria* × *ananassa*	The mutant variety Shin-Nyohou was officially approved in 1989. Somaclonal mutation by callus formation. The real mutant characteristic were earlier harvest type and high yielding.
Akita Berry	Japan	*Fragaria* × *ananassa*	The mutant variety Akita Berry was officially approved in 1992. Somaclonal mutation by meristem culture. The real mutant characteristic is resistance to black leaf spot disease (*Alternaria alternata*).

(continued)

Table 4.3 (continued)

Mutant variety	Country of origin	Common and scientific name	Improved features
Anteher	Japan	*Fragaria × ananassa*	The mutant variety Anteher was officially approved in 1994. Somaclonal mutation by anther culture. Main improved attribute of mutant variety is early maturity.
Himatsuri	Japan	*Fragaria × ananassa*	The mutant variety Himatsuri was officially approved in 1995. Somaclonal mutation through meristem culture. Main improved attributes of mutant variety are bright red fruit skin and light red flesh colour.
Smile Heart	Japan	*Fragaria × ananassa*	The mutant variety Smile Heart was officially approved in 1991. Somaclonal mutation through tissue culture. Main improved attribute of mutant variety is resistance to phytophthora rot (*Phytophthora nicotinae*).
Green Magic	Japan	*Wasabia japonica*	The mutant variety Green Magic was officially approved in 1997. Somaclonal mutation by meristem culture. Main improved attributes of mutant variety are longer stem, less anthocyanin at leaf stem, light colour at root stem and stronger bitterness.
Amagi-nishiki	Japan	*Wasabia japonica*	The mutant variety Amagi-nishiki was officially approved in 1998. Somaclonal mutation by tissue culture. Main improved attribute of mutant variety is less pithy tissue.
Kishu Kasumi 1 Gou	Japan	*Gypsophila elegans M. Bieb.* (*Annual baby's-breath*)	The mutant variety Kishu Kasumi 1 Gou was officially approved in 2000. Somaclonal mutation through meristem culture. Main improved attributes of mutant variety are more wax on the leaf surface and top of the petal V-shape.
Cocktail Dress	Japan	*Cymbidium* sp.	The mutant variety Cocktail Dress was officially approved in 1997. Somaclonal mutation through tissue culture (mericlone). Main improved attribute of mutant variety is petal colour and shape.
Capli Ice	Japan	*Pelargonium* sp.	The mutant variety Capli Ice was officially approved in 1988. Somaclonal mutation through petal culture.
Capli Luluby	Japan	*Pelargonium* sp.	The mutant variety Capli Luluby was officially approved in 1986. Somaclonal mutation through anther culture.
Gold One	Japan	*Oncidium* sp.	The mutant variety Gold One was officially approved in 1995. Somaclonal mutation by mericlone culture. Main improved attribute of mutant variety is flower colour.

(continued)

Table 4.3 (continued)

Mutant variety	Country of origin	Common and scientific name	Improved features
Sherley Baby White	Japan	*Oncidium* sp.	The mutant variety Sherley Baby White was officially approved in 1997. Somaclonal mutation by mericlone culture. Main improved attribute of mutant variety is flower colour.
Sunny Sky	Japan	*Delphinium* sp.	The mutant variety Sunny Skyn was officially approved in 2005. Somaclonal mutation by tissue culture. Main improved attributes of mutant variety are shorter plant and shorter inflorescence.

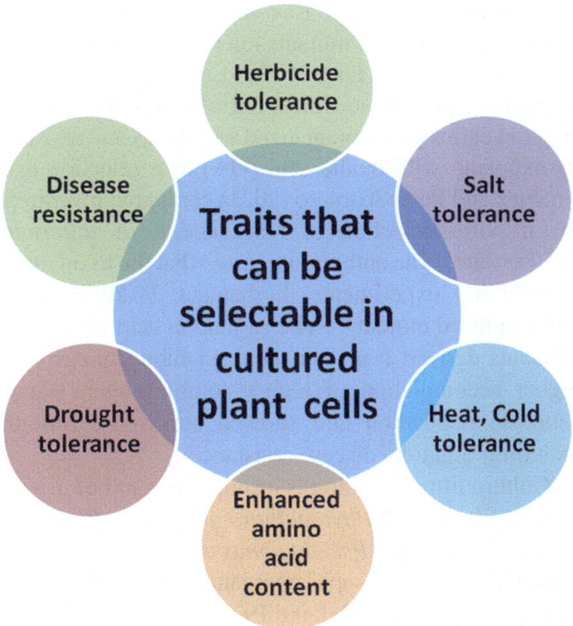

Fig. 4.5 Selectable traits using in vitro cultured plant cells

use of these approaches for selection against plant diseases. One of the approaches employs pathogen itself for selection. The pathogen is allowed to grow in association with the plant cells or tissues. In certain cases, where disease determinants such as toxins are not available, pathogen itself may be used for selection. The approach of using 'pathogen itself' has been used n few plants including tobacco, rice and alfalfa (Miller and Hughes 1980; Palit and Reddy 1990). In potato, discs of pathogen, *Fusarium avenaceum* were used for selection of Fusarium dry root disease resistant in EMS mutagenized potato in vitro plants (Arici et al. 2017). Ghani and

Sharma (2019) used fungal conidial suspension to screen for powdery mildew disease and isolated resistant gerbera plants from gamma irradiated in vitro shoot cultures.

Pathotoxins (in pure or partially pure form), either host-specific or non-specific have been successful for use as selection agents (Gendloff et al. 1987). In other studies, researchers have also considered crude culture filtrates or purified culture filtrate or compounds that closely resemble toxins as selection agents. The *Helminthosporium sacchari* toxin has been very useful for selecting sugarcane (Larkin and Scowcroft 1981). Fusaric acid has found wide application for the selection of Fusarium wilt-resistant plants in several crop plants (Remotti et al. 1997; Matsumoto et al. 1997). Using gamma irradiated in vitro cultured shoots of apple, Saito et al. (2001) screened for resistant mutants for Alternaria blotch disease in the presence of a chemically synthesized AM-toxin I of *Alternaria alternata* (Fr.) Keissler. In date palm (*Phoenix dactylifera* L.), gamma-irradiated embryogenic cell cultures were exposed to Bayoud toxin from the fungus *Fusarium oxysporum* f. sp. albedinis, and several mutants tolerant to Bayoud disease field evaluated for stability of resistance trait (Jain 2012). Liu et al. (2005) developed haploid calli from microspore-derived seedlings for selection of resistance to *Sclerotinia sclerotiorum.* Haploid callus was mutagenized with EMS and resistance was scored in two mutants by exposing with oxalic acid (OA) as a selection agent. In a recent study, Pérez-Jiménez and Pérez-Tornero (2021) compared mycotoxin and culture filtrate methods for in vitro selection of resistance to *A. alternata* in explants of mandarin cv. 'Fortune'. The authors suggested that the toxin method proved to be most reliable in selection experiments. In banana, Wang et al. (2021) successfully isolated EMS-induced mutant ('ReFen 1') upon field based low temperature winter selection of plants derived from mutagenized shoot tip derived cultures. The plants showed higher accumulation of higher soluble sugar, starch accumulation and reduced cellulose deposition and showed superior agronomic traits and cold tolerance including Sigatoka disease resistance.

Culture filtrate method is a simple method of in vitro selection. In several plants, culture filtrates of *Phytophthora infestans*, *Phoma lingam*, *Verticillium dahliae*, *V. alboartrum* and *Pseudomonas solanacearum* have been successfully used for selection in potato, rape, eggplant, hop plant, alfalfa, grapevine, banana, cotton, turmeric and tomato (Behnke 1979; Sacristan 1982; Rotino et al. 1987; Toyoda et al. 1989; Hartman et al. 1984; Binarova et al. 1990: Jayashankar et al. 2000; Okole and Schulz 1997; Ganesan and Jayabalan 2006; Gayatri et al. 2005; Selvapandiyan et al. 1988). Use of double selection system with culture filtrate and host-specific toxin has also been successful for the selection of black sigatoka-resistant banana plants (Okole and Schulz 1997). Saraswathi et al. (2016) treated proliferating shoot buds of cv. Rasthali with EMS and sodium azide and the proliferating explants were then in vitro selected against Fusarium wilt disease by employing fusaric acid and culture filtrate. Three putative resistant mutants were isolated and multiplied in vitro for field evaluation. In conclusion, in vitro selection for disease resistance has become a feasible approach for enabling resistance selection at cellular level; however, extensive investigations are required to translate the research from laboratory and green house level experiments to field level evaluation and varietal development for use in crop improvement programmes.

4.11.2 Selection for Abiotic Stress Tolerance

Plants have to adopt and sustain productivity under environmental stress conditions. Conventional breeding and biotechnological methods have been employed to develop stress-tolerant plants (Suprasanna et al. 2018). Mutants have been developed for tolerance to different abiotic stresses including drought, salinity cold/frost, heavy metals and high temperature. Selection for salt tolerance has been performed by using NaCl in sugarcane (Nikam et al. 2014) and sweet potato (Luan et al. 2007). Use of polyethylene glycol, sorbitol or mannitol has been well established for plant for drought tolerance. For example, banana plants were selected for drought tolerance using sorbitol as an agent to increase osmolality for drought experiments (Bidabadi et al. 2012). In peanuts, Sui et al. (2015) performed in vitro mutagenesis with Pingyangmycin using embryonic leaflets and screened the embryos on Hydroxyproline supplemented medium for the selection of drought-tolerant mutants. Dalvi et al. (2021) exposed EMS-exposed sugarcane callus to polyethylene glycol and reported that this provides good in vitro osmotic selection regime. In sugarcane, EMS mutagenized callus cultures were treated with 28% PEG-6000 to obtain drought-tolerant plants with high adaptation (Khalil et al. 2018).

Salt-tolerant plantlets have been obtained using NaCl as the major selection mediator in flax (Rowland et al. 1989), sugar beet (Freytag et al. 1990), *Brassica juncea* (Kirti et al. 1991), *Nicotiana* sp. (Sumaryati et al. 1992) and rice (Reddy and Vaidyanath 1986; Elshafei et al. 2019). Selection with proline analogues or ABA insensitivity was also performed in developing salt-tolerant plants in *Brassica napus* (Chandler and Thorpe 1986) and *Vigna radiata* (Kumar and Sharma 1989). Gamma ray-induced in vitro mutagenesis followed by in vitro selection for NaCl tolerance has been successfully accomplished in sugarcane (Patade et al. 2008; Nikam et al. 2014) and the mutant clones showed high sugar yield. Hossain et al. (2006) obtained a NaCl-tolerant mutant line of *Chrysanthemum morifolium* Ramat isolated through in vitro mutagenesis with gamma radiation and selection with sodium chloride. In another study, gamma irradiated axillary buds of in vitro strawberry plants were exposed to drought conditions for selecting drought-tolerant plants (Jain 1997). In an endangered medicinal plant, *Picrorhiza kurroa* Royle, callus was mutagenized with EMS and selected for salinity and drought stress tolerance using NaCl and mannitol suggesting that mutant lines can be generated in medicinally important plants (Bisht et al. 2017).

Similarly, selection for resistance to metal stresses has also been studied using higher lethal or sublethal levels of heavy metals as selection agents for in vitro selection (Ashrafzadeh and Leung 2015). Successful in vitro selection are available, for example, aluminum resistance in tomato (Conner and Meredith 1985), manganese resistance in tobacco (Petolino and Collins 1985), zinc resistance in *Haplopappus* (Gilissen and Van Staveren 1986), mercury resistance in Petunia

(Colijn et al. 1979), and cadmium resistance in *Datura* and potato (Ashrafzadeh and Leung 2017). The studies suggest the possibility that such methods can be extended to mutant population for selection against metal stress. Using chemical mutagenesis, several mutants have been isolated having enhanced metal accumulation (Phang et al. 2012). For example, Piotto et al. (2014) described an efficient selection of EMS mutagenized mutant plants with high Cd levels and obtained two cadmium-tolerant mutants of tomato (cv. Micro-Tom).

4.11.3 Selection for Enhanced Nutritional Content

Nutritional quality of crop plants assumes significance and hence plants with higher nutritional content are developed using different breeding and biotechnological methods (Soares et al. 2019). Cereal crops have deficiency of essential amino acids (lysine, threonine and tryptophan) whereas legume crops have deficiency of methionine (Widholm 1974, 1985). Selection in vitro has been useful for screening improved nutritional quality by using amino acid analogues in culture medium (Dix 1986). It has been demonstrated that analogues of an amino acid or higher growth inhibitory levels of amino acids such as lysine and threonine can be made as the selection criteria (Hibberd and Green 1982). Several studies have shown success of the in vitro selection approach in rice, maize, and carrot (Widholm 1985; Hibberd and Green 1982; Miao et al. 1983; Wakasa and Widholm 1987). Rice, normally deficient for lysine, was selected in vitro for resistance to *S*-(2-aminoethyl)-cysteine (AEC) and the regenerated plants showed higher levels of lysine (Schaeffer 1981). Kim et al. (2014) employed gamma-irradiated rice cell lines and selected against AEC and interestingly four AEC-resistant M_3 lines showed higher amino acids contents than the parent cultivar. In another study, 5-methyltryptophan (5MT)-resistant rice mutant lines were developed with high free tryptophan content and increased protein (Kim et al. 2005). Recently, it has been shown that *E. coli* lysine-overproducing strains possessed stress tolerance (Isogai and Takagi 2021) suggesting that lysine over production in plants can similarly be considered as a criterion for stress tolerance.

4.12 Concluding Remarks

Mutation breeding is an important strategy of plant improvement programmes, and it has greatly contributed to global food security and nutrition. Stable and inherited mutations are induced in plants by means of physical or chemical mutagens in diverse seed and vegetatively propagated plants. To date, More than 3400 mutant plant varieties have been released for cultivation in cereals and oilseed crops to orchids and woody perennials, indicating the widespread use of induced mutations in crop improvement. Although mutations can be induced in all plant species, it is important to consider suitability of the plant species, ploidy, propagation type to obtain desired mutant in given time and, efforts needed with available field facilities.

Once the mutant is isolated, depending on its superiority either it can be released as a new variety or it can be used indirectly through crossing to develop high yielding superior varieties.

The culture of totipotent cells has great potential for in vitro mutagenesis and selection at cellular level. In vitro approaches include the use of shoot apical meristems cultures, embryogenic callus, cell suspensions and protoplast cultures. Both physical and chemical mutagens are applied in tissue cultures so as to enhance the recovery of the selected variants. It is now possible to select for tolerance or resistance to different biotic/abiotic stress factors and regenerate plants. For selection, different approaches of culture filtrate, pathotoxins, amino acid analogues, heavy metals and salts can be used. Considerations such as appropriate in vitro culture system(s), high frequency plant regeneration, post-mutagenesis handling of mutant population and evaluation of mutants have to be dealt carefully before embarking on in vitro mutagenesis program. Mutant selection in plant cell and tissue cultures is exciting and offers considerable scope for generating genetic variability for a range of agronomic traits and mutants have been isolated in cereals, fruit crops and pulse crops for disease resistance, salt stress and amino acid enhancement. There is a need for consistent research efforts to create and characterize mutation-induced genetic diversity, multiplication of desirable mutants in order to complement plant biodiversity and large-scale assessment on the impact of induced mutants on various agronomic, yield and nutritional quality characters under normal and challenging environment. The field of in vitro mutagenesis and mutant selection is exciting with immense opportunities to utilize the system to generate useful mutants of interest for use in crop improvement programmes.

References

Ahloowalia BS (1998) *In vitro* techniques and mutagenesis for the improvement of vegetatively propagated plants. In: Jain SM, Brar DS (eds) Somaclonal variation and induced mutations in crop improvement. Kluwer Acad. Publ., Dordrecht, pp 293–309

Ahloowalia B, Maluszynski M, Nichterlein K (2004) Global impact of mutation-derived varieties. Euphytica 135:187–204

Álvarez-Holguín A, Morales-Nieto CR, Avendaño-Arrazate CH, Corrales-Lerma R, Villarreal-Guerrero F, Santellano-Estrada E, Gómez-Simuta Y (2019) Mean lethal dose (LD50) and growth reduction (GR50) due to gamma radiation in Wilman lovegrass (*Eragrostis superba*). Rev Mex Cienc Pecu 10:227–238

Arici SE, Tuncel ZN, Kara A, Çaltili O (2017) *In vitro* mutagenesis and selection of potato mutants resistance to fusarium dry root (*Fusarium Avenaceum*). Comm Appl Biol Sci 82(3):378–385

Ashrafzadeh S, Leung DMW (2015) *In vitro* breeding of heavy metal-resistant plants: a review. Hortic Environ Biotechnol 56:131–136

Ashrafzadeh S, Leung DWM (2017) Novel potato plants with enhanced cadmium resistance and antioxidative defence generated after *in vitro* cell line selection. PLoS One 12(10):e0185621. https://doi.org/10.1371/journal.pone.0185621

Bado S, Forster BP, Nielen S, Ali AM, Lagoda PJ, Till BJ, Laimer M (2015) Plant mutation breeding: current progress and future assessment. In: Janick J (ed) Plant breeding reviews, vol 39. https://doi.org/10.1002/9781119107743.ch2

Behnke M (1979) Selection of potato callus for resistance to culture filtrates of *Phytophthora infestans* and regeneration of resistant plants. Theor Appl Genet 55:69–71

Bidabadi SS, Meon S, Wahab Z, Subramaniam S, Mahmood M (2012) *In vitro* selection and characterization of water stress tolerant lines among ethyl methanesulphonate (EMS) induced variants of banana (*Musa* spp., with AAA genome). Aust J Crop Sci 6:567–575

Binarova P, Nedelnik J, Fellner M, Nedbalkova B (1990) Selection of resistance to filtrates of Fusarium spp. in embryogenic cell suspension culture of *Medicago sativa* L. Plant Cell Tissue Organ Cult 22:191–196

Bisht SS, Bisht AS, Chauhan RS (2017) *In-vitro* mutagenesis induction to improve abiotic stress in tissue cultured plantlet of *Picrorhiza kurroa* Royle ex. Benth: an endangered plant of Western Himalayas, India. Med Aromat Plants 6:1–11

Breseghello F, Coelho ASG (2013) Traditional and modern plant breeding methods with examples in Rice (*Oryza sativa* L.). J Agric Food Chem 61(35):8277–8286

Brown N (2013) Mutagenesis. http://plantbreeding.coe.uga.edu/index.php?title=20._Mutagenesis

Calicioglu O, Flammini A, Bracco S, Bellù L, Sims R (2019) The future challenges of food and agriculture: an integrated analysis of trends and solutions. Sustainability 11(1):222. https://doi.org/10.3390/su11010222

Çelik Ö, Atak Ç (2017) Applications of ionizing radiation in mutation breeding. In: Maghraby AM (ed) New insights on gamma rays. IntechOpen, Rijeka, Croatia. https://doi.org/10.5772/66925

Chaleff RS, Parsons MF (1978) Direct selection *in vitro* for herbicide-resistant mutants of *Nicotiana tobaccum*. Proc Natl Acad Sci U S A 75:5104–5107

Chandler SF, Thorpe TA (1986) Variation from plant tissue cultures: biotechnological application to improving salinity tolerance. Biotechnol Adv 4:117–135

Chang S, Lee U, Hong MJ, Jo YD, Kim JB (2020) High-throughput phenotyping (HTP) data reveal dosage effect at growth stages in Arabidopsis thaliana irradiated by gamma rays. Plants (Basel) 9(5):557. https://doi.org/10.3390/plants9050557

Colijn CM, Kool AJ, Nijkaamp HJJ (1979) An effective chemical mutagenesis procedure for Petunia hybrid cell suspensions. Theor Appl Genet 55:101–106

Conner AJ, Meredith CP (1985) Large scale selection of aluminum-resistant mutants from plant cell culture: expression and inheritance in seedlings. Theor Appl Genet 71:159–165

Dalvi SG, Tawar PN, Suprasanna P, Dixit GB, Theertha Prasad D (2021) EMS-based *in vitro* mutagenesis and mutant screening for smut resistance with agronomic traits in Sugarcane. Sugar Tech 23:854–864

Datta SK (2020) Induced mutations: technological advancement for development of new ornamental varieties. Nucleus 63. https://doi.org/10.1007/s13237-020-00310-7

Dix PJ (1986) Cell line selection. In: Yeomann MM (ed) Plant cell culture technology. Blackwell Science, Oxford, UK, pp 143–201

Duncan DR, Widholm JM (1990) Techniques for selecting mutants from plant tissue cultures. Methods Mol Biol 6:443–453

El Hadrami A, El Idrissi-Tourane A, El Hassni M, Daayf F, El Hadrami I (2005) Toxin-based *in-vitro* selection and its potential application to date palm for resistance to the bayoud Fusarium wilt. C R Biol 328:732–744

Elshafei A, Barakat M, Milad S (2019) Regeneration of rice somaclones tolerant to high level of abscisic acid and their characterization via RAPD markers. Bull Natl Res Cent 43:107

Freisleben RA, Lein A (1944) Möglichkeiten und praktische Durchführung der Mutationszüchtung. Kühn-Arhiv 60:211–222

Freytag AH, Wrather JA, Erichsen AW (1990) Salt tolerance Sugar beet progeny from tissue cultures challenged with multiple salts. Plant Cell Rep 8:647–650

Fróna D, Szenderák J, Harangi-Rákos M (2019) The challenge of feeding the world. Sustainability 11(20). https://doi.org/10.3390/su11205816

Ganapathi TR, Ujjappa KM, Badigannavar A (2016) Characterization of gamma ray induced clones in 'giant cavendish' banana (AAA) for morphological and yield contributing traits. Int J Fruit Sci 16(3):310–322

Ganesan M, Jayabalan N (2006) Isolation of disease-tolerant cotton (*Gossypium hirsutum* L. cv. SVPR 2) plants by screening somatic embryos with fungal culture filtrate. Plant Cell Tissue Organ Cult 87:273–284

Gaul R (1964) Mutations in plant breeding. Radiat But 4:155–232

Gayatri MC, Roopa Darshini V, Kavyashree R (2005) Selection of turmeric callus for tolerant to culture filtrate of *Pythium graminicolum* and regeneration of plants. Plant Cell Tissue Organ Cult 83:33–40

Gendloff EH, Scheffer RP, Somerville SC (1987) An improved bioassay for victorin based on the use of oat protoplasts. Physiol Mol Plant Pathol 31:421–427

Gengenbach BG, Green CE, Donovan CM (1977) Inheritance of selected pathotoxin resistance in maize plants regenerated from cell cultures. Proc Natl Acad Sci U S A 74:5113–5117

Ghani M, Sharma SK (2019) Induction of powdery mildew resistance in gerbera (*Gerbera jamesonii*) through gamma irradiation. Physiol Mol Biol Plants 25(1):159–166

Gilissen LJW, Van Staveren MJ (1986) Zinc-resistant cell lines of *Haplopappus gracilis*. J Plant Physiol 125(1):95–103

Harlan JR (1992) Crops and men, 2nd edn. American Society of Agronomy, Crop Science Society, Madison, WI. 284 pp

Hartman CL, Mc Coy TJ, Knous TR (1984) Selection of alfalfa (*Medicago sativa*) cell lines and regeneration of plants resistant to the toxin(s) produced by *Fusarium oxysporum* f. sp. medicaginis. Plant Sci Lett 34:183–194

Hase Y, Satoh K, Seito H, Oono Y (2020) Genetic consequences of acute/chronic gamma and carbon ion irradiation of *Arabidopsis thaliana*. Front Plant Sci 11:336. https://doi.org/10.3389/fpls.2020.00336

Hibberd KA, Green CE (1982) Inheritance and expression of lysine plus threonine resistance selected in maize tissue cultures. Proc Natl Acad Sci U S A 79:559–563

Holme IB, Gregersen PL, Brinch-Pedersen H (2019) Induced genetic variation in crop plants by random or targeted mutagenesis: convergence and differences. Front Plant Sci 10:1468. https://doi.org/10.3389/fpls.2019.01468

Hossain Z, Mandal AKA, Datta SK, Biswas AK (2006) Isolation of a NaCl-tolerant mutant of *Chrysanthemum morifolium* by gamma radiation: *in vitro* mutagenesis and selection by salt stress. Funct Plant Biol 33(1):91–101

Ibrahim R, Ahmad Z, Salleh S, Abu Hassan A, Ariffin S (2018) Mutation breeding in ornamentals. In: Van Huylenbroeck J (ed) Ornamental crops, handbook of plant breeding, vol 11. Springer International Publishing AG, part of Springer Nature, Cham, pp 175–211

Isogai S, Takagi H (2021) Enhancement of lysine biosynthesis confers high-temperature stress tolerance to *Escherichia coli* cells. Appl Microbiol Biotechnol 105(18):6899–6908

Jain SM (1997) Creation of variability by mutation and tissue culture in improving plants. Acta Hortic 447:69–78

Jain SM (2000) Tissue culture induced variation in crop improvement. Euphytica 118:153–166

Jain SM (2007) Mutation-assisted breeding for improving ornamental plants. Acta Hortic 714: 85–98

Jain SM (2012) *In vitro* mutagenesis for improving date palm (*Phoenix dactylifera* L.). Emir J Food Agric 24(5):400–407

Jain S, Ochatt S, Kulkarni VM, Predieri S (2010) *In vitro* culture for mutant development. Acta Hortic 865:59–68

Jain SM, Till B, Suprasanna P, Roux N (2011) Mutations and cultivar development of banana. In: Pillay M, Tenkouano A (eds) Banana breeding: progress and challenges. CRC Press Taylor & Francis Group, New York, pp 203–217

Jankowicz-Cieslak J, Gössnitzer F, Datta S, Viljoen A, Ingelbrecht I, Till B (2021) Induced mutations for generating bananas resistant to Fusarium wilt tropical race 4. In: Mutation breeding, genetic diversity and crop adaptation to climate change. CABI, Wallingford, UK. https://doi.org/10.1079/9781789249095.0038

Jayashankar S, Li Z, Gray DJ (2000) *In vitro* selection of *Vitis vinifera* Chardonnay with *Elsinoe ampelina* culture filtrate is accompanied by fungal resistance and enhanced secretion of chitinase. Planta 211:200–208

Kaplan RW (1948) Uber Möglichkeiten der Mutationsauslösung in der Pflanzenzüchtung. Zeits. Pflanzenzucht 32:121–131. 1953

Kashtwari M, Wani AA, Dhar MK, Jan S, Kamili AN (2018) Development of an efficient in vitro mutagenesis protocol for genetic improvement of saffron (*Crocus sativus* L.). Physiol Mol Biol Plants 24(5):951–962

Khalil F, Naiyan X, Tayyab M, Pinghua C (2018) Screening of EMS-induced drought-tolerant sugarcane mutants employing physiological, molecular and enzymatic approaches. Agronomy 8:226. https://doi.org/10.3390/agronomy8100226

Kim DS, Lee I, Jang C, Kang S-Y, Park I, Song H, Seo YA (2005) High amino acid accumulating 5-methyltryptophan-resistant rice mutants may include an increased antioxidative response system. Physiol Plant 123:302–313

Kim DS, Lee IS, Jang CS, Lee SJ, Song HS, Lee YI, Seo YW (2014) AEC resistant rice mutants induced by gamma-ray irradiation may include both elevated lysine production and increased activity of stress related enzymes. Plant Sci 167:305–316

Kirti PB, Hadi S, Kumar PA, Chopra VL (1991) Production of sodium-chloride-tolerant Brassica juncea plants by in vitro selection at the somatic embryo level. Theor Appl Genet 83:233–237

Kodym A, Afza R, Forster BP, Ukai Y, Nakagawa H, Mba C (2012) Methodology for physical and chemical mutagenic treatments. Plant mutation breeding and biotechnology. FAO-IAEA, Rome, Italy, pp 135–158

Kranz J (1988) Measuring plant disease. In: Kranz J, Rotem J (eds) Experimental techniques in plant disease epidemiology. Springer-Verlag, New York, pp 35–50

Kulkarni VM, Ganapathi TR, Suprasanna P, Bapat VA (2007) *In vitro* mutagenesis in banana (*Musa* spp.) using gamma irradiation. In: Jain SM, Häggman H (eds) Protocols for micro-propagation of woody trees and fruits. Springer, Dordrecht, pp 543–559

Kumar V, Sharma DR (1989) Isolation and characterization of sodium chloride resistant callus cultures of *Vigna radiata* (L.) Wilczek var. radiata. J Exp Bot 40:143–147

Larkin PJ, Scowcroft WR (1981) Somaclonal variation—a novel source of variability from cell culture for plant improvement. Theor Appl Genet 60:197–214

Leitão JM (2012) Chemical mutagenesis. In: Shu QY, Forster BP, Nakagawa H (eds) Plant mutation breeding and biotechnology. CABI Publishing, Wallingford, pp 135–158

Li Q, Wang B, Yu J, Dou D (2021) Pathogen-informed breeding for crop disease resistance. J Integr Plant Biol 63:305–311

Liu S, Wang H, Zhang J, Fitt BD, Xu Z, Evans N, Liu Y, Yang W, Guo X (2005) *In vitro* mutation and selection of doubled-haploid *Brassica napus* lines with improved resistance to *Sclerotinia sclerotiorum*. Plant Cell Rep 24(3):133–144

Lopez J, Rayas A, Santos A, Medero V, Beovides Y, Basail M (2017) Mutation induction using gamma irradiation and embryogenic cell suspensions in plantain (Musa spp.). In: Jankowicz-Cieslak J, Tai T, Kumlehn J, Till B (eds) Biotechnologies for plant mutation breeding. Springer, Cham, pp 55–71

Loyola-Vargas VM, Ochoa-Alejo N (2018) An introduction to plant tissue culture: advances and perspectives. Methods Mol Biol 1815:3–13

Luan Y, Zhang J, Gao X-R, An L-J (2007) Mutation induced by ethylmethane sulphonate (EMS), *in vitro* screening for salt tolerance and plant regeneration of sweet potato (*Ipomoea batatas* L.). Plant Cell Tissue Organ Cult 88:77–81

Maluszynski M, Ahloowalia BS, Sigurbjörnsson B (1995) Application of in vivo and in vitro mutation techniques for crop improvement. Euphytica 85:303–315

Matsumoto K, Souza LAC, Barbosa ML (1997) *In vitro* selection for Fusarium wilt resistance in banana I. Co-cultivation technique to produce culture filtrate of race 1 *Fusarium oxysporum* f. sp. cubense. Fruits 54(2):97–102

Mba C, Afza R, Bado S, Jain SH (2010) Induced mutagenesis in plants using physical and chemical agents. In: Davey MR, Anthony P (eds) Plant cell culture: essential methods. Wiley, Chichester, UK, pp 111–130
Melsen K, van de Wouw M, Contreras R (2021) Mutation breeding in ornamentals. HortScience 56(10):1154–1165
Miao S, Duncan DR, Widholm JM (1983) Selection of regenerable maize callus cultures resistant to 5-methyl tryptophan, S2aminoethyl-L-cysteine and high levels of L-lysine plus L-threonine. Plant Cell Tissue Organ Cult 14:3–14
Miller OK, Hughes KW (1980) Selection of paraquat resistant variants of tobacco from cell cultures. In Vitro 16:1085–1091
Nabors MW (1990) Environmental stress resistance. In: Dix PJ (ed) Plant cell line selection procedures and applications. VCH Weinheim, New York, pp 167–186
Nayar NM (1969) Considerations on overcoming intrasomatic selection during mutation breeding of vegetatively propagated plants. Theor Appl Genet 39:99–103
Negrutiu I (1990) *In vitro* mutagenesis. In: Dix PJ (ed) Plant cell line selection. VCH Publ, Weinheim, pp 19–38
Nikam AA, Devarumath RM, Shitole MG et al (2014) Gamma radiation, *in vitro* selection for salt (NaCl) tolerance, and characterization of mutants in sugarcane (*Saccharum officinarum* L.). In Vitro Cell Dev Biol Plant 50:766–776
Novak FJ (1991) *In vitro* mutation system for crop improvement. In: Plant mutation breeding for crop improvement. Proc. Int. Symp. on the contribution of plant mutation breeding, vol 2. IAEA, Vienna, Austria, pp 327–342
Novak F, Afza R, Van Duren M, Omar MS (1989) Mutation induction by gamma irradiation of *in vitro* cultured shoot-tips of banana and plantain (*Musa* cvs). Trop Agric (Trinidad) 67:21–28
Okole BN, Schulz FA (1997) Selection of *Mycosphaerella fijiensis*-resistant cell lines from micro-cross sections of banana and plantain. Plant Cell Rep 16(5):339–343
Palit R, Reddy GM (1990) Selection of callus resistant to *Pyricularia oryzae* and regeneration of plants in rice. Indian J Exp Biol 28:928–931
Patade VY, Suprasanna P (2009) An *in vitro* radiation induced mutagenesis-selection system for salinity tolerance in sugarcane. Sugar Tech 11:246–251
Patade VY, Suprasanna P, Bapat VA (2008) Gamma irradiation of embryogenic callus cultures and in vitro selection for salt tolerance in sugarcane (*Saccharum officinarum* L.). Agric Sci China 7(2008):101–105
Pathirana R (2011) Plant mutation breeding in agriculture. CAB Rev Perspect Agric Vet Sci Nutr Nat Resour 6:1–20
Pérez-Jiménez M, Pérez-Tornero O (2021) Comparison of four systems to test the tolerance of 'Fortune' Mandarin tissue cultured plants to *Alternaria alternata*. Plants 10:1321. https://doi.org/10.3390/plants10071321
Petolino JF, Collins GB (1985) Manganese toxicity in tobacco (*Nicotiana tabacum* L.) callus and seedlings. J Plant Physiol 118:139–144
Phang IC, Taylor H, Leung D (2012) Chemical mutagenesis for improving potential of plants to remediate environments with heavy metal contaminants. Curr Top Phytochem 11:47–52
Piotto FA, Tulmann-Neto A, Franco MR, Boaretto LF, Azevedo RA (2014) Rapid screening for selection of heavy metal-tolerant plants. Crop Breed Appl Biotechnol 14:1–7
Predieri S, Di Virgilio N (2007) *In vitro* mutagenesis and mutant multiplication. In: Jain SM, Häggman H (eds) Protocols for micropropagation of woody trees and fruits. Springer, Dordrecht, pp 323–333
Qaim M (2020) Role of new plant breeding technologies for food security and sustainable agricultural development. Appl Econ Perspect Policy 42:129–150
Rai M, Kalia R, Singh R, Gangola MP, Dhawan A (2011) Developing stress tolerant plants through *in vitro* selection—an overview of the recent progress. Environ Exp Bot 71:89–98
Reddy PJ, Vaidyanath K (1986) *In vitro* characterization of salt stress effects and the selection of salt tolerant plants in rice (*Oryza sativa* L.). Theor Appl Genet 71(5):757–760

Remotti PC, Loffler HJM, Vanvlotendoting L (1997) Selection of cell-lines and regeneration of plants resistant to fusaric acid from *Gladiolus grandiflorus* cv. Peter pears. Euphytica 96:237–245

Rines HW, Luke HH (1985) Selection and characterization of toxin-insensitive plants from tissue cultures of oats (*Avena sativa*) susceptible to *Helminthosporium victoriae*. Theor Appl Genet 71:16–21

Rotino GL, Falavigna A, Restaino F (1987) *In vitro* selection of eggplant cells resistant to culture filtrate of *Verticillium dahliae* Kleb. and regeneration of plants. Capsicum Newsl 6:94–95

Roux N, Dolezel J, Swennen R, Zapata-Arias FJ (2001) Effectiveness of three micropropagation techniques to dissociate cytochimeras in Musa spp. Plant Cell Tissue Org Cult 66:189–197

Roux N, Toloza A, Strosse H, Busogoro JP, Jaroslav D (2009) Induction and selection of potentially useful mutants in banana. Acta Hortic 828:315–322

Rowland GG, McHughen A, McOnie C (1989) Field performance at saline affected sites of a somaclonal variant of McGregor flax selected for salt tolerance *in vitro*. Can J Plant Sci 69:49–60

Sacristan MD (1982) Resistance response to *Phoma lingum* of plant regenerated from selected cells and embryogenic cultures of haploid *Brassica napus*. Theor Appl Genet 61:193–200

Saito A, Nakazawa N, Suzuki M (2001) Selection of mutants resistant to *Alternaria* blotch from in vitro-cultured apple shoots irradiated with X- and γ-rays. J Plant Physiol 158:391–400

Saraswathi M, Kannan G, Uma S, Thangavelu R, Backiyarani S (2016) Improvement of banana cv. Rasthali (Silk, AAB) against *Fusarium oxysporum* f.sp. cubense (VCG 0124/5) through induced mutagenesis: Determination of LD50 specific to mutagen, explants, toxins and in vitro and in vivo screening for Fusarium wilt resistance. Indian J Exp Biol 54:345–353

Schaeffer GW (1981) Mutations and cell selections: increased protein from regenerated rice tissue cultures. Environ Exp Bot 21(3–4):333–345

Selvapandiyan A, Mehta AR, Bhatt PN (1988) Cellular breeding approach for development of Fusarium wilt resistant tobacco. Proc Ind Natl Acad Sci B 54:391–394

Soares JC, Santos CS, Carvalho SMP, Pintado MP, Vasconcelos MW (2019) Preserving the nutritional quality of crop plants under a changing climate: importance and strategies. Plant Soil 443:1–26

Souleymane B, Matumelo AR, Kaoutar EA, Enoch S, Stephan N, Abdelbagi MAG, Margit L (2016) *In vitro* methods for mutation induction in potato (*Solanum tuberosum* L.). Afr J Biotechnol 15(39):2132–2145

Sui J, Wang Y, Wang P, Qiao L, Sun S, Hu X et al (2015) Generation of Peanut drought tolerant plants by pingyangmycin-mediated *in vitro* mutagenesis and hydroxyproline-resistance screening. PLoS One 10(3):e0119240. https://doi.org/10.1371/journal.pone.0119240

Sumaryati S, Negrutiu I, Jacobs M (1992) Characterization and regeneration of salt-and water-stress mutants from protoplast culture of *Nicotiana plumbaginifolia* (Viviani). Theor Appl Genet 83: 613–619

Suprasanna P, Nakagawa H (2012) Mutation breeding of vegetatively propagated crops. In: Shu Q-Y, Forster BP, Nakagawa H (eds) Plant mutation breeding and biotechnology. FAO-IAEA & CABI, pp 345–356

Suprasanna P, Negi P (2020) Creating novel genetic variability in sugarcane using induced mutations. IANCAS Bull XV(1):23–29

Suprasanna P, Rao PS (1997) Selection of mutants using plant cell and tissue culture techniques. In: Hemantaranjan A (ed) Advances in plant physiology, vol 1. Scientific Publishers, Jodhpur, India, pp 103–122

Suprasanna P, Jain SM, Ochatt SJ, Kulkarni VM, Predieri S (2012a) Applications of *in vitro* techniques in mutation breeding of vegetatively propagated crops. FAO-IAEA, Rome, Italy

Suprasanna P, Sonawane BV, Patade VY (2012b) *In vitro* mutagenesis and selection: strategies and prospects for crop improvement. Biodivers Biotechnol Bioremed 6:6–14

Suprasanna P, Mirajkar SJ, Bhagwat SG (2015) Induced mutations and crop improvement. In: Bahadur B, Venkat Rajam M, Sahijram L, Krishnamurthy K (eds) Plant biology and biotechnology. Springer, New Delhi, pp 593–617

Suprasanna P, Ghuge SA, Patade VY, Mirajkar SJ, Nikalje GC (2018) Genomic roadmaps for augmenting salinity stress tolerance in crop plants; salinity responses and tolerance in plants, vol 2. Springer International, Cham, pp 189–216

Szarejko I (2012) Haploid mutagenesis. In: Shu QY, Forster BP, Nakagawa H (eds) Plant mutation breeding and biotechnology. FAO-IAEA, Rome, Italy, pp 387–410

Toyoda H, Shimizu K, Chatani K, Kita N, Matsuda Y, Ouchi S (1989) Selection of bacterial wilt-resistant tomato through tissue culture. Plant Cell Rep 8:317–320

Van Harten AM (1998) Mutation breeding: theory and practical applications. Cambridge University Press, Cambridge

Wakasa K, Widholm JM (1987) A 5-methyltryptophan resistant rice mutant, MTR1, selected in tissue-culture. Theor Appl Genet 74:49–54

Wang X, Wang A, Li Y, Xu Y, Wei Q, Wang J, Lin F, Gong D, Liu F, Wang Y, Peng L, Li J (2021) A novel banana mutant "RF 1" (Musa spp. ABB, Pisang Awak Subgroup) for improved agronomic traits and enhanced cold tolerance and disease resistance. Front Plant Sci 12: 730718. https://doi.org/10.3389/fpls.2021.730718

Wanga MA, Shimelis H, Horn LN, Sarsu F (2020) The effect of single and combined use of gamma radiation and ethylmethane sulfonate on early growth parameters in Sorghum. Plants (Basel) 9(7):827. https://doi.org/10.3390/plants9070827

Wenzel G, Foroughi-Wehr B (1994) Isogenic lines and haploids. In: Vasil IK, Thorpe TA (eds) Plant cell and tissue culture. Kluwer Academic Publ, Dordrecht, The Netherlands

Widholm JM (1974) Control of aromatic amino acid biosynthesis in cultured plant tissues: effect of intermediates and aromatic amino acids on free levels. Physiol Plant 30:13–18

Widholm JM (1985) *In vitro* selection with plant cell and tissue cultures-an overview. Iowa State J Res 62(4):587–597

Yamaguchi H (2018) Mutation breeding of ornamental plants using ion beams. Breed Sci 68(1): 71–78

Zia MAB, Bakhsh A, Çalıskan ME (2018) Mutation breeding in potato; endeavors and challenges. J Anim Plant Sci 28(1):177–186

Haploid Mutagenesis: An Old Concept and New Achievements

5

Monika Gajecka and Iwona Szarejko

Abstract

Doubled haploid (DH) technology provides the fastest route to obtain homozygous lines from the heterozygous plant materials. Application of DH system in breeding saves many generations needed to produce pure breeding lines, allows rapid fixing of additive variation and facilitates the selection of desired lines. The same advantages are apparent when DH systems are used for mutant induction and selection. Microspores as haploid cells provide the best material for both physical and chemical mutagenesis, and when the haploid genome is doubled, the mutations are fixed within only one generation. The main benefits of using DH system in mutagenesis are: the possibility to screen for both recessive and dominant mutants in the first generation after mutagenic treatment; the increased selection efficiency of desired mutants due to the gametic versus zygotic segregation ratios in mutagenised populations; the lack of chimerism, and the possibility of applying in vitro selection methods. However, the successful application of haploid mutagenesis depends on the availability of an efficient method of DH production for a species of interest. This chapter provides information on the available methods of DH production and an update on their use in mutant induction and selection. The recent examples of successful application of haploid mutagenesis that allowed to identify valuable phenotypes in Brassica species and barley are reported. We also point out that microspores have recently become the target material for new technologies, such as CRISPR/Cas precise genome editing.

M. Gajecka · I. Szarejko (✉)
Institute of Biology, Biotechnology and Environmental Protection, Faculty of Natural Sciences, University of Silesia in Katowice, Katowice, Poland
e-mail: iwona.szarejko@us.edu.pl

S. Penna, S. M. Jain (eds.), *Mutation Breeding for Sustainable Food Production and Climate Resilience*, https://doi.org/10.1007/978-981-16-9720-3_5

Keywords

Androgenesis · Anther culture · Doubled haploids · Haploid mutagenesis · Induced mutations · In vitro selection · Isolated microspore culture

5.1 Introduction

Haploid is a plant that contains a gametic number of chromosomes (*n*) in somatic cells. In nature, haploids can arise spontaneously from the unfertilised egg cell, however with a very low frequency. For practical purposes, which require a high efficiency of haploid production, haploids can be induced in vivo by interspecific crosses or in vitro by culture of male (androgenesis) or female (gynogenesis) gametophytes. The haploid plant is sterile; therefore, the doubling of chromosomes is needed to obtain the progeny of generatively propagated plants. The chromosome doubling can occur spontaneously or can be induced by the use of anti-microtubular agents such as colchicine or oryzalin (Ahmadi and Ebrahimzadeh 2020). After haploid genome doubling, plants that contain two identical sets of chromosomes are called doubled haploids (DHs). DH technology allows to produce completely homozygous lines within one generation, compared to the conventional breeding that requires seven to eight generations of selfing to achieve homozygosity. Therefore, DH technology provides the shortest route to produce completely homozygous plants from heterozygous materials. Doubled haploidy is predominantly utilised in breeding programmes to speed up the production of genetically stable lines from hybrid materials. Introducing DH technology to breeding programmes resulted in the release of numerous cultivars of cereals, oil crops, and vegetables (Germanà 2011; Tadesse et al. 2012; Dwivedi et al. 2015; Badu et al. 2017). Another crucial area of doubled haploidy application is in genetic research, where DH lines are used to generate mapping populations, especially for analysis of quantitative trait loci (QTL) (reviewed by Tuvesson et al. 2021; Singh and Singh 2015). Alternatively, a QTL mapping population may be obtained by creation of recombinant inbred lines (RIL), which takes six to eight generations (Forster et al. 2007; Ferrie et al. 2014).

Doubled haploid systems give also the beneficial effects in mutagenesis. Haploid cells exposed to mutagenic treatment, followed by chromosome doubling, are the perfect source of homozygous mutants obtained in a shorter time than in the conventional mutagenesis. DH systems increase the probability of identifying a rare mutation event and allow recovering recessive mutations in the first generation after mutagenic treatment. The combining of mutagenesis and double haploidy can speed up the detection of any mutants and subsequent evaluation of lines in field conditions. Furthermore, there is a greater chance of finding a desirable mutant in a smaller population than would be required in conventional breeding programmes (Szarejko and Foster 2007). The utilisation of haploid cells for mutagenic treatment overcomes a problem of M_1 chimerism or a loss of a desired mutant, which may occur in the conventional mutagenesis where multicellular seed embryos are subjected to mutagenic treatment (Ferrie and Möllers 2011). Another significant advantage of using DH systems for mutagenesis is the possibility of in vitro selection with large mutagenised population. Nevertheless, the advantages, that are provided by the utilisation of DH system in mutagenesis, are limited to those species, for which an effective system of DH production has been developed.

The first protocols for large-scale mutagenesis with haploid cells were developed for *Brassica napus* already in the 1980s of the previous century (Beversdorf and Kott 1987; Polsoni et al. 1988; Swanson et al. 1989). Since then, examples of successful application of DH systems in mutation breeding were reported for other species, including cereals (reviewed by Szarejko and Foster 2007; Ferrie and Möllers 2011; Szarejko 2012). In this chapter, we focus on examples of using haploid systems for mutant production/selection in the last 10–15 years, including recent application in targeted mutagenesis. We present the achievements but also discuss the limitations of haploid mutagenesis.

5.2 Methods of Doubled Haploid Production

Since the first haploid plant was obtained in vitro from cultured anthers of *Datura innoxia* (Guha and Maheshwari 1964), various methods of doubled haploid production have been developed. A great number of available protocols for a wide range of species was recently published in Segui-Simarro (2021); however, efficient and reproducible protocols are available for less than 30 species.

Vast majority of DH techniques utilise in vitro cultures, at least at some step of DH production. There are three major approaches to produce doubled haploid plants using in vitro systems: chromosome elimination, androgenesis, and gynogenesis (Maluszynski et al. 2003). There are also methods of haploid induction in vivo applied for some species recalcitrant to in vitro procedures. Generally, for each species, one or two methods are predominantly utilised to produce DH lines, with divergent effectiveness among species.

5.2.1 Crosses with Haploidy-Inducing Lines

Historically, crosses with haploidy-inducing lines were the first system of haploid production utilised in plant breeding. In maize, the ‘Stock6’ haploid inducer line has been applied for induction of maternal haploid formation when used as a pollen donor. This line was identified in the 1950s and, after improvement, gives ca. 10% of seeds containing a haploid embryo, which develops from the unfertilised egg cell. As the central cell of the embryo sac is fertilised, the endosperm, which supports a developing embryo is formed and rescue of haploid embryos to in vitro conditions is not needed. The seeds carrying haploid embryos can be distinguished from those with diploid embryo through the absence/presence of morphological markers (Meng et al. 2021). The haploid-inducer system provides an easy, consistent in vivo method to maize breeding. Nevertheless, the mechanism underlying the induction of haploid by ‘Stock6’ is still unclear thus the introduction of this system to other species failed

(Watts et al. 2018; Kalinowska et al. 2019). The haploidy-inducer line that has the potential to be introduced to other species is a mutant in the *CENH3* gene, identified in *Arabidopsis thaliana*. *CENH3* encodes a H3 histone variant exclusively localised in centromeric nucleosomes. When *cenh3* null mutant was crossed to a wild type, chromosomes from the mutant were eliminated, producing haploid progeny (Ravi and Chan 2010). The lack of CENH3 leads to the improper distribution of chromosomes between daughter cells during mitotic and meiotic divisions. Due to the meiotic nonreduction, haploid cells can be spontaneously converted into fertile diploids (Ravi and Chan 2010). The possibilities and challenges of using CENH3 and other systems for in vivo DH technology were recently discussed by Ren et al. (2017), Watts et al. (2018), and Kalinowska et al. (2019).

Haploid plants were also induced by crossing plants with different ploidy levels. For *Solanum tuberosum* that is a tetraploid ($2n = 4x = 48$), the dihaploid plants ($n = 2x = 24$) are produced by crossing with a diploid species *S. phureja* ($2n = 2x = 24$). In such a crossing, the egg cell is not fertilised by the *S. phureja* sperm, and dihaploid embryo is induced ($n = 2x = 24$). What is more, in this cross one of the haploid central cells ($n = 2x = 24$) is fertilised, which results in the development of triploid endosperm ($3n = 3x = 36$) (Watts et al. 2018).

5.2.2 Wide Hybridisation Followed by Chromosome Elimination

For the first time, the recovery of haploid plants resulting from wide crosses followed by chromosome elimination of one parent was described by Kasha and Kao (1970) in barley. After crossing the cultivated barley (*Hordeum vulgare* L.) used as a female parent with the wild relative *H. bulbosum*, a hybrid zygote was formed, but during subsequent divisions of embryo cells, the chromosomes of *H. bulbosum* were lost, and the haploid embryo with only one set of chromosomes belonging to *H. vulgare* was developed. It was proposed that difference in the timing of mitosis or asynchronous synthesis of nucleoproteins were the main sources of uniparental chromosome elimination (Ishii et al. 2016). At this point, the endosperm failed to develop and the embryo had to be rescued through in vitro culture to regenerate the haploid plant. This system has been successfully adapted for other crop species such as wheat (*Tritucim durum* and *T. arestivum*) (reviewed by Niu et al. 2014), oat (*Avena sativa*) (Marcińska et al. 2013; Ishii 2017) and triticale (Wędzony et al. 1998; Ishii et al. 2010). It was tested for many other species although with varying success (for review, see Ishii et al. 2016).

In wheat, the haploid formation is trigged by the pollination with maize pollen, however to increase the rate of regenerable embryos, the application of plant growth regulators (auxins, cytokinins, or gibberellic acid) into tillers and/or florets is required after pollination (Niu et al. 2014). The proper stage of stigma development is crucial for the high frequency of fertilisation and the highest rate was observed when the stigma was at the feathery stage in wheat. Also, external factors such as temperature, light, and photoperiod significantly influence the frequency of haploid embryos recovery. The great advantage of wide hybridisation as a method of DH production is the absence of gametoclonal variation that occasionally occurs during in vitro culture. Another advantage is a very low genotype dependence, while a requirement for chromosome doubling is the main limitation of this system (Niu et al. 2014).

5.2.3 In Vitro-Based Systems: Gynogenesis

Gynogenesis is a process of haploid plant regeneration from female gametophyte under in vitro conditions. Ovaries, ovules or even floret buds can be used as explants for in vitro culture. The haploid embryos may develop from the unfertilised egg cells, antipodals or synergids of the embryo sac. The crucial step that limits the efficiency of this method is the choice of the proper developmental stage with responsive ovules. Other factors affecting the effectiveness of gynogenesis are: the genotype of donor plant, donor plant growth conditions, the composition of media used for in vitro culture, including plant growth regulators and culture condition (Chen et al. 2011). The gynogenesis-derived embryos regenerate only haploid plants that always require the application of chromosome doubling procedure to obtain fertile doubled haploid plants.

Gynogenesis is characterised as a method of low efficiency, with a low frequency of haploid recovery, thus it requires a high number of female gametic cells per one inflorescence. It is the method of choice when other systems fail to be optimised for DH production. Nevertheless, this method was successfully introduced in breeding programmes and genetic research of *Allium cepa* (Khan et al. 2020) and other Amaryllidaceae species, *Beta vulgaris* (Pazuki et al. 2018; Sohrabi et al. 2021) and some Cucurbitaceae species (Dong et al. 2016). Since many external factors affect the effectiveness of gynogenesis, some efforts were made to increase the frequency of haploid embryo recovery (Pazuki et al. 2018; Ozsan et al. 2017; Grigolava et al. 2021; Sohrabi et al. 2021). Some researchers reported a positive effect of stress factors such as cold or heat applied to inflorescences, flower buds or isolated ovaries before culture (Karakan and Arpaci 2018; Sohrabi et al. 2021).

5.2.4 In Vitro-Based Systems: Androgenesis

During androgenesis, haploid plant develops from microspores that are the immature male gametophytes, the precursors of pollen grains in vivo (Fig. 5.1). Microspores before the second pollen mitotic division, at the particular developmental stage, usually at the mid to late uninucleate, can be switched from gametophytic into sporophytic developmental pathway. Unfortunately, androgenesis is a highly genotype-dependent process and even with optimised protocols, some genotypes are recalcitrant to the androgenetic induction. Besides the developmental stage of microspores and the genotype of donor plants, the effectiveness of androgenesis is highly influenced by a pre-treatment applied before in vitro culture, which is a stress factor that triggers the re-programming of microspore development. The

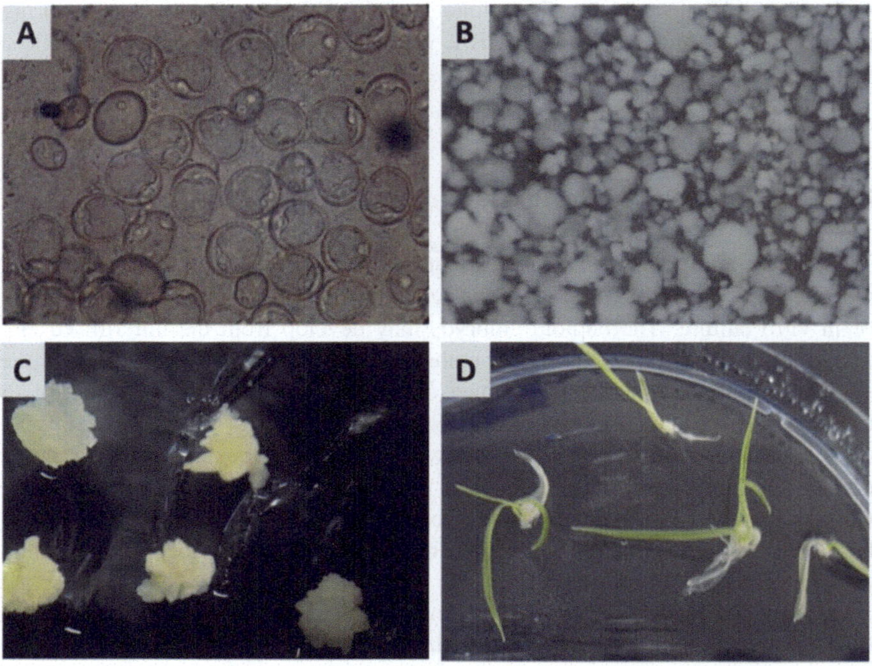

Fig. 5.1 Isolated microspore culture of barley cv. 'Jersey'. (**a**) Freshly isolated microspores. (**b**) Microspore-derived embryos on 21st day of in vitro culture. (**c**) Developed androgenic embryos on 43rd day of in vitro culture. (**d**) Green regenerants in regeneration medium

pre-treatment stress can be applied to floral buds, tillers or excised spikes/panicles as well as anthers or microspores for a few days to several weeks. Many different stress factors were tested; however, the most commonly used are low or high temperature, carbon and nitrogen starvation or osmotic stress (Testillano 2019). The type of stress and time of application should be optimised for a particular species or even cultivar. For the most efficient production of doubled haploids, also the conditions of donor plant growth, as well as in vitro culture media and incubation conditions should be optimised for each species (Ferrie and Caswell 2011). What is an excellent advantage of androgenesis, is the high level of spontaneous diploidisation of haploid genome occurring in in vitro culture of immature male gametophytes. The main process responsible for this phenomenon is nuclear fusion, which takes place after the first mitotic division of uninucleate microspore at the pre-treatment or early induction stage (Kasha et al. 2001). Due to the spontaneous diploidisation, the regenerated plants become doubled haploids, and therefore it is possible to obtain fully homozygous fertile plants directly from androgenic culture, without application of additional treatments (Ahmadi and Ebrahimzadeh 2020).

Contrary to wide hybridisation and gynogenesis, the effectiveness of androgenesis in cereals is highly limited by the regeneration of albino plants (Torp and Andersen 2009; Makowska and Oleszczuk 2014). Recently, Gajecka et al. (2020)

reported that genotype-dependent regeneration of albino plants in androgenic cultures of barley is determined by the plastid differentiation during pollen grain development. The cultivars that contained undifferentiated proplastids in the microspores used for culture initiation were capable of regenerating mostly green plants. On the other side, the cultivars that produced a vast majority of albino regenerants, contained already differentiated amyloplasts in the microspores at the stage of culture initiation. The use of microspores at the stage preceding the mid-to-late microspores utilised commonly for culture initiation, significantly increased the green plants frequency in cultivars that regenerated mostly albinos in standard protocols.

There are two main in vitro culture approaches applied in androgenesis—anthers culture and isolated microspore culture. In vitro culture of anthers is the most widely used technology of doubled haploid production. For initiation of this culture, anthers containing microspores at the mid to late uninucleate or early bi-nucleate stage, after pre-treatment are excised from spikes, panicles or flower buds under sterile conditions, plated onto induction medium and incubated at a dedicated temperature for 4–6 weeks. During the induction phase, the responsive microspores undergo a series of mitotic divisions resulting in the development of microspore-derived structures: embryos or calli, which after transfer onto regeneration medium produce plants. The most valuable approach is the isolated microspore culture, where microspores are mechanically released from spikes, flower buds or florets. Through filtration, the microspore suspension is separated from debris, and by the washing and centrifugation only viable microspores are collected. Microspores are cultured preferably in a liquid medium and incubated under appropriate conditions to induce sporophytic development. It is important to optimise the density of microspores in the medium and provide aeration of microspores by gently shaking during incubation. The advantage of isolated microspore culture is more rapid and synchronous formation of microspore-derived structures, majority of which develop via embryogenesis pathway. Additionally, the isolated microspore culture lacks somatic tissues present in anther culture, thus avoiding development of embryos/calli from non-haploid cells and products of anther degradation that can negatively influence the efficiency of culture. When embryogenesis is the predominant pathway of microspore development, the derived embryos are prone to convert easily into plants on regeneration medium, which results in the high regeneration potential of microspore culture. What is more, when embryos are produced, the derived plants exhibit very low gametoclonal variation compared to plants regenerated from calli.

Androgenesis, primarily through microspore culture, is the most effective method of doubled haploid production. However, the availability of efficient microspore culture protocols is a bottleneck in taking a full advantage of doubled haploid application in breeding and genetics, including mutation projects. When the effective and reproducible protocols of isolated microspores culture are available, haploid mutagenesis is widely exploited with new germplasm produced, especially in *Brassica* species including *B. napus*, *B. rapa*, *B. juncea*, *B. campestris* and *B. carinata* (Ferrie and Möllers 2011).

5.3 Haploid Mutagenesis

When mutagenic treatment is applied to haploid cells, it results in the development of entirely homozygous plants carrying induced mutation in both alleles. Thus, it is possible to identify recessive mutations already in the first generation after mutagenic treatment. Although many routes of doubled haploid formation have been developed, wide hybridisation and gynogenesis have not been employed for mutant production. On the contrary, androgenesis is the most widely used system of DH production in plant mutagenesis (Germanà 2011). An important advantage of haploid mutagenesis is that it can be performed in vitro, which allows to deal with a great number of mutagenised cells within a small area. Microspores, as haploid cells are the best target and most effective system for mutagenesis. Mutagenic treatment can be applied directly to isolated microspores in vitro or to plant organs that contain male gametophytes, such as anthers, spikes, panicles and flower buds before culture initiation. After induction of sporophytic development, followed by the formation of microspore-derived structures, and spontaneous or induced chromosome doubling, the plants carrying mutation in the homozygous state can be developed. Depending on the trait of interest, selection of mutant plants can be performed in vitro already at the embryo stage or in vivo at the haploid/doubled plant level (Ferrie and Möllers 2011). Figure 5.2 presents the utilisation of doubled haploids in mutagenesis and selection.

Doubled haploid technology can also be applied to M_1 plants derived from the mutagenic treatment of dormant seeds. Gametes produced by M_1 plants carrying mutations in the hemizygous stage can serve as explants for DH production, which results in homozygous mutants produced in one generation (Szarejko 2012). Examples of the recent haploid mutagenesis studies, including used mutagens and their doses, explants and mutated traits, are listed in Table 5.1.

5.3.1 Explants Used for Mutagenic Treatment

For species with well-established protocols of microspore embryogenesis, isolated microspores at the uninucleate stage are the best targets for mutagenesis. They can be induced for rapid and synchronous development into embryos followed by regeneration of plants (Ferrie and Caswell 2011). Additionally, at the early stage of in vitro microspore culture, a high frequency of spontaneous chromosome doubling occurs, which results in fertility of completely homozygous regenerants and allows to omit the employment of doubling agents.

There is a narrow time-window when mutagenic treatment should be applied to microspores. Therefore, the timing of the mutagenic treatment is critical for effective mutagenesis, and it should take place before the first pollen mitotic division, thus before in vitro culture initiation. Mutagenic treatment is usually applied to isolated microspores, but also to anthers, flower buds or inflorescences containing microspores at the suitable stage to follow androgenic development.

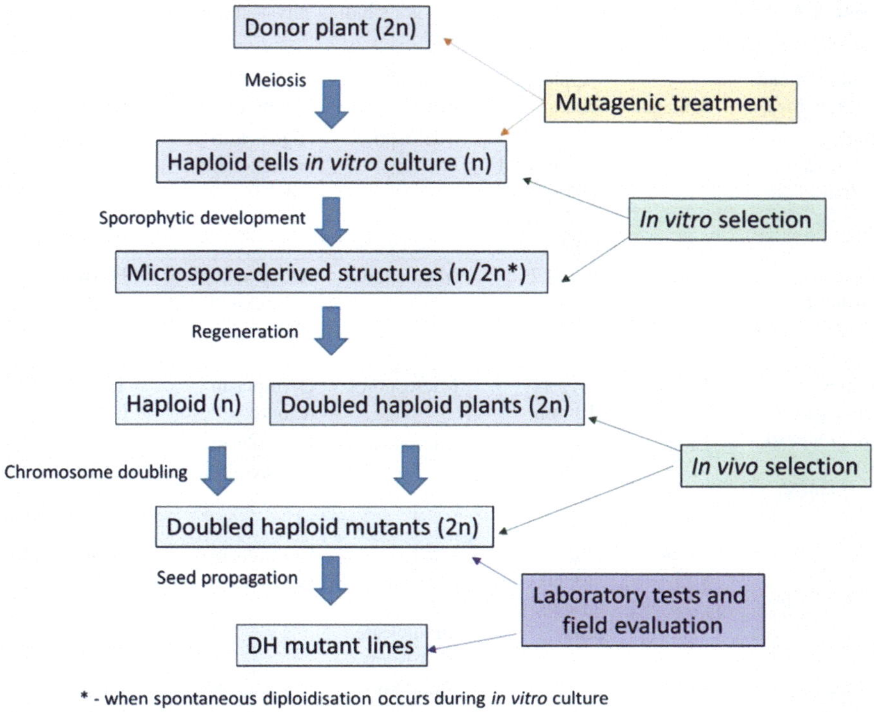

Fig. 5.2 Mutation induction and mutant selection using double haploid systems

Anther culture is used principally as a method of choice when the isolated microspore protocol is not available for a species. In such cases, mutagenic treatment of anthers followed by in vitro culture have been applied, for example, in major cereals, such as barley, rice or wheat (Vagera et al. 2004; Kyin et al. 2005; Plamenov et al. 2013). The mutagenic treatment of microspore-derived embryos of *B. rapa* or calli derived from anther culture of *Oryza sativa* were also reported (Lee and Lee 2002; Daurova et al. 2020). However, the application of mutagenic treatment to the multicellular structure, such as haploid embryo or callus can result in regeneration of chimeric plants that segregate for DH mutants only in the generative progeny of the regenerant.

5.3.2 Mutagens Utilised in Haploid Mutagenesis

The type of mutagenic agent should also be considered before treatment. Both chemical and physical mutagens have been applied for mutation induction in haploid systems (Szarejko and Foster 2007). UV light and gamma rays have been utilised most frequently among physical factors, while alkylating agent EMS (ethyl methanesulfonate) has been applied most often in chemical mutagenesis. UV light

Table 5.1 Recent examples of haploid mutagenesis including physical and chemical treatments, explants and mutated traits

Common and Latin name	Mutagen	Applied doses	Material treated	Mutated trait	References
Barley (*Hordeum vulgare* L.)	EMS; PYM	8, 24, 40 mM EMS; 2, 3 μM PYM; 48 h	Microspores	Higher nitrogen use efficiency (NUE)	Gao et al. (2018)
Brown mustard (*B. juncea* ssp. *juncea*)	EMS	0.1–0.2%; 1.5 h	Microspores	Reduced content of saturated fatty acids	Ferrie et al. (2008)
Chinese cabbage (*B. campestris* ssp. *pekinensis*)	EMS	0.12%; 10 min	Microspores	Germplasm resources	Tang et al. (2018)
	EMS	0.04%, 0.08%, 0.12%; 10 min	Microspores	Germplasm resources	Huang et al. (2016a)
	^{60}Co γ-ray	20, 40, 60 Gy (1.58 Gy/ min)	Buds containing late uninucleate microspores	Germplasm resources	Huang et al. (2014)
Chinese cabbage (*B. rapa* L. ssp. *pekinensis*)	EMS	0.03%, 0.05%, 0.1%; 5–10 min	Floral buds	Germplasm resources	Lu et al. (2016)
Rapeseed/ canola (*B. napus*)	EMS	0.15%; 72 h	Microspore-derived calli	In vitro selection—improved resistance to *Sclerotinia sclerotiorum*	Liu et al. (2005)
	EMS	0.1–0.2%; 1.5 h	Microspores	Reduced content of saturated fatty acids	Ferrie et al. (2008)
	UV	254 nm; 90 s	Microspores	In vitro selection—reduced content of saturated fatty acids	Beaith et al. (2005)
	UV	280 nm; 75–90 s	Microspores	In vitro selection—enhanced cold tolerance	McClinchey and Kott (2008)
Turnip rape (*Brassica rapa* L. ssp. *oleifera*)	EMS	0.1–0.2%; 1.5 h	Microspores	Increased content of oleic acid, reduced content of	Ferrie et al. (2008)

(continued)

Table 5.1 (continued)

Common and Latin name	Mutagen	Applied doses	Material treated	Mutated trait	References
				linolenic acid or saturated fatty acids	
	EMS	4, 8, 12 mM; 1 h	Microspore-derived embryos	Increased oleic acid content	Daurova et al. (2020)

as the non-ionising radiation is relatively easy to apply for in vitro culture, nevertheless considering its low penetration, it should be applied only for culture of single cell layers, such as isolated microspore cultures. Gamma rays irradiation, due to its penetration ability, is the most feasible factor of treatment for all types of plant organs and tissues; however, it is limited to the availability of nearby facilities. Therefore, cells and organs in culture, in contrast to dormant seeds are not a convenient material for irradiation.

The application of radiation has been mostly reported for *Brassica* species due to the optimised protocols of isolated microspore cultures available for many of them. However, some attempts have also been made to introduce the physical treatment to cereals. For example, the effect of gamma rays on androgenic response was examined in anther culture of wheat (Plamenov et al. 2013). The study included irradiation of anthers, calli and combined irradiation of anthers and calli with 5 Gy. As conclusion, the effect of 5 Gy on androgenesis was evaluated as negative for efficiency of cultures derived from irradiated anthers and the reduction of green plant regeneration was mostly visible in the combined treatment group (Plamenov et al. 2013).

Chemical mutagenesis has been exploited for inducing predominantly point mutations. In connection with DH system, the main approach is to treat freshly isolated microspores for a relatively short time with such concentrations of mutagenic agents that ensure not too high negative effect on embryo induction. Chemical mutagens easily dissolve in the culture media, so they can be used directly in the pre-treatment medium before anther and microspore culture initiations. However, such simplification of work requires a thorough cleaning of treated explants to remove residues of the chemicals, which could affect the efficiency of embryogenesis. Microspores are extremely fragile and can be easily damaged during washing and centrifugation.

While chemical mutagen treatment is applied, it is required to optimise the type of mutagen, mutagen concentration and duration of treatment. Due to the high negative impact of mutagens on in vitro culture efficiency, much lower mutagen doses are applied compared to treatment of dormant seeds (Szarejko 2012). Gao et al. (2018) compared the effect of EMS and PYM (antibiotic pingyangmycin) applied at low concentrations on isolated microspore culture of barley. They obtained regenerants from each tested treatment, however the frequency of regeneration depended on the type of used mutagen and its concentration. Authors reported that higher yields of both embryogenic structures and green plants were achieved using EMS treatment, while PYM showed a more harmful effect on microspore viability.

Lu et al. (2016) treated floral buds of five different Chinese cabbage (*Brassica rapa* L. ssp. *pekinensis*) genotypes with EMS in concentrations ranging from 0.03% to 0.2% for 5–15 min. EMS treatment delayed embryo formation in all tested concentrations. The number of developed embryoids decreased with the increasing EMS concentrations and duration of treatment in all genotypes tested. At the highest concentration (0.2%) and duration of treatment longer than 5 min, almost no embryos developed into seedlings. The optimal concentration that resulted in a sufficient number of regenerants depended on the genotype (Lu et al. 2016).

However, the application of mutagens to isolated microspores is not always harmful to embryo development. Sometimes, mutagenic treatment shows a positive effect on the frequency of regeneration of fertile androgenic plants. Huang et al. (2016a) applied 0.04%, 0.08%, or 0.12% EMS to isolated microspores of Chinese cabbage, followed by washing and incubation in induction medium. Despite the lower numbers of developed embryoids per bud, all tested concentrations of EMS remarkably increased the regeneration, rooting and survival rates in comparison to the culture induced from non-treated microspores. The highest number of plants was regenerated from microspores treated with 0.08% EMS. Daurova et al. (2020) showed that the range of EMS concentrations (4, 8, and 12 mM) applied to turnip rape (*Brassica rapa* L. ssp. *oleifera*) microspore-derived embryos (1.5–2.5 mm in size) affected the yield of regenerated and fertile plants. At a concentration of 8 mM EMS, the regeneration of plants was more frequent than at a concentration of 4 mM, while the highest concentration was lethal for treated embryos.

5.3.3 Selection of Desired Phenotypes

The cultures of isolated microspores give the possibility to carry out selection of desired phenotype in vitro; however, the number of features suitable for in vitro selection is limited. Therefore, the most common approach is to select at the later stages of doubled haploid development.

There are many successful applications of selection during in vitro culture, including biotic and abiotic stresses. A selection agent is usually applied during the development of microspore-derived calli or embryos. To select rapeseed (*Brassica napus*) mutants resistant to stem rot disease caused by *Sclerotinia scleotiorum* Liu et al. (2005) treated microspore-derived calli with different concentrations of oxylic acid, a toxin produced by *Sclerotinia scleotiorum*, for 4 weeks. Besides using of toxic compound as the selective agent, it is possible to co-culture the disease-causing bacteria with developing microspore-derived embryos. Zhang and Takahata (1999) applied this approach in order to produce mutants of Chinese cabbage resistant to soft rot disease caused by *Erwinia carotovora*.

Additionally, selection during in vitro culture allows to analyse a wide range of combinations; therefore, it simplifies the choice of a proper compound and its concentration for selection of a trait of interest. In order to select spring canola

(*Brassica napus*), plants with increased cold tolerance (McClinchey and Kott 2008) tested several compounds that play a protective role in oxidative and/or osmotic stress accompanying plant response to cold. The microspore-derived embryos were exposed in culture to such agents as: salicylic acids (SA), jasmonic acid (JA), *p*-fluoro-D,L-phenylalanine (FPA) as well as proline analogues: hydroxyproline (HP), azetidine-2-carboxylate (A2C) and 3,4-dehydro-D,L-proline (DP). It was shown that all compounds, except FPA, could serve as the selection agents for isolating lines with higher tolerance to cold and freezing.

DH plants that are produced in in vitro culture are transferred to soil and grown till maturity. When spontaneous chromosome doubling occurs at low frequency, it is required to introduce colchicine treatment, which causes one more generation to propagate seeds for selection of mutants of interest. The harvested seeds can be used to screen for DH mutants and evaluation of DH lines. Below, we present examples of the recent studies in turnip rape, Chinese cabbage, oilseed rape, brown mustard (*Brassica juncea*) and barley, where many valuable DH mutants were selected in the progeny of microspore-derived plants.

5.4 Examples of Successful Haploid Mutagenesis

The ultimate goal of mutation programmes is to obtain mutants with desired characters. The high efficiency of classical mutagenesis to generate mutations valuable for breeders has been well documented through the official release of more than 3300 mutant cultivars, as indicated in the FAO/IAEA Mutant Varieties Database (https://mvd.iaea.org). Combining mutagenesis with doubled haploidy has been utilised in breeding programmes of those species, for which efficient protocols of DH production have been established. Many DH mutant lines have been employed in breeding programmes of various Brassicaceae species, and some new genotypes were successfully commercialised. Such an example is a rapeseed mutant tolerant to imidazolinone group of herbicides that was released in Canada under the name ‘Smart Canola’ (Swanson et al. 1989; Senior and Bavage 2003). The mutant was produced by the mutagenic treatment of isolated microspores of rapeseed cv. Topas with 20 μM EMS and induction of embryo development in medium containing the imidazolinone herbicide that served as the selection agent. The survived embryos were considered as imidazolinone herbicide-tolerant and regenerated into plants. Two mutants harbouring a mutation in the gene encoding acetohydroxyacid synthase (AHAS) were isolated and crossed to each other. The AHAS enzyme is involved in synthesis of leucine, isoleucine and valine and is the primary target of imidazolinone herbicides. The progeny of the mutant cross exhibited a higher tolerance to AHAS herbicide the parent cv. ‘Topas’ and either mutant alone (Swanson et al. 1989).

5.4.1 New Germplasm for Basic Research and Breeding

Among many features that make DH systems attractive for mutagenesis, the key factor is the rapid fixing of mutations in the complete homozygous state, which makes it possible to screen for both, recessive and dominant mutants as early as M_1 generation and shortens the time required for mutant selection. This advantage is especially important in generating large mutated populations designed for forward genetics studies. Such an example is the population of Chinese cabbage (*Brassica campestris* ssp. *pekinensis*) mutants produced by γ-radiation of floral buds of a doubled haploid line 'FT' developed from variety 'Fukuda 50'. The mutant population has been developed by Huang et al. (2014) and widely used in basic research of Chinese cabbage. Out of 1483 regenerated plants, a total of 492 putative mutants were isolated and phenotyped. The population was the source of DH mutants that exhibited several new phenotypes and their utilisation enhanced the knowledge on the molecular mechanisms underlying plant development. The analysis of selected mutants allowed to identify genes that were associated with male sterility (Huang et al. 2020; Tan et al. 2018, 2019), petal development (Huang et al. 2015, 2016c; Peng et al. 2019), thylakoid development (Huang et al. 2016b) or chlorophyll deficiency (Li et al. 2019).

Another Chinese cabbage mutant population developed for the 'FT' genotype was generated after treatment of isolated microspores with the chemical mutagen ethylmethanesulfonate (EMS). One of the obtained mutants exhibited pale green inner leaves and slower growth compared to the wild type. NGS sequencing enabled to identify a missense mutation in the plastome gene encoding a ribosomal protein S4 (RPS4). The role of RPS4 was proven to be associated with chloroplast rRNA processing (Tang et al. 2018).

Lu et al. (2016) also used EMS for creation of Chinese cabbage mutant population that served as a source of mutations for basic research. The authors applied EMS treatment to flower buds of five genotypes: three breeding lines and two DH lines. In total, 142 mutants with distinct variations in leaf shape, leaf colour, corolla size, flower colour, bolting time and downy mildew resistance were identified among 475 microspore-derived DH lines.

The population of DH mutants derived from EMS-treated microspore cultures was also developed for canola (*B. napus*) breeding line 92-B10. Among regenerated DH plants, the semi-dwarf *ds-1* mutant was identified. This mutant shown a lower position of the first branch and shorter internodes compared to the wild type, which resulted in the significant reduction of plant height (Liu et al. 2010). The *BnRGA* gene that encodes a DELLA protein that functions as a gibberellic acid signalling repressor was associated with the dwarf phenotype. Identification and characterisation of this gene is important in terms of improving lodging resistance and enhancing harvest index in *B. napus*.

5.4.2 Modifications of Fatty Acids Profiles

The typical canola fatty acids profile consists of 61% oleic acid, 21% linoleic acid, 11% α-linolenic acid and 7% saturated fatty acids. Any alteration in the fatty acid profile that increases the proportion of oleic acid and decrease of α-linolenic acid and saturated fatty acids improves the quality of canola oil.

The production of mutants with modified fatty acid composition was the objective of studies conducted by Ferrie et al. (2008). These large-scale studies included 12,300 DH lines of three Brassica species: rapeseed (*B. napus*) 1700 lines, turnip rape (*B. rapa*) 7000 lines and brown mustard (*B. juncea* ssp. *juncea*) 3600 lines that were produced from mutagenised microspores using EMS. Many DH mutant lines of *B. napus* and *B. juncea* were obtained with a reduced level of saturated fatty acids to 5% and 5.4%, respectively. The mutagenic treatment of *B. rapa* turned out to be most successful. The average fatty acid composition of the *B. rapa* 'CV-2' line that was used for mutagenic treatment consisted of 50% oleic acid, 26% linoleic acid, 12% α-linolenic acid and 7% saturated fatty acids. Analysis of fatty acids composition revealed many *B. rapa* DH mutant lines with changed fatty acid composition, among them 69 lines with a reduced level of α-linolenic acid below 8%, 157 lines with low saturated fatty acid content and 197 lines with elevated level of oleic acid above 55%, compared to the donor plants.

An interesting approach was undertaken by Beaith et al. (2005) who generated DH lines of Ethiopian mustard (*Brassica carinata*) with a reduced level of palmitic and stearic acids. In this study, not only mutagenesis but also selection for changed fatty acid composition was conducted in vitro. The isolated microspores were exposed to UV light for 90 s and cultured to induce embryo development at 30 °C for 17 days. Then, the maturation of embryos was conducted at 35 °C for 18 days in order to increase the level of saturated fats artificially. Heat artificially elevated the saturate levels in developing mutant embryos, allowing efficient identification of those with reduced saturates using HPLC fatty acid analysis. Mature embryos, after 35 days of culture, were subjected to evaluation of saturated fatty acids level. To perform the evaluation, one or two cotyledons were excised and subjected to HPLC analysis of fatty acid composition, while embryo axes were transferred to the regeneration medium in order to regenerate plants. The selection at this point reduced the number of plants for further cultivation because only embryos that showed positive changes in fatty acid composition were maintained. As a result, 389 embryos were produced, among which 52% had reduced palmitic acid levels and 46% had lower stearic acid content. The reduced level of palmitic and stearic acids was confirmed in the progeny of DH lines that were grown at standard temperature. What is important, the composition of fatty acids in microspore-derived mature embryos was very similar to the composition of fatty acids in seeds (Beaith et al. 2005).

Average turnip rape fatty acids composition is similar to canola. After mutagenic treatment of haploid embryos, Daurova et al. (2020) obtained mutants of turnip rape cultivars 'Yantarnaya' and 'Zolotistaya' that showed increased content of both, oleic acid by 11.7% and 12.5% and linoleic acid by 1.8% and 6.5%, respectively, with no changes in the content of palmitic and stearic acids, compared to donor plants.

5.4.3 Nitrogen Uptake Modifications

DH mutant lines with important agronomic traits were also produced in barley. Gao et al. (2018), among 356 DH lines derived from mutagenic treatment of malt barley microspores with EMS or PYM, identified five mutants characterised by the higher nitrogen uptake efficiency (NUE) than the NUE of their parent cultivar 'Hua-30'. The obtained DH lines were screened for the number of produced tillers in the field under low and high nitrogen levels and two mutant lines with improved nitrogen uptake efficiency were identified. Due to the excessive use of N fertilisers, production of cultivars with a high nitrogen use efficiency is considered the best strategy for agricultural sustainability.

5.4.4 Disease Resistance

Haploid mutagenesis was also successfully applied to produce genotypes that exhibited higher resistance to pathogens: the bacterial disease soft rot caused by *Erwinia carotovora* and the fungal disease stem rot caused by *Sclerotinia scleotiorum*. Chinese cabbage cv. 'Ho Mei' was used as donor of microspores in the experiment with *E. carotovora*. The microspores were subjected to UV radiation immediately after isolation, followed by the induction of microspores in in vitro culture. Microspore-derived embryos were transferred to the medium containing filtrates of *E. carotovora* for several weeks, and surviving embryos were regenerated. Twelve DH lines among the progeny of microspore-derived plants were described as resistant or moderately resistant to *E. carotovora* (Zhang and Takahata 1999). Different approach was taken by Liu et al. (2005) to produce rapeseed resistant to *S. scleotiorum*. Twenty genotypes that exhibited different levels of resistance to *S. scleotiorum* were used as donor plants to induce isolated microspore culture. Calli produced by eight rapeseed lines were subjected to EMS treatment through supplementation of medium with 0.15% EMS, followed by recovery on EMS-free medium. Then, selection of *S. scleotiorum*-resistant calli was performed during in vitro culture, using twice calli passage to the medium supplemented with oxylic acid, the toxin produced by *S. scleotiorum*. Finally, among 54 DH lines recovered, 2 showed 50% fewer disease indices than the control.

5.4.5 Cold Tolerance

Microspore mutagenesis of *B. napus* has been applied to develop lines with increased tolerance to cold. Five spring canola genotypes were used as donor plants of microspores that were exposed to UV light followed by in vitro culture. The selection was performed in vitro using single cell mutagenised microspores (1–3 days old) by the supplementation of medium with: salicylic acids (SA), jasmonic acid (JA), *p*-fluoro-D,L-phenylalanine (FPA) that play an important role in the frost-induced oxidative stress pathway. Additionally, the proline analogues: hydroxyproline (HP), azetidine-2-carboxylate (A2C) and 3,4-dehydro-D,L-proline (DP) that protect plants from cold-induced osmotic stress were tested. The next step of selection was made when plantlets were 24 and 40 days old by flooding the solid medium with appropriate concentrations of SA, FPA, A2C and DP. Among 329 DH lines derived from microspore culture, 74 showed significantly enhanced tolerance to freezing temperature (−6 °C) and 19 DH mutant lines had a better winter survival rate when compared to the parent lines, without a loss in seed quality and agronomic parameters.

5.5 Haploid Targeted Mutagenesis

Physical, chemical and insertional mutagenesis produce mutations located randomly within the whole genome. Gene editing systems like TALEN (Transcription activator-like effector nuclease) or CRISPR/Cas9 (Clustered, Regularly Interspersed, Palindromic Repeats/CRISPR-associated endonuclease 9) can induce targeted gene knock-outs. These systems are used to introduce double-strand breaks (DSBs) at a particular locus, followed by their repair by host cell repair mechanisms, which occur in plants mostly by the non-homologous end joining (NHEJ) that is error-prone. As a consequence, a few nucleotides can be lost or gained in a targeted gene region, which can result in a loss of function of the gene of interest (for review, see Gaj et al. 2013; Sedeek et al. 2019).

These technologies play an essential role in understanding the function of genes and can be the source of new valuable phenotypes in crop improvement. Designed for a gene of interest, transcriptional cassettes are introduced to plant cells by conventional transformation, mostly using *Agrobacterium* mediated systems and immature plant embryos (Gasparis et al. 2018). To take the advantage of targeted mutagenesis, it is crucial to establish usable and efficient protocols for crop plants. Such procedures, however, are often limited by the amount of time and resources required to produce transgenic plants carrying mutation in the homozygous state. Transgenic plants regenerated through conventional transformation in vitro are usually heterozygous or hemizygous for the introduced change, and homozygous transformants are selected in the next generation. As an alternative, the microspore-based editing system can be introduced for the targeted gene knock-out. Isolated microspores can be transformed by the inoculation with *Agrobacterium*, induced to form embryos and regenerated into plants, that due to the chromosome diploidisation are completely homozygous.

The TALEN technology was the first system of targeted mutagenesis introduced to plant research. Its application was reported for many model and crop species including Arabidopsis, *Brachypodium distachyon*, barley, maize, tobacco, rice, soybean, tomato and wheat (Malzahn et al. 2017). The approach that combines

TALEN with haploidy was successfully adopted for targeted mutagenesis in isolated microspore culture of winter barley cv. 'Igri'. The proper functioning of TALEN in microspore culture was shown by the knock-out of *GFP* gene and the lack of GFP signal in TALEN-modified transgenic lines harbouring the *GFP* gene (Gurushidze et al. 2014).

CRISPR-Cas9 technology has been quickly adopted in plant research, promoting a reverse genetics revolution in gene function analysis and application for agronomic purposes (Malzahn et al. 2017). This system was introduced into microspore culture of barley and wheat, for which the conventional transformation using immature embryo-based system requires a large number of explants to obtain a sufficient number of transformants. In wheat (*Triticum aestivum*), the usefulness of employing microspores as explants for CRISPR-Cas9 editing system was demonstrated by the knock-out of exogenous *DsRed* (red fluorescent protein variant) reporter gene and two endogenous genes: *TaLox2* encoding a lipoxygenase enzyme and *TaUbiL1* encoding a ubiquitin protein (Bhowmik et al. 2018). Two spring wheat cultivars 'Bobwhite' and 'AC Nanda' that are highly responsive in isolated microspore culture were used in this study. The isolated micropores were transformed by electroporation using Neon electroporation system (Thermo Fisher Scientific) with10–20 μg of foreign DNA at the pulse strength of 500 V. It was shown that the electroporation did not affect the regeneration potential of microspores. The reduction in the number of microspores expressing *DsRed* was evidenced by fluorescent microscopy after targeted mutagenesis of *DsRed* gene. Occurrence of mutations in endogenous genes was analysed 48 h after transfection. In all three wheat subgenomes small deletions in *TaLox2*, as well as insertions and deletions in *TaUbiL1* were identified (Bhowmik et al. 2018).

The CRISPR/Cas9 system was also introduced to isolated microspore culture of barley cv. 'Igri'. After microspore isolation and culture for 1 week, the transgene was delivered to microspore-derived multicellular structures by the inoculation with *Agrobacterium tumefaciens* containing the binary vector. The Authors targeted three different genes: *EIF4E* encoding the eukaryotic translation initiation factor 4E, *PDIL5*-1 (*Protein disulphide isomerase like 5-1*), which both are factors responsible for susceptibility to plant viruses, and *QSD1* encoding alanine aminotransferase that controls the seed dormancy in barley. They generated 48 T_0 plants, of which 17 carried indels of 1–22 bp. Among these 17 plants, 9 were chimeric or heterozygous, but 8 plants were completely homozygous in regard to the carried mutations (Hoffie et al. 2021a). In the second experiment, Hoffie et al. (2021b) using the same method obtained mutations in the *EIF4E* that showed small insertions resulting in a frameshift, which caused the loss of function of the encoded protein. The progeny of generated homozygous mutants showed full resistance to mechanical inoculation with barley mild mosaic virus (Hoffie et al. 2021b).

It should be noted that the conventional transformation is often limited to a few responsive cultivars within the species (Harwood 2012). In barley, the cultivar widely used in transformation is spring 'Golden Promise' whereas in wheat 'Fielder' and 'Bobwhite' (Harwood 2012; Bhowmik et al. 2018). When a microspore system is utilised for transformation, the number of responsive cultivars increases, including winter genotypes, which is another advantage of haploid gene editing in cereals. It can be concluded that the targeted haploid mutagenesis is a promising tool for plant research and agricultural purposes.

References

Ahmadi B, Ebrahimzadeh H (2020) *In vitro* androgenesis: spontaneous *vs.* artificial genome doubling and characterization of regenerants. Plant Cell Rep 39:299–316

Badu M, Tripathy B, Sahu GS, Jena KA (2017) Role of doubled haploids in vegetable crop improvement. J Pharmacogn Phytochem 6:384–389

Beaith ME, Fletcher RS, Kott LS (2005) Reduction of saturated fats by mutagenesis and heat selection in *Brassica napus* L. Euphytica 144:1–9

Beversdorf WD, Kott LS (1987) An *in vitro* mutagenesis/selection system for *Brassica napus*. Iowa State J Res 61:435–443

Bhowmik P, Ellison E, Polley B, Bollina V, Kulkarni M, Ghanbarnia K, Song H, Gao C, Voytas DF, Kagale S (2018) Targeted mutagenesis in wheat microspores using CRISPR/Cas9. Sci Rep 8:1–10

Chen JF, Cui L, Malik AA, Mbira KG (2011) *In vitro* haploid and dihaploid production *via* unfertilized ovule culture. Plant Cell Tissue Organ Cult 104:311–319

Daurova A, Daurov D, Volkov D, Karimov A, Abai Z, Raimbek D, Zhapar K, Zhambakin K, Shamekova M (2020) Mutagenic treatment of microspore-derived embryos of turnip rape (*Brassica rapa*) to increase oleic acid content. Plant Breed 139:916–922

Dong Y-Q, Zhao W-X, Li X-H, Liu X-C, Gao N-N, Huang J-H, Wang W-Y, Xu X-L, Tang Z-H (2016) Androgenesis, gynogenesis, and parthenogenesis haploids in cucurbit species. Plant Cell Rep 35:1991–2019

Dwivedi SL, Britt AB, Tripathi L, Sharma S, Upadhyaya HD, Ortiz R (2015) Haploids: constraints and opportunities in plant breeding. Biotechnol Adv 33:812–829

Ferrie AMR, Caswell KL (2011) Isolated microspore culture techniques and recent progress for haploid and doubled haploid plant production. Plant Cell Tissue Organ Cult 104:301–309

Ferrie AMR, Möllers C (2011) Haploids and doubled haploids in Brassica spp. for genetic and genomic research. Plant Cell Tissue Organ Cult 104:375–386

Ferrie AMR, Taylor DC, MacKenzie SL, Rakow G, Raney JP, Keller WA (2008) Microspore mutagenesis of Brassica species for fatty acid modifications: a preliminary evaluation. Plant Breed 127:501–506

Ferrie AMR, Irmen KI, Beattie AD, Rossnagel BG (2014) Isolated microspore culture of oat (*Avena sativa* L.) for the production of doubled haploids: effect of pre-culture and post-culture conditions. Plant Cell Tissue Organ Cult 116:89–96

Forster BP, Heberle-Bors E, Kasha KJ, Touraev A (2007) The resurgence of haploids in higher plants. Trends Plant Sci 12:368–375

Gaj T, Gersbach CA, Barbas CB (2013) ZFN, TALEN, and CRISPR/Cas-based methods for genome engineering. Trends Biotechnol 31:397–405

Gajecka M, Marzec M, Chmielewska B, Jelonek J, Zbieszczyk J, Szarejko I (2020) Plastid differentiation during microgametogenesis determines green plant regeneration in barley microspore culture. Plant Sci 291:110321

Gao R, Guo G, Fang C, Huang S, Chen J, Lu R, Huang J, Fan X, Liu C (2018) Rapid generation of barley mutant lines with high nitrogen uptake efficiency by microspore mutagenesis and field screening. Front Plant Sci 9:1–11

Gasparis S, Kała M, Przyborowski M, Łyżnik LA, Orczyk W, Nadolska-Orczyk A (2018) A simple and efficient CRISPR/Cas9 platform for induction of single and multiple, heritable mutations in barley (*Hordeum vulgare* L.). Plant Methods 14:1–14

Germanà MA (2011) Gametic embryogenesis and haploid technology as valuable support to plant breeding. Plant Cell Rep 30:839–857

Grigolava TR, Vishnyakova AV, Voronina AV, Zubko ON, Monakhos SG (2021) The effects of sugars and growth regulators on embryo- and callusogenesis in isolated ovules culture of beetroot, *Beta vulgaris* L. Caspian. J Environ Sci. https://doi.org/10.22124/cjes.2021.5091

Guha S, Maheshwari SC (1964) Cell division and differentiation of embryos in the pollen grains of Datura *in vitro*. Nature 204:497–497

Gurushidze M, Hensel G, Hiekel S, Schedel S, Valkov V, Kumlehn J (2014) True-breeding targeted gene knock-out in barley using designer TALE-nuclease in haploid cells. PLoS One 9:1–9

Harwood W (2012) Advances and remaining challenges in the transformation of barley and wheat. J Exp Bot 63:1791–1798

Hoffie RE, Otto I, Hisano H, Kumlehn J (2021a) Site-directed mutagenesis in barley using RNA-guided Cas endonucleases during microspore-derived generation of doubled haploids. In: Segui-Simarro JM (ed) Doubled haploid technology. Springer Science + Business Media, New York, pp 199–214

Hoffie RE, Otto I, Perovic D, Budhagatapalli N, Ordon F, Kumlehn J (2021b) Targeted knockout of Eukaryotic Translation Initiation Factor 4E confers bymovirus resistance in winter barley. Front Genome Ed 3:784233

Huang S, Liu Z, Li D, Yao R, Meng Q, Feng H (2014) Screening of Chinese cabbage mutants produced by 60Co γ-ray mutagenesis of isolated microspore cultures. Plant Breed 133:480–488

Huang S, Liu Z, Yao R, Li D, Feng H (2015) Comparative transcriptome analysis of the petal degeneration mutant *pdm* in Chinese cabbage (*Brassica campestris* ssp. *pekinensis*) using RNA-Seq. Mol Gen Genomics 290:1833–1847

Huang S, Liu Z, Li D, Yao R, Feng H (2016a) A new method for generation and screening of Chinese cabbage mutants using isolated microspore culturing and EMS mutagenesis. Euphytica 207:23–33

Huang S, Liu Z, Li D, Yao R, Hou L, Li X, Feng H (2016b) Physiological characterization and comparative transcriptome analysis of a slow-growing reduced-thylakoid mutant of Chinese cabbage (*Brassica campestris* ssp. *pekinensis*). Front Plant Sci 7:1–14

Huang S, Liu Z, Yao R, Li D, Zhang T, Li X, Hou L, Wang Y, Tang X, Feng H (2016c) Candidate gene prediction for a petal degeneration mutant, *pdm*, of the Chinese cabbage (*Brassica campestris* ssp. *pekinensis*) by using fine mapping and transcriptome analysis. Mol Breed 36: 1–10

Huang S, Peng S, Liu Z, Li C, Tan C, Yao R, Li D, Li X, Hou L, Feng H (2020) Investigation of the genes associated with a male sterility mutant (*msm*) in Chinese cabbage (*Brassica campestris* ssp. *pekinensis*) using RNA-Seq. Mol Gen Genomics 295:233–249

Ishii T (2017) Wide hybridization between oat and pearl millet. In: Gasparis S (ed) Oat: methods in molecular biology. Humana, New York, pp 31–42

Ishii T, Ueda T, Tanaka H, Tsujimoto H (2010) Chromosome elimination by wide hybridization between Triticeae or oat plant and pearl millet: pearl millet chromosome dynamics in hybrid embryo cells. Chromosom Res 18:821–831

Ishii T, Karimi-Ashtiyani R, Houben A (2016) Haploidization *via* chromosome elimination: means and mechanisms. Annu Rev Plant Biol 67:421–438

Kalinowska K, Chamas S, Unkel K, Demidov D, Lermontova I, Dresselhaus T, Kumlehn J, Dunemann F, Houben A (2019) State-of-the-art and novel developments of *in vivo* haploid technologies. Theor Appl Genet 132:593–605

Karakan FY, Arpaci BB (2018) Cold/heat shock pre-treatments for gynogenic haploid embryo induction in a Maryllıdaceae and Cucurbıtaceae: a review. Sci Bull Ser F Biotechnol 22:11–14

Kasha KJ, Kao KN (1970) Hight frequency haploid production in barley (*Hordeum vulgare* L.). Nature 228:726–734

Kasha KJ, Hu TC, Oro R, Simion E, Shi YS (2001) Nuclear fusion leads to chromosome doubling during mannitol pretreatment of barley (*Hordeum vulgare* L.) microspores. J Exp Bot 52:1227–1238

Khan PSSV, Vijayalakshmi G, Raja MM, Naik ML, Germana MA, Terry RG (2020) Doubled haploid production in onion (*Allium cepa* L.): from gynogenesis to chromosome doubling. Plant Cell Tissue Organ Cult 142:1–22

Kyin SM, Khine OA, Khin S (2005) Development of a short duration upland rice mutant line through anther culture of gamma irradiated plants. MBNL & R 1:13–14

Lee JH, Lee SY (2002) Selection of stable mutants from cultured rice anthers treated with ethyl methane sulfonic acid. Plant Cell Tissue Organ Cult 71:165–171

Li X, Huang S, Liu Z, Hou L, Feng H (2019) Mutation in *EMB1923* gene promoter is associated with chlorophyll deficiency in Chinese cabbage (*Brassica campestris* ssp. *pekinensis*). Physiol Plant 166:909–920

Liu S, Wang H, Zhang J, Fitt BDL, Xu Z, Evans N, Liu Y, Yang W, Guo X (2005) *In vitro* mutation and selection of doubled-haploid *Brassica napus* lines with improved resistance to *Sclerotinia sclerotiorum*. Plant Cell Rep 24:133–144

Liu C, Wang J, Huang T, Wang F, Yuan F, Cheng X, Zhang Y, Shi S, Wu J, Liu K (2010) A missense mutation in the VHYNP motif of a DELLA protein causes a semi-dwarf mutant phenotype in *Brassica napus*. Theor Appl Genet 121:249–258

Lu Y, Dai S, Gu A, Liu M, Wang Y, Luo S, Zhao Y, Wang S, Xuan S, Chen X, Li X, Bonnema G, Zhao J, Shen S (2016) Microspore induced doubled haploids production from ethyl methanesulfonate (EMS) soaked flower buds is an efficient strategy for mutagenesis in Chinese cabbage. Front Plant Sci 7:1–9

Makowska K, Oleszczuk S (2014) Albinism in barley androgenesis. Plant Cell Rep 33:385–392

Maluszynski M, Kasha KJ, Forster BP, Szarejko I (2003) Doubled haploid production in crop plants. A manual. Kluwer Academic, Dordrecht, pp 1–480

Malzahn A, Lowder L, Qi Y (2017) Plant genome editing with TALEN and CRISPR. Cell Biosci 7: 1–18

Marcińska I, Nowakowska A, Skrzypek E, Czyczyło-Mysza I (2013) Production of double haploids in oat (*Avena sativa* L.) by pollination with maize (*Zea mays* L.). Cent Eur J Biol 8:306–313

McClinchey SL, Kott LS (2008) Production of mutants with high cold tolerance in spring canola (*Brassica napus*). Euphytica 162:51–67

Meng D, Liu C, Chen S, Jin W (2021) Haploid induction and its application in maize breeding. Mol Breed 41:20

Niu Z, Jiang A, Hammad WA, Oladzadabbsabadi A, Xu SS, Mergoum M, Elias EM (2014) Review of doubled haploid production in durum and common wheat through wheat × maize hybridization. Plant Breed 133:313–320

Ozsan T, Gozen V, Onus AN (2017) Cucumber gynogenesis: effects of 8 different media on embryo and plant formation. Int J Agric Innov Res 6:419–422

Pazuki A, Aflaki F, Gürel S, Ergü A, Gürel E (2018) Production of doubled haploids in sugar beet (*Beta vulgaris*): an efficient method by a multivariate experiment. Plant Cell Tissue Organ Cult 132:85–97

Peng S, Huang S, Liu Z, Feng H (2019) Mutation of *ACX1*, a jasmonic acid biosynthetic enzyme, leads to petal degeneration in Chinese cabbage (*Brassica campestris* ssp. *pekinensis*). Int J Mol Sci 20:2310

Plamenov D, Belchev I, Daskalova N, Spetsov P, Moraliyski T (2013) Application of a low dose of gamma rays in wheat androgenesis. Arch Biol Sci 65:291–296

Polsoni L, Kott LS, Beversdorf WD (1988) Large-scale microspore culture technique for mutation-selection studies in *Brassica napus*. Can J Bot 66:1681–1685

Ravi M, Chan SWL (2010) Haploid plants produced by centromere-mediated genome elimination. Nature 464:615–618

Ren J, Wu P, Trampe B, Tian X, Lubberstedt T, Chen S (2017) Novel technologies in doubled haploid line development. Plant Biotechnol J 15:1361–1370

Sedeek KEM, Mahas A, Mahfouz M (2019) Plant genome engineering for targeted improvement of crop traits. Front Plant Sci 10:1–16

Segui-Simarro JM (ed) (2021) Doubled haploid technology. Humana, New York

Senior IJ, Bavage AD (2003) Comparison of genetically modified and conventionally derived herbicide tolerance in oilseed rape: a case study. Euphytica 132:217–226

Singh BD, Singh AK (2015) Marker-assisted plant breeding: principles and practices. Humana, New York, pp 1–514

Sohrabi S, Abdollahi MR, Mirzaie-Asl A, Koulaei HE, Aghaeezadeh M, Seguí-Simarro JM (2021) A refined method for ovule culture in sugar beet (*Beta vulgaris* L.). Plant Cell Tissue Organ Cult 146:259–267

Swanson EB, Herrgesell MJ, Arnoldo M, Sippell DW, Wong RSC (1989) Microspore mutagenesis and selection: canola plants with field tolerance to the imidazolinones. Theor Appl Genet 78: 525–530

Szarejko I (2012) Haploid mutagenesis. In: Shu QY, Forster BP, Nakagawa H (eds) Plant mutation breeding and biotechnology. CABI, Wallingford, pp 387–410

Szarejko I, Foster BP (2007) Doubled haploidy and induced mutation. Euphytica 158:359–370

Tadesse W, Inagaki M, Tawkaz S, Baum S, van Ginkel M (2012) Recent advances and application of doubled haploids in wheat breeding. Afr J Biotechnol 11:15484–15492

Tan C, Liu Z, Huang S, Li C, Ren J, Tang X, Liu W, Peng S, Feng S (2018) Pectin methylesterase inhibitor (PMEI) family can be related to male sterility in Chinese cabbage (*Brassica rapa* ssp. *pekinensis*). Mol Gen Genomics 293:343–357

Tan C, Liu Z, Huang S, Feng H (2019) Mapping of the male sterile mutant gene *ftms* in *Brassica rapa* L. ssp. *pekinensis via* BSR-Seq combined with whole-genome resequencing. Theor Appl Genet 132:355–370

Tang X, Wang Y, Zhang Y, Huang S, Liu Z, Fei D, Feng H (2018) A missense mutation of plastid *RPS4* is associated with chlorophyll deficiency in Chinese cabbage (*Brassica campestris* ssp. *pekinensis*). BMC Plant Biol 18:1–11

Testillano PS (2019) Microspore embryogenesis: targeting the determinant factors of stress-induced cell reprogramming for crop improvement. J Exp Bot 70:2965–2978

Torp AM, Andersen SB (2009) Albinism in microspore culture. In: Touraev A, Forster BP, Jain SM (eds) Advances in haploid production in higher plants. Springer, Dordrecht, pp 155–160

Tuvesson SD, Larsson CT, Ordon F (2021) Use of molecular markers for doubled haploid technology: from academia to plant breeding companies. In: Segui-Simarro JM (ed) Doubled haploid technology, Methods in molecular biology, vol 2288. Humana, New York, pp 49–72

Vagera J, Novotny J, Ohnoutkova L (2004) Induced androgenesis *in vitro* in mutated populations of barley, *Hordeum vulgare*. Plant Cell Tissue Organ Cult 77:55–61

Watts A, Kumar V, Raipuria RK, Bhattacharya RC (2018) *In vivo* haploid production in crop plants: methods and challenges. Plant Mol Biol Rep 36:685–694

Wędzony M, Marcińska I, Ponitka A, Ślusarkiewicz-Jarzina A, Wozna J (1998) Production of doubled haploids in triticale (x *Triticosecale* Wittin.) by means of crosses with maize (*Zea mays* L.) using picloram and dicamba. Plant Breed 117:211–215

Zhang FI, Takahata Y (1999) Microspore mutagenesis and in vitro selection for resistance to soft rot disease in Chinese cabbage (*Brassica campestris* L. *ssp. pekinensis*). Breed Sci 49:161–166

Strategies for Screening Induced Mutants for Stress Tolerance

6

Fatma Sarsu, Suprasanna Penna, and Ganesh C. Nikalje

Abstract

Crop plants are exposed to a multitude of biotic and abiotic factors, which necessitates improving tolerance/resistance in plant germplasm using conventional, mutagenesis and biotechnological approaches. Several induced mutants have been improved for tolerance/resistance with other desirable characteristics and mutant varieties have been released for cultivation. As per the FAO/IAEA mutant variety database, so far 248 and 558 mutant varieties have been developed for biotic and abiotic stress tolerance, respectively. Progress in this area has been majorly based on mutation induction and screening of putative mutants using a wide range of selection strategies especially, pre-field screening at seedling and flowering stages based on simple phenotypic responses. Several methodological considerations are now advocated for taking up a suitable in vivo/in vitro screening method for a given plant stress factor. For a specific crop and stress tolerance, it is necessary to establish reliable, rapid and high throughput screening techniques in putative mutant lines. Plant cell and tissue culture based in vitro selection strategies have been established for developing stress-tolerant mutants using selection agents for biotic stress, for example, culture filtrate, pathotoxin, or spore suspension or NaCl or PEG. Disease-resistant lines have been developed in

F. Sarsu (✉)
Joint FAO/IAEA Centre of Nuclear Techniques in Food and Agriculture, International Atomic Energy Agency Vienna, Vienna, Austria

S. Penna
Nuclear Agriculture and Biotechnology Division, Bhabha Atomic Research Centre, Mumbai, India

Amity Centre for Nuclear Biotechnology & Amity Institute of Biotechnology, Amity University, Mumbai, Maharashtra, India

G. C. Nikalje
Seva Sadan's R.K. Talreja College of Arts, Science and Commerce, Mumbai, India

S. Penna, S. M. Jain (eds.), *Mutation Breeding for Sustainable Food Production and Climate Resilience*, https://doi.org/10.1007/978-981-16-9720-3_6

banana, carnation, grapevine, strawberry, pineapple, tomato, alfalfa, barley, chickpea, asparagus, sugarcane, celery, oats and wheat for use in breeding programmes. Methodologies to screen for the mutant traits using different cellular, analytical and molecular tools will have to be extensively adopted in plant populations under stress conditions.

Keywords

Induced mutants · In vivo screening · In vitro selection · Abiotic stress tolerance · Crop plants · Disease resistance

6.1 Introduction

Plant breeding approaches carry out an important task to increase crop productivity under stressed environment; however, crops must have the genetic potential to produce high yields in such environments. Diverse approaches involved in breeding new crop varieties for biotic and abiotic stress tolerance include, creation of new genetic variation through mutation breeding, screening and selection of superior germplasm, identification of useful novel genetic variation and, conventional and accelerated breeding (Suprasanna et al. 2015; Sivasankar et al. 2020). As per the IAEA database (2021), there are over 248 and 558 mutant varieties developed through mutation breeding possess biotic and abiotic stress tolerance, respectively. Abiotic stresses such as drought, salinity, cold/heat stress, nutrient deficiency or toxicity, and flooding limit crop productivity worldwide (Suprasanna et al. 2014; Meena et al. 2017). These stresses are one of the root causes of food insecurity and poverty for millions of people in developing countries. While drought stress has affected more than 70 million hectares of rice-growing land worldwide, salt stress and nutrient stress render more than 100 million hectares of agricultural land uncultivable thereby resulting in low outputs, poor human nutrition and reduced educational and employment opportunities (Shahzad et al. 2021). The responses that agriculture systems worldwide can put in place to cope with the expected impact of abiotic stress and to reduce the impact on food security range from institutional and policy levels to the best management practices and technology advancement. An important opportunity in terms of technology advancement is offered by the genetic improvement of crops that can adapt to the abiotic stress conditions and, to develop new varieties of with a higher yield which are tolerant to abiotic stress factors and with nutritional quality.

Faced with global warming, temperature extremes lead to a number of stresses, consequently imposing pressure on sustaining agricultural productivity (Mickelbart et al. 2015; Pandey et al. 2017). High temperature, drought and soil salinization have a direct damaging effect on crop development and yield; and locations where crops suffer from these stresses have been identified worldwide (Meena et al. 2016). Breeding for stress tolerance in a sustainable manner strongly needs thorough understanding of the genetic and physiological mechanisms of plants (Chaudhary and Sidhu 2021) for selecting and identifying tolerant plant germplasm and breeding lines. In most cases, targeted plant breeding for stress tolerance has been impeded due to the limited knowledge of the physiological tolerance processes and their complex genetic nature, thus resulting in the lack of efficient selection procedures (Gilliham et al. 2017). Selection of appropriate screening techniques for stress tolerance is one of the most crucial steps in the selection and/or development of stress-tolerant varieties. Most crop species are sensitive to biotic and abiotic stresses at all the stages of plant development mostly seed germination, pre/post-flowering as a result of which, their growth and economic yield are substantially reduced under stress. To identify crop germplasm for stress tolerance, it is necessary to develop and adopt appriate screening methods that are simple and reproducible under the targeted environmental conditions.

A major challenge for the rapid development of crop varieties, especially in the context of induced mutations, where large populations are required, is the ability to quickly evaluate the materials, and hence reliable, simple, consistent and efficient screening/phenotyping methods and strategies are most needed. Therefore, managing screening nurseries requires a careful analysis of likely sources of non-genetic variation among plots, replications and repeated experiments, as well as establishing procedures for minimizing these factors. Several protocols, techniques, which can be used in the field or greenhouse/screenhouse, rainout shelter or controlled/non-controlled environments have recently been described by scientists (Aslam 2008; Bado et al. 2016; Sarsu et al. 2018, 2021). The early-drought screening using the wooden box technique has proved useful for several crops including cowpea (Singh et al. 1999). Another important tool is the rainout shelter facility that can significantly improve the precision of drought screening. The drought screening under rainout shelter, though reliable, has limitations of space and that crops cannot be grown year after year in the same place so to overcome the latter problem, movable rainout shelters have also been designed. In this chapter, we have described different strategies and methodologies for screening stress tolerance in crop plants, which are essential and applicable for mutant population in crop plants.

6.2 Strategies and Techniques for Improving Crop Efficiency Against Stresses

Major strategies for breeding of new crop varieties include, screening and selection of the existing germplasm, conventional selection and other breeding programmes applied effectively in improving crops against biotic and abiotic stresses. However, extent and rate of progress in improving stress tolerance in crops through conventional breeding programme is limited. This is due to complex mechanism of abiotic stress tolerance controlled by the expression of several minor genes, and techniques employed for selecting tolerant plants are time consumable and consequently expensive. Using mutation techniques in combination with biotechnologies provide powerful technology to improve mechanisms of stress tolerance. These findings emphasized that nuclear techniques help solve practical problems of economic significance to developing countries and significantly contribute to food and nutritional security in those countries (Spencer-Lopes et al. 2018). Identification of morpho-physiological traits, effective of mechanisms and processes that reflect the tolerance, should have a high-priority step in stress tolerance researches (Pandey et al. 2017). An applicable screening trait for stress tolerance should meet the following criteria (Badigannavar et al. 2018):

1. A strong link with stable yield especially grain yield in the target stress environment.
2. A high level of heritability.
3. The expression of tolerance should be quantifiable with acceptable replication.

The traits associated with some promising stress-tolerant germplasm are listed below. Most of these characteristics that appear to enhance crop stress resistance are the manifestation of several individual mechanisms and are most likely under complex genetic control.

1. Morphological characteristics: germination and early vigour, leaf area maintenance, root and shoot growth rate and developmental elasticity. Early growth vigour is a crucial factor in stress tolerance as it allows establishment of a root system more effective in extracting water during later stress periods. Tillering characters in cereals could be useful for stresses as well.
2. Phenology: The suitable crop duration is a compromise of various factors, including season length, yield potential and the timing of when drought stress occurs. For example, using an escape mechanism, the development of short-duration genotypes can help mitigate the effects of terminal drought.
3. Drought avoidance: Adaptation of root architecture, for example, a large and/or deep root system can be useful in greater extraction of available soil moisture. Smaller leaf area could reduce the transpirational water loss under water-deficits conditions.
4. Transpiration efficiency (TE): Depending on the crop, there is a large scope for the genetic improvement of the efficiency of crop water use under dryland conditions. Research has also shown that TE and carbon isotope discrimination in leaf (Δ) are well correlated in several crop species, suggesting a possibility of using Δ as a rapid, non-destructive tool for selection of TE.
5. Delayed senescence or stay-green is considered as a useful trait for plant adaptation to post-flowering drought stress, particularly in environments in which the crop depends largely on stored soil moisture for grain filling. Stay-green trait can also be useful for salinity tolerance.
6. Nitrogen fixation: Improvement of symbiotic and associative N_2 fixation stress tolerance will contribute in raising the crop's productivity and system sustainability in low fertility soils.
7. Crop yield under multi-location field-testing under stress condition: The ultimate evaluation method for stress tolerance improvement is the establishment of multi-location field-testing sites for the evaluation of crop yield under the target environment conditions.
8. Stress tolerance index.

To sum up, it is recommended to initiate mutation breeding to support crop improvement programmes for stress tolerance, following an integrated strategy based on the research priorities described below (Spencer-Lopes et al. 2018).

6.2.1 Important Considerations on Handling Mutated Populations and Mutant Lines

1. Perform radio-sensitivity tests for new species and cultivars before large-scale field experiments are conducted. Use simple tests for seedling emergence and growth reduction for establishing a critical dose of mutagen.
2. Use 2–3 doses for field treatments. For breeding purpose, the doses used in field conditions should not give more than 30% seedling growth reduction in laboratory tests.
3. Calculate the size of M_1 generation taking into account the possible somatic effects of mutagen (survival and fertility reduction) and the required size of M_2 generation.
4. Harvest M_1 plants individually and grow M_2 generation on head/row or plant/row basis. The size of M_2 generation should reach 10,000–50,000 M_2 plants, depending on the character to be improved, available selection methods and experimental conditions. Visual selection and combined appropriate screening procedures (e.g. to expose biotic/abiotic stress) are the most effective and efficient method for identifying mutant phenotypes. Mechanical/physical selection can be used very efficiently for seed size, shape of plant, lodging, weight, etc. The other methods chemicals, biochemical, physiological, physico-chemicals screening procedures needed selection of some mutants. Do not select for yield, quality, response to abiotic stresses, etc. in M_2 generation. Perform selection in M_3, using individual progenies of M_2 plants. Screening and selection in M_2 can be performed only for obvious characters such as semi-dwarfness or earliness. Selection can start in M_2 generation especially for biotic stress.
5. Doubled haploids may be useful to speed up production of pure mutant lines combine mutation techniques and crop breeding programmes. Use M_1 generation for DH production, DH system also in conjunction with crossbreeding, including crosses with mutants or mutant varieties.
6. M_3 generations is raised from all M_2 plants as plant progeny rows and M_3 mutants are considered homozygous for the traits. Selection of M_3 is done on a row or plot basis of homozygous plants, screening can be done agronomically for very important traits such as yield, resistance/tolerance to abiotic stresses. Progeny tests are essential for the identification of mutant lines which will be useful for further progeny tests which may be necessary to stabilize potentially useful mutants. Depending on the generation in which the selection was performed, the homozygosity test can be done in M_3 generation or M_4. Breeders are able to develop new varieties in a short breeding cycle since homozygous lines can be obtained in the M_4 generation (four crop cycles) at the latest whereas it would

Table 6.1 Methodological considerations for handling mutant population for screening abiotic stress tolerance

Mutant generation	Method
M_1 generation (Mutant population)	Grow plants in isolation or bagging (Possible to have anther culture for DH production)
M_2 generation (Mutant Population)	Screening against biotic stress. Identified mutants and harvest seed mutated plants
M_3 generation (Mutant Lines)	Screening for abiotic stress Continue identification of mutants, selection for desired traits; homozygosity test can be started
M_4 generation (Advanced Mutant Lines—AML)	First agronomic evaluation; homozygosity test Propagation of promising lines
M_4 up to M_8 generation—AML	Identifying desired mutants, multiplication trials of stable mutant and recombinant lines
M_9 generation (Candidate Mutant Variety)	Official testing and release of mutant variety

take at least 8–10 generations to produce a similar line through crossbreeding (Shu et al. 2012).

7. As of M_5 generation, application of multi-locational trials of stable mutants and recombinant lines, and official testing and release of mutant variety/varieties.

A general methodological based consideration of generation-wise handling of mutant population is presented below (Table 6.1). Selection of induced mutations is more difficult than mutation induction; generally three types of screening can be employed in M_2 and subsequent generations, visual, mechanical/physical and other methods.

6.3 Screening for Abiotic Stress Tolerance

Climate change manifests into occurrences of extreme weather events, especially drought, heat stress, floods, salinity, etc. which result in significant yield losses, and thus cause a threat to food and nutritional security (Mbow et al. 2019). Enhancing crop production for sustainable food security is an increasing issue as the effects of climate change on agriculture are adversely affecting the productivity of existing local varieties. The breeding of stress-tolerant varieties is critical to produce more yield and increase agricultural productivity in the era of climate change (FAO 2010; Beddington et al. 2011; Dhankher and Foyer 2018; Raza et al. 2019). Soil salinization, especially in highly valuable irrigated lands is a major worldwide land degradation problem that has brought about severe soil quality decline, crop productivity losses and overall reduced environmental sustainability. In view of the several limitations of the classical soil and water management measures, focus is on the identification of plant genotypes better adapted to these abiotic (salinity, drought and heat) stresses and the development of new cultivars (Bado et al. 2016; Badigannavar

Fig. 6.1 Screening for seedling stage heat stress in rice: Mutant lines with no heat stress applied (Control) and, with heat stress applied (Treatment)

et al. 2018). These varieties with improved tolerance to such stresses will provide added value and a competitive market advantage, which in turn will contribute to increased farm income and improved human welfare (Mba et al. 2007).

The quantification of stress tolerance has serious difficulties. Direct selection in the field is difficult because uncontrollable environmental factors adversely affect the precision and repeatability of such trials. There is no exact reliable field screening technique that could be used and selection for stress tolerance using phenotypic measurements requires specialized personnel and extensive investments in field nurseries or greenhouse facilities. Thereby, researchers have been continuously investigating for more effective approaches for improving crop plants for complex traits such as salinity (Mba et al. 2007; Bado et al. 2016). Several pre-field screening methods have been developed (Bado et al. 2016; Sarsu et al. 2018) for robust pre-field screening at seedling and flowering stages based on simple phenotypic responses. These protocols may facilitate breeders to screen for improved heat and salinity stress tolerance in cereals breeding programmes. Based on the use of a hydroponics system and/or pot experiments in a glasshouse in combination with a controlled growth chamber where the heat stress treatment is applied, these protocols are designed to be effective, low cost, reproducible and user-friendly (Figs. 6.1, 6.2 and 6.3).

Another successful pre-field screening method has been developed by Singh et al. (1999) for drought tolerance in cowpea using the wooden box technique. The method is based on imposing the stress 15 days after the termination of watering, where all the seedlings of the two susceptible lines will be completely dead and tolerant lines only will survive. Breeding for salt tolerance requires an understanding

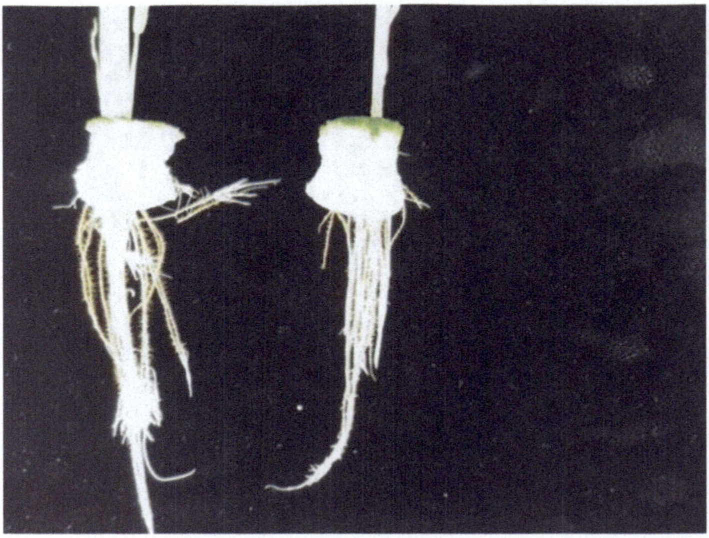

Fig. 6.2 Heat stress effect on seedling root length and weight in rice

Fig. 6.3 Heat stress screening. Heat stress treatment given on rice plants at reproductive stage (in the walk-in-growth chamber)

of the genetic and physiological mechanisms and for phenotyping salt-tolerant germplasm, efficient selection procedures are required. Bado et al. (2016) described effective, low-cost and user-friendly protocols for the pre-field selection of salt-tolerant mutants in cereal crops using hydroponic system (Figs. 6.4 and 6.5).

Fig. 6.4 Wooden box method of screening cowpea plants for drought stress in rainout shelter

Fig. 6.5 Selected drought-tolerant cowpea mutants in the field

Screening of mutants for stress tolerance can also be done using certain physio-biochemical indicators. These include, photosynthesis rate based on CO_2 assimilation, stomatal conductance, chlorophyll fluorescence and various electrophysiological characteristics (Chaudhary et al. 2019; Khadka et al. 2020). Mutant phenotyping can be useful in the analysis of stress tolerance to further pin-down the large population and/or elucidate the tolerance mechanisms. Several phenotyping methods have been applied conventionally; however, they are cumbersome and often destructive. The advanced high-throughput imaging technologies (Kim et al. 2020) enable acquiring quantitative data on plant growth and stress tolerance under treatment

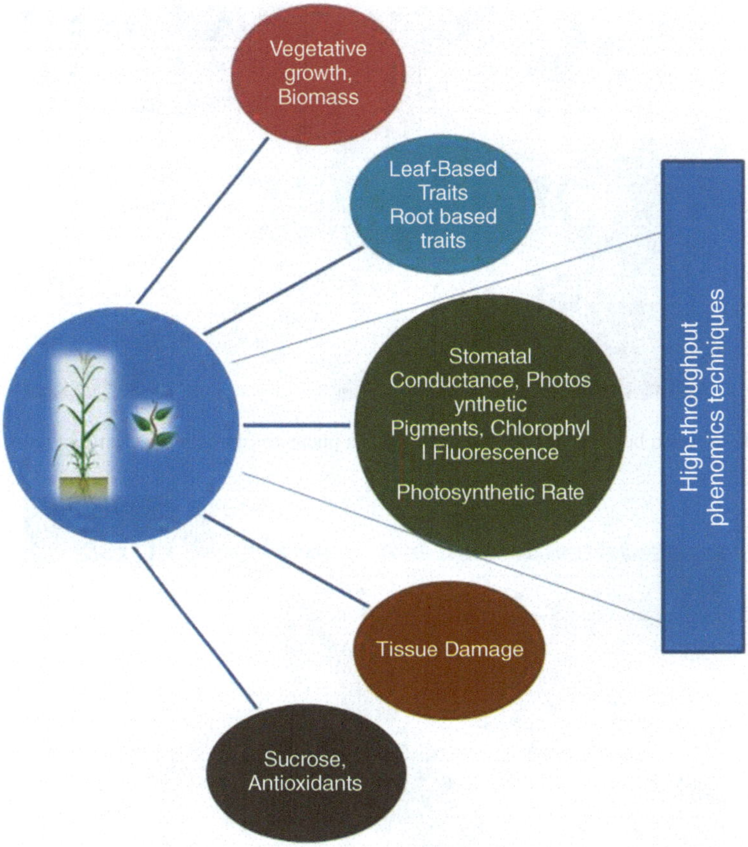

Fig. 6.6 Plant-based screening using different physio-biochemical indices

conditions by taking a series of images of the same plant in spatial and temporal mode, at different wavelengths. Phenomics, the new generation research platform envisages such applications which are now routinely applied to quantify traits related to salt and drought tolerance in a number of crop plants (Yang et al. 2020; Joshi et al. 2021). Chlorophyll fluorescence and thermal imaging are easy and non-destructive rapid methods for monitoring plant stress in the greenhouse and field conditions for collecting data on stomatal and photosynthesis-related parameters (Maxwell and Johnson 2000; Malaspina et al. 2014). Studies on in vivo chlorophyll fluorescence have been useful for screening biotic and abiotic stress tolerance (Baker and Rosenqvist 2004; Kalaji et al. 2016; Perez-Bueno et al. 2019) (Fig. 6.6).

6.4 Considerations for Abiotic Stress Screening

Since the abiotic stress is complex, the problem could be addressed with using and integrated approach physiological dissection of the tolerant traits and molecular genetic tools, together with mutation breeding and agronomical practices that lead to better conservation and utilization of soil moisture as well as enhancing crop genotypes with the environment (Badigannavar et al. 2018). The following points will have to be considered to breed abiotic stress-tolerant crops.

- An improved knowledge and experience of probable soil moisture availability is necessary (using a neutron probe) to further utilize especially for drought tolerance.
- It is recommended that the strategies for abiotic stress tolerance crop improvement focus on not only crop survival, but also on its efficient use in crop establishment, growth, and maximum biomass and seed yield.
- Physiological mechanisms involved in stress tolerance approaches such as the effect of sugars, potassium and calcium nutrition status, scavenging of reactive oxygen species, cell wall biosynthesis and modification should be thoroughly examined using physiological methods and molecular techniques including induced mutation techniques (Meena et al. 2016).
- With respect to salinity, interdisciplinary research on the importance for tolerance of growth processes in the shoot (tillering, leaf number and internode length), their cellular base, cell wall structure and others should be emphasized.
- It is also recommended to monitor salt movement and accumulation in the root zone using isotopes and tracers in selected sites. This enables appropriate and successful plant improvement and study of the plant's impact in soil conditions.
- The genetic enhancement of root systems to make them more effective in water extraction would seem a high-priority effort for rainfed crops. Dissection of root traits and development of a screening system relevant to field conditions are therefore needed, parallel with extensive genotyping and search for molecular markers.
- Other promising integrated traits for improving stress resistance and crop water productivity include harvest index, stay-green (sorghum) and transpiration used efficiency.
- The best approach is needed for combining the relevant stress resistance traits in the major crops of interest, which requires a better knowledge of the physiological mechanisms involved in stress resistance and their genetic control.

6.5 Screening for Biotic Stress Tolerance

Plant diseases and pests cause yield and quality losses in crop production in nearly all agricultural areas around the world. Many wild/original plant genetic resources vital for the present and future agricultural development are threatened by genetic erosion due to the spread of modern commercial agriculture. This, combined with the climate change has increased the vulnerability of existing crops to pests, pathogens and environmental stresses. The most effective approach to prevent disease/pests outbreak is to cultivate resistant varieties. Therefore, improvement of crop production along with pest and disease resistance is one of the main goals of a mutation breeding programme. Using mutation breeding, traits for yield, quality, taste and disease and pest resistance have been improved in staple crops such as cereals, legumes, banana and cassava.

Breeding new varieties which are tolerant to biotic stress factors during their growth and development is one of the solutions to have a sustainable and secure food production. Therefore, it is absolutely necessary to establish reliable, rapid and high throughput screening techniques for disease resistance in putative mutant lines. Combination of traditional breeding methods (hybridization, selection and introduction) into biotechnology and, in vitro mutagenesis can provide excellent approach to breeding for biotic stresses, covering all major biotic challenges to agriculture and food production (Oladosu et al. 2016). Using mutation techniques in combination with in vitro technologies can also be very useful for efficient screening of crop germplasm for quality characters, including nutrition and process characters and tolerance to stress affecting quality in crop plants (Jain and Suprasanna 2011). Biotic stresses are mostly simply inherited (qualitative) traits that are often controlled by one or few genes thereby considerable breeding progress has been made for these traits especially using molecular approaches (Bigini et al. 2021). There seems to be much scope for improving stress tolerance characters, using QTLs and molecular breeding techniques, aided by conventional breeding including mutation breeding to significantly improve the ability of the crops to withstand stress in defined biotic stresses.

Selection of plants for disease resistance/tolerance requires basic knowledge about the biology of the causal agent and its relationship with the host plant. This information is necessary for the development of suitable methods of screening and selecting for resistance (Russell 1978). There is a broad range of different methodological approaches available to screen for resistant mutant genotypes and to select plants with improved resistance (Table 6.2, TNAU 2013; Kiraly et al. 1974; Dhingra and Sinclair 1986; Lebeda 1986; Trigiano et al. 2004; Singh and Singh 2005; Dölfors et al. 2022), among which early in vivo and in vitro screening offers considerable advantages (Svabova and Lebeda 2005).

In cabbage, three early screening methods were developed to assess resistance to yellow disease (*Fusarium oxysporum* f. conglutinans) in 20–30 day-old seedlings (Yamagishi et al. 1985) of the three methods viz., root-clipping, soil-inoculation and ditch-inoculation method, it was suggested that the third method is suitable to test a large number of plants at a time. Pande et al. (2010) reviewed several protocols for field, greenhouse, growth room and laboratory screening techniques for chickpea diseases viz. Ascochyta blight, Botrytis grey mould, Fusarium wilt, dry root rot, collar rot, black root rot and stem rot. Authors suggested Cut twig screening method to be very simple and less time consuming. In case of powdery mildew, a simple, reliable method for assessing reaction on excised mung bean leaves has been developed for screening mutant germplasm (Reddy et al. 1987). Drizou et al.

Table 6.2 Methods of screening for disease resistance

Method	Mandate	Related method(s)	Remarks
Artificial inoculation	The artificial inoculation is required to impose uniform disease exposure for selection	• For efficient screening, adequate amount of inoculum is required • Spraying the spore suspension under high pressure spraying • Introduction of spore suspension into the plant organ surface or into the intercellular air spaces (of a leaf, for example) • Immersion of seedlings in a spore suspension before transplanting them into fields	Spore—suspension in sterile water is often used; in some cases, dry spore powder can be used since spores get damaged quickly if they are wet. Depending upon the convenience and crop materials the procedure can vary from one disease to another and from one crop to another. TNAU (2013)
Field screening	Screening segregating mutant plant population for resistance under natural pathogen infestation condition or artificial conditions	The mutant populations/ lines to be evaluated or the segregating population to be screened for resistance may be raised in fields under natural pathogen infestation condition (natural endemic areas) or in areas where, increasing field infestation can be enhanced for screening	Highly susceptible lines can be maintained as 'spreader rows' along with test plant rows
Greenhouse screening or fibre glass screen cages	Screening mutant population/lines in greenhouses by providing conditions conductive for artificial infestation is more rapid reliable than field screening	The mutant populations/ lines to be evaluated or the segregating population to be screened for resistance may be raised in greenhouse conditions under artificial inculcation conditions	Screening mutant plant population in greenhouses is more rapid reliable than field screening. Depending on the crop and disease/pest, the stage of the crop, it should be possible to decide on possible disease symptoms for screening
Laboratory screening	Screening mutants by providing conditions conductive for artificial pathogen inoculation	Screening for resistance using plant tips or leaf discs and allowing forced or free choice feeding by insects as in the case of lucerne weevil	

(continued)

Table 6.2 (continued)

Method	Mandate	Related method(s)	Remarks
Bioassay techniques	Screening mutants by using bioassays based on disease associated reactions	Resistance screening is based on application of bioassay techniques to monitor disease related biochemical, molecular responses	By periodic observations, larval survival, larval growth and percent pupation are recorded. Data on fecundity and longevity of emerging adults are obtained to screen resistant types
	Measurement of pathogen growth on plant tissue prior to appearance of symptoms	Real-time assay to quantify pathogen growth in live tissue using fluorescent protein tags	Zhang et al. (2021)

(2017) developed a novel high-throughput screening technique based on different growth systems (nutrient media plates, compost trays, light expanded clay aggregate trays and hydroponic pouch and wick system) for screening oilseed rape germplasm resistant to root diseases such as *Rhizoctonia solani* (Kühn) AG 2-1 and it was suggested that use of light expanded clay aggregate trays could be useful for rapid screening of germplasm prior to field evaluation. Recently, Dölfors et al. (2022) reported a protocol wherein pre-grown seedlings of sugarbeet and Arabidopsis were grown in a perlite-based substrate infested with the fungus *Rhizoctonia* and mixed with soil. The protocol is suitable for root crops and other crops with good consistency of plant infection and reproducibility. Some of the examples where mutant development for disease resistance has been achieved based on selection criteria and mutagen are enlisted in Tables 6.3 and 6.4.

6.6 In Vitro Screening

In vitro techniques have been employed successfully in the selection of resistant plants at cellular level. Progress in this area has been majorly based on all the in vitro-related research covering explant cultures, sources of in vitro variability, screening and selection methods and selection agents besides the potential of plant in vitro cultures for regeneration at higher frequency to obtain sizable population after selection and stability of improved pathogen resistance to the pathogen in the progenies (Fig. 6.7; Svabova and Lebeda 2005). The basis for such a selection is that a causative agent in the disease reaction should be employed in the selection process. The selection agents can be pathogen itself, culture filtrate or pathogen metabolites (Pathotoxins). Culture filtrates (CF) represent an easy and simple method of selection by incorporation into the culture media at appropriate concentrations. For many plant diseases, well-characterized toxins may be unavailable and hence culture filtrates can effectively be used as they have been found to have phytotoxic activity.

Table 6.3 Mutant screening for biotic stress tolerance in different plant species

Mutants of plant species	Resistance to	Selection factor	References
Citrus sinensis	Citrus canker	Artificial inoculation of *Xanthomonas citri*	Junior et al. (2008)
Colocasia esculenta	Phytophthora leaf blight disease	Spraying of spore suspension	Sahoo et al. (2015)
Oryza sativa	Rice blast	Spray-inoculated with spores of *M. oryzae*	Kaur et al. (1975)
Vigna mungo	Stem fly damage	Field observations	Reddy (2009)
Abelmoschus esculentus	Yellow vein mosaic virus disease	Whitefly transmission	Phadvibulya et al. (2009)
Lycopersicon esculentum	Bacterial wilt (BW)	Bacterial suspension	Galvez et al. (2009)
Saccharum officinarum	Red rot	Inoculation of spore suspension	Ali et al. (2007)
Musa paradisiaca	Black sigatoka disease	*Inoculation of conidial solution*	Borja et al. (2007)
Musa spp.	Fusarium wilt	Fusaric acid and culture filtrate	Saraswathi et al. (2016)
Malus domestica	*Alternaria alternata* (Fr.) Keissler	AM-toxin I	Saito et al. (2001)
Sesamum indicum	Phytophthora blight	Field observations	Pathirana (2011)
Musa acuminate	Fusarium wilt	Inoculation of conidial solution	Bhagwat and Duncan (1998)
Oryza sativa	Rice blast	Artificial inoculation at seedling stage	Zhang et al. (2003)
Saccharum species	SCMV disease	Field observations	Zambrano et al. (2003)
	Brown rust	Inoculation of fungal solution	Oloriz et al. (2012a)
	Smut	Smut spore inoculation	Dalvi et al. (2012)
	Red rot	Inoculation of fungal solution	Kaur et al. (2016)
Brassica napus	Alternaria leaf spot	Inoculation of spore suspension	Ahmad et al. (1991)
Camellia sinensis	Blister blight	Inoculation of spore suspension	Gunasekera et al. (2003)
Nicotiana glutinosa	TMV disease	Virus infection	Brederode et al. (1991)
Arabidopsis sp.	Downy mildew	Spore suspensions	Kunz et al. (2008)

For a successful crop improvement strategy, it is imperative that selected mutant lines be evaluated, genetically analysed and integrated into breeding programmes. For example in date palm, embryogenic cell cultures were irradiated with gamma radiation and the regenerated plants transferred to greenhouse were treated with

Table 6.4 Mutant screening for abiotic stress tolerance in different crop varieties

Mutant varieties	Released year and country	Tolerance to	References
Rice DT 10	1989 Vietnam	Cold	Do (2009), Vinh et al. (2009)
Rice VND-95-20	1999 Vietnam	Salinity	Do (2009), Vinh et al. (2009)
Rice NIAB IRRI9	1999 Pakistan	Salt tolerance	Haq (2009)
Rice Binadhan-14	2013—Bangladesh	High temperature	Azad et al. (2021)
Rice Binadhan-19	2017—Bangladesh	Drought tolerance	Azad et al. (2021)
Cotton NIAB Karishma NIAB-111 NIAM 277	1996, 2004–2009 Pakistan	Heat tolerance	FAO/IAEA Mutant Variety Database (2021)
Soybean DT 2008	2008—Vietnam	Drought tolerance	Le and Chung (2021)
Tomato Domi and Maybel	2007, 2015/Cuba	Drought tolerance	Sarsu et al. (2021) and FAO/ IAEA MVD (2021)
Tomato Summer Star	2019—Mauritius	Heat tolerance	Sarsu et al. (2021) and FAO/ IAEA MVD (2021)
Sorghum Pahat	2016—Indonesia	Drought tolerance	Soeranto and Indriatama (2021)
Wheat Binagom-1	2016—Bangladesh	Salt tolerance	Azad et al. (2021)
Groundnut Tafra-1	2018—Sudan	Drought	Elgailani et al. (2021)
Groundnut Binachinabadam-5 Binachinabadam-6 Binachinabadam-7 Binachinabadam-9	2011—Bangladesh	Salt tolerance	Azad et al. (2014)

Bayoud toxin, (isolated from the causal fungus *Fusarium oxysporum* f. sp. Albedinis). Several selected mutants showed tolerance to Bayoud disease under field conditions (Jain 2012).

Lebeda and Švábová (2005, 2010) described in detail the in vitro methodologies for in vitro selection by using pathogen isolates, culture filtrates, toxins and toxin-like compounds. To date, several reports of in vitro selection in diverse plant species are available using wide range of in vitro culture systems (excised organ or part of organ, meristem tip culture, anther or pollen, callus, cell or tissue suspension, protoplast culture) and selection agents (natural isolate of pathogen, modified pathogen culture, culture filtrates, toxins, elicitors). Several successful examples include disease-resistant lines developed in banana, carnation, grapevine, strawberry, pineapple, tomato, alfalfa, barley, chickpea, asparagus, sugarcane, celery, oats and wheat (Lebeda and Švábová 2010). In some of these and other crops, in vitro selection technique has become an integral complement to traditional crop improvement programmes.

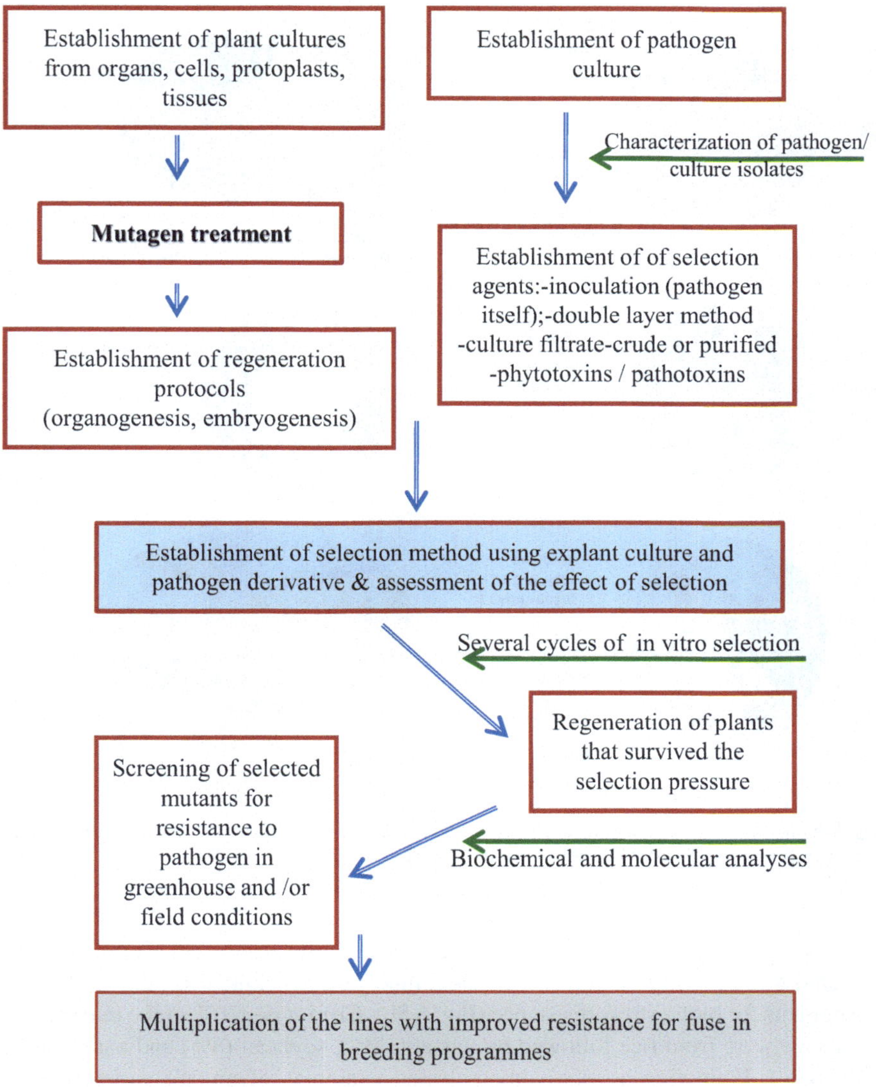

Fig. 6.7 Work-flow for in vitro screening for disease resistance (Lebeda and Švábová 2010)

6.7 Successful Examples of Mutagenesis for Abiotic and Biotic Stress Tolerance

As per FAO-IAEA-MVD (2021) database, 558 mutant varieties were released from 1956 to 2021 for biotic stress resistance. Figure 6.8 presents mutant varieties categorized on the basis of crop, disease, type of mutagen and country. The data

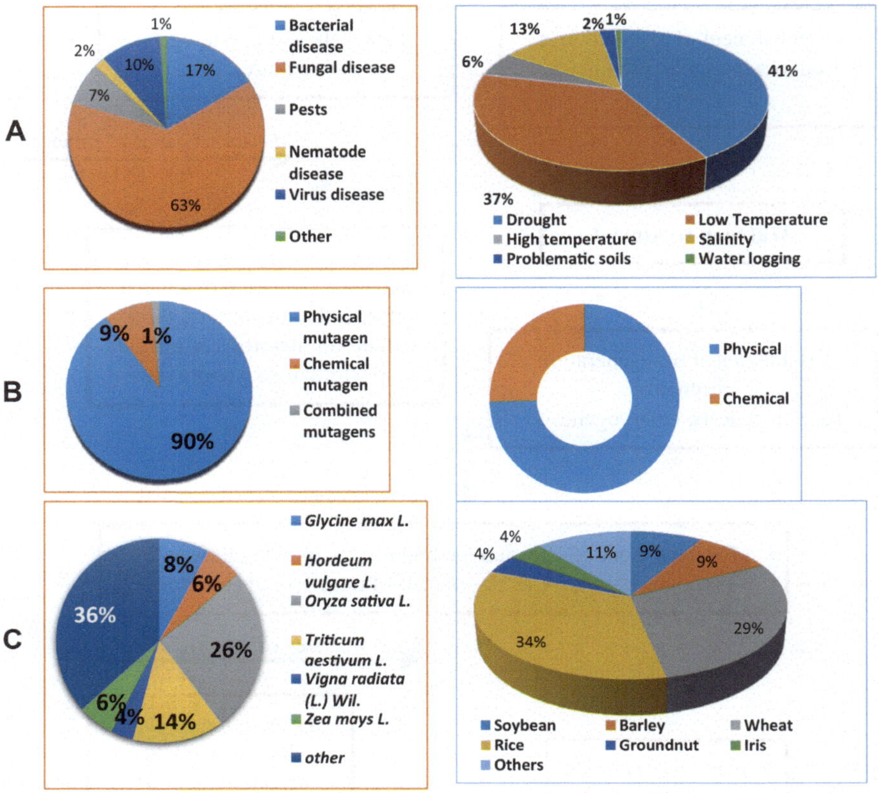

Fig. 6.8 Mutant varieties developed for biotic and abiotic stress tolerance. (**a**) diseases, (**b**) Type of mutagen, (**c**) Crop. (FAO/IAEA MVD database 2021)

showed that resistance to fungal diseases has been most selected (63%) followed by bacterial diseases (Fig. 6.8a). The physical mutagens accounted for almost 90% of the mutants for biotic stress resistance (Fig. 6.8b). Crop-wise, 26% of the mutants for resistance were from rice followed by wheat (14%), soybean (8%) and barley (6%) (Fig. 6.8c). Following are the some selected examples of mutant varieties from different crops specific for biotic stress resistance and other agronomic characters.

In Tomato, bacterial wilt is a major bacterial wilt disease caused by *Ralstonia solanacearum*. In Sri Lanka, the mutant variety 'M 127' derived from 'Manik' variety using Gamma rays (320 Gy) gave significantly higher yield (32.2 t/ha) as compared to check variety T 245 (21.7 t/ha) in both dry and wet seasons. The fruit weight of M 127 was also significantly higher (158.6 g) with red and slightly flattened firm fruits. M 127 was moderately tolerant to bacterial wilt disease (Peiris et al. 2009). Rice is one of the major crops of world. It is highly susceptible to blast, rice yellow mottle virus, bacterial blight diseases. In China, the mutant rice variety 'Zhefu802' derived from Simei No. 2 variety which was induced by Gamma rays, had a short growing period (105–108 days), high yield potential even under poor management and infertile conditions, wide adaptability, high resistance to rice blast and tolerance to cold (Ahloowalia et al. 2004).

Another successful example is from Vietnam, the most outstanding mutant rice variety DT10 has wide adaptability and high yield of 6–8 t/ha (40% increase compared to the average) and especially good tolerance to cold. The variety became very popular and accepted by the farmers in the Northern Vietnam and occupies large areas in the northern provinces, serving as the most suitable rice variety for processing. In Southern Vietnam, mutant rice variety VND 95-20 with high quality and tolerance to salinity has become one of the most planted varieties for exporting rice area in the Mekong Delta (Do 2009; Vinh et al. 2009). Soybean mutant varieties especially DT 2008 contributed to food security in the country (Le and Chung 2021). Tafra-1 is a drought-tolerant peanut mutant variety released in 2018 for Sudanese farmers in areas prone to drought, which can grow with less than 250 mm of rain per year compared to the 350 mm needed for the traditional varieties. It also produces yields which are 11% higher on average than the traditional variety (Abdalla et al. 2018).

In Pakistan, the phenological and physiological investigations revealed heat-tolerant cotton NIAB-878 by maintaining the highest anther dehiscence (82%) and minimum cell injury percentage (39%) along with maximum stomatal conductance (27.7 mmol $CO_2/m^2/s$), transpiration rate (6.89 μmol $H_2O/m^2/s$), net photosynthetic rate (44.6 mmol $CO_2/m^2/s$) and physiological water use efficiency (6.81 mmol CO_2/ μmol H_2O) under the prevailing high temperatures. Additionally, the mutant rice variety NIAB IRRI 9 is a non-aromatic, fine grain, salt-tolerant and high-yielding mutant variety registered in 1999 and has contributed additional incomes to farmers during 2004–2005 (Haq 2009; Hussain et al. 2021).

Three mutant tomato varieties were released in Mauritius with high yielding viz. Summer King, Summer Star and Rising Star, the summer star have heat stress tolerance. These varieties were distributed to farmers and seed multiplication continued to reach more farmers (Sarsu et al. 2021; MVD 2021). Another success in tomato is from Cuba through the development and release of Domi and Maybel mutant varieties for drought tolerance and high yield which are contributing to economic gain for Cuban Farmers (Sarsu et al. 2021; FAO/IAEA MVD 2021).

In chickpea, Ascochyta blight (AB) is a major fungal disease which causes 100% yield loss (Pande et al. 2005). In India, the four high yielding and Ascochyta blight and wilt disease-resistant chickpea mutant varieties Pusa—408 (Ajay), Pusa—413 (Atul), Pusa—417 (Girnar) and Pusa—547 of chickpea, developed at I.A.R.I., New Delhi, are the first examples of direct use of induced micro-mutants in a legume crop. Beside high yield performance under late sown crop, chickpea mutant variety Pusa—547, released in 2006 for farmer's cultivation, has attractive bold seeds, thin testa and good cooking quality (Kharkwal et al. 1986, 2004, 2005, 2008).

Barley ranks fourth most important crop all over world. It is susceptible to Common Root Rot, Spot Blotch, Net Blotch, Stagonospora Leaf Blotch, Septoria Leaf Blotch, Scald, Leaf Rust, Powdery Mildew, Bacterial Blight, Scab or Head Blight, Barley Yellow Dwarf Virus and Loose Smut. In Germany, Trumpf, the best-known barley mutant cultivar obtained after crossing with cultivar Diamant occupied more than 70% of the barley planting area in Germany. The mutant had a yield increased by 15% and better disease resistance. Used extensively in crossbreeding, Trumpf became incorporated into many barley breeding programmes in a large number of countries (Harten 1998). Mutation breeding has been very successfully used in breeding barley, the fourth most important food crop in terms of area in Peru. Centenario, a barley mutant with high yield (37% over the parent cultivar), earliness (18 days), higher protein (10.3%), better test weight and resistance to yellow rust, was released in 2006, is replacing the traditional cultivars of the central highlands of Peru and contributes significantly to the food security of the country (Pando et al. 2008).

Lupine is susceptible to Brown Leaf Spot and Root Rot caused by *Pleiochaeta setosa*. The two lupine mutants (Tanjil-AZ-33 and Tanjil-AZ-55) are highly tolerant, six times more tolerant to metribuzin herbicide than the original parental cultivar Tanjil. This mutant Tanjil-AZ-33 is the most tolerant germplasm in narrow-leafed lupine. Both mutants also maintain the high yield and resistance to the disease anthracnose as cv Tanjil. These facts indicate that the mutation process has created tolerance to metribuzin in Tanjil but has not altered Tanjil's yield capacity and anthracnose resistance. Induced mutation proved to be an effective tool in lupine improvement (Si et al. 2008).

Due to wheat rust almost 14% of wheat grain yield losses was found in Great Plains of North America. Mutant variety, Stadler, a high-yielding wheat mutant released in Missouri (USA) had early maturity, resistance to races of leaf rust and loose smut, as well as better lodging resistance. Almost 90% of wheat varieties are highly susceptible to black stem rust (Ug99). Due to the coordinated efforts made by IAEA, three mutant cultivars for black stem rust (Ug99) resistance have been developed in Kenya (Forster 2014).

In commercial crops like banana and sugarcane, induced mutations have played a significant role in the development of disease-resistant mutants/mutant varieties (Jain 2010; Oloriz et al. 2012b). In sugarcane, mutagenesis with NaN_3 and gamma irradiation resulted in brown rust resistance in mutants which also exhibited superior agronomic traits (Oloriz et al. 2012b). The exposure of sugarcane cultivar CoC671 to EMS resulted in a TC2628 mutant which was highly resistant to smut disease with high sucrose content (23.33%) than parent (21.39%) at maturity (Dalvi et al. 2018).

6.8 Conclusions

Crop plants are often threatened by biotic and abiotic stresses which challenge productivity. Improvement of plant stress tolerance in a sustainable manner strongly needs breeding efforts and knowledge of the genetic and physiological mechanisms. Towards the development of tolerant/resistant crop varieties, induced mutations have played a significant role in the generation novel genetic variability for various traits and trait-associated secondary responses that contribute to resistance phenomena. FAO/IAEA mutant database catalogues about 558 mutants and 248 biotic and abiotic stresses tolerant mutant varieties in different crop plants. Interest in plant mutation breeding is growing as mutants isolated and evaluated for several agronomic traits including biotic stress are grown and contribute to food security in different countries, providing good economic impact. In this regard, it is necessary to expand mutant population-lines and rapidly evaluate the large mutant population by using efficient screening/phenotyping methods. Various in vivo and in vitro methods for screening mutant germplasm are now available. Although mutagenesis and selection processes can be improved by such methods, mutant screening can benefit more by using molecular and/or high-throughput approaches.

References

Abdalla EA, Ahmed TE, Bakhit OA, Gamar YA, Elshaikh SA, Mohammed YE, Sulaiman AA, Mardi HG (2018) Tafra-1, a groundnut mutant with end-of-season drought tolerance for the low rain fall dry lands of North Kordofan State. The 2nd National Variety Release Committee, Elobeid, North Kordofan State, Elobeid, Sudan

Ahloowalia BS, Maluszynski M, Nichterlein K (2004) Global impact of mutation-derived varieties. Euphytica 135:187–204

Ahmad I, Day JP, MacDonald MV, Ingram DS (1991) Haploid culture and UV mutagenesis in rapid-cycling Brassica napus for the generation of resistance to chlorsulfuron and *Alternaria brassicicola*. Ann Bot 67(6):521–519

Ali A, Naz S, Alam S, Javed I (2007) In vitro induced mutation for screening of red rot (*Colletotrichum falcatum*) resistance in sugarcane (*Saccharum officinarum*). Pak J Bot 39: 1979–1994

Aslam Z (2008) Salt tolerance and screening techniques for sat tolerance in plants. In: Sustainable crop production on salt-affected land, Proc. Int. Conf., Faisalabad, Pakistan, 4–6 December 2006. Saline Agriculture Research Centre, UAF, Pakistan

Azad MAK et al (2014) Enhancing abiotic stress tolerance in groundnut through induced mutation. In: Tomlekova NB, Kozgar MI, Wani MR (eds) Mutagenesis: exploring genetic diversity of crops. Academic Publishers, Wageningen, Netherlands, pp 329–344

Azad MAK, Yasmine F, Kamruzzaman M, Rani MH, Begum HA (2021) Development of climate adaptable/resilient crop varieties through induced mutation. In: Sivasankar S, Ellis N, Jankuloski L, Ingelbrecht I (eds) Mutation breeding, genetic diversity and crop adaptation to climate change. CAB International, Wallingford, UK, pp 157–171. https://doi.org/10.1079/9781789249095.000

Badigannavar A, Niaba T, Oliveira AC, Li G, Vaksmann M, Viana VV, Ganapathi TR, Sarsu F (2018) Physiological, genetic and molecular basis of drought resilience in sorghum [Sorghum bicolor (L.) Moench]. Indian J Plant Physiol 23(2018):670–688

Bado S, Forster BP, Ghanim AMA, Jankowicz-Cieslak J, Berthold G, Liu L (2016) Protocols for pre-field screening of mutants for salt tolerance in rice, wheat and barley. Springer International. ISBNs 978-3-31-926588-9, 978-3-31-926590-2

Baker NR, Rosenqvist E (2004) Applications of chlorophyll fluorescence can improve crop production strategies: an examination of future possibilities. J Exp Bot 55:1607–1621

Beddington J, Asaduzzaman M, Fernandez A, Clark M, Guillou M, Jahn M, Erda L, Mamo T, Van Bo N, Nobre CA, Scholes R, Sharma R, Wakhungu J (2011) Achieving food security in the face of climate change: Summary for policy makers from the Commission on Sustainable Agriculture and Climate Change. CGIAR Research Program on Climate Change, Agriculture and Food Security (CCAFS), Copenhagen, Denmark

Bhagwat B, Duncan EJ (1998) Mutation breeding of banana cv Highgate (Musa spp., AAA Group) for tolerance to *Fusarium oxysporum* f. sp. cubense using chemical mutagens. Sci Hort 73:11–22

Bigini V, Camerlengo F, Botticella E, Sestili F, Savatin DV (2021) Biotechnological resources to increase disease-resistance by improving plant immunity: a sustainable approach to save cereal crop production. Plants 10:1146. https://doi.org/10.3390/plants10061146

Borja WOR, Sotomayor I, Ivan G, Vera D, Cedeño M, Castillo B, Tanaka A, Hase Y, Sekozawa Y, Sugaya S, Gemma H (2007) Alteration of resistance to black Sigatoka (*Mycosphaerella fijiensis* Morelet) in banana by in vitro irradiation using carbon ion-beam. Plant Biotechnol 24:349–353

Brederode FT, Linthorst HJ, Bol JF (1991) Differential induction of acquired resistance and PR gene expression in tobacco by virus infection, ethephon treatment, UV light and wounding. Plant Mol Biol 17(6):1117–1125

Chaudhary S, Sidhu GPS (2021) Climate change regulated abiotic stress mechanisms in plants: a comprehensive review. Plant Cell Rep. https://doi.org/10.1007/s00299-021-02759-5

Chaudhary J, Alisha A, Bhatt V, Chandanshive S, Kumar N, Mir Z, Kumar A, Yadav SK, Shivaraj SM, Sonah H, Deshmukh R (2019) Mutation breeding in tomato: advances, applicability and challenges. Plants 8(5):128. https://doi.org/10.3390/plants8050128

Dalvi SG, Vasckar VC, Yadav A, Tawar PN, Dixit G, Prsad T, Deshmukh RB (2012) Screening of promising sugarcane somaclones for agronomic traits and smut resistance using PCR amplification of inter transcriped region (ITS) of *Sporisorium scitaminae*. Sugar Tech 14:68–75

Dalvi S, Prasad DT, Dixit G, Suprasanna P, Tawar P (2018) EMS induced in vitro mutagenesis for abiotic and biotic stress tolerance, and isolation of morphologically and biochemically distinct phenotypes in sugarcane. In: FAO-IAEA International symposium on plant mutation breeding and biotechnology, 27–31 August 2018, Vienna, Austria. Abstract no-IAEA-CN-263-166, p 33

Dhankher OP, Foyer CH (2018) Climate resilient crops for improving global food security and safety. Plant Cell Environ 41(5):877–884

Dhingra OD, Sinclair JB (1986) Basic plant pathology methods. CRC Press, Boca Raton, FL, p 355

Do KT (2009) Socio-economic impact of mutant rice varieties in Southern Vietnam. In: Shu QY (ed) Symposium proceeding. Induced Plant Mutations in the Genomic Era. ISBN-978-82-5-106324-8, I0956e.pdf (fao.org)

Dölfors F, Holmquist L, Tzelepis G, Dixelius C (2022) Rhizoctonia solani infection assay of young sugar beet and Arabidopsis plantlets. Bio-protocol 12(2):e4300. https://doi.org/10.21769/BioProtoc.4300

Drizou F, Neil G, Toby B, Rumiana R (2017) Development of high-throughput methods to screen disease caused by Rhizoctonia solani AG 2-1 in oilseed rape. Plant Methods 13. https://doi.org/10.1186/s13007-017-0195-1

Elgailani A, Ahmed T, Bakhit O, Gamar Y, Elshaikh S, Mohammed Y, Sulaiman A, Mardi H (2021) Groundnut mutants with end-of-season drought tolerance for the marginal dry of North Kordofon State, Sudan. In: Sivasankar S, Ellis N, Jankuloski L, Ingelbrecht I (eds) Mutation breeding, genetic diversity and crop adaptation to climate change. CAB International, Wallingford, pp 243–257. https://doi.org/10.1079/9781789249095.000

FAO/IAEA Mutant Variety Database (2021) MVD – Home. https://mvd.iaea.org

Food and Agriculture Organization of the United Nations (2010) Global survey of agricultural mitigation projects, Mitigation of climate change in agriculture series No. 1. Food and Agriculture Organization of the United Nations, Rome, Italy

Forster BP (2014) The development of "EldoNgano I": the world's first Ug99 resistant mutant wheat variety. IAEA Bull 55:18–11

Galvez HF, Canama AO, Colle MG, Quilloy RB, de Vera ML, Bituin JL, Tiongco RL, Frankie RB, Hautea DM (2009) Development of deletion mutants of tomato Hawaii 7996 for resistance to *Ralstonia solanacearum* and other important traits by targeted EMS and irradiation mutagenesis. University Library, University of the Philippines at Los Baños, Los Baños

Gilliham M, Able JA, Roy SJ (2017) Translating knowledge about abiotic stress tolerance to breeding programmes. Plant J 90:898–917. https://doi.org/10.1111/tpj.13456

Gunasekera TS, Paul ND, Ayres PG (2003) The effects of ultraviolet-B (UV-B: 290–320 nm) radiation on blister blight disease of tea (*Camellia sinensis*). Plant Pathol 46(2):179–185

Haq MA (2009) Development of mutant varieties of crop plants at NIAB and the impact on agricultural production in Pakistan. In: Shu QY (ed) Symposium proceeding. Induced Plant Mutations in the Genomic Era. ISBN-978-82-5-106324-8, I0956e.pdf (fao.org)

Harten AM (1998) Mutation breeding: theory and practical applications. Cambridge University Press, Cambridge

Hussain M, Jankuloski L, Rahman MH, Malek M, Islam MDK, Raheemi MR, Dana J, Lwin KM, Ahmad F, Rizwan M, Talha GM, Asif M, Ali S (2021) Improving sustainable cotton production through enhanced resilience to climate change using mutation breeding. In: Sivasankar S, Ellis N, Jankuloski L, Ingelbrecht I (eds) Mutation breeding, genetic diversity and crop adaptation to climate change. cabi.org. https://doi.org/10.1079/9781789249095.0000

Jain SM (2010) In vitro mutagenesis in banana (*Musa sp*) improvement. Acta Hortic 879:605–614

Jain SM (2012) Date palm biotechnology: current status and prospective - an overview. Emir J Food Agric 24(5):386–399

Jain SM, Suprasanna P (2011) Induced mutations for enhancing nutrition for food production. Gene Conserv 40:201–215

Joshi S, Thoday-Kennedy E, Daetwyler HD, Hayden M, Spangenberg G, Kant S (2021) High-throughput phenotyping to dissect genotypic differences in safflower for drought tolerance. PLoS One 16(7):e0254908. https://doi.org/10.1371/journal.pone.0254908

Junior BJ, Latado RR, Neto TA (2008) Resistance of mutants of sweet orange induced by gamma-rays to citrus canker (*Xanthomonas citri* subsp citri) under artificial inoculation (IAEA-CN--167). International Atomic Energy Agency (IAEA), Vienna, pp 112–113

Kalaji HM, Jajoo A, Oukarroum A, Brestic M, Zivcak M, Izabela AS, Cetner MD, Łukasik I, Goltsev V, Ladle RJ (2016) Chlorophyll a fluorescence as a tool to monitor physiological status of plants under abiotic stress conditions. Acta Physiol Plant 38:102. https://doi.org/10.1007/s11738-016-2113-y

Kaur S, Padmanabhan SY, Kaur P (1975) Induction of resistance to blast disease (Pyricularia oryzae) in the high yielding variety, Ratna (IR 8 x TKM 6). In: Induced mutations against plant diseases, Proceedings of a symposium, Vienna, 31 Jan. 4 Feb. 1977. International Atomic Energy Agency (IAEA), pp 147–156

Kaur M, Thind KS, Sanghera GS, Kumar R, Kashyap L (2016) Gamma rays induced variability for economic traits, quality and red rot resistance in sugarcane (*Saccharum spp.*). Int J Sci Environ Technol 5(2):355–365

Khadka K, Earl HJ, Raizada MN, Navabi A (2020) A physio-morphological trait-based approach for breeding drought tolerant wheat. Front Plant Sci 11:715. https://doi.org/10.3389/fpls.2020.00715

Kharkwal MC, Jain HK, Sharma B (1986) Induced mutations for improvement of chickpea, lentil, pea and cowpea. In: Proc. FAO/IAEA workshop on improvement of grain legume production using induced mutations, 1–5 July, Pullman, Washington (USA), 1988. IAEA, Vienna, Austria, pp 89–109

Kharkwal MC, Pandey RN, Pawar SE (2004) Mutation breeding for crop improvement. In: Jain HK, Kharkwal MC (eds) Plant breeding – Mendelian to molecular approaches. Narosa Publishing House, New Delhi, pp 601–645

Kharkwal MC, Nagar JP, Kala YK (2005) 547 - A high yielding chickpea (*Cicer arietinum* L.) mutant variety for late sown conditions of North Western Plains Zone of India. Indian J Genet 65:229–230

Kharkwal MC, Gopalakrishna T, Pawar SE, Ahsanul HM (2008) Mutation breeding for improvement of food legumes. In: Kharkwal MC (ed) Food legumes for nutritional security and sustainable agriculture, vol. 1. Proc. Fourth International Food Legumes Research Conference (IFLRC-IV), Oct. 18–22, New Delhi, India. Indian Society of Genetics and Plant Breeding, New Delhi, India, pp 194–221

Kim SL, Kim N, Lee H, Lee E, Cheon KS, Kim M, Baek JH, Choi I, Ji H, Yoon IS, Jung KH, Kwon TR, Kim KH (2020) High-throughput phenotyping platform for analysing drought tolerance in rice. Planta 252:38. https://doi.org/10.1007/s00425-020-03436-9

Kiraly Z, Klement Z, Solymosy F, Vörös J (1974) Methods in plant pathology. AkadémiaiKiadó, Budapest, Hungary, p 509

Kunz BA, Dando PK, Grice DM, Mohr PG, Schenk PM, Cahill DM (2008) UV-induced DNA damage promotes resistance to the biotrophic pathogen Hyaloperonospora parasitica in Arabidopsis. Plant Physiol 148(2):1021–1031

Le TD, Chung TBP (2021) Soybean breeding through induced mutation in Vietnam. In: Sivasankar S, Ellis N, Jankuloski L, Ingelbrecht I (eds) Mutation breeding, genetic diversity and crop adaptation to climate change. CAB International, Wallingford, pp 40–46. https://doi.org/10.1079/9781789249095.000

Lebeda A (ed) (1986) Methods of testing vegetable crops for resistance to plant pathogens. Olomouc, Czech Republic, VHJ Sempra, VŠÚZ (Research and Breeding Institute for Vegetable Crops), p 286

Lebeda A, Švábová L (2005) *In vitro* selection for improved plant resistance to toxin-producing pathogens. J Phytopathol 153:52–64

Lebeda A, Švábová L (2010) In vitro screening methods for assessing plant disease resistance. In: Mass screening techniques for selecting crops resistant to diseases. IAEA, Austria, pp 201–235

Malaspina P, Giordani P, Faimali M, Garaventa F, Modenesi P (2014) Assessing photosynthetic biomarkers in lichen transplants exposed under different light regimes. Ecol Indic 43:126–131. https://doi.org/10.1016/j.ecolind.2014.02.034

Maxwell K, Johnson GN (2000) Chlorophyll fluorescence—a practical guide. J Exp Bot 51:659–668

Mba C, Rownak A, Jain SM, Gregorios GB, Zapata-Arias FJ (2007) Induced mutations for enhancing salinity tolerance. In: Jenks MA et al (eds) Rice, Advances in molecular breeding toward drought and salt tolerant crops, pp 413–454

Mbow CC, Rosenzweig LG, Barioni TG, Herrero BM, Krishnapillai M, Liwenga E, Pradhan P, Rivera-Ferre MG, Sapkota T, Tubiello FN, Xu Y (2019) Food security. In: Shukla PR, Skea J, CalvoBuendia E, Masson-Delmotte V, Pörtner H-O, Roberts DC, Zhai P, Slade R, Connors S, van Diemen R, Ferrat M, Haughey E, Luz S, Neogi S, Pathak M, Petzold J, Portugal Pereira J, Vyas P, Huntley E, Kissick K, Belkacemi M, Malley J (eds) Climate change and land: an IPCC special report on climate change, desertification, land degradation, sustainable land management, food security, and greenhouse gas fluxes in terrestrial ecosystems. IPCC, Geneva

Meena HP, Bainsla NK, Yadav DK (2016) Breeding for abiotic stress tolerance in crop plants, recent advances in plant stress physiology. Daya Publishing House, New Delhi

Meena KK, Sorty AM, Bitla UM, Choudhary K, Gupta P, Pareek A, Singh DP, Prabha R, Sahu PK, Gupta VK, Singh HB, Krishanani KK, Minhas PS (2017) Abiotic stress responses and microbe-mediated mitigation in plants: the omics strategies. Front Plant Sci 8:172. https://doi.org/10.3389/fpls.2017.00172

Mickelbart M, Hasegawa P, Bailey-Serres J (2015) Genetic mechanisms of abiotic stress tolerance that translate to crop yield stability. Nat Rev Genet 16:237–251. https://doi.org/10.1038/nrg3901

Oladosu Y, Rafii MY, Abdullah N, Hussin G, Ramli A, Rahim HA, Miah G, Usman M (2016) Principle and application of plant mutagenesis in crop improvement: a review. Biotechnol Biotechnol Equip 30(1):1–16

Oloriz MI, Gil V, Rojas L, Portal O, Izquierdo Y, Jiménez E, Höfte M (2012a) Sugarcane genes differentially expressed in response to *Puccinia melanocephala* infection: identification and transcript profiling. Plant Cell Rep 31(5):955–969. https://doi.org/10.1007/s00299-011-1216-6

Oloriz MI, Gil V, Rojas L, Veitía N, Höfte M, Jiménez E (2012b) Selection and characterisation of sugarcane mutants with improved resistance to brown rust obtained by induced mutation. Crop Pasture Sci 62(12):1037–1044

Pande S, Siddique KHM, Kishore GK, Bayaa B, Gaur PM, Gowda CLL, Bretag TW, Crouch JH (2005) Ascochyta blight of chickpea (Cicer arietinum L.): a review of biology, pathogenicity, and disease management. Aust J Agric Res 56:317–332

Pande S, Rao JN, Sharma M, Pathak M, Stevenson P (2010) Mass-screening techniques for the early selection of disease resistance in chickpea (*Cicer arietinum*). In: Mass screening techniques for selecting crops resistant to diseases. IAEA, Austria, pp 201–235

Pandey P, Irulappan V, Bagavathiannan MV, Senthil-Kumar M (2017) Impact of combined abiotic and biotic stresses on plant growth and avenues for crop improvement by exploiting physio-morphological traits. Front Plant Sci 188:537. https://doi.org/10.3389/fpls.2017.00537

Pando LG, Eguiluz A, Jimenez J, Falconi J (2008) Barley (*Hordeun vulgare*) and kiwicha (*Amaranthus caudatus*) improvement by mutation induction in Peru. Abstract. In: Book of abstracts, FAO/IAEA international symposium on induced mutations in plants, 12–15 Aug, Vienna, Austria, p 141

Pathirana R (2011) Plant mutation breeding in agriculture. CAB Rev Perspect Agric Vet Sci Nutr Nat Resour 6:1–20. https://doi.org/10.1079/pavsnnr20116032

Peiris R, Wickramasinghe TK, Indrasena SP (2009) M 127 – a promising tomato variety developed through induced mutation technique. In: Induced plant mutations in the genomics era. Proceedings of an international joint FAO/IAEA symposium, 2008. International Atomic Energy Agency, Vienna, Austria, pp 379–380

Perez-Bueno ML, Pineda M, Barón M (2019) Phenotyping plant responses to biotic stress by chlorophyll fluorescence imaging. Front Plant Sci 10:1135. https://doi.org/10.3389/fpls.2019.01135

Phadvibulya V, Boonsirichai K, Adthalungrong A, Srithongchai W (2009) Selection for resistance to yellow vein mosaic virus disease of okra by induced mutation. In: Shu QY (ed) Induced plant mutations in the genomics era. Food and Agriculture Organization of the United Nations, Rome, pp 349–351

Raza A, Razzaq A, Mehmood SS, Zou X, Zhang X, Lv Y, Xu J (2019) Impact of climate change on crops adaptation and strategies to tackle its outcome: a review. Plants (Basel) 8(2):34. https://doi.org/10.3390/plants8020034

Reddy KS (2009) A new mutant for yellow mosaic virus resistance in mungbean variety Sml-668 by recurrent gamma rays irradiation. In: Shu QY (ed) Induced plant mutations in the genomics era. Food and Agriculture Organization of the United Nations, Rome, pp 361–336

Reddy KS, Pawar SE, Bhatia CR (1987) Screening for Powdery mildew (*Erysiphe polygoni* DC) resistance in mung bean (*Vigna radiata*. L) using excised leaves. Indian Acad Sci (Plant Sci) 97: 365–369

Russell GE (1978) Plant breeding for pest and disease resistance. Butterworths, London, UK, p 485

Sahoo BB, Kole PC, Sahoo MR (2015) Effects of γ irradiation on leaf blight disease of some taro (*Colocasia esculenta* L. Schott) genotypes. Int J Bioresour Stress Manage 6(1):7–14

Saito A, Nakazawa N, Suzuki M (2001) Selection of mutants resistant to *Alternaria* blotch from *in vitro*-cultured apple shoots irradiated with X- and γ-rays. J Plant Physiol 158(3):391–400

Saraswathi M, Kannan G, Uma S, Thangavelu R, Backiyarani S (2016) Improvement of banana cv. Rasthali (Silk, AAB) against *Fusarium oxysporum* f. sp. cubense (VCG 0124/5) through induced mutagenesis: determination of LD50 specific to mutagen, explants, toxins and in vitro and in vivo screening for Fusarium wilt resistance. Indian J Exp Biol 54:345–353

Sarsu F, Ghanim AMA, Das P, Bahuguna RN, Kusolwa PM, Ashraf M, Singla-Pareek SL, Pareek A, Forster BP, Ingelbrecht I (2018) Pre-field screening protocols for heat-tolerant mutants in rice. Springer, Cham. ISBN 978-3-319-77338-4, https://www.springer.com/gp/book/9783319773377

Sarsu F, Koffi IB, Jankuloski L (2021) Contribution of induced mutation in crops to global food security. ACI Avances En Ciencias E Ingenierías 12(3):10. https://doi.org/10.18272/nv12i3.2031

Shahzad A, Ullah S, Dar AA, Sardar MF, Mehmood T, Tufail MA, Shakoor A, Haris M (2021) Nexus on climate change: agriculture and possible solution to cope future climate change stresses. Environ Sci Pollut Res Int 28(12):14211–14232

Shu QY, Forster BP, Nakagawa H (2012) Plant mutation breeding and biotechnology. CABI, Wallingford

Si P, Buirchell B, Sweetingham M (2008) Induced mutation in narrow-leafed lupin improvement: an example of herbicide tolerance. Abstract. In: Book of abstracts, FAO/IAEA international symposium on induced mutations in plants, 12–15 Aug, Vienna, Austria, p 14

Singh DP, Singh A (2005) Disease and insect resistance in plants. Science Publishers Inc., Enfield, NH, p 417

Singh BB, Mai-Kodomi Y, Terao T (1999) A simple screening method for drought tolerance in cowpea. Indian J Genet Plant Breed 59:211–220

Sivasankar S, Heng LH, Kang S (2020) Agriculture: improving crop production. https://doi.org/10.1016/B978-0-12-409548-9.12323-1

Soeranto SH, Indriatama WM (2021) Mutation breeding of sorghum to support. In: Sivasankar S, Ellis N, Jankuloski L, Ingelbrecht I (eds) Mutation breeding, genetic diversity and crop adaptation to climate change. CAB International, Wallingford, pp 120–126. https://doi.org/10.1079/9781789249095.000

Spencer-Lopes MM, Forster BP, Jankuloski L (2018) Manual on mutation breeding. ISBN: 978-92-5-130526-3. http://www.fao.org

Suprasanna P, Mirajkar SJ, Patade VY, Jain SM, Tomlekova NB, Kozgar MI, Wani MR (2014) Induced mutagenesis for improving plant abiotic stress tolerance. Wageningen Acad Publ, Wageningen

Suprasanna P, Mirajkar SJ, Bhagwat SG (2015) Induced mutations and crop improvement. Plant diversity, organization, function and improvement, Plant biology and biotechnology. Springer, India, pp 593–617

Svabova L, Lebeda A (2005) In vitro selection for improved plant resistance to toxin-producing pathogens. J Phytopathol 153:52–64

TNAU (2013) Crop improvement: screening methods for specific traits - A. Screening techniques for pest and disease resistance. http://www.agritech.tnau.ac.in/crop_improvement/crop_imprv_screen.html

Trigiano RN, Windham MT, Windham AS (2004) Plant pathology. Concepts and laboratory exercises. CRC Press, Boca Raton, FL, p 413

Vinh MQ, Thinh DK, Bang DT, At DH, Ham LH (2009) Current status and research directions of induced mutation application to seed crops improvement in Vietnam, 2009. In: Shu QY (ed) Symposium proceeding. Induced Plant Mutations in the Genomic Era. ISBN-978-82-5-106324-8, I0956e.pdf (fao.org)

Yamagishi H, Ashizawa M, Yui S (1985) Screening techniques of disease resistance in cruciferae vegetables. Jpn Agr Res Q 19:185–189

Yang W, Feng H, Zhang X, Zhang J, Doonan JH, Batchelor WD, Xiong L, Yan J (2020) Crop phenomics and high-throughput phenotyping: past decades, current challenges, and future perspectives. Mol Plant 13:187–214

Zambrano AY, Demey JR, Fuchs M, González V, Rea R, De Sousa O, Gutierrez Z (2003) Selection of sugarcane plants resistant to SCMV. Plant Sci 165(1):221–225

Zhang MX, Xu JL, Luo RT, Shi D, Li ZK (2003) Genetic analysis and breeding use of blast resistance in a japonica rice mutant R917. Euphytica 130:71–76

Zhang C, Mansfeld BN, Lin Y-C, Grumet R (2021) Quantitative high-throughput, real-time bioassay for plant pathogen growth in vivo. Front Plant Sci 12:637190. https://doi.org/10.3389/fpls.2021.637190

Induced Mutagenesis for Developing Climate Resilience in Plants

7

Mithlesh Kumar, Kirti Rani, and Suprasanna Penna

Abstract

Climate change has an impact on global agricultural productivity. Increased food demand and significant reduction in crop yields are projected to worsen as a result of global warming. This scenario has prompted plant scientists to produce climate change resilient crops that can endure a wide range of conditions, including drought, heat, cold, salt, flood, submergence, diseases, insect and pests, and thus boost yield. Understanding mutations and their exploitation in plant improvement has aided in the discovery of the genetic, physiological, and biochemical basis of mutation induction and mutant characterization. As a result, creating genetic variability through mutations has become one of the most essential techniques for crop improvement. Random and targeted mutations have yielded successful genetic changes. Techniques have also been developed to identify mutations and the molecular control of DNA damage repair systems in the plant genome. Consequently, the genetic diversity of accessible germplasm is being expanded with novel alleles influencing essential agronomic traits being discovered and new crop cultivars with greater resistance to biotic and abiotic stresses and superior nutritional quality are developed. In light of this, the present chapter elaborates on climate resilience and approaches of induced mutagenesis

M. Kumar (✉)
AICRN on Potential crops, ARS Mandor Jodhpur Agriculture University, Jodhpur, Rajasthan, India

K. Rani
Crop Improvement Unit, ICAR-Directorate of Groundnut Research, Junagadh, Gujarat, India

S. Penna
Nuclear Agriculture and Biotechnology Division, Bhabha Atomic Research Centre, Mumbai, India

Amity Centre for Nuclear Biotechnology & Amity Institute of Biotechnology, Amity University, Mumbai, Maharashtra, India

S. Penna, S. M. Jain (eds.), *Mutation Breeding for Sustainable Food Production and Climate Resilience*, https://doi.org/10.1007/978-981-16-9720-3_7

for the development of climate resilience in crop varieties, which are expected to play an increasingly important role in global food security.

Keywords

Classical mutagenesis · Climate change · Stress tolerance · Reverse genetics · Random mutations · Targeted mutations · TILLING · Insertional mutagenesis · Gene editing techniques

7.1 Introduction

Climate change is currently posing a severe threat to global agriculture, which is expected to result in lower agricultural output. Reduced productivity results in rising food costs and increased global food insecurity (FAO 2014), and this situation, if it continues, will result in even higher food prices, and famine in some cases. Climate resilience is considered as an essential core of sustainable agriculture. Resilience is generally framed in such way so as to combine both aspects of anticipation and preparedness to develop adaptiveness to thrive under the impacts of climate changes. Within a given ecosystem, resilience can be built up by taking into vulnerabilities of exposure and responses of the organisms, and this could involve strategies of climate mitigation and adaptation (Fig. 7.1). There are several strategies for adopting climate resilient measures which take into plant and environmental conditions.

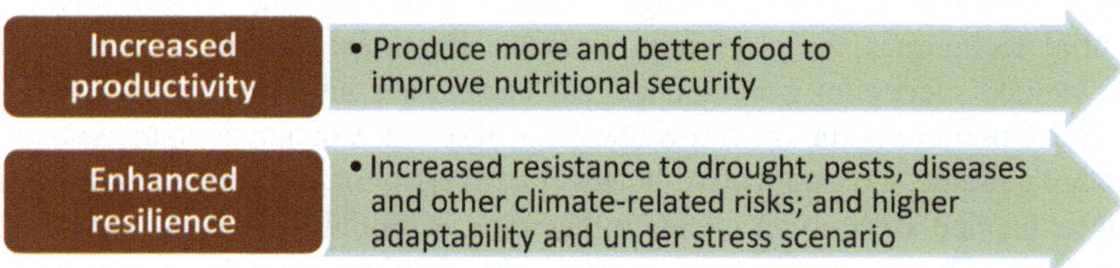

Traits	Options			
Heat Tolerance	Novel genetic variability to develop new crop varieties	Heat tolerant crops or crop varieties	Short duration varieties	
Drought		Drought tolerant crops	Crops with low water requirements	Increase water-use efficiency
Salinity		Use of saline resistant varieties	halophyte crops	
Pests/Diseases		Resistant crop varieties	Association between Beneficial fungi and bacteria	

Fig. 7.1 Climate resilient breeding: strategy and framework

Global warming, changes in rainfall patterns, and other extreme weather events could all play a role in food insecurity, with the changing climate trend leading to increased pathogen and insect attacks. Furthermore, increased CO_2 levels are expected to lower the nutritional content of many crops, while some crops may become toxic as a result of changes in their tissue's chemical makeup (Dwivedi et al. 2013). As a result, increasing crop resilience to climate change is a critical component of ensuring food and nutritional security, which could be accomplished through creation of novel genetic variation. New concerns, such as climate change and human population expansion, are now posing a significant danger to global food production. Local temperatures and terrestrial ecosystems will shift, posing a hazard to human livelihoods in many circumstances. Climate change will have a greater negative impact on food production, environmental services, and rural livelihoods in developing countries. These are confronted with pressing need to increase food security, eliminate poverty, and provide a sufficient standard of living for ever-increasing populations. Furthermore, substantial percentages of the population in poor nations, as high as 80%, rely on agriculture for their livelihoods, making them more vulnerable to climate change. Furthermore, there is no longer any potential for expanding arable land, which is gradually being lost to other human development activities. Rapid industrialization is taking a severe toll on the environment, including ozone layer depletion, acid rain, irregular weather patterns, and global warming. The unpredictable rainfall pattern could result in a water deficit or an increase in flooding, both of which would have a negative impact on food production and raise food prices.

7.2 Climate Change and Associated Plant Functional Traits

Crop plants have evolved to adapt to environmental stresses, and there are several plant responses and traits ranging from physiological, morphological, cellular, biochemical to molecular alterations (Raza et al. 2019; Suprasanna et al. 2021). Diverse changes involving physiological features related to plant growth, development include morphological attributes, regulation of stomata, photosynthesis, transpiration, water balance, nutrient homeostasis, etc. Many of these "functional traits" are now considered as indicators of climate change-associated plant responses (Zhang and Li 2019). In this context, it is often viewed that understanding of these traits in relation to the environmental setting is essential to evolve strategies for developing crops with better resilience to climate change-associated stresses (Kole et al. 2015; Milanović et al. 2020). Plant functional traits act in response to environmental stimuli not only in terms of their individual response, but the traits can signify the responsive plant population. One of the traits, the leaf related morphological, physiological response profile has been used since the tissue exhibits plasticity as well as differential responses in diverse environmental stresses (Ahrens et al. 2020). In recent years, there has been a great emphasis on exploring different functional traits related to whole plant, canopy, leaf, root, and other tissues. Technological breakthroughs in phenomics and image analysis have enabled to predict the

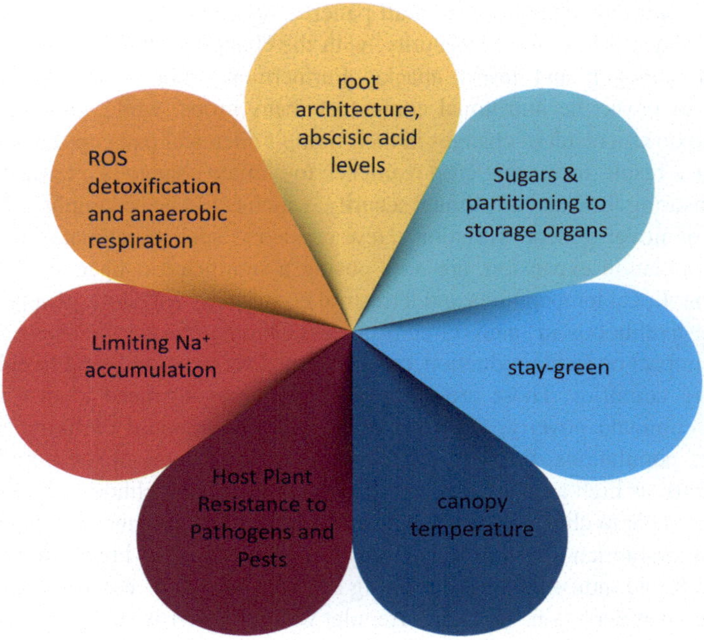

Fig. 7.2 Climate change-associated plant functional traits

impact of global climate change by simulation studies under controlled conditions (Tardieu et al. 2017; Bohra et al. 2021). With the advances in metabolomics, interest has begun into investigating plant metabolism-related responses as indicators of stress responses (Bueno and Lopes 2020).

Plant scientists use different strategies for plant improvement, such as conventional breeding, hybridization, and recombination followed by selection, molecular breeding, and genetic engineering to develop genetically modified and genetically edited crops to ensure that food supply is sustained to feed the increasing population (Jain 2010; Suprasanna et al. 2015; Qaim 2020). Novel plant breeding technologies have also been developed to understand and design crop plants having better adaptability to climate change-driven stresses (Suprasanna et al. 2021). In this regard, genetic variants or induced mutants for diverse phenological and physiological traits can be explored to understand and develop crops with better stress adaptation (Snowdon et al. 2021). Figure 7.2 has illustrated the different climate change-associated plant functional traits including host plant resistance to pathogens and pests, drought-adaptive traits, root architecture, stay-green traits, canopy temperature, and limiting Na^+ accumulation. Methodologies to screen for these traits using different cellular, analytical and molecular tools will have to be extensively adopted in plant populations under stress scenario. Additionally, the harmful impacts of climate change necessitate appropriate strategies for ensuring food security. Since we often depend on only 12 crops for all of our 75% of plant-based food requirements, there is a need to explore new, novel germplasm among the existing plant species and exploit other 10,000 edible plants through precision domestication (Fernie and Yan 2019). In this regard, innovative, technological advancements will have to emerge, for example, random mutagenesis or precise targeted mutagenesis that will become useful for creating useful genetic variation (Jung and Till 2021). In this chapter, we have presented an account of classical and advanced induced mutagenesis tools and their potential applications for the development of climate resilient plant varieties.

7.3 Classical Mutagenesis and Plant Breeding

Plant breeding is a field that focuses on the production of new plant types in a systematic and continued manner. It takes advantage of genetic variation among individuals within a plant species and combines desired traits to create new and improved varieties. Plant breeding relies on genetic variation, and new variation is critical for introducing novel features into breeding programs. When a certain genetic feature is not immediately available for crossing into breeding materials, the genetic variation in a crop species can be increased by alternative techniques. Moreover, crop breeders' thorough screening in the crop domestication process and subsequent breeding of elite cultivars has resulted in a significant loss of natural genetic variation. As a result, further genetic erosion has become a bottleneck to improve crops (Sikora et al. 2011). As a result, induced mutations to extend the genetic base of crops has become useful for creating novel genetic variation for use in crop improvement programs (Mousavi-Derazmahalleh et al. 2019). The goal of induced mutations is to increase the frequency of mutations in order to select appropriate variants for plant breeding. The mutation frequency rate of spontaneous mutations is rather low and difficult for plant breeders to exploit. Physical (e.g., X-rays, gamma rays, fast neutrons) and chemical (e.g., ethyl methanesulfonate, methyl methanesulfonate, sodium azide) mutagen treatments of both seed and vegetatively propagated plants cause novel genetic variations. The mutagen treatment breaks the nuclear DNA, and new mutations are induced randomly and heritably during the DNA repair mechanism (Jain 2010). The mutants with desirable phenotypes are then searched in the populations generated through physical and chemical mutagenesis, and the genetic basis for those characteristics is understood through mutant characterization (Li et al. 2020). Physical mutagenesis can increase the natural mutation rate by 10^6-fold and routinely employed to generate heritable genetic variability in the production of novel agricultural cultivars for the past century, producing billions of dollars in the process (Mousavi-Derazmahalleh et al. 2019). Chemical mutagens often affect single nucleotide pairs, with the extent of mutation depending on the tissue targeted, time, and degree of exposure. Physical mutagens such as gamma rays and neutrons often cause large-scale DNA deletion and chromosomal structure alterations (Jankowicz-Cieslak et al. 2017). The goal of mutagenesis breeding is to cause maximal genetic variation with maximum viability (Sikora et al. 2011). Because large-scale modifications in chromosome architecture frequently have substantial negative repercussions, crop breeders are particularly interested in mutations at single nucleotide pairs (Singh et al. 2014).

In crop plants, for example, several mutant cultivars in cereals, legumes (pulses), oilseeds, forage/fodder, vegetables, fruits, fiber crops, medicinal plants with better productivity, earliness, abiotic and biotic stress tolerance have been produced using mutagenesis techniques (FAO/IAEA-MVD 2021; https://mvd.iaea.org/) (Table 7.1; Fig. 7.3). Mutant germplasm will be of significant interest to be viewed as a genetic resource for understanding climate change-associated plant functional traits and also exploiting such available mutant germplasm for cultivation in the environmentally challenging conditions.

7.4 TILLING (Targeting Induced Local Lesions in Genomes)

The mutational changes enable plant breeders to select for useful mutants for the development of climate resilient crops. When one or a few characteristics of an outstanding cultivar are to be modified, a specific advantage of mutation induction is also the possibility of exploring unselected genetic variation for associated traits. Despite the benefit of useful mutations, the mutagenesis procedures cause hundreds, if not thousands, of random mutations, all of which have unknown effects. Meanwhile, developments in genome sequencing, bioinformatics, and HTP (high throughput phenotyping) technologies, as well as recent advances in reverse genetics and gene discovery tools, have sparked renewed interest in the use of chemically induced mutations for crop enhancement (Fruzangohar et al. 2019). Chemically induced-mutagenesis-based crop breeding has been transformed by the introduction of TILLING (targeting induced local lesions in genomes) technology. TILLING is a reverse genetics technology that identifies a series of single base pair allelic changes within a gene of interest by combining standard chemical mutagenesis with a high-throughput mismatch detection technique (Chen et al. 2014). This method produces a large range of mutant alleles, is quick and easy to automate, and may be used on any organism that can be mutated chemically (McCallum et al. 2004). TILLING has been successfully used for the improvement of abiotic stress tolerance like salt tolerance for genes *OsAKT1*, *OsHKT6*, *OsNSCC2*, *OsHAK11*, and *OsSOS1* in rice (Hwang et al. 2016); drought tolerance for genes *PLD* in peanut (Guo et al. 2015); *Lis1* and *NAC7* in sorghum (Nida et al. 2016); and low leaf temperature phenotype gene *slac1* in rice (Kusumi et al. 2012). The biotic stresses like powdery mildew resistance for gene *TaMlo* in wheat; fungal immunity for gene *HvHox1* in barley (Gottwald et al. 2009); resistance to northern corn leaf blight for the genes *ZmWAK-RLK1* and *ZmWAK-RLK2* in maize (Severune et al. 2015); and disease resistance genes *EDR1* and *NPR1* in barley (Hu 2012). Although TILLING populations have

Table 7.1 Mutant varieties for stress tolerance released through classical mutagenesis (from 1990 to till date; https://mvd.iaea.org/, accessed on 17th August, 2021)

Species name	Mutant name	Mutagen	Stress/trait	Reg. Year	Country
Cereals					
Oryza sativa L.	A-20	0.015% MNH treatment	Early maturity and high tolerance to salinity	1990	Viet Nam
Oryza sativa L.	Zhefu 802	Gamma irradiation	Higher rice blast resistance, higher yield, early maturity, good grain quality	1990	China
Oryza sativa L.	Norin PL 12	Gamma irradiation (200 Gy)	Thermosensitive genetic male sterility which express male sterility under high temperature above 30°	1991	Japan
Oryza sativa L.	Shua 92	Gamma irradiation	Resistance to salinity and high yield	1993	Pakistan
Oryza sativa L.	DT17	–	High yield and tolerance to salinity	1999	Viet Nam
Oryza sativa L.	VND 95-20	Gamma irradiation (200 Gy)	Salinity tolerance	1999	Viet Nam
Oryza sativa L.	VND99-3	Gamma irradiation (200 Gy)	Salinity tolerance	2004	Viet Nam
Oryza sativa L.	GINES	Proton irradiation (20 Gy)	Tolerance to salinity and good quality	2007	Cuba
Oryza sativa L.	Wonhaebyeo	Gamma irradiation of callus (50 Gy)	Tolerance to salinity	2007	Korea, Republic of
Oryza sativa L.	Rc272 (Sahod Ulan 2)	Gamma irradiation (250 Gy)	Drought tolerance	2013	Philippines
Oryza sativa L.	Rc346 (Sahod Ulan 11)	Gamma irradiation (250 Gy)	Drought tolerance	2013	Philippines
Oryza sativa L.	Sinar 2	Gamma irradiation	High yield, higher aromatic value, and higher disease resistance to BLB diseases	2020	Indonesia
Oryza sativa L.	Lampai Sirandah	–	High yield, early maturity, semi dwarf, moderate resistant to BLB diseases, lodging resistant	2020	Indonesia
Triticum aestivum L.	H6765	Gamma irradiation of pollen grain (1.5 Gy)	High yield (17.3% higher), tolerance to salinity and tolerance to drought	2004	China
Triticum aestivum L.	Longfumai 17	–	Good grain quality, drought tolerance, and high yield	2007	China

(continued)

Table 7.1 (continued)

Species name	Mutant name	Mutagen	Stress/trait	Reg. Year	Country
Triticum aestivum L.	Longfumai 19	–	Drought tolerance, high yield, and good adaptability	2010	China
Triticum aestivum L.	Hangmai901	–	Increased yield and drought tolerance	2011	China
Triticum aestivum L.	Binagom-1	Direct use of an induced mutant	Has higher salinity tolerance, higher yield	2016	Bangladesh
Triticum aestivum L.	Darkhan-172	Chemical mutagenesis using sodium azide	Higher yield, early maturity	2018	Mongolia
Hordeum vulgare L.	Phenix	Hybridization with mutant Kharkivsky 99	Improved drought tolerance	2000	Ukraine
Hordeum vulgare L.	Centenario	Gamma irradiation (333 Gy)	Altered maturity, seed production traits	2006	Peru
Zea mays L.	DT-6	Gamma irradiation (200 Gy) and NMU (0.08%)	Early maturity and shortness	1990	Viet Nam
Zea mays L.	DT-8	Gamma irradiation (200 Gy) and NMU (0.08%)	Early maturity and resistance to lodging	1990	Viet Nam
Zea mays L.	P26	Treatment with fast neutrons (7.5 Gy)	Agronomic and botanic traits (combining ability)	2001	Hungary
Zea mays L.	Longfuyu 3	Direct use of an induced mutant	Improved resistance to bacterial diseases	2007	China
Zea mays L.	Kneja 627	Hybridization with mutant (from the cross PCM4658 × Mo17)	Improved grain (seed) yield, late maturity	2009	Bulgaria
Panicum miliaceum L.	Cheget	Hybridization with two chemo mutants	Improved drought tolerance, improved smut resistance	1993	Russia
Sorghum bicolor L.	Fambe	Gamma irradiation (300 Gy)	Resistance to lodging and high grain yield (increased number of grains per panicle)	1998	Mali

Sorghum bicolor L.	Sofin	Gamma irradiation (250 Gy)	Early maturity, dwarfness, and resistance to lodging	1998	Mali
Sorghum bicolor L.	Gnome	Gamma irradiation (300 Gy)	Resistance to lodging and high grain yield	1998	Mali
Sorghum bicolor L.	Sadje	Gamma irradiation (300 Gy)	Early maturity and shortness	1998	Mali
Sorghum bicolor L.	Fambe	Direct use of an induced mutant, gamma irradiation (300 Gy)	Resistance to lodging, high grain yield (increased number of grains per panicle)	1998	Mali
Sorghum bicolor L.	PAHAT	Direct use of an induced mutant, gamma irradiation	High yielding, semi dwarfness, early maturity, grain quality (protein, tannin, starch)	2013	Indonesia
Sorghum bicolor L.	Samurai 1	Direct use of an induced mutant, gamma irradiation (0.3 Other)	High yield, improved food processing quality, improved biomass, lodging resistance, resistance to midrib rot disease, large seed size	2014	Indonesia
Setaria sp.	Lugu 2	Gamma irradiation (400 Gy)	Resistance to bacterial diseases and high yield	1991	China
Setaria sp.	Longgu 29	Fast neutrons	Resistance to bacterial diseases and high yield	1992	China
Setaria sp.	Nunxuan 14	Gamma irradiation (250 Gy)	Drought resistance	1992	China
Setaria sp.	Yugu 6	–	Drought tolerance, high yield, and resistance to lodging	1995	China
Setaria sp.	Fugu 6	Gamma irradiation (250 Gy)	Resistance to drought, high quality, high yield, and wide adaptability	1999	China
Avena sativa L.	Horicon	Thermal neutron	Resistance to crown rust	1990	USA
Avena sativa L.	Ozark	Thermal neutron	Resistance to low temperature and high yield potential	1991	USA
Fagopyrum esculentum Gili	Skorospelaya 86	0.05% of EI treatment	Early maturity and good cooking quality	1990	Russia
Sorghum sudanense (Piper) Stapf	Mironovskaya 8	DMS treatment of seeds	Early maturity, drought tolerance, resistance to lodging and resistance to diseases	1990	Russia

(continued)

Table 7.1 (continued)

Species name	Mutant name	Mutagen	Stress/trait	Reg. Year	Country
Legumes (pulses)					
Vigna mungo L.	Binamash-1	Gamma rays (600 Gy)	Resistance to diseases (yellow mosaic virus and cercospora leaf spot), high yield and early synchronous maturity (80–85 days)	1994	Bangladesh
Vigna radiata (L.) Wil.	Camar	Gamma rays (100 Gy)	Resistance to cerospora leaf spot, resistance to *Uromyces* sp., medium resistance to scrab diseases, high yield and tolerance to salinity and acid soil	1991	Indonesia
Vigna radiata (L.) Wil.	NIAB Mung 92	–	Resistance to diseases (MyMV), early maturity, resistance to grain shattering and large seed size	1992	Pakistan
Vigna radiata (L.) Wil.	Binamoog-2	Developed by hybridization with gamma ray-induced mutant MB-55(4)	Larger seed size, early and synchronous maturity (7–10 days earlier), high yield (16%), tolerant to leaf YMV and cercospora leaf spot	1994	Bangladesh
Vicia faba L.	Tuwaitha	Gamma rays (40 Gy)	Resistance to wilt, high yield, and protein content	1994	Iraq
Vicia faba L.	Babylon	Gamma rays (30 Gy)	Resistance to diseases, high yield, and high protein content	1994	Iraq
Cicer arietinum L.	NIFA-88 (CM-1918)	Gamma rays (100 Gy)	Moderate resistance to Ascochyta rabiei, 2 weeks earlier maturity, high yield (15–20%), higher nitrogen amount fixation	1990	Pakistan
Phaseolus vulgaris L.	CIAT 899	Gamma rays (100 Gy)	Tolerance to salinity and tolerance to P deficit	2007	Tunisia
Lathyrus sativus L.	Bogdan	Gamma rays (250 Gy)	Good resistance to drought and high temperatures, high yield and high protein content	2005	Moldova, Republic of

Oilseeds					
Glycine max	Albisoara	–	Drought tolerant, high protein content, and high yield	2010	Maldova, Republic of
Glycine max	DT 96	Gamma irradiation (180 Gy)	Drought tolerant	2004	Viet Nam
Glycine max	Clavera	–	Increased yield and drought tolerance	2010	Maldova, Republic of
Arachis hypogaea	Huayu 22	–	Increased drought and waterlogging conditions	2003	China
Arachis hypogaea	Huayu 32	Gamma rays	High yield, good quality, medium drought, and waterlogging tolerance	2009	China
Arachis hypogaea	Binachinabadam-5	Gamma rays	Salinity tolerance	2011	Bangladesh
Arachis hypogaea	Binachinabadam-6	Gamma rays	Salinity tolerance	2011	Bangladesh
Arachis hypogaea	Binachinabadam-9	Gamma irradiation (250 Gy)	Salinity tolerance	2014	Bangladesh
Forage/fodder					
Beta vulgaris L.	Hybrid Umanskiipolusakharnyi	Chemical mutagens	White rhizocarp, resistance to mildew and resistance to insects	1990	Russia
Beta vulgaris L.	Tymiryazevskaya 87	Gamma rays and chemical mutagens	High productivity and resistance to diseases	1992	Russia
Morus alba L.	Fuzaofeng	Gamma irradiation (5 Gy)	Improved leaf yield, drought tolerance, resistance to mulberry blight	1992	China
Vegetables					
Brassica oleracea L. var. *acephala*	Veha	Chemical mutagen	Resistance to peronosporosis, bacteriosis, and insects	1990	Russia
Amaranthus sp. L.	Sterk	Chemical mutagen	Drought resistance and medium resistance to low temperature	1992	Russia

(continued)

Table 7.1 (continued)

Species name	Mutant name	Mutagen	Stress/trait	Reg. Year	Country
Capsicum annuum L.	Gornooriahovskakapia	Developed by hybridization with mutant Zlaten medal ms-8 obtained by irradiation with gamma rays (135 Gy)	Early maturity and high yield	1997	Bulgaria
Lycopersicon esculentum M.	MAGINE	Gamma irradiation (300 Gy)	Tolerance to drought and good quality	2007	Cuba
Lycopersicon esculentum M.	Maybel	–	Tolerance to drought	2007	Cuba
Fruits					
Musa sp.	Novaria	Gamma irradiation (60 Gy)	Early maturity and fruit quality	1995	Malaysia
Prunus avium L.	Ferrovia spur	X-ray (4 Gy)	Compact tree habit, early maturity, and high yield	1992	Italy
Rubus idaeus L.	Colocolchnik	Seeds with water solution of 0.025% NEU	Resistance to diseases and to low temperature	1991	Russia
Fiber crops					
Gossypium sp.	Emian 15	Gamma irradiation (300 Gy)	High yield and resistance to bacterial diseases	1991	China
Gossypium sp.	NIAB Karishma	Hybridization with gamma irradiated (300 Gy) variety NIAB-86	Tolerance to high temperature and tolerance to salt	1996	Pakistan
Gossypium sp.	NIAB 999	Hybridization with gamma irradiation of F1 (300 Gy)	Early maturity, resistance to high temperature and high yield	2003	Pakistan
Gossypium sp.	NIAB 111	Hybridization with gamma irradiation of F1 (300 Gy)	Early maturity, resistance to high temperature and high yield	2004	Pakistan
Gossypium sp.	NIAB 846	Hybridization with gamma irradiated (300 Gy) mutant variety NIAB-78	Fiber length, boll size, high yield, resistance to viruses, resistance to high temperatures	2008	Pakistan

Gossypium sp.	NIAB 777	Irradiation of pollen	High yield, good quality, tolerance to high temperatures and resistance to virus diseases	2009	Pakistan
Corchorus capsularis L.	Binadeshipat-2	Sodium azide (NaN_3) with treatment dose 12 mM	High fiber yield, early sowing potential, and white color of fiber	1997	Bangladesh
Corchorus capsularis L.	Xianghuangma 3	Gamma irradiation	Improved plant structure and resistance to virus diseases	1997	China
Linum usitatissimum L.	M-5	DMS (0.05%)	Resistance to lodging, resistance to fusariosis, immune to rust, anthracnose and bacteriosis	1991	Russia
Linum usitatissimum L.	Baltyuchai	NEU (0.012%)	Resistance to lodging and resistance to fusariosis	1991	Russia
Linum usitatissimum L.	DLV 20	EMS treatment	Early maturity, compact growth habit, and higher seed yield	2019	India
Medicinal plants					
Mentha arvensis L.	TN-8	Gamma rays	Improved menthol content (78–82%), resistance to diseases and resistance to pests	1995	Viet Nam

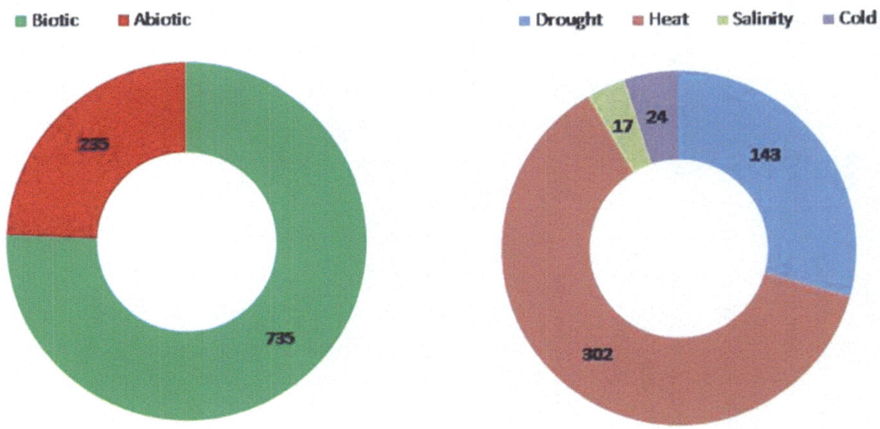

Fig. 7.3 Proportion of mutant varieties developed for biotic and abiotic stress tolerance

traditionally been used for reverse genetic approaches, recent advances in whole-genome sequencing have opened up new avenues for the identification of mutants in candidate genes, and the availability of sequence data from the entire mutant population that will allow TILLING populations to be used for forward genetic methodologies such as starting from a known genotype (Bettgenhaeuser and Krattinger 2019). The efficacy of TILLING technology has been demonstrated, for example, in the cloning of the wheat gene *Stb6* that gives pathogen resistance to the fungal disease *Septoria tritici* blotch (Saintenac et al. 2018). TILLING technique presents new opportunities for gene cloning, disease resistance and abiotic stress genes in plants in the face of a changing global climate (Bettgenhaeuser and Krattinger 2019).

7.5 Insertional Mutagenesis

Insertional mutagenesis is based on the random insertion of T-DNA (transfer DNA), transposon, and retrotransposon into the genome, resulting in a wide spectrum of mutations (Sallaud et al. 2004). The entire genome sequencing of most crop plants makes insertion detection easier, as does the efficacy of *Agrobacterium* infection methods, which enhance T-DNA insertion in the plant genome (Wang et al. 2013). T-DNA insertions, which can induce loss of function and a direct relation to their biological function, are without a doubt one of the most commonly utilized techniques for gene function identification (Lo et al. 2016). The insertional activity of transposons has been routinely used to generate large-scale alterations in plant genomes. DNA repair can cause up to ten times more mutations when transposons excision and reinsertion sequences, owing to the fact that transposons preferentially insert in sites proximal to genes, resulting in a higher frequency of mutations in genic regions. Because it preferentially inserts into genic rather than intergenic areas, retrotransposon insertion can have an impact on gene function (Miyao et al. 2003). As a result of the retrotransposon *Tos17* insertion in the upstream region of *OsHd1* (Heading date 1), rice blooming time was altered (Hori et al. 2016). Transposable elements played a significant role in crop evolution and continue to do so. The role of insertional mutagenesis in tolerance against the biotic and abiotic stresses providing the functional identification of crop genes involved in many important traits have been described (Table 7.2).

7.6 Targeted Mutagenesis Using Gene Editing Tools

Random mutations are caused by physical and chemical mutagens, which provide a limited mutation frequency in desired/target loci. In contrast, the genome editing systems, such as meganucleases (MN), zinc finger nucleases (ZFN), transcription activator-like effector nucleases (TALENS), and clustered regularly interspaced short palindromic repeat (CRISPR) cause target mutations which can be utilized as an alternative systems for developing biotic and abiotic stress tolerance in crop plants. There are several examples of application of CRISPR and TALEN based genome editing technology for the improvement of specific traits of relevance to climate resilience (Table 7.3). In wheat, TALEN-induced editing of three TaMLO homoeologs which are responsible for heritable broad-spectrum resistance to powdery mildew, resulted in plants having mutations in the TaMLO-A1 allele for improving powdery mildew disease resistance in wheat and the technology could pave way to improve other polyploid crops (Wang et al. 2014). Osakabe et al. (2016) developed new mutant alleles of OST2 which resulted in alterations in stomatal closing suggesting the application of the method for developing plants with adaptability to different environmental conditions.

7.7 Conclusion and Future Prospectus

Climate change is expected to pose a serious threat to agricultural productivity around the world, putting food security and nutritional security in jeopardy. Crop breeders are always looking for quick and accurate ways to obtain genetically enhanced breeding elites with desirable attributes. Generating new plants with better or desirable characteristics has traditionally relied on arduous and time-consuming breeding processes. The frequency of spontaneous mutations is extremely low, making them challenging to exploit in plant breeding. The goal of mutation induction is to increase the rate of mutation in a short period of time. Technological advancements, especially transgene-based induced mutagenesis and genome editing technologies, have aided the establishment of elite genotypes with long-term climate change resilience. Induced mutations have the advantage of being able to isolate several trait mutants as opposed to conventional transgenic approaches, which only allow for the introduction of a single characteristic into the crop. Further, advancements have been made to enhance the mutation frequency, to characterize

Table 7.2 Application of biological mutagens for biotic and abiotic stress tolerance in plants

Crop	Method	Stress	Gene(s)	Reference
Biotic stresses				
Arabidopsis	t-DNA	*Pseudomonas* resistance	*Cdr1-d*	Xia et al. (2004)
Rice	t-DNA	Blast resistance	*Rac1*	Kim et al. (2012)
Rice	t-DNA	False smut	*UvPRO1*	Lv et al. (2016)
Rice	t-DNA	BLB resistance	*Xa21*	Vo et al. (2018)
Rice	t-DNA	BLB resistance	*rrsRLK*	Yoo et al. (2018)
Rice	t-DNA	Blast and BLB resistance	*SPL35*	Ma et al. (2019)
Abiotic stresses				
Arabidopsis	t-DNA	Cold temperature germination tolerance	*Ctg*	Salaita et al. (2005)
Arabidopsis	t-DNA	Drought tolerance	*edt1*	Yu et al. (2008)
Rice	t-DNA	Drought and osmotic stress tolerance	*DSM2*	Du et al. (2010)
Solanum pennellii	t-DNA	Salt and drought tolerance	*SPss-2, SPds-1*	Atarés et al. (2011)
Rice	t-DNA	Drought tolerance	*OsDERF1*	Wan et al. (2011)
Rice	t-DNA	Fe efficiency	*OsYSL16*	Lee et al. (2012)
Barley	*Copia* retroposon	Drought stress	*HvDME*	Kapazoglou et al. (2013)
Rice	t-DNA	Drought tolerance	*Glutamate-receptor like gene*	Lu et al. (2014)
Rice	*TOS17* retroposon	Tolerance to PEG-generated osmotic stress	*OsGI*	Li et al. (2016)
Rice	*Ac/Ds* transposon	Water use efficiency/ abiotic stress tolerance	*RPL6, RPL23A*	Moin et al. (2016)
Rice	t-DNA	Cold tolerance	*OsCPK17*	Almadanim et al. (2018)
Rice	t-DNA	Chilling stress	*OsWRKY94 + D14*	Chen et al. (2018)
Rice	t-DNA	P use efficiency	*OsAUX1*	Giri et al. (2018)
Rice	t-DNA	Salt tolerance	*osphyB*	Kwon et al. (2018)
Rice	t-DNA	Aluminum tolerance	*WRKY22*	Li et al. (2018a)

(continued)

Table 7.2 (continued)

Crop	Method	Stress	Gene(s)	Reference
Biotic stresses				
Rice	t-DNA	Cu detoxification	*OsWAK11*	Xia et al. (2018)
Rice	t-DNA	Cu toxicity	*OsMSRB5*	Xiao et al. (2018)
Rice	t-DNA	Fe distribution	*OsYSL13*	Zhang et al. (2018a)
Maize	Retroposon	Salinity tolerance	*ZmHKT1*	Zhang et al. (2017a)
Rice	t-DNA	Salt and drought tolerance	*OsPLC4*	Deng et al. (2019)
Rice	t-DNA	Salt stress	*OsDOF15*	Qin et al. (2019)
Rice	t-DNA	N use efficiency	$OsUPS1^{OX}$	Redillas et al. (2019)
Rice	t-DNA	Arsenic accumulation in shoots	*OsABCC7*	Tang et al. (2019)
Rice	t-DNA	Salt stress	*OsHSP40*	Wang et al. (2019)
Rice	t-DNA	Osmotic stress	*SQD2.1*	Zhan et al. (2019)
Rice	t-DNA	Drought and salt tolerance	*OsMADS23*	Li et al. (2021)

genes, to create new cultivars from mutants, and even to create genetically engineered mutant genotypes with multiple insertions of DNA sequences. As a result, the mutant crop varieties developed contributed significantly to global food and nutritional security. In many field crops, for example, various cultivars with better production, abiotic stress tolerance, biotic stress resistance, and nutritional value have been produced. The use of mutation induction techniques has resulted in a significant quantity of allelic diversity, which has been useful in plant breeding, genetics and advanced genomics research. For developing climate resilient crop cultivars, this has prompted application of reverse genetic methods such as TILLING and insertional mutagenesis, and the use of novel gene editing tools like TALENs and CRISPR-Cas9. Genome editing tools have now enabled the mutational modification of target genes and have successfully yielded biotic and abiotic stress tolerant lines. In the coming years, it is expected that crop improvement using conventional and genome editing-based mutagenesis methods will contribute to development of new crop varieties for sustainable agriculture and climate resilience.

Table 7.3 Use of gene editing techniques as mutagens for stress tolerance in plants

Crop	Method	Target gene	Stress	Reference
Rice	TALEN	*OsSWEET14*	BLB resistance	Li et al. (2012)
Bread wheat	TALEN	*TaMLO-A1*, *TaMLO-B1*, *TaMLOD1*	Powdery mildew resistance	Wang et al. (2014)
Potato	TALEN	*Vlnv*	Cold storage shelf life	Clasen et al. (2016)
Rice	TALEN	*LOX3*	Enhanced disease resistance	Ma et al. (2015)
Rice	TALEN	*OsDERF1*	Drought tolerance	Zhang et al. (2016)
Rice	TALEN	*OsSWEET14*	Bacterial leaf blight resistance	Blanvillain-Baufumé et al. (2017)
Rice	CRISPR	*OsERF922*	Blast resistance	Wang et al. (2016)
Rice	CRISPR	*OsPDS*, *OsMPK2*, *OsBADH2*	Multiple abiotic stress tolerance	Shan et al. (2013)
Rice	CRISPR	*OsMPK5*	Abiotic stress tolerance and disease resistance	Xie and Yang (2013)
Rice	CRISPR	*OsMPK2*, *OsPDS*	Multiple stress tolerance	Shan et al. (2014)
Rice	CRISPR	*OsMPK2*, *OsDEP1*	Yield under stress	Shan et al. (2014)
Wheat	CRISPR	*TaMLO-A*	Powdery mildew resistance	Wang et al. (2014)
Rice	CRISPR	*OsDERF1*, *OsPMS3*, *OsEPSPS*, *OsMSH1*, *OsMYB5*	Drought tolerance	Zhang et al. (2014)
N. benthamiana	CRISPR	*ORFs* and the *IR* sequence sDNA of virus	Tomato yellow leaf curl virus (TYLCV) and Merremia mosaic virus (MeMV)	Ali et al. (2015)
N. benthamiana	CRISPR	*BeYDV*	Bean yellow dwarf virus (BeYDV) resistance	Baltes et al. (2015)
A. thaliana/N. benthamiana	CRISPR	*dsDNA* of virus *(A7*, *B7*, and *C3* regions)	Beet severe curly top virus resistance	Ji et al. (2015)
Rice	CRISPR	*OsAOX1a*, *OsAOX1b*, *OsAOX1c*, *OsBEL*	Multiple abiotic stress tolerance	Xu et al. (2015)
Rice	CRISPR	*OsSWEET13*	Bacterial blight disease resistance	Zhou et al. (2015)

(continued)

Table 7.3 (continued)

Crop	Method	Target gene	Stress	Reference
Cucumber	CRISPR	*eIF4E*	Cucumber vein yellowing virus (CVYV), Zucchini yellow mosaic virus (ZYMV), and papaya ring spot mosaic virus type-W (PRSV-W)	Chandrasekaran et al. (2016)
Soybean	CRISPR	*Avr4/6*	Resistance to *Phytophthora sojae*	Fang and Tyler (2016)
Grape	CRISPR	*MLO-7*	Effective against the powdery mildew	Malnoy et al. (2016)
A. thaliana	CRISPR	*OST2 (AHA1)*	Increased stomatal closure in response to abscisic acid (ABA)	Osakabe et al. (2016)
A. thaliana	CRISPR	*IF(iso)4E*	Turnip mosaic virus (TuMV) resistance	Pyott et al. (2016)
A. thaliana	CRISPR	*MIR169a*	Drought tolerance	Zhao et al. (2016)
Rice	CRISPR	*OsHAK-1*	Low caesium accumulation	Cordones et al. (2017)
A. thaliana	CRISPR	*UGT79B2, UGT79B3*	Cold, salt, and drought stress tolerance	Li et al. (2017)
Tomato	CRISPR	*SlMlo1*	Improved resistant against powdery mildew	Nekrasov et al. (2017)
Citrus, Duncan grapefruit	CRISPR	*CsLOB1*	Citrus canker resistance	Peng et al. (2017), Jia et al. (2017)
Rice	CRISPR	*OsAnn3*	Cold stress tolerance	Shen et al. (2017)
Maize	CRISPR	*ARGOS8-v1, ARGOS8-v2*	Increased grain yield under drought stress	Shi et al. (2017)
Tomato	CRISPR	*SlMAPK3*	Drought tolerance	Wang et al. (2017)
Wheat	CRISPR	*TaEDR1*	Improved resistant against powdery mildew	Zhang et al. (2017b)
Soybean	CRISPR	*GmDrb2a, GmDrb2b*	Salt and drought tolerance	Curtin et al. (2018)
Lettuce	CRISPR	*NCED4*	Seed germination thermo-tolerance	Bertier et al. (2018)
Cacao	CRISPR	*TcNPR3*	Resistance to pathogen *Phytophthora tropicalis*	Fister et al. (2018)
Wheat	CRISPR	*TaDREB2, TaERF3*	Abiotic stress tolerance	Kim et al. (2018)
Tomato	CRISPR	*SlCBF1*	Cold tolerance	Li et al. (2018b)
Rice	CRISPR	*SAPK1, SAPK2*	Salinity tolerance	Lou et al. (2018)
Rice	CRISPR	*eIF4G*	Rice *tungro* spherical virus	Macovei et al. (2018)

(continued)

Table 7.3 (continued)

Crop	Method	Target gene	Stress	Reference
Rice	CRISPR	*OsPRX2*	Potassium deficiency tolerance	Mao et al. (2018)
Tomato	CRISPR	*SlJAZ2*	Bacterial speck resistant	Ortigosa et al. (2018)
Rice	CRISPR	*OsNAC14*	Drought tolerance	Shim et al. (2018)
Wheat	CRISPR	*TaLpx-1*, *TaMLO*	Resistance to *Fusarium graminearum* and powdery mildew	Wang et al. (2018a)
Grape	CRISPR	*VvWRKY52*	Resistance to fungal infection (*Botrytis cinerea*)	Wang et al. (2018b)
A. thaliana	CRISPR	*IF(iso)4E*	Turnip mosaic virus (TuMV) resistance	Zhang et al. (2018b)
Cotton	CRISPR	*Gh14-3-3d*	Resistance to cotton *Verticillium* wilt	Zhang et al. (2018c)
Rice	CRISPR	*TMS5*	Generation of a new thermo-sensitive genic male sterile line	Barman et al. (2019)
Rice	CRISPR	*OsNAC041*	Salinity tolerance	Bo et al. (2019)
Rice	CRISPR	*Pi21*, *Xa13*	Resistance to blast and bacterial blight diseases	Li et al. (2019b)
Tomato	CRISPR	*SlNPR1*	Drought tolerance	Li et al. (2019a)
Rice	CRISPR	*OsOTS1*	Salinity tolerance	Sadanandom et al. (2019)
Rice	CRISPR	*OsRR22*	Salinity tolerance	Zhang et al. (2019)
Tomato	CRISPR	*PMR4*	Powdery mildew resistance	Santillán Martínez et al. (2020)
Watermelon	CRISPR	*ClPSK1*	Enhanced resistance to *Fusarium oxysporum* f.sp. *niveum*	Zhang et al. (2020)

References

Ahrens CW, Andrew ME, Mazanec RA, Ruthrof KX, Challis A, Hardy G, Byrne M, Tissue DT, Rymer PD (2020) Plant functional traits differ in adaptability and are predicted to be differentially affected by climate change. Ecol Evol 10:232–248

Ali Z, Abulfaraj A, Idris A, Ali S, Tashkandi M, Mahfouz MM (2015) CRISPR/Cas9-mediated viral interference in plants. Genome Biol 16:238

Almadanim MC, Gonçalves NM, Rosa MTG, Alexandre BM, Cordeiro AM, Rodrigues M, Saibo NJM, Soares CM, Romão CV, Oliveira MM, Abreu IA (2018) The rice cold-responsive calcium-dependent protein kinase *OsCPK17* is regulated by alternative splicing and post-translational modifications. Biochim Biophys Acta, Mol Cell Res 1865:231–246

Atarés A, Moyano E, Morales B, Schleicher P, García-Abellán JO, Antón T, García-Sogo B, Perez-Martin F, Lozano R, Flores FB, Moreno V, Bolarin MC, Pineda B (2011) An insertional mutagenesis programme with an enhancer trap for the identification and tagging of genes involved in abiotic stress tolerance in the tomato wild-related species *Solanum pennellii*. Plant Cell Rep 30(10):1865–1879. https://doi.org/10.1007/s00299-011-1094-y

Baltes NJ, Hummel AW, Konecna E, Cegan R, Bruns AN, Bisaro DM, Voytas DF (2015) Conferring resistance to geminiviruses with the CRISPR-Cas prokaryotic immune system. Nat Plants 1:15145

Barman HN, Sheng Z, Fiaz S, Zhong M, Wu Y, Cai Y, Wang W, Jiao G, Tang S, Wei X, Hu P (2019) Generation of a new thermo-sensitive genic male sterile rice line by targeted mutagenesis of *TMS5* gene through CRISPR/Cas9 system. BMC Plant Biol 19:109. https://doi.org/10.1186/s12870-019-1715-0

Bertier LD, Ron M, Huo H, Bradford KJ, Britt AB, Michelmore RW (2018) High-resolution analysis of the efficiency, heritability, and editing outcomes of CRISPR-Cas9-induced modifications of *NCED4* in lettuce (*Lactuca sativa*). G3 8(5):1513–1521. https://doi.org/10.1534/g3.117.300396

Bettgenhaeuser J, Krattinger SG (2019) Rapid gene cloning in cereals. Theor Appl Genet 132:699–711. https://doi.org/10.1007/s00122-018-3210-7

Blanvillain-Baufumé S, Reschke M, Solé M, Auguy F, Doucoure H, Szurek B, Meynard D, Portefaix M, Cunnac S, Guiderdoni E, Boch J, Koebnik R (2017) Targeted promoter editing for rice resistance to *Xanthomonas oryzae* pv. *oryzae* reveals differential activities for *SWEET 14*-inducing TAL effectors. Plant Biotechnol J 15:306–317

Bo W, Zhaohui Z, Huanhuan Z, Xia W, Binglin L, Lijia Y, Xiangyan H, Deshui Y, Xuelian Z, Chunguo W, Wenqin S, Chengbin C, Yong Z (2019) Targeted mutagenesis of NAC transcription factor gene, *OsNAC041*, leading to salt sensitivity in rice. Rice Sci 26:98–108. https://doi.org/10.1016/j.rsci.2018.12.005

Bohra A, Satheesh Naik SJ, Kumari A, Tiwari A, Joshi R (2021) Integrating phenomics with breeding for climate-smart agriculture. In: Kumar A, Kumar R, Shukla P, Patel HK (eds) Omics technologies for sustainable agriculture and global food security, vol II. Springer, Singapore, pp 1–24

Bueno PCP, Lopes NP (2020) Metabolomics to characterize adaptive and signaling responses in legume crops under abiotic stresses. ACS Omega 5(4):1752–1763

Chandrasekaran J, Brumin M, Wolf D, Leibman D, Klap C, Pearlsman M, Sherman A, Arazi T, Gal-On A (2016) Development of broad virus resistance in non-transgenic cucumber using CRISPR/Cas9 technology. Mol Plant Pathol 17:1140–1153. https://doi.org/10.1111/mpp.12375

Chen L, Hao L, Parry MA, Phillips AL, Hu Y (2014) Progress in TILLING as a tool for functional genomics and improvement of crops. J Integr Plant Biol 56:425443

Chen L, Zhao Y, Xu S, Zhang Z, Xu Y, Zhang J, Chong K (2018) *OsMADS57* together with *OsTB1* coordinates transcription of its target *OsWRKY94* and *D14* to switch its organogenesis to defense for cold adaptation in rice. New Phytol 218(1):219–231. https://doi.org/10.1111/nph.14977

Clasen BM, Stoddard TJ, Luo S et al (2016) Improving cold storage and processing traits in potato through targeted gene knockout. Plant Biotechnol J 14:169–176

Cordones MN, Mohamed S, Tanoi K, Kobayashi NI, Takagi K, Vernet A, Guiderdoni E, Périn C, Sentenac H, Véry A (2017) Production of low-Cs + rice plants by inactivation of the K + transporter OsHAK1 with the CRISPR-Cas system. Plant J 92:43–56. https://doi.org/10.1111/tpj.13632

Curtin SJ, Xiong Y, Michno JM, Campbell BW, Stec AO, Čermák T, Starker C, Voytas DF, Eamens AL, Stupar RM (2018) CRISPR/Cas9 and TALENs generate heritable mutations for genes involved in small RNA processing of *Glycine max* and *Medicago truncatula*. Plant Biotechnol J 16:1125–1137

Deng XJ, Yuan S, Cao HS, Lam SM, Shui G, Hong Y, Wang X (2019) Phosphatidylinositol-hydrolyzing phospholipase C4 modulates rice response to salt and drought. Plant Cell Environ 42:536–548. https://doi.org/10.1111/pce.13437

Du H, Wang N, Cui F, Li X, Xiao J, Xiong L (2010) Characterization of the *beta*-carotene hydroxylase gene *DSM2* conferring drought and oxidative stress resistance by increasing xanthophylls and abscisic acid synthesis in rice. Plant Physiol 154(3):1304–1318. https://doi.org/10.1104/pp.110.163741

Dwivedi SL, Sahrawat KL, Upadhyaya HD, Ortiz R (2013) Food, nutrition and agrobiodiversity under global climate change. Adv Agron 120:1–128. https://doi.org/10.1016/B978-0-12-407686-0.00001-4

Fang Y, Tyler BM (2016) Efficient disruption and replacement of an effector gene in the Oomycete *Phytophthora sojae* using CRISPR/Cas9. Mol Plant Pathol 17:127–139. https://doi.org/10.1111/mpp.12318

FAO (2014) The State of Food Insecurity in the World 2014. Strengthening the enabling environment for food security and nutrition. FAO, Rome

FAO/IAEA-MVD (2021) Food and agriculture organization of the United Nations/International atomic energy agency-mutant variety database. https://mvd.iaea.org/#!Search. Accessed 31 Aug 2021

Fernie AR, Yan J (2019) De novo domestication: an alternative route toward new crops for the future. Mol Plant 12:615–631

Fister AS, Landherr L, Maximova SN (2018) Transient expression of CRISPR/cas9 machinery targeting *TCNPR3* enhances defense response in *Theobroma cacao*. Front Plant Sci 9:268. https://doi.org/10.3389/fpls.2018.00268

Fruzangohar M, Kalashyan E, Kalambettu P, Ens J, Wiebe K, Pozniak CJ, Tricker PJ, Baumann U (2019) Novel informatic tools to support functional annotation of the durum wheat genome. Front Plant Sci 10:1244

Giri J, Bhosale R, Huang G et al (2018) Rice auxin influx carrier *OsAUX1* facilitates root hair elongation in response to low external phosphate. Nat Commun 9:1408. https://doi.org/10.1038/s41467-018-03850-4

Gottwald S, Bauer P, Komatsuda T, Udda Lundqvist U, Stein N (2009) TILLING in the two-rowed barley cultivar 'Barke' reveals preferred sites of functional diversity in the gene HvHox1. BMC Res Notes 2:258

Guo Y, Abernathy B, Zeng Y, Ozias-Akins P (2015) TILLING by sequencing to identify induced mutations in stress resistance genes of peanut (*Arachis hypogaea*). BMC Genomics 16:157. https://doi.org/10.1186/s12864-015-1348-0

Hori K, Matsubara K, Uga Y, Yano M (2016) A novel *Tos17* insertion upstream of *Hd1* alters flowering time in rice. Plant Breed 135:588–592. https://doi.org/10.1111/pbr.12406

Hu X (2012) TILLING-based analysis of disease resistance genes in barley. J Shandong Agric Univ 43(1):1–6

Hwang JE, Jang DS, Lee KJ, Ahn J, Kim SH, Kang S, Kim DS, Kim J (2016) Identification of gamma ray irradiation-induced mutations in membrane transport genes in a rice population by TILLING. Genes Genet Syst 91(5):245–256. https://doi.org/10.1266/ggs.15-00052

Jain SM (2010) Mutagenesis in crop improvement under the climate change. Rom Biotechnol Lett 15:88–106

Jankowicz-Cieslak J, Mba C, Till BJ (2017) Mutagenesis for crop breeding and functional genomics. In: Jankowicz-Cieslak J, Tai TH, Kumlehn J, Till BJ (eds) Biotechnologies for plant mutation breeding. Springer, Cham, Switzerland, pp 3–18

Ji X, Zhang H, Zhang Y, Wang Y, Gao C (2015) Establishing a CRISPR-Cas-like immune system conferring DNA virus resistance in plants. Nat Plants 1:15144. https://doi.org/10.1038/nplants.2015.144

Jia H, Zhang Y, Orbović V, Xu J, White FF, Jones JB, Wang M (2017) Genome editing of the disease susceptibility gene *CsLOB1* in citrus confers resistance to citrus canker. Plant Biotechnol J 15:817–823

Jung C, Till B (2021) Mutagenesis and genome editing in crop improvement: perspectives for the global regulatory landscape. Trends Plant Sci 26:1258–1269

Kapazoglou A, Drosou V, Argiriou A et al (2013) The study of a barley epigenetic regulator, *HvDME*, in seed development and under drought. BMC Plant Biol 13:172. https://doi.org/10.1186/1471-2229-13-172

Kim D, Alptekin B, Budak H (2018) CRISPR/Cas9 genome editing in wheat. Funct Integr Genomics 18:31–41. https://doi.org/10.1007/s10142-017-0572-x

Kim SH, Oikawa T, Kyozuka J et al (2012) The bHLH Rac immunity1 (*RAI1*) is activated by *OsRac1* via *OsMAPK3* and *OsMAPK6* in rice immunity. Plant Cell Physiol 53:740–754

Kole C, Muthamilarasan M, Henry R, Edwards D, Sharma R, Abberton M, Batley J, Bentley A, Blakeney M, Bryant J, Cai H, Cakir M, Cseke LJ, Cockram J, de Oliveira AC, De Pace C, Dempewolf H, Ellison S, Gepts P, Greenland A, Hall A, Hori K, Hughes S, Humphreys MW, Iorizzo M, Ismail AM, Marshall A, Mayes S, Nguyen HT, Ogbonnaya FC, Ortiz R, Paterson AH, Simon PW, Tohme J, Tuberosa R, Valliyodan B, Varshney RK, Wullschleger SD, Yano M, Prasad M (2015) Application of genomics-assisted breeding for generation of climate resilient crops: progress and prospects. Front Plant Sci 6:563. https://doi.org/10.3389/fpls.2015.00563563

Kusumi K, Hirotsuka S, Kumamaru T, Iba K (2012) Increased leaf photosynthesis caused by elevated stomatal conductance in a rice mutant deficient in *SLAC1*, a guard cell anion channel protein. J Exp Bot 63(15):5635–5644

Kwon CT, Song G, Kim SH et al (2018) Functional deficiency of phytochrome B improves salt tolerance in rice. Environ Exp Bot 148:100–108. https://doi.org/10.1016/j.envexpbot.2017.12.020

Lee S, Ryoo N, Jeon JS, Guerinot ML, An G (2012) Activation of rice yellow stripe1-like 16 (*OsYSL16*) enhances iron efficiency. Mol Cells 33:117–126

Li T, Liu B, Spalding MH, Weeks DP, Yang B (2012) High-efficiency TALEN-based gene editing produces disease-resistant rice. Nat Biotechnol 30:390–392. https://doi.org/10.1038/nbt.2199

Li S, Yue W, Wang M, Qiu W, Zhou L, Shou H (2016) Mutation of *OsGIGANTEA* leads to enhanced tolerance to polyethylene glycol-generated osmotic stress in rice. Front Plant Sci 7:465. https://doi.org/10.3389/fpls.2016.00465

Li P, Li YJ, Zhang FJ, Zhang GZ, Jiang XY, Yu HM, Hou BK (2017) The Arabidopsis UDP-glycosyltransferases *UGT79B2* and *UGT79B3*, contribute to cold, salt and drought stress tolerance via modulating anthocyanin accumulation. Plant J 89:85–103. https://doi.org/10.1111/tpj.13324

Li GZ, Wang ZQ, Yokosho K et al (2018a) Transcription factor *WRKY22* promotes aluminum tolerance via activation of *OsFRDL4* expression and enhancement of citrate secretion in rice (*Oryza sativa*). New Phytol 219(1):149–162. https://doi.org/10.1111/nph.15143

Li R, Zhang L, Wang L, Chen L, Zhao R, Sheng J, Shen L (2018b) Reduction of tomato-plant chilling tolerance by CRISPRCas9-mediated SlCBF1 mutagenesis. J Agric Food Chem 66:9042–9051

Li R, Liu C, Zhao R, Wang L, Chen L, Yu W, Zhang S, Sheng J, Shen L (2019a) CRISPR/Cas9-mediated *SlNPR1* mutagenesis reduces tomato plant drought tolerance. BMC Plant Biol 19:38

Li S, Shen L, Hu P, Liu Q, Zhu X, Qian Q, Wang K, Wang Y (2019b) Developing disease-resistant thermosensitive male sterile rice by multiplex gene editing. J Integr Plant Biol. https://doi.org/10.1111/jipb.12774

Li J, Yang J, Li Y, Ma L (2020) Current strategies and advances in wheat biology. Crop J 8:879–891

Li X, Yu B, Wu Q, Min Q, Zeng R, Xie Z, Huang J (2021) *OsMADS23* phosphorylated by *SAPK9* confers drought and salt tolerance by regulating ABA biosynthesis in rice. PLoS Genet 17(8):e1009699. https://doi.org/10.1371/journal.pgen.1009699

Lo SF, Fan MJ, Hsing YI, Chen LJ, Chen S, Wen IC, Liu YL, Chen KT, Jiang MJ, Lin MK, Rao MY (2016) Genetic resources offer efficient tools for rice functional genomics research. Plant Cell Environ 39:998–1013. https://doi.org/10.1111/pce.12632

Lou D, Wang H, Yu D (2018) The sucrose non-fermenting-1-related protein kinases *SAPK1* and *SAPK2* function collaboratively as positive regulators of salt stress tolerance in rice. BMC Plant Biol 18:203

Lu G, Wang X, Liu J, Yu K, Gao Y, Liu H, Wang C, Wang W, Wang G, Liu M, Mao G (2014) Application of T-DNA activation tagging to identify glutamate receptor-like genes that enhance drought tolerance in plants. Plant Cell Rep 33:617–631

Lv B, Zheng L, Liu H, Tang J, Hsiang T, Huang J (2016) Use of random T-DNA mutagenesis in identification of gene *UvPRO1*, a regulator of conidiation, stress response, and virulence in *Ustilaginoidea virens*. Front Microbiol 7:2086. https://doi.org/10.3389/fmicb.2016.02086

Ma J, Wang Y, Ma X, Meng L, Jing R, Wang F, Wang S, Cheng Z, Zhang X, Jiang L, Wang J (2019) Disruption of gene *SPL35*, encoding a novel CUE domain-containing protein, leads to cell death and enhanced disease response in rice. Plant Biotechnol J17:1679–1693. https://doi.org/10.1111/pbi.13093

Ma L, Zhu F, Li Z, Zhang J, Li X, Dong J, Wang T (2015) TALEN-based mutagenesis of lipoxygenase *LOX3* enhances the storage tolerance of rice (*Oryza sativa*) seeds. PLoS One 10:1–16. https://doi.org/10.1371/journal.pone.0143877

Macovei A, Sevilla NR, Cantos C, Jonson GB, Slamet-Loedin I, Cermák T, Voytas DF, Choi IR, Mohanty PC (2018) Novel alleles of rice *eIF4G* generated by CRISPR/Cas9-targeted mutagenesis confer resistance to Rice *tungro* spherical virus. Plant Biotechnol J 16(11):1918–1927. https://doi.org/10.1111/pbi.12927

Malnoy M, Viola R, Jung MH, Koo OJ, Kim S, Kim JS, Velasco R, Nagamangala KC (2016) DNA-free genetically edited grapevine and apple protoplast using CRISPR/Cas9 ribonucleoproteins. Front Plant Sci 7:1904. https://doi.org/10.3389/fpls.2016.01904

Mao X, Zheng Y, Xiao K, Wei Y, Zhu Y, Cai Q, Chen L, Xie H, Zhang J (2018) *OsPRX2* contributes to stomatal closure and improves potassium deficiency tolerance in rice. Biochem Biophys Res Commun 495:461–467. https://doi.org/10.1016/j.bbrc.2017.11.045

McCallum C, Henikoff S, Colbert T (2004) Fred Hutchinson Cancer Research Center, Assignee. Reverse genetic strategy for identifying functional mutations in genes of known sequences. U.-S. Patent Application US 10/240, 456

Milanović M, Knapp S, Pyšek P, Kühn I (2020) Trait–environment relationships of plant species at different stages of the introduction process. NeoBiota 58:55–74

Miyao A, Tanaka K, Murata K et al (2003) Target site specificity of the *Tos17* retrotransposon shows a preference for insertion within genes and against insertion in retrotransposon-rich regions of the genome. Plant Cell 15(8):1771–1780. https://doi.org/10.1105/tpc.012559

Moin M, Bakshi A, Saha A et al (2016) Activation tagging in Indica rice identifies ribosomal proteins as potential targets for manipulation of water-use efficiency and abiotic stress tolerance in plants. Plant Cell Environ 39(11):2440–2459. https://doi.org/10.1111/pce.12796

Mousavi-Derazmahalleh M, Bayer PE, Hane JK et al (2019) Adapting legume crops to climate change using genomic approaches. Plant Cell Environ 42:6–19

Nekrasov V, Wang C, Win J, Lanz C, Weigel D, Kamoun S (2017) Rapid generation of a transgene-free powdery mildew resistant tomato by genome deletion. Sci Rep 7(1):482. https://doi.org/10.1038/s41598-017-00578-x

Nida H, Blum S, Zielinski D, Srivastava DA, Elbaum R, Xin Z, Erlich Y, Fridman E, Shental N (2016) Highly efficient *de novo* mutant identification in a *Sorghum bicolor* TILLING population using the ComSeq approach. Plant J 86:349–359

Ortigosa A, Gimenez-Ibanez S, Leonhardt N, Solano R (2018) Design of a bacterial speck resistant tomato by CRISPR/Cas9-mediated editing of *SlJAZ2*. Plant Biotechnol J 17:665–673

Osakabe Y, Watanabe T, Sugano SS, Ueta R, Ishihara R, Shinozaki K, Osakabe K (2016) Optimization of CRISPR/Cas9 genome editing to modify abiotic stress responses in plants. Sci Rep 6:26685. https://doi.org/10.1038/srep26685

Peng A, Chen S, Lei T, Xu L, He Y, Wu L, Yao L, Zou X (2017) Engineering canker-resistant plants through CRISPR/Cas9-targeted editing of the susceptibility gene *CsLOB1* promoter in citrus. Plant Biotechnol J 15:1509–1519

Pyott DE, Sheehan E, Molnar A (2016) Engineering of CRISPR/Cas9-mediated potyvirus resistance in transgene-free *Arabidopsis* plants. Mol Plant Pathol 17:1276–1288. https://doi.org/10.1111/mpp.12417

Qaim M (2020) Role of new plant breeding technologies for food security and sustainable agricultural development. Appl Econ Perspect Policy 42(2):129–150

Qin H, Wang J, Chen X et al (2019) Rice *OsDOF15* contributes to ethylene-inhibited primary root elongation under salt stress. New Phytol 223:798–813. https://doi.org/10.1111/nph.15824

Raza A, Razzaq A, Mehmood SS, Zou X, Zhang X, Lv Y, Xu J (2019) Impact of climate change on crops adaptation and strategies to tackle its outcome: a review. Plants (Basel) 8(2):34. https://doi.org/10.3390/plants8020034

Redillas MCFR, Bang SW, Lee DK et al (2019) Allantoin accumulation through overexpression of *ureide permease1* improves rice growth under limited nitrogen conditions. Plant Biotechnol J 17:1289–1301. https://doi.org/10.1111/pbi.13054

Sadanandom A, Srivastava AK, Zhang C (2019) Targeted mutagenesis of the *SUMO* protease, overly tolerant to *Salt1* in rice through CRISPR/Cas9-mediated genome editing reveals a major role of this *SUMO* protease in salt tolerance. BioRxiv 2019:555706

Saintenac C, Lee WS, Cambon F, Rudd JJ, King RC, Marande W, Powers SJ, Bergès H, Phillips AL, Uauy C, Hammond-Kosack KE (2018) Wheat receptor-kinase-like protein *Stb6* controls gene-for-gene resistance to fungal pathogen *Zymoseptoria tritici*. Nat Genet 50:368–374. https://doi.org/10.1038/s41588-018-0051-x

Salaita L, Kar RK, Majee M, Downie AB (2005) Identification and characterization of mutants capable of rapid seed germination at 10 C from activation-tagged lines of *Arabidopsis thaliana*. J Exp Bot 56:2059–2069

Sallaud C, Gay C, Larmande P, Bès M, Piffanelli P, Piégu B, Droc G, Regad F, Bourgeois E, Meynard D, Périn C (2004) High throughput T-DNA insertion mutagenesis in rice: a first step towards *in silico* reverse genetics. Plant J 39(3):450–464. https://doi.org/10.1111/j.1365-313X.2004.02145.x

Santillán Martínez MI, Bracuto V, Koseoglou E, Appiano M, Jacobsen E, Visser RG, Wolters AMA, Bai Y (2020) CRISPR/Cas9-targeted mutagenesis of the tomato susceptibility gene *PMR4* for resistance against powdery mildew. BMC Plant Biol 20:284. https://doi.org/10.1186/s12870-020-02497-y

Severune H, Daniela S, Simon GK, Kessel B, Wicker T, Herren G, Fitze MN, Breen J, Presterl T, Ouzunova M, Keller B (2015) The maize disease resistance gene *Htn1* against northern corn leaf blight encodes a wall-associated recepter-like kinase. Proc Natl Acad Sci U S A 112:8780–8785

Shan Q, Wang Y, Li J, Zhang Y, Chen K, Liang Z, Zhang K, Liu J, Xi JJ, Qiu JL, Gao C (2013) Targeted genome modification of crop plants using a CRISPR-Cas system. Nat Biotechnol 31: 686–688

Shan Q, Wang Y, Li J, Gao C (2014) Genome editing in rice and wheat using the CRISPR/Cas system. Nat Protoc 9:2395–2410

Shen C, Que Z, Xia Y, Tang N, Li D, He R, Cao M (2017) Knock out of the annexin gene *OsAnn3* via CRISPR/Cas9-mediated genome editing decreased cold tolerance in rice. J Plant Biol 60: 539–547. https://doi.org/10.1007/s12374-016-0400-1

Shi J, Gao H, Wang H et al (2017) *ARGOS8* variants generated by CRISPR-Cas9 improve maize grain yield under field drought stress conditions. Plant Biotechnol J 15:207–216. https://doi.org/10.1111/pbi.12603

Shim JS, Oh N, Chung PJ et al (2018) Overexpression of *OsNAC14* improves drought tolerance in rice. Front Plant Sci 9:310

Sikora P, Chawade A, Larsson M, Olsson J, Olsson O (2011) Mutagenesis as a tool in plant genetics, functional genomics, and breeding. Int J Plant Genomics 2011:314829. https://doi.org/10.1155/2011/314829

Singh R, Tiwari R, Sharma D, Tiwari V, Sharma I (2014) Mutagenesis for wheat improvement in the genomics era. J Wheat Res 6:120–125

Snowdon RJ, Wittkop B, Chen TW, Stahl A (2021) Crop adaptation to climate change as a consequence of long-term breeding. Theor Appl Genet 134:1613–1623

Suprasanna P, Mirajkar SJ, Bhagwat SG (2015) Induced mutations and crop improvement. In: Bahadur B, Rajam MV, Sahijram L, Krishnamurthy KV (eds) Plant biology and biotechnology. Springer, New York, pp 593–618

Suprasanna P, Negi P, Kumar V (2021) Genomics and epigenetics of plant abiotic stress. In: Fink G (ed) Stress: genetics, epigenetics and genomics. Academic Press, pp 49–56

Tang Z, Chen Y, Miller AJ (2019) The C-type ATP-binding cassette transporter *OsABCC7* is involved in the root-to-shoot translocation of arsenic in rice. Plant Cell Physiol 60(7): 1525–1535. https://doi.org/10.1093/pcp/pcz054

Tardieu F, Cabrera-Bosquet L, Pridmore T, Bennett M (2017) Plant phenomics, from sensors to knowledge. Curr Biol 27(15):R770–R783

Vo KTX, Kim CY, Hoang TV et al (2018) *OsWRKY67* plays a positive role in basal and *XA21*-mediated resistance in rice. Front Plant Sci 8:2220. https://doi.org/10.3389/fpls.2017.02220

Wan L, Zhang J, Zhang H et al (2011) Transcriptional activation of *OsDERF1* in *OsERF3* and *OsAP2-39* negatively modulates ethylene synthesis and drought tolerance in rice. PLoS One 6(9):e25216. https://doi.org/10.1371/journal.pone.0025216

Wang N, Long T, Yao W, Xiong L, Zhang Q, Wu C (2013) Mutant resources for the functional analysis of the rice genome. Mol Plant 6:596–604. https://doi.org/10.1093/mp/sss142

Wang Y, Cheng X, Shan Q, Zhang Y, Liu J, Gao C, Qiu JL (2014) Simultaneous editing of three homoeoalleles in hexaploid bread wheat confers heritable resistance to powdery mildew. Nat Biotechnol 32:947–952

Wang F, Wang C, Liu P et al (2016) Enhanced rice blast resistance by CRISPR/Cas9-targeted mutagenesis of the ERF transcription factor gene *OsERF922*. PLoS One 11:e0154027

Wang L, Chen L, Li R et al (2017) Reduced drought tolerance by CRISPR/Cas9-mediated *SlMAPK3* mutagenesis in tomato plants. J Agric Food Chem 65:8674–8682. https://doi.org/10.1021/acs.jafc.7b02745

Wang W, Pan Q, He F, Akhunova A, Chao S, Trick H, Akhunov E (2018a) Transgenerational CRISPR-Cas9 activity facilitates multiplex gene editing in allopolyploid wheat. CRISPR J 1: 65–74. https://doi.org/10.1089/crispr.2017.0010

Wang X, Tu M, Wang D, Liu J, Li Y, Li Z, Wang Y, Wang X (2018b) CRISPR/Cas9-mediated efficient targeted mutagenesis in grape in the first generation. Plant Biotechnol J 16:844–855. https://doi.org/10.1111/pbi.12832

Wang X, Zhang H, Shao LY, Yan X, Peng H, Ouyang JX, Li SB (2019) Expression and function analysis of a rice *OsHSP40* gene under salt stress. Genes Genom 41(2):175–182. https://doi.org/10.1007/s13258-018-0749-2

Xia Y, Suzuki H, Borevitz J, Blount J, Guo Z, Patel K, Dixon RA, Lamb C (2004) An extracellular aspartic protease functions in *Arabidopsis* disease resistance signaling. EMBO J 23:980–988

Xia Y, Yin S, Zhang K et al (2018) *OsWAK11*, a rice wall-associated kinase, regulates Cu detoxification by alteration the immobilization of Cu in cell walls. Environ Exp Bot 150:99–105

Xiao T, Mi M, Wang C (2018) A methionine-R-sulfoxide reductase, *OsMSRB5*, is required for rice defense against copper toxicity. Environ Exp Bot. https://doi.org/10.1016/j.envexpbot.2018.04.006

Xie K, Yang Y (2013) RNA-guided genome editing in plants using a CRISPR-Cas system. Mol Plant 6:1975–1983

Xu RF, Li H, Qin R, Li J, Qiu C, Yang Y, Ma H, Li L, Wei P, Yang J (2015) Generation of inheritable and “transgene clean” targeted genome-modified rice in later generations using the CRISPR/Cas9 system. Sci Rep 5:11491. https://doi.org/10.1038/srep11491

Yoo Y, Park JC, Cho MH et al (2018) Lack of a cytoplasmic *RLK*, required for ROS homeostasis, induces strong resistance to bacterial leaf blight in rice. Front Plant Sci 9:577. https://doi.org/10.3389/fpls.2018.00577

Yu H, Chen X, Hong YY et al (2008) Activated expression of an *Arabidopsis HD-START* protein confers drought tolerance with improved root system and reduced stomatal density. Plant Cell 20:1134–1151

Zhan X, Shen Q, Chen J et al (2019) Rice sulfoquinovosyltransferase *SQD2.1* mediates flavonoid glycosylation and enhances tolerance to osmotic stress. Plant Cell Environ 42(7):2215–2230. https://doi.org/10.1111/pce.13554

Zhang J, Li P (2019) Response of plant functional traits to climate change. IOP Conf Ser Earth Environ Sci 300:032078

Zhang H, Zhang J, Wei P et al (2014) The CRISPR/Cas9 system produces specific and homozygous targeted gene editing in rice in one generation. Plant Biotechnol J 12:797–807

Zhang H, Guo F, Zhang J et al (2016) TALEN-mediated targeted mutagenesis produces a large variety of heritable mutation in rice. Plant Biotechnol J 14:186–194. https://doi.org/10.1111/pbi.12372

Zhang M, Cao Y, Wang Z (2017a) A retrotransposon in an *HKT1* family sodium transporter causes variation of leaf Na^+ exclusion and salt tolerance in maize. New Phytol 217(3):1161–1176. https://doi.org/10.1111/nph.14882

Zhang Y, Bai Y, Wu G et al (2017b) Simultaneous modification of three homoeologs of *TaEDR1* by genome editing enhances powdery mildew resistance in wheat. Plant J 91:714–724

Zhang C, Shinwari KI, Luo L, Zheng L (2018a) *OsYSL13* is involved in iron distribution in rice. Int J Mol Sci 19(11):3537. https://doi.org/10.3390/ijms19113537

Zhang T, Zheng Q, Yi X et al (2018b) Establishing RNA virus resistance in plants by harnessing CRISPR immune system. Plant Biotechnol J. https://doi.org/10.1111/pbi.12881

Zhang Z, Ge X, Luo X et al (2018c) Simultaneous editing of two copies of *Gh14-3-3d* confers enhanced transgene-clean plant defense against *Verticillium dahliae* in allotetraploid upland cotton. Front Plant Sci 9:842. https://doi.org/10.3389/fpls.2018.00842

Zhang A, Liu Y, Wang F et al (2019) Enhanced rice salinity tolerance via CRISPR/Cas9-targeted mutagenesis of the *OsRR22* gene. Mol Breed 39:47

Zhang M, Liu Q, Yang X et al (2020) CRISPR/Cas9-mediated mutagenesis of *Clpsk1* in watermelon to confer resistance to *Fusarium oxysporum* f.sp. *niveum*. Plant Cell Rep 39:589–595. https://doi.org/10.1007/s00299-020-02516-0

Zhao Y, Zhang C, Liu W et al (2016) An alternative strategy for targeted gene replacement in plants using a dual-sgRNA/Cas9 design. Sci Rep 6:23890. https://doi.org/10.1038/srep23890

Zhou J, Peng Z, Long J et al (2015) Gene targeting by the TAL effector *PthXo2* reveals cryptic resistance gene for bacterial blight of rice. Plant J 82:632–643. https://doi.org/10.1111/tpj.12838

Molecular Markers for Mutant Characterization

8

R. S. Bhat, M. P. Brijesh Patil, I. S. Tilak, and K. Shirasawa

Abstract

Mutants are either obtained spontaneously or artificially induced frequently in various organisms to serve several purposes like expanding the variability, undertaking genetic studies, and adopting them as new genotypes for diverse commercial applications. Identification and development of a mutant is generally accomplished through phenotypic screening and/or molecular analysis at DNA, RNA, or protein levels. Markers are currently being used to identify and characterize the mutants and epimutants against their parents (wild type) and rest of the genotypes. Though morphological and biochemical markers were among the first to be employed, DNA markers became the widely used system for mutant characterization. These markers when used along with a set of conventional and novel methods of mutant analysis would reveal the extent and nature of mutations which help in understanding the phenotypic effects of mutations. With the advancements in DNA sequencing, novel classes of markers (single nucleotide polymorphism, transposable element, copy number variation, and presence/absence variation) are being envisaged. Markers have been successfully employed for mutant characterization, trait mapping, and identification of mutant loci and molecular breeding in various plant systems like wheat, rice, barley, maize, soybean, peanut, and tomato. This chapter reviews the types of mutants (classical, DNA-tagged, gene silencing, gene edited, and deletion) and epimutants at DNA level, and discusses the morphological, biochemical, and

R. S. Bhat (✉) · M. P. Brijesh Patil · I. S. Tilak
Department of Biotechnology, University of Agricultural Sciences, Dharwad, India
e-mail: bhatrs@uasd.in

K. Shirasawa
Department of Frontier Research and Development, Kazusa DNA Research Institute, Chiba, Japan

S. Penna, S. M. Jain (eds.), *Mutation Breeding for Sustainable Food Production and Climate Resilience*, https://doi.org/10.1007/978-981-16-9720-3_8

DNA markers used to detect them. Further, the successful applications of these markers in mutant characterization have been discussed.

Keywords

Mutations · Genetic changes · Epigenetic changes · Molecular markers · Mutant characterization · Transcriptomic analysis · Next-generation mapping techniques

8.1 Introduction

Genes, being constituted by the nucleic acids and carried on the chromosomes, are the units of heredity. They undergo mutation to change from one form to another. Mutations involve sudden heritable changes as termed by Hugo de Vries (1901). These heritable changes can be identified at the chromosomal level and nucleic acid level (see Auerbach 2013). The changes at chromosome level include changes in chromosome number, changes in the entire set of genes, and changes in individual genes, while the molecular changes at nucleic acid level include base substitutions, additions, and deletions. Apart from occurring spontaneously, the mutations can be induced. Mutations are the chief source of genetic variability leading to phenotypic alterations profoundly contributing to biological evolution (Kutschera and Niklas 2004; Schoen and Schultz 2019; Jay et al. 2021) (Fig. 8.1).

Apart from being extensively used in genetics, mutations have been applied for crop improvement. Till date, several mutants have been developed in plants, animals, and microbes, and many have been characterized. With the advancements

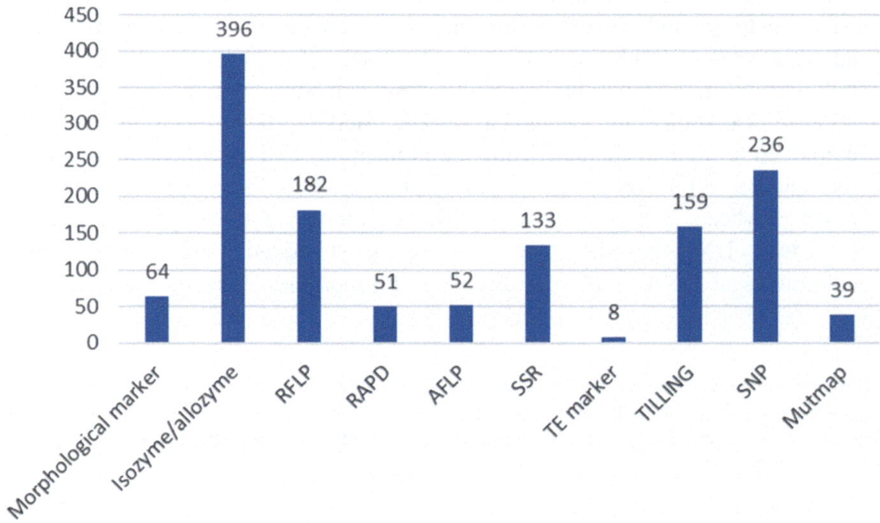

Fig. 8.1 Number of PubMed publications reporting the use of genetic markers in mutant analysis till date

in genetics and genomics, now the characterization of mutants and mutations is employed as an approach for gene discovery. Mutant characterization generally refers to identifying the type and location of genetic changes, determining the mode of inheritance, and recognizing the phenotype due to the genetic changes. Nowadays, the mutants are characterized at the level of phenome, genome, epigenome, transcriptome, proteome, metabolome, ionome, etc. With the advances in DNA sequencing, the genome-wide approach of characterization is becoming the method of choice, however considering its huge demand for the resources, molecular markers are widely used for mutant characterization. Reciprocally, the mutant resources offer an opportunity to develop new markers and marker systems that can be employed for mutant characterization. In this chapter, an effort is made to review the molecular markers that have been successfully employed for identifying and characterizing the mutations in plants.

8.2 Types of Mutations

8.2.1 Classical Mutations

Mutation refers to change in the genetic material and the process by which the changes occur. Thus, the mutations involve sudden and heritable changes that cannot be explained by recombination of pre-existing genetic variability. Such genotypic changes include changes in the chromosome number (euploidy and aneuploidy), gross changes in the structure of chromosomes (deletions/deficiencies, duplications, inversions, and translocations), and changes individual genes (excluding changes in chromosome number or structure) (Gardener et al. 1991). Changes in the structure of chromosomes require breaks (one or more) in the chromosome (single or a set of chromosomes). Based on the first cytological observations of maize chromosome rearrangements by McClintock, four structural changes were identified; (1) deletions/deficiencies (parts of chromosomes lost or deleted), (2) duplications (parts added or duplicated), (3) inversions (sections detached and reunited in reverse order), and (4) translocations (parts of chromosomes detached and join to nonhomologous chromosomes).

Mutations involving single base-pairs are referred to as point mutations which can be found due to substitutions (change of one base-pair to another), duplications, and deletions. Mutations resulting from tautomeric shifts in the DNA involve the replacement of a purine in one strand of DNA with other purine and/or the replacement of a pyrimidine in the complementary strand with the other pyrimidine, which collectively are called transitions. While base-pair substitutions involving the replacement of a purine with a pyrimidine and vice versa are called transversions. Four different transitions and eight different transversions are possible in DNA. Additions and deletions of one or a few base-pairs are collectively called as frameshift mutations since they alter the reading frame of a gene distal to the site of mutation.

Based on the occurrence, the mutations can be classified as spontaneous (those occurring without a known cause) or induced (resulting from the exposure to mutagenic agents). Operationally, it is difficult to discern spontaneous mutations from induced mutations. The effect of mutation on the phenotype might range from minor alterations, which can be detected only through special genetic or biochemical techniques to gross modifications of morphology to lethals.

Any mutation occurring within a given gene will produce a new allele. Because of the degeneracy of the genetic code, the mutant allele coding for the unaltered protein and the phenotype is called as isoallele. Mutant alleles coding for altered gene product (protein) may show modified phenotype, and the mutation resulting into loss of gene product in an essential gene may be lethal.

Mutations may be either recessive or dominant. In diploids (or polyploids), the recessive mutations can be recognized only in their homozygous state, whereas the dominant mutations can be identified both in the heterozygous state and their homozygous state. Most of the mutations that have been identified and studied by geneticists are recessive in nature. It is also important to note that a mutation can continue to exist in heterozygous condition (Branch et al. 2020). Mutations can occur in any cell cycle state of somatic and germinal cells (Schoen and Schultz 2019). Thus, the effect of mutation and the subsequent phenotypic change depends on its nature of dominance, the cell type, and state of the cell cycle.

8.2.2 Epimutations

Classical mutations (explained above) involve changes in DNA base sequences. In contrast, the changes in gene function that are mitotically and/or meiotically heritable and that do not involve any change in DNA sequence were named "epimutation" by Holliday (1984). These changes generally entail DNA methylation and histone/chromatin modifications (Noshay and Springer 2021; Shah 2021). Two types of epimutations can be identified: primary and secondary (Horsthemke 2006). The primary epimutations occur in the absence of any DNA sequence change while the secondary epimutations occur secondary to a DNA mutation in a cis- or trans-acting factor (Oey and Whitelaw 2014). Epimutations can also be recognized as germ line and somatic. Germ line epimutations being derived from the germ line would be present in all of the tissues (constitutive) of an individual, while the somatic epimutations arise in the somatic cells (Hitchins and Ward 2009). These possibilities explain the differences in zygosity of epimutations. Spontaneous epimutations leading to phenotypic changes in plants are well documented (Johannes and Schmitz 2019). DNA methylation is not always faithfully maintained somatic and germ line cells despite the regulation. As a result, cytosine methylation is sometimes gained or lost in a stochastic fashion due to spontaneous epimutations (Johannes and Schmitz 2019) not only in the somatic cells but also in the germ line cells thereby passing through the gametes to subsequent generations, and giving rise to heritable epigenetic variation. Epimutations can also be induced by biotic (Dowen et al. 2012; Bhat et al. 2019a) and abiotic (Verhoeven et al. 2010) stresses in plants. Profiling the

DNA methylomes of Arabidopsis plants exposed to bacterial pathogen and salicylic acid (SA) hormone revealed numerous differentially methylated regions, many of which were transposon-associated regions (Dowen et al. 2012). Multi-generational drought-induced random epimutations were found at a high proportion among the cluster of drought-responsive genes in rice. Many of such epimutations were maintained in advanced generations, and they improved the drought adaptability of offsprings (Zheng et al. 2017).

8.2.3 Gene-Tagged Mutants

A large number of mutants in various crops have been developed using gene-tagging and trapping methods. They employed T-DNA or transposons for generating the insertional inactivation mutants (loss of gene function), activation tagged mutants (gain of gene function), and trans-activation tagged mutants for functional genomics (see Upadhyaya 2007). Till date, a large number of mutants have been characterized and the genes have been identified (Upadhyaya et al. 2003; Lo et al. 2016). Also, *Ds* insertion mutagenesis has been used as an efficient tool to produce diverse variations for rice breeding (Jiang et al. 2007). However, this method of characterizing the mutants largely do not involve any DNA markers since the genes are discovered by isolating the flanking sequences and further validated by overexpression, gene silencing, reversion, etc. (Pereira 2011).

8.2.4 Gene Silencing Mutants

Gene silencing is a straightforward approach to reduce or knockout expression of a gene with the hope of seeing a phenotype that is suggestive of its function (see Ghosh et al. 2020). Transcriptional gene silencing (TGS) and post-transcriptional gene silencing (PTGS) methods have been demonstrated in several plant species with clear advantages over insertional mutagenesis like not being limited by gene redundancy, free from lethal knockouts, lack of non-tagged mutants, and the ability to target the inserted element to a specific gene. Mutant characterization requires gene expression and phenotypic evaluation (see Curtin et al. 2007).

8.2.5 Gene-Edited Mutants

Targeted mutagenesis through genome-editing technologies (Przybyla and Gilbert 2021; Singh and Shekhawat 2021) allows direct and irreversible mutations through nonhomologous end joining of double-stranded breaks generated by CRISPR–Cas9 to get altered phenotype. Recently developed Cas9 variants, novel RNA-guided nucleases and base-editing systems, and DNA-free CRISPR–Cas9 delivery methods now provide great opportunities for inducing mutations (see Yin et al. 2017; Atkins and Voytas 2020; Porto et al. 2020). DNA sequencing and DNA markers are useful in detecting such gene-edited mutations.

Targeted mutagenesis in the conserved coding regions of AhFAD2 genes was undertaken using TALENs (Wen et al. 2018). Mutation frequencies among *AhFAD2* mutant lines were significantly correlated with oleic acid accumulation. Using CRISPR/Cas9 activity, three mutations were generated; G448A in *AhFAD2A*, and 441_442insA and G451T in *AhFAD2B* leading to high oleic acid content in peanut (Yuan et al. 2019). Applications of gene-editing technologies are being demonstrated in various systems (Sattar et al. 2021; Singh and Shekhawat 2021).

8.2.6 Deletion Mutants

Gene isolation for genomic research (in the absence of genome sequence) employs generating random DNA fragments and sequencing. Unfortunately, this strategy requires a labor-intensive assembling of contiguous sequences. In addition, the redundant DNA segments which are quite frequent in plant genomes make this strategy inefficient in terms of time, effort, and cost. The effective methods typically involve the production of a series of nonrandom nested DNA deletions using DNA sonication, DNase I digestion, exonuclease III digestion, and restriction endonuclease digestion. Several nested deletion strategies utilizing PCR have also been reported (see Dennis and Zylstra 2002). In addition, a number of in vivo or in vitro transposon-mediated methods have been developed that utilize random transposon insertions as a binding site for sequencing primers. An alternative method which rapidly produces deletions in a cloned DNA fragment using a rare-cutting restriction enzyme and a frequent-cutting restriction enzyme was also demonstrated to generate contiguous and colinear cloned DNA fragment (Dennis and Zylstra 2002).

Like gene-tagged mutants and gene silencing mutants, these deletion mutants also do not largely involve any DNA markers but depend on expression analysis. Recently, a CRISPR/Cas9 genome-editing system based on RNA endoribonuclease Csy4 processing to induce high-efficiency and inheritable targeted deletion of transcription factors involved in floral development was reported in Arabidopsis (Liu et al. 2019).

8.3 Characterization of Mutants

Mutations in both somatic and germ lines are retained at a tolerable level in spite of DNA repair mechanisms (photoreactivation, excision repair, and postreplication recombination repair) which are probably universal. These mutations are invaluable to the process of evolution, and they provide the new alleles required for the various types of genetic analysis (Mendel's two-factor crosses to chromosome mapping to studies on genetic structures of populations. Mutations have extensively been used for elucidating the metabolic pathways. These applications of mutations are now being increasingly realized with the use of several advanced molecular tools and techniques. Among them, genetic markers (Amom and Nongdam 2017; Nadeem et al. 2018; Barman and Kundu 2019) are being used for various purposes like mutant characterization, mapping the mutations, fine-mapping, candidate gene discovery, etc. The genetic marker is a gene or DNA sequence that directly controls a trait or shows linkage/association with a trait. They can be classical markers (morphological and biochemical) and DNA/molecular markers. Genetic markers are playing a major role in detecting the mutations (Wu et al. 2012) and evaluating mutant and non-mutant populations (Gupta et al. 1999; Jehan and Lakhanpaul 2006) since 1970s.

8.3.1 Morphological Markers

Morphological markers can visually distinguish phenotypes at the traits like seed structure, flower color, growth habit, and other important agronomic traits. They are easy to use, with no requirement for specific instruments and specialized biochemical and molecular techniques to measure. Wagner et al. (1992) used hypocotyl color, monogerm character, pollen fertility, and stem fasciation as the morphological markers to construct the linkage groups using the mutants in sugar beet. Morphological markers have been successfully employed for linkage map construction and diversity analysis among the mutants of alfalfa (Kiss et al. 1993), rye (Voylokov et al. 1998), banana (Miri et al. 2009), etc. However, the main disadvantages of morphological markers are: they are limited in number, influenced by the plant growth stages and various environmental factors.

8.3.2 Biochemical Markers

With the application of biochemical techniques, identifying the variation in the physical and chemical properties of proteins has been made possible using the gel electrophoresis. Forms of the same protein that show different mobility in gel electrophoresis due to the difference in amino acid sequence and the electric charge are called allozymes. Since allozymes are coded by different alleles, allozymic variation is direct indication of genetic variation. Specifically, two or more forms an enzyme coded by the different alleles of the same gene are called allozymes, and the enzymes that process or catalyze the same reaction but coded by different genes are called as isozymes. Though allozyme analysis has the advantage of being relatively rapid, cost-effective, efficient, and sampling being spread over a variety of presumably independent gene loci, it has the chief disadvantage of relatively low abundance and low level of polymorphism. In many species, the maximum allozymic variation is 20–30%. Another drawback of allozyme electrophoresis is that the bands (alleles) that have the same electric charge and migrate to the same pole in the gel may not be truly allozymic. In an initial effort on biochemical markers, Scandalios and Espiritu (1969) reported two forms of aminopeptidase

(AmP) using conventional zone electrophoresis in *Pisum sativum*. Isozyme analysis was used for characterizing the mutants in maize (Schwartz 1971), tomato (Caruso and Glier 1973; Mattoo and Vickery 1977), rice (Alvarez et al. 2000), etc.

With the advances in proteomics (Ingole et al. 2021), metabolomics (Francisco et al. 2021; Murphy et al. 2021; Shen et al. 2021; Zheng et al. 2021), and ionomics (Murgia and Vigani 2015; Sevanthi et al. 2018), the mutant characterization and gene identification turned out to be more robust. The role of gibberellins (GAs) in germination of Arabidopsis seeds was examined by a proteomic approach, and changes in 46 proteins were detected during germination using two-dimensional (2D) electrophoresis (Gallardo et al. 2002). 2D electrophoresis was used to analyze the proteome of the salt-tolerant mutant (RH8706-49) and the salt-sensitive mutant (H8706-34) of wheat (Huo et al. 2004). With MALDI-TOF-MS analysis, the qualitative and quantitative differences were identified between the two mutants for five chloroplast candidate proteins: H^+-transporting two-sector ATPase, glutamine synthetase 2 precursor, putative 33 kDa oxygen evolving protein of photosystem II and ribulose-1,5-bisphosphate carboxylase/oxygenase small subunit.

8.3.3 DNA Markers

DNA/molecular markers detect differences in DNA sequence arising due to base-pair substitutions, deletions, and additions, and are thus less ambiguous than the morphological and biochemical markers, which depend upon gene expression (Williams et al. 1993). Wessler and Varagona (1985) attempted the molecular analysis of more than 40 mutant alleles of the waxy (*Wx*) locus of maize which were phenotypically unstable due to the insertion of the maize transposable activator (*Ac*) and dissociation (*Ds*) elements. The Wx transcription unit had insertions ranging in size from 150 base-pairs to 6.1 kilobases and deletions resulting in the mutant phenotype. Later, Spell et al. (1988) sequenced the upstream region of Wx transcription unit and found restriction fragment length polymorphism (RFLP). Okagaki and Wessler (1988) cloned the *waxy* gene from maize and rice and found that many restriction sites within the translated exons were conserved. Helentjaris et al. (1986) constructed the first genetic map based on the RFLP markers in maize and tomato. RFLP markers were also employed to construct genetic map using the Arabidopsis mutants (Chang et al. 1988). *Petunia hybrida* line V30 harbors three dihydroflavonol-4-reductase (DFR) genes (A, B, C), and they were mapped by RFLP analysis on three different chromosomes (IV, II, and VI, respectively) (Beld et al. 1989).

Williams et al. (1993) demonstrated mapping of mutant genes using RAPD markers which employed a set of primers corresponding to mapped RAPDs distributed throughout the genome of Arabidopsis on the pools of F_2 plants obtained by crossing homozygous wild type genotype with a mutant. Subsequent advances in marker technology witnessed the use of AFLP (Castiglioni et al. 1998; Qu et al. 1998; Li et al. 2000), SSR (Tang et al. 2001), ISSR (Schwarz-Sommer et al. 2003), DArT (Vipin et al. 2013; Shasidhar et al. 2017), and EST (Kuraparthy et al. 2008) markers for mutant characterization and mapping of mutations. da Silva et al. (2016) employed TRAP and SRAP markers on guarana plant accessions which showed more polymorphism than the RAPD markers. SRAP markers were also useful in capturing the variation among morphologically similar accessions. Utility of TRAP markers to determine indel mutation frequencies induced by gamma ray irradiation was also demonstrated in faba bean (*Vicia faba* L.) by Lee et al. (2019). The TRAP markers distinguished mutant lines and showed association between mutation frequency and gamma doses.

By far the most common type of molecular variation exists at single nucleotide level in terms of substitutions, insertions, and deletions. These polymorphisms collectively are referred to as single nucleotide polymorphisms (SNPs). SNP genotyping can be done either at whole-genome level or reduced genome representation level (atrial genome) to differentiate the mutants from its parents. Till date, a large number of SNPs have been identified and their structural and functional features have been studied (Bhat et al. 2022).

An initial study with SNP-derived cleaved amplified polymorphic sequence (CAPS) analysis could identify the structural variations for the rice semi-dwarfing gene, *sd-1*, the rice "green revolution gene" encoding a mutant enzyme involved in gibberellin synthesis (Monna et al. 2002) between the Dee-Geo-Woo-Gen-type sd-1 mutant and the normal-type variety. Further, PCR-RF-SSCP (PRS), which combines CAPS and single-strand conformation polymorphism (SSCP) was used to detect SNP between wild type *Waxy* gene and its mutant types (Sato and Nishio 2003). SNP detection reactions based on multiplexed primer extension assay called multiplexed SNaP shot assay and matrix-assisted laser desorption/ionization time-of-flight (MALDI-ToF) assay were established for high-throughput SNP genotyping the mutants in Arabidopsis (Torjek et al. 2003). SNPs derived via diversity arrays technology (DArT)seq were used to map liguleless mutant (LM) of *Aegilops tauschii* (Dresvyannikova et al. 2019).

Genotyping by sequencing (GBS)-derived SNPs were used to map the homologous transformation sterility gene (*hts*) in wheat (Yang et al. 2018), Anthocyanin Acyltransferase1 (AAT1) gene in maize (Paulsmeyer et al. 2018), the three-pistil gene (*Pis1*) in wheat (Yang et al. 2017) Breviaristatum-e (*ari-e*) locus in cultivated barley (Liu et al. 2014), etc. where the polymorphic alleles were derived from the mutations.

Genetic mapping using a 9 K SNP genotyping assay and restriction site-associated DNA sequencing (RAD-Seq) on bulked segregants derived from a cross between the susceptible cultivar Columbus, thought to possess the suppressor, and Columbus-NS766, a resistant, near-isogenic line believed to contain a mutant non-suppressor allele introgressed from Canthatch could identify the markers linked to a locus of stem rust resistance in wheat (Pujol et al. 2015).

To expedite the discovery of mutant gene for assessing gene function, the SHOREmap (SHOrtREad map) (Schneeberger et al. 2009), NGM (Next-Generation Mapping) (Austin et al. 2011), MutMap (Abe et al. 2012), MutMap+ (Fekih et al. 2013), MutMap-Gap (Takagi et al. 2013b), and QTL-Seq (Takagi et al. 2013a) methods were developed. In all of these methods, except QTL-Seq, mutation is generated using mutagens and the mutant population is screened for the desired phenotype.

Later, a few other variants of SHOREmap like synteny-based mapping-by-sequencing (Galvao et al. 2012) and SHOREmap (version 3.0) (Sun and Schneeberger 2015) were introduced. SHOREmap, the GATK pipeline, and the samtools pipeline were used to identify the mutations in HASTY locus governing leaf hyponasty in Arabidopsis (Allen et al. 2013). NGM was used to identify three genes involved in cell-wall biology in *Arabidopsis thaliana* (Austin et al. 2011). MutMap-Gap was applied to isolate the blast-resistant gene *Pii* from the rice cv. Hitomebore using mutant lines that have lost *Pii* function (Takagi et al. 2013b). These related techniques were used to identify the mutation in ent-kaurene synthase, a key enzyme involved in gibberellin biosynthesis conferring a non-heading phenotype in Chinese cabbage (*Brassica rapa* L. ssp. pekinensis) (Gao et al. 2020), GWC1 for high grain quality in rice (Guo et al. 2020b), PAMP-triggered immunity in Arabidopsis (Kato et al. 2020), single nucleotide substitution at the 3′-end of SBPase gene involved in Calvin cycle affecting plant growth and grain yield in rice (Li et al. 2020), ZmCLE7 underlying fasciation in maize (Tran et al. 2020), Brnym1, a magnesium-dechelatase protein, causing a stay-green phenotype in an EMS-mutagenized Chinese cabbage (Wang et al. 2020), PINOID regulating floral organ development by modulating auxin transport and interacting with MADS16 in rice (Wu et al. 2020), etc.

Genes underlying mutant phenotypes can be isolated and characterized by combining marker discovery, genetic mapping, and resequencing. However, direct comparison of mutant and wild type genomes could be a straightforward method to identify the mutant loci. NIKS (needle in the k-stack), a reference-free algorithm based on comparing k-mers in whole-genome sequencing data for precise discovery of homozygous mutations was proposed (Nordstrom et al. 2013). NIKS was applied in eight mutants induced in nonreference rice cultivars and also in two mutants of the nonmodel species *Arabis alpina*. In both the species, comparing pooled F_2 plants selected for mutant phenotypes revealed small sets of causal mutations. Thus, NIKS enables forward genetics without requiring segregating populations, genetic maps, and reference sequences possibly in any species. Later, NIKS along with high-resolution melting (HRM) was used to identify regulators of postharvest senescence in EMS-derived mutants of Arabidopsis (Hunter et al. 2018).

A targeting-induced local lesions in genomes (TILLING) population was developed by irradiating rice seeds with X-rays, and the wild type and the early-maturing mutant were subjected to whole-genome resequencing (WGRS) to identify the SNPs. The expression of at least 202 structurally altered genes was changed in the mutant, and functional enrichment analysis of these genes revealed that their molecular functions were related to flower development (Hwang et al. 2014).

Genome-wide survey of artificial mutations induced by ethyl methanesulfonate and gamma rays in tomato indicated that C/G to T/A transitions were predominant in the EMS mutants, while C/G to T/A transitions, A/T to T/A transversions, A/T to G/C transitions, and deletion mutations were equally common in the gamma ray mutants. More than 90% of the mutations were located in intergenic regions, and only 0.2% were deleterious (Shirasawa et al. 2016).

8.3.4 Transcriptome and miRNA Profiling

RNA-Seq (RNA-sequencing) is a technique employed to examine the quantity and sequences of RNA in a sample using next-generation sequencing (NGS). It has become an indispensable tool for transcriptome-wide analysis of differential gene expression, differential splicing of mRNAs and RNA biology (single-cell gene expression, translation, and RNA structure) (Stark et al. 2019). RNA-Seq has been successfully used for characterizing the transcriptome of mutants in comparison to the wild types, thus identifying the underlying genes.

RNA-Seq was used to characterize a late leaf spot (LLS) disease susceptible EMS-derived mutant (M14) from a resistant genotype (Yuanza 9102) of peanut, and to identify the candidate genes under diseased condition. A total of 2219 differentially expressed genes including 1317 up-regulated genes and 902 down-regulated genes were detected. Pathogenesis-related (PR) protein genes were significantly up-regulated, while photosynthesis genes were down-regulated in M14. Moreover, the up-regulated WRKY transcription factors and down-regulated plant hormones related to plant growth were detected in the M14 (Han et al. 2017). RNA-Seq was also used to understand the molecular and genetic basis of plant height using a semi-dwarf peanut mutant and its wild line Fenghua 1 (FH1) at the mature stage. The DEGs were involved in hormone biosynthesis and signaling pathways, cell-wall synthetic and metabolic pathways (Guo et al. 2020a).

Transcriptome analysis and miRNA profile sequencing of seeds from the hydroxyproline (HYP)-tolerant mutant and its parent (Huayu 20) was conducted to elucidate the molecular basis of higher grain size and oil content. Major transcription factors linked to seed development and/or oil biosynthesis were differentially expressed between the genotypes. Moreover, differentially expressed genes related to seed development or oil biosynthesis were also identified. Differentially expressed miRNAs (116) and their target genes playing important role in seed development were identified (Sui et al. 2019).

8.3.5 Transposable Element Markers

Various types of transposable elements together make up a large (50–70%) portion of the plant genome. Largely, two types of TEs have been identified based on the transposition intermediates. Class I TEs transpose through RNA, while class II TEs do so by DNA. This difference results in the increase in the copy number of class I TEs within the genome. Structurally, class I TEs (specifically called retrotransposons) can either have direct long terminal repeats (LTRs) (as in LTR retrotransposons) or can be free from LTRs (as in non-LTR retrotransposons). Class II transposons generally have short terminal inverted repeat (TIR). Transposons have allowed the development of TE markers (Izsvak et al. 1999; Casa et al. 2000; Kumar and Hirochika 2001) because of their ubiquitous and wide distribution in genomes, and high polymorphism at their insertion sites. Also, the transposons have a particular genomic and transcriptional pattern thereby differentially influencing the genome and transcriptome (Marcon et al. 2015), and shaping the organization, stability (Chen and Ni 2006; Lu et al. 2012), and evolution of the genome (Casacuberta and Santiago 2003). Since a large number of mutations and genetic instabilities involve TE polymorphisms, development of TE markers (Flavell et al. 1998; Bhat et al. 2019b; Venkatesh and Nandini 2020) to detect such polymorphisms is very important.

8.3.6 Retrotransposon Markers

Several types of retrotransposon markers have been developed. They are sequence-specific amplification polymorphisms (S-SAP) (Waugh et al. 1997), inter-retrotransposon amplified polymorphism (IRAP), retrotransposon-microsatellite amplified polymorphism (REMAP) (Kalendar et al. 1999), and iPBS (Kalendar et al. 2010). Recently, a high-throughput sequencing platform was developed for an efficient screening of LTR retrotransposon families that show high levels of insertion polymorphism among closely related cultivars (Monden et al. 2014a, d). This approach was tested in strawberry genome to determine 24 LTR retrotransposon families. Among them, several families were experimentally confirmed for their high levels of insertion polymorphism among closely related cultivars. Additionally, a large number of insertion sites for retrotransposon families that showed diverse insertion patterns were identified using high-throughput sequencing (Yamane et al. 2012; Monden et al. 2014b, c, e). A marker system based on TaqMan quantitative PCR (qPCR) combined with retrotransposon-based insertion polymorphism (RBIP) was developed to estimate the dosage of an LTR retrotransposon (*scIvana*) in sugarcane (Metcalfe et al. 2015). Apart from LTR retrotransposons, Short Interspersed Nuclear Elements (SINEs), a non-LTR retrotransposon, was employed for the development Inter-SINE Amplified Polymorphism (ISAP) marker, which was employed for genotype-specific high-resolution fingerprinting in potato (Wenke et al. 2015). Suppression PCR was used for the whole-genome experimental identification of insertion/deletion polymorphisms of interspersed repeats (Mamedov et al. 2005). The same technique was applied for identifying genome-wide polymorphic insertion sites of PHARE1, a retrotransposon from the red bean (*Vigna angularis* (Willd.) Ohwi & H. Ohashi) (unpublished). Since many of the spontaneous and induced mutants involve TE activities, the aforesaid markers can be employed for mutant characterization. In fact, IRAPs were developed to differentiate the mutants from non-mutants in apple (Antonius-Klemola et al. 2006) and citrus (Du et al. 2018). In addition, IRAPs were also used in wheat (Belyayev et al. 2010; Nasri et al. 2013), flax (Smýkal et al. 2011), and *Lallemantia iberica* (Cheraghi et al. 2018), and REMAPs were generated in apple Antonius-Klemola et al. (2006), barley (Kalendar et al. 1999, 2000), wheat (Nasri et al. 2013), *Lallemantia iberica* (Cheraghi et al. 2018), almond (Sorkheh et al. 2017), etc. Sequence-specific amplified polymorphism (S-SAP) was developed through modification of the AFLP method, which has been widely used in several plant species (Waugh et al. 1997; Syed et al. 2005; Lou and Chen 2007; Konovalov et al. 2010; Petit et al. 2010; Melnikova et al. 2012; Du et al. 2018).

iPBS was used in *Tetradium ruticarpum* (Xu et al. 2018) and many Fagaceae species (Coutinho et al. 2018). To screen diverse LTR retrotransposon families at a genome-wide scale, iPBS method was employed using the Illumina HiSeq 2000 sequencing platform to obtain LTR sequences in citrus, apple, and soybean (unpublished data).

8.3.7 DNA Transposon Markers

Among the DNA transposons, miniature inverted-repeat transposable elements (MITEs) are most abundant (1000–15,000 per haploid genome) among the genomes. When compared to the non-autonomous elements of class II TEs, MITEs are relatively small (100–500 bp) and have a preference for insertion into 2–3 bp targets that are rich in A and T residues (Casa et al. 2000). Various families of MITEs have been described (Zhang et al. 2000). *Heartbreaker* (*Hbr*) family was the first one to be considered for marker development (Casa et al. 2000) after it was identified (Johal and Briggs 1992) and isolated from maize (Zhang et al. 2000). Subsequently, the diversity and dynamics of *DcMaster*-like elements of the *PIF/Harbinger* superfamily was studied in carrot (Grzebelus and Simon 2009) and *Medicago truncatula* (Grzebelus et al. 2007b). *Vulmar/VulMITE* were identified in *Chenopodiaceae* subfamily *Betoideae* (Grzebelus et al. 2011). Genomic distribution of MITEs in barley was determined by MITE-AFLP mapping (Takahashi et al. 2006), and the novel MITEs were isolated and analyzed for marker utility (Lyons et al. 2008). TEs were also identified using whole-genome sequencing in banana (Hřibová et al. 2010), peanut (Bertioli et al. 2016), wheat (Wanjugi et al. 2009), foxtail millet (Yadav et al. 2015), etc.

An exhaustive effort has been made in peanut to identify (Patel et al. 2004) and develop *Arachis hypogaea* miniature inverted-repeat transposable element 1 (*AhMITE1*) markers (Bhat et al. 2008; Gowda et al. 2010, 2011; Shirasawa et al. 2012a, b; Gayathri et al. 2018). Before the availability of the genome sequence of peanut (or its progenitors), Shirasawa et al. (2012a), followed a modified method of Nunome et al. (2006), which is generally used to develop SSR markers, to find out the polymorphic sites of *AhMITE1*.

Transposon display (TD) was also attempted for TEs other than MITE. TD for *dTph1* (*Ac/Ds* family) in *Petunia hybrida* (Van den Broeck et al. 1998), *DcMaster* (PIF/Harbinger-like) in carrot (Grzebelus et al. 2007a), *Rim2/Hipa* (CACTA family) in *Oryza* ssp. (Kwon et al. 2005), *nDart* (*hAT* family) in rice (Takagi et al. 2007) etc. have been reported in the past.

Alternatively, transposon sites in genomes could be identified from the whole-genome sequence reads obtained by next-generation sequencing technology. Many tools are available for TE discovery and annotation (Goerner-Potvin and Bourque 2018). In peanut, with the availability of the genome sequences of the diploid progenitors of peanut (Bertioli et al. 2016), efforts were made to identify the genome-wide distribution of *AhMITE1* (Gayathri et al. 2018). For this, a set of diverse genotypes (33) including the genetically unstable peanut mutants which show hyperactivity of *AhMITE1* (Hake et al. 2018) were used to discover the *AhMITE1* insertion polymorphic sites (AIPs). Whole-genome re-sequencing (WGRS) reads from a large number of diverse genotypes were analyzed using the computational method polymorphic TEs and their movement detection (PTEMD) (Kang et al. 2016) for the de novo discovery of AIPs. This tool has been applied in peanut lines (Gayathri et al. 2018) to find hundreds of *AhMITE1* copies over the diploid peanut genomes and to develop *AhMITE1* markers. These AhTE markers showed as high as 20% polymorphism. This effort demonstrated the utility of mutants as a source of TE polymorphism to develop TE markers. These AhTE markers were used to find the differences between TMV 2 and its EMS-derived mutant TMV 2-NLM for the *AhMITE1* insertion polymorphisms and to map important taxonomic and productivity traits in peanut (Hake et al. 2017b; Jadhav et al. 2021). AhTE markers were also employed to identify the marker-trait associations for disease resistance (Kolekar et al. 2016; Shirasawa et al. 2018), quality traits (Hake et al. 2017a), and nutritional traits (Nayak et al. 2020) in peanut. Checking background genome recovery in marker-assisted backcross breeding is an important activity to ensure that the promising backcross line resembles the recurrent parent at all other genomic region except the region being transferred. AhTE markers have been successfully used assessing the background genome recovery in peanut (Yeri and Bhat 2016; Kolekar et al. 2017).

8.3.8 Markers to Detect Epimutations

With the growing understanding on epimutations, techniques to detect DNA methylation and histone modifications are being developed (Albertini and Marconi 2013; Ghosh et al. 2021). The methods which use comprehensive knowledge of an organism's genome sequence are bisulfite modification and chromatin immunoprecipitation (ChIP). Methylation-sensitive amplification polymorphism (MSAP) is a modified amplified fragment length polymorphism (AFLP) technique which does not require genome sequencing. High-performance separation techniques such as high-performance capillary electrophoresis (HPCE) and high-performance liquid chromatography (HPLC) can also be used for detecting the cytosine methylation (Ghosh et al. 2021). metAFLP approach which allows for partition of complex variation into sequence changes, de novo methylation and demethylation was proposed and employed among the regenerants derived tissue culture in triticale (Machczyńska et al. 2014). A computational method AlphaBeta was demonstrated for estimating the precise rate of stochastic events using pedigree-based DNA methylation data as input to study transgenerationally heritable epimutations in clonal or sexually derived mutation accumulation lines, as well as somatic epimutations in long-lived perennials (Shahryary et al. 2020).

Initial method for the identification and quantification of histone post-translational modifications depended on mass spectrometry (Freitas et al. 2004). Later, the techniques like DNase-seq, which is based on nuclease DNase I, and ATAC-seq, which is based on transposase Tn5, have been widely used to identify genomic regions associated with open chromatin. These techniques have played a key role in dissecting the regulatory networks in gene expression in both animal and plant species. Zhao et al. (2020) developed a technique, named MNase hypersensitivity sequencing (MH-seq), to identify genomic regions associated with open chromatin in *Arabidopsis thaliana*. Genomic regions enriched with MH-seq reads were referred as MNase hypersensitive sites (MHSs), and these MHSs overlapped with the majority (~90%) of the open chromatin identified previously by DNase-seq and ATAC-seq. Further, 22% MHSs not covered by DNase-seq or ATAC-seq reads were identified by this technique, and they were referred to as "specific MHSs" (sMHSs).

8.4 Opportunities to Develop New Marker Systems

Various marker systems are in demand for their applications in crop improvement. The development of new marker systems requires identification of genome-wide DNA features. Like SNPs and TE insertion polymorphisms, DNA copy number variation (CNV) is an important source of genetic variation, which has been recognized recently (Sebat et al. 2004). Nearly 12% of the human genome is covered by 1447 CNV (Redon et al. 2006). A new method called CNV-Seq was proposed to detect copy number variation using high-throughput sequencing (Xie and Tammi 2009). Genome-wide analysis in the mutant and wild type genotypes could identify the CNVs in wheat (Diaz et al. 2012; Nilsen et al. 2020), soybean (Bolon et al. 2014; Lemay et al. 2019), maize (Jamann et al. 2014), banana (Datta et al. 2018), etc. Yin et al. (2020) observed the structural variations (SVs like CNVs and PAVs) among the resistant gene analogs (RGAs) in peanut. Differential expression between the resistant and the susceptible genotypes was more pronounced among the RGAs with SV than those RGAs without SVs. Thus, availability of SV-based markers in future may help identifying the candidate genes apart from characterizing the mutants.

In peanut, we made an effort to identify SNPs, insertion sites of *AhMITE1*, copy number variations and DNA methylation sites among the mutants and non-mutant genotypes to assess their scope in developing the new marker systems (Table 8.1) by analyzing the WGRS data of 231 peanut genotypes, including a few mutants, available in the public domain as an effort towards marker development (unpublished data). CNV analysis with a mean window size of 7672 bases identified a large number of CNVs in mutants though non-mutant genotypes showed more CNVs across all the chromosomes. Presence and absence variations (PAVs) were also observed at a few loci. Currently, the efforts are in progress to use CNVs and PAVs for marker development by designing the primers flanking the DNA sequence

Table 8.1 Single nucleotide polymorphism, copy number variation, *AhMITE1* insertion polymorphism, and DNA methylation sites among the peanut genotypes (Hake et al. 2018; Bhat et al. 2019a)

	SNP		CNV		*AhMITE1*		DNA methylation	
Chromosome	Mutant	Non-mutant	Mutant	Non-mutant	Mutant	Non-mutant	Mutant	Non-mutant
Arahy.01	139,343	26,676	5405	12,409	11	22	1433	1313
Arahy.02	49,249	22,023	406	11,935	6	11	2941	2527
Arahy.03	61,709	36,329	827	15,847	21	26	2631	3265
Arahy.04	54,021	28,542	655	14,451	14	21	1957	1707
Arahy.05	39,960	22,944	794	13,217	16	25	1444	1238
Arahy.06	43,593	26,485	94	13,852	17	22	2218	2615
Arahy.07	39,357	20,909	300	9213	8	12	2260	1870
Arahy.08	31,490	16,201	55	7769	14	22	1630	1858
Arahy.09	41,581	26,084	293	12,614	10	18	1928	2182
Arahy.10	43,998	23,094	599	12,980	14	19	2266	2419
Arahy.11	13,756	16,960	753	12,895	11	21	1664	2129
Arahy.12	15,756	16,077	260	12,743	8	17	2839	2897
Arahy.13	20,489	18,923	478	15,012	20	28	3650	3874
Arahy.14	19,068	16,148	6152	18,436	15	22	4305	3246
Arahy.15	15,913	18,091	3260	15,449	16	30	3494	3196
Arahy.16	21,729	17,491	2203	16,088	17	21	2926	3038
Arahy.17	15,080	16,699	1204	12,089	16	23	3037	3347
Arahy.18	14,110	14,947	319	13,389	14	19	2360	2143
Arahy.19	31,283	19,173	290	14,826	18	26	4512	4033
Arahy.20	27,414	16,237	1446	12,900	6	16	3300	3190
Total	738,899	420,033	25,793	268,114	271	420	52,790	52,083

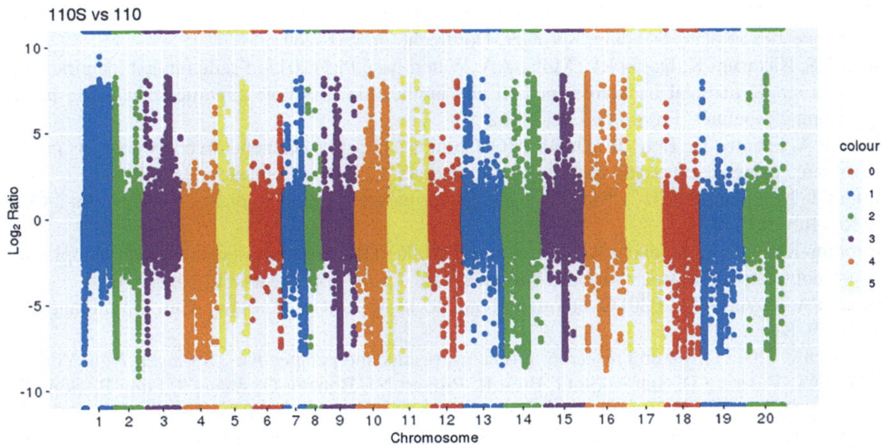

Fig. 8.2 Copy number variation between the mutant and the wild types of peanuts (Hake et al. 2018)

which shows CNV/PAV. Peanut mutants also showed considerably large number of SNPs, *AhMITE1* insertion sites, and DNA methylation sites though the numbers were more in non-mutant genotypes (Table 8.1), indicating the possibility of developing the markers to detect the SNPs, *AhMITE1,* and DNA methylation for various applications (Fig. 8.2).

8.5 Conclusions and Prospects

Beginning with the morphological markers to biochemical markers to DNA markers, they have been successfully used for mutant characterization, and mapping and identification of mutant loci in various plant systems. Considerably large number of genes have been identified from these mutants for various uses. Other categories of mutants obtained with DNA tagging, gene editing, gene silencing, etc. have also contributed extensively for the gene discovery. In spite of these progresses, efforts are underway to use other genomic features (CNVs, PAVs) for marker development. DNA and RNA sequencing and methylome sequencing are being applied in various crops to detect the mutations and epimutations. In the future, the mutant characterization at global levels of genome, transcriptome, and methylome could greatly enhance the efficiency of gene identification for better crop productivity.

References

Abe A, Kosugi S, Yoshida K, Natsume S, Takagi H, Kanzaki H, Matsumura H, Yoshida K, Mitsuoka C, Tamiru M (2012) Genome sequencing reveals agronomically important loci in rice using MutMap. Nat Biotechnol 30(2):174–178

Albertini E, Marconi G (2013) Methylation-based markers. In: Lübberstedt T, Varshney RK (eds) Diagnostics in plant breeding. Springer, Dordrecht, pp 251–264

Allen RS, Nakasugi K, Doran RL, Millar AA, Waterhouse PM (2013) Facile mutant identification via a single parental backcross method and application of whole genome sequencing based mapping pipelines. Front Plant Sci 4:362

Alvarez A, Fuentes JL, Deus JE, Duque MC, Cornide MT (2000) Genetic diversity analysis in rice mutants using isozyme and morphological markers. Cultivos Tropicales 21(4):39–44

Amom T, Nongdam P (2017) The use of molecular marker methods in plants: a review. Int J Curr Res Rev 9(17):1–7

Antonius-Klemola K, Kalendar R, Schulman AH (2006) TRIM retrotransposons occur in apple and are polymorphic between varieties but not sports. Theor Appl Genet 112(6):999–1008

Atkins PA, Voytas DF (2020) Overcoming bottlenecks in plant gene editing. Curr Opin Plant Biol 54:79–84

Auerbach C (2013) Mutation research: problems, results and perspectives. Springer, New York

Austin RS, Vidaurre D, Stamatiou G, Breit R, Provart NJ, Bonetta D, Zhang J, Fung P, Gong Y, Wang PW (2011) Next-generation mapping of Arabidopsis genes. Plant J 67(4):715–725

Barman M, Kundu S (2019) Molecular markers and a new vista in plant breeding: a review. Int J Curr Microbiol Appl Sci 8(12):1921–1929

Beld M, Martin C, Huits H, Stuitje AR, Gerats AG (1989) Flavonoid synthesis in Petunia hybrida: partial characterization of dihydroflavonol-4-reductase genes. Plant Mol Biol 13(5):491–502

Belyayev A, Kalendar R, Brodsky L, Nevo E, Schulman AH, Raskina O (2010) Transposable elements in a marginal plant population: temporal fluctuations provide new insights into genome evolution of wild diploid wheat. Mob DNA 1(1):6

Bertioli DJ, Cannon SB, Froenicke L, Huang G, Farmer AD, Cannon EK, Liu X, Gao D, Clevenger J, Dash S, Ren L, Moretzsohn MC, Shirasawa K, Huang W, Vidigal B, Abernathy B, Chu Y, Niederhuth CE, Umale P, Araujo AC, Kozik A, Do Kim K, Burow MD, Varshney RK, Wang X, Zhang X, Barkley N, Guimaraes PM, Isobe S, Guo B, Liao B, Stalker HT, Schmitz RJ, Scheffler BE, Leal-Bertioli SC, Xun X, Jackson SA, Michelmore R, Ozias-Akins P (2016) The genome sequences of *Arachis duranensis* and *Arachis ipaensis*, the diploid ancestors of cultivated peanut. Nat Genet 48(4):438–446

Bhat RS, Patil VU, Chandrashekar TM, Sujay V, Gowda MVC, Kuruvinashetti MS (2008) Recovering flanking sequence tags of miniature inverted-repeat transposable element by thermal asymmetric interlaced-PCR in peanut. Curr Sci 95(4):452–453

Bhat RS, Rockey J, Shirasawa K, Tilak IS, Brijesh Patil MP, Reddy VB (2019a) DNA methylation and expression analyses reveal epialleles for the foliar disease resistance genes in peanut (*Arachis hypogaea* L.). BMC Res Notes 13(1):20

Bhat RS, Shirasawa K, Monden Y, Yamashita H, Tahara M (2019b) Developing transposable element marker system for molecular breeding. In: Jain M, Garg R (eds) Legume genomics: methods and protocols. Humana, New York, pp 233–252

Bhat RS, Shirasawa K, Chavadi SD (2022) Genome-wide structural and functional features of single nucleotide polymorphisms revealed from the whole genome resequencing of 179 accessions of *Arachis*. Physiol Plant. https://doi.org/10.1111/ppl.13623

Bolon YT, Stec AO, Michno JM, Roessler J, Bhaskar PB, Ries L, Dobbels AA, Campbell BW, Young NP, Anderson JE, Grant DM, Orf JH, Naeve SL, Muehlbauer GJ, Vance CP, Stupar RM (2014) Genome resilience and prevalence of segmental duplications following fast neutron irradiation of soybean. Genetics 198(3):967–981

Branch W, Tallury S, Clevenger J, Schwartz B, Hanna W (2020) Inheritance of a novel heterozygous peanut mutant, 5-small leaflet. Peanut Sci 47(1):33–37

Caruso JL, Glier JH (1973) Oxidative enzymes of the lanceolate mutant in tomato: a reappraisal. Biochem Genet 8(4):351–355

Casa AM, Brouwer C, Nagel A, Wang L, Zhang Q, Kresovich S, Wessler SR (2000) The MITE family *Heartbreaker* (*Hbr*): molecular markers in maize. Proc Natl Acad Sci 97(18): 10083–10089

Casacuberta JM, Santiago N (2003) Plant LTR-retrotransposons and MITEs: control of transposition and impact on the evolution of plant genes and genomes. Gene 311:1–11

Castiglioni P, Pozzi C, Heun M, Terzi V, Muller KJ, Rohde W, Salamini F (1998) An AFLP-based procedure for the efficient mapping of mutations and DNA probes in barley. Genetics 149(4): 2039–2056

Chang C, Bowman JL, DeJohn AW, Lander ES, Meyerowitz EM (1988) Restriction fragment length polymorphism linkage map for *Arabidopsis thaliana*. Proc Natl Acad Sci 85(18): 6856–6860

Chen ZJ, Ni Z (2006) Mechanisms of genomic rearrangements and gene expression changes in plant polyploids. BioEssays 28(3):240–252

Cheraghi A, Rahmani F, Hassanzadeh-Ghorttapeh A (2018) IRAP and REMAP based genetic diversity among varieties of *Lallemantia iberica*. Mol Biol Res Commun 7(3):125–132

Coutinho JP, Carvalho A, Martin A, Lima-Brito J (2018) Molecular characterization of Fagaceae species using inter-primer binding site (iPBS) markers. Mol Biol Rep 45(2):133–142

Curtin SJ, Wang MB, Watson JM, Roffey P, Blanchard CL, Waterhouse PM (2007) RNA silencing and its application in functional genomics. In: Upadhyaya NM (ed) Rice functional genomics—challenges, progress and prospects. Springer Life Sciences, New York, pp 291–332

da Silva EF, de Sousa SB, da Silva GF, Sousa NR, do Nascimento Filho FJ, Hanada RE (2016) TRAP and SRAP markers to find genetic variability in complex polyploid *Paullinia cupana* var. sorbilis. Plant Gene 6:43–47

Datta S, Jankowicz-Cieslak J, Nielen S, Ingelbrecht I, Till BJ (2018) Induction and recovery of copy number variation in banana through gamma irradiation and low-coverage whole-genome sequencing. Plant Biotechnol J 16(9):1644–1653

Dennis JJ, Zylstra GJ (2002) Rapid generation of nested deletions by differential restriction digestion. BioTechniques 33(2):310, 312, 314–310, 312, 315

Diaz A, Zikhali M, Turner AS, Isaac P, Laurie DA (2012) Copy number variation affecting the Photoperiod-B1 and Vernalization-A1 genes is associated with altered flowering time in wheat (*Triticum aestivum*). PLoS One 7(3):e33234

Dowen RH, Pelizzola M, Schmitz RJ, Lister R, Dowen JM, Nery JR, Dixon JE, Ecker JR (2012) Widespread dynamic DNA methylation in response to biotic stress. Proc Natl Acad Sci 109(32): E2183–E2191

Dresvyannikova AE, Watanabe N, Muterko AF, Krasnikov AA, Goncharov NP, Dobrovolskaya OB (2019) Characterization of a dominant mutation for the liguleless trait: *Aegilops tauschii* liguleless (Lg(t)). BMC Plant Biol 19(Suppl 1):55

Du D, Du X, Mattia MR, Wang Y, Yu Q, Huang M, Yu Y, Grosser JW, Gmitter FG (2018) LTR retrotransposons from the *Citrus* x *clementina* genome: characterization and application. Tree Genet Genomes 14(4):43

Fekih R, Takagi H, Tamiru M, Abe A, Natsume S, Yaegashi H, Sharma S, Sharma S, Kanzaki H, Matsumura H, Saitoh H, Mitsuoka C, Utsushi H, Uemura A, Kanzaki E, Kosugi S, Yoshida K, Cano L, Kamoun S, Terauchi R (2013) MutMap+: genetic mapping and mutant identification without crossing in rice. PLoS One 8(7):e68529

Flavell AJ, Knox MR, Pearce SR, Ellis TN (1998) Retrotransposon-based insertion polymorphisms (RBIP) for high throughput marker analysis. Plant J 16(5):643–650

Francisco M, Kliebenstein DJ, Rodriguez VM, Soengas P, Abilleira R, Cartea ME (2021) Fine-mapping identifies *NAD-ME1* as a candidate underlying a major locus controlling temporal variation in primary and specialized metabolism in Arabidopsis. Plant J 106(2):454–467

Freitas MA, Sklenar AR, Parthun MR (2004) Application of mass spectrometry to the identification and quantification of histone post-translational modifications. J Cell Biochem 92(4):691–700

Gallardo K, Job C, Groot SP, Puype M, Demol H, Vandekerckhove J, Job D (2002) Proteomics of Arabidopsis seed germination. A comparative study of wild-type and gibberellin-deficient seeds. Plant Physiol 129(2):823–837

Galvao VC, Nordstrom KJ, Lanz C, Sulz P, Mathieu J, Pose D, Schmid M, Weigel D, Schneeberger K (2012) Synteny-based mapping-by-sequencing enabled by targeted enrichment. Plant J 71(3): 517–526

Gao Y, Huang S, Qu G, Fu W, Zhang M, Liu Z, Feng H (2020) The mutation of ent-kaurene synthase, a key enzyme involved in gibberellin biosynthesis, confers a non-heading phenotype to Chinese cabbage (*Brassica rapa* L. ssp. pekinensis). Hortic Res 7(1):178

Gardener EJ, Simmons MJ, Snustad DP (1991) Principles of genetics. Wiley, New York

Gayathri M, Shirasawa K, Varshney RK, Pandey MK, Bhat RS (2018) Development of new *AhMITE1* markers through genome-wide analysis in peanut (*Arachis hypogaea* L.). BMC Res Notes 11(1):10

Ghosh S, Biswas A, Nanda S, Das S, Basak S, Mandal S, Biswas A, Sen DJ, Mahanti B, Saha D (2020) Gene Silencing is one of the most efficient and promising functional genomics tools. Res J Sci Technol 12(3):222–232

Ghosh A, Igamberdiev AU, Debnath SC (2021) Tissue culture-induced DNA methylation in crop plants: a review. Mol Biol Rep 48:823–841

Goerner-Potvin P, Bourque G (2018) Computational tools to unmask transposable elements. Nat Rev Genet 19:688–704

Gowda MVC, Bhat RS, Motagi BN, Sujay V, Varshakumari, Bhat S (2010) Association of high-frequency origin of late leaf spot resistant mutants with *AhMITE1* transposition in peanut. Plant Breed 129(5):567–569

Gowda MVC, Bhat RS, Sujay V, Kusuma P, Varshakumari, Bhat S, Varshney RK (2011) Characterization of *AhMITE1* transposition and its association with the mutational and evolutionary origin of botanical types in peanut (*Arachis* spp.). Plant Syst Evol 291(3–4):153–158

Grzebelus D, Simon PW (2009) Diversity of *DcMaster*-like elements of the *PIF/Harbinger* superfamily in the carrot genome. Genetica 135(3):347–353

Grzebelus D, Jagosz B, Simon PW (2007a) The DcMaster transposon display maps polymorphic insertion sites in the carrot (*Daucus carota* L.) genome. Gene 390(1):67–74

Grzebelus D, Lasota S, Gambin T, Kucherov G, Gambin A (2007b) Diversity and structure of *PIF/Harbinger*-like elements in the genome of *Medicago truncatula*. BMC Genomics 8(1):409

Grzebelus D, Stawujak K, Mitoraj J, Szklarczyk M (2011) Dynamics of *Vulmar/VulMITE* group of transposable elements in *Chenopodiaceae* subfamily *Betoideae*. Genetica 139(9):1209–1216

Guo F, Ma J, Hou L, Shi S, Sun J, Li G, Zhao C, Xia H, Zhao S, Wang X, Zhao Y (2020a) Transcriptome profiling provides insights into molecular mechanism in peanut semi-dwarf mutant. BMC Genomics 21(1):211

Guo L, Chen W, Tao L, Hu B, Qu G, Tu B, Yuan H, Ma B, Wang Y, Zhu X, Qin P, Li S (2020b) GWC1 is essential for high grain quality in rice. Plant Sci 296:110497

Gupta P, Varshney R, Sharma P, Ramesh B (1999) Molecular markers and their applications in wheat breeding. Plant Breed 118(5):369–390

Hake AA, Shirasawa K, Yadawad A, Nayak SN, Mondal S, Badigannavar AM, Nadaf HL, Gowda MVC, Bhat RS (2017a) Identification of transposable element markers associated with yield and quality traits from a large population of independent mutants in peanut (*Arachis hypogaea* L.). Euphytica 213(12):283

Hake AA, Shirasawa K, Yadawad A, Sukruth M, Patil M, Nayak SN, Lingaraju S, Patil PV, Nadaf HL, Gowda MVC, Bhat RS (2017b) Mapping of important taxonomic and productivity traits using genic and non-genic transposable element markers in peanut (*Arachis hypogaea* L.). PLoS One 12(10):e0186113

Hake AA, Shirasawa K, Yadawad A, Nadaf HL, Gowda MVC, Bhat RS (2018) Genome-wide structural mutations among the lines resulting from genetic instability in peanut (*Arachis hypogaea* L.). Plant Gene 13:1–7

Han S, Liu H, Yan M, Qi F, Wang Y, Sun Z, Huang B, Dong W, Tang F, Zhang X (2017) Differential gene expression in leaf tissues between mutant and wild-type genotypes response to late leaf spot in peanut (*Arachis hypogaea* L.). PLoS One 12(8):e0183428

Helentjaris T, Slocum M, Wright S, Schaefer A, Nienhuis J (1986) Construction of genetic linkage maps in maize and tomato using restriction fragment length polymorphisms. Theor Appl Genet 72(6):761–769

Hitchins MP, Ward RL (2009) Constitutional (germline) MLH1 epimutation as an aetiological mechanism for hereditary non-polyposis colorectal cancer. J Med Genet 46(12):793–802

Holliday R (1984) The significance of DNA methylation in cellular aging. In: Woodhead AD, Blackett AD, Hollaender A (eds) Molecular biology of aging. Plenum, New York, pp 269–283

Horsthemke B (2006) Epimutations in human disease. Curr Top Microbiol Immunol 310:45–59

Hřibová E, Neumann P, Matsumoto T, Roux N, Macas J, Doležel J (2010) Repetitive part of the banana (*Musa acuminata*) genome investigated by low-depth 454 sequencing. BMC Plant Biol 10(1):204

Hunter DA, Jibran R, Dijkwel P, Chagne D, Sullivan K, Kanojia A, Crowhurst R (2018) Identification of postharvest senescence regulators through map-based cloning using detached Arabidopsis inflorescences as a model tissue. Methods Mol Biol 1744:195–220

Huo CM, Zhao BC, Ge RC, Shen YZ, Huang ZJ (2004) Proteomic analysis of the salt tolerance mutant of wheat under salt stress. Yi Chuan Xue Bao 31(12):1408–1414

Hwang SG, Hwang JG, Kim DS, Jang CS (2014) Genome-wide DNA polymorphism and transcriptome analysis of an early-maturing rice mutant. Genetica 142(1):73–85

Ingole KD, Dahale SK, Bhattacharjee S (2021) Proteomic analysis of SUMO1-SUMOylome changes during defense elicitation in Arabidopsis. J Proteome 232:104054

Izsvak Z, Ivics Z, Shimoda N, Mohn D, Okamoto H, Hackett PB (1999) Short inverted-repeat transposable elements in teleost fish and implications for a mechanism of their amplification. J Mol Evol 48(1):13–21

Jadhav MP, Gangurde SS, Hake AA, Yadwad A, Supriya SM, Pattanashetti SK, Gowda MVC, Shirasawa K, Varshney RK, Pandey MK, Bhat RS (2021) Genotyping-by-sequencing based genetic mapping identified major and consistent genomic regions for productivity and quality traits in peanut. Front Plant Sci 12:2034. https://doi.org/10.3389/fpls.2021.668020

Jamann TM, Poland JA, Kolkman JM, Smith LG, Nelson RJ (2014) Unraveling genomic complexity at a quantitative disease resistance locus in maize. Genetics 198(1):333–344

Jay P, Chouteau M, Whibley A, Bastide H, Parrinello H, Llaurens V, Joron M (2021) Mutation load at a mimicry supergene sheds new light on the evolution of inversion polymorphisms. Nat Genet 53(3):288–293

Jehan T, Lakhanpaul S (2006) Single nucleotide polymorphism (SNP)–methods and applications in plant genetics: a review. Indian J Biotechnol 5:435–459

Jiang S-Y, Bachmann D, La H, Ma Z, Venkatesh P, Ramamoorthy R, Ramachandran S (2007) *Ds* insertion mutagenesis as an efficient tool to produce diverse variations for rice breeding. Plant Mol Biol 65(4):385–402

Johal GS, Briggs SP (1992) Reductase activity encoded by the *HM1* disease resistance gene in maize. Science 258(5084):985–987

Johannes F, Schmitz RJ (2019) Spontaneous epimutations in plants. New Phytol 221(3):1253–1259

Kalendar R, Grob T, Regina M, Suoniemi A, Schulman A (1999) IRAP and REMAP: two new retrotransposon-based DNA fingerprinting techniques. Theor Appl Genet 98(5):704–711

Kalendar R, Tanskanen J, Immonen S, Nevo E, Schulman AH (2000) Genome evolution of wild barley (*Hordeum spontaneum*) by *BARE-1* retrotransposon dynamics in response to sharp microclimatic divergence. Proc Natl Acad Sci 97(12):6603–6607

Kalendar R, Antonius K, Smýkal P, Schulman AH (2010) iPBS: a universal method for DNA fingerprinting and retrotransposon isolation. Theor Appl Genet 121(8):1419–1430

Kang H, Zhu D, Lin R, Opiyo SO, Jiang N, Shiu S-H, Wang G-L (2016) A novel method for identifying polymorphic transposable elements via scanning of high-throughput short reads. DNA Res 23(3):241–251

Kato H, Onai K, Abe A, Shimizu M, Takagi H, Tateda C, Utsushi H, Singkarabanit-Ogawa S, Kitakura S, Ono E, Zipfel C, Takano Y, Ishiura M, Terauchi R (2020) Lumi-Map, a real-time

Luciferase Bioluminescence Screen of mutants combined with MutMap, reveals Arabidopsis genes involved in PAMP-triggered immunity. Mol Plant Microbe Interact 33(12):1366–1380

Kiss GB, Csanádi G, Kálmán K, Kaló P, Ökrész L (1993) Construction of a basic genetic map for alfalfa using RFLP, RAPD, isozyme and morphological markers. Mol Gen Genet 238(1–2): 129–137

Kolekar RM, Sujay V, Shirasawa K, Sukruth M, Khedikar YP, Gowda MVC, Pandey MK, Varshney RK, Bhat RS (2016) QTL mapping for late leaf spot and rust resistance using an improved genetic map and extensive phenotypic data on a recombinant inbred line population in peanut (*Arachis hypogaea* L.). Euphytica 209(1):147–156

Kolekar RM, Sukruth M, Shirasawa K, Nadaf HL, Motagi BN, Lingaraju S, Patil PV, Bhat RS (2017) Marker-assisted backcrossing to develop foliar disease resistant genotypes in TMV 2 variety of peanut (*Arachis hypogaea* L.). Plant Breed 136(6):948–953

Konovalov FA, Goncharov NP, Goryunova S, Shaturova A, Proshlyakova T, Kudryavtsev A (2010) Molecular markers based on LTR retrotransposons *BARE-1* and Jeli uncover different strata of evolutionary relationships in diploid wheats. Mol Gen Genomics 283(6):551–563

Kumar A, Hirochika H (2001) Applications of retrotransposons as genetic tools in plant biology. Trends Plant Sci 6(3):127–134

Kuraparthy V, Sood S, Gill BS (2008) Genomic targeting and mapping of tiller inhibition gene (*tin3*) of wheat using ESTs and synteny with rice. Funct Integr Genomics 8(1):33–42

Kutschera U, Niklas KJ (2004) The modern theory of biological evolution: an expanded synthesis. Naturwissenschaften 91(6):255–276

Kwon S-J, Park K-C, Kim J-H, Lee JK, Kim N-S (2005) *Rim 2/Hipa CACTA* transposon display; a new genetic marker technique in *Oryza* species. BMC Genet 6(1):15

Lee MK, Lyu JI, Hong MJ, Kim DG, Kim JM, Kim JB, Eom SH, Ha BK, Kwon SJ (2019) Utility of TRAP markers to determine indel mutation frequencies induced by gamma-ray irradiation of faba bean (*Vicia faba* L.) seeds. Int J Radiat Biol 95(8):1160–1171

Lemay MA, Torkamaneh D, Rigaill G, Boyle B, Stec AO, Stupar RM, Belzile F (2019) Screening populations for copy number variation using genotyping-by-sequencing: a proof of concept using soybean fast neutron mutants. BMC Genomics 20(1):634

Li CY, Zheng HG, Weng ML, Jia JH, Mou TM, Nguyen HT, Wang B (2000) AFLP analysis of photoperiod-sensitive genic male sterile (PGMS) rice mutant lines. Sheng Wu Gong Cheng Xue Bao 16(1):91–95

Li C, Li N, Huang R, Chen C, Guo J, Yang X, Zhang X, Sun C, Deng X, Wang P (2020) A single nucleotide substitution at the 3′-end of SBPase gene involved in Calvin cycle severely affects plant growth and grain yield in rice. BMC Plant Biol 20(1):345

Liu H, Bayer M, Druka A, Russell JR, Hackett CA, Poland J, Ramsay L, Hedley PE, Waugh R (2014) An evaluation of genotyping by sequencing (GBS) to map the Breviaristatum-e (*ari-e*) locus in cultivated barley. BMC Genomics 15:104

Liu Y, Gao Y, Gao Y, Zhang Q (2019) Targeted deletion of floral development genes in Arabidopsis with CRISPR/Cas9 using the RNA endoribonuclease Csy4 processing system. Hortic Res 6(1):1–10

Lo SF, Fan MJ, Hsing YI, Chen LJ, Chen S, Wen IC, Liu YL, Chen KT, Jiang MJ, Lin MK (2016) Genetic resources offer efficient tools for rice functional genomics research. Plant Cell Environ 39(5):998–1013

Lou Q, Chen J (2007) *Ty1-copia* retrotransposon-based SSAP marker development and its potential in the genetic study of cucurbits. Genome 50(9):802–810

Lu C, Chen J, Zhang Y, Hu Q, Su W, Kuang H (2012) Miniature inverted–repeat transposable elements (MITEs) have been accumulated through amplification bursts and play important roles in gene expression and species diversity in *Oryza sativa*. Mol Biol Evol 29(3):1005–1017

Lyons M, Cardle L, Rostoks N, Waugh R, Flavell AJ (2008) Isolation, analysis and marker utility of novel miniature inverted repeat transposable elements from the barley genome. Mol Gen Genomics 280(4):275–285

Machczyńska J, Orłowska R, Zimny J, Bednarek PT (2014) Extended metAFLP approach in studies of tissue culture induced variation (TCIV) in triticale. Mol Breed 34(3):845–854

Mamedov IZ, Arzumanyan ES, Amosova AL, Lebedev YB, Sverdlov ED (2005) Whole-genome experimental identification of insertion/deletion polymorphisms of interspersed repeats by a new general approach. Nucleic Acids Res 33(2):e16

Marcon HS, Domingues DS, Silva JC, Borges RJ, Matioli FF, de Mattos Fontes MR, Marino CL (2015) Transcriptionally active LTR retrotransposons in *Eucalyptus* genus are differentially expressed and insertionally polymorphic. BMC Plant Biol 15(1):198

Mattoo AK, Vickery RS (1977) Subcellular distributions of isoenzymes in fruits of a normal cultivar of tomato and of the rin mutant at two stages of development. Plant Physiol 60(4): 496–498

Melnikova NV, Kudryavtseva AV, Speranskaya AS, Krinitsina AA, Dmitriev AA, Belenikin MS, Upelniek VP, Batrak ER, Kovaleva IS, Kudryavtsev AM (2012) The *FaRE1* LTR-retrotransposon based SSAP markers reveal genetic polymorphism of strawberry (*Fragaria* × *ananassa*) cultivars. J Agric Sci 4(11):111

Metcalfe CJ, Oliveira SG, Gaiarsa JW, Aitken KS, Carneiro MS, Zatti F, Van Sluys M-A (2015) Using quantitative PCR with retrotransposon-based insertion polymorphisms as markers in sugarcane. J Exp Bot 66(14):4239–4250

Miri SM, Mousavi A, Naghavi MR, Mirzaei M, Talaei AR, Naserian KB (2009) Analysis of induced mutants of salinity resistant banana (*Musa acuminata* cv. Dwarf Cavendish) using morphological and molecular markers. Iran J Biotechnol 7(2):86–92

Monden Y, Fujii N, Yamaguchi K, Ikeo K, Nakazawa Y, Waki T, Hirashima K, Uchimura Y, Tahara M (2014a) Efficient screening of long terminal repeat retrotransposons that show high insertion polymorphism via high-throughput sequencing of the primer binding site. Genome 57(5):245–252

Monden Y, Takai T, Tahara M (2014b) Characterization of a novel retrotransposon *TriRe-1* using nullisomic-tetrasomic lines of hexaploid wheat. Sci Rep Fac Agric Okayama Univ 103:21–30

Monden Y, Takai T, Tahara M, Umeno Y, Nakamura R (2014c) High-throughput development of DNA markers for wheat cultivar discrimination based on an active retrotransposon TriRe-1 insertion polymorphism. DNA Polymorphism 22:60–65

Monden Y, Yamaguchi K, Tahara M (2014d) Application of iPBS in high-throughput sequencing for the development of retrotransposon-based molecular markers. Curr Plant Biol 1:40–44

Monden Y, Yamamoto A, Shindo A, Tahara M (2014e) Efficient DNA fingerprinting based on the targeted sequencing of active retrotransposon insertion sites using a bench-top high-throughput sequencing platform. DNA Res 2014:dsu015

Monna L, Kitazawa N, Yoshino R, Suzuki J, Masuda H, Maehara Y, Tanji M, Sato M, Nasu S, Minobe Y (2002) Positional cloning of rice semidwarfing gene, *sd-1*: rice "green revolution gene" encodes a mutant enzyme involved in gibberellin synthesis. DNA Res 9(1):11–17

Murgia I, Vigani G (2015) Analysis of *Arabidopsis thaliana* atfer4-1, atfh and atfer4-1/atfh mutants uncovers frataxin and ferritin contributions to leaf ionome homeostasis. Plant Physiol Biochem 94:65–72

Murphy KM, Edwards J, Louie KB, Bowen BP, Sundaresan V, Northen TR, Zerbe P (2021) Bioactive diterpenoids impact the composition of the root-associated microbiome in maize (*Zea mays*). Sci Rep 11(1):333

Nadeem MA, Nawaz MA, Shahid MQ, Doğan Y, Comertpay G, Yıldız M, Hatipoğlu R, Ahmad F, Alsaleh A, Labhane N (2018) DNA molecular markers in plant breeding: current status and recent advancements in genomic selection and genome editing. Biotechnol Biotechnol Equip 32(2):261–285

Nasri S, Mandoulakani BA, Darvishzadeh R, Bernousi I (2013) Retrotransposon insertional polymorphism in Iranian bread wheat cultivars and breeding lines revealed by IRAP and REMAP markers. Biochem Genet 51(11–12):927–943

Nayak SN, Hebbal V, Bharati P, Nadaf HL, Naidu GK, Bhat RS (2020) Profiling of nutraceuticals and proximates in peanut genotypes differing for seed coat color and seed size. Front Nutr 7:45

Nilsen KT, Walkowiak S, Xiang D, Gao P, Quilichini TD, Willick IR, Byrns B, N'Diaye A, Ens J, Wiebe K, Ruan Y, Cuthbert RD, Craze M, Wallington EJ, Simmonds J, Uauy C, Datla R, Pozniak CJ (2020) Copy number variation of TdDof controls solid-stemmed architecture in wheat. Proc Natl Acad Sci 117(46):28708–28718

Nordstrom KJ, Albani MC, James GV, Gutjahr C, Hartwig B, Turck F, Paszkowski U, Coupland G, Schneeberger K (2013) Mutation identification by direct comparison of whole-genome sequencing data from mutant and wild-type individuals using k-mers. Nat Biotechnol 31(4):325–330

Noshay JM, Springer NM (2021) Stories that can't be told by SNPs; DNA methylation variation in plant populations. Curr Opin Plant Biol 61:101989

Nunome T, Negoro S, Miyatake K, Yamaguchi H, Fukuoka H (2006) A protocol for the construction of microsatellite enriched genomic library. Plant Mol Biol Rep 24(3–4):305

Oey H, Whitelaw E (2014) On the meaning of the word 'epimutation'. Trends Genet 30(12): 519–520

Okagaki RJ, Wessler SR (1988) Comparison of non-mutant and mutant waxy genes in rice and maize. Genetics 120(4):1137–1143

Patel M, Jung S, Moore K, Powell G, Ainsworth C, Abbott A (2004) High-oleate peanut mutants result from a MITE insertion into the *FAD2* gene. Theor Appl Genet 108(8):1492–1502

Paulsmeyer MN, Brown PJ, Juvik JA (2018) Discovery of anthocyanin acyltransferase1 (AAT1) in maize using genotyping-by-sequencing (GBS). G3 (Bethesda) 8(11):3669–3678

Pereira A (2011) Plant reverse genetics: methods and protocols. Humana, New York

Petit M, Guidat C, Daniel J, Denis E, Montoriol E, Bui Q, Lim K, Kovarik A, Leitch A, Grandbastien MA (2010) Mobilization of retrotransposons in synthetic allotetraploid tobacco. New Phytol 186(1):135–147

Porto EM, Komor AC, Slaymaker IM, Yeo GW (2020) Base editing: advances and therapeutic opportunities. Nat Rev Drug Discov 19(12):839–859

Przybyla L, Gilbert LA (2021) A new era in functional genomics screens. Nat Rev Genet 23:89–103

Pujol V, Forrest KL, Zhang P, Rouse MN, Hayden MJ, Huang L, Tabe L, Lagudah E (2015) Identification of a stem rust resistance locus effective against Ug99 on wheat chromosome 7AL using a RAD-Seq approach. Theor Appl Genet 128(7):1397–1405

Qu LJ, Foote TN, Roberts MA, Money TA, Aragon-Alcaide L, Snape JW, Moore G (1998) A simple PCR-based method for scoring the ph1b deletion in wheat. Theor Appl Genet 96(3–4): 371–375

Redon R, Ishikawa S, Fitch KR, Feuk L, Perry GH, Andrews TD, Fiegler H, Shapero MH, Carson AR, Chen W, Cho EK, Dallaire S, Freeman JL, Gonzalez JR, Gratacos M, Huang J, Kalaitzopoulos D, Komura D, MacDonald JR, Marshall CR, Mei R, Montgomery L, Nishimura K, Okamura K, Shen F, Somerville MJ, Tchinda J, Valsesia A, Woodwark C, Yang F, Zhang J, Zerjal T, Zhang J, Armengol L, Conrad DF, Estivill X, Tyler-Smith C, Carter NP, Aburatani H, Lee C, Jones KW, Scherer SW, Hurles ME (2006) Global variation in copy number in the human genome. Nature 444(7118):444–454

Sato Y, Nishio T (2003) Mutation detection in rice waxy mutants by PCR-RF-SSCP. Theor Appl Genet 107(3):560–567

Sattar MN, Iqbal Z, Al-Khayri JM, Jain SM (2021) Induced genetic variations in fruit trees using new breeding tools: food security and climate resilience. Plants (Basel) 10(7):1347

Scandalios JG, Espiritu LG (1969) Mutant aminopeptidases of *Pisum sativum*. I. Developmental genetics and chemical characteristics. Mol Gen Genet 105(2):101–112

Schneeberger K, Ossowski S, Lanz C, Juul T, Petersen AH, Nielsen KL, Jørgensen J-E, Weigel D, Andersen SU (2009) SHOREmap: simultaneous mapping and mutation identification by deep sequencing. Nat Methods 6(8):550–551

Schoen DJ, Schultz ST (2019) Somatic mutation and evolution in plants. Annu Rev Ecol Evol Syst 50:49–73

Schwartz D (1971) Subunit interaction of a temperature-sensitive alcohol dehydrogenase mutant in maize. Genetics 67(4):515–519

Schwarz-Sommer Z, de Andrade Silva E, Berndtgen R, Lonnig WE, Muller A, Nindl I, Stuber K, Wunder J, Saedler H, Gubitz T, Borking A, Golz JF, Ritter E, Hudson A (2003) A linkage map of an F_2 hybrid population of *Antirrhinum majus* and *A. molle*. Genetics 163(2):699–710

Sebat J, Lakshmi B, Troge J, Alexander J, Young J, Lundin P, Månér S, Massa H, Walker M, Chi M (2004) Large-scale copy number polymorphism in the human genome. Science 305(5683): 525–528

Sevanthi AMV, Kandwal P, Kale PB, Prakash C, Ramkumar MK, Yadav N, Mahato AK, Sureshkumar V, Behera M, Deshmukh RK, Jeyaparakash P, Kar MK, Manonmani S, Muthurajan R, Gopala KS, Neelamraju S, Sheshshayee MS, Swain P, Singh AK, Singh NK, Mohapatra T, Sharma RP (2018) Whole genome characterization of a few EMS-induced mutants of upland rice variety Nagina 22 reveals a staggeringly high frequency of SNPs which show high phenotypic plasticity towards the wild-type. Front Plant Sci 9:1179

Shah JM (2021) Epimutations and mutations, nurturing phenotypic diversity. Genetica 150:171–181

Shahryary Y, Symeonidi A, Hazarika RR, Denkena J, Mubeen T, Hofmeister B, van Gurp T, Colome-Tatche M, Verhoeven KJF, Tuskan G, Schmitz RJ, Johannes F (2020) AlphaBeta: computational inference of epimutation rates and spectra from high-throughput DNA methylation data in plants. Genome Biol 21(1):260

Shasidhar Y, Vishwakarma MK, Pandey MK, Janila P, Variath MT, Manohar SS, Nigam SN, Guo B, Varshney RK (2017) Molecular mapping of oil content and fatty acids using dense genetic maps in groundnut (*Arachis hypogaea* L.). Front Plant Sci 8:794

Shen J, Ge D, Song X, Xiao J, Liu X, Che G, Gu R, Wang Z, Cheng Z, Song W, Liu L, Chen J, Han L, Yan L, Liu R, Zhou Z, Zhang X (2021) Roles of CsBRC1-like in leaf and lateral branch development in cucumber. Plant Sci 302:110681

Shirasawa K, Hirakawa H, Tabata S, Hasegawa M, Kiyoshima H, Suzuki S, Sasamoto S, Watanabe A, Fujishiro T, Isobe S (2012a) Characterization of active miniature inverted-repeat transposable elements in the peanut genome. Theor Appl Genet 124(8):1429–1438

Shirasawa K, Koilkonda P, Aoki K, Hirakawa H, Tabata S, Watanabe M, Hasegawa M, Kiyoshima H, Suzuki S, Kuwata C, Naito Y, Kuboyama T, Nakaya A, Sasamoto S, Watanabe A, Kato M, Kawashima K, Kishida Y, Kohara M, Kurabayashi A, Takahashi C, Tsuruoka H, Wada T, Isobe S (2012b) *In silico* polymorphism analysis for the development of simple sequence repeat and transposon markers and construction of linkage map in cultivated peanut. BMC Plant Biol 12(1):80

Shirasawa K, Hirakawa H, Nunome T, Tabata S, Isobe S (2016) Genome-wide survey of artificial mutations induced by ethyl methanesulfonate and gamma rays in tomato. Plant Biotechnol J 14(1):51–60

Shirasawa K, Bhat RS, Khedikar YP, Sujay V, Kolekar RM, Yeri SB, Sukruth M, Cholin S, Asha B, Pandey MK, Varshney RK, Gowda MVC (2018) Sequencing analysis of genetic loci for resistance for late leaf spot and rust in peanut (*Arachis hypogaea* L.). Front Plant Sci 9:1727

Singh BD, Shekhawat NS (2021) Gene editing: principles and applications. Scientific Publishers, Jodhpur, India

Smýkal P, Bačová-Kerteszová N, Kalendar R, Corander J, Schulman AH, Pavelek M (2011) Genetic diversity of cultivated flax (*Linum usitatissimum* L.) germplasm assessed by retrotransposon-based markers. Theor Appl Genet 122(7):1385–1397

Sorkheh K, Dehkordi MK, Ercisli S, Hegedus A, Halász J (2017) Comparison of traditional and new generation DNA markers declares high genetic diversity and differentiated population structure of wild almond species. Sci Rep 7(1):5966

Spell ML, Baran G, Wessler SR (1988) An RFLP adjacent to the maize waxy gene has the structure of a transposable element. Mol Gen Genet 211(2):364–366

Stark R, Grzelak M, Hadfield J (2019) RNA sequencing: the teenage years. Nat Rev Genet 20(11): 631–656

Sui J, Qiao L, Li E, Zhu H, Wang X, Zhao C, Zhang Z, Zhao L, Xu K, Wang W (2019) Transcriptome and miRNA profiling of a hydroxyproline-tolerant peanut mutant with higher grain size and oil contents. Int J Agric Biol 22(6):1331–1337

Sun H, Schneeberger K (2015) SHOREmap v3.0: fast and accurate identification of causal mutations from forward genetic screens. Methods Mol Biol 1284:381–395

Syed N, Sureshsundar S, Wilkinson M, Bhau B, Cavalcanti J, Flavell A (2005) *Ty1-copia* retrotransposon-based SSAP marker development in cashew (*Anacardium occidentale* L.). Theor Appl Genet 110(7):1195–1202

Takagi K, Ishikawa N, Maekawa M, Tsugane K, Iida S (2007) Transposon display for active DNA transposons in rice. Genes Genet Syst 82(2):109–122

Takagi H, Abe A, Yoshida K, Kosugi S, Natsume S, Mitsuoka C, Uemura A, Utsushi H, Tamiru M, Takuno S (2013a) QTL-seq: rapid mapping of quantitative trait loci in rice by whole genome resequencing of DNA from two bulked populations. Plant J 74(1):174–183

Takagi H, Uemura A, Yaegashi H, Tamiru M, Abe A, Mitsuoka C, Utsushi H, Natsume S, Kanzaki H, Matsumura H (2013b) MutMap-Gap: whole-genome resequencing of mutant F_2 progeny bulk combined with *de novo* assembly of gap regions identifies the rice blast resistance gene *Pii*. New Phytol 200(1):276–283

Takahashi H, Akagi H, Mori K, Sato K, Takeda K (2006) Genomic distribution of MITEs in barley determined by MITE-AFLP mapping. Genome 49(12):1616–1620

Tang J, Zeng W, Wang W, Ma B, Liu Y, Li H, Xia H, Li P, Zhu L (2001) Genetic analysis and gene mapping of a rice few-tillering mutant in early backcross populations (*Oryza sativa* L.). Sci China Life Sci 44(6):570–575

Torjek O, Berger D, Meyer RC, Mussig C, Schmid KJ, Rosleff Sorensen T, Weisshaar B, Mitchell-Olds T, Altmann T (2003) Establishment of a high-efficiency SNP-based framework marker set for Arabidopsis. Plant J 36(1):122–140

Tran QH, Bui NH, Kappel C, Dau NTN, Nguyen LT, Tran TT, Khanh TD, Trung KH, Lenhard M, Vi SL (2020) Mapping-by-sequencing via MutMap identifies a mutation in *ZmCLE7* underlying fasciation in a newly developed EMS mutant population in an elite tropical maize inbred. Genes (Basel) 11(3):281

Upadhyaya NM (2007) Rice functional genomics - challenges, progress and prospects. Springer Life Sciences, New York

Upadhyaya NM, Zhu Q-H, Eamens AL, Dennis ES (2003) Rice gene machine: a vehicle for finding functions of cereal genes. Asia Pacific Biotechnol News 6:936–942

Van den Broeck D, Maes T, Sauer M, Zethof J, De Keukeleire P, D'hauw M, Van Montagu M, Gerats T (1998) Transposon display identifies individual transposable elements in high copy number lines. Plant J 13(1):121–129

Venkatesh, Nandini B (2020) Miniature inverted-repeat transposable elements (MITEs), derived insertional polymorphism as a tool of marker systems for molecular plant breeding. Mol Biol Rep 47(4):3155–3167

Verhoeven KJ, Jansen JJ, Van Dijk PJ, Biere A (2010) Stress-induced DNA methylation changes and their heritability in asexual dandelions. New Phytol 185(4):1108–1118

Vipin CA, Luckett DJ, Harper JD, Ash GJ, Kilian A, Ellwood SR, Phan HT, Raman H (2013) Construction of integrated linkage map of a recombinant inbred line population of white lupin (*Lupinus albus* L.). Breed Sci 63(3):292–300

Voylokov A, Korzun V, Börner A (1998) Mapping of three self-fertility mutations in rye (*Secale cereale* L.) using RFLP, isozyme and morphological markers. Theor Appl Genet 97(1–2): 147–153

Vries H d (1901) Die Mutationstheorie. Veit & Co, Leipzig

Wagner H, Weber W, Wricke G (1992) Estimating linkage relationship of isozyme markers and morphological markers in sugar beet (*Beta vulgaris* L.) including families with distorted segregations. Plant Breed 108(2):89–96

Wang N, Zhang Y, Huang S, Liu Z, Li C, Feng H (2020) Defect in *Brnym1*, a magnesium-dechelatase protein, causes a stay-green phenotype in an EMS-mutagenized Chinese cabbage (*Brassica campestris* L. ssp. pekinensis) line. Hortic Res 7(1):8

Wanjugi H, Coleman-Derr D, Huo N, Kianian SF, Luo M-C, Wu J, Anderson O, Gu YQ (2009) Rapid development of PCR-based genome-specific repetitive DNA junction markers in wheat. Genome 52(6):576–587

Waugh R, McLean K, Flavell A, Pearce S, Kumar A, Thomas B, Powell W (1997) Genetic distribution of *Bare–1*-like retrotransposable elements in the barley genome revealed by sequence-specific amplification polymorphisms (S-SAP). Mol Gen Genet 253(6):687–694

Wen S, Liu H, Li X, Chen X, Hong Y, Li H, Lu Q, Liang X (2018) TALEN-mediated targeted mutagenesis of fatty acid desaturase 2 (FAD2) in peanut (*Arachis hypogaea* L.) promotes the accumulation of oleic acid. Plant Mol Biol 97(1–2):177–185

Wenke T, Seibt KM, Döbel T, Muders K, Schmidt T (2015) Inter-SINE amplified polymorphism (ISAP) for rapid and robust plant genotyping. In: Batley J (ed) Plant genotyping. Springer, New York, pp 183–192

Wessler SR, Varagona MJ (1985) Molecular basis of mutations at the waxy locus of maize: correlation with the fine structure genetic map. Proc Natl Acad Sci 82(12):4177–4181

Williams JG, Reiter RS, Young RM, Scolnik PA (1993) Genetic mapping of mutations using phenotypic pools and mapped RAPD markers. Nucleic Acids Res 21(11):2697–2702

Wu D, Shu Q, Li C (2012) Applications of DNA marker techniques in plant mutation research. In: Shu QY et al (eds) Plant mutation breeding and biotechnology. CABI, Wallingford, pp 287–298

Wu HM, Xie DJ, Tang ZS, Shi DQ, Yang WC (2020) PINOID regulates floral organ development by modulating auxin transport and interacts with *MADS16* in rice. Plant Biotechnol J 18(8): 1778–1795

Xie C, Tammi MT (2009) CNV-seq, a new method to detect copy number variation using high-throughput sequencing. BMC Bioinformatics 10(1):80

Xu JY, Zhu Y, Yi Z, Wu G, Xie GY, Qin MJ (2018) Molecular diversity analysis of *Tetradium ruticarpum* (WuZhuYu) in China based on inter-primer binding site (iPBS) markers and inter-simple sequence repeat (ISSR) markers. Chin J Nat Med 16(1):1–9

Yadav CB, Bonthala VS, Muthamilarasan M, Pandey G, Khan Y, Prasad M (2015) Genome-wide development of transposable elements-based markers in foxtail millet and construction of an integrated database. DNA Res 22(1):79–90

Yamane F, Hirashima Y, Shindo A, Tahara M, Fujii N, Ikeo K, Yamashita Y (2012) Cultivar identification markers of common beans based on the retrotransposon insertion site sequences obtained by next-generation DNA sequencing. DNA Testing (DNA Kantei) 4:67–74

Yang Z, Chen Z, Peng Z, Yu Y, Liao M, Wei S (2017) Development of a high-density linkage map and mapping of the three-pistil gene (*Pis1*) in wheat using GBS markers. BMC Genomics 18(1): 567

Yang Q, Yang Z, Tang H, Yu Y, Chen Z, Wei S, Sun Q, Peng Z (2018) High-density genetic map construction and mapping of the homologous transformation sterility gene (*hts*) in wheat using GBS markers. BMC Plant Biol 18(1):301

Yeri SB, Bhat RS (2016) Development of late leaf spot and rust resistant backcross lines in JL 24 variety of groundnut (*Arachis hypogaea* L.). Electron J Plant Breed 7(1):37–41

Yin K, Gao C, Qiu J-L (2017) Progress and prospects in plant genome editing. Nat Plants 3(8):1–6

Yin D, Ji C, Song Q, Zhang W, Zhang X, Zhao K, Chen CY, Wang C, He G, Liang Z, Ma X, Li Z, Tang Y, Wang Y, Li K, Ning L, Zhang H, Zhao K, Li X, Yu H, Lei Y, Wang M, Ma L, Zheng H, Zhang Y, Zhang J, Hu W, Chen ZJ (2020) Comparison of *Arachis monticola* with diploid and cultivated tetraploid genomes reveals asymmetric subgenome evolution and improvement of peanut. Adv Sci 7(4):1901672

Yuan M, Zhu J, Gong L, He L, Lee C, Han S, Chen C, He G (2019) Mutagenesis of *FAD2* genes in peanut with CRISPR/Cas9 based gene editing. BMC Biotechnol 19(1):24

Zhang Q, Arbuckle J, Wessler SR (2000) Recent, extensive and preferential insertion of members of the miniature inverted-repeat transposable element family *Heartbreaker* into genic regions of maize. Proc Natl Acad Sci 97(3):1160–1165

Zhao H, Zhang W, Zhang T, Lin Y, Hu Y, Fang C, Jiang J (2020) Genome-wide MNase hypersensitivity assay unveils distinct classes of open chromatin associated with H3K27me3 and DNA methylation in *Arabidopsis thaliana*. Genome Biol 21(1):24

Zheng X, Chen L, Xia H, Wei H, Lou Q, Li M, Li T, Luo L (2017) Transgenerational epimutations induced by multi-generation drought imposition mediate rice plant's adaptation to drought condition. Sci Rep 7(1):1–13

Zheng Y, Wang P, Chen X, Yue C, Guo Y, Yang J, Sun Y, Ye N (2021) Integrated transcriptomics and metabolomics provide novel insight into changes in specialized metabolites in an albino tea cultivar (*Camellia sinensis* (L.) O. Kuntz). Plant Physiol Biochem 160:27–36

9 Application of TILLING as a Reverse Genetics Tool to Discover Mutation in Plants Genomes for Crop Improvement

M. I. Siddique, A. Younis, M. A. Gururani, and J. Venkatesh

Abstract

The deployment of mutant resources for plant functional genomics is ever increasing with the availability of genome information of economically important crops. Targeting Induced Local Lesions IN Genomes (TILLING) is a reverse genetics approach for high-throughput discovery of induced mutations in the desired gene(s) from a mutant population developed through mutagenesis. TILLING accelerates trait discovery and crop genetic improvement. TILLING strategies for several crops have been developed for the identification of the desired mutants. The combined use of TILLING and the high-throughput sequencing technologies can overcome the bottlenecks of traditional TILLING approaches and facilitates the rapid discovery of useful mutations. Several important mutants related to economically important traits have been identified in agronomic and horticultural crops by employing TILLING by sequencing (TbyS) and, thus, have sizeable potential in crop improvement. In this chapter, we highlight significant developments in the use of TILLING strategies along with the application of high-throughput techniques, such as high-resolution

M. I. Siddique
Vegetable Research Division, National Institute of Horticulture and Herbal Science, Rural Development Administration, Jeonju, Republic of Korea

A. Younis
Institute of Horticultural Sciences, University of Agriculture, Faisalabad, Pakistan

M. A. Gururani
Department of Biology, College of Science, UAE University, Al Ain, United Arab Emirates

J. Venkatesh (✉)
Department of Plant Science and Plant Genomics and Breeding Institute, Seoul National University, Seoul, South Korea
e-mail: jvs15@snu.ac.kr

S. Penna, S. M. Jain (eds.), *Mutation Breeding for Sustainable Food Production and Climate Resilience*, https://doi.org/10.1007/978-981-16-9720-3_9

melting (HRM) and next-generation sequencing (NGS) in the discovery of mutations and further discuss the challenges and prospects of TILLING in modern crop breeding.

Keywords

Genome editing · TILLING · Reverse genetics · Mutation breeding · TILLING by sequencing

9.1 Introduction

Forward and reverse genetics approaches are two classical approaches often used to discover the underlying mechanisms of biological processes and systems (Robinson and Parkin 2009). Basically, both approaches include analysis of mutants or plants with altered phenotypes to reveal the genetic association between genotype and phenotype in order to determine the gene function underlying the phenotype. Forward genetics is a typical approach used for the identification of the genetic alterations or mutations responsible for the altered or abnormal phenotypes. It is a phenotype-driven approach and does not rely on any a prior assumption about the molecular basis of phenotypes. The classical forward genetic approach involves random mutagenesis and subsequent isolation of defective genetic mutants for a specific biological process or phenotype. The genetic variations responsible for the varied phenotype can be identified by employing map-based cloning, genome-wide association studies, and resequencing-based approaches. The forward genetic approaches have been effectively used for cloning various genes in different crop species (Keller et al. 2005; Komatsuda et al. 2007; Desiderio et al. 2016; Szurman-Zubrzycka et al. 2018; Liu et al. 2016). Reverse genetics is a classical tool that has been successfully employed in plant biology to infer causal relationships between phenotype and genotype. Unlike forward genetic approaches, reverse genetic approaches aim to identify the function of a known gene through phenotypic analysis of an organism's impaired or altered gene function. Commonly used reverse genetic approaches include targeted gene disruption/replacement (knockout), gene silencing (knock-down), insertional mutagenesis, targeting induced local lesions in genomes (TILLING), and genome-editing approaches (Caldwell et al. 2004; Gilchrist and Haughn 2010; Desiderio et al. 2016). Typically, reverse genetics encompasses selecting a target gene, creating mutants of the selected gene, and examining the associated phenotypes to reveal the function of the gene of interest. Mutants can also be identified in natural plant populations, or mutant populations developed by various approaches, including chemical and physical mutagenesis, T-DNA- and transposon-based insertional mutagenesis, RNAi-mediated knock-down, or genome-editing approaches (Martienssen 1998; Ostergaard and Yanofsky 2004; Robinson and Parkin 2009; Altpeter et al. 2016). Mutant populations developed through the aforementioned approaches can be used for both forward and reverse genetic studies. In this chapter, we reviewed the main concepts, recent research trends, and importance of TILLING analyses in plants. In particular, we focused on the utilization of TILLING for crop improvement.

9.2 Targeting Induced Local Lesions IN Genomes (TILLING)

Large-scale DNA sequencing approaches have radically changed the way many biological problems are addressed. The traditional genetic analysis of a gene starts with a phenotype (forward genetics), where most of the genes are known from their sequence, but the mechanism underpinning the phenotypes and identity of causative genes is unknown, which has paved the way for the rapid reverse genetic approaches. Thus, reverse genetic approaches have become an important tool for many researchers, and new technological advancements are in great demand (Nagy et al. 2003). Nevertheless, reverse-genetic approaches are not similarly applicable for all organisms. TILLING is the conventional strategy, applicable for all organisms which involve traditional mutagenesis and high-throughput screening for induced mutations. Furthermore, reverse genetics approaches provide indispensable resources to study the desirable plant features and the identification of sequence variations of the genes with known function. There are yet voluminous predicted genes in fully sequenced plant genomes with unknown function or phenotype that can be discovered through experimental evidence (Desiderio et al. 2016). TILLING (Targeting Induced Local Lesions IN Genomes) term was coined by McCallum et al. (2000) for large-scale mutation discovery in *Arabidopsis thaliana* mutant population targeting the *CMT1* gene, which encodes a DNA methyltransferase homolog with a chromodomain, termed as "chromomethylase." TILLING does not involve transgenic modifications. It is an attractive approach for functional genomics and is also an ideal pre-breeding tool for the development of improved crop varieties that can overcome public concerns as they are not regulated as genetically modified organism (GMOs). TILLING can be applied in both prokaryote and eukaryote organisms of any genome size, the number of chromosomes, and ploidy level. Furthermore, TILLING method does not require a complete genome sequence but the minimum information of the DNA sequence of the gene(s) of interest. It is based on conventional PCR followed by mutation detection by any suitable method according to the accessibility of facilities and equipment, as described in the following sections. A TILLING mutagenized population can, however, be used as an effective forward and reverse genetic tools for the identification of economically important genes.

TILLING integrates conventional mutagenesis with high-throughput mutation detection (Colbert et al. 2001; McCallum et al. 2000). It allows the generation of novel allelic series of mutations in candidate genes in the mutagenized populations. This approach combines the use of synthetic chemical or physical mutagenesis techniques with sequence-specific mutation discovery processes. As described above, TILLING has the benefit of genome-wide mutation discovery and broader applicability for the pursuit of point mutations in any organism. Many TILLING populations have been constructed for numerous crop species using diverse mutagens (Table 9.1). The TILLING technology can also provide a collection of

Table 9.1 Overview of TILLING in different crops

Species	Ploidy	Mutagen	Mutation frequency (1/kb)	Target traits	Target gene	Mutation detection technology	References
Rice	2*x*	Az-MNU	1/265		*Os1433*, *OsBZIP*, *OsCALS8R*, *OsDREB*, *OsEXTE*	LI-COR	Till et al. (2007)
Rice	2*x*	MNU	1/135	Leaf emergence	*OsAHP1*, *OsSAD1*, *PLA1*	CE	Suzuki et al. (2008)
Rice	2*x*	Gamma rays	1/6190		–	Agarose gel	Sato et al. (2006)
Rice	2*x*	DEB EMS	1/1000		–	LI-COR	Wu et al. (2005)
Rice	2*x*	EMS	1/293	Phytic acid metabolism	*AtIPK1*, *AtIPK2β*, *AtMRP5*, *AtITPK1*, *AtITPK4*, *AtMIK*, *At5g60760*	TILLING by sequencing	Kim and Tai (2014)
Wheat	2*x*	EMS	1/92	Waxy and Lignin	*COMT1*, *HCT2*, *4CL1*	Agarose gel	Rawat et al. (2012)
Wheat	4*x*	EMS	1/40	Starch quality	*GBSSI*	LI-COR	Slade et al. (2005)
Wheat	4*x*	EMS	1/51	Starch quality	*SBEIIa*, *SBEIIb*, *WKS1*, *WKS2*	PAGE, LI-COR	Uauy et al. (2009)
Wheat	6*x*	EMS	1/24	Starch quality	*Granule-bound starch synthase I*, *GBSSI*	LI-COR	Slade et al. (2005)
Wheat	6*x*	EMS	1/38	Starch quality	*SBEIIa*, *SBEIIb*, *WKS1*, *WKS2*	PAGE, LI-COR	Uauy et al. (2009)
Wheat	6*x*	EMS	1/37 ~ 1/23	Starch quality and grain hardness	*Wx-A1*, *Wx-D1*, *Pina*, *Pinb*	Agarose gel	Dong et al. (2009a)
Wheat	6*x*	EMS	1/47 ~ 1/34	Spike development	*Ppd-D1*, *Rubisco activase A*, *Rubisco activase B*	Agarose gel, PAGE	Chen et al. (2012)
Wheat	6*x*	HII	1/84	Disease resistance	*TaPFT1*, *PFT1*	TaqMan qPCR	Fitzgerald et al. (2010)

Wheat	6*x*	EMS	–	Starch biosynthesis	*Sgp-1*, *Wx*	LI-COR	Sestili et al. (2010)
Wheat	4*x*	EMS	1/77	Carotenoid metabolism	*Lycopene epsilon-cyclase*, *ε*-LCY, Lycopene *beta-cyclase*, *β-LCY*	CEL I-Agarose gel, dHPLC	Colasuonno et al. (2016)
Wheat	4*x*	EMS	–	Plant height	*Rht-B1*	Exome sequencing	Mo et al. (2018)
Wheat	4*x*	EMS	–	Thousand grain weight	*TaGW2*	TILLING by Sequencing, CEL-1-PAGE	Wang et al. (2018)
Wheat	4*x*	EMS	–	Gluten content	*lys3a*, *BPBF*, *WPBFs*	TILLING by Electrophoresis	Moehs et al. (2018)
Wheat	4*x*	EMS	1/26	Resistance against powdery mildew	*TaMlo-A1*, *TaMlo-B1*, *TaMlo-D1*	HRM	Acevedo-Garcia et al. (2017)
Wheat	4*x*	EMS	1/35,000	–	*TaGA20ox1*	TILLING by sequencing	King et al. (2015)
Wheat	4*x*	EMS	1/24	Quality of starch	*SBEIIa*	LI-COR	Slade et al. (2012)
Wheat	4*x*	EMS	1/84	Quality of starch	*SBEIIa*	LI-COR, HRM	Botticella et al. (2011)
Wheat	4*x*	EMS	–	Kernel hardness and starch	*Pinb*, *waxy*, *Agp2*, *SSIIa*	LI-COR	Li et al. (2017)
Maize	2*x*	EMS	1/500	Chromomethylase	*DMT102*	LI-COR	Till et al. (2004)
Barley	2*x*	EMS	1/1000	Floral organ regulation	*Hin-a*, *HvFor1*	dHPLC	Caldwell et al. (2004)
Barley	2*x*	EMS	1/500	Row type morphology and immunity to fungal disease	*HvHox1*	LI-COR	Gottwald et al. (2009)
Barley	2*x*	NaN3	1/374		*HvCO1*, *Rpg1*, *eIF4E*, *NR*	LI-COR	

(continued)

Table 9.1 (continued)

Species	Ploidy	Mutagen	Mutation frequency (1/kb)	Target traits	Target gene	Mutation detection technology	References
				Virus resistance and immunity to fungal disease			Talamè et al. (2008)
Barley	2*x*	NaN3	1/374	Starch metabolism	*BMY1*, *GBSSI*, *LDA1*, *SSI*, *SSIIa*	CEL-I-Agarose gel	Sparla et al. (2014)
Oat	6*x*	EMS	1/40 ~ 1/20	Increased digestibility and improved food quality	*AsPAL1*, *AsCslF6*	MALDI-TOF	Chawade et al. (2010)
Sorghum	2*x*	EMS	1/526	Forage digestibility	*COMT*	LI-COR	Xin et al. (2008)
Tomato	2*x*	EMS	1/574 ~ 1/322	Shelf life	*RIN*, *Gr*, *Rab11a*, *Exp1*, *PG*, *Lcy-b*, *Lcy-e*	LI-COR	Minoia et al. (2016)
Tomato	2*x*	EMS	1/737	Proline biosynthesis	*ProDH*	CE, HRM	Gady et al. (2009)
Tomato	2*x*	EMS		Virus resistance	*eIF4E*	Keypoint technology/NGS	Rigola et al. (2009)
Tomato	2*x*	EMS	1/574	Virus resistance	*eIF4E1*	LI-COR	Piron et al. (2010)
Tomato	2*x*	EMS	1/1237	Shelf life	*Sletr1-1*, *Sletr1-2*	LI-COR	Okabe et al. (2012)
Tomato	2*x*	EMS	1/1710 ~ 1/737	Delayed fruit ripening and shelf life	*SlETR1*, *SlETR2*, *SlETR3*, *SlETR4*, *SlETR5*, *SlETR6*	LI-COR	Okabe et al. (2011)
Tomato	2*x*	EMS	1/367	Carotenoids and folate metabolism	*ADCL1*, *CCD4A*, *CRTISO*, *GGH3*, *NCED1*, *PAP3*, *PHYF*, *PSY1*, *SPA3*, *SPA3LIKE*	TILLING by sequencing	Gupta et al. (2017)

Pepper	2*x*	EMS	–	Virus resistance	*eIF4E*	Capillary gel electrophoresis	Siddique et al. (2020)
Pea	2*x*	EMS	1/669	Gibberellin metabolism	*At1g11190*, *At1g68290*, *At4g21590*, *At4g21585*, *At4g21600*	LI-COR	Triques et al. (2007)
Sunflower	2*x*	EMS	1/475	Fatty acid biosynthetic pathway and downy	*kasII*, *kasIII*, *fad2-1*, *AY490791*	LI-COR	Sabetta et al. (2011)
Arabidopsis	2*x*	EMS	1/300	–	–	LI-COR	Greene et al. (2003)
Arabidopsis	2*x*	EMS	1/170	–	–	LI-COR	Till et al. (2003)
Brassica rapa	2*x*	EMS	1/60	DNA methylation	*BraA.RPL.a*, *BraA.RPL.b*, *BraA.RPL.c*, *BraA.IND.a*, *BraA.MET1.a*, *BraA.MET1.b*	CE (ABI3730)	Stephenson et al. (2010)
Brassica oleracea	2*x*	EMS	1/447	Wax biosynthesis and dwarf stature	*PgiC*, *COR314-TM2*, *Methyltransferase*, *CEO1*, *ANNAT1*, *LEA14*, *NDPK2*	LI-COR	Himelblau et al. (2009)
Melon	2*x*	EMS	1/573	Fruit quality	*CmACO1*	LI-COR	Dahmani-Mardas et al. (2010)
Melon	2*x*	EMS	1/1500	Disease resistance, fruit quality	*Cm-PDS*, *Cm-eIF4E*, *Cm-eIFI(iso)4E*, *m-ACO1*, *Cm-NOR*, *Cm-DET1*, *Cm-DHS*	LI-COR	Gonzalez et al. (2011)
Soybean	4*x*	NMU	1/140	–	–	LI-COR	Cooper et al. (2008)
Soybean	4*x*	EMS	1/176 ~ 1/126	Stearic acid content	*GmSACPD-A*, *GmSACPD-B*, *GmSACPD-D*	TILLING by sequencing	Lakhssassi et al. (2020)
Soybean	4*x*	NMU	–	Sucrose level in seed	*AFFINOSE SYNTHASE2*, *RAFFINOSE SYNTHASE3*	TILLING by sequencing	Thapa et al. (2019)
Rapeseed	4*x*	EMS	1/130 ~ 1/41	Oil quality	*FAE1*	LI-COR	Wang et al. (2008)

(continued)

Table 9.1 (continued)

Species	Ploidy	Mutagen	Mutation frequency (1/kb)	Target traits	Target gene	Mutation detection technology	References
Potato	4*x*	EMS	1/91	Starch quality	*waxy*	Direct sequencing	Muth et al. (2008)
Peanut	4*x*	EMS	1/967	Seed quality	*Ara h 1*, *Ara h 2*, *AhFAD2A*, *AhFAD2B*	LI-COR	Knoll et al. (2011)
Tef	4*x*	EMS	1/115 ~ 1/370	Plant height	*dw3*, *rht1*	NGS (Roche 454)	Zhu et al. (2012)

alleles that can offer a range of phenotypic asset and enable structural and functional studies. Moreover, mutations causing stop codon or amino acid replacements in the encoded proteins have the capability to modify the specific action of enzymes by providing alleles with varying strength; for example, a mutation might lead to an enzyme with only 10% of wild-type activity. This may cause a major reduction in the end product but maintains a basic level needed for normal plant growth and development. Thus, mutagenesis can offer useful novel insights into the gene function in a species under investigation.

9.2.1 TILLING Process

Producing a suitable mutagenized population and TILLING process requires consideration of several important factors, such as the target population genetic structure, the choice of mutagen, the mode of sampling, DNA quality, and the pooling strategy. The genetic structure of the target population can be simple for highly homozygous, and inbreeding species and factors, such as ploidy level, heterozygosity, and dioecy complicate the TILLING process (Draper et al. 2004; Slade and Knauf 2005; Wienholds et al. 2003).

Mutagenesis is generally performed by soaking seeds in the suitable mutagen. Alternatively, pollen can be treated with mutagens, which is the method of choice in maize (Till et al. 2004). Mutant individuals (the M1 population) derived from mutagenized seeds are not suitable for TILLING since they are chimeric for mutations; therefore, seeds from M1 are planted to produce the M2 generation. M2 individual DNA samples are pooled using an appropriate pooling strategy, followed by PCR amplification of the target gene sequence (Fig. 9.1). DNA containing point mutations were detected by denaturing and re-annealing of the PCR products. This process results in the formation of heteroduplex DNA in which one strand drives from the mutant and the other strand originate from the wild-type PCR product. A mismatch formed at the mutation can be recognized by mismatch-specific endonucleases, such as CEL I (Yang et al. 2000) or ENDO1 (Triques et al. 2008). These enzymes cleave the heteroduplex DNA at the mismatched sites. The cleavage products can be resolved by gel electrophoresis, usually, sequencing-type denaturing polyacrylamide gels using an LI-COR DNA analyzer (PAGE).

9.3 Mutagenesis Agents

The first prerequisite for the TILLING approaches is to construct a mutant population by applying appropriate mutagens (Fig. 9.1). Mutagens can be either physical energy or chemical elements that induce a range of genetic variations, including DNA point mutations, indels, and larger chromosomal aberrations, and thus in the genomes of treated organisms, novel allelic series that can specify a range of phenotypes can be created. Defining the ultimate dosage of mutagen is crucial for optimal mutagenesis; at suboptimal levels, a huge number of plants will be needed to

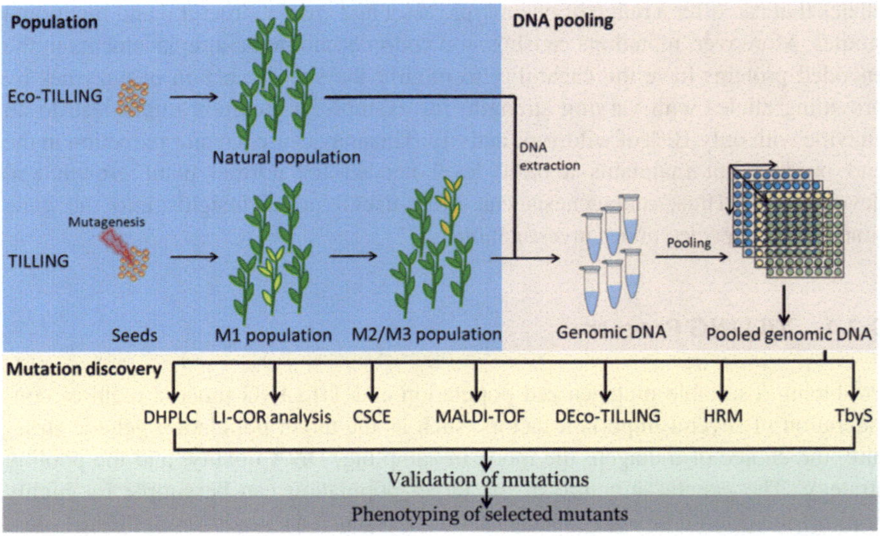

Fig. 9.1 Overview of TILLING approaches. Generally, seeds are treated with either physical or chemical mutagen to induce genetic variation, and then seeds are sown to germinate. The resulting M1 population is chimeric for mutations. Therefore, each M1 is selfed to create the M2 population. M2 DNA is extracted from leaf tissue, and M3 seeds are stored (Occasionally, M3 population can be used for TILLING). DNA samples are pooled to increase throughput. Mutation detection can be performed by any of the known mutation detection approaches, such as DHPLC, LI-COR, CSCE, DEco-TILLING, HRM, and TbyS. Mutations were validated through experiments by analyzing the related phenotypes of mutants

identify mutations in a gene of interest, and high dosage levels could cause seed or plant mortality, loss of viability, and sterility problems. It is critical to identify an optimal dosage because different species and varieties respond differently to mutagen dosage in terms of lethality (Mba et al. 2010; Till et al. 2007). Observation of differences in mutation density between mutagenized organisms can be due to differences in the cytotoxic response to the mutagen and restoration of injuries induced by the mutagen. The density of induced mutations in the genome is the main factor for measuring the effectiveness and rate of mutation discovery. The density of mutation has been assessed for several mutagenized populations (Table 9.1). The ploidy level of a target mutagenized species can be another factor that may contribute to the different mutation densities. Due to the redundancy of several of their genes, polyploids may endure a greater number of mutations per genome as homologous sequences compensate for the loss of gene function. The maximum mutation density was found in hexaploid wheat (1 mutation/24 kb), followed by tetraploid wheat (1 mutation/40 kb), Arabidopsis (1 mutation/300 kb), maize (1 mutation/500 kb), and rice and barley (1 mutation/1 Mb) (Greene et al. 2003; Caldwell et al. 2004; Till et al. 2004; Wu et al. 2005; Slade et al. 2005).

9.3.1 Chemical Agents

Chemical mutagenesis induces irreversible high-density mutations, including point mutations and minor deletions, and can also cause chromosomal breaks. Among the chemical mutagens, alkylating agents are the most frequently used in plant mutant population construction (Till et al. 2006; Siddique et al. 2020). These are base-altering compounds that modify the arrangement and properties of the bases. Alkylating agents trigger all type of point alteration, such as transversions (A↔C; G↔T), transitions (A↔G; C↔T), small base omissions, which may induce frameshifts in coding sequences, as well as chromosome ruptures (Maluszynski et al. 2003; Desiderio et al. 2016). Various alkylating substances have been used for TILLING projects, such as ethyl methanesulfonate (EMS), *N*-methyl-*N*-nitrosourea (MNU), *N*-ethyl-*N*-nitrosourea (ENU), methyl nitrosourea (MNH), and sodium azide. EMS is one of the chemical mutagens most commonly used in TILLING experiments (Table 9.1). EMS induces random point mutations at a high frequency, some of which can create novel stop codons in genes of interest. EMS attaches the alkyl group to the hydrogen-bonding oxygen of guanine, which produces *O*-6 alkylguanine that pairs up with T (instead of C) and causes G/C to A/T base conversion in DNA. In Arabidopsis, wheat, and maize, up to 98–99% of all EMS-induced mutations are ascribed to this kind of DNA base conversion (Greene et al. 2003; Till et al. 2003, 2004; Slade et al. 2005). Whereas in rice, barley, and pepper, G/C to A/T transitions were observed up to 70% (Caldwell et al. 2004; Till et al. 2007; Siddique et al. 2020). Point mutations induced by the EMS are usually randomly distributed across the genome and are less destructive compared with the chromosomal aberrations generally induced by the physical mutagens. Furthermore, a high degree of mutation saturation can be attained in a mutant population, which enables the investigation of gene function at a genomic level. ENU usually produces *O*6-ethylguanine alterations, but some researches revealed frequent transversions (A:T/T:A) in the genome, perhaps due to the outcome of *O*4-ethylthymidine modifications (Jansen et al. 1995; Skopek et al. 1992). *N*-methyl-*N*-nitrosourea (MNU, MNH) generates merely G/C to A/T transitions. Studies in rice and soybean indicated, MNU treatment resulted in approximately 90% GC–AT base modifications (Cooper et al. 2008; Suzuki et al. 2008). The mutation rate achieved with MNU in soybean (1 mutation/140 kb) was comparable to the mutation rate obtained with EMS treatment, whereas in rice the MNU mutant population (1 mutation/135 kb) had twofold higher mutation rate than that of an EMS mutant population (1 mutation/294 kb) (Till et al. 2007; Cooper et al. 2008; Suzuki et al. 2008). These studies evidently proved that the chemical reagents are crucial for mutation density attained. Sodium azide is a chemical substance that is usually used as a mutagenic agent in some microbes and plants, such as *Escherichia coli*, *Saccharomyces cerevisiae*, *Brachypodium distachyon*, *Oryza sativa*, and *Hordium vulgare* (Dalmais et al. 2013; Talamè et al. 2008). Various species have been tested with different combinations of concentration and treatment time with this mutagen. For example, in the barley TILLING experiment, 10 mM concentration of sodium azide was found to be ideal. The mutation rate achieved was 1 mutation/374 kb, which was three times greater than the mutation frequency obtained with EMS treatment (1 mutation/1 Mb) (Caldwell et al. 2004; Talamè et al. 2008).

It has been observed that most insertional mutagenesis applications generate only loss-of-function alleles, whereas chemical mutagenesis can produce either loss-of-function (hypomorphic phenotypes) or gain-of-function mutations (hypermorphic phenotypes) (Desiderio et al. 2016). The hypomorphic alleles might be highly suitable for devising wild-type gene function. In contrast, hypermorphic alleles are more expected to be dominant and are highly possible to produce an observable phenotype for either trivial or important genes. Generally, the incidence of missense mutations is much higher than the nonsense mutations; however, the majority of them may not cause visible phenotype changes (Greene et al. 2003).

9.3.2 Physical Agents

In earlier mutation breeding projects, physical mutagens, usually ionizing radiations, have been used broadly for inducing hereditary mutations, and about 70% of mutant cultivars were bred through the application of physical mutagenesis (reviewed by Oladosu et al. 2016). X-rays were the initial agents used to induce mutations. Other physical mutagens, such as fast neutrons and gamma irradiation are also used to mutagenize the organisms in functional genomics. In the case of radiation-induced mutations, the frequency, rate, and spectrum of mutations rely on several factors, involving radiation form and dosage, linear energy transfer (LET; i.e., the volume of energy transferred to the test materials), and plant tissue nature and condition (Jo and Kim 2019). Thus, an extensive delineation of the mutagenic effects for a particular irradiation treatment is essential for breeding and genomics studies. Gamma irradiation can induce about 30% of small indels of few base pairs length, suggesting it as a powerful tool for knocking out genes (Sato et al. 2006). In rice, exposure of seeds with 500 Gy of gamma rays induced single nucleotide substitutions and small deletions (2–4 bp) with the mutation rate of 1 mutation/6.2 Mb (Sato et al. 2006). Fast neutrons are a type of strong-energy radiation, which predominantly induces deletions of several hundred to thousands of base pairs in plant genomes. A TILLING process that utilizes fast neutrons for inducing mutations was named as delete-a-gene TILLING (Li et al. 2001). Fast neutron induces a range of diverse deletions. However, one of the potential drawbacks of using deletion mutants generated through the use of neutron mutagenesis is that some genomic deletions can affect more than one gene. Furthermore, large genomic deletions are poorly transmitted through male gametes (Li et al. 2001).

Physical and chemical mutagens vary in their efficiency in inducing genetic mutations as well as in the range of mutations induced. Physical mutagens generally cause chromosomal abbreviation or rearrangements to a greater degree and rigorously decrease viability compared to chemical mutagens (Wu et al. 2005). However, chemical mutagens are easy to use, easily available; no requirement of expensive irradiation chambers and can facilitate a very high mutation frequency and thus gained wider popularity in mutation breeding programs.

9.4 DNA Extraction and Pooling

Once an appropriate mutant population has been constructed, then good-quality DNA requires to be extracted, and the DNA samples are needed to be at the same concentrations for pooling to be performed. Generally, for diploid organisms, pooled DNA can include up to eight individuals to permit successful uncovering of mutations (Fig. 9.1). Further increase in pool sizes, the sensitivity of mutation identification reduces due to the decreased fraction of heteroduplex (hybrids between mutant and wild-type DNA) compared with homoduplexes. Accordingly, the ploidy level, heterozygosity, and the volume of naturally occurring SNPs, optimal pooling for a test population should be ascertained experimentally. Two basic pooling methods have been used frequently including one-dimensional and two-dimensional methods. In a one-dimensional pooling strategy, DNA from eight individuals is generally mixed to achieve the desired DNA pooling. If a mutation is detected in a pool containing eight individuals, then every member of that pool is screened separately to identify the individual line exhibiting the mutation. In the two-dimensional pooling strategy, eightfold pools are organized in a bidimensional array so that, detection of the same mutations in two pools identifies the mutant individual with a unique row and column address; therefore, you need to exclude the screening of individual samples as performed for the one-dimensional scheme. Moreover, since two-dimensional pooling implicates screening each individual with double coverage, possible false-negative and false-positive errors are decreased at the preliminary screening phase. Another pooling strategy is three-dimensional pooling, in which samples are pooled in *X*, *Y*, and *Z* dimensions (Reddy et al. 2012). This approach is usually combined with next-generation sequencing (NGS)-based TILLING approaches (Tsai et al. 2011; Reddy et al. 2012). The three-dimensional pooling scheme permits the direct examination of individual mutant sample for the molecular identification of mutation and avoid the additional sequencing steps required for mutation discovery (Tsai et al. 2011). Regardless of the approach used, it is important that all DNA samples are equivalent and normalized with respect to the concentration; otherwise, distinctive induced mutations may remain unidentified regardless of pooling schemes.

9.5 Mutation Detection Methods for TILLING

An easy, basic, and simple method to detect mutations in a mutagenized population is direct Sanger sequencing of the target gene from individual samples. This method has been used to screen a mutant population of wheat (Feiz et al. 2009). Nonetheless, compared to other approaches for mutation identification, direct Sanger sequencing is expensive, not very effective, and thus not regularly used. Therefore, as described above, most approaches for mutation detection depend on DNA pools in a range of between 4 and 8 individuals (8–16 alleles) from the mutant population. Once the DNA pools have been constructed, mutation detection can be followed by any of the methods as described below.

9.5.1 Denaturing High-Performance Liquid Chromatography (DHPLC)

The initial TILLING technique was performed using a commercial denaturing HPLC referred to as Denaturing High-Performance Liquid Chromatography (DHPLC) (McCallum et al. 2000). This is a powerful approach originally developed to examine a mutant population of barley (Caldwell et al. 2004). This approach allows the discovery of point mutation, such as single-base substitutions as well as small insertions and/or deletions in a huge number of samples (Xiao and Oefner 2001) and relies on heteroduplex formation between mutated and wild-type DNA strands due to mismatches in the DNA sequence. Heteroduplex molecules can be detected by mixing, denaturing, and re-annealing the PCR products, followed by the sample separation by reverse-phase liquid chromatography on a unique column matrix with heat denaturation of DNA strands (Keller et al. 2001). However, this approach would not scale up easily (Colasuonno et al. 2016), this led to the development of alternative approaches.

9.5.2 Endonuclease Cleavage Followed by Electrophoresis

An enzymatic mismatch cleavage approach was successfully adapted for high-throughput mutant screening (Oleykowski et al. 1998) much before the discovery of TILLING. Henikoff and coworkers, in the year 2001, developed a "high-throughput TILLING" approach that combined eightfold sample pooling, enzymatic mismatch cleavage assay, single base-pair resolution denaturing polyacrylamide gel electrophoresis, and laser-based fluorescence detection analysis using LI-COR gene analyzers (Colbert et al. 2001). Since then, many TILLING studies to date have commonly employed the LI-COR system for the identification of mutations. In this approach, the target DNA of around 1.0–1.5 kb is end tagged by PCR using fluorescently labeled primers with the IRDye 700 (forward) and IRDye 800 (reverse) dyes. Subsequently, the PCR samples are denatured and re-annealed to produce heteroduplexes between mutant and wild-type amplicons. Mismatches are then cleaved by a single strand-specific nuclease, generally CEL 1 nuclease from celery stalks (*Apium graveolens*), or ENDO1 from Arabidopsis. Cleaved products are then refined, normally on Sephadex plates, fractionated according to size by denaturing polyacrylamide gel electrophoresis, and envisioned by fluorescence detection by means of the LI-COR gene analyzers (Till et al. 2006). To date, the LI-COR system has been employed successfully in several crops, such as Arabidopsis, wheat, melon, and pepper (Henikoff et al. 2004; Dahmani-Mardas et al. 2010; Sestili et al. 2010; Siddique et al. 2020).

9.5.3 Alternate Approaches to LI-COR Screening

TILLING systems commonly use LI-COR gene analyzers for mutation detection, which use fluorescently labeled primers for amplification and are relatively expensive. An economical and cost-effective substitute to the LI-COR system has been demonstrated with the use of conventional agarose or polyacrylamide electrophoresis. The procedure is similar to the mismatch-cleave system as mentioned above; however in this system, labeled primers are not used, and thus expensive fluorescently labeled primers are not required. The ethidium bromide staining is utilized to visualize the DNA fragments after separation on agarose or polyacrylamide gel. This technique has proved to have nearly comparable sensitivity as that of LI-COR systems and can be the optimal method for inexpensive TILLING (Raghavan et al. 2007; Uauy et al. 2009).

Other potential economic alternates are agarose-based mutation detection methods using SYBR Green I dye (Sato et al. 2006) or fine agarose gels (<4 mm) containing ethidium bromide (Dong et al. 2009a). However, these methods may have less mutation detection sensitivity compared to an LI-COR gel analyzer system. Another approach named as "DEco-TILLING" (double-stranded EcoTILLING) originally used for SNP detection in chum salmon (*Oncorhynchus keta*) is a rapid and inexpensive technique that includes digestion of both strands of DNA using CEL 1 enzyme similar to the EcoTILLING. Unlike EcoTILLING, in which the cleavage products need to be purified to remove unincorporated dye-labeled primers and dye label prior to loading samples on a gel agarose gel (Garvin and Gharrett 2007). Another alternative method, Endonucleolytic Mutation Analysis by Internal Labelling (EMAIL), was originally used for mismatch finding in rice starch synthase IIa encoding gene. In this approach, internal amplicon labeling is performed using fluorescently labeled deoxynucleotides instead of 5′-end labeling, which enhances CEL nucleases exonucleolytic activity, causing the removal of the dye-labeled terminal nucleotides and thus limited signal strength (Cross et al. 2008).

However, it should be noted that procedures utilizing specific mismatch cleavage enzymes have certain shortcomings since these enzymes do not distinguish all types of mismatches and they also have less sensitivity in sensing an allele of the DNA or yield gels with a high background produced by nonspecific cleavage (Kurowska et al. 2011; Desiderio et al. 2016).

9.5.4 Conformation Sensitive Capillary Electrophoresis (CSCE)

CSCE is an approach used for non-enzymatic mutation discovery based on differential migration of DNA heteroduplexes (Gady et al. 2009). In this approach, DNA pools of amplified product from mutant and wild-type alleles are allowed to anneal randomly to form several duplex species, including homoduplexes from the annealing of mutant/mutant or wild-type/wild-type fragments, and heteroduplexes from mutant/wild-type fragments. Mismatches that are formed are resolved and identified by capillary electrophoresis on an ABI 3130xl Genetic Analyzer. The DNA heteroduplexes and homoduplexes migrate at differential speeds during electrophoresis in 3130xl Genetic Analyzer, thus facilitating the detection of DNA pools carrying a mutation within the targeted gene fragment (Gady et al. 2009).

The CSCE depend on direct assessment of the PCR product from mutant DNA pools and therefore circumvent some of the laborious and time-consuming steps require in CEL 1 or gel-based systems. One of the limitations of the CSCE is the shorter target fragment lengths of 200–500 bp are optimal (Davies et al. 2006) compared to fragments up to 1.5 kb for the LI-COR-based systems (Gady et al. 2009). CSCE for mutation detection approach has been used in tomato for identification of SNPs between *Solanum lycopersicum* and *S. pimpinellifolium* (Gady et al. 2009).

9.5.5 Matrix-Assisted Laser Desorption/Ionization-Time of Flight (MALDI-TOF)

Matrix-assisted laser desorption/ionization-time of flight mass spectrometry (MALDI-TOF) is a technique in which samples are ionized, followed by the transfer to the mass analyzer to record their mass to charge ratio and then analyzed by using specially developed software. In TILLING using the MALDI-TOF approach, samples were amplified in a standard PCR-reaction using the gene-specific primers followed by MALDI-TOF analysis (Chawade et al. 2010). The MALDI-TOF generated mass spectra were compared to the mass peaks of the wild-type spectra, and the samples showing a noticeable deviation from wild-type were sequenced to confirm the mutations. This system was utilized to discover mutations in a hexaploid oat TILLING population (Chawade et al. 2010). This method has been described as a worthy alternative to other screening methods, especially due to its low cost and capability of high-throughput screening (Chawade et al. 2010).

9.5.6 High-Resolution Melt (HRM)

High-resolution melt (HRM) curve analysis discerns mutations based on dissimilarities in melting curves of heteroduplexes and homoduplexes. HRM has been reported as a technique for high-throughput screening in TILLING experiments (Dong et al. 2009b; Gady et al. 2009; Jeong et al. 2012). The system circumvents all post-PCR handling of products and can differentiate heterozygous and homozygous mutations among samples. The optimum amplicon size is smaller compared to those used in LI-COR and other gel-based systems, making this method most appropriate for genes with small exons.

9.5.7 NGS-Based TILLING Approaches

NGS technologies have emerged to overcome some of the difficulties associated with the traditional TILLING procedures, such as the sample pooling with greater than eight individuals. NGS approaches have simplified the procedure of linking mutations detected in TILLING populations to the pertinent phenotypes (Fig. 9.1). By leveraging NGS technologies, it is now feasible to sequence candidate amplicons in multiple parallel reactions (SCAMPR) for detecting mutations in target genes in cost-effective ways comparable to current TILLING strategies. TbyS has been performed in Arabidopsis, wheat, rice, tobacco, and tomato using Illumina sequencing of genes amplified from the multidimensional pooled DNA samples (Rigola et al. 2009; Tsai et al. 2011; Reddy et al. 2012). Each pool of PCR products was barcoded individually by the ligation of specific DNA adapters allowing accumulation and processing of manifold amplicons. The sequenced samples were deconvoluted using bioinformatics tools, which can sort sequence reads based on the adapter sequence. Mutations are identified by comparison against the wild-type reference sequence; thus TbyS is devoid of enzymatic cleavage or PCR-based melting systems. It was shown that for grain crops, such as wheat and rice, TbyS could be very useful to find mutations in target genes using mutated populations (Tsai et al. 2011).

Another approach, called MutMap which also depends on NGS-based whole-genome resequencing to detect the causative mutation that is underlying a phenotype spotted in the mutant population (Fekih et al. 2013). In the MutMap+ approach, pools of genomic DNA of phenotypically distinct individuals obtained from a mutant line are subjected to whole-genome sequencing followed by the alignment of reads to the reference genome and SNP-calling. An SNP-index was determined for all the SNPs induced by mutagenesis. Genomic regions displaying a distinctive SNP-index peak were linked to locations of the underlying mutations accountable for the altered phenotype (Fekih et al. 2013).

A variant of TbyS strategy, exome-capture technology (Henry et al. 2014), in which a particular part of genes (exons) have been sequenced in order to examine the expressions of specific proteins that absolutely affect the phenotypes of the plants (Henry et al. 2014). This approach allows sequencing the coding region of mRNA for proteins. It circumvents the sequencing of non-coding regions of genes (introns) that generally do not participate in protein function (Uauy et al. 2017). This approach has been utilized in various grain crops, such as rice and wheat (Henry et al. 2014; Uauy et al. 2017).

9.6 Identification and Evaluation of the Individual Mutant

Once a DNA pool having a tentative mutation has been determined, it is required to identify a DNA sample containing mutations (Fig. 9.1). When a one-dimensional pooling scheme has been applied, in the second phase of TILLING, only pools of wild-type DNA and mutant DNA need to be analyzed to verify the mutated individuals before resequencing the mutant allele. Sequence assessment against the wild-type allele will endorse an induced mutation and define its nature. It will also determine whether it aligns with the known mechanism of action of the mutagen applied. The connection between induced mutation and phenotype can be evaluated by sibling scrutiny of heterozygous M3 families directly. Another way is to perform selfing of families of homozygous M3 plants, followed by backcross to the wild-type parent. Practically, there is little need for widespread backcrossing or further population expansion to verify the association between mutation and phenotype, especially when several independent alleles have been discovered. Moreover, it is not required to eliminate background mutations to evaluate potential phenotypic impact though the construction of isogenic lines as they may be needed for other purposes. However, in polyploid species like wheat, the condition is slightly different in which introgression of mutant alleles of all homologs through conventional breeding is normally required to achieve the required phenotype. Furthermore, recurrent backcrossing to the wild-type or an adapted variety to remove the unwanted background mutations also is required (Slade et al. 2005, 2012; Uauy et al. 2009).

9.7 Bioinformatics Tools

Bioinformatics tools and packages are used in the TILLING approach from the earlier stages, at the time of amplicon selection, till the final step and when the identity of alleles in question is realized in terms of their influence on protein function. The analysis of the amplicon is a decisive stage in the TILLING procedure. In a CEL 1 or gel-based system, the designed primers should be highly specific to the target gene to achieve the optimum sensitivity of mutation identification. For polyploid species, such as wheat, this involves designing of homolog-specific primers (Slade et al. 2005, 2012; Uauy et al. 2009). The primers should bind and amplify the exonic regions of the gene, largely those that are significant for protein function analysis or where the probability of finding deleterious mutations is highest (McCallum et al. 2000). The web-based tool CODDLE (Codons Optimized to Discover Deleterious Lesions) has been frequently used to anticipate the ideal region for TILLING. The primer design can also be performed using the web-based program Primer3 (http://primer3.sourceforge.net/) through recommended parameters as described in the previous study (Till et al. 2006). The designing of gene-specific primer and the condition for amplification and optimization are critical, and there is a strong association between amplicon quality and the success of TILLING experiment. After the alleles have been detected and categorized, the SIFT (Sorting Intolerant from Tolerant; http://sift.jcvi.org/) program can be employed to predict the consequence of mutations on protein function (Ng and Henikoff 2003). Mutations that have a SIFT score value less than 0.05 was considered to be deleterious. In addition, the bioinformatics tools SOPMA (Self-Optimized Method for secondary structure prediction with Alignment; http://npsa-pbil.ibcp.fr/cgi-bin/npsa_automat.pl?page=npsa_sopma.html) and SAS (Sequence Annotated by Structure; http://www.ebi.ac.uk/thornton-srv/databases/sas/) can be used to estimate the hypothetical influence of mutations on the structure of the secondary protein. In silico analysis, using ASD (Alternative Splicing Database; http://www.ebi.ac.uk/asd/) can also be implemented for mutations existing in the non-coding zones with respect to supposed alternative splicing events.

For the analysis of the TbyS results, several tools have been used. Among them, MAPS (mutations and polymorphisms surveyor) is a common tool for assessing polyploid genomes since it uses each sample as a control for others, and it allows differentiation between allelic alternatives as per the deviation among homologous and induced mutations (Henry et al. 2014). However, further improvements in the use of bioinformatics tools employed for the TbyS datasets are required for precise results. TbyS technology has ample prospects for future applications. The bioinformatics programs to estimate the possible effect of the observed mutations can be used as a guide for making a strategy for the specific mutants to be analyzed; however, they may need to be validated through experiments by analyzing the related phenotypes of each mutant (Desiderio et al. 2016).

9.8 Modified TILLING Approaches

9.8.1 EcoTILLING

EcoTILLING denotes the modification of TILLING to examine natural populations to discover natural single nucleotide polymorphism associated with significant phenotypic traits (Comai et al. 2004). This approach provides a rapid and precise method to identify a wide range of haplotype varieties in the gene of interest in the group of accessions or cultivar collections. All the methodologies used in the TILLING process for mutation detection can be utilized in EcoTILLING. The level of pooling is smaller in EcoTILLING, compared to TILLING, because the occurrence of variations prevailing in natural populations is greater than those in artificially mutagenized populations. Hence, DNA pooling in EcoTILLING is performed by mixing the DNA of the examined material and the control genotype in a 1:1 ratio (Chen et al. 2011; Comai et al. 2004). EcoTILLING allows the fast discovery of variations in a large number of individuals. It could be more cost-effective by the grouping of haplotypes and then performing sequencing of one representative from each haplotype group. The procedure can be applied to any organism when the sequence information of the gene(s) of interest is accessible. It offers an invaluable tool for mining of SNPs in plant populations, uncovering variants for the trait of interests, or determining the regulatory element controlled by any gene, and also the identification of DNA polymorphisms in individuals with high genetic similarity (Desiderio et al. 2016).

EcoTILLING has been applied to study a variety of crops, including Arabidopsis, wheat, rice, and sorghum to discover new allelic variations for imminent molecular breeding programs (Jawher et al. 2018; Wang et al. 2008). For example, in an EcoTILLING study, 192 different kabuli and desi chickpea accessions were assessed to detect polymorphisms for 1133 transcription factors linked to seed weight (Bajaj et al. 2016). Concerning the abiotic and biotic stress, EcoTILLING accelerates the process of identifying naturally occurring resistance alleles, which can be used in breeding programs and improve elite cultivars. Different researches have shown the effectiveness of this approach to identify alleles that could enhance resistance to various plant-related stresses (Desiderio et al. 2016). For example, after identifying polymorphisms using EcoTILLING, association analyses were performed to categorize allelic variations underlying drought resistance in rice (Kadaru et al. 2006). Similarly, EcoTILLING was applied in barley to discover an allelic variation within the powdery mildew resistance genes *mlo* and *Mla* (Mejlhede et al. 2006). In another study, this technique was effectively applied to study natural allelic variation for resistance to Melon Necrotic Spot Virus disease (MNSV) in diverse *Cucumis* species (Nieto et al. 2007). An overview of previous studies using EcoTILLING is summarized in (Table 9.2).

9.8.2 Individualized TILLING (iTILLING)

TILLING approach has been applied with minor modifications, which have proven to be effective. A modified TILLING method, that lessens costs and the duration required for mutation examination, was developed in Arabidopsis and named as "individualized TILLING" (iTILLING) (Bush and Krysan 2010). The iTILLING approach offers that M2 seeds from the whole population are collected in bulk, so no classification of M2 progenies or DNA samples is needed. The M2 individuals are grown on agar plugs of 96-well spin plates, and the plant tissue for genomic DNA isolation is collected directly from plates using the Ice Cap method (Krysan 2004; Clark and Krysan 2007). The high-resolution melt curve (HRM) analysis of the PCR amplicon is then conducted to identify mutations without using enzymatic cleavage and gel electrophoresis. After the detection of mutations, the matching seedling with the coveted genotype can be transplanted from the 96-well plate to soil or growing media for seed collection to delineate its phenotype and maintain the individuals and produce further generations. The iTILLING approach is suitable for the detection of mutations at a comparatively small number of loci since the screening individuals used are grown for a short span, and only plants containing mutations that can be specified in a rather short time could be grown to reach maturity and harvest seeds for storage.

9.8.3 VeggieTILLING

Another approach is called VeggieTILLING which represents the application of TILLING in vegetatively propagated crops (apple, potato, strawberry). Natural vegetative propagation happens through diverse mechanisms involving the generation of rhizomes and stolons. To develop TILLING platform for vegetatively propagated plants, the first step is to generate a mutagenized population with an appropriate density of induced mutation and removal of chimeric sectors (Mba et al.

Table 9.2 Important mutants identified by Eco-TILLING in various crops

Species	Ploidy	Trait	Gene name	Methodology	References
Arabidopsis	2*x*	–	*DMMT2, DRM1C7, PIF2, AtWR*	LI-COR	Comai et al. (2004)
Wheat	6*x*	Vernalization	*VRN-A1*	CJE/Agarose	Liang et al. (2011)
Wheat	6*x*	Kernel hardness	*Pin a, Pin b*	LI-COR	Ma et al. (2017)
Rice	2*x*	–	–	LI-COR/UL	Rakshit et al. (2007)
Rice	2*x*	–	*MYB1, TPP, ADF*	Agarose gel	Raghavan et al. (2007)
Rice	2*x*	Boron toxicity		CJE/Agarose gel	Ochiai et al. (2011)
Rice	2*x*	Salt tolerance	*OSCP17*	CJE/Agarose gel	Negrão et al. (2011)
Rice	2*x*	Drought tolerance	–	LI-COR	Yu et al. (2012)
Rice	2*x*	Salt resistant	*osCPK17, osRMC, osNHX1*	LI-COR	Negrao et al. (2013)
Rice	2*x*	–	*osHKTI;5, SalT*		
Rice	2*x*	Starch synthesis	*GBSSI, SSI, SSIIa, SSIIIa*	EcoTbyS	Raja et al. (2017)
Rice	2*x*	–	*SBEIa, SBEIIb*		
Barley	2*x*	Chlorophyll protein	*Lhcb1*	LI-COR	Xia et al. (2012)
Barley	2*x*	Heat shock protein	*HSP17.8*	LI-COR	Xia et al. (2013)
Brassica spp.	2*x*	Erucic acid content	*FAE1-A8, FAE1-C3*	LI-COR	Wang et al. (2010)
Brassica spp.	4*x*	Organelle genome	*accD, matK, rbcL, atp6*	LI-COR	Zeng et al. (2012)
Brassica spp.	–	Chloroplast DNA	*Chloroplast DNA*	EcoTbyS	Qiao et al. (2016)
Tomato	2*x*	Folate biosynthesis	*GCH1, ADCS, ADCL1, ADCL2, FPGSp, FPGSm, GGH1, GGH2, GGH3*	LI-COR	Upadhyaya et al. (2017)
Tomato	2*x*	Plant development	*ACS2, CoP1, CYC-B, MSH2, NAC-NOR, PHoT1, PHYA, PHYB, PSY1*	LI-COR	Mohan et al. (2016)
Cotton	4*x*	Sucrose synthesis	*GhSus1At, GhSus1Dt, GhSus3At, GhSus4Dt, GhSus5Dt, GhSus6At*	LI-COR	Zeng et al. (2016)

(continued)

Table 9.2 (continued)

Species	Ploidy	Trait	Gene name	Methodology	References
Pepper	2*x*	Virus resistance	*eIF4E, I1F(iso)4E, eIF(iso)4G, eIF4G*	LI-COR	Ibiza et al. (2010)
Black poplar	2*x*	Lignin biosynthesis	*CAD4, HCT1, C3H3, CCR7, 4CL3*	TbyS	Marroni et al. (2011)
Rice	2*x*	Seed protein	*alk, waxy*	LI-COR	Kadaru et al. (2006)
Olive	2*x*	Fatty acid enzyme	*fad7*	LI-COR	Sabetta et al. (2013)
Jatropha	2*x*	Oil and stress tolerance	*AF, DQ66, EUO6, EFO3, DQ98, EU10, EU22, EU39, EU23, EU21, SUSY11, DQ15*	LI-COR	Maghuly et al. (2015)
Chickpea	2*x*	Seed weight	–	Agarose gel	Bajaj et al. (2016)

2009). The procedure of elimination of chimeras involves successive rounds of tissue culture. Previous studies indicate that chimeras can be significantly reduced or stabilized at the third vegetative cycle (Jain et al. 2011).

9.9 Application of TILLING in Crop Improvement

9.9.1 TILLING for Disease-Resistance Traits

TILLING populations have been used for the purposes of crop improvement by applying high-throughput mutation screening. Particularly for disease resistance, the technique can be used to validate candidate genes for the disease resistance (Desiderio et al. 2016). The powdery mildew disease caused by *Blumeria graminis f.* sp. *tritici* is a major problem in wheat cultivation. It has been reported that natural and induced loss of function in the barley *Mlo* gene confers durable resistance. In wheat, TILLING was employed to detect loss-of-function alleles for the *TaMlo* gene, which is found to be an ortholog of the barley *Mlo* gene. Through TILLING, 16 mutations that caused amino acid alterations were discovered in three *TaMlo* homologs which were proven to affect powdery mildew susceptibility and were used for development of powdery mildew resistant lines. These resistant lines signify an important move towards the development of commercial non-transgenic wheat varieties resistant to powdery-mildew (Acevedo-Garcia et al. 2017; Irshad et al. 2020).

TILLING in the hexaploid wheat line "UC1041+Gpc-B1/Yr36" (carrying the *Gpc-B1* gene for high grain protein content, and the *Yr36* gene for partial stripe rust-resistance) identified mutations in the single copy number genes, *Kinase Start 1* (*WKS1*) and *WKS2* located on chromosome arm 6BS. By TILLING six mutations in the *WKS1* gene causing an alteration in conserved amino acids were identified, whereas three mutations causing premature stop codons in the *WKS2* gene were identified for stripe rust resistance (Fu et al. 2009). The *WKS2* mutants displayed considerable resistance compared with controls suggesting that TILLING method can be used to identify and breed resistant cultivar in wheat (Fu et al. 2009). Similarly, in maize, TILLING populations were examined for two genes, *ZmWAK-RLK1* and *ZmWAK-RLK2* conferring resistance to northern corn leaf blight. A total of seven mutant lines were identified carrying amino acid substitutions in these two genes. The analysis showed that the mutants of *ZmWAK-RLK2* (*RLK2b*, *RLK2d*, and *RLK2e*) did not show any change in susceptibility compared to wild-type. Whereas the mutants of *ZmWAK-RLK1* (*RLK1b*, *RLK1d*, and *RLK1f*) exhibited high susceptibility compared to wild-type. One *ZmWAK-RLK1* mutant, *RLK1e*, showed comparable disease resistance to wild-type (Hurni et al. 2015).

TILLING was also performed in a mutant population of the tomato for potyvirus resistance gene, *eIF4E* isoforms. TILLING was accomplished by screening a TILLING population for mutations in the five translation initiation factors, *eIF4E1*, *eIF4E2*, *eIF(iso)4E*, *eIF4G*, and *eIF(iso)4G* (Piron et al. 2010). In this study, a splicing mutant of *eIF4E1* was discovered conferring resistance to two strains of Potato virus Y and Pepper Mottle Virus (Piron et al. 2010). Similarly, TILLING was performed in 1300 EMS-mutagenized pepper M2 lines and discovered the nine mutations in the *eIF4E1* (Siddique et al. 2020). TbyS was used in tomato to target the *eIF4E* gene, and two mutants were discovered carrying mutations in the *eIF4E* gene. Furthermore, the *eIF4E* gene was also screened using 92 accessions of tomato to discover natural polymorphisms using TbyS, and six haplotypes were identified (Rigola et al. 2009).

A TILLING-based method was also used for the positional cloning of the soybean gene, *Rhg4* harboring resistance to the soybean cyst nematode (SCN) *Heterodera glycines* (Liu et al. 2012). Previous molecular mapping work detected QTLs for resistance for SCN on chromosomes 18 (*rhg1*) and 8 (*Rhg4*) in the cultivar "Forrest." A molecular cloning procedure designated two genes in the *Rhg4* vicinity coding for a subtilisin-like protease (SUB1) and a serine hydroxyl methyltransferase (SHMT), respectively (Liu et al. 2012). A population of 1920 M2 families mutagenized by EMS containing cultivar "Forrest" (harboring *rhg1* and *Rhg4* resistance) genetic background were examined for mutations in the *SHMT* gene, which identified two mutants causing missense mutations at amino acid positions 61 (E61K) and 125 (M125I). Both mutants were highly susceptible to SCN (Liu et al. 2012). In the barley mutant population, two genes (*NPR1* and *EDR1*) involved in plant defense mechanism through the salicylic acid (SA) signaling pathway were examined using TILLING approach. Two missense mutations were discovered among the five total mutations (Hu et al. 2012). Thus these studies demonstrated that TILLING is a suitable technology to discover new resistance alleles for economically important plant diseases and could be utilized as a new genetic resource for breeding programs. Details of other TILLING studies for disease resistance are listed in Table 9.1.

9.9.2 TILLING for Abiotic Stress Tolerance Traits

In rice, TILLING was performed using an M2 mutant population consisting of 961 lines with the aim of screening nine target genes that show vital roles in membrane transport and the control of the salt-tolerance mechanism. Forty-one mutants comprising SNPs were identified in the target genes, including *OsAKT1*, *OsHKT6*, *OsNSCC2*, *OsCAX2*, *OsHKA11*, *OsP5CS1*, *OsSOS1*, *OsNHX1*, and *OsNAC60*. The mutation rate was one mutation per 492 kb, and the mutation percentage per total sequence was 0.67%. Among these 41 mutants, 9 showed mutations in the exon region, and 7 of them were highly salt-tolerant (Hwang et al. 2017). In barley a mutant population was constructed applying a chemical mutagen (sodium azide). Screening of mutant alleles through TILLING revealed a substantial role of the target genes in waterlogging and DNA repair mechanism in barley (Stolarek et al. 2015; Mendiondo et al. 2016).

In a legume crop *Medicago truncatula*, identification of a leucine-rich repeat RLK gene, *Srlk*, which is responsible for salt stress tolerance, was confirmed through TILLING approach. Two *Srlk*-TILLING mutants in *M. truncatula* were used to validate the role of LLR-RLKs related to salt stress tolerance (de Lorenzo et al. 2009). Recently, lipoxygenase (*LOX*) genes were analyzed by exploiting TbyS, and these genes exhibited greater expression in peanut roots (Guo et al. 2015). A total of 782 individuals of an M2 population were used to screen, which resulted in the discovery of four missense variations in *AhLOX*. Likewise, the phospholipase gene, *AhPLD*, which plays an imperative role in stress and drought tolerance, was assessed, and three missense mutations were discovered.

9.9.3 TILLING for Plant Architecture

In barley, two large EMS mutagenized populations were used for the mutation discovery (Caldwell et al. 2004). Two target genes *Hordeum vulgare Floral Organ Regulator-1* (*HvFor1*) and *Hordoindoline-a* (*Hin-a*) were assessed, and out of 10 SNPs detected, 6 of them were missense mutations (Caldwell et al. 2004). Phenotyping of the M3 generations revealed that 20% of the individuals had noticeable mutant phenotypes (Caldwell et al. 2004). In another study, the barley *HvD14* gene, which is involved in strigolactone signaling, was studied. TILLING was employed to examine 6912 M2 individuals from the mutant population named HorTILLUS, which identified seven mutations of the *HvD14* gene (Marzec et al. 2016). Among them, one mutation named *hvd14.d*, was semi-dwarf and exhibited a considerably greater number of tillers compared to wild-type (Sebastian). This mutation changed the amino acid "glycine" at position 193 to glutamic acid (Marzec et al. 2016). An EMS-mutagenized oat population was constructed to discover mutants through TILLING (Chawade et al. 2010). The M2 plants were at first evaluated by visual observation for numerous phenotypes containing a range of early-flowering to late-flowering, dwarfs to tall, and variations in leaf physiology and chlorosis. Phloroglucinol/HCl staining using M3 seeds from 1824 M2 mutants uncovered several potential lignin mutants. These mutants were further confirmed by experiments using a quantitative approach (Chawade et al. 2010).

9.9.4 TILLING for Yield and Quality-Related Traits

The identification of the genes controlling the yield-related traits in economically important crops and development of novel genomic resources and tools offers unique opportunities for improving productivity and yield. In wheat, many significant genes have been discovered using mutant populations. For example, in noodle manufacturing, fractional waxy wheat varieties are preferred. The *SBEIIa* (*starch branching enzymes of class II*) gene has been studied in various studies, and overexpression or suppression of this gene influences amylose content. In wheat, 246 allelic sequences were discovered in the *SBEIIa* gene homologs through TILLING (Slade et al. 2012). Among these allelic sequences, 84 missense, 3 nonsense, and 5 splice junction mutations were detected, with a mutation frequency of one mutation per 40 kb (Slade et al. 2012). Wheat *TaAGP.L-B1* gene, encoding the AGPase large subunit plays a vital role in starch biosynthesis in wheat endosperm and also promotes carbon metabolism and photosynthesis. Using TILLING approach, Guo et al. (2017) identified four major missense mutations in the *TaAGP.L-B1* alleles. Out of these four, an allele with a nonsynonymous mutation influences grain starch content (Guo et al. 2017). Similarly, in wheat starch synthesis gene, *TaSSIV* which plays an important role in granule development during starch production 54 mutations were identified. One of the nonsense mutations and three of the missense mutations were found to be affect the protein function (Guo et al. 2017). TILLING was also performed to identify mutations in rice *SBEI* gene, which plays a crucial role in amylose synthesis (Kim et al. 2018). A total of 37 mutations were recognized using a chemically induced mutant population of cultivar Nipponbare; among the 37 mutants discovered, one mutant M-4936 was found to be characterized with drastically decreased grain thickness and width (Kim et al. 2018). In peanut TILLING identified three silent and six missense mutations in *Ara h1.02*, *Ara h2.02*, and *Arah1.01* genes, which are responsible for the regulation of the ratio of linoleic acid contents in seeds. Mutants carrying these mutations were successively employed to discover the functions of mutated genes (Guo et al. 2015).

In rice, to investigate the stomatal limitations to photosynthesis, mutant population was screened for stomatal anion channel protein gene, *SLAC1* which regulates stomatal closure in response to environmental [CO_2] (Kusumi et al. 2012). TILLING identified four mutations of the *SLAC1*, one mutation designated as "*slac1*," was characterized with a lower leaf temperature phenotype and high stomatal conductance exhibiting high photosynthesis level (Kusumi et al. 2012). In barley, the "Sebastian" cultivar was mutagenized to develop a population comprising 9600 M2 mutant lines. Thirty-two genes related to plant growth and development was pursued in a population screening. This screening revealed 382 mutations containing a mutation rate of 1 mutation per 477 kb. Among the 67 identified mutations, 71% detected as missense, and others were nonsense (Szurman-Zubrzycka et al. 2018). In oat, using the TILLING approach, 520 oat mutant lines were assessed for alterations in seed lignin contents. The lignin content of the examined population differs from 20 to 63 g/kg. The lignin content in the mutant lines was enhanced significantly, and the quality was also improved compared to wild-type (Chawade et al. 2010).

In sorghum, few enzymes play a role in synthesizing, and catabolizing cyanogenic glycosides (CG), which adversely affect humans and animals because of poisonousness upon discharge of hydrogen cyanide (HCN). Dhurrinase1 and dhurrinase2 are enzymes that regulate the release of HCN in sorghum. In sorghum, EMS-mutated 1000 M2 families were screened using TbyS technology (Krothapalli et al. 2013). A single point mutation was discovered and revealed to cause premature stop codon in the coding region of the gene encoding dhurrinase2 enzyme, thus influencing the dhurrin catabolic pathway of the sorghum mutants. In line with this, the mutant is characterized by an altered phenotype of less HCN accumulation in leaves compared to wild-type plants (Krothapalli et al. 2013). TbyS has also been applied in polyploid species, such as camelina, to improve the nutritional value (increased protein content and less oil content) by targeting the *FUSCA3* gene (Kashtwari et al. 2019).

TILLING was also performed to find the mutation in the genes associated with the tomato shelf-life enhancement, indirectly affecting the yield. For instance, unique mutations in the *ETHYLENE RECEPTOR 1*, the *sletr1-1,* and *sletr1-2* alleles were identified by TILLING, and found to be involved in fruit shelf-life increase. Thus, these alleles could be exploited as an ultimate breeding tool to develop tomato cultivars with an increased shelf-life (Okabe et al. 2011).

9.10 TILLING Versus Genome Editing

Recently, genome-editing technologies have developed rapidly for targeted editing of the plant genomes. Unlike TILLING, which generally creates random knockout mutants, the key advantage of genome editing is that it can produce a range of mutations in a specific region in the target genes in a wide range of plants, including diploid or polyploids. This can both produce functional alleles and substitute mutant alleles (Petolino et al. 2016). CRISPR/Cas9 technology has been successfully used in a number of crops to induce mutations with detectable mutation rates (Jiang et al. 2013; Zhang et al. 2015; Xu et al. 2017; Yoon et al. 2020). For instance, in rice and pepper, CRISPR/Cas9 has been used to improve resistance against blast disease and potyviral disease by disrupting the ethylene-responsive factor (ERF) transcription factor and *eIF4E1* genes, respectively (Wang et al. 2016; Yoon et al. 2020). The main challenge to integrate these genome-editing technologies and TILLING is to obtain mutants with the required phenotypic response as well as reduce the off-target variations. Both approaches can be employed to generate allelic variants in the same gene; nevertheless, the result can be different since CRISPR/Cas9 uses only a few targets, whereas TILLING generates genome-wide mutations. For the genetic improvement of polyploids, it is essential to recognize the allelic multiplicity and homologous genes that govern the phenotypes of diverse traits. Recent studies have determined that both approaches are imperative tools to analyze these traits (Wang et al. 2018; Avni et al. 2017). For example, the CRISPR/Cas9 and TILLING technologies were applied to induce mutation in all homologs of the *TaGW2* gene in the wheat cultivars Bobwhite and Paragon, respectively. The results revealed that the *TaGW2* gene homologs negatively affected the size of grain and weight for thousand grains (Wang et al. 2018). Plants containing single-copy nonsense mutations in different genomes revealed different levels of grain weight and size, and the weight of a thousand grain enhanced by an average of 5.5% in edited lines and 5.3% in TILLING mutants. In several combinations, the double homolog mutants exhibited greater phenotypic effects than the individual single-genome mutants. The double mutants showed on average 10.5% (TILLING mutants) and 12.1% (edited lines) greater weight for thousand grains compared to wild-type lines. The maximum increase in grain weight was revealed for triple mutants with 20.7% and 16.3% increase in weight for thousand grains, respectively, for TILLING and edited lines (Wang et al. 2018). This pattern evidently determines that both modern technologies have exclusive advantages and disadvantages, and their integrated use may improve the efficiency of generation/development of desired mutants for crop improvement. It can be anticipated that TILLING and genome editing can be a powerful complement in plant science for crop improvement.

9.11 Conclusions and Future Prospective

Recent interest in genetic dissection of genes involved in the control of economically important traits, such as biotic and abiotic stress tolerance and yield-related traits, has promoted the development of high-efficiency TILLING techniques. Since recently only, modern technologies emerged, including next-generation sequencing, genome-wide association analysis, genomic selection, and new plant breeding technologies, TILLING as a reverse genetic approach has been advanced to expedite trait discovery. With the advancements made in the NGS technologies, genome sequencing become much cheaper and the genome sequences for several plant species are now available. The availability of high-density linkage maps along with the accessibility of the reference genomes can immensely facilitate the detection of candidate genes for the trait of interest. Once candidate genes are predicted, validation of these genes can be achieved by the identification of loss-of-function mutants using the TILLING, EcoTILLING, and improved TILLING methods. Currently, available TbyS procedures are enabling resourceful mutation discovery. Furthermore, the application of multidimensional pooling with increased likelihood threshold is permitting to uncover mutation with related base alteration and having genetic effects (Tsai et al. 2011).

Identification of candidate genes underlying mutant phenotypes of important traits would further benefit from appropriate phenotyping strategies. High-throughput phenotyping, also denoted as phenomics, is a recent and most interesting research area in biology. Progressive plant phenomics technologies comply with a high-throughput examination of phenotypes linked to the genetic variants

(Douchkov et al. 2014), thus refining the discovery of phenotypes linked to a certain trait in mutants detected by TILLING or TILLING-associated approaches.

Advancement in genome-editing technology will further circumvent or reduce off-target mutations by CRISPR/Cas9. The CRISPR/Cas9 specificity can be enhanced by using the Cas9 nickase or Cas9-FokI fusion proteins approaches, and SpCas9-HF or the Cas12a nuclease, which both significantly mitigate the number of possible off-target sequences knockouts (Bortesi and Fischer 2015; Kleinstiver et al. 2016; Strohkendl et al. 2018). Although the emerging CRISPR/Cas9 and other gene-editing approaches to modify plant genomes have increasingly become popular (Yin et al. 2017) and can be an alternate for TILLING approach for some applications. The decision of the European Union to subject CRISPR/Cas9-edited crops to regulatory requirements as similar to transgenic crops (Callaway 2018) indicate that crops developed by TILLING may have easier access to the market than the genome-edited crops at least in some countries.

References

Acevedo-Garcia J, Spencer D, Thieron H, Reinstädler A, Hammond-Kosack K, Phillips AL, Panstruga R (2017) *mlo*-based powdery mildew resistance in hexaploid bread wheat generated by a non-transgenic TILLING approach. Plant Biotechnol J 15:367

Altpeter F, Springer NM, Bartley LE, Blechl AE, Brutnell TP, Citovsky V, Conrad LJ, Gelvin SB, Jackson DP, Kausch AP, Lemaux PG (2016) Advancing crop transformation in the era of genome editing. Plant Cell 7:1510–1520

Avni R, Nave M, Barad O, Baruch K, Twardziok SO, Gundlach H, Hale I, Mascher M, Spannagl M, Wiebe K (2017) Wild emmer genome architecture and diversity elucidate wheat evolution and domestication. Science 357:83–97

Bajaj D, Srivastava R, Nath M, Tripathi S, Bharadwaj C, Upadhyaya HD, Tyagi AK, Parida SK (2016) EcoTILLING-based association mapping efficiently delineates functionally relevant natural allelic variants of candidate genes governing agronomic traits in chickpea. Front Plant Sci 7:450

Bortesi L, Fischer R (2015) The CRISPR/Cas9 system for plant genome editing and beyond. Biotechnol Adv 33:41–52

Botticella E, Sestili F, Hernandez-Lopez A, Phillips A, Lafiandra D (2011) High resolution melting analysis for the detection of EMS induced mutations in wheat *SbeIIa* genes. BMC Plant Biol 11: 156

Bush SM, Krysan PJ (2010) iTILLING: a personalized approach to the identification of induced mutations in Arabidopsis. Plant Physiol 154:25–35

Caldwell DG, McCallum N, Shaw P, Muehlbauer GJ, Marshall DF, Waugh R (2004) A structured mutant population for forward and reverse genetics in barley (*Hordeum vulgare* L.). Plant J 40: 143–150

Callaway E (2018) CRISPR plants now subject to tough GM laws in European Union. Nature 560: 16

Chawade A, Sikora P, Brautigam M, Larsson M, Vivekanand V, Nakash MA, Chen T, Olsson O (2010) Development and characterization of an oat TILLING-population and identification of mutations in lignin and beta-glucan biosynthesis genes. BMC Plant Biol 10:86

Chen L, Wang SQ, Hu YG (2011) Detection of SNPs in the *VRN-A1* gene of common wheat (*Triticum aestivum* L.) by a modified Ecotilling method using agarose gel electrophoresis. Aust J Crop Sci 5:321–329

Chen L, Huang LZ, Min DH, Phillips A, Wang SQ, Madgwick PJ, Parry MAJ, Hu YG (2012) Development and characterization of a new TILLING population of common bread wheat (*Triticum aestivum* L.). PLoS One 7:e41570

Clark KA, Krysan PJ (2007) Protocol: an improved high-throughput method for generating tissue samples in 96-well format for plant genotyping (Ice-Cap 2.0). Plant Methods 3:8

Colasuonno P, Incerti O, Lozito ML, Simeone R, Gadaleta A, Blanco A (2016) DHPLC technology for high-throughput detection of mutations in a durum wheat TILLING population. BMC Genet 17:43

Colbert T, Till BJ, Tompa R, Reynolds S, Steine MN, Yeung AT, McCallum CM, Comai L, Henikoff S (2001) High-throughput screening for induced point mutations. Plant Physiol 126: 480–484

Comai L, Young K, Till BJ, Reynolds SH, Greene EA, Codomo CA (2004) Efficient discovery of DNA polymorphisms in natural populations by Ecotilling. Plant J 37:778–786

Cooper JL, Till BJ, Laport RG, Darlow MC, Kleffner JM, Jamai A, El-Mellouki T, Liu S, Ritchie R, Nielsen N, Bilyeu KD, Meksem K, Comai L, Henikoff S (2008) TILLING to detect induced mutations in soybean. BMC Plant Biol 8:9

Cross MJ, Waters DLE, Lee LS, Henry RJ (2008) Endonucleolytic mutation analysis by internal labeling (EMAIL). Electrophoresis 29:1291–1301

Dahmani-Mardas F, Troadec C, Boualem A, Leveque S, Alsadon AA, Aldoss AA, Dogimont C, Bendahmane A (2010) Engineering melon plants with improved fruit shelf life using the TILLING approach. PLoS One 5:e15776

Dalmais M, Antelme S, Ho-Yue-Kuang S, Wang Y, Darracq O (2013) A TILLING platform for functional genomics in *Brachypodium distachyon*. PLoS One 8:e65503

Davies H, Dicks E, Stephens P, Cox C, Teague J, Greenman C, Bignell G, O'Meara S, Edkins S, Parker A (2006) High throughput DNA sequence variant detection by conformation sensitive capillary electrophoresis and automated peak comparison. Genomics 87:427–432

de Lorenzo L, Merchan F, Laporte P, Thompson R, Clarke J, Sousa C, Crespi M (2009) A novel plant leucine-rich repeat receptor kinase regulates the response of *Medicago truncatula* roots to salt stress. Plant Cell 21:668–680

Desiderio F, Torp AM, Valè G, Rasmussen SK (2016) Tilling in plant disease control: applications and perspectives. In: Collinge DB (ed) Biotechnology for plant disease control. Wiley, New York, pp 365–384

Dong C, Dalton-Morgan J, Vincent K, Sharp P (2009a) A modified TILLING method for wheat breeding. Plant Genome 2:39

Dong C, Vincent K, Sharp P (2009b) Simultaneous mutation detection of three homoeologous genes in wheat by high resolution melting analysis and Mutation Surveyor®. BMC Plant Biol 9: 143

Douchkov D, Baumb T, Ihlowc A, Schweizer P, Seiffert U (2014) Microphenomics for interactions of barley with fungal pathogens. In: Tuberosa R, Graner A, Frison A (eds) Genomics of plant genetic resources, vol 2. Springer, Heidelberg, pp 123–148

Draper BW, McCallum CM, Stout JL, Slade AJ, Moens CB (2004) A high-throughput method for identifying N-ethyl-N-nitrosourea (ENU)-induced point mutations in zebrafish. Methods Cell Biol 77:91–112

Feiz L, Martin JM, Giroux MJ (2009) Creation and functional analysis of new *Puroindoline* alleles in *Triticum aestivum*. Theor Appl Genet 118:247–257

Fekih R, Takagi H, Tamiru M, Abe A, Natsume S, Yaegashi H, Sharma S, Sharma S, Kanzaki H, Matsumura H, Saitoh H, Mitsuoka C, Utsushi H, Uemura A, Kanzaki E, Kosugi S, Yoshida K, Cano L, Kamoun S, Terauchi R (2013) MutMap+: genetic mapping and mutant identification without crossing in rice. PLoS One 8:e68529

Fitzgerald TL, Kazan K, Li Z, Morell MK, Manners JM (2010) A high-throughput method for the detection of homologous gene deletions in hexaploid wheat. BMC Plant Biol 10:264

Fu D, Uauy C, Distelfeld A, Blechl A, Epstein L, Chen X, Sela H, Fahima T, Dubcovsky J (2009) A kinase-START gene confers temperature-dependent resistance to wheat stripe rust. Science 323: 1357–1360

Gady AL, Hermans FW, Van de Wal MH, van Loo EN, Visser RG, Bachem CW (2009) Implementation of two high throughput techniques in a novel application: detecting point mutations in large EMS mutated plant populations. Plant Methods 5:13

Garvin MR, Gharrett AJ (2007) Deco-TILLING: an inexpensive method for single nucleotide polymorphism discovery that reduces ascertainment bias. Mol Ecol Notes 7:735–746

Gilchrist E, Haughn G (2010) Reverse genetics techniques: engineering loss and gain of gene function in plants. Brief Funct Genomics 9:103

Gonzalez M, Xu M, Esteras C, Roig C, Monforte AJ, Troadec C, Pujol M, Nuez F, Bendahmane A, Garcia-Mas J, Pico B (2011) Towards a TILLING platform for functional genomics in Piel de Sapo melons. BMC Res Notes 4:289

Gottwald S, Bauer P, Komatsuda T, Lundqvist U, Stein N (2009) TILLING in the two-rowed barley cultivar 'Barke' reveals preferred sites of functional diversity in the gene *HvHox1*. BMC Res Notes 2:258

Greene EA, Codomo CA, Taylor NE, Henikoff JG, Till BJ (2003) Spectrum of chemically induced mutations from a large-scale reverse-genetic screen in *Arabidopsis*. Genetics 164:731–740

Guo Y, Abernathy B, Zeng Y, Ozias-Akins P (2015) TILLING by sequencing to identify induced mutations in stress resistance genes of peanut (*Arachis hypogaea*). BMC Genomics 16:157

Guo H, Yan Z, Li X, Xie Y, Xiong H, Liu Y, Zhao L, Gu J, Zhao S, Liu L (2017) Development of a high-efficient mutation resource with phenotypic variation in hexaploid winter wheat and identification of novel alleles in the *TaAGP.L-B1* gene. Front Plant Sci 8:1404

Gupta P, Reddaiah B, Salava H, Upadhyaya P, Tyagi K, Sarma S, Datta S, Malhotra B, Thomas S, Sunkum A, Devulapalli S, Till BJ, Sreelakshmi Y, Sharma R (2017) Next-generation sequencing (NGS)-based identification of induced mutations in a doubly mutagenized tomato (*Solanum lycopersicum*) population. Plant J 92:495–508

Henikoff S, Till BJ, Comai L (2004) TILLING. Traditional mutagenesis meets functional genomics. Plant Physiol 135:630–636

Henry IM, Nagalakshmi U, Lieberman MC, Ngo KJ, Krasileva KV, Vasquez-Gross H, Akhunova A, Akhunov E, Dubcovsky J, Tai TH (2014) Efficient genome-wide detection and cataloging of EMS-induced mutations using exome capture and next-generation sequencing. Plant Cell 26:1382–1397

Himelblau E, Gilchrist EJ, Buono K, Bizzell C, Mentzer L, Vogelzang R, Osborn T, Amasino RM, Parkin IA, Haughn GW (2009) Forward and reverse genetics of rapid-cycling *Brassica oleracea*. Theor Appl Genet 118:953–961

Hu X, Qi X, Lv B, Jj W, Dl F (2012) TILLING-based analysis of disease resistance genes in barley. J Shandong Agric Univ (Natural Science Edition) 1:002

Hurni S, Scheuermann D, Krattinger SG, Kessel B, Wicker T, Herren G, Fitze MN, Breen J, Presterl T, Ouzunova M, Keller B (2015) The maize disease resistance gene *Htn1* against northern corn leaf blight encodes a wall-associated receptor-like kinase. Proc Natl Acad Sci U S A 112:8780–8785

Hwang JE, Jang DS, Lee KJ, Ahn JW, Kim SH, Kang SY, Kim DS, Kim JB (2017) Identification of gamma ray irradiation-induced mutations in membrane transport genes in a rice population by TILLING. Genes Genet Syst 91:245–256

Ibiza VP, Cañizares J, Nuez F (2010) EcoTILLING in *Capsicum* species: searching for new virus resistances. BMC Genomics 11:631

Irshad A, Guo H, Zhang S, Liu L (2020) TILLING in cereal crops for allele expansion and mutation detection by using modern sequencing technologies. Agronomy 3:405

Jain S, Till B, Suprasanna P, Roux N (2011) Mutations and cultivar development of banana. In: Pillay M, Tenkouano A (eds) Banana breeding, progress and challenges. CRC Press, Boca Raton, FL, pp 203–217

Jansen JG, Vrieling H, Vanteijlingen CMM, Mohn GR, Tates AD, vanZeeland AA (1995) Marked differences in the role of O-6-alkylguanine in HPRT mutagenesis in T-lymphocytes of rats exposed in-vivo to ethylmethanesulfonate, N-(2-hydroxyethyl)-N-nitro-sourea, or N-ethyl-N-nitrosourea. Cancer Res 55:1875–1882

Jawher M, Arabi MIE, MirAli N, Till BJ (2018) Efficient discovery of single-nucleotide variations in Cochliobolus sativus vegetative compatibility groups by EcoTILLING. J Plant Biochem Physiol 6:2

Jeong HJ, Kwon JK, Pandeya D, Hwang J, Hoang NH, Bae JH, Kang BC (2012) A survey of natural and ethyl methane sulfonate-induced variations of *eIF4E* using high-resolution melting analysis in Capsicum. Mol Breed 29:349–360

Jiang W, Zhou H, Bi H, Fromm M, Yang B, Weeks DP (2013) Demonstration of CRISPR/Cas9/sgRNA-mediated targeted gene modification in *Arabidopsis*, tobacco, sorghum and rice. Nucleic Acids Res 41:e188

Jo YD, Kim JB (2019) Frequency and spectrum of radiation-induced mutations revealed by whole-genome sequencing analyses of plants. Quantum Beam Sci 3:7

Kadaru SB, Yadav AS, Fjellstrom RG, Oard JH (2006) Alternative EcoTILLING protocol for rapid, cost-effective single-nucleotide polymorphism discovery and genotyping in rice (*Oryza sativa* L.). Plant Mol Biol Rep 24:3–22

Kashtwari M, Wani AA, Rather RN (2019) TILLING: an alternative path for crop improvement. J Crop Improv 33:83–109

Keller G, Hartmann A, Mueller J, Hofler H (2001) Denaturing high pressure liquid chromatography (DHPLC) for the analysis of somatic *p53* mutations. Lab Invest 81:1735–1737

Keller B, Feuillet C, Yahiaoui N (2005) Map-based isolation of disease resistance genes from bread wheat: cloning in a supersize genome. Genet Res 85:93–100

Kim SI, Tai TH (2014) Identification of novel rice low phytic acid mutations via TILLING by sequencing. Mol Breed 34:1717–1729

Kim H, Yoon MR, Chun A, Tai TH (2018) Identification of novel mutations in the *rice starch branching enzyme I* gene via TILLING by sequencing. Euphytica 214:94

King R, Bird N, Ramirez-Gonzalez R, Coghill JA, Patil A, Hassani-Pak K, Uauy C, Phillips AL (2015) Mutation scanning in wheat by exon capture and next-generation sequencing. PLoS One 10:e0137549

Kleinstiver BP, Pattanayak V, Prew MS, Tsai SQ, Nguyen NT, Joung JK (2016) High-fidelity CRISPR–Cas9 nucleases with no detectable genome-wide off-target effects. Nature 529:490–495

Knoll JE, Ramos ML, Zeng Y, Holbrook CC, Chow M, Chen S, Maleki S, Bhattacharya A, Ozias-Akins P (2011) TILLING for allergen reduction and improvement of quality traits in peanut (*Arachis hypogaea* L.). BMC Plant Biol 11:81

Komatsuda T, Pourkheirandish M, He C, Azhaguvel O, Kanamori H, Perovic D, Stein N, Graner A, Wicker T, Tagiri A, Lundqvist U, Fujimura T, Matsuoka M, Matsumoto T, Yano M (2007) Six-rowed barley originated from a mutation in a homeodomain-leucine zipper I-class homeobox gene. Proc Natl Acad Sci U S A 104:1424–1429

Krothapalli K, Buescher EM, Li X, Brown E, Chapple C, Dilkes BP, Tuinstra MR (2013) Forward genetics by genome sequencing reveals that rapid cyanide release deters insect herbivory of *Sorghum bicolor*. Genetics 195:309–318

Krysan P (2004) Ice-Cap. A high-throughput method for capturing plant tissue samples for genotype analysis. Plant Physiol 135:1162–1169

Kurowska M, Daszkowskagolec A, Gruszka D, Marzec M, Szurman M, Szarejko I, Maluszynski M (2011) TILLING—a shortcut in functional genomics. J Appl Genet 52:371

Kusumi K, Hirotsuka S, Kumamaru T, Iba K (2012) Increased leaf photosynthesis caused by elevated stomatal conductance in a rice mutant deficient in SLAC1, a guard cell anion channel protein. J Exp Bot 63:5635–5644

Lakhssassi N, Zhou Z, Liu S, Piya S, Cullen MA, El Baze A, Knizia D, Patil GB, Badad O, Embaby MG, Meksem J, Lakhssassi A, AbuGhazaleh A, Hewezi T, Meksem K (2020) Soybean

TILLING-by-Sequencing+ reveals the role of novel GmSACPD members in unsaturated fatty acid biosynthesis while maintaining healthy nodules. J Exp Bot 71:6969–6987

Li X, Song Y, Century K, Straight S, Ronald P, Dong X, Lassner M, Zhang Y (2001) A fast neutron deletion mutagenesis-based reverse genetics system for plants. Plant J 27:235–242

Li W, Guo H, Wang Y, Xie Y, Zhao L, Gu J, Zhao S, Zhao B, Wang G, Liu L (2017) Identification of novel alleles induced by EMS-mutagenesis in key genes of kernel hardness and starch biosynthesis in wheat by TILLING. Genes Genomics 39:387–395

Liang C, Wang SQ, Hu YG (2011) Detection of SNPs in the VRN-A1 gene of common wheat (*Triticum aestivum* L.) by a modified EcoTILLING method using agarose gel electrophoresis. Aust J Crop Sci 5:318–326

Liu S, Kandoth PK, Warren SD, Yeckel G, Heinz R, Alden J, Yang C, Jamai A, El-Mellouki T, Juvale PS, Hill J, Baum TJ, Cianzio S, Whitham SA, Korkin D, Mitchum MG, Meksem K (2012) A soybean cyst nematode resistance gene points to a new mechanism of plant resistance to pathogens. Nature 492:256–260

Liu L, Venkatesh J, Jo YD, Koeda S, Hosokawa M, Kang JH, Goritschnig S, Kang BC (2016) Fine mapping and identification of candidate genes for the *sy-2* locus in a temperature-sensitive chili pepper (*Capsicum chinense*). Theor Appl Genet 8:1541–1556

Ma X, Sajjad M, Wang J, Yang W, Sun J, Li X, Zhang A, Liu D (2017) Diversity, distribution of Puroindoline genes and their effect on kernel hardness in a diverse panel of Chinese wheat germplasm. BMC Plant Biol 17:158

Maghuly F, Jankowicz-Cieslak J, Pabinger S, Till BJ, Laimer M (2015) Geographic origin is not supported by the genetic variability found in a large living collection of *Jatropha curcas* with accessions from three continents. Biotechnol J 10:536–551

Maluszynski M, Szarejko I, Maluszynska J (2003) Mutation techniques. In: Thomas B, Murphy DJ, Murray BG (eds) Encyclopedia of applied plant sciences. Elsevier Academic Press, San Diego, pp 186–201

Marroni F, Pinosio S, Di Centa E, Jurman I, Boerjan W, Felice N, Cattonaro F, Morgante M (2011) Large-scale detection of rare variants via pooled multiplexed next-generation sequencing: towards next-generation EcoTILLING. Plant J 67:736–745

Martienssen RA (1998) Functional genomics: probing plant gene function and expression with transposons. Proc Natl Acad Sci U S A 95:2021–2026

Marzec M, Gruszka D, Tylec P, Szarejko I (2016) Identification and functional analysis of the *HvD14* gene involved in strigolactone signaling in *Hordeum vulgare*. Physiol Plant 158:341–355

Mba C, Afza R, Jankowcz-Cieslak J, Bado S, Matijevic M, Huynh O, Till B (2009) Enhancing genetic diversity through induced mutagenesis in vegetatively propagated plants. In: Qy S (ed) Induced plant mutations in the genomic era. Food and Agriculture Organization of the United Nations, Rome, pp 293–296

Mba C, Afza R, Bado S, Jain S (2010) Induced mutagenesis in plants using physical and chemical agents. In: Davey M, Anthony P (eds) Plant cell culture: essential methods. Wiley, New York, pp 111–130

McCallum CM, Comai L, Greene EA, Henikoff S (2000) Targeting Induced Local Lesions IN Genomes (TILLING) for plant functional genomics. Plant Physiol 123:439–442

Mejlhede N, Kyjovska Z, Backes G, Burhenne K, Rasmussen SK, Jahoor A (2006) EcoTILLING for the identification of allelic variation in the powdery mildew resistance genes *mlo* and *Mla* of barley. Plant Breed 125:461–467

Mendiondo GM, Gibbs DJ, Szurman-Zubrzycka M, Korn A, Marquez J, Szarejko I, Maluszynski M, King J, Axcell B, Smart K (2016) Enhanced waterlogging tolerance in barley by manipulation of expression of the N-end rule pathway E3 ligase PROTEOLYSIS 6. Plant Biotechnol J 14:40–50

Minoia S, Boualem A, Marcel F, Troadec C, Quemener B, Cellini F, Petrozza A, Vigouroux J, Lahaye M, Carriero F (2016) Induced mutations in tomato *SlExp1* alter cell wall metabolism and delay fruit softening. Plant Sci 242:195–202

Mo Y, Howell T, Vasquez-Gross H, de Haro LA, Dubcovsky J, Pearce S (2018) Mapping causal mutations by exome sequencing in a wheat TILLING population: a tall mutant case study. Mol Genet Genomics 293:463–477

Moehs CP, Austill WJ, Holm A, Large TA, Loeffler D, Mullenberg J, Schnable PS, Skinner W, van Boxtel J, Wu L (2018) Development of reduced gluten wheat enabled by determination of the genetic basis of the *lys3a* low hordein barley mutant. Plant Physiol 179:1692–1703

Mohan V, Gupta S, Thomas S, Mickey H, Charakana C, Chauhan VS, Sharma K, Kumar R, Tyagi K, Sarma S, Gupta SK (2016) Tomato fruits show wide phenomic diversity but fruit developmental genes show low genomic diversity. PLoS One 11:e0152907

Muth J, Hartje S, Twyman RM, Hofferbert HR, Tacke E, Prufer D (2008) Precision breeding for novel starch variants in potato. Plant Biotechnol J 6:576–584

Nagy A, Perrimon N, Sandmeyer S, Plasterk R (2003) Tailoring the genome: the power of genetic approaches. Nat Genet 33:276–284

Negrão S, Almadanim C, Pires I, McNally KL, Oliveira MM (2011) Use of EcoTILLING to identify natural allelic variants of rice candidate genes involved in salinity tolerance. Plant Genet Resour 9:300–304

Negrao S, Almadanim MC, Pires IS, Abreu IA, Maroco J, Courtois B, Gregorio GB, McNally KL, Oliveira MM (2013) New allelic variants found in key rice salt-tolerance genes: an association study. Plant Biotechnol J 11:87–100

Ng PC, Henikoff S (2003) SIFT: predicting amino acid changes that affect protein function. Nucleic Acids Res 31:3812–3814

Nieto C, Piron F, Dalmais M, Marco CF, Moriones E, Gomez-Guillamon ML, Truniger V, Gomez P, Garcia-Mas J, Aranda MA, Bendahmane A (2007) EcoTILLING for the identification of allelic variants of melon *eIF4E*, a factor that controls virus susceptibility. BMC Plant Biol 7: 34

Ochiai K, Shimizu A, Okumoto Y, Fujiwara T, Matoh T (2011) Suppression of a NAC-like transcription factor gene improves boron-toxicity tolerance in rice. Plant Physiol 156:1457–1463

Okabe Y, Asamizu E, Saito T, Matsukura C, Ariizumi T, Brès C, Rothan C, Mizoguchi T, Ezura H (2011) Tomato TILLING technology: development of a reverse genetics tool for the efficient isolation of mutants from Micro-Tom mutant libraries. Plant Cell Physiol 52:1994–2005

Okabe Y, Asamizu E, Ariizumi T, Shirasawa K, Tabata S, Ezura H (2012) Availability of Micro-Tom mutant library combined with TILLING in molecular breeding of tomato fruit shelf-life. Breed Sci 62:202–208

Oladosu Y, Rafii MY, Abdullah N, Hussin G, Ramli A, Rahim HA, Miah G, Usman M (2016) Principle and application of plant mutagenesis in crop improvement: a review. Biotechnol Biotechnol Equip 30:1–16

Oleykowski CA, Bronson Mullins CR, Godwin AK, Yeung AT (1998) Mutation detection using a novel plant endonuclease. Nucleic Acids Res 26:4597–4602

Ostergaard L, Yanofsky MF (2004) Establishing gene function by mutagenesis in *Arabidopsis thaliana*. Plant J 39:682–696

Petolino JF, Srivastava V, Daniell H (2016) Editing plant genomes: a new era of crop improvement. Plant Biotechnol J 14:435–436

Piron F, Nicolai M, Minoia S, Piednoir E, Moretti A, Salgues A, Zamir D, Caranta C, Bendahmane A (2010) An induced mutation in tomato *eIF4E* leads to immunity to two potyviruses. PLoS One 5:e11313

Qiao J, Cai M, Yan G, Wang N, Li F, Chen B, Gao G, Xu K, Li J, Wu X (2016) High-throughput multiplex cpDNA resequencing clarifies the genetic diversity and genetic relationships among *Brassica napus*, *Brassica rapa* and *Brassica oleracea*. Plant Biotechnol J 14:409–418

Raghavan C, Naredo MEB, Wang H, Atienza G, Liu B, Qiu F, McNally KL, Leung H (2007) Rapid method for detecting SNPs on agarose gels and its application in candidate gene mapping. Mol Breed 19:87–101

Raja RB, Agasimani S, Jaiswal S, Thiruvengadam V, Sabariappan R, Chibbar RN, Ram SG (2017) EcoTILLING by sequencing reveals polymorphisms in genes encoding starch synthases that are associated with low glycemic response in rice. BMC Plant Biol 17:13

Rakshit S, Rakshit A, Matsumura H, Takahashi Y, Hasegawa Y, Ito A, Ishii T, Miyashita NT, Terauchi R (2007) Large-scale DNA polymorphism study of *Oryza sativa* and *O. rufipogon* reveals the origin and divergence of Asian rice. Theor Appl Genet 114:731–743

Rawat N, Sehgal SK, Joshi A, Rothe N, Wilson DL, McGraw N, Vadlani PV, Li W, Gill BS (2012) A diploid wheat TILLING resource for wheat functional genomics. BMC Plant Biol 12:205

Reddy TV, Dwivedi S, Sharma NK (2012) Development of TILLING by sequencing platform towards enhanced leaf yield in tobacco. Ind Crop Prod 40:324–335

Rigola D, van Oeveren J, Janssen A, Bonné A, Schneiders H, van der Poel HJ, van Orsouw NJ, Hogers RC, de Both MT, van Eijk MJ (2009) High-throughput detection of induced mutations and natural variation using KeyPoint™ technology. PLoS One 4:e4761

Robinson SJ, Parkin IA (2009) Bridging the gene-to-function knowledge gap through functional genomics. Methods Mol Biol 513:153–173

Sabetta W, Alba V, Blanco A, Montemurro C (2011) sunTILL: a TILLING resource for gene function analysis in sunflower. Plant Methods 7:20

Sabetta W, Blanco A, Zelasco S, Lombardo L, Perri E, Mangini G, Montemurro C (2013) *Fad7* gene identification and fatty acids phenotypic variation in an olive collection by EcoTILLING and sequencing approaches. Plant Physiol Biochem 69:1–8

Sato Y, Shirasawa K, Takahashi Y, Nishimura M, Nishio T (2006) Mutant selection from progeny of gamma-ray-irradiated rice by DNA heteroduplex cleavage using Brassica petiole extract. Breed Sci 56:179–183

Sestili F, Botticella E, Bedo Z, Phillips A, Lafiandra D (2010) Production of novel allelic variation for genes involved in starch biosynthesis through mutagenesis. Mol Breed 25:145

Siddique MI, Back S, Lee JH, Jo J, Jang S, Han K, Venkatesh J, Kwon JK, Jo YD, Kang BC (2020) Development and characterization of an ethyl methane sulfonate (EMS) induced mutant population in *Capsicum annuum* L. Plants 9:396

Skopek TR, Walker VE, Cochrane JE, Craft TR, Cariello NF (1992) Mutational spectrum at the Hprt locus in splenic T cells of B6C3F1 mice exposed to N-ethyl-N-nitrosourea. Proc Natl Acad Sci U S A 89:7866–7870

Slade AJ, Knauf VC (2005) TILLING moves beyond functional genomics into crop improvement. Transgenic Res 14:109–115

Slade AJ, Fuerstenberg SI, Loeffler D, Steine MN, Facciotti D (2005) A reverse genetic, nontransgenic approach to wheat crop improvement by TILLING. Nat Biotechnol 23:75–81

Slade AJ, McGuire C, Loeffler D, Mullenberg J, Skinner W, Fazio G, Holm A, Brandt KM, Steine MN, Goodstal JF (2012) Development of high amylose wheat through TILLING. BMC Plant Biol 12:10

Sparla F, Falini G, Botticella E, Pirone C, Talamè V, Bovina R, Salvi S, Tuberosa R, Sestili F, Trost P (2014) New starch phenotypes produced by TILLING in barley. PLoS One 9:e107779

Stephenson P, Baker D, Girin T, Perez A, Amoah S, King GJ, Ostergaard L (2010) A rich TILLING resource for studying gene function in *Brassica rapa*. BMC Plant Biol 10:62

Stolarek M, Gruszka D, Braszewska-Zalewska A, Maluszynski M (2015) Alleles of newly identified barley gene *HvPARP3* exhibit changes in efficiency of DNA repair. DNA Repair 28:116–130

Strohkendl I, Saifuddin FA, Rybarski JR, Finkelstein IJ, Russell R (2018) Kinetic basis for DNA target specificity of CRISPR-Cas12a. Mol Cell 71:816–824

Suzuki T, Eiguchi M, Kumamaru T, Satoh H, Matsusaka H, Moriguchi K, Nagato Y, Kurata N (2008) MNU-induced mutant pools and high performance TILLING enable finding of any gene mutation in rice. Mol Genet Genomics 279:213–223

Szurman-Zubrzycka ME, Zbieszczyk J, Marzec M, Jelonek J, Chmielewska B, Kurowska MM, Krok M, Daszkowska-Golec A, Guzy-Wrobelska J, Gruszka D (2018) HorTILLUS—a rich and renewable source of induced mutations for forward/reverse genetics and pre breeding programs in barley (*Hordeum vulgare* L.). Front Plant Sci 9:216

Talamè V, Bovina R, Sanguineti MC, Tuberosa R, Lundqvist U, Salvi S (2008) TILLMore, a resource for the discovery of chemically induced mutants in barley. Plant Biotechnol J 6:477–485

Thapa R, Carrero-Colón M, Rainey KM, Hudson K (2019) TILLING by sequencing: a successful approach to identify rare alleles in soybean populations. Genes (Basel) 10:1003

Till BJ, Reynolds SH, Greene EA, Codomo CA, Enns LC, Johnson JE, Burtner C, Odden AR, Young K, Taylor NE, Henikoff JG, Comai L, Henikoff S (2003) Large-scale discovery of induced point mutations with high-throughput TILLING. Genome Res 13:524–530

Till BJ, Reynolds SH, Weil C, Springer N, Burtner C, Young K, Bowers E, Codomo CA, Enns LC, Odden AR, Greene EA, Comai L, Henikoff S (2004) Discovery of induced point mutations in maize genes by TILLING. BMC Plant Biol 4:12

Till BJ, Zerr T, Comai L, Henikoff S (2006) A protocol for TILLING and Ecotilling in plants and animals. Nat Protoc 1:2465–2477

Till BJ, Cooper J, Tai TH, Colowit P, Greene EA, Henikoff S, Comai L (2007) Discovery of chemically induced mutations in rice by TILLING. BMC Plant Biol 7:19

Triques K, Sturbois B, Gallais S, Dalmais M, Chauvin S, Clepet C, Aubourg S, Rameau C, Caboche M, Bendahmane A (2007) Characterization of *Arabidopsis thaliana* mismatch specific endonucleases: application to mutation discovery by TILLING in pea. Plant J 51:1116–1125

Triques K, Piednoir E, Dalmais M, Schmidt J, Le Signor C, Sharkey M, Caboche M, Sturbois B, Bendahmane A (2008) Mutation detection using ENDO1: application to disease diagnostics in humans and TILLING and Eco-TILLING in plants. BMC Mol Biol 1:9

Tsai H, Howell T, Nitcher R, Missirian V, Watson B, Ngo KJ, Lieberman M, Fass J, Uauy C, Tran RK, Khan AA, Filkov V, Tai TH, Dubcovsky J, Comai L (2011) Discovery of rare mutations in populations: TILLING by sequencing. Plant Physiol 156:1257–1268

Uauy C, Paraiso F, Colasuonno P, Tran RK, Tsai H, Berardi S, Comai L, Dubcovsky J (2009) A modified TILLING approach to detect induced mutations in tetraploid and hexaploid wheat. BMC Plant Biol 9:115

Uauy C, Wulff BB, Dubcovsky J (2017) Combining traditional mutagenesis with new high-throughput sequencing and genome editing to reveal hidden variation in polyploid wheat. Annu Rev Genet 51:435–454

Upadhyaya P, Tyagi K, Sarma S, Tamboli V, Sreelakshmi Y, Sharma R (2017) Natural variation in folate levels among tomato (*Solanum lycopersicum*) accessions. Food Chem 217:610–619

Wang N, Wang YJ, Tian F, King GJ, Zhang CY, Long Y, Shi L, Meng JL (2008) A functional genomics resource for *Brassica napus*: development of an EMS mutagenized population and discovery of *FAE1* point mutations by TILLING. New Phytol 180:751–765

Wang N, Shi L, Tian F, Ning H, Wu X, Long Y, Meng J (2010) Assessment of *FAE1* polymorphisms in three Brassica species using EcoTILLING and their association with differences in seed erucic acid contents. BMC Plant Biol 10:137

Wang F, Wang C, Liu P, Lei C, Hao W, Gao Y, Liu YG, Zhao K (2016) Enhanced rice blast resistance by CRISPR/Cas9-targeted mutagenesis of the ERF transcription factor gene OsERF922. PLoS One 11:e0154027

Wang W, Simmonds J, Pan Q, Davidson D, He F, Battal A, Akhunova A, Trick HN, Uauy C, Akhunov E (2018) Gene editing and mutagenesis reveal inter-cultivar differences and additivity in the contribution of *TaGW2* homoeologues to grain size and weight in wheat. Theor Appl Genet 131:2463–2475

Wienholds E, van Eeden F, Kosters M, Mudde J, Plasterk RH, Cuppen E (2003) Efficient target-selected mutagenesis in zebrafish. Genome Res 13:2700–2707

Wu JL, Wu C, Lei C, Baraoidan M, Bordeos A, Madamba MR, Ramos-Pamplona M, Mauleon R, Portugal A, Ulat VJ, Bruskiewich R, Wang G, Leach J, Khush G, Leung H (2005) Chemical- and irradiation-induced mutants of indica rice IR64 for forward and reverse genetics. Plant Mol Biol 59:85–97

Xia Y, Ning Z, Bai G, Li R, Yan G, Siddique KH, Baum M, Guo P (2012) Allelic variations of a light harvesting chlorophyll a/b-binding protein gene (*Lhcb1*) associated with agronomic traits in barley. PLoS One 7:e37573

Xia Y, Li R, Ning Z, Bai G, Siddique KH, Yan G, Baum M, Varshney RK, Guo P (2013) Single nucleotide polymorphisms in *HSP17.8* and their association with agronomic traits in barley. PLoS One 8:e56816

Xiao W, Oefner PJ (2001) Denaturing high-performance liquid chromatography: a review. Hum Mutat 17:439–474

Xin Z, Wang ML, Barkley NA, Burow G, Franks C, Pederson G, Burke J (2008) Applying genotyping (TILLING) and phenotyping analyses to elucidate gene function in a chemically induced sorghum mutant population. BMC Plant Biol 8:103

Xu Y, Li P, Zou C, Lu Y, Xie C, Zhang X, Prasanna BM, Olsen MS (2017) Enhancing genetic gain in the era of molecular breeding. J Exp Bot 68:2641–2666

Yang B, Wen X, Kodali NS, Oleykowski CA, Miller CG, Kulinski J, Besack D, Yeung JA, Kowalski D, Yeung AT (2000) Purification, cloning, and characterization of the CEL I nuclease. Biochemistry 39:3533–3541

Yin K, Gao C, Qiu JL (2017) Progress and prospects in plant genome editing. Nat Plants 3:1–6

Yoon YJ, Venkatesh J, Lee JH, Kim J, Lee HE, Kim DS, Kang BC (2020) Genome editing of *eIF4E1* in tomato confers resistance to Pepper Mottle Virus. Front Plant Sci J 24:1098

Yu S, Liao F, Wang F, Wen W, Li J, Mei H, Luo L (2012) Identification of rice transcription factors associated with drought tolerance using the EcoTILLING method. PLoS One 7:e30765

Zeng CL, Wang GY, Wang JB, Yan GX, Chen BY, Xu K, Li J, Gao GZ, Wu XM, Zhao B, Liu L (2012) High-throughput discovery of chloroplast and mitochondrial DNA polymorphisms in *Brassicaceae* species by ORG-EcoTILLING. PLoS One 7:e47284

Zeng YD, Sun JL, Bu SH, Deng KS, Tao T, Zhang YM, Zhang TZ, Du XM, Zhou BL (2016) EcoTILLING revealed SNPs in *GhSus* genes that are associated with fiber- and seed-related traits in upland cotton. Sci Rep 6:29250

Zhang H, Xu R, Xie C, Huang C, Liao H, Xu Y, Wang J, Li WX (2015) Large-scale evaluation of maize germplasm for low-phosphorus tolerance. PLoS One 10:e0124212

Zhu Q, Smith SM, Ayele M, Yang L, Jogi A, Chaluvadi SR, Bennetzen JL (2012) High-throughput discovery of mutations in *tef* semi-dwarfing genes by next-generation sequencing analysis. Genetics 192:819–829

Plant Mutagenesis Tools for Precision Breeding: Conventional CRISPR/Cas9 Tools and Beyond

10

Rahul Mahadev Shelake, Dibyajyoti Pramanik, Rajesh Ramdas Waghunde, and Jae-Yean Kim

Abstract

Plant breeding facilitates the selection of plant populations with the desired trait from the genetic pool. A set of expressed or non-expressed genetic combinations are generally evolved due to spontaneous or induced mutations in the genetic pool of particular plant species. Genetic modification is the basis of diversity that allows domestication alongside adaptation in changing environmental conditions, i.e., from wild to cultivated species. Plant phenotype-based selective breeding has been an efficient way used in the past. Also, induction of random mutations by physical or chemical agents and further selection for desired traits from the mutant library has produced several plant varieties. However, conventional mutagenesis combined with selective breeding has some drawbacks. It makes the random mutations; the process is laborious and slower; it comes with the risk of losing beneficial traits during breeding. In recent times, genome-editing (GE) tools comprising molecular genetic scissors are tailored to precisely target the desired location in the plant genome. The most used GE tool is the clustered regularly interspaced short palindromic repeat-associated (CRISPR/Cas) endonuclease system. The CRISPR-based tools are part of a new technology called new plant breeding techniques (NBTs) that accelerate plant breeding. Diverse CRISPR-based tools have been optimized to achieve expected goals in NBTs

R. M. Shelake and D. Pramanik contributed equally to this work.

R. M. Shelake (✉) · D. Pramanik · J.-Y. Kim
Division of Applied Life Science (BK21 Four Program), Plant Molecular Biology and Biotechnology Research Center, Gyeongsang National University, Jinju, South Korea
e-mail: rahulms@gnu.ac.kr; kimjy@gnu.ac.kr

R. R. Waghunde
Department of Plant Pathology, College of Agriculture, Navsari Agricultural University, Bharuch, India

S. Penna, S. M. Jain (eds.), *Mutation Breeding for Sustainable Food Production and Climate Resilience*, https://doi.org/10.1007/978-981-16-9720-3_10

like higher yield, tolerance to biotic and abiotic stresses, prevention of post-harvest losses, value addition, and novel traits-in-demand by farmers and consumers. This chapter summarizes an overview of the CRISPR/Cas system, CRISPR-based tools for plant breeding, and recent applications of CRISPR technology in agriculture.

Keywords

CRISPR/Cas · Crop improvement · Genetic engineering · Mutation · New plant breeding technologies

10.1 Introduction

Since the beginning of civilization, humans have continuously opted to find novel ways to improve the quality or yield of plant and animal species through domestication and natural or selective breeding. Plant breeding deals with the science of manual selection of plants with desired traits. In early times, plant varieties were selected from a genetic pool generated by recombination and spontaneous mutations (Capdeville et al. 2021). However, natural or selective breeding is time-consuming and limits the attainment of desired traits. Also, there is a high risk of losing some crucial traits during the traditional breeding process. Induction of random mutations followed by selective breeding became a regular plant breeding mode in the twentieth century. Plant scientists used mutagenesis tools to introduce the mutations in the plant genome. The genome-editing (GE)-based plant breeding (NBTs) are a set of targeted mutagenesis and GE methods to accelerate the trait improvement in crop species. The adoption of CRISPR/Cas for GE in 2012 has transformed the field of genetic engineering across all kingdoms (Jinek et al. 2012). The power of the CRISPR/Cas system to target and deliver the GE components at the desired locus in the complex genome enables the alternation of DNA, RNA, and ultimately protein (Pramanik et al. 2021a).

The plant genome engineering tools can be categorized into three groups based on CRISPR/Cas timeline: conventional (physical and chemical mutagens); pre-CRISPR/Cas (PCR-based tools, transposon insertions, T-DNA insertions, RNA interference, synthetic meganucleases like ZNFs and TALENs); and CRISPR/Cas-based tools (Shelake et al. 2019a). Since the first proofs-of-concept of CRISPR/Cas application for plant genome editing in 2013 (Li et al. 2013; Nekrasov et al. 2013; Shan et al. 2013), various CRISPR-based tools have been designed and were successfully applied in different model plants and crop species. This chapter provides an overview of traditional mutagenesis tools, the historical development of CRISPR technology, the basics of CRISPR/Cas components, recent

CRISPR-based tools for NBTs, and applications of CRISPR-based tools in crop improvement.

10.2 Conventional and Pre-CRISPR Mutagenesis Tools

The mutation-causing agents, also known as mutagens, are categorized as physical and chemical mutagens. Some popularly used physical mutagens include high-energy ion beams, X-rays, gamma rays, and neutrons. Particular radiation doses can cause double-stranded breaks (DSBs) in DNA due to photochemical damage triggering indel mutations in the genome (Bado et al. 2015). Several chemical mutagens, such as alkylating agents, base analogs, nitrous acid, hydroxylamine, and acridine dyes, have been used in crops (Oladosu et al. 2016). Physical and chemical mutagens were effectively used in traditional mutation breeding in the past and continue to play a part in the current crop variety development (Suprasanna et al. 2015). However, these tools produce random mutations without any user-control, mutagens are harmful to human health and the environment, and need standardization for each plant species (Bado et al. 2015).

In the mid-1990s, polymerase chain reaction (PCR) and PCR-based screening has revolutionized the field of plant breeding. A PCR-based modified TILLING tool (Targeting Induced Local Lesion IN Genomes) was adopted in the early 2000s that mediate the screening of only a gene of interest in the mutant library. TILLING methods need to screen many plants to get the desired genotype and cause undesirable mutations. Transposable genetic elements were repurposed as a method for mutation breeding in several crop species but limited success. *Agrobacterium*-mediated introduction of transfer DNA (T-DNA) into plant genome is manipulated as a useful tool for the production of T-DNA insertional mutant lines. RNA interference (RNAi) is a promising genome engineering method that includes the suppression of gene expression by RNAi via obstruction of the mRNA of chosen targets (Dalakouras et al. 2020). However, the RNAi approach needs the continuous expression of RNAi machinery, and a stable transgenic system requires the approval of genetically modified (GM) regulatory authorities. Therefore, alternative GM-free tools like synthetic meganucleases are an attractive option for achieving GE plants without continuous transgene expression. CRISPR/Cas-based tools are the most popular meganucleases among the three of them (the other two include ZNFs, zinc-finger nucleases, TALENs, transcription activator-like effector nucleases). Owing to several beneficial features, CRISPR/Cas-based tools are adopted for various applications in the last decade (Fig. 10.1). The CRISPR tools used for simple knockout/knockin generation, homologous recombination (HR)-based mutations at a specific locus, base substitution in DNA and RNA, directed evolution of the desired genomic region, gene tagging, DNA/RNA imaging, pathogen detection, barcoding, modulation of epigenetic marks, transcription regulation, and lineage tracing, etc. (summarized in Shelake et al. 2019a, b; Pramanik et al. 2021a). Various aspects of CRISPR technology are discussed in the following sections.

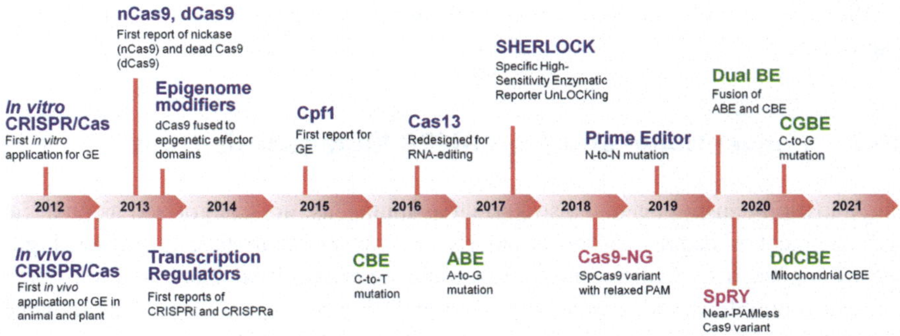

Fig. 10.1 Milestones in development of CRISPR-based tools

10.3 Basics of CRISPR/Cas-Based Tool

The CRISPR/Cas is a part of the immune system reported in bacteria and archaea. The native CRISPR/Cas system elements from bacteria were repurposed for targeted GE (Jinek et al. 2012). The essential two components of the CRISPR/Cas-mediated GE tool comprise Cas nuclease protein and an RNA molecule (Fig. 10.2). The structure of predominantly used *Streptococcus pyogenes* Cas9 (SpCas9) protein has two endonuclease domains, HNH and RuvC (Nishimasu et al. 2014). RNA molecule, also known as single guide RNA (sgRNA), is a fusion of CRISPR RNA (crRNA) and a *trans*-activating RNA (tracrRNA). Some important terms related to the CRISPR/Cas9-based tool are shown in Fig. 10.2 and summarized in the following section.

CRISPR RNA (crRNA)—Short RNA fragment (about 20 bp for SpCas9, 23 bp for Cas12a) at the 5′-end of the engineered RNA scaffold.

Single guide RNA (sgRNA)—Fusion of crRNA and a *trans*-activating RNA (tracrRNA).

Target strand—The crRNA binding DNA strand of the target region.

Non-target strand—The non-crRNA binding DNA strand of the target region.

HNH domain—Cas9 domain that cleaves the target DNA strand bound to the crRNA.

RuvC domain—Cas9 domain that cuts non-target DNA strand unbound to the crRNA.

Nickase Cas9 (nCas9)—The mutant version of Cas9 generated by a point mutation in one of the catalytic domains of active Cas9, i.e., the RuvCI (D10A) and HNH (H840A) domains, that cleave only the target and the non-target strand, respectively.

Dead Cas9 (dCas9)—A double mutant version of Cas9 (D10A and H840A) with no catalytic activity but still recognize and bind the target region together with sgRNA.

Protospacer adjacent motif (PAM)—Short (typically 2–6 bp) recognition sequence present on non-target DNA strand in the target region.

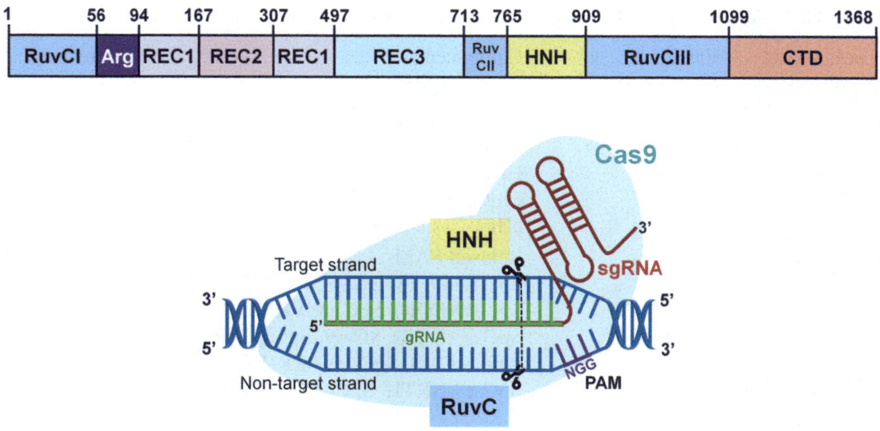

Fig. 10.2 Overview of CRISPR/Cas9 system. Description about components provided in the main text

Another common Cas9 ortholog is Cas12a (previously called Cpf1), which requires only a crRNA (Zetsche et al. 2015). Cas12a mostly recognizes T-rich PAM site (Cas9 recognizes G-rich PAM site), and cleavage generate staggered ends (Cas9 produces blunt ends). Point mutations have developed several Cas9 and Cas12a variants either for relaxed PAM or higher precision activity (Table 10.1).

10.4 CRISPR-Based Tools for Genetic Engineering of Plants

Many CRISPR-based tools have been developed, which allowed desired editing in the targeted genomic site and beyond (Table 10.1). The fusion of various enzymes to the nCas9 or dCas9 has been exploited as a platform to deliver the CRISPR-tool-cargo at the genomic site of interest. Here, the CRISPR-based tools and their applications are discussed based on the type of edited outcomes in GE experiments.

10.4.1 Epigenetic Modification Tools

The genome of plants and higher eukaryotes is highly organized into chromosomes. Chromatin in the chromosome consists of DNA and protein. The packaging and unfolding of DNA with histone and non-histone proteins in chromatin is strongly associated with gene expression or silencing. A set of epigenetic changes driven by genome–environment interactions leads to epigenetic memory development even without modifying the DNA code itself (Pramanik et al. 2021a). Epigenetic regulation governs gene expression by modulating the chromatin structure, and those changes can be inherited to the next generation. Consequently, chromatin-modifying

Table 10.1 Available CRISPR-based tools (adapted from Pramanik et al. 2021a)

Tool	Version	Components	Feature
Cas9	Cas9 Cas9 variants Cas9 orthologs (PAM)	Cas9, sgRNA	Knockout/ knockin
		Higher specificity- eSpCas9(1.0), eSpCas9(1.1), SpCas9-HF1, HeFSpCas9, HypaCas9, evoCas9, HiFiCas9, Sniper Cas9, eHF1-Cas9, eHypaCas9	
		Relaxed PAM- SpCas9(D1135E), SpCas9(VQR), SpCas9(EQR), SpCas9(VRER), SpCas9(QQR), xCas9, SpCas9-NG, iSpmacCas9, SpCas9-NRRH, SpCas9-NRTH, SpCas9-NRCH, Cas9-Sc++, HiFi-Sc++, SpG, SpY, CasΦ	
		Orthologs- SpCas9 (NGG), StCas9 (NNAGAAW), NmeCas9 (NNNNGMTT), SaCas9 (NNNRRT), FnCas9 (NGG), CjCas9 (NNNVRYAC), ScCas9 (NNG), and CasX (TTCN), ten Cas12f orthologs (5′T- or C-rich), SmacCas9 with a variant iSpyMac (NAAN), whereas N: any nucleotide, R:A/G, M:A/C, W:A/T, V:G/C/A, R:A/G, Y:C/T	
Cpf1	Cpf1 Cpf1 variants	Cpf1, crRNA	Knockout/ knockin
		AsCpf1(RR), AsCpf1(RVR), enAsCpf1, LbCpf1(RR), LbCpf1 (RVR), ttLbCas12a, FnCpf1(RR), FnCpf1(RVR), MbCpf1(RR), and MbCpf1(RVR)	
	Cpf1 orthologs (PAM)	FnCpf1 (TTV/TTTV/KYTV), LbCpf1 (TTTV), AsCpf1 (TTTV), and MbCpf1 (TTV/TTTV)	
Cas13	Cas13a	Cas13a, sgRNA	RNA targeting
	Cas13b	Cas13b, sgRNA	
	Cas13d	Cas13d, sgRNA	
Imaging	DNA	dCas9 fused with a fluorescent marker, sgRNA	DNA imaging
	RNA	RNA-targeting dCas9 fused with GFP, sgRNA	RNA imaging
Base Editor	Cytidine DNA Base Editor (CBE)	nCas9 fused with different cytidine deaminases and uracil glycosylase inhibitor (UGI), sgRNA	C-to-T base editing in DNA sequence
		CBE versions- Target-AID (pmCDA1), rAPOBEC1-UGI (BE3), SaBE3, SaKKH-BE3, VQR-BE3, EQR-BE3, VRER-	

(continued)

Table 10.1 (continued)

Tool	Version	Components	Feature
		BE3YE1-BE3, YE2-BE3, EE-BE3, YEE-BE3, nCDA1-BE3, cCDA1-BE3, YE1-BE4, YE2-BE4, YEE-BE4, EE-BE4, R33A + K34A-BE4, YE1-BE4-CP1028, YE1-BE4-NG, AALN-BE4, BE4-PLUS, hAPOBEC3A, BE4, BE4max, AncBE4max, evoAPOBEC1-BE4max, evoFERNY, evoCDA1-BE4max, hyeA3A-BE4max, A3Bctd-VHM-BE3 and A3Bctd-KKR-BE3, YE1-BE3-FNLS, APOBEC3G	
	DddA-derived cytosine base editors (DdCBEs)	DddA (an interbacterial cytidine deaminase working on dsDNA) fused with TALEN effectors	C-to-T in mitochondria
	C-to-G base editor (CGBE)	nCas9 fused with CBE variants and uracil DNA *N*-glycosylase	C-to-G
	Adenine DNA Base Editor (ABE)	nCas9, fused with engineered tRNA adenosine deaminase, sgRNA	A-to-G base editing in DNA sequence
		ABE versions- ABE7.10 (TadA-TadA*), ABEmax, miniABEmax, ABE8e, 20 ABE8 versions	
	Adenine RNA editor (REPAIR)[a]	dCas13 fused with adenine deaminase domain of ADAR2, sgRNA	A-to-Inosine (G) in RNA
	Cytidine RNA editor (RESCUE)[b]	dCas13 fused with engineered adenine deaminase domain of ADAR2[c], sgRNA	C-to-U in RNA sequence
Directed evolution tools	Dual base editor (ABE-CBE)	nCas9 fused with ABE and CBE	Targeted random mutagenesis
	EvolvR[d]	nCas9 fused with engineered DNA polymerase, sgRNA	
	ROS mutator[d]	nCas9 fused with ROS-producing enzyme	
	Target-G[d]	dCas9 fused with engineered DNA glycosylase, sgRNA	
Prime Editor	Prime Editor (PE)	nCas9 fused with engineered reverse transcriptase, prime editing guide RNA (pegRNA)	GE, base change
Targeted Deletor	AFID[e]	Fusion of human APOBEC3A (A3A), uracil DNA-glucosidase and apurinic or apyrimidinic site lyase	Multi-nucleotide deletion
Transcription regulators	CRISPR interference (CRISPRi)	dCas9 fused with KRAB-repressive domain, sgRNA	Transcription repression
	CRISPR activator (CRISPRa)	dCas9 or split dCas9 fused with effectors, sgRNA Effector set—VPR (VP64, P65, and Rta)	Transcription activation

(continued)

Table 10.1 (continued)

Tool	Version	Components	Feature
		Effector set—Synergistic activation mediator, SAM (MS2, p65, VP64, HSF1) Effector set—SunTag	
Epigenetic modifier	Demethylase	dCas9 fused to 10–11 translocation (TET) proteins, dCas9-LSD1	Epigenetic modification
	Methylase	dCas9-DNMT3A, dCas9-MQ1, dCas9-PRDM9, dCas9-KRAB, dCas9-DOT1L	
	Deacetylase	dCas9-HDAC3	
	Acetylase	dCas9-P300	
Pathogen detection	SHERLOCK[f]	Cas13a, reporter (quenched fluorescent RNA), sgRNA	RNA or DNA detection
DNA integrator	Tn7 transposon[d]	Cas 6, Cas7, Cas8, transposon complex (tnsA/B/C, TniQ), sgRNA	DNA integration
	Cas12k-based DNA integrase[d]	Cas12k, TnsB, TnsC, TniQ, sgRNA	

[a] REPAIR: RNA Editing for Programmable A to I Replacement
[b] RESCUE: RNA Editing for Specific C-to-U Exchange
[c] ADAR2: Adenosine Deaminase Acting on RNA type 2
[d] Not yet tested in plant systems
[e] APOBEC-Cas9 fusion-induced deletion system
[f] SHERLOCK: Specific High-Sensitivity Enzymatic Reporter Unlocking

enzyme fused to Cas protein used to design epigenetic modifiers (EMs) for CRISPR-mediated epigenetic modification.

The EM tools can induce epigenetic changes at targeted loci, facilitating reading, writing, and erasing epigenetic signatures (Pulecio et al. 2017; Adli 2018). Several EM tools have been characterized, focusing on targeted genetic methylation modifications and histone modifications in higher eukaryotes (Fig. 10.3). The dCas9-TET1 tool was successfully applied to remove the DNA methylation site and activate the silent gene (Liu et al. 2016). The dCas9-based CLOuD9 tool was used to change the chromosomal organization in the target region in the human genome (Morgan et al. 2017). Recently, a dCas9-DRM tool was engineered to modulate the activities of FWA and SUPERMAN (SUP) promoters for functional studies of underlying genes (Papikian et al. 2019). Histone modifiers have also been designed to enable precise modulation of epigenetic marks present in targeted locus-wrapping histones. For instance, dCas9 fused with EMs [histone demethylases (LSD1) (Kearns et al. 2015) and methyltransferases (SMYD3, PRDM9, and DOT1L) (Kim et al. 2015; Cano-Rodriguez et al. 2016), acetyltransferases (p300) (Hilton et al. 2015), and deacetylases (HDAC3) (Kwon et al. 2017)] for histone modification and gene regulation in human cells. Lately, acetyltransferases (p300)

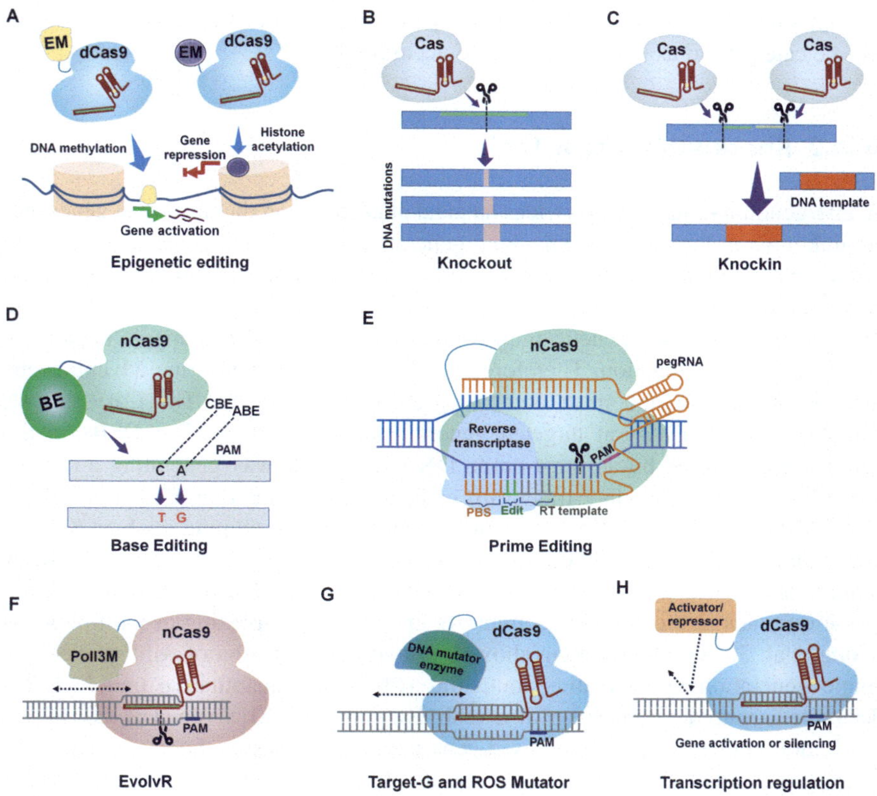

Fig. 10.3 CRISPR-based toolbox for plant engineering. (**a**) Schematic illustration of epigenetic modifier tools that generate epigenetic modifications in the DNA and histones. DNA methylation, histone acetylation modifications can either activate or repress gene function. (**b**) CRISPR/Cas9-mediated GE at target genomic site producing a double-stranded DNA break and different pattern of mutations. (**c**) Targeted gene replacement using CRISPR tools. (**d**) Schematic illustration of base editing of the target site by C-to-T (CBE) or A-to-G (ABE) substitution. (**e**) The prime editing system consists of a nickase Cas9 (nCsa9) fused to a reverse transcriptase (RT) enzyme, and prime editing guide RNA (pegRNA) carrying primer binding site (PBS). The pegRNA also functions as a template for RT that further introduces mutations to the target locus. (**f**) The EvolvR consists of two components: enCas9 (enhanced version of nCas9) fused to a mutant form of DNA polymerase I with an error-prone activity. (**g**) The dCas9-DNA glycosylase-based Target-G or ROS (reactive oxygen species)-producing enzyme-based ROS mutator tool can generate mutations near the sgRNA target site. (**h**) CRISPR-based activator and repression systems designed to switch on/off the target gene expression

and H3K9 methyltransferase (KRYPTONITE) were successfully applied in *Arabidopsis thaliana* to alter flowering time targeting *FT* locus (Lee et al. 2019).

Overall, the use of EM tools will provide new information about the basics of developmental biology and epigenetic aspects for fine-tuning the crop species with improved traits. One of the critical elements of EM tool application is the necessity of continual transgene expression. Thus, plant varieties developed with EM tools will be categorized as GM and regulated accordingly.

10.4.2 Knockout/Knockin Tools

Whole-genome sequencing provides an ideal platform to predict the novel genes that are unknown in the past, and knockout generation of predicted genes facilitates the further characterization of additional features. Knockin/knockout mutant population generated by precise GE serves as an excellent source for breeding to avoid linkage drag. Another way for genetic improvement is to delete the elements negatively affecting the gene function. Therefore, knockout of genes conferring advantageous traits is the easiest and most common CRISPR/Cas system. The Cas9-sgRNA complex induces DSBs at targeted genomic loci (Fig. 10.3). The DSBs in the genome are repaired by an error-prone DNA repair pathway generating a diverse type of mutations, including frameshift or premature stop codon (Gaj et al. 2013). The PAM constraint for gRNA selection is relaxed by developing evolved SpCas9 and mining of Cas9/Cpf1 orthologs (Table 10.1). Similarly, multiple gRNA sequences can be designed as a single CRISPR array to enable simultaneous editing of various sites in the genome, commonly known as multiplexing or multiplex GE (Cong et al. 2013). Multiplexing allows de novo domestication of wild species in a single generation by recreating mutations (SNPs and quantitative trait loci) for desirable agronomic traits (Li et al. 2018b; Zsögön et al. 2018; Yu et al. 2021).

Homologous recombination-based gene targeting possesses the vast potential to generate plant material with improved or novel traits. In nature, homology-directed repair (HDR) is a less preferable pathway for DNA repair; thus, genetic engineering approaches met with low success in the pre-CRISPR era. Many reports showed lower efficiency in the case of HR-mediated gene transfer. The CRISPR/Cas system has demonstrated the potential to improve HDR efficiency by creating precise DSBs and increased gene-targeting efficiency in several crops (Table 10.2).

10.4.3 Point Mutation Tools

Many agronomic traits are conferred by single-nucleotide substitutions (SNP). CRISPR/Cas-mediated DSB formation usually introduces indels at the targeted genomic sites. Therefore, base editors (BE) generated by the fusion of deaminase enzymes to the nCas9 or dCas9 are more suitable for SNPs or novel allele generation.

10.4.3.1 DNA and RNA Base Editors

DNA base editors (BEs) were developed fusing a Cas nuclease (dCas9/nCas9) and a base-modification enzyme, generally a deaminase. Two types of BEs are extensively studied for diverse applications in prokaryotes and eukaryotes, including plants (Huang et al. 2021). The first type comprises cytosine base editors (CBEs) that

Table 10.2 Representative examples of trait improvements in crop species achieved using major CRISPR-based tools

Plant	Target	Application/trait	Reference
Epigenetic tools			
Rice	*OsVAL*	Modification on gibberellin regulation in rice	Xie et al. (2018)
Rice	*LC2*	Increase in tiller number	Lee et al. (2020)
Rice	*WX1*	Higher amylopectin	Zeng et al. (2020)
Knockout/knockin tools			
Rice	*Large gelation*	Improve yield	Wang et al. (2017b)
Soybean	*GmFT2a*	Flowering phenotype	Cai et al. (2018)
Potato	*S-locus (S-RNase)*	Self-compatibility	Enciso-Rodriguez et al. (2019)
Tomato	*PMR4*	Powdery mildew resistant	Santillán Martínez et al. (2020)
Rice	*PDS*, *MPK2*, *BADH2*	Abiotic stress tolerance	Shan et al. (2013)
Rice	*MPK5*	Abiotic stress tolerance	Xie and Yang (2013)
Rice	*DERF1*, *PMS3*, *EPSPs*, *MSH1*, *MYB5*	Drought tolerance	Zhang et al. (2014)
Rice	*AOX1a*, *AOX1b*, *AOX1c*, *OsBEL*	Abiotic stress tolerance	Xu et al. (2015)
Rice	*OsERF922*	Enhanced blast resistance	Wang et al. (2016)
Mushroom	*ppo1*	Non-browning mushroom	Waltz (2016)
Cucumber	*eIF4E1*	Multiple virus resistance	Chandrasekaran et al. (2016)
Tomato	*Slmlo1*	Powdery mildew resistance	Nekrasov et al. (2017)
Rice	*OsHAK1*	Reduced Cesium uptake	Nieves-Cordones et al. (2017)
Rice	*OsPRX2*	Tolerance to potassium deficiency	Mao et al. (2018)
Wheat	*sgAlpha-1*, *sgAlpha-2*	Low-gluten or gluten-free wheat	Sánchez-León et al. (2018)
Potato	*StPPO*	Non-browning potato	González et al. (2019)
Tomato	*SlHAK20*	Loss of salt tolerance	Wang et al. (2020b)

(continued)

Table 10.2 (continued)

Plant	Target	Application/trait	Reference
Rice	*RAV2*	Salt tolerance	Duan et al. (2016)
Citrus	*CsLOB1*	Canker disease resistance	Jia et al. (2017)
Barley	*HvPAPhy_a*	Grain quality	Holme et al. (2017)
Maize	*ARGOS8*	Drought stress resistance	Shi et al. (2017)
Orange	*CsLOB1*	Canker disease resistance	Peng et al. (2017)
Rice	*SWEET11*, *SWEET13* and *SWEET14*	Bacterial blight disease resistance	Oliva et al. (2019)
Rice	*Xa13*	Blight-resistant rice	Li et al. (2020a)
Maize	*WX1*	Higher amylopectin	Gao et al. (2020)
Rice	*WX1*	Higher amylopectin	Huang et al. (2020)
Pumelo	*LOB1* promoter	Canker disease resistance	Jia and Wang (2020)
Rice	*Waxy*	Reduced amylose content	Ma et al. (2015)
Rice	*TMS5*	Rapid breeding of hybrids	Zhou et al. (2016)
Rapeseed	*FAD*	Improved fatty acid	Jiang et al. (2017)
Rice	*Sc*	Rescues male sterility	Shen et al. (2017)
Rice	*Sa*	Overcomes hybrid incompatibility	Xie et al. (2017a)
Rice	*S1*	Overcomes hybrid incompatibility	Xie et al. (2017b)
Tomato	*RIN*	RIN function in fruits ripening	Ito et al. (2020)
Lettuce	*GGP1*, *GGP2*	Oxidation stress tolerance	Zhang et al. (2018)
Tomato	*GGP1*	Improved vitamin C	Li et al. (2018b)
Rice	*OsSWEET14*	Bacterial blight resistance	Wang et al. (2020a)
Tomato	*SlPelo*, *SlMlo1*	Viral and fungal disease resistance	Pramanik et al. (2021b)
Cytidine Base editors			
Rice	*wx*	Soft rice	Xu et al. (2021a)
Rice	*NRT1.1B*	Nitrogen use efficiency	Lu and Zhu (2017)
Oilseed	*ALS*, *RGA*, *IAA7*	Herbicide resistance, decreased plant height	Cheng et al. (2021)

(continued)

Table 10.2 (continued)

Plant	Target	Application/trait	Reference
Rice	*ALS, FTIP1e*	Herbicide resistance	Shimatani et al. (2017)
Potato	*ALS*	Herbicide resistance	Veillet et al. (2019)
Wheat	*TaALS*	Herbicide resistance	Zong et al. (2017)
Wheat	*TaMTL*	Haploid induction	
Adenine Base editors			
Rice	*ACC, ALS, DEP1, CDC48, NRT1.1B*	Proof-of-concept, herbicide resistance	Li et al. (2018a)
Wheat	*TaDEP1, TaGW2*	Proof-of-concept	
Brassica	*ALS, PDS*	Late flowering	Kang et al. (2018)
Rice	*MPK6, MPK13, SERK2, WRKY45, Tms9-1*	Proof-of-concept	
Rice	*OsPDS*	Proof-of-concept	Ren et al. (2021)
Prime Editor			
Rice	*ALS*	Herbicide resistance	Butt et al. (2020)
Rice	*EPSPS*	Glyphosate resistance	Li et al. (2020c)
Rice	*OsALS, OsACC*	Proof-of-concept	Xu et al. (2020b)
Rice	*OsPDS, OsACC1, OsWx*	Proof-of-concept	Xu et al. (2020a)
Maize	*ZmALS1, ZmALS2*	Proof-of-concept, herbicide resistance	Jiang et al. (2020)
Potato	*StALS1*	Proof-of-concept	Veillet et al. (2020)
Tomato	*SlGA, SlALS2, SlPDS1*	Proof-of-concept	Lu et al. (2020)
Rice	*OsALS, OsKO2, OsDEP1, and OsPDS*	Proof-of-concept	Tang et al. (2020)
Rice	*OsALS, OsCDC48, OsDEP1, OsEPSPS, OsGAPDH, OsLDMAR*	Proof-of-concept	Lin et al. (2020)
Wheat	*TaGW2, TaUbi10, TaLOX2, TaNLO, TaDME, TaGASR7*	Proof-of-concept	
Direction evolution tools			
Populus	Multiple genes	Database of CRISPR mutant	Zhou et al. (2015)
Tomato	*SlCLV3*	Fruit size variation	Rodríguez-Leal et al. (2017)
Rice	*ALS*	Herbicide resistance	Kuang et al. (2020)
Rice	*ACC*	Herbicide resistance	Li et al. (2020b)
Strawberry	*FvebZIPs1.1*	Fine-tuning sugar content	Xing et al. (2020)
Transcription regulation tools			
Rice	*miR159b*	Modulation of the plant transcriptome	Tang et al. (2017)
Rice	*ER1*	Increase transcription	Li et al. (2017)

(continued)

Table 10.2 (continued)

Plant	Target	Application/trait	Reference
Tobacco	*DFR*, *AN2*	Gene regulation	Selma et al. (2019)
Rice	*WOX11* and *YUC1*	Increased transcription	Gong et al. (2020)

convert a C:G into a T:A (Komor et al. 2016). CBEs consists of three components: a Cas enzyme (nCas9 or dCas9), a cytidine deaminase, and uracil glycosylase inhibitor (UGI). Secondly, adenine base editors (ABEs) convert A:T to G:C (Gaudelli et al. 2017). ABEs typically have two components: a Cas enzyme (nCas9 or dCas9) and evolved adenosine deaminase. The fusion of CBE and ABE was also designed as a dual base editor (CABE) for allele generation and discussed in another section about diversification of targeted loci.

The first CBE system was developed as BE1 by fusing rat cytidine deaminase (APOBEC1) with dCas9 to facilitate C-to-T base substitution (Komor et al. 2016). BE1 can act on single-stranded DNA (ssDNA) as a substrate but is inefficient on double-stranded DNA (dsDNA). Specific ssDNA recognition by BE1 is helpful to reduce the off-target activity. The second CBE system (BE2) developed from BE1 after replacing dCas9 by nCas9 to produce nick at the non-edited strand. Furthermore, UDG inhibitor (UGI) was added to the BE2 system to generate BE3 (CBE-UGI) and BE4 (CBE-2xUGI) to improve editing efficiency in vivo. The CBE activity is observed in a core location inside the gRNA-binding region called the activity window or editing window. In the case of first-generation CBEs (BE1 to BE4), ~5-bp editing window was observed in the gRNA region from 1 to 5 bp distal to the PAM site. Similarly, target-AID (activation-induced cytidine deaminase) system was developed where cytidine deaminase from sea lamprey (PmCDA1) fused with nCas9-UGI employed to achieve efficient C-to-T base editing (Nishida et al. 2016). The CBE toolkit is recently expanded using different versions of evolved CBEs and Cas9 orthologs (Table 10.1). Two recently reported CBE versions using the human APOBEC3B provide higher CBE activity with lower off-target effects (Jin et al. 2020).

The second class of BE system contains ABE, which is developed using a mutant version of tRNA adenosine deaminase enzyme TadA from *Escherichia coli*. TadA mutant fused with dcas9 to produced TadA*-dCas9 that can act on ssDNA and efficiently allowed conversion of A-to-inosine (I) that further transforms from A-to-G (Gaudelli et al. 2017). nCas9 was replaced with dCas9 to generate TadA*-nCas9 system. The ABE7.10 system was developed combining with newly evolved TadA*, native TadA, and nCas9 (TadA-TadA*-nCas9) that showed significantly improved ABE activity in mammalian cells. ABE7.10 system was further modified, and new variants were reported. For example, ABE8 and ABE8e showed the highest A-to-G conversion in mammalian cells (Gaudelli et al. 2020; Richter et al. 2020). The ABE8 fusion with SpRY (relaxed PAM variant of SpCas9) shown improved ABE activity in rice (Xu et al. 2021b).

While CBE and ABE permit the transition from C-to-T and A-to-G, respectively, the latest C-to-G editor (CGBE) was reported to efficiently generate the transversion [C-to-A in *E. coli* and C-to-G in human cells (Zhao et al. 2021)] [C-to-G in human cells (Kurt et al. 2021)]. In CGBE, uracil DNA glycosylase (UDG) fused to a CBE (No UGI) produces an abasic (AP) site, which is repaired by endogenous machinery into resultant transversion mutation. However, current versions of CGBE can edit the base in site-dependent manner and exact mechanism is not known. In this regard, the CGBE application in plants is expected to generate transversions but yet to be confirmed by experimental evidence. Bacterial deaminase DddA-derived CBE is repurposed for C-to-T conversion in mitochondrial dsDNA (Mok et al. 2020). Therefore, plant organelle GE would be an exciting aspect for plant breeding applications in the near future.

Apart from the DNA BE, CRISPR systems were repurposed for RNA targeting and base conversion in RNA molecules in vivo. The ADAR family enzyme, adenosine deaminases, catalyzes hydrolytic adenosine deamination, transforming adenosine (A) to inosine (I). The catalytically dead Cas13b enzyme (dPspCas13b) fused to the deamination domain (DD) of an ADAR (ADARDD) to create an RNA-guided editor named RNA Editing for Programmable A-to-I Replacement (REPAIR) (Cox et al. 2017). Similarly, ADAR2 were evolved to a cytidine deaminase that fused with dCas13 to develop programmable RNA Editing for Specific C-to-U Exchange (RESCUE) (Abudayyeh et al. 2019).

10.4.3.2 Prime Editor

Prime editing (PE) is the latest CRISPR-based technology that generates desired point mutations at the targeted site. The PE tool consists of a prime editing gRNA (pegRNA) and a fusion protein comprising nCas9, and an engineered reverse transcriptase (RT) enzyme (Anzalone et al. 2019). The PE tool is described as a "search-and-replace" GE tool that can induce base substitutions in more extended regions. A series of PE1, PE2, PE3, PE3b systems were developed. However, low editing efficiency was observed in animals and plants. Several attempts were reported to improve PE efficiency. For instance, the M-MLV RT domain was replaced with the CaMV RT (RT-CaMV) from cauliflower mosaic virus or a retron-derived RT (RT-retron) from *E. coli* (Lin et al. 2020). However, similar to earlier studies, lower editing efficiency is still a significant obstacle for PE application in plants. Recently, an improved level of pegRNAs expression was reported to enhance PE efficiency in maize (Jiang et al. 2020).

Even though the PE technique is promising, lower editing efficacy needs to be improvized by additional efforts. Also, present BE systems show higher efficiency for transition mutations than PE. A diverse ABE/CBE toolbox offers more options for user-defined mutations in the targeted region in the genome.

10.4.4 CRISPR Tools for Directed Evolution

Many favorable phenotypes of crops are produced by multiple SNPs present at the same loci. The CRISPR/Cas systems were engineered to edit such multiple nucleotide bases. In this regard, the dual BE system is an efficient tool for simultaneous C-to-T and A-to-G substitutions at the targeted site. Recently, several dual BE systems were engineered using different ABE and CBE versions. For example, Synchronous Programmable Adenine and Cytosine Editor (SPACE), which consists of APOBEC3A cytidine deaminases and Target-AID (Grünewald et al. 2020), A & C-BE system designed using BE3 with ABE7.10 (Zhang et al. 2020), and ACBE system containing Target-AID and ABE-7.10 (Xie et al. 2020).

Another CRISPR-based tool, EvolvR, is effectively employed for targeted mutagenesis in the bacterial cell that offers a different way of creating novel genetic variants of the desired locus (Halperin et al. 2018). The EvolvR comprises two components: enCas9 (enhanced version of nCas9) fused to a mutant form of DNA polymerase I possessing an error-prone activity. EvolvR is yet to be tested in plants. Still, a unique feature of introducing all 12 possible mutations (A-to-G, A-to-C, A-to-T, C-to-T, C-to-A, C-to-G, G-A, G-T, G-to-C, G-to-A, G-to-C, G-to-T) has immense potential in plant breeding applications. Additionally, tools like Target-G (Nishida and Kondo 2017) and ROS mutator (Kim and Shelake 2021) need optimization for the directed evolution of desired loci in plant GE. One of the targets in plant systems is the promoter region. The transcription factors recognize the *cis*-regulatory elements present in the promoter sequence that dictate the spatiotemporal gene expression pattern. Mutations in regulatory elements in the promoter region changes the levels of gene expression, creating phenotypic differences.

10.4.5 Transcriptional Regulation Tools

As discussed earlier, DNA mutation-based protein engineering is not the only way to attain the anticipated plant phenotype. In addition to epigenetic modifications, modulation of gene expression allows fine-tuning of plant traits governed by spatiotemporal expression levels of targeted genes. Recruitment of synthetic artificial transcription factors (ATFs) mediates the controlled expression of the desired gene (Shrestha et al. 2018). The ATF fusion with dCas9 activates or suppresses the gene expression, therefore, referred to as CRISPRa and CRISPRi, respectively (Table 10.1). Several types of effector domains have been tested alone or in combination for dosage-dependent regulation of transcriptional activity. Some examples include the dCas9 fused with KRAB-repressive domain (Gilbert et al. 2014), VPR activation domain (VP64, P65, and Rta) (Chavez et al. 2015), synergistic activation mediators (MS2, p65, VP64, HSF1) (Konermann et al. 2015), and SunTag activation system (Tanenbaum et al. 2014).

10.5 Agricultural Applications

Climate change is a significant threat to crop productivity. Plant scientists need to use novel ways to develop crop varieties that withstand under changing environmental conditions. CRISPR-based tools are the most popular GE tools for fundamental and applied plant research (Pramanik et al. 2021a). Diverse CRISPR-based tools provide alternatives to achieve expected goals in NBTs like higher yield, tolerance to biotic and abiotic stresses, prevention of post-harvest losses, value addition, and novel traits-in-demand by consumers. Therefore, in a practical scenario, NBT-mediated improved crop productivity by targeted gene modification is beneficial to the farmers and consumers.

The CRISPR-based tools are extensively applied for basic and applied research in the broad range of the plant species, such as rice, oilseed, barley, strawberry, cucumber, tomato, tobacco, maize, soybean, wheat, citrus, and so on. Several phenotypic traits were modulated using knockout experiments to increase yield and quality, biotic and abiotic stress resistance, improve metabolites content (Table 10.2). Some examples of HR-based gene targeting consists of potato (Butler et al. 2016), tomato (Čermák et al. 2015), rice (Wang et al. 2017a), wheat (Gil-Humanes et al. 2017), and cassava (Hummel et al. 2018). Additionally, the insertion of important trait carrying alleles was introduced into the target genome allowed to produce higher anthocyanin in tomato (Čermák et al. 2015), herbicide-resistant soybean (Li et al. 2015), maize (Svitashev et al. 2015), and rice (Sun et al. 2016) varieties. Applications of various CRISPR-based tools are summarized in Table 10.1. Some of the tools are used only for validation in plant systems and within reach to find its application in NBTs for trait improvement.

10.6 Conclusion and Future Perspective

The application of NBTs, especially CRISPR-based tools, is transforming the field of plant breeding. Nonetheless, precision editing demands a comprehensive understanding of the functioning of genome and proteome in plant life. Although off-target activities (mutations at unintended locations in the genome) is a major concern in human, it can be circumvented by backcrossing and segregation in plants (Kadam et al. 2018). In this century, climate change may cause drastic environmental changes. The changing environmental conditions may create new challenges for agriculture and human health. Hence, tools like CRISPR/Cas are boon to agriculture and food production. Also, regulation of GE crops should be science-based to allow plant breeders to apply NBTs worldwide without any barriers.

Acknowledgments This work was supported by the National Research Foundation of Korea (grants NRF 2021R1A5A8029490, 2021R1I1A3057067, 2020R1A6A1A03044344, 2022R1A2C3010331); the Program for New Plant Breeding Techniques (NBT, grant PJ016867022022), Rural Development Administration, Korea.

References

Abudayyeh OO, Gootenberg JS, Franklin B, Koob J, Kellner MJ, Ladha A et al (2019) A cytosine deaminase for programmable single-base RNA editing. Science 365:382–386

Adli M (2018) The CRISPR tool kit for genome editing and beyond. Nat Commun 9:1911

Anzalone AV, Randolph PB, Davis JR, Sousa AA, Koblan LW, Levy JM et al (2019) Search-and-replace genome editing without double-strand breaks or donor DNA. Nature 576:149–157

Bado S, Forster BP, Nielen S, Ali AM, Lagoda PJL, Till BJ et al (2015) Plant mutation breeding: current progress and future assessment. In: Janick J (ed) Plant breeding reviews, vol 39. Wiley, Hoboken, NJ, pp 23–88

Butler NM, Baltes NJ, Voytas DF, Douches DS (2016) Geminivirus-mediated genome editing in potato (*Solanum tuberosum* L.) using sequence-specific nucleases. Front Plant Sci 7:1045

Butt H, Rao GS, Sedeek K, Aman R, Kamel R, Mahfouz M et al (2020) Engineering herbicide resistance via prime editing in rice. Plant Biotechnol J 18:2370–2372

Cai Y, Chen L, Sun S, Wu C, Yao W, Jiang B et al (2018) CRISPR/Cas9-mediated deletion of large genomic fragments in soybean. Int J Mol Sci 19:3835

Cano-Rodriguez D, Gjaltema RAF, Jilderda LJ, Jellema P, Dokter-Fokkens J, Ruiters MHJ et al (2016) Writing of H3K4Me3 overcomes epigenetic silencing in a sustained but context dependent manner. Nat Commun 7:12284

Capdeville N, Merker L, Schindele P, Puchta H (2021) Sophisticated CRISPR/Cas tools for fine-tuning plant performance. J Plant Physiol 257:153332

Čermák T, Baltes NJ, Čegan R, Zhang Y, Voytas DF (2015) High-frequency, precise modification of the tomato genome. Genome Biol 16:232

Chandrasekaran J, Brumin M, Wolf D, Leibman D, Klap C, Pearlsman M et al (2016) Development of broad virus resistance in non-transgenic cucumber using CRISPR/Cas9 technology. Mol Plant Pathol 17:1140–1153

Chavez A, Scheiman J, Vora S, Pruitt BW, Tuttle M, Iyer EPR et al (2015) Highly efficient Cas9-mediated transcriptional programming. Nat Methods 12:326–328

Cheng H, Hao M, Ding B, Mei D, Wang W, Wang H et al (2021) Base editing with high efficiency in allotetraploid oilseed rape by A3A-PBE system. Plant Biotechnol J 19:87–97

Cong L, Ran FA, Cox D, Lin S, Barretto R, Habib N et al (2013) Multiplex genome engineering using CRISPR/Cas systems. Science 339:819–823

Cox DBT, Gootenberg JS, Abudayyeh OO, Franklin B, Kellner MJ, Joung J et al (2017) RNA editing with CRISPR-Cas13. Science 358:1019–1027

Dalakouras A, Wassenegger M, Dadami E, Ganopoulos I, Pappas ML, Papadopoulou K et al (2020) Genetically modified organism-free RNA interference: exogenous application of RNA molecules in plants. Plant Physiol 182:38–50

Duan Y-B, Li J, Qin R-Y, Xu R-F, Li H, Yang Y-C et al (2016) Identification of a regulatory element responsible for salt induction of rice OsRAV2 through ex situ and in situ promoter analysis. Plant Mol Biol 90:49–62

Enciso-Rodriguez F, Manrique-Carpintero NC, Nadakuduti SS, Buell CR, Zarka D, Douches D et al (2019) Overcoming self-incompatibility in diploid potato using CRISPR-Cas9. Front Plant Sci 10:376

Gaj T, Gersbach CA, Barbas CF (2013) ZFN, TALEN, and CRISPR/Cas-based methods for genome engineering. Trends Biotechnol 31:397–405

Gao H, Gadlage MJ, Lafitte HR, Lenderts B, Yang M, Schroder M et al (2020) Superior field performance of waxy corn engineered using CRISPR-Cas9. Nat Biotechnol 38:579–581

Gaudelli NM, Komor AC, Rees HA, Rees HA, Packer MS, Badran AH et al (2017) Programmable base editing of A·T to G·C in genomic DNA without DNA cleavage. Nature 551:464–471

Gaudelli NM, Lam DK, Rees HA, Solá-Esteves NM, Barrera LA, Born DA et al (2020) Directed evolution of adenine base editors with increased activity and therapeutic application. Nat Biotechnol 38:892–900

Gilbert LA, Horlbeck MA, Adamson B, Villalta JE, Chen Y, Whitehead EH et al (2014) Genome-Scale CRISPR-mediated control of gene repression and activation. Cell 159:647–661
Gil-Humanes J, Wang Y, Liang Z, Shan Q, Ozuna CV, Sánchez-León S et al (2017) High-efficiency gene targeting in hexaploid wheat using DNA replicons and CRISPR/Cas9. Plant J 89:1251–1262
Gong X, Zhang T, Xing J, Wang R, Zhao Y (2020) Positional effects on efficiency of CRISPR/Cas9-based transcriptional activation in rice plants. aBIOTECH 1:1–5
González MN, Massa GA, Andersson M, Turesson H, Olsson N, Fält A-S et al (2019) Reduced enzymatic browning in potato tubers by specific editing of a polyphenol oxidase gene via ribonucleoprotein complexes delivery of the CRISPR/Cas9 system. Front Plant Sci 10:1649
Grünewald J, Zhou R, Lareau CA, Garcia SP, Iyer S, Miller BR et al (2020) A dual-deaminase CRISPR base editor enables concurrent adenine and cytosine editing. Nat Biotechnol 38:861–864
Halperin SO, Tou CJ, Wong EB, Modavi C, Schaffer DV, Dueber JE (2018) CRISPR-guided DNA polymerases enable diversification of all nucleotides in a tunable window. Nature 560:248–252
Hilton IB, D'Ippolito AM, Vockley CM, Thakore PI, Crawford GE, Reddy TE et al (2015) Epigenome editing by a CRISPR-Cas9-based acetyltransferase activates genes from promoters and enhancers. Nat Biotechnol 33:510–517
Holme IB, Wendt T, Gil-Humanes J, Deleuran LC, Starker CG, Voytas DF et al (2017) Evaluation of the mature grain phytase candidate HvPAPhy_a gene in barley (Hordeum vulgare L.) using CRISPR/Cas9 and TALENs. Plant Mol Biol 95:111–121
Huang L, Li Q, Zhang C, Chu R, Gu Z, Tan H et al (2020) Creating novel Wx alleles with fine-tuned amylose levels and improved grain quality in rice by promoter editing using CRISPR/Cas9 system. Plant Biotechnol J 18:2164–2166
Huang TP, Newby GA, Liu DR (2021) Precision genome editing using cytosine and adenine base editors in mammalian cells. Nat Protoc 18:1–40
Hummel AW, Chauhan RD, Cermak T, Mutka AM, Vijayaraghavan A, Boyher A et al (2018) Allele exchange at the EPSPS locus confers glyphosate tolerance in cassava. Plant Biotechnol J 16:1275–1282
Ito Y, Sekiyama Y, Nakayama H, Nishizawa-Yokoi A, Endo M, Shima Y et al (2020) Allelic mutations in the ripening-inhibitor locus generate extensive variation in tomato ripening. Plant Physiol 183:80–95
Jia H, Wang N (2020) Generation of homozygous canker-resistant citrus in the T0 generation using CRISPR-SpCas9p. Plant Biotechnol J 18:1990–1992
Jia H, Zhang Y, Orbović V, Xu J, White FF, Jones JB et al (2017) Genome editing of the disease susceptibility gene CsLOB1 in citrus confers resistance to citrus canker. Plant Biotechnol J 15:817–823
Jiang WZ, Henry IM, Lynagh PG, Comai L, Cahoon EB, Weeks DP (2017) Significant enhancement of fatty acid composition in seeds of the allohexaploid, Camelina sativa, using CRISPR/Cas9 gene editing. Plant Biotechnol J 15:648–657
Jiang Y-Y, Chai Y-P, Lu M-H, Han X-L, Lin Q, Zhang Y et al (2020) Prime editing efficiently generates W542L and S621I double mutations in two ALS genes in maize. Genome Biol 21:257
Jin S, Fei H, Zhu Z, Luo Y, Liu J, Gao S et al (2020) Rationally designed APOBEC3B cytosine base editors with improved specificity. Mol Cell 79:728–740.e6
Jinek M, Chylinski K, Fonfara I, Hauer MI, Doudna JA, Charpentier E et al (2012) A programmable dual-RNA-guided DNA endonuclease in adaptive bacterial immunity. Science 337:816–821
Kadam US, Shelake RM, Chavhan RL, Suprasanna P (2018) Concerns regarding 'off-target' activity of genome editing endonucleases. Plant Physiol Biochem 131:22–30
Kang B-C, Yun J-Y, Kim S-T, Shin Y-J, Ryu J, Choi M et al (2018) Precision genome engineering through adenine base editing in plants. Nat Plants 4:427–431
Kearns NA, Pham H, Tabak B, Genga RM, Silverstein NJ, Garber M et al (2015) Functional annotation of native enhancers with a Cas9–histone demethylase fusion. Nat Methods 12:401–403

Kim J-Y, Shelake RM (2021) Method for inducing reactive oxygen species-mediated base mutation of target gene. Korean Patent No. 1022080310000. Appl. No. 1020180096931. https://doi.org/10.8080/1020180096931?urlappend=en

Kim J-M, Kim K, Schmidt T, Punj V, Tucker H, Rice JC et al (2015) Cooperation between SMYD3 and PC4 drives a distinct transcriptional program in cancer cells. Nucleic Acids Res 43:8868–8883

Komor AC, Kim YB, Packer MS, Zuris JA, Liu DR (2016) Programmable editing of a target base in genomic DNA without double-stranded DNA cleavage. Nature 533:420–424

Konermann S, Brigham MD, Trevino AE, Joung J, Abudayyeh OO, Barcena C et al (2015) Genome-scale transcriptional activation by an engineered CRISPR-Cas9 complex. Nature 517:583–588

Kuang Y, Li S, Ren B, Yan F, Spetz C, Li X et al (2020) Base-editing-mediated artificial evolution of OsALS1 in planta to develop novel herbicide-tolerant rice germplasms. Mol Plant 13:565–572

Kurt IC, Zhou R, Iyer S, Garcia SP, Miller BR, Langner LM et al (2021) CRISPR C-to-G base editors for inducing targeted DNA transversions in human cells. Nat Biotechnol 39:41–46

Kwon DY, Zhao Y-T, Lamonica JM, Zhou Z (2017) Locus-specific histone deacetylation using a synthetic CRISPR-Cas9-based HDAC. Nat Commun 8:15315

Lee JE, Neumann M, Duro DI, Schmid M (2019) CRISPR-based tools for targeted transcriptional and epigenetic regulation in plants. PLoS One 14:e0222778

Lee JH, Mazarei M, Pfotenhauer AC, Dorrough AB, Poindexter MR, Hewezi T et al (2020) Epigenetic footprints of CRISPR/Cas9-mediated genome editing in plants. Front Plant Sci 10:1720

Li J-F, Norville JE, Aach J, McCormack M, Zhang D, Bush J et al (2013) Multiplex and homologous recombination–mediated genome editing in Arabidopsis and Nicotiana benthamiana using guide RNA and Cas9. Nat Biotechnol 31:688–691

Li Z, Liu Z-B, Xing A, Moon BP, Koellhoffer JP, Huang L et al (2015) Cas9-guide RNA directed genome editing in soybean. Plant Physiol 169:960–970

Li Z, Zhang D, Xiong X, Yan B, Xie W, Sheen J et al (2017) A potent Cas9-derived gene activator for plant and mammalian cells. Nat Plants 3:930–936

Li C, Zong Y, Wang Y, Zhang D, Song Q, Zhang R et al (2018a) Expanded base editing in rice and wheat using a Cas9-adenosine deaminase fusion. Genome Biol 19:59

Li T, Yang X, Yu Y, Si X, Zhai X, Zhang H et al (2018b) Domestication of wild tomato is accelerated by genome editing. Nat Biotechnol 36:1160–1163

Li C, Li W, Zhou Z, Chen H, Xie C, Lin Y (2020a) A new rice breeding method: CRISPR/Cas9 system editing of the Xa13 promoter to cultivate transgene-free bacterial blight-resistant rice. Plant Biotechnol J 18:313–315

Li C, Zhang R, Meng X, Chen S, Zong Y, Lu C et al (2020b) Targeted, random mutagenesis of plant genes with dual cytosine and adenine base editors. Nat Biotechnol 38:875–882

Li H, Li J, Chen J, Yan L, Xia L (2020c) Precise modifications of both exogenous and endogenous genes in rice by prime editing. Mol Plant 13:671–674

Lin Q, Zong Y, Xue C, Wang S, Jin S, Zhu Z et al (2020) Prime genome editing in rice and wheat. Nat Biotechnol 38:582–585

Liu XS, Wu H, Ji X, Stelzer Y, Wu X, Czauderna S (2016) Editing DNA methylation in the mammalian genome. Cell 167:233–247.e17

Lu Y, Zhu J-K (2017) Precise editing of a target base in the rice genome using a modified CRISPR/Cas9 system. Mol Plant 10:523–525

Lu Y, Tian Y, Shen R, Yao Q, Zhong D, Zhang X et al (2020) Precise genome modification in tomato using an improved prime editing system. Plant Biotechnol J 19:415–417

Ma X, Zhang Q, Zhu Q, Liu W, Chen Y, Qiu R et al (2015) A robust CRISPR/Cas9 system for convenient, high-efficiency multiplex genome editing in monocot and dicot plants. Mol Plant 8:1274–1284

Mao X, Zheng Y, Xiao K, Wei Y, Zhu Y, Cai Q et al (2018) OsPRX2 contributes to stomatal closure and improves potassium deficiency tolerance in rice. Biochem Biophys Res Commun 495:461–467

Mok BY, de Moraes MH, Zeng J, Bosch DE, Kotrys AV, Raguram A et al (2020) A bacterial cytidine deaminase toxin enables CRISPR-free mitochondrial base editing. Nature 583:631–637

Morgan SL, Mariano NC, Bermudez A, Arruda NL, Wu F, Luo Y et al (2017) Manipulation of nuclear architecture through CRISPR-mediated chromosomal looping. Nat Commun 8:15993

Nekrasov V, Staskawicz B, Weigel D, Jones JDG, Kamoun S (2013) Targeted mutagenesis in the model plant Nicotiana benthamiana using Cas9 RNA-guided endonuclease. Nat Biotechnol 31: 691–693

Nekrasov V, Wang C, Win J, Lanz C, Weigel D, Kamoun S (2017) Rapid generation of a transgene-free powdery mildew resistant tomato by genome deletion. Sci Rep 7:482

Nieves-Cordones M, Mohamed S, Tanoi K, Kobayashi NI, Takagi K, Vernet A et al (2017) Production of low-Cs^+ rice plants by inactivation of the K^+ transporter OsHAK1 with the CRISPR-Cas system. Plant J 92:43–56

Nishida K, and Kondo A. (2017) Method for modifying genome sequence to introduce specific mutation to targeted DNA sequence by base-removal reaction, and molecular complex used therein. U.S. Patent Appl. No. 15/523,939

Nishida K, Arazoe T, Yachie N, Banno S, Kakimoto M, Tabata M et al (2016) Targeted nucleotide editing using hybrid prokaryotic and vertebrate adaptive immune systems. Science 353:aaf8729

Nishimasu H, Ran FA, Hsu PD, Konermann S, Shehata SI, Dohmae N et al (2014) Crystal structure of Cas9 in complex with guide RNA and target DNA. Cell 156:935–949

Oladosu Y, Rafii MY, Abdullah N, Hussin G, Ramli A, Rahim HA et al (2016) Principle and application of plant mutagenesis in crop improvement: a review. Biotechnol Biotechnol Equip 30:1–16

Oliva R, Ji C, Atienza-Grande G, Huguet-Tapia JC, Perez-Quintero A, Li T et al (2019) Broad-spectrum resistance to bacterial blight in rice using genome editing. Nat Biotechnol 37:1344–1350. https://doi.org/10.1038/s41587-019-0267-z

Papikian A, Liu W, Gallego-Bartolomé J, Jacobsen SE (2019) Site-specific manipulation of Arabidopsis loci using CRISPR-Cas9 SunTag systems. Nat Commun 10:729

Peng A, Chen S, Lei T, Xu L, He Y, Wu L et al (2017) Engineering canker-resistant plants through CRISPR/Cas9-targeted editing of the susceptibility gene CsLOB1 promoter in citrus. Plant Biotechnol J 15:1509–1519

Pramanik D, Shelake RM, Kim MJ, Kim J-Y (2021a) CRISPR-mediated engineering across the central dogma in plant biology for basic research and crop improvement. Mol Plant 14:127–150

Pramanik D, Shelake RM, Park J, Kim MJ, Hwang I, Park Y et al (2021b) CRISPR/Cas9-mediated generation of pathogen-resistant tomato against *Tomato yellow leaf curl virus* and powdery mildew. Int J Mol Sci 22:1878

Pulecio J, Verma N, Mejía-Ramírez E, Huangfu D, Raya A (2017) CRISPR/Cas9-based engineering of the epigenome. Cell Stem Cell 21:431–447

Ren Q, Sretenovic S, Liu S, Tang X, Huang L, He Y et al (2021) PAM-less plant genome editing using a CRISPR-SpRY toolbox. Nat Plants 7:25–33

Richter MF, Zhao KT, Eton E, Lapinaite A, Newby GA, Thuronyi BW et al (2020) Phage-assisted evolution of an adenine base editor with improved Cas domain compatibility and activity. Nat Biotechnol 38:883–891

Rodríguez-Leal D, Lemmon ZH, Man J, Bartlett ME, Lippman ZB (2017) Engineering quantitative trait variation for crop improvement by genome editing. Cell 171:470–480.e8

Sánchez-León S, Gil-Humanes J, Ozuna CV, Giménez MJ, Sousa C, Voytas DF et al (2018) Low-gluten, nontransgenic wheat engineered with CRISPR/Cas9. Plant Biotechnol J 16:902–910

Santillán Martínez MI, Bracuto V, Koseoglou E, Appiano M, Jacobsen E, Visser RGF et al (2020) CRISPR/Cas9-targeted mutagenesis of the tomato susceptibility gene *PMR4* for resistance against powdery mildew. BMC Plant Biol 20:284

Selma S, Bernabé-Orts JM, Vazquez-Vilar M, Diego-Martin B, Ajenjo M, Garcia-Carpintero V et al (2019) Strong gene activation in plants with genome-wide specificity using a new orthogonal CRISPR/Cas9-based programmable transcriptional activator. Plant Biotechnol J 17:1703–1705

Shan Q, Wang Y, Li J, Zhang Y, Chen K, Liang Z et al (2013) Targeted genome modification of crop plants using a CRISPR-Cas system. Nat Biotechnol 31:686–688

Shelake RM, Pramanik D, Kim J-Y (2019a) Evolution of plant mutagenesis tools: a shifting paradigm from random to targeted genome editing. Plant Biotechnol Rep 13:423–445

Shelake RM, Pramanik D, Kim J-Y (2019b) Exploration of plant-microbe interactions for sustainable agriculture in CRISPR era. Microorganisms 7:269

Shen R, Wang L, Liu X, Wu J, Jin W, Zhao X et al (2017) Genomic structural variation-mediated allelic suppression causes hybrid male sterility in rice. Nat Commun 8:1310

Shi J, Gao H, Wang H, Lafitte HR, Archibald RL, Yang M et al (2017) ARGOS8 variants generated by CRISPR-Cas9 improve maize grain yield under field drought stress conditions. Plant Biotechnol J 15:207–216

Shimatani Z, Kashojiya S, Takayama M, Terada R, Arazoe T, Ishii H et al (2017) Targeted base editing in rice and tomato using a CRISPR-Cas9 cytidine deaminase fusion. Nat Biotechnol 35: 441–443

Shrestha A, Khan A, Dey N (2018) cis–trans engineering: advances and perspectives on customized transcriptional regulation in plants. Mol Plant 11:886–898

Sun Y, Zhang X, Wu C, He Y, Ma Y, Hou H et al (2016) Engineering herbicide-resistant rice plants through CRISPR/Cas9-mediated homologous recombination of acetolactate synthase. Mol Plant 9:628–631

Suprasanna P, Mirajkar SJ, Bhagwat SG (2015) Induced mutations and crop improvement. In: Bahadur B, Rajam MV, Sahijram L, Krishnamurthy KV (eds) Plant biology and biotechnology. Springer India, New Delhi, pp 593–617

Svitashev S, Young JK, Schwartz C, Gao H, Falco SC, Cigan AM (2015) Targeted mutagenesis, precise gene editing, and site-specific gene insertion in maize using Cas9 and guide RNA. Plant Physiol 169:931–945

Tanenbaum ME, Gilbert LA, Qi LS, Weissman JS, Vale RD (2014) A protein-tagging system for signal amplification in gene expression and fluorescence imaging. Cell 159:635–646

Tang X, Lowder LG, Zhang T, Malzahn AA, Zheng X, Voytas DF et al (2017) A CRISPR-Cpf1 system for efficient genome editing and transcriptional repression in plants. Nat Plants 3:17018

Tang X, Sretenovic S, Ren Q, Jia X, Li M, Fan T et al (2020) Plant prime editors enable precise gene editing in rice cells. Mol Plant 13:667–670

Veillet F, Chauvin L, Kermarrec M-P, Sevestre F, Merrer M, Terret Z et al (2019) The Solanum tuberosum GBSSI gene: a target for assessing gene and base editing in tetraploid potato. Plant Cell Rep 38:1065–1080

Veillet F, Kermarrec M-P, Chauvin L, Guyon-Debast A, Chauvin JE, Gallois J-L (2020) Prime editing is achievable in the tetraploid potato but needs improvement. bioRxiv 2020:1–7

Waltz E (2016) Gene-edited CRISPR mushroom escapes US regulation. Nature 532:293

Wang F, Wang C, Liu P, Lei C, Hao W, Gao Y et al (2016) Enhanced rice blast resistance by CRISPR/Cas9-targeted mutagenesis of the ERF transcription factor gene OsERF922. PLoS One 11:e0154027

Wang M, Lu Y, Botella JR, Mao Y, Hua K, Zhu J-K (2017a) Gene targeting by homology-directed repair in rice using a geminivirus-based CRISPR/Cas9 system. Mol Plant 10:1007–1010

Wang Y, Geng L, Yuan M, Wei J, Jin C, Li M et al (2017b) Deletion of a target gene in Indica rice via CRISPR/Cas9. Plant Cell Rep 36:1333–1343

Wang S, Zong Y, Lin Q, Zhang H, Chai Z, Zhang D et al (2020a) Precise, predictable multi-nucleotide deletions in rice and wheat using APOBEC-Cas9. Nat Biotechnol 38:1460–1465

Wang Z, Hong Y, Zhu G, Li Y, Niu Q, Yao J et al (2020b) Loss of salt tolerance during tomato domestication conferred by variation in a Na^+/K^+ transporter. EMBO J 39:e103256

Xie K, Yang Y (2013) RNA-guided genome editing in plants using a CRISPR-Cas system. Mol Plant 6:1975–1983

Xie Y, Niu B, Long Y, Li G, Tang J, Zhang Y et al (2017a) Suppression or knockout of SaF/SaM overcomes the Sa-mediated hybrid male sterility in rice. J Integr Plant Biol 59:669–679

Xie Y, Xu P, Huang J, Ma S, Xie X, Tao D (2017b) Interspecific hybrid sterility in rice is mediated by OgTPR1 at the S1 locus encoding a peptidase-like protein. Mol Plant 10:1137–1140

Xie Y, Zhang Y, Han J, Luo J, Li G, Huang J et al (2018) The intronic cis element SE1 recruits trans-acting repressor complexes to repress the expression of ELONGATED UPPERMOST INTERNODE1 in rice. Mol Plant 11:720–735

Xie J, Huang X, Wang X, Gou S, Liang Y, Chen F et al (2020) ACBE, a new base editor for simultaneous C-to-T and A-to-G substitutions in mammalian systems. BMC Biol 18:131

Xing S, Chen K, Zhu H, Zhang R, Zhang H, Li B et al (2020) Fine-tuning sugar content in strawberry. Genome Biol 21:230

Xu R-F, Li H, Qin R-Y, Li J, Qiu C-H, Yang Y-C et al (2015) Generation of inheritable and "transgene clean" targeted genome-modified rice in later generations using the CRISPR/Cas9 system. Sci Rep 5:11491

Xu R, Li J, Liu X, Shan T, Qin R, Wei P (2020a) Development of plant prime-editing systems for precise genome editing. Plant Commun 1:100043

Xu W, Zhang C, Yang Y, Zhao S, Kang G, He X et al (2020b) Versatile nucleotides substitution in plant using an improved prime editing system. Mol Plant 13:675–678

Xu Y, Lin Q, Li X, Zhao S, Kang G, He X et al (2021a) Fine-tuning the amylose content of rice by precise base editing of the Wx gene. Plant Biotechnol J 19:11–13

Xu Z, Kuang Y, Ren B, Yan D, Yan F, Spetz C et al (2021b) SpRY greatly expands the genome editing scope in rice with highly flexible PAM recognition. Genome Biol 22:6

Yu H, Lin T, Meng X, Du H, Zhang J, Liu G et al (2021) A route to de novo domestication of wild allotetraploid rice. Cell 184:1–15

Zeng D, Liu T, Ma X, Wang B, Zheng Z, Zhang Y et al (2020) Quantitative regulation of Waxy expression by CRISPR/Cas9-based promoter and 5′UTR-intron editing improves grain quality in rice. Plant Biotechnol J 18:2385–2387

Zetsche B, Gootenberg JS, Abudayyeh OO, Slaymaker IM, Makarova KS, Essletzbichler P et al (2015) Cpf1 is a single RNA-guided endonuclease of a class 2 CRISPR-Cas system. Cell 163: 759–771

Zhang H, Zhang J, Wei P, Zhang B, Gou F, Feng Z et al (2014) The CRISPR/Cas9 system produces specific and homozygous targeted gene editing in rice in one generation. Plant Biotechnol J 12: 797–807

Zhang H, Si X, Ji X, Fan R, Liu J, Chen K et al (2018) Genome editing of upstream open reading frames enables translational control in plants. Nat Biotechnol 36:894–898

Zhang X, Zhu B, Chen L, Xie L, Yu W, Wang Y et al (2020) Dual base editor catalyzes both cytosine and adenine base conversions in human cells. Nat Biotechnol 38:856–860

Zhao D, Li J, Li S, Xin X, Hu M, Price MA et al (2021) Glycosylase base editors enable C-to-A and C-to-G base changes. Nat Biotechnol 39:35–40

Zhou X, Jacobs TB, Xue L-J, Harding SA, Tsai C-J (2015) Exploiting SNPs for biallelic CRISPR mutations in the outcrossing woody perennial Populus reveals 4-coumarate:CoA ligase specificity and redundancy. New Phytol 208:298–301

Zhou H, He M, Li J, Chen L, Huang Z, Zheng S et al (2016) Development of commercial thermo-sensitive genic male sterile rice accelerates hybrid rice breeding using the CRISPR/Cas9-mediated TMS5 editing system. Sci Rep 6:37395

Zong Y, Wang Y, Li C, Zhang R, Chen K, Ran Y et al (2017) Precise base editing in rice, wheat and maize with a Cas9-cytidine deaminase fusion. Nat Biotechnol 35:438–440

Zsögön A, Čermák T, Naves ER, Notini MM, Edel KH, Weinl S et al (2018) De novo domestication of wild tomato using genome editing. Nat Biotechnol 36:1211–1216

Crop Improvement Through Induced Genetic Diversity and Mutation Breeding: Challenges and Opportunities

11

Sobhana Sivasankar

Abstract

Spontaneous and induced mutations form the bases for genetic diversity and evolution. Induced mutations in plants have been successfully used for the improvement of crop species over the last several decades and remain an important component of crop-improvement programs in Asia, and increasingly in Africa and Latin America. Induced mutations create rapid genetic diversity, which when combined with present-day genomics and speed-breeding technologies such as doubled haploidy, pave the way for accelerated genetic gain in crop plants. The FAO/IAEA Mutant Variety Database currently holds records of over 3200 mutant crop varieties released for cultivation across 220 plant species, the majority being in rice. Close to 2000 of these have been released in Asia. In the genomic era, the molecular bases of the phenotypes of mutants can be more easily elucidated than ever before with cost- and time-efficient sequencing technologies combined with precision phenotyping and bioinformatics. Beyond routine crop improvement, the combination of induced mutations and current omics technologies constitute a powerful functional genomics platform that identifies molecular markers for forward breeding and also candidate genes and sequences for gene editing.

Keywords

Mutation breeding · Genetic diversity · Crop improvement · Genomics

S. Sivasankar (✉)
Plant Breeding and Genetics Section, Joint FAO/IAEA Centre of Nuclear Techniques in Food and Agriculture, Vienna International Centre, Vienna, Austria
e-mail: s.sivasankar@iaea.org; sobhana.sivasankar@fao.org

S. Penna, S. M. Jain (eds.), *Mutation Breeding for Sustainable Food Production and Climate Resilience*, https://doi.org/10.1007/978-981-16-9720-3_11

11.1 Introduction

Induced genetic diversity and breeding with promising mutations for farmer-preferred traits in agronomically sound genetic backgrounds have paved the way for crop improvement globally for over seven decades now. Plant transgenic technologies debuting in the late eighties and early nineties led to the development and use of continuously evolving methods in plant transformation, DNA manipulation and transgene integration, the most recent being the CRISPR-Cas mediated site-specific gene/genome editing, in crop improvement. Simultaneously, rapid evolution in genomics and computational biology is supporting increased precision and efficiency in breeding selections through marker-assisted breeding and genomic selections. While transgenic technologies and molecular breeding remain in heavy use in the improvement of major crops, namely, maize, soybean, rice, wheat, and barley in developed countries, especially in the seed/biotechnology industry, improvement of many crops of importance to the diet diversity of the global population follow traditional breeding. This is especially so in developing countries with limited technology and/or infrastructure and rudimentary or absent regulatory structures, yet where many species of underutilized crops are central to the food and nutritional security of the local population.

Increasing the productivity of crops is critical to enable self-sufficiency in food especially in developing countries. Technical and infrastructure capacities for systematic plant breeding research remain insufficient in developing countries, consequent to which the productivity of crops remain low. Climate change adds a further dimension to this challenge by triggering crop losses from droughts and warming global temperatures and causing biodiversity loss. Increasing temperatures and uncertainties in the intensity, frequency, and distribution of precipitation adversely affect crop production. These adverse effects are mainly due to the inability of existing crop varieties to cope with these climate anomalies, and because of increasing incidence and severity of pests and diseases, some of which are becoming transnational. Changing climate also affects prevailing biodiversity and the geographic distribution of crops. Without adaptation of crop varieties to the impacts of climate change, crop production will suffer serious setbacks, leading to food insecurity and the breakdown of food systems, especially in developing nations. Sustainable increases in crop productivity requires continuous improvement of the expression of crop genetic capacity, and this is the focus of plant breeding. Improved genetics constitute the most fundamental and sustainable solution for food security and crop adaptation to climate change as it intrinsically improves crop performance or stabilizes it under stress.

Induced genetic diversity and plant mutation breeding remains the fastest approach to introduce new genetic diversity that can address both simple traits like disease resistance or grain quality, and complex traits like yield or tolerance to climate extremes. Cost-efficient, high-throughput next-generation genotyping facilitates combined with mutation breeding enables accelerated genetic gain and also held establish genetic associations and facilitate understanding the molecular bases of the genetic variability.

11.2 Induced Genetic Diversity for Crop Improvement

Induced mutations create more rapid changes in the genome than that normally occurs spontaneously in nature. Nuclear technologies offer a tried-and-proven avenue to safely and easily induce mutations. Since the first demonstration that radiation causes mutations in fruit flies (Muller 1927) and in the crop plants maize and barley (Stadler 1928), heritable positive genetic changes, and their associated phenotypes have been routinely selected for the development of improved crop varieties. In crop improvement, the ability of induced mutations to create rapid evolutionary change, and the combination with "speed-breeding" technologies such as doubled haploidy, shuttle breeding, and marker-assisted selection, can accelerate the pace of developing improved crop varieties.

The Joint FAO/IAEA Division of Nuclear Techniques in Food and Agriculture was established in 1964 with the mandate of technology development and capacity building in FAO/IAEA Member States (countries) in the field of induced genetic diversity and plant mutation breeding. Since then, the Joint FAO/IAEA Centre, as it is currently known, has remained the global leader in the application of irradiation in plant mutation breeding and crop improvement and provided capacity building for inducing genetic variation and developing improved varieties in more than 100 countries. Gamma rays and X-rays are the predominant mutagen sources used, and the Joint Centre Laboratories in Seibersdorf, Austria, provides seed irradiation services to Member States as well. The technical support of the Joint Centres on mutation induction, precision selection, and generation advancement for variety development along with molecular marker discovery and other efficiency-enhancing techniques are also being incorporated.

The Joint Centre delivers its overall goal of crop improvement for sustainable agricultural production systems through mutation breeding using nuclear and associated technologies. This is achieved using a fluid pipeline that includes three distinct operational areas, namely, mutation induction for novel genetic diversity, precision phenotyping and selection, and speed-breeding technologies. This pipeline, as it stands currently, ends with the development and release of improved varieties in FAO/IAEA Member States, and misses a crucial final step, namely, enabling access of the improved varieties to farmers. This is an especially important step in nations where formal seed systems do not exist. Therefore, efforts are now under way to incorporate seed system elements at the end of the current mutation breeding pipeline to facilitate seed access to farmers.

Induced genetic diversity fortifies plant germplasm pools and enables faster genetic gain in all crops, and especially so in vegetative crops where the genetic base is narrow. As such, it is a highly productive means for the improvement of both simple and complex crop traits in developing countries when coupled with access to induction or mutation facilities and technical support from the Joint FAO/IAEA Centre. Testament to this is the >3300 mutant varieties from >70 countries and >220 plant species represented in the Mutant Variety Database (mvd.iaea.org). While mutation induction in annual seed crops have a long history, the technique requires considerable improvement in vegetatively propagated crops where the potential for mutation breeding is significant (Suprasanna et al. 2012). Improved techniques in micropropagation and mutagenesis are important for generating chimera-free populations that can be used to select traits of interest which remain stable.

Currently, the induction of mutation is primarily through gamma ray irradiation in inbred and outbred annual seed crops, and occasionally in vegetatively propagated crops. Use of new mutagen sources including ion beam, e-beam, and cosmic rays or space breeding are of increasing relevance for accelerated variety development in FAO/IAEA Member States, together with functional genomics for marker-assisted breeding and gene editing.

Selection of improved mutant lines involves the development of physiological, biochemical, or molecular screening protocols that enable the identification of mutant lines which show specific improvement in targeted, farmer-preferred traits. These protocols are implemented in the laboratory, greenhouse or field, depending on the nature of the trait. A good part of the crops for which improvement is sought through plant mutation breeding by FAO/IAEA Member States in sub-Saharan Africa are crops involved in subsistence, smallholder agriculture, where women are the main farmers. Improvement of these crops, most of them dryland cereals and legumes, is of critical importance to women farmers. Traits of national and regional interest routinely include increased yield, improved quality, resistance to insect pests, diseases, or parasitic weeds, and resistance to extreme temperatures and drought. In the genomic era, precise selection using molecular markers is becoming increasingly cost- and time-efficient.

A variety of speed-breeding technologies are of focus to reduce the breeding time required to develop new crop varieties from selected mutant lines. These include doubled haploidy, rapid cycling, and molecular marker-assisted selection. Mutation breeding is advantageous over traditional breeding in that it creates and utilizes a much larger genetic diversity than that used in traditional breeding. However, the time required for the development of a new variety in both cases is normally 7–10 years, or more, depending on the plant species. Speed-breeding technologies, when combined, can significantly reduce this time, sometimes even to half.

11.3 Examples of Recent Outcomes in Crop Improvement

Through mutation breeding and crop improvement initiatives, the Joint FAO/IAEA Centre helped Member States build capacities to ensure sustainable agricultural practices, improve crop production, and tackle environmental stresses. In the last decade alone, this support has facilitated the development of more than 230 mutant varieties in 44 different crop species that were officially released in 26 Member States. Furthermore, thousands of developed mutant lines have been contributing to national breeding programs in Member States, as well as to global biodiversity. Over the past decade, the Joint Centre contributed to several impactful achievements in Member States, a few examples are presented.

In Viet Nam, since 2012, scientists have developed 7 rice mutant varieties with high yield and drought tolerance, and in total 20 new mutant rice varieties have been released, which are currently grown by more than 300,000 farmers. This includes Mutant Khang Dan, one of the best-known rice varieties in the northern and central provinces of Viet Nam. Furthermore, mutant soybean varieties are also popular, now occupying about 50% of the area dedicated to this crop. Pakistan has released four cotton mutant varieties since 2013 which have been rapidly adopted by farmers due to their ability to withstand high temperatures and heavy rains and resist critical pests and diseases, while maintaining fiber yield. Of the 3.1 million hectares planted with cotton in the country, between 15% and 25% are planted with mutant varieties, a number expected to increase to 30–40% very soon.

Mutation breeding made an important contribution to Peruvian agriculture through improved nutrition, increased yields, and enhanced livelihoods, as well as increased biodiversity. The new improved barley and amaranth varieties are adaptable to harsh climatic conditions in high altitudes. The mutant barley variety, Centenario II, now covers 18% of the dedicated barley growing area and is estimated to contribute US$ 6.6 million annually to the national economy. The mutant amaranth variety, Centenario, currently covers 47% of the area dedicated to this crop.

Plant mutation breeding led to the development and release of two new mutant wheat varieties in less than 5 years that are resistant to the fungus, Ug99, responsible for the most damaging wheat black stem rust disease. This unprecedented rapid development of resistant varieties positioned Kenya as a leading nation in the fight against Ug99, with the first-ever release, in 2014, of 54 tons of seed to cultivate over 500 hectares, for the benefit of 1000 framers. The varieties retained their resistance to Ug99 in the following years and the demand for seeds remains high.

11.4 Mutations in the Study of Plant Biology

In the present genomic era, the molecular bases of mutations are elucidated more easily than ever before with cost- and time-efficient sequencing technologies, combined with precise phenotyping. Such information constitutes the foundation for molecular breeding in its different connotations. Beyond routine crop improvement, induced mutations in combination with current genomic technologies constitute a powerful and cost-efficient functional genomics platform to identify molecular markers for forward breeding, and candidate genes for biotechnology.

Spontaneous or induced mutations have been the major entry points for elaborating biochemical pathways, describing morphological variants, characterizing physiological phenomena, and understanding signaling pathways very early in plant biology. Developmental mutants in seed plants and the use of mutants to analyze the photosynthetic pathway in higher plants and algae were elegantly described in the 1980s, before the advent of the genomic era (Marx 1983; Somerville 1986). *Arabidopsis thaliana* remains the predominant model plant species for these experiments, wherein genetic associations based on mutations or quantitative trait loci led to early cloning of genes, further discovery of gene networks and interactions through protein–protein interaction techniques such as the yeast two-hybrid system, and extension of gene cloning to other plant species based on sequence homology. A good description of the early efforts in this area prior to the publication of the Arabidopsis genome in 2000 is available from the meeting report of the Seventh International Conference on Arabidopsis Research (Somerville and Somerville 1996). While spontaneous mutations in the dwarfing genes of wheat Norin 10 and rice Dee-gee-woo-gen drove the 1960s Green Revolution, it was not until 1999 and 2002 with the advent of genomics that the causal mutations were identified to be in the *Reduced height* (*Rht*) and the *semidwarf1* (*sd1*) genes in these two crops, respectively (Hedden 2003).

The use of transferred DNA (T-DNA) tagged mutant populations for cloning of plant genes was first reported in 1989, and it utilized the ability of the T-DNA of the bacteria, *Agrobacterium tumefaciens*, to insert into the plant genome to generate insertion mutant populations and the border sequence of the inserted T-DNA as the tag to facilitate gene cloning (Koncz et al. 1989). This was the earliest report of a functional genomics approach in higher plants that allowed the association of a phenotype with its causal gene(s). Subsequently, activation-tagged mutant populations generated first in Arabidopsis began to be used in forward genetics approaches for identifying the gene related to a phenotype (Hayashi et al. 1992). This utilized transcriptional enhancers inserted randomly into the plant genome alongside of T-DNA to produce dominant mutations, screening for phenotypes of interest in the resulting mutant population, followed by the cloning of the gene associated with the phenotype. A genome-wide insertional mutant population of Arabidopsis with T-DNA to facilitate functional genomics explorations was first reported in 2003 (Alonso et al. 2003).

Spontaneous mutations, especially those arising from transposon insertions, have also been used in forward genetics approaches to establish gene-to-phenotype associations. Thus, the cloning of the maize *Anther ear1* (*An1*) gene resulted from the observation of the phenotype of perfect flowers on normally pistillate ears in a transposon insertion mutagenesis population of the *Mutator* transposable element system (Bensen et al. 1995). The identity of the sequence was established through its amplification from independent *Mu*-induced *an1* alleles using one primer from the terminal inverted repeat of Mu and the other from the putative *An1* clone. A systematic reverse genetics platform named Trait Utilization System for Corn or TUSC established at the seed company, then Pioneer Hi-Bred International Inc., in 1995 facilitated the use of a maize *Mutator* population by public sector maize researchers for functional genomics studies where mutations in known gene sequences could be used to explore the nature of the phenotype (Meely and Briggs 1995). Advances in molecular biology technologies enabled the description in the year 2000 of the reverse genetics tool of Targeting induced local lesions IN genomes (TILLING) for functional genomics in an Arabidopsis mutant population resulting from chemical mutagenesis using ethyl methanesulfonate (McCallum et al. 2000). It used denaturing high-performance liquid chromatography to determine base pair changes by heteroduplex analysis.

Increasingly cost- and time-efficient sequencing technologies, bioinformatics, and Artificial Intelligence now facilitate faster and more precise identification of molecular variants underlying mutations, and their deployment for marker-assisted breeding and gene editing. Tested and proven analytical techniques of the present decade such as MutMap, MutMap+, MutMap Gap, and MutRenSeq allows efficient and precise retrieval of variant information from mutants. These, along with genome-wide association studies, next-generation sequencing with bulk segregant analyses, and crop pan-genomes all assume increasing significance when applied to mutant germplasm already proven to be effective for traits of interest to the farmer. While the MutMap approach, first described in 2012, facilitated the identification of a mutation in *OsRR22*, a response regulator gene, as conferring the *hst1* salinity tolerance in rice in 2015, MutMap+ was used to report the identification of a rice starch branching enzyme responsible for cooked rice texture in 2018. In 2016, MutRenSeq was first reported for combining mutagenesis with targeted cloning of disease resistance genes. Most recently in Nov 2020, the barley pan-genome revealed the hidden legacy of mutation breeding through tracing a single mutational event that removed the entire *NUD* gene sequence as the basis of all naked barley varieties cultivated to date.

11.5 Future Outlook

Innovative, cost- and time-efficient technologies of the genomic era offers scope for significant process efficiencies in crop improvement through plant mutation breeding. The utilization of techniques including molecular marker-assisted selection, doubled haploidy, rapid cycling, and shuttle breeding can accelerate the breeding process to reduce the time required for the development of new varieties, potentially, by half. Mutation induction and associated technologies can also constitute functional genomics platform(s) to identify causative genes for crop traits that can in turn be used in non-traditional breeding approaches. The use of hormesis levels of irradiation is also emerging as an increasingly sought technology to enhance seed germination in forest trees, which normally have long periods of dormancy, to strengthen afforestation and reforestation efforts.

In the near-term future, it is of importance to retain and strengthen the focus on a few critical areas. Foremost is the emphasis on crop adaptation to climate change using nuclear techniques to develop new varieties with tolerance to drought and heat, including the use of environmental classification, and process efficiencies in multi-environmental testing. Combating the increasing intensities and frequencies of transboundary plant diseases, and pests (including parasitic weeds) through the development of novel genetic diversity, screening methodologies and pathogen race identification is another important area. Acceleration of breeding cycles through the development of speed-breeding technologies including doubled haploidy, shuttle breeding, rapid cycling, and molecular marker-assisted selection for target crops should remain a continuing priority for enhancing the rate of genetic gain. Yet another area of continuing emphasis is widening the scope of plant mutation breeding through its adaptation to vegetatively propagated crops and to normally recalcitrant seeds of perennials and tree crops. Finally, enabling farmers' access to improved crop varieties by fostering seed system models in pilot tests or field implementation, and upscaling, along with wider deployment of tools, technologies, and knowledge is critical for impact on food security.

References

Alonso JM, Anna NS, Thomas JL et al (2003) Genome-wide insertional mutagenesis of *Arabidopsis thaliana*. Science 301:653–657

Bensen RJ, Johal GS, Crane VC, Tossberg JT, Schnable PS, Meeley RB, Briggs SP (1995) Cloning and characterization of the maize *An1* gene. Plant Cell 7:75–84

Hayashi H, Czaja I, Lubenow H, Schell J, Walden R (1992) Activation of a plant gene by T-DNA tagging: auxin-independent growth in vitro. Science 258:1350–1353

Hedden P (2003) The genes of the Green Revolution. Trends Genet 19(1):5–9

Koncz C, Mayerhofer R, Kocnz-Kalman ZS, Körber H, Redei GP, Schell J (1989) High-frequency T-DNA-mediated gene tagging in plants. Proc Natl Acad Sci U S A 86:8467–8471

Marx GA (1983) Developmental mutants of some annual seed plants. Annu Rev Plant Physiol 34: 389–417

McCallum CM, Comai L, Greene EA, Henikoff S (2000) Targeting induced local lesions IN genomes (TILLING) for plant functional genomics. Plant Physiol 123(2):439–442

Meely RB, Briggs SP (1995) Reverse genetics for maize. Maize Genet Coop Newsl 69:67

Muller HJ (1927) Artificial transmutation of the gene. Science 66(1699):84–87

Somerville CR (1986) Analysis of photosynthesis with mutants of higher plants and algae. Annu Rev Plant Physiol 37:467–507

Somerville S, Somerville C (1996) Arabidopsis at 7: still growing like a weed. Plant Cell 8(11): 1917–1933

Stadler LJ (1928) Mutations in barley induced by X-rays and radium. Science 68:186–187

Suprasanna P, Jain SM, Ochatt SJ, Kulkarni VM, Predieri S (2012) Application of *in vitro* techniques in mutation breeding of vegetatively propagated crops. In: Plant mutation breeding and biotechnology. Food and Agriculture Organization of the United Nations, Rome, pp 371–385

Induced Mutations for Development of New Cultivars and Molecular Analysis of Genes in Japan

12

Hitoshi Nakagawa

Abstract

Mutation breeding was accelerated in Japan following the construction of the Gamma Field in 1960. There are 332 direct-use mutant cultivars generated through irradiation by gamma rays, X-rays, ion beams, and other methods of generating induced mutations. Approximately 79% of these mutations are radiation induced. There are 450 indirect-use mutant cultivars including 405 mutant cultivars of rice, of which nearly 50% were descendants of cv. Reimei. The economic impact of mutant cultivars developed by induced mutations is very huge. Some useful mutations are discussed for protein content, amylose content, giant embryo and non-shattering characteristics in rice, and radiosensitivity, fatty acid composition, and super-nodulation traits in soybean. Beneficial mutations have been uncovered in Japanese pear, chrysanthemum, etc. through various unique induction and screening techniques. The identification of the characteristics and deletion size generated by gamma rays and ion beams revealed that the deletion patterns are mostly similar. Furthermore, genetic studies using induced mutations for phytochrome, aluminum tolerance, and epicuticular wax and a Mendel's gene have also been conducted. Mutation breeding is of great interest and serves as a useful technology for isolation of genes and for elucidation of the molecular mechanisms underlying gene functions and metabolic pathways, as well as is a very useful aid to conventional breeding for developing superior cultivars. The advantages of gamma rays and other methods for mutation

H. Nakagawa (✉)
Hamamatsu Photonics K.K., Central Research Laboratory, Hamamatsu, Shizuoka, Japan

Institute of Radiation Breeding, National Institute of Agrobiological Sciences, Tsukuba, Ibaraki, Japan
e-mail: hitoshi.nakagawa@crl.hpk.co.jp

S. Penna, S. M. Jain (eds.), *Mutation Breeding for Sustainable Food Production and Climate Resilience*, https://doi.org/10.1007/978-981-16-9720-3_12

induction including the more precise genome editing technology are also discussed.

Keywords

Gamma Field Symposia · Gamma ray · Ion beams · Mendel's gene · Mutation breeding · Radiosensitivity · Reimei · Rice · Soybean · Yasuo Ukai

12.1 Introduction

In Japan, the pioneering works on induced mutation in rice were conducted in the second quarter of the 1930s by Imai (1934, 1935) using traditional X-ray irradiation. In the RIKEN (Institute of Physical and Chemical Research; https://www.riken.jp/en/), many mutations were induced in broad bean (Nakaidzumi and Murai 1938) and other crops using 2.8 MeV deuterons from their own cyclotron. This cyclotron, being the second in the world following the first cyclotron in Prof. E. O. Lawrence's laboratory at the University of California, Berkeley, was developed with the support of Prof. Laurence's laboratory (Kim 2006). However, this cyclotron was destroyed and submerged into the Tokyo Bay, along with four other cyclotrons at the RIKEN, Kyoto University, and Osaka University, by the order of the General Head Quarters (GHQ: Supreme Commander for the Allied Powers) in 1945, just after the end of the WWII, despite the request of many Japanese scientists to keep these cyclotrons for the purpose of advancing biological and medical research towards promoting the health of the Japanese people in the postwar era (Nishina 1947). After 1950, when isotopes could be used freely without permission of the GHQ, radiation breeding especially using gamma ray was extensively attempted (Watanabe 1981). With the release of atomic energy for peaceful uses, many scientists were interested in inducing mutations in plants and utility of induced mutants both for genetic improvement of crop plants and for genetic analysis of the mutation in Japan.

The first Gamma Field, which was the largest facility in the world at that time, was constructed at Brookhaven National Laboratory in 1949 (Yamakawa and Sparrow 1965). However, it was not used primarily for the goal of mutation breeding in crops, but rather for identifying the effects of radiation on the natural vegetations and farmers' fields when a nuclear war was envisioned during the Cold War period. In Japan, the Gamma Field (Fig. 12.1) was constructed in the Institute of Radiation Breeding (IRB) (http://www.naro.affrc.go.jp/english/laboratory/nics/research/division_of_radiation_breeding/index.html) (now affiliated to the National Agriculture and Food Research Organization (NARO)), Ohmiya (which is now Hitachiōmiya city), Ibaraki, in 1961. This Gamma Field remains the largest gamma-ray irradiation field in the world, primarily for inducing mutations in crop breeding programs and molecular studies in plants; however, the operation of the Gamma Field has been suspended since May 30, 2019. The Gamma Field has a 100 m radius with an irradiation tower installed with an 88.8 TBq of ^{60}Co source at the center (Fig. 12.1). The field is surrounded by an 8 m high-shielding dike (Fig. 12.2). Prior to the

Fig. 12.1 Gamma Field of IRB (Courtesy of IRB, NARO). Yellow line: Restricted area

Fig. 12.2 Irradiation Tower of the Gamma Field. (Courtesy of IRB, NARO)

commencement of the Gamma Field operations, its structural characteristics and gamma-ray exposure rate inside the facility were measured in detail (Kawahara 1967). Following the Gamma Field, the Gamma Greenhouse (a regular octagonal greenhouse with a 7 m radius installed with a 4.81 TBq ^{137}Cs source: now

Fig. 12.3 A shielded building of the Gamma Room inside the restricted area. (The photo taken by the author with the permission of IRB, NARO)

abandoned) and the Gamma Room (a shielded building installed with a 44.4 TBq ^{60}Co source: Fig. 12.3) were constructed inside the restricted area of the Gamma Field in 1964 and 1967, respectively, thereby accelerating mutation breeding through gamma-ray irradiation in Japan (Yamaguchi 2001).

In the New York Times (Broad 2007), late Dr. Pierre J.L. Lagoda of the Joint FAO/IAEA, Vienna, Austria, was quoted with this saying about the mutation breeding using radiation: "I'm doing the same thing. I'm not doing anything different from what nature does. I'm not using anything that was not in the genetic material itself ... Spontaneous mutations are the motor of evolution. We are mimicking nature in this. We're concentrating time and space for the breeder so he can do the job in his lifetime. We concentrate how often mutants appear—going through 10,000 to one million—to select just the right one ... It's not a panacea. It's not the solution. But it's a very efficient tool that helps us reduce the breeding time."

The concept of the Gamma Field is completely same as the idea of Dr. Lagoda. The facility just serves to artificially induce mutations at a higher frequency than natural mutation. The dosage of radiation at the nearest point of the field (10 m from the center) is ca. 2 Gy/day. This irradiation dosage is estimated to be about 300,000 times that of natural background radiation rate. Then, the plants placed at the nearest point are, in other words, treated to a 1000 years of total accumulated normal background rates of radiation every day. The lowest dosage is ca. 0.01 Gy/day, that is, ca. 2000 times that of background radiation at the farthest point (100 m from the center). Therefore, optimum places within the Gamma Field (that is, optimum dose (Gy/day) of chronic gamma-ray irradiation) suitable for various fruit and tea trees and annual and perennial crops have been identified for effective induction of mutations (Shu et al. 2012). One can say that evolution of crops in the Gamma Field can be highly accelerated, although the resulting effects might not be completely similar to the real 1000 years in nature because crop plants may have at most 1000 generations in 1000 years.

Without prior knowledge of genes or mechanisms of mutations, radiation breeding could facilitate the development of many useful mutant cultivars and materials to meet the needs of both farmers and researchers. In 2005, the genome sequence of the entire 12 chromosomes of rice was identified (International Rice Genome Sequence Project 2005), and genetic studies based on the sequence information became one of the most powerful tools for advancing rice molecular genetic research and breeding.

In this report, many mutant cultivars developed in Japan and their socioeconomic impacts, as well as molecular research of some induced mutations, which contribute to the food production and to the elucidation of the genomic mechanisms underlying the mutations, are discussed.

12.2 Mutation Breeding and Released Cultivars in Japan

A search (on May 10, 2022) of the number of developed mutant varieties in the IAEA database (http://mvgs.iaea.org/AboutMutantVarities.aspx) ranked China as the topmost country with 817 mutant varieties released (Table 12.1). Japan is placed in the second rank with 500 varieties released, followed by India in the third place that accounted for 345 mutant varieties among a total of 3402 varieties documented worldwide (Table 12.1). Furthermore, other Asian countries, including Bangladesh, Vietnam, Indonesia, and Korea, are among the top 15 countries documented in the database for the release of mutant varieties. This scenario demonstrates the popularity of mutation breeding as a highly utilized technique in Asian countries for the release of important crop varieties. However, it is important to realize that the total number of real mutant varieties might far exceed these documented totals, because

Table 12.1 The number of mutant varieties of top 15 countries in the Joint FAO/IAEA Mutant Variety Database (10 May 2022)

Ranking	Country	Number
1	China	817
2	Japan	500
3	India	345
4	Russian Fed.	216
5	Netherlands	176
6	Germany	171
7	USA	139
8	Bangladesh	78
9	Bulgaria	76
10	Pakistan	59
11	Vietnam	58
12	Indonesia	52
13	Korea	40
14	Canada	40
15	France	39

all the mutant varieties might not yet have been registered to this database by the breeders who developed the mutant varieties.

It is worthy of note that the Gamma Green House in Malaysia and the Gamma Phytotron in Korea were newly established in 2008 and 2005, respectively (Nakagawa and Kato 2017).

12.2.1 The Number of Cultivars Developed by Mutation Breeding

Figure 12.4 indicates the total number of direct-use and indirect-use mutant cultivars officially registered in each 5-year period from 1960 to 2020 in Japan. The number of direct-use cultivars in each period had been increasing until 1996–2000 (65 cultivars registered in 5 years). The number, however, has become smaller, with 50 cultivars registered in 2001–2005, 61 in 2006–2010, and 34 in 2011–2015. The total number of direct-use mutant cultivars came to 332 in June 2020.

The number of indirect-use cultivars, which are descended from the direct-use cultivars through hybridization and comprised mostly rice, has been increasing with 103 cultivars registered in 2011–2015 period. The total number of these indirect-use mutant cultivars was 450 in June 2020. This number of indirect-use cultivars could be increased if useful direct-use mutant cultivars, such as cv. Reimei, that will be discussed in this report, are developed by breeders, because breeders utilize these elite cultivars as the maternal line to develop new cultivars. Developing a true number for indirect-use mutant cultivars is difficult because there is disagreement regarding how many generations of ancestors should be traced back to the original mutant cultivar. In this report, however, the author utilized not more than five generations in a cultivar family history. No attempt was made to identify whether the original mutant genes are existing in the mutant cultivars or not.

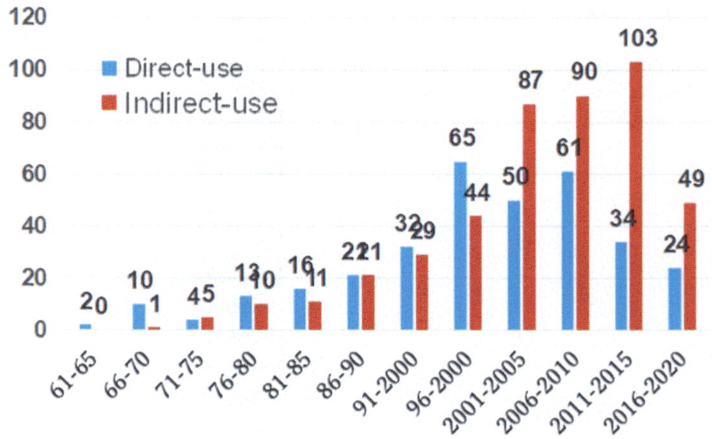

Fig. 12.4 Number of cultivars developed by mutation breeding in each 5-year period from 1961 to 2020. Total number of direct-use cultivars is 332 while that of indirect-use cultivars is 450

Table 12.2 The number of registered direct-use mutant cultivars of 79 crops developed by radiation, gamma rays, and those irradiated in the Institute of Radiation Breeding (IRB)

	Cultivar[a]	Radiation	Gamma rays	IRB[b]
Crops	332	258	186	136
Rice	46	23	22	21
Wheat	4	2	2	0
Barley	5	4	3	0
Soybean	19	18	17	9
Barnyard millet	5	5	4	4
Job's tears	2	2	2	2
Buckwheat	3	3	3	3
Tartary buckwheat	3	2	1	1
Sugarcane	1	1	1	1
Burdock	5	5	5	4
Apple	2	2	2	2
Japanese pear	3	3	3	3
Peach	2	2	2	2
Loquat	1	1	1	1
Enoki mushroom	2	2	2	2
Chrysanthemum	67	62	40	39
Rose	10	7	7	6
Carnation	15	11	3	2
Cytisus	8	8	8	8
Russell prairie gentian	2	2	2	2
Zoysia grass	6	5	3	3
Others	121	88	53	21

[a] Total number of mutant cultivars developed by radiation, chemicals (excluding colchicine treatment), and somaclonal variation
[b] Cultivars irradiated in the Institute of Radiation Breeding (IRB), Japan

In Japan, a total 332 direct-use mutant cultivars of 79 species have been registered (Table 12.2; Fig. 12.5). Figure 12.5 exhibits that 79% of these mutant cultivars were radiation induced, including gamma ray (56.0%), ion beams (13.3%), X-ray (8.4%), and other radiation (ultraviolet (UV) radiation with no precise irradiation: 1.5%). The mutant cultivars induced via somaclonal variation and chemical mutagen (colchicine treatment was not included) are 13.9% and 6.9%, respectively.

Table 12.2 also provides the number of officially registered direct-use mutant cultivars of some crops developed by radiation. Those mutant cultivars induced through gamma-ray irradiation at the IRB are also counted separately in this table. Presently, 46 rice (*Oryza sativa* L.) mutant cultivars, 19 soybean (*Glycine max* (L.) Merrill), 4 wheat (*Triticum aestivum* L.), 5 barley (*Hordeum vulgare* L.), etc. are officially registered in Japan. For ornamental plants, 67 chrysanthemum (*Chrysanthemum* spp.) mutant cultivars, 10 rose (*Rosa* spp.), etc. were registered, too. It is noteworthy that 136 cultivars (ca. 41%) have been induced in the facilities of IRB.

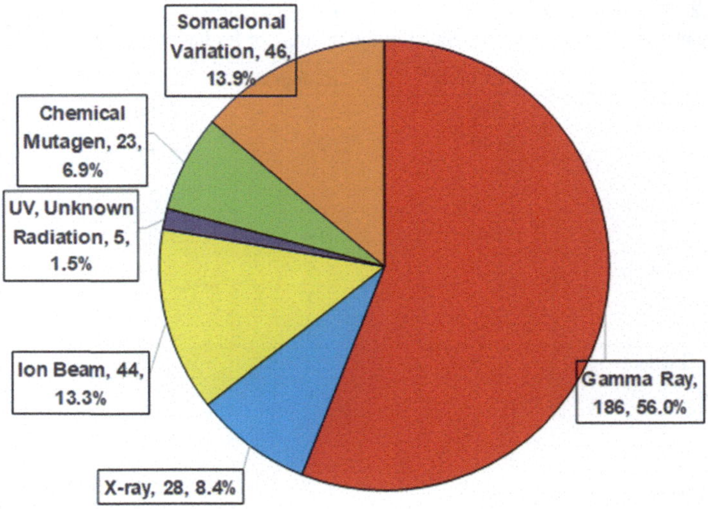

Fig. 12.5 Numbers and percentages of total 332 direct-use cultivars developed by mutation breeding using various methods in Japan (2020). Radiation to callus is not included in somaclonal variation, but in radiation, and chemical mutagen does not include colchicine treatment

These data clearly indicated the contributions of the IRB and its extensive history of use of the facilities in Japan and demonstrated that mutation breeding via radiation, such as gamma-ray and ion beam irradiation, offers the most effective and extraordinary tool for breeders to successfully develop commercially useful cultivars.

The total number of indirect-use mutant cultivars stood at 450 in 2020, which included rice (405), soybean (16), barley (9), wheat (7), tomato (3), lettuce (4), eggplant (1), Japanese lawn grass (*Zoysia japonica* Steud) (1), mat rush (*Juncus effusus* L. var. *decipiens* Buchen.) (2), and Job's tears (*Coix lacryma-jobi* L. var. *ma-yuen* Stapf) (1) (Table 12.3). It is notable again that unique cultivars with useful mutations would be utilized by breeders as breeding materials of new cultivars, which have these unique mutant characteristics or genes, and the mutated gene is propagated into new indirect-use cultivars, again and again.

12.2.2 The Economic Impact of Mutant Cultivars in Japan

Figure 12.6 illustrates the number of mutant cultivars including both direct-use and indirect-use mutant rice cultivars in the farmers' field in Japan from 1960 to 2015 (Nakagawa 2008; Nakagawa and Kato 2017). All these mutant cultivars were derived primarily from gamma-ray irradiations. The number has been increasing since 1960, when the cv. Reimei was released, with a total of 99 mutant cultivars including 2 direct-use and 97 indirect-use cultivars cultivated in 2005. However, it became difficult to obtain the same kind of data after 2006, because the data of cultivated area of all cultivars has not been published from the Ministry of

Table 12.3 The number of indirect-use mutant cultivars in Japan (2020)

Rice	Wheat	Barley	Soybean	Rush	Tomato	Eggplant	Lettuce	Coix	Zoysia	Total
405	7	9	16	2	3	1	4	1	2	450

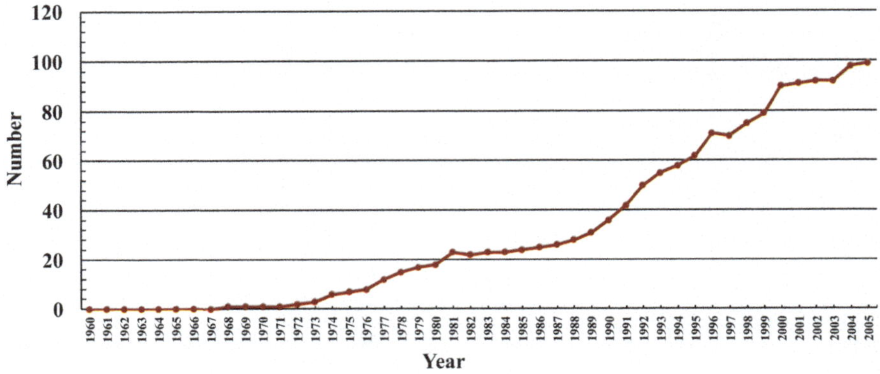

Fig. 12.6 Total number of direct-use and indirect-use mutant rice cultivars mostly developed through gamma-ray irradiation, cultivated during 1960–2005 in Japan. (Modified from Nakagawa 2008)

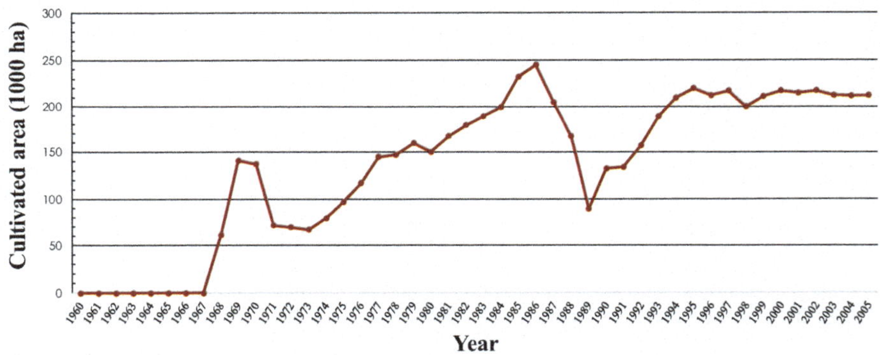

Fig. 12.7 Total cultivated areas of direct-use and indirect-use mutant rice cultivars mostly developed through gamma-ray irradiation and cultivated in Japan during 1960–2005. (Modified from Nakagawa 2008)

Agriculture, Forestry, and Fisheries (MAFF). The data of Fig. 12.7 was also missing after 2006, for the same reason.

Figure 12.7 shows the increase of total area of cultivation of the abovementioned mutant cultivars during 1960 and 2005. That is, the cultivated area was increased after the release of cv. Reimei for the farmers in 1968, that is, a kind of Green Revolution in Japan, as discussed below. The peak of total area reached 250,000 ha in 1986 and remained constant at around 200,000 ha from 1994 to 2005. The mutant cultivars covered a total area of 210,692 ha, that is, 12.4% of the total rice-cultivated area (1,702,000 ha) in 2005 (Nakagawa 2008; Nakagawa and Kato 2017).

It is remarkable that the top 20 cultivated paddy rice cultivars in Japan in 2019 (Rice Stable Supply Support Organization (Beikoku Kikou) 2020: https://www.komenet.jp/pdf/R01sakutuke.pdf) listed 13 indirect-use mutation cultivars, and the total cultivated area covered 261,482 ha of Japanese paddy field (17.8% of total

Table 12.4 Cultivated areas of the top-ranked mutant rice cultivars in 2019 (Rice Stable Supply Support Organization 2020)

Cultivar (Ranking[a])	Area (ha)[b]	%	Origin[c]
1. Haenuki (6)	41,132	2.8	R
2. Masshigura (7)	32,318	2.2	R
3. Kinuhikari (8)	30,849	2.1	H
4. Asahi-no-yume (9)	24,973	1.7	M
5. Kinumusume (11)	22,035	1.5	H
6. Koshi-ibuki (12)	20,566	1.4	R
7. Tsuya-hime (13)	17,628	1.2	H
8. Yume-tsukushi (14)	14,690	1.0	H
9. Fusakogane (15)	13,221	0.9	M
10. Tsugaru-roman (16)	11,752	0.8	R
11. Aichi-no-kaori (17)	11,752	0.8	M
12. Sai-no-kagayaki (18)	10,283	0.7	R, M
13. Ten-no-tsubu (19)	10,283	0.7	R, H
Total	261,482		
All paddy field	1,469,000		
%	17.8		

Cultivation areas are estimated by using data from Kome-net (https://www.komenet.jp/data/jishuchousa/hinshitu/)

[a] Ranking based on top 20 cultivars with the highest cultivated areas

[b] The areas are calculated from the %

[c] R: cv. Reimei; H: Hokuriku No. 100 or cv. Kinuhikari; M: cv. Mineasahi

1,469,000 ha of paddy field) (Table 12.4). These indirect-use mutant cultivar linages are primarily derived from direct-use mutant cultivars: (1) cv. Reimei (Futsuhara 1968), (2) Hokuriku No. 100 (a parental line of cv. Kinuhikari (Koga et al. 1989)), and (3) cv. Mineasahi (Morimoto et al. 1980) (Table 12.4). Furthermore, 193 cultivars (47.8%) among the total 404 indirect-use mutant rice cultivars in 2020 were descendants of the cv. Reimei. The cv. Haenuki (Kyoya et al. 2002) and cv. Masshigura (Mikami et al. 2007), which are descendants of cv. Reimei (both the fourth generation), are ranked sixth and seventh among all rice cultivars and ranked first and second most popular mutant cultivars, respectively. The cv. Kinuhikari, that is developed from a hybridization between a gamma ray-induced semidwarf lodging-tolerant mutant line "Hokuriku 100" (that was induced from the cv. Koshihikari) (Samoto and Kanai 1975) and a semidwarf cv. IR8 (Koga et al. 1989), was ranked as the third most widely cultivated mutant cultivar with shorter stems. The cv. Kinuhikari and its descendants covered an area of more than 95,103 ha (5.8%) of all the paddy fields in 2019 (Beikoku Kikou 2020).

The cultivated areas of total mutant cultivars are more than these data because the data include only the top 20 cultivars. We can say that the economic impacts of mutant cultivars of rice are huge. Especially, the roles of cv. Reimei, cv. Kinuhikari, and cv. Mineasahi are highly important. The amount of total income of rice farmers who cultivated mutant rice cultivars was ca. 250 billion Yen (2.34 billion USD) in 2005 (Nakagawa 2008).

Table 12.5 Cultivated areas (ha) of soybean mutant cultivars and income of farmers from the production in 1997, 2001, 2005, 2011, 2015, and 2017 in Japan

Cultivar	1997	2001	2005	2011	2015	2017
Raiden[a]	80	8				
Wase-suzunari[a]	120					
Mura-yutaka[a]	3507	5910	2466	1403	1021	1039
Kosuzu[a]	498	863	576	194	111	39
Ichihime[a]		35	130			
Akita-midori[a]		8	87		40	
Nanbu-shirome[b]	1246	1550	1534	2132	1423	610
Tomoyutaka[b]	2					
Suzunone[b]	10	50				
Eru-star[b]			447			
Suzu-sayaka[b]			10	65	73	
Suzu-kaori[b]				56	45	
Ryuhou[b]	1150	7050	8033	10,548	10,295	9616
Tsuya-homare[b]				1	1	
Kinusayaka[b]					100	96
Kyo-shirotanba[b]					4	
Total	6613	15,474	13,283	14,398	13,112	11,400
Total soybean area (× 1000)	83.2	143.9	134.0	136.7	136.7	121.3
%	7.95	10.75	9.91	8.79	10.53	9.40
Income of farmers (Mil. USD)	ca. 20	ca. 59	ca. 52	–	22.6 (115.7 with subsidy)	–

[a] Direct-use mutant cultivars
[b] Indirect-use mutant cultivars

Nineteen direct-use soybean mutant cultivars have been registered in Japan (Table 12.2). Firstly, cv. Raiden and cv. Raikou with induced early-maturing characteristics were developed through gamma-ray irradiation in the IRB by Tohoku Agricultural Experiment Station (now Tohoku Agricultural Research Center, NARO) in 1960 (Laboratory of Soybean Breeding 1970). The other improved characteristics through gamma-ray irradiation were late maturity, white hilum, semidwarf, etc., which are discussed in this report. The cv. Mura-yutaka with white hilum mutation was induced with X-ray in Saga University (Nakamura et al. 1991). Recently, a soybean cv. Akita-honoka was developed for edamame (a boiled green soybean), a traditional food source with unique utilization in Japan, via somaclonal variation and officially registered (Sato et al. 2015). The other 17 mutant cultivars were also induced by gamma rays. Of the 11 indirect-use mutant soybean cultivars (Table 12.3), cv. Ryuhou (Nakamura et al. 1996) and the other 3 cultivars are descendants of cv. Raiden with an induced mutation for early maturity. Table 12.5 shows the change of cultivated areas and farmers' income from the cultivation of soybean mutant cultivars in 1997, 2001, 2005, 2011, 2015, and 2017 (http://www.maff.go.jp/j/seisan/ryutu/daizu/d_data/attach/pdf/index-19.pdf;

http://www.jsapa.or.jp/daizu/date28/28-all.pdf). The cultivated area of direct-use mutant cv. Mura-yutaka was 5910 ha in 2001 and 1039 ha in 2017. The cultivated area of indirect-use mutant, cv. Ryuhou, ranked fourth of all the soybean cultivars and covered 9616 ha. The cv. Nanbu-shirome, which was developed from the hybridization between cv. Raiden and cv. Kitami-nagaha (Matsumoto et al. 1979), covered an area of 2132 ha in 2011 but declined to 610 ha in 2017. As for rice, the soybean mutant cultivars also hugely impacted the economy. The total cultivated area of soybean mutant cultivars covered 11,400 ha, which means that 9.40% of the total cultivated area (121,300 ha) was planted with mutant cultivars in the farmers' soybean field in 2017 (Nakagawa and Kato 2017). Therefore, total farmers' crude income for the cultivation of mutant cultivars was estimated to be 11.6 billion Yen (ca. 116 million USD) in 2015 (Nakagawa and Kato 2017).

The economic impact of mutation induction, as well as the other peaceful uses of radiation, such as nuclear power plants, industrial usage (semiconductor and radial tire production, etc.), and medical application (radiation therapy, diagnosis, etc.), was discussed by the Ministry of Education, Culture, Sports, Science and Technology (MEXT) with the support of Japan Atomic Energy Research Institute (JAERI) (Japan Atomic Energy Agency 2007; Tagawa et al. 2002; Kume et al. 2002). In the summary of these two reports regarding the peaceful uses of radiation, out of a total 8850 billion Yen of the economic size (market size), nuclear power generation represented 54% and radiation use was 46%. Of the total radiation use of 4111 billion Yen, industrial use was 56%, medical use was 37%, and agricultural use, that is, mutation breeding and food irradiation (that is, gamma-ray irradiation only applied to potato to stop germination is allowed in Japan), was 7%. The economic impact of agricultural uses, estimated at 278 billion Yen (mainly from mutant rice cultivars), is not relatively large, but it has a unique point different from the other radiation impact. Radiation uses for industrial and medical uses end only with the irradiation. However, mutant cultivars or genes induced through radiation can be replicating and increasing in farmers' fields; that is, economic impacts of these mutated genes are growing year by year.

12.3 Gamma Field Symposium and Selected Reports of Gamma Field Symposia

The first Gamma Field Symposium was held on the theme "Radiation injury and somatic mutation" in 1962 at the Conference Room of the IRB, where 56 researchers from national agricultural experiment stations, national universities, and seed companies gathered for the exchange of information and discussions in the new research area of mutation breeding. The last 53rd symposium having the theme "Genes and mutations that fight against environment" was held in 2014. The author of this report was the final speaker at the symposium and presented a review titled "A history of radiation breeding for 50 years." Parts of this presentation were published in a new Bulletin of NARO, Crop Science (Nakagawa and Kato 2017). Following the symposium, Gamma Field Symposia as a proceeding of each symposium (from

No. 1 to 50) were published (http://www.naro.affrc.go.jp/archive/nias/newsletter/#symposia or http://www.naro.affrc.go.jp/archive/nias/eng/public/index2.html).

The 50 volumes of the proceedings include Japanese summary and/or some questions or comments from the audience and answers from the lecturers in Japanese.

12.3.1 Some Recommended Reports of Gamma Field Symposia

One of the greatest pioneers of mutation breeding in Japan is Dr. Yasuo Ukai, Emeritus Professor of the University of Tokyo. This leader in mutation breeding passed away on 20 November 2019. As a statistical geneticist, he produced many informative textbooks on plant breeding and molecular mapping and analysis, such as "Genetic analysis at the genomic level: Map and QTL" (Ukai 2000) and "Genetic analysis of quantitative traits" (Ukai 2002). Prof. Yasuo Ukai was also a pioneer at the dawn of the molecular breeding era in Japan when few researchers attempted to apply molecular technologies to plant breeding. His textbooks also contributed to understanding the outcomes of mutation breeding. As a young man, he was a scientist at the newly established IRB and contributed immensely to the mutation research in the Gamma Field, as well as the management of Gamma Field Symposium.

In 2008, he gave the author a note on his personal 20 selected and recommended reports (including his reports) in the Gamma Field Symposia. The contents, authors, and volume numbers are shown in Table 12.6. From this list, some reports are briefly explained from the author's point of view.

12.3.1.1 Development of cv. Reimei Rice Through a Gamma-Ray Irradiation

The first mutant rice cultivar is cv. Reimei, as the author already mentioned in this report. "Reimei" means "dawn" in Japanese. This name represents the first radiation-induced mutant cultivar that ushered in the potential of use of gamma rays for breeding programs in Japan. The primary useful character induced by the mutation was reduced plant height, that is, semidwarf. The semidwarf traits are one of the most frequently induced through gamma-ray irradiation and are easily identified at first glance in the irradiated population in the field. The cv. Reimei was successfully registered in 1966 (Futsuhara 1968). The cv. Reimei exhibited a shorter culm by 15 cm relative to the original cv. Fujiminori, and it was observed that cv. Reimei possessed a mutation of the *sd-1* locus (Ashikari et al. 2002), similar to the mutation of cv. IR8, through cv. Dee-geo-woo-gen. The cv. IR8, developed in the International Rice Research Institute (IRRI), became a very famous cultivar that contributed to the rice "Green Revolution" (Dalrymple 1986). The semidwarf trait is a primary key character of the high-yielding ability in cv. Reimei, resulting in the highest yield recorded in Japan for the cultivar in 1967 (Futsuhara 1968). The cv. Reimei was cultivated on 1201 ha in 1966, but was extended to 141,047 ha in 1969, making it the fifth largest cultivated area in the year (http://ineweb.narcc.affrc.go.jp/). As the total

Table 12.6 Lists of 20 recommended reports of Gamma Field Symposia

Contents	Authors, reference	Vol.
1. Dose distribution of the Gamma Field	Kawahara (1967)	6
2. Radiosensitivity of soybean	Takagi (1969)	8
3. History of mutation breeding of seed crops	Yamashita (1981)	20
4. Characteristics of mutation in barley	Ukai (1983)	22
5. Screening of resistant stems by using fungal toxin in fruit trees	Sanada (1986)	25
6. Induced mutations for powdery mildew and YMV resistance in barley	Yamaguchi et al. (1988)	27
7. Recurrent mutation breeding for outcrossing crops (Italian ryegrass)	Ukai (1990)	29
8. Analysis of seed storage protein mutants in rice	Iida (1995)	34
9. Mutation breeding of chrysanthemum (tissue culture and radiation)	Nagatomi et al. (1996)	35
10. Induction of resistance for Alternaria blotch in apple	Yoshioka et al. (2000)	39
11. Development of low-protein rice for kidney disease patients	Nishimura (2006)	45
12. Molecular analysis of gamma-ray-induced mutation in rice	Morita (2008)	47
13. Development of Reimei rice with gamma-ray irradiation	Futsuhara (1968)	7
14. Translocation induction in interspecific hybrid to transfer resistant gene	Mukade et al. (1970)	9
15. Acute gamma-ray irradiation to mulberry tree	Katagiri (1973)	12
16. Induction of cytoplasmic male sterility in sugar beet	Kinoshita (1980)	19
17. Somaclonal mutations in tissue culture in rice	Oono et al. (1984)	23
18. Gene regulation at the waxy locus in rice	Sano (1985)	24
19. Mutation induction by ion beams in Arabidopsis	Tanaka (1999)	38
20. A mutable slender glume gene in rice	Okumoto et al. (1999)	38

cultivated areas came to 886,188 ha between 1961 and 2017, cv. Reimei became the most successful rice cultivar in Japan.

It is notable that 193 cultivars (47.7%) were descendants of the cv. Reimei among the 405 indirect-use mutant rice cultivars in 2020. This demonstrates that this mutant cultivar possessed unique and agronomically useful semidwarf mutations that have been utilized by many rice breeders to develop new cultivars with this character, as discussed above.

12.3.1.2 Radiosensitivity of Soybean

The theme of the eighth Gamma Field Symposium was "Genetical control of radiosensitivity" and radiosensitivity of *E. coli* (Kato 1969), silkworm (*Bombyx mori*; Tazima and Murakami 1969), Sugi tree (*Cryptomeria japonica* S. Don; Ohba and Murai 1969), and yeast (*Saccharomyces cerevisiae*; Nakai 1969); genetic analysis of radiosensitivity in fruit fly (*Drosophila melanogaster*; Ogaki et al. 1969), varietal difference in some crops (Ukai and Yamashita 1969), gamma ray-sensitive gene in soybean (Takagi 1969), and metabolism in radiosensitive and resistant soybean cultivars (Naito and Yamaguchi 1969) were presented and discussed.

A detailed study was conducted to identify the radiosensitivity of various plant species in the Gamma Field at Brookhaven National Laboratories. Sparrow and Evans (1961) reported that white trillium (*Trillium grandiflorum* (Michx.) Salisb.; $2n = 10$) exhibited 260 times higher sensitivity than the least radiosensitive Arabidopsis (*Arabidopsis thaliana* (L.) Heynh.; $2n = 10$). Among the food crops, broad bean (*Vicia faba* L.), rye (*Secale cereal* L.), and wheat (*Triticum aestivum* L.) exhibited more than ten times the sensitivity to gamma-ray irradiation than tea (*Camellia sinensis* (L.) Kuntze), rice (*Oryza sativa* L.), Italian millet (*Setaria italica* P. Beauv.), and okra (*Abelmoschus esculentus* (L.) Moench).

Sparrow and Evans (1961) demonstrated that there is a wide interspecies diversity caused by nuclear volume, chromosome length, and centromere number. Following this research, Yamakawa and Sparrow (1965) reported that the radiosensitivity of growing plants to chronic gamma-ray irradiation, relative to the plants' interphase chromosome volume, showed an inverse correlation on the logarithmic axis as ($\log_{10}Y = 3.03 - 1.16 \log_{10}X$).

For the intraspecific variation, Takagi (1974) treated acute gamma-ray irradiation in the Gamma Room and/or chronic irradiation in the Gamma Field to more than 600 soybean cultivars and wild accessions to evaluate the diversity of radiosensitivity. The tested accessions were classified into radiosensitive and radioresistant groups by measuring the RD_{50} of the root length. As the results of gamma-ray acute seed irradiation, he observed that the radioresistant cv. Tachisuzunari exhibited an RD_{50} value twice as high as the radiosensitive cv. Shinmejiro. Based on the hybridization of radioresistant cv. Tachisuzunari and radiosensitive cv. Shinmejiro and the F_2 segregation, Takagi (1974) identified a single RS_1 gene, which controls resistance to acute gamma-ray irradiation to the seed, and he further demonstrated that the genotype of radioresistance is RS_1RS_1 and RS_1rs_1 and the genotype of radiosensitivity is rs_1rs_1. Next, Takagi (1969) hybridized radioresistant cv. Tachisuzunari and cv. Goishishirobana, and radiosensitive cv. Shinmejiro with each other, and identified a second single major gene that controls radiosensitivity to a chronic gamma-ray irradiation from the germination to the maturity stage in the Gamma Field. Based on the results of hybridizations, he proved that the F_2 of the cross between cv. Tachisuzunari and cv. Goishishirobana segregated into 3 (resistant to chronic irradiation):1 (susceptible) in accord with Mendelian inheritance. Based on the analysis of the results, Takagi (1969) identified a single RS_2 gene, which controls resistance to chronic gamma-ray irradiation. Therefore, he further proposed that cv. Tachisuzunari's genotype is $RS_1RS_1RS_2RS_2$, which exhibited resistance to both acute and chronic irradiation; cv. Goishishirobana's is $RS_1RS_1rs_2rs_2$, which exhibited resistance to acute irradiation but susceptibility to chronic irradiation; and cv. Shinmejiro's is $rs_1rs_1rs_2rs_2$, which indicated that it is susceptible to both acute and chronic irradiation.

The optimum acute and chronic gamma-ray irradiation dosages to different soybean cultivars, as well as other crops identified in Japan, were listed in detail in the Appendix 2 and 3 of "Plant mutation breeding and biotechnology" (Shu et al.

Table 12.7 Mutagens and dosages treated to induce various mutations in soybean (Nakagawa et al. 2011a)

Mutagen	Treatment	Mutation	References
X-ray	100–200 Gy	Early maturity, germination under low temperature, many stems, large seed, hilum color	Zacharis (1956)
	75–200 Gy	Plant height, maturity, fatty acid content, protein content	Rawling et al. (1958), Williams and Hanway (1961)
Gamma ray	29–111 Gy	Chlorophyll	Takagi (1968)
	29–69 Gy	High protein content	Hiraiwa et al. (1976)
	100–200 Gy	Early maturity, large seed	Gotoh (1964)
	150–300 Gy	Resistance to rust	Tsai et al. (1974)
	200–300 Gy	Morphological character	Constantin et al. (1976)
Neutron	5.6×10^{12} N_{th}/cm^2	Plant height, maturity, seed weight	Rawling et al. (1958)
	2.4×10^{13} N_{th}/cm^2	Fatty acid content, protein content	Williams and Hanway (1961)
	15–25 Gy	Morphological character	Constantin et al. (1976)
EI	0.008–0.02%	Hilum color	Takagi and Hiraiwa (1980)
	0.5%	Protein content, methionine, and threonine content	Kaizuma (1975)
EMS	0.025–0.050% (8 h)	Morphological character	Klobe et al. (1973)
	0.1–0.2%	Plant height, maturity, leaf color and shape	Nguyen (1974)

2012). Some optimum doses of various radiation sources and chemical mutagens are provided in Table 12.7.

12.3.1.3 Gene Regulation at the Waxy Locus in Rice (Low Amylose Content)

Stickiness is one of the most important quality characteristics of cooked rice for the Japanese people. People of Thailand, Myanmar, and Southern China prefer the sticky rice, too. Amylose contents are closely related to the stickiness of cooked rice and ranges from 0 (waxy: glutinous: extremely sticky when cooked) to higher than 20 (non-waxy: not sticky), and *indica*-type rice cultivars exhibit high value. In Japan, glutinous rice with 0% of amylose content is cooked as "okowa" and "mochi" for festivals or a new year's special meal. Non-glutinous rice is typically used in daily cooking and exhibits ca. 17% of amylose content.

The *Waxy* (*Wx*) gene expression is controlled by various genetic elements, and the regulatory changes in the *Wx* gene expression play an important role in the diversity of grain amylose content rather than the changes in the structure of *Wx* gene product (Sano 1985).

This *Wx* locus was mapped to chromosome 6 in rice, and knockout of this *Wx* gene for example through gamma-ray irradiation makes the non-glutinous rice completely glutinous (*waxy*: *wx*) (Iwata and Omura 1971). This *Wx* or *wx* gene was reported to encode granule-bound starch synthesis and a key enzyme in the synthesis of amylose in cereal plants (Nelson and Pan 1995). Several glutinous rice cultivars have been generated by radiation breeding to non-glutinous cultivars through gamma-ray irradiation in the IRB. For example, glutinous cv. Miyuki-mochi was developed from non-glutinous cv. Toyonishiki (Toda 1982), glutinous cv. Fujimi-mochi from non-glutinous cv. Akichikara, and glutinous cv. Odoroki-mochi from non-glutinous cv. Takanari (Imbe et al. 2004a, b).

Another seed starch mutation is called "dull" characteristics, of which amylose content is not zero as the *waxy* (*wx*) rice, but exhibits usually about 10%, which is much lower than non-waxy (*Wx*) ordinary rice in Japan (Okuno et al. 1983). The Japanese consumers primarily prefer the partly sticky rice with 15–17% of amylose content. Okuno et al. (1983) demonstrated that dull mutants induced by ^{32}P (beta-ray) possessed a single recessive gene that is nonallelic to the *wx* alleles by the crossing tests (Okuno et al. 1983). In Japan, the popular dull mutant cultivar is cv. Milky Queen induced through NMU (*N*-methyl-*N*-nitrosourea) treatment to the most popular cv. Koshihikari. The amylose content of cv. Milky Queen is 5–8%, while the original cv. Koshihikari is around 17% (Ise et al. 2001).

12.3.1.4 Analysis of Seed Protein Mutants and Development of Low-Protein Rice

Cooked white Japanese rice has a protein content of ca. 7%, and this protein content is higher for the daily amount of consumption for the patients of kidney disease (Mochizuki 2000). To solve this problem for the patients, Iida et al. (1993) developed a mutant line called "NM67" with a low glutelin content through the ethyleneimine (EI) treatment to cv. Nihon-masari, a cultivar that the Japanese Emperor ceremonially transplants in his special paddy field every year. The low-glutelin cv. LGC-1 was developed after several backcrosses to the original cv. Nihon-masari to remove undesirable characteristics of the "NM67" keeping this low-glutelin gene (Iida et al. 1993).

Interestingly, SDS-polyacrylamide gel electrophoresis (SDS-PAGE) revealed that the seed protein of the cv. LGC-1 had primarily a lower level of easily digestible glutelin through a dominant mutation, although the level of compensatory indigestible prolamin increased. Although the total amount of protein is same, the amount of easily digestible glutelin dramatically decreases in the cooked rice of cv. LGC-1, which makes this cultivar useful as, so to speak, a "low-protein rice," and some clinical tests for patients with kidney disease in the hospital demonstrated that it is useful and effective for the diets of such patients (Mochizuki 2000). One of the characteristics of cv. LGC-1 to be improved is its cooked eating quality. Furthermore, it is known that an additional locus controls the biosynthesis of easily digestible globulin. If the amount of this globulin content is reduced, more useful and efficient diets than cv. LGC-1 will be developed.

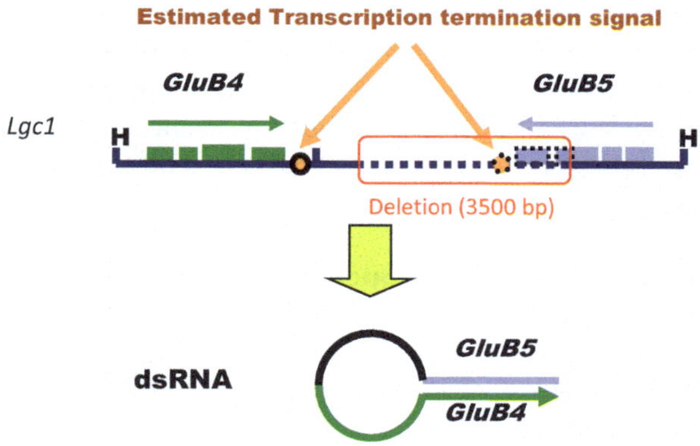

Fig. 12.8 Mechanism of RNAi generation in the low glutelin cv. LGC-1 through a deletion estimated at the transcription termination signal and production of a double-stranded RNA suppressing the glutelin synthesis (Kusaba et al. 2003). (Courtesy of Prof. M. Kusaba)

Nishimura et al. (Nishimura et al. 2005, Nishimura 2006) attempted to induce globulin-free mutants from the cv. Koshihikari with excellent eating quality through gamma-ray irradiations in the IRB. Following the development of the globulin-free Koshihikari line, 89WPKG30-433, they hybridized this mutant line with the cv. LGC-1 and successfully developed two cultivars, cv. LGC-Katsu and cv. LGC-Jun. The glutelin content of these cultivars was similar to the cv. LGC-1, and the globulin content was completely zero. As the good eating quality character from cv. Koshihikari was introduced to these new cultivars, the taste of their cooked rice was highly improved. As a result, total amount of easily digestible protein content of the two cultivars became about 30% of original cv. Nihon-masari and ordinary rice (Nishimura et al. 2005). Because the eating quality is highly improved and easily digestible protein content is significantly reduced to less than that of cv. LGC-1, these two cultivars would greatly contribute to healthy diets in patients with chronic renal failure and the management of dietary proteins in the future.

It is very interesting that the *Lgc-1* (*low glutelin content-1*) gene was applied to produce a high-quality Japanese Ginjo sake, which is basically brewed only from a small central portion of the white grain after milling embryo and seed coat with high protein content. It is particularly important for the brewers to use the grain with reduced protein content for the preparation of Ginjo sake. The milling of the grain leaves approximately 40% of the grain, which provides the basic mash for the brewing. The cv. Shunyou developed through the hybridization of cv. LGC-1 and "Hokuriku 153" line (Uehara et al. 2002), which possesses the *Lgc-1* gene, is effectively used to produce a similar type of sake that omits the extensive milling of the rice grain.

The *Lgc-1* gene of the mutant cv. LGC-1 is, as expressed by the gene symbol, a dominant mutation induced through a chemical mutagen treatment and reduces the amount of glutelin in the grain. Kusaba et al. (2003) discovered that in the *Lgc-1* homozygotes, there is a 3500 bp deletion with an estimated termination signal between two highly identical glutelin genes that forms a tail-to-tail inverted repeat. It is estimated that the generation of a long mRNA, that yields double-stranded RNA (dsRNA) molecule, is a potent inducer of RNA silencing (Fig. 12.8). It is a very unusual case that such a large deletion is generated by chemical mutagens; however, the result of the deletion of a stop codon-generated dsRNA causes the suppression of glutelin synthesis and lowers the glutelin content. This was the first report which demonstrated RNAi as a mutation mechanism in plants (Kusaba et al. 2003). The comparison between this mutation and the other glutelin mutation is discussed again in Sect. 12.5.1.

12.3.1.5 Resistant Induction and Bioassay Screening of Alternaria Disease in Pear and Apple

The cv. Nijisseiki (Japanese pear: *Pyrus serotina* Rehd. var. *culta* Rehd.) was a leading cultivar of green pear and was grown on 28% of the total area under cultivation in 1990 (Sanada et al. 1993). One of the most serious problems of this cultivar was high susceptibility to the pear black spot disease, caused by a fungus, *Alternaria alternata* (Fr.) Keissier (= *Alternaria kikuchiana* Tanaka) (Nishimura et al. 1978). Therefore, Japanese pear growers had to apply treatment of fungicides many times during the growing season for the control of the disease (Sanada et al. 1993). It was challenging, but the induction of resistant mutations conferring resistance to the disease was attempted in the Gamma Field. Small plants of the cv. Nijisseiki were planted at 4 m intervals from 37 to 63 m from a ^{60}Co source in the center of the Gamma Field in the IRB in 1962 (Sanada et al. 1993). The dosages of chronic gamma-ray irradiation treatments were estimated to be 0.3–0.04 Gy/day. Nearly 20 years after the planting, it was in 1981 when they found that a twig of a cv. Nijisseiki tree planted 53 m from the ^{60}CO source exhibited no symptom of the disease, although all the other cv. Nijisseiki trees in the Gamma Field exhibited black spot symptoms. The vegetatively reproduced trees from this single mutant twig with no symptom had all the important characteristics of cv. Nijisseiki with the exception of the resistance to the black spot disease. The vegetatively propagated trees from this single twig were registered and assigned the name cv. Gold Nijisseiki (Sanada et al. 1993) in 1991. This cultivar was also registered under the same name in Australia in 2004 (Certificate Number 2533).

According to Dr. Sanada, a leading breeder of this afore-mentioned cultivar, "The situation of mutation breeding on fruit trees has been severely criticized, because there have been no successful results." It really took 20 years for the identification of a useful mutation and 30 years to register and release cv. Gold Nijisseiki. However, this was a landmark achievement for the Gamma Field and mutation researchers at the IRB.

Though Dr. Sanada is given credit for the successful project, there is an interesting bit of serendipity to the discovery. At that time, technicians were spraying a small amount of fungicide every week to avoid the typical outbreak of black spot disease. This was performed because of the fear of a serious explosion of the disease in the Gamma Field and its spread to other areas. When one researcher took a new position at a new laboratory, his technical staff forgot to spray fungicide for a certain period. This oversight allowed all the pear trees inside the field to become completely infected with the disease and all the leaves had black spots. One technician, observing the alarming outbreak of the black spot disease, noticed one

Fig. 12.9 Bioassay of resistance and susceptibility to the black spot disease using the AK-toxin obtained from the culture medium of the fungus. The first to fifth leaf disc means first (young) to fifth (older) leaf; cv. Chojyuro: highly resistant; cv. Nijisseiki: highly susceptible; cv. Gold Nijisseiki: resistant. (Courtesy of IRB, NARO)

twig on one branch of one tree exhibiting no symptoms. It is this serendipitous event that led to the origin of cv. Gold Nijisseiki. Today, staff retain this tree in the Gamma Field as a monument to this accidental discovery.

Many researchers are critical of mutation breeding, such as the belief that mutation breeders have no control over their treatments and simply wait for adventitious mutants to appear. However, this is not entirely true. The group of the IRB successfully developed an easy and simple but effective screening method to identify resistant plants against this disease. The method is as follows: In a Petri dish, cut leaf discs (cut by 7 mm in diameter) from five leaves (top to bottom) on each branch are placed on the paper towel soaked with the AK-toxin, which is extracted from fungal culture media (Fig. 12.9: Sanada 1988). In a couple of days, susceptible leaves turn black, but resistant leaves stay green. The breeders need to collect leaves from hundreds or thousands of twigs in the irradiated population and apply this screening method to precisely identify mutated resistant twigs.

In this cooperative research for the establishment of this bioassay screening method, Nakashima et al. (1982, 1985) at the Pesticide Research Institute, Kyoto University, identified the chemical structure of the AK-toxin produced by the fungus. Then they used purified AK-toxin molecules for the screening, as they could even artificially synthesize this toxin. It is very important to mention that the cooperation between breeders and biochemist made the establishment of this unique screening method possible. Following the development and application of this screening method, two new resistant mutants cv. Osa-Gold (Masuda et al. 1997,

1998) and cv. Kotobuki Shinsui (Kitagawa et al. 1999) were developed only in 3 and 4 years, respectively.

In addition, a new mutant apple cv. Houiku Indo resistant to another Alternaria disease was developed from gamma-ray irradiation to micropropagated shoots of susceptible cv. Indo using the same bioassay method involving the AM-toxin (Ueno et al. 1977) produced by an Alternaria blotch fungus, *Alternaria mali*, that attacks apple trees (Yoshioka et al. 2000).

Two keys for the success of mutation breeding program of fruit trees are the development of precise screening methods for the selection and patience.

12.3.1.6 Recurrent Mutation Breeding for Outcrossing Crops

Mutation breeding has been effectively applied to self-pollinated species, as well as vegetatively propagated crops. There are widely used methods for the mutation breeding for self-pollinated species, such as the "M_1-spike progeny method" (Stadler 1930; Nybom 1954) and the "one-plant-one-grain method" (Yoshida 1962). These methods, however, could not be effectively applied to obligate allogamous species, because when the M_1 plants with a recessive mutated gene are open pollinated, it is very difficult to obtain homozygous recessive plants and most of the progenies remain as a heterozygous genotype in the M_2 population, so that the plants with a mutated gene will not be identified.

Ukai (1990) developed a unique method to obtain homozygous mutants in outcrossing crop species. This approach was first successfully applied to Italian ryegrass (*Lolium multiflorum* L.), which is widely planted as a cool-season forage grass in winter following maize or sorghum cultivation in Japan. The procedure was named as "crossing-within-spike-progenies method," and the field is shown in Fig. 12.10.

This method is composed of four steps: (1) to take seeds of each spike separately from each plant of a gamma ray-irradiated population, (2) to sow the seeds in a "hill" plot as a spike-progeny, (3) to isolate each "hill" from the others during the flowering time and to allow the open pollination of plants within each "hill" by bagging all plants within the "hill" with doubled cheesecloth before flowering, and (4) to take seeds from each "hill" and sow the seeds in "hill" progenies for the screening of mutants. This procedure needs to be repeated each year for six generations with recurrent gamma-ray irradiation to the seed.

The results of the method demonstrated that when the seed was irradiated with 300 Gy of gamma ray, the frequency of chlorophyll mutations was ca. 70.6% per "hill" progeny and 1.87% per plant. While open-pollinated populations with the same methodology without pollen isolation by bagging the plants in the "hill" exhibited chlorophyll mutations of only 10% and 0.12% per progeny and per plant, respectively. This method could be applied to other outcrossing crop species, such as buckwheat and alfalfa with bees inside the bag.

12.3.1.7 Mutation Breeding of Chrysanthemum

Radiation to plants may generate chimeras that are generally undesirable (Prina et al. 2012). However, De Jong and Custers (1986) reported that the mutants induced by

Fig. 12.10 Bagging of plants of each hill plot at the flowering time of Italian ryegrass. (Ukai 1990: Courtesy of IRB, NARO)

in vitro culture of X-ray-irradiated pedicel segments and petal epidermis of chrysanthemum did not appear to generate chimeras. Nagatomi et al. (1996) attempted to apply this method to chrysanthemum and successfully established a method for inducing mutations through gamma-ray irradiations with fewer chimeras. They could achieve the highest mutation rate with wider mutation spectrum by a chronic gamma-ray irradiation in the Gamma Field and following in vitro culture of irradiated flower petals, callus formation, and regeneration of plants from the callus. Induced different petal colors and different flower shapes from cv. Taihei (top right) are shown in Fig. 12.11a, b. Then, Nagatomi et al. (1996) developed six cultivars with remarkable different colors and petal shapes from the original cv. Taihei. Furthermore, they attempted to use ion beams in the same way in the Takasaki Ion Accelerators for Advanced Radiation Application (TIARA), Japan Atomic Energy Agency (JAEA), Takasaki, Gunma, Japan, instead of gamma-ray chronic irradiation (Nakagawa and Kato 2017).

The presence of only a single large flower on a single stem is one of the highly desirable characteristics in chrysanthemum. The most successful mutant cultivars induced through ion-beam irradiation for this purpose are cvs. Arajin (same pronunciation as "Araddin" but the Chinese character's meaning is "New God") and Imajin (same pronunciation as "imagine" in Japanese but the Chinese character's meaning is "Now God"). It was hard for farmers to remove small axillary buds by hand and leave only one top flower to produce a single big flower in one stem. Imakiire et al. (2005) successfully reduced the number of the axillary buds by irradiating the leaves of a leading cv. Jimba through ion-beam irradiation in TIARA and following in vitro

Fig. 12.11 Mutation of different flower colors (**a**) and different flower shapes (**b**) irradiated and regenerated from petal in vitro culture of cv. Taihei (circled top right) (Nagatomi et al. 1996: courtesy of IRB, NARO)

culture of the leaves and regeneration of the plant from the callus. Although the two mutant cultivars produced a small number of axillary buds, the number became remarkably less than the original cv. Jimba, and the petal number in a flower became more than cv. Jimba. Following this, the other two new mutant chrysanthemum cultivars with relatively larger flowers, cv. Ryujin and cv. Touma, were developed by the same method through ion-beam irradiation and registered in Japan (http://www.pref.kagoshima.jp/ag11/pop-tech/zenbu/documents/48620_20160318155705-1.pdf).

Recently, Ohmiya et al. (2012) induced yellow petal color mutations from a dominant white petal color plant, cv. Jimba, and hypothesized that the knockout and decreased number of *CmCCD4a* homologs, that are related to the carotenoid synthesis, cause the change of white petal color to light yellow color and eventually to a deep yellow color.

12.3.1.8 Characteristics of Gamma Ray- and Ion Beam-Induced Mutations

Developing the understanding of how mutations occurred at the DNA level has been an interesting and significant area of breeding or genetics, especially when applying reverse genetics, employing such useful techniques as PCR and TILLING. However, when deletions are small, PCR cannot identify mutants, and TILLING is the more effective approach. While TILLING method cannot identify mutants with a large deletion, PCR becomes the more effective approach.

Naito et al. (2005) attempted to induce mutations by gamma-ray and ion-beam irradiation to pollen, identified the mutations in Arabidopsis, and found that both large deletions more than 6 Mbp that were not transmissible to subsequent generations and smaller deletions, in size 1–4 bp, that were largely transmissible were induced. Shikazono et al. (1998) analyzed the carbon-ion beam-induced mutations by PCR analysis and found that about half of the mutations were point mutations while the remaining were large alterations in DNA such as inversions, translocations, and large deletions. Through the induction of mutations in Arabidopsis via carbon-ion-beam irradiation, Kazama et al. (2011) indicated that smaller base substitutions, or small deletions or insertions, could be successfully identified as single nucleotide polymorphism (SNP) detection systems.

Morita (2008) and Morita et al. (2009) attempted the molecular analysis of transmissible gamma ray-induced mutants of fixed mutant lines with typical morphological characteristics in rice, such as chlorophyll deficiency, gibberellin insensitivity, glutelin or globulin deficiency, and glutinous endosperm. They revealed the frequency of different mutations such as deletions, base substitutions, and inversions with variable size differences. Consequently, the 24 transmissible mutations were primarily classified into four groups: (1) a small point deletion, (2) very large deletion, (3) base substitution, and (4) very large inversion, and the ratio was 62.5:16.7:12.5:8.3 (Table 12.8; Morita et al. 2009).

In recent years, it has often been discussed whether the mutations induced via ion beams and by gamma rays are similar or not. To solve this question, mutations induced by gamma ray and carbon-ion beams were examined using whole-genome

Table 12.8 Gamma-ray irradiation-induced deletion, base substitutions, and inversions and the size of mutation (Morita et al. 2009)

Mutation type	Gene	Size (bp)
Small deletion	*CAO* (*cao-g1*)	1
	CAO (*cao-g2*)	3
	CPS (*cps-g1*)	1
	GA3ox (*ga3ox-g1*)	1
	GA3ox (*ga3ox-g2*)	3
	GID1 (*gid1-g1*)	1
	GluA1 (*gluA1-g1*)	1
	GluA2 (*gluA2-g1*)	1
	KAO (*kao-g1*)	4
	KAO (*kao-g2*)	16
	PLA1 (*pla1-g1*)	5
	PLA2 (*pla2-g1*)	5
	Wx (*wx-g1*)	2
	Wx (*wx-g2*)	5
	Wx (*wx-g3*)	6
Large deletion	*GID2* (*gid2-g1*)	42,200
	Glb (*glb1*)	62,800
	GluB4/B5 (*glu1*)	129,700
	Wx (*wx-g4*)	9400
Base substitution	*GluA2* (*gluA2-g2*)	1
	PLA1 (*pla1-g2*)	1
	Wx (*wx-g5*)	1
Inversion	*Wx*	1,284,800
	PLA2	3,208,500

CAO chlorophyll b deficiency, *CPS* ent-Copalyl diphosphate synthase (gibberellin deficiency), *KAO ent*-kaurenoic acid oxidase gene (gibberellin deficiency), *Ga3ox2 GA3-beta-hydroxylase* (gibberellin deficiency), *GID* gibberellin insensitivity; *GluA* and *GluB* glutelin deficiency, *Glb* alpha-globulin deficiency, *PLA* shortened plastochron, *WX* glutinous endosperm

re-sequencing in rice (Li et al. 2019b) and in Arabidopsis (Yang et al. 2019). These studies revealed that many gene substitutions were induced through both gamma-ray and ion-beam irradiations, as well as deletions and insertions (Li et al. 2019b). Yang et al. (2019) found mostly the same trends, too, and they indicated that the genomic mutation profiles induced using carbon-ion beams and gamma ray were same in mutation type and in distribution on the chromosomes. However, the authors consider that the sample size of both researches was small so that more research will be required to identify overall picture and differences between gamma-ray and ion-beam irradiation.

This author considered that the similarity between the mechanisms underlying the generation of extremely large deletions and inversion may be caused by a similar phenomenon. That is, in the generation of large deletions, a large base chain segment with the two point deletions at both ends was cut and missing at a crossing point of the two deletions, while in the case of inversion, the cut point was reversely placed at the same site again with two small point mutations at both ends of inversion. As this research and the following sorghum mutation (Mizuno et al. 2013) in Sect. 12.4.6 demonstrated, inversion is not a rare event in the gamma-ray irradiation.

12.3.1.9 Mutation Induction Through Ion-Beam Irradiation

Tanaka (1999) introduced ion-beam irradiation for the induction of mutations in Arabidopsis in the TIARA (Takasaki Ion accelerators for Advanced Radiation Application) in the JAEA. The TIARA was established in 1993, as the first facility of ion-beam irradiation designed specifically for the application of biological materials. Utilizing the facility, Tanaka et al. (1997) induced a stable flavonoid biosynthesis mutant through carbon-ion-beam irradiation and mapped the new mutant locus *ast* (*anthocyanin spotted testa*) in Arabidopsis. Tanaka (1999) discussed the mutation mechanism and mutation frequency induced by ion beams comparing it with electrons and concluded that the ion beam was 8- to 33-fold higher than similar applications using electrons.

The total number of induced mutant cultivars, mainly ornament flowers, developed at the TIARA is presently 21 with the detailed list provided in the Appendix 1 of the publication (Nakagawa and Kato 2017). Furthermore, cv. Koshihikari Kan No. 1 represents a unique mutant rice cultivar that has the ability to absorb very little cadmium in the paddy field and was developed using ion beam in the TIARA (Abe et al. 2017), which is discussed in Sect. 12.4.5.

Another active ion-beam irradiation facility, named the RI Beam Factory, is located in RIKEN Nishina Center, Wako, Saitama, Japan. The facility and a unique automated irradiation equipment are explained in detail elsewhere (Abe et al. 2012). From this facility, 11 cultivars, mainly ornamental flowers, are listed by Nakagawa and Kato (2017), too.

Within Japan, there are additional ion-beam irradiation facilities. These are the Wakasa Wan Energy Research Center Multipurpose Accelerator System with Synchrotron and Tandem (W-MAST), Fukui, Japan (http://www.werc.or.jp/enenews/pdf/pamphlet_english.pdf), and the National Institutes for Quantum and Radiation Science and Technology, Chiba, Japan (https://www.nirs.qst.go.jp/ENG/index.html), that is mainly focusing on the medical use. Ten cultivars have been released by the W-MAST (Nakagawa and Kato 2017).

Recently, the use of ion-beam irradiation to induce mutations in flower cultivars has been increasing. Cultivars such as cv. Viento Flamingo *Osteospermum* (Iizuka et al. 2011), cv. Viento *Labios osteospermum*, cv. TL585 salvia, cv. San-mari-peach verbena, cv. Sunreniripin (https://patents.justia.com/patents-by-us-classification/PLT/487) and cv. Sunrekodou torenia, cv. Nishina-Zaou, and cv. Nishina-otome cherry flower (https://www.riken.jp/en/news_pubs/research_news/rr/8266/) have been developed.

As shown in Fig. 12.5, the total number of ion beam-induced cultivars reached 44 in 2020 and the contribution of ion-beam irradiation applications will continue to increase due to the growth primarily in ornamental flower applications.

12.3.1.10 Mutable Gene(s) or Transposon

Okumoto et al. (1999) introduced a mutable *slender glume gene* (*slg*) in rice cv. Ginbozu. This mutation was induced through gamma-ray irradiation presumed to act as a single recessive gene. Following the hybridization of this mutant with cv. Ginbozu, they attempted a physical mapping of the locus with a chromosome walking technique using a YAC library (Uehara et al. 1995). Results of this study identified *slg* as a mutable gene.

The existence of gene mutations induced by transposon was first suggested in maize by the identification of the maize *Activator* (*Ac*) (Peschke et al. 1987). The activation of a transposable element system was hypothesized to be induced through tissue culture and somaclonal variation (Larkin and Scowcroft 1981). Hirochika et al. (1992) identified various transposons in the genome of rice. Among these transposons, Hirochika et al. (1996) isolated a useful transposable element, *Tos17* retrotransposon in rice cv. Nipponbare, which genome is totally sequenced in the International Rice Genome Sequencing Project (International Rice Genome Sequencing Project 2005). This *Tos17* retrotransposon is activated by tissue culture and can induce mutations. The possible use of this *Tos17* was proposed as (1) a gene tagging, (2) a functional analysis of genes by the retrotransposon-induced mutants, and (3) a functional analysis by analyzing the target-site sequence of *Tos17*.

A total of ca. 50,000 insertion lines having the *Tos17* retrotransposons were developed for functional analysis of rice genes of cv. Nipponbare, and a large-scale phenotypic data of each mutant line, such as growth behavior, leaf color and shape, culm character (including dwarf), heading date, panicle characters, sterility, and seed characters, are deposited in the "*Tos17* mutant panel Web-based database" with the photos of the mutant plant (Miyao et al. 2007). Anyone can obtain both every phenotype and the flanking sequence data at the web site (https://tos.nias.affrc.go.jp/) and the seeds at the web site (https://www.rgrc.dna.affrc.go.jp/jp/).

Although there are no confirmed released cultivars that are indicated to have mutation by *Tos17* transposons, the use of this system remains very effective and useful for gene targeting and functional analysis of genes, and some interesting genes have been identified by using this technology (Sato et al. 1999; Agrawal et al. 2001; Nonomura et al. 2007; Yoshii et al. 2009).

12.4 Some Interesting Induced Mutations

There are many interesting mutations, which are useful for food and biomass crop production. Some reports of the induced mutations are briefly explained here.

12.4.1 Non-shattering Gene and Induction of Non-shattering Cultivars in Rice

Recently, *indica*-type high-yielding rice genetic resources have been attracting the interests of breeders of forage rice breeding programs to improve biomass productivity by trying to improve total dry matter yield of foliage and/or grains without regarding the taste of the grain, which is the most important characteristic of cooked rice. In the 1980s, the Japanese Government recommended rice farmers to convert their paddy fields into forage crop fields of maize, sorghum, and forage grasses primarily to feed dairy or beef cattle because rice was overproduced in Japan. At that time, the Japanese farmers were not familiar with the idea to feed cattle with rice, as rice has always been a staple food for the Japanese people. Recently, however, with the promotion of government subsidies to support livestock farmers who changed their feeding materials from imported cereals and hay to forage rice, there occurred a notable increase in the number of farmers who feed animals with rice foliage and grain. It is noteworthy that for a long time, rice straw has been an important and special fiber-source feed for producing well-known beef cattle with a delicious expensive marbled meat in Japan.

One of the defects of *indica*-type rice to be improved is its seed shattering traits. Konishi et al. (2006) first demonstrated that this seed shattering character is controlled by a single dominant *qSH1* gene, which is a key gene of the abscission layer formation process. There are several mutations related to the abscission layer formation such as *SHAT1* (Zhou et al. 2012) and *OsSh1* (Lin et al. 2012). When these dominant genes can be knocked out, non-shattering rice plants that cannot form an abscission layer between the pedicel and spikelet at the base of the rice seed would be developed.

Kato et al. (2006) induced a mutation at the locus in an already developed forage rice cv. Moretsu through an acute gamma-ray irradiation to the seeds in the Gamma Room, IRB, and developed and released a non-shattering mutant cv. Minami-yutaka for rice silage production. Sakai et al. (2013) attempted to induce a mutation at the same locus in a high-yielding *indica*-type waxy cv. Taporuri, a plant introduction from Taiwan, with the same method in the IRB, and a non-shattering mutant cv. Ruri-aoba for the same purpose was successfully developed. Furthermore, a moderate-shattering cv. Oonari was developed from a high-yielding but easy-shattering *indica*-type cv. Takanari using the same method at the IRB (Imbe et al. 2004a, b). The development of these mutant cultivars indicated the relative ease of inducing non-shattering mutations through acute gamma-ray irradiation at the Gamma Room.

Recently, Li et al. (2019a) directly identified these mutations in the cv. Minami-yutaka (Li et al. 2020) and cv. Oonari through whole-genome resequencing. Then, Li et al. (2020) demonstrated that the non-shattering cv. Minami-yutaka induced through gamma-ray irradiation possessed a 13 bp deletion eliciting defective splicing in exon 3 of the *OsSh1* gene, which is one of the candidate genes for the control of seed shattering (Lin et al. 2012). Furthermore, Li et al. (2020) based on CRISPR/Cas9 gene editing technology identified that a knockout mutation of the *OsSh1* underlies non-shattering in rice.

12.4.2 Giant Embryo in Rice

The rice embryo contains large amounts of proteins, fats, and vitamins of vital importance for human health. When the rice embryo grows following the soaking of seed in water, gamma-aminobutyric acid (GABA), transformed from glutamic acid, is accumulated in the grain (Saikusa et al. 1994a, b). Defatted rice embryos rich in GABA are useful as a functional food for the control of blood pressure and its sedative qualities related to insomnia and autonomic disorder (Okada et al. 2000). The GABA-accumulated brown rice of ordinary rice cultivars is already on the market as a healthy brown rice. Consequently, Satoh (1981) developed two mutant lines, EM40 and GM15-34, which exhibited giant embryo induced by *N*-nitroso-*N*-methylurethane (MNU) treatment of cv. Kinmaze. The embryo of the two lines is 3–4 times larger in volume than that of original cultivar. Satoh and Iwata (1990) identified this giant embryo character of the EM40 to be under the control of only a single recessive gene (*ge*) located on chromosome 7. This giant embryo (*GE*) rice gene is essential for controlling the embryo/endosperm size balance. Nagasawa et al. (2013) identified the function of *GE*, which encodes *CYP78A13*, a gene that controls cell size in the embryo and cell death in the endosperm. Complementation analysis of *CYP78A13* and *GE* demonstrated that complementation of *ge* mutation in either embryo or endosperm could not restore the wild-type embryo/endosperm ratio.

Nemoto et al. (2001) developed the first giant embryo mutant cv. Haiminori for the purpose of cooked brown rice by crossing EM40 and cv. Akenohoshi in 1990. The embryo of cv. Haiminori is 3–4 times in volume that of ordinary Japanese rice. The amount of GABA produced in the embryo of cv. Haiminori is approximately four times greater than that of the cv. Nipponbare, which is the material of the Rice Genome Project in Japan, after soaking in water for 4 h. This new cultivar is used for the commercial production of "healthy rice cake." One of the serious defects of these giant embryo cultivars is a low germination rate and the subsequent weak seedling vigor; therefore, the establishment rate of the seedlings is occasionally only 40–50% (Nemoto et al. 2001). The genetic and functional relationship between the giant embryo and poor seedling establishment will be a most important problem to be solved for the stable cultivation of giant embryo rice cultivars. Formerly, the breeders of giant embryo rice intended to utilize the grain of the giant embryo mutants as a raw material of rice bran oil production, as the nutritional value of cooked brown rice attracted the attention of consumers who opt for a healthier alternative rice to ordinary white rice.

Recently, Sakata et al. (2016) attempted to induce giant embryo mutant lines by NMU treatment in cv. Mizuhochikara for increasing the fatty acid composition in the grain focusing on altering the triacylglycerol (TAG) content in rice bran oil to enhance its various health benefits. They could obtain four giant embryo lines exhibiting agronomically similar phenotype. Then, they found that two mutant lines have a new locus that is nonallelic to ge at chromosome 3, while the other two mutant lines have a mutation at the reported locus allelic to ge at chromosome 7 (Sakata et al. 2016).

12.4.3 Fatty Acid Composition, Lipoxygenase Lacking, and Glycinin Rich in Soybean

12.4.3.1 Fatty Acid Composition

Oil crop produces its unique oil with a specific fatty acid composition. There exists an extremely large diversity of fatty acid composition that ranges within and across the oil crops (Takagi and Anai 2006). For example, linolenic acid content is higher in soybean (8.0%), but is not found in maize, traditional sunflower, and safflower. The biosynthesis of fatty acids, common to all plant species, involves a carbon elongation process from palmitic acid (carbon number 16: C–C double bond number zero) to stearic acid (18:0), which is derived from an unsaturated reaction process: from stearic acid (18:0) through oleic acid (18:1) and linoleic acid (18:2) to linolenic acid (18:3) (Nakagawa et al. 2011a).

Another carbon elongation reaction is unique in rapeseed species. This is the formation of oleic acid (18:1) through eicosenoic acid (20:1) to erucic acid (22:1) (Nakagawa et al. 2011a). The fatty acid content of the species and cultivars is the result of different activities of these enzymes, which is important for the biosynthesis pathway, and the number of genes that code for each enzyme in each step related to the process of carbon elongation and the unsaturated reaction pathways.

Soybean is the second most widely used edible oil resource in the world, after oil palm (https://www.bctj.jp/2017/wp-content/uploads/2021/02/palmWP2019.pdf). For the improvement of fatty acid composition of the seed, target genes, which are known to modify the fatty acid composition of the other oil crops, were mostly identified in the germplasm and incorporated by the various breeding methods. White et al. (1961) reported that contents of both linoleic acid and linolenic acid are controlled by polygenes through the identification of intervarietal variation of fatty acid composition in 251 soybean cultivars and varieties. Alternatively, Takagi et al. (1986) suggested that both oleic acid and linolenic acid contents were controlled by only two and one gene(s), respectively. Then, Takagi et al. (1986) concluded that the effects of each gene were small in determining the levels of oleic and linolenic acid, although environmental effects played a much larger role in the level of expression. Takagi's research group initiated the mutation induction through X-ray irradiation in cv. Bay and identified 46 mutant lines from 12,266 M_2 population. These mutants included J3 (low palmitic acid: 6.1%) (Takagi et al. 1995), J10 (high palmitic acid: 17.2%) (Takagi et al. 1995), M25 (high stearic acid: 21.2%) (Rahman et al. 1995), M23 (high oleic acid: 48.6%) (Rahman et al. 1994), M5 (low linolenic acid: 4.9%) (Takagi et al. 1990), and B739 (high linolenic acid: 18.4%) (Takagi et al. 1989). Furthermore, a lot of lines with a wider range of fatty acid composition have been developed by the hybridization among these mutants (Rahman et al. 2001, 2003). These results proved that the artificial induction of the mutants is a useful approach for enhancing the fatty acid diversity in soybean.

Many genes controlling fatty acid contents, such as low palmitic acid, high palmitic acid, high stearic acid, high oleic acid, low linolenic acid, and high linolenic acid, have been identified from the hybridization and genetic analyses (Nakagawa et al. 2011a). These enzymes of fatty acid biosynthesis are encoded by *Gm βKASII*, *GmSte*, *GmFAD2*, and *GmFAD3* gene family. Genetic analyses of fatty acid composition mutants suggest that several genes would be involved in each step of the pathway for the fatty acid synthesis. Anai et al. (2005) attempted an extensive molecular analysis on some of these mutations and identified some new genes.

To successfully select the allele-specific genotypes in the breeding program, the use of molecular markers related to these gene families can be more effective and efficient than traditional phenotypic selections. It is very useful to have multiple sources of alleles for each candidate gene's isoform in the selection of a species that exhibit a superior fatty acid composition (Takagi et al. 1998; Anai et al. 2005; Takagi and Anai 2006; Nakagawa et al. 2011a).

12.4.3.2 Lipoxygenase Lacking

Being an enzyme, lipoxygenase generates a beany flavor of soybean and the processed products. When the soybean seed is soaked, the lipoxygenases oxidize the unsaturated fatty acids of the seed to produce aldehyde compounds such as *n*-hexanal (Fig. 12.12). Usually, a heat treatment is used to reduce this beany flavor, but the flavor is not completely eliminated. This flavor, once produced, becomes one of the less desirable flavors of soybean products. Now, establishment of processing methods has been attempted to remove this beany flavor from the processed soybean meal in the food industry.

Soybean seed contains three lipoxygenases, L-1, L-2, and L-3 (Arai et al. 1970). In the 1980s, mutant varieties which lack lipoxygenase were isolated from germplasm (Hildebrand and Hymowitz 1981; Kitamura et al. 1983). These early works identified lines lacking both the L-1 and L-3. Following this, the L-2 and L-3 lacking lines were developed by hybridization between single L1, L2, and L3 lacking lines (Kitamura et al. 1985). However, they failed to develop either L-1 and L-3 lacking

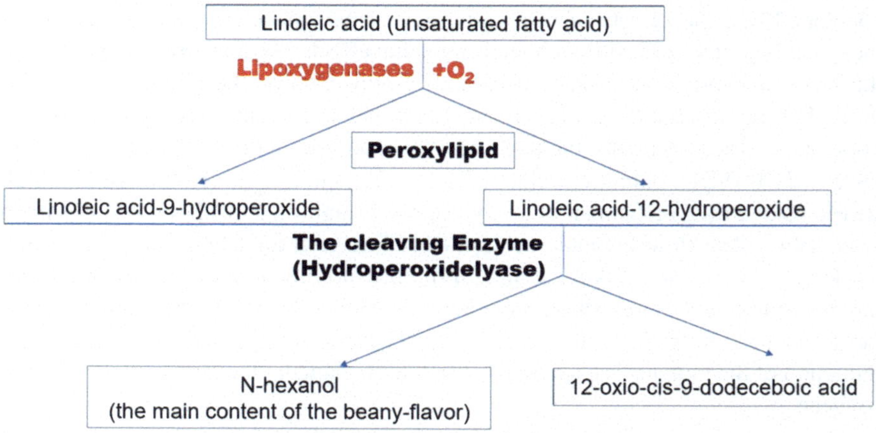

Fig. 12.12 The pathway of generation of beany flavor by the lipoxygenases. (Based on Anai et al. 2005)

line or all-lacking plants. It was suggested that the tight linkage of the two loci may make it difficult to produce L-1 and L3 lipoxygenase-deficient plants (Hajika et al. 1991). Therefore, mutation breeding through a gamma-ray irradiation was attempted to induce all-lipoxygenase-lacking soybean mutants.

Hajika et al. (1991) irradiated the F_2 seeds of the hybridization between a line Kankei 2 (later named as Kanto 102: lacking L-1 and L-3) and a line Kankei 1 (later officially registered as cv. Yumeyutaka (Kitamura et al. 1992; lacking L-2 and L-3) with 100 Gy or 150 Gy of gamma rays for 2 days in the Gamma Room at the IRB in 1989. They screened 1813 M_3 seeds using the SDS-PAGE and obtained only one mutant seed lacking all L-1, L-2, and L-3 (Nakagawa et al. 2011a). From this mutant line, the cv. Ichihime with all-lipoxygenase-lacking characteristics was released in 1995 (Hajika et al. 2002). Following this remarkable achievement, cv. Eru-Star (Takahashi et al. 2003a) and cv. Suzusayaka (Yumoto et al. 2006) with all-lipoxygenase-lacking soybean cultivars have been generated through hybridizations between normal soybean cultivars and the all-lipoxygenase-lacking mutant in Japan (Hajika et al. 2005).

Recently, a lipoxygenase-free soybean mutant was developed using CRISPR-Cas9 technology (Jinek et al. 2012; Wang et al. 2020). This outcome provides an additional example whereby mutant induction of complete gene systems can produce useful and desired products. Once particular mutations and the underlying molecular mechanisms are identified, genome editing technologies can be applied towards the generation of new cultivars using CRISPR-Cas9 technologies.

12.4.3.3 Glycinin Rich and Low Allergenicity

Glycinin (11S globulin) and β-conglycinin (7S globulin) are the major components of soybean seed proteins which make up ca. 70% of the total seed proteins in the seed (Thanh and Shibasaki 1976). Takahashi et al. (1996) demonstrated that the content of sulfur-containing amino acid, such as methionine and cystine, in glycinin is 3–4 times larger than that of β-conglycinin. Glycinin exhibits a better processing property in texturized and filmed soy foods, as well as in tofu gels, than β-conglycinin (Fukushima 1991), composed of α-, α′-, and β-subunits (Thanh and Shibasaki 1976). Kitamura and Kaizuma (1981) identified cv. Keburi, that is characterized by the absence of α′-subunit, and cv. Mo-shi-dou (Gong 503), thereby exhibiting low levels of the α- and β-subunits, in the soybean germplasm maintained in Japan. Ogawa et al. (1989) developed a few lines from these accessions with these spontaneous soybean mutations and demonstrated that it was possible to increase the glycinin amount by decreasing the β-conglycinin amount. An interesting character of a breeding line, Kari-kei 434, for the breeder was the lack of the α′-subunit and decrease in the α- and β-subunits of β-conglycinin level. However, neither α-less nor β-less germplasm were identified in the germplasm. Therefore, Takahashi et al. (1996) attempted an acute gamma-ray irradiation to the seed of the line Kari-kei to knock out the α-subunit gene at the IRB. Consequently, they identified five mutant seeds which exhibited a loss of the α-subunit in addition to the α′-subunit using the SDS-PAGE in M_2 and M_3. The α-subunit was proved to be controlled by a single gene through the crossing studies (Takahashi et al. 1996). A new glycinin-rich

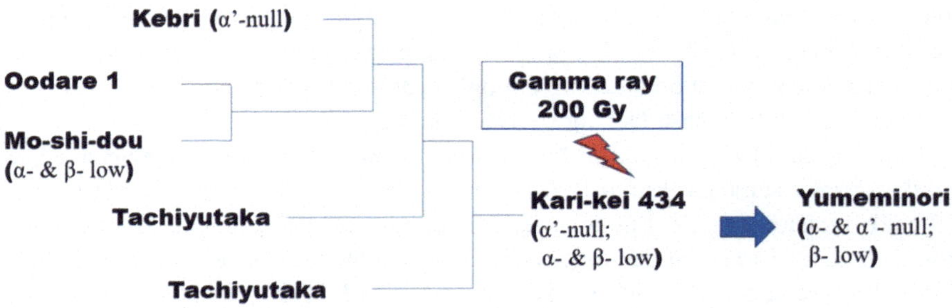

Fig. 12.13 Development of glycinin-rich mutant cv. Yumeminori and the gamma-ray irradiation processes. (Based on Takahashi et al. 2004)

mutant cv. Yumeminori was developed from this induced mutation and commercially released as a first glycinin-rich soybean in 2001 (Fig. 12.13; Takahashi et al. 2004; Nakagawa et al. 2011a).

Following the studies on the quality of the seed proteins of the cv. Yumeminori, Takahashi et al. (2004) found that this cultivar possesses hypoallergenic property, because the α-subunit, which was deleted in the cv. Yumeminori, is a major allergen in soybean (Ogawa et al. 1991), and Gly m Bd 30K, that is also one of the major allergens, was removed from the soybean seed (Samoto et al. 1996).

Later, Ishikawa et al. (2006) clearly identified that the deficiency of the α-subunit in a line, α-null (1) soybean, was caused by a stop codon, which was generated by an insertion of four bases into *CG-2* gene encoding this α-subunit.

12.4.4 Super-Nodulation

Super-nodulation characteristics mean production of a larger number of root nodules in the root of leguminous plants. Formerly, super-nodulating mutants were thought to be an interesting new germplasm, which might improve the productivity of soybean with higher nitrogen-fixation ability. Such materials would also be useful for elucidating molecular mechanisms of Rhizobium-plant interactions. Akao and Kouchi (1992) induced and isolated a super-nodulating mutant line, En6500, from 2800 M_2 of 0.5% ethyl methanesulfonate (EMS) treatment to the seed of cv. Enrei. Then, Kokubun and Akao (1994) identified that this En6500 plant produced a larger number of nodules and that this character was controlled by a single recessive gene. However, the growth and yield of this mutant line were less than those of the original cv. Enrei. Therefore, it was presumed that the inferior growth of the line was caused by the high consumption of carbohydrates to produce more nodules and a reduced nitrogen-use efficiency (Takahashi et al. 1995; Oya and Ishii 1999).

Takahashi et al. (2003b) attempted additional backcrossing of the En6500 to the original cv. Enrei to obtain a plant with both this super-nodulation phenotypes and good agronomic performance in the backcross populations and developed and released a super-nodulating cv. Kanto 100 with a yield similar to cv. Enrei.

The super-nodulating mutant gene has a potential impact of providing insights into Rhizobium-plant interaction mechanism and breeding for higher nitrogen fixation in soybean and other leguminous crop species.

12.4.5 Mendel's Gene

It is well known that Gregor Mendel revealed the law of heredity by using seven traits such as cotyledon color, seed shape, pod color, pod shape, flower color, flower position, and stem length in garden pea (*Pisum sativum* L.) (Mendel 1866). Genetic backgrounds of the two characteristics were identified by using natural mutations in pea: the seed shape (dominant: smooth versus recessive: wrinkled) is controlled by *R* gene that corresponds to a gene of natural mutation of starch-branching enzyme (Bhattacharyya et al. 1990); the stem length (dominant: tall versus recessive: short) is controlled by *Le* gene, which is a gibberellin 3β-hydroxylase gene (Lester et al. 1997). Sato et al. (2007) identified the molecular background of cotyledon color (dominant: yellow versus recessive: green) of the pea seed using a stay-green rice mutant of cv. Nipponbare induced through carbon-ion-beam irradiation (220 MeV) at the TIARA, JAEA. That is, they revealed that Mendel's green cotyledon gene is same as a single recessive rice stay-green gene (*ossgr*) of the rice stay-green mutation (the dominant gene is *OsSGR*) and the gene of green cotyledon color in pea is *pssgr* (the dominant gene is *PsSGR*). In addition, they proved that the *OsSGR* gene in rice was highly expressed in the senescent stage of leaves and activated the chlorophyll degradation pathway through the translational or posttranslational regulation. Pea plant, that has a homozygous recessive mutation (known as *i*), maintains green cotyledon color during maturation of the seed and senescence of leaves. Sato et al. (2007) asserted that tight linkage existed between the *I* locus and stay-green gene (*SGR*) found in rice based on molecular analyses, that the *I* gene encodes *PsSGR* in pea, and that *PsSGR* is involved in chlorophyll degradation by a functional analysis of *pssgr* mutants in pea and *ossgr* mutant in rice (Fig. 12.14).

12.4.6 Low-Cadmium Rice

Cadmium (Cd) is one of the environmental pollutants, and this metal exhibits serous toxicities for human beings after a long-term exposure. In some paddy fields near the sphalerite mine in Japan such as Kamioka mine, Cd content in the soil was elevated and Cd-associated health problems have been documented in the Jinzu River basin, Toyama (Aoshima 1987). The limit of maximum Cd content of edible rice grain is 0.4 mg/kg (Codex Alimentarius 2008).

To avoid the risk of Cd poisoning, a rice cv. Koshihikari Kan 1 gou, that accumulates dramatically low Cd in the grain, was induced from cv. Koshihikari through ion-beam irradiations at the TIARA, JAEA (Ishikawa et al. 2012). The sequence analysis and functional characterization of the mutants and the locus in the original cultivar revealed that the mutant has a single base pair deletion that

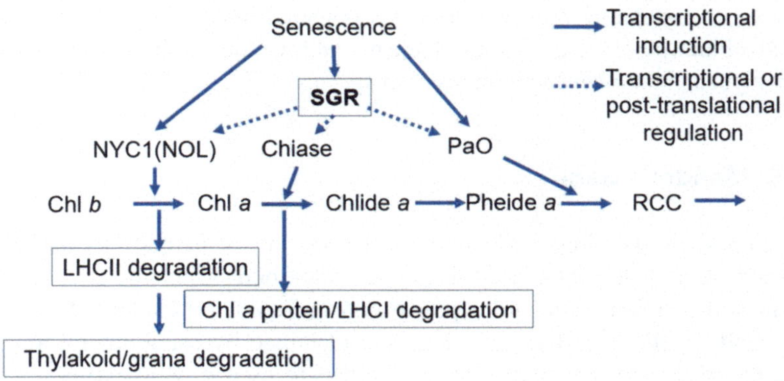

Fig. 12.14 Model of SGR function during leaf senescence (Based on Sato et al. 2007). NYC1, NOL, Chlase, and PaO: the enzymes in the chlorophyll-degrading pathway. The Chl *b* is converted into Chl *a* and metabolized in the Chl *a*-degrading pathway to the red chlorophyll catabolite (RCC). The first step is catalyzed by the Chl *b* reductase NYC1. Following the Chl *b* degradation, the LHCII and thylakoid/grana degrade in the senescent leaf. Another Chl *b* reductase (NOL) may have a minor role in the Chl *b* degradation during leaf senescence (Kusaba et al. 2007). Senescence signals induce the expression of NYC1, NOL, SGR, and PaO (dashed arrows). The SGR is primarily involved in Chl *b* degradation and the LHCII degradation via translational/posttranslational regulation of NYC1 (dashed arrows). The PaO activity is regulated by SGR at the translational/posttranslational process

generates a frameshift mutation at a gene of a transporter protein, *OsNARMP5*, which is known to transport manganese (Mn) ion and located on the seventh chromosome (Ishikawa et al. 2012). This mutant gene *osnramp5-1* ceases the activity of Cd transportation causing the Cd absorption of roots in the mutant line to be highly suppressed. With the mutant line, the Cd contents of the rice grain (white or brown) exhibited less than 0.05 mg Cd/kg compared with 1.73 mg Cd/kg in the original cv. Koshihikari (Ishikawa et al. 2012).

If the rice plants are cultivated aerobically in the paddy fields, absorption of another toxic element, arsenic (As), is also inhibited, although absorption of Cd is accelerated under this condition (Arao et al. 2009). When this low-Cd rice cultivar was cultivated under aerobic paddy field conditions, absorption of both Cd and As was greatly reduced and a low-Cd-low-As rice grain was efficiently produced (Ishikawa et al. 2016). Field studies indicated that the grain of these mutant plants has nearly non-detectable levels of Cd, when cultivated even in highly Cd-contaminated paddy fields.

Because this mutant gene *osnramp5-1* is useful and effective for producing low-Cd rice, this low-Cd mutant gene was first introduced into four different popular cultivars and is being introduced into more than 100 cultivars, adapted to climatically different areas by using marker-assisted selection (Ishikawa 2020).

Ishikawa (2020) emphasized that genome editing technologies would be very useful and easy to generate new low-Cd rice cultivars, because the locus and the gene sequence have already been identified. However, Ishikawa (2020) stated in the report that he does not intend to use these technologies so far, as the Japanese public acceptance for genome editing food was not clearly spelt out.

This research also implies an interesting relationship between mutation breeding and genome editing technologies. Although genome editing, especially CRISPR/Cas9 technology (Cong et al. 2013), is a powerful tool to induce deletions and insertions at a very specific point in the genome of the targeted plant, mutation breeding can induce any targeted phenotypes without knowing the genome sequence and/or the genes of the targeted phenotype. Therefore, it is reasonable to assume that researchers should first induce mutations and then apply genome editing technologies to generate the same mutation in different cultivars at a more rapid and efficient rate.

12.4.7 Epicuticular Wax-Free Mutation of Sorghum

Sorghum (*Sorghum bicolor* (L.) Moench.) generates white bloom or epicuticular wax on the stem and leaf surfaces. Jordan et al. (1983) reported that this wax deposition is a dominant trait (*Bm*) and bloomless (*bm*) mutants do not produce wax on the plant surface. The important role of this wax is estimated to contribute to drought tolerance by reducing cuticular transpiration from the leaf surface (Blum 1975) and to protection from ultraviolet radiation from the sun (Jordan et al. 1983). Furthermore, Jenks et al. (1994) reported that reduction of this wax promotes susceptibility to a fungal disease, known as northern corn leaf blight, caused by *Exserohilum turcicum* (Pass) K.J. Leonard & Suggs, due to the increased epidermal conductance to water vapor that facilitates *Exserohilum turcicum* spore germination and infection. However, sorghum plants possessing the *bm* mutation have distinct advantages such as a resistance to greenbug (Nakagawa et al. 2011b; Peiretti et al. 1980; Weibel and Starks 1986) and sheath blight (Kasuga et al. 2001).

Recently, sweet sorghum has been identified as a promising high-yielding crop for bioethanol production by using both stem sugars and stem lignocellulose through fermentation in Japan (Nakagawa et al. 2013). When sweet sorghum breeding for biofuel feedstocks is conducted, resistance to these biotic stresses has great advantages for low-input and sustainable biofuel production systems.

Seeds of sweet sorghum variety, Italian, were irradiated with a total dosage of 400 Gy of gamma ray for 20 h, in the Gamma Room, IRB, and two *bloomless* (no epicuticular wax: *bm*) mutant plants, 2I-400-2-11-1 and 2I-400-2-11-2, were isolated in the M_2 (Mizuno et al. 2013). These two mutants probably derived from one M_1 plant exhibited the same characteristics with the original Italian, except for the non-waxy trait. The F_2 genetic analysis from the hybridization of the waxy and non-waxy plants indicated an approximate 3:1 segregation ratio, which suggested that *bm* (no epicuticular wax) character was regulated by a single recessive nuclear gene (Nakagawa et al. 2011b; Mizuno et al. 2013).

Mizuno et al. (2013) conducted an RNA sequencing analysis for the wild-type plants exhibiting epicuticular wax (*Bm* phenotype) and mutant plants (*bm* phenotype). The total RNA from the leaf sheath of each plant was converted to cDNA for

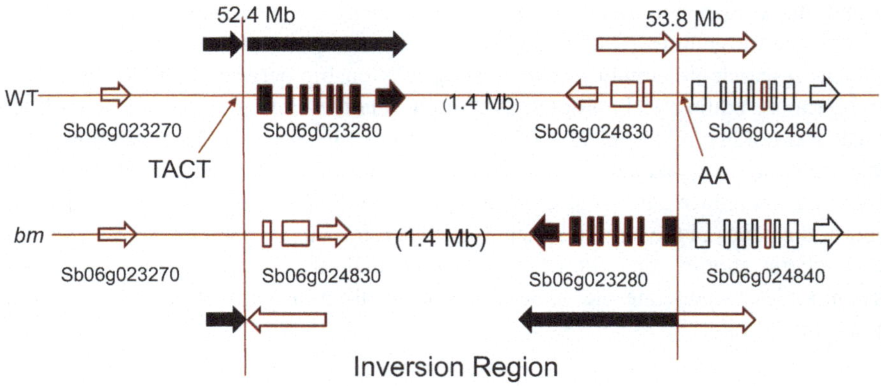

Fig. 12.15 Genomic inversion in the *bm* mutant. The size of an inversion is ca. 1.4 Mb (Based on Mizuno et al. 2013). The arrows indicate the cloned sequence of the wild-type (WT) and the *bm* mutant genome. Four bases (TACT) or two bases (AA) were deleted in the *bm* genome. This inversion causes the exchange of the upstream region of the *Sb06g023280/WBC11* gene (black arrow) with the downstream sequence. (Courtesy of Dr. H. Mizuno, NARO)

massive parallel sequencing in an Illumina Genome Analyzer, and differentially expressed genes were identified. The results of this study revealed that one gene from the 31 downregulated genes was mostly same as the ABC transporter, which regulates wax secretion in Arabidopsis (Bird et al. 2007; Panikashvili et al. 2007).

It is interesting that this induced *bm* mutant carries a 1.4 Mb genomic inversion near the promoter region of Sb06g023280 with small base pair deletions at both ends (Fig. 12.15). Mizuno et al. (2013) revealed that this inversion mutation inhibited wax secretion in the bloomless sorghum, although the epicuticular wax was perfectly synthesized inside the cells.

12.5 Achievement of Biological Research on Mutations Induced by Gamma Ray

There are interesting biological researches on the characteristics of mutations, which were induced via chronic and acute gamma-ray irradiations. Some interesting mutants were induced through chronic gamma-ray irradiations in the Gamma Field. Comparison of chronic and acute irradiation including advantages and disadvantages of chronic gamma-ray irradiation and some reports of the topics are briefly explained here.

12.5.1 Different Sizes and Locations of Deletions Generate Different Kinds of Phenotypes

Through the evolution of plants, genes are often tandemly duplicated and generate a functional redundancy. Sequencing data of rice genome revealed that more than 14% of the genes exhibited this kind of tandem array (International Rice Genome Sequence Project 2005; Arabidopsis Genome Initiative 2000). It is an important point that the analysis of these gene functions by inducing double mutants through chemical mutagens and transposon treatments is extremely difficult because it would be very difficult to knock out two tandemly repeated genes at once.

Glutelin is one of the digestible seed storage proteins. There are reversely and tandemly repeated double loci (Fig. 12.8), which code for mRNA for the glutelin production. Iida et al. (1993, 1997) induced and isolated many mutants which exhibited low glutelin contents using SDS-PAGE. Molecular analyses of these mutants which possess low glutelin contents have revealed that the different sizes and locations of deletions generate very different mutant phenotypes.

As already discussed in Sect. 12.3.1.4, the *Low glutelin content* (*Lgc-1*) is dominant and reduces glutelin content in the grain of rice. Although it is a very rare case, Kusaba et al. (2003) induced this mutation of 3500 bp deletion by chemical mutagenesis with ethyleneimine (EI). The deletion is considered to include a terminator signal of two highly similar glutelin genes with a tail-to-tail inverted repeat, so that a long mRNA that includes these two RNAs with a tail-to-tail inverted repeat is generated to form a double-stranded RNA molecule. It is known that a double-stranded RNA molecule becomes a potent inducer of RNA silencing (Figs. 12.8 and 12.16; Kusaba et al. 2003). The cv. Lgc-1 exhibited high suppression of glutelin synthesis and a concomitant low glutelin content.

Another low-glutelin mutation, *glu1*, that is induced through gamma-ray irradiation in rice, lost an acidic subunit of glutelin. Morita et al. (2007) identified that the

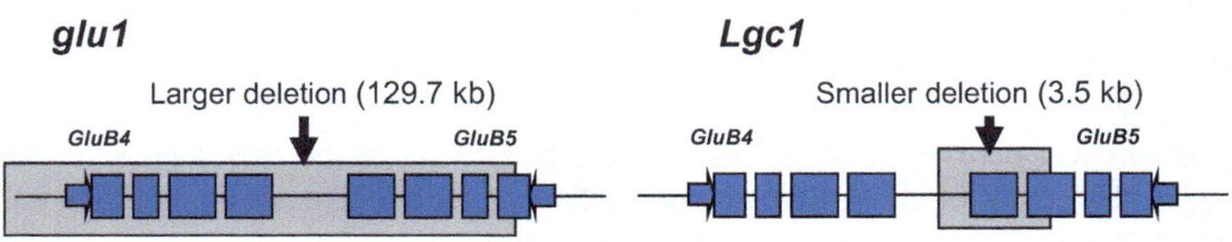

	Phenotyp	Dominant or Recessive	Mechanism
glu1	Deletion of 1 acidic subunit	Recessive	Knock-out of 2 genes
Lgc1	Total glutelin contents lower	Dominant	Suppresses glutelin genes via **RNAi.**

Fig. 12.16 Differences in the phenotypes, dominant or recessive character, and mechanism of *glu1* and *Lgc1* gene with different sizes and locations of deletion in the same region of *GluB4* and *GluB5* (*glu1*: Morita et al. (2007); *Lgc1*: Kusaba et al. (2003)). (Courtesy of Dr. R. Morita, RIKEN)

glu1 possesses a 129,700 bp deletion involving both the tandemly repeated *GluB5* and *GluB4* genes, which silences their expression completely. Therefore, the phenotype of the *glu1* is a complete loss of the acidic subunit of glutelin, and the *glu1* acts as a recessive gene for low glutelin content in the grains (Fig. 12.16).

12.5.2 Useful Mutations Induced with Acute or Chronic Gamma-Ray Irradiation

Shu et al. (2012) listed optimum doses of acute and chronic gamma-ray irradiation to induce mutations in various crop plants (Appendices 2 and 3).

Differences among the dosage effects have been discussed between acute irradiation (mainly to the dormant seed in the Gamma Room: e.g., totally 100–1000 Gy for 2 days) and chronic irradiation (mainly the whole growing period, i.e., from the seedling stage to harvest time through the reproductive stage) at the optimum length from the ^{60}Co radiation source in the Gamma Field, less than 2–3 Gy/day depending on the planting points. The fundamental differences between the treatments are the following: (1) dose rate: acute irradiation is conducted with a high dosage for a short period, and chronic irradiation is conducted with a lower dosage for longer period, more than several years for perennial plants; and (2) irradiated parts and stages: acute irradiation is often conducted to the dormant seed or vegetative parts, but only at certain stages of plant growth such as booting stages: however, chronic irradiation is conducted all through the growing stages when the cell division of the plant is active (from germination to the mature stage), and even to a reproductive stage. Tanaka and Sekiguchi (1966) reported that the highest chlorophyll mutation rate was achieved by chronic gamma-ray irradiation to growing rice plants especially at the meiosis stages. Chronic irradiation was seen to be more efficient at the seed germinating stage, while the lowest frequencies of mutation were observed using acute irradiation to the dormant seeds. However, prior to the construction of the Gamma Fields, the use of the acute irradiation to the seed was a common but inefficient approach. The insufficiency of this method resides with the low survival rate and high sterility of recovered mutant lines (Niwa and El-Keredy 1967). It is estimated that direct physiological damages of acute gamma-ray irradiation to the irradiated seed are higher than those of chronic irradiation in the Gamma Field, even though the total radiation dose is similar (Niwa and El-Keredy 1967).

The Gamma Phytotron was established in Korea in 2005 and the Gamma Green House, the size of which is approximately five times larger than that of the Gamma Greenhouse located at the IRB, Japan, was established in Malaysia in 2008 (Nakagawa and Kato 2017). Both facilities are for mutation induction by a chronic gamma-ray irradiation to growing plants or in vitro culture of medicinal crops, ornamental crops, and tropical crops. As discussed in the following parts of this report, chronic irradiation is a useful tool to generate mutant genome resources that are applied to molecular analysis as well as conventional breeding.

One of the serious disadvantages of such facilities is that management of the facilities is hampered by high costs and a staff having a radiation protection supervision to monitor the safe operation of the facility. However, irrespective of

Fig. 12.17 A shelf between the ^{60}Co source of the irradiation tower and the nearest field, where the rice plants in pots were placed at ca. 5–3 m from the source in the Gamma Field. (Courtesy of IRB, NARO)

the cost and unit maintenance, a breeder who desires to develop new cultivars with certain characteristics can obtain his or her targeted mutants from selected populations and have the material irradiated locally. If success is achieved, the cost of the breeding laboratory for the breeding program is far less expensive than applying GM technology, or even the genome editing technologies.

Another disadvantage or limitation regarding the management of a Gamma Field, as well as all radiation facilities in Japan or elsewhere, is that the facility must commit to the governmental regulation regarding dose limitation that is approved by the Recommendations of the International Commission on Radiological Protection (ICRP) (ICRP 2007). The tendency of the recommended limitation is to utilize the lower limits of radiation treatments based on the latest scientific information. To meet the Commission's regulation without expanding the size of a Gamma Field, the only option is to lower the total irradiation dose in a day, that is, to shorten the irradiation time per day. When the Gamma Field was established in 1961, irradiation was conducted for 20 h per day, from 9:00 AM to 5:00 AM of the next morning for 5 days of the week from Monday to Friday. However, following the dose limit regulation, the Gamma Field is now operated for 8 h per day, from 12:00 to 20:00 for 5 days of the week.

As with the induction rates of mutation in rice with lower radiosensitivity, a shelf was constructed between the ^{60}Co source and the nearest field, where the rice plants are in pots or directly water cultured (ca. 5–3 m from the source) in a water pool in the Gamma Field (Figs. 12.2 and 12.17). Using this shelf, higher dosage is available such as 6 Gy/day (Takano et al. 2005) and 20 Gy/day (Ma et al. 2005) in the Gamma Field, which are explained in the following section.

12.5.2.1 Phytochrome

Takano et al. (2005) identified the relationship between phytochrome A (*phyA*), B (*phyB*), and C (*phyC*) by isolating the mutants in rice and by developing all possible combinations of the double mutants. They characterized their phenotypes mainly for the response to photoperiod and to different light qualities such as red and far-red light. Seedlings of the *phy B* and the *phyB phyC* double mutants partly lost the sensitivity to continuous red light but exhibited significant de-etiolation or greening responses. However, the *phyA phyB* double mutants exhibited no responses to red light. These results demonstrated that the *phyA* and the *phyB* redundantly act to control the de-etiolation under red-light irradiation. Furthermore, they proved that mutations either in the *phyB* or in the *phyC* locus provide a cause of moderate early flowering under long-day photoperiod condition, but monogenic *phyA* mutations give little effect on flowering time. Interestingly, the *phyA* mutation in combination with the *phyB* or the *phyC* mutation exhibited a dramatic early-flowering phenotype.

In this research, *Tos17* mutant panels explained in Sect. 12.3.1.10 played an important role in isolating the *phyA* and the *phyC* mutants, which were efficiently screened by PCR using *PHYA* and *Tos17*-specific primers (Takano et al. 2001). However, the *phyB* mutants were obtained through chronic gamma-ray irradiation with dose rates ranging between 3 and 6 Gy/day (Takano et al. 2005).

12.5.2.2 Aluminum Tolerance

Ma et al. (2005) isolated a highly aluminum (Al)-sensitive mutant from a highly resistant rice cv. Koshihikari following the use of chronic gamma-ray irradiation. The highly Al-sensitive mutant gene they generated is called *als1*. The *ALS*, an Al-tolerant gene, encodes a phloem-localized ABC transporter-like protein that has been cloned in Arabidopsis (Larsen et al. 2005). In this research, M_1 plants were irradiated in the Gamma Field at a dose under 20 Gy/day for 8 days from 7 days before heading to 2 days after heading. A sophisticated screening method was then applied by measuring each elongated root for 24 h both with Al and without Al solution (Ma et al. 2005).

In the absence of an Al solution, the Al-sensitive mutant exhibited the same phenotype as the wild type. The Al-sensitive mutant also exhibited poorer root growth in acid soil (Ma et al. 2005). Genetic analysis of F_2 population of the hybridization between this mutant and the wild type revealed that the Al-resistant plants and Al-sensitive plants segregated in a 3:1 ratio, which indicated that the high sensitivity to Al is under the control of a single recessive *als1* gene. This gene was mapped to the long arm of chromosome 6.

12.6 Conclusions

A.M. van Harten (1998) gave his opinion in a book of "Mutation Breeding—Theory and Practical Application": "An explanation for the decreasing interest in mutation breeding, at least in most 'developed' countries, may be that during the past two decades attention has become more and more directed towards studying the possibilities offered to plant breeding by various new molecular technologies ... As results of these developments mutation breeding seems to have lost part of its previous attraction for young researchers."

This author believes that mutation breeding using radiation is a very powerful technology for induction of mutations for crop breeding and molecular research, isolation of genes, and elucidation of molecular mechanisms and metabolic pathways in every crop. The record has also shown as in this review that mutation induction has been an extremely useful aid to conventional breeding for developing superior cultivars and breeding materials.

We have discussed the mutation rate of various treatments and level of mutation induction many times. The author, as a plant breeder, understands that the mutation induction rate of a chemical mutagen is much higher than that of gamma-ray irradiation because of its mutation induction mechanism. In that light of the above, chemical mutagenesis is a very useful tool for molecular research that investigates the underlying molecular mechanisms of gene expression or physiological metabolism. However, one of the problems of chemical mutagens for breeding applications is the high mutation rate. Since a treated plant could have many mutations including the targeted mutation, this might not be a clean and precise approach, as it makes the development of a single mutation isoline virtually impossible. On the contrary, as the mutation rate from irradiation is relatively lower than chemical mutagenesis, it could be concluded that only one or two mutations are mostly induced through an optimum dosage of gamma-ray irradiation in a single plant, when a targeted mutant is found in the field. Therefore, the selection and subsequent breeding process using these mutants induced by radiation are much easier than those generated by chemical mutagens for breeders.

Recently, mutation breeding has often been compared with genome editing technology. The author agrees that the genome editing technology is superb because it can produce deletions and insertion of DNA(s) in a targeted locus precisely and at a high frequency. The deletion-induction is like mutation breeding, and the insertion-induction is like genetic modification. However, genome editing technologies are useless if breeders do not know the genome sequence of the species or the gene (s) which controls the targeted phenotype, as well as without total genome sequencing data of the crops. These advantages of having complete genomic sequences are available to a limited number of species. The genome sequences of so-called orphan crops are not generally available or complete.

The author believes that based on the achievements of traditional or classic methods, novel fields of science and molecular technologies have been developed, and it is hoped that mutation breeding remains accessible as a useful technique in the field of plant breeding and genetics, especially by incorporating emerging knowledge and technologies.

During the course of writing this review, information was received that the operation of the Gamma Field in Japan was terminated and the ^{60}Co source was removed from the irradiation tower on May 30, 2019, due to the increasing difficulty to fund and maintain the aging facility that was damaged by the Great East Japan Earthquake in 2011. The Gamma Room at the facility, however, is still in operation.

Though the Gamma Field operations in Japan were terminated, there remain other chronic irradiation facilities in Korea and Malaysia as described in this report. The author hopes that this review would support the operation of these facilities.

Acknowledgments I dedicate this report to two great researchers, Emeritus prof. Yasuo Ukai and Dr. Pierre J.L. Lagoda of the Joint FAO/IAEA, who have inspired the author and cultivated larger vision regarding the capacity of mutation breeding.

The author would like to express his sincere thanks to Emer. Prof. Yutaka Takagi, Saga Univ.; Prof. Makoto Kusaba, Hiroshima Univ.; Dr. Ryouhei Morita, RIKEN; Dr. Makoto Takano; Dr. Hiroshi Kato; Dr. Makita Hajika; Dr. Hiroshi Mizuno; Dr. Yutaka Sato; Dr. Shigeki Nagatomi, NARO; as well as the Institute of Radiation Breeding, NARO, for providing figures and/or informative comments for this report. The author identified the number of mutant varieties in the world from the FAO/IAEA Mutant Variety Database (http://mvgs.iaea.org/AboutMutantVarities.aspx). The author identified the linage of rice cultivars in the useful database of "Rice Varieties and Characteristics Database" provided by the Institute of Crop Sciences, NARO (https://ineweb.narcc.affrc.go.jp/). The author identified the cultivated area of rice cultivars from https://www.komenet.jp/pdf/H30sakutuke.pdf supplied by the Rice Stable Supply Support Organization (Beikoku Kikou) and those of soybean from http://www.maff.go.jp/j/seisan/ryutu/daizu/d_data/attach/pdf/index-19.pdf by the Ministry of Agriculture, Forestry and Fisheries and http://www.jsapa.or.jp/daizu/date28/28-all.pdf by Japan Speciality Agriculture Products Association (JSAPA). The author would like to express his thanks to these sites.

The author would like to express his sincere thanks to Dr. Benjamin Ubi, Director of Biotechnology Research & Development Centre and Professor of Ebonyi State University for his critical reading of the manuscript and advices.

Conflicts of Interest The author declares no conflict of interest.

References

Abe T, Ryuto H, Fukunishi N (2012) Ion beam radiation mutagenesis. In: Shu QY, Forster BP, Nakagawa H (eds) Plant mutation breeding and biotechnology. CAVI, Oxfordshire, UK, pp 99–106

Abe T, Kuramata M, Igura M, Arao T, Makino T, Sunohara Y, Kuroki M, Ishikawa S (2017) "Koshihikari Kan No. 1", a new rice variety with nearly cadmium-free in grains. Breed Res 19: 109–115. (In Japanese)

Agrawal GK, Yamazaki M, Kobayashi M, Hirochika R, Miyao A, Hirochika H (2001) Screening of the rice viviparous mutants generated by endogenous *Tos17* insertion: tagging of a zeaxanthin epoxidase gene and a novel *OsTATC* gene. Plant Physiol 125:1248–1257

Akao S, Kouchi H (1992) A supernodulating mutant isolated from soybean cultivar Enrei. Soil Sci Plant Nutr 38:183–187

Anai T, Yamada T, Kinoshita T, Rahman SM, Takagi Y (2005) Identification of corresponding genes for three low-α-linolenic acid mutants and elucidation of their contribution to fatty acid biosynthesis in soybean seed. Plant Sci 168:1615–1623

Aoshima K (1987) Epidemiology of renal tubular dysfunction in the inhabitants of a cadmium-polluted area in the Jinzu River basin in Toyama Prefecture. Tohoku J Exp Med 152:151–172

Arabidopsis Genome Initiative (2000) Analysis of the genome sequencing of the flowering plant *Arabidopsis thaliana*. Nature 408:796–815

Arai S, Noguchi M, Kaji M, Kato H, Fujimaki M (1970) n-Hexanal and some volatile alcohols. Their distribution in row soybean tissue and formation in crude soy protein concentrate by lipoxygenase. Agric Biol Chem 34:1420–1423

Arao T, Kawasaki A, Baba K, Mori S, Matsumoto S (2009) Effects of water management on cadmium and arsenic accumulation and dimethylarsinic acid concentrations in Japanese rice. Environ Sci Technol 43:9361–9367

Ashikari M, Sasaki A, Ueguchi-Tanaka M, Itoh H, Nishimura A, Datta S, Ishiyama K, Saito T, Kobayashi M, Khush GS, Kitano H, Matsuoka M (2002) Loss-of-function of a rice gibberellin biosynthetic gene, *GA20 oxidase* (*GA20ox-2*), led to the rice 'Green Revolution'. Breed Sci 52: 143–150

Bhattacharyya MK, Smith AM, Ellis THN, Hedley C, Martin C (1990) The wrinkled-seed character of pea described by Mendel is caused by a transposon-like insertion in a gene encoding starch-branching enzyme. Cell 60:115–122

Bird D, Beisson F, Brigham A, Shin J, Greer S, Jetter R, Kunst L, Wu XW, Yephremov A, Samuel L (2007) Characterization of Arabidopsis ABCG11/WBC11, an ATP binding cassette (ABC) transporter that is required for cuticular lipid secretion. Plant J 52:485–498

Blum A (1975) Effect of the *Bm* gene on epicuticular wax and water relations of *Sorghum bicolor* (L.) Moench. Isr J Bot 24:50–51

Broad WJ (2007) Useful mutants, bred with radiation. The New York Times. https://www.nytimes.com/2007/08/28/science/28crop.html

Codex Alimentarius (2008) CODEX STAN 193-1995, Codex general standard for contaminants and toxins in foods and feed. www.codexalimentarius.net/download/standards/17/CXS_193e.pdf. Accessed 30 Jun 2012

Cong L, Ran FA, Cox D, Lin S, Barretto R, Habib N, Hsu PD, Wu X, Jiang W, Marraffini LA, Zhang F (2013) Multiplex genome engineering using CRISPR/Cas systems. Science 339:819–823

Constantin MJ, Klobe WD, Skold LN (1976) Effects of physical and chemical mutagens on survival, growth and seed yield of soybean. Crop Sci 16:49–52

Dalrymple DG (1986) Development and spread of high-yielding rice varieties in developing countries. Bur. Sci. Tech., Agency for Int Dev, Washington, DC. 116 pp

De Jong J, Custers JMB (1986) Induced changes in growth and flowering of chrysanthemum after irradiation and *in vitro* culture of pedicels and petal epidermis. Euphytica 35:137–148

Fukushima D (1991) Recent progress of soybean protein foods: chemistry, technology, and nutrition. Food Rev Intl 7:323–351

Futsuhara Y (1968) Breeding of a new rice variety Reimei by gamma-ray irradiation. Gamma Field Symp 7:87–109

Gotoh K (1964) Mutation breeding in soybeans and common beans. Gamma Field Symp 3:74–89

Hajika M, Igita K, Kitamura K (1991) A line lacking all the seed lipoxygenase isozymes in soybean (*Glycine max* (L.) Merrill) induced by gamma-ray irradiation. Japan J Breed 41:507–509

Hajika M, Takahashi M, Igita K, Sakai S, Nakazawa Y (2002) A new soybean variety 'Ichihime'. Bull Nat Agric Res Cent Kyushu Okinawa Reg 40:79–93. (In Japanese with English summary)

Hajika M, Igita K, Nakazawa Y (2005) Induction of a soybean (*Glycine Max* (L.) Merrill) line lacking all seed lipoxygenase isozymes. JARQ 29:73–76

Hildebrand DF, Hymowitz T (1981) Two soybean genotypes lacking lipoxygenase-1. J Am Oil Chem Soc 58:583–586

Hiraiwa S, Tanaka S, Nakamura S (1976) Induction of mutants with higher protein content in soybean. In: Evaluation of seed protein alteration by mutation breeding. IAEA, Vienna, p 185

Hirochika H, Fukuchi A, Kikuchi F (1992) Retrotransposon families in rice. Mol Gen Genet 233: 209–216

Hirochika H, Sugimoto K, Otsuki Y, Tsugawa H, Kanda M (1996) Retrotransposons of rice involved in mutations induced by tissue culture. Proc Natl Acad Sci U S A 93:7783–7788

ICRP (2007) The 2007 recommendations of the International Commission of Radiological Protection. ICRP Publication 103. Ann ICRP 37(2–4):1–332

Iida S (1995) Screening and genetic analysis of rice seed storage protein mutants. Gamma Field Symp 34:31–48

Iida S, Amano E, Nishio T (1993) A rice (*Oryza sativa* L.) mutant having a low content of glutelin and a high content of prolamin. Theor Appl Genet 87:374–378

Iida S, Kusaba M, Nishio T (1997) Mutants lacking glutelin subunits in rice: mapping and combination of mutated glutelin genes. Theor Appl Genet 94:177–183

Iizuka M, Kimura Y, Okada T, Hase Y, Tanaka A, Sekiguchi M (2011) A new cultivar "Viento Flamingo" of *Osteospermum*. Bull Gunma Agric Res Cent 8:89–92. (In Japanese with English summary)

Imai Y (1934) The effect of X-rays on the production of sterile rice. Japan J Genet 10:233–236

Imai Y (1935) Chlorophyll deficiencies in *Oryza sativa* induced by X-rays. Japan J Genet 11:157–161

Imakiire S, Nagayoshi S, Koriyama K, Ueno K (2005) Breeding process and characteristics of standard type chrysanthemum cultivar "Araddin". Bull Kagoshima Agric Expt Stn 34:15–19. (In Japanese with English summary)

Imbe T, Horisue N, Akama Y, Nakane A, Ando I, Suto M, Hata T, Ise K, Numaguchi K, Furutachi H, Takadate M (2004a) "Odorokimochi" - a multipurpose high-yield glutinous rice variety. Bull Natl Inst Crop Sci 6:19–35. (In Japanese with English summary)

Imbe T, Akama Y, Nakane A, Hata T, Ise K, Ando I, Uchiyamada H, Nakagawa N, Furutachi H, Horisue N, Noto M, Fujita Y, Kimura K, Mori K, Takayanagi K, Uehara Y, Ishizaka S, Nakagahara M, Yamada T, Koga Y (2004b) Development of a multipurpose high-yielding rice variety "Takanari". Bull Natl Inst Crop Sci 5:35–51. (In Japanese with English summary)

International Rice Genome Sequence Project (2005) The map-based sequence of the rice genome. Nature 436:793–800

Ise K, Akama Y, Horisue N, Nakane A, Yokoo M, Ando I, Hata T, Suto M, Numaguchi K, Nemoto H, Furutachi H, Imbe T (2001) "Milky Queen", a new high quality rice cultivar with low amylose content in endosperm. Bull Natl Inst Crop Sci 2:39–61. (In Japanese with English summary)

Ishikawa S (2020) Development of paddy rice cultivar "Koshihikari Kan 1 Gou" which does not absorb cadmium, and technical issues for dissemination. Agric Hortic 95:675–678. (In Japanese: The author translated the title)

Ishikawa G, Takada Y, Nakamura T (2006) A PCR-based method to test for the presence or absence of β-conglycinin α′- and α-subunits in soybean. Mol Breed 17:365–374

Ishikawa S, Ishimaru Y, Igura M, Kuramata M, Abe T, Senoura T, Hase Y, Arao T, Nishizawa NK, Nakanishi H (2012) Ion-beam irradiation, gene identification, and marker-assisted breeding in the development of low-cadmium rice. Proc Natl Acad Sci U S A 109:19166–19171

Ishikawa S, Makino T, Ito M, Harada K, Nakada H, Nishida I, Nishimura M, Tokunaga T, Sjirao K, Yoshizawa C, Matsuyama M, Abe T, Arao T (2016) Low-cadmium rice (*Oryza sativa* L.) cultivar can simultaneously reduce arsenic and cadmium concentration in rice grains. Soil Sci Plant Nutr 62:327–339

Iwata M, Omura T (1971) Linkage analysis of reciprocal translocation method in rice plant (*Oryza sativa* L.). II. Linkage groups corresponding to the chromosome 5, 6, 8, 9, 10 and 11. Sci Bull Fac Agric Kyushu Univ 30:137–153. (In Japanese with English summary)

Japan Atomic Energy Agency (2007) FY2007 Survey on economic scale of radiation utilization report - Summary version. 18th Atomic Energy Commission Material No. 1, 15pp. (In Japanese). http://www.aec.go.jp/jicst/NC/iinkai/teirei/siryo2008/siryo18/siryo1.pdf

Jenks MA, Joly RJ, Peters PJ, Rich PJ, Axtell JD, Ashworth EN (1994) Chemically induced cuticle mutation affecting epidermal conductance to water vapor and disease susceptibility in *Sorghum bicolor* (L.) Moench. Plant Physiol 105:1239–1245

Jinek M, Chylinski K, Fonfara I, Hauer M, Doudna JA, Charpentier E (2012) A programmable dual-RNA-guided DNA endonuclease in adaptive bacterial immunity. Science 337:816–821

Jordan WR, Monk RL, Miller FR, Rosenow DT, Clark LE, Shouse PJ (1983) Environmental physiology of sorghum: I. Environmental and genetic control of epicuticular was load. Crop Sci 23:552–558

Kaizuma N (1975) Studies on the theoretical base of seed protein improvement in soybean. J Fac Agric Iwate Univ 12:159–264. (In Japanese with English summary)

Kasuga S, Inoue N, Kaidai H, Watanabe H (2001) Effects of brown midrib and bloomless genes on the resistance to sheath blight (*Rhizoctonia solani* Kuehn) in sorghum. Grassl Sci 47:256–261

Katagiri K (1973) Radiation damage in winter buds and relation of shoot cutting-back to mutation frequencies and spectra in acutely gamma-irradiated mulberry. Gamma Field Symp 12:63–81

Kato T (1969) High susceptibility of an X-ray sensitive *E. coli* strain to single-strand break by X-rays. Gamma Field Symp 8:1–15

Kato H, Yoshioka H, Iba R, Ueda S, Tsutsumi S, Mitsueda H, Wakasugi K, Nakahara T, Takeda H (2006) A newly bred rice variety "Minamiyutaka" for whole-crop silage. Bull Miyazaki Agric Expt Stn 41:51–60. (In Japanese with English summary)

Kawahara K (1967) Structural characteristics and the exposure rates in the Japan's Gamma Field. Gamma Field Symp 6:1–18

Kazama Y, Hirano T, Saito H, Liu Y, Ohbu S, Hayashi Y, Abe T (2011) Characterization of highly efficient heavy-ion mutagenesis in *Arabidopsis thaliana*. BMC Plant Biol 11:161

Kim DW (2006) Yoshio Nishina and two cyclotrons. Hist Stud Phys Biol Sci 36:243–273

Kinoshita T (1980) Induction of cytoplasmic male sterility by gamma ray and chemical mutagenesis in sugar beets. Gamma Field Symp 19:27–48

Kitagawa K, Nagae M, Uchida M, Inoue K, Murata K, Masuda T, Yoshioka T, Kotobuki K (1999) A new Japanese pear cultivar, 'Kotobuki Shinsui'. Bull Tot-tori Hort Expt Stn 3:1–13. (In Japanese)

Kitamura K, Kaizuma N (1981) Mutant strains with low level of subunits of 7S globulin in soybean seed. Japan J Breed 31:353–359

Kitamura K, Davies CS, Kaizuma N, Nielsen NC (1983) Genetic analysis of a null-allele for lipoxygenase-3 in soybean seeds. Crop Sci 23:924–927

Kitamura K, Kumagai T, Kikuchi A (1985) Inheritance of lipogenase-2 and genetic relationships among genes for lipogenase-1, 2 and -3 isozymes in soybean seeds. Japan J Breed 35:413–420

Kitamura K, Ishimoto M, Kikuchi A, Kaizuma N (1992) A new soybean variety "Yumeyutaka" lacking both L-2 and L-3 of the lipoxygenase isozymes. Japan J Breed 42:905–913. (In Japanese with English summary)

Klobe WD, Constantin MJ, Skold LN (1973) Dose response of soybean seed to various mutagens. In: Agronomy abstracts, vol 8. American Society Agronomy, Madison, WI

Koga Y, Uchiyamada H, Samoto S, Ishizaka S, Fujita Y, Okuno K, Uehara Y, Nakagahra M, Horiuchi H, Miura K, Maruyama K, Yamada T, Yagi T, Mori K (1989) Breeding a new rice cultivar "Kinuhikari". Bull Hokuriku Natl Agric Exp Stn 30:1–24. (In Japanese with English summary)

Kokubun M, Akao S (1994) Inheritance of supernodulation in soybean mutant En6500. Soil Sci Plant Nutr 40:715–718

Konishi S, Izawa T, Lin SY, Ebata K, Fukuta Y, Sasaki T, Yano M (2006) An SNP caused loss of seed shattering during rice domestication. Science 312:1392–1396

Kume T, Amano E, Nakanishi TM (2002) Economic scale of utilization of radiation (II): agriculture comparison between Japan and the U.S.A. J Nucl Sci Technol 39:1106–1113

Kusaba M, Miyahara K, Iida S, Fukuoka H, Takano T, Sassa H, Nishimura M, Nishio T (2003) Low glutelin content1: a dominant mutation that suppresses the glutelin multigene family *via* RNA silencing in rice. Plant Cell 15:455–1467

Kusaba M, Ito H, Morita R, Iida S, Sato Y, Fujimoto M, Kawasaki S, Tanaka R, Hirochika H, Nishimura M, Tanaka A (2007) Rice NON-YELLOW COLORING1 is involved in light-harvesting complex II and grana degradation during leaf senescence. Plant Cell 19:1362–1375

Kyoya M, Taguchi M, Kodama M, Yamamoto T, Sato Y (2002) New rice cultivar "HAENUKI". Bull Akita Pref Agric Expt Stn 42:1–20. (In Japanese with English summary)

Laboratory of Soybean Breeding (1970) The new soybean varieties "Raiden" and "Raiko" induced by gamma-ray irradiation. Bull Touhoku Agric Expt Stn 40:65–105. (In Japanese with English summary)

Larkin PJ, Scowcroft WR (1981) Somaclonal variation – a novel source of variability from cell cultures for plant improvement. Theor Appl Genet 60:197–214

Larsen PB, Geisler MJB, Jones CA, Williams KM, Cancel JD (2005) *ALS3* encodes a phloem-localized ABC transporter-like protein that is required for aluminum tolerance in Arabidopsis. Plant J 41:353–363

Lester DR, Ross JJ, Davies PJ, Reid JB (1997) Mendel's stem length gene (*Le*) encodes a gibberellin 3 β-hydroxylase. Plant Cell 9:1435–1443F

Li F, Numa H, Hara N, Sentoku N, Ishii T, Fukuta Y, Nishimura N, Kato H (2019a) Identification of a locus for seed shattering in rice (*Oryza sativa* L.) by combining bulked segregant analysis with whole-genome sequencing. Mol Breed 39:36

Li F, Shimizu A, Nishio T, Tsutsumi N, Kato H (2019b) Comparison and characterization of mutations induced by gamma-ray and carbon-ion irradiation in rice (*Oryza sativa* L.) using whole-genome resequencing. G3 (Bethesda) 9:3743–3751

Li F, Komatsu A, Ohtake M, Eun H, Shimizu A, Kato H (2020) Direct identification of a mutation in *OsSh1* causing non-shattering in a rice (*Oryza sativa* L.) mutant cultivar using whole-genome resequencing. Sci Rep 10:14936

Lin ZW, Li XR, Shannon LM, Yeh CT, Wang ML, Bai GH, Peng Z, Li JR, Trick HN, Clement TE, Doebley J, Schnable PS, Tuinstra MR, Tessp TT, White F, Yu JM (2012) Parallel domestication of the *Shattering1* gene in cereals. Nat Genet 44:720–724

Ma JF, Nagao S, Huang CF, Nishimura M (2005) Isolation and characterization of a rice mutant hypersensitive to Al. Plant Cell Physiol 46:1054–1061

Masuda T, Yoshioka T, Inoue K, Murata K, Kitagawa K, Tabira H, Yoshida A, Kotobuki K, Sanada T (1997) Selection of mutants resistant to black spot disease by chronic irradiation of gamma-rays in Japanese pear 'Osanijisseiki'. J Japan Hortic Sci 66:85–92

Masuda T, Yoshioka T, Sanada T, Kotobuki K, Nagae M, Uchida M, Inoue K, Murata K, Kitagawa K, Yoshida R (1998) A new Japanese pear cultivar 'Osa Gold', resistant mutant to the black spot disease of Japanese pear (*Pyrus pyrifolia* Nakai) induced by chronic irradiation of gamma-rays. Bull Natl Inst Agrobiol Resour (Japan) 12:1–11. (In Japanese with English summary)

Matsumoto S, Kokubun K, Murakami S, Nagasawa T, Koyama T, Nakamura S, Hashimoto K, Ishikawa M, Miyahara T, Matsumoto S, Konno Z (1979) A new soybean variety "Nanbushirome". Bull Tohoku Natl Agric Exp Stn 59:87–104. (In Japanese with English summary)

Mendel G (1866) Versuche über Plflanzen-hybriden. Verh Naturforsch Ver Brunn 4:3–47. (In German)

Mikami T, Kawamura Y, Yokoyama H, Takadate M, Kobayashi W, Tateyama M, Maeda K, Nakahori T, Oyamada Z (2007) A new rice variety "Masshigura". Bull Aomori Agric For Res Cent 41:23–44. (In Japanese with English summary)

Miyao A, Iwasaki Y, Kitano H, Itoh J, Maekawa M, Murata K, Yatou O, Nagato Y, Hirochika H (2007) A large-scale collection of phenotypic data describing an insertional mutant population to facilitate functional analysis of rice genes. Plant Mol Biol 63:625–635

Mizuno H, Kawahigashi H, Ogata J, Minami H, Kanamori H, Nakagawa H, Matsumoto T (2013) Genomic inversion caused by gamma irradiation contributes to downregulation of a *WBC11* homolog in bloomless sorghum. Theor Appl Genet 126:1513–1520

Mochizuki T, Hara S (2000) Usefulness of low protein rice in diet therapy in patients with chronic renal failure. Jpn J Nephrol 42:24–29. (In Japanese with English summary)

Morimoto T, Ito T, Tanabe K, Nakamori M, Taniguchi M, Inagaki A, Inoue M, Asai Y, Ito Y, Shumiya A, Fujii K (1980) A new rice variety "Mineasahi". Res Bull Aichi Agric Res Cent 12: 24–36. (In Japanese with English summary)

Morita R (2008) Analysis and application of gamma-ray-induced mutations. Gamma Field Symp 47:41–48

Morita R, Kusaba M, Iida S, Nishio T, Nishimura M (2007) Knockout of glutelin genes which form a tandem array with a high level of homology in rice by gamma irradiation. Genes Genet Syst 82:321–327

Morita R, Kusaba M, Iida S, Yamaguchi H, Nishio T, Nishimura M (2009) Molecular characterization of mutations induced by gamma irradiation in rice. Genes Genet Syst 84:361–370

Mukade K, Kamio M, Hosoda K (1970) The transfer of leaf-rust resistance from rye to wheat by intergeneric addition and translocation. Gamma Field Symp 9:68–87

Nagasawa N, Hibara K, Heppard EP, Vander Velden KA, Luck S, Beatty M, Nagato Y, Sakai H (2013) *GIANT EMBRYO* encodes CYP78A13, required for proper size balance between embryo and endosperm in rice. Plant J 75:592–605

Nagatomi S, Miyahara E, Degi K (1996) Combined effect of gamma irradiation methods and *in vitro* explant sources on mutation induction of flower color in *Chrysanthemum morifolium* RAMAT. Gamma Field Symp 35:51–69

Naito T, Yamaguchi H (1969) Protein synthesis and ribonucleic acid metabolism in soybean embryos after seed irradiation with γ-rays. Gamma Field Symp 8:95–107

Naito K, Kusaba M, Shikazono N, Takano T, Tanaka A, Tanisaka T, Nishimura M (2005) Transmissible and nontransmissible mutations induced by irradiating *Arabidopsis thaliana* pollen with γ-rays and carbon ions. Genetics 169:881–889

Nakagawa H (2008) Mutation breeding, status quo and future. Techno Innov 68:6–12. (In Japanese)

Nakagawa H, Kato H (2017) Induced mutations for food and energy security: challenge of inducing unique mutations for new cultivars and molecular research. Bull NARO Crop Sci 1:33–124. http://www.naro.affrc.go.jp/publicity_report/pub2016_or_later/files/nics_report01_4.pdf

Nakagawa H, Anai T, Okabe A, Takahashi K, Hajika M, Takagi Y (2011a) Mutation breeding of soybean in Japan. In: Khan S, Kozgar MI (eds) Breeding of pulse crops. Kalyani Publishers, India, pp 55–84

Nakagawa H, Okuizumi H, Nishimura M (2011b) Characteristics and inheritance of *bm* mutant induced by gamma-ray irradiation. Breed Res 13(Suppl. 1):111. (In Japanese)

Nakagawa H, Gau M, Okuizumi H (2013) Introduction of genetic resources and breeding of cellulosic crops and forage grasses. Agric Hortic 88:754–768. (In Japanese)

Nakai S (1969) The studies of radiation-sensitive mutants in yeast. Gamma Field Symp 8:16–31

Nakaidzumi M, Murai K (1938) Effects of Be-D radiations upon *Vicia faba*. Nature 142:534–535

Nakamura D, Yokoh H, Hirota Y, Nonaka K, Shigyo T, Oh-ishi H, Shigetomi O, Takagi Y, Kishikawa H (1991) A new soybean cultivar "Murayutaka". Bull Saga Agric Expt Stn 27:21–42. (In Japanese with English summary)

Nakamura S, Yumoto S, Takahashi K, Tabuchi K, Adachi T, Kikuchi A, Kowata M, Banba H, Takahashi N, Okabe A, Watanabe I, Nagasawa T, Murakami S, Hashimoto K, Sakai S, Igita K (1996) A new soybean variety "Ryuhou". Bull Tohoku Natl Agric Exp Stn 91:1–11. (In Japanese with English abstract)

Nakashima T, Ueno T, Fukami H (1982) Structure elucidation of AK-toxins, host specific phytotoxic metabolites produced by *Alternaria kikuchiana* Tanaka. Tetrahedron Lett 23:4469–4472

Nakashima T, Ueno T, Fukami H (1985) Isolation and structure of AK-toxin I and II. Host-specific phytotoxic metabolites produced by *Alternaria alternata* Japanese phytotype. Agric Biol Chem 49:807–815

Nelson O, Pan D (1995) Starch synthesis in maize endosperms. Annu Rev Plant Physiol Plant Mol Biol 46:475–496

Nemoto H, Iida S, Maeda H, Ishii T, Nakagawa N, Hoshino T, Sakai M, Okamoto M, Shinoda H, Yoshida T (2001) A new rice cultivar with giant embryo, "Haiminori". Bull Chugoku Natl Agric Expt Stn 22:25–40. (In Japanese with English summary)

Nguyen VM (1974) Mutation studies with soybeans. Mutat Breed Newsl 3:10

Nishimura M (2006) Utilization of seed protein composition mutants in rice breeding. Gamma Field Symp 45:61–66

Nishimura S, Sugihara M, Kohmoto K, Otani H (1978) Two different phases in pathogenicity of the Alternaria pathogen causing black spot disease of Japanese pear. J Fac Agric Tottori Univ 13:1–10

Nishimura M, Kusaba M, Miyahara K, Nishio T, Iida S, Imbe T, Sato H (2005) New rice varieties with low levels of easy-to-digest protein, 'LGC-Katsu' and 'LGC-Jun'. Breed Sci 55:103–105

Nishina Y (1947) A Japanese scientist describes the destruction of his cyclotrons. Bull At Sci 3: 145–167

Niwa M, El-Keredy MS (1967) Differential sensitivities of rice plant at early growth stages to acute and chronic gamma-ray. Gamma Field Symp 6:47–69

Nonomura K, Morohoshi A, Nakano M, Eiguchi M, Miyao A, Hirochika H, Kurata N (2007) A germ cell-specific gene of the *ARGONAUTE* family is essential for the progression of premeiotic mitosis and meiosis during sporogenesis in rice. Plant Cell 19:2583–2594

Nybom N (1954) Mutation type in barley. Acta Agric Scand 4:430–456

Ogaki M, Nakashima-Tanaka E, Matsunaga S (1969) Genetic control of radiosensitivity in *Drosophila melanogaster*. Gamma Field Symp 8:32–52

Ogawa T, Tayama E, Kitamura K, Kaizuma N (1989) Genetic improvement of seed storage proteins using three variant alleles of 7S globulin subunits in soybean. Jpn J Breed 39:137–147

Ogawa T, Bando N, Tsuji H, Okajima H, Nishikawa K, Sasaoka K (1991) Investigation of the IgE-binding proteins in soybean by immunoblotting with the sera of the patients with atopic dermatitis. J Nutr Sci Vitaminol 37:555–565

Ohba K, Murai M (1969) Varietal difference in somatic mutability of sugi, *Cryptomeria japonica* D. Don. Gamma Field Symp 8:108–120

Ohmiya A, Toyoda T, Watanabe H, Emoto K, Hase Y, Yoshioka S (2012) Mechanism behind petal color mutation induced by heavy-ion-beam irradiation of recalcitrant Chrysanthemum cultivar. J Jpn Hortic Sci 81:269–274

Okada T, Sugishita T, Murakami T, Murai H, Saikusa T, Horino T, Onoda A, Kajimoto O, Takahashi R, Takahashi T (2000) Effect of defatted rice germ enriched with GABA for sleeplessness depression, autonomic disorder by oral administration. Nippon Shokuhin Kagaku Kogaku Kaishi 47:596–603. (In Japanese)

Okumoto Y, Yamagata H, Horibata A, Teraishi M, Tanisaka T (1999) A mutable slender glume gene in rice (*Oryza sativa* L.). Gamma Field Symp 38:29–41

Okuno K, Fuwa H, Yano M (1983) A new mutant gene lowering amylose content in endosperm starch of rice, *Oryza sativa* L. Jpn J Breed 33:387–394

Oono K, Okuno K, Kawai T (1984) High frequency of somaclonal mutations in callus culture of rice. Gamma Field Symp 23:71–95

Oya T, Ishii R (1999) Growth characteristics and nitrogen use efficiency in a supernodulating soybean mutant, En6500, derived from the variety of Enrei. Jpn J Crop Sci 68:278–282

Panikashvili D, Savaldi-Goldstein S, Mandel T, Yifhar T, Franke RB, Hofer R, Schreiber L, Chory J, Aharoni A (2007) The Arabidopsis *DESPERADO/AtWBC11* transporter is required for cutin and wax secretion. Plant Physiol 145:1345–1360

Peiretti RA, Amini I, Weibel DE, Starks KJ, Mcnew RW (1980) Relationship of "bloomless" (*bm bm*) sorghum to greenbug resistance. Crop Sci 20:173–176

Peschke VM, Phillips RL, Gengenbach BG (1987) Discovery of transportable element activity among progeny of tissue culture-derived maize plants. Science 238:804–807

Prina AR, Landau AM, Pacheco MG (2012) Chimeras and mutant gene transmission (C15). In: Shu QY, Forster BP, Nakagawa H (eds) Plant mutation breeding and biotechnology. CAVI, Oxfordshire, UK, pp 181–189

Rahman SM, Takagi Y, Kubota K, Miyamoto K, Kawakita T (1994) High oleic acid mutant in soybean induced by X-ray irradiation. Biosci Biotech Biochem 58:1070–1072

Rahman SM, Takagi Y, Miyamoto K, Kawakita T (1995) High stearic acid soybean mutant induced by X-ray irradiation. Biosci Biotech Biochem 59:922–923

Rahman SM, Kinoshita T, Anai T, Takagi Y (2001) Combining ability in loci for high oleic and low linolenic acids in soybean. Crop Sci 41:26–29

Rahman SM, Anai T, Kinoshita T, Takagi Y (2003) A novel soybean germplasm with elevated saturated fatty acids. Crop Sci 43:527–531

Rawling JO, Hanway DG, Gardner CO (1958) Variation in quantitative characters of soybeans after seed irradiation. Agron J 50:524–528

Rice Stable Supply Support Organization (Beikoku Kikou) (2020) The cropping trend of paddy rice produced in 2019. https://www.komenet.jp/pdf/R01sakutuke.pdf. (In Japanese)

Saikusa T, Horino T, Mori Y (1994a) Accumulation of γ-Aminobutyric Acid (GABA) in the rice germ during water soaking. Biosci Biotech Bioch 58:2291–2292

Saikusa T, Horino T, Mori Y (1994b) Distribution of free amino acids in the rice kernel fractions and the effect of water soaking on the distribution. J Agric Food Chem 42:1122–1125

Sakai M, Nakano H, Okamoto M, Tamura K, Kaiji R, Tamura Y, Kataoka T, Mizobuchi R (2013) "Ruriaoba", a new rice forage variety for double harvesting. Bull Natl Agric Res Cent Kyushu Okinawa Reg 60:1–12. (In Japanese with English summary)

Sakata M, Seno M, Matsusaka H, Takahashi K, Nakamura Y, Yamagata Y, Angeles ER, Mochizuki T, Kumamaru T, Sato M, Enomoto A, Tashiro K, Kuhara S, Satoh H, Yoshimura A (2016) Development and evaluation of rice giant embryo mutants for high oil content originated from a high-yielding cultivar 'Mizuhochikara'. Breed Sci 66:425–433

Samoto S, Kanai D (1975) Studies on the mutation breeding in rice. I. Short stiff mutations induced by gamma-ray irradiation to the rice variety Koshihikari. Jpn J Breed 25:1–7. (In Japanese with English summary)

Samoto M, Takahashi K, Fukuda Y, Nakamura S, Kawamura Y (1996) Substantially complete removal of the 34kD allergenic soybean protein, Gly m Bd 30K, from soy milk of a mutant lacking the α- and α′-subunit of conglycinin. Biosci Biotech Biochem 60:1911–1913

Sanada T (1986) Induced mutation breeding in fruit trees: resistant mutant to black spot disease of Japanese pear. Gamma Field Symp 25:87–108

Sanada T (1988) Selection of resistant mutants to black spot disease of Japanese pear by using host specific toxin. Jpn J Breed 38:198–204

Sanada T, Kotobuki K, Nishida M, Fujita H, Ikeda T (1993) A new Japanese pear cultivar 'Gold Nijisseiki', resistant to black spot disease of Japanese pear. Jpn J Breed 43:455–461. (In Japanese with English summary)

Sano Y (1985) Gene regulation at the waxy locus in rice. Gamma Field Symp 24:63–79

Sato Y, Sentoku N, Miura Y, Hirochika H, Kitano H, Matsuoka M (1999) Loss-of-function mutations in the rice homeobox gene *OSH15* affect the architecture of internodes resulting in dwarf plants. EMBO J 15:992–1002

Sato Y, Morita R, Nishimura M, Yamaguchi H, Kusaba M (2007) Mendel's green cotyledon gene encodes a positive regulator of the chlorophyll-degrading pathway. Proc Natl Acad Sci U S A 104:14169–14174

Sato T, Himori Y, Tsubaki N, Sano H (2015) Breeding of new green soybean varieties "Akinoush-40" and "Akita-honoka". Bull Akita Pref Agric Expt Stn 55:40–58. (In Japanese with English abstract)

Satoh H (1981) New endosperm mutations induced by chemical mutagens in rice, *Oryza sativa*. Jpn J Breed 31:316–326

Satoh H, Iwata N (1990) Linkage analysis in rice. On three mutant loci for endosperm properties, *ge* (*giant embryo*), *du-4* (*dull endosperm-4*) and *flo-1* (*floury endosperm-1*). Jpn J Breed 40(Suppl. 2):268. (In Japanese)

Shikazono N, Yokota Y, Tanaka A, Watanabe H, Tano S (1998) Molecular analysis of carbon ion-induced mutations in *Arabidopsis thaliana*. Genes Genet Syst 73:173–179

Shu QY, Forster BP, Nakagawa H (2012) Plant mutation breeding and biotechnology. CAVI, Oxfordshire, UK. 608pp. http://www.fao.org/3/a-i2388e.pdf

Sparrow AH, Evans HJ (1961) Nuclear factors affecting radiosensitivity I. The influence of nuclear size and structure, chromosome complement, and DNA content. Brookhaven Symp Biol 14:76–100

Stadler LJ (1930) Some genetic effects of X-rays in plants. J Hered 21:3–19

Tagawa S, Kashiwagi M, Kamada T, Sekiguchi M, Hosobuchi K, Tominaga H, Ooka N, Makuuchi K (2002) Economic scale of utilization of radiation (I): industry. J Nucl Sci Technol 39:1002–1007
Takagi Y (1968) Varietal differences of radiosensitivity in soybean. Jpn J Plant Breed 18(Extra 1): 35–36. (In Japanese)
Takagi Y (1969) The second type of gamma-ray sensitive gene rs_2 in soybean *Glycine max* (L.) Merrill. Gamma Field Symp 8:83–94
Takagi Y (1974) Studies on varietal differences of radiosensitivity in soybean. Bull Inst Radiat Breed 3:45–87. (In Japanese)
Takagi Y, Anai T (2006) Development of novel fatty acid composition in soybean oil by induced mutation. Oleoscience 6:195–203. (In Japanese with English abstract)
Takagi Y, Hiraiwa S (1980) Seed-coat color mutants in soybean (*Glycine max* (L.) Merrill) induced by gamma-irradiation and ethylenimine treatment. Bull Fac Agric Saga Univ 48:87–92
Takagi Y, Matsuo T, Kishikawa H (1986) Inheritance of oleic and linolenic acid in soybean seeds. Jpn J Breed 36:163–176. (In Japanese with English summary)
Takagi Y, Hossain ABMM, Yanagita T, Kusaba S (1989) High linolenic acid mutant induced by X-ray irradiation. Jpn J Breed 39:403–409
Takagi Y, Hossain ABMM, Yanagita T, Matsueda T, Murayama A (1990) Linolenic acid content in soybean improved by X-ray irradiation. Agric Biol Chem 54:1735–1738
Takagi Y, Rahman SM, Joo H, Kawakita T (1995) Reduced and elevated palmitic acid mutants in soybean developed by X-ray irradiation. Biosci Biotech Bioch 59:1778–1779
Takagi Y, Rahman SM, Anai T (1998) Construction of novel fatty acid composition in soybean oil by induced mutation. Gamma Field Symp 37:17–28
Takahashi M, Kokubun M, Akao S (1995) Characterization of nitrogen assimilation in a supernodulating soybean mutant En6500. Soil Sci Plant Nutr 41:567–575
Takahashi K, Yumoto S, Nakamura S, Mizuno Y, Kitamura K (1996) Inheritance of the α-subunit deficiency of β-conglycinin in soybean (*Glycine max* (L.) Merrill) line induced by gamma-ray irradiation. Breed Sci 46:251–255
Takahashi M, Matsunaga R, Komatsu K, Hajika M, Sakai S, Igita K, Nakazawa Y (2003a) Breeding of a new soybean cultivar "Eru-Star" and its characteristics. Bull Natl Agric Res Cent Kyusyu Okinawa Reg 42:49–64. (In Japanese)
Takahashi M, Arihara J, Nakayama N, Kokubun M, Shimada S, Takahashi K, Hajika M (2003b) Breeding of supernodulating soybean cultivar "Sakukei 4". Bull Nat Inst Crop Sci 4:17–28
Takahashi K, Shimada S, Shimada H, Takada Y, Sakai T, Kono Y, Adachi T, Tabuchi K, Kikuchi A, Yumoto S, Nakamura S, Ito M, Banba H, Okabe A (2004) A new soybean cultivar "Yumeminori" with low allergenicity and high content of 11S globulin. Bull Natl Agric Res Cent Tohoku Reg 102:23–39. (In Japanese with English summary)
Takano M, Kanegae H, Shinomura T, Miyao A, Hirochika H, Furuya M (2001) Isolation and characterization of rice phytochrome A mutants. Plant Cell 13:521–534
Takano M, Inagaki N, Xie X, Yuzurihara N, Hihara F, Ishizuka T, Yano M, Nishimura M, Miyao A, Hirochika H, Shinomura T (2005) Distinct and cooperative function of phytochromes A, B, and C in the control of deetiolation and flowering in rice. Plant Cell 17:3311–3325
Tanaka A (1999) Mutation induction by ion beams in Arabidopsis. Gamma Field Symp 38:19–28
Tanaka S, Sekiguchi F (1966) Studies on effective irradiation techniques to induce mutations in rice. Jpn J Breed 16:184–190
Tanaka A, Tano S, Chantes T, Yokota Y, Shikazono N, Watanabe H (1997) A new Arabidopsis mutant induced by ion beams affects flavonoid synthesis with spotted pigmentation in testa. Genes Genet Syst 72:141–148
Tazima Y, Murakami A (1969) Analysis of strain differences in radiosensitivity of the silkworm. Gamma Field Symp 8:53–68
Thanh VH, Shibasaki K (1976) Major proteins of soybean seeds: a straightforward fraction and their characterization. J Agric Food Chem 24:1117–1121

Toda M (1982) The breeding of four new mutant varieties by gamma-rays in rice. Gamma Field Symp 21:7–17

Tsai HK, Lu YC, Oka HI (1974) Mutation breeding of soybean for the resistance to rust disease. SABRAO J 6:181

Uehara Y, Inagaki A, Tanoue H, Yasukochi Y, Nagamura Y, Saji S, Otsuki Y, Fujimura T, Kurata N, Minobe Y (1995) Construction and characterization of a rice YAC library for physical mapping. Mol Breed 1:79–89

Uehara Y, Kobayashi A, Ohta H, Shimizu H, Fukui K, Miura K, Otsuki H, Komaki Y, Sasahara H (2002) A new rice variety "Shunyou". Bull Natl Agric Res Cent 1:1–21. (In Japanese with English summary)

Ueno T, Nakashima T, Uemoto M, Fukami H, Lee SN, Izumiya N (1977) Mass spectrometry of *Alternaria mali* toxins and related cyclodepsipeptides. Biomed Mass Spectrum 4:134–142

Ukai Y (1983) Early maturing mutations as germplasm stocks for barley breeding. Gamma Field Symp 22:49–72

Ukai Y (1990) Application of a new method for selection of mutants in a cross-fertilizing species to recurrently mutagen-treated populations of Italian ryegrass. Gamma Field Symp 29:55–69

Ukai Y (2000) Genetic analysis at the genomic level: map and QTL. University of Tokyo Press, Tokyo. 351pp. (In Japanese)

Ukai Y (2002) Genetic analysis of quantitative traits. IGAKU-SHUPPAN, Inc., Tokyo. 355pp. (In Japanese)

Ukai Y, Yamashita A (1969) Varietal differences in radiosensitivity with special reference to different aspects with different crops. Gamma Field Symp 8:69–81

van Harten AM (1998) Mutation breeding - theory and practical breeding. Cambridge University Press, Cambridge, UK. 353pp

Wang J, Kuang H, Zhang Z, Yang Y, Yan L, Zhang M, Song S, Guan Y (2020) Generation of seed lipoxygenase-free soybean using CRISPR-Cas9. Crop J 8:432–439

Watanabe Y (1981) The 20 years history of the Institute of Radiation Breeding. Gamma Field Symp 20:1–10

Weibel DE, Starks KI (1986) Greenbug nonpreference for bloomless sorghum. Crop Sci 26:1151–1153

White HBJ, Quackenbush FW, Probst AH (1961) Occurrence and inheritance of linolenic and linoleic acid in soybean seeds. J Am Oil Chem Soc 38:113–117

Williams JH, Hanway DG (1961) Genetic variation in oil and protein content of soybean induced by seed irradiation. Crop Sci 1:34–36

Yamaguchi I (2001) Forty years of mutation breeding in Japan - research and fruits. Gamma Field Symp 40:1–14

Yamaguchi I, Ukai Y, Yamashita A (1988) Induced mutations for disease resistance in barley. Gamma Field Symp 27:33–46

Yamakawa K, Sparrow AH (1965) Correlation of interphase chromosome volume and reduction of viable seed set by chronic irradiation of 21 cultivated plants during reproductive stage. Radiat Bot 5:557–566

Yamashita A (1981) Twenty years research of chronic gamma-ray irradiation on seed crops. Gamma Field Symp 20:87–102

Yang G, Luo W, Zhang J, Yan X, Du Y, Zhou L, Li W, Wang H, Chen Z, Guo T (2019) Genome-wide comparisons of mutations induced by carbon-ion beam and gamma-rays irradiation in rice *via* resequencing multiple mutants. Front Plant Sci 10:1514

Yoshida Y (1962) Theoretical studies on the methodological procedures of radiation breeding: I. New methods in autogamous plants following seed irradiation. Euphytica 11:95–111

Yoshii M, Shimizu T, Yamazaki M, Higashi T, Miyao A, Hirochika H, Omura T (2009) Disruption of a novel gene for a NAC-domain protein in rice confers resistance to Rice dwarf virus. Plant J 57:615–625

Yoshioka T, Ito Y, Masuda T (2000) Induction of mutant resistant to Alternaria blotch of apple by gamma-ray irradiation. Gamma Field Symp 39:69–79

Yumoto S, Shimada S, Takada Y, Sakai T, Kondo Y, Shimada H, Takahashi K, Adachi T, Tabuchi K et al (2006) A new Tohoku-district soybean cultivar "Suzusayaka" with three lipoxygenase isozyme deletions. Bull Natl Agric Res Cent Tohoku Reg 105:35–48. (In Japanese with English abstract)

Zacharis M (1956) Mutationsversuche an Kuklturpflanzen IV. Rontgenbestrah-lungen der Sojabohne (*Glycine soja* L.). Zuchter 28:321–338. (In German)

Zhou Y, Lu DF, Li CY, Luo JH, Zhu BF, Zhu JJ, Shangguan YY, Wang ZX, Sang T, Zhu B, Han B (2012) Genetic control of seed shattering in rice by the APETALA2 transcription factor *SHATTERING ABORTION1*. Plant Cell 24:1034–1048

Role of Mutation Breeding in Crop Improvement with Special Reference to Indian Subcontinent

13

M. C. Kharkwal

Abstract

The primary source of all genetic variations existing in any organism, including plants and animals, is mutation. Mutation provides the raw material for natural selection by creating variation and is a driving force in evolution. Natural selection operates to bring about evolution of new races and species through the variability created by natural mutations and amplified by subsequent recombination of genes during sexual reproduction. Besides natural mutations that occur spontaneously due to various kinds of radiations and cosmic rays received from the sun and also emitted by several radioactive elements present on the earth, mutation can also be artificially induced by a number of physical mutagenic agents like gamma rays, X-rays and fast neutrons and several types of chemical mutagens like EMS, NMU and EI. The standard technique of creating variability by means of altering genes through induction of mutations by physical or chemical mutagens and using the same effectively through elaborate methods of selection techniques in various generations for improvement of a particular crop species for desired objectives is called mutation breeding. It is frequently practised by plant breeders all over the world for crop improvement. Nearly 3365 mutant varieties belonging to >240 plant species have been developed and officially released by the turn of the twenty-first century. The cumulative number 3365 of released mutant varieties in six continents of the world indicates that Asia tops the regional list with 2052, closely followed by Europe (960) and North America (209). With more than 100 mutant varieties each, China, India, Japan, Russia, the Netherlands, Germany and the USA are the leading countries among approximately more than 80 countries actively engaged in the development and release of mutant varieties. During the last five decades, several countries took up

M. C. Kharkwal (✉)
Division of Genetics, ICAR-Indian Agricultural Research Institute, New Delhi, India

S. Penna, S. M. Jain (eds.), *Mutation Breeding for Sustainable Food Production and Climate Resilience*, https://doi.org/10.1007/978-981-16-9720-3_13

extensive crop improvement programmes through mutation breeding and made spectacular accomplishments in evolving several superior mutant varieties in a large number of important agricultural crop species including cereals, pulses, oilseeds, vegetables, fibres, medicinal plants and ornamentals. Large numbers of these mutant varieties, being cultivated by farmers in millions of acres in several countries, have made significant contribution worth billions of dollars in global food production. A wide range of characters including yield, maturity, quality and tolerance to biotic and abiotic stresses have been improved in the mutant varieties developed so far. Among the seven countries of the Indian subcontinent, India has made a significant contribution in crop improvement through mutation breeding by developing 542 mutant varieties in 85 crop species. The other two countries of the Indian subcontinent, Pakistan and Bangladesh, have also made major contributions through mutation breeding by releasing 79 and 76 mutant varieties of crop plants, respectively.

Keywords

Mutation · Mutation breeding · Mutant varieties · Crop improvement · Radiation · Gamma rays · X-rays

13.1 Introduction

The entire global population until the year 1810 was only 1 billion people, and today it has grown to more than 7.7 billion and is estimated to reach 9.9 billion by 2050, thus placing tremendous pressures on increasing world food production of all kinds (Gustafson et al. 2021). In 2018, world cereal production was estimated to be 2.56 billion tonnes, and thus the world's farmers will need to produce 3 billion tonnes by 2050 to meet burgeoning population projections. Climate-resilient agriculture can improve sustainable world food production, and mutation breeding can play a major role in crop improvement by developing high-yielding, biotic and abiotic stress-tolerant cultivars. The impetus to the Green Revolution that led to a phenomenal increase in the production of agricultural crops, especially food grains, also emanated from the crucial introduction of the spontaneous mutant dwarfing genes of '*Norin-10*' wheat and '*Dee-Gee-Woo-Gen*' rice into breeding programmes culminating in the development of dwarf, non-lodging and fertilizer-responsive wheat and rice varieties, respectively. The Green Revolution witnessed by Indian agriculture has transformed the country's status from a food-importing nation to not only a self-sufficient but also a food-exporting nation. India, the largest country of the Indian subcontinent, has made spectacular achievements in crop improvement and food security in which mutation breeding has also made a major contribution by developing and reporting 542 mutant varieties and occupying the second position after China at the global level. Similar gains in crop improvement have been achieved by the two other neighbouring countries, Pakistan and Bangladesh. Sri

Lanka, a small island country in the Indian subcontinent, has also succeeded in the development and release of four mutant varieties so far.

13.2 Global Scenario of Mutation Breeding in Crop Improvement

The discoveries made by Muller and Stadler 95 years ago were followed by a large number of breeders in several countries to induce and generate genetic variability through various mutagens leading to mutation breeding. Mutation breeding was acknowledged and adopted as an additional tool by plant breeders globally particularly after the establishment of the Joint FAO/IAEA Division of Nuclear Techniques in Food and Agriculture in 1964. Because of the popularity and easy and safe access, majority (75%) of the genetic variability has been induced through radiations, particularly gamma rays. Among the mutant varieties released, the majority are food crops and ornamentals. A Mutant Variety Database (http://mvd.iaea.org) on the mutation-derived varieties developed and released in major crops all over the world is being maintained by the FAO/IAEA. Extensive use of induced mutation breeding in crop improvement programmes globally resulted in the release of >3365 mutant varieties belonging to >240 crops and plant species. The cumulative number of officially released mutant varieties in six continents of the world indicates that most of the mutant varieties are released in Asia (2052), followed by Europe (960) and North America (209) (Table 13.1). The other remaining three continents have also taken up mutation breeding during the past few decades and have released mutant varieties in several crops.

The proportion of released plant mutant varieties at global level also indicates that largest numbers consisted of cereal species (48.02%) (Table 13.2) followed by flowers/ornamentals (21.64%), legumes and pulses (13.85%) and oil crops (10.76).

Several countries including China, India, Japan, Pakistan, Bangladesh, Viet Nam, Thailand, the Netherlands, Germany, Italy, Sweden, the USA and Canada took up extensive crop improvement programmes through mutation breeding. During the last five decades, they have made spectacular accomplishments in evolving several superior mutant varieties in a large number of important agricultural crop species

Table 13.1 Number of officially released mutant varieties across the various continents

	Mutant varieties	
Continents	Number	Percentage
Africa	82	2.44
Asia	2052	60.98
Australia and Pacific	9	0.27
Europe	960	28.53
Latin America	53	1.57
North America	209	6.21
Total	3365	

http://mvd.iaea.org (accessed in March 2022)

Table 13.2 Proportion of released mutant varieties in various crop groups

S. No.	Crop group	Mutant varieties	
		Number	Percentage
1	Cereals	1616	48.02
2	Ornamentals	728	21.64
3	Legumes	466	13.85
4	Oil crops	362	10.76
5	Industrial crops	120	3.57
6	Fruits/nuts	103	3.06
7	Vegetables	93	2.76
8	Others	123	3.65
9	Total	3365	

http://mvd.iaea.org (accessed in March 2022)

including cereals, pulses, oilseeds, vegetables, fruits, fibres and ornamentals. Since the Mutant Variety Database of the IAEA is perhaps not being updated very regularly and frequently, as is evident from the number of mutant varieties released in China (817), India (341) and Pakistan (56) reported in the MVD, China has already released more than 1000 mutant varieties (Liu 2021), India has already reported 542 (Tables 13.3, 13.4 and 13.11) and Pakistan has released 79 mutant varieties (Tariq M. Shah personal communication). The number of mutant varieties released in the world from >240 plant species by early 2022 is expected to be >3500. Majority of the mutant varieties have been released during the last five decades. With more than 100 mutant varieties each, China, India, Japan, the Russian Federation, the Netherlands, Germany and the USA are the leading countries engaged in the development and release of mutant varieties (Table 13.3). Several other countries, particularly in the European and Asian continents, are among the other important countries actively engaged in the development and release of mutant varieties, which have released more than 20 important cultivars each through mutation breeding.

A perusal of the data on specific crops and number of mutant varieties released in the world (Table 13.3) indicates that the top 20 ranks are occupied by some of the most important food and ornamental plant species in the world agriculture, e.g., rice (853), barley (311), chrysanthemum (266), wheat (265), soybean (181), maize (96), groundnut (79), rose (67), common bean (59) and cotton (48). Although an exact estimate of the total area covered by commercially released mutant cultivars in a large number of countries is not readily available, they are being cultivated in millions of hectares and have made a very significant contribution worth billions of dollars in global agriculture addressing food and nutritional security problems in many countries of the world (Ahloowalia et al. 2004; Kharkwal and Shu 2009; Kharkwal 2012, 2017). Most of the released mutant varieties of crop plants belong to seed-propagated species. About 75% of varieties developed and released using induced mutations are of crop plants and 25% are of the ornamentals. Of the total mutant varieties, majority mutant varieties were developed ‘directly’ after mutagenic treatment and selection in the subsequent generations (Table 13.4). The remaining

Table 13.3 Number of mutant cultivars released in top 20 countries and crops of the world

No.	Country	No. of mutants	No.	Crop	No. of mutants
1	China	817	1	Rice	853
2	India	542[a]	2	Barley	316
3	Japan	479	3	Chrysanthemum	266
4	Russian Federation	216	4	Wheat	265
5	Netherlands	176	5	Soybean	181
6	Germany	171	6	Maize	96
7	USA	139	7	Groundnut	79
8	Pakistan	79[b]	8	Rose	67
9	Bulgaria	76	9	Common bean	59
10	Bangladesh	76	10	Cotton	48
11	Viet Nam	58	11	Mung bean	39
12	Canada	40	12	Dahlia	36
13	South Korea	40	13	Durum wheat	31
14	France	39	14	Pea	34
15	Italy	35	15	Sesame	30
16	UK	34	16	Chickpea	27
17	Poland	31	17	Tomato	25
18	Sweden	26	18	Rapeseed	24
19	Guyana	26	19	Oat	23
20	Thailand	24	20	Faba bean	20

http://mvd.iaea.org (accessed in March 2022)
[a] Details in Tables 13.8 and 13.11
[b] Personal communication T.M. Shah, Director, NIAB, Faisalabad, Pakistan

Table 13.4 Methods used in the development of mutant varieties

S. No.	Development methods	Mutant varieties	
		Number	Percentage
1	Direct use of an induced mutant	2137	65.09
2	Crossing with one mutant	457	13.92
3	Crossing with two mutants	160	4.87
4	Crossing with one mutant variety	299	9.10
5	Mutagenic treatment of breeding material F_1, F_2	168	5.11
6	Mutagenic treatment of hybrid F_1	62	1.89
		3283	

http://mvd.iaea.org (accessed in March 2022)

new mutant varieties were developed 'indirectly' through cross-breeding of mutants or already released mutant varieties as sources of desired characters in cross-breeding programmes.

Different mutagens have been used by various breeders to induce genetic variability in various crops. Among the important mutagenic agents used for developing mutant varieties, a great majority were obtained with the use of physical

Table 13.5 Proportion of top ten mutagens used for the development of mutant varieties released

Physical mutagens (2610 mutant vars.)			Chemical mutagens (384 mutant vars.)		
Mutagen	Number	%	Mutagen	Number	%
Gamma rays	1705	65.25	EMS	108	28.12
X-rays	561	21.50	NEU	57	14.84
Gamma rays chronic		3.41	NMU	48	12.50
Fast neutrons		2.22	Colchicine		12.24
thN		1.91	EI		9.37
Aerospace		1.76	DMS		4.95
Laser		1.00	dES		3.65
Neutrons		0.92	NaN_3		2.86
Ion beams		0.77	MNU		1.82
Beta rays		0.61	MNH		1.56

http://mvd.iaea.org (accessed in March 2022)

mutagens like radiations, particularly gamma rays and X-rays. Chemical mutagens have also been used to develop 384 mutant varieties (Table 13.5). Among the chemical mutagens, the most popular ones are EMS, NEU and NMU. Although a large number of physical as well as chemical mutagens have been successfully used to breed mutant varieties all over the world, the proportion of top ten most commonly used physical and chemical mutagenic agents is given in Table 13.5.

A wide range of characters including yield and yield-contributing traits, resistance/tolerance to biotic and abiotic stresses, quality and nutrition traits, and agronomic and botanic traits have been improved in the mutant varieties developed so far (Table 13.6). Besides these general improved characters, a very large number of specific characters like yield, early maturity, growth habit, drought tolerance, salinity tolerance and disease resistance have also been successfully incorporated in a large number of mutant varieties developed and released.

As per the FAO/IAEA Mutant Variety Database (FAO/IAEA 2022), the total number of mutant varieties officially released at the beginning of the year 2022 is 3365 and they belong to >240 plant species. Its distribution and widespread success in various crop groups (Table 13.7) indicate the enormous popularity of mutation breeding in crop improvement programmes being applied in a large number of countries in all the continents of the world (Shu 2009; Sivasankar et al. 2021). The following selected examples of major crops illustrate the role that mutation breeding has already played and continues to play in plant breeding of a large number of crop plants in big as well as very small countries.

Among the crops, mutation breeding in rice has been very successful worldwide, probably due to its diploid nature, self-fertilizing character and enormous attention of many researchers to this most important food crop. The majority of rice mutant varieties have been developed after selection in segregating mutated populations. The economic impact of mutant varieties of rice released with improved characters such as semi-dwarfness, earliness, improved grain yield, disease tolerance and improved grain quality has been reviewed by Rutger (1992) and Maluszynski et al. (1995a, b). In barley, mutant varieties that proved to be superior to best control

Table 13.6 Proportion of improved characters in mutant plant varieties released

S. No.	Characters	Mutant varieties	
		Number	Percentage
1	Agronomic and botanic traits	2989	49.82
2	Quality and nutrition traits	1175	19.59
3	Yield and contributors	1029	17.15
4	Resistance to biotic stresses	558	9.30
5	Tolerance to abiotic stresses	248	4.13
		5999	

http://mvd.iaea.org (accessed in March 2022)

Table 13.7 Crop varieties developed through mutation breeding

Crop group	No. of mutants	Specific major crops and number of mutant varieties released
Cereals	1616	Rice (853), barley (311), wheat (265), maize (96), durum wheat (31), oats (23), millet (20), sorghum (17)
Legumes	489	Soybean (181), groundnut (79), Common bean (59), mung bean (39), pea (34), chickpea (27), faba bean (20), lentil (18), cowpea (16), black gram (9), pigeon pea (7)
Oilseeds	102	*Brassica* sp. (33), sesame (30), linseed (11), sunflower (6), castor (5) and others (17)
Industrial/ cash crops	120	Cotton (48), tobacco (15), sugarcane (13), Jute (12), fibre flax (11), citronella (6), tea (2) and others (13)
Fruits/nuts	103	Sweet cherry (16), apple (13), citrus (12), mulberry (9), peach (7), sour cherry (5), pear (5), banana (4), pomegranate (2), apricot (1), papaya (1), plum (1), grape (1) and others (26)
Vegetables	93	Tomato (25), chilli (16), lettuce (7), sugar beet (7), potato (7), sweet potato (6), onion (6), brinjal (5), cabbage (4), taro (3), cassava (2), okra (2) and others (3)
Flowers/ ornamentals	728	Chrysanthemum (266), rose (67), dahlia (36), *Alstroemeria* sp. (35), streptocarpus (30), carnation (28), begonia (18), bougainvillea (17), portulaca (17), azalea (15), *Hibiscus* sp. (14), tulip (9), canna lilies (8), snapdragon (5), gladiolus (4), lotus (3), polyanthus (2), *Delphinium* sp. (2) and others (156)

http://mvd.iaea.org (accessed in March 2022)

lines for important plant characters have been isolated and released by various countries. Most of the barley mutant cultivars are characterized with dwarf or semi-dwarf stature, low susceptibility to lodging due to stiff straw, higher tillering and better harvest index. For instance, the induced barley mutants further used by Swedish scientists in cross-breeding gave rise to cultivars which performed very well (Gustafsson 1963). Similarly, a Danish dwarf mutant with remarkably high degree of lodging resistance had very good grain yield performance. Some of the important mutant varieties of barley are 'Pallas' (Sweden), 'Balder J' (Finland), 'Diamant' (Czech Republic), 'Trumpf' (Germany), 'Goldspear' (UK), 'Pennrad' &

'Luther' (USA) and 'Betina' (France) (Van Harten 1998). The two barley parent cultivars 'Pallas' and 'Mari' that are in the pedigrees of many cultivars grown today derive from mutation programmes. The economic impact of short-straw mutants and some other mutant types in barley has been of immense economic value in Sweden, Denmark, Germany, France, Spain, the UK and the Czech Republic (Lundqvist 2009, 2021). Commercial success and economic impact of mutant varieties of grapefruit in Texas, USA, and pear in Japan have been highlighted by Nichterlein et al. (2000).

The success of mutation breeding in some of the leading countries and their important achievements in terms of the mutant varieties released in various crops are summarized and presented here continent- and countrywise in the following paragraphs of this chapter.

13.2.1 Asia

More than half of the total number of mutant varieties released in the world are developed in Asia. China, India and Japan are the three leading Asian countries that have released the largest number of mutant varieties and in turn contributed to crop improvement and their national economy and food security. Mutation breeding has also been successfully used in many other Asian countries like Pakistan, Bangladesh, Viet Nam, South Korea, Thailand and Myanmar. While detailed information for India, Pakistan, Bangladesh and Sri Lanka has been given under the Indian subcontinent heading of this chapter elsewhere, brief status of major Asian countries actively engaged in mutation breeding is highlighted in the following paragraphs.

13.2.1.1 China

Rice breeding through induced mutations was started by China in 1960, and the first mutant variety developed was 12-'Zhefu'. The most widely cultivated rice mutant variety in China between 1986 and 1994 was 'Zhefu-802', and its cumulative planted area reached 10.6 million ha during that period. Induced by gamma rays, mutant variety 'Zhefu-802' has a short growing period (105–108 days), high yield potential even under poor management and infertile conditions, wide adaptability, high resistance to rice blast and tolerance to cold. Two other mutant rice varieties, 'Yuanfengzao' and 'Yangdao #6', developed and released before and after 'Zhefu-802', are further mutant varieties that had been grown on annual scales up to 1 million ha. Using a pollen irradiation technique, two new high-quality, high-yield and early-maturity mutant varieties—'Jiahezazhan' and 'Jiafuzhan', resistant to blast and planthopper, as well as endued with a wide adaptability—were developed and are now planted annually on 363,000 ha in Fujian province of China. China has also been successful in breeding soybean varieties like 'Hefeng-25', 'Henong' and 'Tiefeng-18', possessing different excellent traits such as high yield, good grain quality, disease/insect resistance or drought/salt tolerance. China has also developed and released a large number of high-yielding groundnut mutant varieties during the last few decades. The cumulative cultivated area of the more than 35 mutant cultivars released accounts for about 20% of the total area under groundnut in China. The maximum annual dissemination area of mutant varieties is more than 9 million ha, which results in the production of over 1.5 million tons of crops annually, with an estimated value of about US$ 500 million. In China, only three mutant varieties are covering over 30 million ha and earn US$ 4.9 billion to uplift the socio-economic status (Liu et al. 2021).

In the late 1980s, Chinese scientists initiated new mutagenesis approaches, including space radiation with high-energy and heavy-ion beam irradiation used as new effective and alternative ways for crop improvement. China started the use of space mutagenesis for crop improvement in 1987, and Chinese scientists produced giant sweet peppers and improved quality traits in wheat and rice through rare inheritable genetic mutations using satellites and high-altitude balloons for space radiation. Spaceflight-induced mutation, also called space breeding, is a technique using an aerospace environment to generate genetic variation. The space environment is characterized by its mixture of cosmic ray irradiation, microgravity, weak geomagnetic field and super-vacuum and is super clean. China launched and recovered a specific breeding satellite, 'Shijian 8', in 2006 to conduct basic research and for the application of space breeding. Using space satellite, China has produced 41 varieties developed through space-induced mutation breeding of various crop species, viz. rice, wheat, cotton, sesame, pepper, tomato and alfalfa (Liu et al. 2021).

High-yielding Chinese mutant varieties of wheat (Luyuan-502, Yangfumai No. 3, 4 and 5), rice (Ilyou-D069, C Liangyou-266 and Zhefu-802), soybean (Hefeng-25 and Henong), groundnut (Luhua-11) and other crops like cotton, maize and forage legumes are being cultivated in millions of hectares, and their breeders have won National Invention Awards (Liu 2021). By the end of 2018, the total number of officially released mutant varieties had reached more than 1030 mutant varieties covering 46 crop and ornamental species with high-yielding, fine-quality and multi-resistant traits in cereal, oil and vegetable crops (Liu et al. 2021). The maximum annual dissemination area of mutant varieties was more than 9 million ha. The mutant varieties yielded more than 1.5 million tonnes of crop production annually, which was valued at US$ 500 million. China has developed and officially released 14 wheat mutant varieties with improved salt tolerance, higher yielding potential, better quality or enhanced drought resistance. The wheat mutant variety H-6756, characterized by its salt tolerance, produces 17% higher yield than the control variety. Hangmai-247, which was released nationally in 2016, has high and stable yield potential in a large area. The national variety Luyuan-502 is a widely adaptable, sprouting-resistant, lodging-resistant, drought- and main disease-tolerant and high-yielding variety. The average yield of this variety is 8.2 t/ha and the maximum yield is 12.2 t/ha. By 2017, the variety had a cumulative cultivation area of over 3.6 million hectares. Hangmai-2566, Hangmai-287 and Hangmai-501 are new mutant varieties of wheat released in 2018 and contributing to higher grain yield, higher farmers' income and a better environment. The use of induced mutations and mutation breeding for crop improvement has been playing very important roles in food security and found to be a profitable approach in China (Wang 1991).

13.2.1.2 Japan

Japan started research on radiation-induced mutations and mutation breeding during the 1960s, which was accelerated following the construction of the Gamma Field at the Institute of Radiation Breeding, Hitachiōmiya, in Ibaraki. In total, 479 mutant varieties, 318 direct-use mutant cultivars representing 79 species generated and released through irradiation of gamma rays, X-rays, ion beams and chemical mutagenesis and in vitro culture (somaclonal variation), have been registered in Japan. Approximately 79% of these direct-use cultivars were induced by radiation (Nakagawa 2021). The total number of 375 indirect-use mutant cultivars released in Japan included 332 rice, 16 soybean, 8 barley, 6 wheat, 3 tomato, 4 lettuce, 1 eggplant, 2 Japanese lawn grass, 2 mat rush and 1 Job's tears. Recently, the development of mutant cultivars generated by ion beam irradiation has been a growing area of mutation induction in Japan (Nakagawa 2021).

The first gamma-ray irradiation-induced mutant rice cultivar developed and registered in 1966 in Japan was semi-dwarf cv. 'Reimei'. This cultivar exhibits a mutation of the *sd-1* locus, which is the same as the mutation of a miracle rice, cv. 'IR8', through cv. 'Dee-geo-woo-gen', which later contributed to the 'Green Revolution' of rice. The number of mutant cultivars derived from cv. 'Reimei' has been increasing, and 99 mutant cultivars (2 direct-use and 97 indirect-use cultivars) were in cultivation in Japan in 2005. The total cultivated area of mutant cultivars increased after cv. 'Reimei' was released for cultivation. The peak use of induced-mutation cultivars was 250,000 ha in 1986 and slightly exceeded 210,000 ha in 2005 (Nakagawa 2009).

The economic impact of the mutant cultivars, primarily of rice, soybean and pear, is very large, and the roles of cvs. 'Reimei', 'Kinuhikari' and 'Mineasahi' of rice and 'Gold Nijisseiki' of pear are very important in Japan. Useful mutations in soybean such as radiosensitivity, fatty acid composition and super-nodulation have been identified. There have been 16 direct-use mutant cultivars of soybean registered in Japan since cvs. 'Raiden' and 'Raikou' were developed by gamma-ray irradiation in 1960. The Japanese pear mutant variety 'Gold Nijisseiki' registered and released in 1991 is more resistant to black spot disease caused by *Alternaria alternata* than its parent. The economic effect of this pear mutant variety 'Gold Nijisseiki' has been extraordinary. Following its development, two more new pear mutant cvs. 'Osa-Gold' and 'Kotobuki Shinsui' were also developed. The additional annual income by growing these mutant varieties is estimated to be several million US$ (Nakagawa 2021).

13.2.1.3 Viet Nam

Mutation breeding in rice and soybean has been carried out in Viet Nam since the 1980s, and it has made outstanding achievements by releasing 58 mutant varieties till date, predominantly 36 mutant varieties of rice for farmers' cultivation. Rice which represents more than 60% of mutant varieties, besides contributing to the national food security, also provides income and decreases poverty among millions of rural population in Viet Nam. These rice mutant varieties have 10–20% higher yields in comparison to the parent varieties, with lodging resistance, tolerance to acid soils, salinity, tolerance to biotic stress, short duration and better nutritional quality. Due to short duration, mutant rice varieties can be cultivated 2–3 crops per year and avoid early flooding. With resistance to diseases and pests, cultivation can be done with a reduced number of sprayings, which saves costs and protects the environment from excessive use of pesticide (Ahloowalia et al. 2004; Kharkwal and Shu 2009). Though Viet Nam used to import 2–3 million tons of food annually in the decade of 1970–1980, it now exports 4–5 million tons of rice, becoming the world's second-largest exporter of rice. The wide use of dozens of high-yielding mutant varieties contributed substantially to this transformation into food self-sufficiency. For example, the mutant rice variety 'VND-95-20', grown on more than 300,000 ha/year, has become the top variety in southern Viet Nam, both as an export variety and in terms of growing area (Do et al. 2009). Two more improved rice mutant varieties, 'TNDB-100' and 'THDB', with earliness and improved grain yield were released within only 6 years after mutagenic treatment in Mekong Delta of Viet Nam. These varieties grown in millions of hectares have maintained tolerance to acid sulphate soil or soil salinity. The rest of the crops including soybean, chrysanthemum, maize, groundnut, Indian jujube and field mint have also made significant contributions to national food security (Le et al. 2021). Induction and incorporation of desirable traits in ideal soybean mutant varieties such as DT-84, DT-90, DT-99 and DT-2008 have been achieved in Viet Nam. These outstanding mutants incorporate desirable traits like high and stable yield (2.0–3.5 t/ha), good quality, drought tolerance, resistance to diseases, short growth duration (70–100 days), and wide adaptability and suitability for various ecological regions and cropping systems in the whole country. The most outstanding soybean mutant variety, DT-84, occupies over 50% of the total production area and 80% in Central and North Viet Nam. Mutant varieties DT-90 and DT-2008 contributed to increasing soybean productivity from 0.66 t/ha in 1980 to 1.44 t/ha in 2013. Thus, during the past decade, mutation breeding has played an important role in improving production and productivity of rice and soybean varieties in Viet Nam and has become an icon for the success of mutation breeding (Le et al. 2021).

13.2.1.4 Thailand

The contribution of Thailand in mutation breeding is best reflected by the two aromatic *indica*-type mutant varieties of rice, 'RD-6' and 'RD-15', derived from gamma-irradiated progeny of the popular rice variety 'Khao Dawk Mali 105'. RD-6 has glutinous endosperm and retained all other grain traits, including the aroma of the parent variety. RD-15, on the other hand, is non-glutinous and aromatic like the parent, but ripens 10 days earlier than the parent, which is a major advantage for harvesting before the onset of the rainy season. Even 30 years after their release, these two varieties are still grown extensively covering 80% of the rice fields in north-eastern Thailand. Thailand is the largest exporter of aromatic rice to the world market. Thus, the impact of the two rice mutant varieties is far beyond the farm gate with a major contribution to the export earnings (Ahloowalia et al. 2004; Kharkwal and Shu 2009). Besides the highly popular mutant varieties of rice, the first mung bean variety of Thailand, Chai Nat-72 (CN-72), developed through induced mutation as high yielding and tolerant to bean fly was officially released in the year 2000 (Nagampongsai et al. 2009).

13.2.1.5 South Korea

Sesame yield in the Republic of Korea has been increased more than twice (from 283 to 720 kg/ha) due to the development and release of 15 improved determinate type, high-oil-content mutant varieties having phytophthora blight resistance and good cooking quality. These mutants occupied 55% of the national acreage during the last two decades in Korea (Kang et al. 2008; Kharkwal and Shu 2009).

13.2.1.6 Myanmar

Rice mutant variety 'Shwewartun' with improvement in grain quality, seed yield and early maturity was developed after irradiation of 'IR5' seeds and released in Myanmar in 1975. The success of this high-yielding rice mutant variety led to its large-scale planting, and it covered annually more than 0.8 million ha—17% of the 4.8 million ha area under rice in Myanmar (Ahloowalia et al. 2004; Kharkwal and Shu 2009).

13.2.2 Europe

Mutation breeding in Europe has become a significant method in the European plant breeding programmes resulting in the release of 959 mutant varieties (Zakir 2018; Sarsu et al. 2021). Induced mutations have become an inherent component of many current plant varieties in Europe, particularly for barley, durum wheat and oats. Mutation techniques are also widely used in the breeding of flowers and horticulture cultivars. A few examples of European countries are given here.

13.2.2.1 Sweden

Swedish research on induced mutations in barley was initiated by the eminent Swedish geneticists Herman Nilsson-Ehle and Åke Gustafsson at Svalöf. The systematic approach of the work at Svalöf through the joint efforts of the barley breeders can be used as an example of how mutation breeding can be employed in a crop improvement programme (Gustafsson 1963). In the Swedish programme, the use of macro-mutations has proved to be more successful than recurrent mutagenic treatments. A rather large number of mutant cultivars of two-row barley were registered as originals, and 15 Swedish ones were commercially released. Two of these cultivars, 'Pallas', a stiff-straw, lodging-resistant and high-yielding *Erectoides* mutant, and 'Mari', an extremely early photo- and thermo-insensitive mutant, were produced directly by X-irradiation. All other cultivars derive from crosses and backcrosses with the X-ray-induced mutants 'Pallas', 'Sv 44/3' and 'Mari'. Most of the Swedish barley mutant cultivars are characterized with dwarf or semi-dwarf stature, low susceptibility to lodging due to stiff straw, higher tillering and better harvest index (Lundqvist 2009, 2021).

13.2.2.2 Czech Republic

'Diamant', a gamma ray-induced barley mutant cultivar that was 15 cm shorter than the parent cultivar 'Valticky' and had an increased grain yield of around 12%, was officially released in the Czech Republic in 1965. It became so popular that nearly 45% of spring barley in the Czech Republic was planted under either Diamant or mutant cultivars derived from Diamant. The spring barley cultivars that had mutated Diamant's denso gene in their pedigree were grown all over Europe on an area of nearly 3 million ha (Bouma et al. 1991). The high-yielding, short-height barley mutants Diamant and Golden Promise were a major impact on the brewing industry in Europe; they added billions of US dollars to the value of the brewing and malting industry. More than 150 cultivars of malting barley in Europe, North America and Asia were derived from crosses involving Diamant (Ahloowalia et al. 2004; Kharkwal and Shu 2009).

13.2.2.3 Germany

'Trumpf', the best-known barley mutant cultivar obtained after crossing with the Czech mutant cultivar 'Diamant' occupied more than 70% of the barley-planting area in Germany. The mutant had a yield increased by 15% and better disease resistance. Used extensively in cross-breeding, Trumpf became incorporated into many barley breeding programmes in a large number of European countries (Van Harten 1998; Ahloowalia et al. 2004; Kharkwal and Shu 2009).

13.2.2.4 Italy

Release of 11 durum wheat mutant varieties was a great success story of mutation breeding in Italy. Six of them were developed through the direct use of identified mutants and five were developed through cross-breeding. 'Creso' is one of the best durum wheat mutant varieties which was released in 1974 that had a significant impact on Italian economy contributing with US$ 1.9 billion. Creso was grown in about one-third of the total area of durum wheat in Italy. During a period of 10 years, in Italy alone, an extra economic profit of 1.8 billion US dollars was obtained by growing this cultivar. 'Castelporziano' and 'Castelfusano', the other two high-yielding durum wheat mutants, had shorter culms and spike length and better resistance to lodging, but higher numbers of grain per spikelet. Planted in sizable areas, they contributed notably to the national economy of Italy. Both mutants were also used in extensive cross-breeding. Elsewhere in the world, the FAO estimated in 1994 that almost 70% of the durum wheat in Italy was mutant varieties. Many wheat mutant lines developed in Italy were used in the wheat improvement programmes in Austria and Bulgaria for releasing improved wheat varieties (Scarascia-Mugnoza et al. 1991; Ahloowalia et al. 2004; Kharkwal and Shu 2009).

13.2.2.5 Finland

A high-yielding barley mutant released in Finland, 'Balder J', had higher yield, greater drought resistance, better sprouting resistance and greater 1000-kernel weight. Nearly 1 million kg of 'Balder J' seed was sold by Jokioinen Seed Center. Oat stiff-straw mutant cultivar 'Ryhti' occupied up to 41% of the total area of oat in Finland during 1970–1980. Another stiff-straw oat mutant cultivar, 'Puhti', occupied 30% of the oat-planting area in Finland. Many new varieties now grown are derived from crosses with these mutant varieties (Van Harten 1998; Kharkwal and Shu 2009).

13.2.2.6 Bulgaria

For over 50 years, Bulgaria has efficiently applied mutation breeding in crop improvement. The country has released 76 mutant cultivars in 15 different crop species, such as barley (5), wheat (5), durum wheat (9), maize (26), sunflower (3), lentil (4), bean (2), pea (1), chickpea and vetch (2), soybean (5), tomato (6), pepper (4), cotton (2) and tobacco (2). Some of the mutant varieties, e.g. maize mutant hybrid 'Kneja 509' and durum wheat variety 'Gergana', have become the leading varieties occupying up to 50% of the planting area from 1984 to 2000. In durum wheat, mutant varieties have not only covered nearly all the growing areas but also doubled the yield over the past 30 years (Yanev 2006; Kharkwal and Shu 2009).

13.2.3 North America

In North America, the USA is one of the pioneering countries in the exploitation of mutation breeding for crop improvement and has had many extraordinary successes. Significant progress has also been reported from Canada, and more recently in Mexico.

13.2.3.1 The USA

The USA has developed and released 139 induced mutant varieties. In his classic mutation breeding experiments, Gregory (1960) exposed about 200 pounds (90 kg) of peanuts to high doses of X-rays in 1959 and planted one million M_2 plants in 64 acres in the university's crop development fields. This population ultimately resulted in the release of one of the first commercial groundnut mutant var. NC 4x in the USA, which was about 5% better in total production. Mutation breeding was accelerated in the USA with the nation's first semi-dwarf table-rice mutant variety 'Calrose 76' induced by gamma irradiation released in 1977 in California (Rutger 2006). The semi-dwarf gene allele *sd1*, which was induced through gamma-ray mutagenesis, has enabled the American version of the 'Green Revolution' in rice. This gene is transferred by crossing other varieties and led to the development of 22 new rice mutant cultivars that included 11 released in the USA, 9 in Australia and 2 in Egypt (Rutger 2009). A high-yielding wheat mutant variety 'Stadler' released in Missouri had early maturity, resistance to races of leaf rust and loose smut, as well as better lodging resistance. It was once grown on two million acres annually in the

USA (Ahloowalia et al. 2004). 'Luther', a barley mutant, had 20% increased yield, shorter straw, higher tillering and better lodging resistance. About 120,000 acres were planted annually in three states of the USA—a gain of an estimated US$ 1.1 million in 1 year. It was used extensively in cross-breeding, and several mutants were released. 'Pennrad', a high-yielding winter barley mutant released in Pennsylvania, had winter hardiness, early ripening and better lodging resistance. It was grown on about 100,000 ha in the USA. 'Sanilac', a high-yielding Navy pea bean mutant cultivar developed after irradiation with X-rays and released in Michigan, was grown on more than 87,000 ha. Similarly, about 160,000 ha were planted with common bean cultivars 'Gratiot' and 'Sea-way', developed likewise by cross-breeding with a 'Michelite' mutant, an X-ray-induced white bean mutant with altered plant type, which is in the pedigrees of most of the white beans grown in North America (Ahloowalia et al. 2004; Kharkwal and Shu 2009). The peppermint mutant variety, 'Todd's Mitcham', forms bulk of the world's production of mint oil. Two grapefruit varieties, 'Star Ruby' and 'Rio Red', both developed through thermal neutron mutagenesis (Hensz 1991), have become a widely grown variety during the past two decades. Currently grown on 75% of the grapefruit-planting area in Texas, the fruits of both cultivars are sold under the trademark 'Rio Star' (Kharkwal and Shu 2009).

13.2.3.2 Canada

After wheat and barley, rapeseed-canola is Canada's third most important grain export crop. Mutant cultivars with low erucic acid and very low (more than 30 μm/g) glucosinolates ('Double Zero') have been developed and released in Canada (Ahloowalia et al. 2004). The strongest modification of oil composition with induced mutations has been the development and release of linseed cultivars of the 'linola' type in Australia and Canada. Zero is the low-linolenic acid genotype derived by ethyl methanesulphonate (EMS) mutagenesis of the Australian linseed cultivar 'Glenelg' and recombination of two mutated genes (Green and Dribnenki 1996). The rapeseed cultivar 'Regina II' was developed by mutation in Sweden and was released in Canada in 1953. 'Redwood 65' flax was derived from a mutation programme at the University of Saskatchewan (Larter et al. 1965) and is present in the pedigrees of many western Canadian flax cultivars. In 2000, Canada planted canola in 5,564,000 ha and earned US$ 350.5 million, an outstanding contribution to the Canadian economy (Ahloowalia et al. 2004).

13.2.3.3 Mexico

Radiation-induced mutation breeding started in 1974 in Mexico. Two new wheat mutant varieties, 'Centauro' and 'Bajio Plus' derived from 'Salamanca' seeds irradiated at 500 Gy, showed increased yield and tolerance to lodging. Two soybean mutant varieties, 'Hector' and 'Esperanza', were obtained by irradiation of seeds from the variety 'Suaqui 86' at 150 Gy. These new varieties exhibit an increased yield and reduction in dehiscence and lodging, being tolerant to white fly. 'SalCer' is another new soybean variety obtained through irradiation of seeds from the line

ISAEGBM2 at 200 Gy. Its improved traits are higher yields and increased height to first pod (da Cruz et al. (2013).

13.2.4 Latin America

Mutation breeding through irradiation in Latin America was started during the 1960s initially in six countries, viz. Colombia, Peru, Brazil, Guatemala, Uruguay and Costa Rica, followed by Argentina, Chile and Panama in subsequent years. The IAEA and FAO have been working with 18 countries in Latin America and the Caribbean for many years to introduce and strengthen crop mutation breeding. The project used mutation breeding techniques to develop improved crops of economic interest with better tolerance to environmental stresses and better quality. Participating countries have successfully developed a range of new mutant varieties (Sarsu et al. 2021). A few examples of Latin American countries are given here.

13.2.4.1 Argentina

Mutant varieties of rice covered 40% of the rice production area in Argentina during 2016–2017. They were also distributed in other Latin American countries where weedy red rice is a problem. Currently, imidazoline herbicide-resistant varieties are grown on more than 800,000 ha. They have also been very useful to improve abandoned rice fields for being severely occupied by red rice (Livore et al. 2018). 'Puita INTA-CL', a rice mutant with high yield and herbicide resistance, has occupied more than 18% of the rice-growing area in Argentina since then. Also planted in Brazil, Costa Rica, Paraguay and Bolivia, this mutant variety has contributed significantly to these Latin American countries' economies and their food security (Kharkwal and Shu 2009). 'Colorado Irradiado', a groundnut mutant induced by X-rays with high yield and fat content, occupied more than 80% of the groundnut area in Argentina in the 1970s (Ahloowalia et al. 2004).

13.2.4.2 Cuba

Mutation breeding has been successfully applied in Cuba for crop improvement, and tremendous successes have been achieved in a range of food and ornamental crops. Cuba has released 21 mutant varieties so far in crops including hibiscus (3), sugarcane (4), tomato (3), rice (9) and soybean (2) (González et al. 2008a, b, 2009, 2020). The rice variety 'LP-7' has higher yield and is tolerant to salinity. Attempts to obtain a rice mutant variety with good agronomical characteristics and salinity tolerance have been successful in Cuba. The first mutant released from in vitro mutagenesis using proton radiations in Cuba is 'GINES', which shows the best performance under saline conditions and has been successfully introduced in rural areas (PBGN, 2020). Another significant success achieved by Cuba is in tomato in which the very first tomato mutant variety 'Maybel' and the second 'Domi' mutant have shown significantly increased yields and high performance under drought conditions and have been introduced and released in rural areas of different provinces of Cuba (González et al. 2020). Two new improved mutant varieties,

'Giron-50' of tomato and 'Cuvin-22' of soybean, have also been successfully induced and released by Cuba (Gonzalez et al. 2021).

13.2.4.3 Peru

Mutation breeding has been very successfully used in Peru, and a barley mutant variety 'UNA La Molina-95' characterized by its earliness, short stature and production of huskless grains with higher protein content was released in 1995. Another barley mutant variety 'Centenario' with high yield (37% over the parent cultivar), earliness (18 days), better test weight and better quality with higher protein (10.3%), and resistance to yellow rust is grown at altitudes of up to 5000 m above sea level and is replacing the traditional cultivars of the Central Highlands of Peru, extending the range of crops to higher altitudes contributing significantly to the food security of the country (Gomez-Pando et al. 2009; González et al. 2020). The amaranth mutant variety 'Centenario' released in 2006 is cultivated in the Peruvian highlands and exported as a high-value commodity (Gomez-Pando et al. 2009). Centenario (MSA-011), a Kiwicha (*Amaranthus caudatus*) mutant with high yield, earliness (45 days), tolerance to salinity, wide adaptability, better grain colour and size, as well as higher market price, was released in 2006 and has covered 40% of the total Peruvian land dedicated to Kiwicha crops (Gomez-Pando et al. 2009).

13.2.4.4 Brazil

Brazil has been effectively using induced mutations in the improvement of new rice varieties. The rice mutant variety, 'SCS-118 Marques', was developed through application of gamma rays in 2013. This high-yielding lodging-resistant variety has moderate resistance to blast disease, long grains and consumer-preferred cooking quality. The mutant variety 'SCS-121-CL' which shows resistance to herbicide Kifix was released in 2014. 'Clearfield', an IMI herbicide-resistant cultivar of rice improved through mutagenesis, was released for cultivation. Combination of management practices has increased rice yield in Brazil by approximately 50%. Around 1.1 million ha were planted with IMI herbicide-resistant rice mutant variety Clearfield in Brazil (Merotto et al. 2006; Schiocchet et al. 2015; Singh et al. 2017).

13.2.5 Australia

Nine rice mutant varieties—'Amaroo', 'Bogan', 'Echua', 'Harra', 'Illabong', 'Jarrah', 'Langi', 'Millin' and 'Namaga' derived from the famous American rice mutant variety 'Calrose-76'—have been introduced in Australia. Rice mutant variety Amaroo has covered 60–70% of the rice cultivation area of Australia and on average yielded 8.9 t/ha grain with a potential of 13.3 t/ha (Clampett et al. 2001). The highly successful Japanese pear mutant variety 'Gold Nijisseiki' was also registered under the same name in Australia in 2004. Facing the new challenge of developing herbicide-tolerant lupin cultivars, a dominant grain legume crop in Western Australia, chemical mutagenesis has been used to create new tolerance to herbicide. The two lupine mutants (Tanjil-AZ-33 and Tanjil-AZ-55) are highly tolerant, six times more tolerant to metribuzin herbicide than the original parental cultivar Tanjil, and also have high yield and resistance to the disease anthracnose (Kharkwal and Shu 2009). Induced mutation proves to be an effective tool in the improvement of lupine crop in Australia (Si et al. 2009).

13.2.6 Africa

Several African countries have initiated and successfully applied mutation breeding in their crop improvement programmes during the last three decades. The continent has already released 83 mutant cultivars in crop species, such as rice, wheat, maize, sorghum, finger millet, groundnut, soybean, chickpea, cowpea, common bean, sesame, safflower, tomato, banana and cassava. The 16 countries which have succeeded in releasing mutant varieties in one or more crops include Algeria (soybean), Burkina Faso (rice), Congo (soybean and maize), Cote D'Ivoire (rice), Egypt (chickpea, common bean, safflower and sesame), Ghana (cassava), Kenya (cowpea and wheat), Male (sorghum and rice), Mauritius (tomato and oyster mushroom), Nigeria (rice), Senegal (rice), Sudan (banana and groundnut), Tunisia (common bean), Uganda (wheat), Zambia (finger millet and cowpea) and Zimbabwe (cowpea). A few examples of the African countries are given here.

13.2.6.1 Egypt

A total of nine mutant varieties have been released in Egypt, which include five belonging to sesame, two to safflower and one each to chickpea and common bean. Two semi-dwarf mutant rice varieties, 'Giza 176' (1989) and 'Sakha 101' (1997), derived from the famous American rice mutant variety 'Calrose-76' were released in Egypt during the 1990s with the standard yield abruptly increasing from 3.8 to 8.9 t/ha. 'Giza 196' became the leading rice variety with a yield potential of 10 t/ha (Badawi 2001). Induced mutation technique was also successfully applied in edible oil crops, such as sesame and safflower. Five sesame mutant varieties were developed with high yield and good quality during 1992–1996 (Kharkwal and Shu 2009). Two safflower mutant varieties with high yield and oleic fatty acid, resistant to leaf spot and smut, were released in Egypt in 2011 (Sarsu et al. 2021).

13.2.6.2 Ghana

Ghana has been working on induced mutations since 1983 and has produced improved mutant varieties in two crops. In cassava, irradiation of stem cuttings resulted in the production of 'Tek bankye', a mutant variety with excellent cooking and pounding quality and high dry matter content (40%) released in 1997 (Danso et al. 2009). With significant increases in yield, this mutant variety is tolerant to Africa cassava mosaic virus (ACMV), thereby leading to its quick adoption by many farmers in Ghana as it is used to prepare *fufu*, the nation's most popular cassava-based food (Kharkwal and Shu 2009). Similarly, irradiation of vegetative buds of 'Amelonado', 'Trinitario' and 'Upper Amazon' cocoa varieties resulted in the production of a mutant variety resistant to the cocoa swollen shoot virus (Danso et al. 2009).

13.2.6.3 Sudan

Mutation breeding in Sudan was started about three decades ago and covered crops like cotton, sugarcane, sesame, banana, tomato, groundnuts and cereals under stress environment to ensure sustainable food security and well-being of farmers. The outcome of these efforts resulted in the development of a banana variety 'Albeely' released in the year 2003 with 40% higher yield over the existing cultivars, and it also has better crop stand and fruit quality. 'Albeely' is widely cultivated by farmers in banana production areas along the Blue Nile in the south of Wad Madani (Kharkwal and Shu 2009). A drought-tolerant peanut variety, 'Tafra-1', was released in 2018 for Sudanese farmers in drought-prone areas, which led to an increase in the country's exports (Abdalla et al. 2021).

13.2.6.4 Mauritius

Three high-yielding tomato mutant varieties, namely 'Summer King', 'Summer Star' and 'Rising Star', and four oyster mushroom mutant varieties have been released in Mauritius. These varieties were distributed among farmers, and seed multiplication continues to reach more farmers (Sarsu et al. 2021).

13.2.6.5 Namibia

Namibia developed four sorghum and seven cowpea mutant varieties with 10–20% higher yield than local cultivars under drought conditions and pre-released to farmers. The Ministry of Agriculture is facilitating seed multiplication and planning to provide seeds of the newly developed improved mutant varieties to all potential cowpea growers in the country during the ensuing cropping seasons (Horn et al. 2017).

13.3 Mutation Breeding for Crop Improvement in the Indian Subcontinent

The Indian subcontinent includes the countries of Bangladesh, Bhutan, India, Maldives, Nepal, Pakistan and Sri Lanka. With about one-fourth of the world's population, it is the most populous and also the most densely populated geographical region in the world. Overall, it accounts for about 39.5% of Asia's population and is home to a vast array of people needing huge quantities of food to sustain the ever-increasing demands for their food security. The largest country by area and also by population in the Indian subcontinent, India has made spectacular achievements in crop improvement through mutation breeding and occupies the second position in the release of mutant varieties after China at the global level. Similar gains in crop improvement through mutation breeding have been achieved by the two other neighbouring countries—Pakistan and Bangladesh—as they have also released several high-yielding, biotic and abiotic stress-resistant mutant varieties of rice, wheat, chickpea, mung bean, cotton and some other crops for commercial cultivation. Several of these mutant varieties, whose details are given under their respective countries reported at the end of this chapter, grown in millions of hectares have made significant impact on crop improvement and the total production and productivity in these countries. Sri Lanka, a small island country, has also been working in mutation breeding and succeeded in the development and release of four mutant varieties so far. Brief surveys of the progress in crop improvement through mutation breeding and release of mutant varieties in a wide range of crops achieved by India, Pakistan, Bangladesh and Sri Lanka are listed countrywise in the ensuing paragraphs.

13.3.1 Mutation Breeding for Crop Improvement in India

India is one of the few pioneering countries of the world which started working on systematic mutation breeding seriously right from the middle of the twentieth century and was successful in releasing the first ever X-ray-induced drought-resistant cotton mutant variety MA-9 in 1948. A large number of mutant varieties out of the total of 542 mutant varieties belonging to 85 crop species developed and reported in India are under commercial cultivation and have made significant contribution to crop improvement and food and nutritional security of the country. Mutation breeding in India has yielded considerable dividends in enhancing our knowledge on various mutagenesis processes relevant to crop improvement and food security in India (Kharkwal and Shu 2009). Some of the major research centres actively engaged in mutation breeding work in several crops and having contributed substantially to the development and release of a large number of mutant varieties are the Indian Agricultural Research Institute, New Delhi; the National Botanical Research Institute, Lucknow; the Central Institute for Medicinal and Aromatic Plants, Lucknow; the Bhabha Atomic Research Centre, Mumbai; the Tamil Nadu Agricultural University, Coimbatore; Odisha University of Agriculture and Technology, Bhubaneswar; and several others. The mutant varieties developed in India belong to cereals, grain legumes, oilseeds, fibre crops, vegetables and ornamentals. The highest number was of ornamentals (165) followed by cereals (126), grain legumes (85) and oilseeds (74). The rapid progress in the development of mutant varieties is evident from the fact that Kharkwal et al. (2004) enlisted a total of 309 mutant varieties of 56 plant species released in India till 2004. Concerted further efforts in mutation breeding had already resulted in the development and release of 345 improved mutant varieties belonging to 57 crop species by 2017 (Kharkwal 2017). However, the most up-to-date list of 542 mutant varieties belonging to 85 crop species reported in India (Tables 13.3, 13.8 and 13.11) unambiguously demonstrates the significance and importance of use of mutation breeding in crop improvement in India. More than 72% of these 542 mutants have been developed as direct mutants and 12% by crossing with one or more mutants or mutant varieties. The largest number of mutant varieties reported from India belong to rice (76) among cereal crops, chrysanthemum (63) among the ornamentals, and groundnut (33) among the oilseed crops and mung bean (22) among legume crops

Table 13.8 Number of mutant varieties developed and reported in India

S. No.	Latin name	Common name	Type	No. of mutants
1	*Eleusine coracana*	Finger millet	C	10
2	*Hordeum vulgare*	Barley	C	14
3	*Oryza sativa*	Rice	C	76
4	*Panicum maximum*	Guinea grass	C	1
5	*Panicum sumatrense*	Little millet	C	2
6	*Paspalum scrobiculatum*	Kodo millet	C	5
7	*Pennisetum glaucum*	Pearl millet	C	5
8	*Setaria italica*	Foxtail millet	C	1
9	*Sorghum bicolor*	Sorghum	C	6
10	*Triticum aestivum*	Wheat	C	4
11	*Triticum dicoccum*	Emmer wheat	C	2
		Cereals		*126*
12	*Cajanus cajan*	Pigeon pea	L	10
13	*Cicer arietinum*	Chickpea	L	9
14	*Cyamopsis tetragonoloba*	Cluster bean	L	1
15	*Lablab purpureus*	Hyacinth bean	L	2
16	*Lens culinaris*	Lentil	L	4
17	*Macrotyloma uniflorum*	Horse gram	L	1
18	*Phaseolus vulgaris*	Common bean	L	1
19	*Pisum sativum*	Pea	L	1
20	*Trifolium alexandrinum*	Egyptian clover	L	1
21	*Vigna aconitifolia*	Moth bean	L	6
22	*Vigna mungo*	Black gram	L	13
23	*Vigna radiata*	Mung bean	L	22
24	*Vigna unguiculata*	Cowpea	L	14
		Legumes		*85*
25	*Arachis hypogaea*	Groundnut	O	33
26	*Brassica campestris*	Field mustard	O	3
27	*Brassica juncea*	Indian mustard	O	9
28	*Brassica napus*	Rapeseed	O	2
29	*Glycine max*	Soybean	O	12
30	*Helianthus annus*	Sunflower	O	3
31	*Linum usitatissimum*	Linseed	O	2
32	*Ricinus communis*	Castor bean	O	4
33	*Sesamum indicum*	Sesame	O	6
		Oil seeds		*74*
34	*Corchorus capsularis*	White jute	I	6
35	*Corchorus olitorius*	Tossa jute	I	5
36	*Gossypium arboreum*	Desi cotton	I	2
37	*Gossypium herbaceum*	Desi cotton	I	1
38	*Gossypium hirsutum*	American cotton	I	8
39	*Nicotiana tabacum*	Tobacco	I	1
40	*Saccharum officinarum*	Sugarcane	I	9

(continued)

Table 13.8 (continued)

S. No.	Latin name	Common name	Type	No. of mutants
		Industrial crops		*32*
41	*Abelmoschus esculentus*	Okra	V	2
42	*Capsicum annum*	Pepper	V	1
43	*Luffa acutangula*	Ridged gourd	V	1
44	*Momordica charantia*	Bitter gourd	V	1
45	*Solanum lycopersicon*	Tomato	V	4
46	*Solanum melongena*	Brinjal	V	1
47	*Trichosanthes anguina*	Snake gourd	V	1
		Vegetables		*11*
48	*Carica papaya*	Papaya	Fr	1
49	*Citrus reticulata*	Mandarin	Fr	1
		Fruits		*2*
50	*Coriandrum sativum*	Coriander	S	1
51	*Cuminum cyminum*	Cumin	S	1
52	*Curcuma longa*	Turmeric	S	2
53	*Trigonella foenum-graecum*	Fenugreek	S	2
54	*Zingiber officinalis*	Ginger	S	2
		Spices		*8*
55	*Catharanthus roseus*	Periwinkle	M	1
56	*Chamomilla recutita*	German chamomile	M	3
57	*Cymbopogon flexuosus*	Lemongrass	M	5
58	*Cymbopogon winterianus*	Citronella	M	10
59	*Hyoscyamus niger*	Black henbane	M	2
60	*Lantana depressa*	Wild sage	M	3
61	*Matricaria chamomilla*	German chamomile	M	1
62	*Mentha citrate*	Orange mint	M	1
63	*Mentha piperita*	Peppermint	M	3
64	*Mentha spicata*	Spearmint	M	1
65	*Papaver somniferum*	Opium poppy	M	3
66	*Phyllanthus niruri*	Java citronella	M	1
67	*Plantago ovata*	Psyllium	M	3
68	*Rauwolfia serpentina*	Indian snakeroot	M	1
69	*Solanum khasianum*	Khasianum	M	1
		Medicinal and aromatic		*39*
70	*Bougainvillea spectabilis*	Bougainvillea	F	28
71	*Callistephus chinensis*	Chinese aster	F	2
72	*Canna generalis*	Canna lily	F	2
73	*Chrysanthemum* sp.	Chrysanthemum	F	63
74	*Coreopsis*	Tickseed	F	1
75	*Crossandra infundibuliformis*	The fire cracker	F	1
76	*Dahlia* sp.	Dahlia	F	11
77	*Dianthus caryophyllus*	Carnation	F	1
78	*Gladiolus* sp.	Gladiolus	F	4

(continued)

Table 13.8 (continued)

S. No.	Latin name	Common name	Type	No. of mutants
79	*Hibiscus sinensis*	Hibiscus	F	2
80	*Jasminum* sp.	Jasmine	F	1
81	*Morus alba*	Mulberry	F	2
82	*Polyanthus tuberosa*	Tuberose	F	2
83	*Portulaca grandiflora*	Portulaca	F	11
84	*Rosa* sp.	Rose	F	33
85	*Solenostemon rotundifolius*	Coleus	F	1
		Ornamentals		*165*
		Total of all crops		542

Abbreviations: *C* cereals, *F* flowers, *Fr* fruits, *I* industrial crops, *L* legumes, *M* medicinal plants, *O* oilseeds, *S* spices, *V* vegetables

(Table 13.8). Although an exact estimate of the total area covered by commercially developed/released mutant cultivars in different states of India is not readily available, the limited information gathered clearly indicates that several of these are being cultivated in thousands of hectares and have made a very significant contribution to Indian agriculture addressing food and nutritional security problems of the country.

A close examination of the type of mutagens used and the number of mutant cultivars developed and reported in India (Tables 13.3, 13.8 and 13.11) indicates that the largest number of mutant varieties have been induced by physical mutagens (Table 13.9), with gamma rays being the most commonly used and highly successful followed by X-rays. This is largely because of the ease of availability and convenience of safety in handling and transportation of treated material. Although most of the chemical mutagens are very effective, they are highly carcinogenic, their handling with proper safety is a concern, and if high doses are applied, the desired mutations may be accompanied by too many undesired ones (Sybenga 1983). The survey also indicates that major gains have resulted from the use of induced variability when it was fully integrated with crop breeding and more varieties have resulted by using induced mutants in cross-breeding programme rather than processing them directly into mutant varieties (Table 13.9).

13.3.1.1 Mutant Varieties Released for Crop Improvement in India

A large number of mutant varieties of several food crops, ornamental crops and medicinal and aromatic plants have been released for cultivation in different agro-climatic regions in India. The mutant cultivars released in India have contributed immensely to augmenting the efforts of Indian plant breeders in achieving the target of self-sufficiency in food production, nutritional security and strong economic (Kharkwal et al. 1988) growth of the country (Kharkwal 1996, 1998, 2012, 2017; Kharkwal et al. 2001, 2004, 2005; Kharkwal and Shu 2009). Mutation breeding has made significant contribution to increasing the production of rice, groundnut, castor, chickpea, mung bean and urdbean in the Indian subcontinent (Kharkwal et al. 2004; Kharkwal 2017).

Table 13.9 Mutagens used for the development of 542 mutant varieties reported in India

Mutagen/origin	Mutant varieties	
	Number	Percentage
Gamma rays	285	52.50
X-rays	37	6.90
Neutrons	7	1.30
Chemical mutagens	50	9.30
Combined treatment	12	2.20
Crosses	73	13.40
Bud sports	22	4.10
Natural and others	56	10.40

A perusal of the details of mutant cultivars of agricultural crop plants reported or approved for release and cultivation in India provided in Tables 13.8 and 13.11 indicates that the largest number of mutant cultivars have been produced in ornamentals (165) followed closely by cereals (126), legumes (85) and oilseeds (74). The success story of mutation breeding in ornamentals and horticultural plants in India as is evident with 63 mutants of chrysanthemum topping the list of ornamentals followed by roses (33) is because of the fact that India is at the threshold of a boom in the export market in floriculture and horticulture. Among cereals, the leading position with 76 mutant cultivars is occupied by rice. Keeping in view the rarity with which mutations occur and the general deleterious nature of induced mutants, the success and achievements of mutation breeding in the development of commercially released varieties in such a large and wide range of plant species in India are thus significant. This is particularly true in Indian situation where all varieties (except ornamentals) irrespective of the method through which they are developed have to go through the All India Coordinated Trials grid before getting identified/approved for release. It is already known that mutation breeding is expected to make a contribution primarily as an important adjunct to the conventional breeding approach. Among the research institutes actively engaged in mutation breeding in India, the largest number of 67 Indian mutant varieties belonging to 20 major crop species have been developed and released by the ICAR-Indian Agricultural Research Institute (IARI), New Delhi, a leading institute in agricultural research not only in India, but also in the world (Table 13.10).

13.3.1.2 Success Stories of Prominent Mutant Varieties Released in India

Although the world's very first cotton mutant variety MA-9 induced by X-rays, and endowed with drought tolerance, was released by India way back in 1948, concerted efforts for crop improvement through induced mutations in India were initiated during the second half of the 1950s. India, distinctly, is the first in the world to have released radiation-induced mutant varieties in cotton (MA-9), wheat (NP-836), sugarcane (Co-6608), castor (Aruna), chickpea (Pusa-408 Ajay), pearl millet (NHB-5), black gram (Co-4), mung bean (TAP-7), pigeon pea (Co-3), lablab (Co-10), okra (MDU-5), turmeric (Co-1), bitter gourd (MDU-1) and ridge gourd

Table 13.10 Mutant varieties released by the Indian Agricultural Research Institute, New Delhi

S. No.	Crop	Latin name	No.	Names of mutant varieties
1	Rice	*Oryza sativa*	13	PNR-162, PNR-166, PNR-381, PNR-519, PNR-550-1-2 (JD-8), PNR-551-4-20 (JD-6), PNR-555-28 (JD-10), PNR-555-5, PNR-555-5 (JD-3), PNR-570-17, PNR-571
2	Barley	*Hordeum vulgare*	6	DL-253, Karan-3, Karan-4, Karan-15, Karan-201, Karan-265
3	Wheat	*Triticum aestivum*	3	NP 836, Pusa Larma, Sharbati Sonora
4	Emmer wheat	*Triticum dicoccum*	2	COW-2, Nilgiri khapli (HW-1098)
5	Pearl millet	*Pennisetum glaucum*	4	NHB-3, NHB-4, NHB-5, Pusa-46
6	Chickpea	*Cicer arietinum*	4	Pusa-408 (Ajay), Pusa-413 (Atul), Pusa-417 (Girnar), Pusa-547
7	Cowpea	*Vigna unguiculata*	6	V-16 (Amba), V-37 (Shreshtha), V-38 (Swarna), V-240, V-578, V -585 (Sampada)
8	Castor bean	*Ricinus communis*	2	Aruna, Sowbhagya (157-B)
9	Cotton	*Gossypium hirsutum*	2	Pusa Ageti, Rasmi
10	Soybean	*Glycine max*	2	Pusa-97-12, Pusa-12
11	Pigeon pea	*Cajanus cajan*	1	Pusa-855
12	French bean	*Phaseolus vulgaris*	1	Pusa Parvati
13	Field pea	*Pisum sativum*	1	L-116 (Hans)
14	Tomato	*Solanum lycopersicon*	1	Pusa Lal Meruti
15	Papaya	*Carica papaya*	1	Pusa Nanha
16	Chrysanthemum	*Chrysanthemum*	4	Pusa Amol, Pusa Arunodaya, Pusa Centenary, Pusa Kesari
17	Roses	*Rosa* sp.	11	Abhisarika, Chitra, Climbing Sadabahar, Madhosh, Nav Bahar, Pink Sport Montezuma, Pusa Abhishek, Pusa Christian, Pusa Mansij, Pusa Urmil, Striped Christian Dior
18	Bougainvillea	*Bougainvillea* sp.	1	Visaka
19	Gladiolus	*Gladiolus* sp.	1	Swarnima
20	Tickseed	*Coreopsis*	1	Pusa Tara
	Total		67	

(PKM-1). Mutant varieties like Aruna of castor that matures in 120 days compared to 270 days of the parent variety; TAG-24, TG-37A and TG-38 of groundnut; Pusa-408

(Ajay), Pusa-413 (Atul), Pusa-417 (Girnar) and Pusa-547 of chickpea; TARM-1, Co-4 and TMB-37 of mung bean; TAU-1 and Co-4 of black gram; TT-401 of pigeon pea; Jagannath, Savitri, Dharitri, Gayatri, Sattari, Keshari and Nagina-22 of rice; amber-coloured Sharbati Sonora of wheat; DL-253 of barley; Pusa 46 of pearl millet; and Pusa Nanha, a dwarf variety of papaya, are among the most important varieties of economic importance released in India (Kharkwal 1996, 1998; Kharkwal et al. 2001, 2005, Kharkwal and Shu 2009). However, authentic information on the area covered under these cultivars is unfortunately lacking. It is also, however, accepted that some of the mutant cultivars of rice, wheat, castor, groundnut, chickpea, mung bean and urdbean did find wide acceptability and cover a sizable acreage on their release for cultivation in various states and areas of India.

The first ever examples of direct use of induced micro-mutants in a legume crop in the world are the four gamma ray-induced chickpea mutant varieties, Pusa-408 (Ajay), Pusa-413 (Atul), Pusa-417 (Girnar) and Pusa-547 with resistance to Ascochyta blight, *Fusarium* wilt and other diseases and pests developed by the author at the Division of Genetics, ICAR-Indian Agricultural Research Institute (IARI), New Delhi, and released by the Government of India for commercial cultivation by the farmers. The salient features of these four high-yielding direct mutant varieties of chickpea are given in the following paragraphs.

13.3.1.2.1 Pusa-408 (Ajay)

Pusa-408, an induced mutant through 600 Gy gamma rays, has superior average yield of 23 q/ha and a yield potential of 35 q/ha. It is suitable for normal as well as late plantings, under rainfed and irrigated conditions. Pusa-408 has resistance to Ascochyta blight and Fusarium wilt, stunt virus and collar rot. It shows low incidence of pod borer and nematode damage. This mutant variety has been released for commercial cultivation by farmers of Haryana, Punjab, Rajasthan, Delhi, western UP and J&K states of India.

13.3.1.2.2 Pusa-413 (Atul)

Pusa-413, an induced mutant through 600 Gy gamma rays, has superior average yield of 21 q/ha and a yield potential of 35 q/ha. It is suitable for normal as well as late plantings, under rainfed and irrigated conditions. Pusa-413 has resistance to Fusarium wilt and Ascochyta blight, stunt virus, foot rot and root rot. It shows low incidence of pod borer and nematode damage. This mutant variety has been released for commercial cultivation by farmers of the eastern UP, Bihar, Jharkhand, Odisha, West Bengal and Assam states of India.

13.3.1.2.3 Pusa-417 (Girnar)

An induced mutant through 600 Gy gamma rays, Pusa-417 has superior average yield of 21 q/ha and a yield potential of 35 q/ha. It is suitable for normal as well as late plantings, under rainfed and irrigated conditions. Pusa-417 has high resistance to Fusarium wilt and moderate resistance to Ascochyta blight, stunt virus, collar rot, foot rot and root rot. It shows low incidence of pod borer and nematode damage. This mutant variety has been released for commercial cultivation by farmers of Madhya

Pradesh, Rajasthan, Gujarat, Maharashtra and parts of Telangana and Karnataka states of India.

13.3.1.2.4 Pusa-547

An induced mutant through 600 Gy gamma rays, Pusa-547 has superior average yield of 20 q/ha and a yield potential of 35 q/ha. A golden brown bold-seeded (25 g/ 100 seeds) variety, it is especially suitable for late plantings, under rainfed and irrigated conditions. Chickpea mutant variety Pusa-547, the latest variety released in 2006 for commercial cultivation by farmers in Haryana, Punjab, Rajasthan, Delhi, western UP and J&K states of India, has shown high yield performance particularly under late-sown conditions. Based on its performance on farmers' fields, it has become popular in several other chickpea-growing areas like eastern UP, Bihar and Madhya Pradesh. It has very attractive bold, golden colour grains having thin testa and good cooking quality (Kharkwal et al. 2005). A major seed production programme of this high-yielding chickpea mutant variety Pusa-547 for rapid production of breeder seed, foundation seed and certified seed has been taken up in a big way by the State Farms Corporation of India (SFCI) and also the National Seed Corporation (NSC) at their Hisar (Haryana), Suratgarh, Sardargarh and Jaitsar (Rajasthan) farms, and more than 35,000 q of high-quality seeds of this high-yielding bold-seeded mutant variety Pusa-547 have already been produced and distributed to farmers.

13.3.1.3 Other Prominent Mutant Varieties Released in India

Jagannath (BBS-873), a short rice mutant, was isolated and released as 'Jagannath' in 1969 for the coastal districts of Odisha, West Bengal and Andhra Pradesh. It matured in about 155 days and yielded about 50% more than tall parent T-141 in shallow-water conditions and is moderately responsive to nitrogen fertilization. A late-maturing mutant variety with medium slender translucent grains, it has 1000-grain weight of 20 g. It is strongly photosensitive and possesses strong seed dormancy; it has broad-spectrum field resistance to major pests and diseases. It was the leading high-yielding rice mutant variety in Odisha for nearly two decades. The change of wheat seed coat colour from red to amber by gamma radiation resulting in the development of 'Sharbati Sonora' (Swaminathan et al. 1968) is a classical example. In castor, the mutant variety 'Aruna' developed through neutron irradiation matures in 120 days compared to 270 days of the parent cultivar, HC-6 (Ankineedu et al. 1968), which is also a classical example of prominent mutant variety. One of the black gram (urdbean) mutant var. 'TAU-1' released in India has become the most popular variety in Maharashtra, occupying over 70% of the total area under urdbean cultivation in Maharashtra (Pawar et al. 1991). Since 1990, the Maharashtra State Seed Corporation has distributed about 200,000 q of certified seeds of TAU-1 to the farmers, which has resulted in an additional production of about 129,000 q of urdbean annually in Maharashtra. The notional income generated by additional production amounts to Rs. 258 crores annually (Kharkwal and Shu 2009; Kharkwal 2017). Similarly, several mutant varieties of groundnut—TAG-24 and TG-26—released and cultivated in millions of acres in Maharashtra, Gujarat and several other states have contributed tremendously not only towards achieving the targets of the agricultural production of the respective states but also to the Indian agricultural economy as a whole (Kharkwal et al. 2004; Kharkwal and Shu 2009).

The main improved characteristics of Indian mutant varieties developed and released or approved as reported by the breeders and summarized in Table 13.11 indicate that higher yield, altered maturity duration, improved plant type, increased resistance to biotic and abiotic stresses and quality characters, which ultimately led to higher productivity and acceptability of these mutant cultivars, are the most frequent traits in a new variety reported and released for commercial cultivation by farmers and marketing.

13.3.2 Mutation Breeding for Crop Improvement in Pakistan

Pakistan has successfully developed and released 79 mutant varieties in major crop groups like cotton (24), pulses (29), cereals (16), oilseeds (8) and one each of sugarcane and citrus fruit mandarin (Personal communication T.M. Shah). There are three major institutes engaged in mutation breeding in Pakistan. The largest and main centre, Nuclear Institute of Agriculture and Biology (NIAB) based in Faisalabad, has developed and released 50 mutant varieties belonging to 8 crop species till the beginning of 2022. The second main contributor of 19 mutant varieties belonging to 7 crops is the Nuclear Institute of Agriculture (NIA) based in Tando Jam near Hyderabad in Sindh Province. The third centre of mutation breeding in Pakistan is the Nuclear Institute for Food and Agriculture (NIFA) located at Peshawar in Khyber Pakhtunkhwa, which has developed 10 mutant varieties of 3 crops. NIAB has the distinction of developing not only the largest number of mutant varieties but also the first and most important and popular mutant varieties of rice—'Kashmir Basmati', cotton mutant var. 'NIAB-78', chickpea 'CM-72' and mung bean 'NM-28', all four being the first mutant varieties in these crops released in 1983 (Ullah et al. 2020; Razzaq et al. 2021). The mutant varieties of wheat, rice and pulses like chickpea, mung bean and lentils have improved food sustainability in Pakistan, and mutant varieties of oilseed crops and cash crops like cotton have led to positive economic impacts worth billions of US dollars through their export potential. Individual cropwise details of these mutant varieties are listed in Table 13.12.

Cotton is the second most important crop of Pakistan after wheat. Mutation breeding in cotton initiated in 1970 has resulted in the development of 24 early-maturing, high-yielding mutant varieties with desirable fibre quality (Table 13.12). Some of them like NIAB-78, NIAB-86 and NIAB Karishma covering millions of hectares have resulted in a quantum jump in cotton production in Pakistan. NIAB-78 covered about 70.8% of the total cotton area in Punjab in 1988. It spread gradually to Sindh starting from 1992 to 1993 and covered almost all the area of cotton crop and now also covers some of the areas under cotton in Sindh. Within 5 years of release of NIAB-78, its cultivation contributed enormously to the growth of the textile industry in the country and doubled cotton production in Pakistan. It is estimated that during

Table 13.11 Mutant varieties of crop plants developed, reported and approved/released for farmers' cultivation in India

S. No.	Latin name	Common name	Mutant variety	Release year	Institution	Mutagen and dose/parent var.	Main characters induced/improved	MVID No.@ MBNL No.#
1	*Abelmoschus esculentus*	Okra	MDU-1	1985	TNAU, AC&RI, Madurai	DES 0.04%/Pusa Sawani	Higher yield, tolerant to YMV	MVID-333 MBNL-3334
2		Okra	Anjitha	2006	COA, Vellayani, Trivandrum	Gamma rays 300 Gy/ Kiran × *A. manihot*	High yield with YMV resistance	MVID-2770
3	*Arachis hypogaea*	Groundnut	TG-1 (Vikram)	1973	BARC, Mumbai	X-rays 550 Gy/Spanish Improved	Bold kernel (TKW–8800 g), 20% more yield, oil content 48%	MVID-1559 MBNL-11
4		Groundnut	TMV-10	1975	TNAU, Coimbatore	A natural mutant from 'Argentina'/Virginia bunch	High oil 54.4%, and shelling 77.1%	
5		Groundnut	TG-4 (TG-14)	1976	BARC, Mumbai	X-rays 150 Gy/Spanish Improved	Uniform maturity, high yield	MVID-1557 MBNL-12
6		Groundnut	MH-2	1978	HAU, Hisar	Gamma rays/Gujarat Dwarf Mutant	Yield and resistance for tikka disease	MVID-1573 MBNL-32
7		Groundnut	BP-1	1979	BAU, Ranchi	Gamma rays 450 Gy/41-C	Bold kernels, high yield	MVID-1571 MBNL-31
8		Groundnut	BP-2	1979	BAU, Ranchi	Gamma rays 450 Gy/41-C	Bold kernels, high yield	MVID-1572 MBNL-32
9		Groundnut	BG-1	1983	BAU, Ranchi	Gamma rays/41-C Virginia group	High yield, semi-erect and large seed	MVID-2858
10		Groundnut	BG-2	1983	BAU, Ranchi	Gamma rays/41-C Virginia group	High yield, semi-erect and large seed	MVID-2859
11		Groundnut	Co-2	1984	TNAU, Coimbatore	EMS 0.2%/a bunch mutant from POL-1	High yield, oil 51.4%, tolerant to ELS and LLS	MVID-1568 MBNL-26
12		Groundnut	Kaushal (G-201)	1985	CSAU, Kanpur	Natural mutant of bunchy type 28	High shelling, 70%, resistance to LS and rust	
13		Groundnut	TG-17	1985	BARC, Mumbai	X-rays 150 Gy/inter-mutant cross, Spanish Improved	15–20% more yield, high harvest index	MVID-1558 MBNL-12
14		Groundnut	TG-3	1987	BARC, Mumbai	X-rays 150 Gy/Spanish Improved	High yield and earliness	MVID-1556 MBNL-12
15		Groundnut	VRI-2	1989	Vridhachalam Station, TNAU, Coimbatore	EMS 0.2% Pol-1 × JL 24 crossed with a mutant Co 2	Large seed size	MVID-2861

(continued)

Table 13.11 (continued)

S. No.	Latin name	Common name	Mutant variety	Release year	Institution	Mutagen and dose/parent var.	Main characters induced/improved	MVID No.@ MBNL No.#
16		Groundnut	TGS-1 (Somnath)	1991	BARC, Mumbai	Cross: [TG-18A × M-13 (200 Gy gamma rays mutant)]	23% more yield, large seed early and spreading habit	MVID-1576 MBNL-41
17		Groundnut	TAG-24	1992	BARC, Mumbai	Cross (TG-18A × M-13) × (line × TG-9)	Earliness, drought tolerant and 25–50% more yield	MVID-1575 MBNL-41
18		Groundnut	TG-22	1992	BARC, Mumbai	Cross: (Robout 33-1 × TG-17)	30% more yield, fresh seed dormancy	MVID-1579 MBNL-44
19		Groundnut	TKG-19A	1996	BARC, Mumbai	Cross: (TG-17 × TG-1)	12–13% more yield, fresh seed dormancy	MVID-1578 MBNL-44
20		Groundnut	TG-26	1996	BARC, Mumbai	Cross: Involving mutant derivatives (BARC G-1 × TG-23)	Semi-dwarf, early, high harvest index (50%) and 23–39% more yield	MVID-1580 MBNL-44
21		Groundnut	R-9251	1997	UAS, Raichur Karnataka	Gamma rays Crossing with one mutant JL 24 and JLM 1 of TG 23	Early maturity, tolerant to peanut bud necrosis	MVID-2860
22		Groundnut	Mutant 28-2	2002	UAS, Dharwad	EMS/VL-1	Bold kernels, resistant to late leaf spot and thrips	MVID-2857
23		Groundnut	TG-37A	2004	BARC, Mumbai	Inter-mutant cross TG-25 × TG-26	Oil 51%, 26–38% more yield, early, fresh seed dormancy	MVID-2318
24		Groundnut	TPG-41	2004	BARC, Mumbai	Inter-mutant cross TG-28A × TG-22	Large seed (70 g/100 seeds), 49% more yield, fresh seed dormancy	MVID-2268
25		Groundnut	TG-38	2006	BARC, Mumbai	Inter-mutant cross Girnar × TG-26	Erect semi-dwarf plant, tolerance to stem rot and dry root rot	MVID-2559
26		Groundnut	TLG-45	2007	BARC, Mumbai	Inter-mutant cross TAG-24 × TG-19	High pod yield, semi-dwarf plant and good quality	MVID-2840
27		Groundnut	Dh 4-3	2007	UAS, Dharwad	Dharwad early Runner	High yielding, resistant to LLS and rust	
28		Groundnut	TBG-39	2008	BARC, Mumbai	Inter-mutant cross TAG-24 × TG-19	High yield	MVID-2939
29		Groundnut	TG-51	2008	BARC, Mumbai	Inter-mutant cross TAG-24 × TG-19	High yield	MVID-2938
30		Groundnut	TDG-39	2009	BARC, Mumbai	TAG-24 × TG-19	Large seeds, high oleic acid and early maturity	MVID-3335

31		Groundnut	GPBD-5	2010	UAS, Dharwad	Gamma rays 300 Gy/TG-49 × GPBD-4	Large seed, fungal disease resistance	MVID-3424
32		Groundnut	TG-47 (RARS-T1)	2011	BARC, Mumbai	Cross: RG-24 × TG-19	High yield	
33		Groundnut	G-2-52	2013	UAS, Dharwad	Gamma rays 200 Gy/GPBD-4	Bold, high yielding, resistant to LLS, oil quality (55% oleic acid)	
34		Groundnut	Dh-245	2017	UAS, Dharwad	Gamma rays 300 Gy/GPBD-4	High oleic acid (70.5%) and resistant to foliar diseases	
35		Groundnut	Dh-256	2019	UAS, Dharwad	Gamma rays 300 Gy/[(R 2001-2 × GM 4-3-12 (mutant))]	Drought tolerance along with yield advantage	
36	*Bougainvillea* sp.	Bougainvillea	Shubhra	1969	NBRI, Lucknow	Bud sport of 'Mary Palmer'	Ornamental novelty	
37		Bougainvillea	Archana	1973	NBRI, Lucknow	Bud sport of Roseville's Delight	Ornamental novelty	
38		Bougainvillea	Vishaka	1974	IARI, New Delhi	Bud sport of Mrs. H.C. Buck	Leaves variegated at margins	
39		Bougainvillea	Arka Jawaharlal Nehru	1975	IIHR, Bengaluru	Bud sport of Lalbagh	Variegated leaves and bracts	
40		Bougainvillea	Arjuna	1976	NBRI, Lucknow	Gamma rays 5 Gy/Partha	Variegated leaves	MVID-1894 MBNL-15
41		Bougainvillea	Jaya	1977	BARC, Mumbai	Gamma rays 25 Gy/Jayalakshmi	Ornamental novelty	MVID-1895 MBNL-20
42		Bougainvillea	Jayalaxmi Variegata	1977	BARC, Mumbai	Gamma rays 65 Gy/Jayalakshmi	Ornamental novelty	MVID-1896 MBNL-14
43		Bougainvillea	Parthasarthy	1977	NBRI, Lucknow	Bud sport of Partha	Ornamental novelty	
44		Bougainvillea	Silver Top	1978	BARC, Mumbai	Gamma rays 25 Gy + colchicine 0.5%/versicolor	Ornamental novelty	MVID-1898 MBNL-20
45		Bougainvillea	Lady Hudson of Ceylon Variegata	1979	BARC, Mumbai	Gamma rays 10 Gy + 0.5% colchicine/Lady Hudson Ceylon	Ornamental novelty	MVID-1897 MBNL-20
46		Bougainvillea	Shweta	1979	NBRI, Lucknow	Bud sport of 'Trinidad'	Ornamental novelty	
47		Bougainvillea	Poultoni Variegata	1981	BARC, Mumbai	Gamma rays 15 Gy/Poultoni	Variegated leaves	MVID-1892 MBNL-33

(continued)

Table 13.11 (continued)

S. No.	Latin name	Common name	Mutant variety	Release year	Institution	Mutagen and dose/parent var.	Main characters induced/improved	MVID No.@ MBNL No.#
48		Bougainvillea	Surekha	1981	NBRI, Lucknow	Bud sport of 'Scarlet Queen'	Ornamental novelty	
49		Bougainvillea	Suvarna	1981	BARC, Mumbai	Gamma rays 10 Gy/Lady Hudson of Ceylon	Ornamental novelty flower colour	MVID-1893 MBNL-33
50		Bougainvillea	Nirmal	1982	NBRI, Lucknow	Bud sport of Mrs. McClean	Ornamental novelty	
51		Bougainvillea	Manohar Chandra Variegata	1985	NBRI, Lucknow	Bud sport of 'Manohar Chandra'	Ornamental novelty	
52		Bougainvillea	Pallavi	1987	NBRI, Lucknow	Gamma rays 10 Gy, Roseville's Delight	Variegated leaves	MVID-1947 MBNL-31
53		Bougainvillea	Hawaiian Beauty	1990	NBRI, Lucknow	Bud sport of 'Hawaiian White'	Ornamental novelty	
54		Bougainvillea	Los Banos Variegata	1990	NBRI, Lucknow	Gamma rays 10–15 Gy/'Los Banos Beauty'	Leaf colour	MVID-1959 MBNL-37
55		Bougainvillea	Mahara Variegata	1994	NBRI, Lucknow	Gamma rays 10 Gy/'Mahara'	Ornamental novelty	MVID-1976 MBNL-43
56		Bougainvillea	Los Banos Variegata Silver Margin	2002	NBRI, Lucknow	Gamma rays/'Los Banos Beauty'	Variegated leaves	MVID-4462
57		Bougainvillea	Mahara Variegata Abnormal Leaf	2002	NBRI, Lucknow	Gamma radiation/'Mahara'	Abnormal leaf mutant	MVID-4456
58		Bougainvillea	Los Banos Varigeta 'Jayanti'	2006	NBRI, Lucknow	EMS/'Los Banos Beauty'	Variegated leaves	
59		Bougainvillea	Aruna	2008	NBRI, Lucknow	Gamma rays/'Palekar'	Ornamental novelty	
60		Bougainvillea	Pixie Variegata	2009	NBRI, Lucknow	EMS 0.02%/cv. 'Pixie'	Variegated leaves	MVID-3438
61		Bougainvillea	Abhimanyu	2010	NBRI, Lucknow	Bud sport from mutant var. 'Arjuna'	Higher multiplication rate	MVID-4451
62		Bougainvillea	Dr. P V Sane	2011	NBRI, Lucknow	Gamma rays/Dr B P Pal	Ornamental novelty	

63		Bougainvillea	Dr. A P J Abdul Kalam	2015	NBRI, Lucknow	Bud sport from 'Fantasi'	Ornamental novelty	
64	*Brassica campestris*	Field mustard	Parbati (ORT (M) 2-4	2000	OUAT, Odisha	Gamma rays 1000 Gy/BT-4	High yield, earliness, oil 41.5%	
65		Field mustard	Anuradha (ORT(M) 6-2)	2002	OUAT, Odisha	Gamma rays 800 Gy/TS-29	High yield, oil 44.5% CVRC	
66		Field mustard	Sushree ORT (M) 7-2	2012	OUAT, Odisha	EMS 0.3%/TS-29	Oil 42% dwarf, CVRC	
67	*Brassica juncea*	Indian mustard	RLM-198	1977	PAU, Ludhiana	Gamma rays/RL-18	Oil content, tolerant to aphids and blight	MVID-1711 MBNL-7
68		Indian mustard	RLM-514	1984	PAU, Ludhiana	Gamma rays 2000 Gy/RL-18	Yield, bold grain, early and high oil content	MVID-1706 MBNL-17
69		Indian mustard	RLM-619	1984	PAU, Ludhiana	Gamma rays 2000 Gy/RL-18	Tolerant to white rust, downy mildew and aphids, early	
70		Indian mustard	TM-2	1987	BARC, Mumbai	X-rays 750 Gy/RL-9	Early maturity (90–95 days), black seeded, high oil and yield	MVID-1715 MBNL-43
71		Indian mustard	TM-4	1987	BARC, Mumbai	Cross: TM-1 × Varuna	Earliness, yellow seeded, high oil and yield	MVID-1716 MBNL-43
72		Indian mustard	RL-1359	1987	PAU, Ludhiana	Cross: RLM-514 Varuna	High yield, bold seeds, oil content, tolerant to aphids	MVID-1712
73		Indian mustard	TPM-1	2007	BARC, Mumbai	Inter-mutant cross	Yellow seed coat, tolerant to powdery mildew	MVID-3333
74		Indian mustard	TBM-204	2019	BARC, Mumbai BCKV, Mohanpur	Cross: TM-102 × TM-28	High oil content (41%), no lodging and shattering at maturity	
75		Indian mustard	THPM-1	2021	HPKV, Palampur	Cross: TM-102 × TM-28	High yield, yellow seeds and aphid resistance	
76	*Brassica napus*	Rapeseed	NUDB-38	2004	Nagpur Univ. and Dhara VOFC	Gamma rays + EMS/Westar	Early maturity and high yield	MVID-3370
77		Rapeseed	NUDB-26-11	2008	Nagpur Univ. and Dhara VOFC	Gamma rays + EMS/Westar	Zero erucic acid and low glucosinolate	MVID-3350
78	*Cajanus cajan*	Pigeon pea	TAT-5	1984	BARC, Mumbai	Fast neutrons 150 Gy/T-21	Increased seed size (50%) and earliness	MVID-398 MBNL-28

(continued)

Table 13.11 (continued)

S. No.	Latin name	Common name	Mutant variety	Release year	Institution	Mutagen and dose/parent var.	Main characters induced/improved	MVID No.@ MBNL No.#
79		Pigeon pea	Co-3	1985	TNAU, Coimbatore	EMS 0.6%/Co-1	High yield, resistant to root rot, pod borer and determinate	MVID-391 MBNL-29
80		Pigeon pea	T. Vishaka-1 (TT-6)	1985	BARC, Mumbai	Mutant of T-21	Bold seeds, extra early maturity with 15% more yield	MVID-388 MBNL-23
81		Pigeon pea	Co-5	1985	TNAU, Coimbatore	Gamma rays 160 Gy/Co-1	Early, resistant to pod fly, root rot, sterility mosaic	MVID-392 MBNL-29
82		Pigeon pea	TAT-10	1985	BARC, Mumbai	Cross: TT-2 × TT-8	Extra early (115 days) and medium seeds	MVID-399 MBNL-28
83		Pigeon pea	Co-6	1991	TNAU, Coimbatore	Gamma rays 250 Gy/SA-1	Photosensitive type, tolerant to pod borer	
84		Pigeon pea	Pusa-855	1993	IARI, New Delhi	Gamma rays 600 Gy	Earliness and large seed	
85		Pigeon pea	TT-401	2007	BARC, Mumbai	Cross: ICPL 84008 × TT-6	Early maturing, resistant to wilt	MVID-3334
86		Pigeon pea	TJT-501	2009	BARC, Mumbai	Cross: ICPL 84008 × TT-6	Resistant to sterility mosaic and wilt, tolerant to pod borer and pod fly	MVID-3336
87		Pigeon pea	PKV-TARA (TAT-9629)	2013	PDKV, Akola and BARC, Mumbai	Cross: ICPL 84008 × TT-6	Tolerant to Fusarium wilt, sterility mosaic, pod borer and pod fly	
88	*Callistephus chinensis*	Chinese aster	Arka Advika	2021	IIHR, Bengaluru	Spontaneous mutant of Arka Poornima	Semi-double-white flowers having short pseudo-ray florets	
89		Chinese aster	Arka Nirali	2021	IIHR, Bengaluru	Spontaneous mutant of Arka Violet	Semi-double-violet colour flowers having short pseudo-ray florets	
90	*Canna generalis*	Canna lily	Agnisikha	2009	NBRI, Lucknow	Gamma rays/Lucifer	Bicoloured with 20 shades of red-like flame of fire, canary yellow margin	
91		Canna lily	Raktima	2010	NBRI, Lucknow	Gamma rays/Cleopatra	Blood red-coloured flowers and bronzy brown glossy foliage	
92	*Capsicum annuum frutescens*	Pepper	MDU-1	1985	TNAU, AC&RI, Madurai	Gamma rays 300 Gy/K-1	High yielding and compact dwarf type	MVID-2202 MBNL-10
93	*Carica papaya*	Papaya	Pusa Nanha	1987	IARI, New Delhi	Gamma rays 150 Gy/Ranchi	Dwarf plant type	MVID-238 MBNL-30
94	*Catharanthus roseus*	Periwinkle	Dhawal	2005	CIMAP, Lucknow	NEU 0.06%/Nirmal	Higher herbage and anti-cancer alkaloid yield (1.3–1.7%)	

95	*Chamomilla recutita*	Chamomile	Vallary (M-70-1)	1996	CIMAP, Lucknow	Gamma rays 700 Gy/German line	High flower and oil yields, blue oil	
96		Chamomile	CIM-Sammohak	2010	CIMAP, Lucknow	Gamma rays 700 Gy/Vallary	High flower, oil yields and high chamazulene content, dark-blue oil	
97		Chamomile	CIM-Ujjwala	2019	CIMAP, Lucknow	Gamma rays 600 Gy/Vallary	High, big size flower, high oil yields, no chamazulene content	
98	*Chrysanthemum*	Chrysanthemum	Aruna	1974	NBRI, Lucknow	Gamma rays 15 Gy/Undaunted	Flower colour	MVID-1901 MBNL-15
99		Chrysanthemum	Ashankit	1974	NBRI, Lucknow	Gamma rays 15 Gy/Undaunted	Flower colour	MVID-1903 MBNL-15
100		Chrysanthemum	Gairik	1974	NBRI, Lucknow	Gamma rays 10 Gy/Belur Math	Flower colour	MVID-1906 MBNL-15
101		Chrysanthemum	Himani	1974	NBRI, Lucknow	Gamma rays 20 Gy/E-13	Flower colour	MVID-1908 MBNL-15
102		Chrysanthemum	Kansya	1974	NBRI, Lucknow	Gamma rays 15 Gy/Rose Day	Flower colour	MVID-1911 MBNL-15
103		Chrysanthemum	Kapish	1974	NBRI, Lucknow	Gamma rays 20 Gy/E-13	Flower colour	MVID-1912 MBNL-15
104		Chrysanthemum	Kunchita	1974	NBRI, Lucknow	Gamma rays 15 Gy/Undaunted	Flower colour	MVID-1913 MBNL-15
105		Chrysanthemum	Lohita	1974	NBRI, Lucknow	Gamma-rays 20 Gy/E-13	Flower colour	MVID-1914 MBNL-15
106		Chrysanthemum	Manak	1974	NBRI, Lucknow	Gamma rays/Undaunted	Dark-brown flower	
107		Chrysanthemum	Pingal	1974	NBRI, Lucknow	Gamma rays 15 Gy/Pink Casket	Flower colour	MVID-1918 MBNL-15
108		Chrysanthemum	Shukla	1974	NBRI, Lucknow	Gamma rays 15 Gy/Mrs. H Gubby	Flower colour	MVID-1924 MBNL-15
109		Chrysanthemum	Shveta	1974	NBRI, Lucknow	Gamma rays 20 Gy/Fish Tail	Flower colour	MVID-1925 MBNL-15
110		Chrysanthemum	Tamra	1974	NBRI, Lucknow	Gamma rays 15 Gy/Goldie	Flower colour	MVID-1927 MBNL-15
111		Chrysanthemum	Anamika	1975	NBRI, Lucknow	Gamma rays 20 Gy/E-13	Flower colour	MVID-1900 MBNL-15

(continued)

Table 13.11 (continued)

S. No.	Latin name	Common name	Mutant variety	Release year	Institution	Mutagen and dose/parent var.	Main characters induced/improved	MVID No.@ MBNL No.#
112		Chrysanthemum	Asha	1975	NBRI, Lucknow	Gamma rays 15 Gy/Hope	Flower colour	MVID-1902 MBNL-15
113		Chrysanthemum	Basant	1975	NBRI, Lucknow	Gamma rays 10 Gy/Paul	Flower colour	MVID-1904 MBNL-15
114		Chrysanthemum	Jhalar	1975	NBRI, Lucknow	Gamma rays 15 Gy/Undaunted	Flower colour	MVID-1909 MBNL-15
115		Chrysanthemum	Kanak	1975	NBRI, Lucknow	Gamma rays 15 Gy/Undaunted	Flower colour	MVID-1910 MBNL-15
116		Chrysanthemum	Nirbhaya	1975	NBRI, Lucknow	Gamma rays 15 Gy/Undaunted	Flower colour	MVID-1916 MBNL-15
117		Chrysanthemum	Nirbhik	1975	NBRI, Lucknow	Gamma rays 10 Gy/Undaunted	Flower colour	MVID-1917 MBNL-15
118		Chrysanthemum	Shafali	1975	NBRI, Lucknow	Gamma rays 15 Gy/Undaunted	Flower colour	MVID-1923 MBNL-15
119		Chrysanthemum	Swarnim	1975	NBRI, Lucknow	Gamma rays 15 Gy/Undaunted	Flower colour	MVID-1926 MBNL-15
120		Chrysanthemum	Pitaka	1978	NBRI, Lucknow	Gamma rays 15 Gy/Otome-Zakura	Flower colour	MVID-1919 MBNL-14
121		Chrysanthemum	Pitambar	1978	NBRI, Lucknow	Gamma rays 15 Gy/Otome-Zakura	Flower colour	MVID-1920 MBNL-14
122		Chrysanthemum	Purnima	1978	NBRI, Lucknow	Gamma rays 15 Gy/Otome-Zakura	Flower colour	MVID-1921
123		Chrysanthemum	Basanti	1979	NBRI, Lucknow	Gamma rays 15 Gy/E-13	Flower colour	MVID-1905 MBNL-23
124		Chrysanthemum	Hemanti	1979	NBRI, Lucknow	Gamma rays 15 Gy/Megami	Flower colour	MVID-1907 MBNL-16
125		Chrysanthemum	Rohit	1979	NBRI, Lucknow	Gamma rays 20 Gy/Kingsford Smith	Flower colour	MVID-1922 MBNL-16
126		Chrysanthemum	Taruni	1979	NBRI, Lucknow	Gamma rays 20 Gy/Kingsford Smith	Flower colour	MVID-1928 MBNL-17

127		Chrysanthemum	Khumani (Double Korean)	1980	NBRI, Lucknow	Induced mutant	Apricot flower colour	
128		Chrysanthemum	Alankar	1982	NBRI, Lucknow	Gamma rays 15 Gy/D-5	Flower colour	MVID-1899 MBNL-23
129		Chrysanthemum	Kumkum	1982	NBRI, Lucknow	Gamma rays 20–25 Gy/M-71	Flower colour	MVID-1949 MBNL-31
130		Chrysanthemum	Man Bhawan	1982	NBRI, Lucknow	Gamma rays 15 Gy/Flirt	Flower colour	MVID-1915 MBNL-23
131		Chrysanthemum	Cosmonaut	1984	NBRI, Lucknow	Gamma rays 15–25 Gy/Nimrod	Flower colour	MVID-1891 MBNL-26
132		Chrysanthemum	Colchi Bahar	1985	NBRI, Lucknow	Colchicine 0.0625%/Sharad Bahar	Flower colour	MVID-1948 MBNL-31
133		Chrysanthemum	Sheela	1985	NBRI, Lucknow	Gamma rays 20–25 Gy/Himani	Flower colour	MVID-1951 MBNL-31
134		Chrysanthemum	Tulika	1985	NBRI, Lucknow	Gamma rays 15 Gy/M-24	Flower colour	MVID-1952 MBNL-31
135		Chrysanthemum	Agnisikha	1987	NBRI, Lucknow	Gamma rays 15–25 Gy/D-5	Flower colour	MVID-1960 MBNL-37
136		Chrysanthemum	Navneet	1987	NBRI, Lucknow	Gamma rays 15 Gy/Kalyani Mauve	Flower colour	MVID-1961 MBNL-37
137		Chrysanthemum	Shabnam	1987	NBRI, Lucknow	Gamma rays 15 Gy/D-5	Flower colour	MVID-1950 MBNL-31
138		Chrysanthemum	Ratna (Pompon)	1989	NBRI, Lucknow	Cross with a mutant	Red colour	
139		Chrysanthemum	Sonali	1990	NBRI, Lucknow	Gamma rays 20 Gy/Ratna	Flower colour	MVID-1968 MBNL-42
140		Chrysanthemum	Subarn	1990	NBRI, Lucknow	Gamma rays 20 Gy/Flirt	Flower colour	MVID-1967 MBNL-42
141		Chrysanthemum	Jugnu	1991	NBRI, Lucknow	Gamma rays 15–20 Gy/Lalima	Flower colour	MVID-1973 MBNL-43

(continued)

Table 13.11 (continued)

S. No.	Latin name	Common name	Mutant variety	Release year	Institution	Mutagen and dose/parent var.	Main characters induced/improved	MVID No.@ MBNL No.#
142		Chrysanthemum	Arka Yellow Gold	1992	IIHR, Bengaluru	Gamma rays/Flirt	Yellow flowers with brick red colour stripes	
143		Chrysanthemum	Sharad Har	1992	NBRI, Lucknow	Gamma rays 15 Gy/Sharad Mala	Flower colour	MVID-1971 MBNL-43
144		Chrysanthemum	Surekha Yellow	1992	NBRI, Lucknow	Gamma rays 15 Gy/Surekha	Flower colour	MVID-1970 MBNL-42
145		Chrysanthemum	Navneet Yellow	1993	NBRI, Lucknow	Gamma rays 15 Gy/Navneet	Flower colour	MVID-1972 MBNL-43
146		Chrysanthemum	Batik	1994	NBRI, Lucknow	Gamma rays 20 Gy/Flirt	Flower colour	MVID-1975 MBNL-43
147		Chrysanthemum	Raktima	1998	NBRI, Lucknow	Gamma rays 15 Gy/Shyamal	Flower colour	MVID-1977 MBNL-44
148		Chrysanthemum	Arka Usha Kiran	2001	IIHR, Bengaluru	Gamma rays/Kirti	Semi-double-yellow-coloured flowers	
149		Chrysanthemum	Lalima tubular	2003	NBRI, Lucknow	Gamma rays 20 Gy/Lalima	Flower colour and shape	MVID-4461
150		Chrysanthemum	Lalima Head shape mutant	2003	NBRI, Lucknow	Gamma rays 20 Gy/Lalima	Flower colour	
151		Chrysanthemum	Pusa Anmol	2006	IARI, New Delhi	Gamma rays/Ajay	Yellowish pink	
152		Chrysanthemum	Pusa Centenary	2007	IARI, New Delhi	Gamma rays/Thai Chen Queen	Brilliant yellow flower	
153		Chrysanthemum	Pusa Arunodaya	2014	IARI, New Delhi	Gamma rays/Thai Chen Queen	Flower colour	
154		Chrysanthemum	Pusa Kesari	2014	IARI, New Delhi	Gamma rays/Thai Chen Queen	Flower colour	
155		Chrysanthemum	NBRI-Kesar	2015	NBRI, Lucknow	Gamma rays 7.5 Gy/Puja	Bicoloured florets and large-size capitulum	MVID-4450
156		Chrysanthemum	NBRI-CSIR-75	2017	NBRI, Lucknow	Gamma rays/Suneel	Flower colour	
157		Chrysanthemum	NBRI-Asha Kiran	2017	NBRI, Lucknow	Gamma rays/Suneel	Flower colour	
158		Chrysanthemum	NBRI-Himjyoti	2017	NBRI, Lucknow	Gamma rays/Himanshu	Flower colour	

159		Chrysanthemum	NBRI-Pukhraj	2019	NBRI, Lucknow	Gamma rays/Himanshu	Flower colour	
160		Chrysanthemum	NBRI-Shekhar	2020	NBRI, Lucknow	Gamma rays/Suneel	Late-blooming type	
161	*Cicer arietinum*	Chickpea	RS-11	1970	ARS, Durgapura	A natural mutant of var. RS-10	White flowered and resistant to drought	
162		Chickpea	RSG-2 (Kiran)	1984	ARS, Durgapura	Fast neutrons 4.5 10^{12} n/cm^2/RS-10	Erect, earlier maturity, salinity tolerant and high yield	MVID-390 MBNL-26
163		Chickpea	Pusa-408 (Ajay)	1985	IARI, New Delhi	Gamma rays 600 Gy/G-130	High yield, profuse branching, resistant to Ascochyta blight, pod borer and nematode	MVID-393 MBNL-29
164		Chickpea	Pusa-413 (Atul)	1985	IARI, New Delhi	Gamma rays 600 Gy/G-130	High yield, resistant to Fusarium wilt, stunt virus, root rot, pod borer and nematode	MVID-394 MBNL-29
165		Chickpea	Pusa-417 (Girnar)	1985	IARI, New Delhi	Gamma rays 600 Gy/BG-203	High yield, resistant to Fusarium wilt, stunt virus, colour rot, pod borer and nematode	MVID-395 MBNL-29
166		Chickpea	WCG-1 (Sadbhawna)	1997	GBPUA&T, Modipuram	Gamma rays 300 Gy/C-235	Yield, early maturity and robust plant	
167		Chickpea	WCG-2 (Surya)	1999	GBPUA&T, Modipuram	Gamma rays 300 Gy/G-130	Yield, wilt resistance and white flowered	
168		Chickpea	Pusa-547 (BGM-547)	2006	IARI, New Delhi	Gamma rays 600 Gy/BG-256	High yield, bold, attractive coloured grains, thin testa, good cooking quality and disease resistant	MVID-3354
169		Chickpea	GBM-2	2016	UAS, Raichur	Gamma rays 32 Gy/Annigeri-1	Yield, suitable for mechanical harvesting	
170	*Citrus reticulata*	Mandarin	PAU Kinnow-1	2017	PAU Ludhiana	Gamma rays 30 Gy/Kinnow	Seedless, high yield	MVID-4841
171	*Corchorus capsularis*	White jute	Shyamali (JRC-7447)	1978	JARC, Barrackpore	X-rays 250 Gy/JRC-212	Yield and nitrogen fertilizer responsive	MVID-321 MBNL-18
172		White jute	Padma (Hyb-C)	1983	JARC, Barrackpore	Cross: JRC-6165 × JRC-412	Tolerance to waterlogging and less affected by diseases	MVID-335 MBNL-34
173		White jute	Jaydev KC-1	1992	JRS, Kendrapara, Odisha	Gamma rays/JRC-444	High fibre yield SVRC	

(continued)

Table 13.11 (continued)

S. No.	Latin name	Common name	Mutant variety	Release year	Institution	Mutagen and dose/parent var.	Main characters induced/improved	MVID No.@ MBNL No.#
174		White jute	Bidhan Pat-3	2000	BCKV, Kalyani	D-154 × D18 mutant	High fibre yield	
175		White jute	Bidhan Pat-1	2001	BCKV, Kalyani	Gamma rays/D-154	High fibre yield	
176		White jute	Bidhan Pat-2	2001	BCKV, Kalyani	D-154 × D18 mutant	High fibre yield	
177	*Corchorus olitorius*	Tossa jute	IR-1	1978	JARC, Barrackpore	Gamma rays 100 Gy/JRO 632	Plant vigour and fibre yield	MVID-336 MBNL-37/25
178		Tossa jute	Mahadev (TKG-40)	1983	BARC, Mumbai	Cross: 'Virescent' × 'involute leathery' mutant	Higher fibre yield (10–14%) and early flowering	MVID-322 MBNL-23
179		Tossa jute	Savitri (JRO-3690)	1985	GBPUA&T, Pantnagar	Cross 'Tobacco leaf' × 'long internode'	Yield, resistant to aphids and yellow mite	MVID-334 MBNL-33
180		Tossa jute	Rebati (KOM-62)	1992	OUAT, Bhubaneswar	Gamma rays/JRO-878	High fibre yield	
181		Tossa jute	Bidhan Rupali	2002	BCKV, Kalyani	X-rays/JRO-632	High fibre yield	
182	*Coreopsis*	Tickseed	Pusa Tara	2012	IARI, New Delhi	Spontaneous mutant	Flower colour	
183	*Coriandrum sativum*	Coriander	RCr-684	1999	SKN COA, RAU, Jobner	Gamma rays/Rcr-20	Bold seeds, 0.32% volatile oil	
184	*Crossandra infundibuliformis*	The Fire Cracker	Arka Chenna	2019	IIHR, Bengaluru	Gamma rays/Arka Ambara	Medium-sized orange-coloured flowers	
185	*Cuminum cyminum*	Cumin	RZ-223	2007	SKN COA, RAU, Jobner	Mutant of UC-216	Higher resistance to wilt and blight diseases	
186	*Curcuma longa*	Turmeric	Co-1	1983	TNAU, Coimbatore	X-rays 50 Gy/Erode local	Rhizome colour	MVID-2206 MBNL-29
187		Turmeric	BSR-1	1986	TNAU, ARS, Bhavanisagar	X-rays 100 Gy/Erode local	Rhizome colour	MVID-2205 MBNL-29
188	*Cyamopsis tetragonoloba*	Cluster bean	Kanchan Bahar	1996	ARS, Durgapura	EMS 0.6%/Durga Bahar	High yield	
189	*Cymbopogon flexuosus*	Cochin grass	RRL-38	1976	RRL, Jammu	Gamma rays	68–100% more oil	
190		Cochin grass	RRL-57	1976	RRL, Jammu	Gamma rays	68–100% more oil	
191		Cochin grass	RRL-59	1976	RRL, Jammu	Gamma rays	68–100% more oil	
192		Cochin grass	RRL-(B)-14	1976	RRL, Jammu	Gamma rays	68–100% more oil	

193		Cochin grass	LLM-81	1976	RRL, Jammu	Gamma rays	Oil is colourless and only citral-a is exclusively present	
194	*Cymbopogon winterianus*	Citronella	RRL-JOR-3-1970	1976	RRL, Jammu	Gamma rays	39% more oil	
195		Citronella	Bhanumati (OJC-11)	1987	OUAT, Bhubaneswar	X-rays 60 Gy/Subirrsourav	Oil content	MVID-1741 MBNL-35
196		Citronella	Bibhuti (OJC-5)	1987	OUAT, Bhubaneswar	X-rays 90 Gy/Subirrsourav	Oil content	MVID-1739 MBNL-35
197		Citronella	Niranjan (OJC-6)	1987	OUAT, Bhubaneswar	X-rays 90 Gy/Subirrsourav	Oil content	MVID-1740 MBNL-35
198		Citronella	Phullara (OJC-22)	1987	OUAT, Bhubaneswar	X-rays 90 Gy/Subirrsourav	Oil content	MVID-1742 MBNL-35
199		Citronella	Sourav (OJC-3)	1987	OUAT, Bhubaneswar	X-rays 60 Gy/Subirrsourav	Oil content	MVID-1738 MBNL-35
200		Citronella	Subir (OJC-31)	1987	OUAT, Bhubaneswar	X-rays 90 Gy/Subirrsourav	Oil content	MVID-1743 MBNL-35
201		Citronella	Jalpallavi	1998	CIMAP, Lucknow	Spontaneous mutant	Oil content	
202		Citronella	Manjari (M-3-8)	1999	CIMAP, Lucknow	Gamma rays 30 Gy/Manjusha	High herb and oil, low elemol content oil	
203		Citronella	CIM Jeeva	2007	CIMAP, Lucknow	Gamma rays 60 Gy/Manjusha	High herb and oil, rich in citronellal 42.97%, yellow leaf tolerant	
204	*Dahlia* sp.	Dahlia	Bichitra	1978	FPARC, Sindri Dhanbad, Jhar Khand	Gamma rays 20–30 Gy/Kenya	Plant architecture	MVID-1929 MBNL-14
205		Dahlia	Black Beauty	1978	FPARC, Sindri Dhanbad, Jharkhand	Gamma rays 20–30 Gy/Black Out	Plant architecture	MVID-1930 MBNL-14
206		Dahlia	Happiness	1978	FPARC, Sindri Dhanbad, Jharkhand	Gamma rays 20–30 Gy/Croydon Monarch	Plant architecture	MVID-1931 MBNL-14
207		Dahlia	Jayaprakash	1978	FPARC, Sindri Dhanbad, Jharkhand	Gamma rays 20–30 Gy/Croydon Apricot	Plant architecture	MVID-1932 MBNL-14

(continued)

Table 13.11 (continued)

S. No.	Latin name	Common name	Mutant variety	Release year	Institution	Mutagen and dose/parent var.	Main characters induced/improved	MVID No.@ MBNL No.#
208		Dahlia	Jubilee	1978	FPARC, Sindri Dhanbad, Jharkhand	Gamma rays 20–30 Gy/Kenya	Plant architecture	MVID-1933 MBNL-14
209		Dahlia	Jyoti	1978	FPARC, Sindri Dhanbad, Jharkhand	Gamma rays 20–30 Gy/Kenya	Plant architecture	MVID-1934 MBNL-14
210		Dahlia	Netaji	1978	FPARC, Sindri Dhanbad, Jharkhand	Gamma rays 20–30 Gy/Eagle Stone	Plant architecture	MVID-1935 MBNL-14
211		Dahlia	Pearl	1978	FPARC, Sindri Dhanbad, Jharkhand	Gamma rays 20–30 Gy/Eagle Stone	Plant architecture	MVID-1936 MBNL-14
212		Dahlia	Pride of Sindri	1978	FPARC, Sindri Dhanbad, Jharkhand	Gamma rays 20–30 Gy/Kenya	Plant architecture	MVID-1937 MBNL-14
213		Dahlia	Twilight	1978	FPARC, Sindri Dhanbad, Jharkhand	Gamma rays 20–30 Gy/Kenya	Plant architecture	MVID-1938 MBNL-14
214		Dahlia	Vivekananda	1978	FPARC, Sindri Dhanbad, Jharkhand	Gamma rays 20–30 Gy/Croydon Master	Plant architecture	MVID-1939 MBNL-14
215	*Dianthus caryophyllus*	Carnation	Arka Flame	2007	IIHR, Bengaluru	In vitro mutation/IIHRS-1	Red flowers with smooth petal edges, tolerant to Fusarium wilt	
216	*Eleusine coracana*	Finger millet	Hagari-1	1941	Dept of Agri, Karnataka	Mutant from Gidda Aryan	Suitable for kharif tracts under irrigated conditions	
217		Finger millet	Co-3	1942	TNAU, Coimbatore	Mutant from Co-1	Suited for rainfed and irrigated, drought tolerant	
218		Finger millet	Dibya Sinha	1976	CPR, Berhampur	Mutant from AKP-7	Tolerant to blast and stem borer and early maturity	
219		Finger millet	K-6	1978	ARS, Kovilpatti, TNAU, Coimbatore	Natural mutant of local cultivar	Very early maturity (75–80 days)	

220		Finger millet	Dapoli-1	1994	KVK, Dapoli	Selection from mutant No. 50-1	High yield	
221		Finger millet	BM-11-1	1996	OUAT, Berhampur	Mutant from Budha Mandia	High yield	
222		Finger millet	Neelachal	1999	OUAT, Berhampur	EMS/IE 642	High yield, MR to blast, stem borer	
223		Finger millet	Bhairabi (BM-9-1)	1999	OUAT, Berhampur	EMS Mutant from Budha Mandia	High yield, MR to sheath blight, grasshopper, black spot	
224		Finger millet	Suvra (OUAT-2)	1999	OUAT, Berhampur	EMS/Co-9	High yield, white seeds, MR to leaf neck, finger blast	
225		Finger millet	Dapoli-2	2017	KVK, Dapoli	Somaclone/Dapoli 1	High yield	
226	*Gladiolus* sp.	Gladiolus	Arka Shobha	1988	IIHR, Bengaluru	Gamma rays 10 Gy/Wild Rose	Long spikes, florets light pink with creamy white throat	MVID-1958 MBNL-34
227		Gladiolus	Tambari	1991	NBRI, Lucknow	Gamma-rays/Oscar	Flower colour	MVID-1974 MBNL-43
228		Gladiolus	Swarnima	2000	IARI, New Delhi	Spontaneous mutant of Dhanwantari	Coppery yellow	
229		Gladiolus	Subhangini		GBPUAT, Pantnagar	Gamma rays/cv. Fidelio	White flowers	
230	*Glycine max*	Soybean	Birsa Soybean-1	1983	BAU, Ranchi	Natural mutant of 'Sepaya Black'	Erect, dwarf, early, resistance to bacterial pustules	MVID-2845
231		Soybean	VL Soya-1	1985	VPKAS, Almora	Natural mutant of Bragg	Tolerant to Cercospora leaf spot, determinate, black and bold seeds	MVID-2846
232		Soybean	NRC-7	1993	IISR, Indore	Irradiation	High yield	
233		Soybean	Ahilya-1 (NRC-2)	1997	IISR, Indore	Radiation-induced mutant of Bragg	High yield	MVID-2847
234		Soybean	Ahilya-2 (NRC-12)	1997	IISR, Indore	Radiation-induced mutant of Bragg	High yield	MVID-2848
235		Soybean	MACS-450	2000	ARI, Pune	Cross: Bragg × MACS-111	High yield, resistant to leaf spot, bud blight, yellow mosaic and stem fly	MVID-2849
236		Soybean	TAMS-38	2005	BARC, Mumbai and PDKV Akola	Cross: Monetta × PK-472	Earliness and disease resistance, high yield	MVID-3390

(continued)

Table 13.11 (continued)

S. No.	Latin name	Common name	Mutant variety	Release year	Institution	Mutagen and dose/parent var.	Main characters induced/improved	MVID No.@ MBNL No.#
237		Soybean	TAMS-98-21	2007	BARC, Mumbai and PDKV Akola	Gamma rays 250 Gy, JS-80-21	21% higher yield, multiple disease and pest resistance	MVID-2851
238		Soybean	Pusa-97-12	2009	IARI, New Delhi	Mutant of var. DS-74	Determinate, white flower, resistant to YMV SMV and charcoal rot	
239		Soybean	Pusa-12	2015	IARI, New Delhi	Mutant of var. DS-74	Determinate, white flower, yellow seed, resistant to YMV, *Rhizoctonia*	
240		Soybean	AMS-1001	2019	PDKV, Amaravati, Maharashtra	Mutant of Var. JS-93-05	Determinate, purple flower, yellow seed, YMV and root rot resistant	
241		Soybean	AMS-100-39	2021	PDKV, Amaravati	Mutant of var. JS-93-05	Average productivity 2087 kg	
242	*Gossypium arboreum*	Desi cotton	DS-1	1985	HAU, Hisar, Haryana	Gamma rays 200 Gy/G-27	Resistant to jassids, spotted bollworm, pink bollworm and blight	MVID-337 MBNL-42
243	*Gossypium herbaceum*	Desi cotton	DDhc-11	2008	UAS, Dharwad	Selection from Sujay mutant	High yield, high fibre length, resistant to Alternaria blight	
244	*Gossypium hirsutum*	American cotton	M.A.-9	1948	Mysore	X-rays/Co-2	Drought tolerant	MVID-329 MBNL-30
245		American cotton	Indore-2	1950	IPSI, Indore	X-rays/MU-4 (Dhar Kambodia)	High yield	MVID-327 MBNL-30
246		American cotton	Badnawar-1	1961	IPSI, Indore	Cross with Indore-2	High yield	MVID-326 MBNL-30
247		American cotton	Khandwa-2	1971	IPSI, Indore	Cross with Indore-2	High yield	MVID-328 MBNL-30
248		American cotton	Rashmi	1976	IARI, New Delhi	Gamma rays 300 Gy/MCU-5	High yield, superior quality and day neutral	MVID-325 MBNL-16
249		American cotton	Pusa Ageti	1978	IARI, New Delhi	Gamma rays 250 Gy/ Stoneville213	Improved ginning capacity of fibre, early maturity, tolerant to jassids	MVID-324 MBNL-16
250		American cotton	MCU-10	1982	AC&RI, Madurai	Gamma rays 300 Gy/MCU-4	Long staple, drought tolerant, increased spinning capacity	MVID-331 MBNL-29
251		American cotton	MCU-7	1984	AC&RI, TNAU, Madurai	X-rays 800 Gy/L-1143 EE	Early, yield, long staple, resistant to black arm, root rot and leaf blight	MVID-323 MBNL-2

252	*Helianthus*	Sunflower	K-2	1977	TNAU, Coimbatore	Spontaneous mutant of 'Cerenianka'	Oil 38%, dwarf with short duration and suited to late rainfed seasons	
253		Sunflower	Co-3 (SUF-10)	1995	TNAU, Coimbatore	Gamma rays 50 Gy/Co 2	Tall, oil 38.3%, tolerant to jassid, ash weevil, Alternaria, rust	
254		Sunflower	TAS-82	2007	BARC, Mumbai	Gamma rays 100 Gy/Surya	Increased seed and oil yield, black seed coat, tolerance to drought	MVID-3331
255	*Hibiscus* sp.	Hibiscus	Purnima	1979	NBRI, Lucknow	Gamma rays 200 Gy/Alipore Beauty	Variegated smaller leaves	MVID-1946 MBNL-30
256		Hibiscus	Anjali	1987	NBRI, Lucknow	Gamma rays 40 Gy/Alipore Beauty	Light carmine red-colour flower but five petals, single type	MVID-1953 MBNL-31
257	*Hordeum vulgare*	Barley	RDB-1	1974	ARS, Durgapura, Jaipur	Pile neutrons 4.5×10^{12} NP/CM^2/RS-17	Dwarf, early, high yielding, non-lodging and less water requirement	MVID-1019
258		Barley	PL-56	1978	HPKV, Palampur, HP	EMS 0.2%/C-164	High tillering, high yield, bold grains suited for rainfed areas	MBNL-32
259		Barley	RD-103	1978	ARS, Durgapura, Jaipur	Cross: RBD-1 × K-18	Dwarf, stiff straw, high yield and high tillering	MVID-1020 MBNL-26
260		Barley	K-2578	1980	CSAU, Kanpur	Cross: RBD-1 × Vijaya (K-572/11)	Medium tall, high yield potential and long ears	MVID-1052 MBNL-36
261		Barley	DL-253	1981	IARI, New Delhi	Gamma rays 200 Gy + EMS 0.30%/Ratna	High yield, high tillering, resistant to covered and LS and YR	MVID-1018 MBNL-19
262		Barley	RD-137	1981	ARS, Durgapura, Jaipur	Cross: RDB-1 × EB-795	Shortness, medium tall, high yielding and less water requirement	MVID-1058 MBNL-36
263		Barley	Rajkiran (RD-387)	1982	ARS, Durgapura, Jaipur	Cross: RBD-1 × Marocaine-079	Dwarf, erect, profusely tillering and immune to Molya disease	MVID-1021 MBNL-26
264		Barley	Karan-3	1982	IARI, RRS, Karnal	Cross; RDB-1 × EB-7576, Riso Mut.-1508	Semi-dwarf, non-lodging and hull-less amber grains	MVID-1053 MBNL-36
265		Barley	Karan-15	1982	IARI, RRS, Karnal	Cross: RDB-1 × EB-20	Semi-dwarfness, high tillering and high yield	MVID-1055 MBNL-36
266		Barley	Karan-4	1983	IARI, RRS, Karnal	RDB-1 × EB-7576	Semi-dwarf, lodging resistant, hull-less amber grains	MVID-1054 MBNL-36
267		Barley	BH-75	1983	HAU, Hisar, Haryana	Cross RD-150 (RDB-1 × EB-795) × Ahor-131/68	Semi-dwarf, early, profuse tillering, resistant to YR and CCN	MVID-1051 MBNL-36
268		Barley	Karan-201	1984	IARI, RRS, Karnal	Cross: (Azam d1 × 1B-65) × (RDB-1 Riso Mut. 1508)	Semi-dwarf, hull-less grain and high protein (16%)	MVID-1056 MBNL-36

(continued)

Table 13.11 (continued)

S. No.	Latin name	Common name	Mutant variety	Release year	Institution	Mutagen and dose/parent var.	Main characters induced/improved	MVID No.@ MBNL No.#
269		Barley	RD-2035	1988	ARS, Durgapura, Jaipur	Cross: RD-137 (RDB-1 × EB-795) × PL-101	Medium tall, profuse tillering, early maturity and resistant to CCN	MVID-1059 MBNL-36
270		Barley	Karan-265	1989	IARI, RRS, Karnal	Cross: (RDB-1 × EB-7725) × Riso Mut 1508	Dwarf, non-lodging, high tillering, hull-less grains	MVID-1057 MBNL-36
271	*Hyoscyamus niger*	Black henbane	Aela (Y-17)	1987	CIMAP, Lucknow	Gamma rays 400 Gy/inbred wild strain	Yellow flowered, high yield of tropane alkaloid and high biomass	
272		Black henbane	Aekela (Ub)	1997	CIMAP, Lucknow	Gamma rays 200 Gy/inbred wild strain	Un-branched, more biomass, high tropane alkaloid yield	
273	*Jasminum*	Jasmine	Co-2	1991	TNAU, Coimbatore	Gamma rays/Co-1 Pitchi	Bolder pink buds with long corolla tube, white with pink tinge	
274	*Lablab purpureus*	Hyacinth bean	Co-9	1980	TNAU, Coimbatore	Natural mutant from Mandapam lablab	Photoinsensitive type. Pale green and flat pods	
275		Hyacinth bean	Co-10	1983	TNAU, Coimbatore	Gamma rays 240 Gy/Co-6	Photoinsensitive type. Greenish white, tubular and curved pods	MVID-936 MBNL-29
276	*Lantana depressa*	Wild sage	*Lantana depressa* Bicoloured	1986	NBRI, Lucknow	Gamma rays 10–15 Gy/ *L. depressa*	Leaf colour	MVID-1962 MBNL-37
277		Wild sage	*Lantana depressa* Variegata	1986	NBRI, Lucknow	Gamma rays 10 Gy/*L. depressa*	Variegated leaves, yellow and white flower colour	MVID-1954 MBNL-31
278		Wild sage	Niharika	1998	NBRI, Lucknow	Gamma rays 10-15 Gy/ *L. depressa*	Green leaf colour and light-yellow flower colour	MVID-1963 MBNL-37
279	*Lens culinaris*	Lentil	B-177 (S-177)	1984	PORS, Berhampore	Mutant of B-77 (Asha)	Erect and double-seeded pods, released in Bengal	
280		Lentil	S-256 (Ranjan)	1984	PORS, Berhampore	X-ray mutant/B-77	Spreading type, high yield and high protein content (28.6%)	MVID-389 MBNL-20
281		Lentil	Arun (PI-77-12)	1986	RPCAU, Samastipur	A mutant from 7BR-25	Medium bold seeds, tolerant to rust	
282		Lentil	Rajendra Masoor-1	1996	ARI, Dhelwan, Lohianagar	Gamma rays 100 Gy/BR-25	Tolerance to low temperatures, early maturity and good for late sowing	MVID-2352
283	*Linum usitatissimum*	Linseed	TL-99	2019	BARC, Mumbai	RLC-6 × Solin mutant	Low linolenic acid, high yield and oil content	

284		Linseed	DLV-20	2019	UAS, Dharwad	EMS 0.3%/NL-115	Earliness and high yield	MVID-4844
285	*Luffa acutangula*	Ridged gourd	PKM-1	1984	TNAU, AC&RI, Periyakulam	Gamma rays/H-160	High yielding, tolerant to pumpkin beetles, fruit fly and leaf spot	MVID-2929 MBNL-32
286	*Macrotyloma uniflorum*	Horse gram	Paiyur-2	1998	TNAU, Coimbatore	Gamma rays/Co-1	High protein content (19.25%)	
287	*Matricaria chamomilla*	German chamomile	Vallery	1994	CIMAP, Lucknow	Gamma rays 100–1000 Gy/ German Bulk	High chamazulene, high oil and dry herb yield	
288	*Mentha citrata*	Orange mint	Kiran	1988	CIMAP	Gamma rays	Linalool 48%, tall vigorous growth	
289	*Mentha piperita*	Peppermint	Kukrail	1994	CIMAP	Gamma rays	Menthol 34.5%, menthone 27.9%	
290		Peppermint	Tushar	1999	CIMAP	Gamma rays	Menthol 33.3%, menthone 27.3%	
291		Peppermint	Pranjal	2000	CIMAP	Gamma rays	Menthol 52%, menthone 13.5%	
292	*Mentha spicata*	Spearmint	Neera	1992	CIMAP, Lucknow	Irradiation of sucker pieces	High alkaloid content carvone 58%	
293	*Momordica charantia*	Bitter gourd	MDU-1	1984	TNAU, AC&RI, Madurai	Gamma rays 100 Gy/MC-103	Tolerant to pumpkin beetle, fruit fly and leaf spot disease	MVID-2210 MBNL-32
294	*Morus alba*	Mulberry	S-36	1984	CSRTI, Mysore	EMS 0.1–0.6% Berhampore Local	High leaf yield	
295		Mulberry	S-54	1984	CSRTI, Mysore	EMS 0.1–0.6% Berhampore Local	Improved leaf yield and leaf quality	MVID-233 MBNL-33
296	*Nicotiana tabacum*	Tobacco	GSH-3	1979	CTRI, Rajahmundry	Cross: (LTH × M4) × CTRI Special	Special improved leaf quality and high yield	MVID-330 MBNL-30
297	*Oryza sativa*	Rice	GEB-24 (Kichilisamba)	1921	TNAU, Coimbatore	Spontaneous mutant	Fine grain, good quality rice	
298		Rice	ADT-15 (Senkuruvai)	1937	TNAU, Coimbatore	Mutant of ADT-41	Short bold, white, resistant to foot rot	
299		Rice	K-84	1967	Jammu and Kashmir	Gamma rays/Taichung-65	Early maturity, indica grained mutant	MVID-1035 MBNL-29
300		Rice	Vellayani-1	1968	KAU, Kerala	Thermal neutrons/PTB-10	Early maturity	MVID-1040 MBNL-29
301		Rice	IIT-48	1972	IIT, Kharagpur	Ethylene oxide/IR-8	Earliness and improved grain size	MVID-1027
302		Rice	IIT-60	1972	IIT, Kharagpur	EMS 0.5%/IR-8	Earliness, 1 month earlier than IR 8, yield	MVID-1028

(continued)

Table 13.11 (continued)

S. No.	Latin name	Common name	Mutant variety	Release year	Institution	Mutagen and dose/parent var.	Main characters induced/improved	MVID No.@ MBNL No.#
303		Rice	PL-56	1975	PAU, Ludhiana	EMS 0.2%/C-164	High tillering type, high yield, good adaptability	MVID-1036 MBNL-29
304		Rice	HM-95	1975	PAU, Ludhiana	Gamma rays 500 Gy (Jhona 349 × Taichung Native-1)	Semi-dwarf, high yield, high protein (12.3%) and high lysine (4.07%)	MVID-1026 MBNL-4
305		Rice	Jagannath (BSS-873)	1976	OUA&T, Bhubaneswar	X-rays 300 Gy/T-141	Grain size, 1 month earlier than T 141	MVID-1029
306		Rice	Au-1	1976	TNAU, Coimbatore	Gamma rays/IR-8	Earliness, maturity 105 days, tolerant to alkalinity and salinity	MVID-1030 MBNL-29
307		Rice	CNM-25	1979	Rice Res. Sta., Chinsurah, West Bengal	X-rays 300 Gy/IR-8	Early maturity, high yield, increased, long grain and resistant to thrips	MVID-1024 MBNL-18
308		Rice	CNM-31	1979	Rice Res. Sta., Chinsurah, West Bengal	X-rays 300 Gy/IR-8	Early, long grain, high yield and resistant to BLB, BLS, BPH and BS	MVID-1025 MBNL-17
309		Rice	CNM-20	1980	Rice Res. Sta., Chinsurah, West Bengal	X-rays 300 Gy/IR-8	Early maturity, long grain and resistant to BLB, BLS and BPH	MVID-1023 MBNL-18
310		Rice	Indira	1980	NRRI, Cuttack	EMS/Tainan-3	Earliness, tolerant to Blast, BLB and stem borer	MVID-1033 MBNL-29
311		Rice	Lakshmi (CNM-6)	1982	Rice Res. Sta., Chinsurah, West Bengal	X-rays 300 Gy/IR-8	Early maturity, drought tolerant, long grain and high yield	MVID-1022 MBNL-18
312		Rice	Biraj (CNM-539)	1982	West Bengal	X-rays/OC-1393	Better tolerance to submergence and resistance to *Helminthosporium*	MVID-1031 MBNL-29
313		Rice	Mohan (CSR-4)	1983	West Bengal	Gamma rays/IR-8	Salt tolerance, semi-dwarf, resistant to lodging and fertilizer responsive	MVID-1049 MBNL-37
314		Rice	Sattari (CRM-13-3241)	1983	NRRI, Cuttack	Gamma rays (NSJ 200 × Padma)	Early maturity, nitrogen responsive and suitable for direct seeding in rainfed upland areas	MVID-1038 MBNL-29
315		Rice	Savitri (Ponmani)	1983	NRRI, Cuttack, Odisha	Cross of Pankaj × Jagannath	Daylength, tolerant to blast and sheath blight	MVID-1039 MBNL-29

316		Rice	HPU-8020 (IET-5878)	1984	HPAU, Palampur	Gamma rays 200 Gy/Bala	Synchronous tillering and high yield (7.3 t/ha)	MVID-1032 MBNL-29
317		Rice	Prabhavati	1985	MAU, Parbhani	EMS 0.2%/Ambemohar local	Semi-dwarf, stiff straw, resistant to lodging and responsive to nitrogen	MVID-1037 MBNL-29
318		Rice	Keshari (IET-6215)	1985	NRRI, Cuttack, Odisha	X-rays 300 Gy Cross [(T-141 × Kumar) × Jagannath]	Shortness, earliness, resistant to GLH, blast, BLB and green leaf hopper	MVID-1034 MBNL-29
319		Rice	Rasmi (PTB-44)	1985	RARS, Pattambi	Gamma rays 220 Gy/Oorpandy	Awnless, high yield and tolerant to salinity	MVID-1041 MBNL-30
320		Rice	Rajshree (TCA 80-4)	1987	TCA, Dholi, Bihar	Local landrace mutant	Resistant to blast and bacterial leaf blight	
321		Rice	Hari (TR-RNR-21)	1988	BARC, Bombay	Cross IR-8 × TR-5	Shortness and higher yield	MVID-1047 MBNL-34
322		Rice	Intan mutant	1988	UAS, Bengaluru	Ethyleneimine 0.2%/Intan	Early maturing and photoperiod insensitive	MVID-1048 MBNL-35
323		Rice	Padmini	1988	NRRI, Cuttack	Gamma rays/CR-1014	Earliness, medium slender grains, tillering type	MVID-1050 MBNL-37
324		Rice	PNR-162	1988	IARI, New Delhi	Cross: Jaya mut. × Basmati 370	Early maturity (85–115 days), semi-dwarf	MVID-1066 MBNL-42
325			Dharitri (IET-6272)	1989	NRRI, Cuttack	X-rays 300 Gy [(T-141 × Pankaj) × Jagannath]	Semi-dwarf, high yield, short grains, released for eastern states	MVID-2877
326		Rice	Gayatri (IET-8020)	1989	NRRI, Cuttack	X-rays 300 Gy [(T-141 × Pankaj) × Jagannath]	Semi-dwarf, high yield, short grains, resistant to BLB, MR to blast and GM	MVID-2878
327		Rice	PNR-381	1989	IARI, New Delhi	Tainan-3 mutant × Basmati-370	Blast resistance, semi-dwarf, early maturity and dual purpose	MVID-1064 MBNL-42
328		Rice	PNR-166	1989	IARI, New Delhi	Cross: IR-8 mutant × Basmati-370	Synchronous maturity	MVID-1065 MBNL-42
329		Rice	HUR-36	1990	BHU, Varanasi	Gamma rays 150 Gy + 0.04% EMS/Mahsuri	Earliness	MVID-1060 MBNL-42
330		Rice	PNR-519	1990	IARI, New Delhi	Cross: (Tainan-3 mut × Basmati 370) × NR-417-3	Diseases and pest resistance	MVID-1063 MBNL-42
331		Rice	PNR-555-5	1990	IARI, New Delhi	Cross: [(Dular mut. × N-22 mut.) × MTU-17]	Earliness, resistance to diseases and pests	MVID-1061 MBNL-42
332		Rice	PNR-570-17	1990	IARI, New Delhi	Cross: (Gora mut. × MW-10 mut.) × N-22 mutant	Earliness, resistance to diseases and pests	MVID-1062 MBNL-42

(continued)

Table 13.11 (continued)

S. No.	Latin name	Common name	Mutant variety	Release year	Institution	Mutagen and dose/parent var.	Main characters induced/improved	MVID No.@ MBNL No.#
333		Rice	PNR-571	1990	IARI, New Delhi	Cross: MW-10 mutant × PMR-351 (Tainan-3 mut. × Basmati-370)	Semi-dwarfness and early maturity	MVID-1071 MBNL-44
334		Rice	Lunisree (IET-10678)	1992	NRRI, Cuttack	Mutant of Nonasail variety	High yield, long slender grains, drought, disease and pest resistance	MVID-2875
335		Rice	Bipasha	1993	West Bengal	X-rays/Pankaj	Long bold grains, suitable for shallow lowlands	
336		Rice	MDU-4 (ACM-15)	1993	TNAU, Coimbatore	X-rays 300 Gy/Cross: [(T-141 × AC-3836) × Jagannath]	Semi-dwarf, highly tolerant to cold stress, LB, neck infection, MR to BPH, WBPH and BS	MVID-2879
337		Rice	ADT-41 (JJ-92)	1993	TNRRI, Aduthurai, Coimbatore	Mutant of Basmati-370	ELS, white, aromatic rice, Kuruvai season	MVID-2876
338		Rice	Vytila-5 (KAU-655)	1995	KAU, Kerala	Mutant of Mahsuri	Short bold grains	
339		Rice	Gautam	1996	RPCAU, Samastipur	EMS/Rasi	Dwarfness, long bold grains	MVID-2893
340		Rice	Radhi	1997	NRRI, Cuttack	Mutant of Swarnaprabha	Long bold grains, blast and BPH tolerance	MVID-2892
341		Rice	Tapaswini (IET-12168)	1997	NRRI, Cuttack	X-rays 300 Gy/Cross [(T-141 × Mahsuri) × Jagannath]	Tolerance to WBPH, BB, LF and GM	MVID-2881
342		Rice	Malviya Dhan-36	1997	BHU, Varanasi, UP	Mutant of Mahsuri	Semi-tall, early, white fine long slender grains, resistance to diseases	MVID-2884
343		Rice	PNR-550-1-2 (JD-8)	1997	IARI, New Delhi	Cross: Dular mutant × N-22 mutant-17	Semi-dwarfness	MVID-1069 MBNL-44
344		Rice	PNR-551-4-20 (JD-6)	1997	IARI, New Delhi	Cross: Dular mutant × N-22 mutant	Semi-dwarfness	MVID-1070 MBNL-44
345		Rice	PNR-555-28 (JD-10)	1997	IARI, New Delhi	Cross: (Dular mut × N-22 mut.) × (PNR-351 × Mutant-17)	Semi-dwarfness	MVID-1068 MBNL-44
346		Rice	PNR-555-5 (JD-3)	1998	IARI, New Delhi	Cross: (Dular mut. × N-22 mut.) × (PNR-351 × Mut.17)	Yield	MVID-1067 MBNL-44

347		Rice	PNR-546	1998	IARI, New Delhi	Gamma rays/F_2 of (PNR-125-2 × PNR-130-2)	Grain quality	MVID-1072
348		Rice	Remanika (MO-15)	1998	KAU, Kerala	Mutant of MO-1	High yield	MVID-2785
349		Rice	Ramchandi (IET-13354)	1999	NRRI, Cuttack	X-rays 300 Gy/Cross [(T-141 × IR-17494-32-2-2-1) × Jagannath]	Semi-dwarf, white moderate bold grains, photosensitive	MVID-2882
350		Rice	Padmanth (IET-11876)	1999	NRRI, Cuttack	X-rays 300 Gy/Cross [(T-141 × Pankaj Nagoba) × Jagannath]	Long bold grains, released for Assam state	MVID-2883
351		Rice	CRM-49	1999	NRRI, Cuttack	NaN_3 0.001 M/IR-50	Blast resistance	MVID-1073
352		Rice	CRM-51	1999	NRRI, Cuttack	NaN_3 0.001 M/IR-50	Blast resistance	MVID-1074
353		Rice	CRM-53	1999	NRRI, Cuttack	EMS/IR-50	Blast resistance	MVID-1075
354		Rice	PNR-519	2000	IARI, New Delhi	Gamma rays (Tainan-3 mutant × Basmati-370) × PNR-417-3	Semi-dwarf, earliness, resistant to BS, SB, GM, LF and WM	MVID-2888
355		Rice	Early Samba	2000	Andhra Pradesh	Mutant of BPT-5204 (Samba mahsuri)	Dwarf, tolerance to SB, white grains	MVID-2891
356		Rice	Dhanu	2002	RARS, Kayamkulam, Kerala	Gamma rays 220 Gy/PTB-20	Increased grain and straw yield with tolerance to major pests and diseases	MVID-2894
357		Rice	Kaum-20-19-4	2002	KAU, Kerala	Mutant of MO-1	Dwarf, red medium bold grains, resistance to BPH, early maturity	MVID-2887
358		Rice	Anaswara	2006	KAU, Kerala	Gamma rays/PTB-20	A photoperiod-sensitive semi-tall variety for rabi	MVID-2799
359		Rice	Shivam (IET-17868)	2006		CR 314-5-10 (Open Florat mutant)	Semi-dwarf, highly resistanct to blast, GM, SB and LF	MVID-2885
360		Rice	Geetanjali (CRM 2007-1)	2006	NRRI, Cuttack	Gamma rays 100 Gy Basmati-370	Semi-dwarf, early, high yield, basmati grain quality	MVID-3267
361		Rice	RC Maniphau-7 (RCM-9)	2008	Manipur	Mutant of Punshi	Moderately resistant to blast, BLB, SB and GM	

(continued)

Table 13.11 (continued)

S. No.	Latin name	Common name	Mutant variety	Release year	Institution	Mutagen and dose/parent var.	Main characters induced/improved	MVID No.@ MBNL No.#
362		Rice	Chandan (CR Boro Dhan) CRM-898	2008	NRRI, Cuttack	Gamma rays/01 × China-45	Moderately resistant to SB, BPH, blast and BLB. Released for Assam state	
363		Rice	Malviya Sugandh-105 HUR-105	2009	BHU, Varanasi, UP	Mutant of MPR 7-2	HYV, long grain with strong aroma, moderately resistant to blast, brown spot and stem borer	
364		Rice	Malviya Sugandh 4-3 HUR-4-3	2009	BHU, Varanasi, UP	Mutant of Lanjhi	Fine grain with mild aroma, moderately resistant to blast, BLB, and brown plant hopper	
365		Rice	PUBM-8 (IET-19110)	2009	UAS, Bengaluru, Karnataka	Gamma rays/PUB Local	Medium slender white grains, non-lodging	
366		Rice	KHP-12	2011	UAS, Bengaluru, Karnataka	Gamma rays/BKB Local	Medium slender grains without awns, tolerant to blast	
367		Rice	Nagina-22 (N-22)	2011	NRRI, Hyderabad	Gamma rays/Samba Mahsuri	Water stress and heat tolerance	
368		Rice	Pooma Bhog (CR Basna Dhan-902)	2012	NRRI, Cuttack	Mutant of Pusa Basmati-1	Resistant to NB, GM and MR to SR and SB	
369		Rice	TCDM-1	2019	BARC, Bombay and IGKV, Raipur	Gamma rays 300 Gy/Mai Dubraj	High yield, shorter duration and shorter height	MVID-4840
370		Rice	TKR Kolam	2020	BARC, Bombay—DBSKKV, Dapoli	Gamma rays	Slender fine grain, semi-dwarf high-yielding variety	
371		Rice	Vikram TCR	2021	BARC, Bombay and IGKV, Raipur	Gamma rays/Safari-17	30–35 days earlier and 21% more yield, plant height 101–106 cm	
372		Rice	CG Jawaphool	2021	BARC, Bombay and IGKV, Raipur	Gamma rays/Jawaphool	10–15 days earlier than parent var. Jawaphool and 40% higher yield	
373	*Panicum maximum*	Guinea grass	Marathakam	1993	KAU, Vellayani, Kerala	Gamma rays/F.R. 600	It produces 55 t/ha of green fodder and crude protein is 8.2%	MVID-2835
374	*Panicum sumatrense*	Little Millet	Kolab (OLM-36)	2001	OUAT, Odisha	Mutant of SS-81-1	High yield	

375		Little Millet	Sabara (OLM-20)	2003	OUAT, Odisha	Mutant of SS-81-1	High yield	
376	*Papaver somniferum*	Opium poppy	Vivek (BC-28/ 9/4)	1992	CIMAP, Lucknow	Gamma rays 50 Gy/Shweta	Big capsule size and high morphine content	MVID-1440 MBNL-42
377		Opium poppy	Sujata (LL-34)	1999	CIMAP, Lucknow	Gamma rays 200 Gy + 0.4% EMS/Sampada	Opium less, very low alkaloids (non-narcotic, oilseed crop) with high seed protein content	
378		Opium poppy	Abha	2019	NBRI, Lucknow	Gamma rays/NBRI-5	Thebaine 11%, opium dry 37 g/ha, yield 12 q/ha	
379	*Paspalum scrobiculatum*	Kodo Millet	JK-13	2007	Rewa, JNKVV	Selection from mutant JK-76	High yield	
380		Kodo Millet	RK-390-25	2012	Rewa, JNKVV	Mutant of RK-390	High yield	
381		Kodo Millet	Chhattisgarh Kodo-2	2014	Jagdalpur, IGKVV	Mutant/line of Co-3	High yield	
382		Kodo Millet	Jawahar Kodo-137	1016	Rewa, JNKVV	Mutant of RK-390	High yield	
383		Kodo Millet	KMV-543	2019	ZARS, Jagdalpur, IGKVV	Mutant of Co-3	High yield	
384	*Pennisetum glaucum*	Pearl millet	NHB-3	1975	IARI, New Delhi	Cross MS 5071A × J-104 (gamma rays 350 Gy), Tift-23B	Resistance to *Sclerospora graminicola*	MVID-157 MBNL-37
385		Pearl millet	NHB-4	1975	IARI, New Delhi	Cross MS 5071A × K-560-230	*Sclerospora* resistance	MVID-158 MBNL-37
386		Pearl millet	NHB-5	1976	IARI, New Delhi	Gamma rays 350 Gy male sterile inbred line Tift-23A	*Sclerospora* resistance to mildew disease and bold grains	MVID-153 MBNL-11
387		Pearl millet	Pusa-46	1983	IARI, New Delhi	Gamma rays/J-104 × K-559	Mildew and *Pyricularia* resistance	MVID-154 MBNL-23
388		Pearl millet	ICMH-451	1986	ICRISAT, Patancheru	Gamma rays 300 Gy/Tift-23-DB	Downy mildew resistance	MVID-155 MBNL-30
389	*Phaseolus vulgaris*	Common bean	Pusa Parvati	1970	IARI, New Delhi	X-rays/Wax podded	Earliness, resistant to mosaic and powdery mildew, high yield	MVID-1560
390	*Phyllanthus niruri*	Java Citronella	CIM-Jeeva	2007	CIMAP, Lucknow	Gamma rays 60 Gy	High herb and oil, rich in 42.97% citronellal content, yellow leaf	

(continued)

Table 13.11 (continued)

S. No.	Latin name	Common name	Mutant variety	Release year	Institution	Mutagen and dose/parent var.	Main characters induced/improved	MVID No.@ MBNL No.#
391	*Pisum sativum*	Field pea	Hans (L-116)	1980	IARI, New Delhi	Ethyleneimine, P-1163	High yield and semi-erect	MVID-1561 MBNL-15
392	*Plantago ovata*	Psyllium	Niharika (M-20-22)	1998	CIMAP, Lucknow	Gamma rays 200 Gy + 0.2% EMS	Long inflorescence and high seed yield and seed husk yield	
393		Psyllium	Nimisha	1999	CIMAP, Lucknow	Gamma rays	Dark-green leaves and medium long panicles	
394		Psyllium	Mayuri (CIMAP/S-5)	2007	CIMAP, Lucknow	Gamma rays + 0.2%/EBFEA-5	Early-maturing variety, high seed and seed husk yield	
395	*Polyanthus tuberosa*	Tube rose	Rajat Rekha	1974	NBRI, Lucknow	Gamma rays/single flower cv.	Leaf colour mutant	MVID-1889 MBNL-14
396		Tube rose	Swarna Rekha	1974	NBRI, Lucknow	Gamma rays/single flower cv.	Leaf colour mutant	MVID-1890 MBNL-14
397	*Portulaca grandiflora*	Portulaca	Five Petal	1974	BARC, Mumbai	Gamma rays 25 Gy	Flower colour	MVID-43 MBNL-20
398		Portulaca	Jhumka	1974	NBRI, Lucknow	Gamma rays 10 Gy/Karna Pali	Flower colour	MVID-44 MBNL-14
399		Portulaca	Karna Pali	1974	NBRI, Lucknow	Gamma rays 10 Gy/Portulaca double	Double flower	MVID-45 MBNL-14
400		Portulaca	Karna Phul	1974	BARC, Mumbai	Gamma rays 10 Gy/ *P. grandiflora*	Flower colour	MVID-46 MBNL-17
401		Portulaca	Lalita	1974	NBRI, Lucknow	Gamma rays 10 Gy/Vibhuti	Flower colour	MVID-47 MBNL-14
402		Portulaca	Mukta	1974	NBRI, Lucknow	Gamma rays 10 Gy/Portulaca double	Flower colour	MVID-48 MBNL-14
403		Portulaca	Pink colour	1974	BARC, Mumbai	Gamma rays 150 Gy	Flower colour	MVID-49 MBNL-20
404		Portulaca	Ratnam	1974	NBRI, Lucknow	Gamma rays/Portulaca Double	Flower number	MVID-1964 MBNL-37
405		Portulaca	Rosy green	1974	BARC, Mumbai	Gamma rays 100 Gy chronic	Flower colour	MVID-50 MBNL-20

406		Portulaca	Semi-double	1974	BARC, Mumbai	Gamma rays 100 Gy	Altered flower morphology	MVID-51 MBNL-20
407		Portulaca	Vibhuti	1974	NBRI, Lucknow	Gamma rays 40 Gy/Portulaca Double	Flower colour	MVID-52 MBNL-14
408	*Rauwolfia serpentine*	Indian snakeroot	CIM-Sheel	2004	CIMAP, Lucknow	Gamma rays/wild strain	Reserpine content 0.030–0.035% in dry root	
409	*Ricinus communis*	Castor bean	SA-2	1972	TNAU, Coimbatore	A spontaneous mutant of TMV-1	Resistant to drought, higher TKW and higher yield, oil 53%	
410		Castor bean	Aruna	1976	IARI, RS, Hyderabad	Thermal neutrons 1400 rad, HC-6	Extra early (120 days vs. 270 days for mother variety) and high yield	MVID-1707
411		Castor bean	Sowbhagya (157-B)	1978	IARI Reg. Station, Hyderabad	(Aruna × dwarf mut. of HC-6)	Earliness, dwarf, non-shattering, suitable for intercropping and long-duration maturity	MVID-1709 MBNL-11
412		Castor bean	RC-8	1982	UAS, Bengaluru	Gamma rays 400 Gy/RC 1188-54	Earliness, higher TKW and higher yield	MVID-1708 MBNL-11
413	*Rosa* sp.	Rose	Abhisarika	1975	IARI, New Delhi	Gamma rays 75 Gy/Kiss of Fire	Flower colour	MVID-1746 MBNL-26
414		Rose	Angara	1975	NBRI, Lucknow	Gamma rays/Montezuma	Plant architecture, flower colour	MVID-1940 MBNL-14
415		Rose	Climbing Cri Cri	1975	NBRI, Lucknow	Climbing mutant from cv. Cri Cri	Salmon and coral pink flower colour	
416		Rose	Madhosh	1975	IARI, New Delhi	EMS 0.025%/Gulzar	Flower colour	MVID-1941 MBNL-14
417		Rose	Pusa Christian	1975	IARI, New Delhi	Gamma rays 100 Gy/Christian Dior	Flower colour	MVID-1747 MBNL-26
418		Rose	Striped Christian Dior	1975	IARI, New Delhi	Gamma rays 75 Gy/Christiana Dior	Flower colour	MVID-1745 MBNL-26
419		Rose	Winter Holiday Mutant	1975	NBRI, Lucknow	Mutant from cv./Summer Holiday	Rose pink flower colour	
420		Rose	Nav Sadabahar	1980	IARI, New Delhi	Natural Bud Sports/Rose Sadabahar	Flower colour	
421		Rose	Pink Montezuma	1980	NBRI, Lucknow	Natural Bud Sports of Rose Montezuma	Flower colour	

(continued)

Table 13.11 (continued)

S. No.	Latin name	Common name	Mutant variety	Release year	Institution	Mutagen and dose/parent var.	Main characters induced/improved	MVID No.@ MBNL No.#
422		Rose	Pink Sport Montezuma	1980	IARI, New Delhi	Bud sport of Var. Montezuma	Spinal pink flower colour	
423		Rose	Sharada	1983	NBRI, Lucknow	Gamma rays 30 Gy/Queen Elizabeth	Flower colour	MVID-1969 MBNL-42
424		Rose	Striped Contempo	1983	NBRI, Lucknow	Gamma rays 30 Gy/Contempo	Flower colour	MVID-1966 MBNL-37
425		Rose	Saroda	1983	NBRI, Lucknow	Gamma rays 30 Gy/Queen Elizabeth	Flower colour	MVID-1942 MBNL-23
426		Rose	Sukumari	1983	NBRI, Lucknow	Gamma rays 30 Gy/America's Jr. Miss	Flower colour	MVID-1943 MBNL-23
427		Rose	Tangerine Contempo	1983	NBRI, Lucknow	Gamma rays 30 Gy/Contempo	Flower colour	MVID-1944 MBNL-23
428		Rose	Yellow Contempo	1983	NBRI, Lucknow	Gamma rays 30 Gy/Contempo	Flower colour	MVID-1945 MBNL-23
429		Rose	Curio	1986	NBRI, Lucknow	Gamma rays 30 Gy/Imperator	Flower colour	MVID-1956 MBNL-31
430		Rose	Pink Contempo	1986	NBRI, Lucknow	Gamma rays 30 Gy/Contempo	Flower colour	MVID-1955 MBNL-31
431		Rose	Twinkle	1986	NBRI, Lucknow	Gamma rays 30 Gy/Imperator	Flower colour	MVID-1957 MBNL-31
432		Rose	Pink Imperator	1986	NBRI, Lucknow	Gamma rays/Imperator	Pink flower colour	
433		Rose	Windy City Mutant	1986	NBRI, Lucknow	Gamma rays/Windy City	Flower colour	
434		Rose	Light Pink Prize	1989	NBRI, Lucknow	Gamma rays/First Prize	Flower colour	MVID-1965 MBNL-37
435		Rose	Climbing Sadabahar	1991	IARI, New Delhi	Bud sport of Var. Sadabahar	Deep pink flower colour	
436		Rose	Mrinalini Stripe	1991	NBRI, Lucknow	Gamma rays/Mrinalini	Flower colour	
437		Rose	Salmon Beauty Mutant	1992	NBRI, Lucknow	Gamma rays/Salmon Beauty	Peach pink flower colour	

438		Rose	Zorina Pink	1992	NBRI, Lucknow	Gamma rays/Zorina	Pink flower colour	
439		Rose	Mrinalini Light pink Mutant	1992	NBRI, Lucknow	Gamma rays/Mrinalini	Flower colour	
440		Rose	Summer Holiday Pink Mutant	1992	NBRI, Lucknow	Gamma rays/Summer Holiday	Flower colour	
441		Rose	Arka Nishkant	1993	IIHR, Bengaluru	Bud sport of Rosa multiflora	Flower colour	
442		Rose	Chitra	1995	IARI, New Delhi	Natural Bud Sports of Rose var Janina	Flower colour	
443		Rose	Pusa Abhishek	2001	IARI, New Delhi	Natural Bud Sports Rose/Jantar Mantar	Flower colour	
444		Rose	Pusa Urmil	2001	IARI, New Delhi	Natural Bud Sports Rose Jantar Mantar	Flower colour	
445		Rose	Pusa Mansij	2004	IARI, New Delhi	Natural Bud Sports of var. Raja Surendra Singh of Nalagarh	Flower colour	
446	*Saccharum officinarum*	Sugarcane	Co-312	1928	SBI, Coimbatore	Natural mutant	High cane yield	
447		Sugarcane	Co-449	1946	SBI, Coimbatore	Cross with a mutant	Resistant to red rot	
448		Sugarcane	Co-527	1946	SBI, Coimbatore	Cross with a mutant	High cane yield	
449		Sugarcane	Co-775	1950	SBI, Coimbatore	Cross with a mutant	High cane yield	
450		Sugarcane	Co-6608	1966	SBI, Coimbatore	Gamma rays 30-50 Gy/Co-449	Red rot resistance	MVID-371 MBNL-12
451		Sugarcane	Co-997 Mutant	1967	SBI, Coimbatore	Gamma rays 30–50 Gy/Co-997	Red rot resistance	MVID-370 MBNL-12
452		Sugarcane	Co-8153	1981	SBI, Coimbatore	Gamma rays 150 Gy/ (Co-6304 × Co-6806)	Superior juice quality and yield	MVID-372 MBNL-30
453		Sugarcane	Co-85017	1985	SBI, Coimbatore	Gamma rays 150 Gy/Co-740	Adaptability, resistant to *Ustilago scitaminea* and high sucrose %	MVID-368 MBNL-31
454		Sugarcane	Co-85035	1985	SBI, Coimbatore	Gamma rays/Co-740	Earliness, high cane yield, resistance to *Ustilago* and high sucrose %	MVID-369 MBNL-32

(continued)

Table 13.11 (continued)

S. No.	Latin name	Common name	Mutant variety	Release year	Institution	Mutagen and dose/parent var.	Main characters induced/improved	MVID No.@ MBNL No.#
455	*Sesamum indicum*	Sesamum	Kalika (BM 3-7)	1985	OUAT, Bhubaneswar	EMS 0.1% 6 h/Binayak	Semi-dwarf, high yield, tolerant to *Cercospora* leaf spot, stem, root rot	MVID-1710 MBNL-17
456		Sesamum	Uma (OMT 11-6-3)	1990	OUAT, Odisha	Arsenik-Q 10% 8 h/Kanak	Uniform maturity, higher oil content	MVID-1714 MBNL-43
457		Sesamum	Usha (OMT 11-6-5)	1990	OUAT, Odisha	Rhustox 10% 8 h/Kanak	High yield, high oil content	MVID-1713 MBNL-43
458		Sesamum	Nirmala (OS-Sel-164)	2002	OUAT, Odisha	Gamma rays 700 Gy + NG 0.04%	Increased oil content, light brown colour	
459		Sesamum	Prachi (ORM-17)	2002	OUAT, Odisha	Gamma rays 700 Gy + EMS 0.75%	High yield, resistance to major pests and diseases	
460		Sesamum	DSS-9	2009	UAS, Dharwad	Mutant of Phule Til × E-8	Early maturity, white and bold seeds	
461	*Setaria italica*	Foxtail millet	PS-4	1999	GBPUA&T, Pantnagar	EMS 0.2%/SIA-2616	High yield, early maturity and bold seeds with 13–15% protein	
462	*Solanum khasianum*	Khasianum	RRL-20-2	1976	RRL, Jammu	Gamma rays 250 Gy/Dehradun local	High solasodine content (3.1%)	MVID-332 MBNL-7, 13
463	*Solanum lycopersicon*	Tomato	S-12	1969	PAU, Ludhiana	Gamma rays/Sioux	Dwarfness, high yield, nematode resistant	MVID-2204
464		Tomato	Pusa Lal Meruti	1972	IARI, New Delhi	Gamma rays 300 Gy/Meeruti	Uniform fruit ripening	MVID-2203
465		Tomato	PKM-1	1980	TNAU, AC&RI, Periyakulam	Gamma rays 250 Gy/Annangi	Yield, determinate type, good transportability	MVID-2929 MBNL-32
466		Tomato	Co-3 (Marudham)	1984	TNAU, Coimbatore	EMS 0.1%/Co-1	Yield, dwarf, compact, determinate, high Vit C and heat tolerant	MVID-2207 MBNL-29
467	*Solanum melongena*	Brinjal	PKM-1	1985	TNAU, AC&RI, Periyakulam	Gamma rays/Puzhuthikathiri	High yielding and adapted to rainfed area	MVID-2947 MBNL-32
468	*Solenostemon rotundifolius*	Coleus	Suphala	2006	Kerala Agricultural University	A tissue culture mutant derived from local cultivar	High yielding (15.93 t/ha) var. with a duration of 120–140 days	MVID-2834
469		Coleus	M-7-ISIRI	2008	UAS, Bengaluru	Gamma rays/IC-547017	Forskolin: 2.6–2.7 t/acre, moderately resistant to bacterial wilt	

470	*Sorghum bicolor*	Sorghum	Mothi (SPV-141)	1978	TNAU, Coimbatore	IS-6928	Suitable for late kharif and early maturing (105 days)	
471		Sorghum	SPV-126 (CSV-9)	1983	TNAU, Coimbatore	Cross: CS-3541 × CSV-4	Resistant to charcoal rot, downy mildew, head moulds and stem borer	
472		Sorghum	Co-21 (SPV-80)	1985	TNAU, Coimbatore	X-rays 400 Gy CSV-5 [148 (699)]	High yield, tolerance to Striga	MVID-156 MBNL-29
473		Sorghum	Co-31	2014	TNAU, Coimbatore	Gamma ray 400 Gy/Co (FS) 29	Ratoonable, more seed yield, moderately tolerant to drought	
474		Sorghum	CSV-33-MF	2017	TNAU, Coimbatore	EMS 0.2%/CoFS29	High yield	
475		Sorghum	TRJP-1-5	2021	UAS Raichur BARC, Mumbai	Gamma rays 350 Gy/JP-1-5	High yield, resistance to rust, tolerant to charcoal rot	
476	*Trichosanthes anguina*	Snake gourd	PKM-1	1982	TNAU, AC&RI, Periyakulam	H 375	High yield	
477	*Trifolium alexandrinum*	Egyptian clover	BL-22	1984	PAU, Ludhiana	Gamma rays 400 Gy/Mescavi	Lateness, quick growing	MVID-212 MBNL-26
478	*Trigonella foenum-graecum*	Fenugreek	RMt-305	2007	SKN COA, RAU, Jobner	Gamma rays/RMt-1	First determinate type, multipotent, resistance to root knot nematodes	
479		Fenugreek	RMt-361 (UM-361)	2009	SKN COA, RAU, Jobner, Raj	Gamma rays/RMt-1	Bold and attractive seeds, more number of branches and seeds per pod	
480	*Triticum aestivum*	Wheat	NP-836	1961	IARI, New Delhi	X-rays 160 Gy/NP 799	Fully awned, higher yield, medium-early with rust resistance	MVID-1043
481		Wheat	Sharbati Sonora	1969	IARI, New Delhi	UV 2600 A + gamma rays 200 Gy/Sonora 64	Amber grain colour, short straw and early maturity	MVID-1045 MBNL-6
482		Wheat	Pusa Lerma	1971	IARI, New Delhi	Gamma rays/Lerma rojo 64-A	Grain colour and high yield	MVID-1044
483		Wheat	NI-5643	1975	Agri. Res. Station, Niphad, Nasik	Radiation (New Thatcher × NI-284S)	Earliness, amber grain colour, high yield and resistance to rust	MVID-1042 MBNL-19
484	*Triticum dicoccum*	Emmer wheat	COW-2 Samba Wheat	2010	IARI RS, Wellington and TNAU, Coimbatore	Gamma rays 200 Gy/NP-200	Early, semi-dwarf, high yield, quality grains, resistant to black, yellow and brown rusts	MVID-4846
485		Emmer wheat	Nilgiri Khapli (HW-1098)	2012	IARI RS, Wellington, TN	Gamma rays 200 Gy/NP-201	High yield, semi-dwarf, disease resistance, 25,000 ha under cultivation	MVID-4845

(continued)

Table 13.11 (continued)

S. No.	Latin name	Common name	Mutant variety	Release year	Institution	Mutagen and dose/parent var.	Main characters induced/improved	MVID No.@ MBNL No.#
486	*Vigna aconitifolia*	Moth bean	RMO-40	1995	CAZRI, Jodhpur	Gamma rays 400 Gy + EMS 0.6%/Jwala	High yield and earliness by 40 days	MVID-2895
487		Moth bean	RMO-257	1997	CAZRI, Jodhpur	Gamma rays 300 Gy + 0.6% EMS/Jadia	Earliness and high harvest index	
488		Moth bean	FMM-96	1997	ARS, Fatehpur, Rajasthan	Gamma rays + EMS/Jadia	Extra early, short height, synchronous maturity	
489		Moth bean	Maru Moth-1 (IJMM-259)	1998	CAZRI, Jodhpur	Gamma rays 300 Gy/Jadia	Drought resistant and early maturity	
490		Moth bean	CAZRI Moth-1	1999	CAZRI, Jodhpur	Gamma rays 300 Gy/Jadia	High yield, resistant to YMV and plants having high seed protein	
491		Moth bean	Maru Vardan (RMO-225)	1999	CAZRI, Jodhpur	Gamma rays 400 Gy + 0.6% EMS/Jadia	Earliness and short resistance	
492	*Vigna mungo*	Black gram	Co-4	1980	TNAU, Coimbatore	MMS 0.02%/Co-1	Earliness, erect, determinate and photoinsensitive	MVID-397 MBNL-29
493		Black gram	Sarla (B-12-4)	1985	OUAT, Odisha	Natural mutant/T-9	Resistant to MYMV and powdery mildew	
494		Black gram	TAU-1	1985	BARC, Mumbai and PKV, Akola	Cross: U-196 × T-9	Yield, larger seed size and powdery mildew resistant	MVID-400 MBNL-28
495		Black gram	Manikya	1988	GBPUA&T, Pantnagar	Gamma rays 400 Gy/T-9	High yield and YMV resistance	MVID-3357
496		Black gram	TPU-4	1992	BARC, Mumbai, and MPKV, Rahuri	Cross: U-201 × T-9	High yield and tolerant to MYMV	MVID-402 MBNL-42
497		Black gram	TAU-2	1993	BARC, Mumbai and PKV, Akola	Cross: U-196 × T-9	High yield and tolerant to powdery mildew	MVID-2300
498		Black gram	Vamban-2	1996	TNAU, Coimbatore	Spontaneous mutant from type 9	Resistant to yellow mosaic virus	MVID-2856
499		Black gram	TU-94-2	1999	BARC, Mumbai	Cross: TPU-3 × TAU-5	High yield	MVID-2301
500		Black gram	Prasad (B-3-8-8)	2005	OUAT, Odisha	EMS/T-9	High yield, tolerant to PM, YMV	

501		Black gram	Ujala (OBG-17)	2005	OUAT, Bhubaneswar	EMS 0.125%/B 3-8-8	Short duration of 66–70 days, good yield, MR to PM and YMV	
502		Black gram	DU-1	2008	UAS, Dharwad	Irradiated progeny of cross TAU-1 × 169	Bold seeded, high seed protein content (23.6%)	
503		Black gram	TU-40	2013	BARC, Mumbai	Cross: TU-94-2 × *V. mungo* var. *silvestris*	High yield and resistance	
504		Black gram	TRCRU-22	2021	UAS. Raichur	Cross: TAU-1 × KU-96-3	High yield	
505	*Vigna radiata*	Mung bean	Dhauli (TT-9E)	1979	OUA&T, Bhubaneshwar	Mutant of fixed line of cross T-51 local	High yield	MVID-2855
506		Mung bean	Pant Mung-2	1983	GBPUA&T, Pantnagar	Gamma rays 100 Gy/ML-26	Resistance to MYMV, and high yield	MVID-1562 MBNL-23
507		Mung bean	TAP-7	1983	BARC, Mumbai	Mutant of S-8	Earliness, high yield, tolerance to PM and leaf spot	MVID-1563 MBNL-23
508		Mung bean	ML-26-10-3	1983	PAU, Ludhiana	Gamma rays/ML-26	YMV resistance, yield	MVID-401 MBNL-33
509		Mung bean	Co-4	1984	TNAU, Coimbatore	Gamma rays 200 Gy/Co-1	High yield, early maturity and resistance to drought	MVID-1570 MBNL-29
510		Mung bean	TPM-1	1985	BARC, Mumbai MPKV Jalgaon	Mutant of S-8	High yield	
511		Mung bean	KGM-1 (TAP-70)	1990	UAS, Dharwad	Gamma rays 300 Gy/S-8	Tolerant to drought and PM, suited for irrigated summer crop	
512		Mung bean	MUM-2	1992	CCSU, Meerut	EMS 0.2% 6 h/K-851	Yield, tolerance to MYMV and early maturity	MVID-1577 MBNL-43
513		Mung bean	BM-4	1992	ARS, Badnapur	EMS 0.15%/T-44	High yield, early, tolerant to PM and YMV	MVID-2854
514		Mung bean	LGG-407	1993	APAU, Lam	Gamma rays 400 Gy/Pant Mung-2	High yield	MVID-2853
515		Mung bean	LGG-450	1993	APAU, Lam	Gamma rays 400 Gy/Pant Mung-2	High yield	MVID-2852
516		Mung bean	TARM-2	1994	BARC, Mumbai, and PKV, Akola	RUM-5 × TPM-1	High yield and resistance to MYMV	MVID-2294
517		Mung bean	TARM-1	1997	BARC, Mumbai and PKV, Akola	RUM-5 × TPM-1	High yield and resistance to MYMV and PM	MVID-2304

(continued)

Table 13.11 (continued)

S. No.	Latin name	Common name	Mutant variety	Release year	Institution	Mutagen and dose/parent var.	Main characters induced/improved	MVID No.@ MBNL No.#
518		Mung bean	TARM-18	1997	BARC, Mumbai, and PKV, Akola	PDM-5 × RUM-5	High yield and resistance to MYMV	MVID-2303
519		Mung bean	PBM-1	2000	PAU, Ludhiana	Gamma rays 300 Gy/ML-131	Yield and tolerance to MYMV	
520		Mung bean	Kamadeva (OUM 11-5)	2002	OUAT, Bhubaneswar	EMS/Dhauli	Early, high yield, shining green seeds MR to PM, MYMV, CLS	
521		Mung bean	Durga (OBGG-52)	2002	OUAT, Bhubaneswar	EMS 0.375%/K-851	High yield, medium maturity, resistant to MV, CLS, leaf curl	
522		Mung bean	TMB-37	2005	BARC, Mumbai	Cross: Kopergaon × TARM-2B	High yield and virus resistance	MVID-2934
523		Mung bean	TJM-3	2007	BARC, Mumbai	Cross: Kopergaon × TARM-1B	Early maturity, resistance to YMV, PM and Rhizoctonia root rot	MVID-2936
524		Mung bean	TM-96-2	2007	BARC, Mumbai	Cross: Kopergaon × TARM-2B	Resistance to PM, *Corynespora* leaf spot (CLS)	MVID-2935
525		Mung bean	TM-2000-2	2010	BARC, Mumbai	TARM-1 × JL-781	Resistance to PM and suitable for rice fallows	MVID-3337
526		Mung bean	DGG-7	2018	UAS, Dharwad	Mutant of Selection 04	Shining green medium bold seeds, higher number of clusters/plant	
527	*Vigna unguiculata*	Cowpea	V-16 (Amba)	1984	IARI, New Delhi	DMS/Pusa Phalguni	High yield, resistant to fungal and bacterial diseases	MVID-1564 MBNL-25
528		Cowpea	V-37 (Shreshtha)	1984	IARI, New Delhi	DMS/Pusa Phalguni	High yield and suitable as green fodder	MVID-1565 MBNL-25
529		Cowpea	V-38 (Swarna)	1984	IARI, New Delhi	DMS/Pusa Phalguni	High yield, earlier maturity and resistant to diseases	MVID-1566 MBNL-25
530		Cowpea	V-240 (Rambha)	1984	IARI, New Delhi	DMS/Pusa Phalguni	High yield, resistant to fungal, bacterial and viral diseases	MVID-1567 MBNL-25
531		Cowpea	Gujarat Cowpea-1	1984	GAU, Anand	Barsati mutant	High yield, earliness and resistant to root knot nematode	MVID-2933
532		Cowpea	Co-5	1986	TNAU, Coimbatore	Gamma rays 300 Gy/Co-1	Nutritional, forage cowpea, high yield	MVID-1569 MBNL-29
533		Cowpea	Cowpea-88	1990	PAU, Ludhiana	F1 seed irradiation (Cowpea-74 × H-2)	High grain and fodder yield, highly resistant to MYMV, *Anthracnose*	MVID-1574 MBNL-37

534		Cowpea	V-585 (Sampada)	1995	IARI, New Delhi	DMSO 0.8%	YMV resistant, white seeds	
535		Cowpea	Co (CP)-7	2002	TNAU, Coimbatore	Gamma rays 200 Gy/Co-4	High yield and good quality	MVID-2874
536		Cowpea	V-578	2004	IARI, New Delhi	DMSO 0.8%	YMV resistance, erect and bold	
537		Cowpea	TCM-148-1	2006	BARC, Mumbai	Cross between mutant lines	Yield and resistance	
538		Cowpea	TRC-77-4 (Kalleshwari)	2007	BARC, Mumbai	Mutant of V-130	High yield, suitable for rice-based cropping system in Chhattisgarh	MVID-2771
539		Cowpea	KM-5	2010	UAS, Bengaluru	Gamma rays/KBC-2	High grain yield and tolerant to rust disease	
540		Cowpea	TC-901	2018	BARC, Mumbai	Gamma rays 250 Gy/EC-394763	High yielding, early, synchronous maturity, resistant to CMV	MVID-4835
541	*Zingiber officinalis*	Ginger	Suravi	1991	HARS, OUAT, Pottangi, Odisha	Gamma rays 150 Gy/Rudrapur Local	High rhizome yield	
542		Ginger	Sourabh	2017	HARS, OUAT, Pottangi, Odisha	Chemical mutagen/Rio de Janeiro	High rhizome yield	

[a] MVID No. = ID number of the mutant variety in the IAEA Mutant Variety Database
[b] MBNL No. = Issue number of the IAEA Mutation Breeding Newsletter

the 10 years following its release, it contributed more than US$ 3 billion in cotton production.

So far, Pakistan has released 24 mutant varieties of cotton, out of which the NIAB has developed 19 cotton cultivars through the use of induced mutation, including the famous cultivars NIAB-78, NIAB-Karishma, NIAB-111 and, most recently, the high-yielding and fine-fibre cotton varieties NIAB-Kiran, NIAB-878B, NIAB-545 and NIAB-1048. Area under NIAB-Karishma increased from 17.4% in 1997–1998 to 28.4% in 1998–1999 and more than 40% in 1999–2000. NIAB Kiran covered an area of 292,000 ha (2019–2020) in Pakistan (Hussain et al. 2021). The first induced mutant variety of cotton 'NIAB 78' has assisted transformation of cotton production and stimulated a wider impact on agriculture in Pakistan. NIAB-78 had early maturity and higher yield, showed wider adaptability and eventually covered 80% of the cotton area in Punjab and Sindh provinces (Ahloowalia et al. 2004). By introduction of this variety, cotton production increased from 654,000 to 2,613,000 tons during 1991 and 1992. This mutant variety remained in farmers' fields over a very long period (1983–2000) and led to an increase in cotton production from 1,024,000 to 7,276,600 tons per year in Pakistan (Haq 2009). However, huge yield gap differences still exist among the top three cotton-producing countries of the region (China 1484 kg/ha, Pakistan 700 kg/ha and India 529 kg/ha). These yield differences are being further aggravated due to changing climate associated with higher temperature stresses (Rahman et al. 2018; Razzaq et al. 2021). The cotton mutant varieties developed through the use of induced mutation at the NIAB have shown enhanced resilience against high temperatures under field conditions and also showed significant variation in their root length leading to higher yields and better productivity (Hussain et al. 2021). Cotton of course is a cash crop, important for the textile industry in the region, which has made it a priority for the NIAB's mutation breeding activities. Right now, of the 3.1 million hectares planted with cotton, nearly 25% area is planted with mutant varieties, a number expected to increase to 30–40% in 2018–2019. Three mutant varieties released in 2013, 2016 and 2017 have been well accepted by the farmers because of their ability to withstand high temperatures and heavy rains, resistance to pests and diseases and capacity to sustain yields in this time of climate change while also producing a very-high-quality fibre that brings a higher price than standard varieties at the market. The improved characteristics of the mutant varieties like NIAB-78 and NIAB-86 along with their wide adaptability in different agro-climatic areas leading to quantum jump in cotton production have not only improved the agriculture sector's output and, in turn, national food security, but also contributed to the economy of the country. The cotton mutant variety NIAB-78 alone created an additional income of US$ 486 million to often poor growers between 1986 and 2004. The NIAB estimates that the 50 NIAB mutant varieties had an economic impact on the national economy that, as of April 2018, amounted to US$ 6 billion (Hussain et al. 2021).

Besides the large number of high-yielding mutant varieties of cotton released, the NIAB, Faisalabad, has developed and released six famous mutant varieties of rice (Table 13.12). It released Pakistan's first rice mutant variety 'Kashmir Basmati' from Basmati 370. The basmati types are contributing billions of dollars by export to their

Table 13.12 Brief of 79[a] mutant varieties of 11 crops released in Pakistan

S. No.	Crop and No.	Institute	No.	Mutant varieties
1	Cotton (24)	NIAB, Faisalabad	19	NIAB-78, NIAB-86, NIAB-26N, NIAB-Karishma, NIAB-999, NIAB-111, NIAB-846, NIAB-777, NIAB-852 NIAB-Kiran, NIAB-112, IR-NIAB-824, NIAB-2008, NIAB-878B, NIAB-545, NIAB-1048, NIAB-1011, NIAB-SANAB-M, NIAB-898
		NIA, Tando Jam	5	Chandi-95, Sohni, Sadori, NIA-Ufaq, NIA-Noori
2	Rice (13)	NIAB, Faisalabad	6	Kashmir Basmati, NIAB-IRRI-9, NIAB-2013, NIAB Basmati Rice-2016, Noor Basmati, NIAB-Super
		NIA, Tando Jam	7	Shadab, Shua-92, Khushboo-95, Sarshar, Mehak, Shandar, NIA-Mehran
3	Wheat (3)	NIA, Tando Jam	3	Jauhar-78, Soghat-90, Kiran-95
4	Chickpea (10)	NIAB, Faisalabad	6	CM-72, CM-88, CM-98, CM-2000 (Kabuli), CM-2008 (Kabuli), NIAB-CM-104
		NIFA, Peshawar	4	NIFA-88, NIFA-95, Hassan-2K (Kabuli), NIFA-2005
5	Mung bean (17)	NIAB, Faisalabad	13	NM-28, NM-121-25, NM-19-19, NM-20-21, NM-13-1, NM-51, NM-54, NM-92, NM-98, NIAB Mung-2006, NM-2011, NIAB Mung-2016, Abbas Mung
		NIFA, Peshawar	3	Ramzan, NIFA Mung-2017, NIFA Mung-2019
		NIA, Tando Jam	1	AEM-96
6	Lentil (2)	NIAB, Faisalabad	1	NIAB Masoor-2006
		NIA, Tando Jam	1	NIA-Masoor-05
7	Sesame (3)	NIAB, Faisalabad	3	NIAB Sesame-2016, NAIB Pearl, NIAB-Millennium
8	Brassica (4)	NIFA, Peshawar	3	Absin-95, NIFA Raya, NIFA Saroon T-20
		NIA, Tando Jam	1	NIA-Surhan
9	Castor (1)	NIAB, Faisalabad	1	NAIB Gold
10	Mandarin (1)	NIAB, Faisalabad	1	NIAB Kinnow
11	Sugarcane (1)	NIA, Tando Jam	1	NIA-2011
Total			79[a]	

Total—50 mutant vars. of NIAB, Faisalabad; 19 mutant vars. of NIA, Tando Jam; and 10 mutant vars. of NIFA, Peshawar

[a] Personal communication from T. M. Shah, Director, NIAB, Faisalabad, Pakistan

national economy. NIAB-Super and NIAB-IRRI-9 are fine-grain, salt-tolerant mutant varieties of rice released by the NIAB. Area in Punjab province of NIAB-IRRI-9 in coarse group is about 23.8% on an average basis. It fetches higher price due to fine and translucent grain features. Export share of NIAB-IRRI-9 in non-basmati group from 2010 to 2015 is 3.56% having a worth of US$ 42.5 million (3.68%). The Nuclear Institute of Agriculture (NIA), Tando Jam, has released three famous wheat mutant varieties Jauhar-78, Soghat-90 and Kiran-95, including its first mutant wheat variety Jauhar-78 with salinity tolerance and shattering resistance and Kiran-95 endowed with better grain quality and tolerance to salinity and rusts.

In chickpea, the NIAB has released six high-yielding and disease-resistant mutant varieties of desi and kabuli types (Table 13.12), the most popular and widely cultivated being CM-72, CM-88 and CM-98 (Haq 2009). The Nuclear Institute for Food and Agriculture (NIFA), Peshawar, has released four chickpea mutant varieties including Hassan-2K, a kabuli mutant variety (Hassan et al. 2001). Out of 17 mung bean mutant varieties released by Pakistan, the prominent ones NM-92, NM-98 and NM-2006 are being cultivated in more than 70% area of mung bean acreage in Punjab province and improved mung bean varieties occupy almost 90% of the mung bean area in Pakistan (Ali et al. 1997, 2010). NIAB Mung 2011 covered an area of 102,410 ha in the year 2015–2016, i.e. 70% area of mung bean (Razzaq et al. 2021). Mutation breeding efforts in lentil in Pakistan have culminated in the development and release of a high-yielding and disease-resistant mutant variety NIAB Masoor-2006 and one NIA-Masoor-05 by NIA, Tando Jam (Ali et al. 2010).

The high-yielding, high-oil-content and early-maturing mutant varieties of oilseed crops, Abasin-95 and NIFA Raya (Table 13.12), are the first ever mutant varieties of rapeseed and mustard, respectively, released in Pakistan (Syed et al. 2009). Both Abasin-95 and NIFA Raya are occupying several million acres and giving additional income of several million US dollars to farmers. Two sesame varieties released in 2016 and 2017 have doubled the yield of traditional sesame varieties and are more suitable for modern cultivation techniques. The mutant mandarin variety, NIAB Kinnow, released in 2017, has an increased yield of more than 30% and reduced seed count from around 50 to just 3–5 seeds per fruit, which makes it more valuable and popular for export.

13.3.3 Mutation Breeding for Crop Improvement in Bangladesh

The Bangladesh Institute of Nuclear Agriculture (BINA), Mymensingh, located in northern Bangladesh, is the major centre of mutation breeding. Mutation breeding at the BINA has resulted in the release of 76 mutant varieties belonging to 13 crop species (http://mvd.iaea.org). The BINA has released 28 mutant varieties of pulses, 23 of oilseeds, 12 of rice and 4 of tomato (Table 13.13). Some of the popular mutant varieties that have been widely adopted by farmers in Bangladesh for large-scale cultivation include Binasail, Iratom-24, Binadhan-7, Binadhan-14 and Binadhan-19 of rice; 'Hyprosola', a high-yielding and high-protein mutant variety of chickpea (Sheikh et al. 1982; Shaikh et al. 1995); Binachinabadam-4 of groundnut;

Table 13.13 Mutant varieties released by the Bangladesh Institute of Nuclear Agriculture, Mymensingh in Bangladesh

S. No.	Crop	No.	Mutant varieties
1	Rice	12	Iratom-24, Binasail, Binadhan-4, Binadhan-5, Binadhan-6, Binadhan-7, Binadhan-9, Binadhan-13, Binadhan-14, Binadhan-18, Binadhan-19, Binadhan-20
2	Mustard	10	Agrani, Safal, Binasarisha-3, Binasarisha-4, Binasarisha-5, Binasarisha-6, Binasarisha-7, Binasarisha-8, Binasarisha-9, Shambal BAU-M/248
3	Chickpea	10	Hyprosola, Binasola-2, Binasola-3, Binasola-4, Binasola-5, Binasola-6, Binasola-7, Binasola-8, Binasola-9, Binasola-10
4	Mung bean	9	Binamoog-1, Binamoog-2, Binamoog-3, Binamoog-4, Binamoog-5, Binamoog-6, Binamoog-7, Binamoog-8, Binamoog-9
5	Groundnut	9	Binachinabadam-1, Binachinabadam-2, Binachinabadam-3, Binachinabadam-4, Binachinabadam-5, Binachinabadam-6, Binachinabadam-7, Binachinabadam-9, Binachinabadam-10
6	Lentil	8	Binamasur-1, Binamasur-2, Binamasur-3, Binamasur-5, Binamasur-6, Binamasur-8, Binamasur-9, Binamasur-11
7	Jute	5	Atompet-28, Atompet-36, Atompet-38, Binadeshipat-2, Binapatshak-1
8	Sesame	4	Binatil-1, Binatil-2, Binatil-3, Binatil-4
9	Tomato	4	Bahar, Binatomato-2, Binatomato-3, Binatomato-13
10	Black gram	2	Binamash-1
11	Onion	2	Binapiaz-1, Binapiaz-2
12	Wheat	1	Binagom-1
13	Lathyrus	1	Binakhesari-1
	Total	76	

http://mvd.iaea.org (accessed in March 2022)

Binamoog-2 and Binamoog-5 of mung bean; Binamash-1 of black gram; and Safal and Agrani mutant varieties of mustard. All these mutant varieties planted and cultivated in large areas have contributed substantially towards food security in Bangladesh. A bold kabuli-type high-yielding chickpea mutant variety 'Binasola-9', with cream seed coat colour and larger seed size, maturing in 115–125 days, tolerant to root rot and Botrytis grey mould, showing greater tolerance to pod borer insect-pest infestation than other mutants and check varieties and also suitable for farmers in drought-prone areas in Bangladesh, has been released for commercial cultivation during 2017 (Begum 2021).

Climate change, drought, high/low temperatures, salinity, river and flash floods, tidal surges, cyclones, hailstorms, etc. are recurrent phenomena in Bangladesh that seriously hamper crop production and productivity. To cope with these natural disasters, agricultural research in Bangladesh requires concerted efforts in the development of climate-resilient crop varieties. Therefore, genetic improvement of important traits in major crops has been successfully attempted in Bangladesh through mutation breeding. Drought is an important stress phenomenon in Bangladesh that greatly hampers crop production. For crop production to cope with the problems driven by climate change, such as salinity, drought and extreme temperatures, the Bangladesh Institute of Nuclear Agriculture (BINA) released a late 'Boro' rice variety, 'Binadhan-14', in 2013 which is tolerant to high temperature, has short maturity duration (105–115 days) and gives an average yield of 6.9 t/ha. Another variety, 'Binadhan-19', developed by irradiating the seeds of 'NERICA-10' rice with carbon-ion beams was released by Bangladesh in 2017 as a drought-tolerant, short-duration (95–105 days) and high-yielding (average 4.0 t/ha) variety for the '*Aus*' growing season (Table 13.13). The BINA developed a salt-tolerant wheat variety, 'Binagom-1', by selecting from a segregating population, obtained from the NIAB, Pakistan. Released in 2016, this variety can tolerate salinity (up to 12 dS/m) and produces an average yield of 2.8 t/ha. Apart from these, the BINA developed four salt-tolerant groundnut varieties 'Binachinabadam-5', 'Binachinabadam-6', 'Binachinabadam-7' and 'Binachinabadam-9' by irradiation with gamma rays. All these four varieties can tolerate salinity (up to 8 dS/m) from flowering to maturity and produce yields at 1.8–3.4 t/ha under saline soil conditions. These climate-resilient varieties are playing a significant role in food security and enhancing the nutritional status of the people of Bangladesh (Azad et al. 2021).

13.3.4 Mutation Breeding for Crop Improvement in Sri Lanka

A small island country, Sri Lanka is located to the south-east of the Indian subcontinent. In the island nation of Sri Lanka, the Department of Agriculture initiated mutation breeding in the late 1960s with the introduction of a cobalt-60 source to the country and has successfully developed and released four mutant varieties till date. It released the very first mutant variety of rice 'MI-273' in 1971. This success led to the production of direct and indirect mutant varieties of groundnut ANK-G-1 in 1995 and sesame 'ANK-S2' (Pathirana et al. 2000), and thereafter in tomato, a mutant variety 'Lanka Cherry' was released in 2010 (Pathirana 2011). Improved mutant lines were either used to develop new varieties or conserved at the Plant Genetic Resources Centre (PGRC) for future use. Through the IAEA's Technical Cooperation programme, IAEA has assisted in establishing the Sri Lanka National Centre for Nuclear Agriculture by providing a new cobalt-60 source, to expand the country's mutation breeding activities (Parasuraman et al. 2021).

13.4 Future Scope

Mutation breeding is expected to continue making significant contribution to plant breeding primarily as an important adjunct to the conventional and molecular breeding approaches in future also. The direct use of mutations is a very valuable supplementary approach to plant breeding, particularly when it is desired to improve one or two easily identifiable characters in an otherwise well-adapted variety.

However, in the long run, the use of induced mutations in various cross-combinations in hybridization programmes is probably more important than the direct use of mutants. A survey of the achievements of mutation breeding and related literature during the past eight decades of its history (Kharkwal and Shu 2009) suggests that major gains have resulted from the use of induced variability when it was fully integrated with conventional crop breeding programmes. This is evident from the fact that many more mutant varieties have resulted by using induced mutants in crossing programmes in comparison to those, which resulted through the direct release of mutants. During recent decades, tremendous progress has been made in the research of plant molecular biology and biotechnology, particularly plant genomics. As a result, we are witnessing new impulses in plant mutation research for crop improvement, from fundamental studies of mutagenesis to reverse genetics. Breeders are now aware of the newer potentialities and far-reaching implications of induced mutation for crop improvement and are able to use it with more sophistication and efficiency than before.

Induced mutagenesis is also gaining importance in plant molecular biology as a tool to identify and isolate genes and to study their structure and function. During the past three decades, mutation techniques for crop improvement have also been integrated with other molecular technologies such as molecular marker and high-throughput mutation screening techniques of genomics. Recent progress in the sequencing of entire genomes has led to the identification of many genes in model species. A future challenge will be to understand the functions of all these genes. In reverse genetics, radiation, chemical and transposon mutagenesis are used to discover the unknown functions of genes by knocking out genes and assigning known DNA sequences to the mutant phenotypes. Mutation breeding for crop improvement is therefore entering into a new era of 'molecular mutation breeding', and several ultramodern techniques like high hydrostatic pressure, ion beam technology, Targeting Induced Local Lesions in Genomes, endonucleolytic mutation analysis by internal labelling and space breeding technology are going to play an important role in revolutionizing mutation breeding in the next few decades leading to 'mutagenomics'. The detailed description of these techniques can be found in the work of Shu et al. (2012) as due to page limitations the same is beyond the scope of this chapter. Based on the review of the success of mutation breeding as is evident from the release of more than 3500 mutant cultivars belonging to more than 240 crop species from nearly 100 countries of the world till date, it can be hoped that mutation breeding will continue to play a significant role in crop improvement and solving world food security in the coming years and decades (Kharkwal and Shu 2009; Kharkwal 2017).

13.5 Conclusion

Crop domestication and improvement for agriculture over the past 12,000 years have been driven by spontaneous mutation combined with selection for desirable crop traits such as yield, plant and inflorescence architecture and non-shattering of seeds.

Mutations occurring in the genome, either spontaneous or induced, constitute the basis for this genetic diversity so central to crop improvement. Induced mutations have been used to create novel genetic diversity in plants since as early as 1928, and systematic mutation breeding has been used for crop improvement for about 90 years. Mutation breeding has played very important roles in elite crop variety development in the past, and will also continue to play in the future, to secure the global food supply and food security. More recently, the increasing threat posed by climate change on biodiversity loss, crop productivity and, consequently, food and nutrition security and farmers' livelihoods demands the generation of novel genetic variation for the development of improved varieties that can adapt to adverse climate events. Therefore, the role of mutation breeding in enhancing genetic diversity and crop improvement will become even more important in the coming decades in Asia, particularly in the Indian subcontinent. In the future, mutation induction approaches will be closely associated with biotechnology platforms, such as novel techniques of high-throughput genotyping and phenotyping for the introduction of new genetic alleles for climate-resilient traits like heat and cold tolerance and higher water and nutrient efficiency.

A brief description of the 542 mutant crop cultivars collected from various research institutes and agricultural universities and reported or approved and/or released as induced mutants or from the use of such mutants in cross-breeding or natural and spontaneous mutants, including bud sports, belonging to 85 plant species for cultivation in India is given in Table 13.11.

References

Abdalla E et al (2021) Groundnut mutants with end-of-season drought tolerance for the marginal dry lands of North Kordofan State, Sudan. In: Sivasankar S et al (eds) Mutation breeding, genetic diversity and crop adaptation to climate change. CABI, Wallingford, UK, pp 243–257

Ahloowalia B et al (2004) Global impact of mutation-derived varieties. Euphytica 135:187–204

Ali M et al (1997) The mungbean green revolution in Pakistan. Technical Bulletin No. 24. Shanhua, Taiwan, AVRDC ROC. 66 p

Ali JAF et al (2010) Genetic manipulation of lentil through induced mutations. Pak J Bot 42(5): 3449–3455

Ankineedu G et al (1968) Effect of fast neutrons and gamma-rays on castor. Indian J Genet 28:31–39

Azad MAK et al (2021) Development of climate-adaptable/resilient crop varieties through induced mutation. In: Sivasankar S et al (eds) Mutation breeding, genetic diversity and crop adaptation to climate change. CABI, Wallingford, pp 157–171

Badawi AT (2001) Yield gap and productivity decline in rice production in Egypt. In: Yield gap and productivity decline in rice production. International Rice Commission, FAO, Rome, pp 429–441

Begum SN (2021) Development of the first Kabuli type chickpea mutant variety in Bangladesh. In: Sivasankar S et al (eds) Mutation breeding, genetic diversity and crop adaptation to climate change. CABI, Wallingford, pp 203–208

Bouma J et al (1991) Importance and application of the mutant 'Diamant' in spring barley breeding. In: Plant mutation breeding for crop improvement, vol 1. IAEA, Vienna, pp 127–133

Clampett WS et al (2001) Major achievements in closing yield gap of rice between research and farmers in Australia. In: Yield gap and productivity decline in rice production. International Rice Commission, FAO, Rome, pp 411–428

da Cruz RP et al (2013) Avoiding damage and achieving cold tolerance in rice plants. Food Energy Secur 2(2):96–119

Danso KE et al (2009) Application of induced mutation techniques in Ghana: impact, challenges and the future. In: Induced plant mutations in the genomics era. FAO, Rome, pp 270–272

Do KT et al (2009) Socio-economic impacts of mutant rice varieties in southern Viet Nam. In: Induced plant mutations in the genomics era. FAO, Rome, pp 65–68

FAO/IAEA (2022) Mutant Variety Database. Accessed Mar 2022

Gomez-Pando L et al (2009) Barley (*Hordeum vulgare*) and Kiwicha (*Amaranthus caudatus*) improvement by mutation induction in Peru. In: Induced plant mutations in the genomics era. FAO, Rome, pp 230–232

González MC et al (2008a) Salinity tolerant mutant obtained from protons radiations. Abstract p. 56. In: Book of abstracts, FAO/IAEA international symposium on induced mutations in plants, 12–15 Aug 2008, Vienna, Austria

González MC et al (2008b) Development of drought tolerant tomato varieties through induced mutations in Cuba. Abstract p. 56. In: Book of abstracts, FAO/IAEA international symposium on induced mutations in plants, 12–15 Aug 2008, Vienna, Austria

González MC et al (2009) Development of salinity-tolerant rice varieties using biotechnological and nuclear techniques. In: Induced plant mutations in the genomics era. FAO, Rome, pp 138–140

González MC et al (2020) Girón 50, Nuevo Mutante de Tomate (*Solanum lycopersicon* L.) Tolerante a la Sequía y Altas Temperaturas. II Simposio Latinoamericano de Apliciones Nucleares en Agricultura, 56 March 2020, Quito, Ecuador

Gonzalez MC et al (2021) Evaluation of mutant rice genotypes for tolerance to high temperature. Aust J Crop Sci 15(8):43–49

Green AG, Dribnenki JCP (1996) Breeding and development of Linola™ (low linolenic flax). In: FAO-Proc. 3rd Intern. Rax Breeding Research Group, France. FAO, Rome, pp 145–150

Gregory WC (1960) The peanut NC 4x, a milestone in crop breeding. Crops Soils 12(8):12–13

Gustafson JP et al (2021) World food supply: problems and prospects. In: Sivasankar S et al (eds) Mutation breeding, genetic diversity and crop adaptation to climate change. CABI, Wallingford, UK, pp 3–9

Gustafsson A (1963) Productive mutations induced in barley by ionizing radiations and chemical mutagens. Hereditas 50:211–263

Haq MA (2009) Development of mutant varieties of crop plants at NIAB and the impact on agricultural production in Pakistan. In: Induced plant mutations in the genomics era. FAO, Rome, pp 61–64

Hassan S et al (2001) Gamma ray induced high Yielding Kabuli type chickpea mutant variety "Hassan-2K". Pak J Bot 33:703–707

Hensz RA (1991) Mutation breeding of grapefruit (*Citrus parodist* Macf.). In: Plant mutation breeding for crop improvement, vol I. SM-311. IAEA, Vienna, pp 533–536

Horn LH et al (2017) Genotype-by-environment interaction for grain yield among novel cowpea (*Vigna unguiculata* L.) selections derived by gamma irradiation. Crop J 8. https://doi.org/10.1016/j.cj.2017.10.002

Hussain M et al (2021) Improving sustainable cotton production through enhanced resilience to climate change using mutation breeding. In: Sivasankar S et al (eds) Mutation breeding, genetic diversity and crop adaptation to climate change. CABI, Wallingford, UK, pp 145–156

Kang CW et al (2008) Improvement of sesame crop through induced mutations in Korea. Abstract p. 142. In: Book of abstracts, FAO/IAEA international symposium on induced mutations in plants, 12–15 Aug 2008, Vienna, Austria

Kharkwal MC (1996) Accomplishments of mutation breeding in crop improvement in India. In: Sachdev MS et al (eds) Isotopes & radiations in agriculture and environment research. Indian Society for Nuclear Techniques in Agriculture and Biology, New Delhi, pp 196–218

Kharkwal MC (1998) Induced mutations for improvement of protein in chickpea (*Cicer arietinum* L.). Indian J Genet 58(1):61–68

Kharkwal MC (2012) A brief history of plant mutagenesis. Plant mutation breeding and biotechnology. CABI, Wallingford, UK, pp 21–30

Kharkwal MC (2017) Mutation breeding for crop improvement. Geography You 17(102):26–32

Kharkwal MC, Shu QY (2009) Role of induced mutations in world food security. In: Induced plant mutations in genomics era. FAO, Rome, pp 33–38

Kharkwal MC et al (1988) Induced mutations for improvement of chickpea, lentil, pea and cowpea. In: Proc. FAO/IAEA workshop on improvement of grain legume production using induced mutations, 1–5 July, 1986, Pullman, Washington (USA). IAEA, Vienna, Austria, pp 89–109

Kharkwal MC et al (2001) Seventy five years of research on induced mutations with special reference to Crop improvement in India. In: Ramamurty et al (eds) Proceedings of NAARRI International conference on applications of radioisotopes and radiation technology in the 21st century, December 12-14, 2001, Mumbai, pp 230–235

Kharkwal MC et al (2004) Mutation breeding in crop improvement. In: Jain HK, Kharkwal MC (eds) Plant breeding – Mendelian to molecular approaches. Narosa Publishing House, New Delhi, India, pp 601–645

Kharkwal MC et al (2005) Pusa 547, a high yielding chickpea (*Cicer arietinum* L.) mutant variety for late sown conditions of North Western Plains Zone of India. Indian J Genet 65:229–230

Larter EN et al (1965) Redwood 65, an improved flax variety. Can J Plant Sci 45(5):515–516. https://doi.org/10.4141/cjps65-100

Le TD et al (2021) Soybean breeding through induced mutation in Viet Nam. In: Sivasankar S et al (eds) Mutation breeding, genetic diversity and crop adaptation to climate change. CABI, Wallingford, UK, pp 40–46

Liu L (2021) New breakthroughs in plant mutation breeding prominent among top ten scientific research advances in China in 2019. FAO/IAEA Plant Breed Genet Newsl 46:20–22

Liu L et al (2021) New mutation techniques for crop improvement in China. In: Sivasankar S et al (eds) Mutation breeding, genetic diversity and crop adaptation to climate change. CABI, Wallingford, UK, pp 47–52

Livore AB et al (2018) The success of IMI tolerant rice varieties in Latin America. In: FAO/IAEA International symposium on plant mutation breeding and biotechnology, 27–31 August 2018, Vienna, Austria

Lundqvist U (2009) Eighty years of Scandinavian barley mutation genetics and breeding. In: Induced plant mutations in the genomics era. FAO, Rome, pp 39–43

Lundqvist U (2021) Scandinavian mutation research during the past 90 years – a historical review. In: Sivasankar S et al (eds) Mutation breeding, genetic diversity and crop adaptation to climate change. CABI, Wallingford, UK, pp 10–23

Maluszynski M et al (1995a) Application of *in vivo* and *in vitro* mutation techniques for crop improvement. Euphytica 85(1):303–315

Maluszynski M et al (1995b) Mutation techniques in plant breeding. In: Proc. IAEA/FAO Symp., on Induced mutations and molecular techniques for crop improvement, Vienna, June 19–23, 1995, pp 489–504

Merotto A et al (2006) Evolutionary and social consequences of introgression of non-transgenic herbicide resistance from rice to weedy rice in Brazil. Evol Appl 9:837–846

Nagampongsai et al (2009) Current status of mungbean and the use of mutation breeding in Thailand. In: Induced plant mutations in the genomics era. FAO, Rome, pp 355–357

Nakagawa H (2009) Induced mutations in plant breeding and biological researches in Japan. In: Induced plant mutations in the genomics era. FAO, Rome, pp 48–54

Nakagawa H (2021) History of mutation breeding and molecular research using induced mutations in Japan. In: Sivasankar S et al (eds) Mutation breeding, genetic diversity and crop adaptation to climate change. CABI, Wallingford, UK, pp 24–39

Nichterlein K et al (2000) Achievements and trends of using induced mutations in crop improvement. In: Proc., DAE-BRNS Symp., 6–8 December 2000, Mumbai, India, pp 27–35

Parasuraman M et al (2021) Application of mutation breeding techniques in the development of green crop varieties in Sri Lanka: the way forward. In: Sivasankar S et al (eds) Mutation breeding, genetic diversity and crop adaptation to climate change. CABI, Wallingford, UK, pp 76–82

Pathirana R (2011) Plant mutation breeding in agriculture. CAB Rev 6(0320):1–20

Pathirana R et al (2000) Development and release of gamma ray induced sesame mutant ANK-S2 in Sri Lanka. Trop Agric Res Ext 3(1):19–24

Pawar SE et al (1991) Use of induced mutations in the breeding of pulse crops. In: Proc. IAEA Symp. On plant mutation breeding for crop improvement, 18–22 June 1990, vol 1. IAEA, Vienna, pp 413–418

Razzaq A et al (2021) Cotton germplasm improvement and progress in Pakistan. J Cotton Res 4:1. https://doi.org/10.1186/s42397-020-00077-x

Rahman HM et al (2018) Multi-model projections of future climate and climate change impacts uncertainty assessment for cotton production in Pakistan. Agric For Meteorol 253–254:94–113

Rutger JN (1992) Impact of mutation breeding in rice. A review, Mutation breeding review, vol 8. IAEA, Vienna

Rutger JN (2006) Thirty years of induction, evaluation and integration of useful mutants in rice genetics and breeding. IAEA Mutat Breed Rep 1(2):4–13

Rutger JN (2009) The induced SD1 mutant and other useful mutant genes in modern rice varieties. In: Induced plant mutations in the genomics era. FAO, Rome, pp 44–47

Sarsu F et al (2021) Contribution of induced mutation in crops to global food security. Communicate 12(22):2–11. https://doi.org/10.18272/aci.v12i3.2031

Scarascia-Mugnoza GT et al (1991) Mutation breeding programme for durum wheat (*Triticum turgidum* spp. *durum* Desf). In: Proceedings of the symposium on Plant mutation breeding for crop improvement, 18–22 June 1990, Vienna, Austria

Schiocchet MA et al (2015) SCS121 CL: rice cultivar resistant to herbicides of imidazolinone chemical group. Crop Breed Appl Biotechnol 15(4). https://doi.org/10.1590/198470332015v15n4c47

Shaikh MAQ et al (1995) A disease resistant and high yielding blackgram variety, BINA Mash-1, derived through induced mutation. In: Proc. IAEA/FAO Symp., on Induced mutations and molecular techniques for crop improvement, Vienna, 19–23 June 1995, pp 635–637

Sheikh MAQ et al (1982) A high-yielding and high-protein mutant of chickpea (*Cicer arietinum* L.) derived through mutation breeding. Environ Exp Bot 22:483–389

Shu QY (ed) (2009) Induced plant mutations in the genomics era. FAO/IAEA, Rome, Italy. 458 pp

Shu QY et al (2012) Plant mutation breeding and biotechnology. CABI FAO, Oxfordshire, UK

Si P et al (2009) Induced mutation in narrow-leafed lupin improvement: an example of herbicide tolerance. In: Induced plant mutations in the genomics era. FAO, Rome, pp 77–79

Singh V et al (2017) Rice production in the Americas. In: Rice production worldwide. Springer, New York, pp 137–168

Sivasankar S et al (eds) (2021) Mutation breeding, genetic diversity and crop adaptation to climate change. CABI, Wallingford, UK, p 535

Sybenga J (1983) Genetic manipulation in plant breeding: somatic versus generative. Theor Appl Genet 66:179–201

Syed AS et al (2009) Development of improved varieties of rapeseed and mustard through *in vivo* mutagenesis and hybridization in Pakistan. In: Induced plant mutations in the genomics era. FAO, Rome, pp 376–378

Ullah A et al (2020) Pulses production in Pakistan: status, constraints and opportunities. Int J Plant Prod 14:549–569. https://doi.org/10.1007/s42106-020-00108-2

Van Harten AM (1998) Mutation breeding: theory and practical applications. Cambridge University Press, Cambridge
Wang LQ (1991) Induced mutation for crop improvement in China. In: Proc. IAEA Symp. on Plant mutation breeding for crop improvement, 18–22 June 1990. IAEA, Vienna, pp 9–32. Review on mutation breeding for improvement of food legumes. 14-05-2020. www.researchgate.net
Yanev A (2006) Mutant durum wheat varieties developed in Bulgaria. Plant Mutat Rep 1(2)
Zakir M (2018) Mutation breeding and its application in crop improvement under current environmental situations for biotic and abiotic stresses. Int J Res Stud Agric Sci 4(4):1–10

Success of Mutation Breeding of Sorghum to Support Food Security in Indonesia

14

Soeranto Human, Wijaya M. Indriatama, and Sihono

Abstract

To support food security in the country, Indonesia has launched a national program of food diversification meaning that people should eat locally available food stuff resources other than rice. Indonesia has a great potency to produce food from the available crop diversity in order to support food security. In dryland agroecosystem, Indonesia needs to develop crops that require less agricultural inputs such as water and fertilizer and that have have good adaptability, nutrition, and economic values, and the choice fell on sorghum (*Sorghum bicolor*). Attempts to increase sorghum genetic variation were done by mutation breeding, and by selecting desired mutant genotypes, some promising mutant lines have been developed. Significant adverse impacts of climate change have appeared in some agricultural regions in Indonesia, such as prolonged drought problem in the eastern regions. To face the worsening conditions brought about by climate change and variability, some sorghum mutant varieties having better adaptability, productivity, and quality have been released and developed further by stakeholders. Sorghum cultivation in Indonesia has given significantly positive impacts on mitigating climate change and supporting food diversification program for maintaining food security in the country. Sorghum has also promoted economic growth in the rural areas impacted by climate change.

S. Human (✉) · Sihono
Research Organization of Nuclear Technology (ORTN), National Research and Innovation Agency of Indonesia (BRIN), Jakarta, Indonesia
e-mail: soer001@brin.go.id

W. M. Indriatama
Research Center for Isotopes and Radiation Application, Research Organization of Nuclear Technology (BATAN), Indonesian Research and Innovation Agency (BRIN), Jakarta, Indonesia

S. Penna, S. M. Jain (eds.), *Mutation Breeding for Sustainable Food Production and Climate Resilience*, https://doi.org/10.1007/978-981-16-9720-3_14

Keywords

Sorghum · Mutation breeding · Climate change · Food diversity and security

14.1 Introduction

Crop improvement program is usually conducted through a plant breeding. The improvement itself is focused on the genetics of crop's desirable traits mostly related to crop productivity and/or quality. Plant breeding activities are usually started by increasing the genetic variation so that the breeders can do selection of superior genotypes for further evaluation before releasing them as new varieties that can be grown widely or commercially by stakeholders. There are many ways of increasing genetic variation such as through introduction, collection of available local genotypes, hybridization, mutation, and modern biotechnology. Many researches on modern biotechnology have recently produced transgenic plants, which products are called as genetically modified organism (GMO). For some countries, however, public acceptances of the GMO products are still debatable because of the biosafety and food safety issues. It seems to be different from those products produced from mutation techniques.

Mutation is a sudden, random, and heritable change of genetic material that can happen in the level of genome, chromosome, or gene (DNA). Mutation can happen in a plant naturally due to cosmic rays of the sunshine radiation, but the mutation rate is very low. Mutation can be induced by using a mutagenic agent. There are two types of mutagenic agent, namely chemical or alkylating agent (such as EMS and dES) and physical or ionizing radiation (such as gamma rays, fast neutron, ion beam). Plant mutation breeding is the process of exposing plant reproductive materials such as seeds, tubers, cuttings, anthers, and tissues to a mutagenic agent with the objective of fastening the increase of genetic variation, doing selection of desirable mutated genotypes, and then screening and testing the superior improved plants in productivity and quality before finally submitting for the official release as a new mutant variety. Other biotechnological tools can actually be used to help plant breeder accelerate the breeding process of releasing new varieties with desired traits such as through molecular assisted selection (MAS).

Once released, the mutant variety is safe for human beings because the mutagenic radiation does not reside inside the plant, and also mutation breeding does not involve gene modification but rather uses a plant's own genetic resources and mimics the natural process of spontaneous mutation. Globally, there are more than 3200 mutant varieties developed through mutation breeding, and information of the released mutant varieties is now available at the IAEA Mutant Variety Database. Research collaboration on mutation breeding in Indonesia has so far been supported by IAEA through some technical cooperation (TC) projects and also through Forum for Nuclear Cooperation in Asia (FNCA) projects. Results of the projects have given positive social and economic impacts to some participating member states including that of mutation breeding of sorghum in Indonesia (King et al. 2020). Information and experience of the success in mutation breeding of sorghum in Indonesia are discussed and presented in this chapter.

14.2 Mutation Breeding in Indonesia

In Indonesia, research on plant mutation breeding is mainly conducted at the Center for Isotopes and Radiation Application (CIRA) under the National Nuclear Energy Agency of Indonesia (BATAN). The available infrastructure and facilities include gamma irradiators, plant breeding laboratory, tissue culture and biotechnology facilities, mutant seed germplasm storage, greenhouse, mutant nursery facilities, experimental field, processing unit, and supporting equipment. The objectives of mutation breeding are mainly to support national program on food security with better nutrition for human health and livelihoods.

Food crop production, particularly in Java Island, is now facing problem with shrinking growing areas due to land utilization for nonagricultural purposes such as those of industries, roads, and properties. Expanding agricultural field to outside Java is facing problems with soil fertility and infrastructure, and also adverse impacts of climate change have been observed recently. Therefore, future plant breeding programs are directed to develop climate-smart agriculture through developing crops that can adapt and produce well in the impacted areas with less agricultural inputs, especially outside Java Island.

Plant mutation breeding has so far played a very important role in crop varietal improvement program in Indonesia. According to BATAN Report (2020), Indonesia has released many mutant varieties of different crops including rice (25), soybean (12), sorghum (3), mung bean (2), groundnut (1), tropical wheat (1), cotton (1), and banana (1). All these mutant varieties have been grown and developed further by stakeholders including farmers and some private companies throughout Indonesia. The socioeconomic impact of mutant varieties had been studied and analyzed for rice; for example, farmers who grew rice mutant variety from BATAN increased their income of IDR 4,111,998 per ha per harvest season. This was about 49% higher than using traditional local variety (Suryati et al. 2019). It is not exaggeration to mention that rice mutant varieties have significantly contributed to self-sufficient rice demand in Indonesia. However, Indonesia has so far been very much dependent on rice as staple food. Rice is regarded as a high-input crop because of requiring more water (irrigation), fertilizer, and other inputs. In order to support food security, the Government of Indonesia has launched a food diversity program, meaning that we should develop food sources other than rice. We found that sorghum is very suitable and potential for hot and dryland, low-input crop and its grain can be used as a source of foodstuff.

14.3 Breeding of Mutation of Sorghum

Sorghum is not of Indonesian origin, so the available genetic variability of the plants is low; therefore, a sorghum breeding program is needed to increase the genetic variation for further improved varieties (Human 2015a). In some areas, sorghum has been recognized as a source of food, fodder, and fuel. However, sorghum is still considered a minor crop and its cultivation is limited, mostly cultivated by local farmers in a particular region. Sorghum (*Sorghum bicolor*) is a cereal crop that is commonly grown in hot and dry (arid or semiarid) regions and is considered a multipurpose crop due to its use as food, feed, or fuel. From an agronomic point of view, sorghum is known as a low-input crop and is very efficient in the use of fertilizers and sunlight (Suminar et al. 2017; Aznur et al. 2017). Sorghum grain contains carbohydrates that can be used as a food source, and its stem and leaves (stover) can be used as animal feed. In addition, there is another type of sorghum, the stem of which contains liquid sugar known as sweet sorghum, and it is usually used as a raw material for the production of liquid sugar (syrup) with a low glycemic index (GI), so its products have become popular with people suffering from diabetes (Human 2015a). The glycemic index indicates the effect of a food on a person's blood glucose level (also called blood sugar). Still in many countries, sweet sorghum is commonly used as a raw material for the production of bioethanol (as a renewable bioenergy) through the fermentation process of its sugar juice.

Sorghum can grow and adapt well in Indonesia, especially in the drought-prone areas of the eastern part of the country, but still with low productivity and quality. Farmers in the provinces of Nusa Tenggara Barat (NTB) and Nusa Tenggara Timur (NTT) have grown local varieties of sorghum for a long time and use it as a staple food since other food crops such as rice and maize do not grow well there during the dry season. Local sorghum varieties mostly have undesirable characteristics such as low productivity, late maturity, tall plant height which is prone to lodging, and low-quality dark-colored grains as an indicator of high tannin content. The dark color of sorghum grains has been studied and was controlled not only by additive dominant genes but also by epistatic genes with moderate heritability, which means that the improvement of grain color can be done through a plant breeding program (Trikoesoemaningtyas et al. 2017). The aim of sorghum breeding is to improve agronomic traits including plant height, growth length, and grain yield and quality and to adapt plants to withstand adverse conditions caused by climate change, such as prolonged drought.

In Indonesia, the improvement of sorghum varieties using mutation techniques was carried out at the Center for Isotopes and Radiation Applications (CIRA), National Atomic Energy Agency (BATAN). The main equipment available at this center is a gamma irradiator with an installed cobalt-60 source that can be used to irradiate breeding materials to increase the genetic variation of plants (Fig. 14.1). The plant materials used were seeds of the Zhengzu variety, which was imported from China through the IAEA Regional Project RAS5040. Sorghum seeds were treated with gamma radiation at doses of 0, 100, 200, 300, 400, 500, 600, 700, 800, 900, and 100 Gy, respectively, and then grown on sandy soil in a greenhouse for a radiosensitivity study in the M1 generation. Seedling growth was measured for each irradiation treatment, and optimal irradiation doses were determined using lethal dose (LD) values. The highest plant genetic variation in M2 is usually found between LD-20 and LD-50 values, as reported in the wheat mutation breeding program (Nur et al. 2014). LD values were determined using best curve fitting software, which is based on the function of measured lethality (Y) at given irradiation doses (X) as shown in Fig. 14.2, and the variation of plants in M2 is shown in

Fig. 14.1 Gamma irradiator with installed cobalt-60 source available at BATAN for irradiating breeding materials in mutation breeding program

Fig. 14.2 Screening methods for drought tolerance: (**a**) indirect screening and (**b**) direct screening

Fig. 14.3. It was found that optimal doses of gamma radiation for sorghum mutation breeding purposes were about 200–450 Gy (Fig. 14.2). The mutant population derived from these optimal irradiation doses yielded the highest genetic variation (as measured by statistical variances) in M2. Visual selection of mutants was performed by comparison with the original parent and then continued by seed propagation in the next generation.

Screening of plants for drought tolerance was initiated in M2 using indirect screening in the greenhouse and continued with direct screening in the field (Fig. 14.2). Indirect screening was performed at the seedling stage by growing seedlings in medium containing polyethylene glycol (PEG) in a greenhouse. PEG is a chemical that can regulate the amount of water absorbed by a plant's root (Van

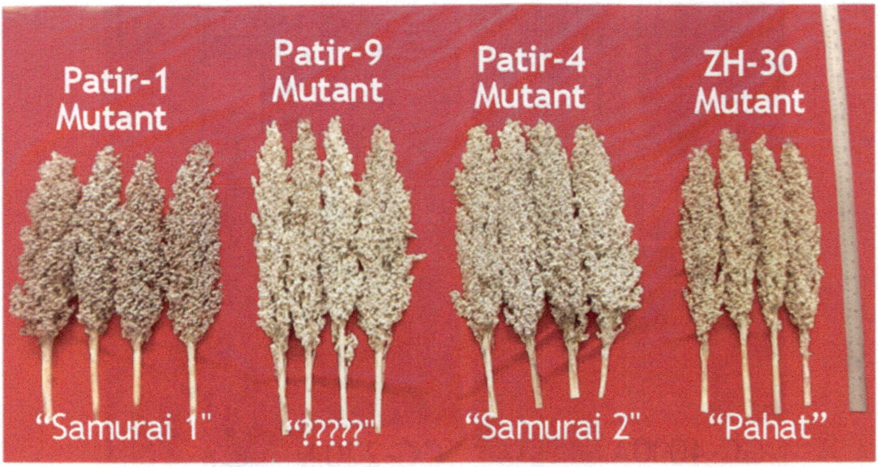

Fig. 14.3 The released mutant varieties of sorghum: Pahat, Samurai-1, and Samurai-2

den Berg and Zeng 2006). The higher the concentration of PEG in the growth medium, the less water the root can absorb. A concentration of 25% PEG in solution is equivalent to a critical drought exposure in the field with an impact on permanent leaf wilt in cereals. The use of the PEG method to screen for drought tolerance has also been reported in the sorghum mutation breeding program (Human 2015b; Human and Sihono 2010). The selected tolerant seedlings were then transplanted to the field for multiplying their seeds, and further direct screening was conducted in the field during dry season in the drought-prone areas. Subsequently, yield trials were evaluated among drought-tolerant plants in the field. Sorghum mutation breeding program followed that of self-pollinated crop, which directed to the formation of pure or homozygous line of the selected mutants (Richards 1997; Van Harten 1998). Homogeneity test and stability of the mutant lines were studied in the M5 (Human et al. 2011), and those of homogenous lines were denoted as the promising mutant lines. Yield multilocation trials and grain quality analysis were conducted for the promising mutant lines before they were proposed for submitting for the official release to the *Varietal Release Team* which is under the Ministry of Agriculture. Sorghum grain quality analysis was conducted at the Center for Chemistry, the Indonesian Academy of Science (LIPI) at Serpong, Banten Province, Indonesia.

Fifteen promising drought-tolerant mutants were obtained in M5 and evaluated together with their parental and national control varieties in multilocation trials. Agronomic data were collected, analyzed, and reported by the Varietal Release Team, and finally three promising mutant lines ZH-30, PATIR-1, and PATIR-4 were accepted by the Ministry of Agriculture for release as new mutant sorghum varieties. The ZH-30 mutant line was released in 2013 and was named Pahat, while the PATIR-1 and PATIR-4 mutant lines were released in 2014 under the names Samurai-1 and Samurai-2. The head performance of these mutant varieties is shown in Fig. 14.3. In the description, Pahat and Samurai-2 were recommended for grain sorghum for food use, while Samurai-1 was sweet sorghum for use as liquid sugar (syrup) or was further processed to make bioethanol (as bioenergy).

Productivity and grain quality data for these sorghum mutant varieties are shown in Table 14.1. All mutant varieties had higher grain productivity than the parental variety Zhengzu and the national control variety Mandau. The grain of the mutant varieties contained sufficient carbohydrates as a food source. However, judging by tannin content alone, Pahat and Samurai-2 met the food requirements. According to Indonesia's Food Regulation, the tannin content of food should be less than 0.025%. The Pahat variety is the most promising for food use because it has good carbohydrates, high protein and fiber content, and low tannin content. The mutated variety Pahat is agronomically early (89 days) and semidwarf (148 cm) in height, making it easy to harvest and tolerant of lodging against strong winds. As one of the negative impacts of climate change is adverse weather conditions such as strong wind (tornado) in some regions, the semidwarf Pahat variety can be used to mitigate climate change. The mutated sorghum variety Pahat is therefore widely cultivated in Indonesia, especially to support food security in drought-prone areas in the eastern part of the country, which is very strongly affected by climate change. Its short growth period (only 3 months) also provides an advantage in the cultivation of the Pahat mutant variety with respect to long-term drought avoidance during the dry season.

Samurai-1 is the only mutant variety with the highest sugar or brix content (12%) in stem juice, and this variety can be classified as sweet sorghum. Despite the high productivity, Samurai-1 grain is not recommended to be used as food due to the high tannin content (0.030%); therefore, the Samurai-1 variety is recommended mainly as a source for the production of liquid sugar (syrup) or another process through fermentation for the production of bioethanol (as bioenergy) industry. The Indonesian Government has recently promoted the use of bioenergy as a renewable energy source, and many companies are now engaged in this type of business.

14.4 Dissemination, Economic, Social, and Environmental Impacts

Based on the dissemination of breeding seeds of mutant varieties produced by BATAN and their distribution to some stakeholders in 2017/2018, the cultivation area of mutant sorghum varieties in Indonesia was estimated to be 800,000 ha in total. Growers or stakeholders included farmers, private companies, local government offices, and some universities. In fact, the commercialization of these mutant sorghum varieties started in 2017 by a private company named PT Sedana Panen Sejahtera (Fig. 14.4) and also PTPN XII, which widely cultivated the Pahat mutant variety in Banyuwangi District, East Java Province (Fig. 14.5). In addition to the company's business, the cultivation of the mutant variety was also intended to increase the productivity of land affected by climate change (drought-prone areas) and ensure food security in Indonesia, especially during the prolonged dry season. Especially for sweet sorghum, the Ministry of Energy and Mineral Resources of

Table 14.1 Grain productivity and quality of the released sorghum mutant varieties

No.	Sorghum variety	Analyzed components						
		Productivity (t/ha)	Carbohydrate (%)	Protein (%)	Fat (%)	Fiber (%)	Tannin (%)	Brix (%)
1	Samurai-1	6.1	73.59	12.55	2.56	2.20	0.030	12
2	Samurai-2	6.4	75.53	12.07	2.80	1.53	0.016	8
3	Pahat	5.8	72.86	12.80	2.42	2.21	0.012	5
4	Zhengzu (parent)	4.2	70.63	10.71	2.59	1.39	0.016	5
5	Mandau (national variety)	5.1	75.40	12.73	2.97	1.22	0.013	6

Fig. 14.4 Some commercial sorghum products sold in the market in Indonesia

Fig. 14.5 Sorghum mutant variety of Pahat grown widely by the private company of PTPN XII in Banyuwangi District of East Java Province during dry season

Indonesia (ESDM) cultivated a mutated variety of sweet sorghum Samurai-1 in Merauke District, Irian Province, for the production of bioethanol industry with a capacity of 40 kg of litter per day. Located in eastern Indonesia, Irian Province is one province that has been adversely affected by climate change.

In a country where rice is the main staple food and where the population was unfamiliar with this new crop, the commercialization of sorghum focused on emphasizing the nutritional added value of the crop. Sorghum grains are high in fiber, iron, protein, calcium, and beneficial polyphenols, but low in fat and cholesterol. In addition, sorghum does not contain gluten and has a low glycemic index, so it is especially suitable for people suffering from diabetes and related diseases (Human 2015a).

In addition to nutritional value, mutant sorghum varieties have proven to be early maturing, high yielding, and drought resistant, making them ideal for dry-season cultivation. This means that they have great potential to increase the marginal productivity of land and support economic growth, especially in drought-prone areas where arable land lies fallow and cannot grow other types of food crops (such as those mostly found in the eastern part of Indonesia). Mutant sorghum varieties have been certified by the Department of Agriculture to have a grain yield of approximately 30% higher than nonmutant varieties. This feature, together with the ability to grow and sell sorghum during the dry season, has the potential to increase farmers' income by an average of 20–30%.

In addition to their potential to enhance economic development through their agronomic properties, these new mutant sorghum varieties hold promise for supporting the country's efforts to reduce dependence on rice and thus ensure future food security. The past decade has seen a diversification of food consumption, which was the highest priority for the country. This is taken into account in the Strategic Plan of the Ministry of Agriculture (2015–2019). The goal of crop mutation breeding is usually to improve the productivity and quality of crops compared to the original variety. Once released, the mutant varieties may be adopted by farmers for their higher yields and better prices. However, by growing mutated varieties of sorghum, farmers can improve their income through soil productivity during the dry season, as mutated varieties are more drought tolerant and high yielding and have good grain quality. Therefore, these mutant sorghum varieties contributed to the success of the food diversification program to promote food security in Indonesia.

Mutated sorghum varieties also have several environmental benefits. They are drought tolerant and therefore use water efficiently. They are also resistant to major diseases, so they require less fertilizer and pesticides. It is estimated that the use of irrigation and pesticides can be significantly reduced in sorghum production by up to 20%. In addition, sorghum is highly efficient in its rate of photosynthesis. This means it produces more biomass that can be recycled into the soil, which helps maintain soil fertility and promotes more sustainable farming practices. Sorghum stover (stem and leaves) can also be used for feeding ruminants, especially for those new mutants with green leaves (Fig. 14.6).

Fig. 14.6 Dwarf and early-maturing sorghum mutants with stay-green leaves at BATAN

14.5 National and International Collaboration

In carrying out mutation breeding research, BATAN has collaborated with many counterparts, either national or international partnerships. National counterparts include the Ministry of Agriculture, farming communities, private companies, and some universities. Together with the Ministry of Agriculture, BATAN supported a national food diversification program aimed at increasing food resilience through sorghum mutation breeding research. Until now, Indonesia has been very dependent on rice as a staple food. Rice is considered a high-input crop because it requires more water (irrigation), fertilizers, and other inputs. In order to promote food security, the Indonesian Government has launched a food diversity program, which means that we should develop food sources other than rice. We found that sorghum is very suitable for dryland in Indonesia, is a low-input crop, and can be used not only as food but also as animal feed. BATAN has so far carried out mutation breeding of sorghum and released some mutant varieties that were developed by some commercial private companies to produce sorghum food products. Since the release of the mutant varieties, sorghum has developed very well and its various food products have become available in the local market. For example, when making a traditional dish called Tumpeng, people usually use rice as the main ingredient, but now a variation has been made using sorghum grains (Fig. 14.7).

The success of sorghum mutation breeding has also been recognized with the Ministry of Agriculture's Food and Agriculture Innovation Award in 2015, as well as the 2016 Indonesian President's 2016 Agrajan Satyalancana Pembangunanor Award and Agricultural Development Award.

Fig. 14.7 Indonesian traditional food "*Tumpeng*" made from sorghum grains

International counterparts include the Joint FAO/IAEA Program, the Forum for Nuclear Cooperation in Asia (FNCA), the Japan Society for the Promotion of Science (JSPS), and the Asia-Pacific Economic Cooperation (APEC). In collaboration with the FAO/IAEA Joint Program, BATAN has been involved in many technical cooperation (TC) projects, including National TC, Regional TC, and Coordinated Research Program (CRP). These projects supported capacity building for the country's sorghum mutation breeding program. Senior researchers attended scientific exchange meetings, while other junior researchers benefited from attending regional training. As part of the regional project, Indonesia itself hosted some of these scientific capacity-building activities, such as training on mutant screening for abiotic stresses and molecular approaches for selecting desirable green traits in crops under the RAS5077 project. Indonesia has also published scientific articles on sorghum mutation breeding, such as in Atom Indonesia journal, Radioisotopes journal, and the IAEA Plant Breeding and Genetics newsletter.

Thanks to its successful achievements, BATAN received the 2014 and 2021 Outstanding Achievement in Plant Mutation Breeding Awards from the Joint FAO/IAEA Program (Fig. 14.8). Also in 2017, BATAN was appointed as an IAEA Collaborating Center for Plant Mutation Breeding, whose activities include collaborating mutation breeding research and organizing various types of IAEA activities including professional meetings, scientific visits, and internships (Fig. 14.9). Regarding mutation breeding, participants interested in joining the training courses come from Asian and African countries such as Malaysia, Burkina Faso, Malawi, Mozambique, Tanzania, and Togo (Fig. 14.10). The duration of training can vary from 2 to 6 months depending on their wishes and budget availability.

Fig. 14.8 Outstanding achievement award from the Joint FAO/IAEA Program

Fig. 14.9 BATAN appointed as the IAEA Collaborating Center for Plant Mutation Breeding

Fig. 14.10 IAEA training course on sorghum mutation breeding at BATAN, participants from Malaysia, Malawi, and Togo

14.6 Conclusions

Sorghum is a potential food crop for Indonesia because it can support the national food diversification program to maintain food security. However, sorghum is not of Indonesian origin and the available genetic variation is limited. Attempts to increase genetic variation were made through mutational breeding by treating seeds of the Zhengzu variety (originating in China) with gamma radiation. Optimum doses of gamma radiation for sorghum mutation breeding were found to be about 200–450 Gy. The mutant population derived from these irradiation doses yielded the highest genetic variation in M2. In M5, 15 promising mutant lines were evaluated together with parental and national control varieties in multilocus trials. Agronomic data were collected, analyzed, and reported to the Variety Release Team under the Ministry of Agriculture. Finally, three promising mutant lines ZH-30, PATIR-1, and PATIR-4 were accepted for official release. The ZH-30 mutant line was released in 2013 under the name Pahat, while the PATIR-1 and PATIR-4 mutant lines were released in 2014 under the names Samurai-1 and Samurai-2. These mutant varieties have been widely cultivated by stakeholders in Indonesia, estimated at about 800,000 ha in 2018. Sorghum cultivation has a significant impact on mitigating climate change, helping to maintain sustainable agricultural practices, supporting the food diversification program, and enhancing food security in Indonesia.

Acknowledgments I would like to thank BATAN for providing research budget to the sorghum mutation breeding program. Also, I would like to thank IAEA for supporting sorghum mutation breeding through some IAEA TC projects including INS5030, INS5039, RAS5040, RAS5045, RAS5056, RAS5070, RAS5077, RAS5079, and CRPs.

References

Aznur F, Suwarto, Purnawati H (2017) Efisiensi penggunaan cahaya matahari dan partisi karbohihrat tanaman sorgum pada berbagai tingkat pemupukan nitrogen. J Agron Indonesia 45(3):278–284. ISSN 2085-2916 [in Indonesian]

BATAN Report (2020) Pemuliaan tanaman dengan teknik mutasi radiasi. https://drive.google.com/file/d/1cmWUx-Qrz-yPoVSlV6i1D6p7VwwgxsCD/view. In Indonesian language

Human S (2015a) Sorghum mutant varieties: from research to market. Joint FAO/IAEA Program, Plant Breeding and Genetics Newsletter, No. 34, January 2015

Human S (2015b) Sorghum mutation breeding for improving tolerance to abiotic stresses brought about by climate change. Joint FAO/IAEA Programme. FAO/IAEA Website. http://www-naweb.iaea.org/nafa/news/pbg-sorghum-mutation.html

Human S, Sihono (2010) Sorghum breeding for improved drought tolerance using induced mutation with gamma irradiation. J Agron Indonesia 38(2) ISSN 2085-2916

Human S, Andreani S, Sihono, Indriatama WM (2011) Stability test for sorghum mutant lines derived from induced mutations with gamma-ray irradiation. Atom Indonesia J 37(3) ISSN 1226-1568

King J, McKegg K, Arau A, Schiff A, Garcia Aisa M (2020) Social and Economic Impact Assessment of the RCA Programme: Mutation breeding case study. International Atomic Energy Agency, Vienna, Austria

Nur A, Human S, Trikoesoemaningtyas (2014) Keragaman genetik gandum populasi mutan M3 di agroekosistem tropis. Jurnal Aplikasi Isotop dan Radiasi 10(1) ISSN 1907-0322. [In Indonesian]

Richards AJ (1997) Plant breeding system, 2nd edn. Chapman & Hall, London. ISBN 0-412-57440-3

Suminar R, Suwarto, Purnawati H (2017) Pertumbuhan dan hasil sorgum di tanah latosol dengan aplikasi dosis pupuk nitrogen dan fosfor yang berbeda. J Agron Indonesia 45(3):271–277. ISSN 2085-2916. [In Indonesian]

Suryati N, Amin Z, Andry A, Humaidi E (2019) Pendapatan petani padi varietas hasil iradiasi Badan Tenaga Nuklir. Jurnal Penelitian Terapan 19(3)

Trikoesoemaningtyas, Wirnas D, Saragih EL, Rini EP, Sari M, Arwiyah S, Sopandie D (2017) Kendali genetik karakter morfologi dan agronomi pada tiga populasi sorgum (*Sorghum bicolor* (L.) Moench) populations. J Agron Indonesia 45(3):271–277. ISSN 2085-2916. [In Indonesian]

Van den Berg L, Zeng YJ (2006) Response of South African indigenous grass species to drought stress induced by polyethylene glycol (PEG). S Afr J Bot 72(2)

Van Harten AM (1998) Mutation breeding: theory and practical applications. Cambridge University Press, Cambridge. ISBN: 0-521-47074-9

Potential of Mutation Breeding in Genetic Improvement of Pulse Crops

15

Souframanien Jegadeesan and Dhanasekar Punniyamoorthy

Abstract

Pulse crops are the primary source of dietary proteins for the predominantly vegetarian population of the country and complement the carbohydrate-rich cereals in ensuring nutritional security. Unlike conventional breeding, mutation breeding through induced mutagenesis presents a powerful tool for creating variability and offers scope for selection and genetic improvement of pulse crops with narrow genetic base. Through mutation breeding, 331 pulse crop varieties have been released worldwide, of which the highest of 122 have been released from India (excluding legumes like soybean and groundnut). In India, maximum number of mutant pulse varieties have been released in mung bean (34) followed by black gram (18), cowpea (16), chickpea (12), pigeon pea (12), moth bean (11), Lablab (6), lentils (4), horse gram (6), peas (2) and common bean (1). Preponderance of gamma ray-induced mutant pulse crop varieties exemplifies the potential of the physical mutagen in varietal development. Apart from broadening the genetic base of pulse crops, induced mutagenesis is increasingly being used as a tool for perusing functional genomics and, in conjunction with new breeding techniques, offers scope for inducing site-specific targeted mutations. In this chapter, we discuss the potential of mutation breeding in genetic improvement of some important pulse crops.

Keywords

Mutation breeding · Pulse crops · Mutagens · Gamma rays · Mutagenesis · Variability · Mutant varieties · Gamma rays

S. Jegadeesan (✉) · D. Punniyamoorthy
Nuclear Agriculture and Biotechnology Division, Bhabha Atomic Research Centre, Mumbai, Maharashtra, India
e-mail: souf@barc.gov.in; sekar@barc.gov.in

S. Penna, S. M. Jain (eds.), *Mutation Breeding for Sustainable Food Production and Climate Resilience*, https://doi.org/10.1007/978-981-16-9720-3_15

15.1 Introduction

Pulse crops are legume crops harvested exclusively for the dry seeds and exclude those which are used primarily as vegetables (green beans, green peas) or for oil extraction (soybeans and peanuts) or as forages (clovers, alfalfa). These fabaceous crops are indispensable elements of Indian agriculture as they are the primary source of dietary proteins for the predominantly vegetarian population of the country. They also form the integral component of the subsistence farming and contribute to sustained income of resource-poor farmers apart from ensuring nutritional security. The important pulses cultivated in India include chickpea (*Cicer arietinum*), pigeon pea (*Cajanus cajan*), mung bean (*Vigna radiata*), black gram (*Vigna mungo*), lentils (*Lens esculenta*) and various beans. Some relatively hardy pulses capable of withstanding low-moisture and high-temperature stresses and commonly referred to as arid legumes include cowpea (*Vigna unguiculata*), moth bean (*Vigna aconitifolia*), cluster bean (*Cyamopsis tetragonoloba*) and horse gram (*Macrotyloma uniflorum*).

As the primary contributor of dietary proteins, pulses or legumes contribute substantially to alleviating protein malnutrition. Rich in lysine and tryptophan amino acids, legumes complement the sulphur-rich amino acids (cysteine and methionine) of cereals in providing a wholesome protein, and therefore, a mixed diet of pulses and cereals commands superior biological value. Pulses are also important in soil amelioration owing to their innate nitrogen-fixing ability, and their comparatively hardy nature enables them to perform relatively better even under poor resources and variable climate. Among the food legumes, chickpea is the major Indian legume contributing 43.3% to the country's total legume production followed by pigeon pea (15.3%), black gram (10.6%) and mung bean (8.6%). In 2018–2019, India imported 15.1 million tonnes of edible oils and 2.6 million tonnes of legumes worth 10.5 billion USD (DACandFW 2019). Annual consumption of these is expected to increase further with increased urbanization, higher disposable incomes and burgeoning population, necessitating more imports. Despite their high economic importance, national productivity levels compare poorly with the rest of the world. Many of the issues attributed to lower productivity were narrow genetic base, lack of tolerance to biotic and abiotic stresses, lack of quality seeds of improved varieties and restriction of these crops on marginal areas with poor inputs. The low productivity further aggravates the widening demand-supply gap entailing import of pulses leading to great economic loss to the exchequer.

Induced mutagenesis is one of the important tools to induce genetic variability, which is the basis for any crop improvement programme. Mutations are the ultimate sources of genetic variation and provide the raw material upon which other factors of evolution act. Mutagenesis is predominantly carried out not only in autogamous diploids like legumes or allopolyploid crops and ornamentals, but also to some extent in allogamous crops in which, however, more methodological difficulties need to be conquered. Mutagenesis in plants can be achieved by using physical (ionizing and non-ionizing radiations) or chemical agents. However, several practical problems with chemical mutagens have been identified, which include soaking of seeds, penetration of the relevant target cells, safety of handling, and disposal among many others (Micke et al. 1990). Among the radiation-based methods, gamma irradiation is known to be the most effective in inducing a wide range of mutations (Bado et al. 2015), due to their deeper penetration power (Mba et al. 2012) and less destructive nature (Sikora et al. 2011). By induced mutations, 3365 mutant varieties were registered in more than 240 different crop and plant species. Among these, the largest number of varieties released is in China—817, followed by Japan—479 and India—415 (updated by adding 74 pulse crop varieties). As per the Mutant Variety Database of the IAEA, globally, 244 mutant pulse crop varieties have been released in 17 species (IAEA-MVD 2021). From various authentic online resources, we have now updated the total number to 331. By developing 122 pulse mutant varieties in 11 species, India (Table 15.1) stands first followed by Pakistan and the Russian Federation (30 each), Bangladesh (29), the USA (26) and Poland (23). In this chapter, we discuss the potential of mutation breeding in varietal development and the impact of mutant varieties in some important pulse crops of relevance to India, especially chickpea, pigeon pea, mung bean, black gram and cowpea.

15.2 Genetic Bottlenecks

The appalling genetic potential of pulses could be primarily ascribed to their inherent narrow genetic base that constrains their genetic improvement in making superior selections. The domestication events capture only a small fraction of the genetic diversity (Ladizinsky 1985), and subsequently, the gene flow between the domesticated cultivars and its wild ancestors is restricted by the breeding system (Cooper et al. 2001). In pulse crops like chickpea, crop introduction followed by natural selection has resulted in *desi* and *kabuli* ecotypes (Kumar et al. 2004). The small population sizes emanating from either novel introductions or crop losses could consequently lead to random genetic drift and eventually loss of genetic diversity. The lack of diversity could also have stemmed from previous breeding practices. For instance, in pulses, the lineage of released varieties showed predominance of a small number of parents or high degree of relatedness between parents. In pigeon pea, 57 parental genotypes were used to develop 47 varieties through hybridization, while in urdbean, a total of 70 varieties were released involving only 26 parental genotypes. The genotypes 'T 1' and 'T 190' were involved in 34% of the pigeon pea varieties, while 'T 9' and 'T 1' appeared in 64% and 35% of urdbean and mung bean varieties, respectively (Kumar et al. 2004). Such extensive and repetitive use of superior genotypes with common ancestors delineated the narrow genetic base of the released varieties, and therefore, broadening the genetic base is vital for sustainable production in major pulse crops.

Table 15.1 List of mutant pulse crop varieties released in India for commercial cultivation

S. no.	Variety name	Year of release	Mutagen/dose	Direct/derived	Improved trait
	Black gram (18)				
1	Co 4	1978	MMS (0.02%)	Mutant of Co-1	Tolerant to leaf crinkle virus, tip blight, PM and less susceptible to stem fly and pod borer
2	TAU 1	1985	Gamma rays	Derived (T-9 × UM-196*)	High yield, resistance to PM and larger seed size
3	Sarala (B-12-4)	1985		Mutant of T-9	MYMV resistant
4	TPU-4	1992	Gamma rays	Derived (UM-201* × T-9)	High yield and high seed weight
5	TAU-2	1992	Gamma rays	Derived (T-9 × UM-196*)	Large seeds
6	TU-94-2	1999	Gamma rays	Derived (TPU-3 × TAU-5*)	High yield and resistance to YMV disease
7	DU-1	2007	Gamma rays (200 Gy)	Irradiation of F_1 TAU-1* × No. 169	High grain yield and resistance to insects
8	Vamban 2	1997	Spontaneous mutant	Mutant of T-9	High yield, early maturity with tolerance or resistance to MYMV
9	Manikya	1988	Gamma rays (400 Gy)	Mutant of T-9	High yield, early maturity, large seed size
10	PKV Udid-15 (AKU-15)	2001	–	Derived (TPU-4* × DU-4)	Tolerant to PM and dry root rot
11	VBN (bg) 4	2003	MMS	Derived (CO 4* × PDU 102)	Resistant to YMV
12	GU-1	2004	–	Derived (RBU-28 × TPU-4*)	Moderately resistant to PM and *Cercospora* leaf spot
13	TU-40	2013	Gamma rays	Derived (TU94-2* × *V. mungo* var. *silvestris*)	Resistant to YMV and PM, suitable for rice fallows
14	DBGV-5	2014	–	Derived (TAU-1* × LBG-20)	Moderately resistant to PM
15	Indira Urd (Pratham)	2016	–	Derived (PDU-1 × TU-94-2*)	Resistant to PM up to podding stage

16	PDKV Blackgold (AKU 10-1)	2016	–	Derived (TAU-1* × AKU 18-1)	Tolerant to MYMV and PM
17	Prasad (B 3-8-8)	2005	–	Mutant of Type-9	Resistant to YMV
18	Ujala (OBG-17)	2005	–	Mutant of B 3-8-8	Moderately resistant to YMV, *Cercospora* leaf spot and PM
	Chickpea (12)				
19	Kiran (RSG-2)	1984	Fast neutrons, 4.5×10^{12} n/cm	Mutant of RS-10	Erect plant type, increased pod number, high yield, early and salt tolerant
20	Pusa 408 (Ajay)	1985	Gamma rays (600 Gy)	Mutant of G-130	High yield, blight resistant, semi-erect, 140–155 days to maturity and improved plant architecture
21	Pusa 413 (Atul)	1985	Gamma rays (600 Gy)	Mutant of G-130	High yield, wilt resistance, resistance to stunt virus, foot root, root rot, semi-erect, 2 seeds per pod, 130–140 days to maturity and plant architecture
22	Pusa 417 (Girnar)	1985	Gamma rays (600 Gy)	Mutant of BG-203	High yield, profuse branching, high pod number, 110–130 days to maturity, wilt resistance, low pod borer and nematode damage
23	WCG1 (Sadbhavana)	1996	–	Mutant of C235	Resistant to dry rot and foot rot, moderate resistance to stunt, wilt/root rot and pod borer, protein content 23.7%
24	WCG2 (Surya)	1999	–	Mutant of G130	Resistant to rot, tolerant to stunt and dry root rot
25	WCG10 (Pant G-10)	2001	–	Mutant of G130	Resistant to root rot, moderately resistant to stunt virus, wilt and dry root rot
26	RS11	1968	–	White-flowered mutant of RS10	Suitable for irrigated conditions
27	WCG3 vallabh Kallar chana 1	2008	–	Mutant of C235	Suitable for saline soils, bold seeded, protein content 22.8%
28	BGM 547	2005	Gamma rays (300 Gy)	Mutant of BG 256	High yield, bold grain size, attractive golden brown colour and moderate resistance to wilt, root rot, stunt and Heliothis

(continued)

Table 15.1 (continued)

S. no.	Variety name	Year of release	Mutagen/dose	Direct/derived	Improved trait
29	Pusa 547	2006	Gamma rays (300 Gy)	Mutant of BG 256	High yield, good cooking quality, tolerance to *Fusarium* wilt, stunt virus and root rot
30	PKV Harita	2012	–	Derived (121-1* × AKG-46)	Green seed coat, semi-spreading
	Cowpea (16)				
31	V16 (Amba)	1981	DMS	Mutant of Pusa Phalguni	High yield, resistance to fungal and bacterial diseases
32	V37 (Shreshtha)	1981	DMS	Mutant of Pusa Phalguni	High yield, luxuriant vegetative growth
33	V38 (Swarna)	1981	DMS	Mutant of Pusa Phalguni	High yield, early maturity, synchronous flowering, better quality pods and grains, resistance to bacterial blight, *Macrophomina* and PM
34	V240	1984	DMS	Mutant of Pusa Phalguni	High yield, resistance to fungal, viral and bacterial diseases, tolerant to moisture stress
35	GC-1	1984	–	Derived	Yield, early maturity, root knot resistance
36	Co-5	1986	Gamma rays	Mutant of Co-1	High yield, suitable for forage
37	Cowpea-88	1992	Gamma rays	Irradiation of F1 (COWPEA-74 × H-2)	Large seed size, resistant to YMV and anthracnose
38	KBC-2	1999	–	Mutant of V-16	Resistant to rust
39	PUSA SAMPADA (V-585)	1999	–	Mutant of Pusa Phalguni	Erect, bushy, multiple disease resistant
40	COCP 702 [CoVu 702 and CO(CP) 7]	2002	Gamma rays (200 Gy)	Mutant of Co-4	High yield and good quality
41	Co (FC) 8	2004	Gamma rays	Derived (Co-5* × N331)	High green fodder, indeterminate type of growth, resistant to cowpea YMV and root rot
42	V578 (Pusa-578)	2004	Gamma rays	Mutant of EC 170578	Resistant to YMV, determinate

43	TRC77-4 (Kalleshwari)	2007	Gamma rays (200 Gy)	Mutant of V-130	Semi-dwarf, high yield, suitable for rice fallows, dual purpose
44	Co-9	2016	Gamma rays	Derived (CO 5* × Bundel Lobia 2)	Higher protein content and reduced fibre confer increased digestibility, palatability and intake rate; MR to YMV and resistant to major pests
45	TC-901	2018	Gamma rays (250 Gy)	Mutant of EC394763	Semi-determinate mutant, high yielding, increased seed size, pod size and pod number, resistance to cowpea mosaic and root rot diseases
46	DCS-6	–	–	Mutant of KBC-1	–
	Mung bean (34)				
46	Dhauli (TT9E)	1979	–	Derived	High yield, early maturity with tolerance or resistance to YMV
47	Co 4	1982	Gamma rays (200 Gy)	Mutant of Co-1	High yield, early maturity and resistance to drought
48	ML 26-10-3 (Pant Moong 2)	1983	Gamma rays (100 Gy)	Mutant of ML-26	Resistant to YMV and high yield
49	TAP-7	1983	Gamma rays	Mutant of S-8	Early maturity (5–7 days), resistance to mildew and leaf spot, higher yield (23% over Kopargoan)
50	MUM-2	1992	EMS (0.2%)	Mutant of K-851	Dwarf and shining green and medium-large seeds
51	BM 4	1992	EMS	Mutant of T-44	High yield, early maturity with tolerance or resistance to MYMV
52	LGG410	1994	–	Mutant of ML 26-10-3	Resistant to YMV
53	LGG 450 (Pushkara)	1995	Gamma rays (400 Gy)	Mutant of Pant Moong 2	High yield, early maturity with tolerance or resistance to MYMV, PM
54	LGG-407 (LAM-407)	1995	Gamma rays (400 Gy)	Mutant of Pant Moong 2	High yield, early maturity with tolerance or resistance to MYMV, PM
55	TARM-2	1994	Gamma rays	Derived (RUM 5* × TPM-1)	High yield, medium late maturity and resistance to PM disease

(continued)

Table 15.1 (continued)

S. no.	Variety name	Year of release	Mutagen/dose	Direct/derived	Improved trait
56	TARM-18	1996	Gamma rays	Derived (PDM54 × TARM2*)	High yield and resistance to PM disease
57	TARM-1	1997	Gamma rays	Derived (RUM-5* × TPM-1)	High yield, resistance to PM disease and medium maturity
58	K1	1998	Gamma rays	Derived (Co4* × ML 65)	Tolerant to drought
59	PBM-1	1998	–	Mutant of ML 131	Resistant to YMV
60	OUM 11-5 (Kamadeva)	2002	–	Mutant of Dhauli	High yielding, moderately resistant to MYMV and CLS
61	RMG-492	2003	–	Mutant of RMG-62	Resistant to web blight and CLS, moderately resistant to YMV and *M. incognita*
62	Meha (IPM99-125)	2004	Gamma rays	Derived (Pant Mung 2* × AMP36)	Small shining seeds, resistant to MYMV
63	TMB-37	2005	Gamma rays	Derived (Kopargaon × TARM-2*)	Early maturing (55–57 days), YMV resistant, high yielding and large seed size
64	TM-96-2	2007	Gamma rays	Derived (Kopargaon × TARM-2*)	PM disease resistance, early maturity, resistance to *Corynespora* leaf spot and suitable for rice fallow cultivation
65	TJM-3	2007	Gamma rays	Derived (Kopargaon × TARM-2*)	YMV resistance, resistance to PM, *Rhizoctonia* root rot disease resistance, early maturity and large seeds
66	Pant Moong 6 (UPM 02-17)	2008	Gamma rays	Derived (Pant Moong 2* × AMP36)	Resistant to MYMV, BLS, CLS and leaf crinkle
67	VBN(Gg)3	2009	Gamma rays	Derived (K1* × Vellore local)	Resistant to MYMV, PM
68	IPM02-3	2009	Gamma rays	Derived (Pusa Vishal × IPM99-125*)	Resistant to MYMV
69	IPM 02-14 (Shreya)	2011	Gamma rays	Derived (IPM99125* × Pusa Bold 2)	Resistant to MYMV and leaf crinkle

70	TM 2000-2 (Pairy mung)	2010	Gamma rays	Derived (JL781 × TARM-2*)	Resistant to PM, synchronous podding and maturity, suitable for rice fallows
71	PKV AKM 4 (AKM-9904)	2011	EMS	Derived (BM4* × PS7)	–
72	DGGV-3	2013	–	Derived (GG4 × TM98-50*)	–
73	DGGV-2	2014	–	Derived (Chinamung × TM98-50*)	Moderately resistant to PM, moderately tolerant to Apion beetle
74	DGGS4	2014	–	Derived (selection 4 × TM 98-50*)	
75	MSJ118 (Keshwanand Mung 2)	2016	–	Mutant of K-851	Moderately resistant of MYMV
76	Gujarat Mung-7 (GM-7)	2018	–	Meha × GM 4	Resistant to YMV
77	Tripura Mung 1 (TRCM131)	2018	–	SPS 5 × IPM 99-125	Resistant to MYMV
78	Pant M 9 (PM09-11)	2019	–	Derivative (PM5 × Bina Mung*)	Resistant to MYMV
79	IPM512-1 (Soorya)	2020	–	Derived (IPM99-125* × Co-5)	Resistant to MYMV, *Cercospora* leaf spot and anthracnose
	Pigeon pea (12)				
81	Co 3	1977	EMS (0.6%)	Mutant of Co-1	High yield, bold seeded, higher shelling, field dormancy for 15–20 days
82	Trombay Vishakha-1	1983	Fast neutrons	Mutant of T-21	Increased seed size with all desirable traits of parent variety T-21
83	Co 5	1984	Gamma rays (160 Gy)	Mutant of Co-1	Early maturity, photoperiod insensitivity and drought tolerance
84	TAT 5	1984	Fast neutrons	Mutant of T-21	Increased seed size (50%), high TGW and early maturity (140 days)

(continued)

Table 15.1 (continued)

S. no.	Variety name	Year of release	Mutagen/dose	Direct/derived	Improved trait
85	TAT 10	1984	Fast neutrons (25 Gy)	Derived (TT2* × TT9*)	Medium-large-size grain and extra early maturity (115–120)
86	Co-6	1991	Gamma rays (250 Gy)	Mutant of SA-1	Photo-insensitive, tolerant to pod borer
87	COPH-2	1997	Gamma rays	Derived (male sterile line of Co-5*)	Hybrid suitable for both irrigated and rainfed conditions
88	CO(RG)-7	2004	Fast neutrons	Derived (ICP 8863 × AL 101) × (PA 1218 × TT6*)	Early, tolerant to wilt, sterility mosaic and *Phytophthora* blight
89	TT-401	2007	Fast neutrons	Derived (ICPL84008 × TT-6*)	High yield, tolerance to *Fusarium* wilt, pod borer and pod fly damage
90	TJT-501	2009	Fast neutrons	Derived (ICPL84008 × TT-6*)	Resistant to *Fusarium* wilt and tolerant to *Phytophthora* blight
91	PKV-TARA	2013	Fast neutrons	Derived (ICPL84008 × TT-6*)	Tolerant to *Fusarium* wilt and sterility mosaic diseases
92	Co-9	2019	Gamma rays	Derived (CO 6* × IC 525427)	Bold seed (100-seed weight 9.9 g), moderately resistant to wilt and SMD diseases, moderately resistant to *Maruca* and pod fly
	Common bean (1)				
93	Pusa Parvati	1970	X-rays (70 Gy)	Mutant of EC1906	Early maturity (40–45 days), bushy type with attractive round meaty light-green pods, high yield (45% more)
	Lentil (4)				
94	B-256 (Ranjan)	1981	X-rays	Mutant of B-77	High yield and spreading type
95	Arun (PL 77-2)	1986	–	–	Spreading type, high yielding, tolerant to wilt and *Ascochyta* blight

96	Rajendra Masoor 1	1996	Gamma rays (100 Gy)	Direct	Tolerance to low temperatures, early maturity and good for late sowing
97	Malaviya Vishwanath (HUL-57)	2005		Mutant of HUL-11	Resistant to wilt and tolerant to rust
	Moth bean (11)				
98	RMO-40	1994	Gamma rays (400 Gy) + 0.1% EMS	Mutant of Jwala Moth	Extra early maturing, tolerance to drought, short stature, non-spreading variety with synchronous maturity
99	CAZRI Moth –3 (CZM-99)	2005	–	Mutant of RMO-40	Resistant to YMV and root rot diseases, drought tolerant
100	RMO-423	2004	Gamma rays (600 Gy)	Mutant of RMO-40	Tolerant to pests and diseases, early and high yielding
101	CAZRI MOTH-1 (CZM-79)	1999	Gamma rays (400 Gy) + 0.6% EMS	Mutant of Jadia cultivar	Resistant to drought and heat
102	CAZRI MOTH-2 (CZM-45)	2006	Gamma rays (400 Gy) + 0.1% EMS	Derivative (Jadia × RMO-40*)	Tolerant to YMV and root rot
103	MARU VARDHAN (RMO-225)	1999	Gamma rays (400 Gy) + 0.6% EMS	Mutant of Jadia cultivar	Early and high yielding
104	FMM-96	1997	Gamma rays (300 Gy)	Mutant of Jadia cultivar	Extra early, highly tolerant to yellow mosaic, stem blight and red leaf spot
105	JMM-259 (Maru-Moth-1)	1988	Chemical mutagenesis	Mutant of Jadia cultivar	Tolerant to YMV
106	RMO-257	2005	Gamma rays (300 Gy) + 0.6% EMS	Mutant of Jadia cultivar	Early maturing, spreading type, suitable for rainfed areas

(continued)

Table 15.1 (continued)

S. no.	Variety name	Year of release	Mutagen/dose	Direct/derived	Improved trait
107	Marudhar (RMO-2251)	2018	Gamma rays	Direct mutant	Early maturing, compact plant type
108	Maru Bahar (RMO-435)	2002	Gamma rays (600 Gy)	Mutant of RMO-40	Early maturing, semi-spreading
	Pea (2)				
109	Hans-2	1979	EI	Mutant of P-1163	Early maturity, high yield and better seed quality
110	JP-179	1987	–	Derived (LOCAL BROWN BARRIS × LINE 6588-1 × MUTANT 46 C)	PM resistant
	Lablab (6)				
111	CO-9	1980	Spontaneous mutant	Mutant of Co-6	Bush variety
112	CO-10	1983	Gamma rays (240 Gy)	Mutant of CO-6	Bush variety
113	CO-11	1990	Spontaneous mutant	Derived (CO 9* × CO 4)	Early maturing
114	CO-12	1991	Spontaneous mutant	Derived (CO 9* × CO 4)	Erect, bushy and tendency to form tendrils
115	CO-13	1997	Spontaneous mutant	Derived (CO 9* × Florikifield)	Erect, bushy and tendency to form tendrils
116	CO(GB)14	2007	Spontaneous mutant	Derived (CO 9* × CO 4)	Dwarf, bushy without tendrils
	Horse gram (6)				
117	Cridaharsha CRGH-19	2014	Gamma rays	Mutant of K42	Tolerant to PM, anthracnose and whiteflies, high protein content

118	CRIDALATHA (CRHG-4)	2010	Gamma rays	Mutant of Hyderabad Local	Tolerant to YMV, PM, anthracnose and mites
119	CRIDA-18R	2009	Gamma rays	Mutant of K42	Early, tolerant to YMV, PM and mites
120	Paiyur-2	1998	Gamma rays (390 Gy)	Mutant of Co-1	Medium growth, tolerant to mosaic
121	BIRSA Kulthi 2	–	–	Mutant of BIRSA Kulthi 1	–
122	BIRSA Kulthi 3	–	–	Mutant of BIRSA Kulthi 1	–

PM powdery mildew, *YMV* yellow mosaic virus, *MYMV* mung bean yellow mosaic virus, *CLS Cercospora* leaf spot
Note: *: mutant, –: not available

15.3 Approaches for Broadening the Genetic Base

The major objective of any crop improvement programme involves development of biotic and abiotic stress-tolerant high-yielding crop varieties, the success of which critically depends on the extent of genetic variation available for selection. For broadening the narrow genetic base of legumes, it is paramount to exploit new and diverse sources of variation. Hybridization, recombination and mutation (spontaneous and induced) contribute to creating variability. More variable landraces, exotics and wild relatives that are stockpiles of novel genes have been exploited through wide hybridizations and pre-breeding efforts in some pulse crops (Di Vito et al. 1996). However, their utilization in crop improvement is largely limited owing to reproductive isolation, genetic incompatibilities and linkage drags. Among all the breeding methodologies adopted, mutation breeding has become a proven tool for creating variation within a crop variety and offers the possibility of inducing novel attributes that either cannot be found in nature or have been lost during evolution.

15.4 Mutation Breeding

Mutation breeding is the process of inducing mutations (mutagenesis) to generate variability and utilizing this variability to detect variants/mutants with desirable traits for breeding better crop varieties. More like the conventional breeding and unlike the genetically modified organisms (GMOs), no regulatory restrictions are imposed on mutation breeding (Parry et al. 2009). Mutation breeding is all about mutation induction (exposing seeds or any plant propagules to chemical(s) or radiation (s) primarily to induce variability for the trait of interest), mutation detection (mutant screening from a large mutated population) and mutant confirmation (putative screened-in mutants are evaluated for their true breeding nature under controlled or replicated environments through progeny testing) for crop improvement. Selecting a mutagen and its optimal dose for irradiating any genotype in any plant species is an important step in mutation breeding programme. Radio-sensitivity assays for physical mutagens are necessitated to ascertain the optimal dose for inducing useful mutations with least possible unintended damage. In comparison to cereals, the success saga of mutation breeding in grain legume improvement is relatively limited, albeit the latter housing majority of autogamous crop species and mutagenesis being predominantly suitable for autogamous diploids. There is a tremendous scope to augment the application of induced mutations in genetic improvement of pulses.

15.5 History of Mutation Breeding in Pulses

The application of mutation breeding in pulses commenced post-demonstration of mutation induction by X-rays in barley by Stadler (1928). The first legume variety through mutation breeding was released in groundnut (N.C. 4-X) in the USA way back in 1959 (Gregory 1955) using X-rays. Since then, nearly 408 mutant varieties of legume crops belonging to 24 different crop species have been released for cultivation worldwide (Kharkwal et al. 2005). In India, a total of 122 pulse crop varieties (excluding peanuts and soybean) as enlisted in Table 15.1 have been developed through mutation breeding. Crop improvement using classical mutagenesis is now well standardized, and as a result, new methods of radiation treatment, as well as chemical agents with mutagenic properties, are serving as invaluable tools for augmentation of the genetic variation in pulse crops. The useful mutations of spontaneous or induced origin are being exploited in crop improvement.

15.6 Spontaneous Mutations

Spontaneous mutations arise naturally from a variety of sources, including errors in DNA replication, spontaneous lesions, transposable genetic elements or as a result of background radiations, without any deliberate mutagen intervention. Such mutants are treated akin to induced mutants in varietal development. The Lablab variety 'Co-9', supposedly the first spontaneous mutant pulse variety, was later on used to derive a series of mutant varieties ['Co-11', 'Co-12', 'Co-13', 'Co(GB)14'] through hybridization with other genotypes. In black gram, a spontaneous mutation resulted in the direct release of mutant variety 'Vamban-2' in 1997. In pulses, only 2% (7) of the mutant varieties have been developed by exploiting spontaneous mutants (Tables 15.1 and 15.3; Figs. 15.1, 15.2 and 15.3).

15.7 Induced Mutations

Majority of the pulse mutant varieties developed across the world are through induced mutagenesis. Both physical and chemical mutagens have been used apart from a combination of both. The data of released mutant varieties shows that 54% (180) of them have been developed through physical mutagens, 20% (65) through chemical mutagens and 2% (8) through a combination of both (Table 15.2; Figs. 15.1, 15.2 and 15.3). Among the physical mutagens, gamma rays have been primarily used to release maximum varieties (141) followed by X-rays (29). Fast neutrons have been used exclusively in India to release 8 mutant varieties in pulses. Among the chemical mutagens, maximum varieties were developed through EMS (32), followed by EI (7), DMS (4), NMU (3) and MMS (2) (Table 15.3).

15.7.1 Physical Mutagenesis

Physical mutagens, namely UV-rays, X-rays, gamma rays, beta particles, and fast and thermal neutrons, have been primarily used for the induction of mutation in various crops including pulses (Saima Mir et al. 2020). Physical mutagens are more commonly preferred by breeders as compared to the chemical mutagens. The history

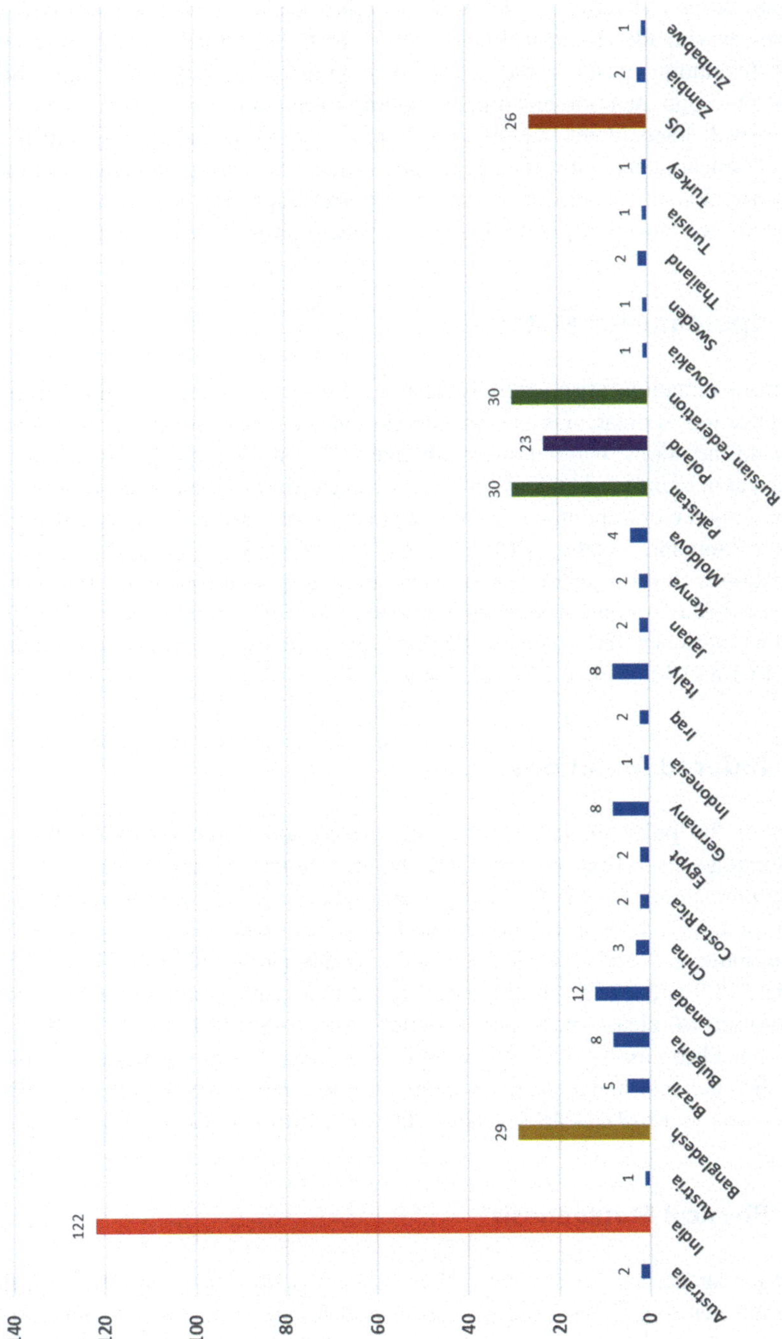

Fig. 15.1 Contribution of different countries in the development of mutant pulse crop varieties

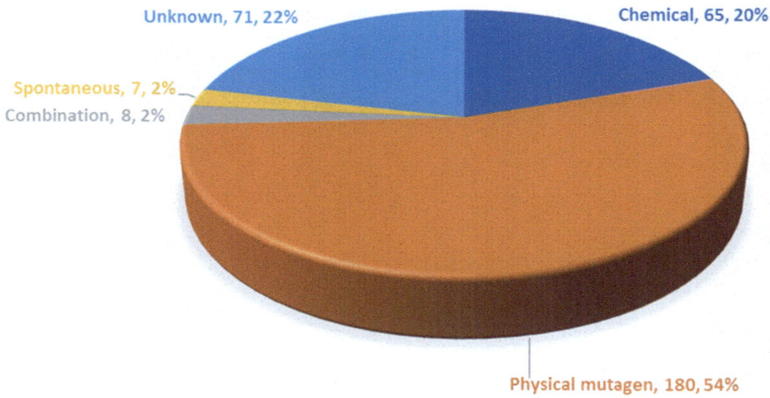

Fig. 15.2 Use of different mutagens in the release of pulse crop varieties worldwide

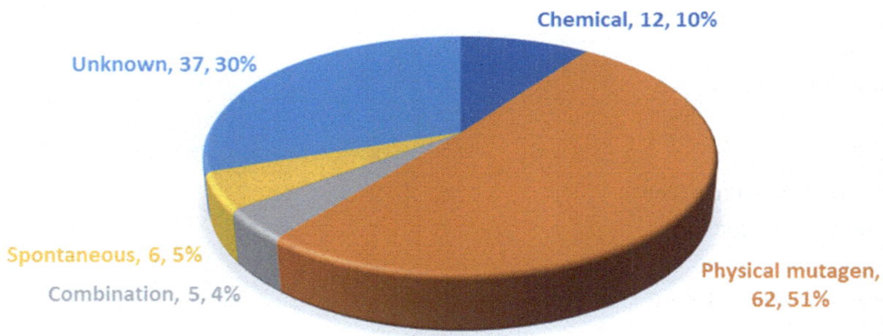

Fig. 15.3 Use of different mutagens in the release of pulse crop varieties in India

Table 15.2 Contribution of different mutagen types to the release of mutant pulse crop varieties

Mutagen category	Number of mutant pulse crop varieties		
	India	Rest of world	Worldwide
Chemical	12	53	65
Physical mutagen	62	118	180
Combination (physical + chemical)	5	3	8
Spontaneous mutation	6	1	7
Unknown	37	34	71
Total	122	209	331

of mutant varieties in pulses showed that the initial phase of mutation breeding in early 1970s was largely based on X-rays, which shifted to fast neutrons in 1980s and were widely replaced with gamma rays thereafter. A rapid glance through Table 15.2 shows that the majority of mutant varieties were developed through gamma rays. Fast neutrons have predominantly been used in crops like chickpea and pigeon pea

Table 15.3 Contribution of different mutagens to the release of mutant pulse crop varieties

Mutagen	Rest of world	India	Worldwide
EMS	28	4	32
DMS	0	4	4
MMS	0	2	2
NMU	3	0	3
EI	6	1	7
MNH	1	0	1
dES	1	0	1
NEU	1	0	1
NEU + NMU	1	0	1
NEU + NMU + DES + DMS + EI	1	0	1
NMH	1	0	1
NMU + EI	1	0	1
NMU + EI + DMS	1	0	1
Datura extract	1	0	1
Unknown chemical mutagen	7	1	8
Unknown physical mutagen	1	0	1
Gamma rays	88	52	140
X-rays	27	2	29
Physical + chemical mutagen	3	5	8
Spontaneous mutation	1	6	7
Recurrent (gamma)	1	0	1
Recurrent (gamma + X-rays)	1	0	1
Fast neutrons	0	8	8
Not known	34	37	71
Total	209	122	331

for varietal improvement. Physical mutagens induce deletions and chromosomal rearrangements unlike the chemical mutagens that largely induce point mutations.

New effective methods with higher mutation induction ability are highly desired for the success of the mutation breeding. Apart from the conventional electromagnetic radiations, like X-rays and γ-rays, electron beam is now looked upon as an alternative source of energy to induce mutations. Electron beams in contrast to gamma rays are generated using particle accelerators that can be switched on or off (Park and Vestal 2002). Electron beams have very high absorbed dose rate (Zhu et al. 2008) and low penetrating power in comparison to γ-rays. Also, pursuant to their charge and mass, electron beams are limited by their kinetic energy and exhibit higher linear energy transfer (Nishant et al. 2020). In pulses like black gram, electron beams have been observed to induce higher mutation frequency and wider spectrum of mutations than γ-rays (Souframanien et al. 2016). However, Eswaramoorthy et al. (2021) reported gamma rays and electron beam to induce higher biological damage than EMS in cowpea.

The use of ion beams for mutagenesis has gained momentum presently to generate novel mutants as they have been shown to present higher relative biological effectiveness (RBE) and induce both single- and double-strand breaks in DNA. Ion beams have been reported to produce higher mutation frequency and induce wider spectrum of mutants. It is plausible that ion beams induce a limited amount of large and irreparable DNA damage, resulting in the production of a null mutation that shows a new mutant phenotype. They are also known to induce minimum number of mutations compared to other mutagens (Tanaka et al. 2010). China and Japan are pioneers in utilizing ion beams in producing a large number of mutant varieties (Nakagawa 2009; Wu et al. 2005), especially in flowers and ornamentals. Low-energy ion beams have also been used to produce improved crop varieties in rice (Kitamura et al. 2006). Their utility in pulse improvement is yet to be exploited. In cowpea, proton beam showed lower LD_{50} and RD_{50} in comparison to gamma rays. Antioxidant enzymes were found to react more to proton beams, while gamma rays had greater effect on chlorophyll content (Kang et al. 2020).

15.7.2 Chemical Mutagenesis

The discovery of the highly mutagenic nature of some chemicals like mustard gas (Auerbach and Robson 1946) led to the unearthing of a number of chemical agents that increased the frequency of artificially induced point mutations and were categorized as base analogues, acridine dyes, nitrous acid, hydroxylamines, etc. The most commonly used chemical mutagens include methyl methanesulphonate (MMS), ethyl methanesulphonate (EMS), diethyl sulphate (DES), methyl nitrosourea (MNH), ethyl nitrosourea (ENH) and ethyleneimine (EI), all of which belong to a special class of alkylating agents. All these chemicals react with DNA by alkylating the phosphate groups, and also the purine and pyrimidine bases, or react with guanine or thymine by adding an ethyl group which causes the DNA replication machinery to recognize the modified base as an adenine or cytosine, respectively, causing GC-to-AT base pair transitions (Till et al. 2004, 2007). The dose of a chemical mutagen mainly depends upon the concentration, duration of treatment and temperature during treatment. The strong carcinogenic nature of these chemical mutagens and the difficulties and precautions to be observed while handling and disposal of these chemicals generally discourage their use by the breeders. A range of modifying factors like pre-soaking, pH of solution, metallic ions, carrier agents, post-washing, post-drying and storage of seeds greatly influence the efficacy of mutagenesis. In varietal development of pulses, the chemical mutagens like MMS and DMS have been widely used in black gram and cowpea, respectively. EMS has been the choice in the varietal development of mung bean, pigeon pea and moth bean, while EI has been successful in peas. Apart from this, a variety of other chemical mutagens including sodium azide, hydrazine hydrate (Wani et al. 2018), caffeine (Shahwar et al. 2020) and heavy metals like lead and cadmium nitrate (Shahwar et al. 2019) have been used in pulses for induction of mutations.

15.7.3 Space Mutagenesis

A new emerging field of mutagenesis known as space mutagenesis involves induction of mutation in space where a clean environment with strong cosmic radiation, near-zero gravity and weak geomagnetic fields under vacuum provide conducive atmosphere for inducing a novel spectrum of mutations (Liu et al. 2009). Induction of mutation in seeds is carried out in space using returnable satellites, space shuttles or high-altitude balloons and proves advantageous over gamma rays in producing lower physiological damage. China is the pioneer country to have exploited space mutagenesis since 1987 and have developed more than 66 varieties in various crop species like rice, wheat, cotton, sesame, pepper, tomato and alfalfa (Liu et al. 2009). In mung bean, China has released two lodging-resistant dwarf varieties, Zhonglyu No. 8 and Zhonglyu No. 12, through space mutation (Wang and Cheng 2018a, b). Long-pod mutant in mung bean (Qiu et al. 1998) and large-grain mutation in adzuki bean (Shi et al. 2000) are some of the breakthrough mutations induced in pulses through space mutagenesis.

15.8 Optimal Dose and Genotypic Sensitivity

The success of any mutation breeding programme is largely dependent on the selection of appropriate dose for maximizing recovery of useful mutations apart from other factors. The varying sensitivities of different crops or species and even different genotypes within a species to mutagenic treatments (Olasupo et al. 2016) necessitate the need to identify crop-wise or genotype-wise optimal dose. Treatment with optimal doses of mutagen maximizes recovery of useful mutations with least unsolicited damages with above-optimal dose leading to increased mortality while below-optimal doses result in low mutation frequencies. The usual strategy to choose doses based on LD_{50} values (50% lethality) is often argued to be disadvantageous owing to the fact that the desirable mutations could be either lost or disregarded due to plant mortality or poor agronomic performance (Oldach 2011). Therefore, doses causing less than 50% lethality, e.g. LD_{20} or LD_{30}, with 80–70% survivability or less than 50% growth reduction (GR_{50}) are understandably more desirable in autogamous plants like legumes. Maluszynski et al. (2009) also stressed the importance of low doses for mutation induction in high-quality genetic backgrounds.

Radio-sensitivity assays for physical mutagens are carried out with a series of doses to identify the LD_{50} or GR_{50} values prior to determination of optimal dose. Based on the previous experiments, the doses for mutation breeding in different pulse crops have been standardized for gamma rays at Bhabha Atomic Research Centre, Mumbai, India. A dose range of 300–400 Gy is effective in case of mung bean and urdbean, whereas in pigeon pea a dose range of 100–200 Gy is helpful. In cowpea, 200–300 Gy has been found to be efficient, while in chickpea 300–400 Gy has been effective in inducing useful mutations. In recent times, radio-sensitivity assays of different crops for electron beam have been done and LD_{50} for different crops has also been ascertained (mung bean: 500 Gy; urdbean: 400 Gy; chickpea: 300 Gy; and cowpea: 270 Gy) (Souframanien et al. 2020). Radio-sensitivity has also been found to vary with seed testa texture, thickness and seed weight. In cowpea, the LD_{50} for gamma rays with respect to seedling survival was low in accessions with rough seed testa (149–357 Gy), while it was higher in accessions with smooth testa (449–620 Gy). Low-dose (100 Gy) irradiation was observed to increase the plant vigour in M_1 seedlings, while progressive reductions were detected at higher doses (Olasupo et al. 2016). In soybean, the optimal dose for proton beam irradiation was reported to be between 250 and 300 Gy (Im et al. 2017). For inducing morphological mutations, EMS doses in the range of 0.01–0.8% have been stated to be beneficial. In cowpea, the LD_{50} value for EMS has been determined to be 50 mM (Yaseen et al. 2020).

Various factors like native plant species, target tissue, water content in the target tissue, temperature, oxygen level and dose rate influence the consequence of induced mutations. Water content of seeds critically affects mutation frequency, and seeds with less than 12–14% moisture usually yield higher mutation frequency. Similarly, for chemical mutagenesis, pre-soaking of seeds in water for 8–12 h and post-treatment drying are very critical. Likewise, the handling of different mutant generations also plays a key role in mutation breeding. The mutagen-treated seeds, referred to as the first generation (M_1), do not express the often-recessive heterozygous mutations. The predominance of lethality in this generation also affects the M_2 population size and hence the mutant recovery. The attaining of homozygosity of the mutant allele usually in M_2 generation makes it very critical as most of the mutations would be visually detectable for selection.

15.9 Mutagenic Effectiveness and Efficiency

The optimal dose of mutagen is dependent on the effectiveness as well as efficiency of the mutagens. Effectiveness refers to the number of mutations per unit dose, while efficiency pertains to the ratio of specific desirable mutations to undesirable effects like height reduction, chromosomal breakages, sterility or lethality. The rate of biological damages like the seedling injury and pollen sterility increases with increasing mutagenic doses in comparison to mutation frequency, and hence, efficiency of mutagens at lower doses or concentrations is presumably high (Konzak et al. 1965). Various contrasting reports on the efficiency and effectiveness of different mutagens could be attributed primarily to genotypic differences. Gamma rays were found to be more efficient than EMS in inducing chlorophyll mutations in black gram and soybean (Vanniarajan et al. 1996; Khan and Tyagi 2010; Deepalakshmi and Kumar 2003). In mung bean, MNNG, EMS and HA were more effective than gamma rays, while gamma rays were more efficient than chemical mutagens (Grover and Virk 1984). In chickpea, HZ and SA were more effective than EMS (Khan et al. 2005). Dhanavel et al. (2008) reported decrease in effectiveness with increasing concentrations of EMS, DES and SA in cowpea. Kharkwal (1998a), while studying the comparative effectiveness and efficiency of gamma rays, fast neutrons, NMU and EMS in chickpea, established that NMU was the most potent mutagen, EMS was the least efficient and gamma rays were the least effective.

15.10 Mutant Screening

In mutation breeding, the larger the M_2 population, the higher the probability of success rate. M_2 population has to be critically observed at different stages of growth and development to isolate desirable mutants. Morphological mutants could be identified visually by screening large population. For identification of disease-resistant mutants, it becomes imperative to grow the M_2 population in hotspots, or alternatively, they could be screened under artificial epiphytotic conditions like infector rows or spraying of inoculum. Often, certain complex mutation events including quantitative traits segregate at higher generations like M_3 or M_4, and therefore, screening should be suitably delayed to later generations. The agronomic performance of these mutants frequently needs improvement if found otherwise by repeated back-crossing or through suitable hybridization programmes (Kharkwal et al. 1988; Rheenen et al. 2003; Shah et al. 2006).

15.11 High-Throughput Mutation Detection and Screening Techniques

DNA marker techniques are increasingly being used in plant mutation breeding and genetics for tracing the pedigree of induced mutants and tagging important mutations. Markers tightly linked with mutant traits can be used for marker-assisted selection (MAS), pyramiding and cloning of mutant genes. Several innovative approaches have been introduced that discriminate the mutations in the mutagenized population. Targeting Induced Local Lesions in Genomes (TILLING), a novel, reverse genetics approach, combines advantages of point mutations provided by chemical mutagenesis, with advantages of high-throughput PCR-based mutant screening (McCallum et al. 2000) for identifying allelic variants. With the advent of next-generation sequencing (NGS), several remarkable techniques like MutMap (mapping-by-sequencing) and MutChromSeq (helps to assort the desired genes in shortest time span) are helpful in identifying basic changes induced through mutagenesis (Sánchez-Martín et al. 2016). The prior information on gene sequences enables the assessment of gene function and identification of mutations through RNAi, TILLING and ECOTILLING techniques (Till et al. 2007). Techniques including SHOREmap (SHOrtREad map), next-generation mapping (NGM), MutMap, MutMap+, Mut-Map-Gap and QTL-Seq have been demonstrated successfully for mapping of causal mutation sites wherein prior information of gene sequences is lacking (Schneeberger et al. 2009; Austin et al. 2011; Abe et al. 2012; Fekih et al. 2013; Takagi et al. 2013a, b). Song et al. (2017) mapped genes controlling cotyledon colour using the yellow and green cotyledon colour mutants in soybean by employing QTL-Seq approach.

15.12 Types of Mutations

15.12.1 Chlorophyll Mutations

Chlorophyll mutations are primarily used to evaluate the genetic effects of various mutagens. They are the telltale signs of irradiation that can be visibly and readily observed in M_2 generation. Chlorophyll mutation frequency in M_2 has been reported to be the most dependable index for evaluating the genetic effects of mutagens (Kharkwal 1998b; Waghmare and Mehra 2001). Gustafsson (1940) grouped chlorophyll mutations into albina, xantha, viridis, chlorina, striata, tigrina and maculata classes. In pulse crops, a wide range of mutations like chlorina, virescens, viridis, flavoviridis, albo-viridis, chlorina-terminalis, chlorina-virescens, albovirescens, chlorotica, albina, xantha, coppery and variegated have been reported (Goyal and Khan 2010; Bhosale and Hallale 2011). Chemical mutagens have been found to induce chlorophyll mutations at higher frequencies in comparison to physical mutagens in chickpea (Kharkwal 1998b) and grass pea (Waghmare and Mehra 2001). In black gram, albinos were found to be the least frequent, while chlorina and xantha were recorded maximum at higher doses (Hemavathy and Ravidnran 2005). In cowpea, a dose-dependent increase in the spectrum and frequency of chlorophyll mutations was observed irrespective of whether mutagens were applied singly or in combination. Singly, EMS was found to be more efficient than gamma rays in inducing chlorophyll mutations, while the combination of EMS and gamma rays was the most efficient of all (Bind et al. 2016).

15.12.2 Mutations Affecting Morphological and Other Quantitative Traits

Mutations affecting the morphology of the plants are visually observable and have been widely reported across the pulse crops. They encompass a wide range of mutations affecting the size, shape, texture, colour, number, stature and arrangement of various plant parts such as leaves, pods, seeds, flowers, stems, branches and peduncles. Many of these mutations affecting quantitative traits are polygenic and could be the pleiotropic manifestations of mutated genes or chromosomal aberrations. Morphological mutations affecting yield-attributing traits are of primary concern to the breeders as they ultimately contribute to yield enhancement. Wide array of morphological mutants affecting various plant parts in different pulse crops like chickpea, mung bean, black gram, cowpea, soybean, moth bean, lentils and grass pea have been isolated following irradiation at 100–500 Gy gamma rays and by using different chemical mutagens. Contrasting plant statures like dwarf mutants and tall mutants have also been reported in these pulse crops (for back references, refer to Rafiq Wani et al. 2014; Kumar et al. 2019). Certain morphological mutants like flower colour gain attention as they could serve as genetic markers in different breeding programmes (Datta and Sengupta 2002; Atta et al. 2003). Bushy mutants with increased yield and high photosynthetic efficiency were identified in cowpea

(Gnanamurthy and Dhanavel 2014) and mung bean (Naik et al. 2002). Leaf mutations pertaining to colour, leaflet size and number, and shape and length of leaflets and rachis have been reported in a number of pulse crops including black gram (Bhosale and Hallale 2011), lentil and faba bean (Shahwar et al. 2019), though most of them were sterile or poor yielders. Mutations affecting pods and seeds have been reported in various pulse crops like lobed and hairy-pod mutants in black gram (Bhosale and Hallale 2011) and long-pod and large-seeded mutants in mung bean (Arulbalachandran and Mullainathan 2009), chickpea (Shah et al. 2011), lentil (Amin et al. 2015) and faba bean (Khursheed et al. 2019). Mutations affecting testa colour have been reported widely in these pulse crops. Gaafar et al. (2016) isolated a black-coat-colour mutant from a white-seeded parent in cowpea, which segregated for colour, eye pattern and seed texture. Nodulation mutants like Nod− (no nodules), nod+/− (few or no nodules), fix− (ineffective nodulation), nod++ (supernodulation or hypernodulation), nts (nitrate-tolerant symbiosis), nitrate-tolerant nodulation, ethylene resistant (insensitivity) and *Mycorrhiza* resistant (myc), and nitrate reductase deficient (nar) have been identified in various legume crops (reviewed in Suresh and Kumar 2020) including a dwarf cowpea with higher nitrogen-fixing ability (Anjana and Thimmaiah 2002).

Role of mutation breeding in increasing the genetic variability for quantitative traits in various crop plants has been proved beyond doubt (reviewed in Goyal and Khan 2010). Number of branches and pods, yield, pod length, plant height, etc. have been improved due to the effect of mutagens in various pulses. Negative effects have also been reported in black gram, wherein plant height, pods per plant and yield were negatively affected by gamma ray treatment (Manapure et al. 1998). In pulse crops, mutants for high yield, high protein content, early maturity, root nodulation, erect plant type, determinate growth and compact growth habit have been extensively identified (Rafiq Wani et al. 2014). In cowpea, a large-seed-size mutant (>30%) was isolated upon exposure to low-dose gamma irradiation (50 Gy) (Gaafar et al. 2016), and high-yielding mutant lines were identified following sodium azide and gamma ray treatment (Raina et al. 2020). Mutants with significant increase in pod lengths have been induced using gamma rays and chemical mutagens like EMS, HZ and SA in black gram, mung bean, cowpea, chickpea and cluster bean (Wani et al. 2011, 2018; Barshile 2006; Sharma and Singh 1992; Singh and Agarwal 1986), while a large-seeded mutant with vigorous growth and more leaves and pods was induced by gamma rays and EMS in black gram (Singh et al. 2000). Mutagenic agents with higher efficiency induce more multiple mutations (more than one mutation in a single plant (Sharma 1969)), and such mutations may accumulate several desirable characters within one plant. Also, the frequency of morphological mutations has been found to increase with increasing doses of the mutagen (Thakur and Sethi 1995). Multiple mutations have been reported earlier by Kharkwal (1999) in chickpea, Appa Rao et al. (1975) in black gram, Odeigah et al. (1998) in cowpea and Auti and Apparao (2009) in mung bean. Therefore, both physical and chemical mutagens could be effectively used to induce mutations for morphological and quantitative traits in pulse crops.

15.13 Mutants in Cross-Breeding

The agronomically poor morphological mutants could be exploited in hybridization programmes to introgress some of the useful traits into elite, high-yielding genetic backgrounds towards varietal development. Similarly, micro-mutants, with hitherto invisible mutations, are also useful in developing new crop varieties. Gaul (1964) emphasized that micro-mutations are useful in plant breeding as they occur more frequently than macro-mutations and often do not adversely affect the viability in comparison to macro-mutations. Morphological mutants play a pivotal role in restructuring the crop ideotype and contribute immensely towards varietal development. Usually, the mutants by itself may not be appropriate for release as varieties, but have the potential to be more productive, when used in cross-breeding (Pawar et al. 2000). Moreover, negative traits of the pleiotropic spectrum could be eliminated to some extent from the positive ones, by transferring it into a specific genotypic background (Sidorova 1981). For example, a black gram variety, No. 55, was irradiated with gamma rays (150–750 Gy) and fast neutrons (20–60 Gy) to obtain mutants having variation for numerous morphological traits. The large-seed mutants, 'UM 196' (dark-green leaf mutant) and 'UM 201', were used in hybridization with the elite cultivar 'T-9' for developing high-yielding black gram varieties 'TAU-1', 'TAU-2' and 'TPU-4' (Pawar and Manjaya 1996). So far, 122 pulse crop varieties have been developed through mutation breeding and released for cultivation in India. Fifty of these varieties are the derivatives of mutants developed through cross-breeding with a mutant.

15.14 Mutation Breeding for Yield and Varietal Development in Pulses

Micro-mutations are quite useful in improving quantitatively inherited traits (such as yield) as they do not interfere much with genotypic and phenotypic architecture of the crop while enhancing the genetic variability. The significance of micro-mutations in evolution (Baur 1924) was recognized well before the importance of physical and chemical mutagens in inducing mutations for polygenic characters was elucidated (Gregory 1955). In pulses, numerous efforts have been made to study the mutagenic effect of various mutagens in inducing mutants with high yield potential and other desirable traits. Both physical and chemical mutagens, alone or in combinations, have been used in pulses for generating variability in yield-attributing quantitative traits. Gupta and Swaminathan (1967) suggested identification of promising mutants in M_2 based on mean and variance. The efficiency of early-generation (M_2) selection in mutation breeding experiments has been reported in crops like lentil (Solanki and Sharma 2002) and green gram (Tickoo and Chandra 1999). The coefficients of variation for all polygenic traits were of higher magnitude in M_3 generation. In rice bean, the lower doses (300 Gy in gamma rays and 0.50% in EMS) were found to be the most effective in generating variability and three traits, viz., pods per cluster, seeds per pod and pod length that showed high heritability coupled with high genetic advance. This indicated that breeding for these traits could be achieved by phenotypic selection (Patial et al. 2017). The role of mutation breeding in yield improvement towards varietal development in some of the important pulse crops is discussed. Realizing the significance of legumes in improving nutrition and livelihood of poor farmers, there is a need to breed new crop cultivars with a broad genetic base, capable of withstanding frequent climatic fluctuations, as well as with resistance/tolerance to biotic and abiotic stresses.

15.14.1 Chickpea

Among the pulses, chickpea (*Cicer arietinum* L.) is one of the most widely grown legume crops ranking first in area and production in India and contributes 65% of the global production. Mutation breeding has significantly contributed to the genetic improvement of chickpea, and 34 varieties have been developed worldwide using mutation breeding, of which 12 are of Indian origin (Table 15.1). The foremost variety 'Plovdiv-8' was released in Bulgaria in 1979 followed by the early- and high-yielding variety 'M699 (Hyprosola)' in 1981 by Bangladesh. Gamma ray is the most commonly used mutagen with dose ranging from 100 to 450 Gy to be most successful in varietal development (Table 15.1). In addition, fast neutrons and EMS have been employed for developing mutant varieties. India, Bangladesh and Pakistan are the prime contributors, while Bulgaria, Egypt and Turkey have released one variety each. Besides high yield performance under late-sown conditions, chickpea mutant variety 'Pusa—547' has attractive bold seeds, thin testa and good cooking quality (Kharkwal et al. 2005). The green testa variety 'PKV Harita' developed by Dr. PDKV, Akola, has been developed through hybridization with a green-seeded mutant '121-1'. Majority of the mutant varieties are direct mutants (Table 15.1). Kabuli chickpeas are more sensitive to mutagens than the desi types (Sharma and Kharkwal 1982). Gamma rays and EMS were more effective in mutation induction in large-seeded and small-seeded types, respectively (Pathania and Sood 2006). Several workers have so far reported encouraging results on the induction of useful quantitative variability in chickpea (Kozgar et al. 2012; Khan et al. 2004; Canci et al. 2004; Cagirgan and Toker 2004). The additive gene effects and high estimates of heritability in yield and yield components in mutant populations are useful and would enable breeders to exercise selection based on phenotypic performance (Navnath and Subhash 2015; Amri-Tiliouine et al. 2018).

15.14.2 Pigeon Pea

Pigeon pea is the second most important pulse crop of India, and nearly 70% of global area under pigeon pea cultivation is in India. All of the 12 pigeon pea mutant varieties have been developed from India (Table 15.1). The first variety 'Co-3' was developed by TNAU in the year 1977 through EMS treatment. Thereafter, in 1983, 'TT 6' (Trombay Vishakha 1), a large-seeded variety developed from 'Type-21' using fast neutrons as mutagen, was released from BARC, Mumbai. In 1984, inter-mutant cross involving two fast-neutron-induced mutants resulted in the release of an extra-early-maturing (115–120 days) variety 'TAT-10'. Later on, by involving 'TT-6' in hybridization with 'ICPL84008', three highly successful varieties 'TT-401', 'TJT-501' and 'PKV-TARA' were released in 2005, 2007 and 2013, respectively. 'TT-6' was also involved in a four-way cross resulting in 'CO(RG)-7' variety in 2004. The 'CO-5' which is an early, photo-insensitive variety was released in the year 1984 and was later used to derive a male sterile line involved in the development of 'COPH-2' hybrid in 1997. The gamma ray-induced mutant variety 'Co-6' was hybridized to derive 'Co-9' variety in 2019. Excepting 'CO-3', all the mutant varieties have been evolved directly or indirectly through physical mutagens.

15.14.3 Mung Bean

Induced mutations have played a determinative role in the genetic improvement of mung bean. A variety of reports suggest that mutation has a great potential in genetic improvement of mung bean including yield. The mutagens like gamma rays, electron beam, EMS and epichlorohydrin have been reported to induce mutations affecting yield-attributing traits (Singh et al. 2001; Khan and Goyal 2009; Dhole and Reddy 2018a). At 0.1% EMS concentration in green gram, there is an enhancement in crop phenology and yield (Arulbalachandran and Mullainathan 2009). In mung bean, nearly about 65 varieties have been developed through mutation breeding, of which more than half (34 varieties) have been released from India (Table 15.1). Nearly 50% of the varieties have been developed through hybridization involving a mutant. Five varieties have been developed subsequent to the irradiation of F_1 hybrid seeds. One-fourth (8 varieties) of the varieties released in India have been developed by BARC, Mumbai. The other major contributors include Pakistan, Bangladesh, Thailand and Indonesia. Gamma rays have been exclusively used in the dose range of 100–600 Gy for successful development of mung bean mutant varieties.

15.14.4 Urdbean

In black gram, the survey yielded a number of reports wherein mutation has been observed to affect a wide range of traits. Mutations affecting plant architecture such as dwarf and tall plant types, leaf size and shape, increased branch number, branch position, stem girth, early and late flowering, increased pod number, pod and seed colour, and increased primary and lateral root length have been induced through EMS and gamma rays (Dhasarathan et al. 2017; Souframanien et al. 2016; Usharani and Kumar 2015; Makeen et al. 2013; Kumar et al. 2009; Sinha and Bharati 1990). Significant levels of variability were induced for yield and other yield-attributing traits following gamma rays and EMS treatments (Arulbalachandran and Mullainathan 2009; Vanniarajan et al. 1996; Hepziba and Subramanian 1994;

Kundu and Singh 1982). Nearly about 19 varieties have been released through mutation breeding in black gram, and all the varieties except one have been released from India. Five of the 18 mutant varieties in India have been developed by BARC, Mumbai. Gamma rays and MMS have been successfully exploited in the varietal development. The BARC-bred variety 'TAU-1' way back in 1985 has wreaked revolution in black gram cultivation in India. During its peak cultivation, more than 90% of the black gram area in Maharashtra state was occupied by this variety. Most of the varieties possess resistance against yellow mosaic virus and powdery mildew diseases. The variety 'TU-40' released in 2013 by BARC, Mumbai, is suitable for rice fallow niches and has the potential to bring additional area under black gram cultivation.

15.14.5 Cowpea

Cowpea is one of the important arid legumes widely cultivated in the arid and semi-arid regions of the world, especially in the Asian and African continents. Being relatively hardy, it assumes importance in the present context of climate change. Ethyl methanesulphonate (EMS) and gamma rays have been frequently used to create variations among different agro-morphological traits of cowpea (Girija et al. 2013; Singh et al. 2013; Girija and Dhanavel 2009). Lower and intermediate doses of EMS (0.25% and 0.30%) and gamma rays (200 and 300 Gy) were effective for inducing variability and isolating promising mutants (Nair and Mehta 2014). Selection for quantitative traits in the M_3 generation proved promising (Raina et al. 2020). Nair and Mehta (2014) reported high GCV coupled with high PCV in many of the mutant traits in both M_2 and M_3 generations. India has contributed 16 mutant varieties, while 6 varieties have been developed from other parts of the world. The initial varieties from India in early 1980s were primarily induced through the chemical mutagen, DMS. Thereinafter, almost all the varieties are the resultants of gamma irradiation. IARI, New Delhi; TNAU, Coimbatore; and BARC, Mumbai, are the major contributors from India. The variety 'TRC77-4' ('Kalleshwari') is suitable for rice fallows, while BARC's other variety 'TC-901' released in 2018 has the distinction of being the country's first summer-suitable federal cowpea variety (Dhanasekar and Reddy 2018). Costa Rica and African countries like Kenya, Zimbabwe and Zambia are the other global contributors for mutant varieties in this crop. Gamma rays in a dose range of 100–600 Gy were found successful in varietal development.

15.15 Mutation Breeding for Pest and Disease Resistance in Pulses

Mutation breeding is one of the best alternatives when gene(s) for resistance to a particular disease or stress cannot be found in the available gene pool. Unlike the transgenics, mutation breeding causes in situ alteration in the existing genome and creates allelic variability and opportunities for selection. In chickpea, mutations for resistance against *Ascochyta* blight (Omar and Singh 1995), the most widespread and dreaded disease and pod borer (Shaikh 1983), the most destructive pest, have been induced. The Indian Agricultural Research Institute (IARI), New Delhi, and the Nuclear Institute for Agriculture and Biology (NIAB), Faisalabad, Pakistan, have released many disease-resistant mutant varieties of chickpea. The first *Ascochyta* blight-resistant chickpea mutant cultivar 'CM-72' was released in Pakistan in 1983. Subsequently, new mutant cultivars 'CM-88' and 'CM-98' were developed with multiple resistance against *Ascochyta* blight and *Fusarium* wilt using gamma rays (Haq et al. 1999). Currently, these disease-resistant mutant cultivars are widely cultivated in over 350,000 ha that constitute more than 30% of chickpea area in Pakistan (Kumar et al. 2019). The IARI-bred chickpea variety 'BGM-547' released in 2005 has moderate resistance against wilt and root rot complex.

In mung bean, a series of mutant cultivars with resistance to yellow mosaic virus (YMV) and early maturity like 'MUM 2', 'BM 4', 'LGG 407', 'LGG 450', 'CO 4', 'Dhauli' (TT9E), 'Pant Mung-1' and 'TAP-7' (Ahloowalia et al. 2014) have been released in India. As seen in Table 15.1, most of the mutant cultivars released post-2000 are resistant to MYMV. The most popular mutant varieties 'TMB-37' and 'SML 668' are endowed with synchronous maturity and resistance to YMV. The mutant varieties from IIPR, Kanpur, like 'IPM99-125', 'IPM02-03' and 'IPM 02-14' have good resistance against MYMV. The latest mutant variety 'IPM512-1' released in 2020 has multiple resistance against MYMV, *Cercospora* leaf spot and anthracnose diseases. In Pakistan, more than ten mutant cultivars of mung bean with induced resistance against YMV and *Cercospora* leaf spot diseases have been released. The first powdery mildew (PM)-resistant mutant mung bean variety ('TARM-2') in the whole of Asia was developed from Bhabha Atomic Research Centre (BARC), Mumbai, through induced mutagenesis (Pandey et al. 2005), and subsequently, a series of PM-resistant mung bean varieties were evolved by utilizing PM-resistant mutant (Reddy et al. 2008).

In black gram, the YMV-resistant mutant cultivars 'Vamban 2' and 'Sarla B-14-4' were derived from 'T 9' which is susceptible to YMV. Besides resistance to YMV, both mutants are drought tolerant, early maturing (70 days) and high yielding (Dixit et al. 2000). At BARC, Mumbai, a number of YMV-resistant mutants have been isolated and successfully used in hybridization leading to development of YMV-resistant varieties (Pawar et al. 2000). The dark-green leaf mutant 'UM 196' (mutant of CV. 'No. 55') was crossed with 'T9' to develop yellow mosaic disease-resistant variety 'TU94-2' (Souframanien and Reddy 2018). The highly resistant mutant lines to MYMV, viz. 'ACM-014-021', 'ACM-015-025', 'ACM-014-006', 'ACM-015-022', 'ACM-015-023', 'ACM-014-019', 'ACM-014-007', 'ACM-015-017', 'ACM-014-003' and 'VBN (Bg) 4', were found to possess reasonably higher total phenol, phytic acid content and lower total sugar content than the susceptible mutant lines (Vanniarajan et al. 2021).

In cowpea, BARC developed a gamma ray-induced mutant variety 'TC-901' in 2018 with resistance against cowpea mosaic and root rot diseases. Mutants with resistance to cowpea aphid-borne mosaic virus and leaf crinkle diseases have also been identified in cowpea (Dhanasekar and Reddy 2015a) and are currently being utilized in the crossing programme towards varietal development. Munyinda and Kanenga (2018) also released a cowpea variety 'Lukusuzi' with tolerance to *Ascochyta* and *Cercospora* and with medium tolerance to cowpea aphid-borne mosaic virus in Zambia.

The progress in breeding for resistance to insect pests in pulse crops has been limited mainly due to non-availability of sources with high level of resistance and lack of effective screening techniques (Gaur and Chaturvedi 2004). BARC has received accolades for doing pioneering research towards development of black gram genotypes resistant to the serious storage pest, namely bruchids (*Callosobruchus maculatus*) (Souframanien and Gopalakrishna 2007). Trombay wild urdbean (*Vigna mungo* var. *silvestris*) collected from the Trombay hills was identified to possess resistance to bruchids and has been registered (INGR10133) as potential genetic donor for bruchid resistance. By involving these genotypes in the mutation breeding programme at BARC, a number of mutant derivatives have been developed and are being tested for bruchid resistance under the All India Coordinated Research Programme (AICRP) of the ICAR (Gopalaswamy et al. 2016). Munyinda and Kanenga (2018) identified cowpea mutant following 150 Gy gamma irradiation with tolerance to aphids and bruchids, which was later released as a mutant variety 'Lunkhwakwa' in Zambia.

15.16 Mutation Breeding for Abiotic Stress Tolerance

Abiotic stresses are highly variable, and resistance breeding for abiotic stress tolerance has not been highly successful like that of biotic stresses. Nevertheless, mutation breeding has contributed to the development of mutants with tolerance against abiotic stresses in pulses. Mutants with tolerance against acidity and water stress have been identified in lentil (Lal and Tomer 2009). In cowpea, gamma ray (150 Gy)-induced mutant lines of 'Moussa Local' ('Moussa M51-4P10' and 'Moussa M43-20P14') exhibited better stress tolerance and produced higher yield under water-stress conditions (Gnankambary et al. 2020). Mutant variety 'Lunkhwakwa' with tolerance against aluminium toxicity has also been developed in cowpea (Munyinda and Kanenga 2018). The mutant cowpea varieties 'Lunkhwakwa' and 'Lukusuzi' were also found to be drought tolerant. A new mutant cowpea variety 'CBC-5' was released in 2017 that has proved promising to farmers in drought-prone areas of Zimbabwe (Dixit and Slavchev 2018). Shah et al. (2010) identified an early-type mutant in 'Pb2000' having high palmitic acid, stearic acid, oleic acid and linoleic acid, which supported their role in tolerance against frost in chickpea. Salt-tolerant and slow-transpiring mutants have also been identified in different cultivars of chickpea (Joshi Saha et al. 2018). In mung bean, mutants with early and synchronous maturity, long roots and drought tolerance were identified (Dhole and Reddy 2018a).

15.17 Mutation Breeding for Improved Nutrition in Pulses

For a simultaneous increase of mineral elements along with yield and its attributing characters, induced mutagenesis could play a determinative role. Mutants with altered seed mineral profiles have been identified in pea (Wang et al. 2003). High-yielding pulse mutants coupled with high protein contents have been reported by various workers (Naik et al. 2002; Kharkwal 1998c; Misra et al. 1973). Gottschalk (1986) suggested that it is possible to analyse the seed proteins of mutants, quantitatively and qualitatively, which had been selected with regard to other useful traits but not with regard to improved seed proteins. Protein profiling in the isolated mutants was reported by Auti and Apparao (2009) in black gram and Barshile et al. (2009) in chickpea. Mutation treatments involving gamma rays increased the softness of seeds, which in turn improved the cooking quality in chickpea (Graham et al. 2002). Surface-active proteins (globulins) and a polysaccharide (arabinogalactan) determine the soft, spongy texture in idli batter made out of black gram seeds, and Vanniarajan et al. (2021) have identified six mutant lines that have implications in developing superior black gram varieties with better batter quality (Vanniarajan et al. 2021).

Mutation breeding has also contributed substantially to reducing the anti-nutritional factors in pulse crops. Phytic acid (PA) is known to conjugate phosphorus and other essential elements like iron and make it unavailable to organisms that feed on seeds rich in PA. Mutations affecting PA content have been identified, and low PA mutants have been isolated in mung bean (Dhole and Reddy 2018b), chickpea (Misra et al. 2017), black gram and cowpea (Dhanasekar and Reddy 2017). These mutants are being exploited in varietal development programme so as to reduce the PA content to reasonable levels without affecting the physiological balance. Mutants with low flatulence causing raffinose family oligosaccharides (RFOs) have been identified in black gram (Souframanien et al. 2014) and cowpea (Dhanasekar and Reddy 2015b) that could be potential donors for developing varieties with low RFOs. Low RFOs containing black gram mutant 'TU43-1' also exhibited the lowest verbascose and stachyose contents. Mutant ('TU1-820-1-5') with high verbascose, but low stachyose and raffinose, was also recorded (Souframanien et al. 2014). A cowpea mutant 'TC501' possessing very low total RFOs (4.27 mmol/100 g) has been identified. Mutants with high RFOs in cowpea have been found to possess long roots and could play a prospective role in better performance potential under water-stress conditions (Dhanasekar and Reddy 2017). Low tannin-containing mutants have also been identified in cowpea (Dhanasekar and Reddy 2012) that could be used to develop pro-nutritious varieties.

Genetic biofortification for improved Fe and Zn with moderate PA can be a better breeding target in chickpea. Genotypes with consistently high iron and/or Zn content with low/moderate PA were identified (Joshi Saha and Reddy 2014; Misra et al. 2017). In addition, two large-seed-size mutants ($M_{2:3}$ seeds) were found to contain significantly higher Fe content (15.6 and 12 mg/100 g, respectively) than the control (range 8.1–10.2 mg/100 g). Advanced progenies ($M_{3:4}$ seeds) of these two mutants also showed significantly higher Fe content (11.09 and 10.9 mg/100 g, respectively) than the control (7.08 mg/100 g) (Misra et al. 2018).

15.18 Impact of Mutant Pulse Crop Varieties in India

The mutant pulse crop varieties with their superior yield attributes are widely adopted by the farmers across the country. Using these varieties, many of the farmers have harvested record yields resulting in increased on-farm income. The mutant variety 'TAU-1' (developed in collaboration with Dr. PDKV, Akola) in black gram released way back in 1985 is still very popular among the Maharashtra farmers. This variety is extensively grown throughout Maharashtra and occupies more than 50% of the area under black gram. The variety 'TJT-501' (developed in collaboration with JNKVV, Jabalpur) occupies almost 40–50% of the area under pigeon pea in Madhya Pradesh and is one of the topmost varieties receiving on an average 12% of national breeder seed indent (Chauhan et al. 2016). In mung bean, the variety TMB-37 is widely preferred by farmers throughout the country for its earliness and suitability for summer cultivation though it was originally released for North East Plain Zone. This variety has been readopted and re-released in the state of Punjab. Likewise, the pigeon pea variety 'TT-401' initially released for central zone has been found promising in southern states like Tamil Nadu, Andhra Pradesh and Karnataka, where it has given record yield of 2500 kg/ha. The mung bean varieties 'TM96-2' (in collaboration with ANGRAU, Lam, AP) and 'TM 2000-2' (in collaboration with IGKVV, Raipur) with resistance against powdery mildew and *Corynespora* leaf spot diseases occupy large areas under rice fallows in Andhra Pradesh and Chhattisgarh, respectively. The recent black gram variety 'TU-40' is gaining popularity in the southern states. The farmers of Maharashtra cultivating 'PVK-TARA' pigeon pea are reaping high yields especially under drip irrigation. The recent summer-suitable cowpea variety 'TC901' is expected to bring additional areas under cultivation. Thus, the mutant varieties of pulse crops have contributed substantially to increasing the productivity and ensuring nutritional security of the country.

15.19 New Breeding Techniques (Targeted Mutagenesis)

New breeding techniques or NBTs are described as precision breeding and involve techniques for incorporating site-specific targeted mutations in the genome with greater accuracy and less off-targeted mutations (Holme et al. 2019). These include techniques like zinc finger nuclease (ZFN) technology, oligonucleotide-directed mutagenesis (ODM), cisgenesis and intragenesis, grafting on GM rootstock, RNA-dependent DNA methylation, agro-infiltration 'sensu stricto' and reverse breeding. The site-directed nuclease (SDN) is often employed to create a site-specific mutation in the plant genome. In addition, TALEN and CRISPR/Cas-based SDN have been introduced, and the latter is now extensively being used

(Holme et al. 2019). The conjoint use of classical mutation breeding and NBT mutations can be employed implicitly in the modification of plant attributes. The main advantage of NBT over the classical mutation technique is its precision and specificity that could be utilized to find robust mutation sites without the unwanted genetic changes. The conjoint use of NBT for creating site-specific mutations and the traditional mutation techniques for retrieval of the desired mutations could resultantly lead to the development of mutant varieties with desired mutations (Jankowicz-Cieslak et al. 2017).

15.20 Conclusion and Future Prospects

The narrow genetic base of pulses impedes the development of new varieties capable of breaking the yield plateau through conventional breeding. Induced mutagenesis with its breeding strategies offers a viable and potential tool for creating variability and offers scope for genetic improvement of pulse crops for both quantitative and qualitative traits within a relatively shorter time than traditional breeding. Induced mutagenesis is a powerful tool for creating allelic variants for traits of interest and empowers breeders to potentially develop resistance against biotic and abiotic stresses for which the existing gene pools are of no aid. The impact of pulse crop varieties bred through mutation breeding globally demonstrates the utility of induced mutagenesis as a flexible and practicable approach for genetic improvement of any crop including legumes provided that appropriate objectives and selection/screening methods are followed accordingly. Induced mutagenesis, though primarily was used to broaden the genetic base of pulse crops, is now becoming widespread as a tool for deciphering functional genomics. Mutants are effectively being used for studying gene expressions and gene regulations and assigning gene functions. High-throughput phenotyping coupled with next-generation sequencing methods has accelerated mutant identification process both through forward and reverse genetic approaches. The accuracy and analyses of large structural variations resulting from induced mutagenesis will be further improved by long-read sequencing technologies (e.g. PacBio sequencing and Oxford Nanopore sequencing). Additionally, conjunction of mutational and high-throughput DNA sequencing through new breeding techniques will maximize precision breeding for inducing site-specific targeted mutations in the genome with greater accuracy and lesser unintended mutations.

References

Abe A, Kosugi S, Yoshida K et al (2012) Genome sequencing reveals agronomically important loci in rice using MutMap. Nat Biotechnol 30:174–178. https://doi.org/10.1038/nbt.2095

Ahloowalia BS, Maluszynski M, Nichterlein K (2014) Global impact of mutation-derived varieties. Euphytica 135(2):187–204

Amin R, Laskar RA, Khan S (2015) Assessment of genetic response and character association for yield and yield components in Lentil (*Lens culinaris* L.) population developed through chemical mutagenesis. Cogent Food Agric 1:1000715

Amri-Tiliouine W, Laouar M, Abdelguerfi A et al (2018) Genetic variability induced by gamma rays and preliminary results of low-cost TILLING on M_2 generation of chickpea (*Cicer arietinum* L.). Front Plant Sci 9:1568. https://doi.org/10.3389/fpls.2018.01568

Anjana G, Thimmaiah SK (2002) Evaluation of dwarf mutant of cowpea (*Vigna unguiculata* L. Walp.) developed through gamma irradiation for nitrogen fixation characters. J Nucl Agric Biol 31:94–98

Appa Rao S, Rao P, Jana MK (1975) New plant type in black gram. Curr Sci 44:679–680

Arulbalachandran D, Mullainathan L (2009) Changes on quantitative traits of blackgram (*Vigna mungo* (L.) Hepper) induced by EMS in M_2 generation. J Phytol 1(4):230–235

Atta BM, Ahsan HM, Shah TM et al (2003) Induced flower colour mutation in chickpea. Int Chickpea Pigeonpea Newsl 10:6–7

Auerbach C, Robson JM (1946) Chemical production of mutations. Nature 157:302

Austin RS, Vidaurre D, Stamatiou G et al (2011) Next-generation mapping of *Arabidopsis* genes. Plant J 67(4):715–725. https://doi.org/10.1111/j.1365-313X.2011.04619.x

Auti SG, Apparao BJ (2009) Induced mutagenesis in mungbean (*Vigna radiata* (L.) Wilczek). In: Shu QY (ed) Induced plant mutations in the genomics era. Food and Agriculture Organization of the United Nation, Rome, pp 97–100. http://cropgenebank.sgrp.cgiar.org/. Accessed 30 Sept 2011

Bado S, Forster BP, Nielen S et al (2015) Plant mutation breeding: current progress and future assessment. Plant Breed Rev 39:23–88. https://doi.org/10.1002/9781119107743.ch02

Barshile (2006) Induction of genetic variability in *Cicer arietinum* employing sodium azide, ethylmethane sulphonate and gamma radiations. Ph.D thesis, University of Pune, India

Barshile JD, Auti SG, Apparao BJ (2009) Genetic enhancement of chickpea through induced mutagenesis. J Food Legum 22:26–29

Baur E (1924) Untersuchungen über das Wesen, die Entstehung und die Vererbung von Rassenunterschieden bei Antirrhinum majus. Borntraeger, Leipzig

Bhosale UP, Hallale BV (2011) Gamma radiation induced mutations in black gram (*Vigna mungo* (L.) Hepper). Asian J Plant Sci Res 1(2):96–100

Bind D, Dwivedi VK, Singh SK (2016) Induction of chlorophyll mutations through physical and chemical mutagenesis in cowpea [*Vigna unguiculata* (L.) Walp]. Int J Adv Res 4:49–53

Cagirgan MI, Toker C (2004) Selection of chickpea mutants for cold tolerance and *Ascochyta* blight resistance. Int Chickpea Pigeonpea Newsl 11:14–16

Canci H, Toker C, Cagirgan MI (2004) Preliminary screening for nodulation in chickpea mutants. Int Chickpea Pigeonpea Newsl 11:13–14

Chauhan JS, Singh BB, Gupta S (2016) Enhancing pulses production in India through improving seed and variety replacement rates. Indian J Genet 76(4):410

Cooper HD, Spillane C, Hodgkin T (2001) Broadening the genetic base of crops: an overview. In: Cooper HD, Spillane C, Hodakin T (eds) Broadening the genetic base of crop production. CAB International, Oxon, UK, pp 1–23

DACandFW (2019) Agricultural statistics at a glance. Ministry of Agriculture and Farmers Welfare, Department of Agriculture Cooperation and Farmers Welfare, Directorate of Economics and Statistics, New Delhi. http://agricoop.nic.in/agristatistics.htm

Datta AK, Sengupta K (2002) Induced viable macromutations in coriander (*Coriandrum sativum* L.). J Genet Plant Breed 48(3):303–304

Deepalakshmi AJ, Kumar CRA (2003) Efficiency and effectiveness of physical and chemical mutagens in urdbean (*Vigna mungo* (L.) Hepper). Madras Agric J 90:485–489

Dhanasekar P, Reddy KS (2012) Variation for seed protein and tannin content in elite lines of cowpea. In: Abstracts of DAE-BRNS Life sciences symposium on Trends in plant, agriculture and food sciences, Dec 17–19, Mumbai, p 94

Dhanasekar P, Reddy KS (2015a) Serological screening of cowpea genotypes for resistance against cowpea aphid borne mosaic virus using DAS-ELISA. Asian J Plant Pathol 9(2):83–90

Dhanasekar P, Reddy KS (2015b) Genotypic variation for raffinose family oligosaccharides in cowpea. In: Proceedings of the national conference on pulses: Challenges and opportunities

under changing climatic scenario, September 20–October 1, 2014. ISPRD, JNKVV, Jabalpur, pp 316–320
Dhanasekar P, Reddy KS (2017) Role of seed raffinose family oligosaccharides and phytic acid in better performance potential of cowpea under water stress conditions. J Basic Appl Plant Sci 1(1):106
Dhanasekar P, Reddy KS (2018) Development of Trombay cowpea variety 'TC-901' amenable for summer cultivation by induced mutagenesis. In: FAO/IAEA International symposium on Plant mutation breeding and biotechnology held during 27–31 August at Vienna, Austria. IAEA-CN-263-170
Dhanavel D, Pavadai P, Mullainathan L et al (2008) Effectiveness and efficiency of chemical mutagens in cowpea (*Vigna unguiculata* (L.) Walp.). Afr J Biotechnol 7:4116–4117
Dhasarathan M, Geetha S, Meenakshiganesan N et al (2017) Studies on effectiveness and efficiency of gamma rays in black gram (*Vigna mungo* (L.) hepper). Bioscan 12(2):1155–1159
Dhole VJ, Reddy KS (2018a) Genetic analysis and variability studies in mutants induced through electron beam and gamma rays in mungbean (*Vigna radiata* L. Wilczek). Electron J Plant Breed 9(1):304–312
Dhole VJ, Reddy KS (2018b) DAE-BRNS Life Sciences Symposium on "Frontiers in sustainable agriculture". DAE Convention Centre, Anushakti Nagar, Mumbai, p 15
Di Vito M, Singh KB, Greco N et al (1996) Sources of resistance to cyst nematode in cultivated and wild *Cicer* species. Genet Resour Crop Evol 43:103–107
Dixit A, Slavchev S (2018) New mutant cowpea variety helps Zimbabwe's farmers in drought prone areas. IAEA Bull:20–21
Dixit GP, Tripathi DP, Chandra S et al (2000) MULLaRP crops: varieties developed during the last fifty years. All India Coordinated Research Project on MULLaRP (ICAR), Indian Institute of Pulses Research, Kanpur, India
Eswaramoorthy V, Kandasamy T, Thiyagarajan K et al (2021) Effectiveness and efficiency of electron beam in comparison with gamma rays and ethyl methane sulfonate mutagens in cowpea. Appl Radiat Isot 171:109640. https://doi.org/10.1016/j.apradiso.2021.109640
Fekih R, Takagi H, Tamiru M et al (2013) MutMap+: genetic mapping and mutant identification without crossing in rice. PLoS One 8:1–10. https://doi.org/10.1371/journal.pone.0068529
Gaafar RM, Hamouda M, Badr A (2016) Seed coat color, weight and eye pattern inheritance in gamma-rays induced cowpea M_2-mutant line. J Genet Eng Biotechnol 14:61–68. https://doi.org/10.1016/j.jgeb.2015.12.005
Gaul H (1964) Mutations in plant breeding. Radiat Bot 4:155–232
Gaur PM, Chaturvedi SK (2004) Genetic options for managing biotic stresses in pulse crops. In: Masood A, Singh BB, Kumar S, Dhar V (eds) Pulses in new perspective. Indian Society of Pulses Research and Development, IIPR, Kanpur
Girija M, Dhanavel D (2009) Mutagenic effectiveness and efficiency of gamma rays, ethyl methane sulphonate and their combined treatments in cowpea [*Vigna unguiculata* (L.) Walp]. Global J Mol Sci 4:68–75
Girija M, Dhanavel D, Gnanamurthy S (2013) Gamma rays and EMS induced flower color and seed mutants in cowpea [*Vigna unguiculata* (L.) Walp]. Adv Appl Sci Res 4:134–139
Gnanamurthy S, Dhanavel D (2014) Effect of EMS on induced morphological mutants and chromosomal variation in Cowpea (*Vigna unguiculata* (L.) Walp). Int Lett Nat Sci 17:33–43
Gnankambary K, Sawadogo N, Diéni Z et al (2020, 2020) Assessment of cowpea (*Vigna unguiculata* (L.) Walp.) mutant lines for drought tolerance. Int J Agron. https://doi.org/10.1155/2020/8823498
Gopalaswamy SVS, Sesha Mahalakshmi M, Souframanien J (2016) Evaluation of certain blackgram varieties for resistance to pulse bruchid, *Callosobruchus maculatus* (F.). J Res ANGRAU 44(1 & 2):8–13
Gottschalk W (1986) Experimental mutagenesis in plant breeding. In: Prasad AB (ed) Mutagenesis: basic and applied. Print House, Lucknow, India, pp 81–96

Goyal S, Khan S (2010) Induced mutagenesis in urdbean (*Vigna mungo* L. Hepper): a review. Int J Bot 6:194–206. https://doi.org/10.3923/ijb.2010.194.206

Graham JA, Panozzo JF, Lim PC et al (2002) Effects of gamma irradiation on physical and chemical properties of chickpeas (*Cicer arietinum*). J Sci Food Agric 82(14):1599–1605

Gregory WC (1955) X-ray breeding of peanut (*Arachis hypogaea* L.). Agron J 47:396–399

Grover SI, Virk GS (1984) Induced chlorophyll mutants in mungbean (*Vigna radiata* (L.) Wilczek). Acta Bot Indica 14:170–180

Gupta N, Swaminathan MS (1967) An induced sphaerococcoid mutant in *Triticum dicoccum*. Curr Sci 36(1):19

Gustafsson A (1940) A mutation system of the chlorophyll apparatus. Lund Univ Arsskr NF Avd 236:1–40

Haq MA, Sadiq M, Hassan M (1999) CM 98 (CM 31-1/85): a very high yielding, disease resistant mutant variety of chickpea. Int Chickpea Pigeonpea Newsl 6:9–10

Hemavathy AT, Ravidnran GR (2005) Mutagenic effects of gamma rays on frequency and spectrum of chlorophyll mutations in urdbean (*Vigna mungo* (L.) Hepper). Madras Agric J 92:325–327

Hepziba SJ, Subramanian M (1994) Gamma ray induced variability in blackgram (*Vigna mungo* (L.) Heper). Ann Agric Res 15:512–512

Holme IB, Gregersen PL, Brinch-Pedersen H (2019) Induced genetic variation in crop plants by random or targeted mutagenesis: convergence and differences. Front Plant Sci 10:1468

IAEA-MVD (2021) Accessed online at https://nucleus.iaea.org/sites/mvd/SitePages/Home.aspx

Im J, Kim WJ, Kim SH, Ha BK (2017) Effects of proton beam irradiation on seed germination and growth of soybean (*Glycine max* L. Merr.). J Korean Phys Soc 71:752–757. https://doi.org/10.3938/jkps.71.752

Jankowicz-Cieslak J, Tai TH, Kumlehn J et al (2017) Biotechnologies for plant mutation breeding: protocols. Springer Nature, Cham

Joshi Saha A, Reddy KS (2014) Diversity study in chickpea (*Cicer arietinum* L.) using REMAP, SSR and ISSR markers. J Food Legum 27:281–288

Joshi Saha A, Misra G, Reddy KS (2018) Mutation breeding in chickpea (*Cicer arietinum* L.) for increased genetic variability, improved quality and agronomic traits. In: FAO/IAEA International symposium on Plant mutation breeding and Biotechnology, 27–31 August 2018, Vienna, Austria

Kang R, Seo E, Kim G et al (2020) Radio sensitivity of cowpea plants after gamma-ray and proton-beam irradiation. Plant Breed Biotechnol 8:281–292. https://doi.org/10.9787/PBB.2020.8.3.281

Khan S, Goyal S (2009) Improvement of mungbean varieties through induced mutations. African J Plant Sci 3:174–180

Khan MH, Tyagi SD (2010) Studies on effectiveness and efficiency of gamma rays, EMS and their combination in soybean (*Glycine max* (L.) Merrill). J Plant Breed Crop Sci 2:55–58

Khan S, Wani MR, Parveen K (2004) Induced genetic variability for quantitative traits in *Vigna radiata* (L.) Wilczek. Pak J Bot 36(4):845–850

Khan S, Wani MR, Bhat M et al (2005) Induced chlorophyll mutations in chickpea (*Cicer arietinum* L.). Int J Agric Biol 7:764–767

Kharkwal MC (1998a) Induced mutations in chickpea (*Cicer arietinum* L.) I. Comparative mutagenic effectiveness and efficiency of physical and chemical mutagens. Indian J Genet 58(2): 159–167

Kharkwal MC (1998b) Induced mutations in chickpea (*Cicer arietinum* L.) II. Frequency and spectrum of chlorophyll mutations. Indian J Genet 58(4):465–474

Kharkwal MC (1998c) Induced mutations for improvement of protein in chickpea (*Cicer arietinum* L.). Indian J Genet 58(1):61–68

Kharkwal MC (1999) Induced mutations in chickpea (*Cicer arietinum* L.). III. Frequency and spectrum of viable mutations. Indian J Genet 59(4):451–464

Kharkwal MC, Jain HK, Sharma B (1988) Induced mutations for improvement of chickpea, lentil, pea and cowpea. In: Proc. FAO/IAEA workshop on Improvement of grain legume production using induced mutations. IAEA, Pullman, Washington, pp 89–109

Kharkwal MC, Nagar JP, Kala YK (2005) BGM 547 – a high yielding chickpea (*Cicer arietinum* L.) mutant variety for late sown conditions of North Western Plains Zone of India. Indian J Genet 65:229–230

Khursheed S, Raina A, Parveen K et al (2019) Induced phenotypic diversity in the mutagenized populations of faba bean using physical and chemical mutagenesis. J Saudi Soc Agric Sci 18(2): 113–119. https://doi.org/10.1016/j.jssas.2017.03.001

Kitamura H, Mori M, Sato D et al (2006) Carbon ion beam breeding of rice suitable for low nitrogen input. TIARA Annu Rep 2004:100–101

Konzak CF, Nilan RA, Wagner J et al (1965) Efficient chemical mutagenesis. Rad Bot 5 (Suppl):49–70

Kozgar MI, Khan S, Wani MR (2012) Variability and correlations studies for total iron and manganese contents of chickpea (*Cicer arietinum* L.) high yielding mutants. Am J Food Technol 7(7):437–444

Kumar S, Gupta S, Chandra S et al (2004) How wide is the Genetic Base of Pulse Crops? In: Ali M et al (eds) Pulses in new perspective. Proceedings of the national symposium on crop diversification and natural resources management, 20–22 December 2003. Indian Society of Pulses Research and Development, Indian Institute of Pulses Research, Kanpur, pp 211–221

Kumar V, Sharma AK, Singh VP, Kumar M (2009) Characterization of pre-breeding genetic stocks of urdbean (*Vigna mungo* (L.) Hepper) induced through mutagenesis. In: Shu QY (ed) Induced plant mutations in the genomics era. Food and Agriculture Organization of the United Nation, Rome, pp 391–394

Kumar S, Katna G, Sharma N (2019) Mutation breeding in chickpea. Adv Plants Agric Res 9:355–362. https://doi.org/10.15406/APAR.2019.09.00448

Kundu SK, Singh DP (1982) Gamma ray induced variability for quantitative characters in blackgram (*Vigna mungo* (L.) Hepper). Madras Agric J 69:644–646

Ladizinsky O (1985) Founder effect in crop-plant evolution. Econ Bot 39:191–199

Lal JP, Tomer AK (2009) Genetic enhancement of lentil (*Lens culinaris* Medikus) for drought tolerance through induced mutations. In: Shu QY (ed) Induced plant mutations in the genomics era. Proceedings of an International Joint FAO/IAEA Symposium. Food and Agriculture Organization of the United Nations, Rome, pp 151–154

Liu LX, Guo HJ, Zhao LS et al (2009) Achievements and perspectives of crop space breeding in China. In: Shu QY (ed) Induced plant mutations in genomics era. IAEA, Vienna, Austria, pp 213–215

Makeen K, Suresh BG, Lavanya GR et al (2013) Study of chlorophyll and macromutations induced by gamma rays and sodium azide in urd bean (*Vigna mungo* L. Hepper). Afr J Agric Res 8(47): 5958–5961

Maluszynski M, Szarejko I, Bhatia CR et al (2009) Methodologies for generating variability. In: Ceccarelli S, Guimar EP, Weltzien E (eds) Plant breeding and farmer participation. Food and Agriculture Organization (FAO), Rome. http://www.fao.org/docrep/012/i1070e/i1070e00.htm

Manapure P, Patil S, Bhoyarand HB et al (1998) Induced mutagenic changes in quantitative characters of blackgram (*Vigna mungo* (L.) Hepper). J Soils Crops 8:204–206

Mba C, Afza R, Shu QY (2012) Mutagenic radiations: X-rays, ionizing particles, and ultraviolet. In: Shu QY, Forster BP, Nakagawa H (eds) Plant mutation breeding and biotechnology. Joint FAO/IAEA Division of Nuclear Techniques in Food and Agriculture International Atomic Energy Agency, Vienna, Austria, pp 83–90. https://doi.org/10.1079/9781780640853.0083

McCallum CM, Comai L, Greene EA et al (2000) Targeted screening for induced mutations. Nat Biotechnol 18:455–457

Micke B, Donini B, Maluszynski M (1990) Induced mutation for crop improvement. Mutat Breed Rev 7:1–4

Misra RN, Jachuck PJ, Sampath S (1973) Induced mutations for practical utility in rice. In: International symposium on use of isotopes and radiations in agriculture and animal husbandry, New Delhi, p 1971

Misra G, Joshi Saha A, Reddy KS (2017) Distribution and genetic analysis of phytic acid content in chickpea (*Cicer arietinum* L.). In: PulseSym2017, 2–4 Dec. Indian Society of Pulses research and Development at IIPR, Kanpur, UP

Misra G, Joshi Saha A, Reddy KS (2018) Mutation breeding for genetic variability and improved quality traits in chickpea (*Cicer arietinum* L.). In: DAE-BRNS Life Sciences symposium 2018 on Frontiers in Sustainable Agriculture, April 26–28, 2018. Anushakti Nagar, Mumbai, DAE Convention Centre, p 51

Munyinda K, Kanenga K (2018) Creating variability in cowpea for adaptation and value addition through induced mutation. In: FAO/IAEA International symposium on plant mutation breeding and biotechnology

Naik BS, Singh B, Kole C (2002) A promising mungbean (*Vigna radiata* (L.) Wilczek) genotype with high protein content and seed yield. Indian J Genet 62(4):342–344

Nair R, Mehta AK (2014) Induced mutagenesis in cowpea [*Vigna unguiculata* (L.) Walp] var. Arka Garima. Indian J Agric Res 48(4):247–257

Nakagawa H (2009) Induced mutations in plant breeding and biological researches in Japan. Crops 242(188):48

Navnath GK, Subhash B (2015) Genetic variability induced by ethyl methane sulphonate and sodium azide on seed characters in chickpea (*Cicer arietinum* L.). Int J Recent Sci Res 6:6676–6679

Nishant C, Singh A, Aswal DK (2020) Electron beam induced tailoring of electrical characteristics of organic semiconductor films. Chem Africa 3:571–592. https://doi.org/10.1007/s42250-020-00168-x

Odeigah PGC, Osanyinpejo AO, Myers GO (1998) Induced mutations in cowpea, *Vigna unguiculata* (Leguminosae). Rev Biol Trop 46(3):579–586

Olasupo FO, Olumuyiwa Ilori C, Forster BP, Bado S (2016) Mutagenic effects of gamma radiation on eight accessions of cowpea (*Vigna unguiculata* [L.] Walp.). Am J Plant Sci 07:339–351. https://doi.org/10.4236/ajps.2016.72034

Oldach KH (2011) Mutagenesis. In: Pratap A, Kumar J (eds) Biology and breeding of food legumes. CABI, Wallingford, pp 208–219

Omar M, Singh KB (1995) Development of early mutants with resistance to *Ascochyta* blight or leaf minor. Int Chickpea Pigeonpea Newsl 2:10–11

Pandey RN, Reddy KS, Dhanasekar P (2005) Induced mutations in mungbean: a review. J Arid Legum 2(2):314–324

Park D, Vestal TA (2002) Electron Beam 21st century food technology. http://oaktrust.library.tamu.edu/bitstream/handle/1969.1/87187/pdf_1751.pdf

Parry MAJ, Madgwick PJ, Bayon C et al (2009) Mutation discovery for crop improvement. J Exp Bot 60(10):2817–2825

Pathania A, Sood BC (2006) Spectrum and frequency of mutations induced by gamma rays and EMS in chickpea (*Cicer arietinum* L.). Res Crops 7(2):453–457

Patial M, Thakur SR, Singh KP et al (2017) Frequency and spectrum of chlorophyll mutations and induced variability in ricebean (*Vigna umbellata* Thunb, Ohwi and Ohashi). Legum Res 40(1): 39–46

Pawar SE, Manjaya JG (1996) Use of induced mutations for developing improved varieties of blackgram. In: Proceeding of 2nd International Crop Science Congress on crop Science Congress on Crop productivity and sustainability-shaping the future, (ICSCCPSF96), New Delhi, p 258

Pawar SE, Manjaya JG, Souframanien J et al (2000) Genetic improvement of blackgram using induced mutations. In: Paper presented in DAE-BRNS Symposium on the use of Nuclear Techniques in crop improvement Dec 6–8, 2000, Mumbai, pp 170–174

Qiu F, Li JG, Weng ML et al (1998) Molecular analysis of long-pod mutant line of mung bean gene rated by space mutagenesis. Sci Agric Sin 31(6):1–5

Rafiq Wani M, Imran Kozgar M, Tomlekova N et al (2014) Mutation breeding: a novel technique for genetic improvement of pulse crops particularly chickpea (*Cicer arietinum* L.). In: Improvement of crops in the era of climatic changes. Springer, New York, pp 217–248

Raina A, Laskar RA, Tantray YR et al (2020) Characterization of induced high yielding cowpea mutant lines using physiological, biochemical and molecular markers. Sci Rep 10:1–22. https://doi.org/10.1038/s41598-020-60601-6

Reddy KS, Dhanasekar P, Dhole VJ (2008) A review on powdery mildew disease resistance in mungbean. J Food Legum 21(3):151–155

Rheenen HAV, Murthy AK, Rao BV et al (2003) Inheritance of albinism in chickpea. Legum Res 26:254–258

Saima Mir A, Maria M, Muhammad S, Mahboob Ali S (2020) Potential of mutation breeding to sustain food security. In: Genetic diversity [Working Title]. IntechOpen, Rijeka, Croatia

Sánchez-Martín J, Steuernagel B, Ghosh S et al (2016) Rapid gene isolation in barley and wheat by mutant chromosome sequencing. Genome Biol 17(1):221

Schneeberger K, Ossowski S, Lanz C et al (2009) SHOREmap: simultaneous mapping and mutation identification by deep sequencing. Nat Methods 6:550–551. https://doi.org/10.1038/nmeth0809-550

Shah TM, Mirza JI, Haq MA et al (2006) Induced genetic variability in chickpea (*Cicer arietinum* L.) I. Frequency and spectrum of chlorophyll mutations. Pak J Bot 38(4):1217–1226

Shah TM, Atta BM, Alam SS et al (2010) High yielding kabuli mutant chickpea (*Cicer arietinum* l.) variety "CM 2008". Pak J Bot 42(5):3533–3545

Shah TM, Atta BM, Mirza JI et al (2011) Induced genetic variability in chickpea (*Cicer arietinum* L.) III. Frequency of morphological mutations. Pak J Bot 43:2039–2043

Shahwar D, Ansari MYK, Choudhary S (2019) Induction of phenotypic diversity in mutagenized population of lentil (*Lens culinaris* Medik) by using heavy metal. Heliyon 5:e01722. https://doi.org/10.1016/j.heliyon.2019.e01722

Shahwar D, Khan Z, Ansari MYK (2020) Evaluation of mutagenized lentil populations by caffeine and EMS for exploration of agronomic traits and mutant phenotyping. Ecol Genet Genomics 14. https://doi.org/10.1016/j.egg.2019.100049

Shaikh MAQ (1983) Genetic improvement of food legumes in developing countries by mutation induction. In: Summerfield RJ (ed) World crops: cool season food legumes. Kluwer Academic, Dordrecht

Sharma B (1969) Chemically induced mutations in cowpea (*Vigna sinensis* L. Savi). Curr Sci 38: 520–521

Sharma B, Kharkwal MC (1982) Induced mutations in grain legumes. In: Induced mutations for the improvement of grain legumes production II. IAEA-TECDOC-260. International Atomic Energy Agency, Vienna, Austria, pp 59–64

Sharma R, Singh VP (1992) The influence of altered moisture levels in seeds on the induction of mutations with EMS and gamma rays in mungbean. J Indian Botanical Soc 71:125–128

Shi J, Fan Q, Wang L et al (2000) Induction of large grain mutation in adzuki bean (*Phaseolus angularis* Wight) by space environmental condition. Acta Agric Nucleatae Sin 14(2):93–98

Sidorova KK (1981) Influence of genotypic background on the expressivity of mutant genes of pea. Pulse Crops Newsl 1(3):23–24

Sikora P, Chawade A, Larsson M et al (2011) Mutagenesis as a tool in plant genetics, functional genomics and breeding. Int J Plant Genomics 2011:314829. https://doi.org/10.1155/2011/314829

Singh VP, Agarwal S (1986) Induced high yielding mutants in cluster bean. Indian J Agric Sci 56(10):595–600

Singh VP, Singh M, Lal JP (2000) Gamma ray and EMS induced genetic variability for quantitative traits in urdbean (*Vigna mungo* L.) Hepper. Indian J Genet 60(1):89–96

Singh G, Sareen PK, Saharan RP et al (2001) Induced variability in mungbean [*Vigna radiata* (L.) Wilczek]. Indian J Genet 61(3):281–282

Singh DP, Sharma SP, Lal M et al (2013) Induction of genetic variability for polygenic traits through physical and chemical mutagens in cowpea [*Vigna unguiculata* (L.) Walp]. Legum Res 36:10–14

Sinha RP, Bharati RC (1990) Variability in mutant populations of urdbean (*Vigna mungo* (L.) Hepper). J Nucl Agric Biol 19(1):44–46

Solanki IS, Sharma B (2002) Induced polygenic variability in different groups of mutagenic damage in lentil (*Lens culinaris* Medik.). Indian J Genet 62(2):135–139

Song J, Li Z, Liu Z et al (2017) Next-generation sequencing from bulked-segregant analysis accelerates the simultaneous identification of two qualitative genes in soybean. Front Plant Sci 8:1–11. https://doi.org/10.3389/fpls.2017.00919

Souframanien J, Gopalakrishna T (2007) Source for bruchid resistance and its inheritance in Trombay wild blackgram [*Vigna mungo* (L.) *silvestris*]. J Food Legum 20:19–21

Souframanien J, Reddy KS (2018) Radiation induced mutations in genetic enhancement and development of new crop varieties in blackgram (*Vigna mungo* (L.) Hepper). In: FAO/IAEA International symposium on Plant mutation breeding and biotechnology held during 27–31, August at Vienna, Austria

Souframanien J, Roja G, Gopalakrishna T (2014) Genetic variation in raffinose family oligosaccharides and sucrose content in black gram [*Vigna mungo* L. (Hepper)]. J Food Legum 27:37–41

Souframanien J, Reddy KS, Petwal VC et al (2016) Comparative effectiveness and efficiency of electron beam and 60Co γ-rays in induction of mutations in black gram [*Vigna mungo* (L.) Hepper]. J Food Legum 29(1):1–6

Souframanien J, Saha AJ, Dhole VJ et al (2020) Genetic improvement of pulse crops through induced mutation and biotechnological approaches. IANCAS Bull XV:71–80

Stadler LJ (1928) Mutations in barley induced by x-rays and radium. Science 68:186–187

Suresh N, Kumar B (2020) Review on Mutation breeding in legumes and nodulation mutants of different legumes. Int J Curr Microbiol Appl Sci 2020(Special Issue-11):362–373

Takagi H, Abe A, Yoshida K et al (2013a) QTL-Seq: rapid mapping of quantitative trait loci in rice by whole-genome resequencing of DNA from two bulked populations. Plant J 74:174–183. https://doi.org/10.1111/tpj.12105

Takagi H, Uemura A, Yaegashi H et al (2013b) MutMap-Gap: whole-genome re-sequencing of mutant F2 progeny bulk combined with de novo assembly of gap regions identifies the rice blast resistance gene Pii. New Phytol 200:276–283. https://doi.org/10.1111/nph.12369

Tanaka A, Shikazono N, Hase Y (2010) Studies on biological effects of ion beams on lethality, molecular nature of mutation, mutation rate, and spectrum of mutation phenotype for mutation breeding in higher plants. J Radiat Res 51:223–233. https://doi.org/10.1269/jrr.09143

Thakur JR, Sethi GS (1995) Comparative mutagenicity of gamma rays, ethylmethane sulphonate and sodium azide in barley (*Hordeum vulgare* L.). Crop Res 9:350–357

Tickoo JL, Chandra N (1999) Mutagen induced polygenic variability in mungbean [*V. radiata* (L) Wilczek]. Indian J Genet Plant Breed 59(2):193–201

Till BJ, Reynolds SH, Weil C et al (2004) Discovery of induced point mutations in maize genes by TILLING. BMC Plant Biol 4:1–8

Till BJ, Cooper J, Tai TH et al (2007) Discovery of chemically induced mutation in rice by TILLING. BMC Plant Biol 7:19. https://doi.org/10.1186/1471-2229-7-19

Usharani KS, Kumar CR (2015) Mutagenic efficiency and effectiveness of gamma rays and EMS and their combination in inducing chlorophyll mutations in M_2 generation of Urdbean (*Vigna mungo* (L.) Hepper). Electron J Plant Breed 6(1):210–217

Vanniarajan C, Vivekanandan P, Ramalingam J (1996) Induced variability for quantitative characters in blackgram. Crop Res 11:341–346

Vanniarajan C, Chandirakala R, Souframanien J (2021) Genetic and biochemical characterization of electron beam and gamma ray induced mutants for superior battering quality trait and MYMV in urdbean [*Vigna mungo* (L.)]. Legum Res. https://doi.org/10.18805/LR-4583

Waghmare VN, Mehra RB (2001) Induced chlorophyll mutations, mutagenic effectiveness and efficiency in *Lathyrus sativus* L. Indian J Genet 61:53–56

Wang S, Cheng X (2018a) A new variety Zhonglyu No.8 from space mutation induction. In: FAO/IAEA International symposium on plant mutation breeding and biotechnology, 27–-31 August 2018. IAEA, Vienna. (Contribution ID: 218)

Wang L, Cheng X (2018b) A new variety Zhonglyu No.12 from space mutation. In: FAO/IAEA International symposium on Plant mutation breeding and biotechnology, 27–31 August 2018. IAEA, Vienna. (Contribution ID: 219)

Wang TL, Domoney C, Hedley CL et al (2003) Can we improve the nutritional quality of legume seeds? Plant Physiol 131(3):886–891

Wani MR, Khan S, Kozgar MI, Goyal S (2011) Induction of morphological mutants in mungbean (*Vigna radiata* (L.) Wilczek) through chemical mutagens. Nucleus 48(3):243–247

Wani MR, Dar AR, Tak A et al (2018) Chemo-induced pod and seed mutants in mungbean (*Vigna radiata* L. Wilczek). SAARC J Agric 15:57–67. https://doi.org/10.3329/sja.v15i2.35161

Wu YJ, Zhang Y, Yu W et al (2005) The progress of the research and application of ion implantation biotechnology in China. In: Solid state phenomena, vol 107. Trans Tech Publications Ltd, Bäch, Switzerland, pp 37–42

Yaseen MS, Naga Parimala Kumar SK, Prabakaran S, Suresh S (2020) Ethyl methane sulphonate (EMS) induced mutations in M1 generation of cowpea (*Vigna unguiculata* (L.) Walp). J Pharmacogn Phytochem 9:1545

Zhu H, Xu J, Li S et al (2008) Effect of high- energy-pulse-electron beam radiation on biomolecules. Sci China Ser B Chem 51:86–91

Advances in Mutation Breeding of Groundnut (*Arachis hypogaea* L.)

16

Anand M. Badigannavar and Suvendu Mondal

Abstract

Induced mutagenesis finds a significant place in crop enhancement methodologies for bringing genetic variability in desired genetic backgrounds. Mutation breeding along with recombination breeding has developed >3300 mutant varieties in various crop species globally. In groundnut, radiation and chemical mutagenesis have been extensively employed for genetic improvement of vegetative, reproductive, agronomical, biochemical and physiological traits. Consequently, these mutant traits were instrumental in delivering 112 suitable productive cultivars in this allotetraploid leguminous crop. Induced mutants and their derived varieties acted not only as basic genetic pool for evolving desirable varieties but also for understanding various functions at biochemical and molecular levels. Numerous farmers, traders and exporters have benefitted by cultivating groundnut mutant varieties in many countries. Recent advances in genomics have facilitated to utilize molecular tools like gene editing, TILLING and mutagenomics for developing desired and improved traits in groundnut, addressing famers' concerns, consumer preference and industrial needs.

Keywords

Induced mutagenesis · Gamma rays · EMS · Recombination breeding · Mutant varieties · TILLING · Mutagenomics · Gene editing

A. M. Badigannavar (✉) · S. Mondal
Nuclear Agriculture and Biotechnology Division, Bhabha Atomic Research Centre, Mumbai, India
e-mail: anandmb@barc.gov.in

S. Penna, S. M. Jain (eds.), *Mutation Breeding for Sustainable Food Production and Climate Resilience*, https://doi.org/10.1007/978-981-16-9720-3_16

16.1 Introduction

Groundnut (*Arachis hypogaea* L.) is an important edible oilseed, food and feed crop cultivated globally on 27.52 million ha with a production of 45.52 million metric tonnes (MMT), with a productivity of 1654 kg/ha during 2016–2020 period (FAOSTAT 2020). Groundnut is the 2nd most important grain legume crop, 4th most important oilseed crop and 13th most important food crop and is currently cultivated in over 80 countries. Among the groundnut-growing countries, China has the highest production (16.70 MMT) followed by India (7.50 MMT), Nigeria (3.15 MMT) and the USA (2.68 MMT). It covers around 19.23% of the cultivated area with 27.33% contribution to the total oilseed production in India (ICAR-IASRI 2019). It is distinct from the rest of the legume species by having above-ground flower and below-ground fruit. Fruit development involves aerial elongation, below-ground pod expansion, seed filling and finally seed drying.

Groundnut is mainly consumed for cooking oil, protein, minerals and vitamins in many nations and contributes considerably to food security in turn alleviating poverty. Around 50% of the total global groundnut produce is utilized for oil purpose; 35% is for food or confectionary purpose, and the remaining 15% is for seed and animal feed (Birthal et al. 2010). Groundnut oil is primarily used for cooking and manufacture of soap and margarine. Seeds are consumed as raw or roasted nuts, in making confectioneries or butter. Tender pods are also eaten as vegetable. Protein-rich de-fatted cake is a nutritious livestock feed. Microbe-processed groundnut shells and cakes are used to manufacture commercially essential enzymes.

Groundnuts have recently been considered as a functional food. Several studies revealed that its consumption has distinct advantages on human health. Diets enriched with groundnut and its butter reduced the risk of heart disease, cancer, total cholesterol, bad cholesterol and triglycerides without affecting beneficial cholesterol due to the presence of monounsaturated fatty acid, resveratrol, beta-sitosterol, vitamin E, folic acid and fibre (Francisco and Resurreccion 2008). It contains nearly half of the minerals and vitamins necessary for normal body growth and maintenance. Additionally, groundnut skins or its extracts are identified as an antioxidant, antimicrobial agent, functional food and animal feed component (Toomer 2020). Due to its satiety value, it prolongs hunger and is helpful in dieting for body weight maintenance. Children gained body weight and height with groundnut protein fortifications.

Genetic variability for a wide array of characters is of utmost necessity for plant breeding success. In nature, mutations, though their frequency is very low, are the main source of variability. Breeder with the help of ionizing radiations and some mutagenic chemicals enhances the mutation frequency. Mutation breeding involves induction of genetic variability and employing induced variability either directly or in cross-breeding. With the specific objectives and targeting, improvement of one or two traits in a well-adapted variety is fundamental to success in mutation breeding at many institutes including ours. Sustained efforts in mutation breeding for many decades resulted in commercial release of >3300 crop varieties in different species worldwide (IAEA 2021).

Cultivated groundnut (*A. hypogaea* L.) exhibits narrow genetic base, regardless of having considerable agronomical, morphological and physiological variability because of its restricted gene flow due to ploidy barrier, self-pollination and monophyletic origin (Holbrook and Stalker 2003). Due to paucity of genetic variability, groundnut is vulnerable to many biotic and abiotic stresses. Breeding for better pod yield continues to be the primary objective in any of groundnut improvement activities. Induced mutagenesis is one of the convenient and desirable approach for broadening genetic variability to overcome the limitations associated with a narrow genetic basis and could bring specific improvement without significantly affecting other traits in groundnut (Patil and Mouli 1979). Extent of induced variability is dependent on mutagen type, studied character and genetic background (Murty et al. 2004; Nadaf et al. 2009; Badigannavar et al. 2020). Induced mutants not only serve as parental material in crop breeding but are also genetic resources for functional genomics. In groundnut, effective application of induced mutagenesis along with cross-breeding has led to the generation of a wide spectrum of mutants for plant type, early maturity, pod type and size, seed size and composition, testa colour and tolerance to both biotic and abiotic stresses in groundnut (Patil and Mouli 1979; Mouli et al. 1989a; Kale et al. 2000a; Mondal et al. 2007; Gowda et al. 2010; Kavera et al. 2013; Yu et al. 2019; Badigannavar et al. 2020; Brown et al. 2021). This review envisages experiences and recent developments in mutation breeding in groundnut.

16.2 Mutagens

In groundnut mutation breeding experiments, the objectives are to induce mutants with better pod yield, ideal plant type, desirable pod traits, high shelling percentage, large seed, moderate seed dormancy, early maturity, high oil content, improved seed nutritional traits and tolerance to diseases, salinity, moisture stress, heat, etc. Many of the physical and chemical mutagens were effectively employed to induce desirable mutants, in turn leading to mutant varieties (Table 16.1). In early years, mainly X-rays and later gamma rays were used for induced mutagenesis. Additionally, fast neutrons, electron beam and heavy-ion beam have been used for mutation induction in groundnut. Electron beam irradiation was standardized using linear accelerator (10 MeV) for low-dose (50–1000 Gy) application for groundnut in India (Mondal et al. 2017). Based on the electron beam treatment of seeds of five groundnut genotypes, GR_{50} (50% growth reduction) values ranged from 108 Gy in TG 51 (most radio-sensitive) to 270 Gy in TAG 24 (most radio-tolerant). For gamma rays, corresponding values were 295 Gy and 385 Gy, respectively. Another ionizing radiation with a high linear energy transfer (LET), fast-neutron irradiation, can bring secondary ionization as well as gene mutations in plant cell, resulting in stable mutant traits (Wang et al. 2015). Heavy-ion beams, due to their greater LET and relative biological effect (RBE) as compared to gamma rays or X-rays, induce greater mutation rate, in turn generating a wide range of mutation types. X-rays

Table 16.1 Different mutagens used for induced mutagenesis in groundnut

Mutagen	Dose/concentration	References
Physical mutagens		
X-rays	10–35 kR	Gregory (1955), Bora et al. (1961), Patil and Bora (1961), Patil (1968), Menon et al. (1970), Patil and Mouli (1977)
Gamma rays	100–600 Gy	Mouli and Patil (1976), Reddy et al. (1977), Mouli and Kale (1982b), Pathirana et al. (1998), Branch (2002), Liu et al. (2004), Badigannavar and Murty (2007), Mondal et al. (2007), Nadaf et al. (2009), Hassan and Anes (2015), Badigannavar et al. (2020), Fu et al. (2021)
Electron beam	150–250 Gy	Mondal et al. (2017)
Fast neutron	10–18 Gy	Wang et al. (2015)
Neutron beam	2.5×10^{12}–5×10^{13} n/cm^2 s	Shivraj et al. (1962), Shivraj and Ramana Rao (1963)
Heavy-ion beam (C, N)	100 Gy	Cabanos et al. (2012)
Laser		Liu et al. (2004), Lin et al. (2005)
Chemical mutagens		
Ethyl methanesulphonate (EMS)	0.01–0.5%	Ashri and Levy (1976), Prasad et al. (1984), Sivaram et al. (1989), Gowda et al. (1996), Mathur et al. (2000), Nadaf et al. (2009), Wang et al. (2011b), Chen et al. (2020)
Diethyl sulphate (DES)	0.01–0.2%	Ashri and Goldin (1965), Ashri and Levy (1976), Mathur et al. (2000)
EMS + DES	0.01–0.1% + 0.01–0.1%	Mathur et al. (2000)
Ethidium bromide	500 ppm	Levy and Ashri (1978)
Acriflavine	1.15 mM	Ashri et al. (1977)
Sodium azide	3 mM; 0.41% for 2 h	Prasad et al. (1985), Mathur et al. (1998), Mondal et al. (2007), Wang et al. (2002, 2011a), Nkuna et al. (2021)
Nitroso methyl urea	0.02%	Prasad et al. (1984)
Pingyangmycin	4 mg/L	Sui et al. (2015)

yield low-LET radiation (0.2 and 5 keV/μm) resulting in few ionizations. Hence, most probably, low-LET radiation will mainly cause easily repairable single-strand DNA breaks. On the contrary, heavy ions and fast neutrons having higher LET induce clusters of double-strand breaks in the DNA, which are difficult to repair (Shikazono et al. 2005; Cabanos et al. 2012). The effective dose for gamma rays ranges from 200 to 350 Gy, which is close to GR_{50} depending on the genotype and radio-sensitivity factors (Badigannavar and Murty 2007; Nadaf et al. 2009). Decreased germination and increased seedling mortality, pollen sterility, and morphological and cytological anomalies were reported with the increasing doses of gamma rays or concentrations of ethyl methanesulphonate (EMS) (Gowda et al. 1996; Badigannavar and Murty 2007; Mondal et al. 2007). Mutagenic effectiveness was higher in gamma rays than the EMS, while with an increased rate of dose/concentration, mutagenic efficiency was decreased for lethality and pollen sterility and increased for injury (Manjunath et al. 2020). With an increase in doses of fast neutrons, there was decrease in the frequency of somatic embryo formation and plantlet regeneration (Wang et al. 2015). Mathur et al. (1998) standardized the LD_{50} dose of 0.41% for a 2-h treatment for sodium azide in groundnut. Subsequently, Mondal et al. (2007) found that the combination of sodium azide and gamma rays was ideal for developing groundnut mutants with greater genetic variability.

Seeds (M_0) of groundnut cultivars, mutants, selections, advanced lines or hybrids are usually treated with various mutagens (Murty et al. 2004). Subsequently, treated seeds (M_1) are sown in the field. Usually, desirable mutants are selected from M_2 generation onwards, and their breeding behaviour is confirmed in the next generations. Since then, usual breeding methods like pedigree method, bulk method or single-seed descent for agronomic evaluation have been followed. The induced mutants are exploited in recombination breeding by crossing mutant with mutant, mutant derivative, breeding line or cultivar and mutant derivative with cultivar (Fig. 16.1).

16.3 Cytogenetic Aberrations

Groundnut is an allotetraploid ($2n = 4x = 40$, AABB) legume crop belonging to the Fabaceae family. Radiations are known to induce chromosomal aberrations in the M_1 generation. The meiotic aberrations due to X-ray irradiation of groundnut cv. Spanish Improved were irregular development of spindles and their activity, abnormal association of chromosomes and their separation at anaphase, occurrence of bridges and fragments, failure of cytokinesis, formation of polyads and anomalies in the development of pollen grains (Bora et al. 1961). Further, many reciprocal translocations leading to the formation of chain and ring multivalents, inversions resulting in fragments and bridges at anaphase I and II and persistent anaphase I bridges at telophase II were reported (Patil and Bora 1961). In M_4 progenies, Patil (1968) observed trisomics, tetrasomics, cytomixis, long chromosomes with disturbed coiling mechanism and 15–18 chromosomes in pollen mother cells. Ashri et al. (1977) induced three trisomic mutations by treating developing embryos with acriflavine. X-ray irradiation of inter-specific (*A. hypogaea* × *A. monticola*) hybrid showed occurrence of monosomics (Menon et al. 1970). Asynaptic chromosomes with very low frequency of bivalents at diakinesis were observed in the X-ray-induced dwarf mutant (Patil and Mouli 1977). An altered chromosomal organization in meiotic prophase and early metaphase stages, a spherical dot instead of normal round centromere in meiotic metaphase and the displaced spindle fibres in most stages of meiosis were observed (Hassan and Anes 2015). Many chromosomal aberration types like earlier separated mitosis metaphase, polypolarity division,

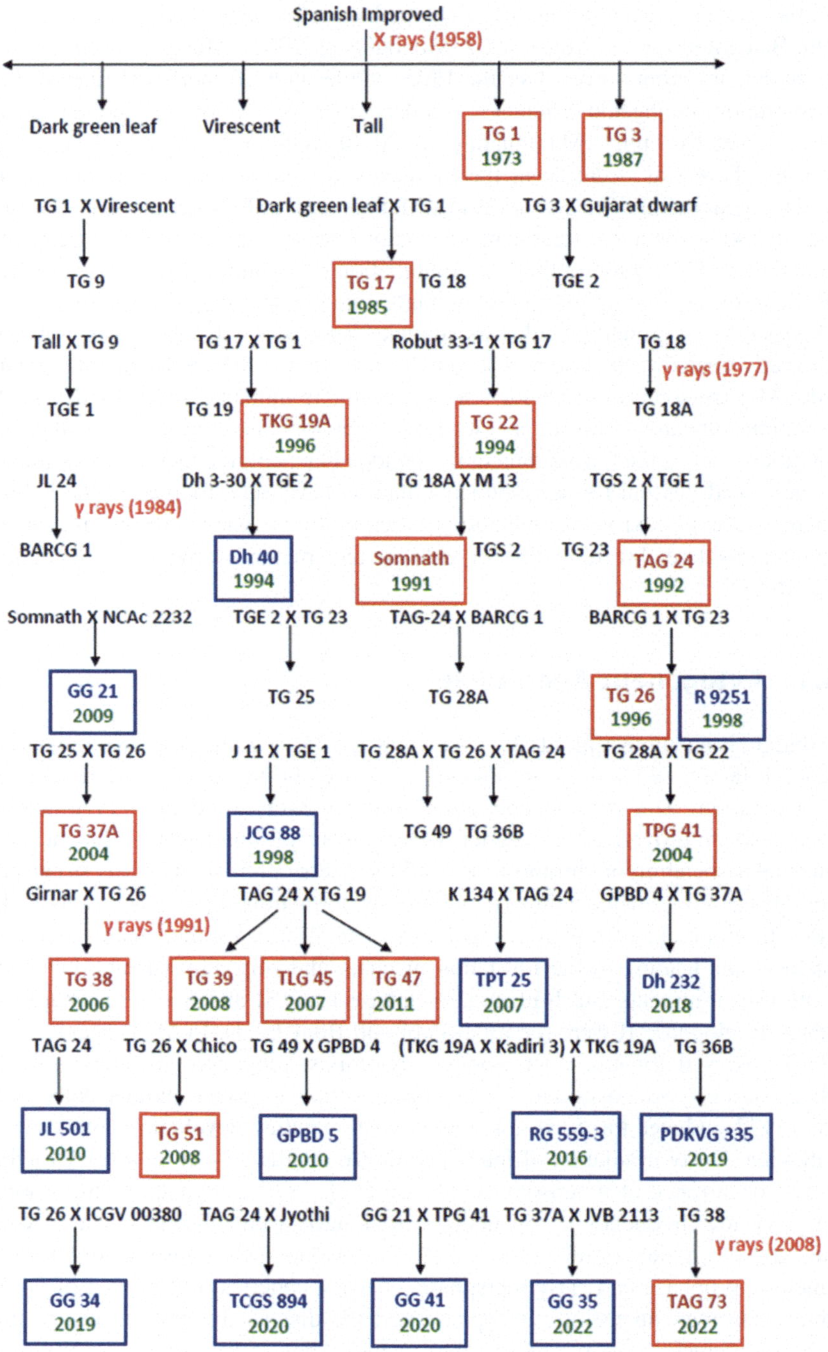

Fig. 16.1 Evolution of groundnut mutant and mutant-derived varieties developed at Bhabha Atomic Research Centre, Mumbai, India (red square), and their usage as parents for mutant variety development by state agricultural universities (blue square), which are released and notified across the country

sticky, non-equal division and nucleus protrusion into cytoplasm were also found due to laser irradiation (Lin et al. 2005).

Recently, sequential genomic in situ hybridization (GISH) and fluorescence in situ hybridization (FISH) were employed to study chromosomal variations in Chinese variety, Silihong, and in 70 gamma ray-induced M_1 plants with two multiplex probe cocktails, total genomic DNAs of *A. ipaensis* and *A. duranensis*, and 45S and 5S rDNAs as the probes (Fu et al. 2021). In 14 M_1 plants, 8 monosomic chromosomes, 1 deletion and 17 translocations were identified. In one of the M_1 plants, one reciprocal translocation between chromosomes 1A and 3B was also observed. In these chromosomal changes, nine translocations were observed between non-homologous chromosomes and eight translocations between homoeologous chromosomes. Higher number of translocations were seen in chromosomes 1, 3 and 5. Such chromosomal variations in groundnut could be employed for gene mapping, radiation hybrid mapping and translocation or deletion mapping.

16.4 Mutations for Seed Size

In groundnut, large seeds attract greater consumer and market preference for confectionery and value addition purpose, which may get better price in domestic and international markets. Groundnut growers often face problems with the large-seed varieties like late maturity, low shelling outturn, less proportion of sound mature kernels, longer seed dormancy and lower yields. Induced mutagenesis was successfully employed for the development of desirable high-yielding large-seed varieties with earliness and optimum seed dormancy, making them fit for diverse cropping patterns. In late 1950s, X-ray treatment of cv. Spanish Improved induced a large-seed mutant, which was later released as TG 1 based on its superior yield performance (Patil 1975) (Fig. 16.1). Inter-mutant cross between TG 1 and dark-leaf mutant resulted in extreme *fastigiata*, high-yielding variety, TG 17 (Patil 1977). Subsequently, by including TG 1 and TG 17 in crosses, TG 22 with medium-large seeds and TKG 19A and Somnath with large-seed varieties were developed and released (Mouli et al. 1989b, 1990). Further, a large-seed sister line of TKG 19A, TG 19, was recombined with TAG-24 to evolve three high-yielding large-seed varieties, TG 39, TLG 45 and RARST-1 (TG 47) with 115–120 days to maturity, 72–75% shelling outturn, 70–80 g HKW (hundred kernel weight) and higher proportion of large seeds (Kale et al. 2008, 2010; Badigannavar et al. 2012). Towards diversification of large-seed genetic base, gamma ray mutagenesis of TAG 24 (Badigannavar and Murty 2007), TG 38 (Badigannavar et al. 2020) and TG 66 (Mondal et al. 2007) and electron beam mutagenesis of TG 26 (Mondal et al. 2017) were carried out periodically and several high-yielding large-seed (HKW: 70–80 g) mutants (TG 77, TG 78, TG 79, TG 89, TG 90) with 110–115 days to maturity were isolated, which are performing better in national and state evaluation trials. Similarly in Sri Lanka, Pathirana et al. (1998) isolated three high-yielding gamma ray-induced mutants with large seeds. Branch (2002) selected four mutants having larger pod and seed and greater percentage of extra-large kernels than parent through gamma ray mutagenesis. Field performance of one mutant showed 23% higher pod yield than parent. Further, sodium azide treatment of cv. L7-1 induced mutants with a greater change in pod and seed size and weight (Wang et al. 2002). Joshua and Bhatia (1983) attributed increased seed size in groundnut to the increased cotyledonary cell volume by retaining similar cell number within the unit area.

16.5 Mutations for Early Maturity

Early-maturing groundnut genotypes facilitate in escaping end-season drought and end-season rains and fit into different cropping systems like paddy fallows, residual moisture situation and post-potato crop. Early maturity also makes farmers to harvest early and bring the produce early to the market, which in turn fetch better price. Breeding for earliness in groundnut is complicated due to its indeterminate nature and subterranean cumulative process of pod maturation. By exercising selection pressure for better pod yield, higher shelling outturn and greater HKW from early harvest, early-maturing (95 days) mutant derivatives TGE-1 and TGE-2 were evolved by recombining mutants (Patil et al. 1982; Mouli and Kale 1982a) (Fig. 16.1). The maturity was further brought down to 90 days with desirable traits using induced mutagenesis (Mouli and Kale 1989). Early maturity and high yielding ability were brought together in TAG-24 (Patil et al. 1995). Further, earliness from Chico was combined with better pod yield of TG 26, leading to the development of a new variety, TG 51, having 90 days to maturity, greater pod yield and better shelling (Kale et al. 2009). Gamma ray mutagenesis of TG 51 along with polyethylene glycol (PEG)-based screening for moisture-stress tolerance has identified drought-tolerant, early-maturing mutant, TG 84, which has shown better performance for seed yield in national evaluation trials.

16.6 Mutations for Subspecific Traits

In *Arachis* genera, 81 species including the cultivated groundnut, *A. hypogaea*, have been reported (Valls et al. 2013). *A. hypogaea* is categorized into subspecies, *fastigiata* and *hypogaea*, which are further classified based on morphological features into botanical varieties. Accordingly, ssp. *fastigiata* is classified into var. *aequatoriana*, var. *fastigiata* (Valencia), var. *peruviana* and var. *vulgaris* (Spanish), while ssp. *hypogaea* into var. *hirsute* (Peruvian runner) and var. *hypogaea* (Virginia runner). Consistent generation of mutations and recombination in groundnut resulted in interchanges for intra-specific and inter-specific characters among genotypes shown in Tables 16.2 and 16.3.

Table 16.2 Studies on mutations between groundnut botanical varieties

From		*To*			
	Variety	Spanish bunch (var. *vulgaris*)	Valencia (var. *fastigiata*)	Virginia bunch (var. *hypogaea*)	Virginia runner (var. *hypogaea*)
	Spanish bunch (var. *vulgaris*)	–	7	1, 7, 8	2, 3, 6, 7
	Valencia (var. *fastigiata*)	7	–	2, 7	7
	Virginia bunch (var. *hypogaea*)	4, 5, 7	–	–	7
	Virginia runner (var. *hypogaea*)	3, 7	7	7	–

References: 1. Patil (1966); 2. Reddy et al. (1977); 3. Levy and Ashri (1978); 4. Mouli and Kale (1982a); 5. Mouli et al. (1986); 6. Hussein et al. (1991); 7. Gowda et al. (1996); 8. Mondal et al. (2007)

Table 16.3 Mutant or mutant derivatives showing mutations at subspecies levels in groundnut

Traits	Mutant or mutant derivatives	References
fastigiata genotypes with *hypogaea* traits	BARCG 1, BARCG 2, TG 7, TG 8, TG 9, TG 10, TG 12, TG 13, TG 13A, TG 16, TG 17, TG 18A, TKG 19A, TG 22, TG 23, TAG 24, TG 25, TG 26, TG 27, TG 28, TG 28A, TG 40, TPG 41, TG 42, TG 44, TG 45, TG 46, TG 47, TG 48, TG 49, TG 77, TG 78, TG 79, TG 89, TG 90	Mouli and Kale (1982b), Mouli and Kale (1989), Mouli et al. (1989a, 1990), Kale et al. (2000b), Badigannavar and Murty (2007), Mondal et al. (2007, 2017), Badigannavar et al. (2020)
hypogaea genotypes with *fastigiata* traits	TG 1A, TG 18, TG 39, TG 43	Kale et al. (2000b)
Genotypes having *hypogaea* spreading habit and sequential flowering	TGS-1 (Somnath), TGS-2 and TGS-3	Mouli et al. (1989b)
Genotypes with one or more modified characters absent in *fastigiata* and *hypogaea*	TGE-1, *sl*, *sl-imp* mutants, TG 18A-84, TG 21	Mouli and Kale (1982a), Patil and Mouli (1984)

16.7 Mutations for Trait Association

Many of the groundnut traits are associated and inherited together in succeeding generations. Such associations are disintegrated due to induced mutagenesis resulting in favourable recombinants. The usual negative correlation between seed size and oil content or shelling outturn or maturity was disassociated through induced mutants or inter-mutant crosses, leading to genotypes with large seeds having high oil content, early maturity and increased shelling outturn (Patil 1973; Mondal et al. 2017). High water-use efficiency and high harvest index, otherwise negatively correlated, were recorded in inter-mutant derivatives, TAG-24 and Somnath, in multi-location trials (Murty et al. 2004).

16.8 Mutations for Biotic and Abiotic Stress Tolerance

In groundnut, biotic and abiotic stresses severely affect the seed productivity and quality. The induced mutants and their derivatives tolerant to diseases, pests, drought, acid soil, heat and cold have been reported in groundnut (Table 16.4). Taxonomically important groundnut genotype, Dharwad Early Runner (DER) upon successive EMS treatments, resulted in induced mutants (mutants 1-45, 1-110, 28-2)

Table 16.4 Mutant or mutant derivatives of groundnut showing tolerance to biotic and abiotic stresses

Trait	Tolerant mutant or mutant derivatives	References
Late leaf spot	Dh 232, GPBD 5, 1-45, 1-110, Mutant 28-2, PBS 30107, PBS 30108	Mathur et al. (2000), Motagi et al. (1996, 2022)
Early leaf spot	ICGV 76, GG 13	Basu (2002)
Rust	Dh 232, GG 13, GPBD 5, PBS 30108, PBS 30138	Mathur et al. (2000), Basu (2002), Motagi et al. (2022)
Collar rot	TG 37A	NRCG (2000)
Peanut bud necrosis disease	TAG 24, TG 37A, R 9251	Patil et al. (1995), Giriraj and Itnal (1999), NRCG (2000)
Peanut stem necrosis disease	TKG 19A	NRCG (2000)
Spodoptera litura	Mutant 28-1, 28-2, 45, 110-1	Rajendra Prasad et al. (1998)
Thrips	Mutant 28-1, 28-2, 45, 110, 172	Rajendra Prasad et al. (1998)
Bruchid beetle	TSP 60, PBS 30001	Mathur et al. (2000)
Phosphorus efficiency	PBS 30016, PBS 30026	Mathur et al. (2000)
Acid soils	ICGV 76, TKG 19A	Annual report (2001)
Salinity	Binachinabadam-5, Binachinabadam-6	Azad et al. (2014)
Drought	Dh 256, GG 35, PBS 30008, PBS 30022, PBS 30023, PBS 30109, Tafra-1, TG 37A, TG 39, TPT 25	Mathur et al. (2000), Kale et al. (2004a, 2010), Abdalla et al. (2018), Motagi et al. (2022)
Heat	PBS 30109, PBS 30138, TG 18	Mathur et al. (2000), Anonymous (2002)
Cold	TAG 24	Dave and Mitra (2000)

with late leaf spot (LLS) resistance (Motagi et al. 1996). Further, some of these mutants showed resistance to *Spodoptera litura* (mutants 28-1, 28-2, 45, 110-1) and to thrips (mutants 28-1, 28-2, 45, 110, 172) (Rajendra Prasad et al. 1998). EMS mutagenesis of mutant VL1 having rust resistance and LLS susceptibility produced morphologically similar mutants with LLS resistance and rust susceptibility (Gowda et al. 2010). Origin of resistant mutant was later related to the insertion of *A. hypogaea* miniature inverted-repeat transposable element (AhMITE1) at FST1-linked site. Subsequent spontaneous mutation in these mutants produced variants which were LLS and rust susceptible. Origin of such variants was linked to the excision of AhMITE1 from FST1-linked site, demonstrating its main role in high-frequency origin of LLS-resistant mutants in groundnut. Towards developing aflatoxigenic resistant fungal mutants in groundnut, Azzam et al. (2007) isolated mutants with significant decrease in occurrence of *Aspergillus flavus* and *A. parasiticus* and aflatoxin in comparison to parent, Giza 5. These mutants compared to Giza 5 had the lowest content of aflatoxin B1 and/or B2 under soil infestation with aflatoxin-producing fungi and were found free from aflatoxin contamination under field conditions.

Salinity, being an important abiotic stress, impacts various growth stages of groundnut including final yield. It is ideal to develop salinity-tolerant groundnut genotypes to minimize yield losses due to salinity problem. So far in groundnut, tangible results have not been achieved due to lack of simple and suitable screening method for salinity tolerance in large populations. Towards this aspect, a simple screening protocol for larger segregating or mutant populations for radicle growth was developed (Badigannavar et al. 2007). Considerable genetic variability for germination under 400 mM NaCl stress was noted in established gamma ray mutants of TAG 24. In six of the TAG 24 mutants, greater radicle growth was noted by screening at 125 and 150 mM NaCl. Further, three mutants at 125 mM and one mutant at 150 mM NaCl had better tolerance due to longer radicle length and lesser radicle reduction. Further, TPG 41, a large-seed variety, was irradiated with gamma rays to induce genetic variability for radicle growth under salinity (Badigannavar et al. 2007). By screening large number of seeds at 100 mM NaCl in the M_3 followed by screening plant-wise in the M_4 and M_5, 91 true breeding mutants with salinity tolerance for radicle growth were obtained. Ahmed and Mohamed (2009) identified three salinity-tolerant mutants obtained through gamma ray and sodium azide treatment when they were evaluated with soil and irrigation water salinity (ECe) having 13.15 dS/m and 6.5 dS/m, respectively, in M_5 and M_6 generations. Mutants produced higher seed and pod yield per plant and more number of pods and seeds per plant than their parents. Azad et al. (2014) studied groundnut mutants in the Bangladesh coastal belt under saline field conditions. It was found that the high-yielding mutants under rainfed condition mostly took a lesser period to close the stomata in spite of their higher stomatal and lower cuticular transpiration rates. Of these, two mutants having comparatively better yield than 'Dacca-1' were commercialized as salt-tolerant varieties by names 'Binachinabadam-5' and 'Binachinabadam-6'.

In vitro mutagenesis of Huayu 20 seeds with pingyangmycin and screening on a medium having hydroxyproline was carried out to obtain drought-tolerant mutants (Sui et al. 2015). The seedlings of M_3 individuals were screened for drought stress. The activities of superoxide dismutase and peroxidase were substantively raised in eight progenies of hydroxyproline-tolerant, regenerated plants than parent. Additionally, under drought stress, few M_3 progenies gave more pods compared to parent. Abdalla et al. (2018) evaluated groundnut mutants for end-season drought tolerance in Sudan. By using the rainout shelter, 25 days' terminal drought was imposed after 60 days from planting. These terminal drought-survived mutants were further evaluated for yield performance under rainfed situation. The mutant Barberton-B-30-3 gave a mean pod yield of 1024 kg/ha as compared to 926 kg in check variety over 12 seasons. Further, the stability and GGE biplot analysis also showed that Barberton-B-30-3 was more stable by producing better yield in both low and high rainfall seasons and was officially released as 'Tafra-1' in 2018.

16.9 Mutations for Physiological Traits

Though the physiological traits are genetically quantitative in nature, few studies have shown mutations for such traits in groundnut. Chlorophyll synthesis was studied in wild-type and virescent mutant leaves by continuous illumination of dark-grown seedlings (Benedict and Ketring 1972). Chlorophyll synthesis in the virescent leaves showed a 72-h lag period compared to the wild-type leaves before the onset of rapid chlorophyll accumulation. The development of chloroplast grana, protein synthesis and activity of many enzymes of the malate dehydrogenase, phosphoenolpyruvate carboxylase and reductive pentose phosphate cycle decreased in the virescent leaves during the lag phase of chlorophyll accumulation. Later, Alberte et al. (1976) compared the photosynthetic activity and chloroplast lamellar system of wild-type and virescent leaves. The leaves in mutant have shown 42% reduction in chlorophyll, a reduction in the number of photosystem I reaction centre, a higher chlorophyll a/b ratio and a change in proportions of the two chlorophyll-protein complexes of the chloroplast. In addition, the mutant had 1.5 times larger photosynthetic unit size than the wild type. Groundnut mutants, TG 1 and TG 16, showed higher photosynthetic rate than their parent Spanish Improved. Among the two, TG 1 was found more efficient. The translocation of photosynthates to nodules was also greater in TG 1. Comparatively higher nitrogen fixation and dark CO_2 fixation capacity in nodules were reported in the same mutants (Lodha et al. 1983, 1985).

16.10 Mutations for Seed Biochemical Traits

Induced mutagenesis has played a vital role in altering seed nutritional traits like oil content, fatty acid, protein, minerals and vitamins. Seed oil content was increased by 2% from 49.5% in the mutants of groundnut cv. JL-24 (Mouli et al. 1987). Inter-mutant crosses resulted in high oil (55%) genotypes with a 5% increase (Patil 1973). On the contrary, a mutation also brought down oil content from 46% to 38% (Mouli and Kale 1991). Doo et al. (2008) observed 43.2–53.5% oil among gamma ray mutants compared to 47.8% in parent Shinnamkwang. Similarly, oil content was enhanced by 4.3–6.1% in seven gamma ray mutants of TAG 24 along with better seed and oil yield (Badigannavar and Mondal 2009). With an in vitro mutagenesis of Huayu 20 seeds with pingyangmycin, mutant seeds had >57% oil as compared to 53.7% in parent (Sui et al. 2015). Subsequently, three mutants with oil contents 57.7%, 61.1% and 59.3% were released as new groundnut varieties, Yuhua 4, Yuhua 9 and Yuhua 14, respectively (Wang et al. 2020).

Groundnut seeds with higher oleic acid have shown greater shelf life, and their consumption has several advantages to human health (Kris-Etherton et al. 2001; Moreira et al. 2014). Mutations affecting modified oleic and linoleic acid (O/L) ratios were obtained in induced TG mutant and mutant derivatives (3.16–3.32) as compared to parent, Spanish Improved (1.02) (Sharma et al. 1981, 1985). Mondal and Badigannavar (2010) isolated gamma ray mutant TGM 71 from large-seed variety, TPG 41, with higher oleic acid (68.7%) and lower linoleic acid (13.6%), in turn enhancing O/L ratio to 5.0 as compared to its parent (61.9%, 19.3%, 3.2, respectively). Another mutant, TGM 192M obtained from gamma rays and sodium azide treatment, showed better oleic acid (70.7%) and lesser linoleic acid (12.8%) than its parent, TFDRG 5 (55.3%; 26.0%), which in turn increased the O/L ratio to 5.5 from 2.0 (Mondal et al. 2011). Further, this high oleate mutant was improved for pod yield by hybridizing it with the varieties, TPG 41 and TG 51, and directed selection resulted in eight high-yielding advanced selections with 70–75% oleic acid (Mondal et al. 2018; Badigannavar et al. 2020). In addition to gamma rays, EMS mutagenesis also generated a narrow-leaf high-oleate mutant (68.6%) from TMV 2 (38.5%) (Prasad et al. 1984; Mondal and Badigannavar 2013). An inter-specific disease-resistant variety, GPBD-4, with both EMS and gamma ray mutagenesis resulted in several mutants with >70% oleic acid as compared to 50.7% in parent (Kavera et al. 2013). Recently, a normal oleic cv., Fuhua 12, was successively treated with EMS in two rounds, and mutants having 76.9–83.9% oleic acid were isolated in contrast to 35.9% in their parent (Yu et al. 2019).

Induced mutations have contributed to protein enhancement in groundnut. Induced mutants, TG 1 (27.68%) and TG 16 (31.64%), were reported to have higher protein compared to parent, Spanish Improved (13.59%) (Sharma et al. 1981, 1985). These mutants also had higher contents of histidine, lysine, phenylalanine, proline and tryptophan and lower contents of methionine, serine and threonine. The limiting amino acids were tryptophan in Spanish Improved; threonine in TG 8, TG 9 and TG 17; and valine in TG 18. Further, the essential amino acid content per seed excepting methionine, valine and threonine was higher in all the mutants. Doo et al. (2008) found 23.3–31.7% protein in gamma ray mutants compared to 26.8% in parent Shinnamkwang. An EMS-derived mutant having 28.6% protein was isolated as compared to 17.7% in its parent (Wang et al. 2013). A high-seed-protein mutant TGM 206 (31%) with selective increase in conarachin fraction was identified from gamma ray mutagenesis of TG 66 (26%) (Mondal and Badigannavar 2016). In an in vitro mutagenesis of Huayu 20 with pingyangmycin followed by screening on a medium with hydroxyproline, mutant seeds contained higher protein (>30%) than parent (26.6%) (Sui et al. 2015). Some of the TG varieties showed enhanced sucrose content. Large-seed mutant derivative like TKG-19A contained higher sucrose, lower raffinose and stachyose and desirable nutritional traits for 'table purpose' (Gadgil and Mitra 1982, 1983). Sucrose content among gamma ray mutants ranged from 2.6% to 6.2% compared to 4.5% in parent Shinnamkwang (Doo et al. 2008).

Heavy-ion (C or N) beam irradiation was employed to develop 17 knockout hypoallergenic groundnut mutants from the Japanese Nakateyutaka variety (Cabanos et al. 2012). Of these, eight mutants lacked either one of the two isoforms of Ara h 2 or other nine mutants lacked one of the isoforms of Ara h 3. Wan et al. (2016) noted lower level of anthocyanin, lignin and proanthocyanidin and higher level of melanin in mutants having seed coat crack as compared to wild type. In white-seed-coat mutant, lower anthocyanins and higher isoflavones and flavones were recorded as compared to its parent (Wan et al. 2020).

16.11 Mutation Breeding for Climate Resilience in Groundnut

Climate change due to addition of greenhouse gases to the atmosphere brings more frequent high/low temperatures, droughts, floods, cyclones, hailstorms and persistent sea-level rise, which are expected to pose a severe threat to agriculture, biodiversity and human society (Karavolias et al. 2021). Climate change along with abiotic stresses would also increase disease and pest incidence by exposing crops to enhanced biotic pressure. To cope with these disasters and to tackle various biotic and abiotic stresses, it is crucial to breed crop varieties with climate resilience. Mutation breeding has the potential to evolve climate-resilient genotypes in various crops including groundnut, in turn moderating the effects of climate change on agriculture.

Groundnut crop encounters frequent early, mid-season and terminal drought stress especially in rainfed-based cultivation, which is the major area in African and Asian countries. Groundnut mutant derivatives like TAG 24, TG 37A and TG 51 with semi-dwarf type, early vigour, better harvest index and assimilate partitioning were reported to have enhanced moisture stress tolerance, and their earliness assists them to escape end-season droughts (Murty et al. 2004; Badigannavar et al. 2020). Recently, Abdalla et al. (2018) developed gamma ray-induced end-of-season drought-tolerant groundnut mutants in Sudan. Another mutant-derived variety, Dh 256, having mid-season drought tolerance has recorded better pod yield and higher relative water content compared to check, G 2-52, under water-stress condition (Motagi et al. 2022).

On the other hand, groundnut crop has been experiencing excessive rains at the pod-filling/maturity stage in recent times due to shift in rainfall pattern. Such excessive humid conditions are highly congenial for groundnut to get greater prevalence of rust and leaf spot diseases affecting seed yield and quality considerably. Towards this aspect, several late leaf spot-resistant groundnut mutants were isolated through EMS mutagenesis (Motagi et al. 1996; Gowda et al. 2010).

In recent times, larger groundnut area in India suffered from in situ seed germination due to end-season rains spoiling almost ready-to-harvest produce. Varieties with optimum period of fresh seed dormancy would rescue the rain-induced seed germination and spoilage. Hussein et al. (1991) induced gamma ray mutants from non-dormant parent, Early Bunch, with 6–7 weeks of seed dormancy. Trombay groundnut mutants or their derivatives like TG 1, TKG 19A, TG 22, TG 26 and TPG 41 with their 3–5 weeks' fresh seed dormancy are ideal to curtail the yield losses due to in situ seed germination (Murty et al. 2004; Badigannavar et al. 2020).

16.12 Mutant Varieties

In groundnut, mutation breeding has played a significant role in enhancing productivity, which has been the main objective in majority of the country's breeding programmes. Selection was practised consistently for yield components like pod number, seed size, shell thickness and branch number in different mutant generations. Generally, most productive and widely adapted popular cultivar was subjected for induced mutagenesis to derive mutant varieties. Usually, mutants have been detected in the M_2 on a single-plant basis (Donini et al. 1984). In many breeding experiments including at Trombay, mutants selected for academic interest sometimes even with inferior agronomic traits had also served as good parents for evolving agronomically superior breeding lines (Patil and Mouli 1979). Based on available literature, around 112 groundnut varieties involving 46 direct mutants and 66 mutant derivatives have been released globally, and Bangladesh, China, India and the USA are leading countries (Table 16.5).

Radiation experiment in groundnut was initiated way back in 1949 in the USA (Gregory 1955). A high-yielding, disease-resistant mutant was developed by irradiating cv. Virginia Bunch with X-rays and subsequent selection. Comparable with the best varieties grown, the mutant was released as NC 4x in 1959, the first groundnut mutant variety in the USA (Gregory 1960). GA-T2636M is another induced mutant with a high O/L ratio derived from gamma irradiation of 'Georgia Runner' (Branch 2000). By involving GA-T2636M in crosses, two varieties, Georgia Hi-O/L and Georgia-02C, and a breeding line, GA 942004, were developed. Further, through GA 942004 in back crosses with Georgia Green, Georgia-09B was developed. Subsequently, by incorporating high oleate trait from Georgia-02C, Georgia-09B and Georgia Hi-O/L, another seven high-yielding varieties (Georgia-08V, Georgia-10T, Georgia-12Y, Georgia-13M, Georgia-14N, Georgia-17SP, Georgia-18RU) were developed in the USA (Table 16.5) (Brown et al. 2021). Besides, high O/L cultivar C458 (Flavor Runner 458) was selected from EMS-treated Florunner. In Argentina, an X-ray-induced mutant, Colorado Irradiado, having high yield and oil content became a successful cultivar in the late 1970s (Livore et al. 2018).

Table 16.5 Groundnut mutant or mutant-derived varieties developed in different countries

Country/trait	Direct mutant varieties	Mutant-derived varieties
Argentina (2)[a]		
High yield	Colorado Irradiado	
Large seed	Virginia No. 3	
Bangladesh (9)		
High yield	Binachinabadam-1, Binachinabadam-2, Binachinabadam-4, BINA Chinabadam-10	
Large seed	Binachinabadam-3	
Salinity tolerance	Binachinabadam-5, Binachinabadam-6, Binachinabadam-7, Binachinabadam-9	
China (40)		
High yield	Fu 21, Huayu 16, Lainong 10, Xianghuasheng 4, Yuhua 5, Yuhua 7	8130, 78961, Huayu 22, Huayu 32, Lu 8130, Luhua 13, P 12, Yangxuan 1, Yangxuan 58, Yueyou 22, Yueyou 33, Yueyou 551, Yueyou 551-6, Yueyou 551-38, Yueyou 551-116, Zhonghua 16
Large seed	Huayu 40	Huayu 9610, Yueyou 169, Yueyou 187, Yueyou 187-93
Early maturity	Ganhua 1, Luhua 6	Luhua 15, Xianghua 1
Low temperature resistance	Changhua 4	
Large seed	Luhua 7	
Resistance to *Aspergillus flavus*	Fu 22	
Rust resistance		Shanyou 27, Yueyou 223
Drought tolerance	Luhua 11	
High oil	Yuhua 4, Yuhua 9, Yuhua 14	
India (41)		
High yield	CO 2, G 2-52, TG 3, TG 38, TAG 73	BSG 0912, Co (Gn)-5, Dh 40, Dh 86, Dharani, JCG 88, GG 21, GG 34, GG 41, PDKVG 335, RG 559-3, TG 51, TCGS 894
Large seed	BG 1, BG 2, TG 1	RARS-T-1 (TG 47), Somnath, TLG 45, VRI 2
Drought tolerance		Dh 256, GG 35, TG 37A, TPT 25

(continued)

Table 16.5 (continued)

Country/trait	Direct mutant varieties	Mutant-derived varieties
Seed dormancy		R 9251, TKG 19A, TG 22, TG 26
High harvest index		JL 501, TG 17, TAG 24
Late leaf spot and rust resistance		GPBD 5, Dh 232
High oleic acid	Dh 245	TG 39, TPG 41
Malaysia (2)		
Early leaf spot resistance	KARISMA Serene	
High vitamin A and invert sugar	KARISMA Sweet	
Myanmar (1)		
Early maturity	Sin Padetha 1	
Pakistan (1)		
High yield, red testa	Golden	
Sri Lanka (1)		
High yield, early maturity	ANK-G1 (Tissa)	
Sudan (1)		
Drought tolerance	Tafra-1	
USA (12)		
High yield	NC 4x	
High oleic acid	C 458	Georgia Hi-O/L, Georgia-02C, Georgia-08V, Georgia-09B, Georgia-10T, Georgia-12Y, Georgia-13M, Georgia-14N, Georgia-17SP, Georgia-18RU
Viet Nam (2)		
High yield	B 5000, DT 332	
Total (112)	46	66

[a] Total number of mutant or mutant-derived varieties of the country

In India, groundnut improvement through radiation-based mutation breeding was initiated during 1957–1958 at the Bhabha Atomic Research Centre, Trombay, Mumbai, which evolved a robust mutant gene pool having distinct mutants and mutant derivatives for developing varieties in subsequent years (Patil and Mouli 1979). In total, 41 varieties were developed by mutation breeding and have been

Fig. 16.2 Some groundnut mutant or mutant-derived varieties in India: TAG 24, TG 37A, TG 39, TLG 45 and TG 51 (left to right)

released for farmer cultivation in India since 1973 (Table 16.5). Of these, nine are direct induced mutants and rest are all mutant derivatives. TG 1 (Trombay groundnut 1) was the first induced large-seed mutant variety developed in 1973 by X-ray irradiation of Spanish Improved with alterations from var. *vulgaris* to var. *hypogaea* (Patil 1966) (Fig. 16.1). TG 3, being another direct mutant variety, also showed superior yield performance with greater number of tertiary branches (Patil 1966). Interaction of two mutant genomes resulted in increase in oil content in TG 9 (Patil 1973), reduced number of branches (extreme *fastigiata* type) and increased pod yield in TG 17 (Patil 1977). TG 1 along with TG 17 resulted in large-seed variety, TKG 19A. Two early-maturing selections, TGE 1 and TGE 2, were evolved as a result of genomic blend of three mutants and mutant genome in the background of Gujarat dwarf mutant, respectively. Under the different genomic backgrounds, suitable mutant derivatives, Somnath (TGS 1) and TG 22, were developed to become subsequently promising cultivars. Genomic blend of four mutants of Spanish Improved under the M 13 background developed TAG 24 (Patil et al. 1995), TG 39 (Kale et al. 2010), TLG 45 (Kale et al. 2008) and RARST-1 (TG 47) (Badigannavar et al. 2012) (Fig. 16.2). Similarly, one mutant of JL 24 and four mutants of Spanish Improved along with M 13 resulted in yet another successful cv. TG 26 (Kale et al. 1997). The genomic constitution of TG 26 was expanded with natural mutant Gujarat dwarf, Girnar 1, and early genetic stock Chico, respectively, to evolve three varieties TG 37A, TG 38 and TG 51 (Kale et al. 2004a, 2007, 2009) (Fig. 16.2). Likewise, TPG 41 was developed from one mutant of JL 24 and four mutants of Spanish Improved under Robut 33-1 and M 13 backgrounds (Kale et al. 2004b). Recently, a gamma ray mutant of TG 38, TAG 73, has been released. The entire genetic base for TG germplasm was constituted by seven parents, viz., Chico, Girnar, Gujarat dwarf, JL 24, M 13, Robut 33-1 and Spanish Improved (Fig. 16.1), with induced mutants of Spanish Improved being the main contributors to TG evolution (Badigannavar et al. 2002). TG varieties were also one or the other parent in the development and release of another 14 varieties in various states (Fig. 16.1) (Badigannavar et al. 2020). EMS mutagenesis of GPBD 4 also contributed to the development of a promising variety G 2-52 (Nadaf et al. 2009) and Dh 245 (Nadaf et al. 2017) and its derivatives, GPBD 5 and Dh 232 (Motagi et al. 2022).

In China, mutation breeding was initiated in the early 1960s and applied successfully to develop improved groundnut varieties. As a result of intensive mutation research in many institutes, around 40 mutant or mutant-derived varieties have been released, which include 16 direct mutant varieties (Table 16.5). In the 1960s, groundnut mutant Fushi was developed by irradiating cv. Shitouqi with beta rays from ^{32}P. By using Fushi in cross-breeding, a first variety Yueyou 22 was developed in 1968 followed by Yueyou 33 in 1971. Later, Yueyou 22 contributed as parent in developing another six varieties (Fu 21, Yuexuan 58, Yueyou 551, Yueyou 551-6, Yueyou 551-38, Yueyou 551-116) between 1972 and 1986. Most of the groundnut varieties planted in Guangdong and South China provinces have a pedigree of mutant Fushi. The popular, widely grown cv. Huayu 22 in northern China was developed by gamma rays and hybridization (Wu et al. 2006). Mutation breeding has also been successfully adopted for groundnut improvement in Uganda (Busolo-Bulafu 1991) and in Egypt (Ahmed and Mohamed 2009).

16.13 Significance and Coverage of Groundnut Mutant and Mutant-Derived Varieties

In China, numerous groundnut mutant varieties have been cultivated on larger area bringing remarkable economic, social and ecological benefit and sustaining crop production of the country (Liu et al. 2004). The cumulative area of Chinese mutant cultivars covers about 20% of the total area under groundnut (Qiu et al. 1997). Series of 'Yueyou' varieties have reached the accumulated growing areas of more than 4 million ha. Cultivation of Yueyou 22 reached an area of 130,000 ha. Fu 21, a gamma ray mutant of Yueyou 22, and Yueyou 33, a selection from Yueyou 22, covered an area of 11,000 ha and 20,000 ha, respectively. Yueyou 551 evolved by crossing Yueyou 22 with Yueyou 431 was grown extensively on 222,000 ha in Guangdong province. By involving Yueyou 551 in cross-breeding, five more varieties, Yangxuan 1, Yuexuan 58, Yueyou 551-6, Yueyou 551-38 and Yueyou 551-116, were developed, and of these, Yueyou 551-116 became the most prominent variety with 130,000 ha area coverage. Fushi-derived varieties had been grown on 3 million ha in South China, and groundnut production was enhanced to the tune of 500,000 tonnes with 500 million Yuan to the farmers (Jiang and Zhou 1987, 1988). In Argentina, an X-ray-induced mutant variety, Colorado Irradiado, occupied around 80% of the groundnut area (262,000 ha) in the 1970s (Livore et al. 2018).

In India, mutation breeding has contributed notably for the development of large-seed varieties with early maturity (TG 39, TPG 41, TLG 45, TG 47), which are suitable for table purpose and export under bold types. Seeds of the varieties like TG 37A, TG 38 and TG 51 are preferred for the export under Java type. Semi-dwarf habit, high harvest index and better partitioning in TAG 24, TG 26 and TG 47 permit pegs to enter the soil early to have better and uniformly developed pods. The compact plants of TAG 24, TG 26, TG 38 and TG 47 facilitate farmers for high-density planting for attaining greater productivity. Fresh seed dormancy in TKG 19A, TG 22, TG 26 and TPG 41 prevents in situ seed germination of matured crop

Fig. 16.3 Field view of groundnut variety, TG 39 (left) along with farmer (right)

due to end-season rains. Hence, this character is very beneficial under current changing climatic conditions, wherein prolonged end-season rains are often experienced. Early maturity in TAG 24, TG 26 and TG 51 is helpful to escape end-season drought and to have groundnut crop in paddy fallows, residual moisture situation and post-potato crop. Hence, farmers are able to earn greater income due to better price in the market. Drought tolerance in TAG 24, TG 37A and TG 39 makes them suitable for cultivation in water-limited areas. High oleic acid (60%) in TG 39 and TPG 41 imparts better oil shelf life and health benefits. TAG 24 is used as the national check variety and TG 37A and TG 51 as zonal check varieties in national evaluation trials.

Widespread coverage of these TG varieties in major groundnut-growing Indian states has benefitted thousands of farmers, traders and exporters. Farmers have been experiencing the better yielding ability of TG varieties by harvesting the record yields (>7000 kg/ha) apart from superior yields (3000–4000 kg/ha) obtained in many parts of the country. Two progressive farmers from Maharashtra state harvested record yields of 9280 kg/ha dry pods and 10,175 kg by cultivating TAG 24 and 9487 and 10,542 kg by growing TG 26 under suitable agro-ecology such as balanced nutrition, uninterrupted but controlled irrigation in summer environment and disease-free conditions (Kale et al. 2002). Another progressive farmer from Andhra Pradesh obtained a record yield of 7800 kg pods/ha by cultivating TG 51 under irrigated conditions (DGR 2010). Of late, large-scale field demonstrations of TG 37A in Maharashtra during 2019 and 2020 showed superiority of 37% in pod yields and 71% in net returns over local variety, SB XI, to the farmers. Similarly, TG 37A also recorded 28% and 24% greater pod yield and net returns, respectively, in Tamil Nadu. Recently, TG 39 has earned price advantage to the Gujarat farmers by securing the highest market price (Fig. 16.3). Annual breeder seed demand for mutant and mutant-derived varieties from the various states in India ranged from 10% to 30% of the total demand for all the varieties put together with an average of 18% (>145 tonnes) in the last two decades.

16.14 TILLING in Groundnut

Targeting Induced Local Lesions in Genomes (TILLING) is a potent reverse genetic tool to detect mutants in high-throughput and gene sequence-assisted manner. The technique is equally applicable for all traits if the gene sequence is known. The universality of TILLING made it popular in many crop species for detecting mutants for agronomic, biochemical and stress-related traits. It involves PCR-based amplification of the target genes, followed by heteroduplex formations and cleavage by CEL1 nuclease or other endonucleases to identify single-nucleotide or small insertion/deletion mutations. The technique has been applied in maize (Till et al. 2004), rice (Till et al. 2007), soybean (Cooper et al. 2008) as well as groundnut (Knoll et al. 2011; Guo et al. 2015). It was a challenge to develop allergen-free groundnut for human consumption. Knoll et al. (2011) addressed this problem by mutation induction and screening the mutant population through TILLING approach. Before the genome sequence availability, *Ara h2.01* and *Ara h2.02* were cloned from cDNA library by employing nested PCR. LICOR-gel-based screening of 3420 M_2 plants from EMS-mutagenized population of Tifrunner and nine SNPs in *Ara h 2.01*, five in *Ara h 2.02* and four in *Ara h 1.01* were identified. In *Ara h 1.01*, a 'C-to-T transition' at 593 nucleotide position was silent, but the other three SNPs were predicted to induce amino acid changes: R333W, P405L and E437K. The non-synonymous mutation R333W at position 333 situated within epitope 12 of Ara h 1.01 (Shin et al. 1998). Only one mutation (a premature stop codon) was confirmed in *Ara h 1.02* that resulted in a truncated protein (Knoll et al. 2011). TILLING was also used to identify a frameshift mutation in *AhFAD2B* (Knoll et al. 2011). The overall mutation frequency for EMS was found to be 1 SNP/967 kb in the above six genes in groundnut.

Subsequently, the researchers investigated mutations for stress-related genes by using a subset of 768 M_2 plants from the above mutant population of Tifrunner through TILLING by sequencing (Guo et al. 2015). Phospholipase D (*AhPLD*) and lipoxygenase 7 (*AhLOX 7*) genes were upregulated in pod in response to drought and infection to *Aspergillus* spp., respectively. Using the above mutant population, four missense mutations in *AhLOX7* and three missense mutations in *AhPLD* were detected. Further, one missense and two silent mutations for *Ara h 1.01*, three silent and five missense mutations in *Ara h 1.02*, one silent mutation in *Ara h 2.02* and one missense mutation in *AhFAD2B* were also identified by following TILLING by sequencing in this subset mutant population (Guo et al. 2015). It was revealed that the SNP detection frequency for single-copy genes was 1 SNP/344 kb and 1 SNP/3028 kb for multiple-copy genes in groundnut. Recently, Karaman et al. (2021) analysed EMS-induced mutant population by TILLING. The mutation in *ahFAD2B* resulted in change from serine to threonine and from glycine to aspartic acid in *ahFAD2A*. The estimated overall mutation rate was 1 mutation in every 2139 kb. The mutation frequencies were also 1/317 kb for *ahFAD2A* in 0.4% EMS and 1/466 kb for *ahFAD2B* in 1.2% EMS treatments. Taking its genome size of 2800 Mbp (2800 × 1000 kbp), at least 2800 plants must be screened to get a mutation in any gene through EMS mutagenesis.

16.15 Molecular Characterizations of Mutants Through Target Gene-Based Approach

In groundnut, genes for oil accumulation in seed were studied by examining transcriptome of developing seeds of a normal oil cultivar, Huayu 22 (49.5%), and its high-oil-EMS mutant, O1 (60.9%), at 40 days and 47 days after flowering (DAF) using GeneFishing technology (Tang et al. 2013). After cloning and sequencing of distinguishable differentially expressed bands, 27 unigenes from 40 DAF and 13 from 47 DAF were identified. Homology search revealed that 20 unigenes (17 from 40 DAF, 3 from 47 DAF) were highly homologous to known gene sequences for oil biosynthesis, energy metabolism, signal transduction and stress response. Of them, three differentially expressed genes encoding thioredoxin, oleosin and transaldolase were further confirmed by real-time quantitative PCR.

By using induced mutagenesis with gamma rays and sodium azide in groundnut, a high-oleic-acid (70%) mutant was isolated and later characterized through target gene sequencing approach (Mondal et al. 2011). Sequencing of mutated *ahFAD2A* gene detected two non-synonymous mutations in the coding region. Multiple sequence alignments of the AhFAD2B gene from Huayu 22 (wild type) and high-oleate mutant revealed a C281T transition in the coding region causing an I94T (isoleucine to threonine at 94th amino acid position) substitution in the oleoyl-PC desaturase (Wang et al. 2011a). A Virginia-type high-oleate (>60%) mutant (E2-4-83-12) was isolated from an EMS treatment of LF2 (44.2% oleic acid). Cloning and sequencing of *AhFAD2*B from LF2 and E2-4-83-12 identified a novel mutation, C313T, in the coding region causing an H105Y non-synonymous substitution in the first histidine box of the FAD2B protein (Fang et al. 2012). Two stable high-oleic-acid (>70%) mutant lines, viz. GM6-1 and GM4-3, isolated from GPBD 4 were utilized for characterization of *AhFAD2*B gene (Nadaf et al. 2017). Cloning and sequencing of FAD2B gene from GPBD 4, GM6-1 and GM4-3 revealed two novel mutations (A1085G and G1111A) in GM 6-1 and single transition (G1111A) in GM 4-3. A CAPS (bF19/R1, *Mob*II enzyme) marker and two SNP markers (bF19/GM6-1-GM4-3 and bF19/GM6-1) that could differentiate the two mutants were also developed (Nadaf et al. 2017). Sequence comparison of the high-oleate EMS mutant and parent Fuhua 12 revealed a G448A substitution in *AhFAD2A* and an A insertion (441_442insA) in *AhFAD2B* genes, which together contributed to high oleate in mutant (Yu et al. 2019). Zhuang et al. (2019) developed mutants Min6-A from EMS treatment of Minhua 6 and Min8-A from gamma ray irradiation of Minhua 8 with ~80% oleic acid. Sequencing identified mutations in *AhFAD2A* (dysfunction mutation on AH09G33970 at 114,779,221 bp of Chr09) and *AhFAD2B* (frameshift on AH19G43590 at 154464257 bp of Chr19), which conferred high oleate trait. Recently, Nkuna et al. (2021) identified a high-oleic Virginia-type groundnut mutant from sodium azide mutagenesis. Sequencing of the mutated and wild-type *FAD2A/FAD2B* genes detected two mutations, viz. the G448A mutation in *FAD2A* resulting in an amino acid change of D150N and G558A in *FAD2B*, causing a stop codon and premature termination of protein synthesis.

16.16 Mutagenomics for Characterization of Mutants

Mutagenomics research in groundnut was initiated by Yu et al. (2015) through the usage of GeneFishing technology. Using this technique, 40 differential transcripts were identified between high-protein mutant and its parent. Of them, three unique genes with nutrient reservoir activity, protein disulphide oxidoreductase activity and ATPase/transporter activity, respectively, were related to high-protein phenotype in the mutant.

Next-generation sequencing is a well-established, versatile genomics platform with many applications in plant biology research. This technique along with mutants and mutant-derived population is now being used in many mapping techniques like MutMap (Abe et al. 2012), MutMap-Gap (Takagi et al. 2013a) and QTL-seq (Takagi et al. 2013b) to know the nature of mutations and to identify mutant gene. Testa colour in groundnut is an important quality attribute that determines consumer preference. Using RNA sequencing approach, Wan et al. (2016) studied the differential transcript behaviour of a brown testa mutant with crack seed coat (pscb) and wild-type seed coat harvested at 20, 40 and 60 DAF. By analysing gene expression patterns and sequences of 62 differentially expressed genes (DEGs) between mutant and wild type, three candidate genes, namely, c36498_g1 (CCoAOMT1; Caffeoyl-CoA *O*-methyltransferase 1), c40902_g2 (kinesin) and c33560_g1 (MYB3), were suggested to be responsible for seed coat cracking and brown colour phenotype. Later, the same group of scientists worked towards identification of DEGs on a small pod width mutant (pw) compared to its parent by using RNA-sequencing of developing pod samples at 20, 40 and 60 DAF. A genome-wide comparative analysis of expression profiles revealed 260 DEGs across all three stages, and two candidate genes, c26901_g1 (CAD; cinnamyl alcohol dehydrogenase) and c37339_g1 (ACS; 1-aminocyclopropane-1-carboxylate synthase), responsible for pod width were identified (Wan et al. 2017). Recently, an integrative analysis of transcriptomes, metabolomes and histocytology of white-seed-coat (wsc) mutant revealed that the mutant gene influenced the flavonoid biosynthesis in testa as well as suberin formation, glycolysis, TCA cycle and amino acid metabolism. Further, common DEGs' analysis in the above three pod development stages detected three testa-specific expressed candidate genes *Araip.M7RY3*, *Aradu.R8PMF* and *Araip.MHR6K* that were likely responsible for the white testa in the mutant (Wan et al. 2020).

The same transcriptomic approach was used to characterize a susceptible late leaf spot EMS mutant M14 of a resistant cultivar Yuanza 9102 (Han et al. 2017). RNA-seq analysis in the leaf tissues of M14 and Yuanza 9102 under pathogen challenge showed 2219 DEGs including 1317 up-regulated genes and 902 down-regulated genes. DEG analysis revealed up-regulation of inducible pathogenesis-related genes and down-regulation of genes related to photosynthesis in susceptible M14 mutant. The study suggested that the up-regulation of WRKY transcription factor along with depression of chloroplast genes and plant hormone-related genes for plant growth happened in response to fungal infection in susceptible mutant that resulted in reduction of photosynthesis and phytohormones and led to symptom formation.

In another high-throughput transcriptome study on gamma ray-induced semi-dwarf mutant 2 (sdm2), DEGs were found to be involved in cell wall synthesis and metabolic pathways and hormone biosynthesis and signalling pathways (Guo et al. 2020). The expression of several genes in BR biosynthesis and signalling was found to be considerably down-regulated in leaf and stem of sdm2 as compared to parent. Further genes in cell wall synthesis and metabolic pathways, which are related to cell elongation, were downregulated in the stem of sdm2 mutant. In the study of muta-epigenomics, Bhat et al. (2020) identified differentially methylated sites in the genome of an EMS mutant TMV2-NLM and its parent, TMV 2. This narrow leaf mutant differed in 240 methylation sites in A genome and 401 methylation sites in B genome. Of the 641 differentially methylated sites, only 45 were found in 37 genes among which 8 had differential expression between mutant and parent. This study first demonstrated the role of EMS mutagenesis in DNA methylation in groundnut.

16.17 Gene Editing for Site-Directed Mutagenesis

Recently, gene editing studies based on zinc finger nucleases (ZFN), single-stranded oligonucleotides, transcription activator-like effector nucleases (TALENs) and clustered regularly interspaced short palindromic repeats (CRISPR/Cas9) have been reported in several polyploid crops (Kaur et al. 2018; Arndell et al. 2019). The site-directed approach also allows selection of desirable plants by delivery of biodegradable Cas9/sgRNA ribonucleoprotein complexes (no foreign DNA) into plant cells, where they are acted transiently but allow for efficient gene editing (Woo et al. 2015). Such a system has been demonstrated in *Camelina sativa*, an emerging oil seed plant, wherein oleic acid was increased from 16% to >50% with a concomitant decrease in linoleic acid and linolenic acid (Jiang et al. 2017). These innovative systems creating precise mutations will be useful in enhancing oleate content even in groundnut.

Much before the availability of complete genome sequence of tetraploid groundnut, scientists have explored the possibility to further improve the oil quality by changing the oleic/linoleic (O/L) ratio in seed oil. Huang et al. (2008) generated RNAi lines against *FAD2* gene in groundnut and isolated transgenic plants with higher O/L ratio than the wild type. After the detection of sequence variation in natural and induced high-oleic mutants, researchers started to manipulate these quality characters through site-directed mutagenesis approach. Wen et al. (2018) attempted TALEN-mediated gene editing in both *FAD2A* and *FAD2B* genes of cv. YueYou no. 7. Oleic acid level in transgenic seeds was increased up to 90.45% with the concomitant reduction of linoleic acid to 7.42%. This is the first successful report on the use of TALEN in targeted mutagenesis of groundnut. However, TALEN technique has its inherent constraints of target-specific TALE protein construction in vector and low mutation frequency of T1 plants. Most of the site-directed approach depends on transformation efficiency, which is still low in groundnut compared to other crops. However, the genome editing efficiency in groundnut can be enhanced by other potential sequence-specific nucleases, including ZFNs, CRISPR/Cas9 and CRISPR/Cpf1 systems (Lowder et al. 2016). CRISPR/Cas9-mediated site-directed gene editing was used to mutate *FAD2* gene by both protoplast and hairy root transformation methods in groundnut (Yuan et al. 2019). They had identified the hotspot site (called as sgRNA6 site) within *FAD2* that showed higher frequency of 'G448A' and 'G451T' transition in *AhFAD2A* and *AhFAD2B*, respectively. In addition, '441_442 insertion A' in FAD2B similar to natural mutation in F435 was also identified in this study. The mutation 'G451T' in *AhFAD2B* was novel and generated a premature stop codon. Such site-specific mutagenesis using CRISPR-Cas9 is equally important to know the function of uncharacterized gene in groundnut. On this direction, Shu et al. (2020) used this technology in groundnut and found that *AhNFR5* gene played a major role in nodule formation. Application of site-directed mutagenesis technique in studying the role of genes or creating valuable traits in groundnut has enough scope in post-genomic era due to the availability of information on gene families, copy number and sequence divergence within homologues. Such information will help to design more specific sgRNA construct, which will selectively mutate the homologue in polyploid crop like groundnut.

16.18 Conclusions and Prospects

One of the impediments in the exploitation of plant genetic resources was transferring specific genes into 'ideal desirable genetic backgrounds' (FAO 1998). Convincingly, a judicious blend of mutation and recombination breeding efforts in groundnut was successful in bringing mutant genes into ideal genetic backgrounds. Consequently, many distinct groundnut varieties were evolved with improved agronomic and nutritional traits along with stress tolerance as exemplified by groundnut breeding efforts undertaken in many countries. Besides these, mutation breeding was also helpful to rectify some of the undesirable features associated with desirable characters. Many of these mutant varieties have been extensively cultivated by the farming community and have occupied considerable groundnut area in several countries. Post-genomic era has come up with ample information on gene families, copy number and sequence divergence within homologues in cultivated groundnut genome. Such information will help to design more specific sgRNA construct, which will selectively mutate the homologue in this polyploid crop for better nutritional value, abiotic and biotic resistance and allergen- and aflatoxin-free food in groundnut.

References

Abdalla E, Bakhit O, Elsheikh S (2018) Groundnut mutants with end-season drought tolerance for the marginal drylands of north Kord Ofan state, Sudan. In: FAO/IAEA International symposium on plant mutation breeding and biotechnology. IAEA, Vienna, p IAEA-CN-263-196

Abe A, Kosugi S, Yoshida K, Natsume S, Takagi H, Kanzaki H, Matsumura H, Yoshida K, Mitsuoka C, Tamiru M, Innas H, Canos L, Kamoun S, Terauchi R (2012) Genome sequencing reveals agronomically important loci in rice using MutMap. Nat Biotechnol 30:174–178

Ahmed MSH, Mohamed SMS (2009) Improvement of groundnut (*Arachis hypogaea* L.) productivity under saline condition through mutation induction. World J Agric Sci 5:680–685

Alberte RS, Hesketh JD, Kirby JS (1976) Comparisons of photosynthetic activity and lamellar characteristics of virescent and normal green peanut leaves. Z Pflanzenphysiol 77(2):152–159

Annual Report (2001) National Research Centre for Groundnut, Junagadh, India. pp 43–47

Anonymous (2002) Development of thermo-tolerant lines in groundnut: an approach based on temperature induction response technique (TIR). In: Progress report. BRNS, Bhabha Atomic Research Centre, Mumbai, India, pp 1–20

Arndell T, Sharma N, Langridge P, Baumann U, Watson-Haigh NS, Whitford R (2019) gRNA validation for wheat genome editing with the CRISPR-Cas9 system. BMC Biotechnol 19(1): 1–12

Ashri A, Goldin E (1965) The mutagenic activity of diethyl sulfate in peanuts. Radiat Bot 5:431–441

Ashri A, Levy A (1976) Natural and induced plasmon variation affecting growth habit in peanuts, *A. hypogaea*. In: Filer K (ed) Experimental mutagenesis in plants. Bulgaria, Bulgarian Acad Sci Sofia, pp 417–430

Ashri A, Offenbach R, Cahaner A, Levy A (1977) Transmission of acriflavine-induced trisomic mutants affecting branching pattern in peanuts, *Arachis hypogaea* L. Z Pflanzenzüchtg 79:210–218

Azad MAK, Hamid MA, Yasmine F (2014) Enhancing abiotic stress tolerance in groundnut through induced mutation. In: Tomlekova NB, Kozgar MI, Wani MR (eds) Mutagenesis: exploring genetic diversity of crops. Wageningen Academic Publishers, The Netherlands

Azzam CR, Azer SA, Khaleifa MMA, Abol-Ela MF (2007) Characterization of peanut mutants and molecular markers associated with resistance to pod rot diseases and aflatoxin contamination by RAPD and ISSR. Arab J Biotech 10(2):301–320

Badigannavar AM, Mondal S (2009) Genetic enhancement of groundnut (*Arachis hypogaea* L.) for high oil content through gamma ray mutagenesis. In: Shu QY (ed) Induced plant mutations in the genomics era. FAO, Rome, pp 101–103

Badigannavar AM, Murty GSS (2007) Genetic enhancement of groundnut through gamma ray induced mutagenesis. Plant Mutat Rep 1(3):16–21

Badigannavar AM, Kale DM, Murty GSS (2002) Genetic base and diversity in groundnut genotypes. Plant Breed 121:348–353

Badigannavar AM, Mondal S, Murty GSS (2007) Induction of salt tolerance for radicle growth in groundnut through gamma ray mutagenesis. BARC Newslett 285:226–236

Badigannavar AM, Mondal S, D'Souza SF (2012) Development and deployment of Trombay groundnut (*Arachis hypogaea* L.) varieties with desirable agronomic features. In: DAE-BRNS Life Sci Symp, p 60

Badigannavar AM, Mondal S, Bhad PG (2020) Radiation based induced mutagenesis in Trombay groundnuts: developments and accomplishments. IANCAS Bull XV 1:45–49

Basu MS (2002) Characterization of released groundnut (*Arachis hypogaea* L) cultivars. National Research Centre for Groundnut, Junagadh, India, pp 1–254

Benedict CR, Ketring DL (1972) Nuclear gene affecting greening in virescent peanut leaves. Plant Physiol 49(6):972–976

Bhat RS, Rockey J, Shirasawa K, Tilak IS, Patil MPB, Lachagam VBR (2020) DNA methylation and expression analyses reveal epialleles for the foliar disease resistance genes in peanut (*Arachis hypogaea* L.). BMC Res Notes 13:20

Birthal PS, Parthasarathy Rao P, Nigam SN, Bantilan CS, Bhagavatulu S (2010) Groundnut and soybean economies in Asia: facts, trends and outlook. ICRISAT, Patancheru, India, pp 22–26

Bora KC, Patil SH, Subbaiah KC (1961) X-ray and neutron induced meiotic irregularities in plants with special reference to *Arachis hypogaea* and *Plantago ovata*. In: Proc effects of ionizing radiations on seeds. IAEA, Vienna, pp 203–216

Branch WD (2000) Registration of 'Georgia Hi-O/L' peanut. Crop Sci 40:1823–1824.

Branch WD (2002) Variability among advanced gamma-irradiation induced large-seeded mutant breeding lines in the 'Georgia Browne' peanut cultivar. Plant Breed 121:275–277

Brown N, Branch WD, Johnson M, Wallace J (2021) Genetic diversity assessment of Georgia peanut cultivars developed during ninety years of breeding. Plant Genome 2021:e20141. https://doi.org/10.1002/tpg2.20141

Busolo-Bulafu (1991) Utilization of induced mutations for groundnut breeding in Uganda. Mutat Breed Newslett 30:4

Cabanos CS, Katayama H, Urabe H, Kuwata C, Murota Y, Abe T, Okumoto Y, Maruyama N (2012) Heavy-ion beam irradiation is an effective technique for reducing major allergens in peanut seeds. Mol Breed 30:1037–1044

Chen T, Huang L, Wang M, Huang Y, Zeng R, Wang X, Wang L, Wan S, Zhang L (2020) Ethyl methyl sulfonate-induced mutagenesis and its effects on peanut agronomic, yield and quality traits. Agronomy 10:655

Cooper JL, Till BJ, Laport RG, Darlow MC, Kleffner JM, Jamai A, El-Mellouki T, Liu S, Ritchie R, Nielsen N, Bilyeu KD, Meksem K, Comai L, Henikoff S (2008) TILLING to detect induced mutations in soybean. BMC Plant Biol 8:9

Dave RS, Mitra RK (2000) Partial characterization of cDNA clone of a low temperature induced gene from groundnut. Int Arachis Newslett 20:36–37

Directorate of Groundnut Research (DGR) (2010) Andhra farmer scores a record yield with Trombay groundnuts. DGR Newslett IX(3–4):1

Donini B, Kawai T, Micke A (1984) Spectrum of mutant characters utilized in developing improved cultivars. In: Selection in mutation breeding. IAEA, Vienna, pp 7–31

Doo HS, Cheong YK, Park KH (2008) Variations in the chemical compositions of peanut mutants induced by gamma radiation. Korean J Breed Sci 40:113–118

Fang CQ, Wang CT, Wang PW, Tang YY, Wang XZ, Cui FG, Yu SL (2012) Identification of a novel mutation in FAD2B from a peanut EMS mutant with elevated oleate content. J Oleo Sci 61(3):143–148

FAO (1998) The state of the World's Plant Genetic Resources for Food and Agriculture. FAO, Rome

FAOSTAT (2020). http://www.fao.org/faostat/en/#data/QC

Francisco MDL, Resurreccion AVA (2008) Functional components in peanuts. Crit Rev Food Sci Nutr 48:715–746

Fu L, Wang Q, Li L, Lang T, Guo J, Wang S, Sun Z, Han S, Huang B, Dong W, Zhang X, Du P (2021) Physical mapping of repetitive oligonucleotides facilitates the establishment of a genome map-based karyotype to identify chromosomal variations in peanut. BMC Plant Biol 21:107

Gadgil JD, Mitra R (1982) Sucrose, raffinose and stachyose concentration in some Trombay groundnut cultures derived from mutation breeding. Mutat Breed Newslett 20:6–7

Gadgil JD, Mitra R (1983) Chemical composition of seeds in induced groundnut mutants and their derivatives. Indian J Agric Sci 53:295–298

Giriraj K, Itnal CJ (eds) (1999) Oilseeds scenario in India. Highlights of oilseeds research in north Karnataka. University of Agricultural Sciences, Dharwad, India, pp 1–64

Gowda MVC, Nadaf HL, Sheshagiri R (1996) The role of mutation in intraspecific differentiation of groundnut (*Arachis hypogaea* L.). Euphytica 90:105–113

Gowda MVC, Bhat RS, Motagi BN, Sujay V, Kumari V, Sujatha B (2010) Association of high-frequency origin of late leaf spot resistant mutants with AhMITE1 transposition in peanut. Plant Breed 129:567–569

Gregory WC (1955) X-ray breeding of peanuts (*Arachis hypogaea* L.). Agric J 47:396–399

Gregory WC (1960) The peanut NC 4x, a milestone in crop breeding. Crops Soils 12(8):12–13

Guo Y, Abernathy B, Zeng Y, Ozias-Atkin P (2015) TILLING by sequencing to identify induced mutations in stress resistance genes of peanut (*Arachis hypogaea*). BMC Genomics 16:157

Guo F, Ma J, Hou L, Shi S, Sun J, Li G, Zhao C, Xia H, Zhao S, Wang X, Zhao Y (2020) Transcriptome profiling provides insights into molecular mechanism in Peanut semi-dwarf mutant. BMC Genomics 21:211

Han S, Liu H, Yan M, Qi F, Wang Y, Sun Z, Huang B, Dong W, Tang F, Zhang X, He G (2017) Differential gene expression in leaf tissues between mutant and wild-type genotypes response to late leaf spot in peanut (*Arachis hypogaea* L.). PLoS One 12(8):e0183428

Hassan IB, Anes UC (2015) Cytogenetic and morphological effects of gamma radiation induced mutation in groundnut varieties in Nigeria. J Adv Biol Basic Res 01:44–46

Holbrook CC, Stalker HT (2003) Peanut breeding and genetic resources. Plant Breed Rev 22:297–356

Huang BY, Zhang XY, Miao LJ, Yan Z, Hai Y, Yi ML, Xu J, Chen ZK (2008) RNAi transformation of AhFAD2 gene and fatty acid analysis of transgenic seeds. Chinese J Oil Crop Sci 30(3): 290–293

Hussein HA, El-Sharkawy AM, Hussein EH, Sorour WA (1991) Mutation breeding experiments in peanuts (*Arachis hypogaea* L.). In: Plant mutation breeding for crop improvement, vol 2. IAEA, Vienna, pp 199–206

IAEA (2021) Mutant variety database [Online]. http://mvd.iaea.org/. Accessed 1 Aug 2021

ICAR-IASRI (2019) Agricultural research data book. Indian Agricultural Statistics Research Institute (IASRI), New Delhi, India, pp 1–366

Jiang X, Zhou Y (1987) The value of the radiation induced mutant Fushi in peanut cross breeding. Mutat Breed Newslett 30:2–3

Jiang X, Zhou Y (1988) "Fushi" - excellent mutant germplasm for peanut improvement. Acta Agric Nucleata Sinica 2:147–153

Jiang WZ, Henry IM, Lynagh PG, Comai L, Cahoon EB, Weeks DB (2017) Significant enhancement of fatty acid composition in seeds of the allohexaploid, *Camelina sativa*, using CRISPR/Cas9 gene editing. Plant Biotechnol J 15(5):648–657

Joshua DC, Bhatia CR (1983) Increased cotyledonary cell size in induced large seed mutants in three grain legumes. Environ Exp Bot 23:175–181

Kale DM, Mouli C, Murty GSS, Rao MVP (1997) Development of a new groundnut variety, TG-26 by using induced mutations in cross breeding. Mutat Breed Newslett 43:25–27

Kale DM, Badigannavar AM, Murty GSS (2000a) Forty years of mutation research in groundnut at Trombay: accomplishments and potentials. In: Proc DAE-BRNS Symp Use of nuclear and molecular techniques in crop improvement. Bhabha Atomic Research Centre, Mumbai, India, pp 184–193

Kale DM, Badigannavar AM, Murty GSS (2000b) Development of new large pod Trombay Groundnut (*Arachis hypogaea*) selections. Indian J Agric Sci 70:365–369

Kale DM, Badigannavar AM, Murty GSS (2002) High yields with ideal ideotypes of groundnut varieties TAG 24 and TG 26. Int Arachis Newslett 22:13–15

Kale DM, Murty GSS, Badigannavar AM (2004a) TG-37A a new Trombay groundnut variety with wide adaptation. Int Arachis Newslett 24:19–20

Kale DM, Murty GSS, Badigannavar AM (2004b) TPG-41 a new large-seeded groundnut variety released in India. Int Arachis Newslett 24:21–22

Kale DM, Murty GSS, Badigannavar AM (2007) New Trombay groundnut variety TG-38 suitable for the residual moisture situation in India. J SAT Agric Res 3:1–2

Kale DM, Murty GSS, Badigannavar AM, Makane VG, Toprope VN, Shirshikar SP, Jangawad NP (2008) TLG 45, a large seed groundnut variety for Marathwada region of Maharashtra in India. J SAT Agric Res 6:1–2

Kale DM, Murty GSS, Badigannavar AM, Dhal JK (2009) New Trombay groundnut variety TG 51 for commercial cultivation in India. J SAT Agric Res 7:1–2

Kale DM, Murty GSS, Badigannavar AM, Gupta PC, Bhanushali TB, Sain RS, Chaudhary BR, Kumawat SM, Kumar S, Parameshwarappa KG, Malligawad LH, Kenchanagoudar PV, Bentur MG (2010) A new large-seeded groundnut variety TG 39 for Rajasthan and Karnataka states in India. J SAT Agric Res 8:1–3

Karaman K, Kizil S, Başak M, Uzun B, Yol E (2021) Development of EMS-induced mutagenized groundnut population and discovery of point mutations in the ahFAD2 and Ara h 1 genes by TILLING. J Oleo Sci 70:1631–1640

Karavolias NG, Horner W, Abugu MN, Evanega SN (2021) Application of gene editing for climate change in agriculture. Front Sustain Food Syst 5:685801. https://doi.org/10.3389/fsufs.2021.685801

Kaur N, Alok A, Kaur N, Pandey P, Awasthi P, Tiwari S (2018) CRISPR/Cas9-mediated efficient editing in phytoene desaturase (PDS) demonstrates precise manipulation in banana cv. Rasthali genome. Funct Integr Genomics 18(1):89–99

Kavera, Nadaf HL, Hanchinal RR (2013) Genetic improvement for oil quality through induced mutagenesis in groundnut (*Arachis hypogaea* L.). Indian J Genet 73:378–385

Knoll J, Ramos M, Zeng Y, Holbrook C, Chow M, Chen S, Maleki S, Bhattacharya A, Ozias-Akins P (2011) TILLING for allergen reduction and improvement of quality traits in peanut (*Arachis hypogaea* L.). BMC Plant Biol 11:81

Kris-Etherton MP, Pearson AT, Ying W, Hargrove LR, Moriarty K, Fishell V, Etherton DT (2001) High monounsaturated fatty acid diets lower both plasma cholesterol and triacylglycerol concentrations. Am Soc Clin Nutr 70:1009–1015

Levy A, Ashri A (1978) Induced plasmon mutations affecting the growth habit of peanuts, *A. hypogaea* L. Mutat Res 51:347–360

Lin O, Chen R, Lin T, Li B, Zhuang W (2005) Cytological effect of laser-irradiated peanut seeds and induction of mutation. In: Proc. SPIE 5630, Optics in health care and biomedical optics: diagnostics and treatment II. https://doi.org/10.1117/12.572813

Liu L, Van Zanten L, Shu QY, Maluszynski M (2004) Officially released mutant varieties in China. Mutat Breed Rev 14:1–64

Livore AB, Landau A, Prina AR (2018) The success of IMI tolerant rice varieties in Latin America. In: FAO/IAEA international symposium on plant mutation breeding and biotechnology. IAEA, Vienna, p IAEA-CN-263-131

Lodha ML, Johari RP, Sharma ND, Mehta SL (1983) Nitrogen fixation in relation to dark CO_2 fixation in developing nodules of *Arachis hypogaea* L. mutants. Indian J Exp Biol 21:629–632

Lodha ML, Johari RP, Sharma ND, Mehta SL (1985) Photosynthesis and translocation rate in *Arachis hypogaea* L. mutants. Biochem Physiol Pflanz 180:337–343

Lowder L, Malzahn A, Qi Y (2016) Rapid evolution of manifold CRISPR systems for plant genome editing. Front Plant Sci 7:1683

Manjunath NG, Saravanan S, Sushmitha R, Arumugam Pillai M, Sheela J, Shoba D (2020) Mutagenic efficiency and effectiveness of gamma rays and EMS in groundnut (*Arachis hypogaea* L.). Electronic. J Plant Breed 11(3):875–880

Mathur RK, Manivel P, Gor HK, Chikani BM (1998) Fixing the sublethal dose for sodium azide on groundnut. Int Arachis Newslett 18:16–17

Mathur RK, Manivel P, Samdur MY, Gor HK, Chikani BM (2000) Creation of genetic variability through mutation breeding in groundnut. In: DAE-BRNS symposium. Bhabha atomic Research Centre, Mumbai, India

Menon PM, Raman VS, Krishnaswami S (1970) An induced monosomic in groundnut. Madras Agric J 57:80–82

Mondal S, Badigannavar AM (2010) Induction of genetic variability for fatty acid composition in a large-seeded groundnut variety through induced mutagenesis. J SAT Agric Res 8:1–3
Mondal S, Badigannavar AM (2013) A narrow leaf groundnut mutant, TMV2-NLM has a G to A mutation in *AhFAD*2A gene for high oleate trait. Indian J Genet 73:105–109
Mondal S, Badigannavar AM (2016) Development of high protein groundnut *(Arachis hypogaea* L.) mutant through induced mutagenesis. J Oilseeds Res 33:98–101
Mondal S, Badigannavar AM, Kale DM, Murty GSS (2007) Induction of genetic variability in a disease resistant groundnut breeding line. BARC Newslett 285:237–247
Mondal S, Badigannavar AM, D'Souza SF (2011) Induced variability for fatty acid profile and molecular characterization of high oleate mutant in cultivated groundnut (*Arachis hypogaea* L.). Plant Breed 130:242–247
Mondal S, Petwal VC, Badigannavar AM, Bhad PG, Verma VP, Goswami SG, Dwivedi J (2017) Electron beam irradiation revealed genetic differences in radio-sensitivity and generated mutants in groundnut (*Arachis hypogaea* L.). Appl Radiat Isotopes 122:78–83
Mondal S, Nazareth J, Bhad PG, Badigannavar AM (2018) Isolation of high oleate recombinants in peanut by Near Infra-Red Spectroscopy and confirmation with allele specific polymerase chain reaction marker. J Am Oil Chemists Soc 95:113–121
Moreira ARD, Boroni MAP, Macedo VS, Bressan J, Alfenas BCG, Mattes R, Costa MB (2014) High oleic peanuts: new perspective to attenuate glucose homeostasis disruption and inflammation related obesity. Obesity 22:1981–1988
Motagi BN, Gowda MVC, Sheshagiri R (1996) Mutants resistant to foliar diseases in groundnut. Curr Sci 71:582–584
Motagi BN, Bhat RS, Pujer S, Nayak SN, Janila P, Pandey MK, Varshney RK, Bera SK, Pal KK, Mondal S, Badigannavar AM, Nagaraju P, Yenagi BS, Sugandhi RS, Nimbal A, Goudar I, Roopa U, Nadaf HL, Gowda MVC (2022) Genetic enhancement of groundnut: current status and future prospects. In: Gosal SS, Wani SH (eds) Accelerated plant breeding, vol 4. https://doi.org/10.1007/978-3-030-81107-5_3
Mouli C, Kale DM (1982a) An early maturing groundnut with foliaceous stipule marker. Curr Sci 51:132–134
Mouli C, Kale DM (1982b) Gamma ray induced Spanish bunch mutant with large pod groundnut. Oleagineux 37:583–588
Mouli C, Kale DM (1989) Early maturing mutants in groundnut cultivar, Phule Pragati (JL-24). Curr Sci 58:690–692
Mouli C, Kale DM (1991) Hard seed character in groundnut. In: National Symp genetics and biotechnology for crop improvement. ICAR-Indian Institute of Rice Research, Hyderabad, India, pp 91–94
Mouli C, Patil SH (1976) Gamma ray induced mutant with suppressed branches in the peanut. J Hered 67:322–324
Mouli C, Kale DM, Patil SH (1986) Inheritance of sequential flowering pattern in the mutants of groundnut cultivar, Robut-33-1. Curr Sci 55:1185–1187
Mouli C, Kale DM, Murty GSS (1987) Induced mutants of groundnut cultivar, Phule-Pragati. Curr Sci 56:1237–1238
Mouli C, Kale DM, Patil SH (1989a) Mutation research on groundnut in India. In: Farook SA, Khan IA (eds) Recent advances in genetics and cytogenetics, ICAR-Indian Institute of Rice Research Hyderabad, India 141–153
Mouli C, Kale DM, Patil SH (1989b) Somnath: a high yielding Virginia runner variety of groundnut. Groundnut News 1:5
Mouli C, Kale DM, Patil SH (1990) TG-19A: a large seeded Spanish groundnut. Groundnut News 2:3
Murty GSS, Badigannavar AM, Mondal S, Kale DM (2004) Research and impact of groundnut mutation breeding in India. In: Basu MS, Singh NB (eds) Groundnut research in India. National Research Centre for Groundnut, Junagadh, India, pp 57–69

Nadaf HL, Kaveri SB, Madhusudan K, Motagi BN (2009) Induced genetic variability for yield and yield components in peanut (*Arachis hypogaea* L.). In: Shu QY (ed) Induced plant mutations in the genomics era. FAO, Rome, pp 346–348

Nadaf HL, Biradar K, Murthy GSS, Krishnaraj PU, Bhat RS, Pasha MA, Yerimani AS (2017) Novel mutations in oleoyl-PC desaturase (ahFAD2B) identified from new high oleic mutants induced by gamma rays in peanut. Crop Sci 57:2538–2546

Nkuna RT, Wang CT, Wang XZ, Tang YY, Wang ZW, Zhang JC (2021) Sodium azide induced high-oleic peanut (*Arachis hypogaea* L.) mutant of Virginia type. Genet Resour Crop Evol 68: 1759–1767

NRCG (2000) Progress rep kharif groundnut workshop. University of Agricultural Sciences, Dharwad, India, pp 9–10

Pathirana R, Weerasena LA, Jayamanna (1998) Induced mutations for improvement of groundnut and mungbean. In: Improvement of grain legume production using induced mutations. IAEA, Vienna, pp 1–544

Patil SH (1966) Mutations induced in groundnut by X-rays. Indian J Genet 26A:334–348

Patil SH (1968) Cytogenetics of X-ray induced in aneuploids in *Arachis hypogaea* L. Canadian J Genet Cytol 10:545–550

Patil SH (1973) Trombay groundnut selections for increased oil content and their comparative yield performances. Indian J Agric Sci 43:370–376

Patil SH (1975) A new groundnut variety for export. Nuclear India 13:7–8

Patil SH (1977) Trombay groundnuts: a home in Gujarat for TG-17. Nuclear India 16:3

Patil SH, Bora KC (1961) Meiotic abnormalities induced by X-rays in *Arachis hypogaea* L. Indian J Genet 21:59–67

Patil SH, Mouli C (1977) X-ray induced asynaptic mutant in groundnut. Indian J Exp Biol 15:521–524

Patil SH, Mouli C (1979) Mutation research on groundnut in India. In: Proc symp the role of induced mutation in crop improvement. ICAR-Indian Institute of Rice Research, Hyderabad, India, pp 221–241

Patil SH, Mouli C (1984) Preferential segregation of two allelic mutations for small leaf character in groundnut. Theor Appl Genet 67:327–332

Patil SH, Mouli C, Kale DM (1982) Progress report on varietal improvement in groundnut at BARC. In: 2nd RC meeting. FAO/IAEA, Vienna

Patil SH, Kale DM, Deshmukh SN, Fulzele GR, Weginwar BG (1995) Semi-dwarf, early maturing and high yielding new groundnut variety, TAG-24. J Oilseed Res 12:254–257

Prasad MVR, Kaul S, Jain HK (1984) Induced mutants of peanut (*Arachis hypogaea* L.) for canopy and pod bearing characters. Indian J Genet 44:25–34

Prasad MVR, Mamede FBF, DaSilva FP (1985) Mutational improvement of peanut. Pesq Agropec Bras Brasília 20:439–445

Qiu Q, Li Z, Shen F, Wang C, Miao H (1997) Peanut breeding through mutation techniques in China. Mutat Breed Newslett 43:6–7

Rajendra Prasad MN, Gowda MVC, Patil RK (1998) Screening groundnut mutants for resistance to *Spodoptera litura* and thrips. Int Arachis Newslett 18:29–30

Reddy PS, Reddi MV, Raju BT, Ali SM (1977) Creation of genetic variability by recourse to irradiation in groundnut (*Arachis hypogaea* L.). Oleagineux 32:59–62

Sharma ND, Mehta SL, Patil SH, Eggum BO (1981) Oil and protein quality of groundnut mutants. Qual Plant Foods Hum Nutr 31:85–90

Sharma ND, Santha IM, Patil SH, Mehta SL (1985) Fatty acid and amino acid composition of groundnut mutants. Qual Plant Foods Hum Nutr 35:3–8

Shikazono N, Suzuki C, Kitamura S, Watanabe H, Tano S, Tanaka A (2005) Analysis of mutations induced by carbon ions in *Arabidopsis thaliana*. J Exp Bot 56:587–596

Shin DS, Compadre CM, Maleki S, Kopper RA, Sampson H, Huang SK, Burks AW, Bannon GA (1998) Biochemical and structural analysis of the IgE binding sites on Ara h 1, an abundant and highly allergenic peanut protein. J Biol Chem 273:13753–13759

Shivraj A, Ramana Rao BV (1963) A note on the effects of fast neutrons and gamma rays on groundnut. Indian Oilseed J 7:156–157

Shivraj A, Gangaprasada RN, Ramana Rao BV, Razvi HA (1962) Effects of fast neutrons on seeds of groundnut. Indian Oilseeds J 11:24–30

Shu H, Luo Z, Peng Z, Wang J (2020) The application of CRISPR/Cas9 in hairy roots to explore the functions of AhNFR1 and AhNFR5 genes during peanut nodulation. BMC Plant Biol 20:417

Sivaram MR, Rathinaswamy R, Appadurai R, Sivasubramaniam S (1989) CO 2 a high yielding mutant bunch groundnut. Madras Agric J 76:424–426

Sui J, Wang Y, Wang P, Qiao L, Sun S, Hu X, Chen J, Wang J (2015) Generation of peanut drought tolerant plants by pingyangmycin-mediated in vitro mutagenesis and hydroxyproline-resistance screening. PLoS One 10(3):e0119240

Takagi H, Uemura A, Yaegashi H, Tamiru M, Abe A, Matsuoka C, Utsushi H, Natsume S, Kanzaki H, Matsumura H, Saitoh H, Yoshida K, Cano LM, Kamoun S, Terauchi R (2013a) MutMap-Gap: whole-genome resequencing of mutant F_2 progeny bulk combined with de novo assembly of gap regions identifies the rice blast resistance gene pii. New Phytol 200:276–283

Takagi H, Abe A, Yoshida K, Kosugi S, Natsume S, Mitsuoka C, Uemura A, Utsushi H, Tamiru M, Takuno S, Innan H, Cano LM, Kamoun S, Terauchi R (2013b) QTL-seq: rapid mapping of quantitative trait loci in rice by whole genome resequencing of DNA from two bulked populations. Plant J 74:174–183

Tang YY, Wang XZ, Wu Q, Fang CQ, Guani SY, Yang W, Wang CT, Wang PW (2013) Identification of differentially expressed genes from developing seeds of a normal oil peanut cultivar and its high oil EMS mutant. Res Crops 14(2):511–516

Till BJ, Reynolds SH, Weil C, Springer N, Burtner C, Young K, Bowers E, Codomo CA, Enns LC, Odden AR, Greene EA, Comai L, Henikoff S (2004) Discovery of induced point mutations in maize genes by TILLING. BMC Plant Biol 4:12

Till BJ, Cooper J, Tai TH, Colowit P, Greene EA, Henikoff S, Comai L (2007) Discovery of chemically induced mutations in rice by TILLING. BMC Plant Biol 7:19

Toomer OT (2020) A comprehensive review of the value-added uses of peanut (*Arachis hypogaea*) skins and by-products. Crit Rev Food Sci Nutr 60(2):341

Valls JFM, Costa LC, Custodio AR (2013) A novel trifoliolate species of *Arachis* (Fabaceae) and further comments on the taxonomic section trierectoides. Bonplandia 22:91–97

Wan L, Li B, Pandey MK, Wu Y, Lei Y, Yan L, Dai X, Jiang H, Zhang J, Wei G, Varshney RK, Liao B (2016) Transcriptome analysis of a new peanut seed coat mutant for the physiological regulatory mechanism involved in seed coat cracking and pigmentation. Front Plant Sci 7:1491

Wan L, Li B, Lei Y, Yan L, Ren X, Chen Y, Dai X, Jiang H, Zhang J, Guo W, Chen A, Liao B (2017) Mutant transcriptome sequencing provides insights into pod development in peanut (*Arachis hypogaea* L.). Front Plant Sci 8:1900

Wan L, Lei Y, Yan L, Liu Y, Pandey MK, Wan X, Varshney RK, Fang J, Liao B (2020) Transcriptome and metabolome reveal redirection of flavonoids in a white testa peanut mutant. BMC Plant Biol 20:161

Wang CT, Yang XD, Chen DX, Zhang JC, Xu JZ, Yang WQ (2002) Production of extra large-podded and small podded peanut mutants following chemical mutagen treatment. J Peanut Sci 31(4):5–8

Wang CT, Tang YY, Wang XZ, Zhang SW, Li GJ, Zhang JC, Yu SL (2011a) Sodium azide mutagenesis resulted in a peanut plant with elevated oleate content. Electron J Biotechnol 14(2). https://doi.org/10.2225/vol14-issue-2-fulltext-4

Wang CN, Wang XZ, Tang YY, Zhang JC, Chen DX, Xu JZ, Yang XD, Song GS, Cui FG (2011b) Huayu 40, a groundnut cultivar developed through EMS mutagenesis. J SAT Agric Res 9:1–3

Wang CT, Zhang JC, Tang YY, Guan SY, Wang XZ, Wu Q, Shan L, Zhu LG, Su JW, Yu ST (eds) (2013) Peanut genetic improvement. Shanghai Science and Technology Press, Shanghai. 531 pp

Wang JS, Sui JM, Xie YD, Guo HJ, Qiao LX, Zhao LL, Yu SL, Liu LX (2015) Generation of peanut mutants by fast neutron irradiation combined with in vitro culture. J Radiat Res 56:437–445

Wang J, Shi L, Liu Y, Zhao M, Wang X, Qiao L, Sui J, Li G, Zhu H, Yu S (2020) Development of peanut varieties with high oil content by in vitro mutagenesis and screening. J Integr Agric 19(12):2974–2982

Wen S, Liu H, Li X, Chen X, Hong Y, Li H, Lu Q, Liang X (2018) TALEN-mediated targeted mutagenesis of fatty acid desaturase 2 (FAD2) in peanut (*Arachis hypogaea* L.) promotes the accumulation of oleic acid. Plant Mol Biol 97:177–185

Woo J, Kim J, Kwon S, Claudia C, Cho SW, Kim H, Kim S, Kim S, Choe S, Kim J (2015) DNA-free genome editing in plants with preassembled CRISPR-Cas9 ribonucleoproteins. Nat Biotechnol 33:1162–1164

Wu LR, Chen J, Shi YQ, Miao HR, Qi W, Chen XD, Hu WG (2006) Breeding of Huayu 22 by ^{60}Co γ-rays mutagenesis combined with hybridization. J Nucl Agric Sci 20:309–311

Yu ST, Yu HB, Yu GQ, Zhao LR, Sun HX, Tang YY, Wang XZ, Wu Q, Sun XQ, Wang CT (2015) Isolation of differentially expressed genes from developing seeds of a high-protein peanut mutant and its wild type using Genefishing™ technology. In: Zhang TC, Nakajima M (eds) Advances in applied biotechnology, Lecture notes in electrical engineering, vol 332. https://doi.org/10.1007/978-3-662-45657-6_5

Yu ST, Wang CT, Wang H, Yu GQ, Shi PX, Ren L, Yu HB, Du ZB (2019) Two-step chemical mutagen treatment to convert Fuhua 12, a normal oleic Spanish peanut cultivar, into its high oleic version. IOP Conf Ser Earth Environ Sci 346:012039

Yuan M, Zhu J, Gong L, He L, Lee C, Hans S, Chen C, He G (2019) Mutagenesis of FAD2 genes in peanut with CRISPR/Cas9 based gene editing. BMC Plant Biotechnol 19:24

Zhuang W, Chen H, Yang M, Wang J, Pandey MK, Zhang C, Chang W, Zhang L, Zhang X, Tang R, Garg V, Wang X, Tang H, Chow C, Wang J, Deng Y, Wang D, Khan AW, Yang Q, Cai T, Bajaj P, Wu K, Guo B, Zhang X, Li J, Liang F, Hu J, Liao B, Liu S, Chitikineni A, Yan S, Zheng Y, Shan S, Liu Q, Xie D, Wang Z, Khan SA, Ali N, Zhao C, Li X, Luo Z, Zhang S, Zhuang R-R, Peng Z, Wang S, Mamadou G, Zhuang Y, Zhao Z, Yu W, Xiong F, Quan W, Yuan M, Li Y, Zou H, Xia H, Zha L, Fan J, Yu J, Xie W, Yuan J, Chen K, Zhao S, Chu W, Chen Y, Sun P, Meng F, Zhuo T, Zhao Y, Li C, He G, Zhao Y, Wang C, Kavikishor PB, Pan R, Paterson AH, Wang X, Ming R, Varshney RK (2019) The *Arachis hypogaea* genome elucidates legume karyotypes, polyploidy evolution and crop domestication. Nat Genet 51:865–876

Mutation Breeding for Sustainable Food Production in Latin America and the Caribbean Under Climate Change

17

María Caridad González Cepero, Luz Gómez Pando, Alexander De Andrade, Juan Felix Arguello Delgado, Héctor Nakayama Nakashima, Alejandra Landau, Alberto Prina, Mayra Rodríguez García, Aura Suchini Farfan, and Eulogio de la Cruz Torres

Abstract

The region of Latin America and the Caribbean is characterized by the great diversity of crops and environmental conditions where the crops coexist under the effects of different abiotic and biotic stresses, which impact unfavorably the food and nutrition security. This vulnerability increases by the effect of climate change

M. C. G. Cepero (✉)
National Institute of Agricultural Science (INCA), Mayabeque, Cuba

L. G. Pando
Agrarian National University La Molina, Lima, Perú

A. De Andrade
Empresa de Pesquisa Agropecuária e Extensão Rural de Santa Catarina (Epagri), Estação Experimental de Itajaí, Itajaí, SC, Brazil

J. F. A. Delgado
National University of Costa Rica, Heredia, Costa Rica

H. N. Nakashima
Multidisciplinary Center of Technological Investigations, Asuncion National University, San Lorenzo, Paraguay

A. Landau · A. Prina
Genetic Institute Edwal A. Favret, Buenos Aires, Argentina

M. R. García
Genetic and Engineering and Biotechnology Center (CIGB), Havana, Cuba

A. S. Farfan
Science and Agricultural Technology Institute (ICTA), Guatemala, Mexico

E. de la Cruz Torres
National Institute of Nuclear Investigations (ININ), Mexico City, Mexico

S. Penna, S. M. Jain (eds.), *Mutation Breeding for Sustainable Food Production and Climate Resilience*, https://doi.org/10.1007/978-981-16-9720-3_17

due to increase in minimal and maximal temperatures, changes in rainfall pattern that increased the intensity of drought, and rise of sea level affecting coastal areas by intrusion of salty water. To alleviate these constraints, several programs of genetic improvement have been developed with the support of the International Atomic Energy Agency (IAEA) directed to obtain new high-yielding varieties under different stressful conditions. From these projects, several mutants have been obtained which are used directly in the production areas because of their good productive potential or as parental in breeding programs. These advances have been achieved in crops such as rice (*Oryza sativa* L.), tomato (*Solanum lycopersicum* L.), common bean (*Phaseolus vulgaris* L.), soybean (*Glycine max* Merrill), quinoa (*C. quinoa* Willd), and *Hibiscus sabdariffa* L. Rice is the crop in which the largest number of mutants have been obtained, and more Latin American countries are involved in rice improvement. In this chapter, we have presented an overview of the progress in mutation breeding in the Latin American and the Caribbean region with successful examples.

Keywords

Mutant · Rice · Tomato · Soybean · Common bean · Quinoa · Barley · Sweet potato

17.1 Introduction

Latin America is the center of origin of several economically important species, but the food production is affected by a number of biotic and abiotic stresses and the climatic alterations associated with global warming, which in the past years have increased the vulnerability of the agricultural sector. The main impacts of climate change on agriculture are related with the increase of differences between minimum and maximum temperature, increased rates of evaporation, change as in rainfall patterns (in amount, spatial and temporal distributions), increased intensity and frequency of extreme events (floods and droughts), and intrusion of salty water. These factors cause the reduction in agricultural yields and limit the production of foods being necessary for the development of breeding programs to obtain varieties tolerant to these different stresses. In this regard, mutation induction techniques have demonstrated their efficiency because new, novel genetic variation can be generated in the available germplasm, and they can shorten the breeding and improvement cycles.

In the 1960s, some Latin American countries began to apply the induction of mutations in their breeding programs, and thanks to IAEA regional projects, 18 countries of the region have used these techniques for genetic improvement of several crops. Some of these countries have been able to register mutants that are directly employed in the production of food and as parental material in programs of plant breeding.

17.1.1 Argentina

In Argentina, the Institute of Genetics "Ewald A. Favret" has a long tradition in applying mutation induction techniques in crop plants, using physical and chemical mutagens, and mutator genes, to increase the naturally occurring mutation rates. These works made available some mutants that have been useful for basic research and plant breeding (Prina et al. 2011; Landau et al. 2021). The peanut variety *Colorado Irradiado*, released in 1972, was the first achievement regarding the improvement in productivity and food quality by applying mutation induction and was isolated after X-ray treatments (200 Gy) applied on seeds of *Colorado de Cordoba cv.* The mutant variety showed higher yield and oil content (Kharkwal and Shu 2009).

As a result of the cooperation between the INTA Rice Breeding Program and IGEAF, imidazolinone (IMI)-tolerant mutants were isolated after mutagenic treatments. The IMI-tolerant varieties were tolerant to the infestation by weedy rice, the most important weed constraint for cultivated rice (Livore et al. 2018). Treatments were applied by soaking seed samples from three rice cultivars (IRGA 417, El Paso, and Don Juan), and selection was carried in M_2 plants by treating with a mix of two imidazolinones (imazapyr and imazamox). Five different alleles of the AHAS gene, encoding the acetohydroxyacid synthase enzyme, were isolated, and two of them were patented (Livore et al. 2010, 2011). The first imidazolinone-tolerant variety in Argentina, *PUITA INTA CL*, was released in 2005. It carried the mutant AHAS allele ala122thr. Afterwards, other three varieties (*GURI INTA CL* in 2011, *ÑU POTY INTA CL* in 2014, and *MEMBY PORA INTA CL* in 2017) with improved yield and better quality were released.

These IMI-tolerant varieties covered 70% of the irrigated area in Latin America and in Brazil, the main rice producer of the region, as they reached over 800,000 ha per year. Moreover, varieties carrying the ala122thr allele have also been released in Italy, Greece, Romania, and Portugal. In conclusion, rice varieties carrying the mutant allele 122 greatly contributed to improving productivity and quality, and, besides, they played an important role in recovering rice fields severely invaded by weedy rice. It is obviously a partial victory in the continuous fight against weeds, which requires a sustainable weed management.

Other achievements were obtained in citrus by a joint work between IGEAF and EEA INTA Bella Vista. One was the lemon mutant *cv Eureka 22 INTA*, in which fruit yield and quality were improved. The mutant was obtained from X-ray treatments (10 Gy) applied on *Frost Eureka cv.* A second mutant, obtained by X-ray treatment (20 Gy) on cv. *Valencia Late*, was orange mutant *cv Valencia 2 INTA*, in which fruit set and quality were improved. Both mutants were released in 1987 (Joint FAO/IAEA Division of Nuclear Techniques in Food and Agriculture 1999).

17.1.2 Brazil

The application of the plant-induced mutation in Brazil occurred systematically and continuously since 1964, using rice as a model plant (Ando 1968). Several studies were carried out on the potential effects of the induced mutation in different plant species with chemical and physical induced mutation applied in seeds (rice, maize, beans, tobacco, wheat, and soybeans), in vegetatively propagated plants (black pepper, citrus, apple, and ornamental plants) and in vitro (banana, tobacco, pineapple, black pepper, and some ornamental species).

In Brazil, the Agricultural Research and Rural Extension Company of Santa Catarina (Epagri) with the support of the Center of Nuclear Energy in Agriculture (CENA/USP) has applied the "induced mutation in rice breeding" program to increase genetic variability. The work has enabled the development of rice lines and cultivars with promising characteristics: productivity, cycle, resistance to iron toxicity, quality, resistance to diseases, architecture, tolerance to cold, and resistance to herbicides (Yokoiama et al. 1996; Bacha et al. 2001; Ishiy et al. 2005; Raimondi et al. 2011; Schiocchet et al. 2014; Andrade et al. 2018). The potential to apply the induced mutation to plant breeding in Brazil was demonstrated in a productivity test with rice mutants from different countries, conducted in Sao Paulo (Ando et al. 1971).

17.1.3 Costa Rica

At the beginning of 1980s, Dr. Willy Navarro began a mutation breeding program in rice with the objective to increase the genetic variability and to obtain new varieties that had resistance to diseases, grain quality, and high yield, and the Camago 8 variety was planted in 30% of the rice area under cultivation (Ahloowalia et al. 2004). This variety showed tolerance to *Pyricularia oryzae*, good grain quality, and high yielding. In 2016, the National University of Costa Rica started a mutation breeding program in rice to obtain drought-tolerant rice varietiese.

17.1.4 Chile

In Chile, mutation breeding programs were developed in wheat and barley by using ^{60}Co gamma rays, and it was possible to obtain the wheat variety Caroline in 1981 and the barley variety Leo-INIA/CCU in 1990. The wheat variety Caroline was developed by irradiation of seeds with 100 Gy of ^{60}Co. The main attributes of this variety were high yield, high protein content, resistance to diseases, and good quality. The barley mutant Leo-INIA/CCU was developed by hybridization with mutant variety Riso Mutant 1508. The main improved attributes of this spring barley mutant variety were early maturity, plant height (80–90 cm), high yield (6 t/ha), HLW (69 kg), and high malting quality.

17.1.5 Cuba

The application of nuclear techniques began in Cuba in the 1970s. The first studies were on the evaluation of different types of mutagenic agents and the study of the types and the frequency of the radio mutants in several crops (Cornide and Kotvics 1974; Castillo 1981; Pérez Talavera and Labrada 1988). In Cuba, the induction of mutations has been used in breeding programs of rice (*Oryza sativa* L.), tomato (*Solanum lycopersicum* L.), sugarcane (*Saccharum* spp.), banana (*Musa* sp.), bean (*Phaseolus vulgaris* L.), soybean (*Glycine max* Merrill), and flower of Jamaica (*Hibiscus sabdariffa* L.) among others. These breeding programs have led to the development of eight varieties of rice that showed improved different attributes such as tolerance to salinity, drought, resistance to *Steneotarsonemus spinki* Smiley, as well as high yield and good grain quality; four drought-tolerant tomato varieties; three varieties of soybean characterized for their high productive potential, one of those which is tolerant to nematodes; three varieties of flower of Jamaica of high productive potential; three varieties of sugarcane tolerant to the virus of mosaic (VMCA); and a variety of banana of low bearing (Table 17.1).

17.1.6 Mexico

The mutation induction program using ionizing radiation began in 1974, at the National Institute of Nuclear Research (ININ), and several varieties have been released in different crops (wheat, barley, soybean, and rice). Overall, in a period of 47 years, ten mutant varieties have been registered in Mexico. The ININ and the School of Postgraduates developed the first mutant varieties. The wheat varieties Centaur and Sandbank Bonus have, among its main attributes, resistance to lodging and high yield. In the Field Experimental Northwest of the National Institute of Forest and Agricultural Investigations, the registered mutant varieties included soybean varieties Nainari (previously well known as Héctor) and Esperanza having high yield and resistance to the white fly (*Bemisia argentifolii* Bellows & Perring); in the High Agricultural School of the State of Guerrero (CSAEGRO), a soybean variety SALCER was registered in 2009, and in 2020 the varieties Valente and Albino. The first one is used for animal and human consumption and the last two for human consumption in dry tropical climates. The National Institute of Forest and Agricultural Investigations (INIFAP) developed rice varieties Morelos A 2010 and Morelos A 2010 for high yield, and regarding ornamental species, the INIFAP has registered the ornamental species the Christmas Eve variety Juanita having compact growth for cultivation in gravel.

17.1.7 Peru

Genetic improvement work using the mutation induction technique in Peru began in 1978 in the Cereals and Native Grains Program of the National Agrarian University

Table 17.1 Mutants released in Cuba

Crops	Scientific name	Name of varieties	Registration year	Main characteristics
Rice	*Oryza sativa* L.	IACuba 21	1995	Grain quality
		IACuba 22	1995	Grain quality
		IACuba 23	1995	Grain quality
		IACuba 27	2001	Grain quality
		IACuba 28	2001	Grain quality
		INCA LP-7	1998	Salt tolerance
		INCA LP-10	1998	High yield
		GINES	2007	Drought tolerance
		Guillemar LP 19	2014	Drought tolerance
		Jose LP20	2015	Drought tolerance
Tomato	*Solanum lycopersicum* L.	Maybel	2007	Drought tolerance
		Magine	2007	Drought tolerance
		Domi	2015	Drought tolerance
		Giron 50	2021	Drought tolerance
Sugarcane	*Saccharum officinarum* L.	CCe 10582	1990	High yield
		CCe 283	1993	High yield
		CCe 483	1993	High yield
		CCe 183	1993	High yield
Flower of Jamaica	*Hibiscus sabdariffa* L.	Dogo	2013	High yield
		Ana Delia	2013	High yield
		Benito	2013	High yield
Soybean	*Glycine max* Merrill	INCASoy 35	2004	High yield
		INCASoy 36	2008	High yield
		CUVIN 22	2021	Black grain color

Source: IAEA (2015). IAEA Mutant Database. Vienna: International Atomic Energy Agency

under the direction of Professor Marino Romero Loli. In 1995, the first mutant variety of barley (*Hordeum vulgare*) called UNA-La Molina 95 was released in 1995, developed by irradiation of seeds with gamma rays (300 Gy). The main improved attributes of the mutant variety are early maturity and naked kernels. Genetic improvement work with the induction of mutations was continued to obtain a second variety of barley called Centenario released in 2006. This variety was developed by irradiating seeds with gamma rays (300 Gy), and its main attributes are earliness, higher yield, and quality characters such as better protein content and greater weight of 1000 grains. Mutation induction work has also been carried out on native Peruvian crops such as kiwicha (*Amaranthus caudatus*) and quinoa (*Chenopodium quinoa*), obtaining a variety of *Amaranthus caudatus* called Centenario, released in 2006 and which was developed by irradiation of seeds with gamma rays (400 Gy); its improved characteristics were higher yield, earliness, and weight of 1000 grains. There are advanced mutant lines in *Chenopodium quinoa* in which the yield, growth habit, life cycle, and quality characteristics such as protein content have been improved (Gomez Pando 2014).

17.2 Mutation Breeding in Latin America Facing Climatic Change

The increase in minimum and maximum temperature and the changes in rainfall patterns as well as salinity are the main environmental factors that affect the food production in Latin America. Among the abiotic factors, drought is one of the most devastating, and it has been estimated that more than 50% of the world's arable land will be affected by drought in the year 2050, and some breeding programs have been initiated with the support of the International Atomic Energy Agency (IAEA).

17.2.1 Mutation Breeding in Rice for Salinity and Drought Tolerance

Rice (*Oryza sativa* L.) is one of the most important crops produced worldwide and is the staple food for more than half of the world's population (Yang and Zhang 2006). With respect to yield, rice is the second most produced crop after maize (*Zea mays* L.) and is the second highest crop after wheat with respect to the harvested area. Rice provides more calories than other cereals (Acevedo et al. 2006), and the low rice yields are associated with different types of abiotic and biotic stresses.

Plant mutation breeding has been very successful in rice and has produced around 850 mutant rice cultivars (FAO/IAEA Mutant Variety Database 2019), and some countries of Latin America are working in mutation breeding in rice to obtain rice varieties tolerant to different abiotic stresses (salinity, drought, and high temperature). The International Atomic Energy Agency (IAEA) has supported these investigations through different projects; one of them was the project ARCAL RLA5068 entitled "Improvement of the yield and commercial potential of crops of economic importance for the region of Latin America and the Caribbean," which was recently concluded.

In the National Institute of Agricultural Sciences (INCA) of Cuba, a program of genetic improvement was developed by means of biotechnical and nuclear techniques to obtain new rice varieties with good performance under saline and drought conditions from the Cuban rice varieties Amistad-82 and J-104. The breeding program was developed using a work protocol that included the employment of biotechnical and nuclear techniques for the induction of genetic variability and the selection in vitro (Fig. 17.1), and it was possible to obtain three mutant varieties (Table 17.1; INCA LP-7, INCA LP-10, and Gines LP-18). The varieties INCA LP-7

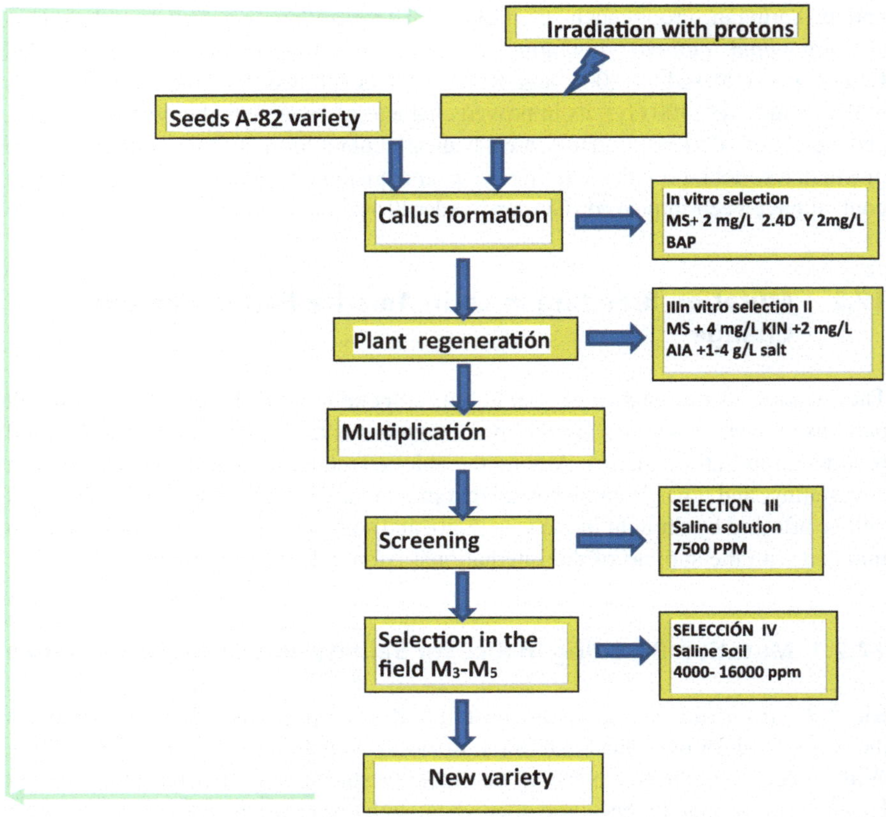

Fig. 17.1 Mutation breeding protocol used in rice for drought and salinity tolerance

and INCA LP-10 were obtained from in vitro culture of Amistad 82 (González 1993). The INCA variety LP-7 is characterized for its tolerance to salinity and resistance to the white Acarus (*Steneotarsonemus spinki* Smiley), while the variety INCA LP-10 is characterized for its tolerance to salinity and high yield. The Guillemar LP-19 and José LP-20 varieties were also developed from crossing with the mutants INCA LP-7 and INCA LP-10, respectively (Cristo et al. 2014, 2015). The variety Gines LP-18 was obtained from the proton beam irradiation of seeds of var. J-104 in the nuclear reactor of DUVNA, and later using in vitro culture protocol, INCA variety LP-7 was developed. This is the first variety of rice obtained using proton beam irradiation (González et al. 2009a, b). The Gines variety stands out for its high productive potential in conditions of low water supplies and the excellent quality of its grain, being highly demanded by the farmers. There is also a group of advanced drought-tolerant lines (Gonzalez and Martínez 2016).

17.2.1.1 Molecular Evaluation of Cuban Rice Mutants

To evaluate the genomic diversity among the mutants and the wild type of variety, amplified fragment length polymorphism (AFLP) technique was used. The four primer combinations tested (E-ACG/M-CAG, E-AGC/M-CAG, E-AGC/M-CAC, E-ACG/M-CTC) produced good polymorphism results. Analyses of the genetic distance between AFLP patterns showed a dendrogram, which depicts the clustering of rice mutants into two groups of individuals as well as shows a clear distinction between mutants and the respective donor variety (Figs. 17.2 and 17.3). Polymorphic AFLP fragments were extracted from polyacrylamide gel and sequenced. Nucleotide sequences were compared with the GenBank nonredundant databases by using the BLAST sequence alignment program. Some AFLP fragment sequences from the mutants were homologues to glutathione *S*-transferase (gst1), auxin-induced protein (pGNT1), and genomic DNA chromosomes 1, 2, 3, 6, and 10.

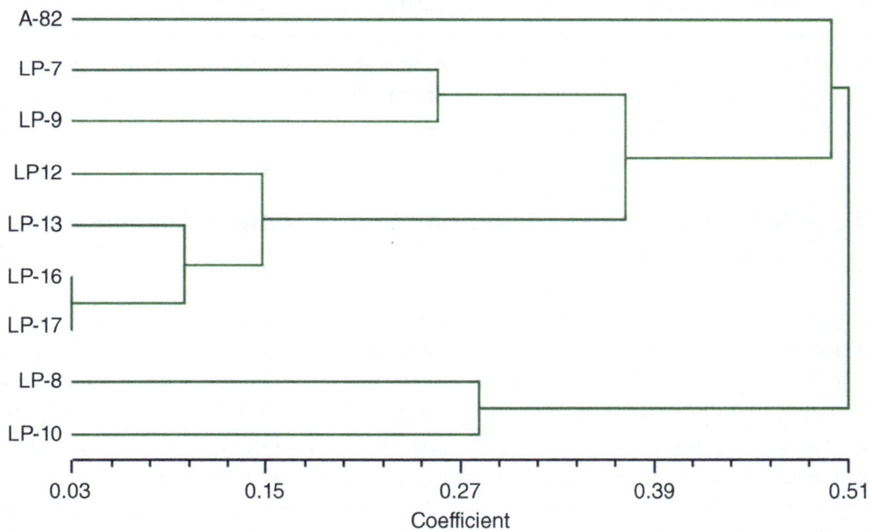

Fig. 17.2 UPGMA cluster analysis of Dice similarity coefficients calculated on the basis of 36 AFLP comparisons among A-82 and eight rice mutants

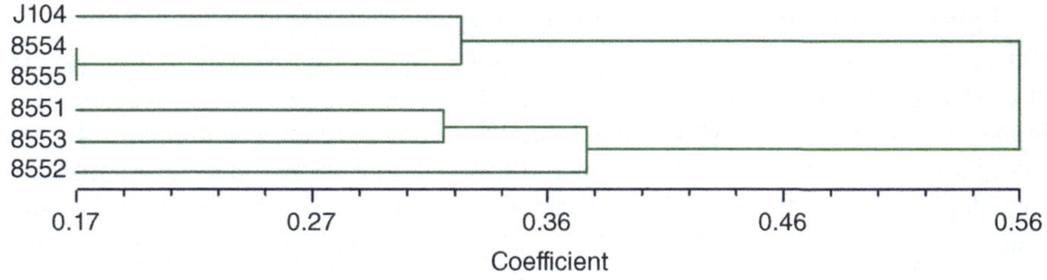

Fig. 17.3 UPGMA cluster analysis of Dice similarity coefficients calculated on the basis of 36 AFLP comparisons among J-104 and five mutant rice genotypes

The enzyme glutathione *S*-transferase is associated with different abiotic stress responses in plants, because a reduced form of glutathione is considered to protect the cell from oxidative damage, based on its redox buffering action and abundance in the cell. It is very interesting to see the association of GST in the tolerant mutant genotypes compared with the susceptible donor variety, suggesting the possible role of this enzyme in salt tolerance.

The National University of Costa Rica has also established a genetic improvement program in rice aimed at obtaining drought-tolerant varieties and developed a protocol for the selection of drought-tolerant genotypes in the greenhouse and field conditions. Seeds M_2 of CR5272 irradiated with 300 Gy were placed on PEG 8000 medium with the osmotic potential of −0.5 MPa in a germination chamber for 14 days and 7 days in a growth room with a photoperiod of 16 h of light. After this period, all selected seedlings were put in a hydroponic system for 50 days for their recovery. The second selection was made in the pre-flowering stage under drought-stress condition (−0.5 MPa) for a period of 27 days, and it was possible to obtain some mutant lines that showed better performance than the parent under drought condition. The studies have demonstrated that mutation induction with ionizing radiation is a valuable tool to obtain drought- and salinity-tolerant rice mutants.

17.2.2 Selection of Heat-Tolerant Mutant of Quinoa (*Chenopodium quinoa* Willd) in Peru

Quinoa (*Chenopodium quinoa* Willd) is a plant species recognized for its high nutritional value and tolerance to drought and salinity and is a valuable alternative to solve future problems for marginal areas and populations with malnutrition problems. Its importance has extended its cultivation to new areas, considering its low tolerance to high temperatures in the coastal region of the Peruvian coast and in other areas of the world where quinoa has been introduced (Mendoza 2013; Bazile et al. 2016). There are various studies that show differences in heat tolerance between varieties of quinoa (Bonifacio 1995). High temperatures cause flower abortion and pollen death in quinoa crops (Jacobsen et al. 2003). Aguilar (1968), cited by Peterson and Murphy (2015), reported that the variety of quinoa subjected to a constant temperature of 32 °C reduced its height, biomass, grain filling, and development time compared to optimal temperatures. Therefore, it has become necessary to develop varieties of quinoa, for the coastal zone, with high productivity and quality for environments with high temperatures. The commercial variety Amarilla Marangani from which the mutant lines were derived by gamma irradiation had a yield of 2771 kg/ha, and it was observed that more than seven mutants exceeded this yield in optimal environments. In summer conditions, only four mutant lines had a higher yield than the Amarilla Marangani parent material. Different stress indices like yield reduction (YR), tolerance (TOL), and stress susceptibility index (SSI) were considered to assess the mutant lines and the parental variety. Mutants M4AM250-70, M4AM250-72, M4AM250-78, M4AM250-85,

M3AM150-302, and M3AM250-580B showed the lowest values for these indices, suggesting higher tolerance to heat stress under the experimental conditions. The highest values for the average productivity index (PM) and geometric mean productivity (PMG) corresponded to the mutant lines M3AM150-1005, M4AM250-70, M4AM150-22, M3AM150-179, and M4AM250-62. Similarly, the highest values of the stress tolerance index (STI) corresponded to the mutant lines M4AM250-78, M4AM250-85, M4AM250-72, M4AM250-70, M3AM250-580B, M3AM150-302, M3AM150-1005, M4AM150-31, and M4AM250-62. Among the mutant lines that showed better performance under optimal and heat-stress conditions were M4AM250-70, M3AM150-302, and M3AM150-1005. On the other hand, mutants M4AM250-85 and M4AM250-78 had acceptable yields considering the yield of quinoa at the national level, equal to 1330 kg/ha (Table 17.2).

17.2.2.1 Mutation Breeding for Drought Tolerance in Soybean (*Glycine max* Merrill) and *Stevia* sp. in Paraguay

Paraguay has begun to use nuclear techniques to generate variability in crops of economic interest, through a program of genetic improvement of crops of the Multidisciplinary Center for Technological Research (CEMIT) of the National University of Asunción. To generate variability and select some drought-tolerant mutant in soybean and *Stevia* sp., gamma ray-irradiated seed of soybean and nodal segments of *Stevia* sp. with X-rays were used. It has been observed that mutants showed phenotypic variability, especially differences in color, shape, and appearance of leaves and plant height, drought tolerance, and development cycle in *Stevia* sp. as well as some mutants in the soybean for drought tolerance (Table 17.3).

17.2.2.2 Mutation Breeding for Drought Tolerance in Sweet Potato (*Ipomoea batatas* L.)

In 2017, internodes of sweet potato varieties CIP 440004 and ICTA Dorado were irradiated with gamma rays ^{60}Co. In the seventh-generation M_1V_7, a mutant was field evaluated in the Vegetable Program for its increased performance. To determine the effect of irradiation on tolerance to water stress in sweet potato varieties, an experiment was carried out in the ICTA Biotechnology Laboratory, using different concentrations of PEG (6, 12, 18, and 24 g/L) in sweet potato seedlings of the varieties CIP code 440004 and ICTA Dorado irradiated with 12.5 Gy. The selection with 12 g/L PEG showed a superior response of tolerance to water stress in plants of CIP code 440004 followed by irradiated plants selected with 18 and 24 g/L PEG. A protocol was thus established for the induction of mutation and for the irradiation and in vitro selection of sweet potato mutants of the varieties CIP 440004 and ICTA Dorado that showed tolerance to drought under in vitro conditions.

17.2.2.3 Mutation Breeding for Drought Tolerance in tomato (*Solanum lycopersicum* L.)

Tomato (*Solanum lycopersicum* L.) is the vegetable of high global importance due to its high demand for superior nutritional qualities and gustatory and varied consumption forms. Fruits contain high content of lycopene, carotene, vitamins, and

Table 17.2 Stress tolerance indices used to determine heat-stress tolerance of advanced quinoa (*Chenopodium quinoa*) mutant lines, La Molina, Peru

Genotype	Grain yield potential (Yp)	Grain yield stress condition (Ys)	% Yield reduction (%YR)	Stress intensity (SI)	Stress susceptibility index (SSI)	Tolerance index (TOL)	Mean productivity index (MP)	Geometric mean productivity index (GMP)	Stress tolerance index (STI)
M4AM150-22	3424.3	1900.0	44.5	0.37	1.203	1524.3	2662.15	2550.72	0.55486
M4MQ150-23	3187.7	1323.3	58.5	0.37	1.581	1864.4	2255.5	2053.85	0.41513
M4AM150-31	2578.3	1714.0	33.5	0.37	0.906	864.3	2146.15	2102.19	0.66478
M4AM250-62	2896.0	1869.7	35.4	0.37	0.958	1026.3	2382.85	2326.94	0.64561
M4AM250-65	2614.7	1387.0	47.0	0.37	1.269	1227.7	2000.85	1904.36	0.530469
M4AM250-69	2546.7	1419.3	44.3	0.37	1.196	1127.4	1983	1901.19	0.55731
M4AM250-70	3083.3	2256.3	26.8	0.37	0.725	827	2669.8	2637.58	0.73178
M4AM250-71	2614.7	1158.3	55.7	0.37	1.505	1456.4	1886.5	1740.29	0.44300
M4AM250-72	2502.7	1880.0	24.9	0.37	0.672	622.7	2191.35	2169.12	0.75118
M4AM250-78	2020.7	1921.0	4.9	0.37	0.133	99.7	1970.85	1970.22	0.95066
M4AM250-85	2391.0	2000.7	16.3	0.37	0.441	390.3	2195.85	2187.16	0.83676
M3AM150-179	3440.0	1693.0	50.8	0.37	1.373	1747	2566.5	2413.28	0.49215

M3AM150-302	2778.7	1973.0	29.0	0.37	0.784	805.7	2375.85	2341.45	0.71004
M3AM150-882	2874.7	1482.3	48.4	0.37	1.309	1392.4	2178.5	2064.26	0.51563
M3AM250-532	2778.3	1398.0	49.7	0.37	1.343	1380.3	2088.15	1970.80	0.50319
M3AM150-561	2843.7	1447.7	49.1	0.37	1.327	1396	2145.7	2029.00	0.50909
M3AM250-711	2330.7	1423.7	38.9	0.37	1.052	907	1877.2	1821.60	0.61085
M3AM150-1338	2607.0	1546.7	40.7	0.37	1.099	1060.3	2076.85	2008.05	0.59329
M3AM250-580B	2585.7	1867.0	27.8	0.37	0.751	718.7	2226.35	2197.16	0.72205
M3AM150-1005	3721.0	2479.7	33.4	0.37	0.902	1241.3	3100.35	3037.60	0.66641
M3AM250-565	3096.3	1571.3	49.3	0.37	1.331	1525	2333.8	2205.72	0.50748
A Maranganí	2771.0	1768.0	36.2	0.37	0.978	1003	2269.5	2213.40	0.63804

Table 17.3 Drought-tolerant lines of soybean and *Stevia* sp. selected in Paraguay

Crops	Mutagenic agent	Characteristic obtained	Number of advanced lines
Glycine max	Gamma ray (^{60}Co)	Drought tolerance	10
Stevia rebaudiana	X-ray	Drought tolerance	5

Fig. 17.4 Participatory selection of tomato mutants

secondary metabolites of great importance to prevent health problems. Tomato mutation breeding program was initiated by seed irradiation of Cuban varieties Amalia and INCA 9-1 using 300 and 500 Gy of gamma rays. The selection was carried out during five cycles in conditions of low water availability and fertilizers, and several variations were observed. It was possible to select some mutant lines of good performance under low water and fertilizer supply. Participatory selection was used in the mutant advanced lines to guarantee the fast introduction of the varieties in the production areas (Fig. 17.4). The mutants Maybel, Magine, Domi, and Giron 50 were registered that showed high productive potential in conditions of low water and fertilizers as well as good quality for their application in the fruit processing industry (González et al. 2009b, 2010). All these results show the importance of induced mutation to obtain new varieties tolerant to biotic and abiotic stresses that contribute to sustainable production of food and reduce the impact of climate change on the agriculture sector of the region.

17.3 Conclusion

In Latin America, several countries have successfully employed the mutation breeding programs with the aim of obtaining varieties tolerant to different types of abiotic and biotic stresses in different crops such as rice, wheat, barley, soybean, tomato, quinoa, potato, sugarcane, and *Hibiscus sabdariffa* among others; however, rice has been the most worked crop. Varieties tolerant to salinity and high and low temperatures have been obtained, as well as varieties with good agronomic behavior in conditions of low water supplies that will contribute to sustainable food production in the face of climate change. The number of countries that have incorporated mutation induction into breeding programs has been increasing to strengthen the work of the Latin American Network of Nuclear Techniques Applied to Agriculture (ATENA) and to continue working on obtaining varieties with environmental stress tolerance, nutritional quality, and good response to the use of biofertilizers to reduce the environmental impact of chemical fertilizers. The mutant varieties have been directly employed in the production of food and as parental material in crop improvement, highlighting the success of mutation breeding in the Latin American and the Caribbean region.

References

Acevedo M, Castrillo W, Belmonte U (2006) Origin, evolution, and diversity of the rice. Trop Agron 56(2):151–170

Ahloowalia BS, Maluszynski M, Nichterlein K (2004) Global impact of mutation-derived varieties. Euphytica 135:187–204

Ando A (1968) Mutation induction in rice by radiation combined with chemical protectants and mutagens. In: Rice breeding with induced mutations, Technical reports series, vol 86. IAEA, Vienna, pp 7–16

Ando A, Banzatto NV, Rocha TR (1971) Ensaio internacional de arroz para mutantes do grupo indica e seleções híbridas. Ciência e Cultura São Paulo 23:157–161

Andrade A, Tulmann Neto A, Tcacenco FA, Marschalek R, Pereira A, Oliveira Neto AM, Scheuermann KK, Wickert E, Noldin JA (2018) Development of rice (*Oryza sativa*) lines resistant to aryloxyphenoxypropionate herbicides through induced mutation with gamma rays. Plant Breed 137:364–369

Bacha RE, Yokoyama S, Ishiy T (2001) Induced mutations as a method of obtaining iron toxicity resistant and high quality rice cultivar. In: Maluszynski M, Kasha KJ (eds) Mutations, in vitro and techniques for environmentally sustainable crop improvement. IAEA, Vienna, pp 175–182

Bazile D, Pulvento C, Verniau A, Al-Nusairi MS, Ba D, Breidy J, Hassan L, Mohammed MI, Mambetov O, Otambekova M, Sepahvand NA, Shams A, Souici D, Miri K, Padulosi S (2016) Worldwide evaluations of quinoa: preliminary results from post international year of quinoa FAO projects in nine countries. Front Plant Sci 7:850. https://doi.org/10.3389/fpls.2016.00850

Bonifacio A (1995). Interspecific and intergeneric hybridization in Chenopod species. Tesis MSc, Provo, Utah Brigham Young University, 150 p

Castillo A (1981) The breeding of rice through mutation induction. Centro Agrícola 8(3):91–103

Cornide MT, Kotvics G (1974) Induction of mutations in tomato (*Lycopersicum esculentum* Mill). Rev CENIC Ser Ciencias Biológicas 5(1):107–118

Cristo E, González MC, Pérez N (2014) Guillemar LP-19. New rice cultivar tolerant to low water and fertilizer supplies for Cuban conditions. Cultivos Tropicales 35(1):85

Cristo E, González MC, Pérez N (2015) José LP-20 rice cultivar tolerant to low water and fertilizer supplies. Cultivos Tropicales 36(2):90

Gomez Pando L (2014) Development of improved varieties of native grains trough radiation-induced mutagenesis. In: Tomlekova NB, Kozgar MI, Wani MR (eds) Mutagenesis: exploring novel genes and pathways. Wageningen Academic Publishers, Wageningen, pp 105–123

González MC (1993) Use of the somaclonal variation in the genetic improvement for the tolerance to the salinity in the cultivation of the rice (*Oryza sativa* L.). Thesis of Doctorate, Instituto Nacional de Ciencias Agrícolas, Dpto. de Genética y Mejoramiento de las Plantas, 100 p

Gonzalez MC, Martínez A (2016) Selection of rice mutants of good agronomic performance under low water supply conditions. Cultivos Tropicales 37(1):102–109

González MC, Pérez N, Cristo E (2009a) GINES: first mutant of rice obtained starting from the irradiation with protons. Cultivos Tropicales 30:3–4

González MC, Mukandama JP, Alí MM, Trujillo D (2009b) Maybel: first tomato variety tolerant to low supplies of water for industrial use obtained by mutation induction. Cultivos Tropicales 30(4):38

González MC, Mukandama JP, Alí MM, Trujillo D (2010) DOMI: tomato mutant of double purpose obtained by induction of mutations with gamma rays of ^{60}Co. Cultivos Tropicales 31(1):86

IAEA (2015) IAEA mutant database. International Atomic Energy Agency, Vienna. http://mvd.iaea.org/. Accessed Jul 2020

Ishiy T, Morel DA, Schiocchet MA, Bacha RE, Ando A, Tulmann Neto A, Knoblauch R (2005) SCS114 Andosan - primeira variedade mutante de arroz irrigado do Brasil. Agropecuária Catarinense Florianópolis 18(3):87–90

Jacobsen SE, Mujica A, Jensen CR (2003) The resistance of quinoa (*Chenopodium quinoa* Willd.) to adverse abiotic factors. Food Rev Int 19:99–109

Joint FAO/IAEA Division of Nuclear Techniques in Food and Agriculture (1999) List of new cultivars. Mutat Breed Newsl (44):31–44

Kharkwal MC, Shu QY (2009) The role of induced mutations in world food security. In: Shu QY (ed) Induced plant mutations in the genomics era. FAO/IAEA, Rome, Italy, pp 33–38. ISBN 978-92-5-106324-8. https://mvd.iaea.org/Colorado. Irradiado Mutant Variety Database, ID 1443

Landau A, Brizuela V, Lencina F, Martínez A, Tcach M, Díaz D, Pacheco MG, Prina AR (2021) Induced mutations in crop plants: mutants of scientific and/or agronomic interest at the Institute of Genetics "Ewald A. Favret". ACI Avances En Ciencias E Ingenierías 12(3):14. https://doi.org/10.18272/aci.v12i3.1928

Livore AB, Prina AR, Singh BK, Ascenzi R, Whitt SR (2010). Herbicide-resistant rice plants, polynucleotides encoding herbicide-resistant acetohydroxyacid synthase large subunit proteins, and methods of use. United States Patent application publication. https://patents.justia.com/inventor/alberto-livore

Livore AB, Prina AR, Birk I, Singh B (2011) Rice plants having increased tolerance to imidazolinone herbicides. European patent application. https://patents.justia.com/inventor/alberto-livore

Livore AB, Landau A, Prina AR (2018) The success of IMI tolerant rice varieties in Latin America. In: FAO/IAEA international symposium on plant mutation breeding and biotechnology. Book of abstracts

Mendoza V (2013) Comparativo de Accesiones de quinua (*Chenpodium quinoa Willd.*) en condiciones de Costa Central. Facultad de Agronomía, Universidad Nacional Agraria La Molina, Lima-Peru. 138p (Thesis)

Pérez Talavera S, Labrada L (1988) Valores de GR_{50} para especies vegetales de interés agrícola en Cuba. Cienc Agr 33:146–149

Peterson AJ, Murphy KM (2015) Quinoa cultivation for temperate North America: considerations and areas for investigation. Improvement and Sustainable Production, Quinoa, p 173

Prina AR, Landau AM, Martínez AE, Pacheco MG, Arias MC, Etchart VJ, Lencina FA, Díaz DG (2011) Fuentes primarias de variabilidad genética en plantas cultivadas. Mutaciones Inducidas y Genes Mutadores. J Basic Appl Genet 22(1) http://ppct.caicyt.gov.ar ppct-caicyt@conicet.gov.ar

Raimondi V, Marschaleck R, Tulmann Neto A (2011) Mutação induzida como fonte de variabilidade para tolerância a temperaturas baixas no estadio de germinação em arroz irrigado. Agropecuária Catarinense Florianópolis 24:75–79

Schiocchet MA, Raimundi V, Tulmann Neto A, Andrade A (2014) SC118 Marques New rice cultivar obtained through induced mutation. Crop Breed Appl Biotechnol Viçosa 14:68–70

Yang J, Zhang J (2006) Grain filling of cereals under soil drying. New Phytol 169:223–236

Yokoiama S, Ishiy T, Schiocchet MA, Tulmann Neto A, Ando A (1996) Mutation breeding for irrigated rice at Empasc, Santa Catarina, Brazil. Use of mutation techniques for improvement of cereals in Latin America, Technical reports series, vol 859. IAEA, Vienna, pp 23–30

18 Mutation Breeding Studies in the Indian Non-basmati Aromatic Rice: Success and Outlook

N. B. Gaikwad and V. A. Bapat

Abstract

Induced mutation is one of the techniques widely used in crop breeding for improvements in desirable traits. Creating variability through mutations is the most significant procedure to be carried out to improve crops. Rice is an important edible cereal and is the staple food for over half of the world's population. Induced mutations have been extensively used in rice breeding programs to generate superior rice varieties. Rice has several genotypes of indigenous landraces having specific multiple characteristics, and aroma is a sought-after quality, always pursued in rice breeding. Aromatic rice varieties are always broadly preferred since they fetch higher revenues to the farmers. This chapter summarizes the work on induced mutations using chemical and physical mutagens in two non-basmati aromatic landraces of rice from India. The stabilities of the identified mutants selected were confirmed and comprehensibly scrutinized in the subsequent four generations in the extensive field trials. Dwarfism, early flowering, non-lodging, and more tillers were the main features of the mutants in both the landraces. Rice landraces, due to their adaptableness to different agroclimatic conditions and exclusive features, are limited to only certain rice-growing areas. Hence, the generated mutants reported in this study would be tested under various agroclimatic zones for demonstrating the utility of these mutants on a wider scale for their integration in rice breeding programs.

N. B. Gaikwad (✉)
Department of Botany, Shivaji University, Kolhapur, India

V. A. Bapat
Department of Biotechnology, Shivaji University, Kolhapur, India

S. Penna, S. M. Jain (eds.), *Mutation Breeding for Sustainable Food Production and Climate Resilience*, https://doi.org/10.1007/978-981-16-9720-3_18

Keywords

Aroma · Breeding · Landraces · Mutations · Rice

18.1 Introduction

Increasing global population, combined with climate changes and striking natural disasters more regularly, is threatening global food security with extraordinary challenges and pressure. It is necessary to produce more food as the world population is expected to extend to 9.7 billion by 2050 (Tripathi et al. 2019). Against this framework, insecurity of global food production poses a hindrance in the pathway of achieving a goal of Zero Hunger, indicated by a latest FAO report (FAO, IFAD, UNICEF, WFP, and WHO 2020). To avoid global food security, it is necessary to design proper crop ideotypes to enhance crop production, which can withstand adverse environmental stress factors, water scarcity, and diseases. In this reference, mutation is a major source of genetic variation and may be used for gene functional analysis. Induced mutations have a long history in both applied and basic aspects of plant research. During the past 50 years, several plant varieties have been developed and released worldwide, either directly from induced mutants or as a result of crossing such mutants with other breeding lines. In the era of plant breeding, conventional breeding approaches including selection based on phenotypes have transformed into mutation approaches to enhance the genetic gain. Additionally, mutation breeding and genetic improvement have been integrated with the updated perceptions of plant tissue culture, DNA markers, genetic engineering, and marker-assisted breeding as expected by more precise background and foreground selection, limitation of breeding cycle, and exploitation and sighting of diverse genetic resources.

Generally, chemical or physical mutations mainly using gamma rays have been employed to generate plant mutations mainly for the plant improvement and for upgrading efficiencies of their desirable properties. Mutation induction, together with mutation detection and mutant stability, is a key element of mutation breeding and has been an important tool for plant breeders for more than 70 years to increase the genetic diversity of plants and derive new mutant lines with value-added characteristics (Oladosu et al. 2016; FAO/IAEA-MVD 2021). The main advantage of mutation breeding is the isolation of one or more desirable, stable mutations in an elite variety without changing the remaining genotype. Furthermore, induced mutations have contributed to the discovery and identification of gene functions in the next genomics era (Sahu et al. 2020). The other listed benefits of mutations are that mutant varieties have proven to be environmentally acceptable, unregulated, and nonhazardous, since there is no difference between artificially induced mutants and naturally spontaneous mutants (Mba 2013). Based on these advantages, mutation breeding has made a significant contribution to the yield, quality, and disease and pest resistance and is free from environmental changes. Mutation breeding programs combined with biotechnologies and advanced omics (genomics, proteomics, transcriptomics, phenomics, metabolomics, etc.) have opened up novel and innovative opportunities for crop improvements. Knowledge from the mutant databases confirms mutagenesis as a significant factor in crop improvement. In view of the advances of induced mutations, in case of rice, it is always a challenge to increase the genetic variability, which could pave the way for advances in the explanation of a genetic, physiological, and biochemical basis of rice traits and generating variability through mutations, which is the utmost tool to increase desirable properties in rice (Wei et al. 2013; Li et al. 2019). Additionally, rice is a versatile plant material for manipulations using in vitro mutagenesis, and in vitro culture not only provides relatively uniform and large populations of cells and tissues in a disease-free situation for the mutagenic treatment but also in a short time makes it possible to screen desirable mutant lines (Jain and Maluszynski 2004; Breseghello 2013; Andrew-Peter-Leon et al. 2021).

Rice (*Oryza sativa* L.) is a staple cereal grain, and one-half of the world population, mainly Asian countries, is dependent upon rice as a key staple food. Rice is the world's second most important cereal crop, and with a production volume of over 755 million metric tons in 2019, China is the world's leading rice producer followed by India and Bangladesh (USDA FAS 2021). Global consumption of rice has seen a slight increase over the last several years. In the 2020/2021 crop year, about 504.3 million metric tons of rice was consumed worldwide. Rice (*Oryza sativa* L.) has two major groups, known by the subspecies names *indica* (*Oryza sativa* ssp. *indica*) and *japonica* (*Oryza sativa* ssp. *japonica*). The indica subspecies is the most widely cultivated form of rice produced universally. The whole-genome sequence studies demonstrated differences of the two subspecies of cultivated rice, and subspecies-specific genes have also been identified in these subspecies (Song et al. 2018). Rice breeding has made an enormous progress and is often referred to as one of the considerable success stories in agriculture. Work on the rice improvement is a continuous research quest, and the rice breeding advances will develop and deploy innovative breeding strategies (Viana et al. 2019). The key research rice breeding areas are breeding for biotic and abiotic adaptive and strong tolerant lines, standardizing rice genotypes for efficient cropping systems, farming practices, and consumer preferences including grain quality, aroma, and nutrition. Simultaneously, other objectives include exactitude phenotyping, germplasm analysis, mutation and molecular breeding, and integrating plant genetic engineering, gene editing, and omics studies in rice breeding programs.

The natural adaptation of rice to different habitats and extensive genotypic and phenotypic diversity and multiplicity has resulted in about 120,000 different accessions. These accessions include traditional rice landraces preserved by local farmers and commercially bred cultivars. The National Bureau of Plant Genetic Resources (NBPGR) in New Delhi has more than 114,724 rice accessions. Landraces derived from the domesticated plant germplasm survive both the abiotic and biotic stress environments. Each landrace has its own specific taste, grain qualities, suitability for milling, cooking, nutritional compositions and other post-harvest procedures, early maturity, adaptability to different soil types, surviving potentialities under biotic and abiotic stresses, and fragrant aromas (Gaur et al.

2016). India is home to many such landraces (Ashraf and Lokanadan 2017), but very few varieties are now being cultivated. Data indicate that about 2000 local landraces are available, and they form about 60% of all rice cultivated in India. Recently, there has been a great interest in inducing genetic variability for agronomic traits so as to bring them into mainstream cultivation (Sao et al. 2021).

18.1.1 Mutation Breeding Research in Rice

Mutation breeding research in rice began several decades back, and currently as per the IAEA Mutant Database, 861 mutant varieties have been registered as mutant varieties released in more than 30 countries, with China and Japan accounting for 37% and 29%, respectively (Fig. 18.1; https://mvd.iaea.org). Other countries to follow in order are India, Indonesia, the USA, Viet Nam, Guyana, Bangladesh, Mali, Pakistan, the Philippines, and Thailand. Among the total mutants, 25 have been aromatic rice types, 271 for disease resistance, 156 dwarf types, and 20 for drought tolerance.

Among the desirable properties of rice, aroma in rice is the most sought-after character fetching always a higher market value. Popularity of aromatic rice is because of globalization, health consciousness, and culinary changes and ever-increasing market demand (Bisht et al. 2018). Rice aroma quality is also exceedingly governed by the farming process along with environmental conditions like temperature, soil type, abiotic stress, water, CO_2, light, salinity, and shading (Mo et al. 2015; Prodhan and Qingyao 2020).

Aromatic rice is a subgroup of rice, which is extensively popular because of the excellent aroma, flavor, specific taste, and superior grain quality. Aromatic rice has

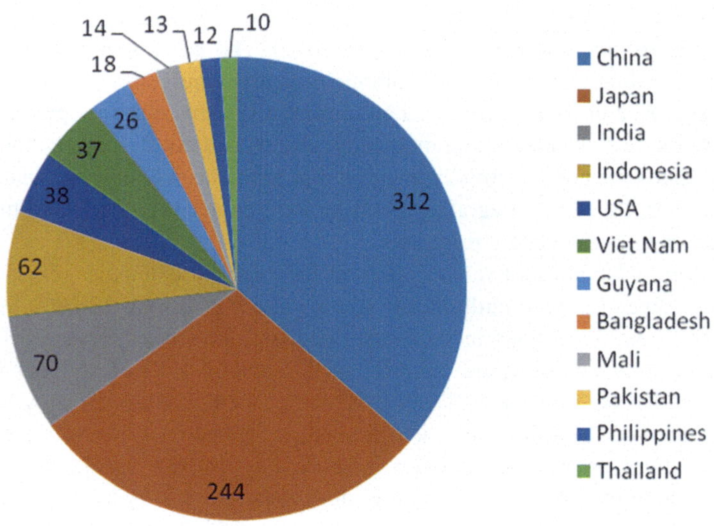

Fig. 18.1 Rice mutant varieties released globally (https://mvd.iaea.org)

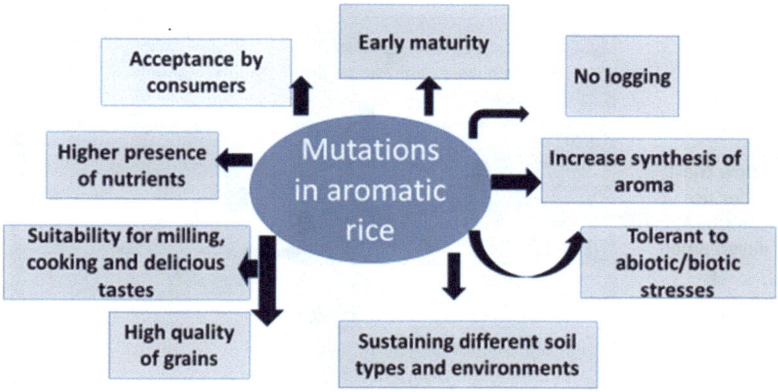

Fig. 18.2 Targeted desirable traits for the improvement of aromatic rice

substantial genetic diversity of germplasm in the Indian subcontinent. Characteristically, these rice varieties are low yielding, have long maturity period, and suffer from various disadvantages such as lodging during maturity, susceptibility to pest and diseases, and stresses. The improvement of aromatic rice has been slower than other nonaromatic rice because of the complex nature of the trait and associated grain value parameters, difficulties related with cross-compatibility with high-yielding nonaromatic rice, and high reliance of expression of quality traits on the environmental factors (Fig. 18.2; Chakravarti et al. 2012; Prodhan and Qingyao 2020). There have been some studies on the induction of mutations for different traits in aromatic rice cultivars (Roy et al. 2018).

Rice aroma is a phenotypical expression of spontaneous recessive mutations of the Badh2 gene and 2-acetyl-1-pyrroline (2AP), which is the principal compound imparting the pleasant aroma to scented rice varieties (Mathure et al. 2011; Wakte et al. 2017). Additional studies demonstrated the presence of alk-2-enals, alka (E)-2,4-dienals, 2-pentylfuran, 2-acetyl-1-pyrroline, and 2-phenylethanol in total aroma profiling of rice. The aroma is caused by the evaporation of 2-acetyl-1-pyrroline. A major gene, *betaine aldehyde dehydrogenase* (*BADH2*), was identified within the *fgr* (fragrance) locus on rice chromosome 8. The presence of dominant *BADH2* allele inhibits the synthesis of 2-acetyl-1-pyrroline (2-AP), a potent flavor component in rice fragrance. The recessive alleles of *BADH2*, resulting from mutations in coding regions, induce 2-AP formation in aromatic rice. Among different *badh2* alleles, an 8 bp deletion in exon 7 is the most frequent and found in all aromatic cultivars, except a few. Although 2-AP is the major aroma compound in aromatic rice, there are several other volatiles in rice whose interaction determines the aroma and flavor of rice grains. Against this background, work on mutation breeding in two non-basmati aromatic local landraces of rice, Ajara Ghansal and Kala Jirga, has been carried out in our laboratory. Both these rice landraces have peculiar characteristics of aromatic grains with superfine grain quality and have a great commercial potential value because of nutritional quality, high iron content, and

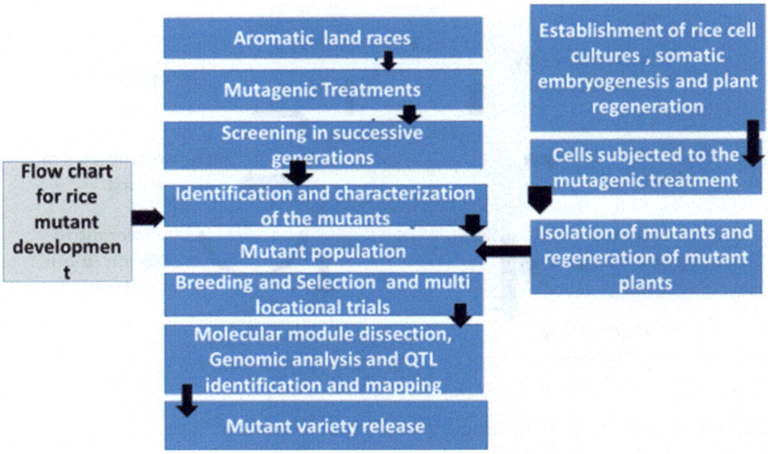

Fig. 18.3 Workflow for induced mutations in aromatic rice landraces

special aroma flavors besides being preferred for consumption especially by old-age patients and lactating mothers for their speedy recovery. In spite of this, these important cultivars are not used for cultivation because of low yield, long maturity period, tallness, and lodging nature. Keeping in view this backdrop, mutation breeding work was undertaken in these two landraces for improvement in agronomic traits. A general workflow for the mutagenesis program for the aromatic rice varieties proposed in our studies is given in Fig. 18.3.

18.2 Mutation Studies in Ajara Ghansal (Non-basmati Aromatic Landrace)

Ajara Ghansal is one of the popular non-basmati aromatic rice landraces and is cultivated only in the Ajara tehsil of Kolhapur district (Desai et al. 2021b). Because of the availability of lateritic soil, high rainfall, and dry and cool climate required for its growth and development of the aroma, it is cultivated at Ajara locality. It is one of the oldest landraces of Kolhapur district and has been cultivated for more than 100 years in Ajara area. Ajara Ghansal rice is famous for its aroma and nutrition properties; due to this, it fetches good returns in the market. In spite of this, farmers are not using this important rice landrace for cultivation because of its low yield, long maturity period, tallness, and lodging nature (Bolake 2013).

Mutagenic treatments of gamma rays (150, 200, and 250 Gy) and ethyl methanesulfonate (EMS; 0.08%, 1.0%, and 1.2%) and sodium azide (SA; 0.002%, 0.004%, and 0.006%) resulted in the reduction in M_1 germination (%) with an increase in the concentration of EMS, whereas in the case of SA and gamma ray treatments, least germination (%) was observed at lower concentration/dose (Desai et al. 2021b). Among all the treatments of EMS, SA, and gamma rays, maximum

Fig. 18.4 Effects of mutagens on various biological parameters of Ajara Ghansal. (**a**) Pollen sterility, (**b** and **c**) chromosomal abnormalities, (**d** and **e**) sectorial mutants, (**f** and **g**) albino, (**h**) xantha, and (**i**) chlorina

pollen sterility was observed in 150 Gy and minimum in 200 Gy gamma ray treatment (Figs. 18.4a and 18.5). Maximum rate of plant survival was reported in EMS 1.0% while the minimum in 200 Gy gamma ray treatment. Chromosomal bridge, sticky metaphase, disturbed metaphase, and scattering of the chromosomes were observed (Fig. 18.4b, c). Highest chromosomal abnormality rate was observed in 1.2% EMS, while the lowest was observed in 150 Gy treatment. Maximum chlorophyll-deficient sectors such as albino, xantha, chlorina, and viridis were observed in M_1 generation in EMS treatment (Fig. 18.4d–I) compared with SA and gamma ray treatments. Among the treatments, pollen sterility was found to be more than 10% as compared to control (Fig. 18.5).

18.2.1 Performance of Mutants

In the M_2 generation, different kinds of mutants such as albino, xantha, chlorina, and viridis were observed. The frequency of chlorophyll mutants varied with variation in concentration/dose of a particular treatment. All the mutagenic treatments with EMS, SA, and gamma rays yielded a total of 201 chlorophyll mutants, out of which albino mutants were maximum followed by viridis, xantha, and chlorine (Fig. 18.6).

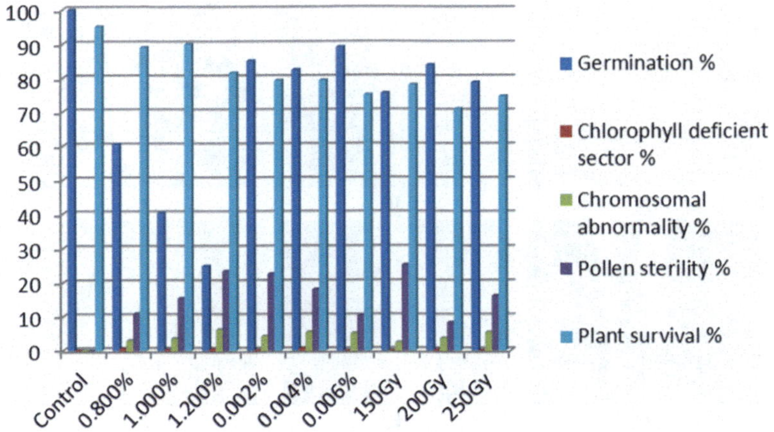

Fig. 18.5 Effects of different mutagens on biological parameters in the M_1 generation of Ajara Ghansal

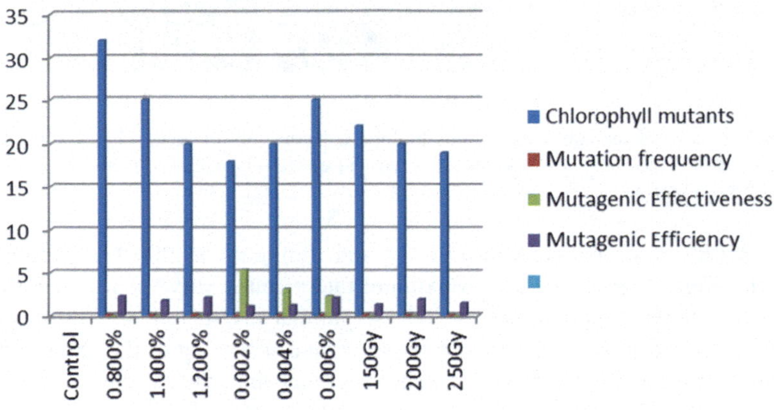

Fig. 18.6 Frequency of chlorophyll mutants, mutagenic effectiveness, and mutagenic efficiency from M_2 generation of Ajara Ghansal

Chlorophyll mutants were reported to be maximum in EMS treatment followed by sodium azide and gamma rays. The mutation frequencies were estimated, and total mutation frequency of chlorophyll mutants (0.839%) was observed. Highest mutation frequency rate was observed in EMS 1.2%, while the lowest mutation frequency rate was observed in SA 0.002%. In case of mutagenic effectiveness, maximum mutagenic effectiveness was observed in SA treatment, while minimum in gamma ray dose (Fig. 18.6). Mutagenic effectiveness declined with an increase in dose/concentration of gamma ray, EMS, and SA treatments. The highest mutagenic efficiency was observed in EMS treatment, whereas minimum efficiency was observed in SA treatment. On the basis of mutagenic efficiency, EMS 0.8%, SA

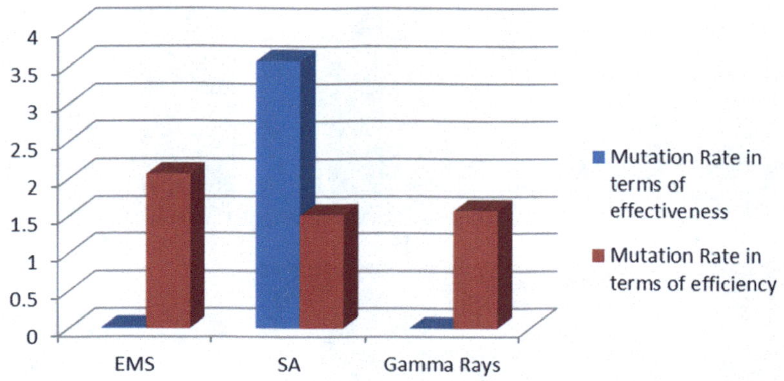

Fig. 18.7 Mutation rate of EMS, SA, and gamma ray treatments in Ajara Ghansal from M_1 and M_2 generation in terms of effectiveness and efficiency

0.006%, and 200 Gy gamma ray treatments were found to be most efficient among all the mutagenic treatments (Fig. 18.6). In this study, the maximum mutation rate in terms of effectiveness was recorded in SA treatment, while the minimum was reported in gamma ray treatment. With regard to mutation rate in terms of efficiency, highest efficiency was recorded under EMS treatment while minimum was reported in SA treatment. Overall EMS treatment reported the highest values of mutation rate when calculated in terms of efficiency (Fig. 18.7) (Desai et al. 2021b).

In Ajara Ghansal, a total of 56 putative mutants were isolated in M_2 generation, which included mutants for dwarfness (16), high tiller number (16), early maturity (15), and lodging resistance (9) (Desai et al. 2021a). In EMS treatment, highest number of mutants (29) were identified followed by gamma ray (16) and SA (11) treatments. EMS (1.0%) treatment was found to be the most effective to induce the highest frequency (0.038%) of dwarf mutants (8) followed by four lodging-resistant mutants with a frequency of 0.019% and two each of early maturing and mutant with higher tiller number at a frequency of 0.009% among all the mutagenic treatments. Overall, EMS (1.0%) was found to be the most effective treatment to induce desirable mutants followed by SA (0.006%) and gamma rays (150 Gy).

In Ajara Ghansal, a total of 16 dwarf mutants were isolated in M_2 generation with plant height ranging from 40 to 95.2 cm. Out of the 16 dwarf mutants, 5 mutants proved to be the most promising, i.e., 4 mutants from EMS and 1 in gamma rays, while the rest of the dwarf mutants had less tiller number and very low grain yield per plant. In M_2 generation, dwarf mutants (5) showed considerable reduction in plant height as compared to the control (Fig. 18.8b). In M_3 generation, all the dwarf mutants expressed true-to-type behavior with significant reduction in plant height (range: 62.8–84.0 cm). All the dwarf mutants flowered significantly late but matured significantly earlier (13–21 days) as compared to control (167 days) (Desai et al. 2021a).

Fig. 18.8 Different morphological mutants of Ajara Ghansal. (**a**) Field view, (**b**–**e**) putative mutants selected from M_2 generation (**b** dwarf, **c** early maturity, **d** lodging resistant, and **e** high yielding), (**f**–**i**) true breeding lines of selected mutants from M_4 generation (**f** dwarf, **g** early maturity, **h** lodging resistant, and **i** high yielding)

Late maturity period is the main hurdle in the large-scale cultivation of Ajara Ghansal; out of 15 early-maturing mutants identified in Ajara Ghansal, only 2 mutants were found promising and reported 25 days' early maturity than control (Fig. 18.8c). Both the mutants expressed true-to-type behavior in M_3 generation with 22–24 days earlier than control (Desai et al. 2021a).

Tallness and lodging are the common phenomena observed in Ajara Ghansal. Nine lodging-resistant mutants were isolated in M_2 generation, out of which two mutants (derived from EMS 1.0% and 150 Gy gamma ray treatments) were promising with resistance to lodging as compared to control (complete lodging) (Fig. 18.8d). In M_3 generation, mutant derived from EMS 1.0% treatment was found to be the most promising and showed significantly higher yield along with higher number of effective tiller per plant and reduction in 1000-seed weight as compared to control (Desai et al. 2021a).

Ajara Ghansal landrace generally is a very less yielder compared with other popular cultivars. In our study, a total of 16 high-tiller-number mutants were noticed in M_2 generation, which showed a range of 58–70 tillers per plant as compared to 19 tillers per plant in the control (Fig. 18.8e). Nine high-tillering mutants showed significant increase in high tiller number, effective tillers, yield per plant, and 1000-seed weight. In M_3 generation, nine high-tillering mutants showed true breeding with significantly high tiller number and effective tillers and reported 1.8- to threefold enhancement in yield per plant compared with control (Desai et al. 2021a).

18.2.2 Stability and Performance of Selected Mutants

18.2.2.1 Dwarf Mutants

Selection of the dwarf mutant plants with a higher number of tiller per plant and good performance in its successive generation has been the main objective of plant breeding programs. In M_4 generation, we have confirmed true-to-type behavior of the dwarf mutant plants obtained from EMS 0.8% and 1.0% and 200 Gy of gamma ray treatments with plant height ranging from 62 to 115 cm compared with control (149 cm). These dwarf mutants showed 24–39 tillers per plant and 22–37 effective tillers per plant compared with control. In the case of yield per plant, dwarf mutants obtained from EMS 0.8% treatment reported 1.5-fold enhancement in the yield and 50% reduction in plant height, while the mutant obtained from 200 Gy gamma ray treatment recorded similar yield with control and 20% reduction in height with 20 days' early maturity (Table 18.1; Fig. 18.8f).

18.2.2.2 Early-Maturity Mutants

Long maturity period is the main problem in the cultivation, so selection of early-maturing mutants with confirmation of its true-to-type behavior has been one of the objectives of the mutation breeding program. From M_3 generation, two early-maturing mutants were selected from EMS 1.0% and 1.2% treatment and were shown to have true-to-type behavior in M_4 generation. An early-maturing mutant was found to mature 16–18 days earlier than control and have 1–1.4-fold higher yield than control. The early-maturing plants showed reduction in plant height by 11–16 cm (Table 18.1; Fig. 18.8g).

18.2.2.3 Lodging-Resistant Mutants

Ajara Ghansal landrace is a tall cultivar (149 cm), and hence, lodging is a common problem at the time of maturity. From M_3 generation, two lodging-resistant mutant plants (isolated from EMS 1.0% and 150 Gy gamma ray treatments) were selected and confirmed for true-to-type behavior at the time of maturity in M_4 generation. Lodging-resistant mutants showed 1.2–1.5-fold enhancement in the yield compared with control (Table 18.1; Fig. 18.8h).

18.2.2.4 High-Yielding Mutants

The nine high-yielding mutants obtained in M_3 generation showed true-to-type behavior in M_4 generation. High-yielding mutants' reported number of tillers per plant and number of effective tillers per plant ranged from 41 to 59 and 38 to

Table 18.1 Performance of true breeding promising mutants of Ajara Ghansal in M_4 generation for important yield traits

S No.	Mutants/ control	Treatment	Days to 50% flowering	Days to 100% flowering	Days to maturity	Plant height (cm)	Panicle length (cm)	No. of tillers/ plant	No. of effective tillers/plant	No. of seeds/ panicle	Yield/ plant (g)	1000-seed weight (g)
1	Ajara Ghansal	Control	118	121	160	149.02	23.85	21.80	20.20	137.68	95.37	21.37
2	DF-1	EMS 0.8%	115	120	155*	77.96*	19.39*	39.20*	37.00*	84.00*	132.69*	21.30
3	DF-2	EMS 1.0%	117	120	154*	62.30*	20.79	33.80*	32.40*	36.12*	46.14*	19.12*
4	DF-3	EMS 1.0%	116	121	155*	70.78*	20.10	24.20	22.40	30.36*	44.84*	18.50*
5	DF-4	EMS 1.0%	120*	125*	155*	78.66*	18.82*	31.20*	29.40*	39.76*	66.96*	19.15*
6	DF-5	200 Gy γ-rays	114	117*	140*	115.22*	19.89*	37.80*	35.40*	72.58*	75.04*	21.90
7	EM-1	EMS 1.0%	108*	111*	142*	138.44*	26.53	28.60*	26.40*	145.82	129.85*	21.49
8	EM-2	EMS 1.2%	109*	113*	144*	133.27*	27.10*	28.40*	27.00*	148.56	96.49	21.60
9	LR-1	EMS 1.0%	119*	123	158	150.94	28.09*	29.80*	27.00*	164.18	110.56*	21.58
10	LR-2	150 Gy γ-rays	111*	115*	157	146.26	27.80*	27.40*	26.20*	180.18*	143.84*	20.80
11	HT-1	EMS 1.0%	114	118	158	147.24	27.97*	59.40*	56.80*	186.72*	225.38*	21.70
12	HT-2	EMS 1.0%	111*	115*	158	147.48	22.63	56.40*	54.60*	187.80*	223.77*	21.38

13	HT-3	EMS 1.2%	114	118	160	146.24	29.33*	54.20*	52.60*	237.60*	162.17*	22.00
14	HT-4	150 Gy γ-rays	111*	115*	158	154.12	29.37*	55.40*	53.00*	243.88*	240.50*	20.39
15	HT-5	150 Gy γ-rays	112	116*	157	141.28*	29.59*	55.60*	53.80*	250.12	235.24*	20.84
16	HT-6	150 Gy γ-rays	111*	115*	156*	145.80	30.13*	54.40*	52.00*	240.56*	225.74*	19.86*
17	HT-7	SA 0.006%	113*	117*	158	139.24*	23.62	59.40*	55.80*	165.64	212.78*	23.32
18	HT-8	SA 0.006%	112*	115*	158	142.68*	22.71	54.20*	52.00*	185.12*	200.77*	23.28
19	HT-9	SA 0.006%	112*	115*	158	141.80	24.85	58.40*	55.80*	224.8*	220.49*	22.81
C.D. at 5%			2.667	2.926	4.690	8.898	2.198	5.942	5.287	40.814	15.194	2.409

DF dwarf mutant, *EM* early-maturing mutant, *LR* lodging-resistant mutant, *HT* high-tiller mutant
*5% level of significance

56 tillers per plant, respectively, compared with 21 and 20 tillers per plant from control. In case of panicle length, it ranged from 22 to 30 cm compared with control (23 cm), and number of seeds per panicle ranged from 161 to 250 seeds per panicle compared with control (137 seeds per panicle). Yield per plant and 1000-seed weight from high-yielding mutant were assessed, and it ranged from 137 to 240 g yield per plant and 19.86 to 23.38 g, respectively, compared with control. Among the high-yielding mutants, mutants obtained from EMS 1.0%, 150 Gy gamma ray, and SA 0.0006% treatments were most promising and reported 2–2.4-fold enhancement in yield compared with control (Table 18.1; Fig. 18.8i). In M_4 generation, stable expression was observed for all the agronomic traits which were selected in M_3 generation.

18.3 Mutation Studies in Kala Jirga (Non-basmati Aromatic Landrace)

Mutagenic treatments with gamma rays (100, 200, and 300 Gy) and ethyl methanesulfonate (0.08%, 1.0%, and 1.2%) and sodium azide (0.002%, 0.004%, and 0.006%) in Kala Jirga showed that maximum pollen sterility percent in M_1 generation was observed in 0.004% SA and minimum in 0.8% EMS treatment, among all the treatments (Fig. 18.9a). The chromosomal abnormalities included

Fig. 18.9 Effects of mutagens on various biological parameters of Kala Jirga. (**a**) Pollen sterility, (**b** and **c**) chromosomal abnormalities, (**d** and **e**) albino, (**f**) xantha, (**g**) chlorina, (**h** and **i**) sectorial mutants

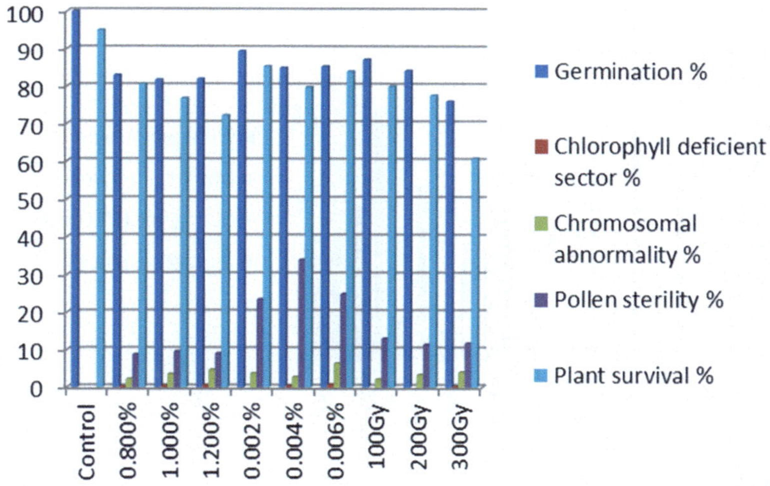

Fig. 18.10 Effects of different mutagens on biological parameters in the M_1 generation of Kala Jirga

anaphase bridge, disturbed metaphase, scattering of the chromosomes, and sticky metaphase (Fig. 18.9b, c). Highest chromosomal abnormality rate was observed in SA 0.006% while the least in gamma ray 100 Gy treatment. Chlorophyll-deficient sectors such as albino, xantha, chlorina, and viridis were observed in M_1 generation where maximum percentage was observed in SA treatment compared with EMS and gamma ray treatments (Fig. 18.9d–i). Mutagenic treatments with gamma rays (100, 200, and 300 Gy) and ethyl methanesulfonate (0.08%, 1.0%, and 1.2%) and sodium azide (0.002%, 0.004%, and 0.006%) in Kala Jirga showed that germination rate decreased with increase in EMS, SA, and gamma rays (Fig. 18.10). In this study, highest plant survival rate was found in SA 0.002% while the lowest in 300 Gy gamma ray treatment (Fig. 18.10).

In the M_2 generation, different kinds of chlorophyll mutants such as albino, xantha, chlorina, and viridis were observed. The frequency of chlorophyll mutants varied with variation in concentration/dose of a particular treatment. A total of 91 chlorophyll mutants were observed in all the treatments of EMS, SA, and gamma rays, out of which chlorina mutants were the highest in number followed by xantha, albino, and viridis (Fig. 18.11). Maximum chlorophyll mutants were reported in EMS treatment followed by gamma rays and sodium azide. The data on mutation frequencies indicated a high total mutation frequency of chlorophyll mutants (2.09%). Maximum mutation frequency rate was observed in SA 0.004% treatment, while the least mutation frequency rate was observed in 100 Gy gamma ray treatment among all the treatments (Fig. 18.11). In case of mutagenic effectiveness, highest mutagenic effectiveness was observed in SA treatment, while lowest in gamma rays. The maximum mutagenic efficiency (18.22) was observed in EMS treatment, whereas minimum efficiency (10.46) was observed in SA treatment. In terms of mutagenic efficiency, EMS 1.0%, SA 0.004%, and 200 Gy gamma rays

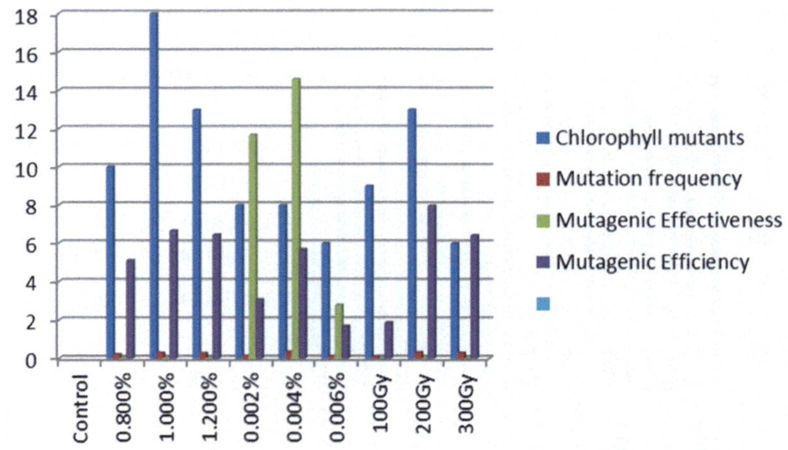

Fig. 18.11 Frequency of chlorophyll mutants, mutagenic effectiveness, and mutagenic efficiency from M_2 generation of Kala Jirga

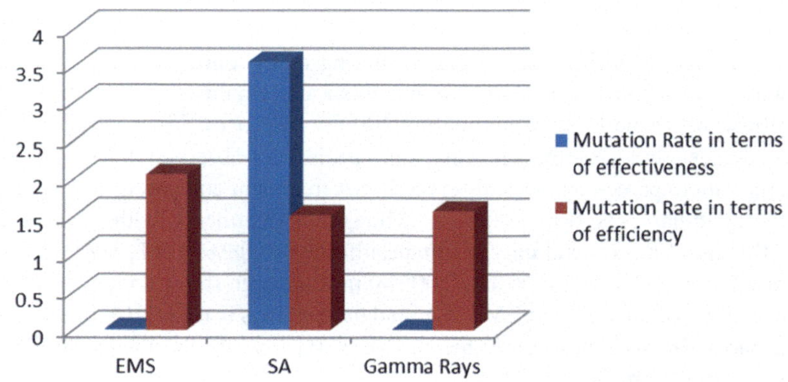

Fig. 18.12 Mutation rate of EMS, SA, and gamma ray treatments in Kala Jirga from M_1 and M_2 generation in terms of effectiveness and efficiency

were found to be most efficient among all the mutagenic treatments (Fig. 18.11). The highest mutation rate (9.677) in terms of effectiveness was recorded in SA treatment, while the minimum (0.0011) was observed in gamma ray treatment. Maximum mutation rate in terms of efficiency was recorded under EMS treatment while minimum was reported in SA treatment. Overall EMS treatment induced the highest mutation rate (6.07) in terms of efficiency (Fig. 18.12) (Desai et al. 2021b).

Fig. 18.13 Different morphological mutants of Kala Jirga. (**a**) Field view, (**b**–**f**) selective putative mutants from M_2 generation (**b** and **c** dwarf, **d** and **e** early maturity, and **f** high yielding), (**g**–**i**) true breeding lines of selected mutants from M_4 generation (**g** dwarf, **h** early maturity, and **i** high yielding)

18.3.1 Performance of Mutants

In Kala Jirga, 24 putative mutants were identified in M_2 generation, which included mutants for dwarf type (8), high tiller number (8), and early maturing (8). Maximum number of mutants (11) were recognized in the EMS treatment followed by SA (8 mutants), while gamma rays noted with 5 mutants.

Development of dwarf plant in Kala Jirga is the major objective of this study. Eight dwarf mutants were isolated; two dwarf mutants derived from 100 and 200 Gy gamma ray treatment were promising with significant 50–70% reduction in plant height compared to control (140 cm) (Fig. 18.13b, c). Remaining dwarf mutants had less tiller numbers and very low grain yield per plant. These two mutants exhibited true breeding behavior in M_3 generation. Height of dwarf mutants was reduced from 40% to 60% as compared to control. Both the mutants performed significantly better than the control in terms of number of tillers per plant, number of effective tillers per plant, grain yield per plant, and 1000-seed weight and yielded 1.15- to 1.8-fold higher yield than control (Desai et al. 2021a).

Early-maturing mutants are enviable in Kala Jirga landrace, which not only minimize the crop duration but also save the time and cost of cultivation. We have identified eight early-maturing mutants in Kala Jirga derived from EMS and SA treatments, which matured 6–12 days earlier than the control in M_2 generation

(Fig. 18.13d, e). In M_3 generation, all the eight mutants matured significantly earlier than the control. Only mutant derived from EMS 0.8% treatment matured 8 days earlier than the control while rest of the mutants were 4–5 days earlier than the control (Desai et al. 2021a).

Enhancement in the yield of Kala Jirga was targeted in this study, and high tillering is one of the most important yield-contributing traits. In M_2 generation, eight mutants with increased tiller number were identified from different treatments of EMS, SA, and gamma rays, which showed a range of 27–32 tillers per plant (compared to 9.4 tillers in the control) and enhancement in yield from 1.8- to 2.8-fold. Maximum number of tillers (32 tillers per plants) was reported in mutants derived from 100 Gy gamma ray treatment with 11 effective tillers (Fig. 18.13f), while highest numbers of effective tillers were recorded in mutants induced under 1.0% and 1.2% EMS treatments. In M_3 generation, five mutants expressed true-to-type behavior for tiller number. All the five high-tiller-number mutants reported an increase in the number of effective tiller per plants and yield as compared to the control. Among the five high-yielding mutants, mutants identified in 1.2% EMS treatment reported 1.5-fold higher yield per plant with significantly higher number of seeds per panicle (Desai et al. 2021a).

18.3.2 Stability and Performance of Mutants

18.3.2.1 Dwarf Mutants

Selection of dwarf mutant plants with a higher tiller number and its true breeding behavior in consecutive generation has been the main focus of mutation breeding studies. In M_4 generation, we have checked true-to-type behavior of the dwarf mutant plants obtained from 100 and 200 Gy gamma ray treatments with a plant height ranging from 53 to 93 cm compared with control (141 cm). In case of yield per plant, dwarf mutants obtained from 100 and 200 Gy gamma ray treatment reported 1.3- to 1.4-fold enhancement in the yield and 40–60% reduction in plant height with 4–5 days' early maturity compared with control (Table 18.2; Fig. 18.13g).

18.3.2.2 Early Maturity Mutants

Long maturity period is the main problem in the cultivation of Kala Jirga, so selection of early-maturing mutants with confirmation of its true-to-type behavior is one of our objectives. From M_3 generation, eight early-maturing mutants were selected derived from EMS (0.8%, 1.0%, and 1.2%) and SA (0.004%) treatments, and true-to-type behavior was observed in M_4 generation. All the mutants matured 6–10 days earlier than the control and had 1.2- to 1.5-fold higher yield than control (Table 18.2; Fig. 18.13h).

18.3.2.3 High-Yielding Mutants

Confirmation of true-to-type behavior of high-yielding mutants obtained from M_3 generation in M_4 generation was the main aim. From M_3 generation, eight

Table 18.2 Performance of true breeding promising mutants of Kala Jirga in M_4 generation for important yield traits

S No.	Mutants/ control	Treatment	Days to 50% flowering	Days to 100% flowering	Days to maturity	Plant height (cm)	Panicle length (cm)	No. of TILLERS/ Plant	No. of effective tillers/ plant	No. of seeds/ panicle	Yield/ plant (g)	1000-seed weight (g)*
1	Kala Jirga	Control	120	123	166	141.08	25.34	22.2	20	120.96	83.31	14.81
2	DF-1	100 Gy γ-rays	126	132*	161*	53.32*	19.09	51*	50*	140.04	112.99*	16.04
3	DF-2	200 Gy γ-rays	120	125	162*	93.44*	22.41	28	26	178.08*	106.11*	12.67
4	EM-1	EMS 0.8%	115*	119*	159*	130.00	24.36	31*	29	198.92*	99.43*	16.44
5	EM-2	EMS 0.8%	118	121	156*	136.32	22.58	22	21	175.28*	61.49	17.47
6	EM-3	EMS 0.8%	119	122	158*	141.96	24.66	32*	31*	194.52*	115.51*	19.05*
7	EM-4	EMS 1.0%	120	123	158*	145.74	23.35	31	29	174.24	93.26	17.70
8	EM-5	EMS 1.0%	120	123	158*	140.3	23.61	33*	30*	160.44	114.09*	21.11*
9	EM-6	EMS 1.2%	118	121	160*	144.24	22.94	31	29	174.24	124.64*	21.13*
10	EM-7	SA 0.004%	115*	120	158*	142.8	24.24	33*	30*	161.44	108.31*	21.50*
11	EM-8	SA 0.004%	119	121	159*	136.74	22.23	30	29	145.28	101.38*	19.33
12	HT-1	100 Gy γ-rays	120	122	160*	148.48	24.03	27	26	191.44*	127.21*	14.79

(continued)

Table 18.2 (continued)

S No.	Mutants/ control	Treatment	Days to 50% flowering	Days to 100% flowering	Days to maturity	Plant height (cm)	Panicle length (cm)	No. of TILLERS/ Plant	No. of effective tillers/ plant	No. of seeds/ panicle	Yield/ plant (g)	1000-seed weight (g)*
13	HT-2	EMS 1.0%	119	122	164	138.64	24.33	30	29	199.28*	92.56	15.35
14	HT-3	EMS 1.2%	120	122	162*	143.6	25.83	33*	31*	167.36	103.45*	18.99
15	HT-4	EMS 1.2%	120	122	165	139.82	25.26	30	29	197.84*	127.36*	15.93
16	HT-5	SA 0.002%	119	122	162*	146.8	24.28	30	29	191.68*	101.69*	14.97
C.D. at 5%			0.518	0.602	3.195	8.051	2.817	5.729	4.641	32.784	15.251	3.579

DF dwarf mutant, *EM* early-maturing mutant, *HT* high-tiller mutant
*5% level of significance

high-yielding mutants were used to check the true-to-type behavior in M_4 generation. The high-yielding mutants obtained from M_3 generation showed true-to-type behavior in M_4 generation. These high-yielding mutants showed 27–33 tillers per plant and 26–31 effective tillers per plant compared with 22 and 20 tillers and effective tillers per plant from control. Yield per plant and 1000-seed weight from high-yielding mutants were assessed, which showed 1.1- to 1.5-fold higher yield than control and enhancement in 1000-seed weight by 1–4 g compared with control (Table 18.2; Fig. 18.13i).

From the above observation, it is clear that mutants were confirmed with improvement of desirable traits in Kala Jirga through EMS, SA, and gamma ray treatments. Among these treatments, EMS (0.8%, 1.0%, and 1.2%), SA (0.004%), and 100 Gy gamma ray treatments were the most effective treatments as compared with other mutagenic treatments for the induction as well as development of the desirable mutants.

18.4 Tissue Culture Studies on Ajara Ghansal and Kala Jirga Rice Landraces

Callus and cell cultures were established from both the rice landraces and their mutants, and protocols were standardized for callus and cell suspension cultures followed by plant regeneration. Cell suspensions are generally an ideal material for the in vitro mutagenic treatments (Suprasanna et al. 2012). Simultaneously, cell suspensions were checked for the presence of aroma, and a high and strong aroma was detected in both the landraces. The powdered callus after drying also retained the aroma flavor offering further opportunities for exploiting in vitro cultures in aromatic rice (Fig. 18.14). Additional studies on the precursors' and elicitors' enrichment of the medium for enhancing aroma synthesis and molecular mechanisms deciphering the activity of aroma synthesis genes are underway. The tissue-cultured raised mutant population will be tested at field level for evaluating the performance and stability of desirable traits.

18.5 Conclusions and Future Outlook

Rice has certain specific characteristics and has been well known and widely used for plant functional genomics research due to its small genome size, perfect genome sequences characterized by linking with the sequences of other cereal crops, high-efficiency transformation technology, and abundant germplasm resources (Jiang et al. 2012; Li et al. 2018). Mutation breeding using chemical or physical mutagens has been well accepted and is an attractive option for inducing genetic variability. EMS-based mutation induction has been applied to create a repository of rice mutants for exploring functional validation of genetic mechanisms for different traits including stress tolerance variability (Jiang and Ramachandran 2010). In India, an EMS-mutagenized M_2 population in the background of Nagina 22 genotype was

Fig. 18.14 Functional prospects of using aromatic rice tissue cultures for aroma applications

created for use in mutant screening (Poli et al. 2021). Considering this, induced mutation in conjunction with genomic approaches needs to be strengthened in rice breeding programs. Genomics could be supportive for better understanding of genetic and molecular insights of induced mutations, and additionally, the availability of mutant populations with induced genetic diversity and advent of molecular modules could be helpful in deciphering the advantageous mutant variation. Genes and genomic regions, through introgression and mutant breeding selection, can be employed with the help of genomic selection and functional genomics (Varshney et al. 2021). Validation and identification of genes controlling prime agronomic traits, especially molecular modules, are necessary in the current post-genomic era (Thao and Phan Tran 2016). Sequencing of landraces, wild relatives, and improved mutant cultivars of rice will provide novel genomic variability platforms for future rice breeding. Incorporation of data generated from rice from in vitro mutagenesis studies would further augment and enrich rice germplasm with the elite varieties (Xiaojia et al. 2006; Chen et al. 2013). Although resistance against biotic and abiotic stresses, grain yield, grain quality, flowering time, and plant height are the prime objectives of rice precision molecular breeding, higher nutritional contents and higher aroma synthesis in rice are also other prized traits which have to be targeted in the rice mutation breeding. Identification and pyramiding of superior alleles controlling key traits have to be pursued further. The journey of rice breeding has

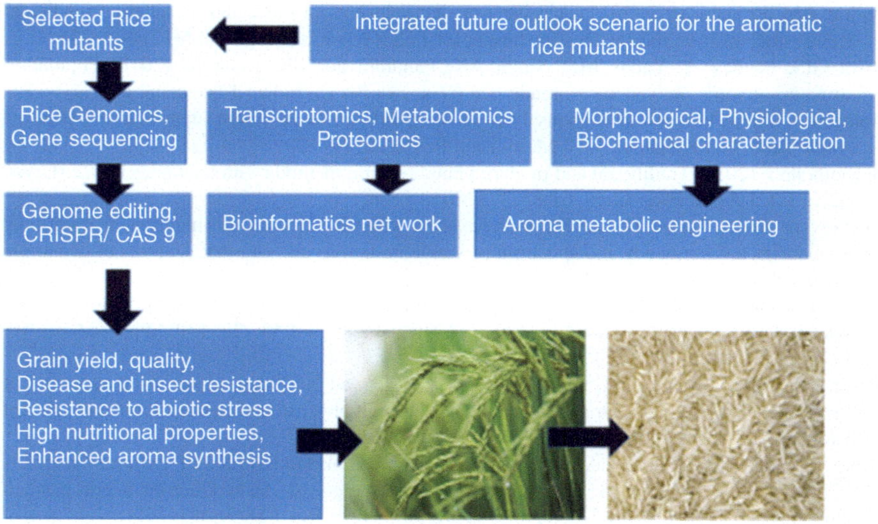

Fig. 18.15 The integrated approaches of induced mutations and molecular studies for improvement of non-basmati aromatic rice cultivars

significantly transcended from the large source concentrated field trials to molecular level with the dawn of molecular breeding, to bring the beneficial changes to the rice plant.

The mutants generated in our laboratory and confirmation of the data from four generation trials on the putative desirable mutants through EMS, SA, and gamma ray treatments can be a valuable addition for the improvement of these two landraces and would be very useful to plant breeders and farmers. These selected mutants can be taken further for appropriate approvals and cultivation on a large scale under agroclimatic surroundings and backgrounds. This is essential to assess the expression and stability of the mutant under specific local environments including biotic and abiotic stresses (Fig. 18.15). The pooled results from several locations and after a proper scrutiny would make these mutant varieties eligible for an authenticated variety release. Simultaneously, mutant plants regenerated from in vitro mutagenic studies may also be tested for their evaluation in the plant breeding programs.

Acknowledgements VAB thanks the National Academy of Sciences, India, for honorary scientist fellowship. We thank Dr. P. Suprasanna for the critical evaluation of the manuscript and providing suggestions for improvement.

References

Andrew-Peter-Leon MT, Ramchander S, Kumar KK, Muthamilarasan MM, Pillai A (2021) Assessment of efficacy of mutagenesis of gamma-irradiation in plant height and days to maturity through expression analysis in rice. PLoS One. https://doi.org/10.1371/journal.pone.0245603

Ashraf AM, Lokanadan S (2017) A review of rice landraces in India and its inherent medicinal values -the nutritive food values for future. Int J Curr Microbiol App Sci 6(12):348–354

Bisht GIS, Pandey A, Yadav MC, Singh AK, Pandraravada SR, Rana JC (2018) Population structure of some indigenous aromatic rice (*Oryza sativa* L). Indian J Biotechnol 17:110–117

Bolake RN (2013) Innovation of ghansal rice farming through Government support, a case study of Ajara takula dist Kolhapur. Int J Multidiscip Res 2(7):119–123

Breseghello F (2013) Traditional and modern plant breeding methods with examples in rice (Oryza sativa L.). J Agric Food Chem 61:8277–8286

Chakravarti SKR, Kumar H, Lal JP, Vishwakarma MK (2012) Induced mutation in traditional aromatic rice – frequency and spectrum of viable mutations and characterizations of economic values. Bioscan 7(4):739–742

Chen YL, Liang HL, Ma XL, Lou SL, Xie YY, Liu ZL et al (2013) An efficient rice mutagenesis system based on suspension-cultured cells. J Integr Plant Biol 55:122–130. https://doi.org/10.1111/jipb.12000

Desai S, Jadhav A, Ramteke A, Dhole V, Bapat V, Gaikwad N (2021a) Genetic improvement of two Indian non-basmati aromatic rice landraces through physical and chemical mutagenesis. Int J Radiat Biol. https://doi.org/10.1080/09553002.2021.1987567

Desai SP, Jadhav AG, Ramteke AR, Dhole VJ, Bapat VA, Gaikwad NB (2021b) Differential effects of gamma rays, EMS, and SA on biological parameters of Ajara Ghansal a non-basmati aromatic rice landrace from Kolhapur, Maharashtra. Agric Res J 58(3):383–389

FAO, IFAD, UNICEF, WFP and WHO (2020) The state of food security and nutrition in the world 2020. Transforming food systems for affordable healthy diets. FAO, Rome. https://doi.org/10.4060/ca9692en

FAO/IAEA-MVD (2021) Food and agriculture organization of the United Nations/International atomic energy agency mutant variety database. https://mvd.iaea.org/#!Search. Accessed 23 Nov 2021

Gaur A, Shabir Wani H, Pandita D, Bharti N, Malav A, Shikari A, Bhat A (2016) Understanding the fragrance in rice. J Rice Res 4:e125. https://doi.org/10.4172/2375-4338.1000e125

Jain SM, Maluszynski M (2004) Induced mutations and biotechnology on improving crops. In: Mujib A, Cho M, Predieri S, Banerjee S (eds) In vitro application in crop improvement: recent progress. IBH, Oxford, India, pp 169–202

Jiang S-Y, Ramachandran S (2010) Assigning biological functions to rice genes by genome annotation, expression analysis and mutagenesis. Biotechnol Lett 32(12):1753–1763

Jiang Y, Cai Z, Xie W, Long T, Yu HQ, Zhang Q (2012) Rice functional genomics research: progress and implications for crop genetic improvement. Biotechnol Adv 30:1059–1070

Li Y, Xiao J, Chen L, Huang X, Cheng Z, Han B, Zhang Q, Wu C (2018) Rice functional genomics research: past decade and future. Mol Plant 11:359–380

Li F, Shimizu A, Nishio T, Tsutsumi N, Kato H (2019) Comparison and characterization of mutations induced by gamma-ray and carbon-ion irradiation in rice (Oryza sativa L.) using whole-genome resequencing. Genes/Genomics/Genetics 9:3743–3751

Mathure S, Shaikh A, Renuka N, Wakte K, Jawali N, Thengane R, Nadaf A (2011) Characterization of aromatic rice (*Oryza sativa* L.) germplasm and correlation between their agronomic and quality traits. Euphytica 179:237–246

Mba C (2013) Induced mutations unleash the potential of plant genetic resources for food and agriculture. Agronomy 3:200–231

Mo ZW, Li W, Pan SG, Fitzgerald TL, Xiao F, Tang YJ, Wang YL, Duan MY, Tian H, Tang XR (2015) Shading during the grain filling period increases 2-acetyl-1-pyrroline content in fragrant rice. Rice 8(1):9

Oladosu Y, Rafii MY, Abdulla N, Hussin G, Ramli A, Rahim HA, Milah G, Usman M (2016) Principles and application of plant mutagenesis in crop improvement, a review. Biotechnol Biotechnol Equip 30(1):1–16. https://doi.org/10.1080/13102828.2015.1087333

Poli Y, Nallamothu V, Hao A, Goud MD, Wang X, Desiraju S, Mangrauthia SK, Jain A (2021) *NH787* EMS mutant of rice variety Nagina22 exhibits higher phosphate use efficiency. Sci Rep 11:9156

Prodhan ZH, Qingyao S (2020) Rice aroma: a natural gift comes with price and the way forward. Rice Sci 27(2):86–100

Roy B, Kumar V, Tulsiram SD, Das BK (2018) Development of high yielding aromatic mutants of rice (*Oryza sativa* L.) from a local aromatic cultivar, Tulaipanji by using gamma radiation. Indian J Genet 78(4):409–416

Sahu PK, Sao R, Mondal S, Vishwakarma G, Gupta SK, Kumar V, Singh S, Sharma D, Das BK (2020) Next generation sequencing based forward genetic approaches for identification and mapping of causal mutations in crop plants: a comprehensive review. Plants 9(1355):1–47. https://doi.org/10.3390/plants9101355

Sao R, Sahu PK, Mondal S, Kumar V, Sharma D, Das BK (2021) Spectrum and frequency of macro and micro mutations induced through gamma rays in traditional rice landraces of Chhattisgarh. Electron J Plant Breed 12(3):693–706

Song S, Tian D, Zhang Z, Hu S, Yu J (2018) Rice genomics: over the past two decades and into the future. Genomics Proteomics Bioinformatics 16:397–404

Suprasanna P, Mohan Jain S, Ochatt J, Kulkarni VM, Predieri S (2012) Applications of *in vitro* techniques in mutation breeding of vegetatively propagated crops. Plant Mutat Breed Biotechnol:371–385

Thao NP, Phan Tran LS (2016) Enhancement of plant productivity in the post-genomics era. Curr Genomics 17(4):295–296. https://doi.org/10.2174/138920291704160607182507

Tripathi AD, Mishra R, Maurya KK, Singh RB, Wilson DW (2019) Estimates for world 1035 population and global food availability for global health. In: Singh RB, Watson RR, Takahashi T (eds) The role of functional food security in global health. Academic, Cambridge, pp 3–24

USDA FAS (2021) United States Department of Agriculture, Foreign Agricultural Service World Agricultural Production, Circular Series WAP 11–21 November 2021

Varshney RK, Bohra A, Yu J, Graner A, Zhang Q, Mark E (2021) Designing future crops: genomics assisted breeding comes of age. Trends Plant Sci 26(6):631–649

Viana VE, Pegoraro C, Busanello C, de Oliveira AC (2019) Mutagenesis in rice: the basis for breeding a new super plant. Front Plant Sci 10:1326

Wakte K, Zanan R, Hinge V, Khandagale K, Nadaf A, Henry R (2017) Thirty-three years of 2-acetyl-1-pyrroline, a principal basmati aroma compound in scented rice (Oryza sativa L.): a status review. J Sci Food Agric 97(2):384–395

Wei FJ, Droc G, Guiderdoni E, Hsing YC (2013) International Consortium of Rice Mutagenesis: resources and beyond. Rice 6(39):1–12

Xiaojia G, Chu Z, Lin Y, Wang S (2006) A tissue culture system for different germplasms of indica rice. Plant Cell Rep 25:392–402

Induced Mutagenesis in Chrysanthemum 19

S. K. Datta

Abstract

Floriculture is now a high-ranking industry in many countries and enriched with scientific knowledge and availability of novel plant materials. New varieties are always in demand in floriculture. Maximum work has been done worldwide on chrysanthemum, a most interesting plant in floriculture trade. The scope of creation of new and novel forms is never-ending in chrysanthemum due to its highly heterozygous genetic makeup. Voluminous work has been carried out on *C. morifolium* Ramat. for its further improvement using a wide range of physical and chemical mutagens. Mutation studies have enriched literature on many important aspects like sensitivity to mutagens, LD_{50} dose, propagules, treatment procedure, mutation event, chimera and its management, in vitro mutation, and desired mutation. All colors were mutable, and more than one color mutation developed in many cultivars. Chrysanthemum developed a number of fascinated mutants like altered flower shape, new appendage like structure on floret, and striped and tubular florets. Mutation developed variability in late-blooming varieties of commercial importance. One can see all types of mutation work starting from classical to modern on this crop and can get a clear picture of technological advancement and its successful application for the development of new varieties. The chapter provides a factual evaluation report on many interesting current topics related to induced mutagenesis work. Chrysanthemum is the only ornamental crop which is considered as a model crop, and one can get all required essential knowledge to design large-scale mutation work on any ornamental crop. Mutation technology package is now enriched with classical and modern techniques for successful development of new and novel varieties.

S. K. Datta (✉)
CSIR-National Botanical Research Institute, Lucknow, India

Bose Institute, Kolkata, India

S. Penna, S. M. Jain (eds.), *Mutation Breeding for Sustainable Food Production and Climate Resilience*, https://doi.org/10.1007/978-981-16-9720-3_19

Keywords

Chrysanthemum · Germplasm · Mutation · Gamma rays · Chimera · In vitro mutagenesis · Mutants

19.1 Introduction

Commercial floriculture is now a very rewarding business and increasing speedily worldwide. Significant growth of this industry has been stimulated in many parts of the world by correct science-based techniques. *Chrysanthemum morifolium* Ramat. is one of the most intriguing ornamental plants in the world. Chrysanthemum is highly appreciated by flower lovers all over the world as it can be grown both as a commercial crop and for exhibition purpose. Diversity in flowers and plant characteristics and broad adjustability to diverse agroecological conditions qualified chrysanthemum to an incomparable position.

Modernization and industrialization of floriculture have created a challenge for the existing varieties and demand for new varieties. Huge amount of work has been done in many countries for improvement of *C. morifolium* Ramat. through induced mutagenesis. A number of physical mutagens (X-ray, gamma rays, fast neutrons, thermal neutrons, radioactive phosphorus, etc.) and chemical mutagens (ethylene imine, ethyl methanesulfonate, and colchicine) have been used for such breeding programs.

Induced mutagenesis work on chrysanthemum is voluminous, and it is not wise to mention all papers in a limited space. Rather, an attempt will be made to discuss some important aspects on the basis of results obtained at the Mutation Breeding Laboratory, Council of Scientific and Industrial Research (CSIR)-National Botanical Research Institute (NBRI), Lucknow, India, in comparison to few other most relevant publications.

In consideration of the importance of induced mutagenesis, CSIR-NBRI started work on different vegetative propagated ornamentals and successfully developed a good number of new promising varieties for the floriculture trade. Voluminous literature has been generated which covers important features like radiosensitivity, selection of material, suitable doses of γ-rays, colchicine treatment, recurrent irradiation, need-based and directive mutations, detection and isolation of mutants, and commercial exploitation of mutants (Datta 2017, 2019b). Attempt has been made in this chapter to focus on different meaningful essential features and scientific progress of mutation methodology, which may be useful as a guideline for extensive mutagenesis work on any ornamental crop.

CSIR-NBRI has developed an appreciable amount of success stories, and the attempt in this direction is continuing. New ornamental/chrysanthemum varieties have created its believability to enhance income and employment generation through boosting production. Vast literature and knowledge generated at CSIR-NBRI are spread in the form of books, bulletins, catalogues, scientific journals, etc. Attempt has been made to highlight the mutation work on chrysanthemum done at

CSIR-NBRI. Careful attempt has been made to focus on every minor detail of technological applications followed at CSIR-NBRI, and important work done at other institutions will be highlighted as and when required.

Most of the mutation breeding works have been carried out with the hexaploid *C. morifolium* Ramat. Chrysanthemum is one of the major ornamental crops, where, perhaps, maximum mutation breeding work has been done worldwide. Chrysanthemum is the only ornamental crop which may be considered as a model crop for induced mutagenesis work. Guidelines about mutation research starting from classical to modern are now reported, which can provide a clear picture of technological advancements and successful creation of new and novel varieties in chrysanthemum (Datta 2020).

19.2 Plant Resources and Methodologies

The autumn-flowering chrysanthemum is a very interesting ornamental plant throughout the world and has acquired enormous recognition as garden flower and as cut flower. Chrysanthemum represents a large polyploid complex ranging from 2*x* to 22*x*, apart from a number of aneuploids. Chrysanthemum is classified into large-flowered and small-flowered type, which are again classified into different categories on the basis of floret shape, size, and position. Most of the mutation breeding works have been carried out with the hexaploid *C. morifolium* Ramat.

Selection criteria for improvement of chrysanthemum varieties are wide considering its genetic diversity and use. A complete picture of conventional and applied mutation technique for creating new and novel chrysanthemum varieties has been shown in Figs. 19.1 and 19.2. All experimental steps have been focused, and most recent technological modifications have been highlighted so that one can follow chrysanthemum as a model experimental plant for mutation experiments. The results and mutants obtained from different experiments are very interesting both from an academic and novelty point of view.

Selection parameters were made on the basis of market potential of varieties like large flowered, small flowered, promising normal season varieties, flower color, flower type, late-blooming varieties, early-blooming varieties, mutant variety, promising color-specific varieties, and induced and spontaneous chimera. The details of large- and small-flowered varieties have been shown in Table 19.1. The other experimental considerations are as follows:

19.2.1 Mutant Genotype

A number of mutant genotypes were selected to study their sensitivity to gamma rays and to induce new mutant characters. The mutant cultivars were “Himani” (small-flowered, white-flower-color mutant of “E-13”), “Sheela” (canary yellow-flower-color mutant of “Himani”), “Man Bhawan” (reddish yellow-flower-color mutant of “Flirt”), “Purnima” (white mutant of “Otome Zakura”), “Colchi Bahar” (terra-cotta

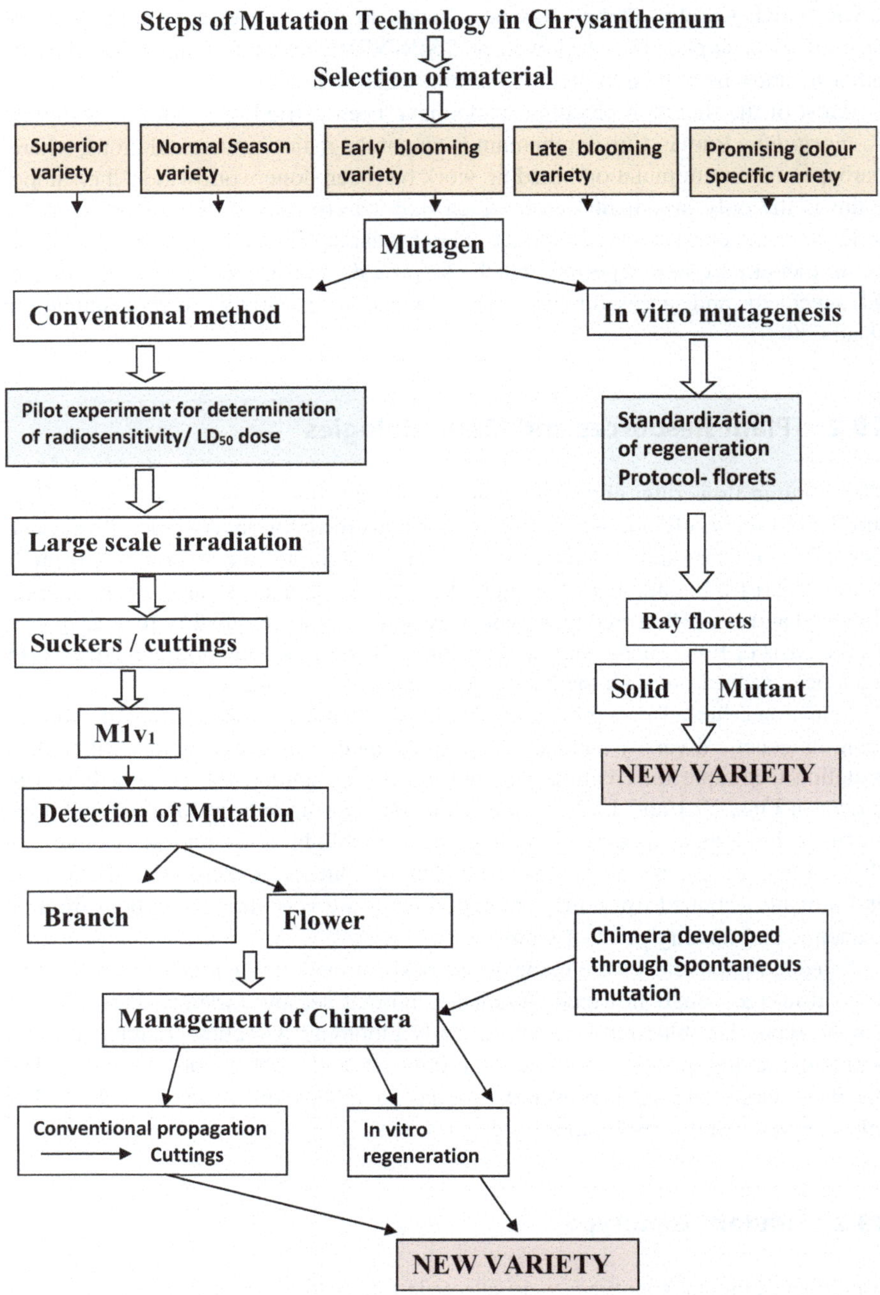

Fig. 19.1 Diagrammatic representation of the application of mutation techniques (traditional and in vitro) in chrysanthemum for the development of new varieties

Fig. 19.2 Practical experiments of mutation techniques on chrysanthemum. (**a**) Cuttings; (**b**) suckers; (**c**) different steps showing how a single mutated cell develops into a chimera; (**d**, **e**) mutated chimeric florets; (**f**, **g**) chimeric mutated branch; (**h**–**k**) isolation of mutant through cuttings from mutated branch; (**h**) mutant "Shabnam"; (**i**) "Purnima"; (**j**) "Colchi Bahar"; (**k**) "Cosmonaut"; (**l**, **m**) management of chimeric florets through in vitro regeneration; (**n**, **o**) in vitro mutation with florets for development of solid mutants; (**p**) directive mutation, white ("Purnima") to yellow; (**q**) in vitro management of chimeric florets developed through spontaneous mutation

red-flower mutant of "Sharad Bahar"), "Alankar" (Spanish orange-flower-color mutant of "D-5"), "Agnisikha" (erythrite red-flower-color mutant of "D-5"), "Shabnam" (flower-shape mutant of "D-5"), and "Navneet" (mutant of "Kalyani Mauve").

19.2.2 Recurrent Irradiation

Rooted cuttings of "Sharad Bahar" (Korean type, purple, early blooming) were exposed to 1.5, 2.0, and 2.5 krad of gamma rays in April 1981. Data were recorded on vegetative and floral characters from treated and control plants, and all plants were propagated in bulk treatment-wise for next vegetative generations. Recurrent irradiation experiment was carried out from these materials again in March 1982. The design and procedures followed for these experiments are as follows: (a) control lot 1981; (b) 1.5 krad 1981; (c) 2.0 krad 1981; (d) 2.5 krad 1981; (e) control lot 1982 maintained from a; (f) cuttings from 1981 1.5 krad; (g) cuttings from 1981 2.0 krad;

Table 19.1 Name of large- and small-flowered varieties of chrysanthemum for experiment

Floret color	Name of variety and type
Large flowered	
Mauve	"Ajina Purple" (ball shaped, incurved), "George Hemming," "Julius Brinas" (reflex, flat floret), "Grape Bowl" (irregular, incurved reflex), "Pink Cloud" (incurving, tubular), "Rose Day," "Hope" (reflex), Spider, "Fish Tail" (intermediate, tubular floret open at tip), "Gallipoli," "H. Townsend," "Kingsford Smith" (incurving or intermediate, tubular), "M-24" (spoon type), "Mrs. H. Gubby," "Meera," "Otome Zakura" (ball shaped), "Pride of Madford" (ball), "Sarojini Naidu," "Silver Cloud," "Thomas W. Pockett," "Undaunted," "M-2," "M-4," "M-9," "M-13," "M-25," "Belur Math" (incurved), "Bharat Ratna," "Mahatma Gandhi," "Ajeets," "Arjuna" (irregular), "Captain Kettle" (incurved), "John Reid," "Red Chief," "Golden Quills," "Goldie," "Paul" (reflex), "Bonaldo," "Alfred Wilson" (reflex), "Miss D. Foxwell," "Red Quills" (quilled, fine tubular floret, very small opening at tip, inner floret incurved)
White	"W-1," "Beatrice May" (reflex, petals irregular type), "Beauty" (incurved, floret tubular with broad open mouth, tip bifid type), "Purnima" (ball type), "White Quills" (quilled), "Fred Conry"
Yellow	"Evening Star" (reflex, outer floret spreading downward and central floret incurved), "Kanchan," "Mrs. Sherrard," "Pitambar" (ball type), "Smith," "Mrs. G. Llyod Wigg," "Mrs. Roger Thompson," Spider, "X"
Small flowered	
Amaranth pink	"Ajay" (decorative/double Korean, early blooming), "Megami" (decorative/double Korean)
Yellow	"Basantika" (decorative pompon), "Bhima" (spider-type floret), "Liliput" (yellow), "Hemanti" (decorative/double Korean)
White	"Cotton Ball" (Pompon), "Lilith" (double Korean), "Nimrod" (Korean type), "Gauri" (decorative pompon, late blooming)
Red	"Colchi Bahar" (decorative type), "Flirt" (double Korean), "Jaya" (double Korean), "Lalkila" (double Korean), "Lalima"
Purple/mauve	"D-5" (decorative), "Fish Tail" (intermediate, tubular floret open at tip), "Shyamal" (double Korean), "Sharad Bahar" (Korean type, early blooming), "Sunil" (double Korean), "M-71," "Puja" (decorative, late blooming), "Khumaini" (double Korean), "E-13" (pompon), "Maghi" (pompon type, late blooming), "Sharad Mala" (double Korean), "Anupam"

(h) cuttings from 1981 2.5 krad; (i) fresh 1.5 krad treatment in 1982; (j) fresh 2.0 krad treatment in 1982; (k) fresh 2.5 krad treatment in 1982; (l) cuttings from 1.5 krad lot of 1981 were collected and treated again with 1.5 krad in 1982; (m) cuttings from 2.0 krad lot of 1981 were collected and treated again with 2.0 krad in 1982; and (n) cuttings from 2.5 krad lot of 1981 were collected and treated again with 2.5 krad in 1982.

19.2.3 Colchicine Treatment

Rooted cuttings of four chrysanthemum cultivars ("D-5": decorative, magnolia purple; "Flirt": double Korean, red; "Lalkila": double Korean, red; and "Sharad

Bahar": Korean type, purple, early blooming) were dipped (2.5 cm) into aqueous solution of 0.0625% and 0.125% colchicine for 5 h. The roots were thoroughly washed in running water before planting.

19.2.4 Early- and Late-Blooming Varieties

There are many attractive flower color varieties in normal season blooming chrysanthemum. But the extent of flower color variation in early- and late-blooming varieties is limited due to limited number of available varieties. For inducing flower color diversification in early- and late-blooming cultivars, rooted cuttings of one early-blooming ("Ajay") and three late-blooming ("Gauri," "Maghi," and "Puja") varieties were exposed to different doses of gamma rays (Shukla and Datta 1993).

19.2.5 Management of Induced Chimera

Rooted cuttings of various promising chrysanthemum cultivars, viz. "Puja," "Purnima," "Colchi Bahar," "D-5," "Shyamal," "Sheela," "Alankar," "Shabnam," "Lalpari," "Lalima," "Navneet," "Puja," "Surekha," "Kundan," "Mother Teresa," "Peet Singer," "Maghi," "Batik," "Flirt," and "Lilith," were exposed to 1.5, 2.0, and 2.5 krad gamma rays and planted in the field up to flowering. For management of chimera, only the plants with mutated chimeric florets were preferred.

19.2.6 Management of Spontaneous Mutation Chimera

In the germplasm collection of CSIR-NBRI, Lucknow, one chrysanthemum cultivar Kasturba Gandhi (large flowered, white) showed some yellow florets due to spontaneous mutation. Shoot buds were regenerated directly from mutated yellow florets applying appropriate culture medium.

19.2.7 In Vitro Mutagenesis

For in vitro mutagenesis, sterilized ray florets of various chrysanthemum cultivars ("Lalima," "Flirt," "Puja," "Maghi," "Sunil," "Madam E. Roger") were exposed to 0.5 and 1.0 krad gamma radiations and regenerated through in vitro culture. Rooted suckers/cuttings (13 cm height) were treated with 1.0, 1.5, 2.0, and 2.5 krad of gamma rays (^{60}Co radiation source). Suckers were treated during March/April, and cuttings were treated during July/August (Fig. 19.2a, b).

19.2.8 General Considerations

19.2.8.1 LD_{50} Dose

Determination of LD_{50} dose prior to conducting experiment is very important for the success of mutation breeding work. Many researchers have reported the appropriate radiation doses to induce somatic mutations in chrysanthemum. Our experimental results determined LD_{50} dose between 1.5 and 2.0 krad for chrysanthemum. Former researchers observed that some cultivars survived even after treatment with 3000r X-rays and determined ideal dose between 2000 and 4000r (Jank 1957; Sheenan and Sagawa 1959). Fujii and Mabuchi (1961) recorded maximum survival after treatment with 2–4 krad gamma rays, and Bowen et al. (1962) noticed only 50% lethality after 4.3 krad treatment. Dowrick and El-Bayoumi (1966a, b) determined 14 krad as the most appropriate dose. Such higher doses have been reported by some other researchers like 25 krad gamma rays by Cawse (1965), 10–12 krad by Yamakawa and Sekiguchi (1968), and 8 krad by Broertjes (1966a, b). Application of such high doses may be due to the use of very low dose rate (1 krad/days, 125–150 rad/ha). The author (Datta 1992, 1994a) determined through repeat experiments 1.5–2.5 krad as the most favorable dose (LD_{50}) to induce mutations in chrysanthemum (suckers and cuttings).

19.2.8.2 Radiosensitivity

Knowledge in radiosensitivity is most imperative for extensive radiation treatment to induce mutations. For determination of radiosensitivity, more than 50 cultivars were selected and influence of a wide range of parameters like flower type, color, size, and shape, chromosome number, chromosomal abnormalities in root meristem, interphase nuclear volume (INV), interphase chromosome volume (ICV), and 2c DNA content were studied after treatment with 1.5, 2.0, and 2.5 krad of gamma rays (Datta 1992, 1994a, b; Banerjee and Datta 1993). Chromosomal abnormalities observed after radiation treatment were clumping, bridges, laggards, late and early separation, and micronuclei. No dose- and cultivar-specific abnormalities were recorded. No positive relationship of radiosensitivity was observed among chromosomal aberrations, INV, ICV, DNA content, and other phenotypic characters. Some cultivars were moderately sensitive, others more sensitive, and others still resistant to mutagens. Induced chromosomal aberrations as the end point of radiosensitivity were studied in root meristems, and it was found that different cultivars were differently sensitive irrespective of flower color, size, and shape. All experimental results clearly indicate that radiosensitivity of chrysanthemum is a genotype-dependent mechanism (Banerjee and Datta 1993; Datta 1992, 1994a).

Survival decreased in all the cultivars with increased treatment dose of gamma rays. Decreased survival and significant reduction in plant height at the time of flowering were recorded in all the cultivars after treatment with increased dose of gamma rays. Higher radiation doses decreased branch, leaf, and flower head numbers. Manifestation of aberrant plant growth was a very striking effect of treatment of suckers/cuttings of the materials under investigation with gamma rays and also colchicine. All the cultivars showed morphological aberrations in leaves, stems,

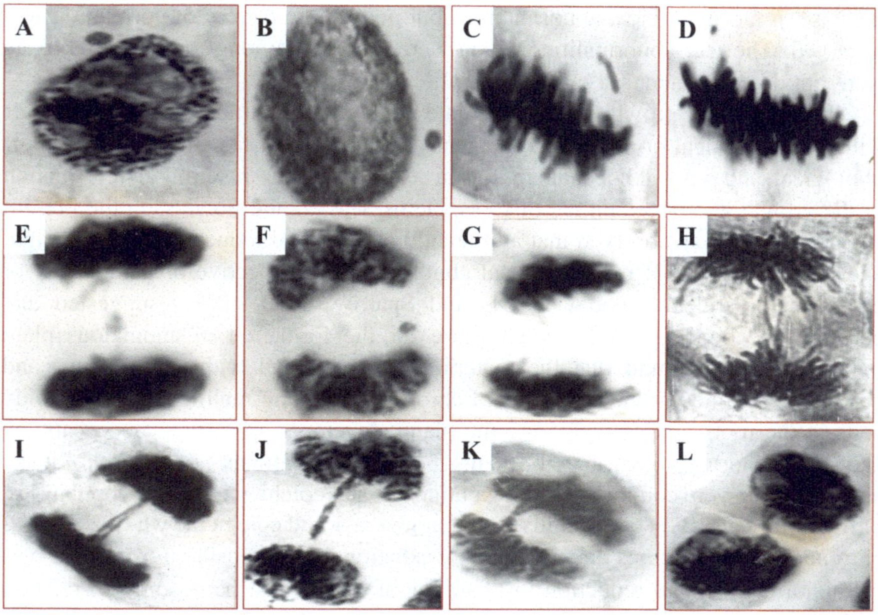

Fig. 19.3 Different types of chromosomal aberrations. (**a**, **b**) Micronuclei; (**c**) early separation; (**d**) fragment; (**e**, **f**) laggards; (**g**) exclusion; (**h**–**l**) different types of bridges at different stages

and flower heads. The leaf abnormalities included changes in shape, size, margin, apex, fission, and fusion of leaves. Floral aberrations were recorded in flower-head shape and size. Asymmetrical development of florets resulted in irregular shape of flower heads, while in some cultivars, smaller flower heads were produced after gamma irradiation. In general, percent of aberrant leaves, plants, and flower heads increased after irradiation and colchicine treatment in all the cultivars. The frequency of plants with morphological aberrations decreased with ageing in most experimental plants. Aberrant cells are eliminated through diplontic discriminating process, and gradually aberrant plants look normal. In some cases, however, the frequency of aberrant leaves per plant decreased after a certain period but increased again indicating that new leaves which developed later were abnormal.

Cytological analysis, with special reference to the impact of gamma rays and colchicine, on chromosomes during root-tip mitosis was studied. Control populations were mostly normal. However, few cells with chromosomal aberrations like clumping, laggards, early separation, bridges, and micronuclei were observed in some of the cultivars. Gamma-irradiated populations showed different types of chromosomal aberrations like clumping, exclusions, laggards, early separation, bridges (chromosomal bridge, sticky bridge), and micronuclei (Fig. 19.3; Datta 1994b), and the frequency of aberrant cells increased in higher doses. Correlation studies of chromosomal aberrations with reduction in plant height and morphological abnormalities yielded interesting results. With increase in exposure, there was increase in chromosomal aberrations and subsequent reduction in plant height and increase in morphological abnormalities, but no specific pattern of correlation was observed. The leaf abnormalities which were recorded among the tested cultivars were not dose, mutagen, or cultivar specific.

Abnormal plant growth is an immediate effect after treatment of plant materials with either physical or chemical mutagens. Treatment induces chromosomal, morphological, and physiological alterations in cells and tissues, resulting in reformation in the growth and development of roots, stems, leaves, and flowers. In the present experiments, it was observed that all the cultivars showed reduction in plant height and different forms of morphological abnormalities irrespective of gamma ray or colchicine treatment. Thoday (1951) and Sparrow et al. (1952) suggested that chromosome breakage is the main cause for the formation of anomalous plant morphology. We found that the materials, when treated with gamma rays and colchicine, resulted in the formation of chromosomal abnormalities and polyploidy cells in the roots, respectively, and the plants developed from the same treatment showed abnormal growth of plants. Different patterns of morphological abnormalities were common to both gamma ray- and colchicine-treated populations. Differential sensitivity of the cultivars to gamma rays and colchicine was recorded. If chromosomal aberrations due to gamma irradiation are responsible for morphological abnormalities and reduction in plant height, there must have some correlation with chromosomal aberrations and morphological changes. The varieties with maximum induced chromosomal aberrations should show maximum morphological abnormalities and reduction in plant height. In the present experiments, the varieties which were most sensitive to gamma rays and colchicine with respect to formation of chromosome aberrations and polyploidy did not show neither maximum morphological abnormalities nor reduction in growth. Therefore, cytological changes due to radiation and colchicine treatment cannot be thought as the agent which directly leads to formation of aberrant plants and reduction in plant height. It is assumed that any alteration in chromosomal status (aberrations or polyploidy) disturbs the route of chemical reactions or mutagens directly act on these chemical pathways, which leads to abnormal growth of plants. It is possible that pathways leading to formation of leaf morphology and plant growth are susceptible differentially to gamma rays and colchicine, and the extent of susceptibility differs from material to material. The author did similar studies and reported same observations on seed-propagated plants using X-ray and colchicine (Datta and Basu 1977). Nonheritable physiological disorder of growth element is the primary reason for the development of unusual plant growth after mutagen treatment. The exact mechanism behind these changes has not yet been authenticated (Datta 1997).

19.2.8.3 Role of Propagule and Time of Irradiation

Chrysanthemum is propagated by both suckers (March/April) and cuttings (July/August). For experiments, rooted cuttings/suckers (13 cm height) were treated with 1, 1.5, 2.0, 2.5, and 3 krad of gamma rays (cobalt-60 gamma radiation source). Treated plants were grown either in pots or in beds. All the standard cultivation methods were applied for proper growth of the experimental plants, and all the flowers were checked to find out mutations during the whole flowering period.

Normally, the somatic mutations in chrysanthemum arise as chimera. The chimeric branches are propagated vegetatively to establish mutant tissues in pure form. Treatment season had definite action in inducing mutation. Plants developed from sucker treatment (March/April) were healthy to a greater extent at flowering time than cutting treatment (July/August). Mutation frequency was higher in cutting irradiation, but the plants were not healthy. On the other hand, sucker-irradiated plants were healthy, but the mutation frequency was less. Separation of mutant sector was not difficult from sucker-irradiated plants, but isolation of mutant tissue was tough from cutting-irradiated plants as the growth of mutant branch was poor. This problem could be solved; that is, chimeric mutants can be established true-to-type using chimera management technique. Secondly, July/August experiments with gamma rays can also be treated with some chemical mutagens as a protection of deleterious effects of gamma rays to encourage healthy growth of plants. The treatment period can be profitably utilized for mutant development considering its benefit and drawback (Datta 1997).

19.2.8.4 Recurrent Irradiation

For recurrent irradiation experiments, the same plant materials are treated in successive generations. In breeding programs, recurrent irradiation has been found to accumulate and expand genetic variability. Rooted cuttings of an early-blooming chrysanthemum cv. "Sharad Bahar" were exposed to 1.5, 2.0, and 2.5 krad of gamma rays. Data were recorded on different morphological aspects from the treated plants, and all the plants were propagated in bulk treatment-wise for next vegetative generations. Recurrent irradiation experiment was carried out with these materials; that is, the materials were again treated with 1.5, 2.0, and 2.5 krad of gamma rays. Different types of effects were noted after gamma irradiation in M_1v_1 and M_1v_2. The reduction in survival, plant height, branch, and leaf number on the one hand and increase in morphological abnormalities and somatic mutation frequency and spectrum on the other hand were more in recurrent irradiated population. Reduction in survival and plant height and induction of different types of morphological abnormalities after gamma ray treatment have been interpreted due to nonheritable physiological disorder of growth substances. It is expected that after radiation treatment, the physiological as well as genetic effects persist in the treated materials. In the first generation, the physiological effects disappear but as a result give over some hidden components within the chromosomes and genes. As a result, the effects are accumulative when experiments are repeated with the same gamma ray doses in successive generations. The result of the experiment lends support to this opinion. From this and other repeat experiments, it may be suggested that recurrent irradiation methods may be utilized in mutation breeding program as this method may create a wide genetic variation than one-time mutation experiment (Datta 1991).

19.2.8.5 Colchi-Mutation

Colchicine has been applied for a long time as a polyploidizing agent, and polyploidy has been developed by colchicine in many important ornamental plants besides agricultural crops. Chromosome doubling and phenotypic changes are the main targets in colchi-ploidy breeding. Very little work has been done on colchicine-induced mutation in vegetatively propagated crops. Experiment was carried out to find out the colchicine-induced mutations in rooted cuttings of chrysanthemum. Rooted cuttings of four chrysanthemum cultivars ("D-5," "Flirt," "Lalkila," and "Sharad Bahar") were dipped (2.5 cm) into aqueous solution of 0.0625% and 0.125% colchicine for 5 h. All the experimental cultivars showed significant decrease in plant height, branch, leaf, and flower-head number and morphological aberrations in leaves and flowers. Higher concentrations showed maximum frequency of aberrations. Time taken for flower bud initiation, first color showing, and full bloom was delayed in colchicine-treated population. Increase in pollen grain sterility and decrease in pollen grain size were significant after colchicine treatment. All experimental cultivars showed mutations in flower color as chimera after colchicine treatment. No mutations could be isolated in the second generation except one mutation in the cultivar "Sharad Bahar." The original color of "Sharad Bahar" was purple, whereas the mutant color was terra-cotta red. The mutant plant has been multiplied, assessed, and released as a new cultivar in the name of "Colchi Bahar" (Datta 1987a, b; Datta and Gupta 1984a, 1987). Colchicine-treated populations of chrysanthemum in this experiment showed exactly the same response as recorded after the treatment of same materials with gamma rays (Datta 1997). This is the first report of colchicine-induced somatic flower color mutation in chrysanthemum. Datta and Gupta (1985) also induced similar flower color mutation for the first time in rose cv. "Contempo" after treatment with colchicine. Colchicine-induced mutations have been reported in *Salvia coccinea* (Haque 1983). Original and colchicine-induced flower color mutant ("Colchi Bahar") cultivars have been critically analyzed and were found to have same chromosome number, interphase nuclear volume, interphase chromosome volume, and DNA content. TLC assay indicated quantitative differences in existing pigments between the two. Present experimental results clearly indicate that colchicine can induce gene mutation (colchi-mutation, C-mutation, Datta 1985a) in place of developing polyploidy. Usually, colchicine experiments are considered ineffective when no polyploidy is formed. Datta (1990a) from his experimental results interpreted that colchicine may be used as a good mutagen although it is famous as a polyploidizing agent.

19.2.8.6 Mutant Genotype (Mutant of a Mutant)

Sensitivity of different chrysanthemum cultivars to gamma rays has been estimated on the basis of different cytomorphological characters. Literature on sensitivity of mutant genotype was scanty. Broertjes et al. (1980) were successful to develop hundreds of mutants by consecutive use of radiation-induced mutants of chrysanthemum cv. "Horim." The author (Datta 1985b, 1997) did a series of experiments to test the radiosensitivity of a number of chrysanthemum mutant genotypes. Rooted cuttings of original cv. "D-5" and its three gamma ray-induced flower color mutants "Alankar" (Spanish orange), "Agnisikha" (erythrite red), and "Shabnam" (flower-shape mutant); "E-13" (mauve) and its mutant "Himani" (white); "Sheela" (canary yellow mutant of mutant "Himani"); and "Flirt" (red) and its mutant "Man Bhawan" (reddish yellow) were exposed to 1.5, 2.0, and 2.5 krad of gamma rays. Original and their respective mutant cultivars showed no noteworthy dissimilarity in ICV, INV, and DNA content. After gamma irradiation, a wide range of morphological (vegetative and floral) anomalies and chromosomal abnormalities were observed in all the original and their mutant cultivars. Percent of cells with chromosomal abnormalities increased with increase in exposure in all the cases. The frequency of induced flower color mutations changed with the cultivars and gamma ray doses. Series of new flower color mutants and at times considerable enhancement in quality and yield were successfully evolved by consecutive application of radiation-induced mutants of chrysanthemum. Reddish yellow flower color mutation was detected in the original cultivar "D-5." Out of the three mutants of "D-5," "Agnisikha" was found to be very stable. No mutation could be detected after irradiation. Yellow flower color mutation in "Alankar" and erythrite red flower color mutation in "Shabnam" were induced. The original cultivar "E-13" was found to be very sensitive to gamma rays with respect to flower color change. A wide range of flower color changes, i.e., erythrite red, white, brownish, and brownish red, were recorded in "E-13." "Himani" produced only canary yellow flower color mutation. "Sheela" was found to be very resistant and no change in flower color could be induced after gamma irradiation. Yellow flower color mutation was induced in the original cultivar "Flirt." "Man Bhawan" developed light-yellow flower color mutation. The concept of developing mutant of a mutant can be exploited in commercial floriculture. The mutant which is well appreciated by the consumer market can be utilized to induce further mutations to increase the new genetic variability for consumers.

19.3 Detection of Mutation (M_1V_1 and Later Vegetative Generations)

Somatic mutations in chrysanthemum are mostly detected in M_1V_1. But repeat experiments and careful observations have proved that mutations that can be detected in M_1V_2, M_1V_3, and later vegetative generations form normal-looking irradiated plants in M_1V_1 (Gupta and Jugran 1978; Das et al. 1974; Datta 1990b; 1994a, b; Usenbaev and Imankulova 1974; Buiatti and Tesi 1968). Probability of recovering solid mutants is more in M_1v_2 and later generations. Results showed that screening for mutations should not be restricted to M_1V_1 only; it should be continued in M_1V_2 and subsequent vegetative generations. Detection of mutation in the M_1V_2 has been explained from the cultural practices of propagation of chrysanthemum. The mutant character of mutated cell is manifested in M_1V_1 under suitable situations. The mutant cells situated at the lower axillary buds persist at dormant state and show their mutant traits when incorporated in vegetative propagation system in M_1V_2.

19.3.1 Mutation in Flower Morphology

Changed flower type is always attractive to consumer, and as such it has good demand in the market. Many novel flower-type mutant varieties have been induced by radiation. Names of a few induced novel changed flower-type mutants in chrysanthemum are worth mentioning, i.e., "Cosmonaut" (Fig. 19.2k; Datta and Gupta 1984b), "Tulika" (Datta et al. 1985), and "Shabnam" (Fig. 19.2h; Datta 1990c). In "Cosmonaut," the flower shape was anemone type, developed from "Nimrod" with Korean-type flower. Florets of "Tulika" look like paintbrush-type (semi-quilled) flower shape developed from "M-24" with spoon-type florets. In "Shabnam," small appendage-like structure is developed at the tip of each floret, which looks like dewdrops. A number of mutations in flower-head types (anemone to Korean type, fringed ray florets, fishtail shaped, etc.) have been reported in different cultivars of chrysanthemum (Gupta and Shukla 1971; Gupta 1979; Heslot 1964; Broertjes 1966a, b). Creation of tubular florets is one of the curious observations in chrysanthemum. It has been noted that cultivars having small tube at the base of each floret can be mutated to the entire tubular floret (Datta 2019a).

19.3.2 Color Mutation

Presence of maximum unit of dominant genes has been reported in early work, and suggested chances of developing recessive mutations are more in pink cultivars, which can be detected in M_1V_1 (Bowen et al. 1962; Jank 1957; Bowen 1965; Broertjes 1966a, b; Dowrick and El-Bayoumi 1966a, b). Development of new color mutations from bronze cultivars has been reported (Jank 1957), but Gupta (1966, 1979) could not induce mutations in white chrysanthemum. Yellow chrysanthemum cultivars were found to be very stable (Jank 1957; Bowen 1965; Broertjes 1966a, b; Gupta 1979). The author did large-scale repeat experiments choosing color-variant varieties. Gamma rays induced yellow sectorial mutation in white-flowered cultivar "Lilith" (Datta and Gupta 1980) and canary yellow mutation in a white mutant ("Himani") of mauve-colored, pompon-type small-flowered chrysanthemum cultivar "E-13" (Datta 1985b; Gupta 1979).

19.3.3 Chlorophyll Variegation

Plants with chlorophyll variegated leaves look very attractive both at blooming time and in off-season. There is a wide range of variegated plants in different ornamentals. However, there is no prominent chlorophyll variegated chrysanthemum varieties. Attempts were made to induce chlorophyll variegated pattern in leaves of chrysanthemum after mutagen treatment. Chlorophyll variegation was induced in leaves of cultivars "Grape Bowl" and "Lilith" with gamma irradiation. In case of "Grape Bowl," only one plant after treatment with 1.5 krad produced five variegated leaves (Gupta and Datta 1978). In case of "Lilith," the percentages of plants with chlorophyll variegation in leaves after exposure of 1.5 and 2.5 krad were 75 and 92.86, respectively. The frequencies of leaves with chlorophyll variegation after treatment with 1.5 and 2.0 krad were 8.28% and 8.76%, respectively (Datta and Gupta 1980). Cultivar "Maghi" had chlorophyll variegation in leaves after exposure to gamma rays. The percentage of plants showing chlorophyll variegation and the percentage of chlorophyll variegated leaves after treatment with 10.5 and 20 Gy were 72.22 and 16.53 and 54.54 and 6.21, respectively. Induction of a huge amount of chlorophyll variegation plants and leaves in "Maghi" is an interesting finding. Such a high rate of chlorophyll variegation has never been observed in earlier experiments with chrysanthemum (Shukla and Datta 1993).

19.3.4 Spectrum of Mutations

Literature review clearly indicates that mutation can be induced in almost all the colors, and some color-specific varieties develop high spectrum of mutations. Induction of more than one flower color mutations has been possible in single promising variety of chrysanthemum ("Undaunted," "E-13," "Otome Zakura," "D-5," "Surekha," "Anupam," "Khumaini," "Kalyani Mauve," "Lalima," Banerji and Datta 1990; Datta and Banerji 1991, 1993, 1995; Gupta 1966, 1979; Datta and Gupta 1981a, b, Datta and Gupta 1980, 1983) and rose ("Contempo"). Gamma radiation successfully induced nine mutations (color and/or shape) in chrysanthemum cultivar "Undaunted," six mutations in "E-13," and five each in "Flirt," "D-5," "Maghi," and rose cultivar "Contempo" (Datta 1997, 2019b).

19.4 Possibilities of Inducing Desired Flower Color Mutation (Directive Mutation)

In induced mutagenesis, work spectrum of flower color mutations is really an attractive topic as many cultivars have developed more than one flower color/shape mutations. These observations inspired irradiation of different cultivars with different color combinations. From the extensive work on such selective cultivars, it has been almost accurately found out that the flower color (pigment composition) of parent cultivar is a meaningful index for new flower color mutation. It has been clearly found out from the repetition of induced mutagenesis experiments with the same and/or different cultivars that white varieties will develop either flower shape mutation or mutation in color (yellow, Fig. 19.2p), red varieties will develop either yellow or red and yellow mixture, and yellow varieties will produce either various hue of yellow or white or a blend of yellow and white. Work on the preparation of a color chart based on the color (pigment composition) of the parent cultivars and the spectrum of induced flower color mutations, observed by the author and reported from other laboratories, is under way. From this color chart, it will be possible to decide beforehand the color combinations of expected mutants. This will help to develop desired flower color mutation (directive mutation) for the floriculture trade by selecting specific parent varieties with specific color (Datta 2019a).

19.5 Demand-Based Experiments

Mutation technique has been very successfully applied in chrysanthemum to develop need-based varieties. For example, many attractive colors/shapes are available in normal season blooming chrysanthemum. Colors are very limited in early- and late-blooming varieties. Choice of colors is very limited to consumers in such cases. Gamma rays induced new flower color mutations in both early-blooming and late-blooming chrysanthemum varieties. Early-blooming variety "Ajay" (amaranth pink) produced dark-pink variety. Late-blooming variety "Gouri" (white) developed two mutations—one cream color floret color and another yellow. Another late-blooming variety "Maghi" (mauve) produced "light mauve," "white," "yellow," and "dark-yellow" floret color mutants (Datta 2015). Ornamental plants are very suitable where most appropriate and commercially viable mutants can be developed by deliberate and appropriate application of technique. To cite an example: Pompon-type large-flowered pastel lilac chrysanthemum "Otome Zakura" is a highly demanded variety in the consumer market. For increasing the income of nurserymen and to increase public choice, complete white mutant variety ("Purnima") was produced by gamma ray treatment in "Otome Zakura." Further, keeping in mind the idea of directive mutation, white mutant ("Purnima") was again treated with gamma rays and a new yellow mutant variety was developed (Fig. 19.2i, p). These directive-induced mutation experimental results have gifted admirable influence in the floriculture industry.

19.6 Bottlenecks

Buds present at the tips of all types of vegetatively propagated propagules are multicellular constituted by a number of moderately autonomous cell layers. Development of chimera after treatment with physical or chemical mutagens is the main bottleneck. Mutation in general is a one-cell event (Fig. 19.2c). Mutagen-induced mutated cells in multicellular organisms are left open for diplontic selection, i.e., struggle to survive among the surrounding normal cells. Diplontic selection restricts the number of mutated plants and mutation spectrum (Datta 1994a, b). Sectorial mutation may appear as a small streak on a petal or whole flower and from a part of a branch to total branch (Fig. 19.2d–g). In mutation breeding experiment, management of such chimeras, i.e., establishment of mutant tissue in pure form, is a very vital operation. In traditional mutation breeding experiments, a large number of chimeric mutants are lost every year.

19.7 Induced Chimera and Management

Common proliferation process can help the establishment of mutant tissue in pure form when a whole branch is mutated. Almost all chrysanthemum mutants have been isolated by conventional cutting methods. CSIR-NBRI, Lucknow, developed a large number of mutant varieties by this method (Fig. 19.2h–k). It is very difficult to isolate mutant tissues from chimeric flower with the help of convenient prevailing multiplication technique. Therefore, in vitro technique has been standardized for direct regeneration from florets of chrysanthemum. Many new flower color/shape mutations, developed through induced mutation and spontaneously, have been established in pure form utilizing direct shoot regeneration technique (Chakrabarty et al. 1999, 2000; Mandal et al. 2000a, b; Datta et al. 2001a, b). The prospect of management of chimeric flower color and creation of novel varieties has been wide through in vitro methods (Fig. 19.21, m). This is a very interesting and almost unexploited field in floriculture.

Gamma ray-treated propagules (cuttings) of different varieties were grown in the field. Chimeric mutated florets were regenerated in standardized basal medium (Chakrabarty et al. 2000; Mandal et al. 2000a, b; Dwivedi et al. 2000; Datta et al. 2001a, b). Few examples are white and yellow chimeric florets that were produced after 1.5 and 2.0 krad gamma rays in cultivar "Maghi" (mauve color), which were regenerated in pure form (Chakrabarty et al. 1999). "Lilith" (white) produced flower color mutation (yellow) in 59 ray florets after gamma irradiation (1.5 and 2.0 krad). The yellow mutant has been regenerated through tissue culture (Dwivedi et al. 2000). 1.5 and 2.0 krad gamma rays induced yellow florets in cultivar "Purnima" (white) (a mutant of "Otome Zakura"—mauve florets) and in "Colchi Bahar" (terra-cotta red florets) (a colchicine-induced mutant of "Sharad Bahar"—purple floret). Mutant florets were established as true-to-explant florets through direct shoot organogenesis (Mandal et al. 2000a, b). Two mutants, one with new yellow-orange florets and another with yellow-orange and flat florets, were induced after gamma irradiation in cultivar "Puja" (red purple with flat spoon-shaped florets). Both the mutants have been established in pure form through in vitro regeneration (Datta et al. 2001a, b). To increase mutation events and to induce solid mutants, in vitro mutagenesis experiments were started in chrysanthemum (Maliga 1984; Ahloowalia 1995; Maluszyuski et al. 1995; Jung-Heliger and Horn 1980; Preil et al. 1983; Huitema et al. 1986, 1989; Jerzy 1990; Jerzy and Zalewska 1996; Schum and Preil 1998; Mishra et al. 2003; Datta and Mandal 2005).

19.8 In Vitro Management of Chimera Developed Through Bud Sprout

It is necessary to mention the role of bud sports/spontaneous somatic mutations in the development of new chrysanthemum cultivars. Wasscher (1956) estimated that sports created approximately 30% new chrysanthemum cultivars. Some varieties like "Sweetheart," "Favourite," and "Indianapolis" need special mention as large number of sports have developed from these varieties (Anonymous 1961; Bowen et al. 1962; Yoder 1967). Spontaneous mutations created many excellent chrysanthemums in different countries, specially in India. Names of some remarkable sport varieties along with their parent varieties (in parenthesis) are mentioned here: "Kasturba Gandhi"—white ("Mahatma Gandhi"—mauve); "Sonar Bangla"—yellow ("Snow Ball"—white); "White Cloud"—white; "Pink

Cloud"—pink; "Sharda"—yellow ("Sharad Shobha"—white); "Queen of Tamluk"—yellow ("Casa Grandi"—white); "R. Venkatraman"—yellow ("S.S. Arnold"—white); "William Turner"—white ("Pink Turner"—pink); "J.S. Lloyd"—yellow ("William Turner"—white); "White Ball" ("Pride of Madford"), etc. These spontaneous mutation varieties were mostly detected as branch chimera and isolated through conventional cutting methods. In chrysanthemum, huge number of floret chimeric mutants develop every year and are lost. Now, in vitro method can be exploited by nurserymen to isolate and establish these natural chimeras as new varieties.

Large white-flowered chrysanthemum "Kasturba Gandhi" has good appreciation in floriculture trade for its appealing shape and size. New variety with yellow floret color has been established in pure form using chimera management technique (Fig. 19.2q) (Chakrabarty et al. 2000). Quite a large number of chrysanthemum germplasms develop floret chimera through spontaneous mutations due to its high heterozygosity every year. Now, all these chimeras can be established as new varieties using in vitro chimera management technology.

19.8.1 In Vitro Mutagenesis

Commonly, plants are grown under field conditions after in vivo treatment with mutagens. In this method, limited land and fund and unfavorable environmental conditions restrict population size. It is well documented that chances of getting mutations are more from a large population. As reported, development of chimera is the main bottleneck in in vivo mutation breeding, and in vitro mutagenesis technique was standardized to increase the number of mutants and to get solid mutants. In vitro technique helped to overthrow the formation of chimera in M_1V_1. In vitro mutagenesis has many benefits over in vivo methods—explants can be uniformly treated with physical or chemical mutagens, treated explants can grow in unchanging cultural environment, and large population can be treated at any time of the year within restricted space.

The author and his colleagues have done significant research work on the manipulation of chimera and in vitro mutagenesis. Five decorative-type chrysanthemum cultivars ("Lalima," "Flirt," "Puja," "Maghi," and "Sunil") were selected as experimental materials. Ray florets were selected as explants and treated with gamma rays after inoculation. All proliferated shoots were transplanted in the field after isolation, rooting, and hardening. It was interesting to note that all gamma-induced flower color and floret shape mutants were solid in nature (Fig. 19.2n, o) (Datta et al. 2005; Mishra et al. 2003; Chakrabarty et al. 1999, 2000; Mandal et al. 2000a, b; Datta et al. 2001a, b, 2005). This novel in vitro mutation technique is most suitable for creating new solid mutant varieties in chrysanthemum for floriculture trade.

Methodical attempt was made for creation of trait-oriented mutation, i.e., NaCl-tolerant chrysanthemum variety following in vitro mutation. Hossain et al. (2006, 2007) developed stable NaCl-tolerant variants through whole plant and callus selection after in vitro treatment of ethyl methanesulfonate (EMS). Experimental data reflects that NaCl-tolerant callus line can be induced when NaCl concentration is increased step-by-step from a low level to cytotoxic level (Hossain et al. 2004).

Mandal and Datta (2005) developed an effective method for somatic embryogenesis from a single cell of chrysanthemum floret for retrieval of single mutated cell.

19.9 Acute and Chronic Irradiation

Plants are distinctly sensitive to acute and chronic gamma radiation. Mutation frequency and spectrum were much more in chronic irradiation than acute dose. Nagatomi and Degi (2009) detected ten times more and non-chimeric mutations when materials were exposed together to chronic and acute radiations. Chrysanthemum propagules (cuttings, petal, and/or bud culture) were found to be 2.5 times more sensitive to chronic radiation than acute dose. This technical procedure favorably created many desirable mutants in ornamentals and other crops (Broertjes 1971; Nagatomi 1991, 1992, 2002; Nagatomi et al. 1993a, b; 1996a, b; Richter and Singleton 1995).

19.10 Ion Beam Technology

Ion beam has been considered as a strong mutagen which can induce a wide spectrum of mutations in comparison to other chemical, physical, and T-DNA-based mutagens. Ion beams have considerably superior linear energy transfer (LET) than other types of radiations (X-rays, gamma rays, and electrons). Helium (He), carbon (C), neon (Ne), and argon (Ar) with 220 MeV C ions are also used in mutagenesis studies. Effects of ion beam have been studied on different plants including few ornamentals (chrysanthemum). The extent of C ion-induced mutation frequency and spectrum was more. C ions have successfully produced complex patterns of colorations, novel mutant phenotype, and a wide spectrum of flower color and shape in chrysanthemum, petunia, rose, Torenia, lotus, cyclamen, etc. (Nagatomi et al. 1996a, b; Okamura et al. 2003, 2006; Ikegami et al. 2005; Ueno et al. 2002; 2004; Yamaguchi et al. 2003; Oka-Kira et al. 2005; Miyazaki et al. 2006; Sugiyama et al. 2008; Tanaka 2009). Ion beams have the potential to be used in induced mutagenesis work on ornamental crop breeding, and also the new basic knowledge generated from these studies can be exploited in biotechnology and molecular biology (Tanaka 2009).

19.11 Annual Chrysanthemum

The seed-propagated annual species of chrysanthemum, unlike the vegetatively multiplied perennials, are not very popular mainly because of the less attractive form, size, and color of their flowers. *C. carinatum* is self-incompatible and has 18 as its diploid chromosome number. *C. carinatum* is very distinct from other annual chrysanthemum for keeled involucral bracts with solitary white or yellow flower heads and sometimes a dark pigmented ring at the base of petals. Induced mutation work on this species is limited. A number of researchers treated seeds of annual chrysanthemum (*C. carinatum*) with X-ray and gamma ray radiations, and interesting cytological and morphological changes in floret forms have been reported. Jain et al. (1961) exposed seeds of *C. carinatum* to an X-ray dose of 15,000r and studied the treated populations in X_1, X_2, X_3, and X_4 generations. Many attractive morphological variations in flower have been noticed in radiation-treated population. They detected a recombinant form combining the tubular shape of the petals with the double condition, which appeared more attractive than either the control or the induced mutant types. The results have suggested that variation in annual chrysanthemum offers considerable scope for selecting and synthesizing new varieties. Rana (1964a, b, c, d) applied 15 krad acute X-ray dose to dry seeds of *C. carinatum* (L.) and observed changed petals (flat to tubular) in X_2 population. He has proposed alike genetic constitution among the tubular phenotypes on the basis of understanding of quantitative inheritance after intercrossing. Rana (1965a) in anther experiment exposed dry seeds of a single-type Chelsea variety of *C. carinatum* to 15 krad X-ray and detected single-type flower heads to double type, and conservation of open-type ray florets instead of the usual pistillate type was also detected. Breeding analysis showed that this mutant condition was determined by a recessive gene. Rana (1965b) further treated dry seeds of *C. carinatum* with X-rays with an acute dose of 15 krad. Four morphological mutants affecting the ray florets (dissected type, tubular type, small-size floret, and apetalous type) were detected in the X_2 population.

19.12 Cause of Flower Color Mutation

It is very clear that mutation technique successfully developed many important new ornamental cultivars including chrysanthemum. Although many new flower color/shape mutants have been developed, reports on the analysis of mutants and their parental varieties are very few. The general question is how these mutations arise. Early reports indicated that flower color changes may develop due to abnormalities in chromosomes, alteration in chromosome number, gene mutation, rearrangement in histogenic layers, and changes at pigment biosynthesis pathway. Exact process of development of these mutations cannot be interpreted with confidence based on all these explanations. The author made extensive comparative cytomorphological, radiosensitivity, anatomical, palynological, and biochemical studies to highlight the appropriate and definite perception in the development of somatic mutation in flower color. Early investigations have reported that flower color change in chrysanthemum is due to change in chromosome number and chromosome fragments (Dowrick 1951, 1952, 1953, 1958; Dowrick and El-Bayoumi 1966a, b; Walker 1955; Ichikawa et al. 1970). The author studied cytological aspects covering karyotype, ICV, INV, and DNA content and observed no change in chromosome number in any of the mutants. No mutant-specific chromosomal abnormalities could be detected. No significant differences in ICV, INV, and DNA content were recorded between control and mutants. Different types of chromosomal aberrations were detected in all the gamma ray-treated populations, but there was no change of flower color/shape in all the plants. This confirms that chromosomal aberrations are not responsible for flower color changes. It may also be noted that the chromosome number of all the present experimental chrysanthemum cultivars was $2n = 54$, but all the varieties were different from each other in flower color and shape. Reports are also available that although 54 or 52 chromosome numbers were present in somatic cells of many control plants, no changed floret color mutants were found in any of the control plants (c.f. Dowrick and El-Bayoumi 1966a, b; Nazeer 1981; Nazeer and Khoshoo 1982, 1983, 1985). Sampson et al. (1958) reported that chromosome number is not necessarily associated with change in flower color. Original and mutant cultivars depicted believably symmetrical karyotypic configurations. There are also differences of opinion that the root tips from which chromosome numbers are determined originate from L-III tissue and flowers from L-I (Bowen 1965; Sampson et al. 1958; Weaver 1963; Stewart and Derman 1970), and therefore it is not wise to connect changed flower color with changed chromosome or chromosomal abnormalities.

Sometimes, rearrangement of cell layers is incorrectly ascribed as a mutation effect (Fujii and Mabuchi 1961; Fuji 1962; Shimotsuma and Sakurai 1962; Yamakawa 1970). Similarly, development of maximum alike mutations and large sectorial or solid mutants are considered rearrangement or some related phenomena (Shapiro and Broertjes 1961). On the basis of present observation and from induced somatic flower color mutation observations of the author from different other ornamentals, it is very clear that although root, shoot, and flowers are developed from different histogenic layers, rearrangements of histogenic layers do not have any specific role in the development of somatic mutations in flower color. Reshuffling of histogenic layer theory cannot explain when there are series of induced flower color mutations from a starting cultivar not only in the first generation but also in the subsequent later generations. Induction of striped (bicolored or spotted) mutant also does not support this opinion. Datta (1994a, b) induced striped flower color mutants in rose cvs. "Contempo," "Mrinalini," and "Imperator." He also induced several leaf chlorophyll variegated mutants. But the pattern of chlorophyll variegation in *Lantana depressa* (Datta 1995) and pattern of stripe (bicolored) in "Contempo" and "Imperator" mutants are variable. Out of several mutants induced by the author in different ornamentals, the variable nature of abovementioned three mutants cannot rule out the possible indication for involvement of transposable elements in these cases.

From a wide range of observations in different ornamentals by the author, it is very clear that reshuffling of histogenic layers is applicable to the development of chlorophyll variegation in leaves. As both the phenomena (flower color mutation and chlorophyll variegation) are found very frequently in different vegetatively propagated ornamentals, different workers are bias to explain that development of new flower color is also due to reshuffling of histogenic layers.

Dowrick (1953) and Bowen et al. (1962) mentioned that chrysanthemum flower color is controlled by qualitative and quantitative differences of three different pigments which are present in upper and lower epidermal cells of ray florets. Fujii and Mabuchi (1961) presumed that new pigments are formed due to mutation in pigment biosynthesis pathway. Kawase and Tsukamoto (1976) classified different chrysanthemum cultivars on the basis of presence or absence of different pigments (anthocyanin, carotenoids, and flavonols). Heslot (1968) on the basis of comparative analysis of pigments in the original and mutant rose cultivars mentioned that mutant colors develop due to quantitative differences of existing one or several pigments present in the control. However, new peonidin pigment was detected in some mutants. The author's analysis of pigments in chrysanthemum detected very clearly that both qualitative and quantitative changes in pigments due to mutation are responsible for changes in flower color. Similar observation has also been reported from the analysis of pigments in a series of induced flower color mutants in rose (Datta 1986; Datta and Gupta 1983a, b). Love and Malone (1967) have clearly indicated from their pigment analysis that color differences between fast neutron-induced mutant and non-mutant Coleus plant are for quantitative difference of one anthocyanin pigment without any change in molecular structure. Fast-neutron irradiation is capable of modifying the system which controls pigment synthesis in Coleus. Datta (1987b) used thin-layer chromatographic technique in mutation breeding, and mutants have been characterized with changes in phenolic compounds. Changes in intensity and number of spots of phenolic compounds were observed in the original chrysanthemum cv. "E-13" and its five gamma ray-induced somatic flower color mutants, indicating that mutants were due to both qualitative and quantitative differences in phenolic compounds (Datta 1994a, b). Datta (1994a, b) from comparative studies of original and mutant varieties reported that gene mutation causes changes in flower color and not change in chromosome number, chromosomal aberrations, and karyomorphology.

19.13 Chrysanthemum Mutants

More than 285 mutants have been evolved, reported, and commercialized worldwide. Information about the name of mutant and parent varieties, mode of origin, countries and year of release, and mutant characters is available in different literature (FAO IAEA Mutant Database, Anonymous 1985, 1988, 1989, 1990, 1992; Broertjes and Van Harten 1978; Datta 2019a). The author prepared passport data of the entire ornamental germplasm and mutant varieties maintained and developed at CSIR-NBRI. Details about some of the promising chrysanthemum mutants developed at CSIR-NBRI, Lucknow (author along with other scientists of Mutation Breeding Laboratory), have been shown in Table 19.2.

Table 19.2 Chrysanthemum mutant varieties developed in India

Cultivar mutant/ [original]	Mutagen	Author/ organization	Mutant character	Reference
Agnisikha [D-5]	Gamma rays (15 Gy)	Datta (1987a, b), CSIR-NBRI	Flower color	Floriculture 9(II), 1988; MBNL, IAEA, Vienna, Issue No. 37, page 22, 1991
Alankar [D-5]	Gamma rays (15 Gy)	CSIR-NBRI	Flower color	Floriculture 9(X, 1), 1983; MBNL, 23: 16
Anamika [E-13]	Gamma rays (15 Gy)	CSIR-NBRI	Flower color	MBNL, Issue No. 15, page 15
Aruna [Ashankit]	Gamma rays (20 Gy)	Gupta (1982), CSIR-NBRI	Flower color	NBG Newsletter 1(4), 1974; MBNL, 15:14
Asha [Hope]	Gamma rays (15 Gy)	CSIR-NBRI	Flower color	MBNL, Issue No. 15, page 15
Ashankit [Undaunted]	Gamma rays (15 Gy)	CSIR-NBRI	Flower color	NBG Newsletter 1(4),1974; MBNL, 14:14
Basant [Paul]	Gamma rays (10 Gy)	Gupta (1979), CSIR-NBRI	Flower color	MBNL Issue No. 15, page 15
Basanti [E-13]	Gamma rays (15 Gy)	CSIR-NBRI	Flower color	NBG Newsletter, VI(3): 22, 1979; MBNL Issue No. 23, page 16
Batik [Flirt]	Gamma rays (20 Gy)	CSIR-NBRI	Striped	NBG Newsletter 1(4), 1974; MBNL, 14:14
Colchi Bahar [Sharad Bahar]	Colchicine (0.0625%)	Datta (1985), CSIR-NBRI	Flower color	The Chrysanthemum 43(1):40,1987; MBNL Issue No. 31, page 12, 1988
Cosmonaut [Nimrod]	Gamma rays (15 Gy)	Datta and Gupta (1984a, b), CSIR-NBRI	Flower shape	NBRI Newsletter XI(4): 29, 1984
Gairik [Belur Math]	Gamma rays (15 Gy)	CSIR-NBRI	Flower color	NBG Newsletter, 1(1),1974; MBNL, Issue No. 31, page 12, 1988
Hemanti [Megami]	Gamma rays (15 Gy)	CSIR-NBRI	Flower color	NBRI Newsletter VI(3): 25, 1979; MBNL Issue No. 16:21
Himani [E-13]	Gamma rays (15 Gy)	CSIR-NBRI	Flower color	NBG Newsletter 1(1), 1974; MNNL Issue No. 15, page 15
Jhalar [Undaunted]	Gamma rays (15 Gy)	Gupta (1975), CSIR-NBRI	Flower shape	MBNL Issue No. 15, page 15

(continued)

Table 19.2 (continued)

Cultivar mutant/ [original]	Mutagen	Author/ organization	Mutant character	Reference
Jugnu [Lalima]	Gamma rays (15 Gy)	Datta and Banerji (1991), CSIR-NBRI	Flower color	NBRI Newsletter 1991
Kanak [Undaunted]	Gamma rays (15 Gy)	CSIR-NBRI	Flower color	NBG Newsletter I(1), 1974; MBNL Issue No. 15, page 15
Kansya [Roseday]	Gamma rays (15 Gy)	CSIR-NBRI	Flower shape	NBG Newsletter I(4), 1974; MBNL Issue No. 15, page 15
Kapish [E-13]	Gamma rays (15 Gy)	CSIR-NBRI	Flower color	NBG Newsletter I(4), 1974; MBNL Issue No. 15, page 15
Kumkum [M-71]	Gamma rays (20 Gy)	Datta and Banerji (1987); CSIR-NBRI	Flower color	NBRI Newsletter XIV (1),1987; MBNL Issue No. 31, 1988
Kunchita [Undaunted]	Gamma rays (15 Gy)	CSIR-NBRI	Flower shape	NBG Newsletter I(4), 1974; MBNL Issue No. 15, page 14
Lohita [E-13]	Gamma rays (15 Gy)	CSIR-NBRI	Flower color	NBG Newsletter I(4), 1974
Man Bhawan [Flirt]	Gamma rays (15 Gy)	CSIR-NBI	Bicolored	NBRI Newsletter X(1), 1983
Navneet [Kalyani Mauve]	Gamma rays (15 Gy)	Datta and Banerji (1987), CSIR-NBRI	Flower color	MBNL, Issue No. 37:24, 1991
Navneet Yellow [Navneet]	Gamma rays (15 Gy)	Datta and Banerji (1993), CSIR-NBRI	Flower color	NBRI Newsletter XX(3): 31, 1993
Nirbhaya [Undaunted]	Gamma rays (15 Gy)	CSIR-NBRI	Flower shape	NBG Newsletter I(4), 1974
Nirbhik [Undaunted]	Gamma rays (15 Gy)	CSIR-NBRI	Flower shape	NBG Newsletter I(1), 1974
Pingal [Pink Casket]	Gamma rays (15 Gy)	CSIR-NBRI	Flower color	NBG Newsletter I(1), 1974
Pitaka [Kansya]	Gamma rays (15 Gy)	CSIR-NBRI	Flower color	NBG Newsletter V(1), 1978

(continued)

Table 19.2 (continued)

Cultivar mutant/ [original]	Mutagen	Author/ organization	Mutant character	Reference
Pitambar [Otome Zakura]	Gamma raya (15 Gy)	CSIR-NBRI	Flower color	NBG Newsletter IV(1):2, 1977
Purnima [Otome Zakura]	Gamma rays (15 Gy)	CSIR-NBRI	Flower color	NBG Newsletter V(1), 1978
Raktima [Shyamal]	Gamma rays (15 Gy)		Flower color	NBRI Newsletter XXIII (3)
Rohit [Kingsford Smith]	Gamma rays (15 Gy)	CSIR-NBRI	Flower color	NBRI Newsletter VI(3), 1979
Shabnam [D-5]	Gamma rays (15 Gy)	Datta (1987a, b), CSIR-NBRI	Flower shape	NBRI Newsletter XIV (1), 1987, MBNL, Issue No. 31, page 17, 1988
Shafali [Undaunted]	Gamma rays (20 Gy)	CSIR-NBRI	Flower color	MBNL Issue No. 31, 1988
Sharad Har [Sharad Mala]	Gamma rays (15 Gy)	Datta (1992), CSIR-NBRI	Flower color	NBRI Newsletter XIX (3), 33, 1992
Sheela [Himani]	Gamma rays (20–25 Gy)	Datta (1985a, b), CSIR-NBRI	Flower color	J.Nuclear Agric.Biol. 14: 131–133, 1985, MBNL Issue No. 31, page 17, 1988
Sheveta [Fish Tail]	Gamma rays (15 Gy)	CSIR-NBRI	Flower color	AICRP Bulletin No. 11: 42, 2001
Surekha Yellow [Surekha]	Gamma rays (15 Gy)	CSIR-NBRI	Flower color	NBRI Newsletter XIV (2):19, 1992
Sonali [Ratna]	Gamma rays (20 Gy)	Datta (1991), CSIR-NBRI	Flower color	NBRI Newsletter XVIII (1), 1991
Subarna [Flirt]	Gamma rays (20 Gy)	Datta (1991), CSIR-NBRI	Flower color	NBRI Newsletter XVIII (1), 1991
Tamra [Goldie]	Gamma rays (20 Gy)	CSIR-NBRI	Flower color	NBG Newsletter I(1), 1974
Taruni [Kingsford Smith]	Gamma rays (25 Gy)	CSIR-NBRI	Flower color	NBRI Newsletter VI(3): 25, 1979
Tulika [M-24]	Gamma rays (15 Gy)	Datta, Banerji and Gupta (1985) CSIR-NBRI	Flower shape	MBNL, Issue No. 31: 17, 1988
Yellow Gold [Flirt]	Gamma rays	IIHR	Flower color	AICRP on Floriculture, Technical Bulletin No. 11: 43

19.14 Present Status of Mutation Research on Chrysanthemum

As mentioned, maximum mutation breeding work has been done on chrysanthemum for its improvement using both physical and chemical mutagens. It is not wise to mention all the works done on chrysanthemum using different mutagens. At a glance, one can see the mutagens used for mutagenesis work on chrysanthemum and names of only few early researchers (in parenthesis) who initiated work as follows: physical mutagens like X-ray (Jank 1957; Jain et al. 1961; Rana 1964a, b, c, d, 1965a; b; Broertjes 1966a, b, 1979; Dowrick and El-Bayoumi 1966a, b; Chan 1966; Crandell et al. 1966; Broertjes et al. 1976; Broertjes et al. 1980, 1983; Preil et al. 1983; Broertjes and Jong 1984; Broertjes and Lock 1985; De Jong and Custers 1986; Huitema et al. 1986), gamma rays (Sheenan and Sagawa 1959; Fujii and Mabuchi 1961; Nybom 1961; Bowen et al. 1962; Poole 1963; Bowen 1965; Cawse 1965, 1966; Das et al. 1974; Dowrick and El-Bayoumi 1966a, b; Fujii and Matsumura 1967; Ichikawa et al. 1970; Jung-Heliger and Horn 1980; Nakajima and Kawara 1967; Weaver 1963; Datta 1988, 1992, 1997, 2000, 2009a, b, 2012, 2014, 2015), fast neutron (Love and Constantin 1965; Broertjes et al. 1980), thermal neutrons (Bowen 1965), and radioactive phosphorus (Drewlow and Widmer 1971) and chemical mutagens like ethylene imine (Bowen 1965), ethyl methanesulfonate (Bowen 1965), and colchicine (Weedle 1941; Datta 1987a, 1990a; Datta and Gupta 1984a) have been used for mutation studies. These are mainly traditional mutation work. One can get all references from these and other recent publications. As a result of these efforts, chrysanthemum got an early lead and substantial gains in some areas related to mutagenesis.

Chrysanthemum is a unique material where researchers have utilized maximum traditional methods of mutation technique for its improvement. It has also motivated scientists to develop in vitro techniques on a need basis for optimum utilization towards the development of new varieties. Presently, mutation means a complete technology package, which covers traditional and in vitro methods. This package is very helpful and has enriched the floriculture industry through the development of need-based varieties like late-blooming varieties, early-blooming varieties, and desired color/type varieties.

Mutagenesis work is still going on in large scale worldwide. If we analyze the work pattern, we will find that scientists have been experimenting traditional mutation technology, over the decades, with new experimental varieties to develop mutant varieties. Many research papers are published each year and just carry out a stereotype model year after year. We have generated all required basic knowledge on traditional mutagenesis in chrysanthemum, but still now every year, we are just filling up the gaps in our experiments with new varieties only. These types of routine mutagenesis work do not increase the quality of research output, and most publications seem to be deficient in innovative components. Many of the published works are simple confirmation of another laboratory's results and therefore have little value in terms of science. This should be a matter of concern to all mutation breeding scientists.

The mutation technology has been enriched on working with chrysanthemum. Traditional mutagenesis technique is now not the proper approach for mutation breeding in chrysanthemum. Present status of mutation technology for chrysanthemum is traditional along with modern in vitro methods. Few laboratories are using in vitro techniques. To meet the demand of the floriculture industry, demand-based research subject should be formulated for induction of new varieties in chrysanthemum considering the whole package of mutation breeding. We can stop some of our routine activities and can train and equip ourselves with essential knowledge of recent scientific infrastructure so that we may plan our activities directly oriented with industry demand. We will have to take deep interest to apply the full package of mutation technology for the futuristic studies on chrysanthemum.

19.15 Conclusion and Prospects

Results clearly indicate that mutation technique is most suitable to develop commercially important novel chrysanthemum through genetic manipulation. Chrysanthemum has been found to be the most suitable ornamental crop where maximum mutation breeding work has been carried out successfully starting from traditional to in vitro techniques (Datta 1988). Different basic parameters for success of induced mutagenesis have been standardized, and one can get a clear picture of technological advancement and its successful application for the development of new varieties. Chrysanthemum is the only ornamental crop which may be considered as a model crop for induced mutagenesis work. The scope of developing new varieties through induced mutagenesis is very high in chrysanthemum for its genetic richness due to highly heterozygous nature.

Application of chemical mutagens is not appropriate for vegetative propagated plant for inadequate suck-up and piercing. This group of plants are very suitable for radiation-induced mutagenesis. Any propagules (seeds, pollen grains, entire plant, rooted or unrooted cuttings, tubers, corms, bulbs, stolons, tissues, etc.) can be exposed to radiation treatment. It has been clearly determined that mutation technique is very relevant for vegetative propagated plants as single or few traits can be mutated without changing the remaining outstanding characters. Mutant cells which survive in diplontic selection can be recognized in the first vegetative generation. Cultural practices play an important role in mutation in vegetatively propagated crops.

Acknowledgements Thanks are due to CSIR-National Botanical Research Institute (NBRI), Lucknow, for providing facilities to work on mutation breeding.

References

Ahloowalia B (1995) *In vitro* mutagenesis for the improvement of vegetatively propagated plants. In: Induced mutations and molecular techniques for crop improvement. Proceedings of a symposium, Vienna, 19–23 June, Jointly organized by IAEA and FAO, pp 531–541

Anonymous (1961) Chrysanthemums. Bull Minist Agric Fish Food London 2:6

Anonymous (1985) Mutation breeding newsletter, vol 26. FAO/IAEA, Vienna

Anonymous (1988) Mutation breeding newsletter, vol 31. FAO/IAEA, Vienna

Anonymous (1989) Mutation breeding newsletter. FAO/IAEA, Vienna, p 34

Anonymous (1990) Mutation breeding newsletter, vol 5. FAO/IAEA, Vienna

Anonymous (1992) Mutation breeding newsletter, vol 39. FAO/IAEA, Vienna

Banerjee BK, Datta SK (1993) Varietal differences in radiosensitivity of garden chrysanthemum. Nucleus 36(3):114–117

Banerji BK, Datta SK (1990) Induction of somatic mutations in chrysanthemum cultivar 'Anupam'. J Nucl Agric Biol 19:252–256

Bowen HJM (1965) Mutations in horticultural chrysanthemum. Radiat Bot Suppl 5:695–700

Bowen HJM, Cawse PA, Dick MJ (1962) The induction of sports in chrysanthemum by gamma irradiation. Radiat Bot 1:297–303

Broertjes C (1966a) Mutation breeding of Chrysanthemum. Euphytica 15:156–162

Broertjes C (1966b) Mutation breeding in Chrysanthemum. Induced mutations and their utilization. Gatersleben. p 357

Broertjes C (1971) Dose-fractionation studies and radiation-induced phenomena in African violet. In: Benson DW, Sparrow AH (eds) . Survival of food crops and livestock in the mission, Oak Ridge, TN, pp 325–342

Broertjes C (1979) The improvement of *Chrysanthemum morifolium* Ramat. by induced mutations. In: Proc. Eucarpia meeting on chrysanthemum, Littlehampton, vol 1978, pp 93–102

Broertjes C, Jong DE (1984) Radiation-induced male-sterility in daisy type of *Chrysanthemum morifolium* Ram. Euphytica 33(1):433–434

Broertjes C, Lock CAM (1985) Radiation-induced low-temperature tolerant solid mutants of *Chrysanthemum morifolium* Ram. Euphytica 34(1):97–103

Broertjes C, Van Harten AM (1978) Application of mutation breeding methods in the improvement of vegetatively propagated crops. Elsevier Scientific Publishing Company, Amsterdam

Broertjes C, Roest S, Bokelmann GS (1976) Mutation breeding of *Chrysanthemum morifolium* Ramat. using *in vivo* and *in vitro* adventitious bud technique. Euphytica 25:11–19

Broertjes C, Koene P, van der Veen JWH (1980) A mutant of a mutant of a mutant of a...: irradiation of progressive radiation-induced mutants in a mutation-breeding programme with *Chrysanthemum morifolium* Ram. Euphytica 29(3):525–530

Broertjes C, Pouwer A, Evers JFM (1983) Kleurenkeus in overvloed. Alleen geelontbreekt nog bij Achimenes. Vakbl Bloemisterij 38(35):34–35

Buiatti M, Tesi R (1968) Gladiolus improvement through radiation induced somatic mutation. Publication No. 1. Laboratorio di Mutagenesi e differeziamento del C.N.R., Pisa, pp 1–6

Cawse PA (1965) Production of Chrysanthemum sports by gamma radiation. Radioisot. Rev. Sheet. Wantage Res. Lab. (A.E.R.E.), Wantage

Cawse PA (1966) Using atomic radiation to produce colour sports in flowers. Commercial Growers 3660:381

Chakrabarty D, Mandal AKA, Datta SK (1999) Management of chimera through direct shoot regeneration from florets of chrysanthemum (*Chrysanthemum morifolium* Ramat). J Hortic Sci Biotechnol 74(3):293–296

Chakrabarty D, Mandal AKA, Datta SK (2000) Retrieval of new coloured chrysanthemum through organogenesis from sectorial chimeras. Curr Sci 78(9):1060–1061

Chan AP (1966) Chrysanthemum and Rose mutations induced by Xrays. Proc Am Soc Hortic Sci 88:613–620

Crandell PC, Clore WJ, Nilan RA (1966) Dr. X and Sparkle. New Chrysanthemums for Washington. Wash. Agric. Exp. Stn., Vancouver, WA. 3pp
Das PK, Ghosh P, Dube S, Dhua SP (1974) Induction of somatic mutations in some vegetatively propagated ornamentals by gamma irradiation. Technology (Coimbatore, India) 11(2,3): 185–188
Datta SK (1985a) Colchi-mutation (C-mutation). Everyman's. Science XIX(3):70–72
Datta SK (1985b) Gamma ray induced mutant of a mutant chrysanthemum. J Nucl Agric Biol 14(4): 131–133
Datta SK (1986) Thin layer chromatographic studies and evolution of somatic mutations in rose. Indian Rose Annu V:41–48
Datta SK (1987a) 'Colchi Bahar'- a new chrysanthemum cultivar evolved by colchi mutation. Chrysanthemum 43(1):40
Datta SK (1987b) Induced changes of phenolic compounds in mutants. New Botanist 14(1–4): 17–19
Datta SK (1988) Chrysanthemum cultivars evolved by induced mutations at National Botanical Research Institute, Lucknow. Chrysanthemum 44(1):72–75
Datta SK (1990a) Colchicine induced mutations in seed and vegetatively propagated plants. J Indian Bot Soc 69:261–266
Datta SK (1990b) Induction and analysis of somatic mutations in garden chrysanthemum. Adv Hortic Forest 31:241–254
Datta SK (1990c) 'Shabnam'- a gamma ray induced new chrysanthemum cultivar with new characters. Chrysanthemum 46(4):168–169
Datta SK (1991) Evaluation of recurrent irradiation on vegetatively propagated ornamental: Chrysanthemum. J Nucl Agric Biol 20(2):81–86
Datta SK (1992) Radiosensitivity of garden chrysanthemum. J Indian Bot Soc 71(I–IV):283–284
Datta SK (1994a) Induction and analysis of somatic mutations in vegetatively propagated ornamental. DSc Thesis, Kanpur University, Kanpur, India
Datta SK (1994b) Cytological interpretation of induced somatic flower colour mutation in garden chrysanthemum. In: Ghosh S (ed) Plant cytogenetics in India. Association for Cell and Chromosome Research, Dept. of Botany, University of Calcutta, Calcutta, pp 107–114
Datta SK (1995) Induced mutation for plant domestication: *Lantana depressa*. Proc Indian Sci Acad B61(1):73–78
Datta SK (1997) Ornamental plants - role of mutation. Daya Publishing House, Delhi. 219pp
Datta SK (2000) Mutation studies on garden Chrysanthemum - a review. Sci Hortic 7:159–199
Datta S (2009a) A report on 36 years of practical work on crop improvement through induced mutagenesis. In: Shu QY (ed) Induced plant mutations in the genomics era. Joint FAO/IAEA Division of Nuclear techniques in Food and Agriculture, International Atomic Energy Agency, Vienna, Austria, pp 253–256
Datta SK (2009b) Role of classical mutagenesis for development of new ornamental varieties. In: Shu QY (ed) Induced plant mutations in the genomics era. Joint FAO/IAEA Division of Nuclear techniques in Food and Agriculture, International Atomic Energy Agency, Vienna, Austria, pp 300–302
Datta SK (2012) Success story of induced mutagenesis for development of new ornamental varieties. Bioremed Biodivers Bioavailabil 6(Special Issue I):15–26. Global Science Books, Invited Review
Datta SK (2014) Induced mutagenesis: basic knowledge for technological success. In: Tomlekova NB, Kozgar ML, Wani MR (eds) Mutagenesis: exploring genetic diversity of crops. Wageningen Academic Publishers, Wageningen, pp 95–137
Datta SK (2015) Indian floriculture: role of CSIR. Regency Publications, A Division of Astral International(P) Ltd, New Delhi
Datta SK (2017) Improvement through induced mutagenesis: ornamental crops. In: Malik CP, Wani SH, Kushwaha HB, Kaur R (eds) Advanced technologies for crop improvement and agricultural productivity. Agrobios (India), Jodhpur, pp 49–86

Datta SK (2019a) Floriculture work at CSIR-National Botanical Research Institute, Lucknow. Sci Cult 85(7–8):274–283
Datta SK (2019b) Determination of radiosensitivity: prerequisite factor for induced mutagenesis. In: Trivedi PC (ed) Harnessing plant biotechnology and physiology to stimulate agricultural growth. Agrobios, Jodhpur, pp 39–54
Datta SK (2020) Induced mutations: technological advancement for development of new ornamental varieties. Nucleus 63:119–129
Datta SK, Banerji BK (1991) Analysis of induced somatic mutations in chrysanthemum. J Indian Bot Soc 70(I–IV):59–62
Datta SK, Banerji BK (1993) Gamma ray induced somatic mutation in chrysanthemum cv. 'Kalyani Mauve'. J Nucl Agric Biol 22(1):19–27
Datta SK, Banerji BK (1995) Improvement of garden chrysanthemum through induced mutation. Flora Fauna 1(1):1–4
Datta SK, Basu RK (1977) Abnormal plant growth in M_1 and C_1 generation of two species of *Trichosanthes*. Trans Bose Res Inst 40(3):63–67
Datta SK, Gupta MN (1980) Effects of gamma irradiation on rooted cuttings of small flower chrysanthemum. New Botanist VII:73–85
Datta SK, Gupta MN (1981a) Cytomorphological, palynological and biochemical studies on control and gamma induced mutant of Chrysanthemum cultivar 'D-5'. Chrysanthemum 34(4): 193–200
Datta SK, Gupta MN (1981b) Cytomorphological, palynological and biochemical studies on control and gamma induced mutant of Chrysanthemum cultivar 'E-13'. SABRAO J 134(2): 136–148
Datta SK, Gupta MN (1983) Somatic flower colour mutation in Chrysanthemum cv. 'D-5'. J Nucl Agric Biol 12(1):22–23
Datta SK, Gupta MN (1983a) Effect of gamma rays on pigment synthesis of rose cv. 'Contempo'. J Nucl Agric Biol 12(3):79–80
Datta SK, Gupta MN (1983b) Thin layer chromatographic and spectrophotometric analysis of flower colour mutations in roses. Am Rose Annu:102–106
Datta SK, Gupta MN (1984a) Effects of colchicine on rooted cuttings of chrysanthemum. Chrysanthemum 40(5):191–194
Datta SK, Gupta MN (1984b) 'Cosmonaut' - a new chrysanthemum cultivar evolved by gamma irradiation. J Nucl Agric Biol 13:140
Datta SK, Gupta M (1985) Treatment of budding eyes with colchicines for induction of somatic mutations in rose. Indian Rose Annu IV:80–83
Datta SK, Gupta MN (1987) Colchicine induced somatic flower color mutation in Chrysanthemum. Bangladesh J Bot 16(1):107–108
Datta SK, Mandal AKA (2005) *In vitro* mutagenesis – a quick method for establishment of solid mutant in chrysanthemum. Curr Sci 88(1):155–158
Datta SK, Banerji BK, Gupta MN (1985) 'Tulika' - a new chrysanthemum cultivar evolved by gamma irradiation. J Nucl Agric Biol 14(4):160
Datta SK, Chakraborty D, Mandal AKA (2001a) Gamma ray induced genetic variation and their manipulation through tissue culture. Plant Breed 120:91–92
Datta SK, Mandal AKA, Saxena M (2001b) Direct organogenesis from ray and disc florets of a newly evolved chlorophyll variegated chrysanthemum (*Chrysanthemum morifolium*). Indian J Agric Sci 71(10):655–657
Datta SK, Mishra P, Mandal AKA (2005) *In vitro* mutagenesis - a quick method for establishment of solid mutant in chrysanthemum. Curr Sci 88(1):155–158
De Jong, Custers JBM (1986) Induced changes in growth and flowering of chrysanthemum after irradiation and in vitro culture of pedicels and petal epidermis. Euphytica 35:137–148
Dowrick GJ (1951) Sporting in Chrysanthemums, Chrysanthemum, pp 152–155
Dowrick GJ (1952) The chromosomes of Chrysanthemum. I. Heredity 6:365–376
Dowrick GJ (1953) The chromosome of Chrysanthemum. III. Heredity 7:219–226

Dowrick GJ (1958) Chromosome numbers and the origin and nature of sports in the garden Chrysanthemum. Natl Chrysanthemum Soc Yearb:60–79

Dowrick GJ, El-Bayoumi A (1966a) The origin of new forms of the garden chrysanthemum. Euphytica 15:32–38

Dowrick GJ, El-Bayoumi A (1966b) The induction of mutations in chrysanthemum using X- and gamma radiation. Euphytica 15:204–210

Drewlow LW, Widmer RE (1971) Effect of radioactive phosphorus and apical bud removal on the incidence of apparent mutation in *Chrysanthemum morifolium* Ramat. J Am Soc Hortic Sci 96(5):633–636

Dwivedi AK, Banerji BK, Chakraborty D, Mandal AKA, Datta SK (2000) Gamma ray induced new flower colour chimera and its management through tissue culture. Indian J Agric Sci 70(12): 853–855

Fuji T (1962) Mutations induced by radiation in vegetatively propagated plants with special reference to flower colour. Gamma Field Symp 1:51–59

Fujii T, Mabuchi T (1961) Irradiation experiments with Chrysanthemum. Seiken Jiho 12:40–44

Fujii T, Matsumura S (1967) Somatic mutations induced by chronic gamma irradiation. Seiken Jiho 19:43–46

Gupta MN (1966) Induction of somatic mutation in some ornamental plants. In: Proc. All India Symp.Hortic, pp 107–114

Gupta MN (1979) Mutation breeding of some vegetatively propagated ornamentals. Prog Plant Res 2:75–92

Gupta MN, Datta SK (1978) Effects of gamma rays on growth, flowering behavior and induction of somatic mutations in chrysanthemum. SABEAO J 10(2):149–161

Gupta MN, Jugran HM (1978) Mutation breeding of Chrysanthemum II. Detection of gamma ray induced somatic mutations in vM2. J Nucl Agric Biol 7(2):50–59

Gupta MN, Shukla R (1971) Mutation breeding of chrysanthemum. I. Production of new cultivars by gamma ray induced somatic mutations in vM1. In: Intnl. Symp. on use of isotopes and radiation in agric. and animal husbandry research, New Delhi, Dec 1971, pp 164–174

Haque MS (1983) Colchicine induced mutants in Salvia coccinea Juss. Geiobios. New Rep 2:149–150

Heslot H (1964) Induction experiment de mutations chezles plantes florales. C R Acad Agri Fr 17: 1281–1308

Heslot H (1968) Mutation research done in 1967 on Barley, Roses and Marigolds, a progress report. Mutation in plant breeding. IAEA, Vienna, pp 153–159

Hossain Z, Mandal AKA, Shukla R, Datta SK (2004) NaCl stress – its chromotoxic effects and antioxidant behaviour in roots of *Chrysanthemum morifolium* Ramat. Plant Sci 166:215–220

Hossain Z, Mandal AKA, Datta SK, Biswas AK (2006) Isolation of a NaCl-tolerant mutant of *Chrysanthemum morifolium* by gamma radiation: *in vitro* mutagenesis and selection by salt stress. Funct Plant Biol 33:91–101

Hossain Z, Mandal AKA, Datta SK, Biswas AK (2007) Development of NaCl-tolerant strain in *Chrysanthemum morifolium* Ramat. through shoot organogenesis of selected callus line. J Biotechnol 129:658–667

Huitema JBM, Cussenhoven GC, Dons JJM, Broertjes C (1986) Induction and selection of low temperature tolerant mutants of *Chrysanthemum* Ramat. Acta Hortic 197:89–96

Huitema JBM, Preil W, Gussenhoven GC, Schineidereit M (1989) Methods or the selection of low-temperature tolerant mutants of *Chrysanthemum morifolium* Ramat. by using irradiated cell suspension cultures. I. Selection of regenerations *in vivo* under suboptimal temperature conditions. Plant Breed 102:140–147

Ichikawa S, Yamakawa K, Sekiguchi F, Tatsuno T (1970) Variation in somatic chromosome number found in radiation-induced mutants in *Chrysanthemum morifolium* Hemsl. Cv. Yellow Delaware and Dalaware. Radiat Bot 10:557–562

Ikegami H et al (2005) Mutation induction through ion beam irradiations in protoplasts of chrysanthemum. Bull Fukuoka Agric Res Cent 24:5–9

Jain HK, Bose AK, Sathpathy D, Sur SC (1961) Mutation studies in annual Chrysanthemum. I. Radiation-induced variation in flower form. Indian J Genet Plant Breed 21:68–74

Jank H (1957) Experimental production of mutations in *Chrysanthemum indicum* by X-rays. (Translation of Jank 1957a). U.K. At. Energy Established, Harwell, 36 pp

Jerzy M (1990) *In vitro* induction of mutation in Chrysanthemum using x- and gamma radiation. Mutat Breed Newsl 35:10

Jerzy M, Zalewska M (1996) Polish cultivars of *Dendranthema grandiflora* Tzvelev and *Gerbera jamesonii* Bolus bred *in vitro* by induced mutations. Mutat Breed Newsl 42:19

Jung-Heliger H, Horn W (1980) Variation nach mutagener Behandlung von Stecklingen and *in vitro*-Kulturen bei Chrysanthemum. Z Pflanzenzlichg 85:185–199

Kawase K, Tsukamoto Y (1976) Studies on flower colour in *Chrysanthemum morifolium* III. Quantitative effects of major pigments on flower colour variation, and measurement of colour qualities of petals with a colour difference meter. J Jpn Soc Hortic 45(1):65–75

Love JE, Constantin MJ (1965) The response of some ornamental plants to fast neutrons. Tenn Farm Home Sci 56:10–12

Love JE, Malone BB (1967) Anthocyanin pigments in mutant and nonmutant Coleus plants. Radiat Bot 7:549–552

Maliga P (1984) Isolation and characterization of mutants in plant cell cultures. Annu Rev Plant Physiol 35:519–552

Maluszyuski M, Ahloowalia BS, Sigurbjornsson B (1995) Application of *in vivo* and *in vitro* mutation techniques for crop improvement. Euphytica 85:303–315

Mandal AKA, Datta SK (2005) Direct somatic embryogenesis and plant regeneration from ray florets of chrysanthemum. Biol Plant 49(1):29–33

Mandal AKA, Chakrabarty D, Datta SK (2000a) Application of *in vitro* techniques in mutation breeding of chrysanthemum. Plant Cell Tissue Organ Cult 60:33–38

Mandal AKA, Chakrabarty D, Datta SK (2000b) *In vitro* isolation of solid novel flower colour mutants from induced chimeric ray florets of chrysanthemum. Euphytica 114:9–12

Mishra P, Datta SK, Chakrabarty D (2003) Mutation in flower colour and shape of *Chrysanthemum morifolium* induced by gamma irradiation. Biol Plant 47(1):153–156

Miyazaki K et al (2006) Flower pigment mutations induced by heavy ion beam irradiation in an interspecific hybrid of Torenia. Plant Biotechnol 23:163–167

Nagatomi S (1991) Enlargement of induced variations by combined method of chronic irradiations with callus culture in sugarcane. Gamma Field Symp 30:87–110

Nagatomi S (1992) Effects of various irradiation methods on mutation induction of regenerants through leaf trimming and tissue culture of pineapple. Technical News 41:1–2

Nagatomi S (2002) Effectiveness of gamma ray chronic irradiation on *in vitro* mutagenesis in crops. In: SEMINAR I: agriculture & biosciences, pp 49–58

Nagatomi S, Degi K (2009) Mutation breeding of Chrysanthemum by gamma field irradiation and *in vitro* culture. In: Shu QY (ed) Induced plant mutations in the genomics era. Food and Agriculture Organization of the United Nations, Rome, pp 258–261

Nagatomi S, Katsumata K, Nojiri C (1993a) Induction of dwarf-type plants derived from *in vitro* cultures by gamma-ray irradiation in Cytisus Genus. Technical News 51:1–2

Nagatomi S, Degi K, Yagaguchi M, Miyahira E, Skamoto M, Takaesu K (1993b) Six mutant varieties of different flower colour induced by floral organ culture of chronically irradiated chrysanthemum plants. Technical News-Inst Radiat Breed 43:1–2

Nagatomi S et al (1996a) Three mutant varieties in Eustoma grandiflorum induced through *in vitro* culture of chronic irradiated plants. Technical News 53:1–2

Nagatomi S et al (1996b) Mutation induction on chrysanthemum plants regenerated from *in vitro* cultured explants irradiated with C ion beam. TIARA Annu Rep 5:50–52

Nakajima K, Kawara K (1967) Induction of mutations in Chrysanthemum by gamma ray irradiation. Hoshasen Ikushujo Nenpo (Annu Reo Natl Inst Radiat Breed) 1965:46–48

Nazeer MA (1981) Karyological studies in two species of Chrysanthemum L. Cell Chrom Newsl 4(2):33–35
Nazeer MA, Khoshoo TN (1982) Cytogenetical evaluation of garden Chrysanthemum. Curr Sci 51(12):583–586
Nazeer MA, Khoshoo TN (1983) Variation chromosome complement of *Chrysanthemum morifolium* complex. Nucleus 26(1):22–24
Nazeer MA, Khoshoo TN (1985) Meiotic variation in *Chrysanthemum morifolium* complex. Nucleus 28(1 & 2):35–41
Nybom N (1961) The use of induced mutations for the improvement of vegetatively propagated plants. In: Symposium mutation and plant breeding, Cornell University, vol 891. NAS-NRC, Ithaca, NY, pp 252–294
Oka-Kira E et al (2005) Klavier (klv), a novel novel-hypernodulation mutant of *Lotus japonicus* affected in vascular tissue organization and floral induction. Plant J 44:505–515
Okamura M et al (2003) Wide variety of flower-color and-shape mutants regenerated from leaf cultures irradiated with ion beams. Nucl Inst Meth Phys Res B206:574–578
Okamura M et al (2006) Advances in mutagenesis in flowers and their industrialization. Floriculture Ornam Plant Biotechnol 1:619–628
Poole RT (1963) Effect of gamma rays on *Chrysanthemum morifolium* "Bluechip" grown at various nutritional levels. Proc Flo State Hort Soc 75:443–447
Preil W, Engelhardt M, Walther F (1983) Breeding of low temperature tolerant poinsettia (*Euphorbia pulcherrima*) and *Chrysanthemum* by means of mutation induction in *in vitro* cultures. Acta Hortic 131:345–351
Rana RS (1964a) Multiple allelism in radiation induced ray floret mutants of annual Chrysanthemum. Curr Sci 33:592–593
Rana RS (1964b) Radiation induced chromosomal arrangements in annual Chrysanthemum. Experientia 20(11):617
Rana RS (1964c) A radiation induced chimera in annual Chrysanthemum. Naturwissemschaften 51: 642–643
Rana RS (1964d) Phenotypic variability of an induced mutant of annual Chrysanthemum. Genetica 35:236–240
Rana RS (1965a) Radiation induced variation in ray floret characteristics on annual Chrysanthemum. Euphytica 14:296–300
Rana RS (1965b) A radiation induced useful mutant of annual Chrysanthemum. Curr Sci 34:58
Richter A, Singleton WR (1995) The effect of chronic gamma radiation on the production of somatic mutations in carnations. Proc Natl Acad Sci U S A 41(5):295–300
Sampson DR, Walker GWR, Hunter AWS, Bragd M (1958) Investigation in the sporting process in greenhouse chrysanthemum. Can J Plant Sci 38:346–356
Schum A, Preil W (1998) Induced mutations in ornamental plants. In: Jain SM, Brar S, Ahloowalia (eds) Somaclonal variation and induced mutations in crop improvement. Springer, pp 333–366
Shapiro S, Broertjes C (1961) Radiation-induced sports in floricultural plants. Brookhaven Natl. Lab. Rep., Upton, NY, pp 1–2
Sheenan TJ, Sagawa Y (1959) The effects of gamma radiation on chrysanthemum and gladiolus. Proc Flo State Hort Soc 72:388–391
Shimotsuma M, Sakurai N (1962) Irradiation experiments with chrysanthemum. Seiken Jiho 14: 106–110
Shukla R, Datta SK (1993) Mutation studies on early and late varieties of garden chrysanthemum. J Nucl Agric Biol 22(3–4):138–142
Sparrow AH, Moses MJ, Dubow R (1952) Relationship between ionizing radiations, chromosomes breakage and certain other nuclear disturbances. Expt Cell Res Suppl 2:245–267
Stewart RN, Derman H (1970) Somatic genetic analysis of the apical layers of chimera sports in Chrysanthemum by experimental production of adventitious shoots. Am J Bot 57:1061–1071
Sugiyama M et al (2008) Biological effects of heavy-ion beam irradiation on Cyclamen. Plant Biotechnol 25:101–104

Tanaka A (2009) Establishment of ion beam technology for breeding. In: Shu QY (ed) Induced plant mutations in the genomics era. Food and Agriculture Organization of the United Nations, Rome, pp 216–219

Thoday JM (1951) The effects of ionizing radiations on the broad bean root. IX. Chromosomes breaks and the lethality of ionizing radiations to the root meristem. Br J Radiol 24:572–576, 622–628

Ueno K et al (2002) Effects of ion beam irradiation on chrysanthemum leaf discs and Sweet potato callus. JAERI Rev 35:44–46

Ueno K et al (2004) Additional improvement of chrysanthemum using ion beam re-irradiation. JAERI Rev 25:53–55

Usenbaev EK, Imankulova KI (1974) Radiation mutants of roses. Proc Int Hort Congr 19:109

Walker GWR (1955) Chromosome numbers in Chrysanthemum sports. In: Dep. Agric. Hortic. Div., Progr. Rep.: 1949–1953, pp 69–70

Wasscher J (1956) The importance of sports in some florist's flower. Euphytica 5:163–170

Weaver GM (1963) The effect of Cesium 137 gamma radiation on plant growth and flower color of greenhouse Chrysanthemum cultivars. Can J Genet Cytol 5:73–82

Weedle C (1941) Two colchicines-induced polyploids of the green house Chrysanthemum and their progeny. Proc Am Soc Hortic Sci 38:658–660

Yamaguchi H et al (2003) Mutation induced with ion beam irradiation in rose. Nucl Instrum Methods Phys Res B206:561–564

Yamakawa K (1970) Radiation-induced mutants of chrysanthemum and their somatic chromosome number. Inst. Radiat. Breed. Techn. Newsl. No. 6

Yamakawa K, Sekiguchi F (1968) Radiation induced internal disbudding as a tool for enlarging mutation sectors. Gamma Field Symp 7:19–39

Yoder BI (1967) Evolution of the Indianapolis family. Grow Circ Newsl 51:1–4

Mutation Breeding Research in Sweet Pepper

20

A. Popatanasov, O. Timina, and N. Tomlekova

A tribute: **Over quarter century passed since the publication of one of the first remarkably detailed and useful reviews on the mutation breeding in pepper from Prof. Stefan Daskalov (Bulgaria). It encompassed half-century mutation breeding research on all varieties of Capsicum spp. used in the agriculture. His passionate long-term dedication to pepper breeding led to the production of many new varieties and many grateful students and followers, who continued his work, which brought the small Bulgaria to the forerunner position in pepper mutation breeding. As continuation of his work, here we have narrowed the focus on the sweet pepper varieties, which are most commonly consumed in the northern hemisphere, but less studied than their hot and pungent relatives.**

Abstract

Sweet pepper is delicious, loved, and an increasingly popular nutritional vegetable that is grown and produced in significant amounts throughout the world and especially in the northern hemisphere. Such wide distribution places the demand on the crop to be able to grow in rather diverse environmental conditions. The

A. Popatanasov
Institute of Neurobiology, Bulgarian Academy of Sciences, Sofia, Bulgaria

O. Timina
Faculty of Natural Sciences and Geography, Transdnistrian State University, Tiraspol, Transdnistria, Moldova

N. Tomlekova (✉)
Maritsa Vegetable Crops Research Institute, Agricultual Academy, Plovdiv, Bulgaria

S. Penna, S. M. Jain (eds.), *Mutation Breeding for Sustainable Food Production and Climate Resilience*, https://doi.org/10.1007/978-981-16-9720-3_20

process of natural adaptation can be rather slow. Besides, the increasing consumption of this crop calls for high-yield varieties. Induced mutagenesis invented in the twentieth century came as a possible solution to these problems. The research on mutation breeding in sweet pepper started in 1940 in India and gradually spread over the world. This review provides a global historical picture of the essence of the works of many researchers who dedicated part of their lives to improve this crop, which we have categorized by the used mutagens to make it easier for the reader.

Keywords

Sweet pepper · *Capsicum annuum* L. · Mutation breeding · Mutant varieties

20.1 Introduction

Sweet pepper (some varieties are also called bell pepper, Cuban, squash, Kapia, etc.), usually taxonomically referred to as *Capsicum annuum* var. *grossum*, originated from Mexico, Central America, Guatemala, and Bulgaria being the secondary center of origin (Bosland 1996; Malik et al. 2011). The Bulgarian pepper germplasm was widely spread to Europe; therefore, in many (especially East European) countries, sweet pepper is also known as Bulgarian pepper (Adhikari et al. 2016; Maryina 2021; Pyshnaya et al. 2012). This spread started in the past (and continues till now) when Bulgarian gardeners brought many of the local varieties to other European countries—Hungary, Austria, Romania, Serbia, Russia, Moldova, Macedonia, etc. Part of them was used and is still used as gene carriers in breeding programs for developing new varieties in these countries (Masheva and Mihov 2009). Sweet pepper is an important crop grown worldwide for its delicious nonpungent fruits (Sharma et al. 2020). More than half of the global production of pepper (*Capsicum* spp.) is accounted to the sweet pepper. However most of the sweet pepper production is consumed in large amounts in the northern hemisphere due to local cuisine specifics and preferences (Penella and Calatayud 2018; Subbiah and Jeyakumar 2009; Bosland 1996). Sweet pepper is rich in vitamins as C, B-complex group, and especially A provitamins as alpha- and beta-carotenes and antioxidants (Dagadkhair 2020); therefore, it can have beneficial effects on human health besides its other nutritional values. The fruits are the most used part of the plants, and often the varieties are named according to their shape and traits, and their modification is one of the main goals of the pepper breeders. Csilléry (2006) did fruit shape division of *C. annuum* into ten types, while Musaev et al. (2013) excluded hot varieties and suggested sweet pepper to be divided into five fruit types: (1) tomato-shaped (example is the variety “Rotund”); (2) bell-shaped (“California Wonder”); (3) cone-shaped (“Kurtovska Kapia 1”); (4) Bulgarian (“Polet”); and (5) cylindrical (“Sivria”; “Italian Sweet”).

Mutations are the heritable changes in the individual genome, which are passed on from the parent to the offspring. If the changes are in the DNA-coding regions, they may be noticed in the phenotype and lead to a new trait either in M_1 generation (if they are dominant) or in M_2 (if they are recessive) (Bado et al. 2015; Yadav et al. 2021). Changes in the coding sequence of DNA occur not only during normal biological processes as errors in DNA replication or crossing-over but also due to reactive compounds in cellular milieu or irradiations from the environment; therefore, during the evolution, the organisms have developed mechanisms for DNA proofreading and repair. When these mechanisms get compromised or overloaded or when changes get unnoticed by them, then mutations occur (Khursheed et al. 2021; Yadav et al. 2021). When exposing the living organisms or cells to mutagens, we can deliberately increase the rate of mutation occurrence.

According to Pathirana (2011), mutation breeding can be described as the purposeful application of mutations in plant breeding. This is a relatively wide definition, which encompasses all types of induced mutation techniques as traditional physical and chemical mutagenesis, newer genome editing/modifying techniques, etc. which are also known as targeted or site-directed mutagenesis (Oladosu et al. 2016; Holme et al. 2019). Most of these techniques were tested and applied in pepper breeding with different success. However, the later ones although being more predictable and productive in some aspects are still the object of ethical and safety concerns due to their similarity with genetic engineering aimed to insert coding sequences from other species into host crop genome known in the popular culture as the GMO techniques (from genome-modified organism), which in reality produce new partially hybrid organisms of sometimes rather distant species as plants and animals, and in some countries these techniques and products are legally restricted (Araki and Ishii 2015; Holme et al. 2019).

Additionally, another hindrance to efficiently use these newer methods in pepper is that many of them rely on plant regeneration techniques and some on genetic transformation ones (Irikova et al. 2011; Rajam et al. 2021). However, unlike other crops from *Solanaceae* family, pepper so far is rather recalcitrant to plant regeneration and genetic transformation; hence, their usage is limited only to laboratory settings (Rajam et al. 2021). Therefore, at the moment, practical application specifically for the sweet pepper breeding has found methods that use seed or gamete material for the mutation-induced treatment. Here, we focus mostly on the former plant material which is still the most preferred and popular one for the mutation breeding regardless of some of the accompanying drawbacks such as the accompanying chimerism.

In order to help the reader evaluate how adopted is a particular mutagen in mutation breeding, we present a slightly modified procedure from Oladosu et al.'s (2016) workflow for developing new mutant variety with mutation breeding. Schematically, it comprises the following steps:

1. Development of treatment protocol for the selected crop (when necessary)
2. Treatment of the plant material with the selected mutagen/s
3. M_1 plant chimera
4. M_2 seed harvesting
5. Mutation screening and selection in M_2

6. Mutation and line selection in M_3 through segregation population
7. Homogeneity test in M_4
8. Further evaluation, description, and registration of a new mutant variety line
9. Hybridization with other varieties, which have some useful traits
10. Sequencing and description at molecular level of the new useful mutation/s in the new mutant variety

As we shall see, most of the studies reach only the first stages of this scheme and fewer reach the final stages.

In the next pages briefly, we shall present the adoption and usefulness of the various mutagens used so far in sweet pepper mutation breeding in a historical and geographical fashion categorized by mutagens. The related studies are grouped as a set of long-term experiments regardless of the time gap to help the reader trace the depths, complexity, and use of the mutation breeding methods. In some occasions, at our discern, we may also include studies with mild or pungent pepper varieties.

20.2 Physical Mutagenesis and Mutagens

Physical mutagenesis was the first artificially induced human-type mutagenesis to be discovered and consequently to be applied in agriculture. Even before its discovery, the inventor of the term mutations, Hugo de Vries, proposed the idea in 1900s that the newly discovered radiation phenomena can cause mutations, and less than three decades later, the first artificially induced mutant in tobacco was introduced (Oladosu et al. 2016). In general, the most commonly used physical mutagens are some type of sufficiently high-energy radiations capable to disrupt/damage the DNA itself or appropriate chemical compounds in the living organism or cells that can damage it and which have sufficient penetration capacity to perform such actions. Radiation was defined as energy travelling through a distance in the form of waves or particles (Oladosu et al. 2016; Spencer-Lopes et al. 2018).

One of the greatest inspirers and promoters of the peaceful application of the high-energy radiations—IAEA—recognizes the following common physical mutagens based on three criteria: (1) deployed energy, (2) penetration capacity, and (3) level of hazard for operators (Spencer-Lopes et al. 2018).

According to this scheme, the physical mutagens are divided into (a) nonionizing radiations, which have less energy than it is necessary to remove electrons from the atom shells, and (b) ionizing radiations, which have sufficient energy to remove electrons from the atom shells. Most commonly used nonionizing radiations in plant breeding are the ultraviolet (UV) rays and laser irradiations. However, the former ones have rather low penetrability, which limits their use (Mba et al. 2012). Ionizing radiations can be divided into (1) non-particulate radiations and (2) particulate radiations. The most commonly used non-particulate radiations are the X-rays and gamma rays. Particulate radiations can be divided further into (1) radiations from nuclear decay (alpha and beta particles and neutrons) and (2) radiations from particle accelerators (such as betatrons, cyclotrons, synchrotrons, linear accelerators)

(Spencer-Lopes et al. 2018). However, the latter division is not fully accepted since some particles gained from nuclear decay can also be produced through the usage of some accelerators (Oladosu et al. 2016). Most commonly used particulate radiations in plant breeding are the neutrons and ion beam ones (Spencer-Lopes et al. 2018; Oladosu et al. 2016).

And finally, a separate type of radiation can be defined, which is of complex origin and composition and requires special preparations and equipment to be applied on crops, and it is known as "cosmic irradiation" (Spencer-Lopes et al. 2018; Schimmerling 2011; Mba 2013).

Physical mutagens in general produce multiple stochastic damages on the hereditary cell's material, which makes the process of mutation selection and isolation laborious and costly, but it is better in this regard than the breeding through natural mutations.

Most of these physical mutagens have been so far tested/applied on sweet pepper varieties with various success and results.

20.2.1 X-Rays

X-rays are electromagnetic radiation originating from electrons and are produced by X-ray machines. Since this radiation is poly-energetic, sometimes filters are used to select/limit the energy range (Spencer-Lopes et al. 2018). Their energy can be absorbed by atoms of the tissue through which they pass, which causes ejection of electrons from their shells, thus ionizing these atoms. The ejected electrons further cause ionizations in the neighbor atoms. The biological effects result from the paths of these ionizations, which are of relatively low or sparse ion density (Smith 1958).

This is the earliest mutagen that was used in plant breeding, and sweet pepper is not an exception from this rule (Stadler 1928). Probably one of the first applications of X-ray irradiation for sweet pepper mutation, breeding was performed in India in 1940 by Raghavan and Venkatasubban (1940). They irradiated with X-ray seeds of Paramakkudy variety and detected six types of mutants in M_1. Among the observed abnormalities were gigantism, dwarfism, exceeding or reduced branching, narrow leaves, pigmentation shifts, etc.

In the USSR, in 1965, Batikyan et al. (1970) evaluated the radiosensitivity to X-rays of dry seeds of sweet pepper variety "Bolgarskiy 079." They irradiated the seeds with X-ray doses from 5 to 300 Gy. They found that doses up to 10 Gy can have a stimulating effect on germination, while the dose of 300 Gy is absolutely lethal.

In 1965, in Bulgaria at the Bulgarian Academy of Sciences, Daskalov (1968) started one of the most successful long-term series of experiments in the sweet pepper mutation breeding. Along with the usage of gamma rays, some of the most stable and appealing elite male sterile mutants were produced, which, thanks to his dedication to pepper research and his generosity, were distributed and used in many countries to facilitate the pepper breeding and research and were used for the development of several mutant varieties and F_1 hybrids (Csilléry 2013; Timina

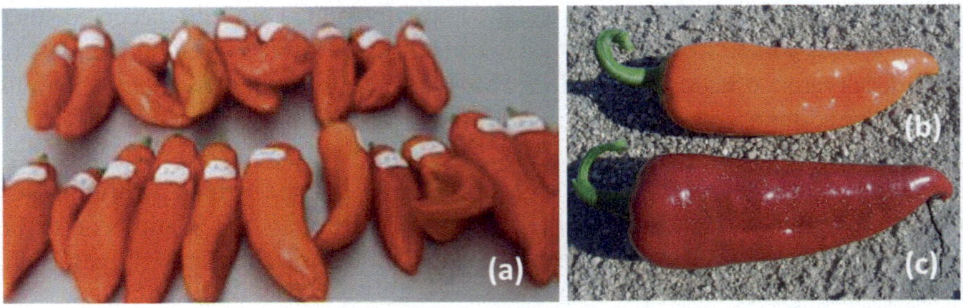

Fig. 20.1 Fruit phenotype of the first high β-carotene mutant variety of sweet pepper in Bulgaria compared to its parent: (**a**, **b**) "Oranzheva kapia": first mutant sweet pepper variety for high β-carotene released by Prof. S. Daskalov; (**c**) "Pazardzhishka kapia 794" (initial variety with red fruits)

et al. 2011, 2014; Timin and Timina 2020; Tomlekova et al. 2008, 2014, 2021). He irradiated with X-rays dry seeds of the highly valued local variety "Pazardzhishka kapia 794." He irradiated seeds with X-rays as follows: 180 kV, 22 mA, and 0.5 mm Cu filter. Initially, he determined the lethal dose for this variety using a dose range of 40–220 Gy and found $LD_{50} = 100$ Gy. So, he used 120 Gy for the mutation breeding experiments, as this dose provides a survival rate of approx. 35% of the treated plants. In M_2 generation from 15,500 plants, he found 8 male sterile mutants. He crossed them with the parent line and got a segregation ratio of 3.28:1 and concluded that the male sterility was caused by a single recessive gene. He further tested the possibility of these mutants to be used for manless cross-pollination for transfer of useful traits from chosen variety by planting them in alternating rows, which strategy proved to be successful.

Later, Daskalov (1973a) irradiated dry seeds of two sweet pepper varieties "Pazardzhishka kapia 794" and "Kalinkov 800/7" with 120 Gy X-rays. He acquired in M_2 generation additional series of male sterile mutants denoted as *ms3*, *ms4*, *ms5*, *ms6*, *ms7*, and *ms8*.

Daskalov and Baralieva (1992) continued the development of orange fruit mutant found in the pool generated in 1965. In 1967, 23 plants were found with orange fruits in progeny no. 357 out of 85. They were subjected to stabilization and refinement beyond M_8 generation. In 1991, it was officially registered as variety "Oranzheva kapia" (see Fig. 20.1). The mutant has green immature fruits, which change to orange instead of red when mature, and has slightly higher yield than the parent. It underwent biochemical evaluation and was found to have 2.0–2.5 times more β-carotene while maintaining the same level of dry matter, sugar, and vitamin C as the parent "Pazardzhishka kapia 794" (Tomlekova et al. 2021).

In a follow-up study, Chalukova et al. (1993) and later Tomlekova et al. (2021) analyzed the disturbance of the carotenoid biosynthetic pathways in the newly registered "Oranzheva kapia" with orange fruits, variety "Pazardzhishka kapia 794" (red fruit), and variety "Kurtovska kapia 1" (red fruit). They made green and mature fruit extracts and processed them with TLC and HPLC. The results showed lack of many of the hydroxy-carotenoids in the orange fruits but presence in the red ones, which the authors assumed to be an indication of participation of various oxidizing agents in the hydroxylation of the beta-carotene and the obtained xanthophylls. They proposed that the mutation hindered the hydroxylation of the beta-carotene to beta-cryptoxanthin, and alpha-carotene to lutein; therefore, the fruits accumulated beta-carotene and alpha-carotene, while the other stages of the carotenoid biosynthetic pathway seemed unchanged. These studies had one-way results in the studies of Tomlekova et al. (2004), Tomlekova et al. (2007a), and Petrov et al. (2013). The other genes sequenced by capsanthin-capsorubin synthase and geranyl-geranyl phosphorylase were not affected by mutation.

Further, the work on pepper mutation breeding was continued (Timina et al. 2011; Tomlekova et al. 2016a). In a follow-up study, the genetic background of the mutant's orange fruit color was identified in depth by using the mutant variety, its initial one "Pazardzhishka kapia 794," and a commonly used variety "Kurtovska kapia 1" (Tomlekova et al. 2016a). The authors narrowed the gene search to two prospective candidate genes encoding β-carotene hydroxylase in pepper $CrtZ_{Chr3}$ and $CrtZ_{Chr6}$ (localized on chromosomes (Chr) 3 and 6, respectively). Using public database, they constructed gene primer pairs of $CrtZ_{Chr3}$ and $CrtZ_{Chr6}$ and a neighbor *Pds* (Chr3) gene. Using simple PCR technique, the authors found that there was no amplification product from the mutant variety for $CrtZ_{Chr3}$ gene although the product from the neighbor *Pds* gene was present; therefore, they assumed that deletion induction of $CrtZ_{Chr3}$ occurred after the gamma-irradiation experiments. Tomlekova et al. (2021) used the varieties and biochemically evaluated in detail the intermediate and end products of the carotenoid biosynthetic grid (lycopene, α-carotene, lutein, β-carotene, β-cryptoxanthin, zeaxanthin). The results revealed that there was inhibition of the α-carotene and β-carotene transformation to the next products of the pathways in the mutant variety, which led to accumulation of these substrates and negligible amounts of the consecutive products, while the profile for the initial genotype was the opposite—presence of end products and negligible substrate levels. The authors further amplified, cloned, and sequenced $CrtZ_{Chr3}$ and $CrtZ_{Chr6}$ genes from all the three varieties. The authors found absence of amplification in the mutant line at all positions, which they attributed to deletion of the gene. To test this hypothesis, they evaluated the expression of this enzyme in the varieties and as a result could not detect any mRNA of $CrtZ_{Chr3}$ in the mutant variety. Later, to evaluate the effects of the orange color mutation on the other fruit biochemical characteristics, Tomlekova et al. compared the red color variety "Albena" with mutant line M38 (a genotype developed by transferring the orange-fruited mutation (*of*) from variety "Oranzheva kapia" and the anthocyanin-free mutation (al) from variety "Albena" into M38) (see Fig. 20.2). The authors found no detrimental effects of the high β-carotene concentrations on mineral elements and other evaluated parameters (Tomlekova et al. 2017).

Very fruitful long-term collaboration was initiated in 1988 as a joint effort between institutes from Bulgaria and Moldova. They started breeding work for the development of new hybrids in order to overcome some of the shortcomings of the obtained mutants and to better suit for the climate in Moldova. For that purpose, they used several Bulgarian mutant lines and varieties as the mutants with male sterility

Fig. 20.2 Bulgarian mutant sweet pepper genotypes with mature fruits: (**a**) M38 with orange color of the mature fruit; (**b**) P31 ("Albena") with red color of the mature fruit

ms-3 and *ms-8*, "Oranzheva kapia 1," etc. The Bulgarian mutant variety "Oranzheva kapia 1," although well suited to the Bulgarian environmental specifics, did not perform well in the Moldavian climate. The canning industry required higher level of β-carotene and vitamins in the pepper fruits in order for sufficient portion of them to result after thermal processing. The researchers developed three mutant lines (L-22, L-29, L-40) from this variety with differing pericarp coloring during the maturation process based on the genes *bc* (orange color of the fruits) and y (yellow fruits) and did extensive crossing between them. They found that these genes had *cis-trans* effect being located on chromosome 6 being far from each other. However, this placed a difficulty on the breeders since two genes were responsible for fruit color and one of them also for high β-carotene. To cope with it, the authors recommended using lines with orange-red coloration developed without the usage of yellow-colored forms (Timina et al. 2014). Other widely used mutants in these collaborative series of experiments were the mutants with male sterility *ms-3* and *ms-8*. They were used for the development of heterosis of F_1 hybrids. For this purpose, the initial genotypes were translated on a sterile background through a complex set of recurrent breeding, multiple backcrossing, etc. In such fashion, new mutant lines were developed as L-5/92 with early ripening trait, which was used for the establishment of new early-ripening varieties (Timina et al. 2014). After a long-term breeding work using the *ms-3* mutation, the variety "Yubileiny Semko" was developed and successfully registered in Russia and Belarus, which had high economic value, yield, multiple disease resistance, and early ripening (Timina et al. 2011, 2014). O. Timin in Transnistria extensively used the *ms-3* mutant to develop varieties. Using backcrossing and recurrent breeding methods, he developed several mutant lines

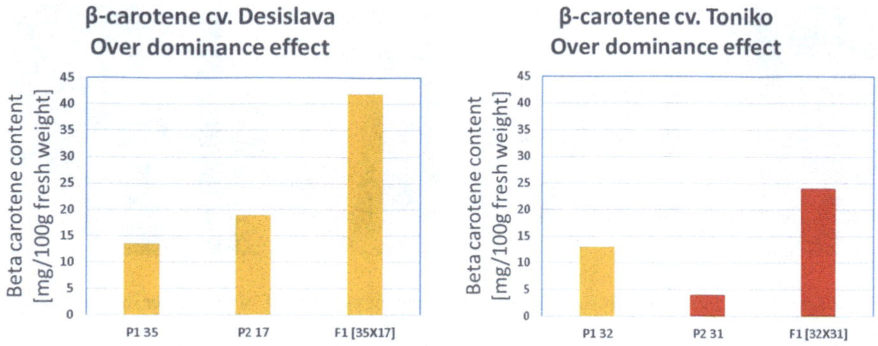

Fig. 20.3 Beta-carotene levels in the fruits of mutant varieties of sweet pepper in Bulgaria

as L-147, L-61/1, and L-46, and later, a new hybrid variety "Vitamin" was registered in Russia with high levels of vitamin C, carotenes, and *Verticillium wilt* resistance (Timin 2005; Tomlekova et al. 2008).

Recently, Timin and Timina (2020) in Russia used the mutant *ms-3* and the local line L-144 to develop a new hybrid variety. After 8 years of complex scheme of crossing, backcrossing, and recurrent breeding, they obtained a new hybrid variety registered in 2019 as "Quinta" (Квинта) in Russia with earliness, high yield, and disease resistance.

In Bulgaria, in an effort to improve the traits of the high β-carotene mutant variety "Oranzheva kapia" through a complex crossing scheme using as parents "Oranzheva kapia," "Zlaten medal 7," and "Albena," Tomlekova et al. developed a new mutant variety named "Desislava." Besides high β-carotene concentrations (Fig. 20.3), it has high yield and improved fruit morphology. It was submitted for registration in Bulgaria (DUS No. 4829/870.2019) and after successful tests was granted certificate in 2020 (BG Fermer 2020). The following year, Tomlekova et al. developed another mutant variety also using as parents "Oranzheva kapia," "Zlaten medal 7," and "Albena," The new variety along with high β-carotene content has added traits as anthocyaninless and bigger red fruits with improved morphology. The variety was submitted for registration in Bulgaria under the name "Toniko" (Fig. 20.4) in 2021 and awaits approval.

In India, Thakur et al. (1980) used another sweet pepper variety "California Wonder" to study the effect of X-rays after their colleagues Raghavan and Venkatasubban started induced mutation experiments with the variety "Paramakkudy" 40 years ago. They irradiated the seeds with five doses of X-rays (50, 100, 150, 200, and 250 Gy). In M_1, they found negative dose-dependent effect on the germination and viability. In M_2 generation, they observed many interesting mutations as tomato-shaped and pointed fruits, flower buds without sepals, male sterility, etc.

In Cuba, Fonseca et al. (2013) treated dry seeds of sweet pepper variety "California Wonder" with X-rays (5, 10, 20, 30 Gy) with work regime 55 kV and 10 mA. In M_1 generation, the authors found an overall increase of the yield in all

Fig. 20.4 Released/under-release high β-carotene mutant varieties of sweet pepper in Bulgaria: (**a**) initial variety “Pazardzhishka kapia 794”; (**b**) cv. “Oranzheva kapia” by prof. S. Daskalov and D. Baralieva, 1992; (**c**) cv. “Zlaten medal 7”; (**d**) cv. “Desislava” (unripe) by prof. N. Tomlekova et al., (BG Fermer 2020); (**e**) cv. “Desislava” (ripe); (**f**) cv. “Toniko” by prof. N. Tomlekova et al., expected release 2022

treated groups with more than twice increase in the 5 Gy-treated group. Additionally, the authors reported increase of the plant height, stem thickness, and root length in the treated groups.

20.2.2 Gamma Rays

Gamma rays are electromagnetic radiation originating from the nuclear decay of unstable isotopes or nuclear reactions. It has more energy per photon and therefore has higher penetrability than X-rays (Spencer-Lopes et al. 2018; Mba 2013). The most commonly used sources are the isotopes cesium-137 (^{137}Cs) and cobalt-60 (^{60}Co) which produce monoenergetic radiation (FAO 1977). Similarly to the X-rays, they eject electrons from the atoms, which absorb them producing ionization. The ejected electrons consequently also produce ionization. Overall, the produced ionization paths are of higher ion density than the one from X-rays due to their higher energy (Smith 1958).

Gamma ray mutagenicity was tested on plants in 1928, a year after the X-rays, but for more than two decades, it was under the shadow of the later mutagen (Stadler 1928). However, nowadays, due to its ease of use, lower cost, etc., this is one of the most commonly used physical mutagens in plant breeding and in sweet pepper one as well.

Probably, one of the first evaluations of this mutagen on pepper was in Japan by Fujii and Matsumura (1959). They irradiated pepper seeds with gamma rays (200, 400, 700, 1000 Gy) from ^{60}Co for 18 h and planted them after 3–10 days. They found slight increase of the germination rate at a dose of 200 Gy, but 40% drop at 400 Gy group and no germination at higher doses. For the tetraploid plants, the decrease was more gradual and 50% rate at 700 Gy, but none at higher doses. As for the seedling survival, the impact was even stronger. For the parameter seedling survival in the aneuploid group, it dropped to zero at higher than 20 Gy doses, while for the tetraploid one, the drop was at doses higher than 40 Gy.

In Bulgaria, Daskalov (1973b) irradiated dry seeds of sweet pepper variety "Zlaten medal 7" with 135 Gy gamma rays. Later, Tomlekova et al. (2008) irradiated the variety "Hebar" with a set of doses of gamma rays ^{60}Co and ethyl methanesulfonate (EMS) being under development. In M_2 generation, Daskalov (1973b) screened 57,000 plants and found 3 male sterile mutants. Further, he performed allelic test of the genes in charge for the male sterility and denoted them as *ms-6*, *ms-7*, and *ms-8*. After a few years of testing, he found that the *ms-8* mutant is most suitable for hybrid seed production due to its stability and independence of climatic conditions or genotype. Later, he used the stabilized Zlaten medal-*ms-8* mutant line as female parent to cross it with the male parent line "GO-250B" with many outstanding traits. The new F_1 hybrid was named "Gornooriahovska kapia F_1," and it has excellent yield and multiple disease resistance (Daskalov 2001; Tomlekova et al. 2014).

Further, Daskalov (1975) irradiated with gamma ray dry seeds of the variety "Zlaten medal 7." As a result, he acquired mutants lacking anthocyanins (*al*). Based on them, a mutant variety with earliness and high yield, more attractive fruit, and better flavor was developed registered under the name "Albena." The work with this mutant continued, the anthocyaninless (*al*) trait was combined with the trait of orange fruit (*of*) from the mentioned earlier X-ray experiments, near-isogenic lines (NILs)/recombinant inbred lines (RILs) (crossed with the corresponding initial genotypes) were acquired up to M_8 generations, further they were crossed by commonly used local varieties, and they are under development (Tomlekova et al. 2009). Consequently, the mutant genes introduced from the mutant pepper genotypes were studied in order to better understand their influence on the concentration of β-carotene in the green and ripe fruits. The biochemical analysis of the fruits revealed that the β-carotene is genotype dependent; that is, the mutants and hybrids with the orange fruit had higher concentration than red fruit, only initial genotype and hybrids. In a follow-up study, Tomlekova et al. (2016b) used the affordable RAPD technique to build variety-specific profiles with a selection of markers in order to differentiate the mutants, hybrids, and initial genotypes in order to facilitate the breeding work. From the tested 16 decamer primers, 12 produced monomorphic profiles in all investigated genotypes, while 3 produced polymorphism in one line and 1 primer between two lines.

Todorova and Daskalov (1997) in 1975 started large-scale induced mutation work in order to develop a line for powdery mildew *Leveillula solanacearum* Gol. since no source for resistance to this disease was available. Dry seeds of the common local "Kurtovska kapia 1" were irradiated with gamma rays (60–120 Gy). In M_2 generation, 65,153 plants and 14,217 control ones were inoculated with the pathogen. After screening, only one resistant mutant was found. This mutant was self-pollinated up to M_8 generation, and in each generation, the plants were inoculated and tested for resistance. Afterwards, the seeds were bulked and performance trials for official registration were started. After 16 years from the start, a variety "Pirin" was officially registered.

Sotirova and Daskalov (1983) used gamma rays to induce mutations for resistance towards *Phytophthora capsici* Leonian. They irradiated dry seeds of the sweet pepper "Kurtovska kapia 1" with doses 60, 80, 100, and 120 Gy. In M_2 generation, the resistant mutants from the groups with 80 and 100 Gy treatments were 3–6 times more than control group, and 10 resistant lines were selected for the development of advanced mutant lines.

In Austria, at IAEA laboratories, chimerism after gamma irradiation in a few crop species including *Capsicum* sp. was studied (Hermelin et al. 1983). Here, Daskalov used for hybridization two sweet pepper lines: (1) Wibault *ms-8* (W-8) (with recessive alleles for male sterility (*ms-8*), lack of anthocyanin (*al*), sulfury white immature fruits (*sw*), and sweet taste of the fruits (*c*)) and (2) line C-3-1 with corresponding dominant genes. These lines were crossed, and the F_1 seeds were irradiated with gamma rays (60, 80, 100 Gy). One thousand two hundred and ten M_1 plants were tested for chimerism. From them, he selected 54 M_1 plants with chlorophyll mutations and observed different chimeric patterns in M_2 generation. Analyzing the results, he suggested that from all the M_1 fruits, those from the main bifurcation should be harvested because their seeds most probably will segregate mutations.

Joshi and Khalatkar (1983) in India irradiated pepper seeds with gamma rays (20, 40, 80, 120, 160, and 200 Gy). In M_2, they evaluated mutants with flower variants as fasciflora, polypetalous mutant, petaloidy of calyx, formation of staminodes, male sterility, and early-flowering types. They observed dose-dependent frequency of occurrence of such type of mutations ranging from 1 to 2.4 per 100 M_2 plants.

In Italy, Cristinzio and Saccardo (1982, 1984) irradiated with gamma rays (70 Gy) seeds of sweet pepper variety "Yolo Wonder," which is susceptible to *Phytophthora capsici*. Afterwards, they inoculated with the pathogen 10,000 M_2 seedlings from 941 progeny M_1. As a result, they acquired 115 plants from 55 progeny M_1 with resistance to *P. capsici*. In a further report, the authors traced the stability of the resistance to M_5 generation. They found one mutant line denoted as 704 to be resistant to all three strains of the pathogen.

In Cuba, researchers (Arias et al. 1998) irradiated seeds of sweet pepper variety "California Wonder" with gamma rays from ^{60}Co in order to determine its radiosensitivity. The treatment was with acute doses between 100 and 800 Gy, at 100 Gy intervals. They found negative dose dependence on seed viability and that the most appropriate interval for induction of mutations was 130–460 Gy.

In India, Sood et al. (2017) used ^{60}Co to irradiate with gamma rays (5, 10, 3, 50, 80, 110, 130, 160, 190, and 220 Gy) seeds of sweet pepper variety "California Wonder." In M_2 generation, they evaluated the frequency of the chlorophyll mutations. They could categorize four types of chlorophyll mutations: chlorina, viridis, xantha, and yellow xantha. The first two were viable mutations, while the second ones were lethal. Also, the authors noted that the viable mutations were more frequent at lower doses while the lethal were more common at higher doses.

Recently, in Nigeria, Adekola and Oluleye (2012) studied mutants of the sweet pepper variety "Tatase" with 240 and 300 Gy gamma ray treatment. In M_4 generation, they evaluated 11 mutant lines and found that most of the mutants were taller and had slightly bigger leaves; however, the yield was either equal or up to twice lower than the wild type.

Later, Abu et al. (2020) continued to evaluate the effect of gamma rays on the same variety. The seeds were irradiated with doses of 50, 100, 150, or 200 Gy. The authors found that the dose of 100 Gy had slightly stimulatory effect on seedling emergence and fruit ripening. Additionally, in M_1 generation on average, the plants from the 150 Gy dose group had more and bigger leaves and longer internodes and were 1.5 times taller than the controls.

20.2.3 Neutrons

Neutrons are neutral particles which are stable only when confined in the atom nucleus; once separated from the nucleus, their lifetime is approx. 15 min (Spencer-Lopes et al. 2018). Due to their neutrality, they have good penetration in tissues and matter. The energy of the neutrons can vary from less than 0.003 eV to over 10 MeV, and based on this, several types are defined. The most commonly used in mutation breeding are the slow thermal neutrons (0.003–0.4 eV) and the fast neutrons (200 keV to 10 MeV) (Spencer-Lopes et al. 2018; Mba et al. 2012). The slow neutrons produce their effects mostly through transmutation reactions when captured by nuclei and consequent radiations or by decay of the new isotopes and their products. The effects of the fast neutrons are mostly due to the collisions with atom nuclei in tissues, mainly hydrogen ones (Smith 1958).

In sweet pepper mutation breeding research, the treatment with fast neutrons is the preferred type of neutron irradiation.

Although it was found that neutrons can cause mutations in 1937, shortly after their discovery (Nagai 1937), it was not until the 1950s when neutrons were applied for plant breeding when accessible facilities to generate neutron irradiation were built and made available for agricultural research (Shapiro 1956). However, even to date, neutron irradiation is less commonly used in sweet pepper breeding than the previous two mutagens; however, some researchers successfully developed new mutant varieties by using them.

In 1974, Saccardo and Ramulu used fast neutrons as mutagen to induce mutations for CMV resistance; they screened M_2 population and found that the frequency of the plants with no symptoms was tenfold higher than the group treated with EMS. Three years later, the same team compared the efficiency of fast neutron vs. gamma radiation to induce mutations for resistance against *Verticillium dahlia*; the results showed 33% higher frequency of M_2-resistant mutant plants in neutron-treated group than the gamma-treated ones (Daskalov 1986; Saccardo and Ramulu 1977).

Todorova and Daskalov (1979) irradiated with fast neutrons seeds of the sweet pepper variety "Albena" and "Kurtovska kapia 1" and screened for resistance against the powdery mildew (*Leveillula solanacearum*), and the rate of the resistant mutants was 0.51%. Based on these mutants, Todorova further developed eight resistant lines, and one of them was officially registered (Todorova 1982).

20.2.4 Ion Beam Irradiation

In this technique, single-type ions are accelerated to particular energy values, and afterwards, the target plant material is irradiated with the beam. In this process, an important parameter is the linear energy transfer (LET) of the ions, which depends on their type and velocity (Abe et al. 2018). LET is related to the resulted mutation rate, but the most effective LET (LETmax) for mutation induction differs for the different species and has to be determined experimentally (Abe et al. 2018). Most of the resulted mutations are small deletions, base changes, etc. (Abe et al. 2018; Honda et al. 2006).

Research with ion beams in pepper mutation breeding is scarce probably due to the cost and the limited availability of appropriate accelerators for agricultural usage and has so far been performed mostly in East Asian countries (Honda et al. 2006; Jo et al. 2016, 2018). Honda and coauthors in Japan used ^{12}C and ^{20}Ne ions with LET 22.7 keV/μm and 64.2 keV/μm to irradiate seeds of variety "California Wonder." They found that doses above 100 Gy are lethal in both ions and doses below 10 Gy have negligible difference in germination parameters compared with controls. Therefore, they used 10 Gy for the mutagenesis experiments and consequently raised 150 M_1 and 200 M_1 plants after ^{20}Ne or ^{12}C ion beam irradiation exposure. As a result, they could identify three types of mutants in the ^{20}Ne-irradiated group, which differed in branching and yellow coloration of the ovary and pericarp. The authors evaluated the inheritance of these traits in M_2 generation and also did crossing with the wild type. They found that all of the explored mutations in the mutants were monogenic and recessive (Honda et al. 2006).

20.2.5 Cosmic Irradiation

Cosmic irradiation is a complex mix, but two components dominate at the altitudes mostly used by human devices on Earth's orbit. The first component is the galactic cosmic ray background radiation. Most of it consists of nuclei of hydrogen (protons) 85% and helium 14%. However, 1% is of heavier particles known as HZE (abbreviated from high (H) atomic number (Z) and energy (E)). The most abundant nuclei part of the HZE mix are C, O, and N. Although they have relatively small atomic number, they have very high energies and can penetrate many centimeters in materials and specifically in tissues; in addition, they are highly charged and therefore are densely ionizing (Schimmerling 2011; Mba 2013).

The second main component is the SPE radiation (from solar particle events), mostly of high-energy protons; however, some of them can have energies of several hundred MeV. However, the effects of SPE radiation can be mitigated, if the space vehicle is at altitudes where Earth's magnetic field is strong enough.

Experiments with HZE show that they can cause multiple chromosomal aberrations in seeds and abnormal rate of plant development (Liu et al. 2008).

In addition to the HZE and SPE radiation, there is another unique factor for space mutagenesis, which is microgravity. Although little is known of its mechanism of action, the experiments show that the rate of aberrations increases with the time the seeds spend in space (Liu et al. 2008).

Typically, the seed material is loaded on returnable satellites and is sent to low Earth orbit (200–400 km from Earth's surface, cycle 90 min) for several days (5–15) at a temperature of 15–40 °C, where the seeds are exposed to microgravity (10–5 g), vacuum (10–5 Pa), radiation, etc. (Xie et al. 2010; Jiyuan et al. 1999).

Although pepper was among the plant crops selected for the first biological experiments in the space during the 1960s (Young 1968), it was not until the end of the twentieth and the beginning of the twenty-first centuries when the research results of the space mutation breeding program of the leader in this field—China—became available to the scientific community. China as the biggest global producer of pepper naturally selected this crop to be among the ones treated with space radiation for breeding purposes (Hu et al. 2010). In 1987, China sent the first high-altitude balloon carrying sweet pepper variety "Dragon 2"; consequently, a new mutant variety was established with fruits weighing 250 g and yield increase by 120% (Jiang et al. 2020; Guo et al. 2004). This success was the prelude of space mutation breeding program of China. So far, all of the officially registered pepper mutant varieties from China in the FAO/IAEA Mutant Variety Database (MVD) are developed through space mutation breeding and 40% of them are sweet pepper ones (IAEA 2021; Maluszynski 2001). The most common traits improved through space mutation breeding are increase of the fruit size and often also the whole plant, increased yield, improved taste, and disease resistance against some pathogens as blight, some viruses, etc. (Guo et al. 2004; Liu et al. 2008; Jiang et al. 2020; BAACVD 2021). Because of the great success, China continues to develop and improve sweet pepper mutants as "satellite 87-2" and L06-30 SP lines (Junmin 1999; Liu et al. 1999; Xie et al. 2010). Additionally, ground-based "space simulation" breeding techniques are in the process of development, and other more technologically developed countries as Russia and the USA were inspired by China's success and initiated space breeding programs, so hopefully more sweet pepper space mutants will emerge in the future (Jiang et al. 2020).

20.2.6 Laser Beam Irradiation

Laser irradiation is considered as a softer physical mutagen since it cannot directly ionize atoms and molecules, and its low-intensity application causes less damage to the tissues and survivability of the laser-irradiated seeds usually is higher than for example from gamma rays (Navrotskaya et al. 2018).

The data show that the mutagenic and cytogenetic effects of laser irradiation depend on the dose, wavelength, metabolic state of the seeds, water content, variety, etc. (Pillai 1998; Navrotskaya et al. 2018). It is assumed that laser mutagenic action is exercised through several mechanisms as conformational changes of macromolecules, DNA damage, induction of radical formation, radiolysis of water, etc. (Balaur 1978; Navrotskaya et al. 2018).

The studies on the mutagenic action in agriculture of the laser irradiation started shortly after it was found in 1969 that coherent light can cause mutations in vitro by Berns et al. and have beneficiary biological effects on seeds by Wilde et al. (Berns et al. 1969; Hernandez et al. 2010). Few monocotyledon crops as wheat, maize, and rice benefited greatly from this technique; however, the application in Solanaceae crops is still scarce (Balaur 1978). One of the possible reasons is the presence of seed shell, which limits and obstructs light penetration.

Application of laser irradiation on sweet pepper variety "Buzau 10" was done by Burnichi et al., in Romania in 2011; as a result, the number of fruits and the yield increased. With similar results, Rankov and Ilieva in Bulgaria (1986) irradiated seeds of sweet pepper variety "Zlaten medal 7." Timina et al. (2001) in Russia irradiated seeds of three lines of sweet pepper and acquired seedlings with earlier growth and better Phytophthora resistance.

20.3 Chemical Mutagenesis and Mutagens

Although for the first time in 1932 Vladimir Saharov (Stroeva 1997; Eiges et al. 2014) in the Soviet Union discovered that some inorganic compounds can cause chemical mutations, it was not until 1946 when the term chemical mutagenesis entered in circulation. This happened with almost joint and independent discoveries of mutagenic action of some organic chemicals on *Drosophila melanogaster* by Charlotte Auerbach, in the UK (with mustard gas) (Auerbach and Robson 1946), and Iosif Rapoport (1946) in the Soviet Union (with carbonyl compounds).

Unlike Auerbach, Rapoport starting from the mid-1930s built and developed a complex algorithm to screen and identify chemical compounds that have strong mutagenic action. Based on it, he discarded the inorganic compounds and focused on selected types of organic ones, and as a result, he discovered over 300 of them (Stroeva 1997; Eiges et al. 2014; Mitrofanov 2001).

Rapoport later established the most influential center of chemical mutagenesis in the USSR, which greatly impacted the plant breeding practices mostly in the socialists' states at that time (which encompassed nearly half of the world). His efforts and the ones of his students and followers gave birth to hundreds of new mutant varieties and thousands of mutant lines including pepper ones by 1990—therefore, he is also known as the father of the chemical mutagenesis in agriculture (Eiges et al. 2014; Pylneva 1997; Stroeva 1997; Leitao 2012).

In comparison with the physical mutagens, the chemical ones have slightly lower degree of freedom of the DNA sites of mutation induction; that is, some of them show preference towards the DNA sequence context as will be presented for some of them below, which can be used by the thoughtful breeder to increase the chance to induce mutations in the DNA area of interest.

However, a major drawback of many of the chemical mutagens is that they have high toxicity and hazard for the users (for example, I. Rapoport got lifetime chronic sickness from them) (Mitrofanov 2001). In addition, many of them are highly reactive and decomposing in water necessitating fresh preparation in each experiment.

Large-scale comparative research of different mutagens from Vietnam mutation breeding program (Tran et al. 2001) reveals:

1. Arrangement of the mutagens by the degree of causing chromosome mutants Gamma ray > NMU > NEU > EI > DMS > DES
2. Arrangement of the mutagens by the rate of morphological/chlorophyll mutations NEU > NMU > EI > gamma ray > DMS

The large pool of chemical compounds has to be organized and classified in order to be better studied and applied and for new ones to be developed. A commonly used leading criterion is their way of action followed by chemical structure similarities. In 2012, Leitao proposed the following groups of chemical mutagens used in agriculture: (1) alkylating agents, (2) nitrous acid and nitric oxide, (3) base analogues and related compounds, (4) antibiotics, (5) intercalating agents, (6) topoisomerase poisons, and (7) other chemical mutagens (Leitao 2012). The usage of mutagens from the groups nitrous acid and nitric oxide, base analogues and related compounds, and topoisomerase poisons is mostly limited to in vitro research, while chemicals from the other groups were used for the treatment of seed and whole organ/plant material too.

20.3.1 Alkylating Agents

Rapoport was the first to discover ethylation as a mutation-inducing process, which consequently was generalized as alkylation and given the name of the most important group chemical mutagens (Rapoport 1948; Bartoshevich 1966; Loveless and Howarth 1959). Alkylation is the substitution of a hydrogen atom by an alkyl group (simple or complex) and also the direct addition of this radical to the molecule (Bartoshevich 1966). Alkylating agents under physiological conditions can alkylate parts of DNA molecule, thus impairing its biological functions besides other molecules.

The group of the alkylating agents is by far the most numerous one. The largest share (over 80%) of mutants from chemical mutagenesis was obtained from treatment with such agents. Alkylating agents can be found in various classes of compounds as aldehydes, urethanes, alkylnitrosoureas, ethyleneimines and ethyleneimides, yprite analogues, epoxides, alkyl methanesulfonates, alkylnitrosoamines, alkylnitrosoamides, alkyl halides, alkyl sulfates, alkyl phosphates, chloroethyl sulfides, chloroethylamines, diazoalkanes, etc. (Bartoshevich 1966; Leitao 2012). By their mode of action, they can be divided into monofunctional, bifunctional, etc. By their mechanism of action, they can be divided into (1) S_1N mechanism—unimolecular nucleophilic substitution; (2) S_2N—bimolecular nucleophilic substitution; and (3) capable of both. The alkylating agents can also be differentiated by their Swain-Scott substrate constant(s) whose low

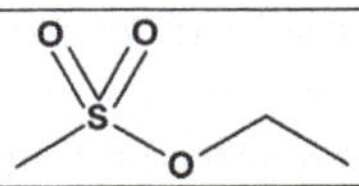	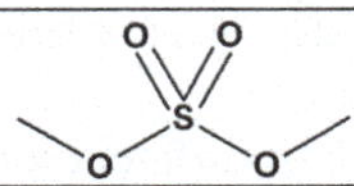	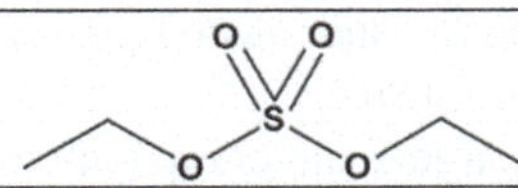
Ethyl methanesulfonate (EMS)	Dimethyl sulfate (DMS)	Diethyl sulfate (DES)
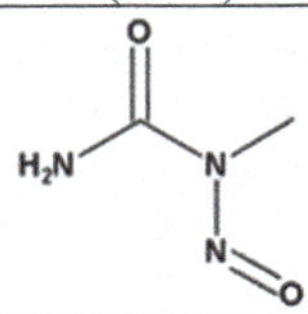	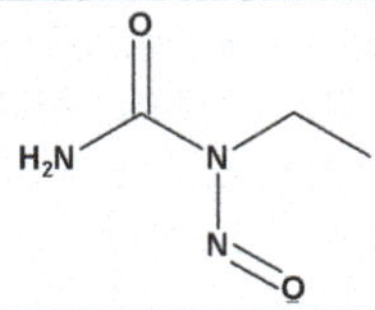	
N-Nitroso-N-methylurea (NMU)	N-Nitroso-N-ethylurea (NEU)	Ethyleneimine (EI)

Fig. 20.5 The most common alkylating agents in sweet pepper mutation breeding

values define high reactivity towards low nucleophilicity centers (for example O-atom sites in DNA) and its high values on the opposite define reactivity towards sites of high nucleophilicity (ring N sites in DNA) (Neale 1976; Bartoshevich 1966).

Although currently thousands of alkylating agents have been discovered, the number of mutagens used in sweet pepper mutation breeding is fairly modest and the most common are presented in Fig. 20.5.

20.3.1.1 Ethyl Methanesulfonate

Over the years, ethyl methanesulfonate (EMS), although not the strongest or most efficient mutagen, has become the most preferred one among plant breeders. One of the reasons for this is its simple mode of application, good penetrability, reproducibility, high mutation frequency, and less problems with the disposal (Shah et al. 2016).

EMS as an alkylating agent can transfer its ethyl group through either unimolecular or bimolecular nucleophilic substitution (Shah et al. 2016). Recently, it was found that EMS has pronounced local sequence context preference of action and specifically targets the guanine residues in the context of RGCG (where R is A or G and the mutated guanine is in italic). EMS forms mostly point mutations of mispairing type probably as a consequence from its strong sequence context bias (Ingelbrecht et al. 2018). The mutations that EMA produces are 93% G:C–A:T transitions and the rest are A:T–G:C transitions, transversions, deletions, etc. (Leitao 2012; Shah et al. 2016).

In 1959, Heslot for the first time demonstrated that EMS has mutagenic activity by applying it on barley (Loveless and Howarth 1959). However, it was not until the 1970s when pepper breeders started to evaluate and apply this mutagen on pepper varieties.

Bansal in India probably pioneered to use EMS on pepper. He treated seeds of variety "NP-46A" (pungent) with EMS (0.15%, 0.30%, 0.45%, 0.60%). In M_3 generation were identified mutants with dark-orange fruits with more than twice-increased β-carotene and mutant with yellow fruits with twice-reduced β-carotene level (Bansal 1969). His results seemed to be rather inspiring since in the following few years more studies with EMS in pepper were published (Zubrzycki and Pahlen 1970, 1973; Pochard 1970; Breuils and Pochard 1975; Saccardo and Ramulu 1977).

Zubrzycki and von der Pahlen in Argentina in a series of experiments treated dry seeds of sweet pepper variety "California Wonder" for 20 h. They acquired mutants with chlorophyll and morphological mutations, and the former ones had higher rate than from X-rays (Zubrzycki and Pahlen 1970). The following year, they reported that EMS surpasses X-rays in the induction of both chlorophyll and morphological mutations in the same sweet pepper variety (Zubrzycki and Pahlen 1973).

Pochard in 1970 treated monoploid material with EMS and could obtain three mutants with male sterility which were denoted as *mr9*, *mc705*, and *mc509*. Later, they were used for the establishment of new hybrids (Pochard 1970; Breuils and Pochard 1975).

In Italy, Restaino (1983) attempted to obtain mutants with resistivity against *Verticillium dahliae* Kleb. using EMS as mutagen. He treated seeds of sweet pepper variety "Friariello" for 13 h at 20 °C with relatively high doses of EMS (0.3%, 0.6%, 0.9%, 1.2%, 1.5%[1]). In M_4 generation, he obtained two dwarf mutant lines denoted as Mutant K 80/78 and Mutant K 80/112. Their plant height was reduced up to 50%, while their fruits maintained their size plus they showed moderate resistance against *Verticillium* wilt. Later in 1985, he officially registered the mutant variety under the name "Friari KS80," which had increased fruit production (15–20%) and tolerance to *Verticillium dahliae* Kleb. (MBNL 1991).

In India, Pillai and Abraham (1996) treated dry seeds of sweet pepper variety "Palmulagu" with EMS (0.25%, 0.50%, 0.75%, 1.0%, 1.25%) for 5 h. Instead in M_2, they did selection in M_3 generation and found three prospective mutants from the middle-dose groups with up to twice higher yield and one with higher vitamin C content. The mutant stability was traced up to M_5 and M_6 generations.

In Mexico, Alcantara et al. (1996) tried to optimize the conditions of seed EMS treatment in pepper. For that purpose, they used different combinations of three variables: EMS concentration (0.5%, 1.0%, 1.5%) in 0.1 M phosphate buffer with pH 7, length of treatment (3, 6, 9 h), and temperature of treatment (5, 10, 15, 20 °C). They presoaked the seeds of sweet pepper variety "Keystone Resistant Giant #3" (which is "California Wonder" type) for 12 h; after the treatment, the seeds were washed with 5 mM sodium thiosulfate for 45 min. From each of the 36 differently treated groups, they planted M_2 populations of 300 individuals each. The group with the highest values in all variables (1.5%, 9 h, 20 °C) showed the highest percentage of mutants in M_2. Some of the mutant types were expressed only at a particular combination of the variable values; for example, campanulate fruits were observed in several treatments with 1.5% EMS with the highest rate at 9 h and 20 °C. The authors also reported many morphological mutants, some of them unique as the ones

[1]NOTE: Actually in the article, the dose range is 3–15%, which probably is a series of typo mistakes, since EMS concentrations over 2.5% are usually lethal.

with single or fused cotyledon. They assumed that pepper is not very sensitive to EMS and proposed that the treatment concentration can be increased even more.

In a follow-up study, Bosland (2002) also treated seeds of the variety "Keystone Resistant Giant #3" with EMS. After 12-h presoaking in water at 24 °C, the seeds were treated with 1.5% EMS (in 0.1 M phosphate buffer with pH 7) for 3 h at 10 °C and washed for 45 min in 0.5% ethyl acetate (pH 7) and for another 45 min in 5 mM sodium thiosulfate (pH 9). In M_2 generation, he acquired mutants with a flaccid (or wilting) phenotype. He did reciprocal hybridization by crossing them with the parent and acquired a ratio of 3.06:1, which is close to the theoretical for a single recessive gene which he denoted as *flc*. The function of this gene is probably related to turgor maintenance and drought stress management.

In Israel, at the Volcani Center were initiated astonishing long-term series of induced mutation-based experiments; Paran et al. (2007) performed a large-scale study with purpose to extend our knowledge of the genetic control of the pepper plant architecture. They applied EMS treatment of seeds of sweet pepper variety "Maor" ("California Wonder" type). A priori, they ran small batch germination test to determine LD15 for this particular variety and found that EMS concentration should be 0.2%, which was used for the larger scale seed treatment. Afterwards, 1800 M_1 plants were grown and from the seeds of their self-pollinated fruits were established 1650 M_2 families. They obtained mutations in 47% of the families and only 4% of them were dominant. The authors found ten mutants with altered sympodial development. Seven of the mutants segregated in crosses with the parent line and had single recessive genes involved in the genetic control of the architecture and were denoted as e-172, 252, 543, 648, 1047, 1463, and 2580. Further, they used the available databases and research reports to search for phenotypic similarities with the data known for other species from *Solanaceae* as tomato and petunia and more distant species as *Arabidopsis*, *Antirrhinum*, pea, etc. The authors found some matches which helped them to narrow the search of candidate genes. These research efforts are part of the long-term goal to develop a collection of pepper mutants in a single genetic background. Such library will serve as infrastructure to study pepper development and relations between phenotypes and gene sequences.

Several in-depth genetic studies (Jeifetz et al. 2011; Cohen et al. 2012; Borovsky et al. 2013, 2020) followed up in order to identify the mutated genes and their role in the control of plant architecture. Jeifetz et al. (2011) used the e-2580 and e-1051, which had dramatically reduced axillary shoot formation and alterations in the sympodial growth habit. The mutants were backcrossed with the wild-type parent. Also, e-2580 was crossed with *Capsicum frutescens* BG 2816 for the generation of F_2 mapping population. In addition, double mutants were developed by crossing e-2580 with the mutant fasciculate (Elitzur et al. 2009) and were identified with molecular markers for the corresponding gene through screening of the F_2 progeny. The authors did anatomic and microscopic evaluation of the mutation and found similarities with tomato *blind* (*bl*) mutation caused by altered *BLIND* (*BL*) gene. They used tomato *BLIND* cDNA to clone and map its pepper homolog *CaBLIND* (*CaBL*), which was sequenced afterwards. The results revealed that *CaBL* consists of three exons and encodes a putative protein composed of 315 amino acids, which in e-2580 is altered with stop codon in the middle through point mutation. The results from the analysis of the double mutants revealed that *CaBL* interacts with *FASCICULATE*, but is independent from *CaREVOLUTA* and *CaLATERALSUPPRESSOR*.

Cohen et al. (2012) used another set of mutants generated by Paran et al. (2007), namely the allelic late-flowering mutants e-252 and e-2537, which had multiple pleiotropic effects on the organization of the sympodial shoot. Here, a similar experimental design and rationale were used as the previous study. In order to map the E-252 mutation, the authors constructed segregating BC1 population by crossing e-252 and *Capsicum chinense* PI 159234. Also, two F_2 double mutants were generated between e-252: CaBL and e-252: *FASCICULATE* for determination of gene interactions. Tomato (*Solanum lycopersicum*) *JOINTLESS* (*j*) seeds were used to facilitate the cloning and mapping of the *CaJOINTLESS* (*CaJ*). The results revealed that *CaJ* encodes protein of 234 amino acids and consists of eight exons. The mutant e-252 had missense mutation (G to A), which causes substitution of glycine to aspartic acid at the middle of the protein. And the other mutant e-2537 had point mutation (C to T) resulting in stop codon in the middle of the protein.

As continuation of the evaluation of the mutants generated by Paran et al. (2007), Borovsky et al. (2013) did in-depth genetic analysis of another mutant, e-172-3, which had orange-colored mature fruits. The authors found that the mutant accumulates mostly β-carotene instead of a complex mix of yellow-red and carotenoids. The authors identified mutation (A to G) in the *β-carotene hydroxylase 2* gene (we named the same gene $CrtZ_{Chr3}$). They crossed the mutant with other orange fruit varieties, but the results were red-fruiting plants, which meant different genes as a cause for the orange color. Further, they evaluated the expression profiles of other enzymes from the carotenoid biosynthesis pathway, and there was increased expression of *PSY*, *CHY2*, *VDE*, and *CCS*, which they attributed to possible feedback regulation mechanisms.

Further few studies of mutants with flowering abnormalities followed (Borovsky et al. 2015), which studied early-flowering mutant after the fourth leaf on the primary stem instead of nine. They used bulked segregant analysis to identify the region of occurred mutation. Comparison with similar region in the relative tomato genome narrowed the search to four prospective genes. After the sequencing of the ORFs of these genes, it was found that the *CaAP2* gene had point mutation which creates premature stop codon, thus producing 70% shorter protein than in the initial genotype.

Later, Mohan et al. (2018) studied late-flowering mutant e-2698. Using similar procedure as the previous study, they identified mutated *CaVIL1* gene (*Arabidopsis* homolog of *VERNALIZATION INSENSITIVE 3-LIKE 1*) in chromosome 5, which had point mutation resulting in stop premature codon, thus losing the VID domain. Further, it was found that the mutation had multiple pleiotropic effects on the shoot and root anatomy. Besides the overall increase in size of the vegetative organs, there were observed distortions in their inner structure accompanied by smaller cellular size, plus weak apical dominance. To reveal the possible interaction with other genes involved in the flowering regulation, double mutants were utilized. This revealed synergistic positive interaction with *CaJ* and *CaFT*. To evaluate possible suppressing effects upon *FLOWERING LOCUS C-LIKE* clustered genes, qRT-PCR was used, and it revealed significant effect of the mutation on their expression levels. Thus, the authors conclude that *CaVIL1* is an important flowering regulator with sophisticated interactions with other promoters and repressors.

Another mutant from the pool generated by Paran et al. (2007) was evaluated recently by Borovsky et al. (2020). They analyzed genetically another late-flowering mutant e-172. The results revealed the presence of missense mutation (G to A) in the *CaFT-LIKE* gene which leads to loss of the third exon, so the expressed protein was nearly half the length of the wild-type one. The authors also tested if *CaFT-LIKE* functions as florigen in tomato; therefore, they created transgenic graft in late-flowering tomato double-mutant uf sft as the florigen receptor using the *CaFT-LIKE* gene. As a result, the non-grafted tomato flowered after approx. 30% more leaves than the grafted one, thus proving the florigen action hypothesis.

In India, Kumar and Gupta (2009) evaluated the EMS effect on the chromosomes of *C. annuum* var. Azad. The seeds were pre-soaked and treated with EMS (0.5%) for 3 h, 5 h, or 7 h. In all treatment schemes, they observed many chromosomal anomalies as bridges, stickiness, and multivalent secondary, associations, laggards, and precocious movement. Predominance of bridges and increased chiasma frequency were observed with the high dose resulting also in phenotype alterations.

In China, Arisha et al. (2014, 2015) performed a series of mutagenicity experiments with sweet pepper variety "B12" and EMS treatment. First, they performed extensive experiments (Arisha et al. 2014) for the determination of optimal conditions of the treatment of this variety. They applied few pre-soaking durations (0 h, 6 h, 12 h, 6 h + 12 h at 28 °C) and ten EMS concentrations (0.25%, 0.50%, 0.75%, 1.00%, 1.25%, 1.50%, 1.75%, 2.00%, 2.25%, 2.50%). The dose effect of EMS on germination percentage was linear only in 12-h pre-soaking group, while the other groups had approx. S-type dependence, and depending on the pre-soaking seed metabolic activation, the LD50 varied from 0.5% to 1.0%. Doses from 1.2% onwards were most damaging in all groups. Later, Arisha et al. (2015) performed larger scale (2000 seeds) treatment of the same variety with 0.6% EMS in 0.1 M phosphate buffer (pH 7) at 20 °C for 12 h and 6 h pre-soaking and 3 h post-washing with water. As a result, only half of the seeds germinated and from them were raised 500 M_2 families with a total of 15,000 individuals. The authors described in detail three most notable mutants: (1) with yellow leaf color, and the leaves contained 62.19% less chlorophyll a and 64.06% less chlorophyll b than the wild type, and (2) dwarf with a very short stature (6 cm), compact internodes, and rough and thick structure. The anatomical evaluation of the leaf blade section of the latter revealed that it has more xylem and collenchyma tissue in the leaf midrib and the leaf blade contained thicker palisade and spongy tissue than the wild type.

In Brazil, Nascimento et al. (2015) treated seeds of pepper (accession UFPB 77.1) with EMS (0.025%, 0.050%, 0.10%, 0.15%, 0.20%, 0.25%) for 3 h or 6 h, after 12-h pre-soaking in water. Here, the seeds were planted in Murashige and Skoog medium (instead in soil or moistened filer paper as the majority of the studies presented here) and were grown there for 50 days under artificial lightening. Authors report of no significant differences for the monitored germination traits; also at acclimation, the plantlets treated with EMS showed a similar trend. However, the authors recorded significant variations in fruits as fruit weight, length, smallest diameter, and fresh matter. Also, they observed reduction of fruit weight and number with dose increase.

In the Republic of Korea, several research groups elaborated on the EMS mutation breeding in pepper in the twenty-first century. Jeong et al. (2012) did a large-scale allelic mining of 248 accessions belonging to 7 *Capsicum* species; in addition, they used the Korean landrace "Yuwol-cho" (low pungent, which sometimes gives nonpungent mutants (Hwang 2014)). Initially, the seeds were pre-soaked for 18 h at 24 °C and then in EMS (0.5%, 1.0%, 1.5%, 2.0%) (0.1 M phosphate buffer, pH 7.0) followed with two washings with 0.5% ethyl acetate (pH 7) and 0.5 mM sodium thiosulfate (pH 9) for 45 min each. From the kill curve, LD15 was determined to be 1.5% for the experiment. They applied high-resolution melting (HRM) method to detect natural variations and EMS-induced mutations. They focused on the first exon of the eIF4E gene (wherein the mutations confer resistance to potyviruses), which they scanned for single polymorphic mutations. As a result, they found five single polymorphic mutations in the eIF4E gene in the EMS-induced mutant population of 1092 M_1 individuals. Further, they determined that 60% of these mutations were of transition type (GC/AC), while the others were transversion ones.

Later, a large M_2 mutant population from the same variety and EMS treatment protocol was developed (Hwang 2014; Hwang et al. 2014). From this pool the authors (Hwang et al. 2014; Hwang 2014) specifically tested for nonpungency 917 M_2 lines (8632 individuals). As a result, six nonpungent mutants were isolated. Further M_3 lines were generated in order to confirm inheritance of nonpungency, and as a result, four mutant individuals were isolated denoted as 221-2-1, 2, 3, and 4 mutants which showed nonpungent phenotypes regardless of their developmental stages.

As a follow-up study, Lee in 2018 did genetic analysis of one of the nonpungent "Yuwol-cho" EMS mutants (number 221-2-1a) to study the capsaicinoid synthesis pathway. Initially, the Pun1 gene was tested, but no mutations in the sequence were found. Therefore, evaluation of the expression of 12 capsaicinoid biosynthesis-related genes was performed. As a result, 7 of these genes had significantly decreased expression: *pAMT*, *BCAT*, *ACL*, *KAS*, *FatA*, *PAL*, and *Pun1*. Further, the author crossed the mutant with the parent and found that the nonpungency in this mutant is controlled by two recessive genes. Afterwards, the author sequenced DNA from samples of the parent and the mutant and found 11 SNPs. The genes were annotated but were not from the capsaicinoid biosynthesis pathway. Additional studies are needed to determine their role in the regulation of the capsaicinoid biosynthesis.

Again in Korea in 2016, Siddique et al. treated with EMS (1.3%) seeds from "Micro-Pep" pepper. Three hundred and fifty M_2 individuals were selected for evaluation and screening for pungency, and 3 nonpungent mutants were identified and will be used for genetic studies.

Recently, Siddique et al. (2020) used EMS mutagen to induce mutations and applied the innovative crop approach TILLING technology (Targeting Induced

Local Lesions in Genomes) to analyze the resulted mutations. They used the same protocol as Jeong et al. in 2012 applied on seeds from "Micro-Pep" pepper. In M_2, they screened 13,000 individuals and described all the phenotypic alterations in 4 main classes and 11 subclasses. Largest percentage (almost 50%) was the alterations in leaf color and morphology. Further, they selected 700 M_2 samples and 1760 M_2 samples from Hwang et al. (2014) originating from the variety "Yuwol-cho" for optimization and processing with the TILLING protocol. As a result, nine EMS-induced mutations were detected in the eIF4E gene, where most of them were GC:AT transitions.

20.3.1.2 *N*-Nitroso-*N*-methylurea (NMU) and *N*-nitroso-*N*-ethylurea (NEU)

In 1962, Rapoport reported the extreme mutagenicity of nitroso-ethylurea (also known as NEU, ENU, ENUA). Nitroso-ethyl- and nitroso-methylureas (also known as NMU, MNU, MNUA) belong to the special group mutagens named by Rapoport "supermutagens." Some comparative tests with other mutagens show that they fully deserve such categorization—the mutagenic effectiveness of NMU is 23 times higher than the one of EMS (Prasad 1972). These agents although chemically similar have different profile of the sites of DNA alkylation. NMU preferably alkylates N7–guanine, thrice less phosphotriesters, etc., while NEU has reverse trend in action—mostly phosphotriesters and five times less N7–guanine (Leitao 2012). Some authors suspect phosphotriesters to be the cause for nucleic acid breakages (Singer 1975). The mutations that NMU produces are almost 100% G:C–A:T transitions, while in NEU their share is 3/4 and the rest is A:T–G:C transitions and transversions (Leitao 2012).

Six years after Rapoport found the extreme mutagens in the USSR at Yerevan State University, several researchers started a decade-long research on the effects of nitroso-alkylureas on the sweet pepper. In a series of experiments, Galukyan (1968a, b, 1969) treated with NMU (0.008%, 0.01%, 0.012%) and NEU (0.0125%, 0.025%, 0.05%) two pepper varieties "Novocherkasskij 35" (sweet) and "Astrakhanskij A-60" (hot) for 18 h. In the first set of experiments, he did cytological evaluation of the mutagenic effects of NMU and NEU. In both varieties, he found that the used mutagens caused chromosomic bridges and fragmentations mostly of singular type and NEU had more pronounced effect over NMU (Galukyan 1968a).

In the second set of experiments, he sowed the treated seeds and in M_1 generation he observed increase in the plant height, yield, and growth rate, and these effects were more pronounced in the hot variety than in the sweet one (Galukyan 1968b). Also, the author noted a high number of dominant mutations from both mutagens. In 1969, as a continuation of these experiments, the author found differences in the response to NMU and NEU mutagens between the two varieties, and no dose dependence of the mutation frequency was observed (Galukyan 1969).

Later again in the same university, Gukasyan et al. (Gukasyan and Akopyan 1974, 1975; Batikyan et al. 1975; Gukasyan and Tumanyan 1976, 1977) performed a series of experiments with NMU on the sweet pepper variety "Yubileynyy 307."

Initially, Gukasyan and Akopyan (1974) evaluated the effect of single and repeated NMU (0.006%, 0.01%, 0.05%) 24-h treatments on the germination, vigor, and cytological changes of the seeds. They found that the lowest dose had stimulating effect on the germination and vigor, while the highest dose had the opposite effect. Additional findings are that both single and repeated treatments cause dose-dependent increase of chromosome rearrangements but more pronounced in the later treatment. The cytogenetic analysis revealed that the repeated treatment diminishes the difference between the mutagen-induced chromosome bridges and fragmentation.

In consequent experimental set, the authors planted the seeds after the same treatment scheme and observed the mutations in the M_1 and M_2 generations (Gukasyan and Akopyan 1975). They found mutants with enlarged fruit and yield increase in M_1 in the repeatedly treated group and chemomorphoses, which showed NMU concentration dependence. In a following study, Gukasyan and Tumanyan (1976) in M_3 generation evaluated biochemically two mutants with large and orange fruits and higher yield. The results showed increase of the vitamin C and carotenes in both mutants.

Further Gukasyan and Akopyan (1975) performed microscopic evaluation of the chloroplast ultrastructure of yellowish mutant from M_2. The results revealed significant reorganization of the chloroplasts—disappearance of the granas and their replacement with lamellae, and also increase of the number and size of liposomes.

In 1976, Gukasyan and Tumanyan evaluated the long-term cytogenetic effects of the mutagen treatment. They found that the chromosome rearrangements persisted even in M_3 generation although in slightly lower degree and also noted the inhibitory effects of the highest NMU dose on few of the germination parameters in M_2 and M_3. Later, Gukasyan and Tumanyan (1977) evaluated another dwarf M_4 mutant. They found anatomical changes in the stem and roots. The stem had increased number of layers of cortex parenchyma with thicker cells, and phloem had doubled thickness. The main root also had significant changes—overall distortion of the arrangements of the structural elements and increased diameter and thickness of xylem vessels.

In Czechoslovakia, Vagera and Havranek (1983) treated seeds of sweet pepper variety "Vesna" with NMU (0.5 mM) pre-soaked for 48 h. In M_2 generation, they inoculated the leaves of the mutants with ringspot strain of cucumber mosaic virus (R-CMV) and screened for resistance against it. M_2 population was of 9000 plants, and 565 mutants had relatively high level of mutations. Thirteen of them (0.16%) showed resistance against R-CMV, and the rest of the mutants had chlorophyll and morphological mutations.

20.3.1.3 Ethyleneimine

Ethyleneimine's (EI) mutagenic action was discovered by Rapoport back in 1940s, and probably this was the first chemical mutagen to surpass "the radiation limit" at that time; that is, it proved that chemical mutagenesis is more efficient than radiation-induced one. It is 53 times more efficient than HCHO and 4.7 than nitrogen mustard. Such powerful mutagens played a pivotal role in the establishment of the chemical mutagenesis as the "traditional" technique in plant breeding practice (Abilev 2012; Mitrofanov 2001). Ethyleneimine is an extremely reactive alkylating agent undergoing a ring-opening reaction with cellular nucleophiles such as guanosine. Ethyleneimine bound to guanine N7 induces one of the highest known rates of depurination (Verschaeve and Kirsch-Volders 1990). It can induce chromosome aberrations, but their frequency can be mitigated by treatments with Cd or Cu nitrates (Ruposhev 1976). Unusual feature of this mutagen is its wavelike kinetics; that is, by varying the posttreatment seed storage time, the mutation and aberration patterns can be changed (Verschaeve and Kirsch-Volders 1990). However, a major drawback of ethyleneimine is its low boiling point and high toxicity, which is a cause for safety concerns that can be a limiter for some researchers.

In the USSR, Galukyan (1968b) treated two pepper varieties "Novocherkasskij 35" (sweet) and "Astrahanskij A-60" (hot) with ethyleneimine (0.008%, 0.01%, 0.02%) for 18 h. He observed in M_1 generation increase in the plant height, yield, and growth rate, and these effects were more pronounced in the hot variety than in the sweet one. However, the mutagen negatively impacted the germination, decreased the plant viability by 9% for "Novocherkasskij 35" and 33% for "Astrakhanskij A-60" variety, and increased slightly the time to flowering, but had negligible effect on the fruiting time. In M_1, six fruit mutants were obtained in the hot variety and one mutant in the sweet pepper but only in the 0.02% group.

In the USSR, Skripnikova (1976) treated dry pepper seeds with 0.05% solution of ethyleneimine and acquired tomato-form mutant. In the USSR, Nushikyan and Martirosyan (1987) treated sweet pepper variety "Nork" and "Lastochka" with 0.05% solution of ethyleneimine for 18 h. In M_5 and M_6 generation, they selected several mutants with higher yield, vitamin C content, and disease resistance. The most prospective mutant line from "Lastochka" in 1991 was officially registered under the name "Nush-51" (MBNL 1994).

20.3.1.4 Dimethyl Sulfate and Diethyl Sulfate

Dimethyl sulfate (DMS) and diethyl sulfate (DES) are monofunctional alkylating agents with similar structure, which attack DNA through S_2N mechanism. However, DES in addition to S_2N is able to alkylate through S_1N mechanism too. Both agents can induce mutations, chromosomal aberrations, and other genetic alterations in the living cells (Hoffmann 1980). Their significant drawbacks are the low boiling temperature, lack of odor (DMS), and short half-life in water, which decreases with temperature increase; this requires frequent change of the solution (Hoffmann 1980; Daskalov 1986).

Rapoport discovered the mutagenicity of these compounds back in 1947 (Auerbach 1973; Hoffmann 1980), but yet their use in agriculture is not very common.

In the USSR, Rubzov and Solomatin (1974) treated dry seeds of three sweet pepper varieties ("Bjala kapia," "Rotund 449," and "Michurinskji Krasnji") with 0.05% DMS solution for 20 h. They acquired mutants with higher yield and bigger fruits.

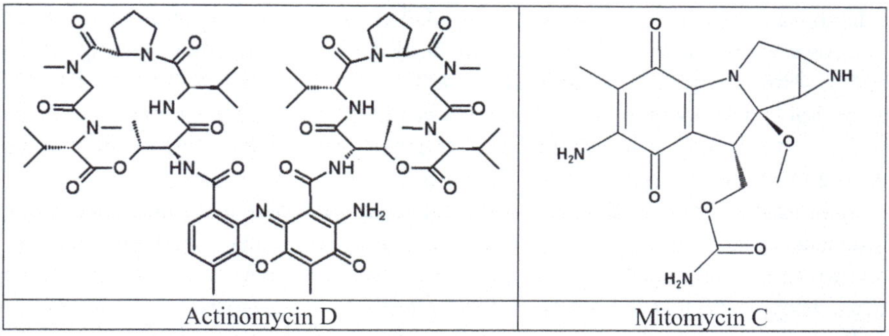

Fig. 20.6 The most common antibiotics in sweet pepper mutation breeding

Skripnikova (1976) treated (0.005%, 0.02%, 0.05%) dry seeds of the variety "Michurinskji 41" with DMS. She acquired mutants with bigger fruits and yield.

20.3.2 Antibiotics

Antibiotics are a structurally diverse group of compounds but functionally united on the basis of their antimicrobial activity (Leitao 2012). Although probably they exercise their mutagenic action through different mechanisms and reactions, a commonly cited result from their action is that they cause various degrees of chromosomic aberrations and there are reports of male sterility too (Oladosu et al. 2016). Commonly used antibiotics in plant breeding (see Fig. 20.6) are actinomycin D, mitomycin C, azaserine, streptonigrin, etc. (Oladosu et al. 2016). In pepper breeding, antibiotics are rarely used as mutagens.

Actinomycin D, first isolated in 1940, is a peptide-containing, pigmented, antibiotic. It can form complexes with DNA and inhibit DNA and RNA synthesis. In addition, there are reports that it can induce DNA breaks and can inhibit repair replication (Vig 1977). However, the extent of these actions can vary significantly between the species, and specific testing is needed for the organism of interest. In 1989, Qihan and Henshaw treated pepper callus culture with actinomycin D. Consequently, they found that DNA synthesis was impaired, which they attributed to the resulting genetic variation in the pepper callus cells.

More interest was paid to mitomycin C (see Fig. 20.6). Mitomycin C, isolated in 1956, is known to selectively bind to DNA and inhibit its synthesis. Further, it was found that mitomycin C cross-links complementary DNA strands in GC-rich regions likely through mono- or bi-alkylation mechanism (for this property, it is also classified as a bifunctional alkylating agent) (Vig 1977; Leitao 2012). In India, Venkatrajam and Subhash (1984) pre-soaked pepper seeds for 24 h and afterwards treated them with 0.01% solution of mitomycin C and compared them to EMS 0.1% treated group. They evaluated the results in up to M_3 generation. Mitomycin C induced sterility with a frequency of 11.39%, which is almost twice higher than the

EMS-treated offspring as revealed in M_2. Other noteworthy mutations are orange fruit color with increased vitamin C content of 0.55%, purple color of unripe fruits 2.04%, semi-sterility 4.51%, spindle-shape fruit 3.75%, etc. They observed in M_2 the desirable multilocular trait with a frequency of 2.84%, which was in concordance from a similar study with mitomycin C (Subhash et al. 1981). In M_3, the authors recorded stability of gigantism and dwarfism.

In the USSR, in 1983, Samovol and Zhuchenko (Samovol and Zhuchenko 1983; Zhuchenko and Samovol 1983) in a series of experiments tested the effect of mitomycin C solutions (0.002–0.05%) on the crossing-over frequency in pepper seeds. They found that mitomycin C moderately decreases crossing-over frequency for the tested gene markers in comparison with other used treatments.

20.3.3 Intercalating Agents

These compounds can reversibly intercalate with double-stranded DNA, but they do not covalently bind to it (Leitao 2012). Some acridine dyes (see Fig. 20.7) used in biochemical research are examples of such agents: ethidium bromide, acridine yellow, etc. (Buiatti and Ragazzini 1966; Levy and Ashiri 1975; Wu and Zhang 1986).

Although it was shown that the classical DNA-stain ethidium bromide is a potent mutagen in higher plants nearly half a century ago, so far mutation breeding research with it has been scarce (Levy and Ashiri 1975; Ingelbrecht et al. 2018). In India, Mehta (1998) studied the effect of ethidium bromide (0.01% and 0.05% solutions) in two red pepper varieties alone or in combination with other chemical mutagens. He pre-soaked the seeds in water for 18 h and afterwards immersed them for 30 min in mutagen solutions. He found no significant changes in M_2, when ethidium bromide was used alone, but when combined with sodium azide, it gave the highest rate of macromutations than the rest tested mutagens.

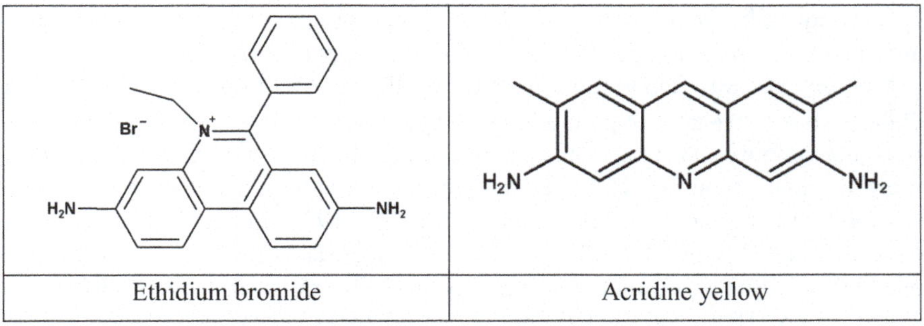

Fig. 20.7 Intercalating agents used in pepper mutation breeding

20.3.4 Other Chemical Mutagens

20.3.4.1 Sodium Azide

Sodium azide (NaN_3) is a highly toxic inorganic compound with relatively common usage in crop breeding. De facto, this is a pro-mutagen; that is, in order to perform its mutagenic action, it has to covalently bind to other compounds. In barley, this mutagenic metabolite was identified as L-azidoalanine (N_3-CH_2-$CH(NH)_2$-COOH) (Khan et al. 2009; Ingelbrecht et al. 2018). Further, it was found that the production of this metabolite is dependent on the enzyme *O*-acetylserine sulfhydrylase. This necessity of sodium azide to be properly metabolized according to some authors may explain the differences in the mutagenic response among the species (Ingelbrecht et al. 2018). In water, sodium azide dissociates partly to hydrogen azide, which is considered to penetrate more easily through the cell membrane. Lowering the pH of the solution significantly increases the dissociation and availability of HN_3; therefore, usually acidic buffer with pH is ≈ 3 (Khan et al. 2009). One of the advantages of the sodium azide is that it causes less chromosome aberrations and therefore it has high mutagenic efficiency when measured as a ratio of mutation yield to chromosome aberrations (Sharma 1994).

Mutagenic action of sodium azide was found in 1947, but for the first time in 1973, it was tested on plants specifically as a mutagen (Nilan et al. 1973). In 1981, a brief report for the mutagenic action of sodium azide on pepper seeds was published by Umalkar et al. (1981). The authors screened the M_2 generation and found that the germination percentage and the vigor increased over the parents.

In 1992, in the Philippines, Longid treated pepper seeds with solutions of sodium azide (0.12, 0.25, 0.50, 0.75 mM) in phosphate buffer at pH 3 for 2 h. In M_1 generation, overall decrease of germination percentage, height, and number of leaves was observed. In M_2 generation, multiple chlorophyll mutations were reported and the most frequent of them was of viridis type followed by chlorina, xantha, and albina.

One of the most extensive researches of sodium azide mutagenicity in pepper was presented in India by Sharma (1994). He tested three varieties with sodium azide (0.01%, 0.02%, 0.03%, 0.04%), EMS (0.1%, 0.2%, 0.3%, 0.4%), and MMS (0.01%, 0.02%, 0.03%, 0.04%) soaked for 6 h and after another 6 h pre-soaked in water. In M_2 generation, he reported that sodium azide had the lowest frequency of chlorophyll mutations and morphologically as well. He noticed that overall the mutants induced with sodium azide had lower plant height in all three varieties compared to other mutagens.

A recent study in Nigeria by Omeke (2021) treated pepper seeds of varieties "Shombo" and "Tatase" with sodium azide (0.01%, 0.02%, 0.03%, 0.04%) for 6 h after another 6-h pre-soaking in water. They evaluated the selected morphological parameters in M_1 generation. The results showed that the M_1 mutants from both varieties had decreased height and such height decrease had dose dependence—the higher the mutagen concentration induced, the lower the plant height. Similar trend was observed with the number of leaves per plant.

20.3.4.2 Caffeine

Caffeine (see Fig. 20.8), mostly popular as a compound of the famous coffee, tea, Coca-Cola drinks, and chocolate, is also known to have mutagenic and cytotoxic effects. It is alkaloid in some plants as coffee and cocoa and is a purine (l,3,7-trimethyl xanthine). Having similarity with some DNA bases, it can impair the replication and repair of DNA and DNA nucleotide precursor pool balances, and under some conditions, it can be incorporated in DNA replacing adenine and guanine. Besides that, it can negatively affect cytokinesis (Haynes and Collins 1984; Kuhlmann et al. 1968).

In Romania, Rosu et al. (2006) germinated dry sweet pepper seed varieties “Cosmin” and “Export” in a solution containing caffeine (0.025%, 0.05%, 0.1%, 0.25%). After the germination, the authors did cytological analysis of the root cells and found overall decrease of the mitotic index and also presence of division aberrations, of which the most common were the anaphases and telophases with chromosomal bridges.

In Brazil, Montes et al. (2014) studied the caffeine effect on the young plant growth in sweet pepper variety “Misano” (“California Wonder” type). The caffeine (2.25 and 9.0 μM) was added to the fertilizer. They measured the level of caffeine accumulation in plants and found it to be more than 300 μM. The evaluation of the caffeine impact showed that there were no significant differences in the monitored morphological parameters during the plant development in two vegetational seasons.

In another recent study in India, Aslam et al. (2017) germinated pepper seeds in caffeine (0.10%, 0.25%, 0.50%, 0.75%, and 1.0%). In M_1 generation, reduction in the germination, survivability, pollen fertility, and root and shoot length was observed with the concentration increase of the caffeine. The evaluation in M_2 generation revealed a slight increase of the size of the mutants compared with the controls at lower concentrations, which trend was reversed at high concentrations. Similar trend was observed with the fruits and yield. There was dose-dependent increase of the chromosome anomalies as changes in chromosome number, structure, base substitution, and deletion. Many and various flower mutations were

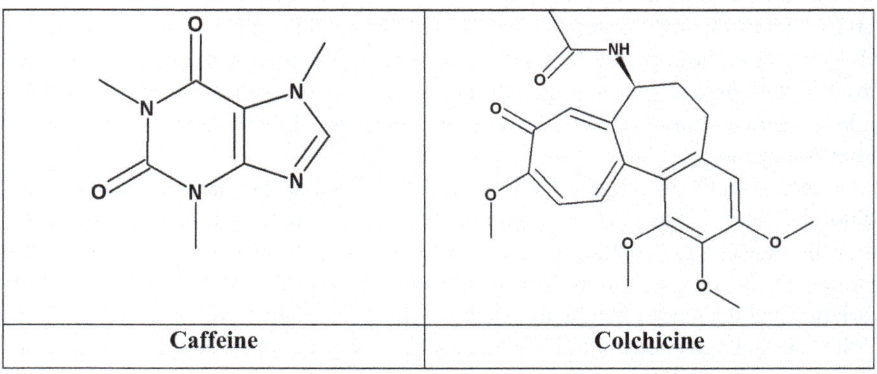

Caffeine	Colchicine

Fig. 20.8 Some miscellaneous mutagens in sweet pepper mutation breeding

observed as changes in number of petals, sepals, anther size, color, etc. Both mutagenic efficiency and effectiveness decreased with the increase of concentrations. Cytological aberrations were lower at the final stages of the cell division. RAPD analysis revealed considerable polymorphism in the observed M_2 mutants; almost half from the acquired 93 bands were polymorphic.

20.3.4.3 Colchicine

Although in strict sense colchicine (see Fig. 20.8) does not directly damage DNA and its main action is on the cell's division apparatus, it still acts on the cellular hereditary material mostly causing polyploidy. Some authors report that it has also mutachromosomic ability in hot pepper (Rao and Kumar 1983). Hence, it is considered mostly as a chemical mutagen *sensu lato* or as a polyploidizing agent (FAO 1977; Ingelbrecht et al. 2018). Many plant breeders try to benefit from its properties since polyploid progeny sometimes is more vigorous, bigger, etc. than the parents.

Colchicine for breeding purposes in sweet pepper is mostly used for treatment of various cell materials as in Hungary by Mitykó and Gémes Juhász (2006); there are some studies that treat seeds and young seedlings in an attempt to obtain polyploid organism.

In India, Rao and Kumar (1983) attempted to obtain from *Capsicum annuum* var. *cerasiformis* (now part of var. *grossum*) (Basu and De 2003) seeds of tetraploid plants by treatment of the seeds with colchicine (0.2%, 0.4%) for 24 and 72 h. They could not get tetraploid plants but observed chromosome aberrations. An year later, the same authors plus Panda obtained full tetraploid and even octoploid plants by treatment of the seedlings with 0.3% colchicine (Panda et al. 1984). The tetraploid plants were bigger than the octoploid ones.

In China, Xie and Deng (2009) treated red bush pepper with colchicine (0.01%, 0.03%, 0.05%) and treatment periods (2, 4, 6 days). They obtained plants with varying levels of mixoploidy. These plants were with thicker and bigger leaves and stems than the controls. The authors concluded that the best treatment combination is 0.05% colchicine with 6-day duration, at which the induction rate could reach 26.1%.

In Indonesia, Tammu et al. (2021) treated dry seeds of mild red pepper ("Katokkon"—bell shaped) with colchicine (0.025%, 0.05%, 0.075%, 0.1%) for 24 h and afterwards washed and planted. Tissues form 2-month-old plants were analyzed with flow cytometry, and the results showed that 50% were diploid ($2n$) and the rest had various levels of mixoploidy ($2n + 4n$) most pronounced at the lowest and highest concentrations of colchicine. The treated plants were 20–30% taller in all groups, while the size of the leaves had mixed variation, but the differences with the controls were within 10% range. The fruits in general were smaller than the controls but were 2–3 times more numerous.

20.4 Mutagenesis with Combination of Mutagens

Besides the so far mentioned application of single mutagens in plant breeding, there are studies which try combining more than one mutagen sometimes of different classes, i.e., physical and chemical ones, in order to explore or boost or mitigate their effects, action, and results. Such scheme is more common among the hot pepper mutation breeding research (Kulshrestha 2000; Kumar and Ponnuswami 2010). Kulshrestha (2000) found that the combination of gamma irradiation and EMS gives the highest mutation rate, but also significantly impairs the chromosomes, but also noted that the different varieties respond differently to it.

In the abovementioned study of Mehta (1998) in India, 11 combinations of two to five different chemical mutagens were used (colchicine, EMS, DMS, DES, SA, ethidium bromide). The author found that overall combined treatment had a boosting effect of mutation gain, but it had mixed effects on plant viability, size, and morphological parameters depending on the particular set of mutagens.

In Bulgaria, Tomlekova et al. (2007b; 2009) applied EMS treatment and gamma irradiation upon seeds of the commonly used sweet pepper variety "Hebar." As a result, anthocyanin-free plants were found at the seedling stage and their phenotype has been further screened. In M_2 generation, mutants with the following traits in the phenotype were selected: lack of anthocyanins, erected fruit, erected fruit in a bunch, and altered fruit shape. The stability of these traits was confirmed till M_4 generation, and they were used for the development of advanced mutant lines.

20.5 Summary of the Achievements and Some Perspectives of the Traditional Mutation Breeding in Sweet Pepper

20.5.1 Registered New Mutant Varieties

So far, many varieties have been released which were developed through mutation breeding. A short list is provided in Table 20.1.

20.5.2 Remarks on the Future of the Traditional Mutation Breeding in Sweet Pepper

The agriculture and food industry are confronted with the need to solve a number of problems associated with global challenges such as food production crisis and declining confidence in food quality and food safety. Today, researchers in the field of sweet pepper production are faced with more complex issues related to low productivity and quality of the local varieties, leading to lack of robust vegetable production and thereafter reliance on import of pepper in many countries.

In this review, we presented the essence of the work of many researchers from all over the globe since the beginning of the mutation breeding research in sweet pepper, which started almost 80 years ago in India. We saw amazing studies

Table 20.1 List of the officially registered new mutant varieties (the list has no claim to be comprehensive although the authors did their best to bring it there)

Variety	Year	Reference	Country	Mutagen	Main improved attributes
"Albena"	1976	Daskalov (1975)	Bulgaria	Gamma, 135 Gy	Early and high yielding, more attractive fruits, better flavor because of lack of anthocyanin
"Krichimski ran"	1972	Daskalov and Milkova (1976)	Bulgaria	X-rays, 120 Gy	Hybrid variety, high yield, early, improved fruit quality
"Lyulin"	1982	Milkova and Daskalov (1983)	Bulgaria	Gamma, 135 Gy	Hybrid variety based on induced male sterility, early maturity, high yield
"Friari KS80"	1985	MBNL (1991)	Italy	0.6% EMS, 13 h	Increased fruit production (15–20%), tolerance to *Verticillium dahliae* Kleb.
"Nush-51"	1991	MBNL (1994)	USSR	0.05% EI, 18 h	High yield and good quality
"Pirin"	1991	Todorova and Daskalov (1997)	Bulgaria	Gamma, 60 Gy	Resistance to powdery mildew, high yield, early
"Oranzheva kapia" (see Fig. 20.1)	1991	Daskalov and Baralieva (1992)	Bulgaria	X-rays, 120 Gy	High content of β-carotene (provitamin A)
"Gornooriahovska kapia"	1997	Daskalov (2001)	Bulgaria	Gamma, 135 Gy	Multiple disease resistance, high yield
"Yubileiny Semko"	1997	Timina (1997), Tomlekova et al. (2008)	Russia	Gamma, 135 Gy	High yield, disease resistance
"Vitamin"	2006	Timina and Timin (2006), Tomlekova et al. (2008)	Russia	Gamma, 135 Gy	High vitamin content and yield, disease resistance

(continued)

Table 20.1 (continued)

Variety	Year	Reference	Country	Mutagen	Main improved attributes
"Yujiao 1"	2002	MVD (2021)	China	Aerospace	Good quality and altered shape
"Yujiao 2"	2006	MVD (2021)	China	Aerospace	Large fruit, good quality and yield, disease resistance
"Oranzheva Krasavitza"	2011	Timina et al. (2011)	Russia	X-rays, 120 Gy	High β-carotene, high content of vitamin C, resistance to *Verticillium* disease
"Quinta"	2019	Timin and Timina (2020)	Russia	Gamma, 135 Gy	Early, high vitamin content and yield, disease resistance
"Desislava" (see Fig. 20.9)	2020	Tomlekova et al. (unpublished)	Bulgaria	X-rays, 120 Gy	Early ripening, high β-carotene, orange fruit, good quality and yield
"Toniko" (see Fig. 20.9)	Submitted for registration	Tomlekova et al. (unpublished)	Bulgaria	X-rays, 120 Gy	High beta-carotene, good quality and yield

which tried to get the utmost from the newest advances and technologies in biology and molecular biology specifically. However, when looking at a global scale, we also saw a dramatic gap in the adoption of these technologies between the economically developed and the developing countries.

So the future of the traditional mutation breeding in sweet pepper should go with at least two goals in mind—how to get the utmost from the ethically safer random mutation wealth that the traditional mutation breeding offers and how to make the mutation breeding much more accessible so that its beneficiary effects at the end can be felt even by the poorest human beings who try to make living from this crop. In regard to the first goal, new technologies have been added to the arsenal of the traditional mutation plant breeders in order to get the utmost from the invested labor, time, and money.

One such technology is the bioinformatics post-processing of the mutant phenotype trait profiles by screening for analogous matches with similar traits with already

Fig. 20.9 Phenotype of the new mutant varieties/candidate varieties of sweet pepper in Bulgaria: (**a**). cv. Desislava—unripe pale yellow fruit; (**b**). cv. Desislava—ripe orange fruit; (**c**). cv. Toniko—ripe red fruits

determined gene involvement from other better studied organisms as *Arabidopsis* as the team of prof. I. Paran applied (Paran et al. 2007). But we are yet to start building sweet pepper functional genomics database as the one started for tomato (Fei et al. 2010), which hopefully will further facilitate such type of analysis.

Less than a decade ago started the whole-genome sequencing of pepper and several reference and annotated genome databases have become available since then (Kim et al. 2014; Popatanasov and Tomlekova 2021). This hopefully will speed up the process of discovery of gene functions and development and better implementation of molecular marker methods, which rely on such knowledge and have proven to be very useful in breeding (Popatanasov and Tomlekova 2021). Our lab along with other ones had taken some steps in the utilization of this wealth as in the analysis and development of molecular markers (Cheng et al. 2016; Popatanasov and Tomlekova 2019; Uncu 2019), mutant gene identification and analysis (Borovsky et al. 2020; Tomlekova et al. 2021), etc. However, there is room for a lot more to be done in order for these achievements to be used routinely and affordably in the sweet pepper breeding practice.

Another technology from which the sweet pepper mutation breeding can greatly benefit is the TILLING, which was successfully adopted and applied in the pilot study by the team of Siddique et al. (2020). The authors used a large pool of two mutated varieties, and TILLING technology could detect and determine nine mutations in one gene. Such results can greatly facilitate and speed up pepper breeders' work. Currently, the cost and complexity of these technologies are prohibitive factors for many pepper researchers however.

However, we are yet to see more newer methods for analysis and selection of mutants applied in other crops to be introduced in sweet pepper as some from the arsenal of forward and reverse genetics like different TILLING modifications, large-scale screening of mutant populations through amplicon sequencing, whole-genome resequencing, etc. (Li et al. 2019; Sashidhar et al. 2020; Taheri et al. 2017). We are yet to see creation of mutant databases from sweet pepper varieties as the ones already available in tomato, barley, etc. (Chaudhary et al. 2019; Szurman-Zubrzycka et al. 2018). One of the reasons that sweet pepper is lagging behind is its few-fold larger genome compared to other major *Solanaceae* crops and its lesser economic importance (Popatanasov and Tomlekova 2021). Hopefully, this will change in the future since there is relatively steady increase of the interest and production demand of this crop and the costs of genomic technologies and services continuously drop (FAO 2021; Salgotra and Zargar 2020).

As for the goal of how to make the mutation breeding problem solver accessible even for the poorest farmers and to tackle their sweet pepper farming challenges and problems, there are some steps already made in this direction. In fact, most of the presented research and use of the mutation breeding in sweet pepper started after the initiative of one of the most influential humanitarian organizations as is the FAO at the UN in collaboration with the IAEA in 1964 (Toker et al. 2007). However, here we see very pronounced unpleasant tendency in the developing countries, where the mutation breeding experiments end after M_2 and even after M_1 generations, which probably is due to insufficient funding. Such scenarios are not favorable neither for the users there nor for the future use of these methods in these countries. The IAEA has taken some steps as establishing facilities for mutagenesis induction with the needed training as in Nigeria (Muhammad et al. 2018).

Development of simpler low-cost methods can further promote the mutation breeding research in countries with limited budgeting. Some of the researchers as S. Daskalov and N. Tomlekova (Chalukova et al. 1993; Tomlekova et al. 2016a, b) addressed this issue and showed that less expensive methods as TLC or molecular markers can produce very useful results and information from the obtained sweet pepper mutants and their hybrids. So, the traditional sweet pepper mutation breeding probably will continue to evolve and develop in the future as it proved to be able to solve some of the challenges that this crop faces in our dynamic world and since it is one of the very few currently ethically acceptable and worldwide adopted accelerated breeding methods in sweet pepper.

References

Abe T, Hayashi Y, Morita R, Ichida H (2018) Ion beam mutagenesis: an innovative and effective method for plant breeding and gene discovery. In: Proceedings of the FAO/IAEA international symposium on plant mutation breeding and biotechnology, 27–31 August 2018, Vienna, Austria, No. IAEA-CN—263-302

Abilev SK (2012) Chemical mutagens and genetic toxicology. Priroda 10:39–46. ISSN 0032-874X. (Russian)

Abu NE, Ojua EO, Udensi OU (2020) Induction of variability in three Nigerian pepper varieties using gamma irradiation. J Exp Agric Int 42(4):111–119

Adekola OF, Oluleye F (2012) Induced morphological variations among Capsicum annum 'Tatase' mutants. Product Agric Technol J 8(2):50–58. ISSN 0794-5213

Adhikari P, Khanal A, Subedi R (2016) Effect of different sources of organic manure on growth and yield of sweet pepper. Adv Plant Agric Res 3(5):111–118

Alcantara TP, Bosland PW, Smith DW (1996) Ethyl methanesulfonate-induced seed mutagenesis of *Capsicum annuum*. J Hered 87(3):239–241

Araki M, Ishii T (2015) Towards social acceptance of plant breeding by genome editing. Trends Plant Sci 20(3):145–149

Arias P, Nunez G, Fernandez R (1998) Radiosensibilidad de la variedad de pimiento California Wonder a los rayos gamma del Co-60. In: Cien-R. Institute de Investigaciones Agropecuarias "Jorge Dimitrov" Bayamo, La Habana, Cuba

Arisha MH, Liang BK, Shah SM, Gong ZH, Li DW (2014) Kill curve analysis and response of first generation *Capsicum annuum* L. B12 cultivar to ethyl methane sulfonate. Genet Mol Res 13: 10049–10061

Arisha MH, Shah SN, Gong ZH, Jing H, Li C, Zhang HX (2015) Ethyl methane sulfonate induced mutations in M_2 generation and physiological variations in M_1 generation of peppers (*Capsicum annuum* L.). Front Plant Sci 6:399

Aslam R, Bhat TM, Choudhary S, Ansari MYK, Shahwar D (2017) Estimation of genetic variability, mutagenic effectiveness and efficiency in M_2 flower mutant lines of *Capsicum annuum* L. treated with caffeine and their analysis through RAPD markers. J King Saud Univ Sci 29(3):274–283

Auerbach C (1973) History of research on chemical mutagenesis. In: Hollaender A (ed) Chemical mutagens. Springer, Boston, MA, pp 1–19

Auerbach C, Robson JM (1946) Chemical production of mutations. Nature 157(3984):302–302

BAACVD (2021) Beijing Academy of Agriculture Crop Variety Database. http://www.agridata.ac.cn:8888/web/default.aspx. Accessed 25 Aug 2021

Bado S, Forster BP, Nielen S, Ali AM, Lagoda PJ, Till BJ, Laimer M (2015) Plant mutation breeding: current progress and future assessment. Plant Breed Rev 39:23–88

Balaur NS (1978) On the mutagenic effect of laser radiation. In: Arystanbekov HA (ed) Problems of photoenergetics of plants. V. Kazakh Agricultural Institute Press, Alma Ata, pp 142–150. (Russian)

Bansal HC (1969) Induced chilli mutants containing high capsaicins and low carotenoids in fruits. In: Proceedings of the symposium on radiations and radiomimetic substances in mutation breeding, food and agriculture. Commission of the Dept. of Atomic Energy, India, pp 285–292

Bartoshevich YE (1966) Molecular mechanisms of alkylation of a biological system. Bull Moscow Soci Naturalists Biol Div 71(5):70–87. (Russian)

Basu SK, De AK (2003) Capsicum: historical and botanical perspectives. In: De AK (ed) Capsicum. CRC Press, New York, pp 21–35

Batikyan GG, Sahakyan TA, Terzyan RT (1970) The effect of pre-sowing X-ray irradiation of seeds on the growth and development of pepper plants (*Capsicum annuum*). Biol J Armenia 23(3): 13–17. (Russian)

Batikyan GG, Gukasyan LA, Sarkisyan SA, Abramyan LK (1975) Ultrastructure of chloroplasts of pepper leaves with chlorophyll changes when exposed to nitrosomethylurea. Biol J Armenia 28(5):12–16. (Russian)

Berns MW, Olson RS, Rounds DE (1969) *In vitro* production of chromosomal lesions with an argon laser microbeam. Nature 221(5175):74–75

BG Fermer (2020) 5 new varieties of vegetables of IZK "Maritsa" are waiting for a patent. BG Fermer News (online), 2 June 2020. https://www.bgfermer.bg/article/8647866. Accessed 25 Aug 2021 (Bulgarian)

Borovsky Y, Tadmor Y, Bar E, Meir A, Lewinsohn E, Paran I (2013) Induced mutation in β-CAROTENE HYDROXYLASE results in accumulation of β-carotene and conversion of red to orange color in pepper fruit. Theor Appl Genet 126(3):557–565

Borovsky Y, Sharma VK, Verbakel H, Paran I (2015) CaAP2 transcription factor is a candidate gene for a flowering repressor and a candidate for controlling natural variation of flowering time in *Capsicum annuum*. Theor Appl Genet 128(6):1073–1082

Borovsky Y, Mohan V, Shabtai S, Paran I (2020) CaFT-LIKE is a flowering promoter in pepper and functions as florigen in tomato. Plant Sci 301:110678

Bosland PW (1996) Capsicums: innovative uses of an ancient crop. In: Janick J (ed) Progress in new crops. ASHS Press, Arlington, VA, pp 479–487

Bosland PW (2002) Inheritance of a novel flaccid mutant in *Capsicum annuum*. J Hered 93(5): 380–382

Breuils G, Pochard E (1975) Essais de fabrication de l'hybride de piment Lamyo-INRA avec utilization d'une stérilité maie génétique (me 509). Ann Amelior Plant (Paris) 25:399–409

Buiatti M, Ragazzini R (1966) The mutagenic effect of acridine orange in tomato (*Lycopersicum esculentum*). Mutat Res 3(4):360–361

Burnichi F, Pârvu G, Danailaguidea S, Petru N, Ristici E, Marian R (2011) Research regarding the influence of laser radiation upon the growth and development of some vegetable species (eggplant and sweet pepper). Rom Biotechnol Lett 16(1):27

Chalukova M, Daskalov S, Lukarska E, Baralieva D (1993) Beta-orange mutant in pepper (*Capsicum annuum* L.). Capsicum Eggplant Newsl 12:57–58

Chaudhary J, Alisha A, Bhatt V, Chandanshive S, Kumar N, Mir Z, Kumar A, Yadav SK, Shivaraj SM, Sonah H, Deshmukh R (2019) Mutation breeding in tomato: advances, applicability and challenges. Plants 8(5):128

Cheng J, Zhao Z, Li B, Qin C, Wu Z, Trejo-Saavedra DL, Luo X, Cui J, Rivera-Bustamante RF, Li S, Hu K (2016) A comprehensive characterization of simple sequence repeats in pepper genomes provides valuable resources for marker development in Capsicum. Sci Rep 6(1):1–12

Cohen O, Borovsky Y, David-Schwartz R, Paran I (2012) CaJOINTLESS is a MADS-box gene involved in suppression of vegetative growth in all shoot meristems in pepper. J Exp Bot 63(13): 4947–4957

Cristinzio G, Saccardo F (1982) Resistance in Capsicum to *Phytophthora capsici* introduced by using gamma-rays. Capsicum Newsl 1:70

Cristinzio G, Saccardo F (1984) Resistance in a mutant of pepper to six isolates of *Phytophthora capsici*. Capsicum Newsl 3:31–32

Csilléry G (2006) Pepper taxonomy and the botanical description of the species. Acta Agron Hun 54(2):151–166

Csilléry G (2013) Introduction to a collection of generative organs pepper mutants. In: Proceedings of the XV EUCARPIA meeting on genetics and breeding of capsicum and eggplant "Breakthroughs in the genetics and breeding of capsicum and eggplant", 2–4 September, 2013, Torino, Italy, p 215

Dagadkhair MR (2020) Health benefits of capsicum (bell pepper). In: Dagadkhair RA, Shinde DB, Shelke SA, Kale (eds) Vegetables & its health benefits. Taurean Publications, New Delhi, pp 11–17

Daskalov S (1968) A male sterile pepper (*C. annuum* L.) mutant. Theor Appl Genet 38(8):370–372. https://doi.org/10.1007/bf00934170

Daskalov S (1973a) Male sterile mutants in sweet pepper (*Capsicum annuum*). Mutat Breed Newsl 3:13

Daskalov S (1973b) Investigations of induced mutants in *Capsicum annuum* L. III. Mutants in the variety Zlaten medal 7. Genet Plant Breed 6:419–429. (Bulgarian)

Daskalov S (1975) Albena: a new mutant cultivar in sweet pepper for early and mid-early field production. Horticulture 12:22

Daskalov S (1986) Mutation breeding in pepper. Mutat Breed Rev 4:1–26

Daskalov S (2001) 'Gornooriahovska Kapia F_1'-a new hybrid pepper variety based on radiation induced male sterility. Mutat Breed Newsl 45:4–5

Daskalov S, Baralieva D (1992) Oranzheva kapiya: a new cv. in sweet pepper with increased beta-carotene (provitamin A) content. Mutat Breed Newsl 39:9–10

Daskalov S, Milkova L (1976) Krichiminski ran - a new heterosis cultivar in sweet pepper. Gradinarstvo 10:22–23. (Bulgarian)

Eiges NS, Volchenko GA, Volchenko SG (2014) "Chemical mutagenesis" and "Phenotypic activation of enzymes" are the largest discoveries of the twentieth century. To the birthday of I. A. Rapoport. Hist Pedagogy Nat Sci 1:55–60. (Russian)

Elitzur T, Nahum H, Borovsky Y, Pekker I, Eshed Y, Paran I (2009) Co-ordinated regulation of flowering time, plant architecture and growth by FASCICULATE: the pepper orthologue of SELF PRUNING. J Exp Bot 60(3):869–880

FAO (1977) Manual on mutation breeding. International Atomic Energy Agency (IAEA), Vienna

FAO (2021) FAOSTAT. https://www.fao.org/faostat/en/#data/QCL. Accessed 1 Aug 2021

Fei Z, Joung JG, Tang X, Zheng Y, Huang M, Lee JM, McQuinn R, Tieman DM, Alba R, Klee HJ, Giovannoni JJ (2010) Tomato Functional Genomics Database: a comprehensive resource and analysis package for tomato functional genomics. Nucleic Acids Res 39(Suppl_1):D1156–D1163

Fonseca AA, Suárez LC, Fernández RR, Prado WE, Labrada YE, Rodríguez AM (2013) Efecto del tratamiento de semillas con bajas dosis de rayos X en plantas de pimiento (*Capsicum annuum* L.). Nucleus 53:14–18

Fujii T, Matsumura S (1959) Radiosensitivity in plants III. Experiments with several polyploid plants. Jpn J Breed 9(4):245–252

Galukyan MG (1968a) Study of the cytological changes in pepper (*Capsicum annuum* L.) under the influence of chemical mutagens. Biol J Armenia 21:59–65. (Armenian)

Galukyan MG (1968b) Experimental mutagenesis in pepper. Communication I. The effect of chemical mutagens – EI, NMU, NEU on pepper in M_1. Biol J Armenia 21(9):53–62. (Armenian)

Galukyan MG (1969) Study of the effect of chemical (EI, NMU and NEU) mutagens on pepper (*Capsicum annuum* L.) in M_2. Biol J Armenia 22(2):72–80. (Armenian)

Gukasyan LA, Akopyan DI (1974) The effectiveness of the action of nitrosomethylurea on pepper seeds. Biol J Armenia 27(2):46–51. (Russian)

Gukasyan LA, Akopyan DI (1975) Mutagenic effect of nitrosomethylurea on *Capsicum annuum* L. Biol J Armenia 28(1):44–50. (Russian)

Gukasyan LA, Tumanyan ER (1976) Cytogenetic study of pepper under the action of nitrosomethyl urea in M_2 and M_3. Sci Notes Yerevan State Univ 1:94–98. (Russian)

Gukasyan LA, Tumanyan ER (1977) Anatomical examination of the stem and root of the under-sized mutant *Capsicum annuum* L. Biol J Armenia 30(6):97–101. (Russian)

Guo Y, Xie L, Wang X, Deng L (2004) Technological innovation of induced breeding of space and selection of new variety on pepper. Acta Agriculturae Nucleatae Sinica 18(4):265–268. ISSN 1000-8551

Haynes RH, Collins JDB (1984) The mutagenic potential of caffeine. In: Dews PB (ed) Caffeine. Springer, Berlin, pp 221–238

Hermelin T, Brunner H, Daskalov S, Nakai H (1983) Chimerism in M_1 plants of *Vicia faba*, *Capsicum annuum* and *Linum usitatissimum*. In: Chimerism in irradiated dicotyledonous plants. IAEA-TECDOC-289, pp 35–42

Hernandez AC, Dominguez PA, Cruz OA, Ivanov R, Carballo CA, Zepeda BR (2010) Laser in agriculture. Int Agrophys 24(4):407–422

Hoffmann GR (1980) Genetic effects of dimethyl sulfate, diethyl sulfate, and related compounds. Mutat Res 75(1):63–129

Holme IB, Gregersen PL, Brinch-Pedersen H (2019) Induced genetic variation in crop plants by random or targeted mutagenesis: convergence and differences. Front Plant Sci 10:1468

Honda I, Kikuchi K, Matsuo S, Fukuda M, Saito H, Ryuto H, Abe T (2006) Heavy-ion-induced mutants in sweet pepper isolated by M_1 plant selection. Euphytica 152(1):61–66

Hu X, Li Y, Gao Y, Luo A, Zhao D, Hu X (2010) Review and prospect of space mutation application in pepper breeding. China Veg 24:14–18

Hwang D (2014) Development of an EMS mutant population and isolation of non-pungent mutants in *Capsicum annuum*. Doctoral dissertation, Seoul National University Graduate School, Seoul, Republic Korea

Hwang D, Jeong HJ, Kwon JK, Kim H, Kang SY, Kang BC (2014) Phenotypic variants among ethyl methanesulfonate M_2 mutant lines in *Capsicum annuum*. Plant Genet Resour 12(S1): S141–S145

IAEA (2021) FAO/IAEA Mutant Variety Database (MVD). https://mvd.iaea.org/#!Search. Accessed 25 Aug 2021

Ingelbrecht I, Jankowicz-Cieslak J, Szurman M, Till BJ, Szarejko I (2018) Chemical mutagenesis. In: Spencer-Lopes MM, Forster BP, Jankuloski L (eds) Manual on mutation breeding. Food and Agriculture Organization of the United Nations (FAO), Rome, Italy, pp 51–82

Irikova T, Grozeva S, Rodeva V (2011) Anther culture in pepper (*Capsicum annuum* L.) *in vitro*. Acta Physiol Plant 33:1559–1570

Jeifetz D, David-Schwartz R, Borovsky Y, Paran I (2011) CaBLIND regulates axillary meristem initiation and transition to flowering in pepper. Planta 234(6):1227–1236

Jeong HJ, Kwon JK, Pandeya D, Hwang J, Hoang NH, Bae JH, Kang BC (2012) A survey of natural and ethyl methane sulfonate-induced variations of eIF4E using high-resolution melting analysis in Capsicum. Mol Breed 29(2):349–360

Jiang L, Zhang B, Chen W, Qiu T, Wang L, Yang S (2020) Space breeding in modern agriculture. Am J Agric Res 5:81. ISSN: 2475-2002

Jiyuan L, Zhenye Q, Yongcheng S, Tianjun Q, Jun H (1999) Seed growth experiments after space flight: the Chinese experience. In: Seminars on the United Nations programme on space applications, vol 10, pp 71–74. ISBN: 9211008107

Jo YD, Kim SH, Hwang JE, Kim YS, Kang HS, Kim SW, Kang SY (2016) Construction of mutation populations by gamma-ray and carbon beam irradiation in chili pepper (*Capsicum annuum* L.). Hortic Environ Biotechnol 57(6):606–614

Jo YD, Kang HS, Ryu J, Kim JB, Kim SH, Kang SY (2018) Development and screening of genetic variants on radiation mutation techniques for breeding and functional genomics study in pepper (No. IAEA-CN—263-108). In: Proceedings of the FAO/IAEA international symposium on plant mutation breeding and biotechnology, 27–31 August 2018, Vienna, Austria

Joshi MM, Khalatkar AS (1983) Effect of neutrons and gamma radiations on the floral apparatus in M_2 generation. Capsicum Eggplant Newsl 2:57–59

Junmin Z (1999) Preliminary results on inheritance and variation of sweet pepper progenies after space flight. Heilongjiang Agric Sci 3. ISSN 1002-2767. (Chinese)

Khan S, Al-Qurainy F, Anwar F (2009) Sodium azide: a chemical mutagen for enhancement of agronomic traits of crop plants. Environ We 4:1–21

Khursheed S, Wani MR, Khan S, Bhat TA (2021) Mutagens, their types and mechanism of action with an emphasis on sodium azide and gamma radiations. In: Bhat TA (ed) Mutagenesis, cytotoxicity and crop improvement: revolutionizing food science. Cambridge Scholars Publishing, Newcastle upon Tyne, pp 1–37

Kim S, Park M, Yeom S, Kim YM, Lee JM, Lee AH, Seo E, Choi J, Cheong K, Kim TK, Lee WG, Oh KS, Bae C, Kim SB, Lee YH, Kim SY, Kim MS, Kang CC, Jo DY, Yang HB, Jeong JH, Kang HW, Kwon JK, Shin C, Lim YJ, Park HJ, Huh HJ, Kim SJ, Kim DB, Cohen O, Paran I, Suh CM, Lee BS, Kim KY, Shin Y, Noh JS, Park J, Seo SY, Kwon YS, Kim AH, Park MJ, Kim JH, Chio SH, Lee SM, Yu Y, Choi DY, Park SB, Deynze A, Ashrafi H, Hill T, Kim TW, Pai SH, Ahn KH, Yeam I, Giovannoni JJ, Rose KJ, Sorensen TW, Lee JS, Kim WR, Choi YI, Choi SB, Lim SJ, Lee HY, Doil C (2014) Genome sequence of the hot pepper provides insights into the evolution of pungency in Capsicum species. Nat Genet 46:270–278

Kuhlmann W, Fromme HG, Heege EM, Ostertag W (1968) The mutagenic action of caffeine in higher organisms. Cancer Res 28(11):2375–2389

Kulshrestha S (2000) Studies on the induction of mutations with physical and chemical mutagens in Capsicum species. Doctoral dissertation, Aligarh Muslim University, Aligarh, India

Kumar G, Gupta P (2009) Induced karyomorphological variations in three phenodeviants of Capsicum annuum L. Turk J Biol 33(2):123–128

Kumar GA, Ponnuswami V (2010) Mutagenic effectiveness and efficiency of gamma rays and ethyl methane sulphonate and their combined treatments in paprika (*Capsicum annuum* L.). Veg Sci 37(2):129–131

Lee J (2018) Genetic analysis of a Nonpungent EMS mutant in pepper (*Capsicum annuum* L.). Doctoral dissertation, Seoul National University Graduate School, Seoul, Republic Korea

Leitao J (2012) Chemical mutagenesis. In: Shu QY, Forster BP, Nakagawa H (eds) Plant mutation breeding and biotechnology. CABI and FAO, Rome, Italy, pp 135–158

Levy A, Ashiri A (1975) Ethidium bromide - an efficient mutagen in higher plants. Mutat Res 28(3): 397–404

Li F, Shimizu A, Nishio T, Tsutsumi N, Kato H (2019) Comparison and characterization of mutations induced by gamma-ray and carbon-ion irradiation in rice (*Oryza sativa* L.) using whole-genome resequencing. G3 9(11):3743–3751

Liu M, Li J, Wang Y, Zhang Z, Xue H, Zhang C, Li G (1999) Preliminary study on peroxidase isoenzyme detection and RAPD molecular verification for sweet pepper 87-2 carried by a recoverable satellite. Acta Agric Nucl Sin 13(5):291–294. (Chinese)

Liu LX, Guo HJ, Zhao L, Gu J, Zhao S (2008) Advances in crop improvement by space mutagenesis in China. In: Proceedings of the 5th International Crop Science Congress (ICSC), Jeju, Republic of Korea, vol 4, p 274

Longid CS (1992) Mutagenic effects of sodium azide on *Capsicum annuum* L. Sci Diliman (Philippines) 4(1):25–42

Loveless A, Howarth S (1959) Mutation of bacteria at high levels of survival by ethyl methane sulphonate. Nature 184(4701):1780–1782

Malik AA, Chattoo MA, Sheemar G, Rashid R (2011) Growth, yield and fruit quality of sweet pepper hybrid SH-SP-5 (*Capsicum annuum* L.) as affected by integration of inorganic fertilizers and organic manures (FYM). J Agric Technol 7(4):1037–1048

Maluszynski M (2001) Officially released mutant varieties-the FAO/IAEA Database. Plant Cell Tissue Organ Cult 65(3):175–177

Maryina T (2021) Lastochka in salad. Saint-Petersburg vedomosti. No. 155 (6993). (Russian)

Masheva S, Mihov M (2009) Trends of the research work in vegetable crops and potatoes in Bulgaria. Acta Hortic 830:45–52. https://doi.org/10.17660/ActaHortic.2009.830.4

Mba C (2013) Induced mutations unleash the potentials of plant genetic resources for food and agriculture. Agronomy 3(1):200–231

Mba C, Afza R, Shu QY (2012) Mutagenic radiations: X-rays, ionizing particles and ultraviolet. In: Shu QY, Forster BP, Nakagawa H (eds) Plant mutation breeding and biotechnology. CABI and FAO, Rome, Italy, pp 83–90

MBNL (1991) List of cultivars. Mutat Breed Newsl 37:22

MBNL (1994) List of new mutant cultivars. Mutat Breed Newsl 41:25

Mehta PJ (1998) Chemical mutagenesis in chillies (*Capsicum annuum* L.). Doctoral dissertation, Gujarat Agricultural University, Anand, India

Milkova L, Daskalov S (1983) Lyulin - a hybrid pepper cultivar for early field production. In: Proceedings of the 5th EUCARPIA meeting of the Capsicum and Eggplant Working Group, 4–7 July, 1983, Plovdiv, Bulgaria, p 202

Mitrofanov VG (2001) Ioseph Abramovich Rapoport - scientist, warrior, citizen. Essays, memoirs, materials. Nauka, Moscow, Russia. (Russian)

Mitykó J, Gémes Juhász A (2006) Improvement in the haploid technique routinely used for breeding sweet and spice peppers in Hungary. Acta Agron Hun 54(2):203–219

Mohan V, Borovsky Y, Kamara I, Zemach H, Paran I (2018) CaVIL1, a plant homeodomain gene that promotes flowering in pepper. Theor Appl Genet 131(12):2639–2649

Montes Ó, Diánez F, Camacho F (2014) Doses of caffeine on the development and performance of pepper crops under greenhouse. Hortic Bras 32:398–403

Muhammad ML, Falusi OA, Adebola MO, Daudu OAY, Abejide DR, Gado AA (2018) Overview of induced mutagenesis in Nigeria. In: Status of mutation breeding of crop plants in Nigeria. Lambert Academic Publishing, Beau Bassin, pp 6–17

Musaev FA, Zaharova OA, Morozova NI (2013) Pepper: botanical characteristics, classification, use in the food industry. FGBOU VPO RGATU Press, Ryazan, Russia. (Russian)

Nagai MA (1937) The production of mutations in *Drosophila melanogaster* with fast neutrons. Doctoral dissertation, William Marsh Rice Institute, Houston, USA

Nascimento KS, Rêgo MM, Nascimento AMM, Rêgo ER (2015) Ethyl methanesulfonate in the generation of genetic variability in Capsicum. In: Proceedings of the XXV International EUCARPIA symposium section ornamentals: crossing borders, 28 June–2 July, 2015, Melle, Belgium, pp 357–363

Navrotskaya LV, Zaginailov VI, Navrotskaya SR (2018) The effect of laser radiation on the seeds of agricultural crops. Int Tech Econ J 1:74–79

Neale S (1976) Mutagenicity of nitrosamides and nitrosamidines in micro-organisms and plants. Mutat Res 32(3–4):229–266

Nilan RA, Sideris EG, Kleinhofs A, Sander C, Konzak CF (1973) Azide—a potent mutagen. Mutat Res 17(1):142–144

Nushikyan VA, Martirosyan GS (1987) Descriptions of some economically useful red pepper mutants produced by the action of chemical mutagens. Chemical mutagenesis in the breeding process. Nauka Press, Moscow, pp 164–167. (Russian)

Oladosu Y, Rafii MY, Abdullah N, Hussin G, Ramli A, Rahim HA, Usman M (2016) Principle and application of plant mutagenesis in crop improvement: a review. Biotechnol Biotechnol Equip 30(1):1–16

Omeke JO (2021) Effect of sodium azide induction on morphological traits of shombo and tatase (*Capsicum annuum* L.). Niger Agric J 52(1):111–117

Panda RC, Kumar OA, Rao KR (1984) Cytomorphology of induced octoploid Chili pepper (*Capsicum annuum* L.). Theor Appl Genet 68(6):567–570

Paran I, Borovsky Y, Nahon S, Cohen O (2007) The use of induced mutations to study shoot architecture in Capsicum. Isr J Plant Sci 55(2):125–131

Pathirana R (2011) Plant mutation breeding in agriculture. CAB Rev Perspect Agric Vet Sci Nutr Nat Resour 6(032):1–20. https://doi.org/10.1079/PAVSNNR20116032

Penella C, Calatayud A (2018) Pepper crop under climate change: grafting as an environmental friendly strategy. In: Rao CS, Shanker AK, Shanker C (eds) Climate resilient agriculture: strategies and perspectives. IntechOpen, London, pp 129–155

Petrov V, Denev I, Draganov M, Timin O, Panchev I, Tomlekova NB (2013) Molecular characterization of advanced mutants for early detection of high β-carotene concentrations in pepper breeding programmes. C R Acad Bulg Sci 66(2):303–310

Pillai UPR (1998) Studies on the effect of laser radiation and other mutagens on plants. Doctoral dissertation, Cochin University of Science and Technology, Cochin, India

Pillai PS, Abraham S (1996) Improvement of fruit characters and yield in sweet pepper by mutation induction. Mutat Breed Newsl 42:17–18

Pochard E (1970) Induction de trois mutations de stérilité mâle chez le poivron rouge (*Capsicum annuum* L.) par traitement mutagène de matériel monoploïde. In: Proceedings of EUCARPIA "La sterilite male chez les plantes hort", Paris, France, pp 93–95

Popatanasov A, Tomlekova NB (2019) Simple sequence repeats in the genome of *Capsicum annuum* L. Acad J Ind Technol 6(1):18–23. ISSN: 13149911. (Bulgarian)

Popatanasov A, Tomlekova NB (2021) Bioinformatics and microsatellite marker development for Solanaceae crops breeding. Acad J Ind Technol 8(1):149–156. ISSN: 13149911

Prasad MVR (1972) A comparison of mutagenic effectiveness and efficiency of gamma rays, EMS, NMU and NG. Indian J Genet 32(3):360–367

Pylneva EV (1997) Development of varieties of agricultural crops. Priroda 1:21–26. ISSN 0032-874X. (Russian)

Pyshnaya ON, Mamedov MI, Pivovarov VF (2012) Pepper breeding. VNIISSOK Publishing House, Moscow, Russia. (Russian)

Qihan X, Henshaw GG (1989) Induction of chromosomal variation of plant cell *in vitro* by treatment of Actinomycin D. Acta Genet Sin 04. ISSN 0379-4172. (Chinese)

Raghavan TS, Venkatasubban KR (1940) Studies in the south Indian chilies. I. A description of the varieties, chromosome numbers and the cytology of some X-rayed derivatives in *Capsicum annuum* Linn. Proc Indian Acad Sci B 12(2):29–46. ISSN 0370-0097

Rajam MV, Nandy S, Pandey R (2021) Biotechnology of red pepper. In: Kishor PBK, Rajam MV, Pullaiah T (eds) Genetically modified crops, vol 2. Springer, Singapore, pp 53–83

Rankov V, Ilieva V (1986) Vegetative and reproductive performances of pepper following laser irradiation at various fertilizer rates. Bulg J Crop Sci 23(4):60–63. ISSN 0568-465X. (Bulgarian)

Rao KR, Kumar OA (1983) Colchicine induced interchanges in chillies (*Capsicum annuum* L.). Experientia 39(1):76–77

Rapoport IA (1946) Carbonyl compounds and the chemical mechanism of mutations. Dokl Akad Nauk SSSR 54(1):65–68. (Russian)

Rapoport IA (1948) Alkylation of the gene molecule. Dokl Akad Nauk SSSR 59(6):1183–1186. (Russian)

Rapoport IA (1962) Mutations in sex chromosome under influence of nitrosoethylurea. Dokl Akad Nauk SSSR 146(6):1418–1421. (Russian)

Restaino F (1983) Dwarf pepper types (*Capsicum annuum* L.) obtained by chemical mutagenesis. Capsicum Newsl 2:54–55

Rosu E, Ieremie I, Rusen FV, Bara I (2006) The influence of caffeine on mitotic division at *Capsicum annuum* L. Analele Stiintifice ale Universitatii "Alexandru Ioan Cuza" din Iasi Sec II a. Genetica si Biologie Moleculara 7(1):159–164

Rubzov M, Solomatin M (1974) Valuable forms of pepper. Kartofel' i Ovoshchi 1:30–31. (Russian)

Ruposhev AR (1976) Modification of the ethylenimine mutagenic effect in *Crepis capillaris*. TSitologiia i Genetika 10(2):111–113. ISSN 0095-4527. (Russian)

Saccardo F, Ramulu KS (1977) Mutagenesis and cross breeding in Capsicum for disease resistance against *Verticillium dahliae*. In: Proceedings of the 3rd Congress EUCARPIA, 5–8 July 1977, Piment, Avignon-Montfaiet, France, pp 161–169

Salgotra RK, Zargar SM (2020) Potential of genetic and genomic resources for genetic improvement of food crops. In: Salgotra RK, Zargar SM (eds) Rediscovery of genetic and genomic resources for future food security. Springer, Singapore, pp 1–23

Samovol AP, Zhuchenko AA (1983) The effect of mutagenic factors on recombination processes in pepper (*C. annuum* L.). Report 1: The induced alteration of segregation type for monogenic marker characters in F_2. Capsicum Newsl 2:40–42

Sashidhar N, Harloff HJ, Jung C (2020) Identification of phytic acid mutants in oilseed rape (*Brassica napus*) by large-scale screening of mutant populations through amplicon sequencing. New Phytol 225(5):2022–2034

Schimmerling W (2011) The space radiation environment: an introduction. https://three-jsc.nasa.gov/concepts/SpaceRadiationEnviron.pdf. Accessed 5 Feb 2011

Shah D, Kamili AN, Wani AA, Nazir N, Sajad N, Khan I, Shah S (2016) Mutagenic action of ethyl methane sulphonate (EMS): a review. J Res Dev (Srinagar) 16:63–68

Shapiro S (1956) The Brookhaven radiations mutation program. In: Proceedings of the conference on radioactive isotopes in agriculture, January 12–14, 1956 at Michigan State University, East Lansing, Michigan, pp 141–150

Sharma N (1994) Induced mutations and improvement in *Capsicum annuum* L. Doctoral dissertation, Aligarh Muslim University, Aligarh, India

Sharma VK, Srivastava A, Mangal M (2020) Recent trends in sweet pepper breeding. In: Gosal SS, Wani SH (eds) Accelerated plant breeding, vol 2. Springer, Cham, pp 417–444

Siddique MI, Han K, Kang BC (2016) Assessment of phenotypic variants and identification of non-pungent mutants among EMS-treated M_2 population (of pepper). Korean J Hortic Sci Technol 34(Suppl. 3):190

Siddique MI, Back S, Lee JH, Jo J, Jang S, Han K, Kang BC (2020) Development and characterization of an ethyl methane sulfonate (EMS) induced mutant population in *Capsicum annuum* L. Plants 9(3):396

Singer B (1975) The chemical effects of Nucleic acid alkylation and their relation to mutagenesis and carcinogenesis. Prog Nucleic Acid Res Mol Biol 15:219–284

Skripnikova ZV (1976) Mutants of sweet pepper induced by chemical mutagens. Genetika 12:37–44. (Russian)

Smith HH (1958) Radiation in the production of useful mutations. Bot Rev 24(1):1–24

Sood S, Jambhulkar S, Singh Y, Gupta N, Sharma S (2017) Induced chlorophyll mutations in bell pepper (*Capsicum annuum* L. var. *grossum*). J Appl Nat Sci 9(2):1085–1089

Sotirova V, Daskalov S (1983) Use of induced mutations in developing pepper forms resistant to Phytophtora capsici Leonian. Capsicum Eggplant Newsl 2:48–50

Spencer-Lopes MM, Jankuloski L, Mukhtar Ali Ghanim A, Matijevic M, Kodym A (2018) Physical mutagenesis. In: Spencer-Lopes MM, Forster BP, Jankuloski L (eds) Manual on mutation breeding. Food and Agriculture Organization of the United Nations (FAO), Rome, Italy, pp 5–50

Stadler LJ (1928) Mutations in barley induced by X-rays and radium. Science 68(1756):186–187

Stroeva OG (1997) Joseph Abramovich Rapoport in the mirror of his works (in Russian). Priroda 1: 3–11. ISSN 0032-874X

Subbiah A, Jeyakumar S (2009) Production and marketing of chillies. Facts You 29(6):1–19

Subhash K, Rajam MV, Sathyanarayana B, Meerabai A (1981) Induction of multilocular ovary in *Capsicum annuum* L. by mitomycin C. Indian J Exp Biol 19(3):301–302

Szurman-Zubrzycka ME, Zbieszczyk J, Marzec M, Jelonek J, Chmielewska B, Kurowska MM, Krok M, Daszkowska-Golec A, Guzy-Wrobelska J, Gruszka D, Gajecka M, Gajewska P, Stolarek M, Tylec P, Sega P, Lip S, Kudełko M, Lorek M, Gorniak-Walas M, Malolepszy A, Podsiadlo N, Szyrajew KP, Keisa A, Mbambo Z, Todorowska E, Gaj M, Nita Z, Orlowska-Job W, Maluszynski M, Szarejko I (2018) HorTILLUS—a rich and renewable source of induced mutations for forward/reverse genetics and pre-breeding programs in barley (*Hordeum vulgare* L.). Front Plant Sci 9:216

Taheri S, Abdullah TL, Jain SM, Sahebi M, Azizi P (2017) TILLING, high-resolution melting (HRM), and next-generation sequencing (NGS) techniques in plant mutation breeding. Mol Breed 37(3):40

Tammu RM, Nuringtyas TR, Daryono BS (2021) Colchicine effects on the ploidy level and morphological characters of Katokkon pepper (*Capsicum annuum* L.) from North Toraja, Indonesia. J Genet Eng Biotechnol 19(1):1–8

Thakur PC, Gill HS, Bhagchandani PM (1980) X-ray induced mutations in sweet pepper. Veg Sci 7(1):118–121

Timin OY (2005) Development of F_1 sweet pepper hybrids with improved biochemical content on a sterile background. Doctoral dissertation, All-Russian Scientific Research Institute of Vegetable Growing, Moscow, Russia. (Russian)

Timin OY, Timina OO (2020) New hybrid of sweet pepper "Quinta". In: Proc. Int. scientific-practical conference "Breeding, seed production and technologies of cultivation of agricultural crops", Tiraspol, Transnistria, pp 156–160. (Russian)

Timina OO (1997) Copyright certificate No. 29969 for sweet pepper variety "Yubileiny Semko"/ 1997. (Russian)

Timina OO, Timin OY (2006) Sweet pepper "Vitamin". Author's certificate No. 42540, given by Federal State Establishment "State Russian Federation Committee of Trail and Protection of Breeding achievements", Russia. (Russian)

Timina OO, Malov AN, Feshchenko VS, Feschenko LV, Rogozhnikova OA, Yusu OB (2001) Induction of photomorphogenetic reaction of sweet pepper under the effect of He–Ne laser light. In: Proceedings of the XI-th EUCARPIA meeting on genetic and breeding of capsicum & egg plant, 9–13 April, 2001, Antalya, Turkey, pp 147–151

Timina OO, Timin OY, Fedorov S, Tomlekova NB (2011) Inheritance of pericarp coloration and β-carotene content in vegetable pepper. Vavilov J Genet Breed 15(3):585–594. ISSN: 1814-554X. (Russian)

Timina OO, Hlebnikov VF, Timin OY, Tomlekova NB (2014) Using mutant resources of Capsicum L. in breeding for productivity, early maturity and quality: methodological recommendations. Publishing house at the Dniester University, Tiraspol. (Russian)

Todorova Y (1982) Pepper resistance to *Leveillula solanacearum* Gol. obtained by experimental mutagenesis. In: Proceedings of the second national symposium of plant immunity, 1982, Plovdiv, Bulgaria, vol 3, pp 91–98

Todorova Y, Daskalov S (1979) Possibilities for the utilization of some mutagenic factors for changing sweet pepper susceptibility to powdery mildew pathogen (*Leveillula solanacearum* Gol f. *capsici* Berg.). Genetika i selektsiia 2:174–179. ISSN: 0016-67661. (Bulgarian)

Todorova Y, Daskalov S (1997) Pirin - a new mutant variety in pepper resistant to powdery mildew *Leveilula solanacearum* Gol. Mutat Breed Newsl 43:28–29

Toker C, Yadav SS, Solanki IS (2007) Mutation breeding. In Lentil, Springer, Dordrecht, pp 209–224

Tomlekova NB, Panchev I, Daskalov S, Todorova V (2004) Analysis of β-carotene hydroxylase activity in pepper (*Capsicum annuum* L.) mutant. In: Proceedings of the international conference on horticulture, post-graduate study system and conditions in Europe, 16–19 November 2004, Lednice, Czech Republic, pp 223–226

Tomlekova NB, Panchev I, Yancheva S, Todorova V, Baudoin JP, Daskalov S (2007a) Established molecular marker in pepper mutants with orange fruit colour and application in crop breeding for high beta-carotene. In: Proceedings of the international scientific conference "Plant germplasm basis of the modern agriculture", 13–14 June 2007, Plovdiv, Bulgaria, pp 379–383

Tomlekova NB, Todorova V, Daskalov S (2007b) Creating variation in pepper (*Capsicum annuum* L.) through induced mutagenesis. Bulg J Crop Sci 44:44–47. ISSN 0568-465X. (Bulgarian)

Tomlekova NB, Timina OO, Timin OY (2008) Achievements and perspectives of sweet pepper breeding towards high beta-carotene. In: Proceedings of the IV Balkan symposium on vegetables and potatoes, 9–12 September, Plovdiv, Bulgaria, pp 205–212

Tomlekova NB, Todorova V, Petkova V, Yancheva S, Nikolova V, Panchev I, Penchev E (2009) Creation and evaluation of induced mutants and valuable tools for pepper breeding programmes. In: Shu QY (ed) Induced plant mutations in the genomics era. Rome, Italy, Food and Agriculture Organization of the United Nations, pp 187–190

Tomlekova NB, Timin O, Arnaoudova Y, Timina O, Wani MR, Kozgar MI (2014) Trends and achievements in F_1 hybrids of sweet pepper utilizing induced male-sterility. In: Tomlekova NB, Kozgar MI, Wani MR (eds) Mutagenesis: exploring genetic diversity of crops. Wageningen Academic Publishers, Wageningen, pp 732–736

Tomlekova NB, Spasova-Apostolova V, Panchev I (2016a) Characterization of mutation affecting a region of the chromosome 3 in orange-coloured and beta-carotene-rich mutant registered as the variety Oranzheva kapiya. C R Acad Bulg Sci 69(2):145–150

Tomlekova NB, Spasova-Apostolova V, Panchev I (2016b) RAPD analysis of Bulgarian pepper induced mutant. C R Acad Bulg Sci 69(6):731–738

Tomlekova NB, White PJ, Thompson JA, Penchev EA, Nielen S (2017) Mutation increasing β-carotene concentrations does not adversely affect concentrations of essential mineral elements in pepper fruit. PLoS One 12(2):e0172180

Tomlekova NB, Spasova-Apostolova V, Pantchev I, Sarsu F (2021) Mutation associated with orange fruit color increases concentrations of β-carotene in a sweet pepper variety (*Capsicum annuum* L.). Foods 10(6):1225

Tran DQ, Nguyen HD, Bui HT, Nguyen VB (2001) Use of physical/chemical mutagens in plant breeding program in Vietnam (No. JAERI-CONF-2001-003). https://inis.iaea.org/search/search.aspx?orig_q=RN:33010119. Accessed 25 Aug 2021

Umalkar GV, Vyawhare MK, Kashikar RM, Kashikar SG (1981) Sodium azide induced mutations for quality seeds in *Capsicum annuum* L. Acta Hortic 111(3):63

Uncu AT (2019) Genome-wide identification of simple sequence repeat (SSR) markers in *Capsicum chinense* Jacq. with high potential for use in pepper introgression breeding. Biologia 74(2): 119–126

Vagera J, Havranek P (1983) Induction of mutants in *Capsicum annuum* L. by N-nitroso-N-methylurea treatment. Capsicum Newsl 2:51–53

Venkatrajam M, Subhash K (1984) Studies on induced mutations in Capsicum by mitomycin C. Mutat Res 138(1):47–54

Verschaeve L, Kirsch-Volders M (1990) Mutagenicity of ethyleneimine. Mutat Res 238(1):39–55

Vig BK (1977) Genetic toxicology of mitomycin C, actinomycins, daunomycin and adriamycin. Mutat Res 39(2):189–238

Wu HN, Zhang SZ (1986) Effect of acridine yellow on development of anthers of *Capsicum frutescens* var. *longum* cultured *in vitro*. Chiang-su nung yeh hseuh pao 2(3):34–49

Xie XL, Deng GZF (2009) Studies on the influence of colchicine on growth and multiploid induction of bush red pepper *Capsicum annuum* L. Seed 10:41–45

Xie L, Guo Y, Wang X, Gao Y, Meng F, Liu L (2010) Molecular detection of genomic DNA variation induced by space environment in sweet pepper (*Capsicum annuum* L.). J Nucl Agric Sci 24(2):254–258

Yadav R, Gorathoki S, Dhakal S, Purnima BC, Shah A, Poudel S (2021) A review on overview role of mutation in plant breeding. Rev Food Agric 2(1):31–34

Young RS (1968) Biological experiments in space. Space Sci Rev 8:665–689

Zhuchenko AA, Samovol AP (1983) The effects of mutagenic factors on recombination processes in pepper. Report 2. The induced alternation of dihybrid segregation for linked and unlinked markers in F_2. Capsicum Newsl 2:43–45

Zubrzycki HM, Pahlen A (1970) Comparación entre los efectos del metanosulfonato de etilo y los Rayos X en la inducción de mutaciones en *Capsicum annuum* L. In: Proceedings of induced mutations and plant improvement - Latin American Study Group Meeting on induced mutations and plant improvement, 16–20 November 1970, Buenos Aires, Argentina, pp 387–396

Zubrzycki HM, Pahlen A (1973) Frequencies of viable mutants in *Capsicum annuum* L. induced with ethylmethane-sulphonate and X-ray treatments. Boletin Genetico (Inst Fitotec Castelar) 8: 23–28s

Induced Mutation Technology for Sugarcane Improvement: Status and Prospects

21

Suprasanna Penna, S. J. Mirajkar, M. V. Purankar, A. A. Nikam, S. G. Dalvi, E. R. Vaidya, and R. M. Devarumath

Abstract

Induced mutation technology is a powerful means to introduce novel genetic variability for use in crop improvement. Induced mutants have been developed in several crop plants and economic use of the mutant varieties has demonstrated their potential across the globe. Sugarcane is an important cash crop cultivated primarily for its distinctive sucrose accumulation ability. Being a vegetatively propagated crop, mutagenesis in sugarcane can be a good means of induction of genetic variation to overcome problems associated with conventional breeding such as longer life cycle, poor fertility, and narrow genetic base. In this regard, in vitro cultures have provided an excellent resource for induced mutagenesis. Strategies of selection for different desirable traits at the cellular level followed by

This article is in memory of our colleague who was passionate and contributed to the sugarcane in vitro mutagenesis work.

S. Penna (✉)
Nuclear Agriculture and Biotechnology Division, Bhabha Atomic Research Centre, Mumbai, India

Amity Centre for Nuclear Biotechnology & Amity Institute of Biotechnology, Amity University, Mumbai, Maharashtra, India

S. J. Mirajkar
School of Nanoscience and Biotechnology, Shivaji University, Kolhapur, India

Plant Tissue Culture Section, Vasantdada Sugar Institute, Pune, India

M. V. Purankar · R. M. Devarumath
Molecular Biology and Genetic Engineering Division, Vasantdada Sugar Institute, Pune, India

S. G. Dalvi
Plant Tissue Culture Section, Vasantdada Sugar Institute, Pune, India

E. R. Vaidya
Pulses Research Unit, Dr. Panjabrao Deshmukh Krishi Vidyapeeth, Akola, India

S. Penna, S. M. Jain (eds.), *Mutation Breeding for Sustainable Food Production and Climate Resilience*, https://doi.org/10.1007/978-981-16-9720-3_21

rigorous ex vitro evaluation and field testing have been successfully adopted for the isolation of sugarcane mutants carrying improved traits. It should also be noted that such mutants for important agronomic traits can be readily adopted by breeders without the regulatory and licensing restrictions that are often applied to transgenic lines. Till date, the Mutant Variety Database (MVD) of FAO-IAEA has listed 13 officially registered mutant varieties of sugarcane developed through induced mutagenesis demonstrating the potential of induced mutagenesis for sugarcane improvement. These induced mutants have been developed for improved traits including disease resistance, abiotic stress tolerance, and agronomic yield and are evaluated for performance and stability. Induced mutations offer a valuable resource for the discovery of genes controlling important traits and understanding the functional basis of the mutant genes. Mutation discovery in polyploids like sugarcane is challenging and currently, TILLING platforms are being developed. In addition, advanced mutagenesis tools based on genome editing have been initiated to target agronomically useful genes. This article outlines the techniques and success of mutagenesis strategies and isolation of desirable mutants for use in sugarcane improvement.

Keywords

Sugarcane · Induced mutation · In vitro cultures · Mutants · Crop improvement

21.1 Introduction

The cornerstone of crop improvement is the induction and sustenance of genetic variability and hence strategies need to be in place in crop breeding programs. In this regard, the use of induced mutations has played a key role in the improvement of superior plant varieties (Jain 2005; Suprasanna et al. 2014, 2015). To the present date, efforts were laid for the improvement of many economically important crops wherein several improved mutant varieties have been released for cultivation, those occupying majority of the cultivated areas across the globe. This is lucid and vivid with more than 3200 mutant varieties of different crops that are registered in the IAEA's Mutant Variety Database (MVD) those are officially released for cultivation (Suprasanna et al. 2015). Among others, radiation-induced mutations have been useful for modification in plant canopy architecture, root traits, oil and protein quality, flowering time, seed size and color, besides the crop yield. Over the past several decades, mutation breeding has remained a viable option to genetically modify existing commercial clones. However, recent developments in terms of induced mutagenesis using in vitro plant cell and tissue cultures offers a feasible mean to generate novel genetic variability which was considered a bottleneck in the classical mutation breeding efforts (Brar and Jain 1998; Kharkwal and Shu 2009; Suprasanna et al. 2015). Current research on induced mutations in these crops is expected to provide a valuable resource of the mutant gene pool for use in direct or indirect breeding programs (Suprasanna et al. 2015).

Induced mutagenesis has become a core approach to introduce desirable genetic variability and to produce new plant types with unique traits. The impeding factors of conventional breeding approaches such as loss of vigor, narrow gene pool, complex genome, poor fertility, and the long breeding and selection cycle, can be resolved through the use of induced mutagenesis (Suprasanna et al. 2011). Often, if the intention is to improve only one or few easily identifiable traits in an otherwise well-adopted variety, then the choice of direct use of mutations becomes a valuable supplementary approach. A major gain of mutation induction in vegetatively propagated crop (VPC) plants is the feasibility to alter one or few traits of an outstanding elite cultivar without disturbing the remaining genetic makeup. The development of improved varieties through the use of induced mutagenesis in the case of VPCs such as sugarcane is often challenged by constraints of heterozygosity, systemic disease, and pest incidence. Moreover, VPCs have certain peculiarities such as the absence of meiotic sieves, the concurrent fixation of deleterious alleles, transmission of pathogens to subsequent generations, and most importantly, the prevalence of chimeras (Mba et al. 2009). One of the means of achieving food security is through enhancing and sustaining biodiversity (Thrupp 2000).

21.2 Outlook of Sugarcane Crop Improvement

Sugarcane (*Saccharum officinarum* L.) is a predominant agro-industrial crop cultivated on over 26.5 million hectares, accounting for approximately about 0.54% of the world's total arable land area (Anon 2022; FAOSTAT 2013). It contributes to over 65% of the world's total sugar production and is also gaining increased attention as a potential source of biofuel (D'Hont et al. 2008; Barnabas et al. 2015). Sugarcane varietal improvement has played a vital role in the overall sustainable growth of sugar industries around the globe. The major objectives of sugarcane breeding programs usually include improved cane yield, sugar content, ratooning ability, and improved disease resistance besides the ability to combat against abiotic stresses (Ming et al. 2006) (Fig. 21.1). During the course of development of existing improved varieties, diverse *Saccharum* species have been used for the incorporation of desirable traits. However, because of the narrow genetic base of *Saccharum* species, enhancement of genetic base has slowed down impeding the pace of development of improved sugarcane varieties and, resistance to biotic or abiotic stresses is still a challenge necessitating the development of new resistant clones (Devarumath et al. 2013; Shrivastava and Srivastava 2016).

Conventional breeding efforts often take a decade or more to release a new improved cultivar and will have to pass through a thorough selection process during which clonal material of selected genotypes is multiplied and clones are then chosen for sucrose content and stress and disease resistance (Parfitt 2005). Improvement in sugar content has high economic value as it increases sugar produced from farms and mills with very little increase in marginal costs through harvesting, cane transport, or milling (Jackson 2005). In an interesting study, Parajuli et al. (2020) attempted to engineer sugarcane through metabolic engineering into oil cane. Since sugarcane is a

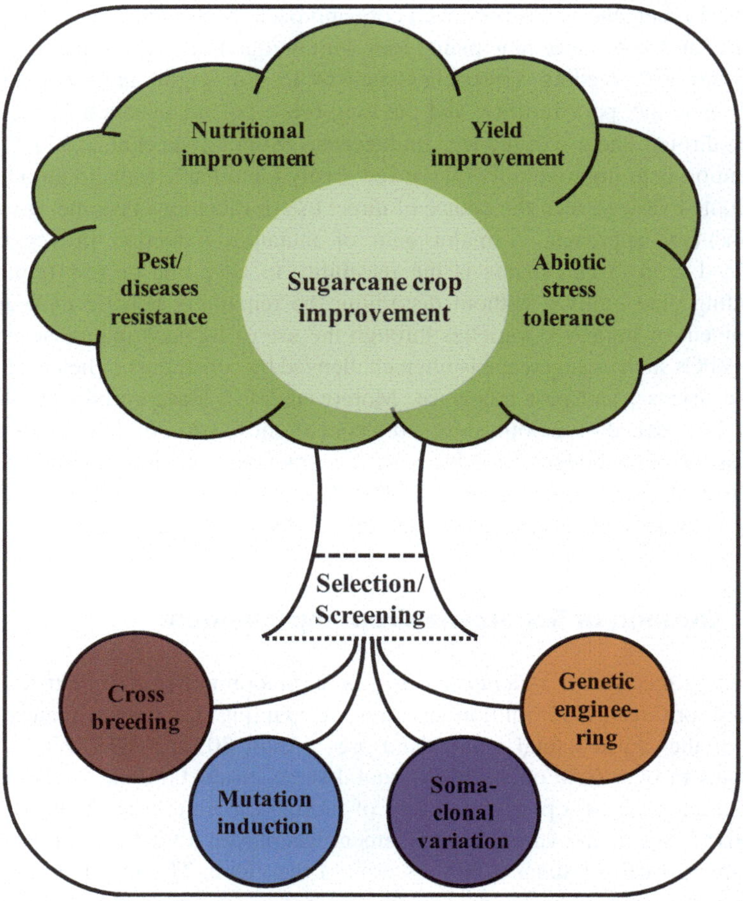

Fig. 21.1 Key methods and approaches employed for sugarcane crop improvement for diverse agronomic traits

high biomass crop, these researchers proposed deflection of carbon flux (sucrose) to oil so as to reap high lipid yield (4.3% triacylglycerol) for biofuel purposes.

21.3 Induced Mutagenesis in Sugarcane

So far, 13 mutant varieties of sugarcane have been registered in the Mutant Variety Database (MVD) those were developed through induced mutagenesis (Table 21.1; FAO-IAEA 2021). These initial attempts demonstrated the successful use of induced mutagenesis in sugarcane improvement. Of these, five mutants were reported from India followed by three each from China and Cuba. Most of these have been developed for improved disease resistance by using gamma radiation of vegetative propagules. Taken together, these attempts have paved a way for the development of

Table 21.1 List of registered induced mutants of sugarcane (Mutant Variety Database (MVD) of FAO/IAEA 2021)

Sr. No.	Mutant variety	Parent variety	Year of release	Country	Target tissue	Mutagen used (dose)	Trait(s) improved
1.	Co6608	Co449	1966	India	Vegetative propagules	Gamma rays (30–50 Gy)	Red rot resistance
2.	Co997 mutant	Co997	1967	India	Vegetative propagules	Gamma rays (30–50 Gy)	Red rot resistance
3.	Nanei	Ni 1	1981	Japan	Vegetative propagules	Chronic gamma rays (420 Gy)	Better tillering, longer and thicker cane, higher yield and sugar
4.	Co8153	Co 6304 × Co 6806	1981	India	Seed	Gamma rays (150 Gy)	Improved juice quality and yield
5.	Co 85017	Co 740	1985	India	–	Gamma rays (150 Gy)	Resistance to *Ustilago scitaminea*, better cane yield and sucrose %
6.	Co 85035	Co 740	1985	India	–	Gamma rays (150 Gy)	Resistance to *Ustilago scitaminea*, better cane yield and sucrose %
7.	Guifu 80-29	Guitang 72-28	1989	China	–	Gamma rays (80 Gy)	Late maturity, high sugar, small stem
8.	CCe 10582	C 87-51	1990	Cuba	–	–	Improved cane yield
9.	Yutangfu 83-5	Yuetang 71-210	1992	China	–	–	–
10.	CCe 183	C 87-51	1993	Cuba	–	–	Resistance to eyespot
11.	CCe 283	C 87-51	1993	Cuba	–	–	Resistance to eyespot
12.	CCe 483	C 87-51	1993	Cuba	–	–	Resistance to eyespot
13.	Guitang 22	Xintaitang 1	2005	China	Vegetative propagules	Gamma rays (80 Gy)	Higher sugar yield, higher tonnage

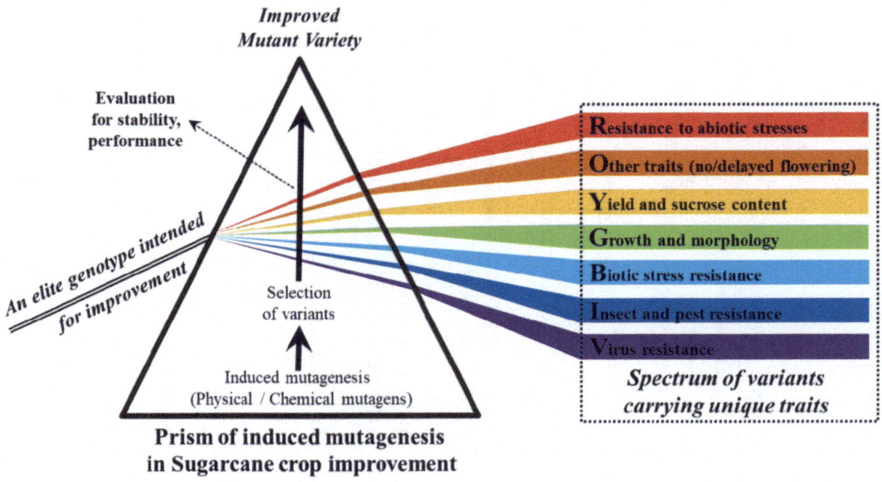

Fig. 21.2 Hypothetical prism model representing scope and potential of induced mutagenesis in sugarcane crop improvement to derive a spectrum of variants (VIBGYOR) carrying unique and valuable traits

a spectrum of variants/mutants bearing unique traits that are difficult and/or not achievable through any other conventional approaches (Fig. 21.2).

21.3.1 Combination of In Vitro Culture and Induced Mutagenesis

In addition to vegetative propagules, sugarcane in vitro culture techniques have provided a great impetus to basic as well as applied research. This has been achieved chiefly through micropropagation of elite clones, production of disease-free propagules, isolation of agronomically superior somaclones, screening methods for biotic and abiotic stress tolerance, and conservation of novel and useful germplasm. Mutagenesis has been achieved for the improvement in diverse traits such as quantitative traits (Kaur et al. 2001), salt and drought tolerance (Patade et al. 2006) and disease resistance (Singh et al. 2008). Jagathesan and Ratnam (1978) isolated a vigorous and fast-growing mutant from gamma-irradiated vegetative buds of variety Co 527 which was able to complete its vegetative growth 2 months prior to that of parent variety while other economic characteristics like sucrose content and juice purity remained unaffected. Selection of mutants for disease resistance through induced mutagenesis has been reported in sugarcane with mutants isolated for improved resistance to smut disease (Siddiqui and Javed 1982; Dalvi et al. 2021), Sugarcane mosaic virus (SCMV) (Zambrano et al. 2003), and red rot disease (Ali et al. 2007; Singh et al. 2008).

Mutagenesis using cultured plant cells and tissues has generated great interest in creating novel genetic variability (Sengar et al. 2011; Suprasanna et al. 2012). Plant cell and tissue cultures offer an exciting resource in in vitro mutant selection for

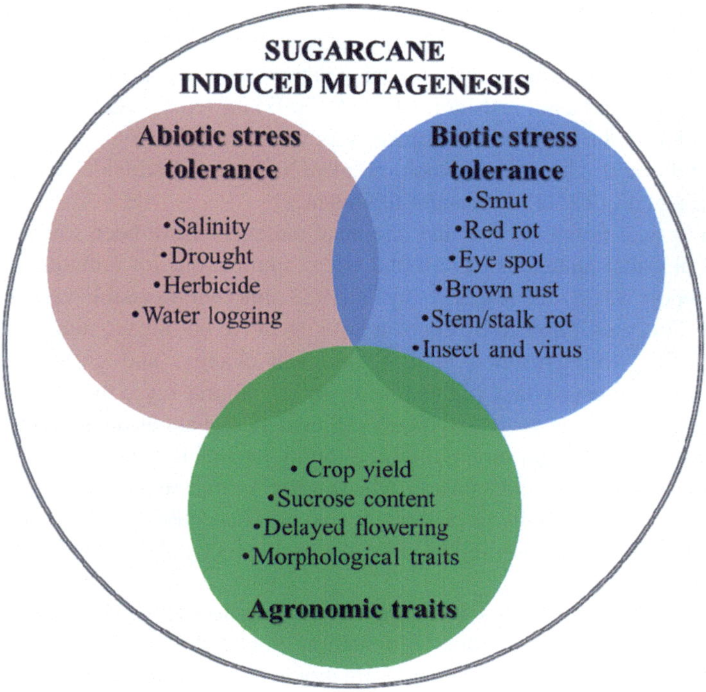

Fig. 21.3 Sugarcane-induced mutagenesis as a source to derive mutants bearing valuable abiotic, abiotic, and agronomic traits

generating genetic variability. However, certain factors such as established and reproducible in vitro plant regeneration protocols, optimization of mutagenic treatments, and efficient screening of the mutagenized populations, etc. play decisive roles in the success of in vitro mutagenesis (Lakshmanan et al. 2005). The in vitro selection methods are relatively inexpensive, simple, and rapid, and can be performed in the laboratory at any time of the year. It is possible to select cell lines resistant to biotic stresses and, tolerant to abiotic stresses with improved nutritional quality. The protocols for somatic embryogenesis are useful for high-volume micropropagation, mutagenesis and in vitro selection and transgenic methods of genetic manipulation (Suprasanna et al. 2010, 2011). Although genetic transformation has been successful, genetically modified (GM) sugarcane is not yet grown commercially partly due to legislation, lack of public acceptance and perceived negative impacts on the environment. Hence, the most feasible approach for the in vitro genetic improvement of sugarcane can be through induced mutagenesis combined with somaclonal variation.

In vitro mutagenesis using embryogenic callus cultures followed by in vitro selection and ex vitro field evaluation of selected lines for agronomic characters has been successful for the isolation of desirable mutants for agronomic and stress tolerance (Patade and Suprasanna 2008, Patade and Suprasanna 2009) (Fig. 21.3).

There has been increasing interest in adopting this strategy for isolation of useful mutants (Table 21.2). Majid et al. (2001) and Samad et al. (2001) developed mutants resistant to red rot, water logging, and delayed flowering; Rodríguez et al. (2001) for wider stem girth and smut resistant; Ali et al. (2007) for red rot, Saif et al. (2001) and Patade et al. (2008a, b) for salt tolerance; Vaidya et al. (2008) for agronomical, cane quality and yield characters, Koch et al. (2012) for herbicide tolerance and Weksanthia et al. (2021) for drought tolerance.

Both physical (gamma rays) and chemical mutagens have been useful in sugarcane, but physical mutagens have found wider applications for mutation induction. The first report was by the use of X-rays in 1927 although no visible variations were observed. This was followed by exposure of vegetative cuttings to neutron radiation (Tysdal 1956) which resulted in morphological changes and growth reduction. Heinz (1973) observed that higher the dose of gamma rays, the lower was the survival, reduction in growth, and increased incidence of abnormal mutants. Historical accounts of studies on mutation induction in sugarcane were detailed by Heinz (1987) and Sreenivasan and Jalaja (1998) suggesting that sugarcane breeders could explore the possibility of mutation breeding to rectify specific defects or to improve a specific desirable trait of highly adapted and genetically balanced sugarcane cultivars.

Different chemical agents like ethyl methanesulfonate (EMS), sodium azide (NaN_3) have been used for mutation induction. Ponce et al. (1998) evaluated three rust-resistant somaclones through NaN_3 treatment of calli and a mutant clone of var. B 4362 obtained from gamma-irradiated calli. The authors observed that the somaclones showed agronomic disadvantages including yield reduction, increased flowering, and smut (*Ustilago scitaminea* Syd) susceptibility. The other mutant clone obtained by an induced mutation in buds was selected as a commercial variety because of its outstanding agronomic traits (Ponce et al. 1998). EMS is well known to produce a large spectrum of mutations. Oloriz et al. (2012) applied 0.003% NaN_3 solution for 30 min on spindles and buds of sugarcane for callus induction and further plantlet regeneration. Direct application of 0.1% NaN_3 solution for 30 min on buds from field-grown plants for planting in plastic bags was also demonstrated. The study led to the successful isolation of five brown rust-resistant mutants out of a total population of 11,167 regenerated plants. The mutagenesis treatment resulted in higher selection frequencies for brown rust resistance along with traits such as sugar yield, internode shape, bud shape, leaf sheath hairiness, outer auricule shape, intensity of flowering, stool growth habit, number of stalks per stool, stalk length, stalk number, and stalk diameter. In vitro mutagenesis using EMS as chemical mutagen resulted in the development of sugarcane mutants bearing different agronomic traits including for disease resistance (Dalvi et al. 2012, 2021). Cha-um et al. (2013) compared six mutant cultivars of sugarcane derived from γ-irradiation and EMS, and commercial cultivars, and found that salt-induced reduction in net photosynthetic rate and growth characters such as shoot length, number of internodes, and internodal length as well as yield traits, was not observed among the mutants suggesting their salt-tolerant nature.

Table 21.2 Developments in sugarcane in vitro mutagenesis for the selection of biotic and abiotic stress tolerance

Sr. No.	Trait (s) improved	Mutagen used	Method used	Selection method	Reference
Resistance to biotic stresses					
1.	Red-rot resistance	γ-rays	Vegetative buds exposed to gamma rays	Vegetative progeny of treated buds resistant to red rot pathogen, *Glomerella tucumanensis*	Rao et al. (1966)
2.	Resistance to SCMV	γ-rays	Calli of susceptible cultivar—gamma irradiated (2 krads)	Manual selection in greenhouse inoculation with strain B of SCMV	Zambrano et al. (2003)
3.	Five brown rust-resistant mutants	NaN_3 and γ-rays	Spindles and buds—submerged in NaN_3 solution (0.003% and 0.1% for 30 min, respectively); Calli irradiated with γ-rays (10 and 20 Gy)	Greenhouse screening using inoculation of uredospore suspension	Oloriz et al. (2012)
4.	Three herbicide (imazapyr) tolerant mutants	EMS	Embryogenic calli—immersed in EMS solution (8–96.6 mM) for 4 h	In vitro selection of calli on 0.025–0.1 μM Imazapyr containing medium; followed by ex vitro selection using spraying of commercial herbicide Arsenal (250 g/L imazapyr)	Koch et al. (2012)
5.	Mutants resistant to *Fusarium sacchari*	EMS	Embryogenic calli—treated with EMS solution (32 mM) for 4 h	In vitro selection using *Fusarium sacchari* culture filtrate; followed by ex vitro selection	Mahlanza et al. (2013)
6.	Two mutants resistant to smut disease	EMS	Calli induction—on MS medium containing EMS (0.8 μM), PEG	Ex vitro screening using smut spore suspension	Dalvi et al. (2012)

(continued)

Table 21.2 (continued)

Sr. No.	Trait (s) improved	Mutagen used	Method used	Selection method	Reference
			(0.125 μM) and 2,4-D (13.57 μM)	treatment for 30 min	
7.	*Fusarium sacchari*-tolerant mutants	EMS	Embryogenic calli—treated with EMS (32 mM) for 4 h	Glasshouse selection by artificial inoculation of *F. sacchari* conidial suspension (10^5 conidia/mL) followed by inoculation with *Eldana saccharina*	Mahlanza et al. (2015)
8.	Mutants moderately resistant to red rot	γ-rays	Embryogenic calli irradiated with γ-rays (0–80 Gy)	Ex vitro selection by artificial inoculation of red rot and wilt fungal spore suspension	Vaidya et al. (unpublished)
9.	Two mutants moderately resistant and one resistant to smut disease	EMS	Calli induction—on MS medium containing EMS (0.8 μM), PEG (0.125 μM) and 2,4-D (13.57 μM)	Ex vitro screening using smut spore suspension treatment for 30 min	Dalvi et al. (2021)
Tolerance to abiotic stresses					
10.	Salinity tolerance	γ-rays	Embryogenic cultures—γ-irradiated (10–50 Gy)	In vitro selection on NaCl (42.8–256.7 mM) containing medium	Suprasanna et al. (2009)
11.	Salinity tolerance	NaN_3	Calli cultured on MS medium containing 0.5% NaN_3	In vitro selection of calli on MS medium containing NaCl (0–75 mM)	Ikram-ul-Haq et al. (2011)
12.	Salinity tolerance	γ-rays	γ-irradiation and EMS	Field screening for salinity tolerance	Cha-Um et al. (2013)
13.	Salinity tolerance	γ-rays	Embryogenic calli—irradiated with γ-rays (0–80 Gy)	In vitro selection of calli on medium containing	Nikam et al. (2014)

(continued)

Table 21.2 (continued)

Sr. No.	Trait (s) improved	Mutagen used	Method used	Selection method	Reference
				NaCl (0–250 mM); followed by ex vitro selection under saline soil conditions (3.55 EC)	
14.	Salinity tolerance	γ-rays	Embryogenic calli—irradiated with γ-rays (0–80 Gy)	In vitro selection of calli on medium containing NaCl (0–250 mM); followed by ex vitro selection under saline soil conditions	Nikam et al. (2015)
15.	Salinity tolerance	EMS	Embryogenic calli—treated with EMS (0.5%) for 0–4 h	In vitro selection on medium containing NaCl (0–200 mM)	Gadakh et al. (2015)

21.3.2 Radiation-Induced Mutagenesis Program in Sugarcane

Radiation-induced mutagenesis program in sugarcane has been widely undertaken and significant success has been achieved using embryogenic callus cultures followed by in vitro selection of mutants and field evaluation. Figures 21.4 and 21.5 illustrate the general methodology and different stages that can be adopted for an in vitro mutagenesis program in sugarcane. Briefly, embryogenic callus is irradiated at different doses of gamma rays to assess the radiosensitivity of the variety being used (Patade et al. 2008a, b). After determining LD_{50}, in vitro cultures are irradiated at LD_{50} and irradiated embryogenic callus cultures are passed through V_1M_1 to V_1M_4 (or V_1M_6 if chimera separation has to be done) and either regenerated into plants or can undergo selection at in vitro level, i.e., exposed to different concentrations of salt (NaCl) or PEG (drought) or pathogen causative agents. The putative mutants will have to be evaluated under field conditions to select the trait of interest and confirm their stability (performance) in further generations.

Our previous studies on radiation-induced mutagenesis have resulted in the generation of several mutants for desirable traits. By using gamma-irradiated embryogenic calli of commercial sugarcane cultivars (CoC-671, Co 86032, and Co 94012), 7500 plantlets were regenerated, hardened, and field planted at the

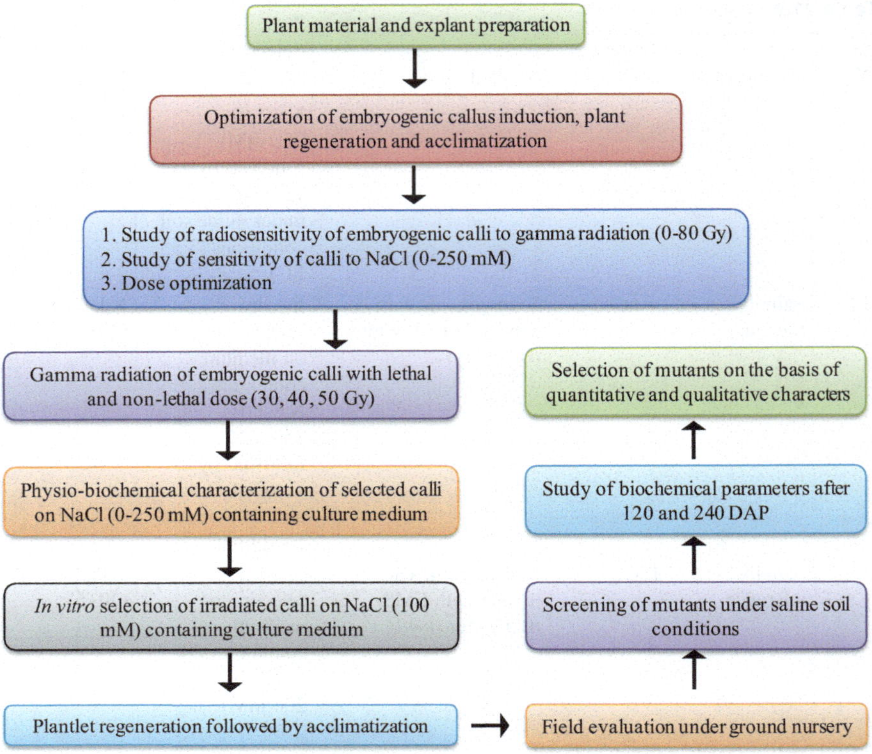

Fig. 21.4 Flow diagram of in vitro mutagenesis using gamma radiation for salt tolerance in sugarcane. (Modified from Nikam et al. 2014)

Sugarcane Research Station, Akola, India. Field observations indicated that there was a spectrum of agronomically desirable mutations from morphological, quality to other yield contributing traits (Fig. 21.6). Mutation spectrum was broader in the case of Co 94012 for morphological characters whereas, cultivar Co 86032 exhibited a wider range of mutations for quality and yield characters. The range of mutations obtained for cane yield attributing characters was 0.09–0.38% for different doses of gamma rays and cultivars. In the case of yield attributing characters, maximum mutations are obtained for cane weight per plant (0.023%) followed by cane diameter and cane height. So far, 22 sugarcane mutant clones were found promising for agronomic and biochemical characteristics (Table 21.3). The results on the reaction of ten sugarcane mutants to red rot showed that three mutants (AKTS-AKTS-03, AKTS-05, AKTS-17) derived from variety Co-86032, one mutant (AKTS-15) of CoC-671 and one mutant (AKTS-18) of Co-94012 showed moderate red rot and wilt disease resistance. Biochemical and molecular profiling analyses were performed in order to ascertain and establish a correlation of mutations and phenotypic influence. Spatial dynamics of sucrose metabolizing enzymes in the different stalk tissues

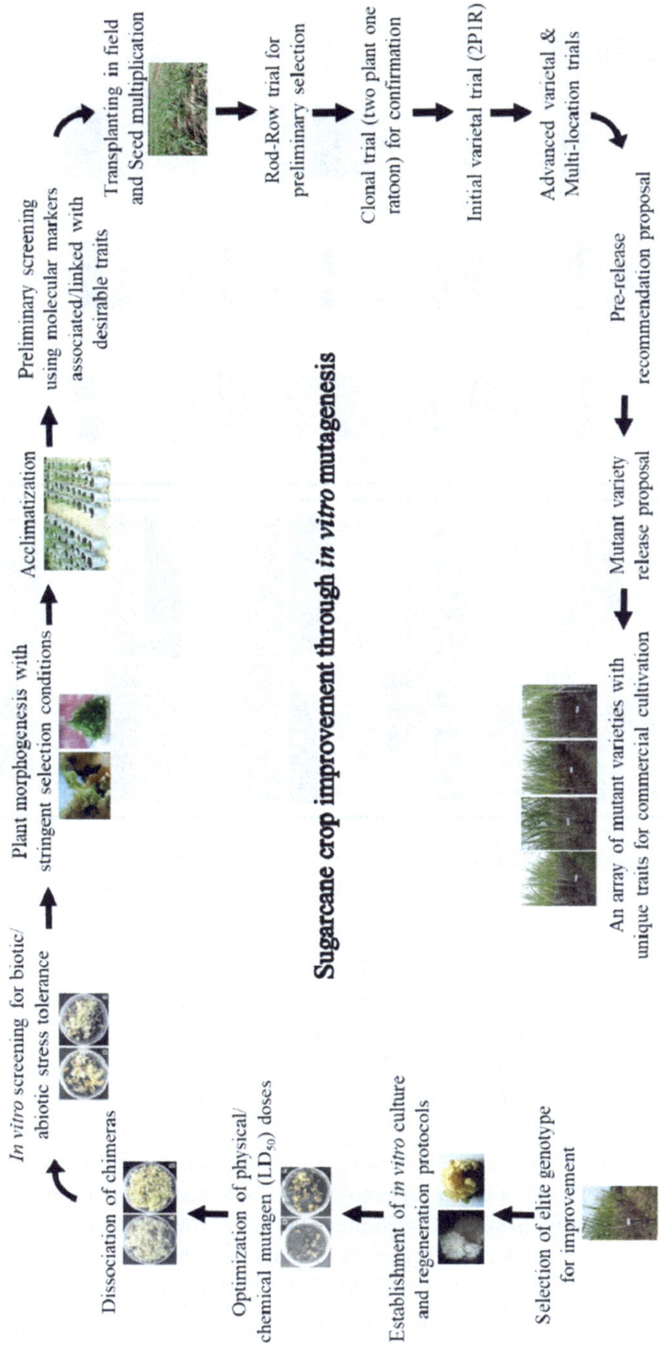

Fig. 21.5 An overview of sugarcane improvement through in vitro induced mutagenesis

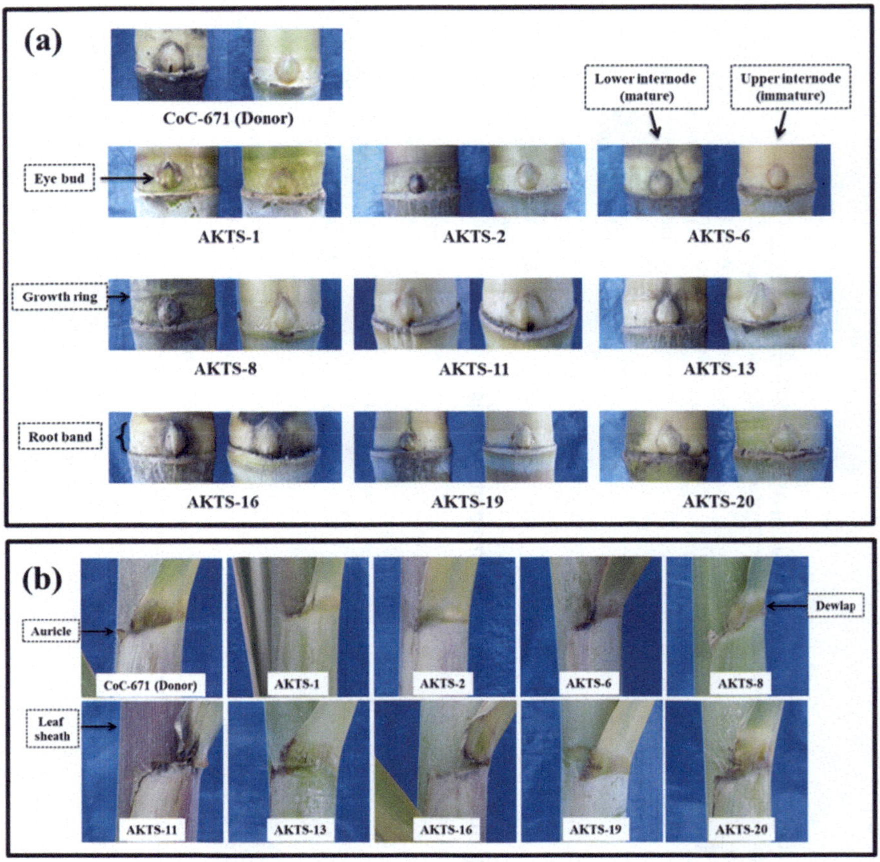

Fig. 21.6 Gamma radiation-induced mutants of cultivar CoC-671 bearing variability in terms of variations in eye bud, growth ring (**a**), auricle, dewlap and leaf sheath color (**b**)

differing in maturity resulted in pinpointing with high sucrose accumulating mutants (AKTS-02 and AKTS-20) having lower soluble acid invertase (SAI) enzyme activities together with increased sucrose synthase (SS) and sucrose phosphate synthase (SPS) activities (Mirajkar et al. 2016).

At the Vasantdada Sugar Institute (VSI), Pune, India, the in vitro selection of gamma irradiated embryonic callus of commercial cultivars Co740, Co86032, CoM 0265 on NaCl containing media was carried out. A total of 790 putative mutant clones have been field screened for agronomic characteristics under normal as well as saline soil conditions. Some of the mutant clones showed phenotypic variation (Figs. 21.7, 21.8, 21.9, and 21.10). Out of which clone M3990, M4160, M4199, M4209, M8733, M8737, M8754, and M8785 developed from Co86032 were selected for their superior agronomic characteristics such as cane height, cane girth, net millable cane (NMC) and commercial cane sugar (CCS). In another

Table 21.3 Induced mutants of sugarcane developed in authors' laboratories[a]

Location	Parent genotype (s) selected for improvement	Improved characters	Trait(s) targeted for introgression through mutagenesis	Method(s) employed for isolation and selection of variants
Dr. Panjabrao Deshmukh Krishi Vidyapeeth (Dr. PDKV), Akola, India	CoC-671, Co-86032, Co-94012	High sugar High yielding	Improvements in sucrose content, agronomic traits and disease resistance	In vitro gamma irradiation of callus culture, in vitro and ex vitro selections
Vasantdada Sugar Institute (VSI), Pune, India	CoM-0265, Co-86032, Co-740	High sugar High yielding	Salinity tolerance, disease resistance and improvement in agronomic traits	In vitro callus culture—gamma irradiation and/or treatment with EMS, in vitro and ex vitro selections
Vasantdada Sugar Institute (VSI), Pune, India	Co-86032	High yielding, disease resistance	Disease resistance, drought tolerance and improvement in agronomic traits	In vitro callus culture—treatment with EMS, in vitro and ex vitro selections

[a] BARC (Mumbai) in collaboration with Dr. PDKV (Akola), and VSI (Pune)

attempt, clone M3489 developed from Co86032 through EMS-induced mutagenesis also showed superior agronomic characteristics. These clones also showed superior performance during consecutive agronomic trials over 3 years at VSI, Pune. Furthermore, the evaluation of salinity-tolerant sugarcane mutant "M4209" was done using transcriptomic and physio-biochemical profiling suggested that salt tolerance was conferred by active transcriptional reprogramming coupled with enhanced photosystem efficiency (Negi et al. 2020). Similarly, on the other hand, CoM0265-derived mutant clones were analyzed for agronomic performance and physio-biochemical analysis under saline field conditions. The clones showed a significant correlation between lower leaf Na^+ content and higher catalase and peroxidase enzyme activities. The lower Na^+ content of the mutant clones in turn showed a negative correlation with improved yield and CCS (t/ha) (Purankar et al. 2022). Mutant clones M8711, M8457, and M8721 developed from CoM 0265 showed higher sugar content at the 12th month than parent CoM 0265 and also showed superior yield at the 12th month on saline soil and subsequent field trials under control soil conditions. In another study using EMS based mutagenesis, Dalvi et al. (2021) isolated several mutants of sugarcane cultivar CoC 671 and evaluated them for smut disease resistance along with agronomic and quality traits viz. early maturity, high sucrose, high cane yield and for quality. Mutants VSI TC 2813 and TC 2819 showed early maturity and high sugar content, and mutants 2813, 2819 were moderately to highly smut resistant.

Fig. 21.7 Field trials of gamma radiation-induced mutants of important sugarcane genotypes (Co86032, Co 740, and CoM 0265) for agronomic and yield performance

a) Co86032 b) M8151 c) M8733 d) M8737 e) M8785

Fig. 21.8 Variations in phenotypic characters observed in gamma radiation-induced mutant clones (**b**–**e**) of variety Co86032 (**a**)

a) CoM-0265 b) M8299 c) M8457 d) M8463 e) M8711 f) M8721

Fig. 21.9 Variations in phenotypic characters observed in gamma radiation-induced mutant clones (**b**–**f**) of CoM 0265 (**a**)

a) CoC 671 b) TC 906 c) TC 2813 d) TC 2819 e) TC 2826 f) TC 2875

Fig. 21.10 EMS-induced mutants (**b**–**f**) of variety CoC 671 (**a**) selected for smut disease resistance and yield

21.4 Mutagenomics and Advances in Post-mutagenesis Analyses

With the rapid advancement in crop genomics and availability of whole genome sequences, the mutation discovery and characterization of induced mutants have greatly benefitted (Suprasanna and Jain 2017). It has been shown in several studies that induced mutations constitute genome alterations, small insertions and deletions (indels), chromosomal aberrations, duplication of genes, and transposable element-mediated insertion/deletion events (Nogue et al. 2016; Negi et al. 2016). Detailed studies of mutants have led to the development of (1) mutant cultivars and (2) a genetic resource for functional genomics studies (Jo and Kim 2019). Advances in genomics technologies including Next-generation sequencing have made great strides in unraveling the nature and characterization of mutations at the whole-genome level and linking of candidate genes with the mutant phenotypes (Oono et al. 2020; Negi et al. 2020). In sugarcane, Negi et al. (2020) used an integrated approach of genomics and physiological, biochemical analyses to understand the salt tolerance trait of the radiation-induced sugarcane mutant line. It was suggested that mutant trait was associated with efficient photosystem and active transcriptional reprogramming of several metabolic processes, transmembrane transport, and photosynthesis leading to yield improvement under saline conditions.

Genomics techniques are now available to detect genetic variation, TILLING (Targeted Induced Local Lesions IN Genomes), ecoTILLING, resequencing, RNAi (RNA interference), mismatch site-specific mutagenesis, homoeologous recombination, forward and reverse genetics via transposable elements, gene replacement, gene addition, and transcriptome modification by mutagenic treatment, aneuploidy, and uniparental chromosome loss (Phillips and Rines 2009). Understanding the mutant gene structure, function, spatial and temporal expression, and genetic regulation can be useful for trait modification aimed at plant improvement. Both TILLING, ecoTILLING have been useful in sugarcane (McIntyre et al. 2006). TILLING (targeting induced local lesions in genomes) technique has transformed mutation research with the provision to isolate specific mutations (MutTILLING) (Wang et al. 2012) especially to explore targeted mutational events in plant species where very limited genomic resources are available, and transformation and forward genetic screens, often not easy. TILLING also offers a distinctive advantage of being a non-genetic manipulation (non-GM) based technology and has direct application especially in developing countries where necessary biosafety regulatory setup is still not in place.

In polyploidy plant species, genes are often represented by multiple homoeologous copies having high sequence identity. This cause additional barrier during usual phenotype-based forward genetic screening processes (Lawrence and Pikaard 2003). Despite this peculiarity, mutation discovery in polyploidy crop plants is challenging because polyploidy genome absorbs mutations at higher densities. TILLING platforms are available for polyploidy (PolyTILLING) crop plants such as Brassica (Gilchrist et al. 2013) and wheat (Uauy et al. 2009; Singh et al. 2015). Hence, PolyTILLING provides well-suited implications for mutation discovery in polyploids and facilitates the identification of large allelic series in target genes involving screening of a relatively small number of individuals (Wang et al. 2012; Chen et al. 2014).

TILLING for vegetatively propagated plants (VeggieTILLING) such as apples, bananas, citrus, cassava, grapevine, hops, and potatoes is important for global economies and also for food security in developing nations. Sugarcane is a yet another unique crop plant which has a high ploidy genome and is very often propagated vegetatively. Tissue culture mutagenesis can also be considered in facultative vegetatively propagated species where seed propagation is possible. Development of TILLING platforms for banana and cassava is being established (Wang et al. 2012). Therefore, under the current scenario of lack of whole genome sequence for sugarcane, attempts of TILLING for mutation discovery are challenging but future endeavors are progressive and scintillating. Once desired mutant alleles are identified and isolated, it can be easily introgressed into different genetic backgrounds or the mutagenesis itself can be developed in advanced genetic material.

21.5 Mutagenesis Through Genome Editing in Sugarcane

Being a polyploid, sugarcane crop has several interesting complexities which often pose difficulties for improvement through conventional breeding methods (Thirugnanasambandam et al. 2018). Techniques of genome editing (GE) have revolutionized the improvement of crop plants, and in sugarcane, there has been a steady progress in optimizing the tool-box for editing traits of agronomic and industrial importance (Meena et al. 2020; Mohan et al. 2022). The technique can enable a speedy mode of introduction of specific mutations on both single and multiple genes precisely (Mohan et al. 2022). Initial successful attempts have been made using TALEN (transcription activator-like effector nuclease) by Jung and Altpeter (2016) to generate target (lignin content) mutations in *COMT* (caffeic acid *O*-methyl transferase) gene upto 99% in 74% of transgenic lines which showed 29–32% reduction in lignin content. Oz et al. (2021) developed CRISPR/Cas9-mediated co-editing method enabling precise targeting of multiple alleles in sugarcane. Authors observed specific nucleotide substitution in the acetolactate synthase gene in the GE lines. In another study, multiallelic editing of magnesium chelatase subunit I was successfully achieved in sugarcane to develop a scorable phenotype based on the CRISPR/Cas9-mediated targeted mutagenesis of magnesium chelatase alleles and reduced chlorophyll content in GE lines (Eid et al. 2021). Using the TALEN-mediated GE method, Kannan et al. (2018) further observed up to 19.7% lignin reduction in the field-grown mutant lines and higher (43.8%) saccharification efficiency. Genome editing mediated mutagenesis has shown considerable interest and more studies are underway to capitalize on the potential of this technology to functionally validate genes of agronomic and industrial interest in sugarcane.

21.6 Conclusions

The isolation and deployment of mutations can be augmented through the use of in vitro cell and tissue culture techniques in vegetatively propagated crops while molecular biological tools can aid in the rapid genotyping of the mutational events. There have been substantial technological developments in the in vitro techniques, mutation induction, screening and utilization of mutants. With the expansion of genome resources and techniques for modifying specific genes, the area of mutagenesis is passing through a phase of resurgence. In addition, induced mutation technology offers a way to avoid biosafety and socio-ethical concerns associated with the use of transgenic methods. Sustained research and awareness about the potentials of mutagenesis will let researchers realize the increasing potential of induced genetic variation in sugarcane crop improvement. Future advancements are indeed necessary for overcoming some of the technical and practical hurdles and perceive the potentials of conventional and advanced mutagenic tools in sugarcane.

Acknowledgments Authors would like to place on record sincere thanks to the Heads of the respective institutions for their support to the sugarcane research program. Thanks are also due to all the students and associates who participated during the course of induced mutagenesis research in sugarcane and contributed to all our accomplishments, and it is a happy moment to say, "*sugarcane research is a sweet success.*"

This article is dedicated to the memory of our colleague Late Dr. Ashok A. Nikam.

References

Ali AA, Naz SS, Iqobal AJ (2007) *In vitro* induced mutation for screening of red rot resistance in sugarcane (*Saccharum officinarum*). Pak J Bot 39(6):1979–1994

Anon (2022). https://www.statista.com/statistics/249679/total-production-of-sugar-worldwide/. Accessed 29 Oct 2022

Barnabas L, Ramadass A, Amalraj RS, Palaniyandi M, Viswanathan R (2015) Sugarcane proteomics: an update on current status, challenges and future prospects. Proteomics 15:1658–1670

Brar DS, Jain SM (1998) Somaclonal variation: mechanism and applications in crop improvement. In: Jain SM, Brar DS, Ahloowalia BS (eds) Somaclonal variation and induced mutations in crop improvement. Kluwer Academic Publishers, Dordrecht, The Netherlands, pp 15–38

Cha-um S, Chantawong S, Mongkolsiriwatana C, Ashraf M, Kirdmanee C (2013) Field screening of sugarcane (*Saccharum* spp.) mutant and commercial genotypes for salt tolerance. Not Bot Horti Agrobo 41(1):286–293

Chen L, Hao L, Parry MAJ, Phillips AL, Hu YG (2014) Progress in TILLING as a tool for functional genomics and improvement of crops. J Integr Plant Biol 56:425–443. https://doi.org/10.1111/jipb.12192

D'Hont A, Souza GM, Menossi M, Vincentz M, Van-Sluys MA, Glaszmann JC, Ulian E (2008) Sugarcane: a major source of sweetness, alcohol and bio-energy. In: Moore PH, Ming R (eds) Genomics of tropical crop plants. Springer, New York, pp 483–513

Dalvi SG, Vasekar VC, Yadav A, Tawar PN, Dixit GB, Theertha Prasad D, Deshmukh RB (2012) Screening of promising sugarcane somaclones for agronomic traits, and smut resistance using

PCR amplification of inter transcribed region (ITS) of *Sporisorium scitaminae*. Sugar Tech 14(1):68–75

Dalvi SG, Tawar PN, Suprasanna P, Dixit GB, Theertha Prasad D (2021) EMS-based *in vitro* mutagenesis and mutant screening for smut resistance with agronomic traits in sugarcane. Sugar Tech 23:854–864

Devarumath RM, Kalwade SB, Kulkarni PA, Sheelavanthmath SS, Suprasanna P (2013) Integrating OMICS approaches in sugarcane improvement. In: Barh D (ed) OMICS applications in crop science. CRC Press, Boca Raton, FL, pp 191–250

Eid A, Mohan C, Sanchez S, Wang D, Altpeter F (2021) Multiallelic, targeted mutagenesis of magnesium chelatase with CRISPR/Cas9 provides a rapidly scorable phenotype in highly polyploid sugarcane. Front Genome Ed 3:654996. https://doi.org/10.3389/fgeed.2021.654996

FAOSTAT (2013). http://faostat3.fao.org/

Gadakh SS, Patel DU, Patil AB (2015) Evaluation of sugarcane (*Saccharum* spp. complex) mutants for yield, yield contributing traits and quality parameters. Int J Adv Biol Res 5(3):220–228

Gilchrist EJ, Sidebottom CHD, Koh CS, MacInnes T, Sharpe AG, Haughn GW (2013) A mutant *Brassica napus* (Canola) population for the identification of new genetic diversity via TILLING and next generation sequencing. PLoS One 8(12):e84303. https://doi.org/10.1371/journal.pone.0084303

Heinz DH (1973) Sugarcane improvement through induced mutations using vegetative propagules and cell culture techniques. In: Induced mutations in vegetatively propagated plants. IAEA, Vienna, pp 53–59

Heinz DJH (1987) Sugarcane improvement through breeding. Elsevier Science Publishers B.V., Amsterdam, The Netherlands, p 603

Ikram-ul-Haq, Memon S, Gill NP, Rajput MT (2011) Regeneration of plantlets under NaCl stress from NaN_3 treated sugarcane explants. Afr J Biotechnol 10(72):16152–15156

Jackson PA (2005) Breeding for improved sugar content in sugarcane. Field Crop Res 92:277–290

Jagathesan D, Ratnam R (1978) A vigorous mutant sugarcane (*Saccharum* sp.) clone Co 527. Theor Appl Genet 51(6):311–313

Jain SM (2005) Major mutation assisted plant breeding programs supported by FAO/IAEA. Plant Cell Tissue Org Cult 82:113–123

Jo YD, Kim J-B (2019) Frequency and spectrum of radiation-induced mutations revealed by whole-genome sequencing analyses of plants. Quantum Beam Sci 3(2):7. https://doi.org/10.3390/qubs3020007

Jung JH, Altpeter F (2016) TALEN mediated targeted mutagenesis of the caffeic acid O-methyltransferase in highly polyploid sugarcane improves cell wall composition for production of bioethanol. Plant Mol Biol 92(1–2):131–142

Kannan B, Jung JH, Moxley GW, Lee SM, Altpeter F (2018) TALEN-mediated targeted mutagenesis of more than 100 COMT copies/alleles in highly polyploid sugarcane improves saccharification efficiency without compromising biomass yield. Plant Biotechnol J 16(4):856–866

Kaur M, Thind KS, Sanghera GS, Kumar R, Kashyap L (2001) Gamma rays induced variability for economic traits, quality and red rot resistance in sugarcane (*Saccharum* spp.). Int J Sci Environ Technol 5(2):355–365

Kharkwal MC, Shu QY (2009) The role of induced mutations in world food security. In: Shu QY (ed) Induced plant mutations in the genomics era. Food and Agriculture Organization of the United Nations, Rome, pp 33–38

Koch AC, Ramgareeb S, Rutherford RS, Snyman SJ, Watt MP (2012) An *in vitro* mutagenesis protocol for the production of sugarcane tolerant to herbicide imazapyr. In Vitro Cell Dev Biol 48:417–427

Lakshmanan P, Geijskes RJ, Aitken KS, Grof CLP, Bonnett GD, Smith GR (2005) Sugarcane biotechnology: the challenges and opportunities. In Vitro Cell Dev Biol 41:345–363

Lawrence RJ, Pikaard CS (2003) Transgene-induced RNA interference: a strategy for overcoming gene redundancy in polyploids to generate loss of function mutations. Plant J 36:114–121

Mahlanza T, Rutherford RS, Snyman SJ, Watt MP (2013) In vitro generation of somaclonal variant plants of sugarcane for tolerance to Fusarium sacchari. Plant Cell Reports 32:249–262

Mahlanza T, Rutherford RS, Snyman SJ, Watt MP (2015) Potential of *Fusarium sacchari*-tolerant mutants in controlling *Eldana saccharina* and borer-associated *Fusarium* stem rot in sugarcane. Eur J Plant Pathol 141:825–837

Majid MA, Shamsuzzaman KM, Howlider MAR, Islam MM (2001) Development of sugarcane mutants with resistance to red rot, water logging and delayed for non-flowering through induced mutations. IAEA-Tecdoc 1227:31–44

Mba C, Afza R, Jankowicz-Cieslak J, Bado S, Matijevic M, Huynh O, Till BJ (2009) Enhancing genetic diversity through induced mutagenesis in vegetatively propagated plants. In: Shu QY (ed) Induced plant mutations in the genomics era. Food and Agriculture Organization of the United Nations, Rome, pp 262–265

McIntyre CL, Jackson M, Cordeiro GM, Amouyal O, Hermann S, Aitken KS, Eliott F, Henry RJ, Casu RE, Bonnett GD (2006) The identification and characterization of alleles of sucrose phosphate synthase gene family III in sugarcane. Mol Breed 18:39–50

Meena MR, Kumar R, Chinnaswamy A, Karuppaiyan R, Kulshreshtha N, Ram B (2020) Current breeding and genomic approaches to enhance the cane and sugar productivity under abiotic stress conditions. 3 Biotech 10(10):440

Ming R, Moore PH, Wu KK, D'Hont A, Glaszmann JC, Tew T (2006) Sugarcane improvement through breeding and biotechnology. In: Janick J (ed) Plant breeding reviews. Wiley, Hoboken, NJ, pp 15–118

Mirajkar SJ, Suprasanna P, Vaidya ER (2016) Spatial distribution and dynamics of sucrose metabolising enzymes in radiation induced mutants of sugarcane. Plant Physiol Biochem 100: 85–93

Mohan C, Easterling M, Yau YY (2022) Gene editing technologies for sugarcane improvement: opportunities and limitations. SugarTech 24:369–385. https://doi.org/10.1007/s12355-021-01045-8

Mutant Variety Database (MVD) FAO-IAEA (2021). https://mvd.iaea.org/#!Home. Accessed 30 Oct 2021

Negi P, Rai AN, Suprasanna P (2016) Moving through the stressed genome: emerging regulatory roles for transposons in plant stress response. Front Plant Sci 7:1448

Negi P, Pandey M, Dorn KM, Nikam AA, Devarumath RM, Srivastava AK, Suprasanna P (2020) Transcriptional reprogramming and enhanced photosynthesis drive inducible salt tolerance in sugarcane mutant line M4209. J Exp Bot 71(19):6159–6173

Nikam AA, Devarumath RM, Shitole MG, Ghole VS, Tawar PN, Suprasanna P (2014) Gamma radiation, *in vitro* selection for salt (NaCl) tolerance, and characterization of mutants in sugarcane (*Saccharum officinarum* L.). In Vitro Cell Dev Biol Plant 50(6):766–776

Nikam AA, Devarumath RM, Ahuja A, Harinath B, Shitole MG, Suprasanna P (2015) Radiation-induced in vitro mutagenesis system for salt tolerance and other agronomic characters in sugarcane (Saccharum officinarum L.). Crop J 3(1):46–56

Nogue F, Mara K, Collonnier C, Casacuberta JM (2016) Genome engineering and plant breeding: impact on trait discovery and development. Plant Cell Rep 35:1475–1486

Oloriz MI, Gil V, Rojas L, Portal O, Izquierdo Y, Jimenez E, Hofte M (2012) Sugarcane genes differentially expressed in response to *Puccinia melanocephala* infection: identification and transcript profiling. Plant Cell Rep 31:955–969

Oono Y, Ichida H, Morita R, Nozawa S, Satoh K, Shimizu A, Abe T, Kato H, Hase Y (2020) Genome sequencing of ion-beam-induced mutants facilitates detection of candidate genes responsible for phenotypes of mutants in rice. Mutat Res 821:111691. https://doi.org/10.1016/j.mrfmmm.2020.111691

Oz MT, Altpeter A, Karan R, Merotto A, Altpeter F (2021) CRISPR/Cas9-mediated multi-allelic gene targeting in sugarcane confers herbicide tolerance. Front Genome Ed 3:673566. https://doi.org/10.3389/fgeed.2021.673566

Parajuli S, Kannan B, Karan R, Sanahuja G, Liu H, Garcia-Ruiz E, Kumar D, Singh V, Zhao H, Long S, Shanklin J, Altpeter F (2020) Towards oil cane: engineering hyperaccumulation of triacylglycerol into sugarcane stems. GCB Bioenergy 12:476–490

Parfitt RC (2005) Release of sugarcane varieties in South Africa. Proc S Afr Sugar Technol Assoc 79:63–71

Patade VY, Suprasanna P (2008) Radiation induced *in vitro* mutagenesis for sugarcane improvement. Sugar Tech 10(1):14–19

Patade VY, Suprasanna P (2009) An *in vitro* radiation induced mutagenesis-selection system for salinity tolerance in sugarcane. Sugar Tech 11(3):246–251

Patade VY, Suprasanna P, Bapat VA, Kulkarni UG (2006) Selection for abiotic (salinity and drought) stress tolerance and molecular characterization of tolerant lines in sugarcane. BARC Newsl 273:244–256

Patade VY, Suprasanna P, Bapat VA (2008a) Effects of salt stress in relation to osmotic adjustment on sugarcane (*Saccharum officinarum* L.) callus cultures. Plant Growth Regul 55:169–173

Patade VY, Suprasanna P, Bapat VA (2008b) Gamma irradiation of embryogenic callus cultures and *in vitro* selection for salt tolerance in sugarcane (*Saccharum officinarum* L.). Agric Sci China 7(9):101–105

Phillips RL, Rines HW (2009) Expanding the boundaries of gene variation for crop improvement. In: Shu QY (ed) Induced plant mutations in the genomics era. Food and Agriculture Organization of the United Nations, Rome, pp 21–26

Ponce JP, Jimenez EG, and Gomez RK (1998) Field performance of selected sugarcane (*Saccharum* spp. hybrids) mutants. *In*: Jain SM, Brar S and Ahloowalia (eds.) Somaclonal variation and induced mutations in crop improvement. Kluwer Academic Publishers, Dordrecht, pp. 447-462

Purankar MV, Nikam AA, Devarumath RM, Suprasanna P (2022) Radiation induced mutagenesis, physio-biochemical profiling and field evaluation of mutants in sugarcane cv. CoM 0265. Int J Radiat Biol 98(7):1261–1276

Rao TJ, Srinivasan KV, Alexander KC (1966) A red-rot resistant mutant of sugarcane induced by gamma irradiation. Proc Indian Acad Sci 64(B):224–230

Rodríguez GL, Orellana Pérez P, García Rodríguez L, Pérez Ponce JN, Veitía Rodríguez N, Bermúdez Caraballoso I, Clavero García J, Romero Quintana C (2001) Use of the mutagenesis to improve the girth of the stem in the somaclone IBP 89–169 of sugarcane. Biotecnología Vegetal 1 & 2:71–75

Saif UR, Rashid M, Asad S, Zafar Y, Washeed RA (2001) Use of radiation and *in vitro* techniques for development of salt tolerant mutants in sugarcane and potato. IAEA-Tecdoc 1227:51–60

Samad MA, Begum S, Majid MA (2001) Somaclonal variation and irradiation in sugarcane calli for selection against red rot, waterlogged conditions and delayed or non-flowering characters. IAEA-Tecdoc 1227:45–50

Sengar RS, Sengar K, Garg SK (2011) Biotechnological approaches for high sugarcane yield. Plant Sci Feed 1(7):101–111

Shrivastava AK, Srivastava S (2016) Diversity of the germplasm of *Saccharum* species and related genera available for use in directed breeding programmes for sugarcane improvement. Curr Sci 111(3):475–482

Siddiqui SH, Javed MA (1982) Mutation breeding in sugarcane (*Saccharum* sp. hybrid) by gamma irradiation of cuttings and tissue culture. In: Induced mutations in vegetatively propagated plants. II. Joint FAO/IAEA Div. of Isotope and Radiation Applications of Atomic Energy for Food and Agricultural Development, Vienna, Austria, pp 155–166

Singh G, Sandhu SK, Meeta M, Singh K, Gill R, Gosal SS (2008) *In vitro* induction and characterization of somaclonal variation for red rot and other agronomic traits in sugarcane. Euphytica 160:35–47

Singh R, Tiwari R, Sharma D, Tiwari V, Sharma I (2015) Variability in yield traits of TILLING population of bread wheat (*Triticum aestivum* L.). J Appl Nat Sci 7(1):443–446

Sreenivasan TV, Jalaja NC (1998) Induced mutations and somaclonal variation in sugarcane. In: Jain SM, Brar DS, Ahloowalia BS (eds) Somaclonal variation and induced mutations in crop improvement. Springer, Netherlands, pp 421–446

Suprasanna P, Jain SM (2017) Mutant resources and mutagenomics in crop plants. Emir J Food Agric 29(9):651–657

Suprasanna P, Patade VY, Vaidya ER, Patil VD (2009) Radiation induced *in vitro* mutagenesis, selection for salt tolerance and characterization in sugarcane. In: Shu QY (ed) Induced plant mutations in the genomics era. Food and Agriculture Organization of the United Nations, Rome, pp 145–147

Suprasanna P, Jain SM, Ochatt SJ, Kulkarni VM, Predieri S (2010) Applications of *in vitro* techniques in mutation breeding of vegetatively propagated crops. In: Shu QY (ed) Plant mutation. IAEA, Vienna, pp 369–383

Suprasanna P, Patade VY, Desai NS, Devarumath RM, Kawar PG, Pagariya MC, Ganapathi A, Manickavasagam M, Babu KH (2011) Biotechnological developments in sugarcane improvement - an overview. Sugar Tech 13(4):322–335

Suprasanna P, Jain SM, Ochatt SJ, Kulkarni VM, Predieri S (2012) Applications of *in vitro* techniques in mutation breeding of vegetatively propagated crops. In: Shu QY, Forster BP, Nakagawa H (eds) Plant mutation breeding and biotechnology. CABI Publishing, Wallingford, pp 371–385

Suprasanna P, Mirajkar SJ, Patade VY, Jain SM (2014) Induced mutagenesis for improving plant abiotic stress tolerance. In: Tomlekova NB, Kozgar MI, Wani MR (eds) Mutagenesis: exploring novel genes and pathways. Wageningen Academic Publishers, Wageningen, pp 347–376

Suprasanna P, Mirajkar SJ, Bhagwat SG (2015) Induced mutations and crop improvement. In: Bahadur B, Venkat Rajam M, Sahijram L, Krishnamurthy KV (eds) Plant biology and biotechnology, Plant diversity, organization, function and improvement, vol I. Springer, India, pp 593–617

Thirugnanasambandam PP, Hoang NV, Henry RJ (2018) The challenge of analyzing the sugarcane genome. Front Plant Sci 9:616. https://doi.org/10.3389/fpls.2018.00616

Thrupp LA (2000) Linking agricultural biodiversity and food security: the valuable role of agro biodiversity for sustainable agriculture. Int Affairs 76:265–281

Tysdal HM (1956) Promising new procedures in sugarcane breeding. Proc Int Soc Sugar Cane Technol 9:618–631

Uauy C, Paraiso F, Colasuonno P, Tran RK, Tsai H, Berardi S, Comai L, Dubcovsky J (2009) A modified TILLING approach to detect induced mutations in tetraploid and hexaploid wheat. BMC Plant Biol 9:115

Vaidya ER, Patade VY, Thakare SV, Suprasanna P, Patil VD (2008) Mutations for morphological characters in sugarcane. PKV Res J 32(2):239–242

Wang TL, Uauy C, Robson F, Till B (2012) TILLING in extremis. Plant Biotechnol J 10:761–772

Weksanthia N, Chaisan T, Wannarat W, Chotchutima S, Jompuk P (2021) Mutagenesis and identification of sugarcane mutants using survival on polyethylene glycol and leaf damage under managed water stress. Int J Agron 2021(9):4387696. https://doi.org/10.1155/2021/4387696

Zambrano AY, Demey JR, Fuchs M, Gonzalez V, Rea R, De Sousa O, Gutierrez Z (2003) Selection of sugarcane mutants resistant to SCMV. Plant Sci 165:221–225

Induced Mutations for Developing New Ornamental Varieties

22

M. Tütüncü, K. Y. Kantoğlu, B. Kunter, and Y. Y. Mendi

Abstract

The ornamental plant industry is a dynamic and diverse sector worldwide. Plant breeders develop a great number of new cultivars each year to increase production and supply market demands of the ornamental plants. Mutation breeding is a highly effective method for creating genetic variability in ornamental plants with desirable characters expected within a given species' genetic scope. A mutation creation is called induced mutagenesis of which results are varied according to mutagens and the type of the technique and it is a random process. The target-selected mutagenesis including the random mutagenesis and selection of mutants at a selected locus belongs to this category. By the present, mutations were induced by treatments with physical and chemical mutagens and sometimes their combination. However, biotechnological approaches such as transposable elements, disrupting the gene through the insertion of a DNA fragment and using molecular techniques to create a mutation at a defined site in a DNA molecule have been used to obtain mutations since the development of recombinant DNA technology. There are about 720 ornamental mutant cultivars developed by mutation breeding studies that have been accelerated with tissue culture techniques since the 1970s. In vitro mutagenesis provides for the isolation of chimeric tissues and the propagation of irradiated tissues in mainly vegetatively

M. Tütüncü (✉)
Horticulture Department, Agriculture Faculty, Ondokuz Mayıs University, Samsun, Türkiye
e-mail: mehmet.tutuncu@omu.edu.tr

K. Y. Kantoğlu · B. Kunter
Turkish Energy Nuclear and Mineral Research Agency, Nuclear Energy Research Institute, Ankara, Türkiye

Y. Y. Mendi
Horticulture Department, Agriculture Faculty, Çukurova University, Adana, Türkiye

S. Penna, S. M. Jain (eds.), *Mutation Breeding for Sustainable Food Production and Climate Resilience*, https://doi.org/10.1007/978-981-16-9720-3_22

propagated plants. In addition, in vitro techniques can be supported to the breeding program before, after, and during the mutagen treatments, allowing the scientist to perform the studies timeless and independently of environmental conditions. In vitro techniques have been commonly used in mutation breeding since in vitro technologies were extensively developed. This chapter presents a brief history of mutation breeding in ornamentals and in vitro mutagenesis strategies in ornamental plants.

Keywords

Physical · Chemical · Mutagens · In vitro · Floriculture

22.1 Introduction

The ornamental plant industry is very active and has many kinds of sectors that comprise cut flowers and cut foliage, flower bulbs, potted flowering as well as foliage plants and bedding plants. Ornamental plant production areas worldwide have increased by 17.75%, reaching over 1.7 million ha in the last 9 years between 2009 and 2017. The production value has increased by 46.44% in the same period, and it reached approximately 65.2 billion Euros. Among the ornamental crop groups, outdoor ornamental plants have the highest production area, with more than 1.9 million ha. In contrast, cut flowers and indoor ornamental plants have the highest production value with 35.5 million euros (Kazaz et al. 2020).

Plant breeding aims to change and improve the genetic structure of plants considering its economic benefits. Plant breeders develop plenty of new flower varieties, altering plant traits such as pigmentation, flower shape, or prolonged shelf life because of market demands each year (Tutuncu et al. 2017; Datta 2020). Genetics became a fundamental science of plant breeding after Gregor Johann Mendel discovered the laws of heredity in the nineteenth century. However, further advancements in plant breeding took place when the hybridization methodology was developed (Oladosua et al. 2016). New genetic variability in plant breeding is formed via hybridization among parental lines in their genomic structure's face of desired traits. The breeders aim to combine these traits from different sources and then selection of individuals is performed considering breeding targets (Schum 2003). Various conventional methods such as selection and hybridization methods are used in plant breeding programs. Selection is the oldest technique among conventional breeding, and it has still been widely used for crop improvement. It is accomplished primarily based on visual appearance. Armed with a mental picture of the target trait, the person conducting the operation visually discriminates among the available variability based on appearance to identify and select desirable plants (Acquaah 2015; Datta 2020).

Variability is highly essential to allow the breeders to select cultivars considering the following criteria: (a) more adapted to environmental changes, (b) more efficient in utilizing nutrients, (c) more tolerant to diseases and pests, and (d) improved in yield and quality (Donini and Sonnino 1998). The variability attributed to genes that encode specific traits and can be transmitted from one generation into the next is described as a genetic or heritable variation. There are three ways in which genetic or heritable variability originates in nature; gene recombination, modifications in chromosome number, and mutations. The significant fact to note is that, rather than wait for them to occur naturally, plant breeders use various techniques and methods to manipulate these three phenomena more and more intensively, as they generate a genetic variation for their breeding programs (Acquaah 2007; Datta 2020; Nelka et al. 2021).

Mutation breeding is a sophisticated process with specific jargon known at the beginning of the mutation studies. A mutation is a term used for the heritable changes that occur suddenly in the DNA in a living cell. These changes in the DNA are not caused by genetic segregation or genetic recombination (Roychowdhury and Tah 2013; Hernández-Muñoz et al. 2019). Mutations have been shown as a way of procreating variations in a variety (Çelik and Atak 2017). Although spontaneous mutation can be found in nature, it is not easy to distinguish. The results of spontaneous mutation and recombination in populations interact with environmental factors as genetic variations (Novak and Brunner 1992). The frequency of spontaneous mutation is shallow (1×10^{-6}) (Suprasanna et al. 2017), and the usage of spontaneous mutations in plant breeding programs is not possible (Lönnig 2005; Oladosua et al. 2016; Suprasanna et al. 2017).

The mutation can also be formed via experimental tools that are termed mutagen. The process of mutation creation is called mutagenesis of which results are varied according to mutagens and the type of the technique. "Induced mutagenesis" and "mutation induction" have common meanings for the process of mutation induction using physical and chemical mutagens, and it is a random process. In addition, target-selected mutagenesis including the random mutagenesis and selection of mutants at a selected locus belongs to this category. Newer biotechnological approaches such as transposable elements, disrupting the gene through the insertion of a DNA fragment (insertional mutagenesis) and using molecular techniques to create a mutation at a defined site in a DNA molecule (site-directed mutagenesis) have been used to obtain mutations since the development of recombinant DNA technology (Forster and Shu 2011; Spencer Lopes et al. 2018). Mutations are broadly classified based on the effects of an organism. The first mutations were applied successfully and observed mutants were described based on their phenotypes. Classification of the mutations at the genotypic level can be mainly divided into intragenic or point mutations (occurring within a gene in the DNA sequence), intergenic or structural mutations within chromosomes (inversions, translocations, duplications, and deletions), and mutations leading to changes in the chromosome number (polyploidy, aneuploidy, and haploidy) (Pathirana 2011). Plant breeders mainly target point mutations (recessive) because more extensive chromosome mutations have several adverse effects like reduced fertility of male and female flower parts. However, reduced fertility in vegetatively propagated plants is not prominent and sometimes is even desired (Kharkwal 2011).

Mutation breeding is one of the techniques in conventional breeding systems that mutagenesis is used during the processes by breeders. In ornamental plant breeding, phenotypic mutations are valuable tools to produce new cultivars for the breeders. Unlike improvement of fruit or vegetable cultivars through mutation breeding techniques, ornamental plants are superior experimental materials for mutation breeding studies because deviant leaf/flower color or shapes can be approved as the economically important traits in ornamental plant markets by the consumer. Additionally, high mutation frequency can be seen in ornamental plants due to their heterozygous nature (Suprasanna et al. 2017). However, vegetable or fruit cultivars having abnormal fruit shapes or colors may not be approved quickly by the markets due to consumers' consumption patterns.

22.2 A Brief Mutation Breeding History of Ornamental Plants

The use of mutant plants for human welfare has been presumably started from the beginning of agriculture. Spontaneous mutation and mutant plants were used to improve new cultivars and enlarge genetic diversity in a population by the breeders for the centuries. The history of plant mutation traced back to 300 BC with reports of mutant crops in China (Muller 1927). Although there are early examples in plant breeding through mutation breeding, spontaneous mutant "morning glory" (*Ipomoea nil*) in Japan in the late seventeenth century is a significant milestone of spontaneous mutations for ornamental plants. In 1741 and following years, Carl Von Linnaeus described various mutants. In the late nineteenth century, while the Dutch scientist Hugo de Vries was experimenting on the "rediscovery" of Mendel's laws of inheritance, variations were found in evening primrose (*Oenothera lamarckiana*) and snapdragon (*Antirrhinum majus*). This variation differed from others in that it did not follow Mendelian patterns of inheritance (3:1), but it was nevertheless heritable. For the first time, mutation as a mechanism of creating variability was identified in 1901. The mechanism of heritable changes was distinct from recombination and segregation (Muller 1927).

At the beginning of the twentieth century, Dutch scientist Hugo de Vries suggested using X-rays for mutation induction after discovering X-rays. This suggestion has opened a new door for mutation breeding studies. X-ray was improved nearly 25 years later to induce mutation in barley (Stadler 1928a) and other plants (Stadler 1928b; Broertjes and Van Harten 1988; Van Harten 1998).

The history of mutation breeding in ornamental plants started with the first observation on mutation studies in evening primrose and anthurium plants. However, this attempt was mainly based on the variations and their heritance. It did not aim to improve any genetic material during the first years of experimental mutation induction studies. The fundamental studies on mutation breeding have led to understanding the practical significance of mutation, especially in vegetatively propagated ornamental plants that are very suitable for applying mutation breeding. A new useful mutation obtained from vegetatively propagated crops can easily be maintained by clonal propagation. The work of Naylor and Johnson on African violet (*Saintpaulia ionanthä*) in 1937, has historical importance in mutation breeding in vegetatively propagated crops. They observed that the adventitious buds were originated from a single (epidermal) cell origin when the petiole base of cut leaves of African violet was cultured. These findings proved to be of great practical significance for mutation breeding, and similar observations were made in other plant species later (Broertjes and Van Harten 1988; Datta 2020).

Tulip cultivar "Faraday" is the first mutant commercial ornamental plant obtained via irradiation method in 1949. The cultivar differed from its parent in terms of altered flower color (Van Harten 2002; Schum 2003). The cooperation of FAO (Food and Agriculture Organization of the United Nations) and IAEA (International Atomic Energy Agency) is another milestone in plant mutation breeding. In 1964, FAO and IAEA created the Joint FAO/IAEA Division as a strategic partnership to mobilize the talents and resources of both organizations and hence to broaden cooperation between their Member States in the peaceful application of nuclear science and technology safely and effectively to provide their communities with more, better, and safer food and agricultural produce while sustaining natural resources (IAEA 2018a; Spencer Lopes et al. 2018). The Mutant Variety Database was also established under the division by FAO and IAEA to document officially released mutant varieties from the member countries worldwide (IAEA 2018b). FAO and IAEA's established division and database allow us to follow effective plant breeding programs using induced mutation and their outcomes.

The first mutant ornamental plant, the tulip cultivar Faraday was officially approved in 1949. It was developed by irradiation with X-rays that improved attribute of mutant variety is flower color. In 1954, the second tulip mutant cultivar Estella Rijnveld developed by irradiation with X-rays, was approved. It was mainly improved the attribute of mutant variety is flower color.

Another mutant population creation can be done by using ion rays. Ion rays, including proton, helium, and more heavily charged particles, accumulate high energy locally than gamma rays or X-rays. The biological effect of ion rays has been investigated in detail with studies since the 1960s, and it has been found that ion rays show higher relative biological effect (RBE), lethality, and cell inactivation compared to low LET (linear energy transfer) radiation such as gamma rays and X-rays (Tanaka et al. 2010). While it was determined that ion rays mainly affect single or double-stranded DNA breaks with low reparability, the precise effects of ion rays on mutation excitation have not been fully clarified with the studies conducted in those years, whether they provide an advantage over gamma rays and chemical mutagen applications. However, as a result of the advancements in the use of ionizing radiation in different areas, studies on the usability of proton, helium, and carbon-based ion beams in plant breeding have been carried out more effectively since 1991. Many countries, especially Japan, have begun to conduct serious research on this subject. Studies have been carried out intensively in many plant species, especially in vitro and in vivo mutation breeding studies in ornamental plants. Many mutant varieties are produced in this way. Ion rays have been used intensively for cut flowers, potted plants, and bedding plant breeding programs for the last two decades (Tanaka et al. 2010). So many mutant varieties were generated

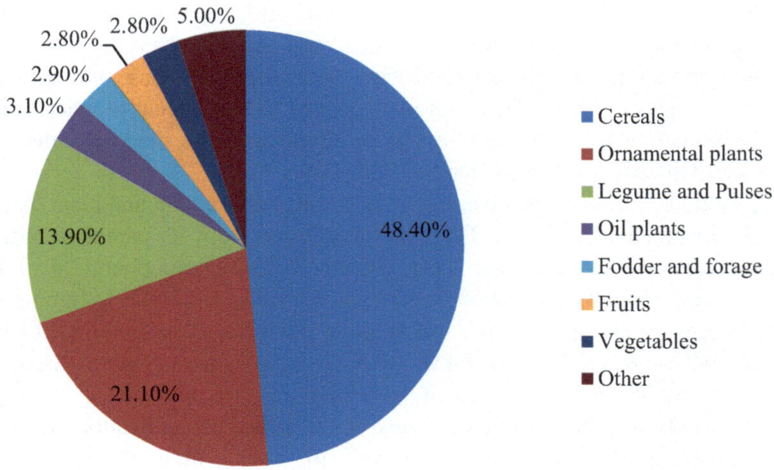

Fig. 22.1 Usage category of mutant varieties listed in IAEA mutant variety database in 2021

by in vitro and in vivo mutation studies by using this kind of irradiation sources in carnation (Okamura et al. 2012), chrysanthemum (Tanaka et al. 2010; Tamaki et al. 2014), rose (Yamaguchi et al. 2003), cyclamen (Ishizaka et al. 2012), *Saintpaulia ionahta* (Zhou et al. 2006), lily, *Limonium* spp., Gypsophila spp., *Gentiana* spp., tulip, *Delphinium* spp., hydrangea, petunia, begonia, *Dendrobium* spp., and *Cymbidium* spp. (Hase et al. 2012; Yamaguchi 2018).

In mutation breeding studies, it is crucial to determine the effective mutation dose (EMD 50) and lethal dose (LD 50) values for the genetic material to be studied, regardless of the plant species to be studied, and to establish the mutant population, especially based on the effective mutation dose. This point, which seems like a small detail for many researchers, has a significant effect on the study's success. If the correct physical or chemical mutagen dose is applied and the breeding work is started, it will be beneficial to increase the diversity of the variation created by ensuring that the mutation frequency is captured at a higher rate. For this reason, it is important to determine the effective mutation dose by trying at least seven doses, depending on the characteristics of the material to be studied (seed, vegetative material, in vitro plantlet, etc.) (Anonymous. 2018; Haspolat et al. 2019; Kantoğlu and Kunter 2021).

According to the "Mutant Variety Database," 3365 officially released mutants in about 225 species are listed (IAEA 2021). Seventy-two member states of the database introduced these mutants, and the usage category of the mutants is illustrated in Fig. 22.1. By 2021, China has the highest number of mutant varieties among countries with more than ten percent of the total mutants, and Japan and India follow it. The rest of them are distributed among other countries (Fig. 22.2). By 2021, there were 720 ornamental mutant varieties documented in more than 20 countries (Fig. 22.3). Holland releases the highest number of ornamental mutants, and it is in parallel with the activity of the countries in the ornamental plant market.

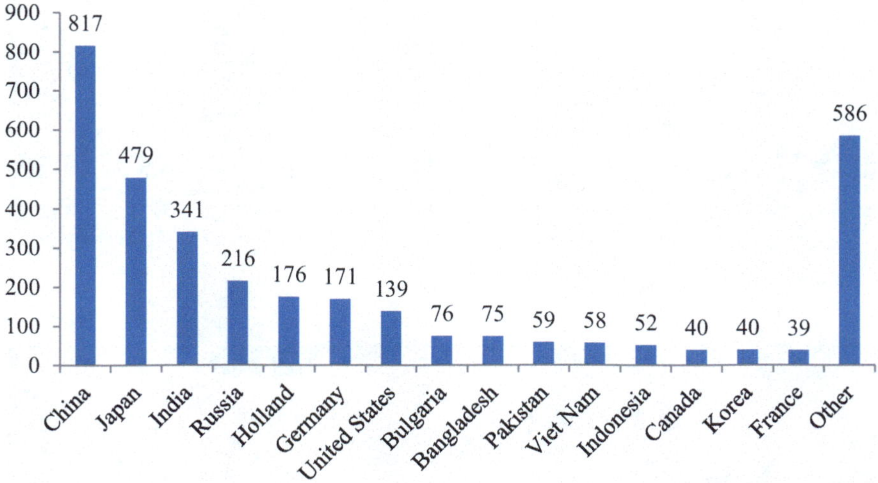

Fig. 22.2 The allocation of mutant varieties among countries is based on the IAEA mutant variety database in 2021

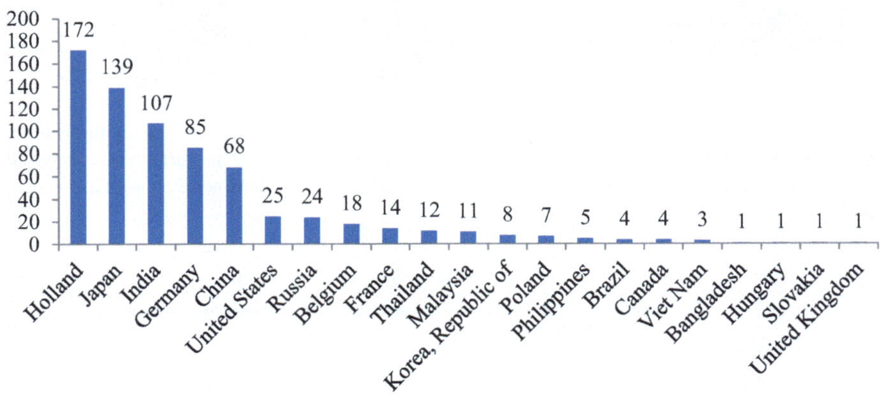

Fig. 22.3 The number of ornamental mutant varieties distributed among the countries (2021)

Chrysanthemum sp. is defined as a model plant for mutation studies due to high susceptibility to mutation breeding and mutational differences in flower morphology can be obtained very easily. On the other hand, it is also an important commercial flower produced with mutation breeding using gamma rays and X-rays (Datta 2014; Yamaguchi 2018). In addition, chrysanthemum is a valuable ornamental plant species in the market due to its natural flower colors and plant habitus (Fig. 22.4). Considering the officially released mutant number among ornamental plant species, it is the most common ornamental plant species (Fig. 22.5).

In 1960, after 4 years of the mutant tulip varieties released, two rose mutants, "Pink Hat" and "Paula," were officially approved. These were the first ornamental mutants of the USA. Both mutants were improved with gamma irradiation, and

Fig. 22.4 "Brandevil" a commercial cultivar of chrysanthemum and mutant types with different flower colors and plant habitus induced by gamma irradiation of original variety. (Courtesy of Dr. K.Y. Kantoğlu and Dr. B. Kunter, Turkish Energy Nuclear and Mineral Research Agency, Nuclear Energy Research Institute, Ankara, Türkiye)

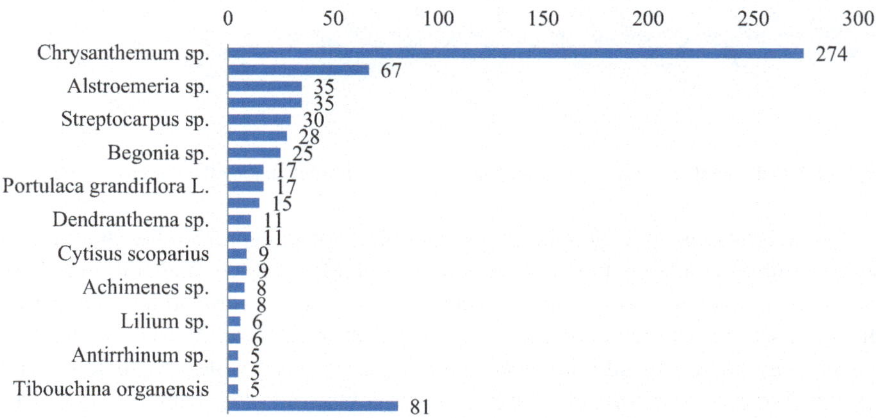

Fig. 22.5 The officially released mutant number of ornamental plant species documented in FAO/IAEA Mutant Variety Database

mainly flower colors of the mutants differed from their parents. In 1961, the mutant variety of Antirrhinum Juliva was obtained in Germany. It was developed by hybridization with X-rays irradiated mutant *A. divaricata*. The first carnation mutant was obtained from irradiated rooted cuttings with gamma rays in the USA. The main improved attributes of mutant variety were fewer ray flowers staying longer after cutting. In 1960, the mutant variety "Izetka Kopenicker Barbarossa Goldkissen" and "Izetka Kopenicker Barbarossa Rotstern" of chrysanthemum, which has the highest number of the mutant variety at present, were developed by treatment with X-rays (10–25 Gy) in Germany (IAEA 2018b).

In the late 1960s, chemical and physical mutagens were started to use for inducing mutant varieties. The streptocarpus was the first species used as a material to induce mutation with both chemical and physical mutagens. The mutant variety "Purple Nymph" was developed by irradiating detached leaves with X-rays and colchicine. The main improved attributes of mutant variety were a giantflower, purple flower color, plant sturdier, and the rest of genotype unchanged. An exciting example of the mutant ornamental plants was the "Raymond Smith" semi-cactus mutant variety developed by irradiating dormant tubers with X-rays (20 Gy) in 1970. The listed mutants were mainly improved in agronomic and some of the plant morphological characteristics, while the mutant variety "Raymond Smith" had a new physiological trait as semi-cactus. In the 1980s, in vitro techniques such as meristem and anther culture were used to start induced mutations in ornamental plants. In 1983, Barton et al. (1983) successfully transferred an exogenous DNA into a tobacco plant using *Agrobacterium tumefaciens* as a vector. It opened the door to transgenic technology that enables the transfer of genes to host plants from any source. However, transformation efficiencies are generally low, and protocols are dependent on species or even cultivar dependent (Donmez et al. 2013). Then, several ornamental plants have been engineered for flower color modifications (Kishi-Kaboshi et al. 2018). Petunia, unable to produce pelargonidin-derivatives as a pigment, was the first genetically modified ornamental plant for flower color. A1 gene from maize transferred into RL01 and intermediate pelargonidin biosynthesis provided by Meyer et al. (1987). Additionally, rose and chrysanthemum do not have violet or blue varieties due to lack of flavonoid 3′,5′-hydroxylase which is a key enzyme in the synthesis of delphinidin, and purple/violet and blue flower in rose and chrysanthemum were produced by transferring the A3′5′GT gene. At present, more than 50 ornamental plants have been genetically modified (GM) via gene transformation but, most of them are not on the market due to a lack of regulatory approval (Boutigny et al. 2020). In the International Service for the Acquisition of Agri-biotech Applications (ISAAA) GM Approval Database, 23 GM events of 19 carnations, 2 petunias, and 2 rose are listed in 2022. In these GM studies, *A. tumefaciens* was used to transfer genes responsible for flower color, herbicide tolerance, and vase life to carnation, petunia, and rose (Table 22.1).

In parallel with the improvement of genetic transformation, somaclonal variation has been used to obtain mutants. In 1986, the mutant variety "CapliLuluby" of geranium was the first officially developed mutant by somaclonal mutation through anther culture. It was followed by "Gold On" a mutant Orchid variety developed by

Table 22.1 Common genes were introduced to ornamental plant species via *A. tumefaciens* and their functions (ISAAA 2022)

Species	Gene	Source	Product	Function
Carnation	dfr	*Petunia hybrida*	Dihydroflavonol-4-reductase (DFR) hydroxylase enzyme	Catalyzes the production of the blue-colored anthocyanin pigment delphinidin and its derivatives
	hfl (f3′5′ h)	*Petunia hybrida*	Flavonoid 3′,5′-hydroxylase (F3′5′ H) enzyme	Catalyzes the production of the blue-colored anthocyanin pigment delphinidin and its derivatives
	surB	*Nicotiana tabacum*	Herbicide-tolerant acetolactate synthase (ALS) enzyme	Confers tolerance to sulfonylurea herbicides and other acetolactate synthase (ALS) inhibiting herbicides
	bp40 (f3′5′ h)	*Viola wittrockiana*	Flavonoid 3′,5′-hydroxylase (F3′5′ H) enzyme	Catalyzes the production of the blue-colored anthocyanin pigment delphinidin and its derivatives
	sfl (f3′5′ h)	Sage (*Salvia splendens*)	Flavonoid 3′,5′-hydroxylase	Involved in the biosynthesis of a group of blue-colored anthocyanins called delphinidins
	dfr-diaca	Carnation (*Dianthus caryophyllus*)	Dihydroflavonol-4-reductase enzyme	Functions in the biosynthesis pathway of the pink/red-colored anthocyanidin 3-*O*-(6-*O*-malylglucoside) pigment in carnations
	cytb5	Petunia (*Petunia hybrida*)	Cytochrome b5	Cyt b5 protein acts as an electron donor to the Cyt P450 enzyme and is required for total activity of the Cyt P450 enzyme Flavonoid 3′,5′ hydroxylase in vivo and the generation of purple/blue flower colors
	acc	*Dianthus caryophyllus*	A modified transcript of 1-amino-cyclopropane-1-carboxylic acid (ACC) synthase gene	Causes reduced synthesis of endogenous ethylene through a gene silencing mechanism and thus delayed senescence and longer vase life
Petunia	uidA	*Escherichia coli*	Beta-D-glucuronidase (GUS) enzyme	Produces blue stain on treated transformed tissue, which allows visual selection

(continued)

Table 22.1 (continued)

Species	Gene	Source	Product	Function
Rose	5AT	*Torenia* sp.	Anthocyanin 5-acyltransferase (5AT) enzyme	Alters the production of a type of anthocyanin called delphinidin
	bp40 (f3′5′ h)	*Viola wittrockiana*	Flavonoid 3′,5′-hydroxylase (F3′5′ H) enzyme	Catalyzes the production of the blue-colored anthocyanin pigment delphinidin and its derivatives

somaclonal mutation by mericlone culture in 1995, the mutant variety Sumi-takara was of chrysanthemum developed by somaclonal mutation through anther culture in 1991, "Cocktail Dress" and "Sherley Baby White" variety of *Cymbidium* sp. and *Oncidium* sp., respectively, in 1997, "Kishu Kasumi 1 Gou" mutant variety of *Gypsophila elegans* M. Bieb. in 2000, and so on. In 2010, "Abhimanyu" variety of *Bougainvillea* sp. is a spontaneous mutant detected as a mericlinal chimera in a very old and injured plant of gamma-ray induced mutant "Arjuna." The spontaneous mutant has been isolated in pure form by the air layering method and released as a new cultivar (Yamaguchi 2018).

The developments in molecular approaches involve to detection mutational events can be characterized by using molecular marker-based tools such as cDNA-amplified fragment length polymorphism (AFLP), single-strand conformational polymorphism (SSCP), serial analysis of gene expression (SAGE), microarray, differential display, TILLING (Suprasanna and Jain 2017). For instance, TILLING is one of the valuable techniques to detect random point mutations in populations created by chemical mutagen such as EMS (Simsek and Kacar 2010). This technique, in 2021, is evaluated as low efficient to discriminate mutations in a complex genome or multiple gene family members. They are now evolved combining with sequence technology such as from TILLING to an optimized high-throughput TILLING by target capture sequencing technology, or TILLING-by-Sequencing+ (TbyS+) (Lakhssassi et al. 2021).

Afterward, development in gene transformation technologies and other genomic approaches lead to precise or targeted genetic modification tools such as engineered homing endonucleases or meganucleases, transcription activator-like effector nucleases (TALENs), and clustered regularly interspersed short palindromic repeats zinc finger nucleases (ZFNs) and CRISPR-Cas9. These novel tools allow efficient and directed modification introduced into cultivated varieties while taking advantage of existing natural genetic variation or through induction of mutations (Suprasanna and Jain 2017). However, the use of these modern tools is minimal due to inadequate optimized gene transformation protocol for the ornamental plant species, little DNA sequence information, and the expensive cost of the procedures compared to the older method used in ornamental plant breeding through mutation induction (Melsen et al. 2021).

By the present, mutations were induced by treatments with physical mutagens (mainly X-rays, gamma rays, and ion rays), chemical mutagens (mainly EMS-ethyl methanesulfonate, DMS-dimethyl sulphate, Colchicine) and sometimes their combination considering listed ornamental mutant varieties at the FAO/IAEA database. These mutant varieties have been modified for traits such as flower shape, flower color, stem length, leaf size, plant height, flowering and harvest time and resistance to edaphic factors.

22.3 The Role of In Vitro Techniques in Mutation Breeding of Ornamental Plants

The mutation induction studies have been experimentally performed over 70 years to produce mutant cultivars by the developing agronomic and botanical characteristics of the plants (Jain 2005). The most common mutagens used are chemical (alkylating agents)-ethyl-nitrosourea (ENH), methyl-nitroso-urea (MNH), sodium azide (SA), ethyl-methane-sulphonate (EMS), and physical agents–X-rays, gamma rays, fast neutrons, ultraviolet, and laser (Jain 2000, 2005). Since induced mutation has begun to study in crops, it has already been evaluated for ornamental crops by in vitro techniques (Broertjes and Van Harten 1988; Oladosua et al. 2016; Pathirana 2011; Datta 2012, 2014, 2020). Ion beams irradiation is a common technique to induce mutation in ornamental plants. Ion beams consist of ion particles that are accelerated by a cyclotron. The resulting high linear energy transfer (LET) radiation has greater biological effects on plants than low LET radiation, such as gamma rays and X-rays. The mutation induction via ion beam irradiation in ornamental plants has been well documented by Yamaguchi (2018).

In the 1970s, the improvements of in vitro technology led scientists to use it in experimental mutation studies to assist mutation breeding programs. In later studies, it has been determined that higher mutation rates can be obtained using in vitro techniques and radiation applications (Suprasanna et al. 2015). At present, these techniques have become powerful tools in breeding ornamentals such as alstroemeria, begonia, chrysanthemum, carnation, dahlia, and streptocarpus. Mutation induction in in vitro, called in vitro mutagenesis, refers to creating mutation by treating explants or in vitro cultures (protoplasts, cells, tissues, and organs) with a mutagen, followed by mutant screening/selection and characterization (Suprasanna et al. 2012). In vitro cultures such as organogenesis, meristem culture, anther culture or ovary culture, and somatic embryogenesis in conjunction with mutation breeding significantly contribute to improving new crops and new varieties (Maluszynski et al. 1995; Çelik and Atak 2017; Abdulhadi et al. 2019).

Initial in vitro mutation studies had focused to use adventive shoot tips on *Streptocarpus* sp. (Broertjes 1969), *Achimenes* sp., and *Saintpaulia* sp. (Broertjes 1972), *Begonia* sp. (Brown and Harney 1974), *Kalanchoe* sp. (Nakornthap 1974), and *Chrysanthemum morifolium* (Broertjes and Roest 1976) species. Using multicellular structures such as shoot tips as explant resources in in vitro mutagenesis leading to form chimeras is a significant problem in plant regeneration after treatments. However, somatic embryogenesis prevents chimeras because it consists of a single cell origin, and a whole mutant plant can be formed from mutagenized somatic embryo cells. On the other hand, the low germination rate of somatic embryos limits commercial production of mutant plants, so the combined use of somatic embryogenesis and organogenesis can be suitable for mutation induction and mass propagation (Suprasanna et al. 2012).

The main advantages of in vitro mutagenesis compared with in vivo are that mutagens can be treated to many homogenous individuals, developmental patterns of explants can be synchronized, in vitro mass selection is possible, and reproducible results can be obtained (Suprasanna et al. 2012). Explants can be uniformly treated with physical and chemical mutagens, be grown in a uniform cultural environment, and conducted with large populations within a limited space and at any time of the year (Datta and Teixeira da Silva 2006; Haspolat et al. 2019).

In addition, the breeding program can be accelerated by in vitro selection, propagation of mutants or mutant candidates and expression of induced mutations in the pure homozygote obtained through microspore, anther, or ovary culture can enhance the rapid recovery of the desired traits. On the other hand, the induced mutation combined with in vitro culture techniques can be the only method of improving an existing cultivar in some vegetatively propagated species (Maluszynski et al. 1995; Su et al. 2019).

22.3.1 The Factors Affecting In Vitro Mutation in Ornamental Plants

The development of highly efficient tissue culture protocols must be optimized to combine mutation breeding. Therefore, several factors such as plant material, in vitro techniques, culture condition, choice of genotype, mutagens affect the processes. By 2018, tissue culture application in ornamental mutants that have been developed by somaclonal mutation through tissue cultures such as anther culture or petal culture alone or irradiation with X-rays and gamma rays directly tissue cultured explants (Table 22.2).

When starting in vitro mutation breeding studies, the factors that will affect the outcome should be considered. Limitations and advantageous factors are presented below.

The advantages of in vitro mutagenesis are:

1. High mutation rate
2. Homogeneous mutation treatments
3. Use of single cell systems and application of selective agents to homogeneous cell population
4. Requirement of less space to handle large population within a short time and keeping the plant material disease-free and dependent on seasons
5. It allows controlling and monitoring all steps of in vitro cycles
6. It offers a broader choice of explants for mutagenic treatment such as root segments, cuttings leaves, apical and axillary buds, anthers, tubers, microspores,

Table 22.2 The mutant varieties developed by induced mutation techniques in combination with tissue culture techniques (data source: Joint FAO/IAEA Mutant Variety Database)

Variety name	Year	Common name	Description	Improvement
Loncerda	1983	Carnation	Irradiation of in vitro culture with chronic gamma rays	Resistance to *Fusarium oxysporum* and flower color change to cherry red
Chaichoompon	1983	Carnation	Irradiation of tissue culture with gamma rays (10 Gy)	White flower color with pink strikes on petal
Baiogiku rainbow yellow	1985	Chrysanthemum	Irradiation with gamma rays. Tissue cultured from petal	Changed flower color from red to yellow
CapliLuluby	1986	Geranium	Somaclonal mutation through anther culture	Changed flower
Golden Cremon	1987	Chrysanthemum	Irradiation of tissue culture with gamma rays (10 Gy)	Golden yellow ray florets with anemone disc florets
Capli Ice	1988	Geranium	Somaclonal mutation through petal culture	
Hae-no-Hatsu-yuki	1990	Chrysanthemum	Irradiation of tissue culture with gamma rays	Yellowish white flower color
Hae-no-Kurenai	1990	Chrysanthemum	Irradiation of tissue culture with gamma rays	Purple, pink flower color
Hae-no-Kirameki	1990	Chrysanthemum	Irradiation of tissue culture with gamma rays	Yellowish orange flower color
Hae-no-Miyarabi	1990	Chrysanthemum	Irradiation of tissue culture with gamma rays	Pale yellowish pink flower color
Hae-no-Yuugure	1990	Chrysanthemum	Irradiation of tissue culture with gamma rays	Altered flower color
Hae-no-Kagayaki	1990	Chrysanthemum	Irradiation of tissue culture with recurrent gamma rays and X-rays	Bright yellow flower color
Ivory Memory	1992	Lily	Irradiation of tissue culture with gamma rays	Pale orange yellow petal color
Mei Wako	1993	Scotch broom	Irradiation of tissue culture with gamma rays	Extremely short plant and altered flower color: pale yellow; brown, deep purple pink

(continued)

Table 22.2 (continued)

Variety name	Year	Common name	Description	Improvement
Mei Hiro	1993	Scotch broom	Irradiation of tissue culture with gamma rays	Improved flower color, outer standard ground color: rainbow; outer wing color: deep reddish brown
Mei Lord	1993	Scotch broom	Irradiation of tissue culture with gamma rays	Brown outer keel color and wide keel
Seto-no-otome	1994	Aster	Irradiation of meristem culture with gamma rays	More ligulose flowers and deep flower color
Mei Eve	1994	Scotch broom	Irradiation of tissue culture with gamma rays	Extremely short plant and flower color: pinkish white, brown reddish purple
Daifura Lavender	1995	Statice	Irradiation of tissue culture with gamma rays	Bright reddish purple sepal color
Gold One	1995	Orchid	Somaclonal mutation by mericlone culture	Improved flower color
Daifura Pink Super	1996	Statice	Irradiation of tissue culture with gamma rays	Bright purple sepal color and a smaller number of floral stalks
White Lineker OW-1	1996	Chrysanthemum	Irradiation of in vitro culture with X-rays	White flower color
Dai-lady Rose	1996	Statice	Irradiation of tissue culture with gamma rays	Vivid yellowish green petal color
Dai-lady White	1996	Statice	Irradiation of tissue culture with gamma rays	Pinkish white sepal color
Yellow Prism	1997	Chrysanthemum	Irradiation of tissue culture with gamma rays	Erect flower petal and bright yellow petal color
Pearl Prism	1997	Chrysanthemum	Somaclonal mutation through anther culture	Erect flower petal and more transparent petal
Cocktail Dress	1997	Boat orchids	Somaclonal mutation through tissue culture (mericlone)	Improved petal color and shape
Sherley Baby White	1997	Orchid	Somaclonal mutation by mericlone culture	Improved flower color
Amazon	1998	Chrysanthemum	Irradiation of tissue culture with gamma rays	Deep orange red flower color
Royal Wedding	1998	Chrysanthemum	Somaclonal mutation through tissue culture	Deep purple pink flower color

(continued)

Table 22.2 (continued)

Variety name	Year	Common name	Description	Improvement
Kopurosu	1998	Japanese lawngrass	Irradiation of meristem culture with X-rays	Better spring vigor, earlier green color of leaf, and no browning in winter
Mei Sum	1998	Scotch broom	Irradiation of in vitro culture with chronic gamma rays	Dwarfness and altered flower color: yellow and deep red
Kirinade Salmon	1999	Carnation	Irradiation of tissue culture with X-rays	Improved flower surface and vivid pink flower color
Garden-Spice carnation Marble (Kirinade Marble)	2000	Carnation	Irradiation of tissue culture with X-rays	Improved flower surface and pinkish white with vivid purple red stripe flower color
Kirimaji Cherry Red (Cherry Red Wave)	2000	Petunia	Irradiation of tissue culture with X-rays	Vivid red monotone flower color
Kishu Kasumi 1 Gou	2000	Annual baby's-breath	Somaclonal mutation through meristem culture	More wax on the leaf surface and top of the petal V-shape
Nazerea Grace White	2001	Abelia	Irradiation of callus with gamma rays (10 Gy)	Improved flower color and form
Nazerea Soft Pink	2001	Abelia	Irradiation of callus with gamma rays (10 Gy)	Improved flower color and form
Chikugo-midori	2001	Mat rush	Somaclonal mutation from meristem culture	More long stems, higher weight of total long stems
Mrs. Elegant	2001	Carnation	Somaclonal mutation by tissue culture	Smaller number of flowers per flower stalk
Etenraku	2001	Chrysanthemum	Somaclonal mutation by tissue culture	Shorter stem
Pretty Wedding	2001	Chrysanthemum	Irradiation of in vitro culture with X-rays	Bicolor outside pink inside white flower color
My Comfort	2001	Creeping bent grass	Somaclonal mutant from tissue culture	Taller plants
Pink Purple	2001	Melastoma	Irradiation of callus with gamma rays	Spot on the leaf and no luster on the leaf
Boh-red	2002	Carnation	Somaclonal mutation through tissue culture	Vivid red flower color
Kirikami Red	2002	Carnation	Somaclonal mutation through tissue culture	Deep orange red surface of flower

(continued)

Table 22.2 (continued)

Variety name	Year	Common name	Description	Improvement
Misty Pink Vital Ion	2002	Carnation	Irradiation of regenerating callus with C ion beam	Altered flower color and petal shape
Red Vital Ion	2002	Carnation	Irradiation of regenerating callus with C ion beam	Bright red flower color
Ion-no-Kouki	2003	Chrysanthemum	Irradiation of in vitro petal culture with 20 Gy C ion beams	Complex with light yellow and pink
Ion-no-Mahou	2003	Chrysanthemum	Irradiation of in vitro culture with Ne ion beams	Altered flower color (light orange on adaxial, dark yellow orange on abaxial side)
Hae-no-Eiokou	2003	Chrysanthemum	Irradiation of in vitro petal culture with 50 Gy gamma rays chronic	Mild orange flower color
Hae-no-Myoujou	2003	Chrysanthemum	Irradiation of in vitro petal culture with 50 Gy gamma rays chronic	Bright yellow flower color
Hae-no-Yumeguruma	2003	Chrysanthemum	Irradiation of bud and petal cultures with 75 Gy gamma rays chronic	Yellowish white color and smaller flower size
Hae-no-Awabeni	2003	Chrysanthemum	Irradiation of in vitro culture (leaf culture) with 30 Gy gamma rays acute	Light pinkish white flower color
Princess Kagawa	2004	Chrysanthemum	Irradiation of in vitro culture with X-rays	Reddish purple stem color and petal color
Yua-red	2005	Carnation	Somaclonal mutation through meristem culture	Vivid red flower color, suspended flower, and large flower
Dark Pink Vital Ion	2005	Carnation	Irradiation of regenerating callus with C ion beam	Altered flower color and petal shape
Sunny Sky	2005	Larkspur	Somaclonal mutation by tissue culture	Shorter plant and shorter inflorescence
Dong-i	2005	Cymbidium orchid	Irradiation of in vitro culture with gamma rays (30 Gy)	Short plant and marginal stripe ion leaves
Emi-akari	2006	Chrysanthemum	Irradiation of in vitro culture with soft X-rays	Irregulate petal color: pale greenish yellow

(continued)

Table 22.2 (continued)

Variety name	Year	Common name	Description	Improvement
ARTIpurple	2011	*Dendranthema grandiflorum* (Ramat.) Kitamura	Irradiation of young in vitro plants	Unspecified
ARTIqueen	2011	*Dendranthema grandiflorum* (Ramat.) Kitamura	Irradiation of young in vitro plants	Unspecified
Yellow Sun	2014	Chrysanthemum	Irradiation of ray floret cultures with ion beams	The smaller flower head size
Majestic Pink	2014	Chrysanthemum	Irradiation of ray floret cultures with ion beams	The shorter plant and light pink flower color
KeenaPearl	2014	Orchid	Protocorm-like bodies (PLBs) of matured self-pollinated seeds exposed to gamma ray at dose 35 Gy	Easy to grow and free blooming with long spike
KeenaPastel	2014	Orchid	Protocorm-like bodies (PLBs) of matured self-pollinated seeds exposed to gamma ray at dose 35 Gy	Long and broad lip with curled at the edges of flower
Golden Eye	2014	Chrysanthemum	Irradiation of ray floret cultures with ion beams	The flower has smaller ray florets which are of incurved type
Cream Marble	2014	Chrysanthemum	Irradiation of nodal cultures with acute gamma (35 Gy) rays	Orangish red flower, yellowish cream/green variegated leaves, medium height plant

ovules, callus, and suspension cells which are more suited to mutation induction techniques as compared to in vivo structures.

Limitations and problems of in vitro mutations are:

1. Problems of regeneration capacity due to cell culture studies do not give positive results for all types
2. Different gene sets that are hidden in cells
3. The inability to select the hereditary material of agronomic importance in the cell culture stage and the loss of the desired material.

Considering all these factors, in vitro mutation studies are very effective in the study's success. The main factor affecting the success of in vitro mutation breeding and other tissue culture techniques is the choice of genotype and explant source. The responses of the selected genotype and explant source vary depending on the regeneration capacity of the explant, mainly dependent on the donor's genetic structure. The regeneration capacity of the explant differs among species cultivars, even in genotypes (Segui-Simarro and Nuez 2008; Tutuncu et al. 2017; Datta 2020; Kantoğlu et al. 2021). Therefore, using a model plant with high regeneration capacity provides high yielded outcomes in vitro.

The use of meristematic, embryogenic tissues or embryos is a more appropriate explant source than differentiated tissue in terms of genetic uniformity. Similar advantages may gain with using explants from a young donor (IAEA 1983).

The mutagen treatments and choice of appropriate mutagen are crucial for mutation induction. In the beginning, the scientists were hopeful that using chemical and physical mutagens would be possible to induce gene-specific mutations. However, it was unrealistic due to the structure of DNA (Maluszynski et al. 1995). Therefore, the choice of mutagen depended on the type of tissue to be treated, type of mutation desired, and availability of mutagen and safety considerations (Harun 2001). In in vitro mutation induction studies, the following parameters were investigated to establish procedures for experiment standardization; dose, pH, chemical or physical properties of the agent, interaction with the culture medium. The aim was to develop methods giving the highest rate of gene mutations with the lowest chromosome and physiological damage (IAEA 1983; Van Harten 2002).

Additionally, the advantages and disadvantages of the selected mutagen should be considered. The physical mutagens (ionizing radiation and UV) have high reproducibility and high penetration in multicellular systems. However, they have a possibly high degree of sterility in plants regenerated from treated culture. Some of the difficulties occurred in reproducibility, which may be overcome by standardizing application methods. On the other hand, chemical mutagens (ENUA, MNUA, EMS, DES) result in predominately point mutation (Abdulhadi et al. 2019), less chromosome damage, high mutation rates, and possibly different mutation spectra compared with physical mutagens. However, penetration of the chemical mutagens isn't stable in multicellular systems. Therefore, some difficulties can occur in standardizing application methods. Most of the chemical mutagens have carcinogenic properties (IAEA 1983).

The somaclonal variation is another way to develop plants in in vitro. The genetic variation in tissue in vitro cultured plants is called somaclonal variation. Since discovering the somaclonal variation, it has been great enthusiasm among plant breeders and raised high hopes of exploiting such variations either directly or by in vitro selection. Somaclonal variation can be observed as the changes in plant qualitative and quantitative characteristics. However, the frequency of the variation and its result on treated plants varied depending on genotype-specific. Somaclonal variation in ornamental plants has been well reviewed by Jain et al. (1998).

The post-treatment handling and selection of mutant genotypes are crucial steps after the in vitro mutagenic treatments. Mutated explants are cultured properly in optimized culture media to continue growth and development to form the whole plant, and then, desired mutant plants are selected. In the period from explants to plant formation, a series of the subculture are needed depending on species, explant type, and the way of plant regeneration. In the first vegetative period (M1V1) of mutated material, screening mutations are not recommended because chimeras may mask mutations. If the functional mutants are discovered at an early stage, the stability of the mutants should be tracked in later generations, up to M1V4 or M1V6 (Suprasanna et al. 2015). The selection of mutants is the final stage in mutagenesis to identify and select desired mutant plants. In ornamental plants, mutant selection based on visual characters such as leaf or flower color and morphological changes can be easily performed growing plants in field or greenhouse conditions (Ibrahim et al. 2018). Additionally, desired traits such as disease resistance or abiotic stress tolerance can also be selected in field conditions comparing a control group. However, in vitro selection provides early selection choice adding an inhibitor or an anti-metabolite into the culture medium and observing mutated cells survival (Suprasanna et al. 2012; Kantoğlu et al. 2021).

22.4 Conclusion and Prospects

The induced mutation was applied to many species from the beginning of the experimental mutation induction studies. About 3365 officially released mutants in 225 species were successfully obtained and listed in FAO/IAEA database. There are about 720 ornamental mutants in the database induced by treatments with physical mutagens (mainly X-rays and gamma rays) and chemical mutagens (mainly EMS, DMS, Colchicine). The improvements of in vitro technology have led to scientists using it in experimental mutation studies to assist mutation breeding program that has been accelerated with the usage of tissue culture techniques since the 1970s. The isolation of chimeric tissues and the development of radioactively mutated tissues in primarily vegetatively propagated plants are both made possible by in vitro mutagenesis. In addition, in vitro techniques can be supported to the breeding program before, after, and during the mutagen treatments, allowing the scientist to perform the studies timeless and independently of environmental conditions. At present, mutation breeding combined with mutation induction has been routinely applied to ornamental plants to supply market demands, improve existing cultivars, and enlarge variation in a genetic pool.

On the other hand, rapid and giant development in recombinant technology provides novel biotechnological techniques such as gene transformation via *Agrobacterium* sp. and genome editing with ZFNs, TALENs, TILLING, kompetitive allele specific PCR (KASP), next generation sequence (NGS), and CRISPR-Cas9 to breed plants toward a specific target. The usage of modern tools in ornamental plants is limited due to inadequate optimized gene transformation protocols for species, little DNA sequence information, and high cost. Besides novel genomic editing techniques, modern tools are combined with classical mutation techniques, and they are efficiently used to screen mutations in a genetic pool. At present, mutation induction in combined with in vitro techniques still contribute to developing new cultivars and modern biotechnological approaches may become more convenient tools in ornamental plant breeding as the development of more protocols specific to species, reducing the cost of process and sequencing genome information in different species near future.

References

Abdulhadi MD, Ayad AO, Kadhim MI (2019) Induced genetic variability in *Chrysanthemum morifolium* single nodes grown *in vitro* using ethyl methane sulphonate. Biochem Cell Arch 19(1):151–156

Acquaah G (2007) Variation: types, origin, and scale. In: Principles of plant genetics and breeding. Blackwell Publishing, Chichester, pp 75–86

Acquaah G (2015) Conventional plant breeding principles and techniques. In: Al-Khayri JM, Jain SM, Johnson DV (eds) Advances in plant breeding strategies: breeding, biotechnology and molecular tools, vol 1. Springer International Publishing, Cham, pp 115–158

Anonymous. (2018) In: Spencer-Lopes MM, Foster BP, Jankuloski L (eds) Manual on mutation breeding, 3rd edn. Food and Agriculture Organization of the United Nations, Rome, 301pp

Barton KA, Binns AN, Matzke AJ, Chilton MD (1983) Regeneration of intact tobacco plants containing full length copies of genetically engineered T-DNA, and transmission of T-DNA to R1 progeny. Cell 32(4):1033–1043

Boutigny AL, Dohin N, Pornin D, Rolland M (2020) Overview and detectability of the genetic modifications in ornamental plants. Hort Res 7(1):1–12

Broertjes C (1969) Mutation breeding of *Streptocarpus*. Euphytica 18:333–339

Broertjes C (1972) Mutation breeding of *Achimenes*. Euphytica 21:48–63

Broertjes C, Roest S (1976) Mutation breeding of *Chrysanthemum morifolium* Ramat, using *in vivo* and *in vitro* adventitious bud technique. Euphytica 22:11–19

Broertjes C, Van Harten AM (1988) Applied mutation breeding for vegetatively propagated crops. Elsevier Science Publishers, Amsterdam, 345p

Brown JL, Harney PM (1974) Induction of non-chimaeral mutants in *Begonia*. Hort Sci 9(3):276

Çelik Ö, Atak Ç (2017) Applications of ionizing radiation in mutation breeding. In: Maghraby AM (ed) New insights on gamma rays. IntechOpen, London, pp 111–132

Datta SK (2012) Success story of induced mutagenesis for development of new ornamental varieties. Bioremediat Biodiver Bioavail 6(Special issue 1):15–26

Datta SK (2014) Induced mutagenesis: basic knowledge for technological success. In: Tomlekova NB, Kozgar MI, Wani MR (eds) Mutagenesis exploring genetic diversity of crops. Wageningen Academic Publishers, Wageningen, pp 97–139. ISBN: 9789086867967

Datta SK (2020) Induced mutations: technological advancement for development of new ornamental varieties. Nucleus 2020(63):119–129

Datta SK, Teixeira da Silva JA (2006) Role of induced mutagenesis for development of new flower color and type in ornamentals. In: Teixeira da Silva JA (ed) Floriculture, ornamental and plant biotechnology advances and topical issue. Global Science Books Ltd., Middlesex, pp 640–645

Donini P, Sonnino A (1998) Induced mutation in plant breeding: current status and future outlook. In: Jain SM, Brar DS, Ahloowalia BS (eds) Somaclonal variation and induced mutations in crop improvement. Kluwer Academic Publishers, Dordrecht, pp 255–291

Donmez D, Simsek O, Izgu T, Aka Kacar Y, Yalcin Mendi Y (2013) Genetic transformation in citrus. Sci World J 2013:491207

Forster BP, Shu QY (2011) Plant mutagenesis in crop improvement: basic terms and applications. In: Shu QY, Forster BP, Nakagawa H (eds) Plant mutation breeding and biotechnology. FAO, Rome, pp 9–20

Harun AR (2001) The effective use of physical and chemical mutagen in the induction of mutation for crop improvement in Malaysia. In: Proceedings of the 8th workshop on plant mutation breeding: effective use of physical/chemical mutagens, 9–13 October 2000, Hanoi, pp 111–122

Hase Y, Akita Y, Kitamura S, Narumi I, Tanaka A (2012) Development of an efficient mutagenesis technique using ion beams: toward more controlled mutation breeding. Plant Biotechnol 29: 193–200

Haspolat G, Senel U, Kantoglu KY, Kunter B, Gunçag N (2019) *In vitro* mutation on chrysanthemums. Acta Hort 1263:261–266

Hernández-Muñoz S, Pedraza-Santos EM, López PA, Gómez-Sanabria JM, Morales-García JL (2019) Mutagenesis in the improvement of ornamental plants. Rev Chapingo Ser Hort 25(3): 151–167

IAEA (1983) IAEA tech doc no: 340. IAEA Publication Series, Vienna

IAEA (Joint FAO/IAEA Division of Nuclear Techniques in Food and Agriculture) (2018a) 50th anniversary: 1964–2014 and beyond joint FAO/IAEA Division of Nuclear Techniques in Food and Agriculture. http://www.naweb.iaea.org/nafa/news/50thAnniversary-JointDivisionFAO-IAEA.html. Accessed 11 Jul 2018

IAEA (Joint FAO/IAEA Division of Nuclear Techniques in Food and Agriculture) (2018b) Plant breeding and genetics. http://www.naweb.iaea.org/nafa/pbg/index.html. Accessed 11 Jul 2018

IAEA (Joint FAO/IAEA Division of Nuclear Techniques in Food and Agriculture) (2021) Mutant variety search. https://mvd.iaea.org/#!Search. Accessed 11 Feb 2021

Ibrahim R, Ahmad Z, Salleh S, Hassan AA, Ariffin S (2018) Mutation breeding in ornamentals. In: Van Huylenbroeck J (ed) Ornamental crops, Handbook of plant breeding, vol 11. Springer, Cham. https://doi.org/10.1007/978-3-319-90698-0_8

International Service for the Acquisition of Agri-biotech Applications (ISAAA) (2022) GM approval database. https://www.isaaa.org/gmapprovaldatabase/default.asp

Ishizaka H, Kondo E, Kameari N (2012) Production of novel flower color mutants from the fragrant cyclamen (*Cyclamen persicum_C. purpurascens*) by ion-beam irradiation. Plant Biotechnol 29: 201–208

Jain SM (2000) A review of induction of mutations in fruits of tropical and subtropical regions. Acta Hort 575:295–302. https://doi.org/10.17660/ActaHortic.2002.575.33

Jain SM (2005) Major mutation-assisted plant breeding programs supported by FAO/IAEA. Plant Cell Tissue Organ Cult 82:113–123. https://doi.org/10.1007/s11240-004-7095-6

Jain SM, Buiatti M, Gimelli F, Saccardo F (1998) Somaclonal variation in improving ornamental plants. In: Jain SM, Brar DS, Ahloowalia BS (eds) Somaclonal variation and induced mutations in crop improvement. Kluwer Academic Publishers, Dordrecht, pp 81–104

Kantoğlu KY, Kunter B (2021) Mutasyonıslahı. In: Mendi NY, Kazaz S (eds) Süs Bitkileri Islahı, Klasikve Biyoteknolojik Yöntemler. Gece Kitaplığı, Ankara, pp 145–202. ISBN: 978-625-7478-51-9 (in Turkish)

Kantoğlu KY, Kuşvuran Ş, Ellialtıoğlu ŞŞ (2021) Doku kültürü yöntemlerinin stres çalışmalarında kullanımı. In: Ellialtığlu ŞŞ, Daşkan HY, Kuşvuran Ş (eds) Sebzelerde Stres Toleransıve Islah Stratejileri. Gece Kitaplığı, Ankara, pp 708–741. ISBN: 978-625-8449-01-3 (in Turkish)

Kazaz S, Kılıç T, Doğan E, Mendi YY, Karagüzel Ö (2020) Süs bitkileri üretiminde mevcut durum ve gelecek. Türkiye Ziraat Mühendisleri IX. Teknik Kongresi Bildiriler Kitabı-1, pp 673–698 (in Turkish)

Kharkwal MC (2011) A brief history of plant mutagenesis. In: Shu QY, Forster BP, Nakagawa H (eds) Plant mutation breeding and biotechnology. FAO, Rome, pp 21–30

Kishi-Kaboshi M, Aida R, Sasaki K (2018) Genome engineering in ornamental plants: current status and future prospects. Plant Physiol Biochem 131:47–52

Lakhssassi N, Zhou Z, Cullen MA, Badad O, El Baze A, Chetto O, Meksem K (2021) TILLING-by-sequencing+ to decipher oil biosynthesis pathway in soybeans: a new and effective platform for high-throughput gene functional analysis. Int J Mol Sci 22(8):4219

Lönnig WE (2005) Mutation breeding, evolution, and the law of recurrent variation. Recent Res Dev Genet Breed 2:45–70

Maluszynski M, Ahloowalia BS, Sigurbjornsson B (1995) Application of *in vivo* and *in vitro* mutation techniques for crop improvement. Euphytica 85:303–315. https://doi.org/10.1007/BF00023960

Melsen K, van de Wouw M, Contreras R (2021) Mutation breeding in ornamentals. Hort Sci 56(10): 1154–1165

Meyer P, Heidmann I, Forkmann G, Saedler H (1987) A new petunia flower colour generated by transformation of a mutant with a maize gene. Nature 330(6149):677–678

Muller HJ (1927) Artificial transmutation of the gene. Science 66(1699):84–87. https://doi.org/10.1126/science.66.1699.84

Nakornthap A (1974) Radiation induced somatic mutations in Kalanchoe (*Kalanchoe laciniata*). Mutat Breed Newsl 3:14

Nelka SAP, Vidanapathirana NP, Dahanayake N, Subasinghe S, Silva TD, Weerasinghe S, Rifnas LM, Rohanadeera H, Madushanka WCMS, Dushane S, Dhanushka TGB, Anuruddhi HIGK (2021) Effect of gamma irradiation on survivability, growth performances and floral characters of *Jasminum officinale* (samanpichcha). J Agro Technol Rural Sci 1(1):24–29

Novak FJ, Brunner H (1992) Plant breeding: Induced mutation technology for crop improvement. IAEA Bull 4:25–33

Okamura M, Umemoto N, Onishi N (2012) Breeding glittering carnations by an efficient mutagenesis System. Plant Biotechnol 29:209–214

Oladosua Y, Rafii MY, Abdullah N, Hussind G, Ramlie A, Rahimf HA, Miaha G, Usmana M (2016) Principle and application of plant mutagenesis in crop improvement: a review. Biotechnol Biotechnol Equip 30(1):1–16

Pathirana R (2011) Plant mutation breeding in agriculture. CAB Rev Perspect Agric Vet Sci Nutr Nat Resour 32(6):1–20. https://doi.org/10.1079/PAVSNNR20116032

Roychowdhury R, Tah J (2013) Mutagenesis—a potential approach for crop improvement. In: Hakeem KR, Ahmad P, Ozturk M (eds) Crop improvement: new approaches and modern techniques. Springer, New York, pp 149–187

Schum A (2003) Mutation breeding in ornamental plants: an efficient breeding method. Acta Hort 612:47–60

Segui-Simarro JM, Nuez F (2008) How microspores transform into haploid embryos: changes associated with embryogenesis induction and microspore derived embryogenesis. Physiol Plant 134:1–12. https://doi.org/10.1111/j.1399-3054.2008.01113.x

Simsek O, Kacar YA (2010) Discovery of mutations with TILLING and ECOTILLING in plant genomes. Sci Res Essays 5(24):3799–3802

Spencer Lopes MM, Forster BP, Mba C, Jankuloski L (2018) Physical mutagenesis. In: Spencer Lopes MM, Forster BP, Jankuloski L (eds) Manual on mutation breeding, 3rd edn. FAO and IAEA, Vienna, pp 15–16

Stadler LJ (1928a) Mutations in barley induced by X-rays and radium. Science 68:186–187. https://doi.org/10.1126/science.68.1756.186

Stadler LJ (1928b) Genetic effects of X-rays on maize. Proc Natl Acad Sci 14:69–75. https://doi.org/10.1073/pnas.14.1.69

Su J, Jiang J, Zhang F, Liu Y, Ding L, Chen S, Chen F (2019) Current achievements and future prospects in the genetic breeding of chrysanthemum: a review. Hort Res 6:109

Suprasanna P, Jain SM (2017) Mutant resources and mutagenomics in crop plants. Emirates J Food Agric 29:651–657

Suprasanna P, Jain SM, Ochatt SJ, Kulkarni VM, Predieri S (2012) Applications of *in vitro* techniques in mutation breeding of vegetatively propagated crops. In: Plant mutation breeding and biotechnology. CABI, Wallingford, pp 371–385

Suprasanna P, Mirajkar SJ, Bhagwat SG (2015) Induced mutations and crop improvement. In: Plant biology and biotechnology. Springer, New Delhi, pp 593–617

Suprasanna P, Ganapathi TR, Ghag SB, Jain SM (2017) Genetic modifications of horticultural plants by induced mutations and transgenic approach. Acta Hort 1187:219–232

Tamaki K, Yamanaka M, Mizutani Y, Hayashi Y, Hirano T, Abe T (2014) Flower color mutants of chrysanthemum obtained using C-ion beam irradiation. RIKEN Accel Prog Rep 47:296

Tanaka A, Shikazono N, Hase Y (2010) Studies on biological effects of ion beams on lethality, molecular nature of mutation, mutation rate, and spectrum of mutation phenotype for mutation breeding in higher plants. J Radiat Res 51:223–233

Tutuncu M, İzgü T, Sevindik B, Mendi YY (2017) *In vitro* haploidy techniques in ornamental plants. Res J Agric Sci 10(1):1–6

Van Harten AM (1998) Mutation breeding theory and practical applications. Cambridge University Press, Cambridge, pp 1–30. ISBN: 0521470749

Van Harten AM (2002) Mutation breeding of vegetatively propagated ornamentals. In: Vainstein A (ed) Breeding for ornamentals: classical and molecular approaches. Kluwer Academic Publishers, Dordrecht, pp 155–127

Yamaguchi H (2018) Mutation breeding of ornamental plants using ion beams. Breed Sci 68:71–78. https://doi.org/10.1270/jsbbs.17086

Yamaguchi H, Nagatomi S, Morishita T, Degi K, Tanaka A, Shikazono N, Hase Y (2003) Mutation induced with ion beam irradiation in rose. Nucl Inst Methods Phys Res B 206:561–564

Zhou LB, Li WJ, Ma S, Dong XC, Yu LX, Li Q, Zhou GM, Gao QX (2006) Effects of ion beam irradiation on adventitious shoot regeneration from in vitro leaf explants of *Saintpaulia ionahta*. Nucl Inst Methods Phys Res B 244:349–353

Improvement of Fruit Crops Through Radiation-Induced Mutations Facing Climate Change

23

S. S. Maan, V. Sharma, and J. S. Brar

Abstract

Genetic improvement of fruits for increasing production and productivity in the era of climate change requires the desirable genetic variation, which is lacking and hampers the breeding of fruit crops. Mutation breeding played a pivotal role in crop improvement of already established genetic backgrounds as well as generating desirable allelic variations for screening against improved horticultural traits and tolerance against abiotic and biotic stresses. According to the Mutant Variety Database (MVD) of fruit crops, mutants of fruit crops were registered with many useful traits affecting plant size, blooming time and fruit ripening, fruit color, self-compatibility, self-thinning, and resistance to pathogens. Radiosensitivity studies prior to starting of large-scale mutagenesis along with the application of in vitro technique significantly increase the mutagenic efficiency and effectiveness. Resultant pre-breeding stock from various radiation-induced mutagenesis programs in different fruit crops produced mutants having improved horticultural traits, and abiotic and biotic stress tolerance. Further, the advent of molecular techniques interplays a significant role in the characterization of genetic variation at early stages, to obtain desirable mutants.

S. S. Maan (✉) · J. S. Brar
Faculty of Applied Sciences and Biotechnology, Shoolini University, Bhajol, Solan (HP), HP, India
e-mail: Sukhjinder-fs@pau.edu

V. Sharma
Faculty of Applied Sciences and Biotechnology, Shoolini University, Bhajol, Solan (HP), Himachal Pradesh, India

S. Penna, S. M. Jain (eds.), *Mutation Breeding for Sustainable Food Production and Climate Resilience*, https://doi.org/10.1007/978-981-16-9720-3_23

Keywords

Climate change · Fruits · Gamma irradiation · Mutation breeding · Physical mutagenesis

23.1 Introduction

Climate change will have profound effects on the production and quality of fruit crops. Most important issues concerning local and global climate change are the rise of temperature, elevated atmospheric CO_2 concentration, and greenhouse gases which will directly or indirectly affect fruit crops, salinity and drought conditions, leading to an impact on soil quality and thereby affecting overall crop production. Floral phenological shifts, changed pollination scenario and population dynamics of pollinators, insect-pest (extra generation and more period of activities of insect-pests), as well as outbreak of new pests and diseases along with increased physiological disorders, inferior fruit qualities, changed suitability and availability of cultivars for current and future production, extreme heat-wave events, frost damage, reduced availability of irrigation water, negative effect on soil due to extreme rainfall and temperature are most important factors which can be associated with the impact of climate change on fruit crops.

Mutation breeding is advantageous over conventional and transgenic breeding because it precludes segregation progenies, and free from regulatory issues while improving the genetic makeup of already established backgrounds by altering or improving one or few traits. Spontaneous mutations have been exploited in past decades for the improvement of different fruit crops, such as in apple, citrus (Liu et al. 2007; Xiao et al. 2009; Latado et al. 2012), pear (Predieri 2001), grapes (Zhao et al. 2015; Walker et al. 2006) etc., simply by selection of superior somatic mutants. A large number of present-day commercial cultivars of fruit plants have their origin as natural bud sports, but these are relatively rare events and the frequency of spontaneous mutation is quite low in nature (Foster and Aranzana 2018). This prevent the breeders to exploit them in a straightforward manner, that is why attempts have been made to accelerate their rate artificially with the mean of certain compounds. Induced mutagenesis has the ability to broaden the genetic background of the elite crop cultivars. Mutagenic compounds are broadly categorized into two main groups (Fig. 23.1).

Irradiation breeding played a prominent role to increase the desirable alleles in the elite lines of various fruit crops. Various mutants of fruit cultivars with a wide range of traits have been isolated and utilized for breeding programs, which are high yield, plant size, blooming time, fruit ripening, fruit color, self-compatibility, self-thinning, and showing resistance to pathogens, drought, and salinity.

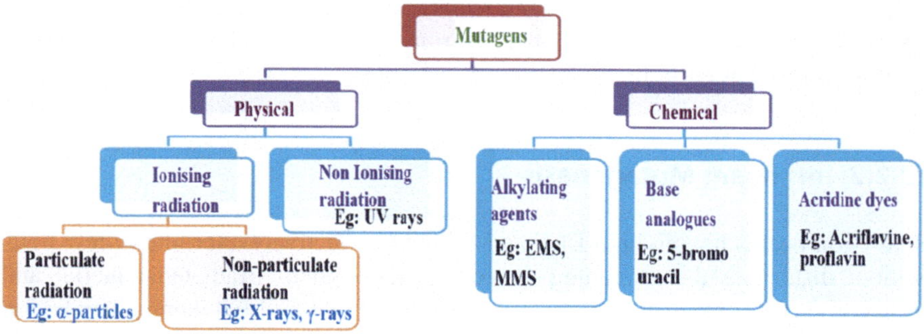

Fig. 23.1 Different types of mutagens used for in vivo and in vitro mutation breeding programs

23.2 Physical Mutagens

Different mutagenic agents such as X-rays, alpha (α), beta (β), and gamma (γ) rays having nature of radioactivity grouped under physical mutagens category serve as sources of radiation for past and on-going present mutational experiments on crop plants (Sharma et al. 2020; Sharma and Thakur 2021). Physical mutagenesis was more preferred over chemical mutagenesis in fruit crops, because of the difficulty in treating the vegetative propagules in a reproducible way with chemicals. Of the ionizing and non-ionizing radiations, the ionizing radiations (X- and γ-rays) have been more efficient and frequently used for mutation induction in fruit plants, due to their nature of scattered ionization in the tissue causing more physiological impact than non-ionizing radiation type (Mba et al. 2010, 2012). The production sources of X- and γ-rays are different, but they are allied to each other with respect to their effect on biological material. However, Gray et al. (1943) reported the lethality order of 1.5:1.0 at equivalent doses of X- and γ-rays. So, it is suggested to be differences lie in their densities, thereby X-rays are more lethal. Recent four decades have witnessed a lessening in the use of X-ray technology, while mutagenesis employing gamma rays as a source has increased significantly owing to its safer use and versatility. Another fact for the overwhelming use of gamma rays in fruit breeding is highest number of fruit crop mutants 46 were registered under the Mutant Varieties Database (MVD) under this category.

23.2.1 Fast Neutron Mutagenesis

Fast neutrons generally cause deletions ranging from a few hundred to a million bases. Fast neutron-induced mutagenesis is an exceptional technique among the other physical mutagenesis tools being employed in crop science in relation to higher impact. Fast neutron was not previously as common as other physical mutagens in fruit crops-mutagenesis, until the release of three Clementina cultivars

(López-García et al. 2012, 2015), and somewhat success in banana (Hautea et al. 2004), while the efficiency of thermal neutrons for apricot and grapefruit improvement is reported as early in 1970–1980.

23.2.2 Ion Beam Mutagenesis

Heavy-ion beams having high LET and RBE can effectively alter one or few traits without disturbing the remaining genetic background of irradiated material and induce a variety of mutagenic effects as compared to low-LET gamma irradiation (Tanaka et al. 2010). Ion-beam mutagenesis technique has gained popularity in Japan and China to produce mutant varieties of field, ornamental, and vegetable crops (Yamaguchi 2018). To the best of our knowledge, few reports pertain to ion-beam mutagenesis methodology on fruit crops with the release of one mutant of Sour cherry in 2011 (Sattar et al. 2021; Anonymous 2021). Twigs of fruit trees smeared with low melting point wax and sealed in plastic foam leaving only the bud-tip, that is to be implanted, are exposed.

23.3 Mutagenic Efficiency, Effectiveness, and Dosimetry

23.3.1 Mutagenic Efficiency and Effectiveness

The efficiency of mutagen employed for experimental mutagenesis is the proportion of certain desired mutagenic alterations relative to biological damage. Mutagenic efficiency reflects the usefulness of mutagen because induced mutagenesis creates favorable mutated alleles accompanied along undesirable effects, sterility, and lethality. In experimental mutagenesis, mutagen yielding multiple mutations is considered as undesirable, as drastic alterations affect plant genetic background. The mutagenic effectiveness is a measure of the frequency of mutations induced by unit mutagenic dose.

The aim of each mutation breeding program should be to obtain a greater number of desirable genetic changes per 100 treated propagules. Factor of effectiveness (FE) was calculated using number of variations and number of treated explants. Frequency of variants depends on genotype and dose of mutagen. The highest FE was obtained for dissected shoot apices of banana at 0.8 krad (Bhagwat and Duncan 1998). In vitro mutagenesis technique has been shown to produce higher frequency of solid mutants on lower dosages of gamma irradiation as compared to high irradiation dosages in banana (Garcia et al. 2002; Karmarkar et al. 2001), grapevine (Kuksova et al. 1997; Patil and Patil 2005), and strawberry (Murti et al. 2013). Similarly, the genotypic specificity for mutation effectiveness was observed in pear (Predieri et al. 1997; Predieri and Zimmerman 2001), strawberry (Weimin et al. 2009), and grape (Dev et al. 2021). Dev (2014) found that the frequency of in vitro mutation was highest in Pusa Navrang (40%) followed by H-76-1 (37.5%) and

Julesky Muscat (37.5%) each. The lowest mutation frequency was noted in Pearl of Casaba (30%).

In vitro mutagenesis techniques significantly improve the effectiveness of mutation induction, especially by handling a large number of regenerated plants (Jain 2000; Jain and Maluszynski 2004), and offer a wide choice of plant material for mutagen treatment, e.g., shoot tips (Broertjes and van Harten 1988), and provide healthy starting plant material (Kulkarni et al. 2007). It appears that the lower dosage of some mutagenic compounds can induce high mutation frequency in the genotypes with highly mutable allelic sites and, further increase in the dosage of these mutagens adds to their frequency.

23.3.2 Dosimetry

The radiosensitivity analysis is the most important requirement for large-scale mutagenesis-based breeding programs that has to be planned in future. It is assayed by growth inhibition, lethality, somatic mutation, and chromosome breakage. Radiosensitivity of higher plants is influenced by a number of physical, chemical, and biological factors. The radiobiological responses are influenced by the cell division stage. Mitotic cells are more sensitive than cells in the G_1 phase, while cells in the S stage and G_2 stage had intermediate sensitivity to resistant. Further, radiosensitivity is also associated with chromosome number, size (Interphase Chromosome Volume), ploidy level, and nuclear factors. Genotype specificity is reported by the number of workers; this is most probably due to the genetic make-up (chromosome morphology, chromosome number, nuclear volume, or DNA content) of the treated material (Bowen 1962). Young or fresh seeds are less susceptible to radiation than aged seeds. Radiosensitivity tends to be decreased in dormant condition, which is due to lesser mitotic activity in this period that resulted in lower nuclear volume. Monoembryonic seeds were more tolerant to radiations than polyembryonic seeds (Spiegel-Roy and Padova 1973; Spiegel-Roy and Vardi 1990).

Radio sensitivity increases if plant propagules were pretreated with chemicals like hydrogen peroxide, potassium cyanide, etc. Environmental conditions modifying the physiological status of the material can subsequently influence irradiation effects. There are different opinions regarding the relative mutagenic efficiency of various doses. Since mutation frequency and radiosensitivity (dosimetry) have been influenced by a number of physical factors. Temperature influences breakage and restitution of chromosomes at the treatment time. Low temperature at the time of X-ray treatment leads to greater aberrations than high temperature. Cellular water content is the prime site for radiation to act, which indirectly effects via the production of hydroxyl and alkoxy radicals, and there is a direct relationship between radiosensitivity and moisture content. So, it would be necessary to equilibrate the seed moisture content prior to ionization. Oxygen content acts as a modifier of radiosensitivity. The presence of oxygen in targeted tissue at the time of irradiation strongly determined the frequency of chromosomal aberrations and is a must for the irradiation to be effective. Dust, fibers, biological agents, and extraneous sources of hormones must be excluded from irradiation setups, to ensure reproducibility of results. Effects of irradiation are modified when the germination of seeds is delayed. Post-irradiation storage for a few weeks results in increased seedling injury and chromosomal aberrations. Mutagenic dose was standardized in various fruit crops for their improvement (Table 23.1).

23.4 In Vitro Mutagenesis

In vitro mutagenesis has several advantages, including an increased mutation rate, homogenous mutagenic treatment, requirement of less space and time for regeneration and evaluation of large mutated populations, and opportunity to keep the plant material disease-free. The main restrictions of fruit crop mutation breeding are the formation of chimera due to the involvement of multicellular cell lines. This enhanced the time frame of the selection process to several generations for dissociation of chimeras, or isolation of non-chimeric mutants. In vitro mutagenesis permitted the use of single cell lines as a plant propagule for irradiation treatment, and this ensures the induction of homo-histonts (non-chimeric mutants). Screening of large mutagenic population to select the variant/mutants before regenerating a whole plant can enhance the success of mutagenesis (Table 23.2).

In vitro mutagenesis applications provide an opportunity for plant breeders to selection of desired mutant in a controlled environment. There are two types of in vitro selection: single step and multistep. In a single-step selection technique, the inhibitor agent is added to the cultures and subcultures, while in multistep technique, the selective agent dose below than the lethal dose is added to culture environment and concentration of selection agent is gradually increased in subcultures. The selected mutant by this method has been defined as more stable than the selected via other methods. Meanwhile, duration of the culture and the selective traits that mutated are the main factors that affect the outcome of in vitro mutagenesis. In fruit crops, in vitro mutagenesis has been successfully performed for abiotic stress and biotic stress tolerance by using different selective agents such as mannitol and polyethylene glycol for drought tolerance (Al-Mousa et al. 2016); NaCl for salt tolerance (Motha et al. 2018; El-Sabagh et al. 2011; Miri et al. 2019); changing the temperature of the cultures to select cold/high-temperature-tolerant plants (Majd et al. 2009), and using artificial inoculum or culture filtrates for selection against particular diseases (Damasco et al. 2019; El-Sayed et al. 2011; Pathirana et al. 2016).

23.5 Impact of Mutation-Assisted Breeding on Fruit-Quality Traits

Negative consequences of global warming at the mid-hill areas resulted in deprivation of required winter chilling units, changed content of bioactive compounds, and non-uniform coloration of sub-tropical fruits. Developing climate-resilient apple genotypes is one of the best strategy suggested for mid-hill areas having warmer

Table 23.1 Standardized dosages of physical mutagens used for in vivo propagules in various fruit crops

Species	Cultivar	Plant material	Dose	References
Grapes	Red Globe	Hardwood cutting	18.49 Gy	Surakshitha et al. (2017)
	Muscat		17.02 Gy	
	Pembe Çekirdeksiz		15.00 Gy	Ekbic et al. (2017)
	Isabella		48.80 Gy	Alyanak (2019)
	Alphonse Lavallée		24.80 Gy	
	Pusa Navrang		10.00 Gy	Dev (2014)
Guava	Safeda	Budsticks	60.00 Gy	Zamir et al. (2009)
	Shweta	Buds	18.25 Gy	Singh et al. (2018)
	Arka Amuyla		25.08 Gy	Maan and Brar (2021)
	Punjab Pink		24.20 Gy	
	Arka Amuyla	Seeds	84.79 Gy	
	Punjab Pink		82.46 Gy	
	Lalit		57.81 Gy	Singh et al. (2021)
	Shweta		57.42 Gy	
Strawberry	Fortuna	Runners	62.00 Gy	El-Oualkadi et al. (2019a)
Amla	Kanchan	Scion	1.00 Gy	Selvi et al. (2007)
Fig	Morocco	Buds	62.00 Gy	El-Oualkadi et al. (2019b)
Mango	Peach	Scion	19.95 Gy	Kumar et al. (2018)
	Bappakai		21.37 Gy	
	Peach	Seed kernel	27.54 Gy	
	Bappakai		23.44 Gy	
Lemon	Fino 49	Budwood	50.00 Gy	Pérez-Jiménez et al. (2020)
Mandarin	Nova		50.00 Gy	
Lime	Bearss		50.00 Gy	
Sour orange		Seeds	156 Gy	
Papaya	Ranchi Local	Pre-soaked seeds	28.35 Gy	Sahu et al. (2019)
	Arka Surya		33.13 Gy	
	Ranchi Local	Seeds immersed in water	24.05 Gy	
	Arka Surya		23.78 Gy	
Carambola			30.00 Gy	Vos et al. (2009)
Cherimoya			30.00 Gy	
Pitanga			45.00 Gy	
Jaboticaba			7.90 Gy	

winters. Mutagenesis breeding has the ability to alter one or few traits of elite genotypes performing already in areas witnessing climate change. Neto et al. (2011) started induced mutation breeding program on apple cv. Fuji for color

Table 23.2 Standardized dosages of physical mutagens used for in vitro propagules in fruit crops for mutagenesis

Species	Cultivar	Plant material	Dose (Gy)	References
Avocado	Duke-7	Zygotic embryos	28.00	Coto et al. (2014)
	Catalina		21.00	
	Hass		27.00	
Lemon	Fino 49	Nodal segments	25.00	Pérez-Jiménez et al. (2020)
	Verna 51			
Lemon		Protoplasts	200.00	Helaly and Hanam El-Hosieny (2011)
Grapes	Red Globe	Nodal segments	18.00	Surakshitha et al. (2017)
Fig	Roxo de Valinhos	Shoots (>2.5 cm)	30.00	Ferreira et al. (2009)
Kiwi	Hort 16A	Callus	30.00	Pathirana et al. (2016)
Mandarin	Limau madu	Embryogenic callus	30.00	Agisimanto et al. (2016)
Banana	Tanduk	Shoot meristem	37.00	Abdulhafiz et al. (2018)
	FHIA-23	Micro shoots	30.00	Qamar et al. (2016)
	Basari	Multiple shoot culture	30.00	Mishra et al. (2007)
	Chakkarakela			
	Rasthali			
Strawberry	Akihime	Buds	177.00	Murti et al. (2013)
	DNKW001		177.00	

improvement in warmer areas of Brazil, and succeeded in selection of two early maturing mutants having superior fruit quality traits (fruiting habit, fruit shape, fruit size, and fruit color). Similarly, 32 mutants performed superiorly to their parent cv. Amasya regarding fruit skin coloring (Atay et al. 2018), whereas Campeanu et al. (2010) reported that some of the mutants showed superior value over parent apple genotypes, rest are inferior for particular traits. Reig and Agustí (2007) reported a new ever-flowering loquat variant, i.e., "Piera" with increased fruit weight, acidity, and hunter coordinates (L^*, a^*, b^*). In the MVD database, most of the mutants of fruit cultivars were registered with improved horticultural traits over well-established parent cultivars, those were lacking particular trait of interest. This success of irradiation breeding initiates the breeders to take up mutagenesis breeding of different fruit crops with the aim of improving fruit-quality traits. Nhat and Chau (2010) reported the superiority of four mutants in terms of fruit weight and fruit quality (TSS) of Papaya cv. "Dai Loan Tim." Kumar et al. (2017) and Pujar et al. (2019) stated the efficiency of lower dosage for fruit attributes (fruit length, fruit girth, fruit weight, and pulp thickness) in papaya, while the central cavity was observed with direct relation in gamma irradiation dosages. Khalil et al. (2011) reported a mutant of "Kinnow" with non-significant depression in physical and physio-chemical fruit characteristics. Similar outcomes of non-significant altered physio-chemical attributes via induced mutagenesis in different citrus species

(Aderdour et al. 2019; Bermejo et al. 2011, 2012; Goldenberg et al. 2014; Arisah and Mariana 2017), whereas, Mariana et al. (2018) reported the mutant of pummelo having increased juice volume. Zamir et al. (2009) reported the most favorable doses for guava were between 0.15 and 0.2 kGy, and it leads to mutations such as: highest fruit size (59 mm) and highest fruit weight (128.38 g). Zheng et al. (2009) developed sweeter taste mutants of Indian jujube. Induced mutagenesis successfully impacted the banana improvement with some superior mutants for bunch related traits, and selection of earlier clones (Ganapathi et al. 2016; Nandariyah et al. 2019; Jamaluddin 1994; Reis et al. 2015). Kunter et al. (2012) reported the superiority of three mutants of cherry for fruit traits and cracking tolerance over its parent cv. 0900 Ziraat. El-Mageid and Al-Kfrawey (2018) reported the chemical composition of mutated seedling of avocado cv. "Duck" by increasing irradiation doses. Liang et al. (2011) reported a large fruited variant of *A. chinensis* with average fruit weight of 138.25 g (the control 115.93 g) with average axis ratio of length to width 1.91 and sugar content 21.23% (the control 1.31% and 14.45%, respectively). Predieri (2002) reported the variants of different pear genotypes having increased sugar content with extended shelf-life. Zanol et al. (2011) reported a different berry skin color mutant (differently colored chimeric sectors) of grapes.

23.6 Impact of Mutation-Assisted Breeding on Seed-Related Traits

Fruit quality of different fruit crops also depends upon seed content of fruit. Breed new cultivars with seedlessness trait and/or low seed content is one of the major objectives in fruit breeding programs, among the different fruit crops, irradiation breeding is well known for creating seedlessness traits in citrus genotypes. Induction of seedless trait in citrus through mutagenesis has been reported since as early as 1935 in several mandarin breeding programs (Hearn 1984; Hensz 1971; Spiegel-Roy 1990; Vardi and Spiegel-Roy 1988; Vardi et al. 1995). A number of seedless or low-seeded clones of different citrus cultivars developed through irradiation such as in "Kinnow" mandarin (Altaf 2007; Khalil et al. 2011; Rattanpal and Sidhu 2015), Pineapple, sweet orange (McCollum and Bowman 2005), "Early Pride" mandarin (Stover et al. 2008), "Murcott" mandarin (Bermejo et al. 2012), "Orri" mandarin (Vardi et al. 2003), "Tango" mandarin (Roose and Williams 2007), "Mambangan" pummelo, "SoE" and "Garut" mandarin (Sutarto et al. 2009). Khalil et al. (2011) developed a sparse-seeded mutant of "Kinnow" (5 ± 3 seeds/fruit) through an exposure of 20 Gy (showed stability up to mV_5). Froneman et al. (1996) via irradiating budwoods of grape fruit and Valencia oranges have a higher frequency of branches with seedless fruits, and one seedless nova tangelo mandarin mutant. Cimen et al. (2019) evaluated the 73 mutants of "Robinson," and the seed number of the mutagenic population ranged between 0 and 3.55 seeds per fruit, and 29% of the evaluated genotypes had been described as low seeded (≤ 1 seed) and two of them were completely seedless. One completely seedless mutant was isolated from the M_1V_2 plants (50 Gy) of citrus cv. "Israeli Villa franca" (usually contained up to

25 seeds), and almost all mutants give fruits with at most one seed (Spiegel-Roy and Vardi 1990). Aderdour et al. (2019) reported that 96% of the irradiated clones (out of which 2% mutants showed an average number of 1.28 seeds) had few seeds as compared to wild-type "Fremont" (4.33 seeds), while 83.33% mutants had a seed number per fruit (SNF) ranged between 2 and 4 seeds, and 8.33% mutants have an SNF ranged between 1 and 2 seeds. Goldenberg et al. (2014) reported that the irradiation reduced average fruit seed number of several mandarin cultivars by 70–92%. Bermejo et al. (2012) reported that the irradiated clones presented lower seed numbers (from 0.23 to 2.47 seeds per fruit), whereas the wild-type "Murcott" showed an average number of 9.03 seeds per fruit. Somsri et al. (2008) found that two (M_1V_4) "Khao Thong Di" pummelo mutants irradiated with gamma rays through chronic irradiation procedure with 9.51 and 32.45 krad have fruits with seedless trait. Also, six (M_1V_3) and five (M_1V_4) tangerine plants treated with acute irradiation procedure by 4–8 krad have fruits with less seeds. Kafa et al. (2015) analyzed the bud-irradiated mutants of "Clementine," "Nova," and "Robinson" mandarin and reported that the budwood irradiation decreased seed number in mutant plants. Montañola et al. (2015) selected 164 seedless selections and 14 individuals from mutagenized population (including hybrids) of lemon and mandarin, respectively. Vardi et al. (2008) conducted an experiment with the aim of inducing complete seedlessness, or a low number of seeds, in mandarin cultivars "Murcott," "Orah," "Yafit," and Michal (local Israeli cultivar), and selected some mutants with up to 40% seedlessness. Mariana et al. (2018) developed a mutant of "Pamelo Nambangan" (40 seeds/fruit) named "Pamindo Agrihorti" through gamma irradiation having 10 seeds/fruit on average. Huang et al. (2017) reported that the seed numbers per fruit in the seedless mutant were significantly decreased, compared to the wild type. Sutarto et al. (2009) reported that the plants obtained from 20 Gy irradiated pummelo budwoods bore completely seedless fruits. In case of mandarins, 11 and 9 fruits checked were seedless (0–5 seeds per fruit) in 20 and 40 Gy gamma irradiation treatment, respectively. Some fruits were completely seedless in "Keprok Soe." Gamma irradiation on mandarin also produced nearly seedless fruits (47 and 9 seeds at 20 and 40 Gy, respectively).

Neto et al. (1998) reported that the average minimum number of seeds per fruit in the "Pera" was 6.63, as compared with 0.86–2.70 in the mutants. Atay et al. (2018) reported the three apple mutants showed markedly low seed numbers. Kumar et al. (2017) reported the mutated seedlings obtained through 20 krad produced minimum number of seeds per fruit. El-Mageid and Al-Kfrawey (2018) recorded the lowest values of seed weight at higher irradiation dosages, i.e., 40 Gy and 30 Gy in "Fuerte" and "Hass" avocado, respectively. Zamir et al. (2009) reported that the highest number of seeds (300) and 100 seed-weight (2.29 g) were recorded at 0.05 kGy while the lowest number of seeds (167) and the lowest 100 seed-weight of 1.56 g was observed in the 0.30 kGy treatments. There was a gradual decrease in the number of seeds and seed weight per fruit with an increase in radiation dose.

23.7 Mutation-Assisted Breeding for Biotic and Abiotic Stress

At present, fruit crops are facing a serious threat from climate change, which is predicted to result in reduced growth, and their productivity. Fruit crops are challenged by different encountered biotic and abiotic stresses, which limit their growth and development, and lead to reduced yield and productivity. Therefore, increasing the resilience of crops to climate change represents a critical component toward ensuring food and nutritional security, which could be achieved through induced mutagenesis strategies.

23.7.1 Abiotic Stress Tolerance

Induced mutagenesis is an efficient tool to address the challenges of increased temperature, heat wave, extended periods of cold or freezing, reduction of water resources (drought) and arable land on the growth, yield and adaptive behavior of fruit crops. Fruit cultivars with tolerance to abiotic stresses such as soil salinity, drought, waterlogging, and climatic adverse conditions showed adapted responses in the era of climate change (Sattar et al. 2021). Supernova mutant cultivar of almond having more tolerance to frost damage is officially registered in 1987, developed with 30 Gy gamma irradiation. Mandarin mutant Hongju 420 is officially registered as having resistance to low temperatures. Radiation breeding in the past yielded some promising mutants of different fruit crops with potential of abiotic stress tolerance. Al-Mousa et al. (2016) reported increased drought tolerance of grape mutants (M_1V_3) using polyethylene glycol (PEG) as selection agent. Fuentes et al. (2009) evaluated the M_1V_3 progenies of different cultivars of avocado using in vitro selective procedures and reported that the mutant showed field performance under severe salinity conditions. Majd et al. (2009) showed that the radiation treatment was able to produce mutant genotypes having cold tolerance in tangerines. In vitro mutagenesis along with NaCl as selection agent for the evaluation of different genotypes of grapes for salinity tolerance yielded some promising results (Motha 2016; Motha et al. 2018). Kepenek (2016) described the methodology of gamma irradiation to the identification of new strawberry cultivars resistant to NaCl within a relatively short period of time. El-Sabagh et al. (2011) observed elevated salinity tolerance response of M_1V_5 generation of apricot cv. Canino. Miri et al. (2019) regenerated the M_1V_3 generation of Dwarf Cavendish, showing tolerance to artificial NaCl stress. In vitro mutagenesis technique along with artificial selection method to generate salinity tolerance has been well established in pineapple (Osei-Kofi et al. 1997). Lokko and Amoatey (2001) derived 54% of the irradiated Sugar Loaf plants survived prolonged drought.

23.7.2 Biotic Stress Tolerance

Climate change affects population dynamics, shift in seasons, cropping patterns, and disease scenarios. Climate change and CO_2 are likely to alter important interactions between horticultural plants, and insect disease. The use of irradiation breeding accompanied with in vitro protocols for evaluation is an efficient technique for regeneration and/or evaluation for biotic stress tolerance mutants. The gamma irradiation-induced "Gold Nijisseiki" pear cultivar with a tolerance to black spot disease was a widely planted variety in Japan and is the classical example of role of irradiation breeding in the development of biotic stress tolerance in already-established genetic backgrounds. Mutants of pear, i.e., Osa Gold and Fuxiangyanghongdli are registered with resistance to black spot disease and bacterial diseases, respectively. Mori-hou-fu 3A and McIntosh 8F-2-32 mutants of apple were officially registered with improved resistance to *Podosphera leucotricha* and *Venturia inaequalis*. Pathirana et al. (2016) used epiphytotic conditions of *Pseudomonas syringae* pv. *actinidiae* to screen 400 putative mutant plants for resistance of kiwi cv. Hort16A under field conditions. In Banana, BMV and BBTV resistant mutant lines had generated through gamma irradiation coupled with in vitro techniques and showed stability for resistance under field trails (El-Sayed et al. 2011; Damasco et al. 2007, 2019). The role played by gamma irradiation in developing Fusarium wilt resistant cultivars with acceptable agronomic traits is well documented (de Beer et al. 2001; Mak et al. 1996; Hwang and Ko 1990; Jain 2010; Tang et al. 2000; Jain and Swennen 2004). Pirama 1 mutant of banana developed with gamma irradiation technology showed tolerance to fusarium wilt disease (TR 4). Rocha et al. (2021) summarized that 4% role is played by in vitro mutagenesis in the obtainment of plants resistant to Fusarium wilt. Smith et al. (2006) reported that the mutant lines of banana developed with 20 Gy have shown a high degree of resistance to Fusarium wilt, as well as equal industry standard to commercial cv. Williams. Several putative mutants tolerant to Bayoud disease in date palm (Jain 2012), and mutants of banana for tolerance to Fusarium wilt, and black Sigatoka were isolated (Jain 2010). NIAB Kinnow mutant of citrus cv. Kinnow showed moderate to high resistance to Citrus canker, scab and wither-tip diseases as well as low incidence to major insect pests. IAC 2014 registered mutant of sweet orange cv. Pera IAC Sweet orange showed greater tolerance to citrus canker (in leaves and fruits). On the basis of biochemical attributes, Kumar et al. (2012) vividly indicated the presence of resistance and tolerance against nematodes in the in vitro-derived mutants of banana. Vos et al. (2009) succeeded in the obtainment of nematode tolerance up to some extent in guava mutants generated through X-ray irradiation. Similarly, Chan (2009) reported that the 1.7% mutants of papaya showed resistance even after the third inoculation of PRSV, and M_3 putative mutants also showed very good resistance to malformed top disease. Litz (2009) performed the irradiation of embryogenic cultures of mango cultivars and used culture filtrate of *Colletotrichum gloeosporioides* Penz as a selection agent for the obtainment of anthracnose tolerance.

23.8 Impact of Mutant Cultivars

Irradiation breeding was a very successful technique in producing fruit cultivars with improved genotypic and desirable phenotypic traits. This mutagenesis technique has not faced intellectual property rights (IPR) and regulatory issues of transgenic breeding. About 89 mutant fruit cultivars belonging to 22 different fruit species have been released in more than 20 countries of the world (Sattar et al. 2021). Irradiation mutagenesis contributed to more than 54 mutant fruit cultivars, out of which 20 mutant fruit cultivars were released by China, Canada released 11, Japan released 9, and India released 2 mutants including PAU Kinnow-1 (citrus), and Pusa nanha (Papaya) (Fig. 23.2). Further, a massive amount of variability at genetic level generated through irradiation breeding is significantly employed for modern genomic studies, and furnish as a basic resource in hybridization programs. This generated mutants and/or variability offer improved quality, disease-resistant, higher productivity, resilient to environmental swings, and a huge impact on boosting up the economic status of different countries. The impact of mutant cultivars on the national economy of different countries was measured in terms of billions of dollars, and additionally improvement of rural economy, nutrition, and helping in sustainable food security. Despite the small number of mutant fruit cultivars, some of them had a significant economic impact. Japanese pear cv. "Gold Nijisseiki" developed through chronic method of gamma irradiation of its parent "Nijisseiki," showed more resistant to black spot disease, has a significant impact on the pear fruit industry of Japan, and according to farmers' group cooperative calculations, this cultivar generated US$ 30 million additional annual incomes. "Osa-Nijisseiki" and "Osa-Gold" mutants of pear with improved disease resistance, and self-compatible nature,

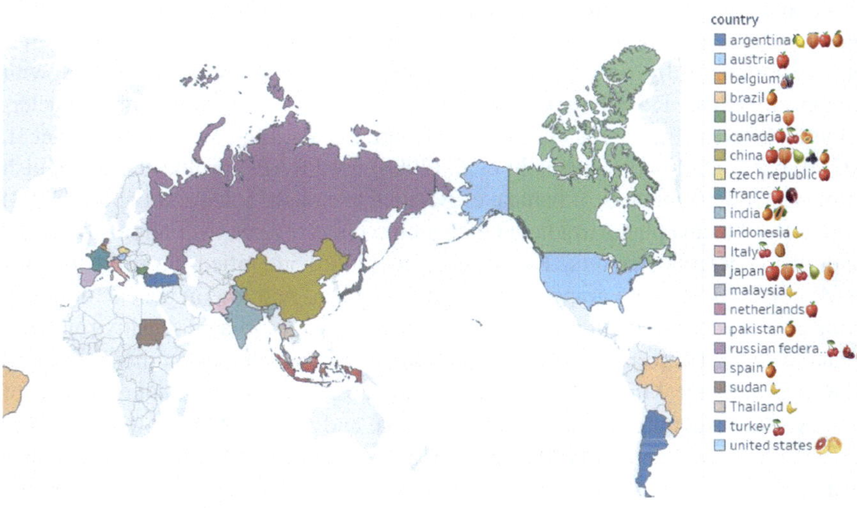

Fig. 23.2 Country-wise distribution of officially released mutant fruit varieties through physical mutagenesis

ultimately reduced the need for pollinizing agents. In Japan, the economic contribution of pear and peach has been estimated a worth of 31.2 million US dollars in 1997. The economic impact of colored variants of grapefruits, i.e., "Star Ruby" and "Rio Red" derived through thermal neutron irradiation is well documented and replaced the commercial cultivars grown in Texas and in other grapefruit growing areas (covered 75% of total area) of USA, and their fruits are sold under the trademark "Rio Star" (Ahloowalia et al. 2004). Data on monetary terms is not much available, but the direct and indirect impact of mutants fruit cultivars is high. MVD-IAEA database depicted that the mutants of commercially important fruit crops developed through irradiation breeding showed improved plant architecture, blooming time, self-compatibility and self-thinning, maturity and fruit ripening, fruit color, quality-related traits besides yield and resistance to various pathogens (Sattar et al. 2021).

23.9 Molecular Tools to Distinguish Mutants

Traditionally, mutated genotypes were distinguished based on morphological markers. The recent development of DNA marker-based technologies has hastened opportunities to utilize by plant breeders to differentiate the mutated progenies and parental lines, and/or characterization of generated variation at the DNA level. Nowadays, in-depth analysis of mutational variations was performed using different high throughput genomic platforms such as single-strand conformational polymorphism (SSCP), cDNA-amplified fragment length polymorphism (AFLP), random amplified DNA polymorphism (RAPD), differential display, microarray, high-resolution melt (HRM) analysis and targeting induced local lesions in genome (TILLING). Among these molecular tools, SSR (simple sequence repeats), RAPD, AFLP, and inter simple sequence repeat (ISSR) have been frequently used in genomic characterization of the fruit crop mutants. Theoretically, mutants differ from their parents by only 5 per million base pairs and frequency varies with mutagens and dose. RAPD markers were used in earlier studies of mutant characterization in different fruit crops (Table 23.3) (Martin et al. 2006; Venkatachalam et al. 2007; Bapat et al. 2007). Ganapathi et al. (2008) used RAPD profiles to distinguish gamma (5–30 Gy)-mutants of banana cv. Giant Cavendish. Sales and Espino (2008) reported that banana plants irradiated with gamma rays revealed variation among the clones using RAPD markers. Khawale et al. (2006) identified 11 solid mutants (36.6%) of Pusa Seedless grape with 7 RAPD primers. A study on genetic diversity amongst 21 induced tolerant mutant clones with one non-irradiated sensitive clone of banana cv. Dwarf Cavendish (AAA) to salinity was conducted using morphological and RAPD markers (Miri et al. 2019). Dev et al. (2021) identified 13 solid mutants of grape cv. Pusa Navrang, H-76-1, Pearl of Csaba, and Julesky Muscat with 6 RAPD primers. Yang and Schmidt (1994) used the RAPD analysis to differentiate mutants developed by X-rays treated cherry leaf plants RAPD assay performed for the identification of 14 in vivo and 1 in vitro lemon mutants in comparison with a known zygotic origin genotype (Deng et al. 1995). Munir et al. (2015) through RAPD analyses indicated that the plantlets subjected to gamma irradiation had a

Table 23.3 Different molecular markers used for the detection of mutations for different traits in various fruit crops

Species	Cultivar	Dosages (Gy)	Generation	Target trait	Marker used	References
Grape	Black Matrouh	0–40	M_1V_3	Drought Tolerance	ISSR	Al-Mousa et al. (2016)
	Salt Creek 140 Ruggeri	0–30		Salinity tolerance	EST SSR	Motha (2016), Motha et al. (2018)
	Pusa Navrang H-76-1 Pearl of Csaba Julesky Muscat		vM_5	Mutant identification	RAPD SSR	Dev et al. (2021)
	Freedom Harmony Ramsey Thompson seedless	0–20		Mutant identification	RAPD	Zayan et al. (2020)
	Pusa Seedless	0–50	vM_3	Mutant identification	RAPD	Khawale et al. (2006)
Banana	Dwarf Cavendish			Salinity tolerance	STMS	Miri et al. (2014)
	Lakatan	5–45	M_1V_4	BBTV resistant	SSR	Damasco et al. (2019)
	Basrai GCTCV-215	0–40	M_1V_4	Mutant identification	ISSR	Khatri et al. (2011)
	Grand-nain	20–60		BMV	ISSR	El-Sayed et al. (2011)
Mandarin	Kinnow	5–20		Mutant identification	SSR	Mallick et al. (2017)
Lemom	Yerli Yuvarlak					Polat et al. (2015)
					SSCP SSR ISSR	Görkem et al. (2020)
Guava	Shweta Lalit	0–125	M_1	Mutant identification	SSR	Singh (2016)
	Shweta	0–30	M_1V_1	Mutant identification	SSR	
Persimmon	Rojo Brillante	15–20			AFLP	Naval et al. (2013)
Avocado	Semil (9 genotypes)	0–50	M_1V_4	Mutant identification	SSR	Avenido et al. (2009)
Pitaya	Zi Honglong				ISSR	Deng et al. (2021)

great genetic variation as compared to the grape cv. Sundar Khani. Zayan et al. (2020) evaluated the genetic variations among mutants of different rootstocks of grape and Thompson seedless by using the RAPD technique, and clustered into two groups; one cluster contained unirradiated five genotypes and irradiated Freedom with 10 Gy, while the second cluster included the rest of irradiated genotypes indicating their higher genetic distinctness between irradiated and unirradiated genotypes. The effectiveness of AFLP markers for monitoring the genetic variation induced by γ-irradiation in persimmon "Rojo Brillante" is described by Naval et al. (2013).

IRAP markers are another class of molecular markers that showed utility in the genetic variability estimation of gamma irradiation-generated lines in banana. Kemal et al. (2018) used IRAP analysis for the genetic characterization of gamma-induced mutants of banana cv. Tanduk, and 10 Gy was suggested as suitable dose for inducing morphological and molecular variation.

ISSR markers are easy to apply, inexpensive in nature, reliable, and more informative than RAPD (Wu et al. 2011; Devarumath et al. 2002). The information obtained from ISSR analysis is more reliable than RAPD to provide supplementary data on the genetic variations of the mutants from the non-overlapping genome regions. Al-Mousa et al. (2016) observed unique bands in mutants developed with 20 Gy and concluded the utility of ISSR primers for initial characterization and selection for drought tolerance in grape. Khatri et al. (2011) elucidated the genetic similarities of mutants of banana cv. Basrai and GCTCV-215 were determined by ISSR analysis and observed a similarity index between 38.5% and 89.8% with 37 total bands produced by 11 primers. Pestana et al. (2011) evaluated genetic dissimilarity among putative "Preciosa" banana mutants generated by gamma-ray irradiation, using morpho-agronomic characteristics and ISSR markers. They found that genetic distances between the putative "Preciosa" mutants varied from 0.21 to 0.66. El-Sayed et al. (2011) used ISSR analysis to find out three BMV-resistant mutants of banana cv. Grand-Nain. Görkem et al. (2020) found the effectiveness of ISSR markers over SSR and SSCP markers for mutant characterization of mandarin and lemon.

SSRs had the choicest marker system in fruit crops due to co-dominant, ease of automation, reproducibility, genomic abundance (average frequency about one every 6.04 kb in plant genomic), and amenable to high throughput, which make them better than other markers for the evaluation of genetic variation. Miri et al. (2014) used microsatellite markers to investigate genetic variability within gamma-irradiated mutants of Cavendish banana for resistance to salinity and observed the absence of two alleles (240 and 280 bp) in irradiated clones, that appeared consistently in non-irradiated susceptible clones. Similarly, Hautea et al. (2004) distinguished the irradiated and non-irradiated clones with SSR primer (Ma 1/3), and detect variation between mutants generated with 40 Gy gamma rays and 3 Gy fast neutron. Dev et al. (2021) identified 13 solid mutants, among the initially selected 36 putative mutants of grape cv. Pusa Navrang, H-76-1, Pearl of Csaba, and Julesky Muscat with 11 SSR primers, and suggested the utility of SSR primers for mutant identification. Mallick (2014) discriminate the mutants of Kinnow and its clones using SSR markers and pointed out the scope of primers showing a high rate of polymorphism toward their further utilization for the characterization of Kinnow variants. Using SSR analysis, Singh (2016) discriminates the mutants of guava cv. Shweta and Lalit, and mutants showed varying degrees of dissimilarity with their respective parents. Polat et al. (2015) identified the genetic differences in the two mutants of Yerli Yuvarlak lemon using SSR markers. Motha (2016) identified eight solid mutants of different grape cultivars with EST-SSRs, and maximum mutants for salinity were identified at 10 Gy dose. The highest number of mutants was observed in Salt Creek and 140 Ruggeri indicating their ability for mutation.

SNPs are new generation markers, mostly biallelic in nature, easy to analyze, PCR-based, reproducible, and amenable to automation and high-throughput genotyping. SSCP technique was originally developed for rapid analysis of mutations (Orita et al. 1989; Hayashi 1991), high precision and success reaches up to 70–95% (Sunnucks et al. 2000). SSCP analysis is mostly single loci and employed for analysis of genetic variation in mutants of mandarin and lemon (Görkem et al. 2020).

23.10 Limitations

Irradiation breeding has the advantage of being simple to implement and does not necessitate complex procedures or a huge laboratory infrastructure. The successfulness lied on the ability to screening of mutant individuals for desired traits.

1. Most of the fruit crops are commercially propagated through vegetative means of propagation, i.e., multicellular tissues. Therefore, after a mutagenic treatment, a few generations are required to dissolving and stabilization of the chimeric sectors before mutant selection.
2. High-throughput mutagenesis in fruit crops is challenging due to big stature, long juvenile phase and absence of pre-selection criteria, heterozygous nature, fruit dropping, and longer maturity period.
3. Screening of larger populations for mutagenesis to select desirable mutation is one of the basic pre-requisites of mutation breeding,
4. Most of the mutations are recessive (Detection in Clonal and polyploidy species difficult), and the frequency of desirable mutants is very low (0.1%). Often mutations produce pleiotropic effects (Mutant × Elite variety).
5. Desirable mutations are commonly associated with undesirable side effects (other mutations, chromosomal aberrations), and mutations in quantitative or polygenic traits are usually in the direction away from the selection history of the parent variety.
6. Identification of micro-mutations, which are more useful to a plant breeder, is usually very difficult.
7. Most of the gamma irradiations mutations are unstable and undesirable.

23.11 Conclusion and Future Perspective

Changing climate is a threat to fruit cultivation leading to a reduction in crop production and quality deterioration. Past success of radiation-induced mutation breeding depicted their potential of resolving the elicited issues of climate change by altering the genome sequences of already established as well as generation of genic variability. Irradiation breeding yielded several superior mutated fruit genotypes with respect to improved productivity, quality parameters, nutraceutical properties, and resistance to insects-pests. Various experimental mutagenesis programs pertaining to the generation of pre-breeding stock, and their molecular characterization could help to shorten the breeding cycle as well as development/identification of novel variations, required to resolve the question of climate change. In vitro mutagenesis has edge over in vivo mutagenesis to shorten the breeding cycle, increase the mutation rate, effectiveness of mutagenic treatments, allow production of homo-histont, selection and evaluation of large number of mutated populations. For the development of climate-resilient fruit crops, radiation-induced in vivo and in vitro mutagenesis is an ideal alternative.

In-depth studies are required in the field of non-directed mutagenesis in vegetatively propagated fruit crops. A pre-selection index at the morphological, physiological, and molecular levels is required to the identification and validation of mutants at early stages. Irradiation with optimal doses and regeneration of irradiated material through plant tissue culture techniques has assumed a new dimension (Ahloowalia and Maluszynski 2001). If cell suspension systems can be developed from haploid plants derived from microspore cultures, the probability to obtain recessive mutants in homozygous conditions would be enhanced many more times. A combination of the existing techniques of anther and microspore culture, cell suspension culture, irradiation of haploid cells, chromosome doubling, and regeneration of doubled haploid plants could be used to obtain the desired genotypes in a short duration. The identification and analysis of mutants are based on the use of molecular techniques of DNA fingerprinting and mapping on PCR-based markers (SSR, RAPD, AFLP). Site-specific insertion of a single base into a targeted gene by using chimeric RNA/DNA oligonucleotides as demonstrated in field crops, and mutant tagging shall bring a new dimension to gene technology. Already, mutations can be linked to changes in DNA sequences for some plant traits, and to establish molecular maps in the structural and functional genomics of crop plants. These in turn would lead to a rapid enhancement of crop yields and quality.

References

Abdulhafiz F, Kayat F, Zakaria S (2018) Gamma irradiation effect on the growth of Musa cv. Tanduk (AAB). Asian J Agric Biol 6(2):135–142

Aderdour T, Handaji N, Brhadda N, Gmira N, Aouad BAE, Mahmoudi K, Ennaciri H, Hmimidi A, Yacoubi R, Mohib M, Benyahia H (2019) Optimization of genetic variability induction by bud irradiation of the Fremont mandarin. Plant Cell Biotechnol Mol Biol 20:778–788

Agisimanto D, Noor NM, Ibrahim R, Mohamad A (2016) Gamma irradiation effect on embryogenic callus growth of Citrus reticulata cv. Limau Madu. Sains Malaysiana 45(3):329–337

Ahloowalia BS, Maluszynski M (2001) Induced mutations - a new paradigm in plant breeding. Euphytica 118:167–173

Ahloowalia BS, Maluszynski M, Nichterlein K (2004) Global impact of mutation-derived varieties. Euphytica 135:187–204

Al-Mousa RN, Hassan NA, Stino RG, Gomaa AH (2016) *In vitro* mutagenesis for increasing drought tolerance and molecular characterization in Grape (*Vitis vinifera* L.) cv. "Black Matrouh". Syrian J Agric Res 3:259–275

Altaf N (2007) A new 'Kinnow' strain with 0-4 seed per fruit. In: Proc international symposium on prospects of horticultural industry in Pakistan. Institute of Horticultural Science, University of Agriculture, Faisalabad, pp 135–138

Alyanak N (2019) 'Alphonse lavallée' and 'Isabella' grape mutation using CO^{60} in variety creating opportunities. MSc thesis, Ordu University, Ordu, Turkey

Anonymous (2021). https://mvd.iaea.org/

Arisah H, Mariana BD (2017) Fruit diversity of SoE mandarin mutant at M1V2 generation resulted from gamma irradiation. Bull Germplasm 23:69–80

Atay AN, Atay E, Lauri P, Kunter B, Kantoglu KY (2018) Phenotyping gamma-ray-induced mutant population of 'Amasya' apple for architectural traits, precocity, floral phenology and fruit characteristics. Sci Hortic 233:195–203

Avenido RA, Galvez HF, Dimaculangan JG, Welgas JN, Frankie RB, Damasco OP (2009) Somatic embryogenesis and embryo culture coupled with gamma irradiation for generating avocado (*Persea americana* Miller) mutants in the Philippines. In: Sangwan R (ed) Induced mutation in tropical fruit trees. IAEA, Vienna, Austria, pp 47–70

Bapat VA, Ganapathi TR, Kulkarni VM, SunilKumar GB, Srinivas L, Suprasanna P (2007) Cellular and molecular approaches for the improvement of banana. In: Singh HP, Uma S (eds) Banana: technological advancements. Association for the Improvement in Production and Utilization of Banana, Natl. Research Center for Banana, Trichy, India, pp 160–176

Bermejo A, Pardo J, Cano A (2011) Influence of gamma irradiation on seedless citrus production: pollen germination and fruit quality. Food Nutr Sci 2:169–180

Bermejo A, Pardo J, Cano A (2012) Murcott seedless: influence of gamma irradiation on citrus production and fruit quality. Span J Agric Res 10:768–777

Bhagwat S, Duncan EJ (1998) Mutation breeding of banana cv. Highgate (*Musa accuminata*, AAA) for tolerance to *Fusarium oxysporum* f.sp. *cubense* using chemical mutagens. Sci Hortic 73:11–22

Bowen HJM (1962) Radiosensitivity in higher plants, and correlations with cell weight and DNA content. Radiat Bot 1:223–228

Broertjes C, Van Harten AM (1988) Applied mutation breeding for vegetatively propagated crops. Elsevier, Amsterdam, pp 345–355

Campeanu G, Neaţa G, Darjanschi G, Stan R (2010) Tree and fruit characteristics of various apple genotypes obtained through mutagenesis. Not Bot Hort Agrobot Cluj 38:248–251

Chan YK (2009) Radiation-induced mutation breeding of papaya. In: Sangwan R (ed) Induced mutation in tropical fruit trees. IAEA, Vienna, Austria, pp 93–100

Cimen B, Yesiloglu T, Incesu M, Yilmaz B, Tuzcu O (2019) Evaluation of Robinson mutant population derived by gamma irradiation. Egypt Univ Ziraat Fak Derg 56:375–381

Coto O, Rodríguez NN, Fuentes JL, Alvarez A, Machado M, Santiago L, Zamora V, Ramos-Leal M (2014) Mutation-based breeding of avocado in Cuba: state of the art. In: Tomlekova NB, Kozgar MI, Wani MR (eds) Mutagenesis: exploring genetic diversity of crops, vol 2014. Wageningen Academic Publishers, pp 265–282

Damasco OP, Dizon TO, Estrella JB, Caymo LS, Guittap EJ, Dela C, Fs JR, Mendoza EMT (2007) Selection and characterization of gamma ray-induced bunchy top virus resistant mutants in banana cv Lakatan. Plant Mutat Rep 1(3):22–30

Damasco OP, Cueva FMD, Descalsota JC, Tayobong RRP (2019) Gamma radiation and *in vitro* induced banana bunchy top virus (BBTV) resistant mutant lines of Banana cv 'Lakatan' (*Musa* sp., AA). Philippine J Sci 149:159–173

de Beer ZC, Severn-Ellis A, Husselman JH (2001) Preliminary assessment of Cavendish banana clones for resistance/tolerance to Fusarium wilt. In: Molina AB, Nik Masdek NH, Liew KW (eds) Banana fusarium wilt management: towards sustainable cultivation, pp 194–200

Deng ZN, Gentile E, Nicolosi E, Domina F, Vardi A, Tribulato E (1995) Identification of *in vitro* and *in vivo* lemon mutants by RAPD markers. J Hortic Sci 70:117–125

Deng R, Fan J, Wang Y, Liu T, Jin J (2021) Mutation induction of EMS and ^{60}Co γ irradiation in vitro cultured seedlings of red pulp pitaya (Stenocereus) and ISSR analyzing of mutant. BMC Plant Biol. https://doi.org/10.21203/rs.3.rs-19273/v1

Dev R (2014) *In vitro* mutagenesis in grape and its validation using molecular markers. PhD dissertation, Division of Fruits and Horticultural Technology ICAR-Indian Agricultural Research Institute, New Delhi

Dev R, Singh SK, Singh R, Singh AK, Patel VB, Alizadeh M, Motha K, Kumar K (2021) Assessment of genetic diversity of grape mutants based on RAPD and SSR markers. Indian J Hortic 78(1):17–24

Devarumath RM, Nandy S, Rani V, Marimuthu S, Muraleedharan N (2002) RAPD, ISSR and RFLP fingerprints as useful markers to evaluate genetic integrity of micropropagated plants of three diploid and triploid elite tea clones representing *Camellia sinensis* (China type) and *C. assamica* ssp. assamica (Assam-India type). Plant Cell Rep 21:166–173

Ekbic BH, Tangolar S, Ekbic E (2017) Mutation induction using ^{60}Co in Pembe cekirdeksiz (*Vitis vinifera* L.) grape cultivar. Akad Ziraat Derg 6:101–106

El-Mageid IS, Al-Kfrawey AAM (2018) Effect of different doses of gamma radiation on avocado buds for produce of new genotypes. Middle East J Agric Res 7:977–985

El-Oualkadi A, Mouhib M, Hajjaj B (2019a) Study of radio-sensitivity of strawberry runners cv. Fortuna under Moroccan conditions. Am J Plant Sci 10:1921–1931

El-Oualkadi A, Mouhib M, Taleb BA, Hajjaj B (2019b) Effect of different gamma radiation doses on the growing of the Achmrar local fig variety *Ficus carica* L. in Morocco. Int J Environ Agric Biotechnol 4:1648–1653

El-Sabagh AS, Barakat MN, Genaidy EA (2011) Towards *in vitro* selection studies for salinity tolerance in Canino apricot cultivar. Effect of gamma irradiation on *in vitro* mutation and selection for salt-tolerance. Adv Hortic Sci 25(4):260–263

El-Sayed E, Mahfouze S, Shaltout A, El-Dougdoug K, Sayed R (2011) Mutation breeding of Banana cv. 'Grand-Nain' for resistance to some banana viruses using biotechnology and physical mutagens. Afr J Plant Sci Biotechnol 5(1):35–40

Ferreira EA, Pasqual M, Neto AT (2009) *In vitro* sensitivity of fig plantlets to gamma rays. Sci Agric (Piracicaba, Braz.) 66:540–542

Foster TM, Aranzana MJ (2018) The far-reaching contributions of bud sport mutants to horticulture and plant biology. Hortic Res 5:44

Froneman LJ, Breedt UJ, Koelemoer PJJ, Van-Rensburg PJJ (1996) Producing seedless Citrus cultivars with gamma irradiation. Proc Int Soc Citricul 1:159–163

Fuentes JL, Santiago L, Rodríguez NN, Arbelo OC, Alvarez A, Valdés Y, Vernhe M, Guerra M, Altanez S, Prieto EF, Velázquez B, Rodríguez JA, Sourd DG, Fuentes VR, Leal MR (2009) Combining zygotic embryo culture and mutation induction to improve salinity tolerance in avocado (*Persea americana* Mill). In: Sangwan R (ed) Induced mutation in tropical fruit trees. IAEA, Vienna, Austria, pp 71–82

Ganapathi TR, Sidha M, Suprasanna P, Ujjappa KM, Bapat VA, D'Souza SF (2008) Field performance and RAPD analysis of gamma irradiated variants of Banana cultivar 'Giant Cavendish' (AAA). Int J Fruit Sci 8(3):147–159

Ganapathi TR, Ujjappa KM, Badigannavar A (2016) Characterization of gamma ray induced clones in 'Giant Cavendish' banana (AAA) for morphological and yield contributing traits. Int J Fruit Sci 16:310–322

Garcia LR, Perej PJ, Bermudez IC, Orellana PP, Veitia NR, Padron YM, Romero CQ (2002) Comparative study of variability produced by induced mutation and tissue culture in banana (*Musa* sp.) cv. 'Grande Naine'. InfoMusa 11(2):4–6

Goldenberg L, Yaniv Y, Porat R, Carmi N (2014) Effects of gamma-irradiation mutagenesis for induction of seedlessness, on the quality of mandarin fruit. Food Nutr Sci 5:943–952

Görkem S, Yıldız AK, İlknur P, Aytül K, Ertuğrul T, Özhan S, Gül S (2020) Identification of genetic diversity among mutant lemon and mandarin varieties using different molecular markers. Turk J Agric Forest 44:1–14

Gray LH, Reed J, Poynter M (1943) The effect of ionizing radiation on broad bean. Br J Radiol 16: 125–128

Hautea DM, Molina GC, Balatero CH, Coronado NB, Perez EB, Alvarez MTH, Canama AO, Akuba RH, Quilloy RB, Frankie RB, Caspillo CS (2004) Analysis of induced mutants of Philippine with molecular markers. In: Jain SM, Swennen R (eds) Banana improvement: cellular, molecular biology and induced mutations. Science Publishers, Inc., USA, pp 45–57

Hayashi K (1991) PCR-SSCP: a simple and sensitive method for detection of mutations in the genomic DNA. PCR Methods Appl 1:34–38

Hearn CJ (1984) Development of seedless orange and grapefruit cultivars through seed irradiation. J Am Soc Hortic Sci 109:270–273

Helaly MNM, Hanam El-Hosieny AMR (2011) Effectiveness of gamma irradiated protoplasts on improving salt tolerance of lemon (Citrus limon L. Burm. F.). Am J Plant Physiol 6(4):190–208

Hensz RA (1971) 'Star Ruby': a new deep red-fleshed grapefruit variety with distinct tree characteristics. J Rio Grande Valley Hortic Soc 25:54–58

Huang J, Wen S, Zhang Y, Zhong Q, Yanga L, Chen L (2017) Abnormal megagametogenesis results in seedlessness of a polyembryonic 'Meiguicheng' orange (*Citrus sinensis*) mutant created with gamma-rays. Sci Hortic 217:73–83

Hwang SC, Ko WH (1990) Selection of improved Cavendish banana mutants resistant to race 4 *Fusarium oxysporum* f. sp. *cubense*. Acta Hortic 275:417–423

Jain SM (2000) Mechanisms of spontaneous and induced mutations in plants. In: Proc. 11th International Congress of Radiation Research. Dublin, Ireland 18–23, July 1999, pp 255–258

Jain SM (2010) *In vitro* mutagenesis in banana (Musa spp.) improvement. Acta Hortic 879:605–614

Jain SM (2012) *In vitro* mutagenesis for improving date palm (*Phoenix dactylifera* L.). Emir J Food Agric 24(5):400–407

Jain SM, Maluszynski M (2004) Induced mutations and biotechnology in improving crops. In: Mujib A, Cho MJ, Predieri S, Banerjee S (eds) *In vitro* application in crop improvement. Science Publishers, London, pp 170–202

Jain SM, Swennen R (2004) Banana improvement: cellular, molecular and mutagenesis approaches. Science Publishers, New Hampshire

Jamaluddin SH (1994) Mutation breeding of banana in Malaysia. In: Jones DR (ed) The improvement and testing of Musa: a global workshop. International Network for the Improvement of Banana and Plantain, Montpellier, pp 228–232

Kafa G, Seday U, Uysal O, Polatoz S (2015) Effect of budwood irradiation on seed number of 'Clementine', 'Nova' and 'Robinson' mandarin mutants. Acta Hortic 1065:539–542

Karmarkar VM, Kulkarni VM, Suprasanna P, Bapat VA, Rao PS (2001) Radio-sensitivity of *in vivo* and *in vitro* cultures of banana cv. Basrai (AAA). Fruits 56:67–74

Kemal FA, Kayat F, Zakaria S (2018) Identification of morphological and molecular variation induced by gamma irradiation on musa cv. Pisang tanduk (AAB). J Biotechnol 15:265–271

Kepenek K (2016) Effects of gamma ray irradiation and NaCl on induced somaclonal variation in Arnavutköy Strawberry cultivar. Acta Phys Polon 130:337–341

Khalil SA, Sattar A, Zamir R (2011) Development of sparse-seeded mutant Kinnow (*Citrus reticulata* Blanco) through budwood irradiation. Afr J Biotechnol 10:14562–14565

Khatri A, Bibi S, Dahot MU, Khan IA, Nizamani GS (2011) *In vitro* mutagenesis in banana and variant screening through ISSR. Pak J Bot 43(5):2427–2431

Khawale RN, Singh SK, Vimala Y (2006) Gamma rays induced in vitro mutagenesis and molecular marker-assisted selection of mutants in grapevine. Acta Hortic 725:643–666

Kuksova VB, Piven NM, Gleba YY (1997) Somaclonal variation and *in vitro* induced mutagenesis in grapevine. Plant Cell Tissue Organ Cult 49:17–27

Kulkarni VM, Ganapathi TR, Suprasanna P, Bapat VA (2007) *In vitro* mutagenesis in banana (Musa spp.) using gamma irradiation. In: Jain SM, Haggman H (eds) Protocols for micropropagation of woody trees and fruits. Springer, New York, pp 138–142

Kumar AR, Kumar N, Poornima K, Soorianathasundaram K (2012) Screening of *in vitro* derived mutants of banana against nematodes. Afr J Biotechnol 11(88):15451–15456

Kumar M, Kumar M, Prakash S, Rao S, Prasad Y, Chand P, Singh MK (2017) Effect of seed treatment with gamma rays on fruit quality of papaya (*Carica papaya* L.). Res Environ Life Sci 10:182–184

Kumar MK, Dinesh MR, Srivastav M, Singh SK, Singh AK (2018) Mutagenic sensitivity of scions and seed kernels of polyembryonic mango cultivars Peach and Bappakai to gamma irradiation. Int J Chem Stud 6:3335–3339

Kunter B, Bas M, Kantoglu Y, Burak M (2012) Mutation breeding of sweet cherry (*Prunus avium* L.) var. 0900 Ziraat. In: Shu QY, Forster BP, Nakagawa H (eds) Plant mutation breeding and biotechnology. CABI Press, Wallingford, UK, pp 453–462

Latado RR, Neto AT, Figueira A (2012) *In vivo* and *in vitro* mutation breeding of citrus. Biorem Biodivers Bioavailab 6:40–45

Liang H, Hu Y, Pang W, Liu W, Yang M (2011) Studies on kiwifruit improvement by multiple top grafting. Acta Hortic 913:365–372

Litz RE (2009) Recovery of mango plants with antrachnose resistance following mutation induction and selection in vitro with the culture filtrate of Colletotrichum gloesporoides Penz. In: Sangwan R (ed) Induced mutation in tropical fruit trees. IAEA, Vienna, Austria, pp 7–16

Liu YZ, Xiong JJ, Deng XX (2007) Difference of a citrus late-ripening mutant (*Citrus sinensis*) from its parental line in sugar and acid metabolism at the fruit ripening stage. Sci China Ser C Life Sci 50:511–517

Lokko Y, Amoatey H (2001) Improvement of pineapple using in vitro and mutation breeding techniques. IAEA, Vienna, pp 25–29

López-García A, Ibanez V, Muñoz JV, Herrero-Ortega A, Talón M (2012) Nero and Clemenverd, two new commercial varieties of clementine. Rural Life 352:18–21

López-García A, Terol J, Tadeo FR, Herrero-Ortega A, Ibañez V, Talón M (2015) Three new cultivars of clementine: 'Clemenverd', 'Nero' and 'Neufina'. Acta Hortic 1065:239–245

Maan SS, Brar JS (2021) Mutagenic sensitivity analysis in guava (*Psidium guajava* L.). Fruits 76(4):181–190

Majd F, Jahangirzadeh E, Vedadi S, Tafti MN, Rastegari J (2009) Mutation induction for improving of tangerine in Iran. In: Sangwan R (ed) Induced mutation in tropical fruit trees. IAEA, Vienna, Austria, pp 41–46

Mak C, Ho YW, Tan YP, Ibrahim R (1996) Novaria-a new banana mutant induced by gamma irradiation. InfoMusa 5:35–36

Mallick M (2014) Characterization of Kinnow mandarin clones and mutants. MSc thesis, ICAR-Indian Agricultural Research Institute

Mallick M, Bhardwaj C, Srivastav M, Sharma N, Awasthi OP (2017) Molecular characterization of Kinnow mandarin clones and mutants using cross genera SSR markers. Indian J Biotechnol 16: 244–249

Mariana BD, Arisah H, Selvawajayanti YM (2018) Seedless fruit pummelo induced by Gamma Ray irradiation: fruit morphological characters and stability evaluation. Biodiversitas 19:706–711

Martin KP, Pachathundikandi SK, Zhang CL, Slater A, Madassery J (2006) RAPD analysis of a variant of banana (*Musa* sp.) cv. Grande Naine and its propagation via shoot tip culture. In Vitro Cell Dev Biol Plant 42(2):188–192

Mba C, Afza R, Bado S, Jain SM (2010) Induced mutagenesis in plants using physical and chemical agents. In: Davey MR, Anthony P (eds) Plant cell culture: essential methods. Wiley, Chichester, pp 111–130

Mba C, Afza R, Shu QY (2012) Mutagenic radiations: X-rays, ionizing particles and ultraviolet. In: Shu Q, Nakagawa H, Forster B (eds) Plant mutation breeding and biotechnology. CABI Press, Wallingford, UK, pp 83–90

McCollum TG, Bowman KD (2005) Fruit and juice quality of seedless Pineapple oranges on four rootstocks. Proc Fla State Hortic Soc 118:1–3

Miri SM, Mousavi A, Naghavi MR, Khiabani BN (2014) Molecular analysis of Musa mutants resistant to salinity by microsatellite markers. Trakia J Sci 2:114–120

Miri SM, Rahimi M, Naserian Khiabani B, Vedadi C (2019) Response of gamma-irradiated banana plants to in vitro and ex vitro salinity stress. Crop Breed J 9(1 & 2):33–44

Mishra PJ, Ganapathi TR, Suprasanna P, Bapat VA (2007) Effect of single and recurrent gamma irradiation on in vitro shoot cultures of banana. Int J Fruit Sci 7:47–57

Montañola MJ, Galaz A, Gambardella M, Mártiz J (2015) New low seeded Mandarin (*Citrus reticulata*) and Lemon (*C. limon*) selections obtained by gamma irradiation. Acta Hortic 1065: 543–548

Motha K (2016) *In vitro* mutagenesis of grape rootstock genotypes and isolation of NaCl tolerant mutants. PhD thesis, ICAR-Indian Agricultural Research Institute

Motha K, Singh SK, Singh AK, Singh R, Srivastav M, Verma MK, Bhardwaj C (2018) Molecular characterization and genetic relationships of some stress tolerant grape rootstock genotypes as revealed by ISSR and SSR markers. Plant Tissue Cult Biotechnol 28(1):77–90

Munir N, Safdar I, Naz S (2015) Effect of induced mutation for varietal improvement in some local grapevine cultivars. J Anim Plant Sci 25(1):234–242

Murti RH, Kim HY, Yeoung YR (2013) Effectiveness of gamma ray irradiation and ethyl methane sulphonate on *in vitro* mutagenesis of strawberry. Afr J Biotechnol 12(30):4803–4812

Nandariyah, Hartati S, Khasanah U (2019) Gamma irradiation on generative growth of Raja Bulu banana (*Musa paradisiaca* L.). In: MV1. Proc 4th International conference on climate change, Yogyakarta, Indonesia

Naval MM, Zuriaga E, Badenes ML (2013) AFLP analysis of mutations induced by gamma irradiation in 'Rojo Brillante' Persimmon. Acta Hortic 996:117–122

Neto AT, Latado RR, Tsai SM, Derbyshire MT, Yemma AF, Filho JAS, Ceravolo L, Rossi AC, Namekata T, Pompeu JP, Figueiredo JO, Pio R, Domingues ET, Santos PC, Boliani A (1998) Mutation breeding in vivo and in vitro in vegetatively propagated crops. Induced mutations in connection with biotechnology for crop improvement in Latin America. Joint FAO/IAEA Division of Nuclear Techniques in Food and Agriculture, Lima, Peru, pp 55–62

Neto AT, Ando A, Figueira A, Latado RR, Santos PC, Correa LS, Peres LEP, Hauagge R, Pulcinelli CE, Ishiy T, Filho AWPF, Camargo CEO (2011) Genetic improvement of crops by mutation techniques in Brazil. In: Plant mutation reports. International Atomic Energy Agency and Food and Agriculture Organization of the United Nations, Vienna, pp 24–37

Nhat HNT, Chau NM (2010) Radiation induced mutation for improving papaya variety in Vietnam. Acta Hortic 851:77–80

Orita M, Suziki Y, Sekiya T, Hayashi K (1989) Rapid and sensitive detection of point mutations and DNA polymorphisms using the polymerase chain reaction. Genomics 5:874–879

Osei-Kofi F, Amoatey HM, Lokko Y (1997) Improvement of pineapple (*Ananas comosus* (L.) Merr.) using biotechnology and mutation breeding techniques. In: Conference: In vitro techniques for selection of radiation-induced mutants adapted to adverse environmental conditions 15–19 April 1996, Cairo, Egypt

Pathirana R, Deroles S, Hoeata K, Montefiori M, Tyson J, Wang T, Datson PM, Hellens RP (2016) Fast-tracking kiwifruit breeding through mutagenesis. Acta Hortic 1127:217–222

Patil SG, Patil VP (2005) Mutation studies in Anab-e-Shahi grape (*Vitis vinifera* L.). Indian J Hortic 62(3):223–226

Pérez-Jiménez M, Tallón CI, Pérez-Tornero O (2020) Inducing mutations in Citrus spp.: sensitivity of different sources of plant material to gamma radiation. Appl Radiat Isotopes 157:1–7

Pestana RKN, Amorim EP, Ferreira CF, Amorim VBO, Oliveira LS, Ledo CAS, Silva SO (2011) Genetic dissimilarity of putative gamma-ray-induced 'Preciosa - AAAB-Pome type' banana (Musa sp) mutants based on multivariate statistical analysis. Genet Mol Res 10:3976–3986

Polat I, Turgutoglu E, Kurt S (2015) Determination of genomic diversity within mutant lemon (*Citrus limon* L.) and Mandatin (*Citrus reticulata*) using molecular markers. Pak J Bot 47(3): 1095–1102

Predieri S (2001) Mutation induction and tissue culture in improving fruits. Plant Cell Tissue Organ Cult 64:185–210

Predieri S (2002) The importance of induced mutations in pear improvement. Acta Hortic 596:161–166

Predieri S, Zimmerman RH (2001) Pear mutagenesis: in vitro treatment with gamma-rays and field selection for productivity and fruit traits. Euphytica 117:217–227

Predieri S, Magli M, Zimmerman RH (1997) Pear mutagenesis: in vitro treatment with gamma-rays and field selection for vegetative form traits. Euphytica 93:227–237

Pujar DU, Vasugi C, Vageeshbabu HS, Honnabyraiah MK, Adiga D, Jayappa J, Kanupriya (2019) Evaluation of mutant progenies for improved morphological, fruit and yield traits. J Pharmacogn Phytochem 8:2324–2334

Qamar M, Qureshi ST, Khan IA, Memon SA, Bano Z, Solangi SK (2016) Influence of gamma radiation on the physiochemical properties of in vitro triploid and tetraploid banana species. Pak J Biotechnol 13(4):237–244

Rattanpal HS, Sidhu GS (2015) Development of low seeded Kinnow through mutation breeding. Agric Res J 52:198–199

Reig C, Agustí M (2007) 'Piera': a new everflowering loquat variety. Acta Hortic 750:165–168

Reis RV, Amorim EP, Ledo CAS, Pestana RKN, Gonçalves ZS, Borém A (2015) Selection of putative Terra Maranhão plantain cultivar mutants obtained by gamma radiation. Genet Mol Res 14(2):4687–4695

Rocha AJ, Soares JMS, Nascimento FS, Santos AS, Amorim VBO, Ferreira CF, Haddad F, Santos-Serejo JA, Amorim EP (2021) Improvements in the resistance of the Banana species to *Fusarium wilt*: a systematic review of methods and perspectives. J Fungi 7:249. https://doi.org/10.3390/jof7040249

Roose ML, Williams TE (2007) Mandarin tree named 'Tango'. US Plant Patent 17863

Sahu P, Dash DK, Lenka J, Dash SN, Tripathy SK (2019) A gamma radiosensitivity study on papaya cv. Ranchi local & Arka Surya. Int J Chem Stud 7:146–153

Sales EK, Espino RRC (2008) Mutation induction in Philippine bananas cv. 'Lakatan' through gamma ray irradiation. In: FAO/IAEA international symposium on induced mutations in plants, 12–15 Aug 2008, Vienna, Austria, p 123

Sattar MN, Iqbal Z, Al-Khayri JM, Jain SM (2021) Induced genetic variations in fruit trees using new breeding tools: food security and climate resilience. Plants 10:1–36

Selvi BS, Ponnuswami V, Kumar N (2007) Radiosensitivity of amla (*Emblica officinalis* Gaertn.) varieties treated with gamma rays. J Hortic Sci 2:108–111

Sharma V, Thakur M (2021) Applicability of SCoT markers for detection of variations in Fusarium yellows resistant lines of ginger (*Zingiber officinale* Rosc.) induced through gamma irradiations. S Afr J Bot 140:454–460. https://doi.org/10.1016/j.sajb.2021.01.021

Sharma V, Thakur M, Tomar M (2020) *In vitro* selection of gamma irradiated shoots of ginger (*Zingiber officinale* Rosc.) against *Fusarium oxysporum* f.sp. *zingiberi* and molecular analysis of the resistant plants. Plant Cell Tissue Org Cult 143:319–330. https://doi.org/10.1007/s11240-020-01919-x

Singh S (2016) Induction and molecular characterization of guava variants generated through mutation breeding. MSc thesis, Punjab Agricultural University, Ludhiana

Singh S, Gill MIS, Arora NK (2018) Induction of variants in guava (*Psidium guajava* L.) through irradiation of budsticks by gamma rays. Agric Res J 55:347–350

Singh S, Gill MIS, Arora NK, Sohi HS (2021) Determination of gamma rays induced mutagenic sensitivity of guava cv. 'Lalit' and 'Shweta' (*Psidium guajava* L.). Agric Res J 58:478–481

Smith MK, Hamill SD, Langdon PW, Giles JE, Doogan VJ, Pegg KG (2006) Towards the development of a Cavendish banana resistant to race 4 of fusarium wilt: gamma irradiation of micropropagated Dwarf parfitt (*Musa* spp., AAA group, Cavendish subgroup). Austral J Exp Agric 46:107–113

Somsri S, Putivoranat M, Kanhom P, Thayamanont P, Meecharoen S, Jompook P (2008) Improvement of tropical and subtropical fruit trees: Tangerine, pummelo in Thailand through induced mutation and biotechnology. Acta Hortic 787:127–140

Spiegel-Roy P (1990) Economic and agricultural impact of mutation breeding in fruit trees. Mut Breed Rev Vienna 5:215–235

Spiegel-Roy P, Padova R (1973) Radiosensitivity of shamouti orange (*Citrus sinensis*) seeds and buds. Radiat Bot 13:105–110

Spiegel-Roy P, Vardi A (1990) Induced mutations in Citrus. Mutat Breed News 35:26–27

Stover EW, McCollum TG, Niedz RP, Bowman KD (2008) Citrus scion breeding at the USDA/ARS U.S. Horticultural Research Laboratory. In: Proc International Society of Citriculture. 11th International Citrus Congress, Vol. I, October 26–30, Wuhan, China, pp 50–54

Sunnucks P, Wilson ACC, Beheregaray LB, Zenger K, French J, Taylor AC (2000) SSCP is not so difficult: the application and utility of single-stranded conformation polymorphism in evolutionary biology and molecular ecology. Mol Ecol 9:1699–1710

Surakshitha NC, Soorianathasundaram K, Ganesan NM (2017) Determination of mutagenic sensitivity of hardwood cuttings of grapes 'Red Globe' and 'Muscat' (*Vitis vinifera* L.) to gamma rays. Sci Hortic 226:152–156

Sutarto I, Agisimanto D, Supriyanto A (2009) Development of promising seedless citrus mutants through gamma irradiation. In: Shu QY (ed) Induced plant mutations in the genomics era. Food and Agriculture Organization of the United Nations, Rome, pp 306–308

Tanaka A, Shikazono N, Hase Y (2010) Studies on biological effects of ion beams on lethality, molecular nature of mutation, mutation rate, and spectrum of mutation phenotype for mutation breeding in higher plants. J Radiat Res 51:223–233

Tang CY, Huang SC, Liu CC (2000) Mutation breeding in bananas: an overview. Chin Soc Hortic Sci 46:251–258

Vardi A, Spiegel-Roy P (1988) A new approach to selection for seedlessness. In: Proc. 6th Citrus Congress International Society of Citriculture, Tel Aviv, Israel, pp 131–134

Vardi A, Elhanati A, Frydman-Shani A, Neumann H, Spiegel-Roy P (1995) New considerations on the choice of irradiation dose rate in Citrus. In: Induce mutation and molecular techniques for crop improvement. International Atomic Energy Agency and Food and Agriculture Organization of the United Nations, Vienna, pp 667–670

Vardi A, Spiegel-Roy P, Frydman-Shani A, Elchanati A, Neumann H (2003) Citrus tree named 'Orri'. US Plant Patent 13616

Vardi A, Levin I, Carmi N (2008) Induction of seedlessness in citrus: from classical techniques to emerging biotechnological approaches. J Am Soc Hortic Sci 133(1):117–126

Venkatachalam L, Venkataramareddy SR, Neelwarne B (2007) Molecular analysis of genetic stability in long-term micropropagated shoots of banana using RAPD and ISSR markers. Electron J Biotechnol 10:1–12

Vos JE, Preez RJD, Froneman I, Hannweg K, Husselman J, Rheeder S (2009) Mutation breeding in South Africa 2003–2004. In: Sangwan R (ed) Induced mutation in tropical fruit trees. IAEA, Vienna, Austria, pp 113–122

Walker AR, Lee E, Robinson SP (2006) Two new grape cultivars, bud sports of Cabernet Sauvignon bearing pale-coloured berries, are the result of deletion of two regulatory genes of the berry colour locus. Plant Mol Biol 62:623–635

Weimin W, Mizhen Z, Zhuangei W, Yaming Q, Ji Y (2009) The study of the irradiation effect on runner plant of strawberry with 60Co-Gamma Ray. In: Lopes-Medina J (ed) Proc. VI. internat. strawberry symposium, Strawberry. Acta Hortic, vol 842, pp 597–600

Wu L, Li M, Yang X, Yang T, Wang J (2011) ISSR analysis of Chlorophytum treated by three kinds of chemical mutagen. J Northeast Agric Univ 18:21–25

Xiao JP, Chen LG, Xie M, Liu HL, Ye WQ (2009) Identification of AFLP fragments linked to seedlessness in Ponkan mandarin (*Citrus reticulata* Blanco) and conversion to SCAR markers. Sci Hortic 121(4):505–510

Yamaguchi H (2018) Mutation breeding of ornamental plants using ion beams. Breed Sci 68:71–78

Yang H, Schmidt H (1994) Selection of mutants from adventitious shoots formed in X-ray treated cherry leaves and differentiation of standard and mutant with RAPDs. Euphytica 77:89–92

Zamir R, Ali N, Shah ST, Mohammad T, Ahmad J (2009) Guava (*Psidium guajava* L) improvement using *in vivo* and *in vitro* induced mutagenesis. In: Sangwan R (ed) Induced mutation in tropical fruit trees. IAEA, Vienna, Austria, pp 101–112

Zanol GC, Cunha J, Brazão JS, Fevereiro PS, Eiras-Dias JE (2011) Identification of a new white-berried grapevine cultivar as a result from bud sport of the portuguese blue black cultivar 'AlfrocheiroPreto'. Acta Hortic 918:673–678

Zayan MA, El-Shereif AR, Gaser ASA, Salama AM (2020) Genetic diversity and relationships among grapevine rootstock mutants through RAPD technique. J Product Dev 25(1):79–99

Zhao Y, Zhaob X, Zhao S, Hanb N (2015) A novel bud sport from the 'Benitaka' table grape cultivar (*Vitis vinifera* L.) improves sugar and anthocyanin accumulation at the berry ripening stage. S Afr J Bot 97:111–116

Zheng HR, Liu HC, Ye ZW, Su MS, Jin YF (2009) Induction of mutation in Jujube (*Zizyphus jujuba* Mill) using tissue culture combined with ^{60}Co gamma rays irradiation. In: Sangwan R (ed) Induced mutation in tropical fruit trees. IAEA, Vienna, Austria, pp 113–122

Induced Mutations for Genetic Improvement of Banana

24

H. Tak, S. Bhakta, S. Negi, and T. R. Ganapathi

Abstract

Banana is one of the most important crops grown in many parts of the world and thus contributes to the economy of many developing countries. However, with changing climatic scenario and the evolution of new pathogenic races, the cultivation of banana crops is under severe threat. Though transgenic approaches have been carried out to address such problems in banana cultivation; however, their limited acceptability by the general public has hindered progress in increasing banana productivity. Most of the elite cultivars of banana are sterile and polyploid in nature despite those efforts toward the genetic improvement of banana has gradually been carried out through induced mutagenesis approaches. This chapter discusses the existing limitations in the genetic improvement of banana and the approaches for mutant development in banana. The chapter further details the various mutants in banana and their improved characteristics observed during analysis and the suitability of different plant materials for different aspects of mutation induction. Despite significant efforts being put into banana improvement by induced mutations, only four improved mutants have till now been registered in the "IAEA Mutant variety database." With the advent of a novel pathogenic race of the dreaded *Fusarium oxysporum* (*Foc*) having the ability to infect all existing cultivated banana varieties, there is a greater need to isolate novel mutants of banana. Concerted efforts by scientists

H. Tak · S. Bhakta · T. R. Ganapathi (✉)
Plant Cell Culture Technology Section, Nuclear Agriculture & Biotechnology Division, Bhabha Atomic Research Center, Mumbai, India

Homi Bhabha National Institute, Mumbai, India

S. Negi
Department of Biotechnology, University of Mumbai, Mumbai, India

S. Penna, S. M. Jain (eds.), *Mutation Breeding for Sustainable Food Production and Climate Resilience*, https://doi.org/10.1007/978-981-16-9720-3_24

worldwide are required for the isolation and identification of mutants having disease resistance, improved stress tolerance, and superior yield.

Keywords

Banana · Mutant · Gamma rays · Induced mutations · Fusarium · Chimerism · Embryogenic cells

24.1 Introduction

Banana and *Musa* spp. plantains are a prominent source of economy from agriculture for many countries. Across tropical and subtropical areas of the globe, banana serves as a source of nutrition and energy for millions of people (FAO 2018). Moreover, banana is considered a cash crop and holds paramount importance for food security in many countries owing to their year-round cultivation and wide acceptance for consumption among people. Banana improvement is a challenging task due to its polyploid nature and vegetative propagation limiting the genotypic variability which can be generated through breeding approaches. Genetic variations in banana are introduced by spontaneous mutations or can be triggered by transgenic approaches or inducing mutations through physical and chemical mutagens (Roux 2004; Jain et al. 2011). Induced mutations can break the barrier in generating genetic diversity in absence of sexual recombination and assortment and its combination with high-throughput techniques for precise identification of molecular changes in mutants can result in rapid genetic improvement of banana.

In nature, banana produces a small number of propagules which limits the separation of desired mutants from chimeras and rapidly increase the time toward achieving the desired goal in induced mutagenesis experiment. Hence, induced mutagenesis in combination with in vitro propagation can rapidly lower the time required for the separation of novel mutants of elite banana cultivars. Induced mutations of embryogenic cell suspension which are derived from somatic embryos can be revolutionary in obtaining novel mutants from banana as they practically circumvent the tedious process of chimera separation and reduces the population size for screening of phenotypic and other agronomic traits in field (Roux et al. 2001; Jain 2010). Generating novel mutants through in vitro mutagenesis in banana will require a comprehensive knowledge of banana tissue culture and regeneration and mutagenesis followed by post-mutagenesis handling of plant material. Chemical mutagens such as ethyl methanesulfonate induce point mutations and hence are easy to identify and have a better effect on generating novel variations in protein functions while physical mutagens such as gamma irradiation can induce an array of mutations ranging from point mutations to variable insertions or deletions which can have a spectrum of effects and are thus tedious to identify and map (Jankowicz-Cieslak and Till 2015). However, only four mutants of banana have been registered as per the mutant variety database and these were derived using gamma irradiation of banana explants suggesting the applicability of physical mutagens in the genetic

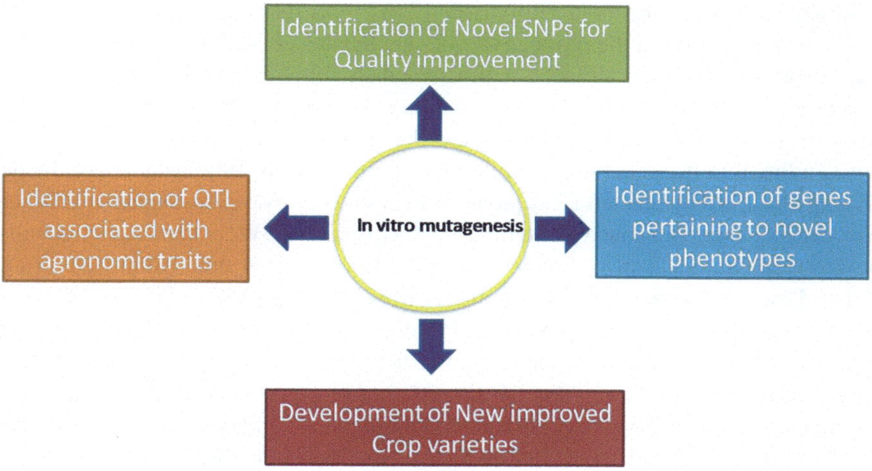

Fig. 24.1 The different facets of in vitro mutagenesis in agriculture research and crop improvement. In vitro mutagenesis is the easiest mutagenesis technique which leads to the development of new crop varieties in vegetatively propagated crops. Different phenotypes are observed in plants after inducing mutation. When in vitro mutagenesis get coupled with advanced sequencing technology it could marshal the identification of new genes and their function corresponding to the phenotype. After mutation induction, small mutations are observed across the genome leading to SNPs. SNPs observed in improved varieties could open a new avenue toward crop improvement research. Varieties developed by in vitro mutagenesis are sometimes agronomically superior than its wild varieties and identification of QTL associated with these agronomic traits could be used for the betterment of the same or related crop species

improvement of banana (MVD 2021). The lower number of improved varieties of banana registered with mutant variety database indicates the hurdles associated with banana improvement such as tissue culture-based propagation, polyploidy, heterozygosity, problems of chimera separation, and requirement of a large field for evaluation which is further aggravated by lower number of scientists associated with banana improvement through mutagenic approaches. The different facets of in vitro mutagenesis in agriculture research and crop improvement are highlighted in Fig. 24.1.

24.2 Problems in Cultivation of Banana

Various stress-causative agents severely affect the productivity of banana worldwide. During its life cycle banana can be targeted by many disease-causing agents belonging to fungi, viruses, bacteria, and nematodes. Among these disease-causing agents, the biggest reduction in banana cultivation has been caused by fungus *Fusarium oxysproum* f. sp. *cubense* (*Foc*) which causes *Fusarium* wilt disease or panama disease. Four races of *Foc* have been classified and race1, race2, and race4 are highly infectious to banana plants and once infected can lead to complete productivity loss in the field (Robinson 1996). *Foc* race1 emergence led to the complete destruction of the productivity of elite banana cultivar *Gros Michel* during the 1950s and was then replaced with Cavendish cultivars which were *Foc* race1 resistant (Dale et al. 2017). However, the recent reports on the emergence of *Foc tropical race4* (*TR4*) which can even infect the now popular Cavendish cultivars has posed a threat to the economy of millions of farmers cultivating banana worldwide (Zheng et al. 2018). Bacteria *Xanthomonas campestris* pv. *musacearum* is another wilt-causing agent in banana plantation and the disease (BBW; banana bacterial wilt) progression can result in a complete loss of productivity in field (Shimwela et al. 2016). Disease of banana originating from viruses such as *banana bunchy top virus* (*BBTV*), *banana streak virus* (*BSV*), and *banana bract mosaic virus* (*BBMV*) also are significant hurdles in the productivity of banana plantation (Kumar et al. 2015). Nematodes such as *Radophilus similis*, *Meloidogyne* spp., and *Helicotylenchus multicinctus* also limit banana cultivation resulting in a significant loss in productivity (Gowen and Queneherve 1990). Stress conditions of abiotic nature such as drought, high salinity and cold are major impeding factors in banana growth and yield, and the conditions are further aggravated by the inherent physiological and phenotypic nature of banana plants such as high transpiration rate due to huge canopy area and narrow root organization (van Asten et al. 2011). Such inherent limitations of banana plants lead to huge water requirement for satisfactory banana cultivation and has been reported to be in the range of 2000–2500 mm of rainfall distributed year-round (Vanhove et al. 2012). There is a dearth of research studies on the loss of productivity after drought exposure to banana plants; however, a significant loss in banana yield was reported in Tamil Nadu after reduced water availability in 2001–2004 (Surendar et al. 2013). Similarly, high-salinity encounters to the roots of banana can reduce plant growth as indicated by a substantial reduction in canopy area and weight of *Musa* cultivar *Calcutta 4* after growth in 100 mM NaCl stress for 21 days (Gomes et al. 2001). Lodging during the fruiting stage due to heavy bunch weight and when plants of tall stature encounter heavy winds is another serious problem in increasing the productivity of banana crop (Haiyan et al. 2018).

24.3 Induced Mutagenesis Studies in Banana

Induced mutagenesis studies have been carried out on seeds and rhizome of diploid banana cultivar *Musa balbisiana Colla* and parthenocarpic cultivar *Gros Michel* and their radiation sensitivity was analyzed (Stotzky et al. 1964; Agarwal 1984). Radiosensitivity of corms of *Musa* cultivar *Gros Michel* has been studied in an attempt to induce resistance toward Sigatoka and panama disease (Azzam and Linden 1965). The use of ethyl methanesulphonate (EMS) to induce novel mutations through mutagenizing seeds of diploid cultivars of banana *Musa acuminata* and *Musa balbisiana* has been attempted in prior studies (Menendez 1973). Banana shoot tips were used to regenerate shoots in vitro and gamma irradiation of explants resulted in plants with better phenotypic characteristics such as improved girth, height, number of suckers, and higher number of hands and fingers (De Guzman et al. 1980; Guzman et al. 1982). One of the earlier studies demonstrated the regeneration of viable shoots from irradiated suckers of banana; however, some of the regenerated plantlets showed growth reduction, chlorosis, and other abnormal characteristics (De Guzman et al. 1976). Shoot apices of different cultivars of *Musa* spp. such as those of dessert banana, plantain, and bluggoe cooking banana exposed to gamma irradiation revealed differences in radiosensitivity, and one early flowering mutant of *Grand Naine* was finally isolated from regenerated plantlets (Novak et al. 1989). Later developments resulted in the standardization of methodology such as explant nature, type and dosage of mutagens, resolving chimera issues, and mass screening of mutagenized population which helped researchers in attempts to isolate novel mutants in banana plants (Roux 2004). Mutagenized population of six cultivars of banana (AAA, AAB, ABB, and BB genotypes) produced by gamma irradiation of in vitro shoot cultures resulted in the establishment of dose regime and identification of some morphological variants in different banana cultivars (Kulkarni et al. 1997). Efforts on genetic improvement of Indian cultivars of banana through radiation-induced mutations of in vitro shoots were initiated at Bhabha Atomic Research Centre, India (Bapat et al. 2007). Dwarf and early flowering mutants were isolated after field trial of gamma irradiated population of important Musa cultivar "*Giant Cavendish*" and were then characterized for DNA polymorphism through RAPD analysis (Ganapathi et al. 2008). An extra dwarf cultivar of *Cavendish* having strong resistance toward *subtropical race4* of *Fusarium oxysporum* was gamma irradiated and one mutant line (DPM25) was isolated showing *Foc race4* resistance and improved agronomic characteristics (Smith et al. 2006). Three different clones of banana (GCTCV-215, Yangambi KM-5, and FHIA-23) were irradiated at different doses of gamma irradiation and responses of in vitro cultures were monitored to determine the optimal dose regime for mutation induction in these varieties (Qamar et al. 2016). GN-60A is an early flowering mutant of banana cultivar *Grand Naine* and was designated as "*Novaria*" obtained after gamma ray treatment and field assessment was performed in Australia, Honduras, Malaysia, and South Africa which demonstrated an earliness of 10 weeks compared to parental clone (Mak et al. 1996). *KlueHom Thong KU1* is a novel mutant isolated after 25 Gy gamma irradiation of parental clone *Hom Thong* in Thailand and is recognized for its large bunch size (Roux 2004). In Sudan, a significantly high-yielding variety of banana cultivar "*Williams*" was developed at "agricultural research corporation" (ARC) of Sudan and was released with the name *Albeely* (Ali et al. 2008). In Indonesia, a novel mutant of banana cultivar, *Ambon Kuning*, with improved characteristics such as *fusarium* wilt resistance and higher vitamin C content was developed by gamma irradiation of in vitro cultures (MVD 2021). Resistance to penetration of *Foc* was developed in *Musa* spp. AAA Group cv. *Highgate* by gamma irradiation and chemical mutagenesis and some of the selected lines showed significantly reduced corm discoloration with no apparent external symptoms (Bhagwat and Duncan 1998a, b). Improvement in agronomic characteristics of a *Foc race4* resistant extra dwarf *Cavendish* cultivar "*Dwarf Parfitt*" was observed in terms of improved height and fruit yield (Smith et al. 1995). Recently through gamma irradiation of multiple shoots, a dwarf mutant of

Table 24.1 List of identified and reported banana mutants

S. No.	Mutant	Parent	Improved trait	Mutagen	Country
1	Novaria	Grand Naine	Early flowering	Gamma irradiation (60 Gy)	Malaysia
2	ALBEELY	Williams	Higher yield	Irradiation	Sudan
3	KlueHom Thong KU1	Hom Thong	Higher yield	Gamma irradiation	Thailand
4	Pirama 1	Ambon Kuning	Fusarium wilt resistance and vitamin C fortified	Gamma irradiation	Indonesia
5	GN35-I to GN35-VIII	Grande Naine	Resistance to *Mycosphaerella fijiensis*	Gamma irradiation	Austria (IAEA)
6	Embul-35 Gy	Embul	Early flowering	Gamma irradiation	Sri Lanka
7	SH-3436-L9	SH-3436	Reduced height	Early flowering	Gamma irradiation

banana cultivar "*Giant Cavendish*" has been isolated at Bhabha Atomic Research Centre, India, and is currently being assessed by multilocation location trials to confirm its utility in preventing yield losses due to lodging (Unpublished data). The list of different novel mutants isolated and identified in banana is provided in Table 24.1.

24.4 Resolving Chimera Occurrence

Utilization of multiple shoot culture or multicellular meristematic culture such as shoot tip generally results in the generation of chimeric plants hence, repeated segregation of chimeric sectors in mutagenized population is required through vegetative propagation. However, the number of multiplication rounds on shoot multiplication medium to successfully segregate chimeric sectors can vary in each experiment. Even then some of the studies have pointed out that the chimeric sectors are successfully resolved after M1V6 generation in in vitro propagated banana plants (Jain et al. 2011; Roux et al. 2001; Jankowicz-Cieslak et al. 2012). Even then, the lack of ease to distinguish mutated cells from wild-type cells, large sample size of mutagenized population, laborious and costly analysis have considerable impact on the choice of explant selection in utility of mutation induction in banana.

24.5 Embryogenic Cells as Explant for Mutation Induction

The utility of cells derived from somatic embryos and therefore having embryogenic potential (embryogenic cells) for mutation inductions has opened the avenues to generate novel mutants free of chimerism issues (Roux et al. 2004). Somatic embryos are derived from single cells and can further produce secondary embryos

Fig. 24.2 Generation of mutagenized population through gamma irradiation of embryogenic cells. (**a**) Somatic embryos are derived from the meristematic tissue such as male flower bud of elite banana cultivars. Somatic embryos are then used to generate embryogenic cell suspension in an appropriate culture medium. (**b**) Embryogenic cells are exposed to gamma rays and then immediately cultured on fresh embryo development medium. (**c**) Embryogenic cells filtered on a glass fiber filter and cultured on embryo development medium. (**d**) Continuous culturing on embryo development medium resulted in the development of somatic embryos which are then converted into shoots. (**e**) Independent shoots emerging from irradiated embryogenic cells are then multiplied individually. (**f**) Different clones of independent lines are then rooted for further analysis. (**g**) The plants are acclimatized in controlled conditions of a glass house before field analysis. The acclimatized mutagenized population are then further screened for improved agronomical characteristics or disease resistance in a field evaluation

which further strengthen the chances of obtaining mutant plants free of chimerism issues (Grapin et al. 1998). The single-cell origin of somatic embryos is further proved by genetic transformation studies on embryogenic cell suspension indicating the generation of transgenic banana plants free of chimeric sectors (Ganapathi et al. 2001). The single-cell origin of somatic embryos in banana was further proved by the failure of the appearance of mixoploid plants from colchicine-treated embryogenic cells (Roux et al. 2003). A schematic description of generating novel mutants through gamma irradiation is depicted in Fig. 24.2.

24.6 Commonly Employed Mutagenic Agents and Their Utility in Banana Improvement

The choice of mutagen for induction of mutations in plant material depends on various factors such as type of explant, species of plant, breeding objectives, screening methods, and ease of availability. Broadly, mutagens have been classified into chemical mutagens and physical mutagens (Ahloowalia 1998). A comprehensive list of categories of mutagens is provided in Table 24.2. However, most of the studies on mutation-based improvement of banana have utilized gamma rays as a source of mutagen (Roux 2004). In most of the experiments on banana plants either using in vitro shoots or embryogenic cells, the optimal dose of gamma irradiation for the generation of mutant plants generally falls in the range of 30–50 Gy (Roux et al. 2003). However, the dearth of studies utilizing different mutagens on genetic improvement of banana has hindered to provide a detailed picture of spectrum of mutants that can be obtained by different mutagenic agents.

Table 24.2 Commonly used mutagens in mutation breeding experiments

S. No.	Type	Class	Effect on genome	Properties
1	Gamma rays	Physical	Point mutations and small deletions	Electromagnetic radiations
2	Fast Neutron	Physical	Translocations, and major deletions	Neutral subatomic particles
3	X-rays	Physical	Single and double strand breaks	Electromagnetic radiations
4	Alpha particles	Physical	Single and double strand breaks	Helium nucleus
5	Beta particles	Physical	DNA lesions	Electron
6	Ion beams	Physical	Chromosomal aberrations	Charged particle beam produced generally produced in accelerator
7	Ethyl methanesulphonate (EMS)	Chemical	DNA damage mostly point mutations	Chemical
8	Sodium azide	Chemical	DNA damage mostly point mutations	Chemical
9	Dimethyl sulfate (DMS)	Chemical	DNA damage mostly point mutations	Chemical
10	Ethyleneimine (EI)	Chemical	DNA damage mostly point mutations	Chemical

24.7 Molecular Analysis of Novel Mutants

Mutations induced by physical and chemical mutagens are random and one of the widely used transgenic-free approaches. Though isolation of mutants with novel characteristics is easy but screening, identification, and mapping of the underlying mutations is a mammoth job. It involves several steps such as (1) visual screening for phenotypic changes, (2) screening using pathogens/abiotic stress for identification of disease-resistant or abiotic stress-resistant mutant, (3) identification of the mutated genes in the chromosome, and (4) validation of the mutation for the mutant trait (Sahu et al. 2020). The last two steps of this process involve extensive molecular analysis. The classical method of identification was through linkage mapping and other physical mapping techniques. Recent advancement in genomic research has developed many DNA- and RNA-based approaches for the identification and mapping of the mutation in a mutant plant. In the following section, we have briefly discussed the recent molecular techniques used for screening and mapping of the mutation in in vitro mutagenesis.

24.8 DNA-Based Markers

DNA-based markers are the most widely used markers for the screening of the mutants in vegetatively propagated crops as they help to identify the presence of a particular locus in a mutant population. The most widely used DNA markers are "Restriction fragment length polymorphism" (RFLP) and "Random amplified polymorphic DNA" (RAPD)s (Suprasanna et al. 2015). Both these markers have their pros and cons as discussed below. RFLP is a codominant marker and requires huge efforts and involves hazardous radioactive labeling for identification and on the other hand, RAPD has reproducibility issues. These problems lead to the development of "Variable number tandem repeats" (VNTRs) dependent fingerprinting in plants. VNTRs are short tandemly repeated sequences present across the genome and they can be identified using PCR-based methods (Gupta et al. 1996). VNTRs have genome specificity and their assays do not depend on hybridization methods. Another molecular marker which is similar to the VNTRs is "Simple Sequence Repeats" (SSRs). SSRs are small repeat sequences present across the genome and they have an advantage over VNTRs as they are present across the genome even in the coding regions while VNTRs are mostly present in the noncoding genes (Zalapa et al. 2012). SSRs can be easily used as tagged sites to identify coding regions in plants with no or little sequence information. The SSR present in coding region is known as expressed sequence tags. AFLP (Amplified fragment length polymorphism) is used as a replacement for RFLP and RAPD as this technique allows selective amplification of restriction enzyme digested fragments. AFLP helps in the detection of marker deletion induced due to external mutagens and it is a promising technique for the screening of induced mutants (Blears et al. 1998). SNPs (Single Nucleotide polymorphism) are the most used maker of recent times and involve the identification of a single nucleotide change in a locus and hence

Table 24.3 Various DNA-based molecular markers useful for identification of mutations

DNA-based molecular markers	Principle	Application
Restriction fragment length polymorphism (RFLP)	Restriction digestion followed by southern hybridization	Comparative mapping, Fingerprinting
Random amplified polymorphic DNA (RAPD)	PCR amplification with random primers	Marker assisted selection
Microsatellites	PCR amplification of small repetitive sequences	Marker assisted selection, Frame work mapping, Fingerprinting
Cleaved amplified polymorphic sequences	Restriction digestion of amplified PCR product	Marker assisted selection, Frame work mapping
Amplified Fragment length polymorphism (AFLP)	Restriction digestion and selective PCR amplification	Marker assisted selection, specific locus identification

SNPs hold great for plants with known genomic information. In SNPs-based identification, the marker density can be measured at kilo base pairs level while in other markers the density is measured at the level of megabases. The higher abundance of SNPs presence in the genome makes it a good candidate for marker-assisted selection, high-resolution gene mapping, and other studies related to the assessment of genetic diversity (Cuenca et al. 2016). However, the mutants can be identified on the basis of various other molecular markers such as CAPs, ISSR, SCAR, STS, etc. A list of various DNA markers for identification of novel mutations is provided in Table 24.3.

24.9 RNA-Based Molecular analysis

Advancement in genome sequencing technology has promoted the mining of sequence information resulting in sequence resources for a huge number of crops. The RNA-based sequencing approach is very convenient in mapping the mutations in the coding region when an altered phenotype is observed and the sequence information is available. It is useful in identifying the possible miss-sense and non-sense mutations, expression changes, and variation in the post-transcriptional processes (Miller et al. 2013). The benefits of this approach are less amount of sequencing data for analysis, analysis of molecular mechanism is easier as it comes with the benefit of quantitative measurement of the change of expression (Weber 2015). Recently, it has been used in understanding the D-genome-specific chromosomal marker in wheat (Nishijima et al. 2018).

24.10 Next Generation Sequencing-Based Approach of Mutation Detection and Mapping

NGS technologies are now one of the most popular approaches for mutation detection and mapping in plant breeding. The easy sequencing technology pipelined with sophisticated and user-accessible computer programming makes it an easy and labor-effective process. These technologies are very useful in identifying molecular markers, transcriptome sequences, and expression analysis. The ability of resequencing of the targeted genomic regions makes it an evident method for the identification of genes involved in the maintenance of economically important agronomic traits in plants (Varshney et al. 2012). The NGS-based techniques have reduced the timing of sequencing in weeks or days while earlier it took years of struggle to sequence a plant genome. The NGS approach coupled with classical mapping approaches helps in the direct mapping of mutation in plants. The most used approaches for mutation identification are whole-genome resequencing and Exome sequencing.

24.11 Whole-Genome Resequencing

This is the easiest way of identifying the mutations in the genome. In this technique, the genome of the mutant plant is sequenced in the NGS platforms and compared with its reference genome to identify the variation. The assay is facilitated as sequence information for the whole genome of most of the crop have been deciphered. Through this approach, it is easier to identify SNPs, insertion deletion mutations, gene losses, and conversions. It is very popular among breeders due to its multifarious application.

24.12 Exome Sequencing

Exome sequencing is the sequencing of cDNAs obtained from transcripts. This technique considers only the coding region of the genome for analysis and thus helps to identify the proteins involved in the phenotypic traits. It is very useful in identifying QTLs as exemplified in sorghum where many QTLs have been identified using this approach (Gelli et al. 2016). Hence, the above-mentioned techniques are the most commonly used technologies for the identification of the causal mutations for a phenotypic trait.

24.13 Recent Advances in Mapping

The field of "induced mutation for crop improvement" is rapidly growing with the advancement in genome sequencing technology. Though in case of banana crop, the advancement of induced mutagenesis has not been routinely applied in comparison

Table 24.4 Advanced techniques for mapping of mutation

Name of the technique	Principles	Scope of application
Exome capture	Only coding region of the genome is analyzed to avoid complexity and repetitive sequences.	Applicable to all plants where genomic information is available.
Needle in the K-stack	Small subsequences of the whole-genome sequence of two related organisms are compared for mapping.	Applicable for all especially for the plant species where reference genome is not available.
Next generation mapping	Casual mutations are detected using reference genome.	Applicable to all plants where genomic information is available
RNA sequencing-based mapping	Used in bulk segregant analysis.	Applicable to all plants where genomic information is available.
Simultaneous Identification of Multiple Causal Mutations	Multiple casual mutations are identified simultaneously without a wild-type reference genome.	No need of genomic information applicable for all plant species.

to cereals like rice, wheat, and barley, advanced genomics technologies can be applied for banana improvement. The most popular advanced gene mapping technologies which could be useful to study the mapping of in vitro mutagenized plants are: (1) Exome capture mapping (Mo et al. 2018), (2) Simultaneous Identification of Multiple Causal Mutations (Yan et al. 2017), and (3) Needle in the k-Stack (Nordström et al. 2013). A list of recent techniques for mapping of mutations is provided in Table 24.4.

24.14 Conclusions and Future Prospective

Generating novel mutants through in vitro mutagenesis in banana is dependent on well-optimized tissue culture, plant regeneration, and mutagenesis followed by post-mutagenesis handling of plant material. In vitro mutagenesis of banana shoots pose issues of chimera formation and this requires a detailed and laborious examination of developed mutants. Isolation of pure mutant lines in banana can be speeded up by using embryogenic cell culture-based induced mutagenesis system. However, the protocols for the development of somatic embryos in most elite cultivars are still to be optimized and further studies in this direction can significantly revolutionize the genetic improvement of banana. Although the efficacy of gamma rays has been proved to be useful, the spectrum of various mutations induced by a combination of physical and chemical mutagens should be explored to isolate useful mutants. Popularization of identified novel mutants in banana should be taken on a priority basis worldwide to educate farmers about the improved traits. Moreover, attempts to characterize the genetic basis of mutations in improved varieties should be undertaken so as to harness the benefits of the same for the improvement of other species. Techniques such as next-generation sequencing, mapping, and transcriptome profiling which have proven useful to identity genetic basis of mutations in many species will have to be studied for their application in banana. Combining such techniques with classical in vitro mutagenesis could lead to the development of disease-resistant, stress-tolerant, and biofortified banana varieties in near future.

Acknowledgments HT, SB, and TRG thank Director, Bio-Sciences group, BARC for his constant encouragement and support. SN thanks the Department of Science and Technology (DST), New Delhi for DST INSPIRE Faculty fellowship.

References

Agarwal PK (1984) Note on effect of ethyl methane sulfonate on *Musa balbisiana* seeds. Indian J Hortic 41:93–94

Ahloowalia BS (1998) *In vitro* techniques and mutagenesis for the improvement of vegetatively propagated plants. In: Jain SM, Brar DS, Ahloowalia BS (eds) Somaclonal variation and induced mutations in crop improvement. Kluwer Academic Publishers, London, pp 293–309

Ali MA, Bakhiet SB, Hamid GA, Ahmed FA, Abdelrahman EA, Elkashif ME (2008) A clone of irradiated banana cultivar 'Williams' with high yield potential (IAEA-CN-167). International Atomic Energy Agency (IAEA), Vienna

Azzam H, Linden DB (1965) Radiation effects on banana corms, Musa sapientum. J Agric Univ Puerto Rico 4:17–21

Bapat VA, Ganapathi TR, Kulkarni VM, Sunil Kumar GB, Srinivas L, Suprasanna P (2007) Cellular and molecular approaches for the improvement of banana. In: Singh HP, Uma S (eds) Banana: technological advancements. Association for the Improvement in Production and Utilization of Banana, Natl. Research Center for Banana, Trichy, India, pp 160–176

Bhagwat B, Duncan EJ (1998a) Mutation breeding of banana cv. Highgate (*Musa acuminata*, AAA Group) for tolerance to *Fusarium oxysporum* f.sp. *cubense* using chemical mutagens. Sci Hort 73:11–22

Bhagwat B, Duncan EJ (1998b) Mutation breeding of Highgate (*Musa acuminata*, AAA) for tolerance to *Fusarium oxysporum* f.sp. *cubense* using gamma irradiation. Euphytica 101:143–150

Blears MJ, De Grandis SA, Lee H, Trevors JT (1998) Amplified fragment length polymorphism (AFLP): a review of the procedure and its applications. J Ind Microbiol Biotechnol 21:99–114

Cuenca J, Aleza P, Garcia-Lor A, Ollitrault P, Navarro L (2016) Fine mapping for identification of citrus alternaria brown spot candidate resistance genes and development of new SNP markers for marker assisted selection. Front Plant Sci 7:1948

Dale J et al (2017) Transgenic Cavendish bananas with resistance to Fusarium wilt tropical race 4. Nat Commun 8:1496

De Guzman EV et al (1976) Banana and coconut in vitro cultures for induced mutation Studies. Improvement of vegetatively propagated plants and tree crops through induced mutations. IAEA, Vienna., TECDOC-194, pp 33–54

De Guzman EV et al (1980) Plantlet regeneration from unirradiated and irradiated banana shoot tip tissues cultured in vitro. Philipp Agric 63:136–140

FAO (2018) Banana facts and figures. FAO, Italy, Rome. http://www.fao.org/economic/est/est-commodities/bananas/bananafacts/en/#.W76y7WgzbIV

Ganapathi TR, Higgs NS, Balint Kurti PJ, Arntzen CJ, May GD, Van Eck JM (2001) Agrobacterium mediated transformation of embryogenic cell suspensions of the banana cultivar Rasthali (AAB). Plant Cell Rep 20:157–162

Ganapathi TR, Sidha M, Suprasanna P, Ujjappa KM, Bapat VA, D'Souza SF (2008) Field performance and RAPD analysis of gamma-irradiated variants of banana cultivar 'giant cavendish' (AAA). Int J Fruit Sci 3:147–159

Gelli M, Mitchell SE, Liu K, Clemente TE, Weeks DP, Zhang C, Holding DR, Dweikat IM (2016) Mapping QTLs and association of differentially expressed gene transcripts for multiple agronomic traits under different nitrogen levels in sorghum. BMC Plant Biol 16:1–18

Gomes EWF, Willadino L, Martins LSS, Cmara TR (2001) The effects of salinity on five banana genotypes (*Musa* spp). In: Horst WJ et al (eds) Plant nutrition: food security and sustainability of agroecosystems through basic and applied research, pp 410–411

Gowen S, Queneherve P (1990) Nematode parasites of bananas, plantains and abaca. In: Luc M, Sikora RA, Bridge J (eds) Plant parasitic nematodes in subtropical and tropical agriculture. CAB International, Wallingford, UK, pp 431–460

Grapin A, Ortiz JL, Domergue R, Babeau J, Monmarson S, Escalant JV, Teisson C, Côte F (1998) Establishment of embryogenic callus and initiation and regeneration of embryogenic cell suspensions from female and male immature flowers of Musa. InfoMusa 7:13–15

Gupta PK, Balyan HS, Sharma PC, Ramesh B (1996) Microsatellites in plants: a new class of molecular markers. Curr Sci 70:45–54

Guzman EV et al (1982) Production of mutants by irradiation of *in vitro* cultured tissues of coconut and banana and their mass propagation by tissue culture technique, Induced mutations in vegetatively propagated plants II. IAEA, Vienna, pp 113–118

Haiyan S, Wei S, Zhengjing W, Shenghe C (2018) Improvement the lodging resistance of Banana (Musa AAA). Biomed J Sci Tech Res. https://doi.org/10.26717/BJSTR.2018.08.001631

Jain SM (2010) *In vitro* mutagenesis in banana (Musa spp.). In: Dubois T (ed) Improvement. Proceedings of IC on banana and plantain in Africa, Acta Horticulture, vol 879, pp 605–614

Jain SM, Till B, Suprasanna P, Roux N (2011) Mutations and cultivar development of banana. In: Pillay M, Tenkouano A (eds) Banana breeding, progress and challenges. CRC Press, Boca Raton, FL, pp 203–217

Jankowicz-Cieslak J, Till B (2015) Forward and reverse genetics in crop breeding. In: Al-Khayri JM, Jain SM, Johnson DV (eds) Advances in plant breeding strategies: breeding, biotechnology and molecular tools. Springer, Berlin, pp 215–240

Jankowicz-Cieslak J, Huynh OA, Brozynska M, Nakitandwe J, Till BJ (2012) Induction, rapid fixation and retention of mutations in vegetatively propagated banana. Plant Biotechnol J 10: 1056–1066

Kulkarni VM, Ganapathi TR, Suprasanna P, Bapat VA, Rao PS (1997) Effect of gamma irradiation on *in vitro* multiple shoot cultures of banana (Musa sp.). J Nucl Agric Biol 26:232–240

Kumar PL, Selvarajan R, Iskra-Caruana ML, Chabannes M, Hanna R (2015) Biology, etiology, and control of virus diseases of banana and plantain. Adv Virus Res 91:229–269

Mak C, Ho YW, Tan YP (1996) NOVARIA - a new banana mutant induced by gamma irradiation. InfoMusa 5:35–36

Menendez T (1973) A note on the effect of ethyl-methanesulphonate on Musa acuminata Seeds. In: Induced mutations in vegetatively propagated plants. IAEA, Vienna, pp 85–90

Miller A, Obholzer N, Arish N, Megason S, Moens C (2013) RNA-seq-based mapping and candidate identification of mutations from forward genetic screens. Genome Res 23

Mo Y, Howell T, Vasquez-Gross H, De Haro LA, Dubcovsky J, Pearce S (2018) Mapping causal mutations by exome sequencing in a wheat TILLING population: a tall mutant case study. Mol Genet Genomics 293(2):463–477

MVD (2021) Mutant variety database. http://mvd.iaea.org

Nishijima R, Yoshida K, Sakaguchi K, Yoshimura S-i, Sato K, Takumi S (2018) RNA sequencing-based bulked segregant analysis facilitates efficient D-genome marker development for a specific chromosomal region of synthetic hexaploid wheat. Int J Mol Sci 19(12):3749

Nordström KJV, Albani MC, James GV, Gutjahr C, Hartwig B, Turck F, Paszkowski U, Coupland G, Schneeberger K (2013) Mutation identification by direct comparison of whole-genome sequencing data from mutant and wild-type individuals using k-mers. Nat Biotechnol 31(4):325–330

Novak F, Afza R, Van Duren M, Omar MS (1989) Mutation induction by gamma irradiation of *in vitro* cultured shoot-tips of banana and plantain (*Musa* cvs). Trop Agric (Trinidad) 67:21–28

Qamar M, Qureshi ST, Khan IA, Memon SA, Bano Z, Solangi SK (2016) Influence of gamma radiation on the physicochemical properties of *in vitro* triploid and tetraploid banana species. Pak Biotechnol 13(4):237–244

Robinson JC (1996) Bananas and plantains. CAB International, Wallingford, UK, p vii & 238

Roux N (2004) Mutation induction in Musa—review. In: Jain SM, Swennen R (eds) Banana improvement: cellular, molecular biology and induced mutation. Science Enfield, pp 23–32

Roux N, Dolezel J, Swennen R, Zapata-Arias FJ (2001) Effectiveness of three micropropagation techniques to dissociate cytochimeras in Musa spp. Plant Cell Tissue Organ Cult 66:189–197

Roux N, Toloza A, Busogoro JP, Panis B, Strosse H, Lepoivre P, Swennen R, Zapata-Arias FJ (2003) Mutagenesis and somaclonal variation to develop new resistance to Mycosphaerella leaf spot diseases on *Musa*. In: Proceedings of 2nd international workshop on mycosphaerella leaf spot diseases of bananas held at San Jose, Costa Rica, 20–23 May 2002, pp 239–250

Roux N, Toloza A, Jaroslav D, Panis B (2004) Usefulness of embryogenic cell suspension cultures for the induction and selection of mutations in Musa spp. In: Banana improvement: cellular molecular biology and induced mutations. Science Publishers Inc., Enfield, pp 33–43

Sahu PK, Sao R, Mondal S, Vishwakarma G, Gupta SK, Kumar V, Singh S, Sharma D, Das BK (2020) Next generation sequencing based forward genetic approaches for identification and mapping of causal mutations in crop plants: a comprehensive review. Plants 9(10)

Shimwela MM, Ploetz RC, Beed RD, Jones JB, Blackburn JK, Mkulila SI, van Bruggen AHC (2016) Banana Xanthomonas wilt continues to spread in Tanzania despite an intensive symptomatic plant removal campaign: an impending socio-economic and ecological disaster. Food Secur 8:939–951

Smith MK, Hamill SD, Longdon PW, Pegg KG (1995) *In vitro* mutation breeding for the development of bananas with resistance to race 4, fusarium wilt (*Fusarium oxysporum* f.sp. *cubense*). In: Final reports of FAO/IAEA Co-ordinated research programme, vol TECDOC-800. IAEA, Vienna, pp 37–44

Smith MK, Hamill SD, Longdon PW, Giles JE, Doogan VJ, Pegg KG (2006) Towards the development of a Cavendish banana resistant to race 4 of fusarium wilt: gamma irradiation of micropropagated dwarf parfitt (Musa spp., AAA group Cavendish sub group). Austral J Exp Agric 46:107–113

Stotzky G et al (1964) Some effects of gamma irradiation on seeds and rhizomes of Musa. Am J Bot 51:724–729

Suprasanna P, Mirajkar SJ, Bhagwat SG (2015) Induced mutations and crop improvement. In: Plant biology and biotechnology. Springer, New York, pp 593–617

Surendar KK, Devi DD, Ravi I, Krishnakumar S, Kumar SR, Velayudham K (2013) Water stress in banana - a review. Bull Environ Pharmacol Life Sci 2:01–18

van Asten PJA, Fermont AM, Taulya G (2011) Drought is a major yield loss factor for rainfed East African highland banana. Agric Water Manag 98:541–552

Vanhove AC, Vermaelen W, Panis B, Swennen R, Carpentier SC (2012) Screening the banana biodiversity for drought tolerance: can an *in vitro* growth model and proteomics be used as a tool to discover tolerant varieties and understand homeostasis. Front Plant Sci 3:176

Varshney RK, Chen W, Li Y, Bharti AK, Saxena RK, Schlueter JA, Donoghue MTA, Azam S, Fan G, Whaley AM (2012) Draft genome sequence of pigeonpea (*Cajanus cajan*), an orphan legume crop of resource-poor farmers. Nat Biotechnol 30(1):83

Weber APM (2015) Discovering new biology through sequencing of RNA. Plant Physiol 169(3): 1524–1531

Yan W, Chen Z, Lu J, Xu C, Xie G, Li Y, Deng XW, He H, Tang X (2017) Simultaneous identification of multiple causal mutations in rice. Front Plant Sci 7:2055

Zalapa JE, Cuevas H, Zhu H, Steffan S, Senalik D, Zeldin E, McCown B, Harbut R, Simon P (2012) Using next-generation sequencing approaches to isolate simple sequence repeat (SSR) loci in the plant sciences. Am J Bot 99(2):193–208

Zheng SJ, Garcia-Bastidas FA et al (2018) New geographical insights of the latest expansion of *Fusarium oxysporum* f. sp. *cubense* tropical race 4 into the Greater Mekong Subregion. Front Plant Sci 9:457

Mutation Breeding in Date Palm (*Phoenix dactylifera* L.) 25

Djamila Yatta El Djouzi, Hafida Khelafi, Shri Mohan Jain, and Jameel M. Al-Khayri

Abstract

Date palm (*Phoenix dactylifera* L.) is one of the oldest fruit crops grown in the arid regions of the Arabian Peninsula, North Africa, and the Middle East. The date palm fruit is pivotal of the economy and the social life of the date palm producing regions. Currently, there are various challenges opposed date palm development, processing, and marketing such as the presence of low-quality cultivars, poor farm management, insect and diseases pest infestations control, deficiencies in harvesting, processing, shortage of qualified and national trained staff and laborers, insufficient research and development activities. To overcome these challenges, there is a need for alternative approach to conventional plant breeding. Genetic enhancement of the date palm by radiation induced mutagenesis represents a major opportunity since it increases the genetic variability to sustain food security. Gamma-irradiated date palm callus regenerated plants resistant to Bayoud toxin isolated from the causal fungus *Fusarium oxysporum* f.sp. *albedinis*. Several selected putative mutants resistant to Bayoud disease have maintained resistance underfield conditions. This chapter presents an overview of induced mutation and their application date palm improvement.

D. Yatta El Djouzi · H. Khelafi
I.N.R.A. Biotechnology and Plant Breeding Division CRP Mehdi Boualem, Alger, Algeria
e-mail: djamila.yatta@inraa.dz; hafida.Khelafi@inraa.dz

S. M. Jain
Department of Agricultural Sciences, PL-27, University of Helsinki, Helsinki, Finland
e-mail: mohan.jain@helsinki.fi

J. M. Al-Khayri (✉)
Department of Agricultural Biotechnology, College of Agriculture and Food Sciences, King Faisal University, Al-Ahsa, Saudi Arabia
e-mail: jkhayri@kfu.edu.sa

S. Penna, S. M. Jain (eds.), *Mutation Breeding for Sustainable Food Production and Climate Resilience*, https://doi.org/10.1007/978-981-16-9720-3_25

Keywords

Breeding · Date palm · In vitro selection · Mutation · Radiation

25.1 Introduction

Date palm (*Phoenix dactylifera* L.), is a monocotyledonous fruit tree belonging to the family Palmaceae (Arecaceae), which inhabit tropical and subtropical habitats where the majority of palms are found. It is an economically important fruit crop widely cultivated in arid and semi-arid regions including the Middle East and North Africa (Al-Khateeb 2006). More than 5000 date palm cultivars are spread worldwide differing in nutritional value, morphological and genetic attributes, although commercial cultivars are limited in number (Abul-Soad et al. 2017). According to FAO statistics (FAOSTAT 2018), the world's largest producer was Egypt ranking first with a total production 1,501,799 metric tons (mt), followed by Iran (1,083,720 mt), Saudi Arabia (1,065,032 mt), Algeria (848,199 mt) and Iraq (6828 mt).

Date palm production faces serious problems such as low yields as well as marketing constraints. These include diseases like Bayoud (*Fusarium oxysporum* f.sp. *albedinis*) and insect pest infestations like red palm weevil (*Rhyncophorus ferrugineus*) and inadequate integrated pest management (Al-Khayri et al. 2018). Furthermore, there are problems as well as deficiencies in harvesting, processing, and marketing practices, shortage of qualified and national trained staff and laborers in addition to insufficient research and development activities (Erskine et al. 2003). The major causes of low date productivity are the large number of the old trees, the existence of many low-quality and undesirable cultivars, lack of sufficient number of offshoots to establish new orchards or even renew the old ones, and the increase in cost of offshoots of good-quality cultivars (Abul-Soad et al. 2017).

Since the spontaneous mutation rate is very slow, induced mutation is crucial to improve the rate of genetic diversity so that breeders can develop the diverse varieties in plant breeding programs (Jatt et al. 2019; Aly et al. 2019).

Mutation breeding is based on the fine art of selecting the "one" genotype, which can bring solution to a given situation, such as tolerance and/or resistance to drought, high salinity content, diseases, and pests. Mutagenesis is method by which sudden heritable changes in the genetic material of an organism induced by mutagens such as chemical, physical or biological agents (Bradshaw 2016; Roychowdhury and Tah 2013). Mutations can be induced at a higher frequency by exposing cells to mutagens (Foster 1991; Miler et al. 2021).

This chapter provides an overview of conventional breeding limitation in date palm and the role of mutation technology. In addition, it presents a case study of mutation induction, date palm in vitro improvement for resistance to Bayouddisease and mutant assessment.

25.1.1 Limitations of Conventional Breeding

The date palm is a diploid, cross-breeding species, where male and female inflorescences develop on separate trees. It is conventionally propagated both asexually (vegetatively) and sexually (seeds). The main advantage is the simplicity of its application and in obtaining a great morphological and genetic variation, thus producing a huge number hydrids and creating a massive reservoir of genetic diversity (Jain 2012). Therefore, this technique is very practical in genetic enhancement programs for the selection of genotypes resistant to abiotic and biotic stresses as well as appreciable characteristics (Jain 2012; Hazzouri et al. 2020; Yatta El Djouzi et al. 2020).

Seed propagation provides progeny comprising about 50% male and 50% female (Yatta et al. 2013; Yatta El Djouzi et al. 2014; Abed et al. 2014). Nevertheless, this method cannot be used to propagate elite cultivars with desirable traits or selected genotypes due to high heterozygous characters of the date palm (Tisserat 1982). The inconvenience of this method is long vegetative phase, 4–8 years for the first inflorescences depending on cultivar and growth conditions. Only after the onset of flowering the gender of the date palm tree can be identified and fruit quality can be assessed following fruit maturation (Al-Khayri 2007; Naik and Al-Khayri 2016). Farmers select among palm trees produced by seeds and produced the best dates. The chosen clones are propagated vegetatively.

The process of vegetative propagation is initiated by the development of axillary buds, arising base of the stem, which gives rise to offshoots (Al-Khayri 2007; Yatta El Djouzi et al. 2014; Behnaz et al. 2018). However, the date palm produces only a limited number of offshoots during its life on an average 10–40 offshoots depending on the cultivar (Naik and Al-Khayri 2018, 2020).

This traditional technique remains slow and inefficient to meet the great demand for the rapid expansion of palm groves. This method is also very slow and takes at least 30 years to obtain a few thousand palms from offshoots plantation, and in addition to that, offshoot constitutes a means of dissemination of diseases such as Bayoud (Jain 2012).

The spontaneous mutations occur at low rate (varying from 10^{-5} to 10^{-8}) which is insufficient to be utilized for enhancing variability in crop improvement (Suprasanna and Jain 2017). Induced mutagenesis using physical and chemical mutagen treatment enhance genetic variability in crop improvement programs (Jain and Suprasanna 2011).

25.1.2 Role of Mutation Breeding

Food security can be ensured by improving crop tolerance to different environmental stresses (Suprasanna and Jain 2017). Traditional plant breeding techniques, mutagenesis, and genetic engineering, have an essential role in harnessing available germplasm resources to increase the genetic variability and to develop improved Crops (Hallerman and Grabau 2016). Induced mutagenesis is very efficient tools for the enhancement of crops (Ahloowalia and Maluszynski 2001; Maluszynski et al. 2004; Miler et al. 2021). It helps to expand genetic variation and crop diversity. It contributes to increasing the potential of phylogenetic resources and allows breeders to create more efficient varieties contributing to food security. For this, a large number of mutagenized crop species have been released to farmers and commercialized, thus indicating the economic value of this approach (Suprasanna and Jain 2017). Genomics tools and bioinformatics could be helpful to select suitable mutagenesis techniques heavy ion beams comparing with physical mutagen can be used to create new cultivars with selected target traits without troubling exist characters of parent cultivar (Suprasanna et al. 2015). Precise confirmation individuals and screening of mutants with desirable traits are most importants points in variation breeding (Suprasanna and Jain 2017). Mutant screen is a way for detecting desired selection criteria in putative mutants. Mutant reevaluation and their confirmation need a large mutated population under a restricted environment.

Mutagenesis technique has advantages to obtain mutants in genomic variation with the trait of interest that may possibly be excellent to cultivate under the global change in climatic conditions (Jain 2010).

The Food and Agriculture Organization of the United Nations/International Atomic Energy Agency—Mutant Variety Database reports on the developed and officially released mutants, a total of 3275 accessions from 225 species (FAO/IAEA-MVD 2021).

25.2 Mutation Sources

25.2.1 Somaclonal Variation

"Somaclonal variation" has been used to refer to genetic variation found in "somaclones" of plants regenerated from any type of tissue cultured in vitro. This concept has been extended to all genotypic and phenotypic alterations occurring during tissue propagation processes. However molecular markers were used to detect variability and screenpositifs variants. They are therefore neutral indicators of genetic variabilty which make it possible to identify polymorphism between species, varieties, populations and even between individuals. Its helps to increase genetic variation and crop diversity. Its contributes to allow breeders to produce more efficient varieties contributing to food security.

In date palm, somaclonal variation and in vitro mutation induction make it possible for select genotypes with new agronomic traits, stress tolerance or quality characteristics (El-Hadrami et al. 2011; Ahloowalia and Maluszynski 2001; Jain 2001; Jain and Maluszynski 2004; Bradshaw 2016), allows the expansion of the date palm genetic base, using micropropagation and tissue culture.

The high rate of desired somaclonal variant can be obtained by introducing selection gents to the tissue propagated in vitro (Jain 2014; Bradshaw 2016). During in vitro tissue propagation, several factors produce genetic variability in these plant tissues. These changes depend on genotype, origin explant, age the culture and number of subculture cycles used; growth regulators concentrations, in vitro multiplication technique type, ploidy level (Jain 2011; Krishna et al. 2016). In date palm, the growth regulators depend on genotypes, and sensitivity to certain auxins often trigger variations during in vitro tissue regeneration (Jain 2012). Cytogenetic Studies have reported different number of chromosome pairs and karyotype also reveals changes structure of chromosomes. Various research in date palm have shown that the chromosome number ranging from $2n = 26, 32, 34, 36$ (Al-Salih et al. 1987; Jatt et al. 2019).

However, the information available on these variants remains unknown, sparse, and inconclusive. It is therefore complex to estimate the percentage of this genetic variability and to verify its impacts on this species (El Hadrami et al. 2005). No comparison between these two in vitro techniques has been made due to the long method of tissue cultures (Bradshaw 2016). Note that some date palm cultivars can produce a high rate of genotype conformity, whereas others are more prone to variation. It has also been observed that certain cultivars propagated by tissue culture present a certain level of recalcitrance; particularly for those with desired production and resistance qualities (Bradshaw 2016).

25.2.2 Induced Mutation

Induced Mutation is used to enhance the probability of creating a favorable genetic variation in plant improvement and the Mutants are generally obtained by mutagenic agents (Mba et al. 2010; Bradshaw 2016; Miler et al. 2021). The concept of the artificial induction of mutations by ionizing radiation began in the twentieth century, and Mutations can be induced at a higher frequency.

It is used successfully for crop improvement and complements conventional methods (Jain 2010; Amin et al. 2015). For this, several types of mutagens are widely used for genetic variability creation for improving crops (Jain 2010, 2012).

Induced mutations involve exposing seeds, stem cuttings, buds, and tubers to mutagens. The success of mutagenesis is dependent on the genetic purity of parental lines. The intensity of the dose, the type, and the concentration of mutagen varies according to plant material used (Suprasanna et al. 2015; Miler et al. 2021).

There are currently numerous chemical agents (e.g., alkylating agents, azide, hydroxylamine, antibiotics, nitrous acid, acridines, and base analogs) capable of inducing mutations in plant species; however, only a limited number have been used in experimental plant mutagenesis and the selection of plant mutants (Wani et al. 2014; Feldman et al. 2017). The consequence of chemical mutagens on plants is often observed as minimally invasive. Alkylating agents, azide, and hydroxylamine react with bases by adding methyl or ethyl groups. The degradation of the alkylated base can go as far as the production of basic sites to cause mutations through DNA replication.

Nitrous acid causes oxidative deamination in which amino groups are converted to ketone groups; cytosine residues are thus converted into uracil which can pair with adenine rather than guanine. Similarly the deamination of adenines to Cytosine rather than Thymine. Antibiotics are caused chromosomal aberrations and cytoplasmic male sterility. Acridines intercalate between the bases of the DNA, creating deformations which can cause deletions or insertions leading to shifts in the reading frame. Base analogs can be incorporated into DNA during replication. Thus bromouracil, an analog of cytosine hybridizes preferentially to Adenine, and 2 aminopurine an analog of adenine pairs with cytosine.

The ratio of mutational changes to unwanted changes is usually higher for chemical mutagens than for physical mutagens (Viana et al. 2019).

Since the discovery that physical radiation (such as gamma rays, X-rays, and UV radiation, as well as particulate radiation, fast and thermal neutrons, α and β particles) have mutagenic properties, interest in the use it has grown increasingly to induce genetic variation in crop species.

The mutagenic dose applied is important in any mutagenesis program. The lethal dose-50 (LD50) gives an inspiration of the optimal mutagenic dose which produces the maximum number of mutations with the minimum risk. An overdose of mutagens causes plants death, while a low dose will show lower mutation. The mutagenic dose depends on the concentration, the time and the temperature during the treatment.

The gamma radiation of radioactive cobalt (^{60}Co) is widely used; the majority of mutant varieties have been developed (Jain 2011; Suprasanna et al. 2015). Radiation technology and colored polyethylene packaging are recommended for shelf-life, sanitary and phytosanitary purposes of palm date (Mohammadzai et al. 2010) and to for improving fruit quality of date palm (El-Beltagi et al. 2019).

In date, X-rays is used to enhance plant growth and also to improve understanding of the physiological responses imposed by irradiation stress (Jain 2011; Al-Enezi et al. 2012). In recent decades, ion beams have been widely used as replacements of gamma and X-rays and neutrons (Jain 2011; Feldman et al. 2017). Radiation causes chromosomal aberrations, lethality the damage caused by ion beams to double-stranded DNA is less repairable due to the removal of DNA fragments of different sizes (Feldman et al. 2017).

In vitro screening of mutants, usually the selection pressure varies and it is better to determine LD50 dose (Jain 2010). Induced mutations in plant breeding changes in the nuclear DNA, cells, organelles, resulting mutations which are interest to breeders to choose useful mutant; In date palm induced mutations is possible at this time due to a dependable plant regeneration system-induced mutations through somatic embryogenesis and organogenesis.

In date palm, there is any publication done on mutation induction, excepting FAO/IAEA Coordinated Research Project on development of date palm mutant varieties resistant to Bayoud disease in North Africa (Jain 2005, 2006). In date palm, mutant plants were selected in the greenhouse using isolated toxin from *Fusarium oxysporum* f.sp. *albedinis* fungus causal agent (Patade and Suprasanna 2008; Jain 2006) and were transferred to infected field for evaluation, resistance confirmation and selection of other agronomic traits, such as yield and date quality. Actually only 1 mutant of cv. Deglet Nour of 14 putative mutants transferred to a disease-infected field was found to be true resistant after 12 years in the field.

For the putative mutants of cv. Teggaza, 11 vitroplants showing resistance to artificial inoculation of the pathogen were transferred in to infested field planting at INRAA Station, Adrar, Algeria. So far, they are growing very well.

25.3 In Vitro Selection

In vitro selection is among the biotechnology techniques frequently used in date palm breeding for tolerance to different stresses, i.e., drought, salinity, cold and diseases and pests (Jain 2011; Nikam et al. 2015; Schaart et al. 2016).

It can shorten the selection time for desirable characters in vitro and can complement field screening (Jain 2012).

Different in vitro culture tools such as suspension cultures and protoplast cultures can be extensively chosen to have a genetic uniformity by adding the selective agents reducing the growth to select tolerant plants.

Mutants have been induced and improved in several plant species (Viana et al. 2019). During in vitro selection, two selections strategies can be used (single-step selection and multi-step selection) (Suprasanna et al. 2012a, b). In the single-step selection, the minimum inhibitory concentration (MIC) is added into the culture medium and cultures are maintained for several subcultures and then resultant surviving tissues are isolated and plants regenerated. In a multi-step selection method, the in vitro cultures exposed to sub-lethal concentration in order to develop and in the following subcultures, a gradual increase in inhibitor level is maintained.

It has been argued, with this process, that selected mutant traits are frequently stable, while variant cells are in stable exposure to improving levels of inhibitor.

25.3.1 Abiotic Stress Agents

Cold stress is the most important abiotic stresses that directly affect crop growth and development. Heat stress has a direct impact on the crop resulting in a loss of yield and quality at harvest. Indeed, Heat stress-induced affect the floral organs and the formation of the fruit. Just as meiosis and the filling phase of the seed are sensitive to the rise in temperature. High temperature stress influences the protein synthesis as well as photosynthetic system dysfunction (Brar and Jain 1998; Akter and Islam 2017; Muhammad et al. 2020). Salt stress affects the reduction in the number of leaves, stomatal conductance, protein synthesis, energy metabolism and photosynthesis. The presence of NaCl in the culture medium limits the supply of the plant with major cations, such as potassium (K^+) and calcium (Ca^{2+}). Bressan et al. (1985) obtained salt-adopted tobacco cells, which were grown for at least 25 generations in 25 g L^{-1} sodium chloride.

Using in vitro tissue cultures to study abiotic stress responses is based on the information that in vitro cultivated cells behave correspondingly to cells of intact plants subjected to abiotic stress situations such as water deficit and salinity stress (Attree et al. 1991).

In vitro culture selection induced by selection pressure alone or associated with mutagenesis is a process to be exploited to enhance genetic variability allowing the creation of new cultivars of cultures (Predieri 2001). This is based on the in vitro culture of plant cells, tissues, or organs on a medium supplemented with selective agents, making it possible to select genotypes one resistant to different abiotic stresses (Pérez-Clemente and Gómez-Cadenas 2012). Drought stress has been caused leaf wilting, a reduction in leaf area, and also raising the rate of photosynthesis. Mainly cases, selection is applied to tissue cultures such as callus, cell suspension, protoplast cultures adding growth inhibitory levels of selection agent in culture medium (Widholm 1972).

In vitro researches in the date palm were conducted to recuperate plants improved tolerance to salinity and drought stress in arid areas. Al Mansoori et al. (2007) examined the effect of sodium chloride (NaCl) on tissue callus issued from immature embryos of local date palm cultivars and they concluded that salinity much affects date palm immature embryos. Al-Mulla et al. (2013) utilized the cvs. Kasab, Barhee and Khalas on in vitro plants to examine their tolerance to salinity under greenhouse conditions. They concluded that both soil salinity and water stress affects date quality and yield. Al-Rokibah et al. (1998) reported a difference in response to salt stress among date palm cultivars grown in Saudi Arabia. Al-Khayri and Al-Bahrany (2004) showed that in response to in vitro drought stress induced by PEG-8000, endogenous free proline content of date palm callus increased gradually in response to increase PEG-concentration. Further research is necessary to optimize in vitro selection and plant regeneration processes to ensure the viability and conserve selected cells totipotency issue from selected cell lines for tolerance to abiotic stress (Al-Khayri and Ibraheem 2014).

Al Kharusi et al. (2017) suggested that there is variation in tolerance to high salinity between different cultivars of date palm. Shoot Na^+ exclusion, photosynthesis, and membrane stability are apparently the main determinants of tolerance and can be used in salinity tolerance selection of cultivars. Heat stress affects the growth of plants and productivity in agricultural crops plants. Climate change is predictable to lead to decrease date palm adaptation. This can be restored by escalating the photosynthesis on a leaf area basis provided that the intercept of solar radiation does not modify significantly. The phytohormone abscisic acid (ABA) plays a main role through plant adaptation to different stresses (high salinity, drought, and extreme temperature) (Jain et al. 1998; Zhu 2002).

25.3.2 Biotic Agents

Biotic stresses in plants are caused by various pathogens such as fungi, bacteria, viruses, etc. diverse agents lead to plant death such as infections caused by fungi and insects. Others affect photosynthesis per leaf area such as virus infections. Selection systems to separate tolerant lines have been planned using screening with culture filtrates, chemicals and toxins.

Culture filtrates (CF) have been successfully used and corresponds a simple process of selection by adding into the culture media at suitable concentrations. Mutants resistant have been screened by using culture filtrate in vitro and disease resistant plants have been regenerated in many crops (Molot et al. 1984; Ahloowalia and Maluszynski 2001). First reported by Carlson (1973), in vitro selection was conducted in tobacco against *Pseudomonas syringae* for disease resistance. Chawla and Wenzel (1987) used callus cultures of barley and of wheat for selection against *Helminthosporium sativum* and the screening with pathotoxins treatments showed in 6–17% of surviving calli. In vitro selection has huge possible for fast production of helpful mutants resistant to biotic stress (Chandra et al. 2010). Callus and regenerated plants were confirmed for resistance by contact to selection agent of *A. alternate*.

Jain (2010) have been reported results on gamma ray irradiation of strawberries grown in vitro using *Phytophthora cactorum* crude extract; where 5% of the plants survived and they were able to resist drought for 5–6 days.

Jain (2006) have been noted that mutant plants of date palm were screened in the greenhouse after infection with bayoud toxin isolated from fungus causal *Fusarium oxysporum* f.sp. *albedinis*.

Bayoud disease is rampant in North Africa, whereas red palm weevil (*Rhynchophorus ferrugineus*) devastates in the Middle East Asia and Mediterranean regions (Jain 2012). In addition, 34 various fungal and Oomycetes have been found to be associated with root diseases of date palm (Saaidi et al. 1981).

25.4 Mutation Induction in Developing Bayoud Resistance

The Bayoud disease, caused by the ascomycete imperfecta fungus *Fusarium oxysporum* special form *albedinis*, is the most serious enemy of Date palm (*Phoenix dactylifera* L.). It has disturbing impact lopsided the ecology of a number of areas and posed serious problems of human, social, and economic problems. It has already destroyed more than ten million palm trees in Morocco and nearly three million in Algeria. The problem is aggravated by pathogen attack on productive cultivars and high commercial values.

Various control strategies are implemented to deal with the disease. However, genetic control remains the most recommended way to reduce or stop the progression of this *Fusarium*. This source of resistance can be found in varieties currently cultivated or in natural populations or induced by gamma irradiation treatment. Confirmation of this resistance must be evaluated by rapid and reliable tests such as the use of toxins or culture filtrates or by artificial inoculations with *F.o.a.* fungus in vitro cultured explants; and mutant evaluation in the glasshouse. Several selected putative mutants were transferred in the field and have maintained tolerance to vascular fusariosis under field conditions. Actually, the resistance is evaluated in contaminated field to obtainBayoudresistant palms, with high-quality dates. Newly developed date palm mutants with Bayoud resistance and high date fruit quality are grown to repopulate the palm groves, devastated by the disease, in the South West

Table 25.1 Date palm mutant development phases and procedural steps of each phase

Phase	Procedural steps numbered in a consecutive order		
Introduction of offshoots for callus induction and irradiation	1. Obtaining actively growing callus	2. Irradiation 60 Cobalt of Embryogenic Cultures (200 embryogenic calluses/Dose)	3. Radio-sensitivity test (Performed on 200 embryogenic callus) (0, 10, 15, 20, 25, 30 Gy) for determining optimal mutagenic dose
Selection in vitro using a culture filtrate or toxin	4. Callus Irradiation (Minimum 1000 callus) with the optimal dose $\leq LD_{50}$	5. Callus proliferation (1: M1V1, 2: M1V2, 3: M1V3, 4: M1V4)	6. In vitro screening of calluses against *F.o.a.* with culture filtrate or fusaric acid.
	7. In vitro selection of somatic embryos against *F.o.a.* with culture filtrate	8. In vitro selection by detached leaves	9. Regeneration of putative mutants
Evaluation of mutants (Confirmation of Resistance in vivo and in field)	10. Transfer of germinated plantlets to the acclimatization phase	11. Greenhouse Evaluation (Treatment of mutant plants with Fusarium toxin and the *F.o.a.* culture filtrate at 35%)	12. Screening and evaluation using flow cytometric and molecular markers
	13. Field evaluation of the mutant plants (Confirmation of resistance in field and selection of other agronomic traits (date quality and others))		
Micropropagation of mutant plants	14. Micropropagation of desired resistant mutant plants exhibiting high date quality		15. Multi-site trials

and center. The use of the mutants obtained in the repopulation of the affected and threatened palm groves by Bayoud would contribute to the protection of the oasis ecosystem, and increase in genetic diversity also they will contribute to the strengthening of the employment policy and to the guarantee food security. The date palm mutant development processes are shown in Table 25.1.

25.4.1 Establishment of Embryogenic Cultures

Somatic embryogenesis is the most efficient way huge scale clonal propagation of date palm. It has several supplementary advantages, such as the production of virus-free plants, the capacity to produce large numbers of plants, facilities for change of genetic material at different levels (national, regional, and international), automation potential, opportunities for synthetic seed production, cryopreservation, and genetic manipulations (Jain 2012; Yatta El Djouzi et al. 2014).

Fig. 25.1 Date palm explants preparation. (**a**) Date palm offshoot isolated from mother plant, (**b**) Removal of leaves and fiber sheets of the out layers, (**c**) Isolated shoot tip, (**d**) Disinfected shoot tip, (**e**) Isolated explants from the shoot tip region, (**f**) Explants cultured on tissue culture medium. (Photos by Al-Khayri JM)

Three date palm cultivars were tested including two cultivars obtained from Adrar oasis, Southwest Algeria (cvs. Taquerbucht and Teggaza), the former is resistant to Bayoud but have low fruit quality, whereas the latter is susceptible but has a good fruit quality. The third cultivar was obtained from Touggourt, Southeast Algeria (cv. Deglet Nour), which is known for high fruit quality but highly susceptible to this disease. Eight healthy offshoots (Fig. 25.1a) weighing 3–5 kg were harvested per cultivar from well-irrigated groves. Outer dry and fibrous leaf sheaths were removed until the inner shoot tip is reached (Fig. 25.1b). Shoot tip region consists of apical meristem bordered by white young leaves surrounding the meristem. The shoot tip region was isolated using a curved knife and trimmed to approximately 10 cm long

Table 25.2 Composition of culture medium used for callus initiation (M1 supplemented 100 mg L^{-1} of 2,4D) and (M2 contained 12.5 mg L^{-1} of Picloram) and for establishment of cell suspension (M3 contained 5 mg L^{-1} of Picloram)

Component	Media designation		
	M1	M2	M3
Murashige and Skoog macroelements salts (Murashige and Skoog 1962)			
Ammonium nitrate NH_4NO_3	1650 mg L^{-1}	1650 mg L^{-1}	1650 mg L^{-1}
Potassium nitrate KNO_3	1900 mg L^{-1}	1900 mg L^{-1}	1900 mg L^{-1}
Potassium phosphate monobasic KH_2PO_4	170 mg L^{-1}	170 mg L^{-1}	170 mg L^{-1}
Calcium chloride $CaCl_2$, $2H_2O$	440 mg L^{-1}	440 mg L^{-1}	440 mg L^{-1}
Magnesium sulphate heptahydrate $MgSO_4{\cdot}7H_2O$	370 mg L^{-1}	370 mg L^{-1}	370 mg L^{-1}
Murashige and Skoog microelements salts (Murashige and Skoog 1962)			
Boric acid H_3BO_3	6.2 mg L^{-1}	6.2 mg L^{-1}	6.2 mg L^{-1}
Cobalt chloride $CoCl_2{\cdot}6H_2O$	0.025 mg L^{-1}	0.025 mg L^{-1}	0.025 mg L^{-1}
Cupricsulphate $CuSO_4{\cdot}5H_2O$	0.025 mg L^{-1}	0.025 mg L^{-1}	0.025 mg L^{-1}
Manganese sulphate $MnSO_4{\cdot}7H_2O$	16.9 mg L^{-1}	16.9 mg L^{-1}	16.9 mg L^{-1}
Molybdic acid sodium salt $Na_2MoO_4{\cdot}2H_2O$	0.25 mg L^{-1}	0.25 mg L^{-1}	0.25 mg L^{-1}
Potassium iodide KI	0.83 mg L^{-1}	0.83 mg L^{-1}	0.83 mg L^{-1}
Zinc sulphate $ZnSO_4{\cdot}7H_2O$	8.6 mg L^{-1}	8.6 mg L^{-1}	8.6 mg L^{-1}
Iron (Fe-EDTA) source			
Ferrous sulphate $FeSO_4{\cdot}7H_2O$	27.8 mg L^{-1}	27.8 mg L^{-1}	27.8 mg L^{-1}
Na_2-EDTA	37.3 mg L^{-1}	37.3 mg L^{-1}	37.3 mg L^{-1}
Vitamins MS (1962)			
Thiamine	1 mg L^{-1}	1 mg L^{-1}	1 mg L^{-1}
myo-Inositol	100 mg L^{-1}	100 mg L^{-1}	100 mg L^{-1}
Plant Growth Regulators (Hormones)			
Dimethylallylamino purine (2iP)	3 mg L^{-1}	1 mg L^{-1}	1 mg L^{-1}
2,4-Dichlorophenoxyacetic acid (2,4-D)	100 mg L^{-1}	–	–
4-Amino-3,4,6 tricloropicolinic acid (Picloram)	–	12.5 mg L^{-1}	5 mg L^{-1}
Others additives			
Monosodium phosphate NaH_2PO_4	170 mg L^{-1}	170 mg L^{-1}	170 mg L^{-1}
Potassium phosphate monobasic KH_2PO_4	100 mg L^{-1}	100 mg L^{-1}	100 mg L^{-1}
Ammonium citrate	200 g L^{-1}	200 g L^{-1}	–
Glutamine	100 mg L^{-1}	100 mg L^{-1}	200 mg L^{-1}
Adenine	40 mg L^{-1}	40 mg L^{-1}	40 mg L^{-1}
Activated charcoal	3 g L^{-1}	0.2 g L^{-1}	–
Polyvinyl pyrrolidones (PVP)	–	–	2 g L^{-1}
Carbohydrates source			
Sucrose	45 g L^{-1}	45 g L^{-1}	45 g L^{-1}
Solidifying agent			
Agar	7 g L^{-1}	7 g L^{-1}	–
pH	5.84	5.84	5.84

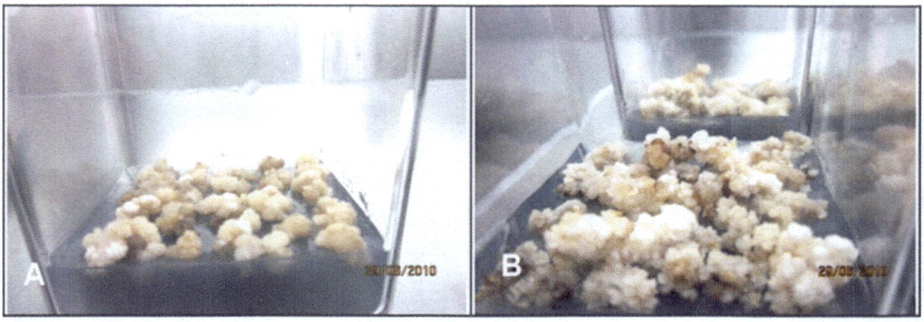

Fig. 25.2 Compact embryogenic calli (**a**) and friable embryogenic calli exhibiting prolific growth after 6 months the shoot tip explants cultured (**b**). (Photos taken in 2010 by Yatta El Djouzi D, Khelafi H and Abed F)

and 5 cm diameter (Fig. 25.1c). This tissue was disinfected in 3 g L^{-1} Benomyl solution for 45 min, and then sterilized for 20 min in 12% sodium hypochlorite (750 mL) solution containing 100 mg L^{-1} potassium permanganate and 2 mL L^{-1} Tween 20. The shoot tip was then rinsed with sterilized distilled water three times, 5 min each (Fig. 25.1d). Under aseptic conditions, a layer of the out-leaf tissue near the shoot tip was detached and meristem region fragmented into 8–12 explants approximately 1 cm^3 (Fig. 25.1e). The shoot tip explants were cultivated on two modified MS media (Murashige and Skoog 1962) designated M1 and M2 (Fig. 25.1f). Their compositions are listed in Table 25.2.

The pH adjusted to 5.7 before autoclaving at 120 °C for 20 min. The cultures were incubated in darkness under a thermoperiod of (28 ± 2 °C for 16 h and 22 ± 2 °C for 8 h) for six to eight subcultures at 5 weeks intervals until callus was produced. Explants which have increased in volume were fragmented into several fractions. The enlargement of the explants was observed between 5 and 9 weeks after their cultivation. As soon as the explants presented small white lumps of a nodular appearance, they were isolated from explant and transferred to new culture medium. The duration of appearance of the first embryogenic strains for the cv. Deglet Nourwas 9 months. The embryogenic calli obtained were of two types; compact and friable (Fig. 25.2a, b).

25.4.2 Irradiation of Embryogenic Cultures

The first step in vitro mutagenesis protocols is to initiate the multiplying calli cultures with high regeneration capacity. Gamma rays are ionizing electromagnetic radiation used on calli culture to induce the mutations. The objective of this work was to test the different doses of cobalt 60 on embryogenic callus of the Deglet Nour and Teggaza cvs. of the date palm to define the threshold of radiosensitivity and determine the optimal dose of irradiation. Embryogenic Calli (Minimum 1000 callus), 1–2 mm in size, were irradiated to diverse levels of gamma rays from a

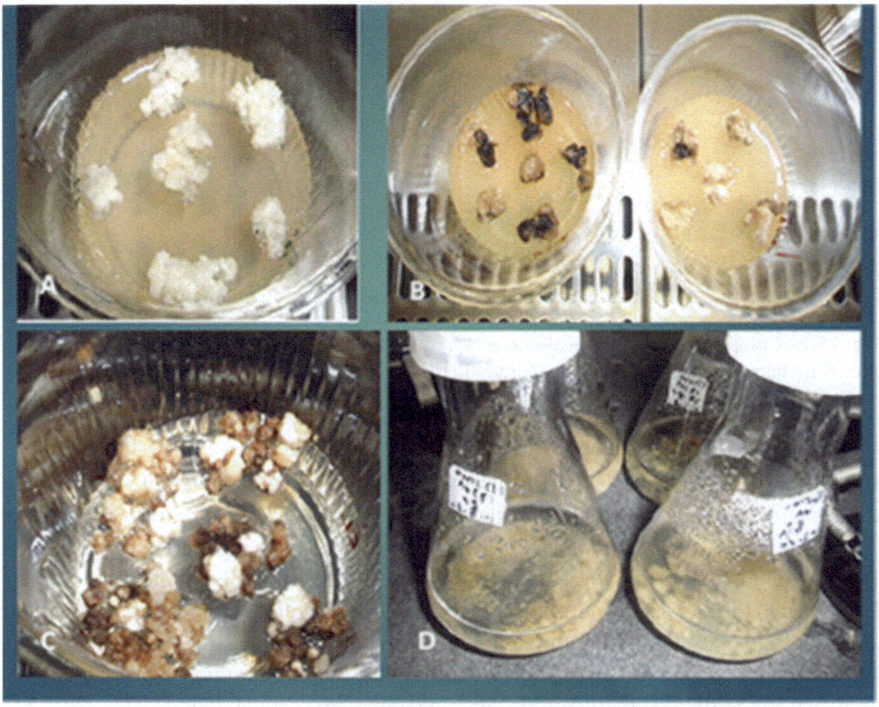

Fig. 25.3 Embryogenic callus irradiated (**a**) embryogenic callus necrosis with living calli in white and dead in brown (**b**), proliferation of survival calli (**c**) and cell suspension cultures (**d**). (Photos by Jain SM)

Cobalt 60 source at dose of 0, 10, 15, 20, 25, and 30 Gy, which was conducted at National Nuclear Center Research (CNRA), Algiers, Algeria. Each treatment consisted of 200 embryogenic calli and 20 embryogenic calli non-irradiated were used as controls. Instantly after irradiation, the embryogenic calli were placed onto fresh solid culture medium (Fig. 25.3a).

After 1 month, the LD_{50} was determined with counting the number and the percentage of survival (white) and death (brown) calli (Fig. 25.3b, c). Cultivars Deglet Nour and Teggaza varied in their percentage survival at various irradiation doses (Fig. 25.4). The percentage survival was constantly elevated in cv. Teggaza than cv. Deglet Nour. The survival rate of both cultivars decreased with dose increases.

These strains were completely necrotic (95.84%) and were found to be incapable of reacting with the exception of one single strain. At the 25 Gy dose, nearly half of the strains do not survive (47.83%). For the 10, 15, and 20 Gy doses, browning of the calluses is reversible and the majority of the strains survive (Table 25.3). We noted, however, an irreversible necrosis of 7 strains at the10 Gy dose, 2 strains at the 15 Gy dose, and 1 strain at a dose of 20 Gy, which respectively represents a mortality rate of 36.85%, 11.12%, 5.56%, 11.12% and 5.5%.

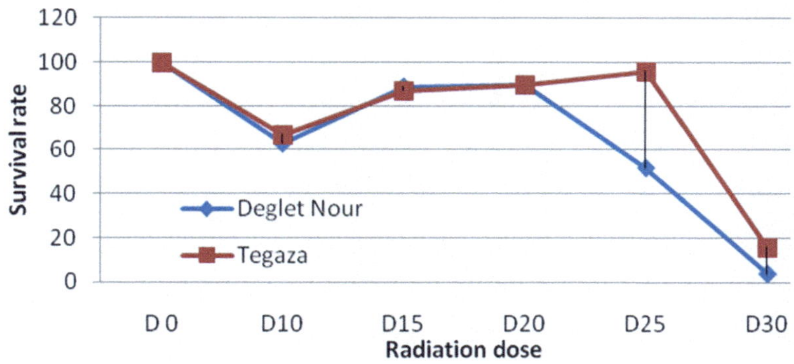

Fig. 25.4 Percentage of survival irradiated embryogenic calli of Deglet Nour and Teggaza after 1 month of culture

25.4.2.1 Gamma Room Cobalt 60 Characteristic

The self-protected irradiator is suitable for processing samples for research and for the calibration of dosimeters. This is type B01 equipment (Transelektro Ltd, Budapest) integrating 24 Cobalt 60 source pencils with an overall activity of around 9604.65 Ci (at the date of irradiation), of a conveying system of the irradiation container. An automatic device allows the irradiation duration to be programmed. The flow dose and the transit reference point dose were determined by the reference dosimeter of Fricke positioned at National Nuclear Center Research (CNRA), Algiers, Algeria. The glass samples were irradiated at doses of, a distance of 20 cm and a dose rate of 11.21 Gy min^{-1} at room temperature. Dosimetry is performed using ionization chambers of varying volumes connected to Fricke Dosimeter. The instrument and ionization chambers are calibrated by the Instrument Calibration Section of the Radiation Biology and Health Physics Branch, against National Research Council of Algeria standards.

25.4.2.2 Post-irradiation

The multiplication of the irradiated calli was carried out on modified MS media of the same composition as those of the induction step and placed in the dark. Subcultures take place every 2 months, the time necessary to observe callus proliferation. Four subcultures were carried out to arrive at the stage which was qualified as M1V4, i.e., 8 months after irradiation (Table 25.4).

The Irradiation of multicellular structures such as seed, meristem tissue or offshoots can lead to chimeras in regenerated plants which would require dissociation by plant multiplication up to the M1V4 generation (Jain 2012, 2010).

In the case of the palm tree, data about the next generation of irradiated sources is not available at this time. In our experiment, we used callus proliferation, i.e., cell divisions or cell proliferation from the irradiated mother cell as a generation. The secondary callus could have been considered a generation, but it was only encountered in the case of the 30 Gy dose, where the necrotic strain.

Table 25.3 Effects of gamma irradiation doses on the embryogenic calli of *Phoenix dactylifera* after one 1 month of culture

Deglet Nour				Teggaza			
Irradiation dose	Number of irradiated calli	Survival number of embryogenic calli	% of embryogenic calli necrosis	Irradiation dose	Number of irradiated calli	Survival number of embryogenic calli	% of embryogenic calli necrosis
0	16	16	0	0	30	30	0
10	19	12	36.85	10	30	20	33.33
15	18	16	11.12	15	30	26	13.33
20	20	18	5.56	20	30	27	10
25	23	12	47.83	25	30	29	3.33
30	24	1	95.84	30	30	5	83.33

Source: Yatta and khelafi (2012); Yatta et al. (2004)

Table 25.4 Proliferation of embryogenic calli from cv. Deglet Nour after fourth subculture

Irradiation dose (Gy)	Number of initial irradiated calli	Number of strains remaining at the first subculture	Number of strains obtained after 4 subcultures	Particularity of the calluses obtained
0	16	16	30	Friable callus + or − vitrified calluses
10	19	12	16	Heterogeneous calluses: vitrified + or − friable + nodular calluses
15	18	16	22	Heterogeneous calluses: vitrified + or − friable + nodular calluses
20	18	17	67	Friable and grainy calluses
25	23	12	25	Friable and grainy calluses
30	24	01	01	Loose-textured secondary calluses

Source: Abed et al. (1999)

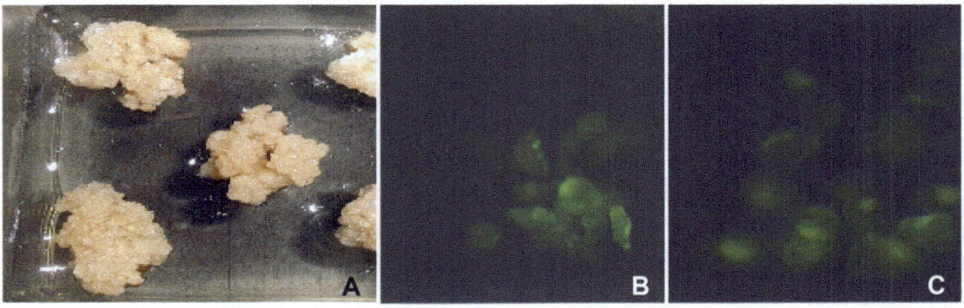

Fig. 25.5 Obtaining irradiated calli from Deglet Nour at stage M1V4 (**a**) Viable Calli of Deglet Nour (**b**) and of Teggaza (**c**). (Photos by Yatta El Djouzi D)

25.4.2.3 Optimal Dose of Mutagen

In our mutagenesis process, first main step is to establish the suitable mutagen doses to achieve best results. Dose 30 Gy can be considered as the lethal dose (LD_{100}). At the 25 Gy dose, half of the strains do not survive and is considered as the LD_{50}. The optimal dose for the irradiation of the strains is estimated at 20 Gy of gamma for Deglet Nour and 25 Gy of gamma irradiation for cultivars Teggaza. Four subcultures were carried out to reach the M1V4 (from stage, i.e., 8 months later irradiation). After treatment with mutagenic, The viability of irradiated calli was assessed by Fluorescein di acetate (FDA) under an inverted microscope (Fig. 25.5a–c). The observation of these cellular masses reveals large cellular aggregates of variable size. The irradiated calluses of more or less compact texture give rise to aggregates of amyliferous cells capable of dividing and mixed with elongated non-amyliferous cells. The increase in the number of cells and aggregates, and the maintenance of a certain level of viability (80–99%) indicate the favorable action of the culture medium used for cell multiplication.

25.4.2.4 Proliferation of Irradiated Callus

The number of subcultures depends mostly on the genotype, LD_{50} dose and on other factors, such as propagation calli rate. After their transfer to germination medium, the well-differentiated embryos germinated and gave seedlings morphologically identical to the control seedlings. Irradiation of embryogenic calli of date palm made it possible to determine the threshold of sensitivity. After 1 month of culture, the embryogenic calli irradiated at a dose 30 Gy were completely necrotic and proved incapable of reacting except for one, which reveals that the toxic dose or close to the lethal (LD_{100}) is from 30 Gy. After the 25 Gy dose, half of the calli do not survive. TheLD_{50} would be this dose. For doses 10 and 15 Gy browning of the callus is reversible and the majority of the calli survive, which shows that the calli are resistant to these doses. It was observed that different sensitivities of the embryogenic calli to the irradiation doses. On the doses 10 and 15 Gy the percentage of mortality varied respectively from 36.85% to 11.2%, whereas for a higher dose 20 Gy 5.56% was recorded (Table 25.4). The proliferating irradiated callus appears to be heterogeneous. These results obtained during the proliferation of irradiated callus have shown that with dose 10 and 15 Gy, the proliferation rate is 1.3, whereas with doses 20 and 25 Gy, the proliferation rate is respectively 36.85% and 11.12%. This leads us to suppose that the irradiated callus exhibited heterogeneity in the stages of cell differentiation. The sensitivity of the callus to decreasing dose can be explained by different cell diversity, or by fact that some calli are in the multiplication phase or in the differentiation phase. Indeed, Devreux et al. (1986) showed that protoplasts with high cell density are less radiosensitive than those at low density. Furthermore their results also demonstrated unambiguously that differentiating callus are more sensitive than those that are only actively multiplying.

25.4.2.5 Histological Analysis of Calli

Irradiated callus fragments were fixed in solution containing in 100 mL, 4 mL 25% glutaraldehyde solution, 50 mL phosphate buffer at pH 7.2, 20 mL of 10% paraformaldehyde solution, 1 g caffeine, and 26 mL distilled water (Schwendiman et al. 1988). The fixation is carried out for 48 h with a passage of 2–3 h under vacuum to facilitate the penetration of the fixer and eliminate the air contained in the tissues. The samples are then rinsed and preserved in alcohol. Dehydration is an operation that is done by successive passages of the calluses in alcohol baths at increasing concentrations. Impregnation of the organs is done by placing the organs in an impregnation mixture composed of 100% alcohol and toluene for 45 min. The samples are then placed in three successive baths of toluene, 1 bath of toluene and paraffin and 1 bath of paraffin. Impregnation of the organs is done by placing the organs in an impregnation mixture composed of 100% alcohol and toluene for 45 min. The samples are then placed in three successive baths of toluene, 1 bath of toluene and paraffin and 1 bath of paraffin. The samples are placed in molds

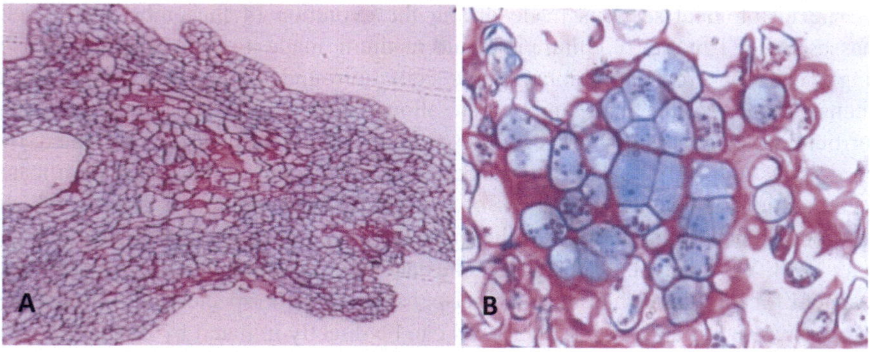

Fig. 25.6 Histology of calli on the floral explants (**a**). Early stage of cell suspension (**b**). (Photos by Yatta El Djouzi D and Abed F)

containing a quantity of approximately 1.5 mL of paraffin. Polymerization takes place at room temperature overnight. The sections are made using a microtome (Leica RM 2125RT) at a thickness of 10 μm and spread in the form of ribbons on glass slides using a brush. The sections are dried on a plate set at 40 °C.

The sections are stained by a double stain with Periodic Acid Schiff (PAS) which stains the polysaccharides of the walls in red, the mucus and the starchy reserves (cellulose, pectic compounds, hemicelluloses, etc.) in purplish pink. Naphthol Blue Black stains the cytoplasm (soluble proteins) light blue, the nucleus and nucleolus dark blue and the protein bodies (protein reserves) dark blue. After 6 months of culture in the Murashige and Skoog (MS) medium containing 100 mg L^{-1} of 2,4-D, the histology shows that after culturing show the presence of compact granular calli which consist of numerous independent spherical globules with a diameter varying between 300 and 500 μm. They are meristematic and well individualized near the vascular tissues. These meristematic cells of the nodules present a nucleus and a dense cytoplasm very rich in protein reserves intensely stained in black by Naphthol Blue Black (Fig. 25.6a). They are surrounded by dense cell layers that are visible on the periphery of the nodules. These cellular areas are the place of numerous divisions and function as a cambium (pseudo cambium) which is at the origin of the growth of nodules. Nodules have three concentric cell layers—a central zone (ZC) formed of meristematic cells, an intermediate zone made up of pseudo-cambial cells in active division, a peripheral zone (ZP) formed of several less active cellular layers. Compact calluses consist of the juxtaposition of more or less voluminous nodules, linked together by inactive cells forming light mucilage which allows the cohesion of the whole. They actively divide between the 6th and 12th month of culture, giving rise to embryogenic stem calluses which are made up of embryogenic cells containing large nuclei very rich in soluble proteins (Fig. 25.6a). These observations are analogous to those already reported for palmaceae such as *Elaeis guineensis* (Chwendiman et al. 1988). Callus proliferation near vascular tissues has also been observed with leaf explants of Cocos nucifera (Buffard-Morel et al. 1992), and in other species such as *Gossypium hirsutum* (Gawel et al. 1986).

The histological sections made during the evolution of the embryogenic cell masses. After 1 month of culture in liquid medium, made it possible to distinguish single cells or groups of embryogenic cells surrounded by a thick outer wall intensely colored in pink, and which allows them to be separated from the surrounding degenerative tissues. These observations were also highlighted by Verdeil et al. (2001). Also, the suspensions are accompanied by a significant accumulation of reserve substances of a starchy and/or lipoprotein nature which are good indicators of tissue development towards embryogenesis (Chwendiman et al. 1988; Verdeil et al. 2001). Unlike the date palm, obtaining a friable callus is not essential for the establishment of cell suspensions in some species. This is the case for cassava and alfalfa. Cell suspensions can be directly initiated by transfer of leaf fragments in liquid medium (Raemakers et al. 1993).

25.4.3 In Vitro Resistance Screening

25.4.3.1 Extraction and Fraction of Fusarium Toxin

The aggressive strain 133 of *Fusarium oxysporum* f.sp. *albedinis* was grown on PDA medium (Potato: 250 g; 20 g: glucose and 15 g agar) for 5 days in the fungal culture was flooded with sterile water to prepare conidial suspension and the concentration was attuned to 10^6 spore mL^{-1}. One millilitre of the conidial suspension was transferred into Erlen Meyer flasks containing 200 mL Czapeck medium (2 g $NaNO_3$, 1 g K_2HPO_4, 0.5 g KCl, 0.5 g $MgSO_4{\cdot}7H_2O$, 0.01 g $FeSO_4$ $7H_2O$, 30 g sucrose, 1000 mL distilled water) and shaken at 200 rpm on a rotary shaker at 25 °C for 10 days. The cultures were filtered with Whatman N°1 filter paper, the filtrate was centrifuged at 4200 g for 20 min, and the supernatant was used for further toxin extraction. The extraction was performed on the *F.o.a.* culture filtrate according to the method of Pringle and Scheffer (1963), modified by El Fakhouri et al. (1996).

Five liters of culture filtrate were evaporated under vacuum below 45 °C pending its initial volume was reduced to 200 mL. An equal volume of methanol was added and the mixture was placed at 4 °C for 48 h. After this time, the precipitated material was removed by filtration, washed two times with a cold methanol/water (1:1, V/V) mixture, and discarded. The resulting filtrate was vacuum-evaporated at 45 °C and the entire volume was reduced to 150 mL. This liquid was poured on a glass column (3 × 40 cm) of Norite mixed with Celite mixed and equilibrated with water. The column was then washed with distilled water until the eluate was completely clear. The absorbed fractions were then eluted by three successive steps of mixture of 10%, 30%, and 50% pyridine in water that generated three fractions: FI, FII, and FIII, respectively (Sedra et al. 1993; El Fakhouri et al. 1996; Sedra and Lazrek 2011). The fraction FII was used for the selection because it showed the most phytotoxic activity during biological assays (Sedra et al. 2008).

The process of infecting a plant with a pathogen begins with the establishment of contact between the two protagonists of the parasitic relationship. The pathogenic agent may use several modes of action to penetrate into plant tissue to overcome mechanical, chemical and/or physiological barriers of the host plant. Pathogenic fungi often damage their host plants by producing toxic metabolites, which causes various symptoms toxins these metabolites could be involved in the pathogenicity (Berestetskiyis 2008). Phytotoxics are classified into two broad groups, those which are nonspecific and those which are host-specific toxins. Some of the toxins have general phytotoxic properties and are active on a wide range of plant species. These are non-host-specific toxins. They contribute to the virulence or development of symptoms of the disease in which they occur, but are not the primary determinants of the host range. Other toxins are host specific and affect only certain plant varieties (Osbourn 2001). Specific toxins play a role in determining the host range of specificity of plant pathogens and can act as a virulence agent for these pathogens. The majority of the special forms of *Fusarium oxysporum* and *Fusarium* species, in general, produce toxins which are derived from primary or secondary metabolism, the nature of which is very diverse: protein (Mussel 1972), terpenic (Casinovi 1972). The first toxic substance, of peptide nature, which was isolated and purified from cultures of *Fusarium oxysporum* f.sp. *lycopesici* is lycomarasmin (Messiaen and Cassini 1968).

The first work carried out on the toxins of *Fusarium oxysporum* f.sp. Surico and Graniti (1977), who isolated fusaric acid and anydro-aspergillomasarin A and B. *F.o. a.*, secretes into a liquid culture media numerous substances like toxins (Mokhlisse 1987; Sedra et al. 1993, 1997); and enzymes (Amraoui et al. 2004). Some of them have been chemically and biologically well characterized, such as fusaric acid and its derivatives which were isolated from culture filtrates (Mokhlisse 1987; Sedra et al. 1993, 1997). Fusaric acid may play an important role in the early stage of date palm infection by *F.o.a.* (Bouizgane et al. 2004; Sedra et al. 2008).

The extraction of secondary metabolites was performed on *F.o.a.* culture filtrates according to the method of Pringle and Scheffer (1963). During extraction, the adsorbed fractions were diluted by three successive steps of mixtures of 10%, 30%, and 50% pyridine in water (v/v) that, respectively, gave three collected fractions labeled FI, FII, and FIII. These fractions were evaporated until dry and a yellowish powder was obtained for FI, while whitish ones were obtained for the FII and FIII fractions. However, the FII (*F.o.a.*) fraction contains the fusaric acid and other toxic sub-fractions (H3, H4, H5 and H6).

25.4.3.2 In Vitro Selection of Irradiated Materials

In vitro screening of irradiated materials is based on the use of purified or crude culture filtrates of the pathogen as a selective agent (culture filtrate from *F.o.a.*). After the radio-sensitivity test, the irradiated calli were irradiated with a 20-Gy dose for Deglet Nourand a 25-Gy dose for Teggaza, then were transferred in the same fresh medium until they reach the M1V4 stage. In our experiment, we used callus proliferation, i.e., cell division or cell proliferation from the irradiated calli as a generation. The secondary callus could have been considered a second generation, but it was only encountered in the case of the 30 Gy dose, where the necrotic embryogenic callus screening of 500 calli using 10% *F.o.a.* filtrate culture. Surviving calli have been selected using a culture medium containing 10% culture filtrates were transferred on liquid medium (Fig. 25.7).

Fig. 25.7 Selection by using filtrate culture at 10% on callus. (Photos by Yatta El Djouzi D)

25.4.3.3 Embryogenic Suspension Cultures

Cell suspension cultures are useful for date palm genetic enhancement based on in vitro selection studies (Jain 2012). A number of researchers successfully obtained in vitro plant regeneration using date palm embryogenic cell suspension cultures (Al-Khayri 2012; Naik and Al-Khayri 2016). These studies proved that cell suspension culture is a prolific source of somatic embryos, suitable for mass propagation of several date palm cultivars.

Embryogenic cultures of "Deglet Nour" and "Teggaza" were transferred in liquid initiation medium in 125 mL Erlenmeyer flasks (40 mL medium per flask) on a rotary shaker (110 rpm), and maintained as suspension cultures and sub cultured into fresh medium at 2-week intervals.

A growth curve was established based on fresh weight of cell suspensions over time. Weighing was carried out regularly before and after each subculture, and estimate the increase of the fresh weight of cell suspensions. The Fig. 25.8 expresses the growth cell suspensions development curve in two cultivars. From the beginning time of suspension culture, the lag phase was indiscriminate and translated into a latency phase, this is not observable. It started during first week of cell culture, between the first and 6 days and embryogenic masses started to isolate from mother callus tissue. The diameter of the particles varied from 2 to 3 mm. From the second week in a liquid medium, the callus weight doubled. The exponential phase, where the highest growth rate occurred, was started 1 week after culture initiation.

After 5 weeks, growth deceleration had occurred and lasted until week 10 and a stationary phase that starts from sixth subculture. At this level, the growth appears nil, this is due to a loss of activity where the multiplication rate is equivalent to that of mortality.

After 3 weeks of the culture, the clusters formed were removed by sieving through a 380 μm sterile sieve to keep only small embryogenic cell suspension. The cultures were subculturing weekly by fresh liquid medium until obtain small pro-embryonic cell aggregates. Embryogenic cultures of Deglet Nour and Teggaza

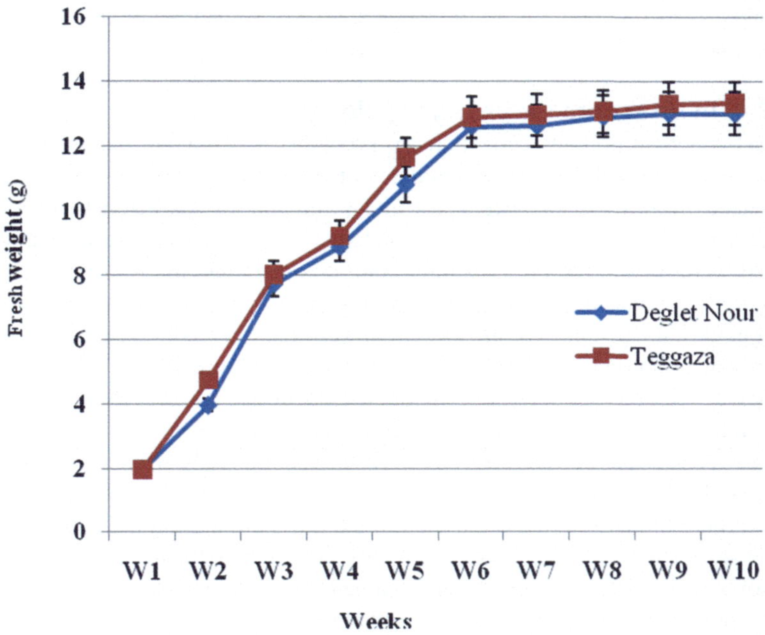

Fig. 25.8 Growth cell suspension evolution in two cultivars Teggaza and Deglet Nour. (Constructed byYatta El Djouzi D)

Fig. 25.9 Somatic embryos (**a**) and pro-embryonic cell aggregates (**b**) obtained in liquid media M3. (Photos by Yatta El Djouzi D and Abed F)

in suspension had similar morphologies, and consisted of small pro-embryonic cell aggregates (heap retained on the filter >380 μm) (Fig. 25.9a). Suspension cultures of Teggaza on the other hand, consisted entirely of pro-embryonic cells and small aggregates of pro-embryonic cells (Fig. 25.9a, b).

In date palm, several researchers successfully obtained in vitro plant regeneration using cell suspension cultures (Bhaskaran and Smith 1992). These studies proved that cell suspension culture is a productive source of somatic embryos, suitable for mass propagation of several date palm cultivars.

25.4.3.4 In Vitro Somatic Embryos Selection

Compact spherical nodules correspond to proembryos. They are maintained under the same culture conditions. These nodules or proembryos evolve into small elongated shapes, varying in size between 7 and 8 mm and are characteristic of somatic date palm embryos. These latter can be isolated. 5200 embryogenic somatic embryos obtained were transferred on a medium culture containing culture filtrates concentration at 35% and 1800 somatic embryos have been selected (Fig. 25.10a–d). The embryogenic somatic embryos are located on germination medium devoid of growth substance at a photoperiod of 16 h. After 6 months of culture, the well-differentiated embryos developed and gave rise to plants. At doses 10 and 15 Gy, the well-differentiated embryos germinated to give green seedlings. Callus is observed which evolves either in roots (46% and 45%) or completely necrosis without any evolution (Table 25.5).

The synchronization of somatic embryo germination is observed for the control cultures as well as for the treated samples. The green plants obtained are not morphologically different from those obtained with the control. However, certain abnormalities were observed during germination in embryos irradiated at different doses. Forty-five percent of germinating somatic embryos gave roots only on doses 10 and 15 Gy, although this phenomenon was not observed at dose 30 Gy. Rooted

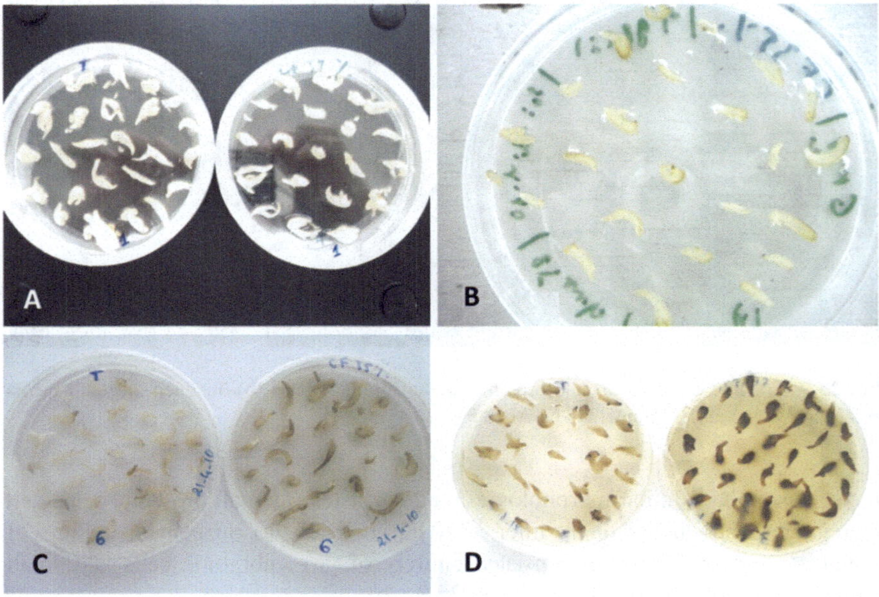

Fig. 25.10 In vitro selection of somatic embryos using the Fusarium culture filtrate at 35% (**a**–**d**). (Photos by Khelafi H, Abed F and Yatta D)

Table 25.5 Effect different irradiation doses on the germination of somatic embryos

Dose (Gy)	Number of somatic embryos germinating	Number of germinated somatic embryos		Green shoots and roots		Cotyledons and roots		Roots only	
		*N***	%***	*N*	%	*N*	%	*N*	%
0	15	10	66.6	9	60	1	66	0	
10	15	08	53.3	9	60	7	46.6	7	46.6
15	22	10	45.5	7	31.8	10	45.4	10	45.5
20	50	20	40	20	40	13	26	0	
25	15	10	66.6	10	66.6	4	26.6	0	
30	01	0		0		0		0	

Table constructed by Khelafi H

N.S.E.G: Number of somatic embryos germinating, N.E.S.G: Number of germinated somatic embryos, P.V.R: Green plants + Root, C.R: Cotyledons + Roots, *N***: Number of germinated somatic embryos, %***: percentage of Number of germinated somatic embryos

fused cotyledons were 45% at dosages 10 and 15 Gy, 26% at dosages 20 and 25 Gy, and 66.6% in the control. The 20 Gy dose was determined to be the optimal irradiation dose for the irradiation of date palm callus.

Similar studies in vitro cultures of date palm in Morocco and Tunisia, on different cultivars, different steps of differentiation, showed that the irradiation dose was around 20 Gy (Jain 2011). Donini (1991) locate the irradiation dose at the Gamma ray close to the LD_{50} dose, where 50% of the irradiated material survived, which confirms the results obtained. The aim of this work is to arrive at determining the radiosensitive dose, and to follow the effect of the irradiation doses on the stages of somatic embryogenesis of Deglet Nour. The lethal dose in which irreversible necrosis of all the embryogenic calli is observed is 30 Gy and represents the LD_{100}.

The LD_{50} dose where half of the irradiated calli survive is 25 Gy. Remaining strains proliferate and the germinating somatic embryos have given rise to whole plants. Following these results, additional studies are to be carried out to improve the process for obtaining homogeneous embryogenic strains and to monitor work aimed at obtaining variants combining date quality and resistance to Bayoud.

25.4.3.5 Regeneration of Putative Mutants

The irradiated and selected calli gradually were transferred into the germination medium to permit the induction of the somatic embryos and their germination. The culture medium was used is the MS base medium (PGRs-free MS4) medium supplemented with 60 g L^{-1} sucrose, 1 mg L^{-1} Thiamine HCl, and 40 mg L^{-1}adenine. The embryogenic strains are maintained in darkness under a thermoperiod of 28 ± 2 °C for 16-h and 22 ± 2 °C for 8 h. The transfer to a new medium is done 2-months intervals. Somatic embryogenesis goes through three stages, namely: somatic embryos induction, somatic embryos into seedlings and development of the root system. The proembryos which are compact spherical nodules are maintained under the same culture conditions. They then evolve into small,

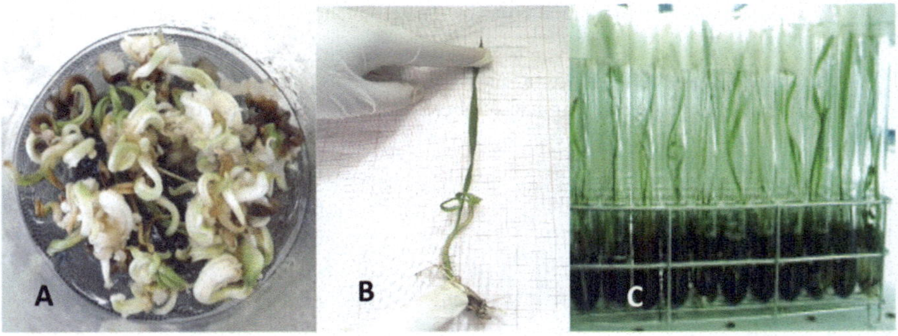

Fig. 25.11 Somatic embryos Maturation and Embryos germination of Deglet Nour (**a**), somatic seedlings (**b**) and seedlings shoot and roots (**c**). (Photos by Yatta El Djouzi D)

elongated shapes and are typical of somatic date palm embryos. The proembryos may appear separate or remain in clumps. The first somatic embryos appeared after 6 months of their transfer.

Embryos evolve and germinate under the same culture conditions. The cotyledon and radical appear. There is a lengthening of the cotyledon which has at its base the cotyledonary slit from which a first leaf emerges (Fig. 25.11a). Embryos evolve and germinate under the same culture conditions (Fig. 25.11b, c).

25.4.3.6 In Vitro Selection by Detached Leaves

Leaves, 20–25 cm in size, were collected from 20 in vitro-plants, issued from irradiated callus, and evaluated for their confrontation to *Fusarium* wilt. Two concentrations, 25 and 50 μg mL^{-1} toxin fractions FII were used. Sterilized water was used as a control. The standard concentration used to screen resistant from sensitive materiel was 50 μg mL^{-1} (Sedra et al. 1993). This concentration can be increased to 100 μg mL^{-1} if large size detached leaves are used. There is a relation between the weight of leaves and the quantity of toxin used (Sedra et al. 1998). The leaves were dried first using cotton, soaked with ethyl alcohol and then with sterile water. They were immersed in test tubes containing 25 mL toxin fraction FII. For each treatment, three leaves per concentration and per vitro-plants were used. For each trial, the treatment is repeated three times. Tubes carrying leaves were maintained under daylight, at an ambient temperature between 22 and 27 °C. The symptoms were observed by measuring the progression of necrosis the petiole every 3 days for 22 days.

25.4.3.7 Test of Artificial Inoculation of Plants

Twenty vitro-plants, previously tested against toxin, were inoculated with 100 mL *F. o.a.* conidia suspension, at the concentration of 10^6 sp mL^{-1}. All tests were carried out in the greenhouse. One Deglet Nourvitro-plant was used as a control, grown in 100 mL water. Weekly assessment was carried out on the mortality of inoculated plantlets. From sensitive plants, re-isolation of the *F.o.a.* from roots was performed on the P.D. A medium (Potato dextrose agar).

25.4.3.8 Effect of Fraction FII of *F.o.a.* Toxin on Detached Leaves of Vitro-Plants

Three days after immersion of the leaves in different concentrations of toxin, fraction FII of *F.o.a.*, browning was observed at the base of the petiole of all treated leaves (Fig. 25.12a–c). This was a wound response to sectioning of the petiole. After 5–6 days, the necrosis progressed in leaves soaked in fraction FII of toxin.

At toxin concentration of 25 μg mL^{-1}, the necrosis varied between 1 and 11 mm in length, whereas it varied between 3 and 40 mm in response to 50 μg mL^{-1} toxin fraction. In the control, necrosis did not exceed 2 mm in length (Table 25.6). The response to the use of 25 and 50 μg mL^{-1} FII fraction of *F.o.a.* toxin on detached leaves of 20 plantlets regenerated from irradiated callus of Deglet Nourcultivar showed that the size of the necrosis increased with increasing dose of the toxin.

The differences in necrosis size varied between 1 and 37 mm in the presence of toxin concentrations 25 and 50 μg mL^{-1}. The necrosis size of vitroplants (M7, M12, M13, M17, and M19) did not exceed 1 mm. Other five vitroplants (M9, M10, M14, M15, and M16) had necrosis size between 2 and 5 mm. The size of detached leaves of eight vitroplants (M1, M2, M4, M6, M8, M11, M18, and M20) varied between 7 and 15 mm. However, necrosis size was 19 mm for M5 and 37 mm for M3 vitroplants.

A correlation between vitroplants noted that showed resistance to artificial inoculation andthe size of the necrosis caused by the treatment of FII fraction of *F.o.a.* toxin on detached leaves (Fig. 25.13). The size of necrosis on detached leaves treated with 25 μg mL^{-1} toxin varied between 1 and 7 mm in vitroplants that had resistance to the fungus, whereas it varied between 3 and 17 mm in response to 50 μg mL^{-1} toxin.

The sensitive date palm vitroplants showing symptoms of *Fusarium* wilt after artificial inoculation had a positive response to the toxin on their detached leaves for both doses. The effect of toxin doses was clearer for the mutants tested. The size of the necrosis varied between 2- and 11-min in the presence of 25 μg $mL^{-1}$and

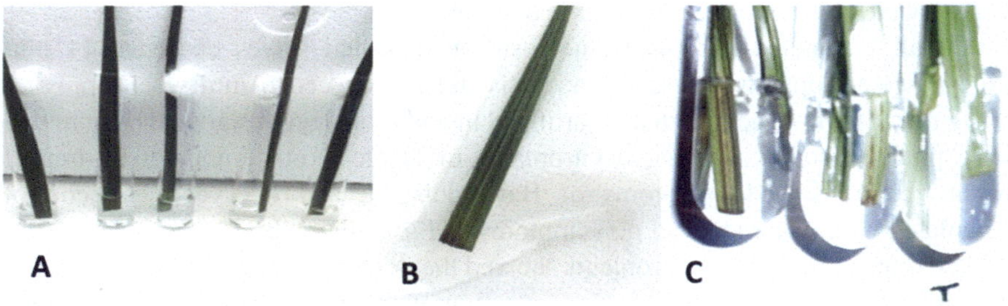

Fig. 25.12 Immersion of leaves (**a**), browning at the base of the petiole (**b**), Progression of necrosis of leaves in the in the solution of FII toxin of *F.o.a.* presence of the fraction FII of *F.o.a.* toxin (**c**). (Photos taken in 2015 by Khelafi H and Yatta El Djouzi D)

Table 25.6 Different responses of date palm mutants to *F.o.a.* based on resistance evaluation test indicated by the size of necrosis developed at the base of the petioles (mm: millimeter) in response to inoculums

	Size of necrosis at the base of the petioles (mm: millimeter)		
Lot	0 μg mL^{-1}	25 μg mL^{-1}	50 μg mL^{-1}
VM1	1	11	20
VM2	1	10	25
M3	1	3	40
VM4	1	1	12
VM5	2	11	30
VM6	1	1	10
VM7	2	12	30
VM8	1	6	17
VM9	1	7	10
VM10	1	3	8
VM11	1	4	11
VM12	2	4	5
VM13	1	5	6
VM14	1	3	6
VM15	2	1	3
VM16	2	6	8
VM17	1	2	3
VM18	1	2	15
VM19	1	5	6
VM20	1	2	15

Constructed by Khelafi H and Yatta El Djouzi D
This evaluation is made on the scale between 0 and 4 cm
1: −1 cm ; 2: <2 cm; 3: +2 cm; 4: 3 cm or larger

between 6 and 40 mm in the presence of 50 μg mL^{-1} toxin. This result is consistent with the work obtained by Asnaghi et al. (2007) and Browne et al. (2005) using detached leaves screening for rusts disease resistance (Xie and Mew 1998). In addition, Twizeyimana et al. (2007) evaluated the detached leaves for selection *Musa* species for resistance to black leaf streak.

For all inoculated vitroplants, necrosis on detached leaves exceeded 11 mm except two vitroplants where the necrosis did not exceed 6 mm when applying 50 μg mL^{-1} toxin. One month after artificial inoculation, leaves started drying in two vitro plants. After 6 months, 9 vitroplants of Deglet Nour among toxin-treated twenty plants sowed symptoms of Bayouddisease and decayed. The control inoculated with water showed no symptoms of *Fusarium* wilt.

The roots of inoculated vitroplants showed the development of fine, curly salmon pink mycelium after their transfer onto a culture medium. This confirmed wild form *F.o.a.*, which is responsible for *fusarium* wilt of date palm.

The most excellent approach to overcome *Fusarium*-wilt difficulty is to obtain resistant palms trees. The mass production of plants using in vitro technique and

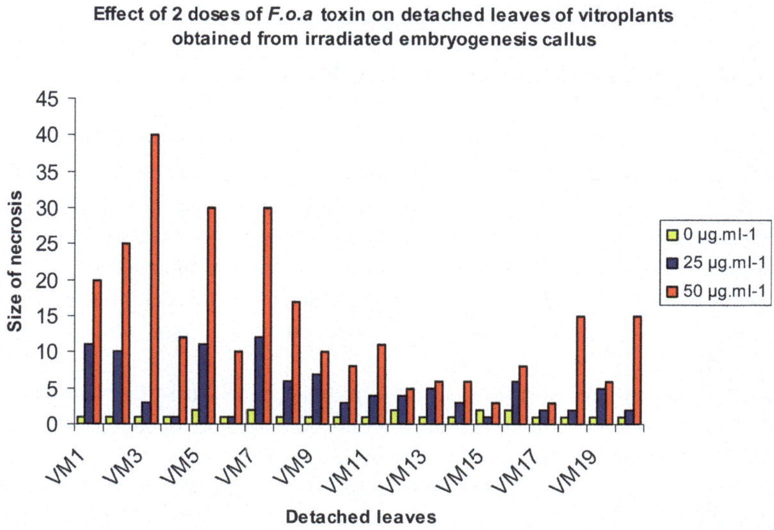

Fig. 25.13 Effect of two doses of filtration of *F.o.a.* toxin on detached leaves of vitroplants obtained from irradiated embryogenic callus. (Constructed by Khelafi H and Yatta El Djouzi D)

screening resistant cultivars was recommended for a long time (Louvet and Toutain 1973; Saaidi et al. 1981). It was noted that fusaric acid isolated from culture filtrate of *Fusarium oxysporum* f.sp. *albedinis* (Amraoui et al. 2005), played an essential role in early phases of date palm contamination by *F.o.a.* (Bouizgane et al. 2004). The selection of mutants is done by detached leaf technique by using FII toxin of *F. o.a.* The pathogen secrets in the medium several toxic substances of various kinds including fusaric acid, anhydrous Aspergillomarasmine-B, and peptidic compounds. These compounds from which, three toxin fractions FI, FII and FIII were determined and the study of their toxicity. Fraction FII showed the most toxic for the date palm (Sedra and Lazrek 2011; Sedra et al. 2008). Amraoui et al. (2005) also tested fusaric acid on date palm leaves to examine its toxic effect with that of the FII (*F.oxysporum* f.sp. *albedinis*) and HPLC—purified fractions. However, they established that fusaric acid and some portions purified from FII (*F. oxysporum* f.sp. *albedinis*), i.e., H (3), H (4) and H (5), were toxic on detached leaves. They showed that FII (*F.o. a.*) fraction contains new toxins, different from enniatins or fusaric acid by their specificity and their elution time in HPLC, respectively. These molecules could act independently or synergistically with fusaric acid to induce the characteristic symptoms of the Bayoud disease.

For these biological tests, several researchers tested whole plants or only a fragment of plant to study the effects of fungal toxins. Molot et al. (1984) used the asparagus plantlets to evaluate the effect of toxins secreted by *Rhizoctonia violacea*. Tomato plantlets were also used to evaluate the effect of toxins secreted by *Pyrenochaeta lycopersici* while Pinon (1984) tested toxins *Hypoxylum mammatum* on poplar leaves.

25.5 Evaluations of Mutants

25.5.1 In Vivo Evaluation of Putative Mutants

Vitro plant acclimatization is an important step after in vitro cultivation. The acclimatization conditions must be well identified to avoid the loss of valuable plant material given the long time the date palm takes to react in vitro culture. The seedlings from the first irradiations and the control seedlings were acclimatized and raised in a greenhouse, in plastic bags containing a mixture of sand and compost in equal volume (1/2). The irradiated seedlings chosen for the acclimatization stage must have a robust structure and a well-developed root system, with at least four roots of a minimum length of 10 cm.

Started drying in two vitroplants. After 6 months, nine vitro-plants among toxin-treated 20 plant sowed symptoms of Bayoud disease and decayed. The control inoculated with water showed no symptoms of *Fusarium* wilt.

The roots of inoculated vitro-plants showed the development of fine, curly salmon pink mycelium (Fig. 25.14) after their transfer onto a culture medium. Fourteen plants of Deglet Nour showed tolerance levels to toxins to the resistant cultivars Taquerbucht. The plants were grown under sheltered shade until reaching the optimal stage for planting in the field.

The infection symptoms development was observed much earlier on detached leaves with the toxin was observed much earlier on detached leaves with the toxin of *F.o.a.* than vitro-plants inoculated with the actual pathogen. The artificial inoculation with liquid form of *F.o.a.* was most efficient in the selection of date palm where concentration was 10^6 sp mL^{-1}.

A similar artificial screening method was used by Ando et al. (1984) to choose plants resistant to fusariosis using soil inoculation with a fungus suspension of 4×10^5 spores mL^{-1}.

Also, the result shows the effect toxic of culture filtrate of the *Fusarium subglutinans* to provide a process of in vitro selection cultivars of pineapple (*Ananas comosus*) for resistance to the disease, as has already been tried in some other host-pathogen system (Ludwig et al. 1992; Song et al. 1994).

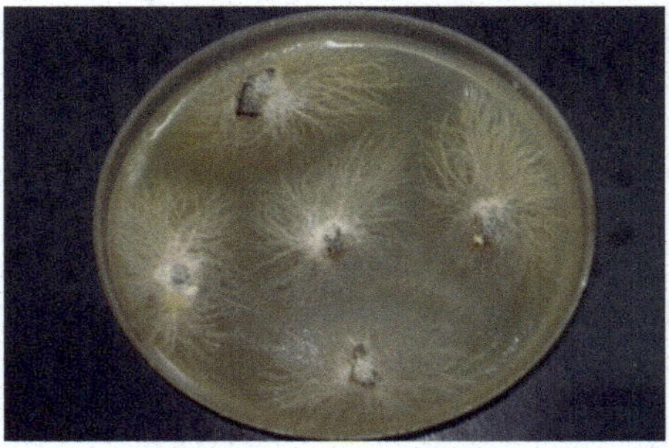

Fig. 25.14 Development of mycelium on PDA medium. (Photo by Khelafi H)

25.5.1.1 Acclimatization of Putative Mutants

In vitro selection of 70 seedlings putative mutant using the *F.o.a.* culture filtrates at 35% (Fig. 25.15a–c).

25.5.1.2 Greenhouse Evaluation

Inspection of the morphological quality of plants showed that the acclimatization step went well and Seedlings issues from the first irradiations and control seedlings grown vigorously. Indeed, the evaluation of the survival of the plants during the first week of acclimatization showed that all Seedlings survived (100% survival). The mutant plantlets selected and healthy adapted in vitro are evaluated in order to select a solid mutant before transfer to the infested field. The screening pressure could be applied 2–3 times to make sure that the preferred putative mutant are resistant (Fig. 25.16a, b). The mutant plantlets were treated with toxin and the *F.o.a.* culture filtrate at 35%. Ten mutant plants were selected from 70 seedlings putative mutant in

Fig. 25.15 Inoculation with *F.o.a.* (**a**, **b**) Selection with respect to the *F.o.a.* and development of cultivars Teggaza (**c**). (Photos by Abed F)

Fig. 25.16 Artificial inoculation test on irradiated tissue culture plants obtained from embryogenic callus. The mutant plants treated with *Fusarium* toxin and the *F.o.a.* culture filtrate at 35% (**a**). The mutant plant infected and mutant plant tolerant after treated with *F.o.a.* culture filtrate (**b**). The mutant plantlets selected tolerant before transfer to the field evaluation (**c**). (Photos by Team date palm INRAA and Jain SM)

the greenhouse, which later on transferred in the contaminated field (Fig. 25.16c). They showed tolerance levels to toxins similar to the resistant cultivars cv. Taquerbucht. The plants were grown under sheltered shade until reaching the optimal stage for planting in the field.

25.5.1.3 Field Evaluation

Transfer plantlets (14 plants showed tolerance levels to toxins to the resistant cultivars Taquerbucht) from irradiated calli, plantlets from non-irradiated calli, (control), and resistant cultivar Taquerbucht (as control) to infested field in order to confirm resistance of mutants, research of interesting agronomic traits and Increase genetic variability of date palm.

However, the field screening of date palm is laborious, expensive, and site specific. Consequently, selection should be increased by the development of mass screening methods at the in vitro level.

The field evaluation of the mutant plants, based on survival rate and other agronomical such as yield and date quality, was carried out in 2008. We choose the Field planting (3 locations in Ghardaia). The controls are sensitive (Deglet Nour), and resistant (Taquerbucht). After 1 year, only 8 plants from 14 plants showed tolerance levels to toxins similar to the resistant cultivars Taquerbucht. We have detected mutant infected by *F.o.a.* after 8 years transplantation in contaminated field and M3 mutant of DN infected with Bayoud, confirmed the sensitivity after isolation of *F.o.a.* from the sampled. The putative mutants of cv. Deglet Nour have been transferred to a disease-infected field 12 years ago. Actually only 1 mutant of Deglet Nour were found to be really resistant after 12 years in Farmer field (Fig. 25.17a–d). This mutant show well development of the tree, emission of four offshoots, and higher bunch with good quality of date palm fruit.

For the putative mutants of cv. Teggaza, in 2017 we have chosen the infested Field planting in south at INRAA Station Adrar (Fig. 25.18a–f). Eleven vitro-plants showing resistance to artificial inoculation of the pathogen were transferred to a plot infested with Bayoud disease also we have transferred the sensitive Controls (Deglet Nour), and resistant Controls (Taquerbucht) in order to study of the behavior of their comportment against *Fusarium oxysporum* f.sp. *albedinis*. This confirmed wild form *F.o.a.*, which is responsible for *Fusarium* wilt of date palm.

Confirm resistance of mutants, research of interesting agronomic traits and Increase genetic variability of date palm.

25.5.2 Molecular Characterization

Molecular approaches were used to understand the mechanisms within the process of regeneration by, genetic stability of plantlets and also to discriminate the genetic variability between the mutant and the mother plants through characterizing the variations at DNA level. Random amplified polymorphic DNA (RAPD), amplified fragment length polymorphism (AFLP), single-strand conformational polymorphism (SSCP), microarray, differential display, targeting induced local lesions in

Fig. 25.17 Only 1 mutant of Deglet Nour were found to be resistant after 10 years in Farmer (**a**, **b**) and mutant M2 fruit of Deglet Nour (**c**). Mutant infected by *F.o.a.* after 8 years transplantation in contaminated field (**d**, **e**) (M3 mutant of DN infected with Bayoud, Confirmation of the sensitivity after isolation of *F.o.a.*). (Photos taken byAbed F and Yatta El Djouzi D)

Fig. 25.18 Transfer in contaminated field of 13 Teggaza mutants selected in vitro and in the greenhouse (**a**, **b**), Teggaza mutants after 1 year (**c**), inoculation technics in field (**d**, **e**) and Teggaza mutants were found to be resistant after 30 months in contaminated field (**f**). (Photos by Yatta El Djouzi D and Abed F)

genome (TILLING) and high-resolution melt (HRM) permi tquick and profoundly analysis of mutational variations (Penna et al. 2012).

Sianipar et al. (2015) have been used RAPD method to identify the genetic variability between the mutant plantlets enhanced from physical mutagens rodent tuber calli. They obtained 69 fragments from 11 mutant plantlets by using 10 RAPD.

ISSR technique was used to discriminate the drought-tolerant sugar beet mutant improved using irradiated shoot tip explants by gamma radiation (Sen and Alikamanoğlu 2012). These authors reported the presence 91 polymorphic bands of 106 PCR fragments with 19 inter simple sequence repeat (ISSR) primers.

In coffee plants, after mutagenic treatment through sodium azide, the genetic variability was determined using RAPD and AFLP markers (Vargas-Segura et al. 2019). They obtained 46 fragments with 12 RAPD primers of which 34 were polymorphic bands (74%). The amplification with AFLP, 6 selective primer combinations revealed detection of 36 polymorphic bands polymorphisms with a percentage of 17.8. The study showed that both NaN_3 and EMS induced genetic variability within the DNA regions amplified with AFLP and RAPD markers.

In date palm, Quenzar et al. (2001) identified two circular plasmid-like DNAs (S and R) in the mitochondria through a PCR-based approach; the presence of R plasmid and absence of S plasmid showing that could be considered as a reliable molecular marker of Bayoud disease resistance. Date palm has been spontaneous mutation induced by mutagen treatments and selected for resistance to Bayoud disease by magnetic fields (Jain 2011).

Molecular tools were used to understand the mechanisms within the process of regeneration, genetic stability of plantlets and also to compare the ploidy level and polymorphism mutant improved and control vitro-plants of the same variety (Figs. 25.19 and 25.20). DNA was isolated form leaf tissues of the plantlet

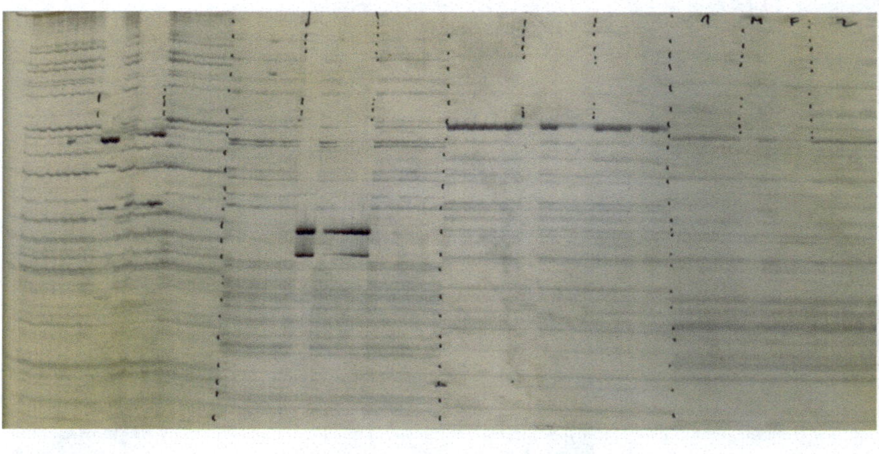

Fig. 25.19 Portion of gel produced with the pairs of primers E-AAC—M-AAG, E-AAG—M-AAG, E-ACG—M-ACG, E-AAC—M-AGT showing the specificity of the AFLP profiles in putative mutant of Tgazza and their control vitro plants. M (putative mutant) and VPC (control vitro plants). (Photo by Yatta El Djouzi D)

Fig. 25.20 Portion of gel produced with the pairs of primers E-AAC—M-AAG, E-AAG—M-AAG, E-ACG—M-ACG, E-AAC—M-AGT showing the specificity of the AFLP profiles in putative mutant of Deglet Nour and their control vitro plants. M1 to M10 (putative mutant) and VPC (control vitro plants). (Photo by Khelafi H)

regenerated by somatic embryogenesis and their mother plants using the mixed alkyltrimethylammonium bromide (MATAB) method (Gawel and Jarret 1991) contained (0.35 M sorbitol, 0.1 M Tris, 1.25 M NaCl, 0.005 M EDTA, 0.5% Na bisulfite, Matab to 4% pH 8). Amplified fragment length polymorphism (AFLP) analysis to be performed in order to verify level polymorphism between irradiated vitroplants of cvs. Deglet Nour, Teggaza and mother plants.

Six primer pairs tested through 17 offered better readability in this study; these are identified as follows: E-AAC—M-AAG, E-AAC—M-AGT, E-AAC—M-ACA, E-AAG—M-AAC, E-AAG—M-AAG, E-AAG—M-ACT. Results of cv. Deglet Nour showed the presence of systematic bands specific to vitroplants control with two pairs of primers E-AAC—M-ACA, E-AAG—M-AAC, and bands specific to irradiated vitroplants. The *mutagen* is a chemical or physical *agent* primers combination selected reveal polymorphic bands with a percentage of 35% for Deglet Nour and 40% for Teggaza.

The electrophoretic profiles showed a number of monomorphic bands equal to 120 between the irradiated and control vitroplants, which corresponds to 65% for Deglet Nour and 10% for Teggaza.

Mutants created by mutagen physic agent at 20, 30 and 40 Gy improved by 20 random primers showed a low polymorphism level compared with the control. Other studies have also shown the efficacy of measure genetic polymorphism using molecular markers for determining mutation procedures in date palm (Saker et al. 2006) and other species such as in peach (Hashmi et al. 2017), soybean (Gesteira et al. 2008; Atak et al. 2004), and rice (Rashid et al. 2009).

However, site-specific mutagenesis with the CRISPR/Cas9 system will allow scientists to analyze the gene expression in date palm genome.

In date palm, CRISPR/Cas9 offers a number of benefits well-suited to modification of dioecious tree crops (Kole et al. 2015), to discover genetic markers for sex determination, to develop flowering, juvenility control and manipulation of fruit ripening, fruit quality, nutritional value and production of biofuel from the date palm wastes (Sabir et al. 2014). As well, CRISPR/Cas9 system can be also explored to curtail phytoplasma diseases. The supplementary applications may help in regulating secondary metabolites production from date palm fruits.

25.5.3 Flow Cytometric Analysis

Flow cytometry protocol was developed for polyploidy detection, dissociating chimeras monitoring of cytochimera dissociation, and chromosomal stability. Moreover, application of DNA flow cytometry is quite straightforward. This technique is based on the use of a specific DNA fluorochrome (propidium iodide) and on the quantification of the relative intensity of the fluorescence of the labeled nuclei. DNA content is correlated with ploidy level. During this experiment, *Zea mays* L. (CE 777) (2C = 5.43 pg of DNA) was used as reference internal standards.

Samples for analysis of the quantity of nuclear DNA are taken from four types of tissues leaves of vitroplants raised from irradiated callus of cv. Teggaza and their (1 cm), vitroplants of cv. Taquerbucht (approx. 300 mg).

These samples were placed in a Petri dish held on an "ice block" containing 1 mL of extraction buffer (Doležel et al. 1997) of the following composition: 15 mM TRIS, 2 mM Na_2EDTA, 0.5 mM spermine 4HCl, 80 mM KCl, 45 mM $MgCl_2$, 30 mM sodium citrate and 40 mM Na_2SO_3 at pH 7.5. They were then chopped very finely using a razor blade. The suspension, which includes isolated nuclei, is then removed using a 3 mL Pasteur pipette and filtered using a 30 μm polyethylene sieve (Partec Celltries). Propidium iodide (IP) at 330 $\mu g\ mL^{-1}$ (IP4170; Sigma) is added to the filtrate then the mixture vortexed and left to stand for 30 min in the dark before analysis by flow cytometry (CyFlow Partec II NOT). Propidium iodide (PI), a nucleic acid intercalating agent, will bind to DNA molecules contained in the cell nucleus.

The DNA assay is carried out using a Partec II PAS (Particle Analyzing System) flow cytometer (Partec GmbH, Münster, Germany) whose laser radiation with a wavelength of between 454 and 515 nm passes through the nucleus and excites by fluorescence the IP bound by the DNA molecule. The quantity of fixed IP is correlated to the quantity of DNA contained in the nuclei and therefore to the ploidy level of the sample. For each sample, 800 nuclei were analyzed on average. The distribution of fluorescence intensities (relative DNA content) obtained by flow cytometry is generally expressed in arbitrary units (C). The 1C value represents the DNA content of a haploid (*n*) chromosome set. To determine the ploidy level, this scale should be calibrated against a reference. In this study, we used *Zea mays* (CE 777) (C = 5.43 pg DNA) as an internal reference standard.

Using flow cytometry, to estimate the nuclear DNA content of *P. dactylifera* L. cv. Teggaza, and analyzed the stability for this factor in regenerated putative mutants. In this case, non-irradiated vitroplant of cv. Teggaza was used as a control for the vitroplants issued from irradiated of cv. Teggaza. The resulting values of the relative Nuclear DNA content of *P. dactylifera* L. cv. Teggazawas 1.68 ± 0.05 pg. However, the significant differences (ca. 60%) in genome size were found in some vitroplants issued from irradiated calli of cv. Teggaza (Fig. 25.18a, c). This is showing that at least some plants could be considered as a mutants induced by irradiation. Molecular approaches were used to understand the mechanisms within the process of regeneration by, genetic stability of plantlets and also to compare the ploidy level and polymorphism in vitro cultures of irradiated vitroplants and no-irradiated control vitroplants of the same variety. The establishment of screening and evaluation using flow cytometric and molecular markers allowed us the selection of date palm lines with good quality and disease resistance.

25.6 Conclusions and Prospects

Date palm plays considerable role socio-cultural economy. Mainly the date palm trees are very old and becoming more vulnerable to various biotic and abiotic constraints. Biotechnological techniques including induced mutation and in vitro selection propose a huge potential for the development of mutants exhibiting resistance to biotic stress in plant breeding programs. However, studies in date palm mutation induction are scarce. Somatic embryogenesis proved to be an ideal approach for mutation-based breeding programs. Selection systems to isolate tolerant lines have been designed using selection for resistance to culture filtrates of the fungal pathogen and toxins in vitro screening of calluses, selection of somatic embryos and regeneration of putative mutants; in vivo evaluation and confirmation of resistance initially in the greenhouse and finally in infested field, and selection of other agronomic useful traits. Flow cytometric and molecular markers were helpful in the selection and confirmation of date palm lines with good quality and Bayoud resistance. They are applied to big numbers of plant cells in culture to establish mutations and the successive regeneration of whole plants offers enormous prospect for dipping the cost of breeding programs.

The next important step is micropropagation of resistant mutants and their multilocation trials for evaluating resistance in contaminated fields to obtain Bayoud resistant high-quality palms. The development of new cultivars with excellent characteristics is a major target of date palm-breeding programs. Other approaches are useful and could be promoted for breeding programs such as genome editing with the CRISPR-Cas9 system, genetic manipulation, and the applications of high-through genome sequencing may develop date palm biotechnology. It possible to repopulate the palm groves already devastated by the disease, in the South West with Teggaza variety and in the center with the Deglet Nour variety. The use of the mutants obtained in the repopulation of the affected and threatened palm groves by Bayoud would contribute to oasis preservation, and increase in genetic diversity also they will contribute to the strengthening of the employment policy and to the guarantee food security.

Acknowledgments This research was financially supported by the National Institute of Agricultural Research (INRA) and FAO/IAEA Coordinated Research Project (RAF/5/035, RAF/5/049, AlG/5/023, AlG/5/024). The technical officers of these projects (Dr. Shri Mohan Jain, Dr. Madeleine Spencer and Dr. Ljupcho Jankuloski) are gratefully acknowledged.

References

Abed F, Saka H, Amara A, Kermiche A (1999) Effet de differentes doses de cobalt 60 sur les cals embryogènes de Deglet nour (*Phoenix dactylifera* L.). Recherche Agronomique 5:73–80

Abed F, Yatta D, Amara B et al (2014) Somatic embryogenesis from offshoot of different cultivars of date palm (Phoenix dactylifera L.) of three regions of southern Algeria. In: Farooq SA, Abed R, Baqir S (eds) Biotechnology and conservation of species from arid regions, vol 1. NOVA publishers, New York, pp 275–286

Abul-Soad AA, Jain SM, Jatoi MA (2017) Biodiversity and conservation of date palm. In: Ahuja MR, Jain SM (eds) Biodiversity and conservation of woody plants. Sustainable development and biodiversity, vol 17. Springer, Cham, pp 313–353

Ahloowalia BS, Maluszynski M (2001) Induced mutations – a new paradigm in plant breeding. Euphytica 118:167–173

Akter N, Islam MR (2017) Heat stress effects and management in wheat. A review. Agron Sustain Dev 37:37

Al Kharusi L, Dekoum VM, Assaha et al (2017) Screening of date palm (*Phoenix dactylifera* L.) cultivars for salinity tolerance. Forests 8:136

Al Mansoori TA et al (2007) Evaluation techniques for salt tolerance in date palm. Acta Hortic 736:301–306

Al-Enezi NA, Al-Bahrany AM, Al-Khayri JM (2012) Effect of x-irradiation on date palm seed germination and seedling growth. Emir J Food Agric 24(5):415–424

Al-Khateeb AA (2006) Role of cytokinen and auxin on the multiplication stage of date palm (*Phoenix dactylifera* L.) cv. Sukry. Biotechnology 5:349–352

Al-Khayri JM (2007) Micropropagation of date palm *Phoenix dactylifera* L. In: Jain SM, Haggman H (eds) Protocols for micropropagation of woody trees and fruits. Springer, Berlin, pp 509–526

Al-Khayri JM (2012) Determination of the date palm cell suspension growth curve, optimum plating efficiency, and influence of liquid medium on somatic embryogenesis. Emir J Food Agric 24(5):444–455

Al-Khayri JM, Al-Bahrany AM (2004) Genotype-dependent in vitro response of date palm (Phoenix dactylifera L.) cultivars to silver nitrate. Sci Hortic 99(2):153–162

Al-Khayri JM, Ibraheem Y (2014) *In vitro* selection of abiotic stress tolerant date palm (*Phoenix dactylifera* L.). Emir J Food Agric 26(11):921–933

Al-Khayri JM, Naik PM, Jain SM, Johnson DV (2018) Advances in date palm (*Phoenix dactylifera* L.) breeding. In: Al-Khayri JM, Jain SM, Johnson DV (eds) Advances in plant breeding strategies: fruits, vol 3. Springer, Cham, pp 727–771

Al-Mulla L, Bhat NR, Khalil M (2013) Salt-tolerance of tissue-cultured date palm cultivars under controlled environment. Int J Food Vet Agric Eng 7:468–471

Al-Rokibah AA, Abdalla MY, El-Fakharani YM (1998) Effect of water salinity on Thielaviopsis paradoxa and growth of date palm seedlings. J King Saud Univ Agric Sci 1:55–63

Al-Salih AA, Al-Najjar NR, Al-Mashhadani AN (1987) A study on the chromosome number of two specific female date palm cultivars. Date Palm J 5:134–143

Aly AA, Eliwa NE, Abd El-Megid MH (2019) Stimulating effect of gamma radiation on some active compounds in eggplant fruits. Egypt J Rad Sci Appl 32:61–73

Amin R, Laskar RA, Khan S (2015) Assessment of genetic response and character association for yield and yield components in lentil (*Lens culinaris* L.) population developed through chemical mutagenesis. Cogent Agric 31(1):1000715

Amraoui H, Sedra MH, Hamdaoui A (2004) Etude des sécrétions protéiques et enzymatiques du *Fusarium oxysporum* f. sp *albedinis:* agent causal de la fusariose vasculaire du palmier dattier. Al Awamia 109(110):47–62

Amraoui H, Lazrek HB, Sedra MH (2005) Chromatographic characterization and phytotoxic activity of *Fusarium oxysporum* f. sp. *albedinis* and saprophytic strain toxins. J Phytopathol 153:203–208

Ando A, Menten JOM, Tulmannn Neto A et al (1984) Obtençao de mutantes resistentes à fusariose en pimento-do-reino (*Peper nigrum* L.). In: Proceedings of the seminario regional sobre Tecnicas nucleares naproducao de plantasagricolaq, Piracicaba. CENA/USP, Sao Paulo, Brazil, pp 47–49

Asnaghi C, D'Hont A, Glaszmannand JC, Rott P (2007) Resistance of sugarcane cultivar R 570 to Puccinia melanocephala isolates from different geographic locations. Plant Dis 85(3):282–286

Atak C, Alikamanoğlu S, Açík L, Canbolat Y (2004) Induced of plastid mutations in soybean plant (*Glycine max* L. *Merrill*) with gamma radiation and determination with RAPD. Mutat Res 556(1–2):34–44

Attree SM, Moore D, Sawhney VK et al (1991) Enhanced maturation and desiccation tolerance of white spruce [*Picea glauca* (Moench) Voss] somatic embryos: effects of a non-plasmolysing water stress and abscisic acid. Ann Bot 68:519–552

Behnaz A, Zarghami R, Salari A (2018) The effects of agno3 and 2ip (n6-(2-isopentenyl) adenine) on different stages of somatic embryogenesis in date palm (*Phoenix dactylifera* L.) (Cv. medjool) Pakistan. J Bot 50:495–502

Berestetskiyis AO (2008) A review of fungal phytotoxins: from basic studies to practical use. Appl Biochem Microbiol 44(5):453–465. https://doi.org/10.1134/S0003683808050013

Bhaskaran S, Smith RH (1992) Somatic embryogenesis in date palm (Phoenix dactylifera L.), Forestry Sciences Chapter book series, vol 44–46. FOSC

Bouizgane B, Brault M, Pennarum AM (2004) Electrophysiological responses to fusaric acid of root hairs from seedlings of date palm susceptible and resistant to *Fusarium oxysrporum* f.sp *albedinis*. J Phytopathol 152:321–324

Bradshaw JE (2016) Mutation breeding plant breeding: past, present and future. Springer International Publishing, Switzerland. ISBN: 978-3-319-23284-3

Brar DS, Jain SM (1998) Somaclonal variation: mechanism and applications in crop improvement. In: Jain SM, Brar DS, Ahloowalia BS (eds) Somaclonal variationand induced mutations in crop improvement. Kluwer Academic Publishers, Dordrecht, The Netherlands, pp 15–38

Bressan RA, Singh NK, Handa AK et al (1985) In: Plant genetics-UCLA symposium on molecular and cellular biology. In: Freeling M (ed) , vol 35. Liss, New York, pp 755–769

Browne RA, Murphy JP, Cooke BM et al (2005) Evaluation of components of *Fusarium* head blight resistance in soft red winter wheat germplasm using detached leaf essay. Plant Dis 89: 404–411

Buffard-Morel J, Verdeil JL, Pannetier C (1992) Embryogenèse somatique du cocotier (Cocos nucifera L.) à partir d'explants foliaires: étude histologique. Can J Bot 70:735–741

Carlson PS (1973) Methionine sulfoximine-resistant mutants of tobacco. Science 180:1366

Casinovi CG (1972) Chemistry of the terpenoid phytotoxins. In: Wood RKS, Ballio A, Graniti A (eds) Phytotoxins in plant diseases. Academic, New York, pp 105–125

Chandra R, Kamle M, Bajpai A, Muthukumar M, Kalim S (2010) In vitro selection: a candidate approach for disease resistance breeding in fruit crops. Asian J Plant Sci 9(8):437–446

Chawla HS, Wenzel G (1987) Regeneration potential of callus from wheat and barley. Arch. Zuchtungforsch., Berlin 17(6):337–343

Chwendiman J, Pannetier C, Michaux-Ferriere N (1988) Histology of somatic embryogenesis from leaf explants of the oil palm *Elaeis guineensis*. Ann Bot 62:43–52

Devreux M, Magnien E, Dalschaert X (1986) Cellules végétales *in vitro* et radiations ionisantes. In: Nuclear techniques and *in vitro* culture for plant improvement. Proceedings of a symposium, Vienna 1985. International Atonic Energy Agency (AIEA/FAO), Vienna, pp 93–100

Doležel J, Greilhuber J, Suda J (1997) Estimation of nuclear DNA content in plants using flow cytometry. Nat Protoc 2(9):2233–2244

Donini B, Mannino P, Ancora A, Sonnino A et al (1991) Mutation breeding programmes for the genetic improvement of vegetatively propagated plants Italy. In: Plant mutation breeding for crop improvement, Proceedings of a Symposium, vol l. IAEA and FAO, Vienna, pp 237–255

El Fakhouri R, Lazrek HB, El Bahraoui E, Rochat H (1996) Preliminary investigation on a peptidic toxin produced by *Fusarium oxysporum* f.sp *albedinis*. Phytopathol Mediterr 35:11–15

El Hadrami H, El Idrissi-Tourane A, El Hasni M et al (2005) Toxin-based *in vitro* selection and its potential application to date palm for resistance to the Bayoud *Fusarium* wilt. Plant Biol Pathol C R Biol 328:732–744

El-Beltagi HS, Aly A, El-Desouky W (2019) Effect of gamma irradiation on some biochemical properties, antioxidant and antimicrobial activities of Sakouti and Bondoky dry dates fruits genotypes. J Rad Res Appl Sci 12:437–446

El-Hadrami A, Daayf F, Elshibli S et al (2011) Somaclonal variation in date palm. In: Mohan S, Al-Khayri JM, Johnson DJ (eds) Date palm biotechnology. Springer, New York, pp 183–203

Erskine W, Moustafa AT, Osman AE et al (2003) Date palm in the GCC countries of the Arabian Peninsula. Euphytica 118:167–173

FAO/IAEA-MVD (2021) Food and Agriculture Organization of the United Nations/International Atomic Energy Agency – mutant variety database. https://mvd.iaea.org/#!Search. Cited 30 Jan 2021

FAOSTAT (2018) Website of the Food and Agriculture Organization of the United Nations. Available online http://www.fao.org/faostat/en/data/QC. Accessed on 1 August 2018

Feldman AB, Leung H, Baraoidan M (2017) Increasing leaf vein density *via* mutagenesis in rice results in an enhanced rate of photosynthesis, smaller cell sizes and can reduce interveinal mesophyll cell number. Front Plant Sci 8:1–10

Foster PL (1991) In vivo mutagenesis. Methods Enzymol 204:114–125

Gawel NL, Jarret RL (1991) A modified CTAB DNA extraction procedure of Musa and Ipomoea. Plant Mol Biol Rep 9:262–266

Gawel NJ, Rao AP, Robacker CD (1986) Somatic embryogenesis from leaf and petiole callus cultures of Gossypium hirsutum L. Plant Cell Rep 5(6):457–459

Gesteira AS, Otoni WC, Barros EG, Moreira MA (2008) RAPD-based detection of genomic instability in soybean plants derived from somatic embryogenesis. Plant Breed 121(3):269–271

Hallerman E, Grabau E (2016) Crop biotechnology: a pivotal moment for global acceptance. Food Energy Secur 5(1):3–17

Hashmi G, Hashmi S, Sen S et al (2017) Polymorphism in heat shock protein gene (hsp70) in entomopathogenic nematodes (rhabditida). J Therm Biol 22(2):143–149

Hazzouri KM, Flowers JM, Nelson D et al (2020) Prospects for the study and improvement of abiotic stress tolerance in date palms in the post-genomics era. Front Plant Sci 11:293. https://doi.org/10.3389/fpls.2020.00293

Jain SM (2001) Tissue culture-derived variation in crop improvement. Euphytica 118:153–161

Jain SM (2005) Major mutation-assisted plant breeding programs supported by FAO/IAEA. Plant Cell Tissue Org Cult 82:113–123

Jain SM (2006) Radiation-induced mutations for developing Bayoud disease resistant date palm in North Africa. In: Zaid A (ed) Proceedings of the international workshop on true-to-typeness of date palm tissue cultured-derived plants. Al Ain, UAE, Plant Tissue Culture Laboratory, UAE University, pp 31–41

Jain SM (2010) Mutagenesis in crop improvement under the climate change. Rom Biotechnol Lett 15(2):88–106

Jain SM (2011) Radiation induced mutations for date palm improvement. In: Mohan S, Al-Khayri JM, Johnson DJ (eds) Date palm biotechnology. Springer, New York, pp 271–286

Jain SM (2012) *In vitro* mutagenesis for improving date palm (*Phoenix dactylifera* L.). Emir J Food Agric 24(5):386–399

Jain SM (2014) *In vitro* culture and mutagenesis in genetic improvement of date palm (*Phoenix dactylifera* L.). In: Proceedings of the fourth symposium on date palm in Saudi Arabia, King Faisal University, Alahsa, 5–8 May 2007, pp 693–705

Jain SM, Maluszynski M (2004) Induced mutations and biotechnology on improving crops. In: Mujib A, Cho M, Predieri S, Banerjee S (eds) *In vitro* applications in crop improvement: recent progress. IBH-Oxford, India, pp 169–202

Jain SM, Suprasanna P (2011) Induced mutations for enhancing nutrition and food production. Geneconserve 40:201–215

Jain SM, Ahloowalia BS, Veilleux RE (1998) Somaclonal variation in crop plants. In: Jain SM, Brar DS, Ahloowalia BS (eds) Somaclonal variation and induced mutations in crop improvement. Kluwer Academic Publishers, Dordrecht, The Netherlands, pp 203–218

Jatt T, Markhand GS, Rayburn L et al (2019) Karyotype variations among date palm (*Phoenix dactylifera* L.) cultivars of Sindh, Pakistan. Sarhad J Agric 35(3):708–718

Kole C, Muthamilarasan M, Henry R et al (2015) Application of genomics-assisted breeding for generation of climate resilient crops: progress and prospects. Front Plant Sci 6:563

Krishna H, Alizadeh M, Singh D et al (2016) Somaclonal variations and their applications in horticultural crops improvement. 3 Biotech 6:54

Louvet JJ, Toutain G (1973) Recherches sur les fusarioses VIII. Nouvelles observations sur la fusariose du dattier et précisions concernant la lutte. Ann Phytopathol 5:35–52

Ludwig AC, Hubstenberger J, Phillips F (1992) Screening of Allium tester lines in vitro with Pyrenochaeta terrestris filtrates. Hort Sci 27:166–168. https://doi.org/10.21273/HORTSCI.27.2.166

Maluszynski M, Szarejko I, Maluszynska J (2004) Mutation techniques. Encyclop Appl Plant Sci 1(3):186–201

Mba C, Rownak A, Souleymane B, Jain SM (2010) Induced mutagenesis in plants using physical and chemical agents. In: Davey MR, Anthony P (eds) Plant cell culture, essential methods. Wiley-Blackwell, Chichester, Sussex). ISBN: 978-0-470-68648-5, pp 111–130

Messiaen CM, Cassini R (1968) Recherches sur les fusarioses. IV. La systématique des *Fusarium*. Ann Epiphytes 19(3):387–454

Miler N, Jedrzejczyk I, Jakubowski S, Winiecki J (2021) Ovaries of Chrysanthemum irradiated with high-energy photons and high-energy electrons can regenerate plants with novel traits. Agronomy 11(6):1111

Mohammadzai IU, Ziarat S, Ihsan I et al (2010) Effect of gamma irradiation, packaging and storage on the nutrients and shelf life of palm dates. J Food Process Preserv 34:622–638

Mokhlisse N (1987) Contribution à l'identification et à l'étude de la toxicité des différents constituants de la toxine sécrétée par le *Fusarium oxysporum* f. sp. *albedinis*. Thèse de 3ême cycle, Faculty des Sciences-Semlalia Marrakech Morocco

Molot PM, Conus M, Cotta J, Ferrière H (1984) Mise en évidence d'une toxine dans les filtrats de culture de *Rhizoctonia violaceae*. Relation entre la sensibilité des plantules d'asperge à cette toxine et aux contaminations par mycélium. Rev Cytol Biol Végét Bot 7:205–217

Muhammad I, Raheem S, Al-Harrasi A et al (2020) Silicon and gibberellins: synergistic function in harnessing ABA signaling and heat stress tolerance in date palm (Phoenix dactylifera L.). Plants (Basel) 9(5):620

Murashige T, Skoog F (1962) A revised medium for rapid growth and bioassays with tobacco tissue cultures. Physiol Plant 15:473–497

Mussel HW (1972) Toxic proteins secreted by cotton isolates of *Verticillium albo-atrum*. In: Wood RKS, Ballio A, Graniti A (eds) Phytotoxins in plant diseases. Academic, New York, pp 443–445

Naik PM, Al-Khayri JM (2016) Somatic embryogenesis of date palm (*Phoenix dactylifera* L.) through cell suspension culture. In: Jain SM (ed) Protocols for in vitro cultures and secondary metabolite analysis of aromatic and medicinal plants, Methods in molecular biology, 2nd edn. Humana, New York, pp 357–366

Naik PM, Al-Khayri JM (2018) Cell suspension culture as a means to produce polyphenols from date palm (*Phoenix dactylifera* L.). Ciênc e Agrotecnologia 42(5):464–473

Naik PM, Al-Khayri JM (2020) Influence of culture parameters on phenolics, flavonoids and antioxidant activity in cell culture extracts of date palm (*Phoenix dactylifera* L.). Erwerbs-Obstbau 62:181–188

Nikam AA, Devarumath RM, Ahuja A et al (2015) Radiation induced *in vitro* mutagenesis system for salt tolerance and other agronomic characters in sugarcane (*Saccharum officinarum* L.). Crop J 3:46–56

Osbourn AE (2001) Tox-boxes, fungal secondary metabolites, and plant disease. Proc Natl Acad Sci U S A 98(25):14187–14188

Patade YV, Suprasanna P (2008) Radiation induced *in vitro* mutagenesis for sugarcane improvement. Sugar Tech 10:14–19

Penna S, Vitthal SB, Yadav PV (2012) *In vitro* mutagenesis and selection in plant tissue cultures and their prospects for crop improvement. Bioremed Biodivers Bioavailab 6:6–14

Pérez-Clemente RM, Gómez-Cadenas A (2012) In vitro tissue culture, a tool for the study and breeding of plants subjected to abiòtic stress conditions. In: Recent advances in plant in vitro culture, pp 91–108

Pinon J (1984) Propriétés biologiques de la toxine d'hypoxylonmammatum parasite des peupliers de la section Leuce. Rev Cytol Biol Végét Bo 7:271–277

Predieri S (2001) Mutation induction, tissue culture in improving fruits. Plant Cell Tissue Organ Cult 64:185–210

Pringle RB, Scheffer RP (1963) Purification of the selective toxin of *Periconia circinata*. Phytopathology 53:785–787

Quenzar B, Trifi M, Bouachrine B, Hartmann C et al (2001) A mitochondrial molecular marker of resistance to Bayoud disease in date palm. Theor Appl Genet 103(2–3):366–370

Raemakers CS, Schavemaker CM, Jacobsen E, Visser RG (1993) Improvements of cyclic somatic embryogenesis of cassava (Manihot esculenta Grantz). Plant Cell Rep 12:226–229

Rashid U, Anwar F, Ansari TM et al (2009) Optimization of alkaline transesterification of rice bran oil for biodiesel production using response surface methodology. J Chem Technol Biotechnol 84:9

Roychowdhury R, Tah J (2013) Mutagenesis a potential approach for crop improvement. In: Hakeem KR, Ahmad P, Ozturk M (eds) Crop improvement: new approaches and modern techniques. Springer, New York, pp 149–187

Saaidi M, Toutain G, Bannerot H, Louvet J (1981) La sélection du palmier dattier (*Phoenix dactylifera* L) pour la résistance au Bayoud. Fruit 4(36):241–249

Sabir JSM, Arasappan D, Bahieldin A et al (2014) Whole mitochondrial and plastid genome SNP analysis of nine date palm cultivars reveals plastid heteroplasmy and close phylogenetic relationships among cultivars. PLoS One 9:e94158. https://doi.org/10.1371/journal.pone.0094158

Saker MM, Adawy SS, Mohamed AA, EL-Itriby HA (2006) Monitoring of cultivar identity in tissue culture-derived date palms using RAPD and AFLP analysis. Biol Plant 50(2):198–204

Schaart JG, van de Wiel CCM, Lotz LAP, Smulders MJM (2016) Opportunities for products of new plant breeding techniques. Trends Plant Sci 21(5):438–449

Schwendiman J, Pannetier C, Michaux-Ferriere N (1988) Histology of somatic embryogenesis from leaf explants of the oil palm Elaeis guineensis. Ann Bot 62:43–52

Sedra MH, Lazrek HB (2011) *Fusarium oxysporum* f. sp. *albedinis* toxin characterization and use for selection of resistant date palm to Bayoud disease. In: Jain SM, Al-Khayri JM, Johnson DV (eds) Date palm biotechnology. Springer, Cham, pp 253–270

Sedra MH, El Fakhouri R, Lazrek HB (1993) Recherche d'une méthode able pour l'évaluation de l'effet des toxines secrétées par *Fusarium oxysporum* f. sp. *albedinis* sur le palmier dattier. Al Awamia 82:89–104

Sedra MH, El Fakhouri R, Lotfi F, Lazrek HB (1997) Activités des toxines secrétées par Fusarium oxysporum f. sp. albedinis, agent causal du bayoud du palmier dattier et d'autres formes spéciales du Fusarium oxysporum. Al Awamia 98:57–65

Sedra MH, Lazrek H, Lot F, Rochat H (1998) Fusarium oxysporum f. sp. albedinis toxin isolation and use for screening of date palm plants for resistance to the bayoud disesase. In: Proceedings of XXV International Horticultural Congress (IHC), 2–7 Aug 1998, Brussel, Belgium

Sedra MH, Lazrek HB, Amraoui H, Nour S (2008) Pathogen toxins of Bayoud disease on date palm: *in vitro* selection and biological and specificity activities. J Plant Pathol 90 (Suppl):53–57

Sen A, Alikamanoğlu S (2012) Analysis of drought-tolerant sugar beet (*Beta vulgaris* L.) mutants induced with gamma radiation using SDS-PAGE and ISSR markers. Mutat Res 738–739:38–44

Sianipar N, Ariandana F, Maarisit W (2015) Detection of gamma-irradiated mutant of rodent tuber (Typhonium Flagelliforme Lodd.) in vitro culture by RAPD molecular marker. Procedia Chem 14:285–294

Song HS, Lim SM, Widholm JM (1994) Selection and regeneration of soybeans resistant to the pathotoxic culture filtrate of *Septoria glycine*. Phytopathology 84:948–951

Suprasanna P, Jain SM (2017) Mutant resources and mutagenomics in crop plants. Emir J Food Agric 29(9):651–657

Suprasanna P, Jain SM, Ochatt SJ et al (2012a) Applications of *in vitro* techniques in mutation breeding of vegetatively propagated crops. In: Shu QY, Forster BP, Nakagawa H (eds) Plant mutation breeding and biotechnology. CABI International, Wallingford, pp 371–385

Suprasanna P, Vitthal SB, Yadav PV (2012b) *In vitro* mutagenesis and selection in plant tissue cultures and their prospects for crop improvement. Bioremed Biodivers Bioavail 6:6–14

Suprasanna P, Mirajkar SJ, Bhagwat SG (2015) Induced mutations and crop improvement. In: Bahadur B, Rajam VM, Sahijram L, Krishnamurthy KV (eds) Plant biology and biotechnology, Plant diversity, organization, function and improvement, vol I. Springer, India, pp 593–617

Surico G, Graniti A (1977) Produzione di tossine da Fusarium oxysporum Schl. f. sp. albedinis. Phytopathol Mediterr 16:30–33

Tisserat B (1982) Factors involved in the production of plantlets from date palm callus cultures. Euphytica 31:201–214. https://doi.org/10.1007/BF00028323

Twizeyimana M, Ojiambo PS, Tenkouano A et al (2007) Rapid screening of *Musa* species for resistance to black leaf streak using *in vitro* plantlets in tubes and detached leaves. Plant Dis 9: 308–314

Vargas-Segura C, López-Gamboa E, Araya-Valverde E et al (2019) Sensitivity of seeds to chemical mutagens, detection of dna polymorphisms and agro-metrical traits in M1 generation of coffee (*Coffea arabica* L.). J Crop Sci Biotechnol 22:451–464

Verdeil JL, Hocher V, Huet C et al (2001) Ultrastructural changes in coconut calli associated with the acquisition of embryogenic competence. Ann Bot 88:9–18

Viana VE, Pegoraro C, Busanello C, Costa de Oliveira A (2019) Mutagenesis in rice: the basis for breeding a new super plant. Front Plant Sci 10:1326

Wani MR, Kozgar MI, Tomlekova N (2014) Mutation breeding: a novel technique for genetic improvement of pulse crops particularly Chickpea (*Cicer arietinum* L.). In: Parvaiz A, Wani MR, Azooz MM, Lam-son PT (eds) Improvement of crops in the era of climatic changes. Springer, New York, pp 217–248

Widholm JM (1972) The use of fluorescein diacetate and phenosafranine for determining viability of cultured plant cells. Stain Techol 47:189–194

Xie GL, Mew TW (1998) A leaf inoculation method for detection of Xanthomonas oryzae Pv. Oryzicola from rice seed. Plant Dis 82:1007–1010

Yatta D, khelafi H (2012) Mise en oeuvre des biotechnologies: cals embryogenes irrradiés, suspensions cellulaires en vue d'amélioration de palmier dattier à la résistance au bayoud. Célébration de la manifestation du 50eme anniversaire de l'indépendance Nationale organisée par la DGRSDT (la Recherche Scientifique et le Développement Technologique). 07 à 20 juillet, 2012 au Palais des expositions des Pins maritimes à Alger

Yatta D, khelafi H, Abed (2004) Effect of gamma rays doses on the embryogenic calli of date palm (Phoenix dactylifera L.). Regional workshop organised by A.I.E.A / F.A.O sur «Cost effective up scale in vitro plant production and long – term storage of mutant plant material» Sfax (Tunisia) 29 November to 3 December 2004

Yatta El Djouzi D, Abed F, Amara B et al (2014) The genomic DNA structure in somatic embryogenesis of date palm (*Phoenix dactylifera* L.) by flow cytometric, RAPD and AFLP analysis, vol 1. NOVA Publishers, New York, pp 241–252

Yatta El Djouzi D, Al-hayri JM, Bouguedoura N (2020) Plant regeneration from cell suspension-derived protoplasts of three date palm cultivars. Propag Ornam Plants 20(1):28–38

Yatta D, Abed F, Amara B et al (2013) Protoplast isolation from cellular suspension of two Algerian cultivars of date palm (Takerbucht and Teggaza). Acta Hortic 994:323–329

Zhu JK (2002) Salt and drought stress signal transduction in plants. Annu Rev Plant Biol 53:247–273

Mutation Breeding in Tropical Root and Tuber Crops

26

A. V. V. Koundinya, Arindam Das, and Vivek Hegde

Abstract

Tropical tuber crops are a unique group of vegetables that can be grown on a wide range of soils from slightly acidic to saline and in climatic conditions from tropical to subtropical regions with limited inputs. But their improvement is limited mostly to clonal selection due to flowering and seed set related inhibitions. Very few of them like sweet potato flower regularly and facilitate the breeding through hybridization. Though cassava flowers in normal tropical conditions, their flowering is irregular and asynchronous. In few crops like arrowroot, Chinese potato and yam bean, the narrow genetic base hinders selection for new or improved traits. Hence, induced mutations could play a greater role in the improvement of these crops. The vegetative propagation nature of these crops except yam bean facilitates the easy fixing of characters. Attempts have been made for the improvement of these crops through induced mutations all over the world. A few varieties have been released in crops like sweet potato, taro, and cassava through mutation breeding. The mutation breeding in tropical root and tuber crops in India was initiated in 1960s at ICAR-Central Tuber Crops Research Institute. Highly six mutant varieties were released in sweet potato all over the world followed by taro (3 varieties) and cassava (2 varieties). In India, the first mutant variety in these crops was released in Chinese potato (*Suphala*).

A. V. V. Koundinya (✉) · V. Hegde
Division of Crop Improvement, ICAR-Central Tuber Crops Research Institute, Thiruvananthapuram, Kerala, India
e-mail: koundinya.avv@icar.gov.in

A. Das
Department of Vegetable Science, Uttara Banga Krishi Viswavidyalaya, Cooch Behar, West Bengal, India

S. Penna, S. M. Jain (eds.), *Mutation Breeding for Sustainable Food Production and Climate Resilience*, https://doi.org/10.1007/978-981-16-9720-3_26

Keywords

Induced mutations · Mutant varieties · Mutation breeding · Tuber crops · Vegetative propagation

26.1 Introduction

By definition "a mutation is a sudden heritable change in the genetic constitution of an organism apart from segregation and recombination." Mutation breeding of crops implies that utilization of physical and chemical mutagens to alter the genetic constitution of plants, thereby performing selection for desirable traits in the subsequent progenies. Since L.J. Stadler used X-ray irradiation in barley in the 1920–1930s, induced mutations have been become an option for plant breeders all over the world to induce new variations or to broaden the existing genetic base of crop species (Beyaz and Yildiz 2017).

Genetic variation is the important prerequisite for effective selection to occur (Koundinya et al. 2013; Koundinya and Dhankhar 2013; Sidhya et al. 2014). The plant breeders mostly aim at the creation of new variations through recombination and segregation for selecting the most superior genotypes for any given trait. They often try to combine two or three desirable mono/oligogenic traits into a single genotype through combination breeding or pooling QTLs of a polygenic trait. This requires selection in the segregating generations that are derived by hybridizing the desirable parents. The method and protocol of hybridization depend upon the breeding system of the crop. Mutation breeding is useful in cases where there exists no variation for a particular trait or the available variation is very narrow. Hence, it is highly useful in crops where the genetic base is narrow and the utilization of available variation is not possible due to hybridization barriers such as sterility, incompatibility, and heterogamy.

The first commercial mutant variety, *Chlorina F1* was developed in Indonesia in Tobacco in 1929 (Solanki et al. 2015). A total of 3347 mutant varieties have been registered with the Mutant Variety Database of the International Atomic Energy Agency as of now. In India, the first mutant variety registered was *M.A.9* in cotton in 1948 and the latest mutant is *Trombay Chhattisgarh Dubraj Mutant-1* in rice in 2019. A total of 341 mutant varieties were registered from India in various crops so far (https://mvd.iaea.org).

26.2 Mutation Breeding in Tropical Root and Tuber Crops

Tropical root and tuber crops are mostly propagated by vegetative means such as cassava by stem cuttings, sweet potato by vine cuttings, yams by tubers, and aroids by corms/cormels. Very few of them like sweet potato flower regularly and facilitate the breeding through hybridization. Though cassava flowers in normal tropical conditions, their flowering is irregular and asynchronous (Hegde et al. 2019).

Moreover, there exists sterility in cassava and yams (Koo and Ruiz 1973). Very rare or nil flowering occurs in aroids like *Colocasia* and elephant foot yam. Often the crosses in these crops result in smaller number of seeds or no seeds. Besides, a highly heterozygous nature and high inbreeding depression result in lacking much breeding efforts in tuber crops. Such difficulties in the introgression of beneficial and desirable genes into one genotype through hybridization minimize the possibility of developing new superior cultivars. Hence, the genetic improvement in these crops is restricted to the collection, evaluation, and clonal selection of superior genotypes. The improvement through simple selection mainly operates on the available genetic variation with less possibility of creation of new variations. So, there is a need to try methods other than hybridization breeding such as mutation breeding, recombinant DNA technology and gene editing tools which can facilitate the creation of new variations and combination of traits. The added advantage of mutation breeding over genetic engineering is that it does not introgress alien genes and does not create negative impact on environment (Kharkwal 2012).

Mutation breeding has certain advantages over conventional hybridization breeding. It requires only one parent and it eliminates the laborious emasculation and pollination. It does not need the source material for the improvement of any particular trait. Since the starting material in mutation breeding is an already adapted and local popular variety, thus generated mutants will probably have the same genetic background except for few mutated genes. Hence, they will be readily adaptable to that particular location and be readily available for cultivation (Suprasanna and Nakagawa 2012). Mutation breeding is highly laborious which demands thorough and judicial screening of a huge number of mutant plants/populations for the selection of promising mutants for further releasing as varieties or using as parents in future breeding programs. Since the mutagens randomly act on the genome, it becomes crucial to identify positive mutants of interest. The differences between hybridization breeding and mutation breeding are presented in Table 26.1.

Any mutation breeding scheme involves the following steps:

1. Induction of mutation either through physical or chemical mutagens
2. Screening of the mutant population for one or a few traits of interest such as earliness, pest and disease resistance, abiotic stress tolerance, improvement in nutritional quality and post-harvest traits
3. Selection and evaluation of putative mutants for stability

Dominant mutants can be identified in the first year/season, i.e., M_1 generation. But the heterozygous dominant will segregate in the M_2 generation upon selfing of M_1. So, homozygous dominant mutants are to be identified by progeny testing. Similarly, homozygous recessive mutants will be seen in the M_2 generation.

Table 26.1 Differences between hybridization breeding and mutation breeding

Hybridization breeding	Mutation breeding
Creation of variation by crossing two genetically dissimilar parents	Creation of variation by treating with mutagens
The purpose is to combine desirable genes from one or more sources	The purpose is to create new variations that are not present in the existing population or to widen the existing variability
Trait specific	Trait non-specific
Variation occurs in the F_2 generation	Variation occurs in M_1 generation
The selection starts from the third year, i.e., F_2 generation after crossing the two homozygous lines.	The selection for dominant mutations starts from the M_1 generation and for recessive mutants in the M_2 generation
Works on the variation within the limit of existing variation	New variations can occur
Time-consuming	Comparatively less time consuming
Specific direction	Directionless
Less laborious	Laborious
The probability of success is more	The probability of success is less, majority mutations are deleterious

Table 26.2 Differences between seed propagated and vegetatively propagated crops for mutation breeding

Seed propagated species	Vegetatively propagated species
Plant material used is seed or pollen or unicellular starting material	Plant material is meristem tissue mostly apical or axillary buds
Next-generation proceeds sexually	Next-generation proceeds vegetatively
Further variation occurs due to selfing, i.e., recombination and segregation	The mutation is the only source of variation
The mutagenic generations are indicated as M_1, M_2, M_3	The mutagenic generations are indicated as M_1V_1, M_1V_2, M_1V_3
No problem with Chimeras	Chimeras are the major problem
Selection starts from M_1 or M_2 generation	Actual selection starts from M_1V_4 after elimination of chimeras
Comparatively consumes less time	Comparatively consumes more time

26.3 Mutation Breeding Methods in Tropical Root and Tuber Crops

Mutation breeding in vegetatively propagated species such as cassava, sweet potato, taro, and elephant foot yam differs from the seed propagated crops (Table 26.2). In seed propagated crops, in general, seeds are used for mutagen treatment but occasionally irradiated pollen is also used for pollination. Since, seeds contain single-cell zygote; whatever changes occur in the genetic constitution through mutagen treatment will be the same in the entire resulting plant body. But in vegetatively

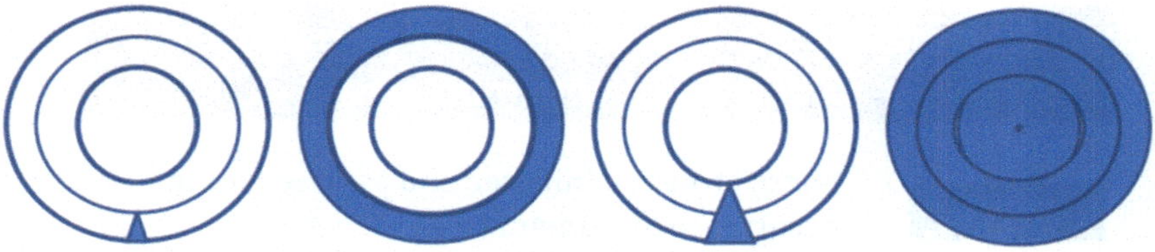

Fig. 26.1 Various types of chimeras in the cross-section of a stem

propagated crops, the meristematic tissues such as apical or axillary buds are used for mutagenic treatment. Since they are multicellular, the different cells in the tissue tend to have different types of mutations. Mutations that occur in meristem will produce a lineage of daughter cells, thereby mutated cells which further develop into large mutated sectors called chimeras (Suprasanna and Nakagawa 2012) which necessitate the thorough examination and elimination of chimeras.

Elimination of chimeras consumes a lot of time of a minimum of three generations. Therefore, the actual selection in vegetatively propagated crops starts after the elimination of chimeras. This makes the identification and propagation of successful solid or homohistant mutants (Fig. 26.1) from the vast population tedious especially in these plants. Periclinal chimeras (Fig. 26.1) are probably stable and they also can be selected if found good. These chimeras can effectively be eliminated by applying the mutagenic treatment to single cells of leaves, stolons, and scales that can form adventitious buds at a later stage (Koo and Ruiz 1973). Sometimes, treating the stem cuttings or tubers before sprouting can also be fruitful for avoiding the chimeras. For this purpose, in crops like cassava and sweet potato, stakes and vine cuttings can be treated with mutagens immediately after harvesting. In crops like yams and aroids, treating the tuber pieces before initiation of sprouting would be beneficial. The senior author observed that treating the cassava stakes after initiation of sprouting resulted in 100% mortality even at lower doses of gamma irradiation. In crops like cassava and sweet potato where fruit and seed set are not much problem, collection of open-pollinated seeds to raise M_2 generation from mutant clones in M_1V_1 would eliminate the chimera problem and also facilitate the more genetic recombination apart from mutagenic changes. However, the biggest advantage in vegetatively propagated crops is once selected, the mutated clone is fixed. The general scheme for mutation breeding in vegetatively propagated crops is given below (Fig. 26.2). In crops like cassava and sweet potato where occasional flowering and seed set is possible, use of irradiated seeds and pollen avoids the problem of chimeras. Abraham (1970) irradiated the seeds and Moh and Alan (1972) used irradiated pollen in cassava to induce variation.

The chimera problems in vegetatively propagated crops can also be resolved effectively by adopting in vitro mutagenesis (Fig. 26.3) techniques to mutate plant material and allowing the regeneration of a large number of plants within a short period. The in vitro mutagenic treatment and subsequent subculturing of plants

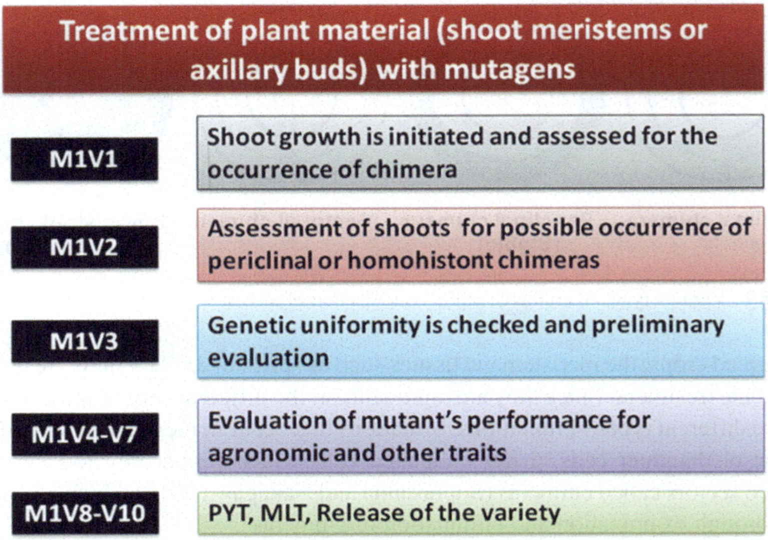

Fig. 26.2 General scheme for mutation breeding in vegetatively propagated crops

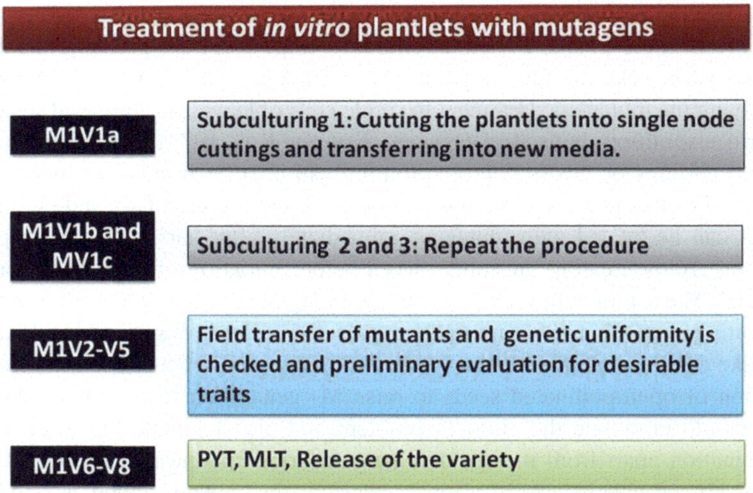

Fig. 26.3 In vitro mutagenesis in vegetatively propagated plants

facilitate the early elimination of chimeras within 9–10 months. Moreover, in vitro mutations facilitate the mutagenic treatment over large populations, selection and cloning of selected variants in a short time on a large scale throughout the year (Suprasanna and Nakagawa 2012).

26.3.1 Cassava

Cassava (*Manihot esculenta* Crantz), belongs to the family *Euphorbiaceae*, is one of the most important tropical root and tuber crops grown across the tropical regions of the world from South America to South East Asia through Sub-Saharan Africa (Koundinya et al. 2018). Cassava was introduced to India during the seventeenth century from Africa through human migrations by the Portuguese. Cassava can be grown on marginal soils with limited inputs and has diverse application in food, feed and industry (Pushpalatha and Gangadharan 2020). Though India occupies 28th position in the world in terms of area under cultivation of cassava, its average productivity is more than double the global average. India occupies the 15th position in total production and 4th position in average productivity in the world. The area under cassava cultivation globally was 27.8 mha resulting in total production and average productivity of 308.5 mt and 12.16 t/ha, respectively, in 2019. The area under cultivation of cassava in India is 16.3 mha with total production and average productivity of and 4.9 mt and 30.5 t/ha in 2019 (FAOSTAT 2021).

Cassava is largely cultivated for human consumption in Kerala and industry in Tamil Nadu states of India. Besides, it is also grown in Andhra Pradesh, Assam, Karnataka, Madhya Pradesh, Pondicherry, Nagaland, Tripura, Mizoram, and the Andaman-Nicobar Islands. Cassava roots are rich in carbohydrates and produce more calories with minimum inputs. The storage roots have edible and commercial importance. They are also an important part of feed in animal husbandry, fish, and poultry farms (Abraham et al. 2006; George et al. 2011). Moreover, the starchy roots also have industrial applications such as extraction of starch and starch-derived products such as sago, textile, alcohol, and high fructose-glucose syrups (Joseph et al. 2004; Yan et al. 2013). The cassava starch is pure as it is free from contamination of proteins and lipids. Nowadays, cassava starch is utilized in the manufacture of biodegradable plastic along with processed products like baby food, vermicelli, chips, and papads (Abraham et al. 2006).

Cassava is being suffered from several production problems like lack of sufficient good quality planting material, less seed multiplication ratio, nutritional deficiencies when grown on marginal soils and Cassava Mosaic Disease (CMD). The starchy roots of cassava contain low protein, vitamin, and mineral nutrients; also contain cyanogenic glucosides which impart a bitter taste. Besides, Post-harvest Physiological Deterioration (PPD) greatly reduces its shelf life (Iglesias et al. 2002). CMD, caused by Cassava Mosaic Virus belongs to the genus Begomovirus of the family *Geminiviridae*, which is one of the major diseases affecting the growth and yield (Asare et al. 2014). In a survey by Jose et al. (2011), cassava mosaic disease incidence in Kerala was estimated at 44.5–96.75%. The tuber yield loss due to CMD was estimated from 10% to 90% with an average of 30% (Thankappan and Chacko 1976). Among other diseases of cassava, Anthracnose caused by *Colletotrichum* species is seen during the rainy season. Root rot is another problem faced by cassava farmers in Africa. In the summer season or dry weather, cassava faces a severe problem from mealybugs and red spider mite. Mealybugs suck the sap from the apical shoots, thereby causing stunting, reduced intermodal length, and

Fig. 26.4 Cassava mosaic disease (**a**), mealybug (**b**) and red spider mite (**c**) infestation

rosette appearance in the top of the shoot. Mealybug infestation in the young plants inhibits the growth (Fig. 26.4). Cassava tubers suffer from vascular blue-lining or blue-streaking, a post-harvest physiological disorder that limits the storage of cassava tubers at ambient conditions. Cassava tubers can be stored for a maximum of up to 2 days without this disorder at ambient room conditions. The estimated losses of fresh cassava roots due to PPD are nearly 1/3rd of total harvest worldwide (Saravanan et al. 2016). Cassava is said to be drought tolerant when compared with other major crops at the cost of tuber yield, dry matter and starch content of tubers (Koundinya et al. 2018, 2021). Water deficit stress at the time of planting adversely affects the sprouting percent and field establishment of plants. Moreover, longer crop duration limits cassava production to one crop per year. Hence, the breeding objectives of cassava include the developing of varieties with the following traits:

- Early bulking and short duration
- High vigor
- Higher Nutrient Use Efficiency (NUE)
- Resistance to Cassava Mosaic Disease (CMD)
- Resistance to mealybug, whitefly, and red spider mite
- Tolerance to abiotic stresses: drought, salinity
- Dual Purpose varieties (Culinary + Processing)
- Tolerance to Post-Harvest Physiological Deterioration (PPD)
- Low HCN content, high dry matter and starch
- Fit for cropping systems

The long cropping season of 10 months along with protogyny and asynchronous flowering is hindering and delaying the conventional breeding in cassava (Hegde et al. 2016, 2019). Moreover, some clones do not produce flowers. Often the crosses result in smaller number of seeds or no seeds. Besides, a highly heterozygous nature and high inbreeding depression (Kawano et al. 1978) result in lacking much breeding efforts in cassava. Hence alternative ways to increase genetic variability are desirable. In such cases, mutation breeding can be an option for genetic improvement. Cassava mutation breeding has been explored in recent decades with few published papers.

26.3.1.1 Mutation Breeding Achievements So Far

The mutation breeding in cassava in India was initiated in the 1960s at ICAR-Central Tuber Crops Research Institute based on the first publication made by Vasudevan and others in 1967. Vasudevan and Jos (1988) suggested that the use of single-node stem cuttings followed by pruning at 5–6 months age facilitates the greater recovery of solid mutants in cassava. Induced mutation breeding in cassava has been resulted in the production of novel genotypes, like high amylase and small granule mutants and tolerance to Cassava Mosaic Disease and Post-harvest Physiological Deterioration. These achievements suggested that mutagenesis could be a viable option for cassava breeding (Yan et al. 2013). So far only two cassava varieties were registered in IAEA Mutant Variety Database namely *Tekbankye* and *Fuxuon 1* (https://mvd.iaea.org/). The mutation breeding progress in cassava reported across the globe so far is presented in Table 26.3.

26.3.2 Sweet Potato

Sweet potato [*Ipomea batatas* (L.) Lam] is one of the important tropical tuber crops that belongs to the family *Convolvulaceae* (Martin and Cabanillas 1966; Ngailo et al. 1991). It is native to South America and a hexaploidy species having a chromosome number of $2n = 6X = 90$ (Srisuwan et al. 2006). Sweet potato is a dicotyledonous plant grown across different agro-climatic zones such as tropical, subtropical, and frost-free temperate climatic conditions. It can grow at altitudes ranging from sea level to 3000 m (Ngailo et al. 2013).

The sweet potato rank seventh among the top ten food crops in the world in terms of volume of production (Clark et al. 2012). It is grown in 108 countries with 66% production concentrated in Asia and 28% in Africa. The global total area under sweet potato cultivation is 10.1 mha with total production and average productivity of 143.8 mt and 11.32 t/ha in 2019. India produced 1.15 mt of sweet potato tubers from 01.1 mha with average productivity of 10.5 t/ha in 2019 (FAOSTAT 2021). In India, sweet potato is grown widely in the subtropical eastern parts of India where it is grown for consumption and is popularly referred to as "Poor man's rich food" providing adequate nutrition at a cheaper cost compared to other staples. In India, the major sweet potato-producing states are Odisha, West Bengal, Maharashtra, and Andhra Pradesh.

Sweet potato storage root has nutrients such as carbohydrates, dietary fiber, vitamins A and C, and minerals. It is a wide spectrum of variability in skin and

Table 26.3 Brief report on mutation breeding achievements in cassava

S. No.	Organization/country	Report	References
1	ICAR-Central Tuber Crops Research Institute, Thiruvananthapuram, India	• A short petiole cassava mutant was isolated from the cultivar H-165 • Chlorophyll mutant had noticeably high HCN content in the leaves and roots	Nayar and Rajendran (1985)
2	ICAR-Central Tuber Crops Research Institute, Thiruvananthapuram, India	• Good peelable (28), medium peelable (20) and poor peelable (10) mutants were observed • Huge variation was observed for proline content 16–61 μg/g during dry period among gamma-irradiated mutants of cassava in cassava • Variation was also observed for CO_2 uptake rate and stomata ratio of upper surface to lower surface among mutants of cassava	Vasudevan et al. (1993)
3	ICAR-Central Tuber Crops Research Institute, Thiruvananthapuram, India	• CMD symptom-free mutants (18) were observed at 4 months after planting among 1040 mutants of M_4	Vasudevan et al. (1995)
4	ICAR-Central Tuber Crops Research Institute, Thiruvananthapuram, India	• Variegated leaf mutant (CAM-61) and non-tuber forming mutant (CAM-3) were identified	Devarajan and Vasudevan (1997)
5	ICAR-Central Tuber Crops Research Institute, Thiruvananthapuram, India	• The highest tuber yield was noticed in the mutant SJ 30–14 (9.7 kg/plant) followed by SJ 15–34 (8.46 kg/plant) which was an 82% and 58.6% improvement over the control, Sree Jaya (5.3 kg/plant) • The mutants SJ 15–14 (41.17%) and SJ 15–27 (40.03%) had more dry matter than control, Sree Jaya (26.76%) • Tuber rind color changed to cream (10 mutants) and yellow (10 mutants) from the original pink color (Fig. 26.5) • Dark cream or yellow color flesh was found in 8 mutants and the remaining had creamish-white color flesh similar to the parent, Sree Jaya (Fig. 26.6) • Huge variation in leaf size, shape, petiole color, petiole length and cassava mosaic	Koundinya et al. (2019, unpublished data)

(continued)

Table 26.3 (continued)

S. No.	Organization/country	Report	References
		disease resistance was observed in the M_2 generation of mutants of Sree Jaya and H-226	
6	Biotechnology and Nuclear Agricultural Research Institute Ghana Atomic Energy Commission, Ghana	• In vitro and in vivo induced mutations in cassava cultivar, *Bosom Nsia* were tried for Cassava Mosaic Disease resistance and isolated four variants resistant to African Cassava Mosaic Virus	Ahiabu and Klu (1997)
7	Department of Crop Science, University of Science and Technology, Kumasi, Ghana	• A mutant cassava cultivar "Tekbankye" (ISU-W) with large-size starch granules, improved cooking quality and resistance to African Cassava Mosaic Virus (ACMV) was developed by radiation treatment (25 Gy) of stakes of *Isunikakiyan*—a local Nigerian cassava cultivar	Asare and Safo-Kantanke (1997), Yan et al. (2013)
8	National Root Crops Research Institute, Nigeria	• Huge variation was created through induced mutagenesis for dry matter, starch, and HCN content in M_1V_2 generation by gamma irradiation of three local cassava cultivars • Identified 14 mutants with low root HCN content and 7 mutants with high dry matter content	Nwachukwu et al. (1997)
9	CIAT, Columbia	• Isolated the mutant lines with smaller starch granules and high amylase content in an M_2 generation by treating the botanical seeds of cassava with gamma radiation	Ceballos et al. (2008)
10	CIAT, Columbia	• Four different root morphotypes were identified viz., small granule and high-amylose starch, "hollow" starch granule, starchless, and tolerance to PPD when 1400 true cassava seeds collected from different full or half-sib families were subjected to irradiation with 200 Gy	Sanchez et al. (2009)
11	University of the Free State, South Africa	• Identified mutants with high (26.8–32.7%) and low amylose contents (11.7–14.0%) four in	Amenorpe (2010)

(continued)

Table 26.3 (continued)

S. No.	Organization/country	Report	References
		each category. • A putative free-sugar mutant was also identified when the stakes of four landraces were treated with 35 Gy radiations	
12	Corporacion Colombiana de Investigacion Agropecuaria (Corpoica), Codazzi (Colombia)	• Mutants with improved granule and paste properties and less post-harvest physiological deterioration were identified in the M_2 population	Tofino et al. (2011)
13	China	• Another mutant cultivar Fuxuan 01 has been developed by irradiating the variety SC124 in China in 2005	Yan et al. (2013)
14	Bogor Agricultural University (IPB), Jl. Meranti Kampus IPB Darmaga Bogor, Indonesia	• Several high-yielding cassava mutant lines (storage root yield >10–20 kg/plant) in the M_1V_1 generation and two mutant lines with high starch content (>39%) were found in cassava	Khumaida et al. (2015)
15	Bogor Agricultural University (IPB), Jl. Meranti Kampus IPB Darmaga Bogor, Indonesia	• Gamma-irradiated 9 putative mutants of two cassava cultivars Jame-Jame and Adira-4 were identified based on weight and number of economic tubers • Out of these nine, six potential mutants namely: V5D1-(2), V5D1-2(2), V5D1-3 (2), V5D1-4(3), V5D2-2(2), and V5D2-6(2) were found to be stable based on the trait namely first branching height, tuber weight, the total number of tubers and the total number of economic tubers	Khumaida et al. (2017)
16	The National University of Singapore, Singapore	• Unfortunately got a 17 and 60 fold reduction in the tuber yield in S14 and S15 mutant lines, respectively • S15 mutants also had a significant reduction as 50% in starch and 30% in amylose content	Joseph et al. (2004)

flesh color of the storage roots starting from white to yellow, orange, and deep purple. The Orange flesh sweet potato is known to have high beta carotene content (Provitamin A) and is widely promoted to combat malnutrition problem in

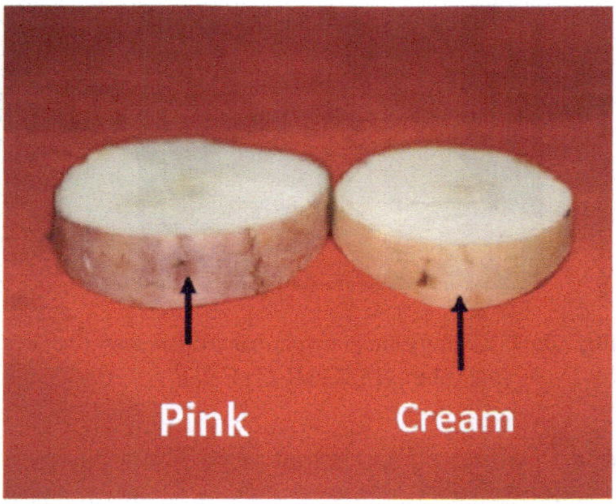

Fig. 26.5 Change in rind color from pink (Control-Sree Jaya) to Cream (Mutants)

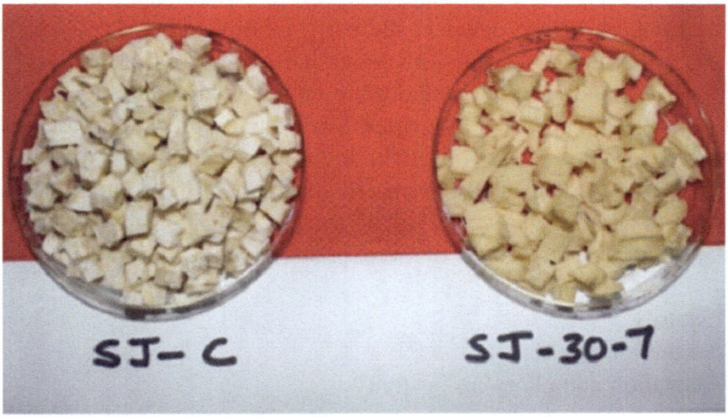

Fig. 26.6 Change in flesh color from Creamish White (Control-Sree Jaya) to Yellow (Mutants)

Sub-Saharan Africa. The purple color sweet potato tubers are rich in anthocyanins which are potent antioxidants. Sweet potato can grow well in marginal soils with a limited nutrient supply and is tolerant to severe weather conditions such as cyclones, drought, and salinity. Sweet potato is known as a famine-relief crop especially in the Kandhamal district of Odisha and it has recently been in the forefront as a health-promoting food crop. Sweet potato roots can be used for diverse applications, such as table use, processed foods like puree, dehydrated functional foods, natural colorants, chips, pasta, noodles, and alcohol. Starchy roots are also used for starch production and animal feed. Sweet potato vines are used as fodder for cattle.

There is a wide variation for sweet potato flowering. Some clones flower profusely, some does not and some produce few flowers (Jones 1980; Huaman 1992). Short days with warm weather are required for the best flowering of sweet potato. Grafting, griddling, and day-length manipulation are required for flowering in temperate regions. Hand-pollinated capsules have fewer seeds (1–2) than naturally

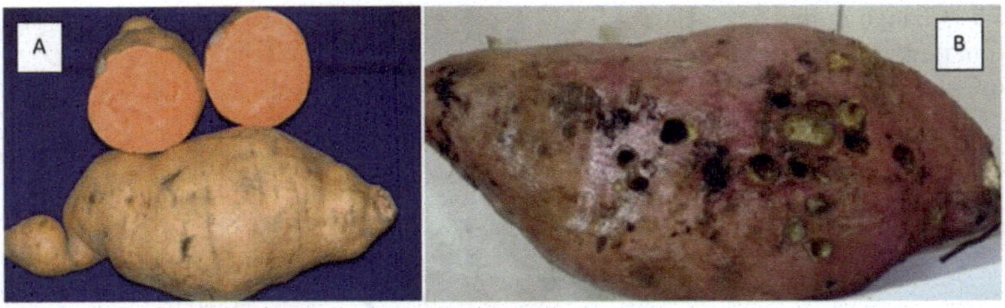

Fig. 26.7 Orange flesh sweet potato (**a**) and sweet potato weevil infestation (**b**). (Photo provided by Dr J. Sureshkumar, ICAR-CTCRI)

open-pollinated seeds that have 3–4 seeds per capsule (Wilson et al. 1989). Very low percent fruit and seed set have been observed in hand-pollinated crosses of sweet potato. This low success rate could be attributed to the self or cross incompatibility contributed by the hexaploidy genome of the crop (Ngailo et al. 2013). Anthesis in sweet potato occurs before dawn in the early morning after 2.00–3.00 AM and flowers start withering after dawn. This necessitates the hybridization work in sweet potato tedious. So, the breeding methods are mostly confined to clonal selection or selection in the open-pollinated progenies. Sweet potato breeding efforts have been made mainly for high storage root yield and quality in India so far. For the past 4–5 years, breeders have actively promoted schemes for developing cultivars suitable for food processing and weevil resistance.

The major breeding objectives of sweet potato include

- Early maturity
- High storage root yield
- Wide adaptation and photo-insensitivity
- Quality storage roots with uniform shape and size
- High tuber weight
- High dry matter/starch content
- High β carotene and anthocyanin content
- Processing and nutritional quality
- Tolerance to drought and salinity stress
- Resistance to sweet potato weevil and viruses

Like other crops, sweet potato also has several production problems. Tuber development in sweet potato is sensitive to photoperiod and other weather parameters. Summer grown sweet potato faces water deficit stress thereby reduction in yields. Water deficit stress conditions immediately after planting affect the establishment of the crop in the field resulting in a lower plant population. Sweet potato weevil (*Cylas formicarius*) is another menace that drastically reduces the marketable tuber yield (Fig. 26.7). In India, it was reported that damage on tubers due to weevil infestation in sweet potato increased at a higher rate of 71% during the summer months from February to May when compared to the rainy season (45%) from June to September (Rajamma 1983; Kyereko et al. 2019). In the temperate world, sweet potato viral diseases caused by the potyvirus complex incur losses from 20% to 40% (Clark et al. 2012). In Africa, Sweet potato viral diseases mainly chlorotic stunt virus and feathery mottle virus cause yield losses from 66% to 90% (Karyeija et al. 1998). Unfortunately, no success has been found so far for the improvement of weevil and other disease resistance traits. There is no known compatible source of resistance exists in the available germplasm. Hence, mutation breeding becomes handy in the creation of variation for weevil resistance, photo-insensitivity and other important traits.

26.3.2.1 Mutation Breeding Achievements So Far

Miller (1935) was the first to report X-ray-induced mutations in sweet potato. The first sweet potato mutant variety was registered with the IAEA mutant variety database in 1986. So far, six mutant varieties of sweet potato were released as per the IAEA mutant variety database. Out of six varieties, five varieties (*91-C3-15, Wanshu-S 367, Yanshu 759, Yanshu 781 and Yushu 5*) were released in China and one variety *Sweet Garden* was released in Japan (https://mvd.iaea.org/). Other important achievements are presented below (Table 26.4).

26.3.3 Yams

Yams are the largest group of tropical tuber crops that belong to the genus *Dioscorea* of the family *Dioscoreaceae*. There are mainly three yams, viz. Greater yam (*Dioscorea alata*), lesser yam (*Dioscorea esculenta*), and white yam (*Dioscorea rotundata*) are being cultivated and consumed largely. There are other yam species like aerial yam (*Dioscorea bulbifera*) which produces aerial bulbils in the leaf axils, *Dioscorea pentaphylla* and *Dioscorea stersicus* often found. Yams are the third most important tropical tuber crops after cassava and sweet potato. Different yam species are said to be originated in different parts of the world. Greater yam or water yam was first cultivated in Assam or Burma (Washaya et al. 2016). Total gross yams production worldwide in 2019 was 74.3 mt from 8.9 mha with average productivity of 9.33 t/ha. Nigeria is the largest producer of yams with a total production of 50 mt followed by Ghana with 8.3 mt. The total production of yams in India in 2018 was 0.75 mt from an area of 0.03 mha with average productivity of 28 t/ha, which was much higher than the global average (FAOSTAT 2021).

Yam tubers are rich in carbohydrates which are a good source of energy. They are also a good source of crude protein, ash, and dietary fiber especially for people who live in rural, marginal, and tribal areas. They also contain minerals like phosphorous, calcium, manganese, potassium, copper, sodium, and zinc (Baah et al. 2009). Some yam clones produce tubers in purple color, which are rich in anthocyanin compounds known as potential antioxidants.

Yam's cultivation also has certain production problems. The growth habit of yams is a vine. It requires staking of vines in the field, which is a laborious and additional burden to the farmer. Moreover, the bulky nature of planting material increases the storage and transportation cost. The crop duration is 10–12 months

Table 26.4 Brief report on mutation breeding achievements in sweet potato

S. No.	Report	Reference
1	• X-ray-induced mutants of the parent *Porto Rico* had creamy white flesh color than the orange-yellow to salmon-colored flesh of the parent. Moreover, induced mutants have more sweetness.	Miller (1935)
2	• X-ray irradiation of the early variety P-30 resulted in the tubers with sport mutations with variegated leaves. • One mutant plant with early maturity and significantly higher yield than the original P-30 variety was found. Another mutant with light-red skin was also found and that was different from the original white.	Cheng (1958)
3	• Twelve mutants exhibited variation in stem and tuber color. The most prominent changes in the stem were long and short, thick and slender stems. Similarly, tuber color variants with yellow-white or purple-red tubers instead of the parental brown-red color was found.	Masima and Sato (1959)
4	• White sectorial mutations were found when roots of three varieties viz., *Unit I Porto Rico, Goldrush,* and *Heartogold* were exposed to gamma irradiation.	Hernandez et al. (1959)
5	• Yield differences were observed among mutated clones. Two clonal irradiation groups had a higher yield than check and one group had a lower yield than the check group.	Poole (1959), Poole and Tanaka (1963)
6	• Mutant clones with short vine length had lower tuber yields and vice versa.	Kukimura and Kouyama (1982)
7	• Adventitious buds were induced after irradiation of hybrid clones of sweet potato with fast neutrons. There was an increase in the dry matter content of the tubers by 1.0–5.5%, storage root yield by 3.6–53.8%, and the number of storage roots per plant by 0.6–4.2%. • These adventitious bud clones had resistances to black rot, root rot, nematodes etc. Three multiple resistance clones Yanshu 568, Yanshu 571-1, 84-C-2 and two good quality clones Yanshu 759 and 781 were selected.	Cui et al. (1993)
8	• Five clonal lines resistant to black rot were obtained from the progenies of irradiated Xu-18. • The mutant line 908-3 (sharp-heart leaf) was obtained from greenhouse-grown cuttings of M_1V_1 which had thick leaves. The sharp-heart leaf character was stable and appeared generation after generation until the time of this report.	Lu et al. (1993)
9	• No significant variation in stem length and pigmentation, tuber skin and flesh colors was observed in the M_1V_2 generation after gamma irradiation of callus in vitro.	Zhen (1993)
10	• Three M_1V_3 mutant clones developed through in vitro mutagenesis showed tolerance to 12 °C but failed in the subsequent test. • A mutant line obtained through irradiation of leaf callus line, HN 892I, displayed salt tolerance in vitro with a medium containing NaCI (10 g/L).	Thinh (1993)

(continued)

Table 26.4 (continued)

S. No.	Report	Reference
	• In the field evaluation of mutants, six early tuberization variants (ET) and condensed formation of tubers (CFT) were identified.	
11	• Variants having differences in leaf shape, leaf vein and petiole color were observed with increasing in vivo irradiation doses. • A high yield and stability mutant clone UPL-SP2 and clones with resistant to sweet potato weevil were identified.	Bader et al. (1993)
12	• Three tuber color (white, rose, and deep red) mutants were recovered from the adventitious roots of sweet potato mutants. • Mutants had maximum sugar content (KLM-22 had 3.25%) and carotene content (SVM-3 had 2687 IU/100 g) than control Kanhangad local.	Vasudevan et al. (1996)
13	• Chlorophyll deficient sectors and patches appeared on M_1V_1 plants. A vine with a chimeric branch was observed in the treated population of "H-4126." • In "S5" gamma irradiation of rooted-cuttings at 20 Gy produced a broad-leaved mutant with changed leaf size and shape. The two higher exposures of gamma rays significantly increased the tuber yield per vine in M_1V_2 generation. • A cordate-leaved mutant was isolated from "Bhadrakalichuvala" rooted-cuttings exposed to 25 Gy.	Bai and Nayar (1997)
14	• Somatic embryogenesis was inhibited when the gamma-irradiated (30–50 Gy) meristems was cultured on induction medium. • Wide differences in yield were observed in field trials of mutants.	Sonnino et al. (1997)
15	• Noticed phenotypic changes, root flesh color changes and improved drought tolerance in vitro gamma-irradiated sweet potato mutant germplasm. • A promising clone in M_1V_2 generation, 91-C3-15 had improved tuber shape and deep red skin color.	Zhen (2001)
16	• Mutant plants were regenerated from the in vitro radiation treated somatic embryos. Variations in the characters such as leaf shape, root skin color and the number of storage roots per plant were observed under field conditions.	Wang et al. (2005)
17	• Successful identification of five root flesh color mutant variants of sweet potato with significantly increased β-carotene content was done through gamma irradiation.	Wang et al. (2007)
18	• "Nongdafu 14"—a sweet potato mutant with significantly higher total carotenoid content, low fiber and better taste than the wild type was developed by treating with the gamma field at 142 Gy.	Wang et al. (2007)
19	• From the putative mutants obtained through in vitro gamma irradiation of nodal explants, 28 mutants were isolated based on storage root size, shape, and number.	Shin et al. (2011)

(continued)

Table 26.4 (continued)

S. No.	Report	Reference
	• The mutant clones' 50 Gy-10 and 50 Gy-8 had comparatively high-amylose content. 50 Gy-23, 50 Gy-26 and 50 Gy-28 had very elevated levels of starch (>30% on a fresh weight basis), while the lines 50 Gy-10, 50 Gy-16, 50 Gy-23, 50 Gy-26 and 50 Gy-28 had a starch content above 30%.	
20	• A significant reduction in the tuber formation was observed in all varieties when treated with 45 Gy radiation	Asare and Akama (2014)

Fig. 26.8 Anthracnose disease in yam

resulting in only one crop per year. Most of the clones are susceptible to anthracnose disease caused by *Colletotrichum gloeosporioides* which is a major production problem in yam worldwide (Fig. 26.8). The disease is more severe during the rainy season. In Nigeria, yam anthracnose causes significant losses in tuber yield ranging from 50 to 90% (Nwadili et al. 2017). Yams can tolerate drought stress up to some extent at the cost of tuber yield. Soil moisture deficiency at the time of sprouting adversely affects the plant population.

The breeding objectives of yams include

- Early maturity and short duration
- Higher tuber yield

- High nutrient use efficiency
- High nutritional value
- High dry matter/starch content
- Reduced vine length/dwarfing nature
- Suitable for cropping systems
- Resistant to anthracnose disease
- Tolerance to abiotic stresses: drought and salinity

Improvement of yams is also suffering from flowering related problems. The majority of the yam clones with desirable traits are sterile or rarely produce inflorescence besides high pollen sterility and low seed set (Koo and Ruiz 1973), but when they do, produce infertile seed (Villamor and Cardinez 2008). Moreover, the dioecious nature of the crop is posing a severe asynchronization problem for hybridization breeding. The attempts of storing pollen in liquid nitrogen to overcome asynchronization barrier did not yield fruitful results. Longer crop duration of 10–12 months prolongs the breeding cycle in yams. Hence, applied mutagenesis could be an option for induction of variation in yams. Few clones of greater yam were found to be resistant to anthracnose disease at ICAR-CTCRI, transfer of resistant genes through conventional breeding is hampered due to non-flowering.

26.3.3.1 Mutation Breeding Achievements So Far

No known mutant variety has been released for cultivation in yams so far. However, a dwarf mutant of 2 m vine length with early maturity was isolated in M_1V_2 generation at ICAR-Central Tuber Crops Research Institute (Vasudevan et al. 1986). Very few attempts have been made for inducing the variation through applied mutagenesis in yams (Table 26.5).

Table 26.5 Brief report on mutation breeding achievements in yams

S. No.	Report	Reference
1	• Treating fresh tubers of *D. alata* resulted in more damage and chimeras than 4 months old tubers. • LD 50 dose for *D. alata* and *D. esculenta* was determined as 3 krad.	Vasudevan and Jos (1992)
2	• Batches of minitubers of local white yam cultivar *Obiaoturugo* was irradiated with gamma rays. LD50 dose was observed as 40 Gy. • Mutants with bushy vegetation and busy with spreading vines were isolated.	Nwachukwu et al. (2009)
3	• Exposure of tubers to UV rays for 24 h increased the tuber size and weight, but exposure for 48 h produced tubers with decreased length and weight. • No significant difference in the yield was observed over treatment duration but violet pigmentation increased.	Villamor and Cardinez (2008)

26.3.4 Aroid Tuber Crops

Aroids are plants that belong to the family *Araceae*. There are two important tropical tuber crops under this family namely Taro (*Colocasia esculenta* (L.) Schott) and Elephant Foot Yam (*Amorphophallus peaoniifolius* (Dennst.) Nicolson). Taro is grown globally in 48 countries in tropical and subtropical regions of the world in 2.1 mha with total production and average productivity of 12.5 mt and 10 t/ha in 2019 (FAOSTAT 2021). Elephant foot yam is grown in very few countries and the data on its global area, production and productivity are not available. In India, the area under taro and elephant foot yam cultivation are 0.04 and 0.03 mha, respectively, in 2018 while total production of 0.6 and 0.8 mt with average productivity of 15.21 and 27.68 t/ha (Byju et al. 2020).

Taro was originated in the Indo-Malayan region. A huge variation/diversity in taro can be found in the North-Eastern states of India. Africa is having the highest area under taro and the largest producer followed by Asia. Taro is having a close association with other tuber-bearing members of the genus, which are equally edible and nutritious. They are Giant Taro (*Alocasia indica*), Swamp Taro (*Cyrtosperma chamissonis*) and tannia (*Xanthosoma sagittifolium*). Besides its starch corms, leaves, and petioles are also edible. Taro starch is used for the preparation of baby food. Corms are also used in various preparations like chips, curry, papad, etc. (Suja et al. 2017). Taro corms have a high energy value of 376–545 kJ/100 g, 24.5–26.5% starch, 1.46–3.80% dietary fiber and low-fat content. Taro corms are also rich in minerals like Magnesium, Iron, Zinc, Copper, and Sodium. It is local staple food in Asian countries such as the Philippines, Indonesia, Malaysia, Bangladesh, India, and China.

Elephant foot yam is a popular cash and remunerative crop of the farmers of Andhra Pradesh, Bihar, Jharkhand, West Bengal, Uttar Pradesh, Tamil Nadu, and Kerala (Ravi et al. 2011). The corms are rich in nutrients and medicinal properties. They contain a moderate amount of protein, calcium, and Vitamin C (Suja et al. 2012). The corms of elephant foot yam produce 324–374 kJ/100 g, have a starch content of 6.72–16.6% and 1.45% dietary fiber. These crops are also used in many ayurvedic treatments. Elephant foot yam corms are used for the treatment of piles.

The cultivation of aroids also faces certain physiological and pathological problems. These crops are sensitive to drought. Water deficit stress immediately after planting adversely affects the sprouting and establishment of the crop. The corms of aroids have anti-nutritional factors like calcium oxalate crystals which cause acridity and itching sensation in the throat upon consumption. However, few edible clones are free from acridity. Dormancy is another physiological problem that prolongs the crop duration in elephant foot yam. The corms of elephant foot yam tend to have dormancy and they take 3–4 months for sprouting (Ravi et al. 2011). The majority of taro clones are susceptible to leaf blight disease caused by *Phytophthora colocasiae* (Fig. 26.9). The infection of leaf blight disease significantly decreases the longevity of taro leaves, thereby reducing the number of functional leaves. Yield reductions due to leaf blight disease are reported up to 50% (Misra et al. 2008). Collar rot caused by *Sclerotium rolfsii* is an important

Fig. 26.9 Leaf blight in taro (**a**), late blight (**b**) and tuber rot (**c**) in elephant foot yam

disease of elephant foot yam often causes huge losses to growers. The infection is most likely to occur in the rainy season followed by warm weather. Late blight is another important disease of elephant foot yam where the leaflets turn black due to necrosis and give a burnt appearance. Physiological or pathological tuber rot in the field and storage is another problem causing huge losses to the farmers (Fig. 26.9).

The breeding objectives of these crops as follows:

- High yield
- Non-acridity
- Lack of dormancy and short duration in elephant foot yam
- Leaf blight resistance in taro
- Collar rot, late blight and tuber rot resistance in elephant foot yam
- Developing varieties suitable for cropping and farming systems
- High dry matter/starch content
- High Processing and nutritional quality
- Tolerance to drought and salinity stress

Akin to other tuber crops, flowering in taro is not common and the majority of clones do not flower under normal conditions. It needs spraying of growth regulators for induction of flowering. Similarly, flowering is scarce in elephant foot yam and the clones do not flower in many cases thereby hindering the improvement through conventional breeding. Even if flowering occurs in some clones of these crops, asynchronization of flowering between male and female clones and protogyny are major problems of hybridization breeding. Therefore, the improvement in these aroid tuber crops is highly limited to clonal selection. Moreover, there exist no known sources of disease resistance and other important traits like lack of dormancy and short duration in elephant foot yam. Hence, mutation breeding can also be the option in these crops for the improvement of these traits.

26.3.4.1 Mutation Breeding Achievements So Far

Three mutant varieties have been registered so far with the IAEA mutant variety database in taro. Out of three, two varieties of *Fukugashira* and *Chiba-maru* were released for cultivation in Japan in 1992 and 2017 respectively. Another variety of *Luyutou 1* was released in 1993 in China (https://mvd.iaea.org/). No mutant variety has been registered so far in elephant foot yam. A review of the literature also suggested that significant mutation breeding attempts have not been made in elephant foot yam so far. Other important contributions in applied mutagenesis in taro are mentioned below (Table 26.6).

A photosynthetically efficient mutant with better performance under shady conditions was isolated at ICAR-Central Tuber Crops Research Institute in tannia (Vasudevan and Jos 1988). An attempt has also been made to induce variation in elephant foot yam which showed that LD50 dosage was around 0.8 krad for corms (Vasudevan and Jos 1991).

Table 26.6 Brief report on mutation breeding achievements in taro

S. No.	Report	Reference
1	• Mutant CM-17 was found to have prolonged storage life up to 6 months after harvest. • Increase in yield (25–30%), decrease in phenols and sugar content and increase in dry matter content was observed among mutants.	Vasudevan and Jos (1988)
2	• In vitro shoot tips of taro were irradiated from 0 to 60 Gy gamma irradiation. • The lethal dose 30 that causes a 30% reduction in growth was found to be 7.65 Gy.	Seetohul et al. (2008)
3	• Seed cormels of 10 diverse taro genotypes were treated 10 Gy γ-irradiation in vitro. • Leaf blight incidence parameters as spot diameter and number of sporangia were low in mutants when compared to the non-irradiated controls.	Sahoo et al. (2015)
4	• Apical shoot cultures of Bogor taro were treated with gamma ray irradiation doses of 0, 5, 10, 15, and 20 Gy. • The influence of gamma irradiation on the number of leaves was found at10 Gy and dose 20 Gy was found to be a lethal dose. • Molecular marker study indicated that gamma irradiation provided 51% diversity when compared with a parent.	Nurilmala et al. (2017)

26.3.5 Chinese Potato

Chinese potato (*Plectranthus rotundifolius* (Poir.) Spreng.) belongs to the family *Lamiaceae* (Labiatae) with a chromosome number of $2n = 64$. Chinese potato is an important tropical tuber crop grown in Africa and Asia (Aculey et al. 2011). Its geographical origin was believed to be East Africa, followed by its distribution to tropical West Africa, Southeast Asia including India, Sri Lanka, Malaysia, and Indonesia (Harlan et al. 1976). In India, it is largely grown in Kerala and Tamil Nadu states.

Chinese potato also known as coleus has a distinct aromatic flavor and delicious taste upon cooking. These tubers are rich in minerals like calcium, iron, and B-complex vitamins such as thiamine (B1), riboflavin (B2) and niacin (B3). Tubers also contain a good amount of ascorbic acid or Vitamin-C (Jayapal et al. 2015). It is also widely used in ayurvedic medicine due to the presence of medicinal properties and is used to reduce the cholesterol level of blood (Sandhya and Vijayalakshmi 2005). It also has anti-tumor and anti-oxidant properties and can be used as a cancer chemoprevention agent.

The yield of Chinese potato is very less compared to other tuber crops. The tuber formation in Chinese potato is found to be sensitive to photoperiod and other weather parameters. Moreover, the tubers of Chinese potato are small, non-uniform and often form in odd shape as clusters at the base of the stem. This makes cleaning and de-skinning time consuming for culinary preparations (Suma et al. 2014). It is a short-day crop and only one crop can be taken from July to November (Radhakrishnan and Abraham 2008). The breeding objectives of this crop include:

- High yield
- Uniform size, the shape of tubers
- Increase in size of tubers
- Good cooking quality and easy peeling
- Increase in mineral nutrient content
- Photo-insensitivity and year-round production

Due to a limited number of available germplasm and the absence of much variability among the available germplasm, the improvement of these important tuber crops did not progress much. Only two cultivars *Sree Dhara* and *Co-1* were developed through selection in this crop so far. Moreover, flowering and fruit set are very scarce, thereby improving through conventional breeding methods is very difficult. Mutation breeding would be fruitful for widening the genetic base and creation of new variations in this underexploited crop.

26.3.5.1 Mutation Breeding Achievements So Far

The first mutant variety in Chinese potato in the world, *Suphala*, was developed and released for cultivation by Kerala Agricultural University, Thrissur, India through

Table 26.7 Brief report on mutation breeding achievements in Chinese potato

S. No.	Report	Reference
1	• The tuber yield of in vitro gamma-irradiated mutants ranged from 15 to 75 g/plant. The number of tubers/plant ranged from 8 to 68. • One commercial mutant clone (TC-9) with a tuber yield of 1 t/ha was identified. • All mutants were found to be photo-insensitive.	Radhakrishnan and Abraham (2008)
2	• The LD 50 dose of gamma irradiation and EMS were found as 40 Gy and 0.4%, respectively. • Two mutants M 131 and M 61 with photoperiod insensitivity and high yield were isolated.	Abraham and Radhakrishnan (2008)

in vitro mutagenesis. The brief report on induced mutations in Chinese potato is mentioned in Table 26.7.

26.3.6 Yam Bean

Yam bean (*Pachyrrhizus erosus* L.) is an important underutilized leguminous tuber crop that belongs to the family Fabaceae with chromosome number $2n = 2X = 22$. This exceptional tuber crop is propagated through sexual reproduction via seeds. It is said to be originated in Mexico and the Central American region. It is grown for fresh tubers to consume as salad due to its watery texture and sweetness in Eastern India. It is mainly cultivated in Bihar, Jharkhand, and West Bengal states of India. It is also grown in South East Asian countries like the Philippines, Indonesia, and Thailand. Other cultivated species in the genus are *P. ahipa* (Wedd.) Parodi and *P. tuberosus* (Lam.) Spreng. Yam bean flour is popularly consumed in Africa. The yam bean flour is rich in carbohydrates and energy consisting of 2.0% reducing sugars and 21% starch. It also contains 5.8% moisture content, 5.7% crude fat, 6.2% crude fiber, and 85% available carbohydrate (Buckman et al. 2018). When it is grown as a tuber crop, pruning of inflorescence promotes tuber development. The seeds of the yam bean contain the toxic alkaloid "Rotenone" whose usage as a botanical pesticide is under investigation. Though the seeds are nutritious and rich in minerals like potassium, phosphorus, calcium, and iron, their consumption is discouraged by the presence of this alkaloid (Pati et al. 2020).

Like other underutilized crops, yam bean improvement is also on a slow path worldwide. Unlike other tuber crops, yam bean flowers regularly and produces fertile viable seed. There are no known hybridization barriers in yam bean and it crosses freely with other *Pachyrrhizous* species. Only two cultivars have been released in this crop so far viz., *Rajendra Misrikand 1* and *Rajendra Misrikand 2* by Bihar Agricultural University, Sabour, India. Compared to other legumes, pest and disease incidence is low in yam bean. However, *Cercospora* leaf spot is a common disease. The breeding objective of yam bean as follows:

- High tuber yield
- Dwarf or bushy growth
- High sugar or carbohydrate content
- Low or no rotenone content in seeds
- High protein and nutritive value in tubers as well as seeds
- Tolerance to biotic and abiotic stresses

26.3.6.1 Mutation Breeding Achievements So Far

Similar to other underexploited crops, available genetic variation is low in yam bean. An experiment on induced mutagenesis in yam bean was done at ICAR-Central Tuber Crops Research Institute, Kerala, India. The seeds of the yam bean were treated with Ethyl Methane Sulfonate ranging from 0.25% to 2.0%. Evaluation of subsequent mutants in M_2 generation resulted in the isolation of a dwarf mutant with a higher tuber yield than the parent. The mutant had similar starch content in tubers and seed rotenone content; slightly high protein content and less sugar content when compared with the parent (Nair 1990).

26.3.7 Other Minor Tuber Crops

There are other minor tuber crops like West Indian arrowroot commonly known as arrowroot (*Maranta arundinacea* L.), native to Mexico, belongs to the family *Marantaceae* and *Tikhur* or starchy Curcuma (*Curcuma angustifolia* Roxb.) of *Zingiberaceae* which have industrial and medicinal properties.

Arrowroot is a perennial herb grown as annual and well known for its starchy rhizomes (Rohandi et al. 2017). The starch in arrowroot is present in a simplified and easily digestible form which makes it an important component of baby food (Asha et al. 2015). Arrowroot biscuits are popular throughout India. Arrowroot rhizomes are nutritive and are also used in the treatment of problems in alimentary canal, pulmonary organs or of the urinary organs (Asha et al. 2015).

Tikhur is grown in hills of Madhya Pradesh, Chhattisgarh, West Bengal, Maharashtra, Tamil Nadu and foot hills of Himalayan ranges (Kumari et al. 2017). The rhizomes of *tikhur* are highly popular diet among the tribal regions. The rhizomes are rich in starch, nutrients and have several health benefits, essential oils, and aroma. The starch powder of *tikhur* is highly nutritious and easily digestible; hence, it is used in the feed of infants, weak children and aged adults. *Tikhur* and its products are popular in Chhattisgarh state of India. The starch is used in the preparation of various sweet meals like *halwa*, *barfi,* and *jalebi*. The herbal drink from *tikhur* is consumed during summer for cooling effect (Shankar et al. 2015). The rhizomes have healing ability of peptic ulcers. They are also used in treatments of dysentery, diarrhea, and colitis, cough and bronchitis (Doble et al. 2011).

The genetic improvement in these two crops is very less so far. Though these two species flower occasionally, there has not been much open pollination and seed set. Arrow root produces white color flowers but seeds are seen rarely (Asha et al. 2015). The available genetic variation is very low in arrowroot. There has not been much variation found among available arrowroot germplasm. Shintu et al. (2016) reported very low 4.0–34.48% genotypic co-efficient variation for various traits among 60 accessions of arrowroot. A single improved variety has not been released so far for cultivation in arrowroot.

Though *tikhur* is native to India, the available genetic variation is low because of nil or low frequent sexual reproduction. It produces flowers in two colors viz., white and light pink (Shankar et al. 2015). Only one variety, *Chhattisgarh Tikhur-1* has been released in *tikhur* for the first time in Chhattisgarh recently. So, mutation breeding is very helpful in these crops to create new variations and to broaden the existing genetic base for the improvement of various traits especially weight and size of rhizomes and starch content.

26.4 Conclusion and Future Aspects

Tropical root and tuber crops are staple foods of masses across the tropical regions of the world. They are not only rich in starch, but also a good source of nutrients and vitamins and providing a balanced diet, especially in rural and tribal areas. These crops are said to be future crops owing to their ability to grow in marginal soils with limited inputs and tolerate adverse weather conditions. But, less attention has been given to the improvement of these crops so far because of their non-breeder friendly traits like vegetative reproduction, little or no flowering, and longer crop durations. It creates a huge scope for the improvement of these crops through mutation breeding. Despite chimeras are problems when vegetative propagules are treated with mutagens, they can effectively be eliminated as discussed in this chapter. A few varieties have been released in crops like sweet potato, taro, and cassava through mutation breeding. Vegetative propagation in these crops helps in the quick fixing of the desirable gene/allele combinations. So, mutation breeding in combination with vegetative propagation enables the speed breeding of these crops. More research is to be undertaken for exploiting the possibility of creation of new variations through applied mutagenesis in these underexploited crops in the coming years.

Acknowledgments The authors are highly thankful to Dr. M.N. Sheela, Head, Division of Crop Improvement for her timely advice on the literature of previous works done at ICAR-Central Tuber Crops Research Institute. The authors also acknowledge the help received from Mr. B.R. Ajeesh and library staff of ICAR-CTCRI in collecting the literature.

References

Abraham A (1970) Breeding work on tapioca (cassava) and few other tropical tuber crops. In: Proceedings of the second international symposium on tropical root and tuber crops, p 76

Abraham M, Radhakrishnan VV (2008) Induced mutations in coleus (*Solenostemon rotundifolius* ((Poir.) J. K. Mortan)—an under-utilized medicinal tuber. In: Proceedings of international symposium on induced mutations in plants, 12–15 August. IAEA and FAO, Vienna, p 111

Abraham K, Edison S, Unnikrishnan M, Sheela MN, Vimala B, Sreekumari MT, Naskar SK (2006) Tuber crop varieties released by the Central Tuber Crops Research Institute. Central Tuber Crops Research Institute, Sreekariyam, Thiruvananthapuram

Aculey K, Quainoo AK, Mahanu G (2011) Feasibility studies on the potential of grafting and budding of Frafra potato (*Solenostemon rotundifolius*). J Hort Forest 3(10):327–332

Ahiabu RKA, Klu GYP (1997) Induced mutation breeding in cassava (*Manihot esculenta* Crantz) cultivar 'BosomNsia' (IAEA-312RC4122). International Atomic Energy Agency (IAEA), p 22

Amenorpe G (2010) Mutation breeding for *in planta* modification of amylase starch in cassava *(Manihot esculenta* Crantz), PhD thesis, University of the Free State, Bloemfontein

Asare PA, Akama CK (2014) Reaction of sweet potato (*Ipomoea batatas* L.) to gamma irradiation. J Appl Sci 14(17):2002–2006

Asare E, Safo-Kantanke O (1997) Improvement of cassava cooking quality through mutation breeding. In: Improvement of basic food crops in Africa through plant breeding, including the use of induced mutations, IAEA-TECDOC-951. IAEA, Vienna, pp 19–24

Asare PA, Galyuon IKA, Asare-Bediako E, Sarfo JK, Tetteh JP (2014) Phenotypic and molecular screening of cassava (*Manihot Esculenta* Crantz) genotypes for resistance to cassava mosaic disease. J Gen Mol Virol 6(2):6–18

Asha KI, Radhika NK, Vineetha B, Devi AA, Sheela MN, Sreekumar J (2015) Diversity analysis of arrowroot (Maranta arundinacea L.) germplasm using ISSR Markers. J Root Crops 41(1):17–24

Baah FD, Maziya-Dixon B, Asiedu R, Oduro I, Ellis WO (2009) Nutritional and biochemical composition of *D. alata* (*Dioscorea* spp.) tubers. J Food Agric Environ 7(2):373–378

Bader RF, Zamora AB, Manguiat PH (1993) Mutation breeding of sweet potato in the Philippines. In: Improvement of root and tuber crops in tropical countries of Asia by induced mutations, IAEA-TECDOC-809. IAEA, Vienna, pp 65–82

Bai DJS, Nayar NK (1997) Sweet potato mutants induced by gamma rays. Mutat Breed Newsl 43: 21

Beyaz R, Yildiz M (2017) The use of gamma irradiation in plant mutation breeding. In: Plant engineering. InTech, London, pp 33–46. https://doi.org/10.5772/intechopen.69974

Buckman ES, Oduro I, Plahar WA, Tortoe C (2018) Determination of the chemical and functional properties of yam bean (*Pachyrhizus erosus* (L.) Urban) flour for food systems. Food Sci Nutr 6: 457–463. https://doi.org/10.1002/fsn3.574

Byju G, Jaganathan D, Koundinya AVV (2020) Quinquennial review team report 2014–2019. ICAR-Central Tuber Crops Research Institute, Sreekariyam, Thiruvananthapuram

Ceballos H, Sanchez T, Denyer K, Tofino AP, Rosero EA, Dufour D, Smith A, Morante N, Perez JC, Fahy B (2008) Induction and identification of a small-granule, high-amylose mutant in cassava (*Manihot esculenta* Crantz). J Agric Food Chem 56:7215–7222

Cheng CF (1958) The effect of radiation on the first irradiated generation of sweet potato and peanut. In: Second United Nations international conference on the peaceful uses of atomic energy A/Conf. 15/P 1131, pp 1–6

Clark CA, Davis JA, Abad JA, Cuellar WJ, Fuentes S, Kreuze JF, Gibson RW, Mukasa SB, Tugume AK, Tairo FD, Valkonen JPT (2012) Sweetpotato viruses: 15 years of progress on understanding and managing complex diseases. Plant Dis 96(2):168–185

Cui GQ, Yang ZC, Lin ZJ, Wang SJ (1993) Mutation breeding in sweet potato with fast neutrons by inducing hypocotyls adventitious buds. In: Improvement of root and tuber crops in tropical countries of Asia by induced mutations, IAEA-TECDOC-809. IAEA, Vienna, pp 9–18

Devarajan V, Vasudevan K (1997) Gamma ray induced modified gene expression in cassava mutants. J Root Crops 23(1):61–62

Doble B, Dwivedi S, Dubey K, Joshi H (2011) Pharmacognostical and antimicrobial activity of leaf of *Curcuma angustifolia* Roxb. Int J Drug Discov Herb Res 1(2):46–49

FAOSTAT (2021). http://www.fao.org/faostat/en/#data/QC

George J, Sureshkumar P, Unnikrishnan M (2011) Description of recommended/released varieties under AICRP on tuber crops 1975–2011. All India Coordinated Research Project on Tuber Crops, Central Tuber Crops Research Institute, Sreekariyam, Thiruvananthapuram

Harlan JR, Dewet JMJ, Stemler ABL (1976) Origins of African plants domestication. De Gruyter Mouton, The Hague. https://doi.org/10.1515/9783110806373

Hegde V, Makeshkumar T, Sheela MN, Chandra CV, Koundinya AVV, Anil SR, Muthuraj R, Darshan S (2016) Production of synthetic seed in cassava (*Manihot esculenta* Crantz). J Root Crops 42(2):3–8

Hegde V, Koundinya AVV, Sheela MN, Chandra CV, Mukherjee A (2019) Storage of cassava pollen for conservation of nuclear genetic diversity and overcoming hybridization barriers. Indian J Hort 76(1):104–111. https://doi.org/10.5958/0974-0112.2019.00015.X

Hernandez TP, Hernandez T Miller JC (1959) Effects of different rates of irradiation on sweet potato roots and tomato seeds. In: Proc. 56th ann. conv. Assoc. South Agr. Workers, Memphis, Tenn., pp 157–158

Huaman Z (1992) Systematic botany and morphology of the sweet potato plant. CIP, Lima, p 25

Iglesias CA, Sanchez T, Yeoh HH (2002) Cyanogens and linamarase activities in storage roots of cassava plants from breeding program. J Food Compos Anal 15:379–387

Jayapal A, Swadija OK, Anju SV (2015) Effect of organic nutrition on quality characters of Chinese potato (*Plectranthus rotundifolius*). J Root Crops 41(1):56–58

Jones A (1980) Sweet potato. Am Soc Agron 46:645–655

Jose A, Makeshkumar T, Edison S (2011) Survey of cassava mosaic disease in Kerala. J Root Crops 37:41–47

Joseph R, Yeoh HH, Loh CS (2004) Induced mutations in cassava using somatic embryos and the identification of mutant plants with altered starch yield and composition. Plant Cell Rep 23:91–98

Karyeija RF, Gibson RW, Valkonen JPT (1998) The significance of sweet potato feathery mottle virus in subsistence sweet potato production in Africa. Plant Dis 82:14–15

Kawano K, Amaya A, Daza P, Rios M (1978) Factors affecting efficiency of hybridization and selection in cassava. Crop Sci 18:373–376

Kharkwal MC (2012) A brief history of plant mutagenesis. In: Shu QY (ed) Plant mutation breeding and biotechnologies. FAO/IAEA, Vienna, pp 21–30

Khumaida N, Ardie SW, Dianasari M, Syukur M (2015) Cassava (*Manihot esculenta* Crantz.) improvement through gamma irradiation. Proc Food Sci 3:27–34

Khumaida N, Ardie SW, Astuti MS (2017) Characterization of irradiation induced mutants of cassava (*Manihot esculenta* Crantz) generated from Jame-jame and Adira-4 genotypes at M1V2 generation. In: Second international conference on sustainable agriculture and food security: a comprehensive approach, KnE Life Sciences, pp 22–28. https://doi.org/10.18502/kls.v2i6.1016

Koo FKS, Ruiz JC (1973) Exploratory induction of solid mutations in yams by γ-irradiation. In: Srb AM (ed) Genes, enzymes, and populations. Plenum Press, New York

Koundinya AVV, Dhankhar SK (2013) Correlation and path analysis of seed yield components in okra *Abelmoschus esculentus* (L.) Moench. Ann Hort 6(1):145–148

Koundinya AVV, Dhankhar SK, Yadav AC, Hegde V (2013) A study on character association and path analysis in okra. Ann Agri-Bio Res 18(2):234–237

Koundinya AVV, Hegde V, Sheela MN, Chandra CV (2018) Evaluation of cassava varieties for drought tolerance. J Root Crops 44(1):70–75

Koundinya AVV, Ajeesh BR, Hegde V, Sheela MN, Mohan C, Asha KI (2021) Genetic parameters, stability and selection of cassava genotypes between rainy and water stress conditions using AMMI, WAAS, BLUP and MTSI. Sci Hort. https://doi.org/10.1016/j.scienta.2021.109949

Kukimura H, Kouyama Y (1982) Studies on mutation breeding in sweet potato (*Ipomoea batatas* (L.) Lam.). In: Improvement of vegetatively propagated crops and tree crops through induced mutations, vol 14. International Atomic Energy Agency, Vienna, pp 199–233

Kumari R, Shrivastava SL, Mishra HN, Meghwal M (2017) Physicochemical and functional properties of *Curcuma angustifolia* (Tikhur)—an underutilized starch. Pharma Innov J 6(7): 114–119

Kyereko WT, Hongbo Z, Amoanimaa-Dede H, Meiwei G, Yeboah A (2019) The major sweet potato weevils; management and control: a review. Entomol Ornithol Herpetol 8:218. https://doi.org/10.35248/2161-0983.8.218

Lu SY, Wu CG, Li WJ, Feng QH (1993) The use of gamma irradiation in vivo and combined use of in vitro culture to select mutant clones in sweet potato. In: Improvement of root and tuber crops in tropical countries of Asia by induced mutations, IAEA-TECDOC-809, vol 3. IAEA, Vienna, pp 19–36

Martin FW, Cabanillas E (1966) Post-pollen-germination barriers to seed set in sweet-potato. Euphytica 15:404–411

Masima I, Sato H (1959) X-ray induced mutations in sweet potato. Jpn J Breed 8:233–237

Miller JC (1935) Further studies of mutations of the Porto Rico sweet potato. J Am Soc Hort Sci 33: 460–465

Misra RS, Sharma K, Mishra AK (2008) Phytophthora leaf blight of taro (*Colocasia esculenta*)—a review. Asian Aust J Plant Sci Biotechnol 2(2):55–63

Moh CC, Alan JJ (1972) Methods for inducing mutations in cassava and the possible use of the mutants. In: Proceedings of the induced mutations in vegetatively propagated plants, Joint FAO/IAEA Division of Atomic Energy in Food and Agriculture, Vienna, 11–15 September, pp 67–73

Nair SG (1990) EHS induced dwarf and high yielding mutant in yam bean (*Pachyrrhizus erosus* L.). Mutat Breed Newsl 36:5–6

Nayar GG, Rajendran PG (1985) Radiation induced mutants in cassava. Mutat Breed Newsl 30:18

Ngailo S, Shimelis H, Sibiya J, Mtunda K (1991) Sweet potato breeding for resistance to sweet potato virus disease and improved yield: progress and challenges. Afr J Agric Res 8(25): 3202–3215. https://doi.org/10.5897/AJAR12.1991

Nurilmala F, Hutagaol RP, Widhyastini I, Widyastuti U, Suharsono (2017) Somaclonal variation induction of Bogor taro (*Colocasia esculenta*) by gamma irradiation. Biodiversitas 18(1):28–33

Nwachukwu EC, Mbanaso ENA, Ene LSO (1997) Improvement of cassava for high dry matter, starch and low cyanogenic glucoside content by mutation induction. In: Improvement of basic food crops in Africa through plant breeding, including the use of induced mutations, IAEA-TECDOC-951. IAEA, Vienna, pp 93–98

Nwachukwu EC, Mbanaso ENA, Nwosu KI (2009) The development of new genotypes of the white yam by mutation induction using yam mini-tubers. In: Shu QY (ed) Induced plant mutations in the genomics era. Food and Agriculture Organization of the United Nations, Rome, pp 309–312

Nwadili CO, Augusto J, Bhattacharjee R, Atehnkeng J, Lopez-Montes A, Onyeka TJ, Lava Kumar PL, Asiedu R, Bandyopadhyay R (2017) Comparative reliability of screening parameters for anthracnose resistance in water yam (*Dioscorea alata*). Plant Dis 101:209–216

Pati K, Kaliyappan R, Chauhan VBS, Bansode V, Nedunchezhiyan M, Hegde V, Koundinya AVV (2020) Phenological growth stages of underutilised crop yam bean (*Pachyrhizus erosus* L. Urban) according to the extended BBCH scale. Ann Appl Biol 117(3):1–7. https://doi.org/10.1111/aab.12637

Poole CF (1959) Improvement in yield of sweet potato clones. Proc Am Soc Hort Sci 73:445–452

Poole CF, Tanaka JS (1963) Transmission of yield variation in irradiated sweet potato clones. Proc Am Soc Hort Sci 80:488–496

Pushpalatha R, Gangadharan B (2020) Is cassava (*Manihot esculenta* Crantz) a climate "smart" crop? A review in the context of bridging future food demand gap. Trop Plant Biol 13:201–211. https://doi.org/10.1007/s12042-020-09255-2

Radhakrishnan VV, Abraham M (2008) *In vitro* mutagenesis for photo insensitivity to tuberisation in coleus (*Solenostemon rotundifolius* (Poir.) J.K. Mortan). In: Proceedings of international symposium on induced mutations in plants, 12–15 August. IAEA and FAO, Vienna, Austria, p 110

Rajamma P (1983) Biology and bionomics of sweet potato weevil Cylas formicarius Fabr. In: Goel S (ed) Symposium on insect ecology & resource management, 2–4 October, Muzaffarnagar, India. Sanatan Dharm College, pp 87–92

Ravi V, Ravindran CS, Suja G, George J, Nedunchezhiyan M, Byju G, Naskar SK (2011) Crop physiology of elephant foot yam (*Amorphophallus paeoniifolius* (Dennst. Nicolson)). Adv Hort Sci 25(1):51–63. https://doi.org/10.13128/ahs-12785

Rohandi A, Budiadi B, Hardiwinoto S, Harmayani E, Sudrajat DJ (2017) Variability in morpho-physiology, tuber yield and starch content of several arrowroot populations in Garut district. AGRIVITA J Agric Sci 39(3):311–323

Sahoo BB, Kole PC, Sahoo MR (2015) Effects of γ irradiation on leaf blight disease of some taro (*Colocasia esculenta* (L.) Schott) genotypes. Int J Bioresour Stress Manag 6(1):007–014

Sanchez T, Roserob A, Tofino AP (2009) Induction and identification of useful mutations for root quality traits in cassava. In: Shu QY (ed) Induced plant mutations in the genomics era. Food and Agriculture Organization of the United Nations, Rome, pp 266–269

Sandhya C, Vijayalakshmi NR (2005) Antioxidant activity of flavanoids from *Solenostemon rotundifolius* in rats fed normal and high fat diets. Food Res Int 38:615–629

Saravanan R, Ravi V, Stephen R, Thajudhin S, George J (2016) Post-harvest physiological deterioration of cassava (*Manihot esculenta*)—a review. Indian J Agric Sci 86(11):1383–1390

Seetohul S, Puchooa D, Ranghoo-Sanmukhiya VM (2008) Genetic improvement of Taro (*Colocasia esculenta var esculenta*) through in-vitro mutagenesis. UoM Res J 13A:79–89

Shankar D, Rao SS, Singh DP, Markam PS, Sahu MK (2015) Assessment of genetic diversity in indigenous genotypes of tikhur (*Curcuma angustifolia* Roxb.). Plant Arch 15(2):743–751

Shin JM, Kim BK, Seo SG, Jeon SB, Kim JS, Jun B, Kang SY, Lee JS, Chung MN, Kim SH (2011) Mutation breeding of sweet potato by gamma-ray radiation. Afr J Agric Res 6(6):1447–1454

Shintu PV, Radhakrishnan VV, Mohanan KV (2016) A study of the genetic variability of West Indian arrowroot (*Maranta arundinacea* L.) in Kerala state of India. Agric Forest Fish 5(5): 186–190

Sidhya P, Koundinya AVV, Pandit MK (2014) Genetic variability, heritability and genetic advance in tomato. Environ Ecol 32(4B):1737–1740

Solanki RK, Gill RK, Verma P, Singh S (2015) Mutation breeding in pulses: an overview. In: Khan S, Kozgar MI (eds) Breeding of pulse crops, 1st edn. Kalyani Publishers, Ludhiana, pp 85–103

Sonnino A, Thinh NT, Santangelo E, Mini P (1997) Genetic improvement of sweet potato through somatic embyrogenesis and *in vitro* induction of mutations. In: Improvement of basic food crops in Africa through plant breeding, including the use of induced mutations, IAEA-TECDOC-951. IAEA, Vienna, pp 131–138

Srisuwan S, Sihachakr D, Siljak-Yakovlev S (2006) The origin and evolution of sweet potato (*Ipomoea batatas* Lam.) and its wild relatives through the cytogenetic approaches. Plant Sci 171:424–433

Suja G, Sundaresan S, John KS, Sreekumar J, Misra RS (2012) Higher yield, profit and soil quality from organic farming of elephant foot yam. Agron Sustain Dev 32:755–764. https://doi.org/10.1007/s13593-011-0058-5

Suja G, Byju G, Jyothi AN, Veena SS, Sreekumar J (2017) Yield, quality and soil health under organic vs conventional farming in taro. Sci Hort 218:334–343

Suma A, Latha M, John KJ, Devi AI, Nair RA, Dwivedi NK (2014) Evaluation of Chinese potato (*Plectranthus rotundifolius* (Poir.) Spreng.) for yield contributing traits. Indian J Plant Genet Resour 27(2):142–148

Suprasanna P, Nakagawa H (2012) Mutation breeding of vegetatively propagated crops. In: Shu QY, Forster BP, Nakagawa H (eds) Plant mutation breeding and biotechnology. CABI, Wallingford, pp 347–358. https://doi.org/10.1079/9781780640853.0347

Thankappan M, Chacko CI (1976) Effect of cassava mosaic on different plant parts and tuber yield in cassava. J Root Crops 2:45–47

Thinh NT (1993) Induction and isolation of mutations in sweet potato through in vitro mutagenesis. In: Improvement of root and tuber crops in tropical countries of Asia by induced mutations, IAEA-TECDOC-809. IAEA, Vienna, pp 53–64

Tofino A, Cabal D, Sanchez T, Ceballos H (2011) Identification of radiation induced mutants of cassava (*Manihot Esculenta* Crantz) using morphological and physicochemical descriptors. Agron Colomb 29(2):361–371

Vasudevan K, Jos JS (1988) Studies on induced mutations in tuber crops. Annual report 1988. ICAR-Central Tuber Crops Research Institute, Thiruvananthapuram, pp 49–51

Vasudevan K, Jos JS (1991) Studies on induced mutations in tuber crops. Annual report 1991. ICAR-Central Tuber Crops Research Institute, Thiruvananthapuram, pp 69–71

Vasudevan K, Jos JS (1992) Gamma ray effects on yams (*D. alata* L. and *D. esculenta* (Lour.) Burk.). J Root Crops 18(1):94–98

Vasudevan K, Unnikrishnan M, Jos JS (1986) Studies on induced mutations in tuber crops. Annual report 1986. ICAR-Central Tuber Crops Research Institute, Thiruvananthapuram, pp 62–65

Vasudevan K, Sundaresan S, Indira P, Jos JS (1993) Induced variation for tuber quality, proline content and photosynthetic rate in cassava mutants. J Root Crops 22(2):35–39

Vasudevan K, Sheela MN, Nair RR (1995) Improvement in Dioscorea, cassava and Chinese potato through induced mutations. Annual report 1995. ICAR-Central Tuber Crops Research Institute, Thiruvananthapuram, pp 50–51

Vasudevan K, Padmaja G, Jayakumar S (1996) Induced tuber colour mutants and their biochemical characters in sweet potato. J Root Crops 22(2):137–138

Villamor CC, Cardinez MT (2008) Application of induced mutation to increase yield of yam (*Dioscorea alata* L.). Philippine J Crop Sci (Philippines):55–56

Wang YP, Liu QC, Hong Z (2005) Mutagenic effects of ion beams on embryogenic cell-aggregates of sweetpotato. Acta Agron Sin 31(4):519–522

Wang Y, Wang F, Zhai H, Liu Q (2007) Production of a useful mutant by chronic irradiation in sweet potato. Sci Hort 111(2):173–178

Washaya S, Mupangwa JF, Muranda E (2016) Nutritional value and utilization of yams (*Dioscorea steriscus*) by residents of Bindura town high density suburbs, Zimbabwe. Adv Agric 7. https://doi.org/10.1155/2016/5934738

Wilson JE, Pole FS, Smith NEJM, Taufatofua P (1989) Sweet potato breeding. IRETA-USP Agro-Facts Publication, Western Samoa, p 39

Yan H, Lu L, Hershey C, Ceballos H, Chen S, Li K (2013) Cassava mutation breeding: current status and trends. Plant Mutat Rep 3(1):37–44

Zhen HR (1993) Induction of mutations in sweet potato by combined use of gamma irradiation and tissue culture. In: Improvement of root and tuber crops in tropical countries of Asia by induced mutations, IAEA-TECDOC-809. IAEA, Vienna, pp 37–52

Zhen HR (2001) In vitro technique for selection of radiation induced mutants of sweet potato. In: *In vitro* techniques for selection of radiation induced mutations adapted to adverse environmental conditions, IAEA-TECDOC-1227. IAEA, Vienna, pp 79–82

The manufacturer's authorised representative in the EU is Springer Nature Customer Service Centre GmbH, Europaplatz 3, 69115 Heidelberg, Germany. If you have any concerns regarding our products, please contact ProductSafety@springernature.com

Printed and bound by CPI Group (UK) Ltd, Croydon, CR0 4YY
06/10/2026
02225252-0019